UNDERSTANDING PHYSICS and PHYSICAL CHEMISTRY USING FORMAL GRAPHS

Eric Vieil

CRC Press
Taylor & Francis Group
Boca Raton London New York

CRC Press is an imprint of the
Taylor & Francis Group, an **informa** business

CRC Press
Taylor & Francis Group
6000 Broken Sound Parkway NW, Suite 300
Boca Raton, FL 33487-2742

First issued in paperback 2019

ISBN-13: 978-1-4200-8612-6 (hbk)
ISBN-13: 978-1-138-38143-8 (pbk)

Library of Congress Cataloging-in-Publication Data

Vieil, Eric.
Understanding physics and physical chemistry using formal graphics / Eric Vieil.
p. cm.
Includes bibliographical references and index.
ISBN 978-1-4200-8612-6 (hardcover : alk. paper)
1. Physics--Graphic methods. 2. Chemistry, Physical and theoretical. 3. Algebra--Graphic methods. 4. Graph theory. I. Title.

QC20.7.G7V54 2012
530--dc23 2011039589

Visit the Taylor & Francis Web site at
http://www.taylorandfrancis.com

and the CRC Press Web site at
http://www.crcpress.com

Contents

Acknowledgments

This book was made possible through fruitful discussions with Professor Rob Hillman from the University of Leicester (UK) and with Dr. Jean-Pierre Badiali from the CNRS Paris (France). I am most grateful for their constructive criticisms and suggestions. I am also indebted to Laurène Surbier and Sylvain Tant, PhD students at the University of Grenoble (France), for their careful reading of the manuscript during the preparation of the book.

Presentation

THE BOOK

The subject of this book is truly original. By using what was originally a purely pedagogical means invented for teaching, that is, the encoding of algebraic equations by graphs, the exploration of physics and physical chemistry reveals common pictures through all disciplines. The hidden structure of the scientific formalism that appears is a source of astonishment and provides efficient simplifications of the representation of physical laws. The transverse view resulting from this approach is enlightening and leads to fruitful thinking about the differences and similarities from mechanics to chemical reactivity in going through electrodynamics and thermodynamics.

More than 80 case studies relevant to the macroscopic world are tackled in the following domains:

Translational mechanics
Rotational mechanics
Electrodynamics
Electric polarization
Magnetism
Newtonian gravitation
Hydrodynamics
Surface energy
Thermics (heat or thermal energy)
Physical chemistry
Corpuscles groups
Chemical reactions

The book is organized according to the transverse structures emerging from the graphs, from simple ones to more elaborated ones, by providing after each series of case studies the theoretical elements necessary for understanding their common features.

PREREQUISITES

A special feature of this book is that it spans over a wide level and range of audiences. The primary audience is graduate students, who will find it a valuable tool for placing their previous knowledge in a coherent framework, connecting items, and allowing a far deeper understanding.

Researchers and specialists will find new proposals for solving classical paradoxes or alternate views on some very fundamental questions in physics.

The book offers engineers new concepts and physical meanings (e.g., by helping them to design physicochemical devices optimizing energy losses).

The simplicity of the visual language and the absence of complicated equations make this textbook easier to access for undergraduate students than any other at an equivalent scientific level. The drawback is that it teaches some new notions that are not yet widely accepted.

Author

Dr. Eric Vieil is a researcher in physical chemistry from the Commissariat à l'énergie atomique et aux énergies alternatives (CEA) (French Alternative Energies and Atomic Energy Commission) at Grenoble, France. He was the head of the Laboratory of Electrochemistry and Physico-Chemistry of Materials and Interfaces (LEPMI) for 10 years. He worked at the CNRS, the Institut National Polytechnique (Grenoble-INP), and the University Joseph Fourier, where he obtained his PhD. Dr. Vieil has more than 80 publications in theoretical and experimental studies on the electrochemical mechanisms of conducting materials.

Most of the contents of this book have been taught to PhD students at the University of Grenoble through the course "Physics and Physical Chemistry without Equations" that started in 2005 at the Doctorate School I-MEP^2 "Ingénierie Matériaux Mécanique Energétique Environnement Procédés Production" (Engineering, Materials, Mechanics, Energetics, Environment, Processes, Production).

Icons Used in This Book

Explanation: For the student or the novice

Remark: Worth noticing, but not essential

Attention: Mistake or misinterpretation to be avoided

See further: Reference to another part or further reading

Restriction: Application to a special case, less general than the main theory

Original: Addition to classical physics and chemistry

Open door: Evocation of a possible track that cannot be followed by lack of space

Open question/Hypothesis: Point that needs to be validated

Companion CD

All Formal Graphs of this book are provided in color bitmap files (png format) in the accompanying CD-ROM.

A software for building simple electric (equivalent) circuits and translating them into Formal Graphs is also provided. See Appendix 5 for further explanations.

CD materials can be found at https://www.routledge.com/9781420086126

Supplementary Resources Disclaimer

Additional resources were previously made available for this title on CD. However, as CD has become a less accessible format, all resources have been moved to a more convenient online download option.

You can find these resources available here: https://www.routledge.com/9781420086126

Please note: Where this title mentions the associated disc, please use the downloadable resources instead.

Website on Formal Graphs

By consulting the website "http://www.vieil.fr/formal_graphs" one may discover supplementary materials: new Formal Graphs, some exercises, links to published articles, rules for coding Formal Graphs in color, correction of (always) possible errors and the Circuit-To-Graph translator of the companion CD as well.

1 Introduction

CONTENTS

1.1 AIM OF THIS BOOK

The purpose of this book is very simple: to provide to the reader a universal toolkit, called Formal Graphs, for understanding a wide range of scientific domains. This toolkit, as any ordinary kit, is composed of two things, some tools and a manual. The tools are graphs that are used for modeling phenomena encountered in physics and chemistry and the manual is a theory based on a classification of objects to be modeled. Universality is the most interesting feature of this toolkit, resulting from the limited number of graphs having the same structure that can be found across all scientific domains in which energy is defined.

A Formal Graph is another way to write equations, not only because one draws relationships instead of writing them, but because its principle differs from ordinary algebra, the most used language for representing relationships, in involving the notion of order, more precisely of topology. One may refer to short presentations already published for an overview (Vieil 2006, 2007). These graphic and topological properties confer to a Formal Graph the remarkable feature of being mainly visual and less abstract than the coded writing of algebra. This greatly facilitates the comprehension of relationships between the pertinent variables describing a physical phenomenon. This new approach is equally suited for advanced and inexperienced readers. Several uses of Formal Graphs are outlined here: as a pedagogical tool, for computing models, for learning basic physics and physical chemistry, and as food for thought in proposing research topics.

1. *Pedagogy*: A first obvious use of Formal Graphs is as a pedagogical tool in the hands of teachers for facilitating access and memorization of several concepts and for presenting alternative viewpoints. However, the simplicity and the power of the tool do not mean that a beginner can use it for learning physics from scratch. At least as long as a sufficient knowledge of experimental facts and processes is not acquired for providing the basic material used in models.
2. *Computing*: The second use is for computing models. Formal Graphs are in fact neural networks, which are easily transposed into algorithms. Neural network-based softwares are widely used for solving many complex and real-world problems in engineering, science, economics, and finance.

3. *Learning*: The third use is for readers having already basic education in at least one scientific domain. This is the same for learning a new foreign language; it is always easier to acquire a second one when one already knows some elements of grammar of a first foreign language. Once the new tool is mastered in one's domain, access to a new scientific domain is immediate: the same graphs and the same objects are directly transposed to another domain. A specialist in electrodynamics, once he has understood the way to graphically represent his equations and algebraic models, can have access to thermodynamics or to chemical reactions without difficulty. (The converse is naturally true.) Indeed, this transversal feature reveals a deeper unity of science than what could be imagined with the help of algebra only.
4. *Research*: The fourth use is for scientists who are interested in questioning the limits of science. In proposing the same graphs for several scientific areas, one meets sometimes some difficulties because the classically used models are less general than in other domains. This is a known problem in translating from one language to another; words do not always have exact correspondents and meanings differ frequently. The generalization brought by the Formal Graph approach leads to the formulation of more general laws in several domains and conversely to restrict some notions in other ones for establishing a common ground valid for every domain. For instance, the well-known relationship between the charge and the potential of an electric capacitor features a linear behavior, that is, to each variation of one of these two variables corresponds a proportional variation of the other. Comparison with other domains, such as physical chemistry, has the effect of proposing a nonlinear relationship instead of a linear one. Obviously, the new relationship may become linear in a restricted range of variables. (This proposal will be demonstrated later, in Chapter 7, but will be used earlier.) On the other hand, the notion of acceleration used in mechanics does not find any correspondent in other domains and must be relegated to an auxiliary role in the understanding of the whole physics. (This will be explained in Chapter 9.)

These are examples of modifications; many others are required when one is searching for a better standardization. It is worth saying that they are not superficial alterations. They raise some fundamental questions that may be the seed of further investigations for many scientists.

1.2 AN IMPERFECT STATE OF SCIENCE

1.2.1 A Partitioned Science

Science can be viewed as a coherent entity when one considers the scientific method, based on observations, experiments, models, and theories, but appears as a strongly partitioned ensemble when one considers its studied objects. The number of scientific domains, as can be evaluated by government agencies through their lists of disciplines, goes over a hundred, depending on the chosen resolution and criteria.

A frequent illusion is to believe that in the beginning of natural science, unity prevailed, as if scientific knowledge has developed from a tiny core growing by successive extensions. This view may be sustained by scientists working in a delimited field who have the feeling that their domain has grown like a tree or a town, encompassing a growing number of phenomena and deepening endlessly the understanding of nature. The existence of specialized disciplines is viewed as a logical and practical consequence of this enormous extension that a single brain cannot hold entirely. The division of scientific work into compartments would be an operational answer in front of the complexity of the task in scientific research. This myth of a golden age, a scientific paradise, lost because of the weakness of the human nature or due to limitations of the human brain, has obviously some religious connotation. It has, as a consequence, certain fatalism among scientists about the possibility of unifying this partitioned science that evokes Babel's tower with its multiplicity of scientific jargons.

In fact, it is the opposite process that occurs in the background when one considers the whole scientific domain. The prescientific situation is a fragmented landscape of many skills, crafts, and myths that have been progressively sorted and gathered as an effect of scientific rationalization.

Let us take the example of a short, but intense, period in which this unification process has been extraordinarily active: the beginning of the nineteenth century. Augustin Fresnel,* in his quest for a greater unity of physics, realized in 1819 the fusion of Newtonian† optics (based on geometry) with wave mechanics. Until the reunification made by Georg Ohm‡ in 1827, with his famous law linking current and potential§ for a conductor, two kinds of electricity were considered: electrostatics, concerned with capacitors and batteries (invented by Alessandro Volta¶ in 1800); and galvanic electricity, concerned with the flow of current (studied by Luigi Galvani** in the late eighteenth century). Electricity and chemistry have been associated with the study of electrolysis by Michael Faraday†† in 1833. Mechanics, hydrodynamics, and thermodynamics have been put on the same footing in terms of contributing to the energy of a system, with the experiments of James Joule‡‡ (1847), establishing equivalence between work and heat, giving birth to the First Principle of Thermodynamics. Later, in 1855, James Clerk Maxwell§§ made the synthesis between electricity and magnetism with the four equations bearing his name and relating electric and electromagnetic fields.

This tendency toward unification is still observable in modern science. Unfortunately, the search for a grand unification, which is very active nowadays, trying to find a common theory encompassing the four fundamental forces (electrodynamical, gravitational, weak, and strong), has not yet produced a satisfying outcome (Smolin 2007). Numerous efforts have been made, and continue to be made, to harmonize theories, spanning from the standardization of units to the generalization of concepts, but the gap between scientific communities seems impossible to bridge. The main difficulty is the natural tendency of every scientific community to develop its own jargon and to define peculiar concepts without connection with other domains. One may point out the drawback of applying specific units under the pretext of convenience. The driving force that makes unity an interesting goal for a scientist (like Fresnel and many others) seems to be weaker than the entropic force of thermal agitation. However, there are structural factors other than human factors that are obstacles to standardization.

1.2.2 Ordinary Algebra has Defaults

Models in natural sciences are based on mathematics. The most used formalism for building models is based on algebraic language, made of symbols and alphabetic characters. This is an extraordinary and powerful tool, allowing encoding of the laws of nature that have been disclosed until now. Algebra is a universal language used in all branches of science; however, other languages exist, such as graphic languages, which will be evoked shortly after, but they are less universal. It may appear surprising to introduce a supplementary graphic language in this context, which is presented as equivalent, or almost. The reason is that algebra is not perfect.

Algebra, in its common version, suffers from a severe drawback: lack of order when dealing with ordinary mathematical objects such as scalar variables. The commutativity of the ordinary

* Augustin Fresnel (1788–1827): French physicist; Paris, France.

† Isaac Newton (1642–1727): English mathematician, physicist, and astronomer (and alchemist); London, UK.

‡ Georg Simon Ohm (1789–1854): German physicist; Cologne and Munich, Germany.

§ Ohm's law was rejected the first time because it associated two domains considered to be separated by definition. Ohm was obliged to abandon his professor's chair in Cologne.

¶ Alessandro Giuseppe Antonio Anastasio Volta (Count) (1745–1827): Italian physicist; Como, Italy.

** Luigi Galvani (1737–1798): Italian physiologist; Padova and Bologna, Italy.

†† Michael Faraday (1791–1867): English physicist and chemist; London, UK.

‡‡ James Prescott Joule (1818–1889): English physicist and brewer; Manchester, UK.

§§ James Clerk Maxwell (1831–1879): Scottish physicist and mathematician; Edinburgh and Cambridge, UK.

multiplication between scalars, which means that the product of two variables A and B can be indifferently written as AB or BA, is a source of disorder in the sense that there is no rule for writing a number of physical laws in a unique form. For instance, Ohm's law relating a current I to a potential difference V through a resistance R of a conductor can be written in four different ways*:

$$V = RI; \quad V = IR; \quad I = \frac{V}{R}; \quad R = \frac{V}{I} \tag{1.1}$$

Commutativity is not restricted to scalar variables, but is also encountered with linear operators. For instance, when the mass M of a body is constant, Newton's second law of motion giving the force F as a function of the body velocity v can be written indifferently, using as an operator the time derivation:

$$F = M\frac{dv}{dt}; \quad F = \frac{dv}{dt}M; \quad F = \frac{dMv}{dt} \tag{1.2}$$

However, when dealing with more structured objects and operations, more rigorous rules are encountered, as for instance with vectors and matrices where the order of multiplication may matter in most cases.

Perhaps, the most annoying drawback is the freedom of composition of variables by combining variables and operators into a single new variable. Anyone is free to define new variables at will, the only condition being that an acceptable symbol can be given to it. No physical meaning is required, except if it corresponds to a deliberate choice.

The consequence is that regularities or analogies between algebraic expressions are harder to detect. The algebraic landscape is so varied that comparisons between domains are made especially difficult (Shive and Weber 1982).

1.2.3 An Incomplete Hierarchy of Objects

The physical world has been progressively understood as being made of objects having various complexities, from quarks and elementary particles to nebulae clusters. The main criterion used for sorting objects along a complexity axis is the object size (leading to the microscopic/macroscopic division of the world), and a secondary criterion is the number of elements that compose an object.

Normally, variables used in physics or chemistry correspond to objects that have a definite degree of complexity or to a class of objects that span a range of complexities. For instance, a position in space is a pertinent variable for a particle or for the center of a star, while distance is a pertinent variable for a system of two particles or twin stars. For a cloud of particles or stars, the average distance is a pertinent variable. Many variables are found under the form of a unique notion, a difference, or a sum (an algebraic average is a sum divided by a number of entities).

The choice of the right form is not always clear. In electrodynamics, sometimes the pertinent variable is a potential; in other cases, it may be a potential difference. Kirchhoff's law† for serial circuits requires a sum of potential differences. The same difficulty arises in mechanics with the velocity, where the position of a referential sometimes needs to be given for considering a difference of velocities. One can multiply examples like these, where the link between objects and variables is not as clear as it should be.

* And even more when one applies conductance, which is the reciprocal of resistance.

† Gustav Robert Kirchhoff (1824–1887): German physicist; Heidelberg and Berlin, Germany.

This may not seem to be a real difficulty for people acquainted with the algebraic language, but it becomes a problem when one wants to translate this into another language. One is obliged to clarify the complexity scale by creating new levels in the object's hierarchy for solving this difficulty.

1.3 IMPROVEMENT THROUGH GRAPHS

1.3.1 Multiple Languages

Algebra is not the only modeling language used in science; several alternative languages exist, most of them bearing the generic naming of *graphic representations*. One of the most important among graphic languages is curve plotting, that is, when the variation of one measured variable is drawn as a function of an imposed variable. It may appear somewhat exaggerated to see these representations as a mathematical language, because one is accustomed to associate mathematics to algebra or geometry mainly, but they are indeed.* (By extension, projections in two-dimensional geometric figures such as surfaces or volumes, maps and pictures are included in this category of curve plotting.) Besides this universal tool, one finds several other graphic languages using graphs, also called networks, which is the representation of links between objects symbolized by nodes or vertices. These languages are more specialized; some are found in peculiar scientific domains, such as electric circuits (junctions are the nodes and components or wires are the links), or chemical formulas (atoms are the nodes and bonds are the links), or Feynman† diagrams in particle physics (nodes are particle encounters and links are the particles themselves) as shown in Figure 1.1.

The concept of graph will be described in more detail in the following section, but the interested reader may refer to the specialized literature (Berge 1962).

1.3.2 Formalism and Graphs

The basic principle of the representation of an algebraic equation by a graph is extremely simple. It is based on the mathematical property that any algebraic equation can be decomposed into elemental equations that express the transformation of one or several variables into a new variable. The category of variable encompasses several mathematical objects representing quantities that are scalars, vectors, or matrices. The transformation of one variable is made by the application of an operator (e.g., $B = A^2$), and the transformation of two variables into a third one by the application of a dyadic operator (e.g., $C = A + B$). An operator implies a large class of mathematical objects that ranges from the simple scalar (that multiplies a variable, thus changing its scale) to more elaborated objects such as matrices, tensors, or any function. By convention, throughout this book, when the

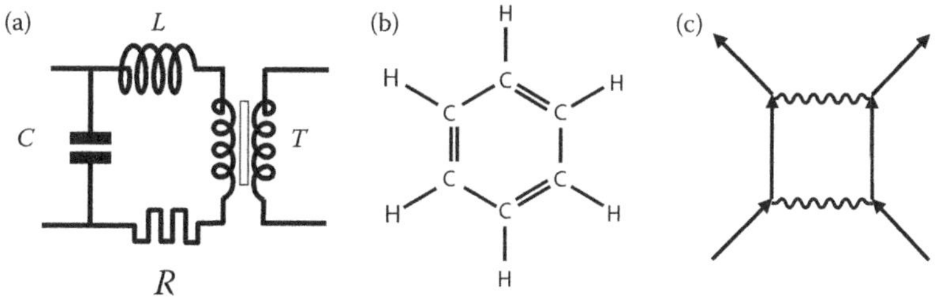

FIGURE 1.1 Examples of graphs used in physics and chemistry: (a) electric circuit, (b) chemical formula, and (c) Feynman diagram.

---

* A curve is the trace of an operator, that is, the set of particular values it takes in given conditions.

† Richard Phillips Feynman (1918–1988): American physicist; New York, USA.

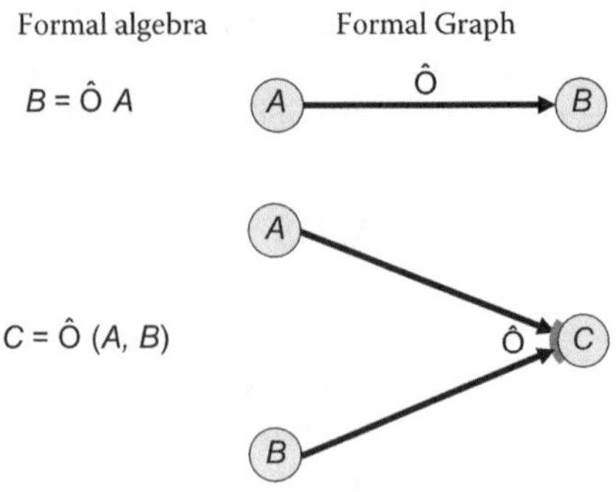

GRAPH 1.1 The principle of transliteration (i.e., correspondence) between algebraic formalism and Formal Graph. *A*, *B*, and *C* are variables linked by an operator. In the case of a monadic operator, its algebraic symbol is placed along the arrow featuring the link, whereas in the case of a dyadic operator, it is placed near the arrow heads.

symbol is not sufficiently explicit for indicating its operator nature, a hat or circumflex (^) is placed over the symbol, with a straight font (e.g., $\hat{\mathrm{H}}$) as opposed to a slanted font used for variables (e.g., *H*).

This basic principle of the representation of an algebraic equation is illustrated in Graph 1.1. Nodes containing the variables are represented by closed geometrical figures, circles or polygons, and links are oriented for indicating the operand variable and the resulting variable (at the arrow end).

Note that by convention, the placement of the operator symbols depends on the monadic/dyadic property of the operator. This allows combining a dyadic operator with two monadic operators as seen in many Formal Graphs (mainly for adding contributions of several links).

The convention for representing a dyadic operator (and, by extension, any operator acting on several variables) is to make the arrows of the links coming from the operand variables arrive at a protruding curve (gray line) at the perimeter of the resulting node. This external piece of curve links all the arrows involved in the same operation, and another arrow can be placed at one of its ends when the order of the operand variable needs to be specified. This symbolism is a reminder that such a graph mimics biology with the axons/synapses/dendrite connections in a neural network. The external gray curve of a node plays the role of a dendrite of a receiver neuron, individual operators are synaptic connections, and graph links are the axons of neurons acting as emitters (see Figure 1.2).

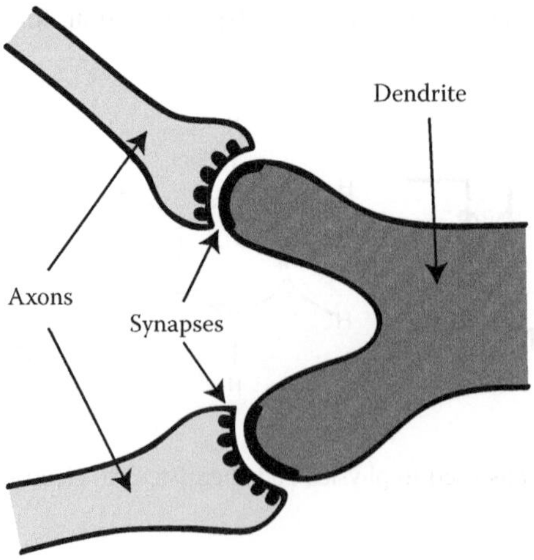

FIGURE 1.2 Connection scheme between neurons through synapses. The dendrite belongs to a receiver neuron and axons are extensions of emitter neurons.

This analogy is far from being fortuitous, since this connection scheme has been used for long in computer simulations of physical processes. Neural networks are classically used for applications where formal analysis is difficult, such as pattern recognition and nonlinear system identification and control (Carling 1992). This common feature between computerized neural networks and Formal Graphs is worth outlining, since it means that a Formal Graph can be directly used for computing the model of system it represents. Many popular computer softwares among the scientific community implement neural network toolboxes.

1.3.3 Comparison of Three Modeling Languages

The algebraic language is compared with two graphic languages, the electric circuit and the Formal Graph languages, taking two slightly different systems, both able to store and to dissipate (by a resistor) energy, but one system storing electrostatic energy (by a capacitor) and the other electromagnetic energy (by an inductor or coil). For a better illustration, two ways of combining storage and dissipation are presented, one in parallel and the other in series.

Electric Circuit: As mentioned earlier, the electric circuit is a graph representation mimicking the disposition of dipolar components that are assembled or "mounted" according to the electric jargon. In Figure 1.3 are sketched the two circuits with indications of the respective potential differences and currents.

The noticeable feature (it is in fact a drawback) of an electric circuit is that it does not provide any explicit information on the behavior of the components. It merely encodes the method, parallel or serial, used for assembling components and helps finding the correct operations for relating currents and potential differences between themselves.

Algebra: The rules for relating circuit variables are called Kirchhoff's laws. Table 1.1 describes the two sets of five algebraic equations used for modeling these circuits.

It is worth outlining that the decomposition into elementary equations is not compulsory when using algebraic equations. Most of the time, equations are more or less combined to form more compact writings, and, in the extreme case, one may find the unique and overall equation of the model written as shown in Table 1.2.

Formal Graph: The two Formal Graphs shown in Graphs 1.2 and 1.3 are exact translations of the two sets of equations shown in Table 1.1 according to the basic transliteration principles described earlier.

The shapes of the nodes are voluntarily chosen among various polygonal shapes in order to distinguish variables in a Formal Graph. When color is available, a distinctive color comes as a supplementary mark for facilitating graph interpretation. These distinctive features will be explained in the subsequent chapter.

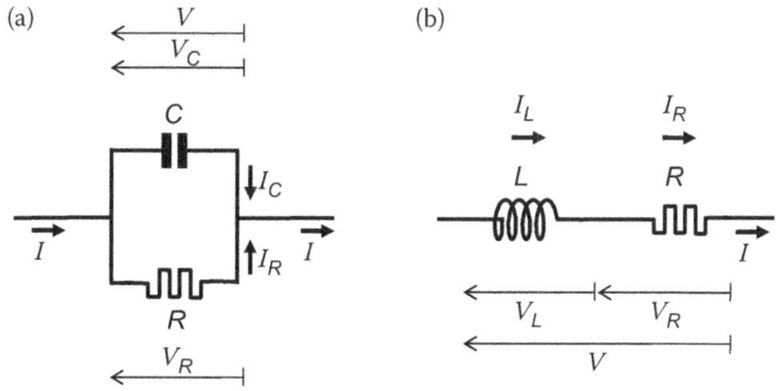

FIGURE 1.3 Electric circuits of a resistor (R) and a capacitor (C) mounted in parallel (a) and of a coil or inductor (L) and a resistor (R) mounted in series (b).

TABLE 1.1
Algebraic Equations Modeling Two Electrical Circuits, One with a Capacitor and a Resistor in Parallel (Left Column), the Other with a Self-Inductance in Series with a Resistor (Right Column)

Parallel RC		Serial RL
	Constitutive Equations	
Ohm's Law: $V_R = R\,I_R$		Ohm's Law: $V_R = R\,I_R$
Capacitance: $Q = C\,V_C$		Self-Inductance: $\Phi_B = L\,I_L$
	Space–Time Equation	
Current–charge: $I_C = \frac{dQ}{dt}$		Induction Law: $V_L = \frac{d\Phi_B}{dt}$
	Mounting Equations	
Kirchhoff's Current Law: $I = I_R + I_C$		Kirchhoff's Voltage Law: $V = V_R + V_L$
Parallel Mounting: $V = V_R = V_C$		Serial Mounting: $I = I_R = I_L$

In the parallel mounting of a resistor, the operator (in this case a scalar) used in Graph 1.2 for representing the behavior of this component is not the resistance R, as in the serial mounting in Graph 1.3, but its reciprocal, which is the conductance G.

$$G = \frac{1}{R} \tag{1.3}$$

The two constitutive equations and the space–time equation of each set are directly translated into corresponding links between two variables, whereas the mounting equation in each set is represented according to a dyadic operator, as explained in the previous section. This operator, according to Kirchhoff's laws, is the addition.

This word-for-word translation proves that these Formal Graphs are perfect working models of the considered systems (as long as one trusts the algebraic model). Naturally, Formal Graphs are not restricted to models that can be represented by electric circuits, even by equivalent circuits working in domains other than electrokinetics. The comparison with electric circuits was just a didactic introduction. Formal Graphs are much more general than models based on dipolar components, whatever their complexity.

Translation Problems: This example does not evidence any advantage of a Formal Graph over algebra. A clear decomposition of the algebraic model into elemental equations, as in Table 1.1, is an excellent tool for modeling systems and can be perfectly used as such. However, there is a

TABLE 1.2
The Same Models as in Table 1.1 but Under the Form of a Unique Algebraic Equation

Parallel RC		Serial RL
	Overall Equation	
$GV + \frac{d}{dt}CV = I$		$RI + \frac{d}{dt}LI = V$

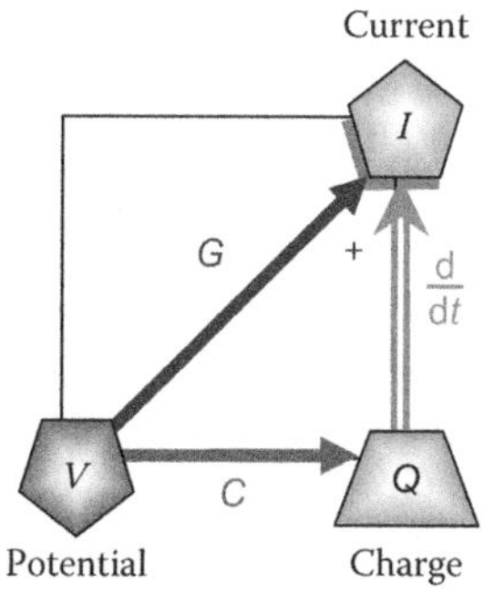

GRAPH 1.2 Formal Graph of a parallel RC circuit.

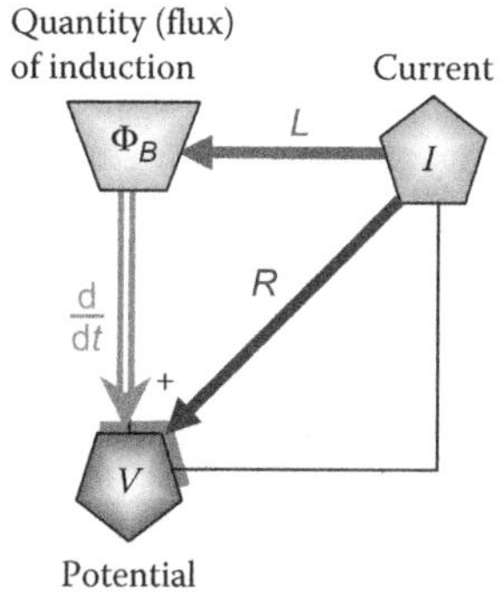

GRAPH 1.3 Formal Graph of a serial RL circuit.

condition for this equality in terms of advantages, which is the existence of a sufficient decomposition. Table 1.1, in fact, has been established with the idea of comparing with the Formal Graphs, in translating "word for word" each graphic element, giving a clear algebraic landscape. The physical meaning, indicated in Table 1.1, comes also from the Formal Graph methodology. Nothing prevents a talented physicist from writing directly these equations without any knowledge of Formal Graphs, but it is not so frequent. In the most used method, one problem in establishing the overall equations is the way of grouping. Table 1.2 could have been written differently, by grouping differently variables and equations, as for instance:

Parallel RC: $$GV + C\frac{\mathrm{d}}{\mathrm{d}t}V = I \tag{1.4}$$

The interesting point is that translation into a Formal Graph of this equation, in the precise form chosen, is impossible because not every algebraic equation can be translated into a Formal Graph. For instance, the grouped Equation 1.4 contains the time derivative of the potential V, which is not written directly in Table 1.1. In fact, it comes from the derivation of the charge Q, and in a second step from the derivation of the product CV. In the grouped equation, as the capacitance is placed on the left of the derivation operator, it means that the capacitance has been assumed to be a constant. This is not exactly equivalent to the model in Table 1.2 and Graph 1.2. However, this is not a sufficient reason for the inability to translate.

The problem comes from the impossibility to represent the time derivative of the potential in a Formal Graph, because other rules, based on physics and not on mathematics, impose limitations of the possible variables in a node and dictate the topology of the graph. In the present case, a first rule says that a capacitance can only link a potential to a charge and a second rule that a current and a charge can only be directly linked by time derivation. The time derivative of a potential is not a node in a Formal Graph because the only possible nodes are variables that feature the thermodynamical state of the system. These rules will be explained in the following chapter; for the moment, it is worth mentioning that they are identical whatever the scientific domain considered.

Forced Generality: An interesting consequence of these constraints is that the Formal Graph imposes a greater generality than pure algebraic reasoning. In forcing the capacitance to stay on the operand* side of the derivation, Formal Graph rules allow, by default, the capacitance to depend on other variables that, in turn, may depend on time. In classical electrostatics, this generality is not a significant advantage since the capacitance of an ideal capacitor is a constant, but there exist some materials in which the behavior is not ideal and the interest of systematically starting from the most general case is clear.

1.3.4 Difficulties?

Are Formal Graphs difficult to master? It would be demagogic to answer yes without warning. However, what is sure is that the difficulty is not the same as in physical mathematics. As will be seen throughout this book, mathematics stays at a simple and general level, without entering into too many technical details. For instance, no space referential, and consequently no components of vectors, will be used in this book. This means that neither vector calculus nor matrix nor tensor calculus will be required. This considerably alleviates the burden of necessary knowledge for understanding physics. Perhaps, the whole physics is not reached, because sometimes details are unavoidable, but a sufficiently wide range is reached for being able to connect many aspects.

Understanding means to connect together, according to the Latin etymology of intelligence, *inter ligare*, to put links between things. This cannot be done by staying too close, in order to be able to discern the main patterns.

The price to pay is the acquisition of new concepts, ten or so, which is a lot at once. New words accompany these new concepts and new ways to use them, to arrange and compose them. Thus, Formal Graphs introduce a new way of thinking physics, and this is a serious difficulty because one has been educated otherwise, in particular without clear hierarchy in complexity scale. (This lack of altitude is the counterpart of investigating in too much detail.) One hopes that the efforts made will be rewarding.

* The operand is the variable on which an operator applies.

2 Nodes of Graphs

CONTENTS

This chapter presents the basic rules that are required for building the nodes of a Formal Graph. Elements are presented in a logical order, beginning with the notion of energy and ending with the graphical coding used in Formal Graphs. Rules for linking the nodes will be introduced in the next chapter.

For beginners this is certainly the most difficult chapter along with the next one because Cartesian* reasoning is quite deprived from examples. Readers who prefer to learn from already known elements may jump to subsequent chapters that are conceived in the opposite way, starting with examples for reaching synthetic conclusions. A glossary at the end (Appendix 1) contains the definitions of the new concepts introduced here.

The essential point in this chapter is the First Principle of Thermodynamics that establishes the conditions for allowing energy to be exchanged or converted. The Second Principle, ruling the behavior of entropy, is more specialized and will be used later.

2.1 ENERGY AND STATE VARIABLES

The seeding idea of the Formal Graphs approach is the primordial concept of energies. The whole theory comes from this concept by defining some principles and properties. This primary status

* René Descartes (1596–1650): French philosopher, mathematician, scientist, and writer; Paris, France and the Netherlands.

makes any definition of energy depending on notions that are external to the body of Formal Graphs itself.

The concept of energy is rather new in human history. It has slowly emerged from the human conscience that something can be exchanged, transformed, or converted, between some materials or objects, accompanying transformations and evolutions of states. An invisible substance, the caloric, was thought as the vector of heat exchange and the notion of force was not very clear until a distinction was made between the *existence* of a force and the *action* of this force. The same word was used with the meaning one uses today—the mechanical tension that stretches a spring or accelerates the movement of a mass—and with the other meaning of *work*. The "living forces," *vis viva* in Latin, were used with the meaning of kinetic energy. The Greek word *energeia*, meaning activity or operation, was used in 1678 by Gottfried von Leibniz[*] for expressing a force in action. One had to wait more than a century to see this notion, instead of *vis viva*, being used by Thomas Young[†] in 1807. The concept of *potential energy* was invented in 1853 by one of the founders of thermodynamics, William Rankine,[‡] as a partner of kinetic energy in the law of conservation of energy, which gradually became a fundamental principle. Thermodynamics developed many other forms of energy around this principle, such as enthalpy, Helmholtz[§] free energy, Gibbs[¶] free energy, grand potential, and so on.

Through this brief and incomplete historical recall, what must be noted is the slow and recent emergence of the modern concept of energy. One must admit that the state of science is far from being perfect and that its development is not completely achieved. Confusions between force and energy, between potential and energy, between *energy-per-particle* and energy still persist in many minds and modern textbooks.

A rigorous definition of energy will progressively appear through the various properties that will be discussed. For the moment, it is sufficient to see energy as a means of exchange that nature uses for realizing transformations or evolutions.

Analogy. A good analogy with energy can be found in an economy with money, which is a means of exchange that humans or economic entities also use for transformations or evolutions (except that the law of money conservation does not really apply).

A first, rather obvious, property of energy is to be contained in something else, which is called a *system*. A system is a part of matter or vacuum, more generally a piece of universe, which eventually can be the universe itself. Space–time is another constituent of the cosmos that contains all systems, enabling conversions and distributions of energy in a system or between systems.

2.1.1 Energy Varieties

Energy is not a single facetted concept but exists under several forms. As mentioned above, kinetic energy, potential energy, work, and heat are instances of energy, to which one may add other forms such as electrostatic, electromagnetic, hydrodynamic, chemical energies, and so on. This is a logical consequence of the ability of energy to be exchanged and to provoke some transformations, in changing from one form to another one.

[*] Gottfried Wilhelm von Leibniz (1646–1716): German philosopher and mathematician; Leipzig and Hanover, Germany.

[†] Thomas Young (1773–1829): English scientist; Cambridge and London, UK.

[‡] William John Macquorn Rankine (1820–1872): Scottish engineer and physicist; Glasgow, UK.

[§] Hermann Ludwig Ferdinand von Helmholtz (1821–1894): German physicist and physiologist; Königsberg and Berlin, Germany.

[¶] Josiah Willard Gibbs (1839–1903): American physicist and chemist; New Haven, Connecticut, USA.

Analogy. A good analogy with energy can be found in an economy with money, which is a means of exchange that humans or economic entities also use for transformations or evolutions (except that the law of money conservation does not really apply).

However, a first difficulty arises with the notion of *energy form*, which is not sufficiently accurate. If the notion of heat is clearly related to only one form, the notion of work may assume several forms, mechanical, hydrodynamical, electrical, etc. Embedding of forms is also seen, such as electrostatics and electromagnetism—two forms belonging to the electrodynamical form.

Due to its huge importance, it is necessary to clarify the concept of energy.

2.1.1.1 Concept of Energy Variety

Every student in the exact sciences has been trained with the notions of heat and work as being the two main forms of energy. This is a quite simplistic view that sees the First Principle of Thermodynamics as restricted to an equivalence between heat and work. In this book, the formulation of the First Principle of Thermodynamics in terms of complementarity between heat and work is abandoned, because it is a much more general principle. This formulation is historically dated (it is a nineteenth-century language) at a time where only two or three energy forms were recognized. (Naturally, one does not abandon the principle itself.) If one still speaks of heat as equivalent to thermal energy, one shall not use the concept of work, because it is too restricted a notion. Also, thermodynamical variations around the notion of energy, such as enthalpy or free energy, are not easily transposable to all domains, so they will not be used outside thermodynamics either.

A clearer definition of energy emerges by introducing the notion of *energy variety*, which replaces the too imprecise notion of form.

What defines an energy variety? There is a simple answer that comes from its essential property to be contained in a system. A first and intuitive approach is to view a container in geometric terms. A system possesses an extent, a size, which may assume various geometric forms. A volume is naturally a container of energy, which is hydrodynamics. A surface contains superficial energy, also called capillary energy. A line contains mechanical energy that is described by various adjectives, including, most commonly, potential, elastic, internal, and static.

A second approach is to view containers as entities, without necessarily referring to a geometric element. An electric charge contains electrodynamical energy. A massive object contains gravitational energy. Energy contained in atoms, molecules, and particles is corpuscular energy (or atomic, molecular, particular energy, respectively).

A more abstract entity is the notion of process. Electric induction contains electromagnetic energy. A movement of an inertial object contains kinetic energy. A chemical reaction contains chemical energy. Sometimes, the container is a very abstract notion, such as the entropy for thermal energy.

2.1.1.2 Variety and Subvarieties

Things would be simple and limpid if a list of containers were sufficient for sorting all energy varieties. Unfortunately, a more subtle degree of classification is necessary because what are thought of as distinct energy varieties, on the sole basis of their different containers, are sometimes so close that a deeper analysis is required. That is the case in mechanics, for example, where potential energy and kinetic energy are recognized as two facets of the same notion called translational mechanics (distinguishing from rotational mechanics). An object may contain both potential and kinetic energies that are able to interact with each other by forward and backward energy exchange, as if they were sharing the same property of complementarity. For instance, a mass attached to a spring end may generate oscillations resulting from such an exchange. This is not the case between all energy containers, as for example, the same mass as in the previous case, but exchanging heat with a container of thermal energy will not generate oscillations.

For taking into account the complementarity of some energy properties and their ability to generate oscillations, the notion of *energy variety* is divided into two *subvarieties* that have some common features. Kinetic and potential energies are therefore considered as two subvarieties of the same energy variety, which is translational mechanics, because of their common characteristic of allowing oscillations. The same division is made in electricity, where electrostatic energy and electromagnetic energy are two subvarieties of the electrodynamical energy variety. An electric circuit made up of a capacitor (filled with charges containing electrostatic energy) and a coil or solenoid (seat of induction process containing electromagnetic energy), mounted in parallel for example, is an electric oscillator.

If the existence of complementary energy subvarieties is well known for long, there is no clear perception among the various scientific disciplines that this dual scheme is a general one and that every subvariety has a corresponding energy subvariety in another energy variety. For example, the electrostatic subvariety in electrodynamics, the potential subvariety in translational mechanics, and the hydrostatic subvariety in hydrodynamics share some common features, one of them being their existence in static regime. On the other hand, electromagnetic and kinetic energy subvarieties (together with a less known subvariety in hydrodynamics, which is the percussive energy) share another common feature, that is, the strong involvement of time in many of their manifestations (i.e., in dynamic processes).

Faced with this lack of perception that there exist two categories of subvarieties sharing some common features, two new concepts were created by adopting known adjectives—*capacitive* and *inductive* energy subvarieties. This is well in accordance with every transverse and synthetic approach to sort and to find out new categories, which naturally need to be named.

Table 2.1 provides a list of most known energy varieties with their subdivision and the corresponding container. For each variety, the inductive subvariety is listed first and the capacitive one after. For some energy varieties, the two subvarieties are known; for some others, this is not the case. This point will be discussed in Section 3.1.5.

Neologisms. It is not always a good idea to introduce new terms; it must be carefully weighted, because any novelty represents a supplementary difficulty in acquiring knowledge. However, this is justified when a stricter meaning is required or when no equivalent word already exists. Unhappily, in a transversal approach one shall meet this situation each time a generalization needs to be made. For instance, there is no name for speaking about the fact that internal, static, potential, static, thermal, chemical, etc., energies belong to the same category, because they have common features that are not shared by kinetic or magnetic energies. Two adjectives were adopted for these subvarieties: *capacitive* and *inductive*. This naming is based on the two electrical properties of a system, called *capacitance* and *inductance*, which allow the system to store electrostatic and electromagnetic energies, respectively. These electrical concepts are generalized to all domains for avoiding the creation of new terms. Notwithstanding, even in trying to limit to the maximum the number of new concepts, it remains a difficulty that is intrinsic to any process aiming at standardizing many notions. It would be marvelous to be able to unify our scientific domains only by keeping existing names.

The list of energy varieties in Table 2.1 is not exhaustive for several reasons. A first reason is the existence of homothetic varieties, that is, varieties in which the container is a multiple of another one. This can be merely a question of the unit used for quantifying the container. An example is the corpuscular energy and the physical chemical energy that are related by the Avogadro* constant.

* Lorenzo Romano Amedeo Carlo Avogadro (1776–1856): Italian scientist; Torino, Italy.

TABLE 2.1
Nonexhaustive List of Energy Varieties, Subvarieties, and Their Containers

Variety	Energy Subvariety	Energy Container	Comments
Rotational mechanics	Rotation kinetic	L (or $\boldsymbol{\sigma}$) Angular momentum	Also called "kinetic momentum"
	Torsion, rotation work	$\boldsymbol{\theta}$, $\boldsymbol{\alpha}$ Angle	Rigorously speaking, it is a vector
Translational mechanics	Kinetic	$\boldsymbol{P}$ Momentum, quantity of movement	Lowercase $\boldsymbol{p}$ is reserved for single particles
	Work, internal, potential, elastic	$\boldsymbol{\ell}$, $\boldsymbol{r}$ Displacement, position, distance, height	Not to be confused with vector coordinates
Superficial energy	Surface kinetic	$\boldsymbol{p}_A$ Superficial impulse	Not used classically
	Surface static	$\boldsymbol{A}$ Area	
Hydrodynamics	Volume kinetic	F Percussion	Not much used
	Hydrostatic	V Volume	
Gravitation	Gravity kinetic	p_M Gravitational impulse	Not much used
	Gravity potential	M_G (or M) Gravitational mass	Identical to the inertial mass M according to Einstein[a]
Electrodynamics	Electromagnetic	$\boldsymbol{\Phi}_B$ Quantity of induction	Also called "induction flux"
	Electrostatic	Q Charge	
Magnetic energy	Magnetism	p_m Magnetic impulse	Not to be confused with a "magnetic charge" (which is not used)
	—		
Electric polarization energy	—		
	Polarization	q_P Polarization charge	
Corpuscles energy	—	p_N Corpuscle impulse	Unknown
	Corpuscular	N (or n) Quantity of corpuscles	Corpuscles are atoms, molecules, particles
Physical chemical energy	—	p_n Substance impulse	Unknown
	Physical chemical	n Substance amount, quantity of moles	n is expressed as the number of moles
Chemical reaction energy	—	p_ξ Reaction impulse	Unknown
	Chemical reaction	ξ Extent, advance	
Thermal energy	—	p_S Entropy impulse	Unknown
	Heat	S Entropy	Entropy is an energy container but temperature T is not

Note: The inductive subvariety is given first, the capacitive one second. This list comprises only subvarieties whose container is made of several entities. They are listed under the variables that quantify their amount, with the most used algebraic symbols. Bold letters are for vectors.

[a] Albert Einstein (1879–1955): German-Swiss American physicist; Bern, Switzerland and Princeton, USA.

This number relates the amount of substance (number of moles) in this latter subvariety to the number of corpuscles.

A second reason stems from the possibility of combining several containers (of the same subvariety but belonging to several subsystems). The chemical energy possessed by a group of reactants is such a combination of the individual energies of reactants, weighted by each stoichiometric coefficient of the reaction. This variety can be called "energy of chemical reaction" but a better name is "chemical reactivity energy" because the notion of chemical energy can be viewed as encompassing all varieties contributing to a chemical substance (bonding, vibrational, rotational, etc.). The number of ways to combine energy subvarieties is in principle unlimited. A physical argument needs to be put forward for legitimating it, but physical systems present such versatility that making an exhaustive list is difficult.

A third reason is the noninclusion of energy varieties that are associated with a single entity, for instance one corpuscle or one elementary charge, or with a wave, for instance one vibration mode or one wavelength. These varieties are not discussed because of a lack of space and also for pedagogical reasons as they can be better comprehended once the collective behavior of entities that make up energy containers is understood. In short, the behavior of a single entity is relevant from the viewpoint of quantum physics, generally associated with a microscopic world, and the collective behavior of several entities is relevant from the viewpoint of macroscopic physics.

"Magnetic charges". Some peculiar energy subvarieties deserve attention because their container is so abstract that it is not really known. This is the case for magnetic energy possessed by magnets. This is an inductive subvariety, like the electromagnetic subvariety in electrodynamics, which was historically believed to be contained in "magnetic charges," as proposed by Charles-Augustin de Coulomb[†] in 1789, on the model of electrostatic charges appearing under the effect of an electric field (polarization charges). These magnetic charges have never been evidenced experimentally, because of a serious misunderstanding about the correct subvariety to look for. The Formal Graph generalization assigns to the magnetic container the same nature as an impulse or a kinetic moment, which corresponds to the quantum physics view as a magnetic moment. It is clear, for taking an analogy, that evidencing a charge or evidencing an electromagnetic induction is not the same task (see case study A4 "Magnetization" in Chapter 4). This illustrates the importance of a good classification. (That is at the beginning of every science.)

Energy and time. Although the existence of different containers, and consequently of different energy varieties, can be intuitively understood, this is not the case for the existence of two subvarieties. Why two and not more? Why is only one really known in certain cases? These are not easy questions and there are no satisfying answers nowadays. Chapter 9 illustrates that these questions are closely related to the notion of time, which is another difficult subject because no clear definition is really given in physics. A fundamental point to outline is that, in the Formal Graph theory, the *notion of energy is independent of any preexistent notion of time*. More precisely, time is defined from energy, in relation with conversion between energy subvarieties. The drawback of the classical approach that assumes the existence of time before any definition of energy or related concepts is that all these notions are entangled. As it will be seen, time is introduced relatively late in the development of the theory. This means that all the notions introduced before must be thought as timeless, as if eternity was the rule, and also valid once time begins to flow, since they are defined independently.

In Table 2.1, some containers are indicated as unknown, corresponding to subvarieties that have not been clearly identified and that bear no name. This merits an explanation that will be given in Section 3.1.5, when dealing with the notion of system property. Their presence in this table is not an affirmation of the existence in a system of the subvariety they are supposed to contain, as no experimental evidence has ever been produced. The reason for mentioning these virtual variables comes from the Mendeleev[†]-type approach chosen: by rationalizing and sorting energy categories, some empty subvarieties appears which raise the question of the existence of the corresponding variables. In fact, it is not because a container is envisaged that it contains necessarily some energy. The Formal Graph theory makes the distinction between the content and the container, this latter being eventually void.

[*] Charles-Augustin de Coulomb (1736–1806): French physicist; Paris, France.

[†] Dmitri Ivanovich Mendeleev (1834–1907): Russian chemist; St. Petersburg, Russia.

It must be outlined that the classification of energy varieties made above is not a simple matter of presentation, for having a satisfactory view on the energy landscape. It has a fundamental and deep meaning that can be described as the key concept allowing the behavior of physical and chemical systems to be understood at a first level: A system may contain several energy varieties in different proportions. This variety composition may depend on various parameters (temperature, pressure, etc.) and may change with their variations.

2.1.2 Some Properties of Energy

The amount of energy in a system is quantified with a variable having the same name and notated here with the script letter $\mathcal{E}$, to avoid confusion with the energy-per-corpuscle variable classically notated E. The SI unit of energy is Joule* (J); 1 J is equivalent to 1 kg m^2 s^{-2}.

2.1.2.1 State Function

The main property of the amount of energy is to define the *state of the system* that contains it. A change in the amount of energy modifies the state of the system. In thermodynamics, this property is called a *state function* and this notion of a *system state* is not defined other than in relation with the energy content. The practical use of this property is that the amount of energy in a given system state does not depend on the way the state has been attained by variation of any parameter. In particular, when variations of two parameters, say x_a and x_b, are required for changing from a first system state to a second one, the order of variations does not matter. When, in a first step, parameter x_a is varied and x_b is maintained at a constant and, in a second step, x_a is maintained at a constant and x_b is varied, the same state is reached at the end of both steps. This has important consequences on mathematically translating the variations of energy that will be detailed shortly after.

States. The notion of *system states* should not be confused with the notion of energy states used by physicists, which corresponds to a discretization of the possible values of an energy-per-corpuscle variable E.

2.1.2.2 Extensivity

The amount of energy follows the extent of the system; that is, the energy amount is an *extensive variable*, as opposed to an intensive variable that is not proportional to the system extent. This means that when two systems are considered together, the total energy is the sum of the two energy amounts:

$$\mathcal{E}(system\ 1+\ system\ 2)=\mathcal{E}(system\ 1)+\mathcal{E}(system\ 2) \tag{2.1}$$

A construction game. Each energy variety is a brick—a cell—that has relationships with other energy varieties and energy may be exchanged or converted between them. With this brick, the task of modeling a physical system becomes analogous to a construction game. The main difficulty is to enumerate the significant energy varieties in a system. Once this step is achieved, establishing relationships between bricks allows prediction of the possible changes in the system.

The main contribution of the Formal Graph theory to physics is to provide this conceptual tool that is extremely simple and powerful, as discussed in this book.

* James Prescott Joule (1818–1889): English physicist and brewer; Manchester, UK.

2.1.2.3 Additivity

The fact that energy exists under several varieties has consequence that the variable quantifying its amount must be additive, because the energy contained in a system is the sum of all existing varieties.

$$\mathcal{E}_{\text{total}} = \mathcal{E}_{\text{chemical}} + \mathcal{E}_{\text{mechanical}} + \mathcal{E}_{\text{electric}} + \cdots = \sum \mathcal{E}_q \tag{2.2}$$

As the addition operation is used for both properties, extensivity and additivity between varieties, the two are often confused. In practice, this is not very important, although the physical meaning is rather different (extensivity refers to the additivity of variables featuring the size of containers). Subscript q used to identify the variety in the above equation is also used to identify related variables.

The subdivision into subvarieties is mathematically written for each energy variety q:

$$\mathcal{E}_q = \mathcal{T}_q + \mathcal{U}_q \tag{2.3}$$

where $\mathcal{T}_q$ stands for the inductive subvariety (kinetic, electromagnetic, etc.) and $\mathcal{U}_q$ for the capacitive subvariety (potential, internal, etc.). As a consequence of Equations 2.2 and 2.3, the total energy $\mathcal{E}$ can be written as the sum of the total inductive energy and the total capacitive energy:

$$\mathcal{E} = \mathcal{T} + \mathcal{U} \tag{2.4}$$

It is important to note that if the total energy is a *state function*, as explained above, this is not the case for any of its subdivisions. Neither $\mathcal{T}$ nor $\mathcal{U}$ is a state function because the knowledge of only one does not allow one to say something about the state of the system.

2.1.3 Entities and Their Energy

It is essential to know whether an energy container is made with one piece, one elementary unit, or is a group with several units. In other words, the question is how to quantify a container.

In fact, there are two answers to this question, containers can be elementary ones or multiunit ones, and one will have to make the distinction all along the development of the theory because it has tremendous consequences.

2.1.3.1 Entities

The elementary container is called an *entity*. The container with several identical entities is called a *collection of entities*. It would have been at liberty to use the words *group* or *set*, but the word *collection* is intentionally chosen because it conveys the notion of collective behavior or common property, which is an elemental feature to include in the notion of container.

2.1.3.2 Energy-per-Entity

This precision allows quantifying a multiunit container by its *number of entities*. Each entity contains the *same amount of energy*, which is called *energy-per-entity*. This constraint of the same amount is very useful for understanding the behavior of containers and does not prevent considering more complex situations by taking into account several collections having different energies-per-entity in the same system. This is a higher level of organization that will be presented in Chapter 3.

2.1.3.3 State Variables

Energy-per-entity and *number of entities* enter in the category of *state variables*. State variables are defined as those variables that it is necessary to know the value and their mutual relations to help determine the amount of energy and therefore the state of the system. They form a pair as each subvariety possesses two of these state variables, also called *conjugate state variables*.

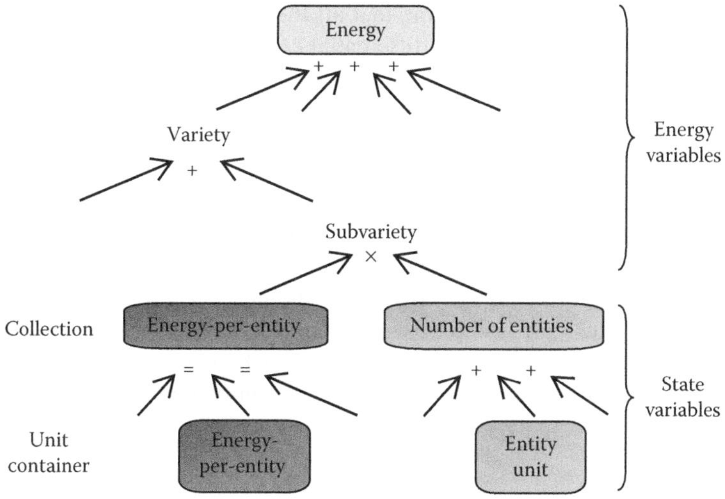

FIGURE 2.1 Arborescence of energy variables and state variables.

Figure 2.1 shows how the various concepts dealing with energy variables and state variables are structurally related.

Analogy. The distinction between *energy* and *energy-per-entity* is of paramount importance. To help grasp this importance, let us take again the economic analogy. The *energy-per-entity* can be compared to the price of a foodstuff, say some eggs, and the *number of entities* to the number of eggs. In this example, the energy corresponds to the budget, the total amount of money, required for buying a certain number of items. Another foodstuff, say some carrots, allows exemplifying the case of non-identical items. In this case, the pertinent entity is not the vegetable itself but its mass. Then, the energy-per-entity is the price per kilogram if the number of entities has the kilogram as a unit.

Now, consider buying two ingredients to make a soup, of carrots and cauliflower, for example. It is obvious for any buyer that he cannot calculate the cost by merely adding the prices per kilogram of the two ingredients and then by multiplying with the total mass of vegetables, unless prices per kilogram are equal, which is a special case. In scientific terms, the energy-per-entity (price per kilogram) is an intensive variable whereas the number of entities (mass) is an extensive one.

When the price per kilogram is independent from the number of kilograms, which is the case when no rebate is made for large quantities, the cost is always equal to the product of the price per kilogram with the number of kilograms. Regarding energy, this independence is not the general rule and the total energy is not merely the product of the energy-per-entity and the number of entities. (This will be discussed in Chapter 13.)

The relationship between energy and its state variables is the subject of the following section.

2.1.4 Energy Variation: The First Principle

Let us symbolize the *energy-per-entity* by y and the *number of entities* by x. Then, let us assume temporally independence between these variables,* so the energy in the considered subvariety is

* This is not the general case.

given by the product yx. Now, let us assume that all these variables are continuous (the case of discrete values will be discussed later). So differential calculus can be used to express their variations and the way they are related. In such a frame, the variation of energy follows Leibniz's rule of derivation of a product of independent variables y and x:

$$\mathrm{d}\mathcal{E}_{subvariety} = y\,\mathrm{d}x + x\,\mathrm{d}y \tag{2.5}$$

In mathematical words, this is called an exact differential equation because its integration can be made without further knowledge by reversing Leibniz's rule and retrieving the product yx as the result. Now, when the two variables are dependent, the previous equation is no longer true. In absence of knowledge of the relationship between them, a hypothesis must be made for being able to relate energy variations to state variable variations. This hypothesis is that the energy variation of a *collection of entities* is only due to the variation of the entity number as if the energy-per-entity could not vary; that is:

$$\delta\mathcal{E}_{subvariety} = y\,\mathrm{d}x \tag{2.6}$$

The energy variation is written, in this case, with a delta (δ) instead of a roman "d" to avoid forgetting that this equation is not an exact differential equation as before. This constitutes in part the First Principle of Thermodynamics upon which relies the whole edifice of the science dealing with collections of entities, because, according to this hypothesis, the number of entities must be able to vary, which is not possible with a single entity. Naturally, the strength of this (partial) principle is that it remains valid when the energy-per-entity varies. Otherwise, it would be a simple consequence of Equation 2.5 obtained by setting $\mathrm{d}y = 0$.

2.1.4.1 First Principle

The full First Principle of Thermodynamics is obtained by adjunction of a supplementary hypothesis that the equation expressing the total energy variation of several varieties or subvarieties is an exact differential equation. In the case of two varieties or subvarieties identified by subscripts a and b, this can be written as:

$$\mathrm{d}\mathcal{E}_{a+b} = \delta\mathcal{E}_a + \delta\mathcal{E}_b \tag{2.7}$$

Physically speaking, this hypothesis is very strong as it means that the energy in a system, which is a state function by definition, can only vary if at least two energy varieties or subvarieties vary simultaneously in the system. The fact that the individual energy variations are not exact differential equations means that they are not themselves state functions, that is, they cannot be integrated alone for finding the energy of the variety or subvariety. But it is the elemental property of energy expressed by this last hypothesis, once grouped with another energy variation, that makes integration possible. In other words, energy in a single subvariety cannot vary. At least two subvarieties, able to vary their amount of energy, in the same variety or in different varieties are necessary. This is the core principle of every science dealing with energy, whatever the organization level in collections or in single entities. From this, all the properties of energy exchanges are derived.

These two hypotheses expressed in Equations 2.6 and 2.7 give the First Principle of Thermodynamics under its differential form:

$$\mathrm{d}\mathcal{E}_{a+b} = y_a\,\mathrm{d}x_a + y_b\,\mathrm{d}x_b \tag{2.8}$$

As explained, this differential formulation only applies to collections of entities and not to elementary units. Would it mean that particle physics is not concerned with the First Principle of Thermodynamics? Not really, and this is the reason why this principle is divided in two parts. The second part, expressed by Equation 2.7, has a universal application. Equation 2.6, used in the first part, must be replaced by an adaptation of Equation 2.5 to the case of a fixed number of entities; that is, by the product of the entity number, equal to a constant, and the energy-per-entity variation:

$$\delta \mathcal{E}_{subvariety} = x \, \mathrm{d}y \tag{2.9}$$

With this adaptation, the First Principle of Thermodynamics becomes valid for single-entity containers.

2.1.5 A First Formal Graph: Differential Graph

The expression in Equation 2.6 of the energy variation for a collection of entities allows one to define energy-per-entity as the partial derivative of the total energy in a system with respect to the number of entities of the considered subvariety, while maintaining at a constant the numbers of entities of all other subvarieties.

$$y_a \overset{def}{=} \left(\frac{\partial \mathcal{E}}{\partial x_a} \right)_{x_b,\, b \neq a} \tag{2.10}$$

Analysis of this definition shows that an operator, the partial derivation with respect to the entity number, applied to a variable, the total energy, gives another variable, the *energy-per-entity*. A more general definition using an operator $\hat{X}_a$ representing the integration operator with respect to the sole variable x_a can be used alternatively for outlining the algebraic structure of the following equation:

$$y_a \overset{def}{=} \hat{X}_a^{-1} \mathcal{E} \tag{2.11}$$

The reciprocal of this operator is used because one does not need integration but derivation. Translation of these algebraic equations into a Formal Graph is straightforward and the resulting graph cannot be simpler than that shown in Graph 2.1, where the two ways to represent the same relationship are drawn.

This double definition may appear as a game with notations and representations; however, it has more profound implications. Our reasoning on the variations of energy and state variables was based on the assumption of continuous variables, for being able to use the differential calculus. The problem is that nothing proves that they are continuous, and the contrary seems more likely. Since the beginning of the nineteenth century, it is known that matter is made up of atoms and, consequently, that variables quantifying many containers are discrete. A number of moles may appear continuous but in fact it is a number of molecules (a discrete variable) divided by the Avogadro constant. A number of electrical charges Q is a discrete variable because any ensemble of charges is composed of elementary charges. It is only because one often deals with large amounts of charges that one may have the illusion that the variable Q is continuous.

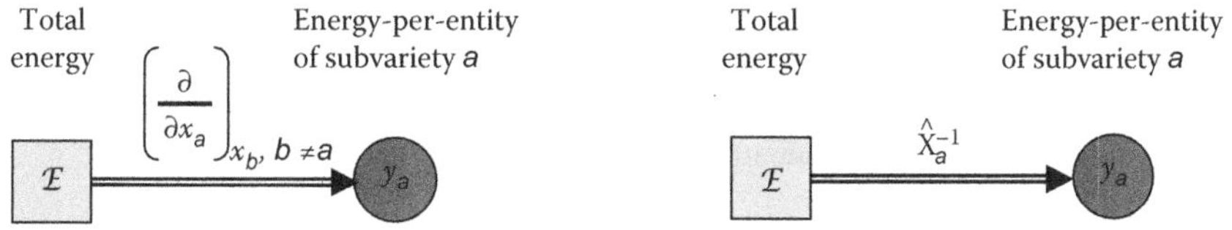

GRAPH 2.1 Formal Graph showing that the energy-per-entity y_a is a partial derivative with respect to the entity number x_a of the total energy $\mathcal{E}$, or more generally is the result of an operator applied to the energy. By convention, nodes with an energy variable are drawn with a square.

Discrete variables. By replacing the partial derivation by an operator, which has to be defined adequately, one generalizes to any structure of containers, the reasoning leading to the definition of the energy-per-entity. The detailed discussion on the shape of this operator is out of the scope of this book and, for the moment, one may return to the initial assumption of continuity for being able to go on with differentials. However, the classical definition of a variation dx of a continuous variable x can be virtually enlarged by admitting that it may also represent the variation of a discrete variable. This is obviously not rigorous and mathematicians may judge this practice condemnable, but it considerably simplifies and generalizes the algebraic language used. Regarding Formal Graphs, nothing changes, they keep the same structure and they adapt to the continuous case or to the discrete one merely by changing the operator linking two nodes. This versatility will be used again, for instance, for modeling space–time properties, which does not need to be continuous in the Formal Graph theory.

2.1.5.1 Differential Graph

The previous Formal Graph is not especially interesting compared to algebraic language, as it brings nothing more, while being more cumbersome and more complex to draw. This relative inferiority can be changed by representing a slightly more complex equation, such as Equation 2.8 expressing algebraically the First Principle of Thermodynamics. The Formal Graph expressing this equation given in Graph 2.2 is built on the same scheme as the previous one and by complementing with the partial derivatives with respect to the other entity number.

The fourth node in Graph 2.2 represents the second derivative of the total energy, with respect to both entity numbers. The fact that the two partial derivatives of the energies-per-entity are equal implies that Equation 2.8 is an exact differential equation. It is indeed the condition to fulfill for satisfying this property.

$$\left(\frac{\partial y_a}{\partial x_b}\right)_a = \left(\frac{\partial y_b}{\partial x_a}\right)_b = \mathcal{E}'' = \frac{\partial^2 \mathcal{E}}{\partial x_a \partial x_b} \tag{2.12}$$

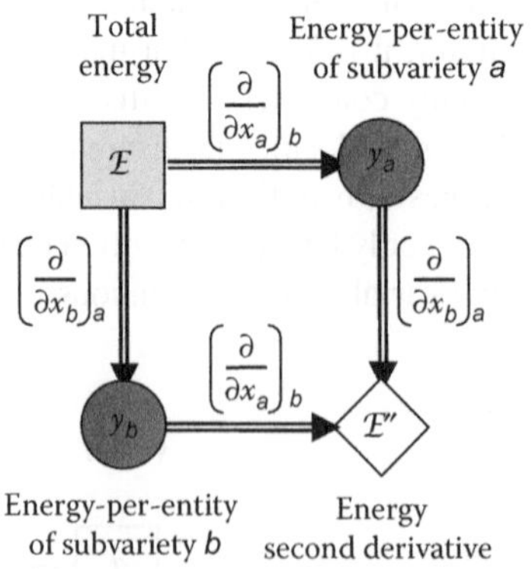

GRAPH 2.2 Differential Formal Graph of the First Principle of Thermodynamics. The graph closure means that the total energy is a state function, because the same result is obtained independently from the order of variations of the entity numbers. Nodes are drawn with different shapes for distinguishing among categories of variables.

Graph closure. In other words, the second derivative is the same whatever the order of partial derivation. This is precisely the mathematical translation of the property of a state function to be independent from the way that has been undertaken for reaching a given system state, as was explained earlier. The Formal Graph above says exactly the same thing in making the two paths coming from the energies-per-entity converge on the same node, which bears the second derivative of the energy. In graph language, we speak about the *closure* of the graph. If the total energy were not a state function, its variation would not be an exact differential equation, and the two paths would arrive on distinct nodes, leaving the graph open. This is an interesting advantage of graphs over a set of equations to provide this criterion of differential path closure for signifying the status of state function of a variable.

Graph types. Note that the two previous Formal Graphs are representatives of a limited class of graphs that are called Differential Formal Graphs. Their characteristic is to relate any type of variables but with differential operators only. In the next chapter, a bigger class is represented by graphs that relate only state variables but with any kind of operators, called Canonical Formal Graphs.

2.1.6 Energies-per-Entity

Once the energy container is identified and quantified by a number of entities, the definition of the corresponding energy-per-entity, also called its *conjugate state variable*, follows Equation 2.10 or 2.11. Table 2.2 provides a list of energies-per-entities corresponding to the containers listed in Table 2.1, with the most used names and symbols.

Contrary to Table 2.1, the list in Table 2.2 has no unknown variables as all virtual subvarieties are endowed with energies-per-entity that are duly recognized by experimentalists. They are known as time-related variables under various names such as "fluxes," "rates," or "flows."

2.1.7 The Canonical Scheme for State Variables

Tables 2.1 and 2.2, giving a list of entity numbers and energies-per-entity, divided each energy variety into two subvarieties, one inductive (kinetic, electromagnetic) and the other capacitive (potential, internal). This means that among entity numbers and energies-per-entity one can distinguish the variables that belong to each subvariety. This distinction is made by defining *four families* of state variables that are individually named and for which a generic symbol is proposed, as shown in Table 2.3.

The classification in Table 2.3 is not recent as it was proposed by Hermann Helmholtz in the mid-nineteenth century under the name of "canonical scheme" (de Broglie 1948). One may find rather odd, after so long a time, that this scheme is not well known among physicists, except among physical chemists who base upon this classification a peculiar modeling tool called *Bond Graphs* (Paynter 1961) or *Thermodynamical Networks* (Oster et al. 1971, 1973).

Translation into a Formal Graph of these four families of state variables is based on the attribution of one node to each variable, in individualizing each node by a specific polygonal shape and a filling color when applicable. For facilitating the correspondence with algebra, the algebraic symbols are placed inside the nodes, and, for pedagogical reasons, the full name is explicitly indicated with a text in the vicinity. These are optional indications, because with the condition that the considered energy variety is given, the sole shape is sufficient for identifying nodes and variables. Graph 2.3 features these conventions with a spatial disposition of nodes that will be used constantly, which also contributes to an easy identification.

TABLE 2.2
Nonexhaustive List of Energy Varieties, Subvarieties, and Energies-per-Entity Corresponding to the Number of Entities given in Table 2.1

Variety	Energy Subvariety	Energy-per-Entity	Comments
Rotational mechanics	Rotation kinetic	$\boldsymbol{\Omega}$ Angular velocity	
	Torsion, rotation work	$\boldsymbol{\tau}$ Torque (or couple)	
Translational mechanics	Kinetic	$\boldsymbol{v}$ Translational velocity	
	Work, internal, potential, elastic	$\boldsymbol{F}$ Force	
Superficial energy	Surface kinetic	$\boldsymbol{f_A}$ Surface expansion velocity	
	Surface static	γ Superficial tension	
Hydrodynamics	Volume kinetic	$\mathcal{Q}$, d Volume flow	Also called "flow rate"
	Hydrostatic	P Pressure	
Gravitation	Gravity kinetic	f_M Mass flow	Also called "mass flow rate" or "mass flux"
	Gravity potential	V_G Gravitational potential	
Electrodynamics	Electromagnetic	I (or i) Electric current	
	Electrostatic	V (or U, $\boldsymbol{\varphi}$) Electric potential (tension)	
Magnetic energy	Magnetism	f_m Magnetic current	Not to be confused with the electric current
	—	e_m Magnetic potential	
Electric polarization energy	—	f_P Polarization current	
	Polarization	e_P Polarization potential	
Corpuscles energy	—	$\mathfrak{I}_N$ Corpuscle flow	
	Corpuscular	E Energy-per-corpuscle	Corpuscles are molecules, atoms, particles
Physical chemical energy	—	$\mathfrak{I}$ Substance flow	Also called "mass flux"
	Physical chemical	μ Chemical potential	
Chemical reaction energy	—	v Reaction rate	
	Chemical reactivity	$\mathcal{A}$ Affinity	Equivalent to the "molar free enthalpy of reaction" $\Delta_r G$
Thermal energy	—	f_S Entropy flow	
	Heat	T Temperature	

Note: The inductive subvariety is given first, the capacitive one second. This list comprises only subvarieties whose container is made of several entities. They are listed under the variables that quantify their value, with the most used algebraic symbols. Bold letters are for vectors.

TABLE 2.3
Four Families of State Variables According to the Canonical Scheme of Helmholtz

Subvariety	Entity Number	Energy-per-Entity
$\mathcal{U}_q$ Capacitive	q Basic quantity	e_q Effort
$\mathcal{T}_q$ Inductive	p_q Impulse	f_q Flow

Time is not requested. It must be recalled that the existence of energy is independent of time, as proven by all static phenomena involving energy that science recognizes. However, many state variables involved in these static phenomena are classically defined as time derivatives of a number of entities. This is the case, among others, for the electric current I, defined as the derivative of the charge with respect to time, or the translational velocity $\boldsymbol{v}$, defined as the derivative of a distance with respect to time. The Formal Graphs approach solves this contradiction by proposing the notion of energy container and in postponing the definition of time by linking it to energy conversions, as explained before.

Notwithstanding this rigorous approach, it remains a serious difficulty for the human mind to accept the idea that a velocity may exist independently from any notion of time. This is rather counterintuitive and constitutes a high level of abstraction. It is easier to imagine a static current that flows indefinitely within a superconductor loop, although it contradicts the common definition. It is an experimental fact that no charges are created or destroyed within the conductor, supposed isolated from any external source of charges, in contradistinction to what should happen if the current were estimated by integration of passing charges in the circuit. This point is detailed in one of the case studies in this book (see case study A3 "Current Loop" in Chapter 4).

The difficulty is greater with the notion of velocity in a mechanical translation. One is so accustomed to think of a movement as associated with a distance or position that evolves concomitantly, that one cannot easily break the direct link between kinetic energy and space and time (see case study C1 "Newton's Second Law" in Chapter 6). However, this is the price to pay for a greater coherence of science and for access to a better understanding of several questions. For example, the famous duality between a particle and a wave, or the problem of localization of a particle in quantum physics, is no longer a problem when one is able to distinguish an energy container (a corpuscle for instance) from any notion of localization in space or along time. For this reason, the notions of *particle* and *corpuscle* are carefully distinguished, this latter being not necessarily endowed with a potential energy container that is a position or a distance.

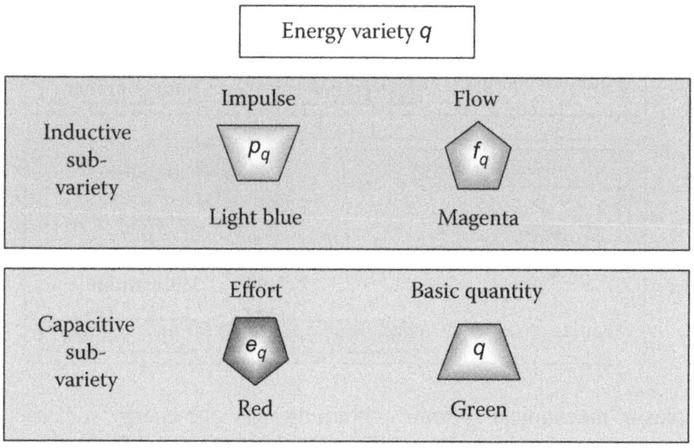

GRAPH 2.3 Convention for representing the four nodes of a Canonical Formal Graph with a specific shape (and color if applicable) for each state variable.

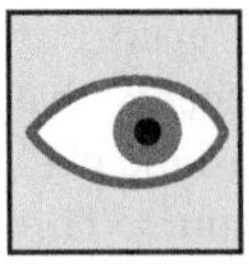

Shapes. Note that four-edge polygons (trapezoids) are chosen for entity numbers, and five-edge polygons (pentagons) for energies-per-entity, in choosing a symmetrical orientation for each subvariety. However, when drawing by hand a Formal Graph, the distinctive shape can be replaced by a circle and the algebraic symbol used for identification purpose.

2.1.7.1 Energy Variations

Adaptation of the energy variation principle for a collection of entities given earlier with Equation 2.6 and these state variables gives for the energy variety identified with subscript q corresponding to the basic quantity:

Capacitive subvariety:

$$\delta \mathcal{U}_q = e_q \, \mathrm{d}q \tag{2.13}$$

Inductive subvariety:

$$\delta \mathcal{T}_q = f_q \, \mathrm{d}p_q \tag{2.14}$$

A condition needs to be imposed for correctly using these equations: For a pair of conjugate state variables (entity number and energy-per-entity in a given energy subvariety) to have finite values does not signify that the corresponding amount of energy, which can be in principle calculated from these equations, has a physical meaning. The condition is that the system must possess the constitutive property allowing the storage of this energy in it.

2.1.7.2 Examples

In Figure 2.2 are given two examples taken from translational mechanics, an elastic solid (spring) elongated at a position ℓ of its end under the action of a force $\boldsymbol{F}$ and a moving solid (massive body) having a velocity $\boldsymbol{v}$ and a momentum (quantity of movement) $\boldsymbol{P}$.

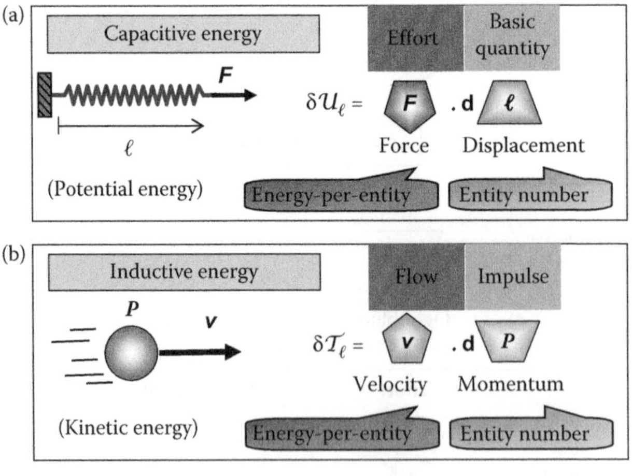

FIGURE 2.2 Examples of mechanical systems containing only one energy subvariety, both belonging to translational mechanics. (a) Potential energy in a spring submitted to a force $\boldsymbol{F}$. (b) Kinetic energy in a moving body with velocity $\boldsymbol{v}$.

Linear case. A mistake is often made in writing the variation of kinetic energy as the product of the momentum by the variation of velocity, in the form of the following unsuitable formula:

$$\delta \mathcal{T}_{\ell} \underset{lin}{=} \boldsymbol{P} \cdot \mathrm{d}v \qquad (2.15)$$

Such a formulation is not mathematically wrong when momentum and velocity are strictly proportional (hence the "*lin*" indication, for *linear*, that is intentionally written under the "equal" sign), as in Newton's theory, but it is false when the inertial mass is not a constant, which is a more general case. Moreover, it induces serious misinterpretation on the physical role of variables.

2.2 IN SHORT

CONTAINERS

1. Energy.
2. Systems: Pieces of universe containing more or less energy.
3. Space–time: Contains systems; allows distribution and conversion of energy.
4. Cosmos: Contains space–time, systems, and energy.

ENERGY PROPERTIES

1. Energy is a *state function*. System state and energy amount are strongly correlated.
2. Energy exists under several *varieties*.
3. Each energy variety is divided into two *subvarieties—capacitive* and *inductive*.
4. Energy is *extensive*. (Energy amount is proportional to system extent.)
5. Energy is *additive*. (Total energy is the sum of energy amounts of various *varieties*.)
6. Energy of one *subvariety* cannot vary alone. At least two *subvarieties* are required.
7. Energy containers are quantified by a *number of entities* for each *subvariety*.
8. Energy contained in an *entity* of a *subvariety* is quantified by an *energy-per-entity*.
9. The variation of energy in a subvariety is given by the product of *energy-per-entity* and variation of *entity number*, provided that the right constitutive property exists.

STATE VARIABLES

1. *Entity numbers* in the *capacitive subvariety* are *basic quantities*.
2. *Entity numbers* in the *inductive subvariety* are *impulses*.
3. *Energies-per-entity* in the *capacitive subvariety* are *efforts*.
4. *Energies-per-entity* in the *inductive subvariety* are *flows*.

FORMAL GRAPH REPRESENTATION OF STATE VARIABLES

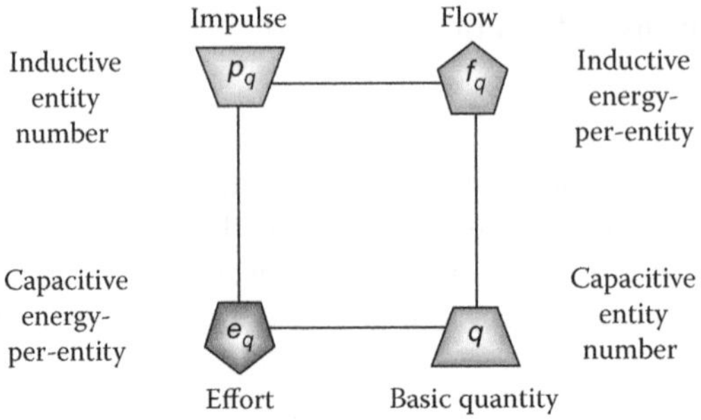

GRAPH 2.4 The conventions for drawing a Formal Graph are the followings: The state variables are placed on the summits of a square. Inductive variables are placed on top and capacitive ones at bottom. Energy-per-entities are in opposed summits.

FIRST PRINCIPLE OF THERMODYNAMICS (DIFFERENTIAL)

If x_a and x_b are the *numbers of entity* of two *collections* in different energy subvarieties, and y_a and y_b are the respective *energies-per-entity*, according to the *First Principle of Thermodynamics*, the total energy variation (in including other eventual varieties) is:

$$d\mathcal{E} = y_a\, dx_a + y_b\, dx_b + \cdots$$

From this differential equation (see Equation 2.8), the definition of an *energy-per-entity* follows (see Equation 2.10), as the partial derivative of the total energy versus the *entity number* of the considered energy subvariety, in maintaining constant any other entity number:

$$y_a \underset{def}{=} \left(\frac{\partial \mathcal{E}}{\partial x_a}\right)_{x_b,\, b \neq a}$$

3 Links and Organization

CONTENTS

A graph is made with two ingredients, nodes and links. Nodes have been introduced in the previous chapter on the foundations of thermodynamics. Knowledge of the values of the variables represented by these nodes allows one to determine the state of the system modeled by the Formal Graph. However, these variables are not independent from each other. This dependence is ensured by the links within the Formal Graph.

Links, contrary to many graphs, are not identical in a Formal Graph. Each of them expresses a property of the system and its context, in distinguishing constitutive properties, featuring the matter of the system (in a broad sense, including vacuum), from space–time properties or from thermodynamical relationships ruling the behavior of energy.

This behavior is not uniform among systems. It depends on the complexity of systems that needs to be carefully analyzed. This can be done by dividing and sorting the complexity of systems in organization levels. This structure will be used as a template for the book, starting from simple organization levels and reaching more elaborated levels.

3.1 SYSTEM CONSTITUTIVE PROPERTIES

A system may contain energy when it possesses the ability to store it, as already stated. This ability takes the form of two constitutive properties, one for each subvariety of energy: *inductance* for the inductive subvariety, and *capacitance* for the capacitive one. These names are borrowed from electrodynamics and generalized to all energy varieties. So, there is a translational mechanical inductance (inertial mass), a rotational mechanical inductance (inertia), a hydrodynamical capacitance (compressibility integrated over the volume), a thermal capacitance (which depends on the specific heat), and so on. In electrodynamics, inductance and capacitance feature components called inductor (or self-inductance) and capacitor, respectively. Electric inductance relates current to the quantity of induction (induction flux) and electric capacitance relates potential to charge.

3.1.1 The Various Properties

3.1.1.1 Capacitance Examples

Here are some examples of generalized capacitances.

Electrodynamics:

$$Q = CV \tag{3.1}$$

where Q represents charge, C capacitance, and V potential difference.

Rotational mechanics:

$$\boldsymbol{\theta} = \frac{1}{C_\theta}\boldsymbol{\tau} \tag{3.2}$$

where $\boldsymbol{\theta}$ represents angle, C_θ torsion constant, and $\boldsymbol{\tau}$ torque.

Translational mechanics:

$$\boldsymbol{\ell} = \frac{1}{k_e}\boldsymbol{F} \tag{3.3}$$

where $\boldsymbol{\ell}$ represents displacement vector, k_e stiffness, and $\boldsymbol{F}$ force. Note that the scalar nature of the stiffness (or spring constant) k_e corresponds to Hooke's* law, which is the limiting linear behavior of more complex laws that may involve matrices or tensors in the general case. The same is for the torsion constant above.

Physicalchemistry:

$$n = n^0 \exp\left(\frac{\mu}{RT}\right) \tag{3.4}$$

where n represents substance amount, n^0 reference substance amount, $\boldsymbol{\mu}$ chemical potential difference, R gas constant, and T temperature.

3.1.1.2 Inductance Examples

Here are some examples of generalized inductances.

* Robert Hooke (1635–1703): British physicist, astronomer, and naturalist; UK.

Electrodynamics:

$$\Phi_B = LI \tag{3.5}$$

where Φ_B represents quantity of induction (induction flux), L self-inductance, and I current.

Rotational mechanics:

$$\boldsymbol{L} = J\boldsymbol{\Omega} \tag{3.6}$$

where $\boldsymbol{L}$ represents angular or kinetic momentum, J rotational inertia, and $\boldsymbol{\Omega}$ angular velocity.

Translational mechanics (Newtonian):

$$\boldsymbol{P} = M\boldsymbol{v} \tag{3.7}$$

where $\boldsymbol{P}$ represents momentum, M inertial mass, and $\boldsymbol{v}$ translational velocity.

> **Symbol case.** Note that uppercase $\boldsymbol{P}$ and M stand for the momentum and the mass of a collection of mechanical entities (body), while lowercase $\boldsymbol{p}$ and m will be used later for the momentum and the mass of one corpuscle (when it is a particle).

Translational mechanics (general):

$$\boldsymbol{P} = \hat{M}\boldsymbol{v} \tag{3.8}$$

where $\boldsymbol{P}$ represents momentum, $\hat{M}$ inertial mass operator, and $\boldsymbol{v}$ translational velocity. According to the theory of special relativity, the inertial mass is a function of the velocity when this latter approaches the maximum velocity c (light velocity). Without entering into details (see case study A1 "Moving Body" in Chapter 4), this function is represented by an operator $\hat{M}$ that becomes equivalent to the scalar mass M for low velocities.

3.1.1.3 Dissipation

A third electric component is the resistor (or resistance), featured by a property called resistance, which relates current to potential, or, by its inverse, conductance. This component, contrary to the inductor and the capacitor, does not store energy but dissipates it into heat. The electric energy that enters a resistor is not lost, but the variety is transformed into the thermal energy variety. Apart from electrodynamics, equivalent dissipation processes are found in all other varieties, under the form of friction in mechanics, chemical reaction, fluid viscosity in hydrodynamics, etc. Conduction is a physical process that transfers a quantity (energy, charges, entropy/heat, molecules, momentum, etc.) through space and that is always a dissipation process. Due to the importance of the physical process, it is preferable to speak about conductance, rather than resistance, for describing the system property allowing dissipation.

3.1.1.4 Conductance Examples

Here are some examples of conductance.

Electrodynamics:

$$I = GV = \frac{1}{R}V \tag{3.9}$$

where I represents current, G conductance, and V potential difference.

Rotational mechanics:

$$\boldsymbol{\Omega} = \frac{1}{k_\theta}\,\boldsymbol{\tau} \tag{3.10}$$

where $\boldsymbol{\Omega}$ represents angular velocity, k_θ rotational friction coefficient, and $\boldsymbol{\tau}$ torque.

Translational mechanics:

$$\boldsymbol{v} = \hat{\mathrm{k}}_f^{-1}\boldsymbol{F} \tag{3.11}$$

where $\boldsymbol{v}$ represents translational velocity, $\hat{\mathrm{k}}_f$ friction coefficient, and $\boldsymbol{F}$ force. The friction coefficient is generally not independent of the velocity; in many media (gases, Newtonian liquids) and in moderate velocity conditions, it is proportional to the velocity, making the force proportional to $\boldsymbol{v}^2$. Apart from these conditions, the friction coefficient is a nonlinear function of the velocity. The general case is represented by an operator $\hat{\mathrm{k}}_f$ which may have different forms.

Physicalchemistry:

$$\Im = \Im^0 \exp\left(\frac{\mu}{RT}\right) \tag{3.12}$$

where $\Im$ represents substance flow (or reaction rate), $\Im^0$ reference substance flow, μ chemical potential difference, R gas constant, and T temperature. Note that this relationship is known as the *Arrhenius* law* and is generally written with a *molar energy of activation* notated E_a in lieu of the chemical potential μ.

Conductance, under a generalized form, is the third constitutive property attributed to a system. There are no other constitutive properties besides inductance, capacitance, and conductance.

3.1.2 Properties as Mathematical Operators

The examples in this section show various forms of mathematical expressions representing system properties, from simple scalars to exponential functions. For taking into account these discrepancies, the use of the operator symbolism is a convenient means of representing various mathematical forms without having to specify them. By using the state variable generalized symbols, the definitions of the three constitutive properties are as follows:

Capacitance:

$$q \overset{def\rightarrow}{=} \hat{\mathrm{C}}_q\, e_q \tag{3.13}$$

Inductance:

$$p_q \overset{def\rightarrow}{=} \hat{\mathrm{L}}_q\, f_q \tag{3.14}$$

Conductance:

$$f_q \overset{def\rightarrow}{=} \hat{\mathrm{G}}_q\, e_q \tag{3.15}$$

* Svante August Arrhenius (1859–1927): Swedish chemist: Stockholm, Sweden.

The symbol $\overset{def\rightarrow}{=}$ is used to indicate that the definition does not concern the left-hand term as usual but the right-hand one.

Differential/integral. Because of the widespread use of alternating current techniques for measuring impedances (admittance and impedance being the generic notions in electrodynamics encompassing these constitutive properties), it is not rare to find as definition of impedance the differential property instead of the integral one, without mentioning the existence of this differential/integral duality. It is important not to follow this reduced point of view and to keep in mind the whole generality of the concept.

3.1.2.1 Constitutive Properties

In Table 3.1 are summarized the three constitutive properties of a system with their specificities.

3.1.3 Formal Graph Representation

Once the graphic convention and the topological disposition for representing state variables are chosen, as explained in Chapter 2, drawing the system constitutive properties for obtaining a Formal Graph is obvious. Graph 3.1 represents a virtual system in which all three possibilities exist for energy to be stored—under the inductive and capacitive forms and to be dissipated. This is not a general case, as only one or two of these possibilities may be effective in a system.

3.1.4 Link Reversibility

An essential feature of the links in a Formal Graph is that they are always reversible, that is, working in two ways, implying that operators linking two state variables can be inverted. Reciprocal system properties not always being in use or even known with genuine names in each energy variety, the naming and the algebraic symbols for the reciprocals of the generalized properties are proposed as follows:

TABLE 3.1
Three Constitutive Properties of a System and Their Role with Respect to Energy

Constitutive Property	Action on Energy	Operand Variable	Resulting Variable
Inductance $\hat{L}_q$	Storage of inductive energy $\mathcal{T}_q$	Flow f_q	Impulse f_q
Capacitance $\hat{C}_q$	Storage of capacitive energy $\mathcal{U}_q$	Effort e_q	Basic quantity q
Conductance $\hat{G}_q$	Dissipation of energy (transformation into heat)	Effort e_q	Flow f_q

Note: The consititutive properties are mathematically represented by operators that apply to a state variable (operand) for producing another state variable.

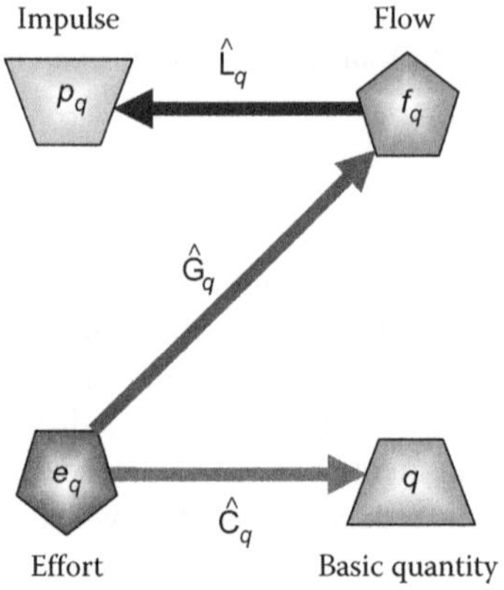

GRAPH 3.1 Generic Formal Graph with the three constitutive properties of a system, in a given energy variety. Only one or two of these properties for an energy variety may exist in a system. This Formal Graph is a canonical one, since only state variables are used. The drawing of oriented links in only one direction must not be taken as impossibility to use the other direction.

Generalized reluctance:

$$\hat{\mathfrak{R}}_q = \hat{L}_q^{-1} \quad (\text{or } \hat{\text{Я}}_q) \quad (\textit{reciprocal of the inductance}) \tag{3.16}$$

Generalized elastance:

$$\hat{€}_q = \hat{C}_q^{-1} \quad (\text{or } \hat{E}_q) \quad (\textit{reciprocal of the capacitance}) \tag{3.17}$$

Nonlinearity is the rule. The discrepancy in forms of constitutive property observed through given examples is a reality seen in different scientific areas that asks for discussion about this lack of homogeneity. The Formal Graph theory proposes a general frame in which most of these discrepancies are explained and even predicted. This will be discussed later in Chapters 4, 6, and 7 once the necessary theoretical elements have been gathered. For the moment, this proposal can be presented as a postulate stating that all constitutive properties of any system are nonlinear functions represented by operators. The linear functions observed in certain cases are asymptotic behaviors resulting from restriction of the validity range for the state variables.

This postulate is based on the observation that the same Formal Graphs are drawn with the same nodes and links in every energy variety and on the necessary coherence between operators belonging to several energy varieties when they contribute to the total energy of a system. This topic will be illustrated for instance with a population of ions, which involves the coupling of electrodynamics with physicalchemistry.

In many domains, system properties obey nonlinear laws and their characterization is often made by choosing a working point around which a small amplitude variation of one state variable is made, generally with a harmonic shape (sinusoidal). This gives access to the *differential property* which is the slope of the tangent at this point to the trace of the operator, called *integral property* curve. The differences between these notions are illustrated in Figure 3.1 where an integral capacitance curve is plotted (with an arbitrary shape) together with its asymptotic linear capacitance and with an example of differential capacitance.

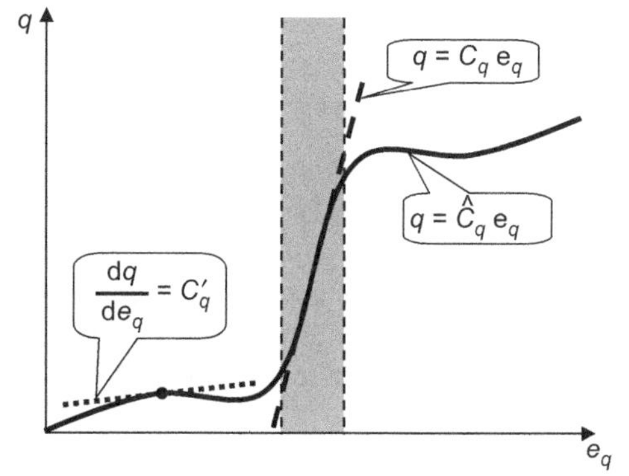

FIGURE 3.1 Plot of the basic quantity q variation as a function of the effort e_q showing three different behaviors: trace of the capacitance operator (solid curve) over the whole range of effort (i.e., integral capacitance), linear capacitance (dashed line) defined in a limited range (gray zone), and differential capacitance (dotted line) defined at a working point (black dot). The shape of the integral capacitance is arbitrary.

Generalized resistance:

$$\hat{\mathrm{R}}_q = \hat{\mathrm{G}}_q^{-1} \quad (\textit{reciprocal of the conductance}) \tag{3.18}$$

(Monetary or cyrillic symbols for two of these reciprocal properties, seldom used, are only proposals.) The existence of reciprocals expresses the symmetry of the relationship between the two state variables. There is no cause or effect notion behind these system properties. Through the direct operator, as defined in Equations 3.13 and 3.14, the system property adapts the *entity number* to the amount of *energy-per-entity* in order to store the adequate amount of energy in the subvariety. Through the reciprocal operator, it is the *energy-per-entity*, which is adapted to the *entity number*, that signifies that entities can be created or destroyed within the subvariety or converted into another subvariety. The same reversibility principle works for the conductance/resistance, except that the two state variables involved are energies-per-entity.

3.1.5 Only Three System Properties

Graph 3.1 shows that all state variables are connected and that three links are sufficient for that. In other words, no supplementary links are necessary; every connection between a pair of state variables can be expressed by composition of the elementary connections described by Graph 3.1. It is therefore not necessary to introduce new system properties, as for instance, the concept of "memristance" which has been suggested for linking a quantity of induction (induction flux) Φ_B to an electric charge Q (Chua 1971) and recently claimed to have been experimentally realized (Tour and He 2008).

Electric memristance:

$$\Phi_B = M_Q\, Q \tag{3.19}$$

In reality, this property should be an operator that is the composition of the other known ones:

Generalized memristance:

$$\hat{\mathrm{M}}_q = \hat{\mathrm{L}}_q\, \hat{\mathrm{G}}_q\, \hat{\mathrm{C}}_q^{-1} \tag{3.20}$$

recalling that the rightmost operator is the elastance, reciprocal of the capacitance. The fact that this property is not a fundamental one does not prevent one from using it for modeling in systems able to store and dissipate both subvarieties of energy, but care must be exercised about its dependence on other system properties.

3.1.6 Link between Energy Storage and System Property

For effectively storing energy, a system must possess a *constitutive property* that is specific to each subvariety. When a system is not endowed with such a property, no corresponding energy can be stored in it.

It must be stressed that such a tight link does not exist between energy storage and values of the associated state variables. The validity of the differential equations governing the variation of energy given in Chapter 2 must in fact be understood as depending on the existence of the system property allowing the storage.

Capacitive subvariety:

$$d\mathcal{U}_q \underset{\hat{C}_q \text{ exists}}{=} e_q \, dq \tag{3.21}$$

Inductive subvariety:

$$\delta\mathcal{T}_q \underset{\hat{L}_q \text{ exists}}{=} f_q \, dp_q \tag{3.22}$$

The point is that the absence of a system property in a system does not mean that the corresponding entities are absent, or even that the associated energies-per-entity are equal to zero. If the two state variables are not connected, the energy in this subvariety cannot vary in the system, and therefore remains constant. This principle is illustrated in Graph 3.2.

This means that it is not a conceptual difficulty to admit the existence of state variables for energy varieties that have never been evidenced, as we have done in Chapter 2, Tables 2.1 and 2.2 listing entity numbers and energies-per-entity respectively.

So, if some subvarieties have not been experimentally identified, this is either because of the nonexistence of systems possessing the right property or naturally because of the experimental difficulty to evidence it. This is owing to the fact that a constant energy has effect only to shift the origin of scale and only a variation of energy can be detected.

In brief, it is more difficult to theoretically explain the nonexistence of some subvarieties and the existence of others than to confer to the systems the responsibility of possessing or not the right property, as shown here.

3.1.7 Reduced Properties versus Global Properties

By definition, a state variable is a global variable, in the sense that it quantifies the totality of the considered system in the same way as energy is counted for a system taken as a whole. Consequently, the system constitutive properties linking these global variables described until now are also global. They

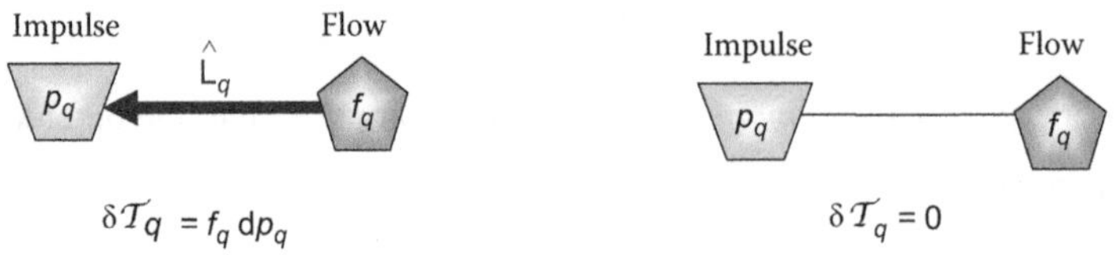

GRAPH 3.2 Two Formal Graphs illustrating the link between existence of a constitutive property in a system and ability to store energy in the considered subvariety. In both cases, the state variables do exist and may possess values.

are not specific from the nature (materials) of the system and they depend on the extent (size and shape) of the system. They are, however, important properties because any characterization of a system is made through its state variables and the system properties are deduced from their measure.

Many systems, if not all when one includes vacuum, are made with matter, which may assume a great number of shapes in space and may be composed of several materials. A global property takes all these features into account, which is not very convenient. When one wants to compare two materials one needs to have systems with identical extents (sizes and shapes), and when the influence of extent is investigated, one needs to have two identical materials. It is therefore interesting to separate these two aspects and this is one of the aims of modeling in science.

3.1.7.1 Three-Dimensional Space

This separation is made by decomposing the extent of a system into geometric elements: line or curve, surface and volume. In restricting this decomposition to three elements, it is assumed that the number of dimensions of space is three and not more. This is not an absolute limitation—one may imagine spaces with *n* dimensions as well—but it is a methodological choice. Before jumping directly into a higher mathematization of physics, it is essential to develop the theory in its simpler form, in order to eventually justify its extension to more complex geometries with solid arguments. Sometimes it is easier to invoke higher dimensions to draw Formal Graphs representing known equations, but the constraint of three dimensions reveals to be more fruitful than the absence of constraint.

3.1.7.2 Localized Variables

Once this spatial decomposition has been stated, modeling the role of space consists of defining local variables that correspond to global state variables. Each local variable results from application of a spatial operator that depends on the geometry of space. The simplest is the Euclidean* space, that is, ordinary space independent from time, also called flat space by opposition with curved space–time, which is the frame of the general relativity.

In addition, spatial operators depend on the sensitivity of space orientation of the operand variable, scalar or vector, and the homogeneity or not of the geometric element. In the simplest case of Euclidean space and homogeneity in this first presentation of the role of space, this operator is a simple division by a length (position along the curve, called curvilinear coordinate), or by an area or by a volume. In other cases, these operators assume generally the forms of gradient, curl (rotational), divergence, etc. All these spatial operators belong to the category of *space–time properties*. Their description will be treated in detail later (Chapter 5).

3.1.7.3 Localization Depths

Thus, three levels of localization are established for each state variable.

The first level is the line, or, more rigorously speaking, a curve. Variables at this level are named *lineic densities*, or fields in certain energy varieties (viz., electrodynamics, gravitation). In case of homogeneous distribution of the state variable in space, the lineic density is the state variable divided by a length.

The second level is the surface. Variables at this level are named *surface (or superficial) densities*. In case of homogeneous distribution of the state variable in space, the surface density is the state variable divided by an area.

The third level is the volume. Variables at this level are named *volume (or volumic) concentrations*. In case of homogeneous distribution of the state variable in space, the *volume concentration* is the state variable divided by a volume.†

---

* Euclid (*ca.* 300): Greek mathematician; Alexandria, Ancient Greece.

† A frequent usage, especially in the Anglo-Saxon world, is to name *density* all these localized variables. As with the Latin world we have at our disposal the alternate word for *concentration*, a helpful distinction can be made by using both terms, allowing one to merely speak about *density* for a surface without mentioning its superficial nature and about *concentration* without adjoining the volume as a qualifier. (The *lineic density* being of less frequent use, it is less annoying to keep it in two words.)

Scalar/vector. Note that when the global variable is a scalar, as in electrodynamics, the variables localized along a curve and on a surface are vectors (in bold symbols) whereas the variable localized in a volume is a scalar. This is the opposite when the global variable is a vector, as in mechanics.

Lost status. Note that localized variables lose the status of state variable (i.e., their sole knowledge is insufficient for determining the amount of energy in the system), and they become intensive variables (i.e., they are not proportional anymore to the system extent if the global variable was extensive).

Table 3.2 summarizes the conventions adopted for handling localized variables, in associating a new parameter, the *depth of localization*, to each level. This quantitative parameter will be used shortly after to determine whether a property is *specific* of the material or not.

3.1.7.4 Reduced Properties

System constitutive properties may also be adapted for linking local variables. However, they cannot be termed as localized properties because they may link variables at different depths of localization. They may relate, for example, a global variable to a volumic one or a lineic density to a surface density. The term *reduced property* is chosen because it better conveys the idea of decrease of the value quantifying a property, which is the case when it is divided by the amount of a geometric element (length, area, or volume).

3.1.7.5 Examples

A first example can be given with the electric conductivity, which is the reduced conductance. This is an operator notated $\hat{\sigma}$ linking the electric field (potential lineic density) $\boldsymbol{E}$ to the current density $\boldsymbol{j}$:

$$\boldsymbol{j} \overset{def\rightarrow}{=} \hat{\sigma} \boldsymbol{E} \tag{3.23}$$

Translation into Formal Graph of this relationship is made in Graph 3.3 together with the conductance defined by Equation 3.9, in assuming that space is isotropic and the matter of the system is homogeneous. More general cases will be treated later.

TABLE 3.2
Depths (or Levels) of Spatial Localization

Depth n_R	Localization	Variable	Generic Symbol	Generic Name
0	Whole system	Global quantity	u	State variable
1	Point of a curve	Global quantity/Length	$\boldsymbol{u}_{/r}$	Lineic density, field strength, gradient, etc.
2	Point of a surface	Global quantity/Area	$\boldsymbol{u}_{/A}$	Density
3	Point of a volume	Global quantity/Volume	$u_{/V}$	Concentration

Note: The generic notation used consists of adjoining a subscript $/r$ or $/A$ or $/V$ to the variable symbol for indicating on which geometric element the variable is localized.

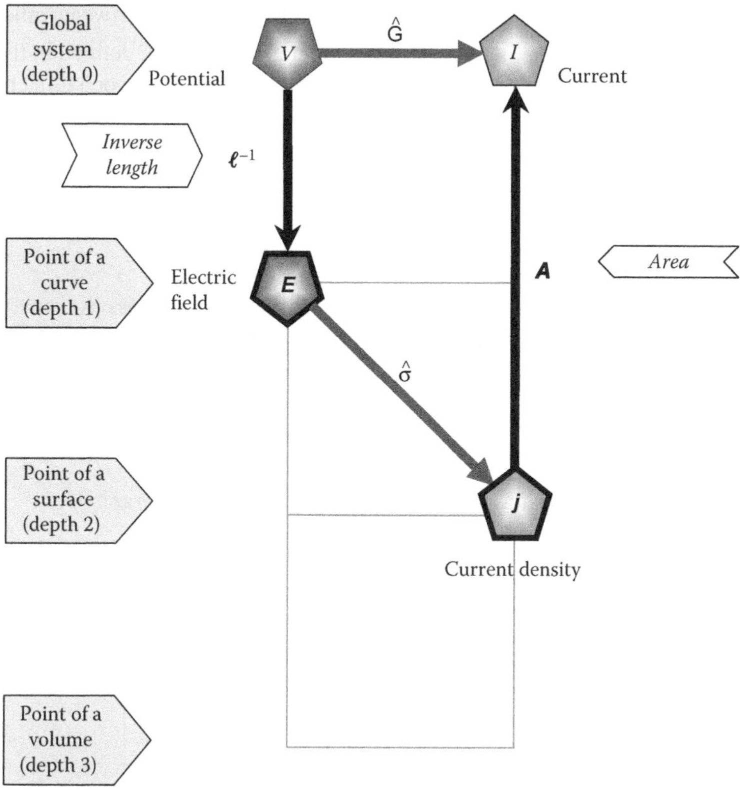

GRAPH 3.3 Formal Graph of the equivalence between the conductance path at the global level and the composed path passing through the conductivity. Properties of space are represented by a length and an area, because the considered system is made with an isotropic and homogeneous matter.

In case of homogeneous distribution of global variables in space, the relationships between localized and global variables are:

$$V \underset{hom}{=} \ell \cdot \boldsymbol{E} \tag{3.24}$$

$$I \underset{hom}{=} \boldsymbol{A} \cdot \boldsymbol{j} \tag{3.25}$$

The indication "*hom*" under the equal sign is a reminder that these relationships are only valid in the special case of a homogeneous distribution. By composing the alternate path starting from the potential node and arriving at the current node, and translating it back to algebra, one obtains the correspondence between the two properties:

$$\hat{G} \underset{hom}{=} \boldsymbol{A}\ \hat{\sigma}\ \ell^{-1} \tag{3.26}$$

In the still more special and limited case of linear conductivity and conductance, one has the well-known relationship

$$G \overset{lin}{\underset{hom}{=}} \sigma \frac{\boldsymbol{A}}{\ell} \tag{3.27}$$

A second example is worth discussing because it illustrates links between other localization depths: In translational mechanics, the inductance is the inertial mass $\hat{M}$, defined in the general case as an operator in Equation 3.8. In one-dimensional systems, such as a string or a thin chain, modeling is done with the lineic* mass $\hat{\mu}_M$ that links the velocity $\boldsymbol{v}$ to the lineic density of the momentum $P_{/r}$ (which is a scalar):

$$P_{/r} \overset{def\rightarrow}{=} \hat{\mu}_M v \quad (3.28)$$

In three-dimensional systems, such as ordinary solids, the volumic mass $\hat{\rho}_M$ is used, which links the velocity $\boldsymbol{v}$ to the concentration of the momentum $\boldsymbol{P}_{/V}$ (which is a vector):

$$\boldsymbol{P}_{/V} \overset{def\rightarrow}{=} \hat{\rho}_M \boldsymbol{v} \quad (3.29)$$

All these three properties are shown in the Formal Graph given in Graph 3.4.

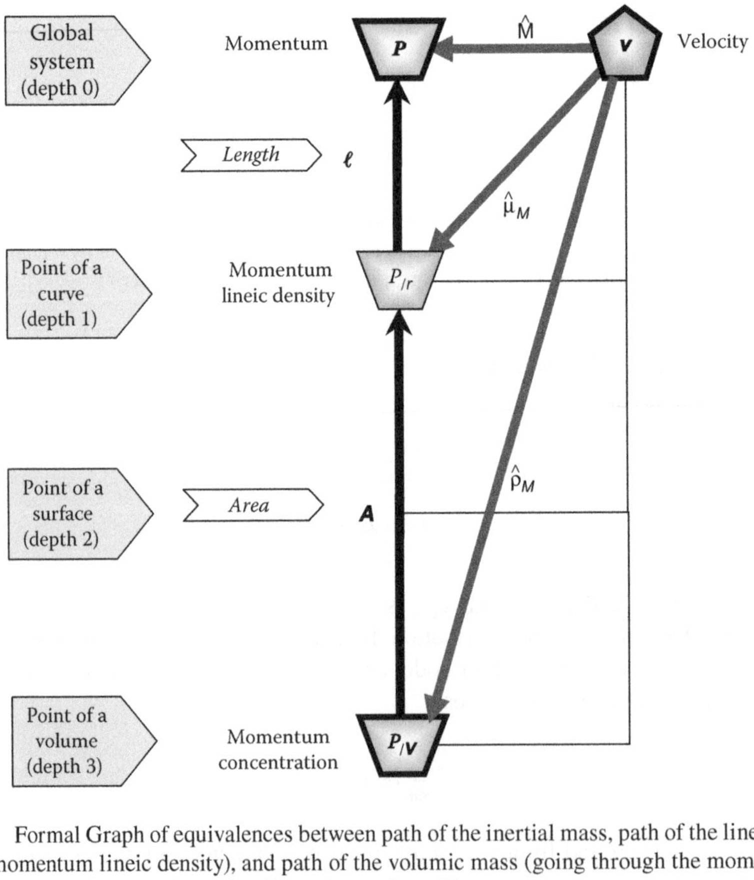

GRAPH 3.4 Formal Graph of equivalences between path of the inertial mass, path of the lineic mass (going through the momentum lineic density), and path of the volumic mass (going through the momentum concentration). The considered system is made with an isotropic and homogeneous matter.

* Often called "linear mass," but it is preferable to reserve the adjective "linear" to an operator or a behavior.

Adopting the same approach as for conductivity that consists in looking for equivalent paths in the graph, one finds the following equalities:

$$\hat{M} \underset{hom}{=} \ell \cdot \hat{\mu}_{M} \underset{hom}{=} \ell \cdot \boldsymbol{A}\, \hat{\rho}_M \tag{3.30}$$

The last equality can be written with volume:

$$\hat{M} \underset{hom}{=} V\, \hat{\rho}_M \tag{3.31}$$

Wrong definitions. It must be stressed that these equalities are not definitions (even in scalar form) of the lineic mass and of the volumic mass, as often said. They are valid only in homogeneous materials. The true definitions are given in Equations 3.28 and 3.29.

3.1.8 Specific Properties

A specific property is a system constitutive property that depends only on the nature of a material and not on its extent (size and shape). The interest to use specific properties in modeling a system is obvious because it allows the prediction of a response to a perturbation knowing the material that constitutes the system, and, conversely, it allows the identification of a material by characterizing some of its specific properties.

Besides this practical interest, another more fundamental reason for looking after specific properties is to put into evidence the existence of *invariant* properties or parameters of our world. Together with the notions of *symmetry* and *conservation*, invariance is one of the fundamental concepts that are the ingredients of our rational understanding of nature.

Specificity Criterion: A Formal Graph clearly evidences the various possibilities of finding equivalent paths of a global property in going down the graph and in passing through reduced properties that can be arbitrarily chosen. In fact, not every reduced property is specific. To know whether a reduced property is specific, there is a criterion based on the number of space dimensions d of the system and on the two depths n_{R1} and n_{R2} of the local variables related by the reduced property. The depth n_R of a local variable is an integer ranging from 0 (global variable) to 3 (variable defined at a point of a volume) quantifying the geometric elements of a system. If the sum of these two depths is equal to the number of space dimensions d of the system, the property is specific:

$$\text{Specificity} \quad \Leftrightarrow \quad n_{R1} + n_{R1} = d \tag{3.32}$$

Thus, the conductivity in Graph 3.3 is specific if the number of dimensions of the system is 3 because the depth of the first local variable, the electric field, is 1 and the depth of the second variable, the current density, is 2. This would not be a specific property if the system had only two dimensions of space (conduction on a surface). The lineic mass in Graph 3.4 is not specific in a three-dimensional system but is specific in a one-dimensional system. At last, the volumic mass in the same graph is specific only if the system has three dimensions. Figure 3.2 shows all the existing possibilities for a reduced property to be specific, for system dimensions ranging from 1 to 3.

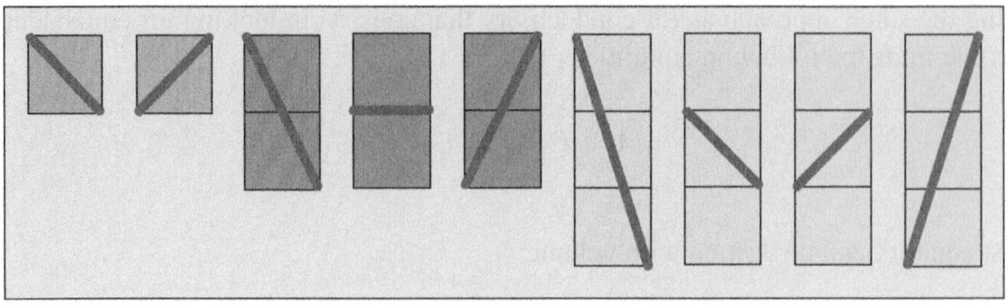

FIGURE 3.2 The nine possibilities for a system constitutive property to be specific, according to the number of space dimensions of the system.

The question remains as to how to choose the right one among specific properties, since there are always several possibilities. There is no simple answer to this question, because it depends on the physics of the phenomenon represented by the reduced property. The Formal Graph theory has no peculiar point of view on this question and a better understanding will emerge from the various case studies given.

Space dependence. The question of independence of a constitutive property from space–time is not as simple as it may appear with the above presentation, which nevertheless follows the classical conception of specificity. In fact, this definition works for simple systems, corresponding to low levels of organization (this topic is tackled hereafter). In higher organization levels, when a system has a variable degree of order—between a perfect crystal (maximum order, anisotropy) and an ideal liquid (minimum order, completely isotropic)—such as a liquid crystal, this degree of order may depend on the respective size of its elemental objects (molecules) compared to the total size of the system. The consequence is that many constitutive properties (viscosity, elastic constant, permittivity, magnetic susceptibility, etc.) can be influenced by this ratio.*

3.2 FORMAL OBJECTS AND ORGANIZATION LEVELS

What is behind a Formal Graph? Which physical system can be really represented? These questions absolutely need to be answered because drawing nodes and links with the only rules previously given can be a game deprived from physical meaning. What was lacking until now is the definition of the physical object represented by a Formal Graph.

Here again, the problem faced is the absence of a clear view on the difficult question of complexity in physics, which is generally tackled in terms of numbers of elements or size of a system. The Formal Graph approach obliges one to clarify the landscape, in distinguishing various degrees of complexity, which are sorted by organization levels. Each level is occupied by a *Formal Object*, which represents a category of systems having common features that are specific to the considered level.

Earlier in this chapter, two types of energy variations were carefully distinguished, according to the grouping or not of entities that constitutes the containers of energy. For a single entity one was

* This can be an explanation of the numerous discrepancies in experimental measures observed between laboratories, each one working with different sample sizes.

obliged to adapt the First Principle of Thermodynamics because the energy variation could not be associated with the variation of the number of entities, though this was possible for a collection of entities. This distinction is described by two Formal Objects called *singleton* and *pole*, the first one representing systems with a single entity and the second one systems with a collection of entities. These two Formal Objects are the very first levels of organization in the complexity scale, at least in the scope of this book.

A good organization scheme for handling increasing physical complexity relies on levels that are not disconnected but that are firmly based on the lower level. Thus, the next level is formed by grouping two poles for building up a *dipole* (or several poles forming a multipole), and a further level is formed by associating several dipoles, always in the same energy variety, for forming a *dipole assembly*, and so on.

3.2.1 The First Organization Levels

Several collections and subvarieties need to be combined for modeling more and more complex systems. Analysis of these elements leads to clearly identify different objects represented by entities and the energy subvarieties. Each Formal Object in a level results not only from combination of Formal Objects of the level just under but also from precise rules governing relationships between variables, allowing a rational progression toward more complexity.

3.2.1.1 Singletons

The first level of organization is represented by pairs of state variables (entity unit–energy-per-entity) belonging to one energy subvariety called *singleton*. By definition, the behavior is not collective, even when several singletons are present in a system. As examples of singletons, one may cite two capacitive singletons: the elementary corpuscle and the elementary vibration, and two inductive ones: the moving particle and the spin of a corpuscle. The difference between capacitive and inductive singletons lies in the number of entities which is by definition one unit in a capacitive singleton and is a variable number in an inductive singleton.

3.2.1.2 Poles

The second level is made up of collections of entities called *poles*. By definition, a pole is a collection of entities of the same energy subvariety, with the same energy-per-entity, which signifies that a collective behavior is required to ensure this condition. Another elemental feature is that poles do not use time at all. A pole belongs to only one energy subvariety and is able to store or to dissipate energy or both. This means that according to the number of system constitutive properties possessed, there exist three fundamental types of poles: capacitive, inductive, conductive, and two mixed types: capacitive + conductive and inductive + conductive. Examples of poles are: a set of electric charges (at the same potential), a moving body (ensemble of particles stuck together), a chemical species (several molecules), a current loop. Rigorously speaking, a pole is considered at the global level of the system (made up only with state variables), but, by extension of the concept for avoiding multiplication of organization levels and types, the case of a space distributed pole is included (with local variables and reduced properties). This will also be the case for the next organization levels.

3.2.1.3 Dipoles

The third level is the association of two poles of the same type to form a *dipole*. A dipole is therefore made of two collections, each one having its own energy-per-entity. Contrary to the pole, the dipole may take time to convert energy (this requires association with another dipole belonging to a different energy subvariety). The same types of dipoles as for the poles are found: three fundamental and two mixed types. Examples of dipoles are: an electric capacitor, two colliding bodies, electrons and holes in a semiconductor, a chemical reaction, etc.

3.2.1.4 Multipoles

The fourth level is similar to the previous one, except that it allows more than two poles to associate in order to form a chain (serial mounting) or a ladder (parallel mounting) or both. Examples of chains are chemical reactions composed of several steps and reactional intermediates, or conduction processes working by exchange or hopping between several sites. Examples of ladders are multi-loops of current (solenoids) in electromagnetism, or magnetic domains in parallel.

3.2.1.5 Dipole Assemblies

The fifth level is the *assembly of dipoles*, made by association of two or three dipoles belonging to different types. Thus, we can have four dipole assemblies: capacitive + inductive, capacitive + conductive, inductive + conductive, and capacitive + inductive + conductive. Examples of dipole assemblies are: an electric LC oscillator, a mass attached to a spring, outflow from a fluid reservoir, a chemical reaction with multiple reactants, etc. Note that the association of two dipoles of the same type does not form an assembly, which requires an association of two different types of dipoles. Two capacitive dipoles, for instance, when energy exchange is possible between them, merely form another capacitive dipole, that combines the properties of both dipoles.

3.2.1.6 Dipole Distributions

The sixth level is represented by *distributions of dipoles* of the same type, but without energy exchange as evoked in the previous level. As will be seen later, a dipole able to store energy (capacitive or inductive) is featured by a *reference energy-per-entity*, also called *energy state* (or *energy level*). This variable is defined as the common value of the energies-per-entity when both numbers of entities are equal. This organization level corresponds to a distribution of these reference energies-per-entity that is featured by a *density of energy states*. This organization level is mainly encountered in solids. Examples are found in crystals, electronic conductors, conducting polymers, etc. As a conductive dipole has no entities, it cannot possess a *reference energy-per-entity*, and consequently cannot participate as such in a distribution, but mixed-type dipoles can.

3.2.2 Multiple Energy Varieties

All these organization levels are defined in the frame of only one energy variety. For taking into account the coexistence of several energy varieties in a system, the concept of *energy coupling* is used. A certain amount of energy, if not all, is shared between coupled varieties to ensure the stability of the system. This allows a minimum amount of energy, or a maximum amount in case of metastability, to be reached by the system. Energy coupling may occur at all organization levels, whatever the energy varieties involved.

3.2.3 Subvarieties and Organization

Lower levels of organization exist, for taking into account the world of particle physics (subparticles, quarks, etc.). Higher levels naturally exist too; they are sufficiently easy to imagine in the functioning of the complexity of physical systems and do not need detailed descriptions. The principle for defining a new level is always the same: an upper level is made by association of elements from a lower level (which may be lower than the level immediately below, as in the case of the dipole distribution).

Table 3.3 gives the organization scheme level by level and for the possible subvarieties involved.

The main observation that can be made about the various Formal Graphs listed in Table 3.3 is that, above the singleton level, all graphs have the same shape, keeping the same square disposition whatever the organization level. This is indeed a remarkable feature of Formal Graphs. Obviously, the variables in the nodes and the system properties in the links differ from one level to a higher one; they are functions of lower-level variables and properties. This conservation of the elementary square shape of a Formal Graph is called the *embedding principle*. It will be justified by explaining the details of the construction of a Formal Graph from lower level ones.

TABLE 3.3
Organization Scheme for One Energy Variety

Organization Level	Capacitive (C)	Inductive (L)	Conductive (G)	Capacitive + Conductive (CG)	Inductive + Conductive (LG)	Capacitive + Inductive + Conductive (CLG)
Dipole distribution						
Dipole assembly						
Multipole						
Dipole						
Pole						
Singleton						

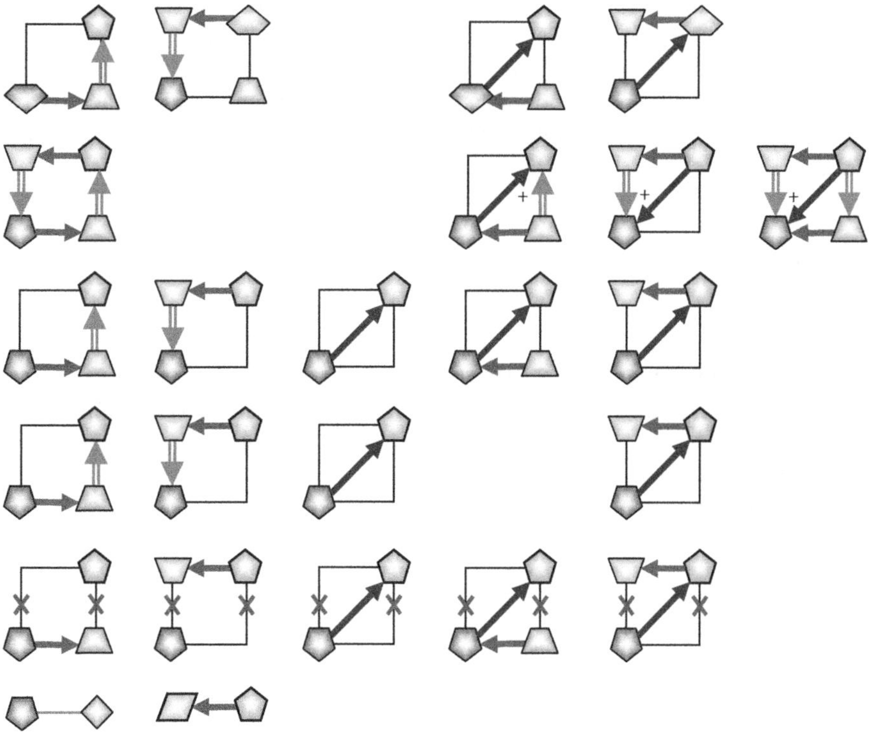

Note: Formal Graphs are sorted by organization level and by energy subvarieties (capacitive and inductive), including the dissipation with the conductive property. In this nonexhaustive list, the simplest organization level is the singleton (bottom) and the most complex is the distribution of dipoles (top).

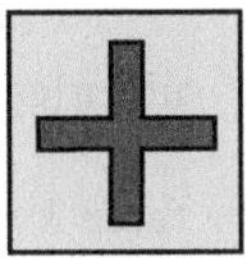

Microscopic–Macroscopic. These concepts of Formal Object and organization levels are new. There is obviously an implicit scale of complexity in classical physics, but levels are not explicitly formalized and generalized to all domains. If they are, they are thought in terms of size, as for the distinction microscopic–macroscopic that delimits the quantum world from the classical one. Here, this classical distinction is made differently between the singletons and the poles, that is to say in terms of individual or collective behavior, whatever the physical size. A large object may behave in a quantum way, provided that its behavior is not collective! (Which is beyond the behavior of a pair or a few objects.)

New names. For new concepts, new names are needed. For the first level, the term *soliton* would have been adequate, but it is already in use for another concept (solitary wave) in physics. It was important not to induce bias by choosing a term already endowed with a physical meaning because of the importance to understand this concept as a totally new one, dissociated, in particular, from the notion of particle. The name *singleton* is borrowed from mathematics, where it means a set containing only one member. It has the advantage of being meaningful in combining the property of singularity with the ending used for many microscopic objects (electron, photon, etc.). The names *pole* and *dipole* are not new, and the preceding drawback of influencing the understanding is not dramatic. They are borrowed from electrostatics or magnetism, because the idea is the same, although with a noticeable difference, which is that a classical electrostatic pole is basically made with one charge, whereas here it is made of several charges (bound together).

3.3 IN SHORT

SYSTEM CONSTITUTIVE PROPERTIES

Property possessed by a system (they are *constitutive* to the system) allowing to store or to dissipate (convert into heat) energy. A property, constitutive or not, links two nodes in a Formal Graph belonging to the same energy subvariety. (Other links that are not constitutive of the system are featuring space–time, independently from the system.) A property is systematically represented by a mathematical operator, which can be reduced to a linear operator (scalar) in some cases.

- Inductance $\hat{L}_q$: Property enabling a system to store *inductive energy* (kinetic, electromagnetic, etc.). An inductance is an operator linking a *flow* to an *impulse*. Its reciprocal is the *reluctance*.
- Capacitance $\hat{C}_q$: Property enabling a system to store *capacitive energy* (internal, potential, elastic, etc.). A capacitance is an operator linking an *effort* to a *basic quantity*. Its reciprocal is the *elastance*.
- Conductance $\hat{G}_q$: Property enabling a system to dissipate energy (i.e., to convert into heat, whatever the *energy subvariety*). A conductance is an operator linking an *effort* to a *flow*. Its reciprocal is the *resistance*.

- Reluctance $\hat{L}_q^{-1}$, ($\hat{\mathfrak{R}}_q$ or $\hat{\mathcal{R}}_q$): Property enabling a system to store *inductive energy*. A reluctance is an operator linking a *flow* to an *impulse*. Its reciprocal is the *inductance*.
- Elastance $\hat{C}_q^{-1}$, ($\hat{E}_q$ or $\widehat{\text{€}}_q$): Property enabling a system to store *capacitive energy*. An elastance is an operator linking a *basic quantity* to an *effort*. Its reciprocal is the *capacitance*.
- Resistance $\hat{G}_q^{-1}$, $\hat{R}_q$: Property enabling a system to dissipate energy. A resistance is an operator linking a *flow* to an *effort*. Its reciprocal is the *resistance*.

FORMAL GRAPH REPRESENTATION OF SYSTEM CONSTITUTIVE PROPERTIES

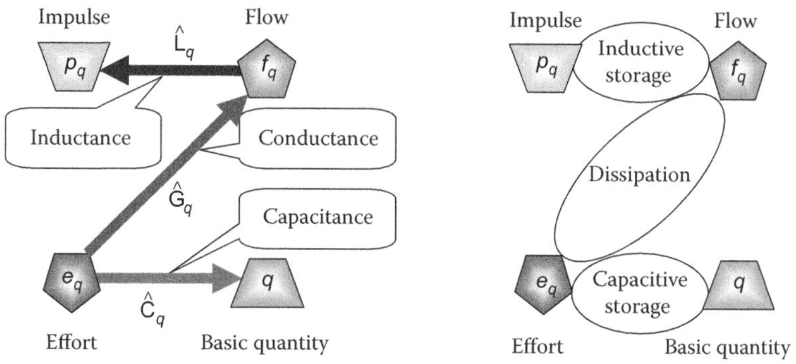

GRAPH 3.5

LOCALIZED VARIABLES

1. State variables are global variables (nonlocalized in a geometric element).
2. At a point of a curve: lineic density, field strength, gradient, etc.
3. At a point of a surface: density.
4. At a point of a volume: concentration.
5. Localized variables cannot be extensive variables.

REDUCED CONSTITUTIVE PROPERTIES

1. A *reduced constitutive property* relates two localized variables or one localized variable to a global variable.
2. To a same system constitutive property at the global level may correspond several *reduced constitutive properties*, depending on the system geometry.
3. A *reduced constitutive property* is specific to the material of the system when the sum of the localization depths of the related variables is equal to the number of spatial dimensions of the system.

ORGANIZATION LEVELS

1. Singleton: Featured by a pair of *energy-per-entity* and *entity number*, in one *energy subvariety*. No collective behavior in a group of *singletons*. The *entity number* is invariable in a capacitive singleton.
2. Pole: Collection of *entities* of the same *subvariety*, in variable *number*, with a common *energy-per-entity*, which requires a collective behavior.
3. Dipole: Association of two *poles* of the same *nature* (same *energy subvariety* and same *system constitutive properties*).
4. Multipole: Several *poles* of the same nature. They may form a chain (in serial), or ladder (in parallel), which is an extension of the notion of *dipole*.
5. Dipole assembly: Association of *dipoles* of the same *energy variety* but different *natures*.
6. Dipole distribution: Ensemble of *dipoles* of the same nature but having different *energies-per-entity* (i.e., distributed over "energy levels").
7. Higher levels may follow if required. (Several *energy varieties* in a system can be associated or *coupled* at all levels.) Lower levels as well (subparticles, quarks, etc.).

4 Poles

CONTENTS

This chapter examines the core of the Formal Graph theory by presenting physical and physical chemical systems that belong to the category of poles that are basic Formal Objects used for building more complex objects.

A pole is among the simplest Formal Objects; some are still simpler (singletons for instance) in the organization levels defined by the Formal Graph theory. It is by far the most important one,

Why begin with poles instead of singletons? A first reason is that many properties of singletons can be best presented once the notions developed in the upper levels in the complexity scale have been understood. A second reason is that singletons are relevant from the viewpoint of quantum physics, which is not reputed for the simplicity of its concepts as a certain number of them are still subject to discussion or varying interpretations. However, it is tempting to begin with the lowest organization level for establishing the properties of the world (limited to the physical one) in climbing the scale of complexity from scratch. This is more or less feasible with the help of the Formal Graphs that allow the emergence of new properties when stepping from one lower level to a higher one. Notwithstanding this appealing idea, to start with poles is a more pedagogical approach.

TABLE 4.1
List of Case Studies of Global Poles

	Name	Energy Variety	Subvariety	Type	Page Number
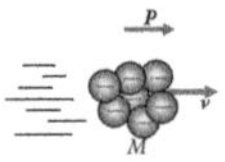	A1: Moving Body	Translational mechanics	Inductive	Fundamental	56
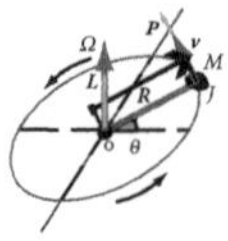	A2: Moment of Inertia	Rotational mechanics	Inductive	Fundamental	59
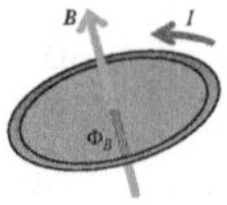	A3: Current Loop	Electrodynamics	Inductive	Fundamental	62
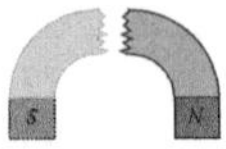	A4: Magnetization	Magnetism	Inductive	Fundamental	65
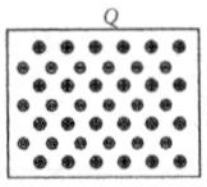	A5: Electric Charges	Electrodynamics	Capacitive	Fundamental	68
A	A6: Chemical Species	Physical chemistry	Capacitive	Fundamental	71
	A7: Group of Corpuscles	Corpuscular energy	Capacitive	Fundamental	75
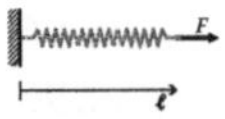	A8: Spring End	Translational mechanics	Capacitive	Fundamental	79
	A9: Thermal Pole	Thermics	Capacitive	Fundamental	82
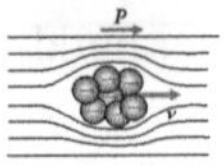	A10: Motion with Friction	Translational mechanics	Inductive	Mixed	85
A →	A11: Reactive Chemical Species	Physical chemistry	Capacitive	Mixed	87

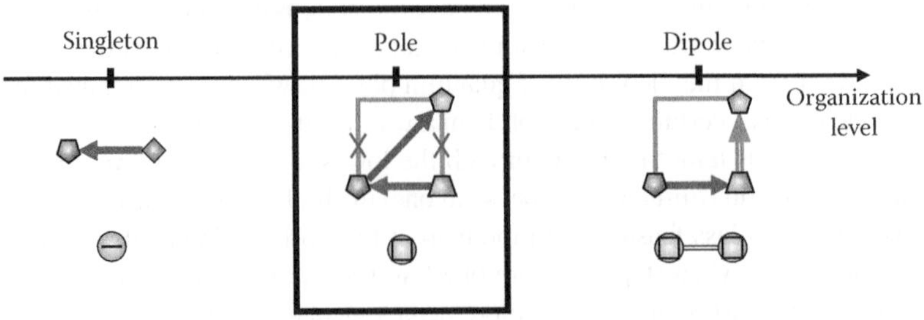

FIGURE 4.1 Limited view of the organization levels along a complexity scale around the Formal Object called a pole.

because it is the universal and elemental brick of the construction game enabling the modeling of the macroscopic world.

The complete definition of the concept of a pole will be given after presenting several examples of poles, of various types and taken from a selection of domains able to give an adequate understanding of the Formal Object.

Table 4.1 lists the case studies of global poles (without space distribution) given in this chapter (see also Figure 4.1).

4.1 THE POLE AS ELEMENTARY COLLECTION

The term "pole" is borrowed from electrostatics and from magnetism, where it generally represents an object that is localized in space. The electric charge or the magnetic moment that quantifies these poles can be as low as the unit or multiple of this unit. Depending on the presence of the electric field or the magnetic field, a pole may or may not contain energy.

Independence of case studies. Unless already skilled in Formal Graphs, there is little chance for the average reader to be familiar with all the domains tackled through the various case studies given. Each case study is written as an independent topic, allowing the reader to browse according to their acquaintance.

Too elementary a level of knowledge? The intrinsic property of a transverse approach is to address a multitude of audiences in many domains. This requires devoting some time for introducing basic notions already known by specialists in the domain.

As every compromise, the knowledge level chosen will appear too low for many readers and too high for others.

This notion is generalized to any ensemble of entities, belonging to any energy variety, which can be localized in a definite region of space, or distributed in space, or even without localization.* The main difference with electrostatics is that a pole is necessarily made of several entities and cannot therefore represent a unit charge.

The essential property of a pole, whose definition will be given at the end of this chapter for pedagogical reasons, is to represent the peculiar behavior of a collection of identical entities. By identical is meant that each entity possesses the same amount of energy as the others, whatever the total amount of energy in the collection and whatever the number of entities!

By extension, a pole may be composed of other poles, under the condition that they cannot exchange energy between them and that they have the same energy-per-entity. For instance, an electrostatic pole can be viewed as a unique collection of identical charges, having the same sign, or as a mixing of positive and negative charges, the total charge being the algebraic sum of these charge amounts. This happens for instance in a population of ions of opposed charge numbers, that is, anions and cations, in the same medium (at the same electric potential), when they are unable to react together. In this case, *composed poles* and subpoles that are more elementary poles are considered. This subject will be exemplified in Section 4.4.

By default, space is not considered in a pole. Because some systems behave independently from space, all the significant variables are state variables that are defined globally for the system. The role of space will be tackled in Chapter 5.

* In the Formal Graph viewpoint, energy and its containers exist before space and time.

4.2 FORMAL GRAPH REPRESENTATION OF A POLE

The graphical representation of a pole without space distribution is quite simple; two kinds are to be considered: a first one when only one system property is present, consisting of three fundamental poles (see Graph 4.1) and a second kind of mixed categories when two system properties are present (see Graph 4.2).

The crosses placed on the vertical links mean that, by definition, a pole has no direct connections in addition to the three system properties. That is a possibility offered by a dipole.

Where does the heat dissipated by a conductance go? The dissipated heat goes to another pole belonging to the thermal energy variety. However, the convention of not representing the associated thermal variety with the conductive pole is adopted for not overloading the drawing of the graph. A deeper reason is because the graphical representation of a conductance has precisely the meaning of an apparent path including this association. This issue will be detailed in Chapter 11, which discusses dissipation, and in Chapter 12, which discusses energy coupling.

As an example of a fundamental pole, Figure 4.2 and Graph 4.3 feature an inductive pole in translational mechanics modeling the movement of an object with an inertial mass.

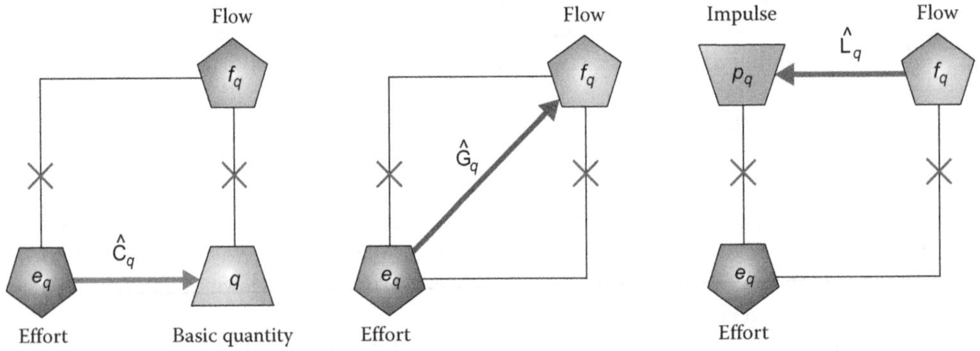

GRAPH 4.1 The three fundamental poles: capacitive (left), conductive (center), and inductive (right) poles.

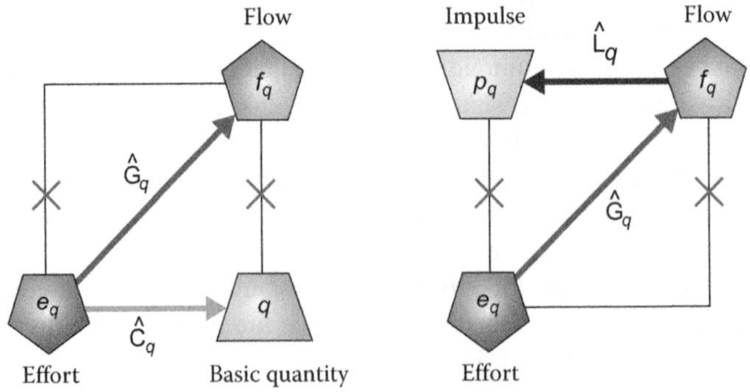

GRAPH 4.2 Capacitive–conductive pole (left) and inductive–conductive pole (right).

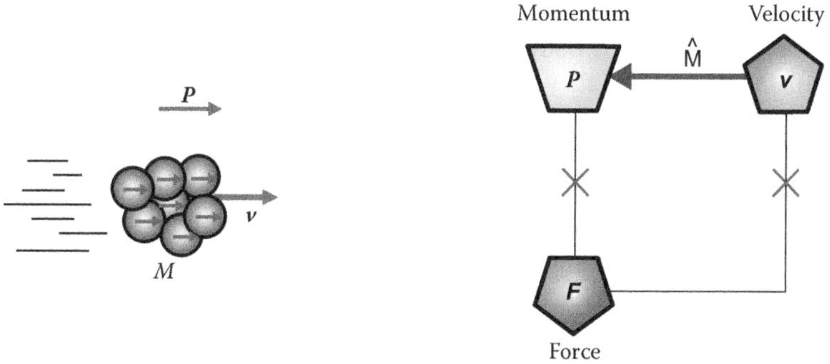

FIGURE 4.2/GRAPH 4.3 A massive object under movement is an inductive pole in the translational mechanics variety.

The inertial mass, which is the inductive system property allowing the object to store kinetic energy, makes the two state variables, the momentum (or quantity of movement) $\boldsymbol{P}$ and the translational velocity $\boldsymbol{v}$, obey the inductive relationship (both state variables are vectors):

$$\boldsymbol{P} = \hat{\mathrm{M}}\,\boldsymbol{v} \tag{4.1}$$

In the general case, the inertial mass is an operator depending on the velocity, which reduces to a simple constant in the Newtonian theory (for low velocities).

It is not compulsory for a pole to possess a system property, meaning that it does not contain energy by itself. The example of a single electrostatic pole illustrates this case in Figure 4.3 and Graph 4.4. The electric charge is eventually composed of several elementary charges and the basic quantity Q is the amount of charge. No capacitance is attributed to this pole because the two state variables, the electric potential V and the electric charge Q, are independent in this case. In electrostatics, a charge placed in a dielectric medium creates some electric potential at a distance but not on itself. Therefore, the potential V is not created by the charge Q but is said to be "seen" by the charge. This potential exists in the medium, possibly created by an external electric field.

However, because the most interesting aspect is the presence of a system constitutive property, all case studies of poles treated in this chapter are endowed with such a property.

The drawing of a Formal Graph is only one part, the graphical one, of the model of a physical system. The other part, the algebraic one, consists of the expressions of the operators used for representing the links of the graph. This part can be void when the model is deprived of system properties, as explained above; however, this is not the general case.

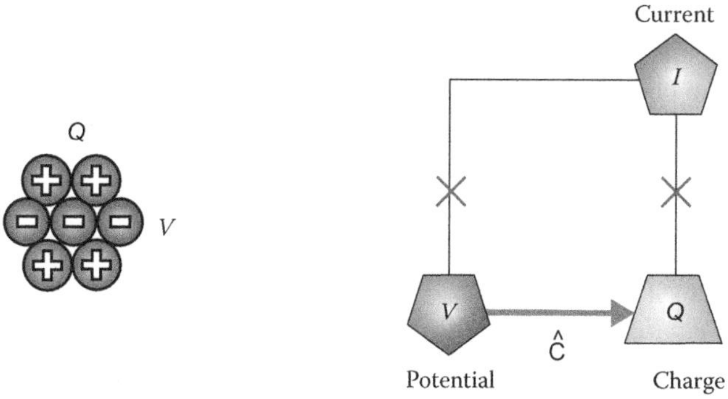

FIGURE 4.3/GRAPH 4.4 A charged body is a capacitive pole in the electrodynamical energy variety.

Why represent isolated nodes? In some Formal Graphs, nodes drawn are not linked to others, because they represent variables that may play a role when necessary, in particular when the pole is connected to other poles. The convention is to show all nodes independent of their effective role.

This point will be discussed in Section 4.5.1 with the notion of a gate variable.

4.3 CASE STUDIES OF POLES

The first case studies are devoted to fundamental poles.

Inductive Poles
- A1: Moving Body in Translational Mechanics
- A2: Moment of Inertia in Rotational Mechanics
- A3: Current Loop in Electrodynamics
- A4: Magnetization in Magnetism

Capacitive Poles
- A5: Electric Charges in Electrodynamics
- A6: Chemical Species in Physical Chemistry
- A7: Group of Corpuscles in Corpuscular Energy
- A8: Spring End in Translational Mechanics
- A9: Thermal Pole in Thermics

The last two case studies are concerned with mixed poles.

Inductive–Conductive Pole
- A10: Motion with Friction in Translational Mechanics

Capacitive–Conductive Pole
- A11: Reactive Chemical Species in Physical Chemistry

4.3.1 Case Study A1: Moving Body — Translational Mechanics

The case study abstract is given in next page.

Depending on the nature and the size of the objects considered, mechanics is divided into several branches: mechanics of fluids, mechanics of solids, quantum mechanics (dealing with particles and waves), etc. A peculiar branch is the *mechanic of the point* when objects have a negligible volume or are sufficiently homogeneous and symmetric for allowing their modeling by a geometrical point.

A distinction is made in the Formal Graph theory between a *body* and a *particle*, in order to correctly model what belongs to the macroscopic world (a body composed of several parts) and to the microscopic world (the elementary particle).

In mechanics, a *body* is an object made up of several elements and its total mass is the sum of the mass of each part. This is in contradistinction with the notion of *particle*, which is an elementary entity endowed with an indissociable mass (without losing the nature of the particle). For distinguishing the two systems, the scheme in the case study abstract shows a body which is a cluster of particles; the variability of the entity number [the momentum (impulse) which is the sum of all momentum associated with every particle] and, in the equations, the variables featuring the impulse (momentum) and the mass are in uppercase for a body and in lowercase for a particle.

An isolated (i.e., not influenced) *body* in movement and endowed with an inertial mass M, constitutes a *pole* in the Formal Graph theory.

Variables: The energy variety to which this system belongs is the mechanical energy of translation, whose basic quantity is the *displacement* ℓ, the impulse is the *momentum* (*quantity of*

A1: Moving Body

Translational Mechanics

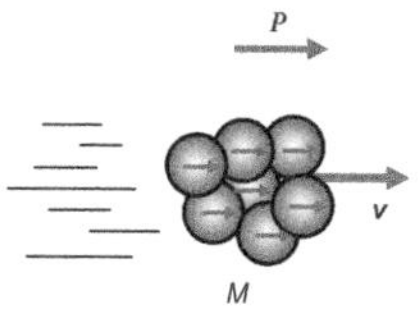

The pole made up by a mechanical object isolated from any influence able to modify its movement, is a fundamental inductive pole.

Pole

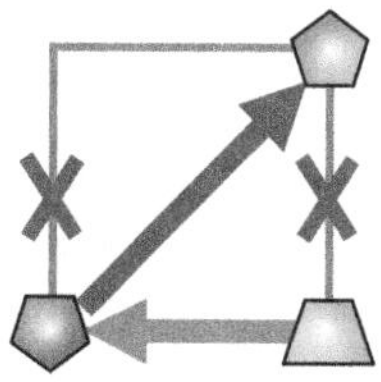

Pole	
✓	Fundamental
	Mixed
	Capacitive
✓	Inductive
	Conductive
✓	Global
	Spatial

Formal Graph:

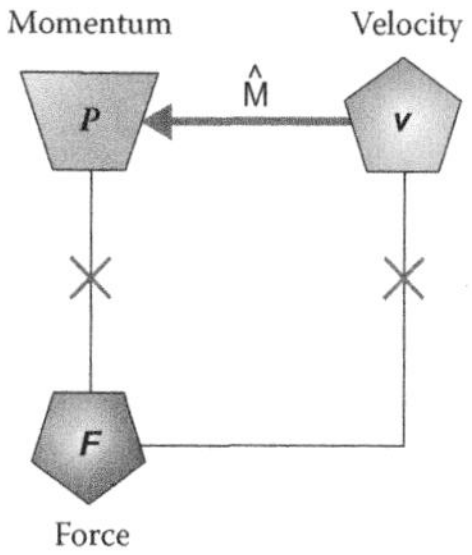

Reluctance operator trace:

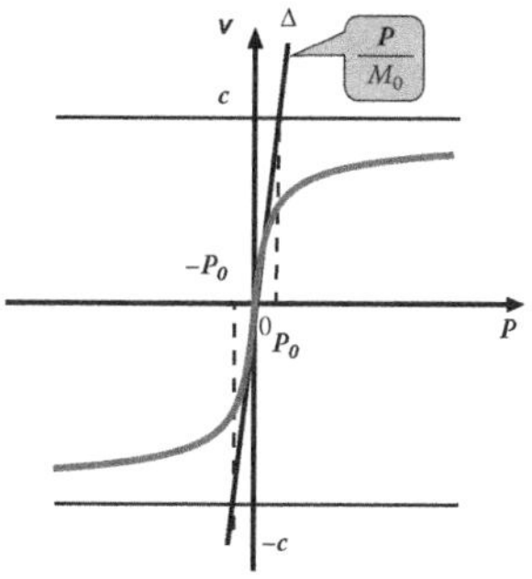

GRAPH 4.5

Variables:

Variety	Translational Mechanics		
Subvariety	Inductive (kinetic)		Capacitive
Category	Entity number	Energy/ entity	Energy/ entity
Family	Impulse	Flow	Effort
Name	Momentum (Quantity of movement)	Velocity	Force
Symbols	$\boldsymbol{P}$, ($\boldsymbol{p}$)	$\mathbf{v}$	$\boldsymbol{F}$
Unit	[N s], [J s m^{-1}]	[m s^{-1}]	[N], [J m^{-1}]

Inductive relationship:

General model

$$\boldsymbol{P} = \hat{M}\mathbf{v}$$

Einstein's model

$$\boldsymbol{P} = M_0 c \frac{\mathbf{v}}{\sqrt{c^2 - v^2}}$$

Newton's model

$$\boldsymbol{P} \underset{lin}{=} M\mathbf{v}$$

System Property	
Nature	Inductance
Level	0
Name	Inertial mass
Symbol	*M*
Unit	[kg]
Specificity	—

Energy variation:

$$\delta \mathcal{T}_\ell \underset{\exists M}{=} \mathbf{v} \cdot \mathbf{d}\boldsymbol{P}$$

movement) $\boldsymbol{P}$ (in Joules per unit of velocity, i.e., J s m^{-1} or, by considering its inductive link, in kg m s^{-1} which is the SI unit). The effort is the *force* and the flow is the *velocity* $\mathbf{v}$. All these state variables are oriented in space and are therefore represented by vectors.

Effort represented. In this Formal Graph (Graph 4.5), the presence of the force—the effort in translational mechanics—merely acts as a reminder that such a pole may become a mixed pole when friction occurs, thus converting kinetic energy into heat. In this case, the frictional force is connected to the velocity by an operator of *resistance*, not shown in this graph that features a fundamental pole (see case study A10 "Motion with Friction").

System Property: The momentum is linked to the velocity by the inductive relation using a translational inductance:

$$\boldsymbol{P} = \hat{\mathsf{M}}\, \boldsymbol{v} \tag{A1.1}$$

This constitutive property is termed *inertial mass* because, historically, it has been thought to oppose the start of movement of an object (which exhibits some inertia). As science progresses, the notion of energy becoming clearer, it has been understood as a means for a system to contain kinetic energy. It is this modern conception that prevails in the Formal Graph theory. The hat (circumflex) over the mass symbol means that it is not a scalar but an operator. Effectively, in the most general case, the inductive relation is not linear. For instance, in Einstein's theory of special relativity, the inductive relation can be written as follows:

$$\boldsymbol{P} = P_0 \frac{\boldsymbol{v}}{\sqrt{c^2 - v^2}} \tag{A1.2}$$

The constant c being the *maximum velocity*, also called *speed of light in free space*, because in vacuum photons propagate at this maximum velocity, the constant P_0 is a scaling coefficient and it is equal to the product of the rest mass M_0 and the velocity of light:

$$P_0 = M_0\, c \tag{A1.3}$$

The curve in the case study abstract represents the inductive relationship by plotting the velocity versus the momentum. This representation is useful for evidencing a feature of the inductive function which is to be odd, that is, $f(-\boldsymbol{P}) = -f(\boldsymbol{P})$. Next, this representation shows that the same sigmoid shape is found again, as for electromagnetic and magnetic inductive relationships (see case studies A3 "Current Loop" and A4 "Magnetization"), although with an inversion of roles between impulses and flows with respect to the present case.

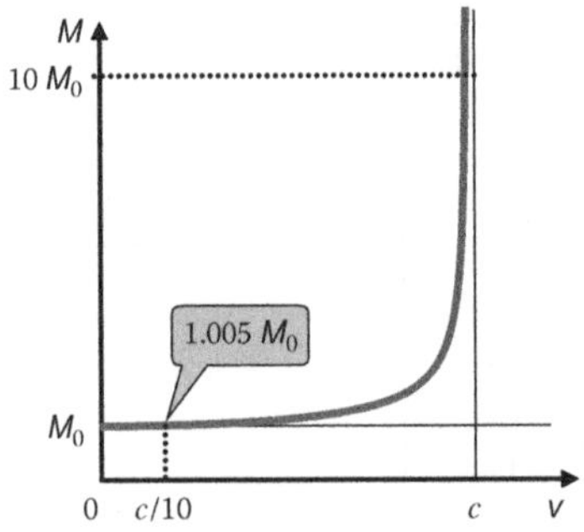

The second plot represents the variation of the mass with the velocity, which is a classical representation based on the formulation of the inertial mass as an eigen-value of the inductive operator (or in a simpler way, as the ratio of the momentum upon the velocity):

$$\hat{\mathsf{M}}\boldsymbol{v} = M\boldsymbol{v} = \frac{M_0\, \boldsymbol{v}}{\sqrt{1 - \dfrac{v^2}{c^2}}} \tag{A1.4}$$

For small velocities compared to the velocity of light, the inductive relationship becomes linear and one returns to the relation predicted by Newton's theory, in assimilating the mass M to the rest mass, as both are equal in that theory.

$$\boldsymbol{P} \underset{lin}{=} M\boldsymbol{v} \tag{A1.5}$$

On the plot in the case study abstract, the straight line Δ of the Newtonian inductive relationship is drawn, and, on the above plot, the validity limit is indicated within 0.5% of Newtonian mechanics which occurs for a velocity equal to approximately one-tenth of the maximum velocity in vacuum.

To write an inductive relation by means of an operator representing the system constitutive property is characteristic of the Formal Graph theory, in which every link, between variables that are placed in the nodes of a graph, is by definition an operation. Therefore, an operator reduced to a scalar will always be considered a peculiar case. Within it, a scalar independent from the variable that it multiplies is a case still more peculiar, corresponding to the "linear" validity domain, systematically indicated by the "*lin*" under the equal sign in relations that are only valid in this restricted domain.

It will appear obvious that this unusual rule is necessary for unifying the various scientific domains since linearity is not the general case. Capacitive relationships are equally submitted to this "operator rule" as the case studies, among others, of electrical (case study A5 "Electric Charges") pole and of physical chemical (case study A6 "Chemical Species") pole demonstrate.

Thus, for the Formal Graph theory, the general case that allows unifying the various energy varieties is that of relativity and not of Newton's theory.

Note that representing an isolated mechanical object by a pole, deprived from any notion of time, corresponds to a uniform movement, which precisely corresponds to the validity condition of special relativity.

Energy: The energy stored by a moving object is inductive energy that is traditionally called *kinetic energy*. Its variation is given by the scalar product

$$\delta \mathcal{T}_\ell \underset{\exists M}{=} \boldsymbol{v} \cdot \mathrm{d}\boldsymbol{P} \tag{A1.6}$$

as the inertial mass expresses the capability of an object to store inductive energy.

4.3.2 Case Study A2: Moment of Inertia — Rotational Mechanics

The case study abstract is given in next page.

Rotational mechanics is a branch of mechanics that is identified with a special energy variety bearing the same name. It includes two energy subvarieties: inductive and capacitive.

The figure in the case study abstract shows a body attached at one end of a rigid rod rotating around a center O placed at the other end. The rotation occurs in a plane represented in perspective in this figure. This system constitutes a pole in rotational mechanics.

Variables: The basic quantity is the *angle of rotation* $\boldsymbol{\theta}$ and the impulse is the *angular momentum* or kinetic moment $\boldsymbol{L}$, also notated $\boldsymbol{\sigma}$. The translational velocity $\boldsymbol{v}$ of a body rotating at a distance $\boldsymbol{R}$ from a center is a tangent to the trajectory in the plane. In the mechanics of rotation, it corresponds to a state variable $\boldsymbol{\Omega}$ from the family of flows, called *angular velocity*. All these variables are oriented in space and therefore represented by vectors. The relationship between angular velocity and translational velocity is expressed by a vector product of the distance (radius of gyration) and the angular velocity:

$$\boldsymbol{v} = \boldsymbol{\Omega} \times \boldsymbol{R} \tag{A2.1}$$

As both vectors $\boldsymbol{R}$ and $\boldsymbol{v}$ are perpendicular and belong to the plane, the vector $\boldsymbol{\Omega}$ stands perpendicular to this plane, pointing upward in case of a positive rotation, as in the figure in the case study abstract and in the following scheme.

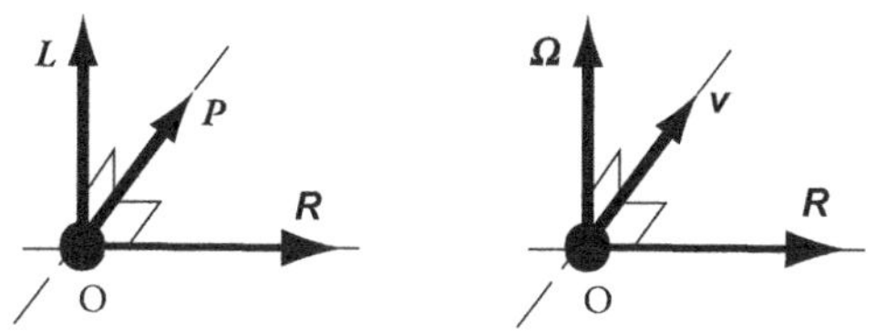

A2: Moment of Inertia **Rotational Mechanics**

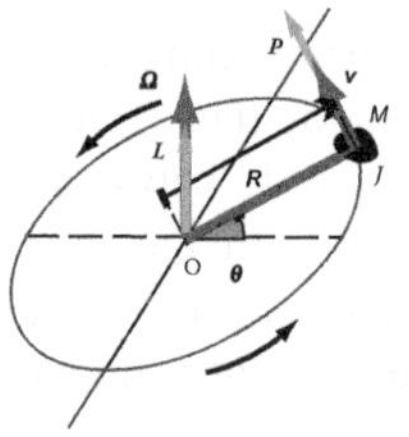

A body rotating around a center O and possessing rotational inertia is a fundamental inductive pole.

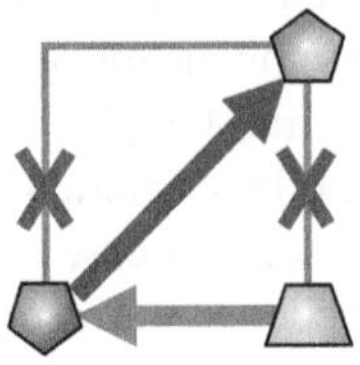

Pole	
✓	Fundamental
	Mixed
	Capacitive
✓	Inductive
	Conductive
✓	Global
	Spatial

Formal Graph:

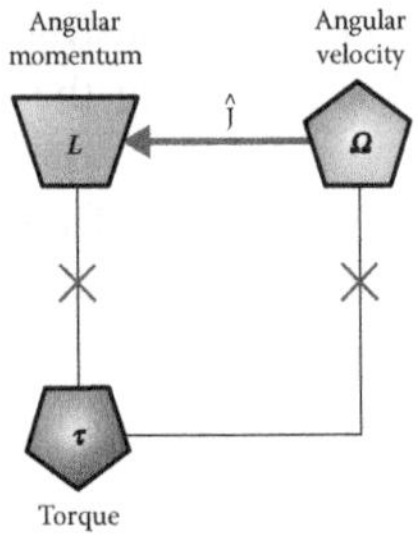

Reluctance operator trace:

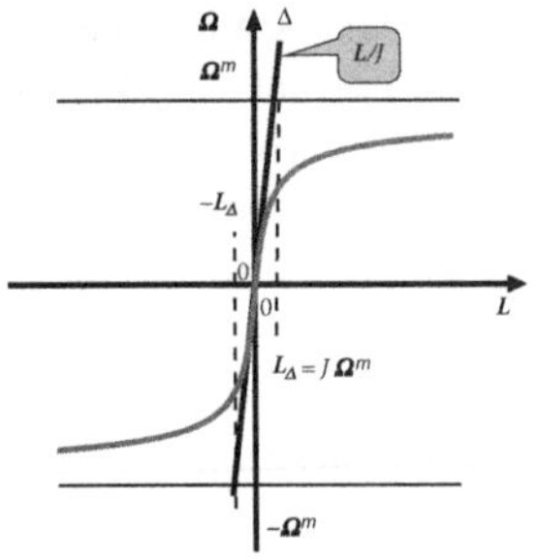

GRAPH 4.6

Variables:

Variety	Rotational Mechanics		
Subvariety	Inductive (kinetic)		Capacitive
Category	Entity number	Energy/ entity	Energy/ entity
Family	Impulse	Flow	Effort
Name	Angular momentum (kinetic moment)	Angular velocity	Torque (couple)
Symbols	$\boldsymbol{L}$, (σ)	Ω	τ
Unit	[N m s rad^{-1}], [J s rad^{-1}]	[rad s^{-1}]	[N m rad^{-1}]

Inductive relationships:

General model:

$$\boldsymbol{L} = \hat{\mathrm{J}}\,\boldsymbol{\Omega}$$

Linear model:

$$\boldsymbol{L} \underset{lin}{=} J\boldsymbol{\Omega}$$

System Property	
Nature	Inductance
Level	0
Name	Rotational inertia
Symbol	J
Unit	[kg m^2 rad^{-2}]
Specificity	—

Energy variation:

$$\delta\mathcal{T}_\theta \underset{\exists J}{=} \boldsymbol{\Omega}\cdot\mathbf{d}\boldsymbol{L}$$

The *angular momentum* (kinetic moment) $\boldsymbol{L}$ is given by the curl product of the radius and the momentum $\boldsymbol{P}$, the impulse in translational mechanics. As illustrated in the above scheme, it has the same orientation as the angular momentum since $\boldsymbol{P}$ and $\boldsymbol{v}$ are colinear vectors.

$$\boldsymbol{L} = \boldsymbol{R} \times \boldsymbol{P} \tag{A2.2}$$

These relationships between translational and rotational mechanics are given here to present the classical definitions, which are not considered as such in the Formal Graph approach (see Chapter 2 "Nodes of Graphs"). In fact, the angular momentum $\boldsymbol{L}$ is defined as the variable (entity number) quantifying the container of inductive energy existing in a rotation, independently of the notion of the radius. The notion of angular velocity $\boldsymbol{\Omega}$ as energy-per-entity is derived from it.

System Property: The angular momentum is linked to the angular velocity by the inductive relation as follows:

$$\boldsymbol{L} = \hat{\mathrm{J}}\,\boldsymbol{\Omega} \tag{A2.3}$$

This constitutive property is termed *rotational inertia* (or *moment of inertia*) because, historically, it has been thought to oppose the start of the rotation of an object (which exhibits some resistance or inertia). The hat (circumflex) over the rotational inertia symbol means that it is not a scalar but an operator. Effectively, in the most general case, the inductive relation is not linear and the rotational inertia is a tensor. If the relativistic model for translational mechanics is relatively amenable, this is not the case in rotation during a translation because of the variation of radius with the velocity at high speed.

The linear case is the most used (Newtonian frame), corresponding to an asymptotic behavior at low velocities, as shown in the plot in the case study abstract.

$$\boldsymbol{L} \underset{lin}{=} J\boldsymbol{\Omega} \tag{A2.4}$$

In this case, the rotational inertia is a scalar, as the two vectors $\boldsymbol{L}$ and $\boldsymbol{\Omega}$ are colinear. As each body is featured in translational mechanics by an inertial mass (mechanical inductance) M (which is also a scalar in the Newtonian frame), relating the velocity $\boldsymbol{v}$ to the momentum $\boldsymbol{P}$, the relationship between the two inductances can be written as:

$$J \underset{lin}{=} MR^2 \tag{A2.5}$$

Energy: When rotational inertia exists in a system, the variation of inductive energy is given by the scalar product of two vector quantities, the angular velocity (the flow) and the angular momentum variation:

$$\delta\mathcal{T}_\theta \underset{\exists J}{=} \boldsymbol{\Omega}\cdot \mathbf{d}\boldsymbol{L} \tag{A2.6}$$

This is a nonexact differential equation because a pole cannot see its energy varying without participation of a dipole.

Coupling between energy varieties. The reference to the state variables of translational mechanics is the traditional way for introducing the state variables of rotational mechanics. This close connection between two energy varieties is tackled in the Formal Graph theory through the notion of *energetic coupling* that will be presented in Chapter 12, after having browsed almost all the Formal Objects based on a single energy variety.

The two relations A2.1 and A2.2 are the algebraic translations of the inductive coupling between rotational and translational mechanics, which are two autonomous energy varieties (i.e., they are defined independently). Case study K5 "Rotating Bodies" in Chapter 13 will develop this notion in more detail.

Frequent confusion. The angular velocity is often called angular frequency.

4.3.3 Case Study A3: Current Loop

Electrodynamics

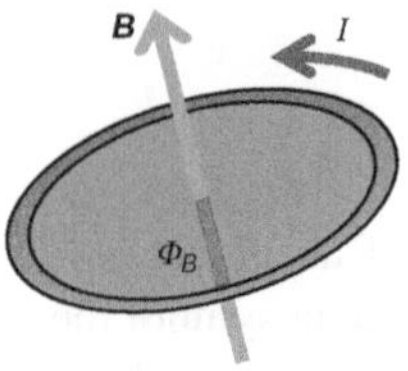

The pole made up by a loop in which flows an electric current isolated from outside (no external source of electric charges) is a fundamental inductive pole.

Pole

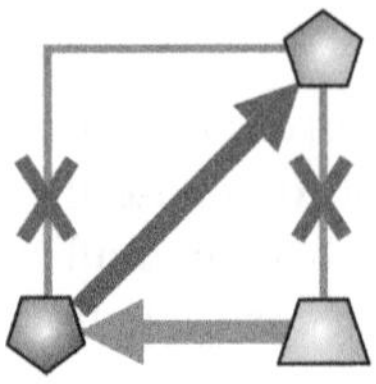

Pole	
✓	Fundamental
	Mixed
	Capacitive
✓	Inductive
	Conductive
✓	Global
	Spatial

Formal Graph:

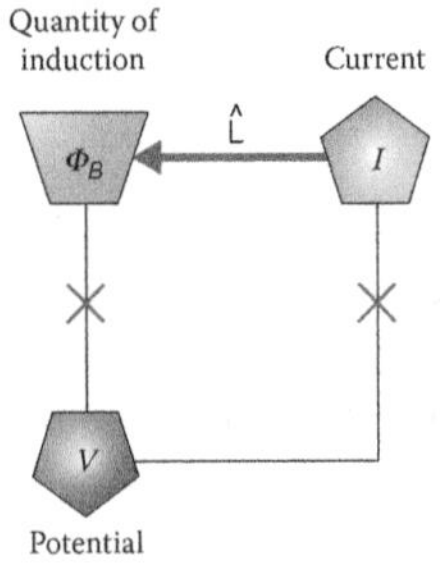

Reluctance operator trace:

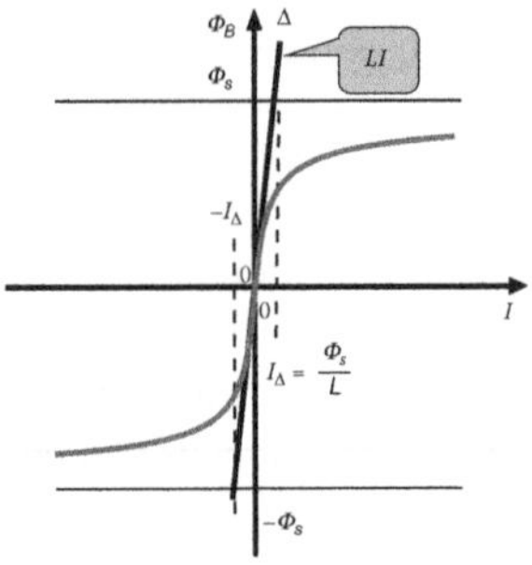

GRAPH 4.7

Variables:

Variety	Electrodynamics		
Subvariety	Inductive (electromagnetic)		Capacitive
Category	Entity number	Energy/entity	Energy/entity
Family	Impulse	Flow	Effort
Name	Quantity (flux) of induction	Current	Potential
Symbols	Φ_B	I, i	V, U, E, φ
Unit	[Wb], [J A^{-1}]	[A]	[V], [J C^{-1}]

Inductive relationship:

General model:

$$\Phi_B = \hat{L}\, I$$

Nonlinear model:

$$\Phi_B = L \frac{I}{\sqrt{1+\frac{I^2}{I_\Delta^2}}}$$

Linear model:

$$\Phi_B \underset{lin}{=} L\, I$$

Energy variation:

$$\delta \mathcal{T}_Q \underset{\exists L}{=} I\, \mathrm{d}\Phi_B$$

A loop, or a turn, formed by an electric conductor along which an electric current I circulates, constitutes a *pole* in the Formal Graph theory.

Variables: The energy variety to which this system belongs is the electric energy, or electrodynamics. Its basic quantity is the *electric charge* Q; its impulse is the *quantity (flux) of electromagnetic induction* Φ_B. The effort is the *electric potential* and the flow is the *electric current* I.

To the circulation of current is associated an electromagnetic induction field $\boldsymbol{B}$ crossing the surface enclosed by the loop. Integration of this vector field on this closed surface S provides the *flux of electromagnetic induction*, which is the *quantity of induction* Φ_B

$$\Phi_B = \oint_S \boldsymbol{B} \cdot \mathbf{d}\boldsymbol{A} \tag{A3.1}$$

Graph: The corresponding Formal Graph is given in the case study abstract.

Representing an isolated electric system by a pole deprived from any notion of time corresponds to a stationary regime. Experimentally, this system consists of a turn of superconducting material cooled down to a low temperature for avoiding losses by the Joule effect that would be caused by a nonzero resistance. The conductor turn is electrically insulated from any external source of electric charges. The circulation of electrons in the material is initiated by imposing a magnetic or electromagnetic field and then removing it. In these conditions, the electromagnetic induction and the current indefinitely remain constant.

System Property: In the Formal Graph theory, an inductance is not necessarily a scalar but an operator

$$\Phi_B = \hat{\mathsf{L}}\, I \qquad \text{(A3.2)}$$

that allows the modeling of the general case. In classical electromagnetism, in vacuum or in materials said to be "linear," the current is linked to the quantity (flux) of induction by a linear inductive relation

$$\Phi_B \underset{lin}{=} \mathsf{L}\, I \qquad \text{(A3.3)}$$

in which the proportionality factor L is the self-inductance. In contrast, in magnetizable materials, particularly ferromagnetic ones, such as iron, cobalt, nickel, and manganese, the electromagnetic inductance is not linear and the electromagnetic induction is limited by a threshold called the *electromagnetic saturation* Φ_s which depends on the nature of the material. A classical model of nonlinear inductive relation (modeling simple ferromagnetic materials such as nickel) is written with two strictly equivalent equations, one being the reciprocal function of the other:

$$I = \frac{1}{\mathsf{L}} \frac{\Phi_s \Phi_B}{\sqrt{\Phi_s^2 - \Phi_B^2}} \qquad \Phi_B = \mathsf{L} \frac{I}{\sqrt{1 + \dfrac{I^2}{I_\Delta^2}}} \qquad \text{(A3.4)}$$

The scaling current is related to the electromagnetic saturation by

$$\Phi_s = \mathsf{L}\, I_\Delta \qquad \text{(A3.5)}$$

The classical curve of the inductive relationship is the one used for expressing the saturation phenomenon during the magnetization of a material under application of a magnetic or electromagnetic field. The inductive relation is an odd function, as both state variables have the same sign, positive as well as negative. This representation shows an obvious similarity with the relativistic variation of the momentum with the velocity of the moving body in translational mechanics given in case study A1.

Inversion! Both inductive relations have the same shape between electromagnetism and mechanics; however, the role of variables is inverted! In Equation 4.4, the current I (which is a flow) is replaced by the momentum $\boldsymbol{P}$ (which is an impulse) and the quantity (flux) of induction Φ_B (which is an impulse) by the velocity $\boldsymbol{v}$ (which is a flow). The same distribution of roles is not always found in an inductive relation from one energy variety to the other. One finds again this same inversion of roles in the capacitive relationship between the thermal energy and the other varieties.

Energy: The energy stored by a current loop is an inductive subvariety and is traditionally called *electromagnetic energy*. It should not be confused with *magnetic energy*, which is the energy contained in magnetizable materials. Its variation is given by the product of the current and the variation of induction quantity (flux)

$$\delta \mathcal{T}_Q \underset{\exists L}{=} I \, \mathrm{d}\Phi_B \tag{A3.6}$$

and not the reverse, that is, the product of the induction quantity (flux) and the variation of current, despite the fact that for a scalar and constant self-inductance, it is equivalent.

Definition of Electromagnetic Induction: From the viewpoint of thermodynamics, and of Formal Graphs, the fundamental variable is not the electromagnetic induction $\boldsymbol{B}$ but its quantity (or flux) of induction Φ_B, which is a state variable, more precisely an entity number in the inductive subvariety of electrodynamical energy. It is therefore more correct (and more physical) to define this variable first by defining it as the quantity of induction, and then to define the electromagnetic induction $\boldsymbol{B}$ as its surface density.

$$\boldsymbol{B} \underset{def}{=} \frac{\mathrm{d}\Phi_B}{\mathrm{d}\boldsymbol{A}} \tag{A3.7}$$

Definition of Current: In the Formal Graph theory, the electric current is the *energy-per-entity* and its definition is therefore given by the partial derivative of the total energy of the system with respect to the electromagnetic quantity (flux) of induction Φ_B, which is the *number of entities*, by maintaining constant the other entities:

$$I \underset{def}{=} \left(\frac{\partial \mathcal{E}}{\partial \Phi_B} \right)_{\substack{\text{other} \\ \text{entities}}} \tag{A3.8}$$

The fact that the amount of charge does not vary in the system (one would observe otherwise a disappearing or saturation and a subsequent stop of the current, which is not experimentally observed) means, without any doubt, that the current cannot be defined in this case as the derivative of charge with respect to time!

$$I \neq \frac{\mathrm{d}Q}{\mathrm{d}t} \tag{A3.9}$$

Nothing forbids one to integrate the current with respect to time and to obtain a quantity that grows linearly with time. This quantity corresponds physically to an *electric charge that has passed* through a point of the circuit, but which is not a number of entities stored in the conducting material (which does not vary). There is no conversion into capacitive energy (electrostatic) because the loop does not possess the capacitive property (the capacitance is null by definition in a purely inductive system) allowing the storage of this subvariety. There is neither transit of charges nor exchange with another system because the loop is isolated from outside.

The notion of passed charge. From the usual definition of a current as being the derivative with respect to the time of a quantity of electric charge, one should have in a closed system, for a stationary current, an increase or decrease of electrical charges as time elapses.

No such observation has been made in an isolated loop with a resistance equal to zero, that is, without losses. (To meet these conditions, a superconducting material at low temperature is taken.)

This paradox can only be removed by distinguishing two kinds of charges: the charge Q contained in the system, which is a state variable as it participates in the capacitive energy in the system, and the passed charge Q_P, counted with the help of a counter put in any place along the circuit, which is not a state variable. This implies a localization of charges in space (in the form of particles) and their displacement with a velocity, which may eventually not be a constant.

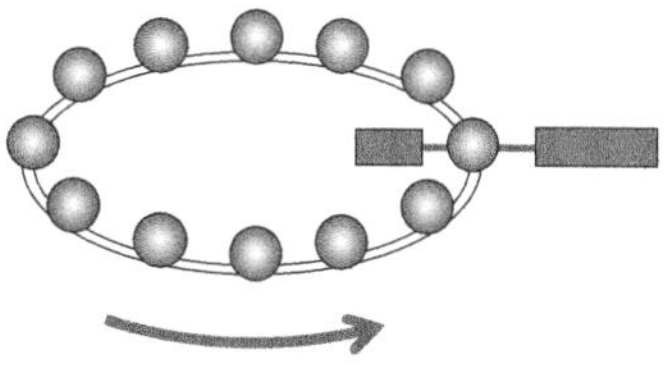

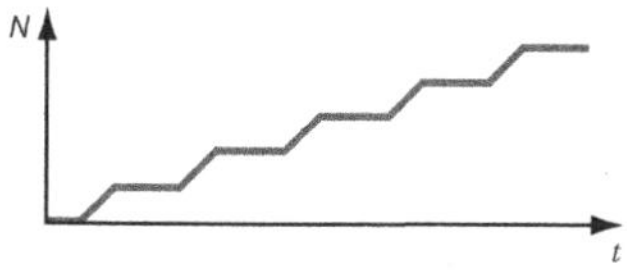

Counting the charges provides a quantity Q_P proportional to the number N of passed particles, which endlessly grows. Furthermore, the individualization of particles makes the growth nonmonotonic (with steps) for the curve of N as a function of times.

So, the classical definition of current must be understood as a derivation of the passed charge Q_P, which normally provides a discontinuous curve (because charges are discrete).

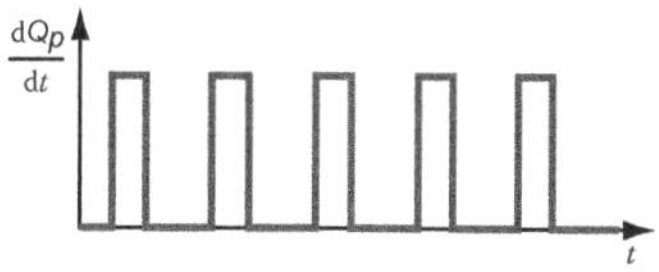

4.3.4 Case Study A4: Magnetization — Magnetism

The case study abstract is given in next page.

A true energy variety. The magnetization of a material is translated in the Formal Graph by a peculiar energy variety, the magnetic energy (energy of magnetization), different from the electrodynamical variety.

More precisely, it is the inductive subvariety of this energy of magnetization that is contained in magnetic materials (magnets) under the form of dipoles.

The comprehension of the force that is exerted between two magnets has been based for long on the strong analogy with electric polarization and electrostatic interaction modeled by Coulomb's law (case study D4 in Chapter 7), in assuming the existence of "magnetic charges" analogous to electric charges (see case study A5 "Electric Charges"). Facing the lack of experimental evidence of the existence of these "magnetic charges," the classical theory has abandoned this analogy and interprets the magnetization of materials in microscopic terms by attributing to the particles that constitute matter (electrons, protons, neutrons) a *magnetic moment* induced by the orbital rotation or the self-rotation of particles. The Formal Graph theory allows the modeling of magnetization by staying at the macroscopic level (without excluding the modern conception at the microscopic level), but without returning to the hypothesis of "magnetic charges," a term that is avoided for not inducing the feeling of a flashback.

Variables: From the Formal Graph viewpoint, the equivalent of the electrostatic charge is the *magnetization impulse* p_m, improperly assimilated in the past to a *magnetic charge*, whose unit is the ampere meter (A m). In the Formal Graph theory, this magnetization impulse is a *number of entities* in the inductive subvariety of *magnetic energy* (or *magnetization energy*, for avoiding confusion with electromagnetic energy). This energy variety is contained in a stable way only in certain materials that are ferromagnetic, magnets being a concrete example, but may temporally appear under the influence of a magnetic field in practically all materials, called diamagnetic in this case.

A4: Magnetization **Magnetism**

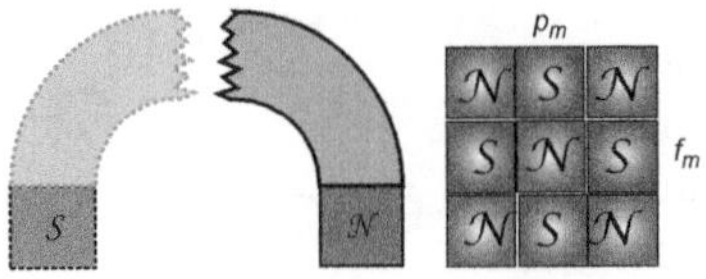

The notion of pole utilized in Formal Graphs is directly borrowed from electrodynamics and magnetism for being generalized to all energy varieties. A nuance is therefore brought in the magnetic domain precisely, in including in this notion a group of magnetic entities (the magnetic impulses and not "magnetic charges") with eventually different signs and in not restricting the notion to a single entity.

Pole

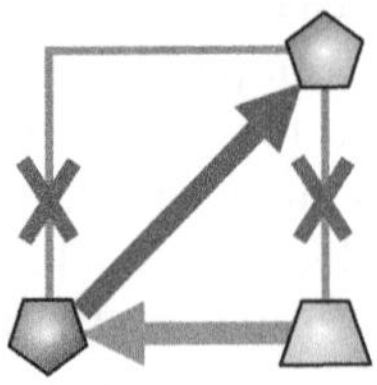

	Pole
✓	Fundamental
	Mixed
	Capacitive
✓	Inductive
	Conductive
✓	Global
	Spatial

Formal Graph:

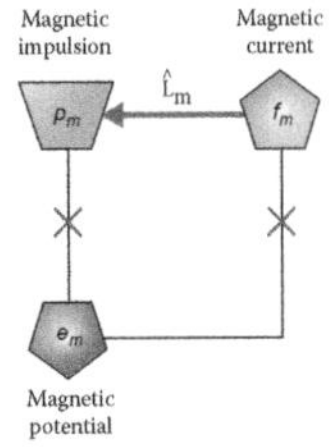

Reluctance operator trace:

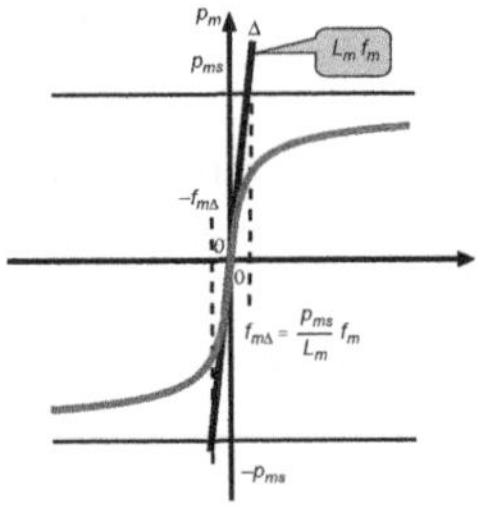

GRAPH 4.8

Variables:

Variety	**Magnetism (Magnetic Energy)**		
Subvariety	Inductive (kinetic)		Capacitive
Category	Entity number	Energy/entity	Energy/entity
Family	Impulse	Flow	Effort
Name	Magnetic impulse	Magnetic current	Magnetic potential
Symbols	p_m	f_m	e_m
Unit	[Wb], [J A^{-1}]	[A]	[V], [J C^{-1}]

Inductive relationship:

General model

$$p_m = \hat{L}_m f_m$$

Nonlinear model

$$p_m = L_m \frac{f_m}{\sqrt{1+\frac{f_m^2}{f_{m\Delta}^2}}}$$

Linear model

$$p_m \underset{lin}{=} L_m f_m$$

System Property	
Nature	Inductance
Level	0
Name	Magnetic inductance
Symbol	L_m
Unit	[H], [J A^{-2}]
Specificity	—

Energy variation:

$$\delta \mathcal{T}_{p_m} \underset{\exists L_m}{=} f_m \, dp_m$$

In the Formal Graph theory, the energy-per-entity is called the *magnetic* or *magnetization current* and it is notated f_m. As for the electric charges, the magnetization impulses exist under two forms of opposite signs, positive and negative, and their groupings are called north and south poles, in reference to the earth's magnetism.

Can a magnetic dipole be split? The attractive (or repulsive) force between two magnetic poles separated by a distance r is given in magnetism by an equation similar to Coulomb's law in electrostatics (see case studies D4 "Coulomb's Law" and D6 "Magnetic Interaction" in Chapter 7).

The experience shows that one cannot practically separate the dipole made with two magnetization impulses of opposite signs for isolating a single magnetic pole. Two opposite poles are always formed in every piece of magnetic material. This is expressed in saying that the magnetic dipoles are *inseparable*, as is the case of two ends of a spring (case study C8 "Spring" in Chapter 6) of a conductor or two faces of an interface. The consideration of a single magnetic pole is therefore purely theoretical for the need of modeling.

Magnetic Pole: A set or a group of magnetic entities *having the same magnetic current* forms a *pole* of the inductive subvariety of magnetic (or magnetization) energy. The magnetic impulses may be a mix of positive (north) and negative (south) entities in unequal proportion for constituting a pole with a nonzero magnetic polarity. Some cohesive forces (belonging to another energy variety, which must be taken into account in the total energy of such a system) are supposed to maintain the whole set clustered. The *total magnetic impulse* is p_m and the *magnetic current* shared by these entities is f_m.

Formal Graph: The Formal Graph relies on these two variables and on the effort, which is the *magnetic potential* e_m, but which is not connected to the *magnetic current* (it can be the case when the system possesses the conductance property, (see case study A11 "Reactive Chemical Species"), which is a mixed capacitive–conductive pole).

System Property: Each individual magnetic entity influences at some distance the others entities by producing a *magnetic field* that creates in every point of space a *magnetic current*. The sum of these currents at a point where the entities are gathered (their total spatial extension is assumed to be reduced to a point, or at least to a space region where the *magnetic current* is a constant) constitutes the current f_m shared by the entities. This current depends on the total magnetic impulse p_m and one traditionally writes the relationship between these two quantities as the linear relationship:

$$p_m \underset{lin}{=} L_m \, f_m \tag{A4.1}$$

The unit of the proportionality factor L_m is the joule per square ampere (J A^{-2}) and represents a *magnetic inductance* (not electromagnetic).

One or several entities? The situation is different when a pole is made up with a single magnetic entity, that is, nondissociable. In this case, the current f_m is not created by this one but must necessarily result from an *external magnetic field* produced by another pole in the vicinity. The two poles are thereby exerting a *mutual influence* on each other. This case does not correspond anymore to a pole but to a dipole (see case study D6 "Magnetic Interaction" in Chapter 7).

Linearity or not? The Formal Graph theory postulates an identity of mathematical forms of system properties for all energy varieties. Thus, the inductive relations are represented by means of inductance operators which allows relation A4.1 to be written under the generalized form

$$p_m = \hat{\mathrm{L}}_m \, f_m \tag{A4.2}$$

In fact, the shape taken by the inductive relationship in this energy variety is the same as for electromagnetism, as shown in case study A3 "Current Loop" (and the same form as for case study A1 "Moving Body").

$$p_m = L_m \frac{f_m}{\sqrt{1 + \dfrac{f_m^2}{f_{m\Delta}^2}}} \tag{A4.3}$$

The scaling magnetic current $f_{m\Delta}$ is the analogous of the scaling I_Δ current seen in the previous case study.

Universally known shape. The sigmoid shape of this inductive relationship is very well known (as "magnetization curve"), especially when illustrating the phenomenon of hysteresis with a back and forth sweep of the magnetic current (or external electromagnetic field) is applied to the material. The parenthood between magnetism and electromagnetism reinforces the idea that, in electromagnetism, the standard behavior of an inductance should not be linear, as asserted by the classical theory, but nonlinear.

Energy: The magnetic energy variation is submitted to the existence of the inductive property

$$\delta \mathcal{T}_{p_m} \underset{\exists L_m}{=} f_m \, dp_m \tag{A4.4}$$

Wrong track. The Formal Graph theory helps to partly understand the difficulty of evidencing the entities of magnetization. This difficulty originates in the abusive assimilation of the magnetic impulse with a "magnetic charge" thought to be similar to the electric charge and therefore having the attributes of a particle. Now, if both are entities, the first one is an impulse and the second one is a basic quantity. They do not belong to the same energy subvariety; the magnetic impulse is inductive while the electric charge is capacitive. To want to isolate a single mechanical impulse of a body at rest (null velocity) is much of an illusion that is well comprehensible; to want to do the same for the magnetization in a material without allowing magnetic current to flow is still trickier!

Unresolved question. The question of the existence of isolated magnetic poles (monopoles) is an active subject of research. Numerous attempts have been made to experimentally isolate magnetic monopoles, without success until now. Theories, notably quantum theories, that predict their existence (in predicting a mass beyond reach by present accelerators of particles) are considered suspiciously while waiting experimental confirmation.

4.3.5 Case Study A5: Electric Charges — Electrodynamics

The case study abstract is given in next page.

A group or a collection of electric charges *at the same potential* forms a pole of the capacitive subvariety of electric energy. The charges may be composed of a mix of positive and negative charges, eventually in unequal proportions. The situation where all the charges are of the same sign is also considered in this case. A supporting lattice or cohesive forces (resulting from the presence of another energy variety, which needs to be taken into account for determining the total energy of such a system) are supposed to maintain the whole group aggregated.

Variables: The total quantity of charge is Q and the electric potential "seen" by these charges is V. The Formal Graph relies on these two variables and on the flow, which is the electric current I, but which here is not linked to the potential (the link exists only if the system possesses the conductance property, cf. case study A11 "Reactive Chemical Species," which is a mixed capacitive–conductive pole).

Energy: The variation of capacitive energy is given by

$$\delta \mathcal{U}_Q \underset{\exists C}{=} V \, dQ \tag{A5.1}$$

A5: Electric Charges **Electrodynamics**

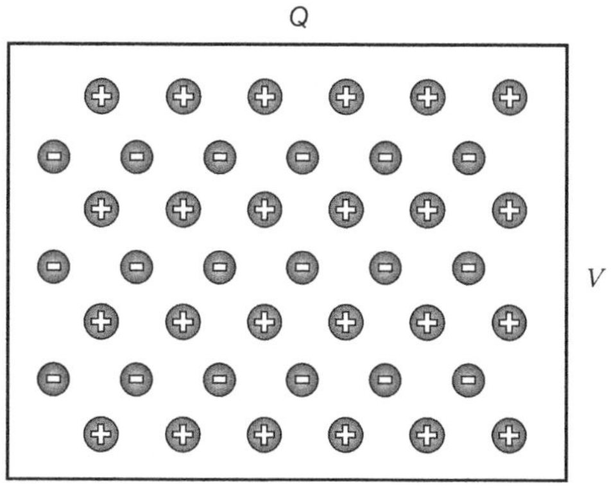

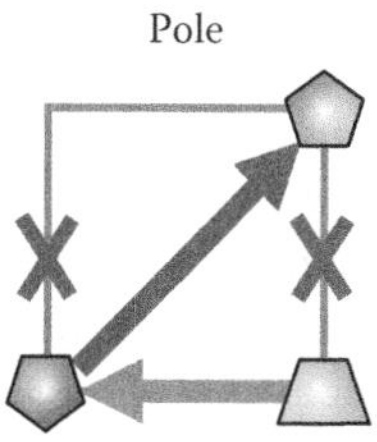

Pole	
✓	Fundamental
	Mixed
✓	Capacitive
	Inductive
	Conductive
✓	Global
	Spatial

Why the plural "charges" instead of a singular? Because when one speaks of a single charge, it has implicitly the meaning of an elementary charge. And, in this case, the number of entities cannot vary for corresponding to the variation of energy, which is not in agreement with the definition of a pole (which is a collection of entities in variable number).

Formal Graph:

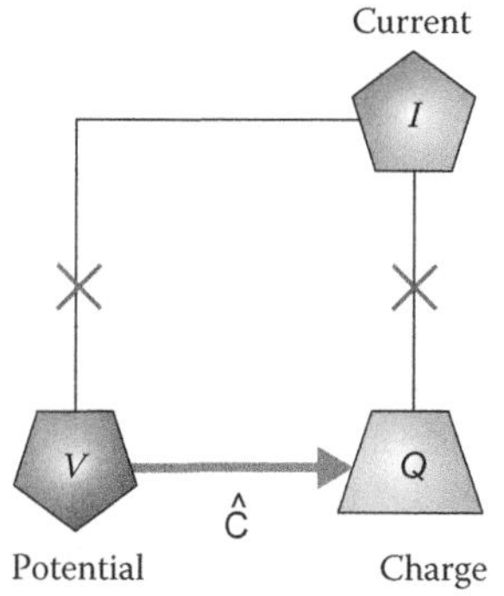

GRAPH 4.9

Capacitance operator trace:

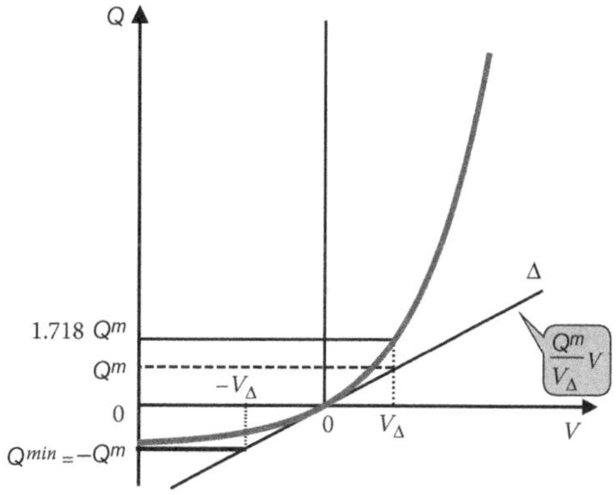

Scaling effort: $V_\Delta = \frac{Q^m}{C}$

Asymptote: $Q = \frac{Q^m}{V_\Delta} V$

Variables:

Variety	Electrodynamics		
Subvariety	Capacitive (potential)		Inductive
Category	Entity number	Energy/ entity	Energy/ entity
Family	Basic quantity	Effort	Flow
Name	Charge	Potential	Current
Symbols	Q	V, U, E, φ	I, i
Unit	[C], [A s]	[V], [J C^{-1}]	[A]

Capacitive relationship:

General model

$$Q = \hat{C} V$$

Nonlinear model

$$\frac{Q + Q^m}{Q^m} = \exp\left(\frac{V}{V_\Delta}\right)$$

Linear model

$$Q \underset{lin}{=} CV$$

System Property	
Nature	Capacitance
Level	0
Name	Capacitance
Symbol	C
Unit	[F], [J C^{-1}]
Specificity	—

Energy variation:

$$\delta \mathcal{U}_Q \underset{\exists C}{=} V \mathrm{d}Q$$

This differential equation expresses the property of energy to vary according to the product of the *energy-per-entity* (the potential) and a variation of the *entity number* (the charge).

System Constitutive Properties: According to electrostatics, each individual charge influences at a distance the other charges in producing an electric field that creates in every point of the space an electric potential. In the present concept of a pole, the sum of these potentials at a point is supposed to be identical for all points and constitutes the potential V in which the charges are covered. This potential V depends of the total charge Q and one traditionally writes the relation between these two quantities as the linear relationship

$$Q \underset{lin}{=} CV \tag{A5.2}$$

The proportionality factor C is called *capacitance* and its unit is the Farad (F).

Linearity or not? The Formal Graph theory postulates an identity between the mathematical forms of system properties for all the energy varieties.* The capacitive relationships are therefore represented by capacitance operators. This changes the writing of relation A5.1 into the generalized form

$$Q = \hat{C}V \tag{A5.3}$$

All these capacitance operators obey the same exponential law as the one in physical chemistry (see Case Study A6 "Chemical Species"), based on the definition of an electrodynamical activity that uses a minimum charge limit Q^{min} as in mechanics in the domain of elasticity of solids where a material cannot be compressed below a certain limit (see case study A8 "Spring End").

$$\frac{Q - Q^{min}}{Q^0 - Q^{min}} = a_Q = \exp\frac{V}{V_\Delta} \tag{A5.4}$$

The scaling potential V_Δ is a quantity classically used for normalizing potentials in many systems as, for instance, in the Shockley† model of an ideal diode (see Chapter 8 "Multipoles"). It plays the same role as the scaling chemical potential μ_Δ in physical chemistry (see the next case study). With a composed pole the reference Q^0 is equal to zero because a null potential entails zero charge (equal amounts of positive and negative charges), whereas in mechanics a null applied force corresponds to the rest length of a solid. (In case of charges with the same sign Q^0 is different from zero; see case study J6 "Ions Distribution" in Chapter 12.) This reference Q^0 is written in the denominator is to allow comparison with the general expression (demonstrated in Chapter 7) in which a reference basic quantity is systematically used. As in electrodynamics the basic quantity can be negative, the minimum charge limit Q^{min} is negative, which is not very convenient for handling algebraic expressions such as the previous one. So, a change of sign is made by using the opposite Q^m of this minimum

$$Q^m = -Q^{min} \tag{A5.5}$$

to write under the more practical form the previous equation

$$\frac{Q + Q^m}{Q^m} = a_Q = \exp\frac{V}{V_\Delta} \tag{A5.6}$$

Another reason for making this change is that when the pole participates in a dipole, this opposite minimum corresponds to the maximum charge of the other pole, which is a positive quantity (see case study C7 "Electric Capacitor" in Chapter 6).

The plot of the charge variations with the potential is given in the case study abstract, with the tangent at the origin Δ playing the role of asymptote for small absolute values of the potential with respect to the scaling potential V_Δ.

---

* This rule relies on the existence of influences between entities, which may not exist in all cases.

† William Bradford Shockley (1910–1989): American physicist; Stanford, California, USA.

Wrong energy expression. In the linear case, it is possible to express the energy variation as the product of the charge and the potential variation. This is contrary to the rule governing energy variations recalled above. It is therefore preferable not to use this peculiar expression.

$$\delta \mathcal{U}_Q \underset{lin}{=} Q\, \mathrm{d}V \tag{A5.7}$$

The concept of a pole. The notion of a pole used in Formal Graphs is directly borrowed from electrodynamics and generalized to all energy varieties. A nuance is brought in the electrostatic domain, precisely, by including in this notion a set of charges, eventually with various signs, and by excluding the case of a solitary electric charge.

Distribution through space. This system is considered as a pole at the global level, without taking into account localization in space. Such a system corresponds, for instance, to an armature of capacitor or to a phase of a charged material (metal, semiconductor, ionic liquid), by assuming that the distribution of charge is homogeneous and that the potential is the same in every point.

The case of charges localized in space is tackled in spatial electric charges and the association of two electric poles for forming a dipole is treated in case study C7 "Electric Capacitor" in Chapter 6.

4.3.6 Case Study A6: Chemical Species — Physical chemistry

The case study abstract is given in next page.

A substance A alone, in a nonreactive medium, constitutes a capacitive *pole* in the Formal Graph theory. This system belongs to the physical chemical energy.

A dipole made of two physical chemical poles is either called a biphasic system, as in the case where the same species is distributed over two media, or called a simple chemical reaction between two reactive species in the same medium (see case study C5 "First-Order Chemical Reaction" in Chapter 6). (The term "simple" here implies the restriction to only two species, because a chemical reaction may involve an unlimited number of species.) When species are distributed in space due to heterogeneity of the medium or due to substance transport or any other cause, the model to consider is the space pole (see case study G3 "Fick's Law of Diffusion" in Chapter 10).

Variables: The basic quantity is the *substance amount n* (or mole number) in moles. The effort is the *chemical potential* μ and the flow is the *substance flow* $\mathfrak{J}$ (see the variables table in the case study abstract). This last state variable would correspond to the reaction rate if the species were able to undertake a reaction (see case study A11 "Reactive Chemical Species").

Formal Graph: In addition to the three state variables, the Formal Graph contains the system constitutive property linking the chemical potential to the substance amount, called physical chemical capacitance.

The system constitutive property used here is the *physical chemical capacitance*, which expresses the capability of a species to store some capacitive energy.

A6: Chemical Species **Physical Chemistry**

A

The pole constituted by a nonreactive chemical species (chemically inert) is a capacitive pole. When the species is reactive, it makes up a mixed (capacitive–conductive) pole.

Pole

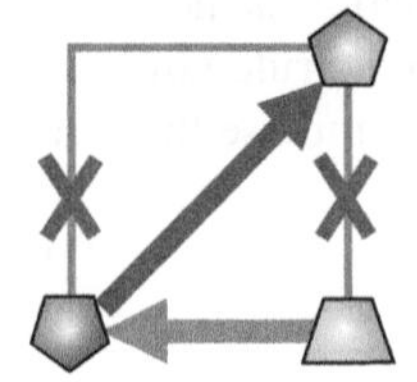

Pole	
✓	Fundamental
	Mixed
✓	Capacitive
	Inductive
	Conductive
✓	Global
	Spatial

Formal Graph:

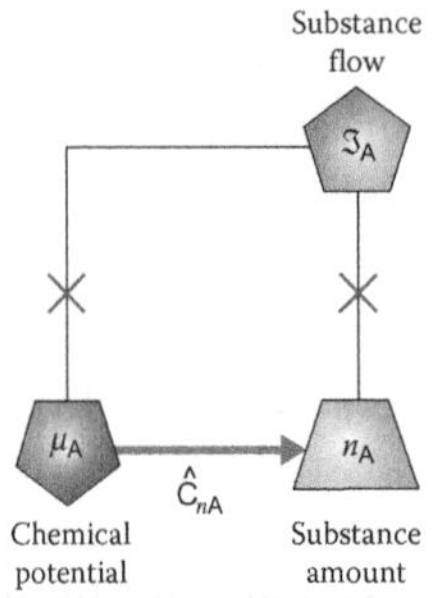

GRAPH 4.10

Capacitance operator trace:

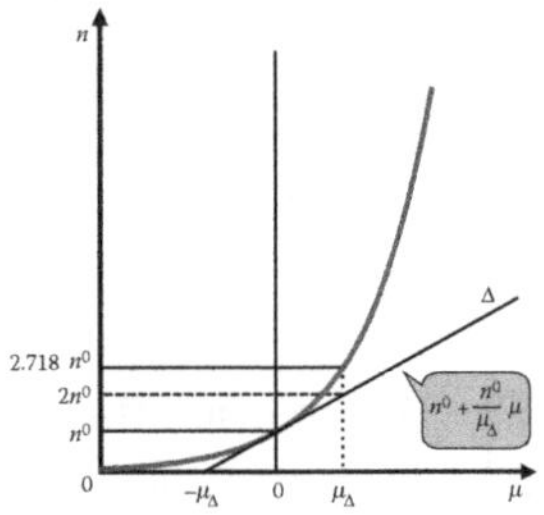

Scaling effort: $\mu_\Delta = RT$

Asymptote: $n = n^\ominus + \dfrac{n^\ominus}{\mu_\Delta}(\mu - \mu^\ominus)$

Variables:

Variety	Physical Chemistry		
Subvariety	Capacitive (potential)		Inductive
Category	Entity number	Energy/entity	Energy/entity
Family	Basic quantity	Effort	Flow
Name	Substance amount	Chemical potential	Substance flow
Symbols	n, N	μ	$\Im, v$
Unit	[mol]	[J mol^{-1}]	[mol s^{-1}]

Capacitive relationship:

$$n = \hat{C}_n \mu$$

General model

$$\frac{n}{n^\ominus} = \exp\left(\frac{\mu - \mu^\ominus}{\mu_\Delta}\right)$$

System Property	
Nature	Capacitance
Level	0
Name	Physical Chemical Capacitance
Symbol	C_n
Unit	[mol J^{-1}]
Specificity	—

Energy variation:

$$\delta \mathcal{U}_n \underset{\exists \hat{C}_n}{=} \mu \, dn$$

One may admit that the interest to represent a variable, which is not effectively used when the species is not reactive, is questionable, but the adopted rule is to represent systematically all the three variables that make a pole.

System Property: The physical chemical capacitive relationship between the chemical potential and the substance amount (mole number) classically relies on the notion of *activity* notated a, which has been introduced by Lewis* (see remark on real substance). The dependence of this activity with the chemical potential has an exponential shape

$$\frac{n}{n^\ominus} = a = exp\left(\frac{\mu - \mu^\ominus}{\mu_\Delta}\right) \tag{A6.1}$$

* Gilbert Newton Lewis (1875–1946): American physical chemist; Massachussets and California, USA.

This expression contains three scaling or reference variables. The first one is n^{θ} (zero-bar superscript), the substance amount when the activity is equal to one, taken by convention equal to 1 mol

$$n^{\ominus} = 1 \text{ mol} \tag{A6.2}$$

The second one is the *standard chemical potential* μ^{θ} ("with a barred zero") which is the chemical potential of 1 mol of the considered substance (it is therefore specific to the substance). The third one is the *scaling chemical potential* μ_{Δ}, which most of the time is equal to

$$\mu_{\Delta} = RT \tag{A6.3}$$

where R is the gas constant and T the temperature. In fact, this proportionality with the temperature expresses a coupling with the thermal energy due to the thermal agitation. In case of an absence of thermal agitation or in case of a presence of additional energy varieties, this proportionality is not verified.

Another formulation consists of choosing an origin of scale equal to 0 for the *chemical potential*

$$n = n^{0} \exp\left(\frac{\mu}{\mu_{\Delta}}\right) \tag{A6.4}$$

with the following correspondence between scaling quantities

$$n^{\ominus} = n^{0} \exp\left(\frac{\mu^{\ominus}}{\mu_{\Delta}}\right) \tag{A6.5}$$

Whatever the retained form, this capacitive relation relies on a unique *capacitance operator* allowing one to write the simple relation

$$n = \hat{\mathsf{C}}_{n} \mu \tag{A6.6}$$

A plot of the variation of substance amount as a function of the chemical potential is shown in the case study abstract.

In making the chemical potential tending toward zero, a straight line Δ is obtained with the equation

$$n - n^{0} = \frac{n^{0}}{\mu_{\Delta}} \mu \tag{A6.7}$$

Such a linearization, which can be done in every point of the exponential curve, is mainly effected for modeling physical chemical systems when they are submitted to low amplitude perturbations around a working point.

Energy: When the physical chemical capacitance exists in the system, the energy variation is

$$\delta \mathcal{U}_{n} \underset{\exists \hat{\mathsf{C}}_{n}}{=} \mu \, \mathrm{d}n \tag{A6.8}$$

Remarks

Pole and stability. To represent a chemical substance with a pole is a way to express the stability of this species, as no evolution is permitted for a solitary pole (a pole has no time).

Universality. An interest of Formal Graphs is to put into evidence a hidden aspect of our formalism in representing with the same graph structure several systems having so many differences, such as a collection of electric charges and a collection of chemical molecules (see case study A5 "Electric Charges").

Corpuscles. It is worth mentioning that when one considers the same system at the atomic scale, that is, in dividing the chemical potential by the Avogadro constant and in multiplying the substance amount by this latter, one has a model of population of molecules, also called the corpuscle group (see case study A7), that obey the same Formal Graph and capacitive relation, but with the *energy-per-corpuscle* E as effort and the *corpuscle number* N as basic quantity. The energy variety is the *corpuscular energy*.

Real substance. There is a tradition in physical chemistry to include into the relation between activity and substance amount an *activity coefficient* γ for taking into account a more complex behavior of the substance, which is called *real substance* in that case. In the Formal Graph viewpoint, the difference between real and ideal substance is attributed to the intervention of another energy variety (for instance, electrodynamical energy in the case of an ion, or superficial energy in the case of a solid, etc.). This more complex system becomes relevant from a model of coupling between energy varieties, which relies on poles as if the substance were ideal. The effect of the coupling is a modification of the chemical potentials (in shifting their origins, so they are not anymore called *standard* chemical potentials but *apparent* ones). In contradistinction, there is no effect on the substance amounts. The present model of a pole is independent from these notions of real/ideal substance, and works in all cases, despite its resemblance to the model of an ideal substance.

Universal exponential shape. A comparison between various domains reveals the same structure of Formal Graphs which points to the fact that identical shapes of relations in these graphs must exist. The most general shape for the capacitive relations is the exponential function, utilized here in the physical chemical energy variety. An exponential function may effectively tend toward a linear asymptote for weak values of its argument, thus allowing the linear laws of electrostatics to be retrieved (see case study A5 "Electric Charges") or the laws of mechanical elasticity to match Hooke's law (see case study A8 "Spring End"). (This is physically justified by a theory of influence.)

This universality relies on the notion of activity, which is generalized to all other energy varieties, in having the meaning of a variable that quantifies the influence between poles.

4.3.7 Case Study A7: Group of Corpuscles

Corpuscular Energy

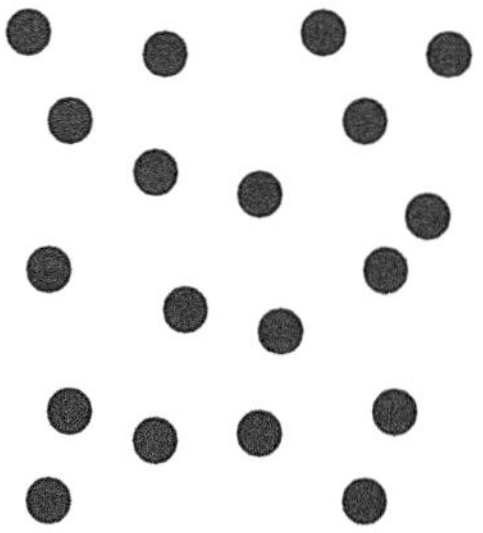

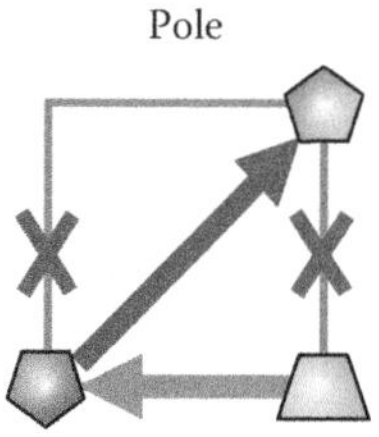

Pole	
✓	Fundamental
	Mixed
✓	Capacitive
	Inductive
	Conductive
✓	Global
	Spatial

A set of identical corpuscles constitutes a pole, belonging to the variety of corpuscular energy.

A collection of nonreactive corpuscles is a purely capacitive pole.

Formal Graph:

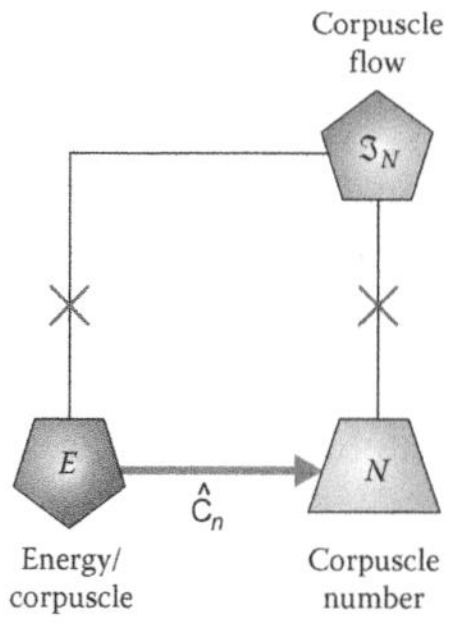

GRAPH 4.11

Capacitance operator trace:

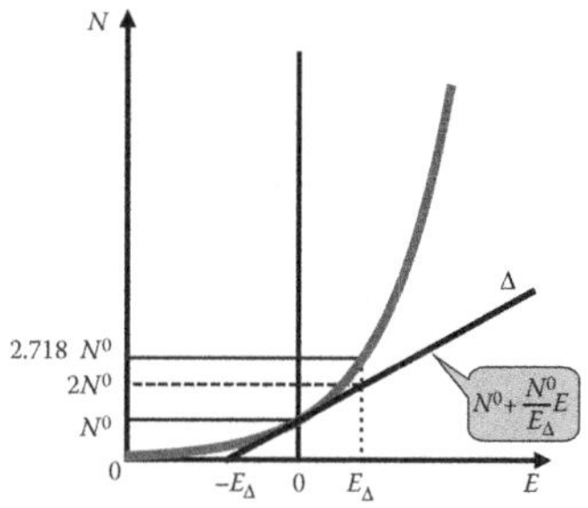

Scaling effort: $E_\Delta = k_B T$

Asymptote: $N - N^0 = \dfrac{N^0}{E_\Delta} E$

Variables:

Variety	Corpuscular Energy		
Subvariety	Capacitive (potential)		Inductive
Category	Entity number	Energy/entity	Energy/entity
Family	Basic quantity 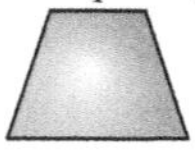	Effort 	Flow 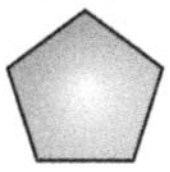
Name	Corpuscle number	Corpuscular energy	Corpuscular flow
Symbols	N, n	E	f_N
Unit	[–], [corpuscle]	[J corpuscle^{-1}]	[s^{-1}]

Capacitive relationship:

General model

$$N = \hat{C}_N E$$

Nonlinear model

$$\frac{N}{N^0} = \exp\left(\frac{E}{E_\Delta}\right)$$

System Property	
Nature	Capacitance
Level	0
Name	Corpuscular capacitance
Symbol	C_N
Unit	[J^{-1}]
Specificity	—

Energy variation:

$$\delta \mathcal{U}_N \underset{\exists C_N}{=} E\,\mathrm{d}N$$

The *corpuscular energy variety* stands for the energy of *collections* of entities considered at the atomic level and not at the molar level. The term "corpuscle" encompasses many kinds of "atomic" objects, including the "particle," the atom, the molecule, the protein, etc., but each one is considered an ensemble of entities having the same energy-per-corpuscle (the effort in this capacitive subvariety), as required by the definition of a collection in the Formal Graph theory. The term corpuscle is used here with a meaning different from that of "particle," which for instance in "particle physics" has a "mechanical" meaning of an object localized in space, to which can be given a momentum out of rest.

Representing corpuscles by a pole expresses their *stability* as no evolution is permitted for a lonely pole (no notion of time in a pole).

Variables: The basic quantity is the *number of corpuscles N*, which in principle has no SI unit because it is customary in the West to consider cardinal numbers (used for counting a countable ensemble) without units. In this work, the unit *corp* is used (to avoid confusing the *energy-per-corpuscle* with *energy*, for instance). The effort is the *energy-per-corpuscle E* (in J $corp^{-1}$) and the flow is the *corpuscle flow* f_N (in corp s^{-1}). This latter state variable would correspond to the rate of disintegration or to the rate of reaction if corpuscles were radioactive or able to undertake a reaction between particles. For understanding the emphasis on the distinction between the two variables *energy-per-corpuscle* and *energy*, see the discussion in Chapter 12, Section 12.2, about the equivalence with the physical chemical energy varieties and on the Avogadro constant.

Graph: Besides the three state variables, the Formal Graph comprises the system constitutive property linking the corpuscular energy to the number of corpuscles called *corpuscular capacitance* C_N. The interest of drawing the corpuscle flow is not perceptible here as this variable is not used for nonreactive corpuscles, but the adopted rule is to represent all the three variables that are constitutive of a pole.

The system constitutive property used here is the *corpuscular capacitance*, analogous to the physical chemical capacitance, which expresses the ability of a collection of corpuscles to store capacitive energy.

System Property: The relationship between the energy-per-corpuscle and the number of corpuscles relies on the notion of *corpuscular activity*, notated a_N, introduced by the Formal Graph theory for standardization. Its dependence on the energy-per-corpuscle has an exponential shape

$$\frac{N}{N^{\ominus}} = a_N = \exp\left(\frac{E - E^{\ominus}}{E_{\Delta}}\right) \tag{A7.1}$$

where N is the standard number of corpuscles corresponding to the unit activity, often taken as equal to 1 corpuscle, and N^{θ} is the standard energy-per-entity corresponding to the standard number of corpuscles (and activity equal to 1). The quantity E_{Δ} is a *scaling energy-per-corpuscle*. When only the thermal agitation acts on the ensemble of corpuscles, this variable is equal to

$$E_{\Delta} = k_B T \tag{A7.2}$$

where k_B is the Boltzmann* constant and T the temperature. When other energy varieties interfere, the expression of the *scaling energy-per-corpuscle* depends on them and the simple proportionality

* Ludwig Boltzmann (1844–1906): Austrian physicist; Vienna, Austria and Leipzig, Germany.

with the temperature is not observed anymore (which happens, for instance, when the free volumes in a dense material are significant or when the superficial energy plays a non-negligible role).

An alternate expression is sometimes useful, based on the number of corpuscles N^0 corresponding to zero energy-per-corpuscle, which is merely related to the standard reference through the standard energy-per-corpuscle

$$\frac{N}{N^0} = \exp\left(\frac{E}{E_\Delta}\right); \qquad \frac{N^\ominus}{N^0} = \exp\left(\frac{E^\ominus}{E_\Delta}\right) \tag{A7.3}$$

This capacitive relation may also be expressed through a *capacitance operator* making the writing simpler:

$$N = \hat{\mathsf{C}}_N\, E \tag{A7.4}$$

The representation of the variation of the corpuscle number as a function of the energy-per-corpuscle is shown below.

The asymptote obtained by making the energy-per-corpuscle tend toward zero is a straight line Δ as shown in Equation A7.5:

$$N - N^0 = \frac{N^0}{E_\Delta} E \tag{A7.5}$$

This linearization corresponds to a negligible thermal agitation, and, notwithstanding its presence, it can be carried out in any point of the exponential curve for modeling the behavior of ensembles of corpuscles when they are subjected to weak perturbations around a functioning point.

Energy: The energy stored by a corpuscle pole is a capacitive subvariety. Its variation, only possible if the capacitive property is owned by the system, is given by the product of the energy-per-entity by the variation of corpuscle number

$$\delta \mathcal{U}_N \underset{\exists C_N}{=} E\,\mathrm{d}N \tag{A7.6}$$

Remarks

Corpuscular dipole. Two populations of different corpuscles able to exchange energy form a dipole. This case is treated in case study C9 "Electrons and Holes" in Chapter 6 with a unique and same energy state. For that type of corpuscles, the *energy-per-corpuscle* is classically called *Fermi level* and the classical model used is the Fermi–Dirac statistics, replaced in the Formal Graph theory by a conservation law applied to both poles, each one being subjected to a capacitive law and being mutually influenced.

The energy of a corpuscle is not energy! Both quantities have the same SI unit, the Joule, but not the same role in thermodynamics: the *energy-per-corpuscle* E is an effort, (it is a *state variable*), whereas the energy $\mathcal{E}$ is a *state function*. To avoid confusion, the J corp^{-1} is given as a unit to the *energy-per-corpuscle* E. For more explanations, see, for instance, the discussion on the equivalence between energy varieties in Chapter 12.

Particle and corpuscle. The distinction between *particle* and *corpuscle* is important in the Formal Graph theory. A *particle* is an object possessing, among other energy varieties, *mechanical* energy, that is, inductive (kinetic) with a momentum and capacitive (potential) with a localization in space (displacement). A *corpuscle* does not necessarily possess mechanical energy. It may contain merely thermal energy or corpuscular energy, for instance. Nevertheless, it may also contain mechanical energy and may therefore become a *particle*. This distinction allows the priority setting on the energy for conceiving and defining an entity and not on the mechanical properties (velocity, position in space) that are not always relevant. This is indeed a cultural break with a long-lasting tradition of pre-eminence of mechanics, but it seriously facilitates the understanding of entangled questions such as the wave–particle duality in *quantum physics* (and not *quantum mechanics*, which in principle acknowledges only *particles*). This topic is briefly treated in Chapter 14 as an example of possibilities of the Formal Graph approach.

Energy states and energetic density. One needs to recall that the studied system is a simple case of a material with a linear distribution of *energy states* (types of sites able to be occupied). The treatment of a material with an irregular distribution of energy states, such as a semiconductor, requires the knowledge of the number of energy states as a function of the energy-per-corpuscle.

The number of corpuscles in each energy state, or *energetic density of corpuscles*, is the product of a *density of energetic states* $\mathcal{D}$, determined by quantum considerations, and the *function of distribution f*, giving the occupancy of each state, which is here equal to the product of the reference number of corpuscles (generally 1 corp) by the activity (which amounts to a more general relation than that shown in Equation A7.1 between activity and basic quantity). The notion of energetic density is applied in physics in a general fashion as the derivative of a variable with respect to the energy-per-corpuscle. Algebraically, all this can be written as

$$\mathrm{d}N = \mathcal{D} N^0 \, a_N \, \mathrm{d}E \qquad \text{(A7.7)}$$

The total number of corpuscles is then the integral over every (allowed) energy-per-corpuscle of this *energetic density of corpuscles*. The analytic calculus is not always possible (it depends on the mathematical form of the density of energy states $\mathcal{D}$, which itself depends on the number of dimensions of space).

Beyond Boltzmann. A comparison between various domains shows the same structure of Formal Graphs which leads to assessing an identity among the forms of relations in these graphs. The most general form for the capacitive relations is the exponential function used for the physical chemical energy variety. This function can effectively tend toward an asymptotic straight line for small values of its argument, thus allowing one to return to the linear laws of electrostatics or of mechanics.

This universality rests on the notion of physical chemical activity (see chemical species) which is generalized to all energy varieties, in bearing the meaning of a variable quantifying the influence between poles.

Classically, the dependence of a number of objects with their energy-per-object is modeled with statistics. When objects are independent of each other (more exactly, are subjected to independent probabilities of presence) and in sufficiently large numbers for allowing the approximation of factorials by exponentials, this dependence is modeled by Maxwell–Boltzmann statistics.

This statistic utilizes the same exponential function as the one used by the Formal Graph theory but leaves no freedom for the scaling energy-per-object, which is always proportional to the temperature via the Boltzmann constant (because an implicit hypothesis is the influence of the thermal agitation only). The higher versatility of the Formal Graphs allows any interaction and therefore any model. This possibility is indispensable for certain materials containing, for instance, important free volumes or a strong superficial energy.

4.3.8 Case Study A8: Spring End — Translational Mechanics

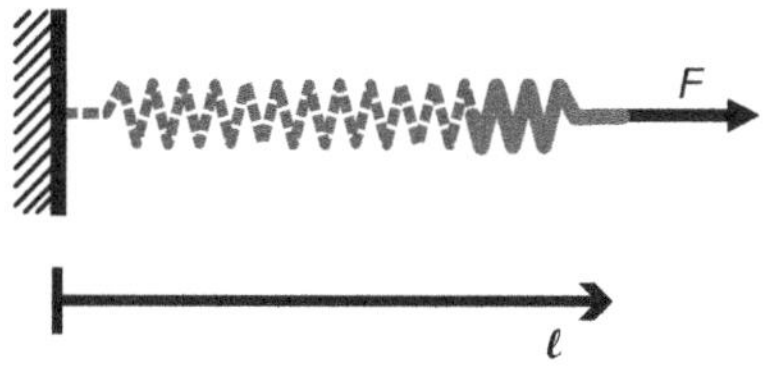

The pole made up by one end of an elastic solid, isolated and without friction, is a fundamental capacitive pole.

Pole

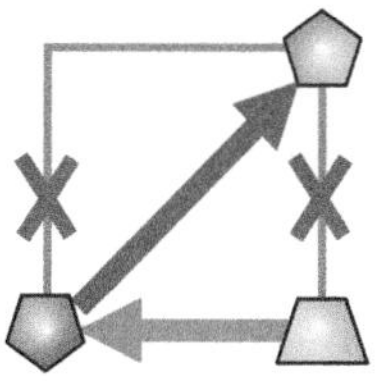

Pole	
✓	Fundamental
	Mixed
✓	Capacitive
	Inductive
	Conductive
✓	Global
	Spatial

Formal Graph:

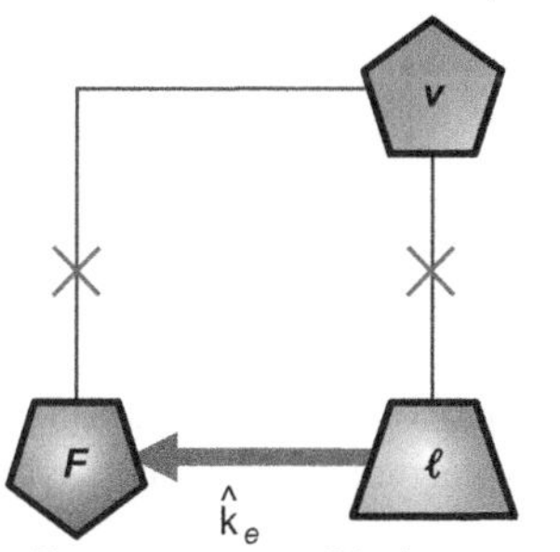

GRAPH 4.12

Capacitance operator trace:

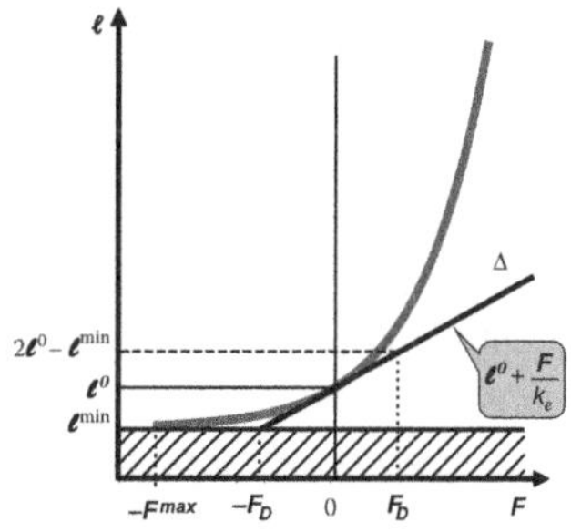

Scaling effort: $F_\Delta = k_e\,(\ell^0 - \ell^{min})$

Asymptote: $\ell = \ell^0 + \dfrac{F}{k_e}$

Variables:

Variety	Translational Mechanics		
Subvariety	Capacitive (potential)		Inductive
Category	Entity number	Energy/entity	Energy/ entity
Family	Basic quantity 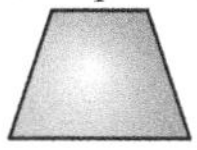	Effort 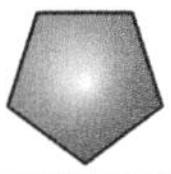	Flow 
Name	Displacement	Force	Velocity
Symbols	ℓ	F	v
Unit	[m]	[N], [J m^{-1}]	[m s^{-1}]

Capacitive relationship:

General model:

$$F = \hat{k}_e\,\ell$$

Nonlinear model:

$$\frac{\ell - \ell^0}{\ell^0 - \ell^{min}} = \exp\frac{F}{F_\Delta}$$

Hooke's law:

$$F \underset{lin}{=} k_e\,(\ell - \ell^0)$$

System Property	
Nature	Capacitance
Level	0
Name	Elasticity
Symbol	k_e
Unit	[N m^{-1}]
Specificity	—

Energy variation:

$$\delta\mathcal{U}_\ell \underset{\exists k_e}{=} F \cdot d\ell$$

A spring, with one fixed end and another free end, gives an approximate idea of a pole in the domain of elasticity of solids. Ideally, only one end of the spring should be considered. Obviously, this is not practically feasible and this is a purely virtual concept. The only real system has two ends, making a dipole, or no ends at all in the case of a tore (e.g., rubber band).

Nevertheless, it is not because a dual system (dipole) is not separable that it is forbidden to consider the properties of each part.

Variables: In the capacitive subvariety of energy in translational mechanics, the basic quantity (number of entities) is the *displacement*, or *length* ℓ, which in the case of a spring is known as *elongation*. The effort (energy-per-entity) is the *force* $\boldsymbol{F}$. Both are quantities oriented in space and therefore represented by vectors. A third state variable, the flow (here the velocity), is only used in case of dissipation due to internal friction, which is not envisaged here.

System Constitutive Property: The free end of the spring obeys the capacitive law linking its displacement (elongation) to the force exerted on the spring

$$\boldsymbol{F} = \hat{\mathrm{k}}_e\, \ell \tag{A8.1}$$

The operator in this capacitive relation above is the *elastance*, the reciprocal of the capacitance. The Formal Graph theory predicts an exponential form for the function representing the capacitance operator, which is identical for all energy varieties.

Standard law. The exponential law proposed as the standard model of capacitive relationship works for ideal solids, that is, containing only one energy variety and homogeneous. When other energy varieties (superficial energy for instance) perturb the ideal scheme, more complex laws, based on this simple one, must be used. This is the case for most solids

A peculiarity of an elastic solid in comparison with other systems that are containers of capacitive energy (see case study A6 "Chemical Species" or A5 "Electric Charges") is to have a lower limit for the basic quantity: This limit is the minimum displacement corresponding to the maximum possible compression without destroying the solid. The model is based on a generalization of the notion of *activity* (see case study A6 "Chemical Species") whose dependency on the effort has an exponential form and the algebraic expression of its relationship with the displacement is as follows:

$$\frac{\ell - \ell^{\min}}{\ell^0 - \ell^{\min}} = a_\ell = exp\,\frac{\boldsymbol{F}}{\boldsymbol{F}_\Delta} \tag{A8.2}$$

This model utilizes two characteristic lengths for the spring, the displacement ℓ^0 of the system at rest ($\boldsymbol{F} = \boldsymbol{0}$), without application of a force, and the minimum displacement $\ell^{\min}$ below which the spring cannot be compressed. The activity must be understood as the ratio between the basic quantity that is effectively available ("active") for influencing some other poles and a reference quantity. For the force, a *scaling force* $\boldsymbol{F}_\Delta$ is used; its expression is written with the eigenvalue of the elastance operator, notated k_e

$$\boldsymbol{F}_\Delta = k_e\,(\ell^0 - \ell^{\min}) \tag{A8.3}$$

For small elongations, the capacitive relation becomes linear and is referred to as Hooke's law. This law is featured by straight line Δ on the plot in the case study abstract and the algebraic expression is

$$\boldsymbol{F} \underset{lin}{=} k_e\,(\ell - \ell^0) \tag{A8.4}$$

In this peculiar case, the elastance is a scalar independent from the displacement, and the proportionality factor k_e is called *rigidity* or *elasticity* in translational mechanics, and in this case study, it is called the *spring constant.*

Energy: The variation of capacitive energy, also called internal energy of the spring, is given by

$$\delta \mathcal{U}_\ell \underset{\exists k_e}{=} \boldsymbol{F} \cdot \mathbf{d}\ell \tag{A8.5}$$

In this domain, the capacitive energy is called *work,* meaning a mechanical energy that can be utilized.*

In the linear case, it is possible to express the energy variation as the product of the length by the force variation.

$$\delta \mathcal{U}_\ell \underset{lin}{=} \ell \cdot \mathbf{d}\boldsymbol{F} \tag{A8.6}$$

This is far from the general case and it is preferable not to use this peculiar expression.

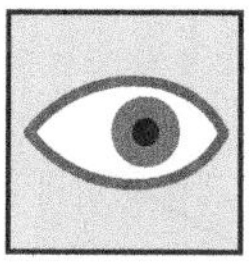

An isolated pole cannot see its energy changed without contribution of an external effort (thus becoming an open system, i.e., nonisolated). Here, it corresponds to the intervention of an external force, for instance, the force on the other end of the spring. Note that a full spring, with its two ends, constitutes a dipole.

Role of space. We have considered one end of the spring as representative of a pole in basing our reasoning on the idea that a dipole is necessarily made of two poles. This concept is sufficient for modeling a number of systems, such as homogeneous bodies like the ideal spring. However, this global approach is not suitable for all cases because it is an oversimplification. In fact, a spring is a bit more complex because it has an extension in space, featured by its length, which means that this longitudinal system is composed of a great number of elementary pieces, each one behaving as a pole. However, we could simplify the model because a chain of identical poles is equivalent to a single dipole, whatever the number of elements.

This behavior is characteristic of what is called a *space pole*, which depends on the properties of space. New variables and new system properties that are defined at a space-localized level (curve, surface, or volume) help model such a system. An example of a space pole is given in the same topic of elasticity with case study B1 "Elastic Element" in Chapter 3.

The modeling of some systems requires entering into the details of space distribution by using space poles, such as the vibrating string, and this is especially the case when an inhomogeneous structure prevents the use of simple models. This topic is treated in elasticity of solids.

* "Work" has too restrictive a meaning, so this term will not be used anymore in this text. As will be progressively apparent, it is interesting to go beyond the mechanical concept of energy by replacing "work" by one among the energy subvarieties.

4.3.9 Case Study A9: Thermal Pole

Thermics

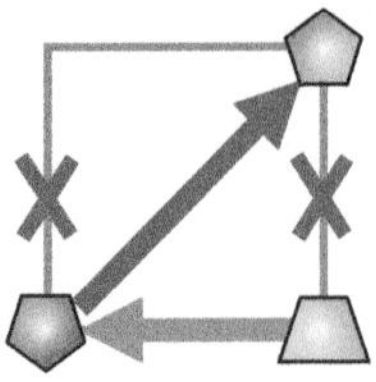

Pole	
✓	Fundamental
	Mixed
✓	Capacitive
	Inductive
	Conductive
✓	Global
	Spatial

Formal Graph:

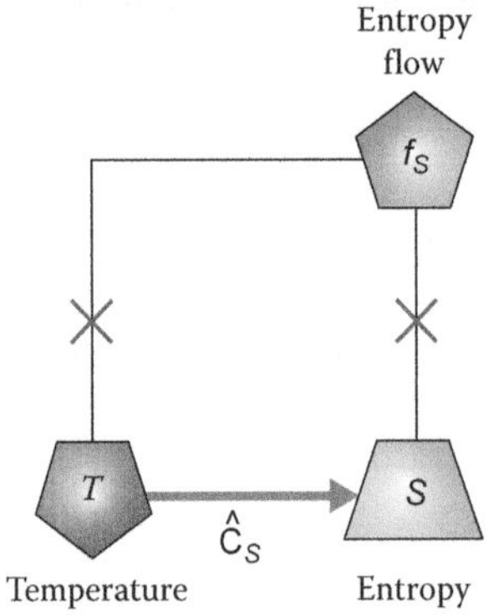

GRAPH 4.13

Elastance operator trace:

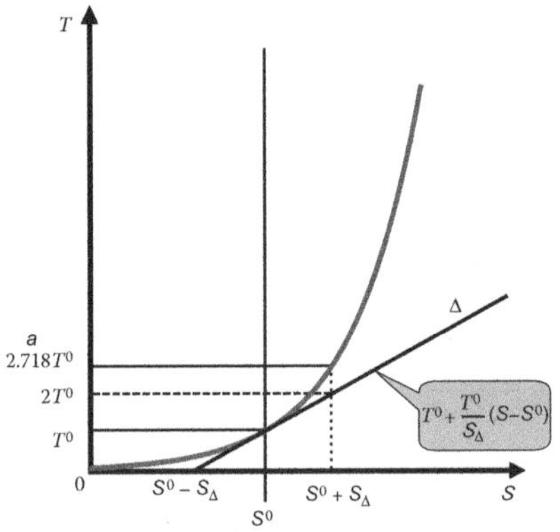

Scaling entropy: $S_\Delta = k_B$

Asymptote: $T - T^0 = \frac{T^0}{S_\Delta}\left(S - S^0\right)$

Variables:

Variety	Thermics		
Subvariety	Capacitive (potential)		Inductive
Category	Entity number	Energy/entity	Energy/entity
Family	Basic quantity 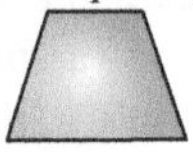	Effort 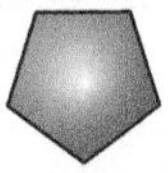	Flow 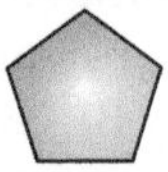
Name	Entropy	Temperature	Entropic flow
Symbols	S	T	f_S
Unit	[J K^{-1}]	[K]	[J K^{-1} s^{-1}]

Capacitive relationship:

$$S = \hat{C}_S\, T$$

Ideal pole

$$S = S^0 + S_\Delta ln\left(\frac{T}{T^0}\right)$$

System Property	
Nature	Capacitance
Level	0
Name	Thermal Capacitance
Symbol	C_S
Unit	[J K^{-2}]
Specificity	—

Energy variation:

$$\delta \mathcal{U}_S \underset{\exists \hat{C}_S}{=} T\, dS$$

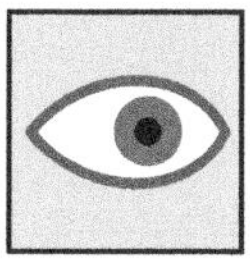

Designations. Thermal energy is also called "heat" and sometimes "thermics," which has mainly a technological meaning but which is used here in a broader sense because there is no better word. "Thermodynamics" would be the correct word but it is universally used for the science dealing with all the energy varieties and their relationships.

A thermal pole is difficult to apprehend. It has no perceptible existence alone, and is easier to envisage when associated in a pair comprising a dipole. For instance, the well-known process of heat transfer (see case study H2 "Fourier Equation" in Chapter 11) is an exchange of heat between two thermal poles. However, the coupling with another energy variety provides many examples of well-known physical systems. The process of dissipation, for instance, consists of the association of a thermal pole with a pole belonging to another energy variety, being able to convert (or to degrade) this energy into heat (see Chapter 11). The coupling with an energy variety in which the entities are objects that can be counted and can move, such as molecules or corpuscles, is known as *thermal agitation*. Notwithstanding this difficulty of conceiving a system only filled with pure heat, a thermal pole is a concept that allows modeling the ability of a system to store some capacitive energy under the form of heat.

Variables: In this energy variety, the basic quantity is not the temperature, as it is often believed, but the *entropy* S, which is a number of entropic entities containing energy-per-entity (effort) that is the *temperature* T. This latter is an intensive variable whereas the entropy is an extensive one. The flow is the *entropy flow* f_S, also called entropy production rate.

System Property: The general capacitive relationship between the temperature and the entropy is written with the help of a thermal capacitance operator

$$S = \hat{C}_S\,T \tag{A9.1}$$

Capacitance. The system constitutive property used here is the *thermal capacitance*, which expresses the capability of a system to store some capacitive energy in the form of heat.

A peculiar specificity of the thermal energy compared to other energy varieties is that the capacitive relationship is not exactly the same. It differs mainly in the respective roles of the two capacitive state variables, basic quantity and effort, which are symmetrically swapped. For instance, the simplest model that corresponds to an ideal thermal pole uses the same exponential function as in the other energy varieties, but in a reversed way. In other words, this is the reciprocal relationship, called "elastive relationship," that is written with the exponential function instead of the capacitive one in all other varieties.

$$T = T^0 exp\left(\frac{S - S^0}{S_\Delta}\right) \tag{A9.2}$$

Normally (i.e., in other energy varieties), the argument of the exponential function is proportional to the effort and the result is the basic quantity. Such an exception is found in the inductive subvariety where a pole in translational mechanics (see case study A1 "Moving Body") has its inductive relationship described by the same quadratic function, but with inverted state variables with respect to other energy varieties such as electromagnetism and magnetism (see case studies A3 "Current Loop" and A4 "Magnetization"). Although the preferred form of this relationship in this theory is the exponential function, another alternative using the logarithmic function is widely used:

$$S = S^0 + S_\Delta ln\left(\frac{T}{T^0}\right) \tag{A9.3}$$

Three parameters are used in these relations. T^0 is a scaling or reference temperature that corresponds to the value of the temperature when the entropy is equal to the reference entropy S^0. As the exponential function allows infinity of reference couples, the choice of a peculiar couple is left to the modeler. The third parameter is the scaling entropy S_Δ. Its value depends on the choice of the temperature scale, and, in the case of the absolute scale (with units in Kelvin), this parameter is equal to the Boltzmann constant:

$$S_\Delta = k_B \tag{A9.4}$$

A plot of the variation in temperature as a function of the entropy is shown in the case study abstract.

In making the entropy tend toward the reference entropy S^0, a straight line Δ is obtained with the following equation:

$$T - T^0 = \frac{T^0}{S_\Delta}\left(S - S^0\right) \tag{A9.5}$$

Such a linearization, which can be done in every point of the exponential curve, is mainly effected for modeling thermal systems when they are submitted to low amplitude perturbations around a working point.

Energy: The energy $\mathcal{U}_S$ stored by a collection of entropic containers is a capacitive subvariety that is traditionally called *heat* or thermal energy and classically notated Q. Its variation is given by the product of the temperature (the energy-per-entity) and the variation of entropy (the basic quantity), provided the thermal capacitance exists in the system:

$$\delta\mathcal{U}_S \underset{\exists \hat{C}_S}{=} T\,dS \tag{A9.6}$$

This is one of the very fundamental laws of physics leading to the First Principle of Thermodynamics, when associated with other energy variations.

Similarities of structures. An interest of Formal Graphs is to put into evidence a hidden aspect of formalism in representing with the same graph structure several systems having so many differences, such as a collection of electric charges and a collection of chemical molecules (see case study A5 "Electric Charges").

Similarities of operator shapes. A comparison between various domains reveals the same structure of Formal Graphs which points to the fact that identical shapes of relations in these graphs must exist. The most general shape for the capacitive relations is the exponential function, utilized here in the thermal energy variety. However, the thermal energy is a notorious exception as the role of the state variables is inverted (meaning that the capacitive relationship is a logarithmic function instead of an exponential function).

4.3.10 Case Study A10: Motion with Friction — Translational Mechanics

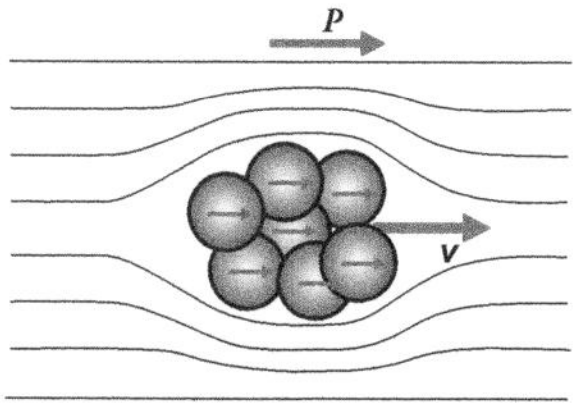

The pole made up of a mechanical object isolated from any interaction capable of modifying its movement, except from friction with its environment, is a mixed inductive–conductive pole

Pole

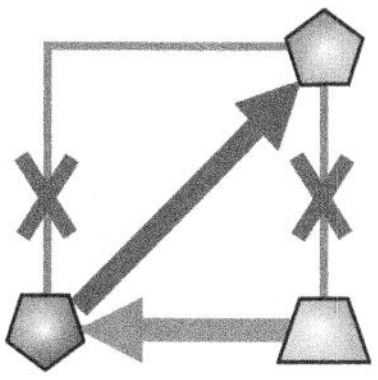

Pole	
	Fundamental
✓	Mixed
	Capacitive
✓	Inductive
✓	Conductive
✓	Global
	Spatial

Formal Graph:

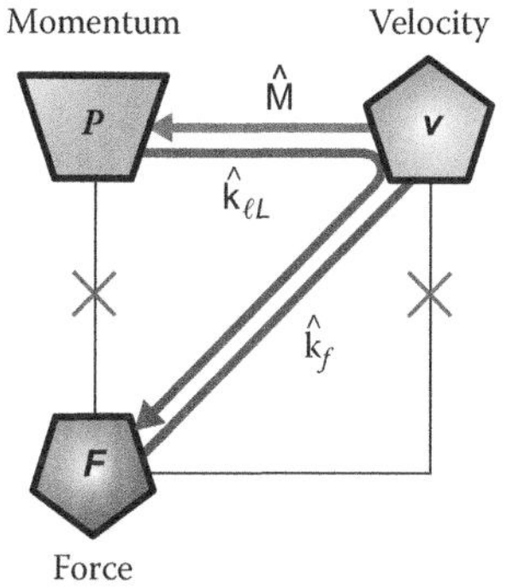

GRAPH 4.14

Variables:

Variety	Translation Mechanics		
Subvariety	Inductive (kinetic)		Capacitive
Category	Entity number	Energy/ entity	Energy/ entity
Family	Impulse 	Flow 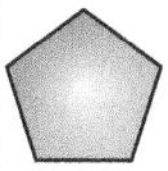	Effort 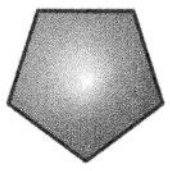
Name	Momentum (quantity of movement)	Velocity	Force
Symbols	P, (p)	$\mathbf{v}$	$\mathbf{F}$
Unit	[N s], [J s m^{-1}]	[m s^{-1}]	[N], [J m^{-1}]

Inductive relationship:

$$P = \hat{M}\mathbf{v}$$

Conductive relationship:

$$F = \hat{k}_f \mathbf{v}$$

Combination:

$$F = \hat{k}_{\ell LG} P$$

$$\hat{k}_{\ell LG} = \hat{k}_f \hat{M}^{-1}$$

System Properties		
Nature	Inductance	Resistance
Level	0	0
Name	Inertial mass	Friction "Coefficient"
Symbol	M	k_f
Unit	[kg]	[N s m^{-1}]
Specificity	Inductance	Resistance

Energy variation:

$$\delta \mathcal{T}_\ell \underset{\exists M}{=} \mathbf{v} \cdot \mathbf{d}P$$

The movement of a body that is slowed down by friction due to resistance of the medium is a topic belonging to *mechanics of the point* as well as to *fluid mechanics*. Here, the case study will be treated in the frame of translational mechanics. In the Formal Graph theory, this system is modeled with the help of a *mixed pole*, because two constitutive properties of the system are required, an inductance (representing the inertial mass) and a conductance (representing the friction process). The case without friction has been treated in case study A1 "Moving Body."

Variables: The energy variety to which this system belongs is the translational mechanic energy, the basic quantity of which is the *displacement* $\boldsymbol{\ell}$, the impulse is the *momentum* or *quantity of movement* $\boldsymbol{P}$. The effort is the *force* $\boldsymbol{F}$ and the flow is the *velocity* $\boldsymbol{v}$. All these state variables are oriented in space and are consequently represented by vectors.

In mechanics, a *body* is an object made up of several parts and its mass is the sum of the masses of each part. This in opposition with the notion of *particle* which is an elementary entity possessing an indissociable mass (without losing the nature of the particle). To distinguish the two systems, the scheme in the case study abstract shows a cluster of particles for heavily insisting on the variability of the entity number (the momentum, or quantity of movement, of the body is the sum of the quantities of movement attributed to each particle). In all equations, the variables featuring the quantity of movement and the mass are in uppercase for a body and in lowercase for a particle.

Formal Graph: An *isolated moving body* with an inductance (the inertial mass) and with a conductance (modeling the friction) constitutes a *mixed pole* in the Formal Graph theory.

Inductive Property: The momentum (quantity of movement) $\boldsymbol{P}$ is linked to the velocity by the inductive relation

$$\boldsymbol{P} = \hat{\mathrm{M}}\,\boldsymbol{v} \tag{A10.1}$$

The hat on the mass means that this inductance is not a scalar but an operator. Actually, in the general case (special relativity), the inductive relation is not linear, whereas in the special case of Newton's theory of movement (velocities lower than $c/10$), the inertial mass is a scalar.

Inertia and friction. The inertial mass expresses the ability of an object to store some inductive energy, here kinetic energy.

Friction can convert kinetic energy into heat. The force of friction is related to the velocity by an operator of friction resistance.

Conductive Property: The friction of the body on the fluid, or on the medium, in which it moves, is modeled by a relation between the velocity and the force of friction:

$$\boldsymbol{F} = \hat{\mathrm{k}}_f\,\boldsymbol{v} \tag{A10.2}$$

An operator of friction resistance, or viscous resistance, is used for linking these two state variables, because the relation is linear only in a validity domain restricted to low velocities (Newtonian fluid). In a gas, such as air, the dependence of the force is most often expressed by a quadratic function of the velocity; however, many different dependences are modeled in practice. The existence of friction signifies a dissipation of the kinetic energy into heat.

A combination of the two relations A10.1 and A10.2, or composition of the two paths in the Formal Graph, provides a third operator of viscous inertia which appears as a "velocity constant" analogous to those "constants" met in chemical kinetics (see the next case study A11 "Reactive Chemical Species"), or more generally in relaxation of dynamic processes

$$\hat{\mathrm{k}}_{\ell LG} = \hat{\mathrm{k}}_f\,\hat{\mathrm{M}}^{-1} \tag{A10.3}$$

This operator allows one to write the direct relation between momentum (quantity of movement) and force:

$$\boldsymbol{F} = \hat{\mathrm{k}}_{\ell LG}\,\boldsymbol{P} \tag{A10.4}$$

Naturally, the reciprocal operator corresponds to a relaxation "time constant," in other words to a characteristic time, estimating the duration of the move when the body is isolated from any external force. When this pole is associated to a thermal pole, conversion of energy into heat is possible, implying, according to the Formal Graph conception of time, that time plays a role of *conversion advance*. When all inductive (kinetic) energy is converted, the body comes to rest, as a result of the effect of the friction. And obviously, the external medium represented by the thermal pole, receives the equivalent thermal energy during the relaxation.

$$\hat{\tau}_{\ell LG} = \hat{M}\,\hat{k}_f^{-1} \tag{A10.5}$$

Energy: Although converted into heat when a possibility of energy exchange exists, energy is initially stored in totality under the inductive form in this system (As explained above, the conversion duration can be estimated with the help of the "viscous time constant" given the previous equation.) The energy of a moving object is traditionally called *kinetic energy*. Its variation is given by the product of the velocity and the momentum variation

$$\delta \mathcal{T}_\ell \underset{\exists M}{=} \boldsymbol{v} \cdot \mathbf{d}\boldsymbol{P} \tag{A10.6}$$

and not in the reverse order, which is only valid in the restricted range of a linear inductive relationship (Newtonian frame).

Remarks

Potentiality. As time is lacking in a pole, no possibility of evolution is left to an isolated body. This means that this pole only models a possibility, a potentiality of friction during movement, in the same way as a reactive species represents a possibility of reaction, which will become effective once a second system (another pole) is connected to this one.

Body, corpuscle, or particle. It must be recalled that a distinction is made in the Formal Graph theory between a *body* and a *particle*, for being able to model correctly what is relevant from the macroscopic world (the body) and from the microscopic world (the elementary particle). A body is necessarily made up of several particles (atoms).

Another distinction is made between *particle* and *corpuscle*. This latter is a generic name for all particles, molecules, atoms, etc. that may or may not possess mechanical properties. A particle does have mechanical properties, such as a momentum or a position in space (displacement). A corpuscle may not have position (i.e., has no capacitive energy in translational mechanics).

4.3.11 Case Study A11: Reactive Chemical Species — Physical Chemistry

The case study abstract is given in next page.

A chemical species A alone, able to react with another chemical substance but in the absence of this latter, constitutes a *mixed pole* (capacitive–conductive) in the Formal Graph theory. The case of reaction corresponds to a mixed capacitive–conductive dipole, which is treated in case study C5 "First Order Chemical Reaction" in Chapter 6.

A11: Reactive Chemical Species **Physical Chemistry**

$$\mathrm{A} \xrightarrow{k} \cdots$$

The pole made up by a chemical species potentially reactive but isolated from any other substance able to react with it, is a mixed pole, of the capacitive–conductive type. For modeling a reaction between two species, a dipole is required.

Pole

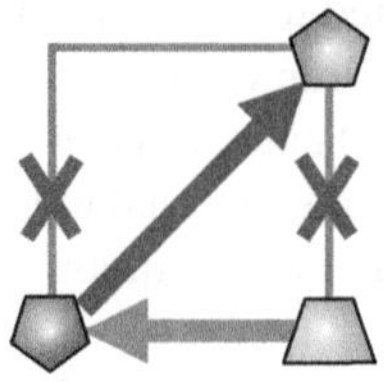

Pole	
	Fundamental
✓	Mixed
✓	Capacitive
	Inductive
✓	Conductive
✓	Global
	Spatial

Formal Graph:

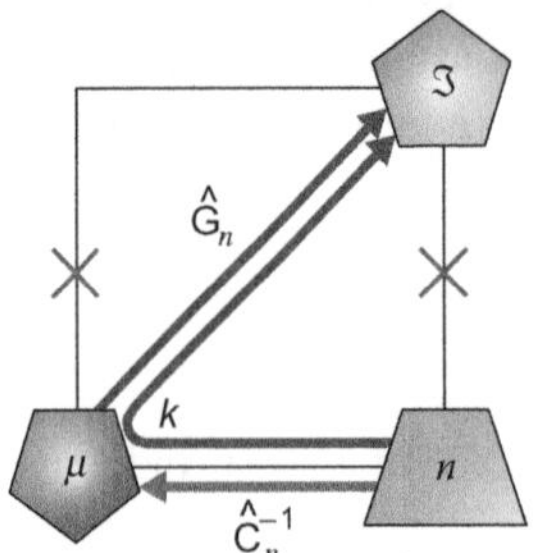

Capacitance operator trace

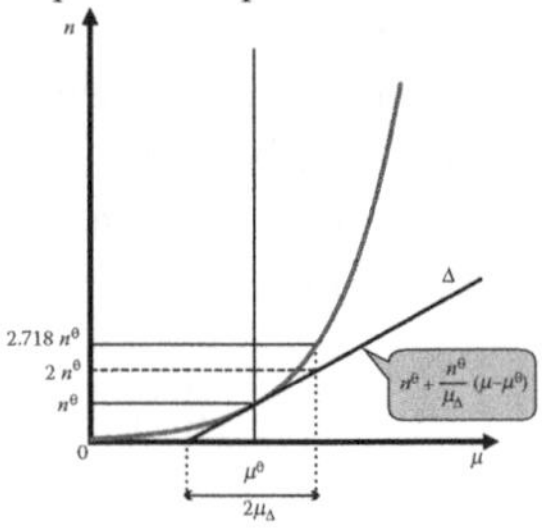

Conductance operator trace

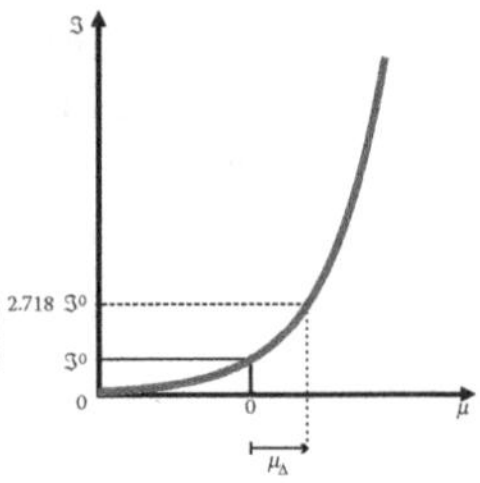

Scaling effort: $\mu_\Delta = RT$

Asymptote: $n = n^\circ + \dfrac{n^\circ}{\mu_\Delta}\left(\mu - \mu^\circ\right)$

GRAPH 4.15

Variables:

Variety	Physical Chemistry		
Subvariety	Capacitive (potential)		Inductive
Category	Entity number	Energy/ entity	Energy/ entity
Family	Basic quantity	Effort	Flow
Name	Substance amount	Chemical potential	Substance flow
Symbols	n, N	μ	$\Im$, v
Unit	[mol]	[J mol^{-1}]	[mol s^{-1}]

Capacitive relationship:

$$\frac{n}{n^\circ} = \exp\left(\frac{\mu - \mu^\circ}{\mu_\Delta}\right)$$

Conductive relationship:

$$\frac{\Im}{\Im^0} = \exp\frac{\mu}{\mu_\Delta}$$

Combination:

$$\Im = k\,n$$

$$k = \frac{\Im^0}{n^\circ}\exp\left(\frac{\mu^\circ}{\mu_\Delta}\right)$$

System Properties		
Nature	Capacitance	Conductance
Level	0	0
Name	Physical chemical capacitance	Physical chemical conductance
Symbol	C_n	G_n
Unit	[mol^2 J^{-1}]	[mol^2 J^{-1} S^{-1}]
Specificity	—	—

Energy variation:

$$\delta\mathcal{U}_n \underset{\exists \hat{C}_n}{=} \mu\, dn$$

The representation of a chemical substance by a pole expresses the stability of this species in the absence of reacting partner since evolution is not permitted for a pole alone (no time in a pole). On the contrary, the formation of a dipole by associating another pole representing another chemical species (or the same species separated from the first one by an interface) allows a temporal evolution of substance amounts, because of a chemical reaction (or an exchange across the interface; see case study C6 "Physical Chemical Interface" in Chapter 6).

Variables: This system belongs to the physical chemical energy, in which the basic quantity is the *substance amount n* (or mole number) with unit in moles. The effort is the *chemical potential* μ and the flow is the *substance flow* $\Im$ (see the table of variables in the case study abstract). The name *substance flow* comes from the span of behaviors that the physical chemical energy has, which is wider than the chemical reactivity. Mass transfer (convection, diffusion, etc.) is a phenomenon belonging to this energy variety (see case study G3 "Fick's Law of Diffusion" in Chapter 10), which thereby justifies the naming.

Dynamics: The *substance flow* corresponds here to the reaction rate when the species undertakes a reaction. When the hypothesis of a first-order kinetic is made, the reaction rate is related to the substance amount through a coefficient called "rate constant" or "specific rate of reaction" k.

$$\mathfrak{I} = k\,n \tag{A11.1}$$

This kinetic equation is represented on the Formal Graph by a composed path taking first the elastance property of the system and then the conductance property. The algebraic translation uses the composition of operators

$$k = \hat{G}_n \hat{C}_n^{-1} \tag{A11.2}$$

System Constitutive Properties: The capacitive relationship is the same as for a nonreactive chemical species (see case study A6 "Chemical Species"); it relies on the notion of *activity*, which depends on the chemical potential through an exponential function

$$\frac{n}{n^{\theta}} = a = \exp\left(\frac{\mu - \mu^{\theta}}{\mu_{\Delta}}\right) \tag{A11.3}$$

Three variables are used as scaling factors, the *standard chemical potential* μ^{θ}, corresponding to the *reference substance amount* n^{θ} (and a unit activity) and the *scaling chemical potential* μ_{Δ}, whose value is infinite in the absence of coupling with another energy variety and finite in the presence of coupling. Coupling with the thermal variety, reflecting the thermal agitation of a population of molecules, makes this potential equal to the product RT.

The conductive relationship is established in the Formal Graph theory on the same model, from a theory of exchange by flow between two poles. It involves as parameters, in addition to the *scaling chemical potential* μ_{Δ}, a reference flow $\mathfrak{I}^0$ corresponding to the value of chemical potential equal to zero. The concept that replaces the activity in the capacitive relationship is an "exchange activity" b

$$\frac{\mathfrak{I}}{\mathfrak{I}^0} = b = \exp\frac{\mu}{\mu_{\Delta}} \tag{A11.4}$$

This equation represents a general model of conduction by exchange of basic quantities through flows (these are the vectors of the exchange), which applies to all varieties of energy. (This is not the only possible model of conduction, but it is the simplest.) A theory of conduction will be proposed in Chapter 8 when studying multipoles.

The plots of the curves of the substance amount versus the chemical potential (capacitive relationship A11.3) and of the substance flow versus the chemical potential (conductive relationship A11.4) are given in the case study abstract.

However, depending on the areas, the concept of conduction is seen differently or not at all. This is the case of chemical kinetics that does not decompose rate constants into properties of capacitance and conductance as do the Formal Graphs. To return to the classical theory, one needs to recombine the processes of conduction and capacitive storage in a single kinetic process.

By grouping the reference variables into an *intrinsic rate constant* k^0

$$k^0 = \frac{\mathfrak{I}^0}{n^{\theta}} \tag{A11.5}$$

and by combining Equations A11.1 and A11.4 one finds an expression for the *reaction rate constant* k

$$k = k^0 \exp\left(\frac{\mu^\theta}{\mu_\Delta}\right) \tag{A11.6}$$

This resembles a classical expression called *Arrhenius** *law* (1889), established on an empirical basis,

$$k = A\exp\left(-\frac{E_a}{RT}\right) \tag{A11.7}$$

where A, in purely mathematical terms, is the "*pre-exponential factor*" and E_a is the *molar energy of activation*. The pre-exponential factor corresponds to the *intrinsic rate constant* k^0 and the molar activation energy is identified with the opposite of the *standard chemical potential* μ^θ. The denominator of the argument of the exponential is the product of the *gas constant* R and the temperature T(K), which equals the *scaling chemical potential* μ_Δ

$$\mu_\Delta = RT \tag{A11.8}$$

In the Formal Graph theory, this relationship reflects a coupling between the physical chemical and the thermal energy varieties (without the intervention of another variety, which would give another term for the scaling chemical potential μ_Δ leading therefore to a law other than the Arrhenius law).

Another approach, based on the theory of the transition state (by Wigner,† Eyring,‡ Polanyi,§ and Evans¶ in 1930; see Laidler and King 1998), provides a more rigorous model based on quantum considerations on the transition frequencies of a reaction intermediate, which is the following:

$$k = \frac{k_B T}{h}\exp\left(-\frac{\Delta_r G_{\neq}}{RT}\right) \tag{A11.9}$$

The quantity that appears in the argument of the exponential at the numerator is the *molar free enthalpy of reaction* of the reaction intermediate that is identified with the *molar activation energy* E_a and with the opposite of the *standard chemical potential* μ^θ. The "pre-exponential" term is the intrinsic rate constant k^0 that is equal to the scaling chemical potential μ_Δ divided by the Avogadro constant N_A and by the Planck constant h or what amounts to the same, by dividing numerator and denominator by N_A to keep only the Boltzmann constant k_B (the values of these constants are given in Appendix 2).

The Formal Graph theory retrieves and merely explains this expression of the intrinsic rate constant k^0 in terms of coupling at a level of organization lower than the poles, the "singletons," which correspond to quantum objects. In Chapter 14, is given a Formal Graph modeling the *intrinsic rate constant* k^0 by means of *singletons* as an example of the possibilities of the Formal Graph theory in quantum physics.

* Svante August Arrhenius (1859–1927): Swedish chemist; Stockholm, Sweden.

† Eugene Paul Wigner (1902–1995): Hungarian-American physicist and mathematician; Princeton, New Jersey, USA.

‡ Henry Eyring (1901–1981): American chemist; Salt Lake City, Utah, USA.

§ Michael Polanyi (1891–1976): Hungarian-British physical chemist; Manchester, UK.

¶ Meredith Gwynne Evans (1904–1952): British physical chemist; Manchester, UK.

Energy: When the physical chemical capacitance exists in the system, the energy variation is

$$\delta \mathcal{U}_n \underset{\exists \hat{C}_n}{=} \mu \, dn \qquad \text{(A11.10)}$$

Alternate Formal Graph: The possibility to model a reactive species or a chemical reaction in the physical chemical energy variety is restricted to first-order reactions. Higher-order reactions require another energy variety especially devoted to the modeling of any kind of reaction called *chemical reaction energy*. The state variables in this energy variety are the *reaction extent* ξ as basic quantity, the *affinity* $\mathcal{A}$ as effort, and the *reaction rate* $\mathcal{V}$ as flow.

Variety	**Chemical Reaction**		
Subvariety	Capacitive (Potential)		Inductive
Category	Entity number	Energy/entity	Energy/entity
Family	Basic quantity	Effort	Flow
	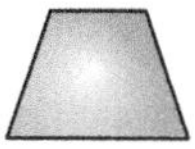	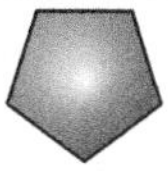	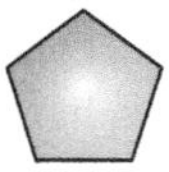
Name	Extent	Affinity	Reaction rate
Symbol	ξ	$\mathcal{A}$	$\mathcal{V}$
Unit	[mol]	[J mol^{-1}]	[mol s^{-1}]

In fact, this energy variety is always associated with the physical chemical one through an equivalence relationship. It consists of linking homologous state variables by coefficients having the same value, corresponding to the order of reaction. Here, in Graph 4.16, is shown the equivalence for a first-order reaction between two poles.

The physical chemical pole is a purely capacitive one, whereas the chemical reaction pole is a purely conductive one. The path representing the specific rate k is the combination of operators

$$k = \hat{G}_\xi 1 \hat{C}_n^{-1} \qquad \text{(A11.11)}$$

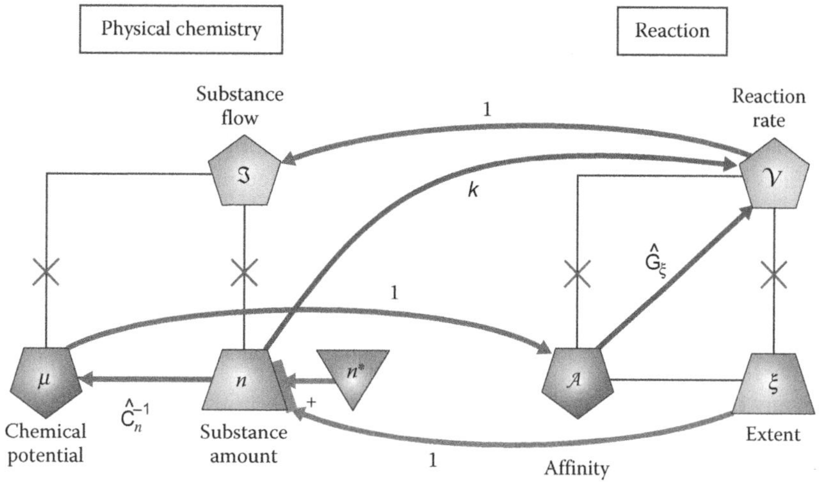

GRAPH 4.16 Equivalence between physical chemical and chemical reaction energies when the reaction order is one.

As the coefficients 1 make all corresponding state variables identical, there is no real interest in using this more complicated graph compared to the one with the sole physical chemical energy given first. However, this alternate representation is shown for pedagogical reasons in allowing comparison with graphs of *n*th-order reactions.

Remarks

Stability. To represent a chemical substance with a pole is a way to express the stability of this species as no evolution is permitted for a solitary pole (a pole has no time).

Structure similarities. An interest of Formal Graphs is to put into evidence a hidden aspect of our formalism in representing with the same graph structure several systems having so many differences, such as a collection of electric charges and a collection of chemical molecules (see case study A5 "Electric Charges").

Similarities of operator shapes. A comparison between various domains reveals the same structure of Formal Graphs which points to the fact that identical shapes of relations in these graphs must exist. The most general shape for the capacitive relations is the exponential function, utilized here in the physical chemical energy variety. An exponential function may effectively tend toward a linear asymptote for weak values of its argument, thus allowing the linear laws of electrostatics to be retrieved (see case study A5 "Electric Charges") or the laws of mechanical elasticity to match Hooke's law (see case study A8 "Spring End").

4.4 COMPOSITION OF POLES

As already mentioned, a collection can be composed of smaller collections, meaning that several poles can be grouped for making up a new pole. Naturally, all subpoles must be of the same nature, all fundamental (inductive, capacitive, or conductive) or all mixed with the same mixing of constitutive properties.

The essential rule to respect is the uniqueness of the value of energies-per-entity among all subpoles, which means an absence of energy exchange between them. Practically, it means that interactions are forbidden between subpoles, or if interactions exist due to other energy (sub)varieties in the system, they must not modify this equality between energies-per-entity. Otherwise, it would make up Formal Objects of a higher organization level (dipoles or multipoles). For instance, several pieces of a solid body may move altogether in translation because they are tied by the rigid structure of the material making the body, so every part of the body moves at the same velocity. It is clear that the interaction through the material structure does not influence the inductive energy-per-entity of any element of the body because this interaction (elasticity) belongs to the capacitive subvariety.

All case studies given previously in this chapter can be completed by describing actual systems involving a composition of poles simply by increasing the number of subsystems that are relevant of the modeling by a pole. We limit the discussion of this extension to two representative cases, one inductive and the second capacitive (which can be fundamental as well as mixed).

4.4.1 Composition of Inductive Poles in Rotational Mechanics

Case study A2 showed a pole consisting of a rotating body in a plane around a center, featured by an inductance J called *inertia* and relating the angular velocity $\boldsymbol{\Omega}$ (flow) to the angular momentum $\boldsymbol{L}$ (impulse).

Figure 4.4 shows two systems, viewed perpendicularly to their rotation plane, a first one of two bodies attached at the ends of a rigid rod rotating around a center O placed along the rod (on the left), a second one of three bodies, each one attached at one end of a rigid rod, the other ends being placed at the center O of rotation (on the right).

This second system of three bodies helps justify the use of a pole composition for modeling such systems, which may comprise any number of rotating bodies, all rigidly secured for keeping the shape of the system. Indeed, the sole consideration of two bodies may naturally induce the recourse to a dipole model (see Chapter 5), which does not allow modeling a third body.

In the following treatment, only the two bodies system is modeled, extension to many bodies being trivial.

The variation of inductive energy (or *kinetic energy* in the classical language) is the sum of variations of inductive energy of each body, each variation being given (provided the existence of the respective inertia):

$$\delta T_\theta = \boldsymbol{\Omega}_1 \cdot \mathbf{d}\boldsymbol{L}_1 + \boldsymbol{\Omega}_2 \cdot \mathbf{d}\boldsymbol{L}_2 = \boldsymbol{\Omega} \cdot \mathbf{d}\boldsymbol{L} \tag{4.2}$$

The constraint of indeformability of bodies is expressed by equating the angular velocities

$$\boldsymbol{\Omega} = \boldsymbol{\Omega}_1 = \boldsymbol{\Omega}_2 \tag{4.3}$$

The equality of angular velocities is translated by equating the sum of variations of dipolar angular momenta to the variation of the composed pole momentum

$$\mathbf{d}\boldsymbol{L}_1 + \mathbf{d}\boldsymbol{L}_2 = \mathbf{d}\boldsymbol{L} \tag{4.4}$$

and by integration (no integration constants are required because the resulting entity number merely counts the number of all entities).

$$\boldsymbol{L}_1 + \boldsymbol{L}_2 = \boldsymbol{L} \tag{4.5}$$

All these relations are translated into the Formal Graph shown in Graph 4.17, which utilizes two poles to combine them in a single pole.

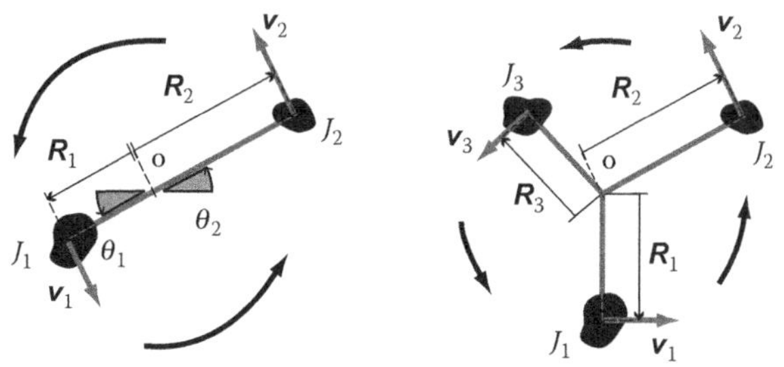

FIGURE 4.4 Devices composed of two (left scheme) and three (right scheme) bodies bound to their rotation axis by rigid rods.

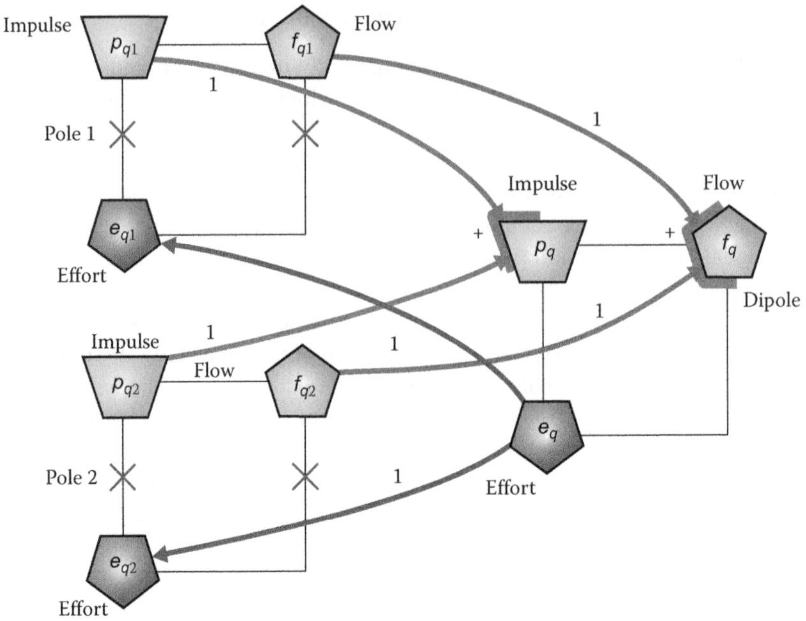

GRAPH 4.17 Formal Graph modeling the composition of two inductive poles for forming a composed pole.

The Formal Graph above is a *graph of connection*: It represents, on the one hand, on the left, the two Formal Graphs of individual poles, and, on the other hand, on the right, the Formal Graph resulting from the composition of these two poles. The relationships between the individual polar state variables and those of the composed pole are represented by connections featured by arrowed arcs, each connection being endowed with an *identity* operator. In the simplest case of linear conductances, the sum of polar torques would represent the total torque, which would be translated into an operation of addition on the effort node of the composed pole, similarly to angular momenta.

This system is tackled again in Chapter 13 in case study K5 "Rotating Bodies" as an example of multiple couplings between energy varieties.

4.4.2 Composition of Capacitive Poles in Electrodynamics

Case study A5 "Electric Charges" was in fact an ensemble of two poles as two kinds of charge, positive and negative, were mixed in the same medium. What follows is therefore a more detailed discussion of this system, in specifying that the model may equally apply to charges with the same sign.

The variation of total capacitive energy for a system with two capacitive poles is given by

$$\delta \mathcal{U}_Q = V_1 \, \mathrm{d}Q_1 + V_2 \, \mathrm{d}Q_2 = V \, \mathrm{d}Q \tag{4.6}$$

where V is the potential of the composed pole, which is the same for all individual poles

$$V_1 = V_2 = V \tag{4.7}$$

Consequently, the variation of the total charge Q is

$$\mathrm{d}Q_1 + \mathrm{d}Q_2 = \mathrm{d}Q \tag{4.8}$$

and by integration, without integration constant, because we need to count all charges with this total charge Q,

$$Q_1 + Q_2 = Q \tag{4.9}$$

The connection Formal Graph in Graph 4.18 relates the two individual poles to the composed pole. Extension to a number N of poles is straightforward:

$$V = V_k; \quad k = 1, N \tag{4.10}$$

$$Q = \sum_{k=1}^{N} Q_k \tag{4.11}$$

When each pole is identical with a charge Q_e and their count is z, the charge of the composed pole is the product of these two quantities:

$$Q = z\, Q_e \tag{4.12}$$

The choice of the letter z instead of N in this relation is for illustrating in an explicit way the case of multicharged atoms or ions where z is a number without dimension (but bearing a sign) called the *charge number*. This example helps to figure out the number N in a more general acceptation of a number of poles eventually multiplied by a weight for taking into account differences in the way of counting entities (including difference in system units).

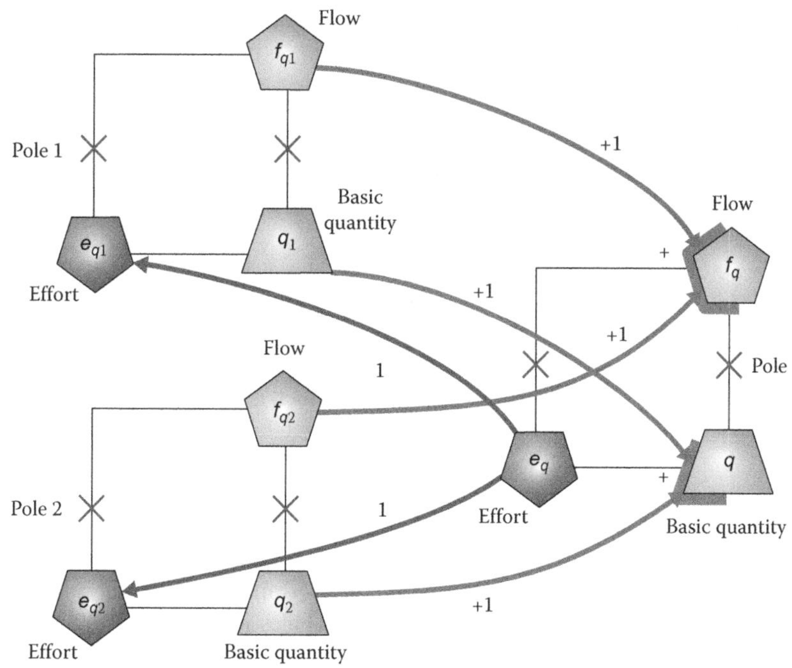

GRAPH 4.18 Formal Graph modeling the composition of two capacitive poles for forming a composed pole.

4.4.3 Generalization

From the above examples, the generalization is straightforward in writing two sets of relations between state variables according to the energy subvariety involved:

Composition of inductive poles:

$$f_q = f_{qk}; \quad k = 1, N \tag{4.13}$$

$$p_q = \sum_{k=1}^{N} p_{qk} \tag{4.14}$$

Composition of capacitive poles:

$$e_q = e_{qk}; \quad k = 1, N \tag{4.15}$$

$$q = \sum_{k=1}^{N} q_k \tag{4.16}$$

A special and interesting case occurs when all subpoles are strictly identical. Each pole containing the same number of entities, the summations in the previous equations are replaced by ordinary multiplications:

$$p_q = N p_{qk}; \quad k = 1, N \tag{4.17}$$

$$q = N q_k; \quad k = 1, N \tag{4.18}$$

For instance, a wheel consisting of N identical pieces disposed at the same distance from the center (same inertia), sees its angular momentum given by N times the angular momentum of each piece.

It must be outlined that a composed pole remains a pole unable to vary its energy alone as no exchange of energy is allowed between subpoles. It means that the sum of individual energy variations remains a nonexact differential.

$$\delta \mathcal{T}_q = \sum_{k=1}^{N} f_{qk} dp_{qk} \tag{4.19}$$

$$\delta \mathcal{U}_q = \sum_{k=1}^{N} e_{qk} dq_k \tag{4.20}$$

Note that when an exchange of energy is permitted between subpoles, the energy variation becomes an exact differential and the system becomes a *multipole*, which is a significant change in object complexity as this new Formal Object is classified at a higher level than dipoles (see Chapter 8).

4.4.4 From Singletons to Poles

By definition, a pole is a collection of entities, all having the same energy-per-entity. Consequently, the previous principle of composition of several poles for forming a bigger pole cannot be reversed to a decomposition principle of a pole into a set of smaller poles without caution. Indeed, the condition for the decomposition is never to reach the number of one entity for one of the poles.

Notwithstanding, the decomposition into elementary units is not impossible, but leaves the concept of pole to adopt that of singleton.

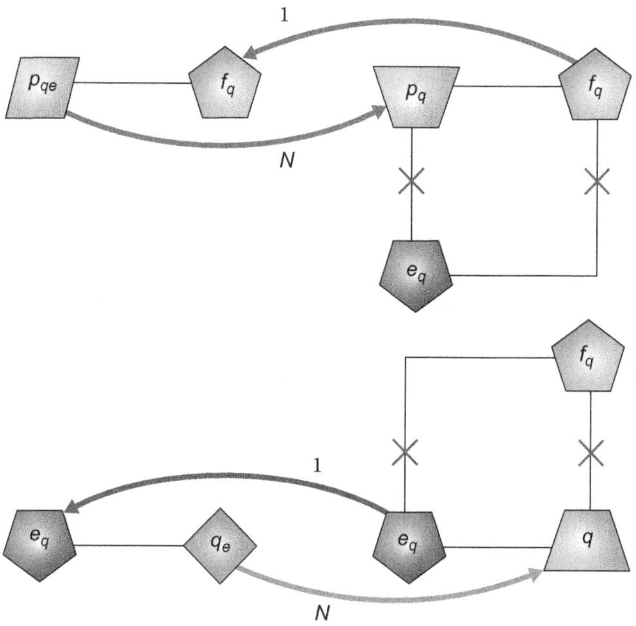

GRAPH 4.19 The relationships between singletons and poles are represented by connections Formal Graphs. The case of inductive energy subvariety is shown on the top and the capacitive one on the bottom.

A singleton is a Formal Object of a lower degree of organization in the complexity scale. In short, a singleton is analogous to a pole but with a single entity instead of a collection. In contradistinction with the pole, it has only one variable in the category of energies-per-entity, which is that a capacitive singleton has an effort but no flow and an inductive singleton has a flow but no effort associated. Only two state variables of the same energy subvariety (a pair entity number, energy-per-entity), are used for modeling a singleton. Algebraically, the relations concerning entities are written by using as multiplier the number of singletons N forming the pole:

$$q = N q_e \tag{4.21}$$

$$p_q = N p_{qe} \tag{4.22}$$

Graphically, connection graphs are drawn with this multiplier for linking entities and the unit coefficient for energies-per-entity as they are equal between singletons and poles (Graph 4.19).

Special shapes for the nodes of single entities are used for distinguishing with collections.

When an entity is identified with a corpuscle, or is associated to a corpuscle, which is the entity of the energy variety called *corpuscular energy*, the number of singletons N forming a pole is replaced by the entity number N of corpuscles. (This is not a crucial distinction and the difference between symbols can be removed if the context permits).

Collective versus individual. The rule, saying that the limit of the decomposition of a pole into smaller ones is the unit number of entities, is not absolute. In fact, this is not a numerical question; it is a physical question about the behavior of entities. Several singletons considered together may not form a pole if their behavior is independent. The notion of pole relies fundamentally on the existence of a collective behavior of entities.

Beyond the macroscopic and nonrelativistic world. The subject of quantum physics and of relativity is not studied in this book; however, a few examples of application of the concept of a singleton are given in Chapter 14 for opening the Formal Graph approach to these domains.

4.5 DEFINITION OF A POLE AND ITS VARIABLES

Before defining a pole, a brief clarification of the exact role of each variable in a pole needs to be done.

4.5.1 State Variables, Conjugate Variables, and Gate Variables

Until now, only one kind of variable has been considered, the state variable, for featuring a pole. A state variable is normally defined in thermodynamics as a variable whose knowledge, with other state variables, allows determining the amount of energy in a system.

From the various examples given, it is clear that in a pole some variables do not participate in this amount of energy, such as a current in an electrostatic pole.

The amount of energy, in a given energy subvariety, is quantified by a pair of state variables, the *entity number* and the *energy-per-entity*. This latter, called the *conjugate energy-per-entity* with respect to the entity, is defined as the partial derivative of energy with respect to the entity number (cf. Chapter 3). In addition to these two state variables, a pole is endowed with a second *energy-per-entity*, belonging to the other category of energy subvariety, as observed with all given case studies.

The reason could not be deduced from the case studies, because all poles were shown alone, without connection with other poles or other Formal Objects (dipoles). The reason will appear in Chapter 6 in showing how a dipole is formed and connected to its poles. Anticipating the way these connections can be made, one merely asserts that these external connections need to link some nodes used as ports or gates, allowing a circulation of energy between Formal Objects. The variables represented by these nodes are always energies-per-entity called *gate variables*. Rigorously speaking, they should not have the status of state variable as their value is not used for determining the amount of energy in the Formal Object. Unless they are simultaneously conjugated to their entity number, in which case they bear the two statuses, conjugate state variable and gate variable. However, these gate variables are still termed state variables but in adding the adjectives conjugate and nonconjugate for distinguishing from true state variables.

Graph 4.20 illustrates these various roles for a fundamental pole belonging to the capacitive type.

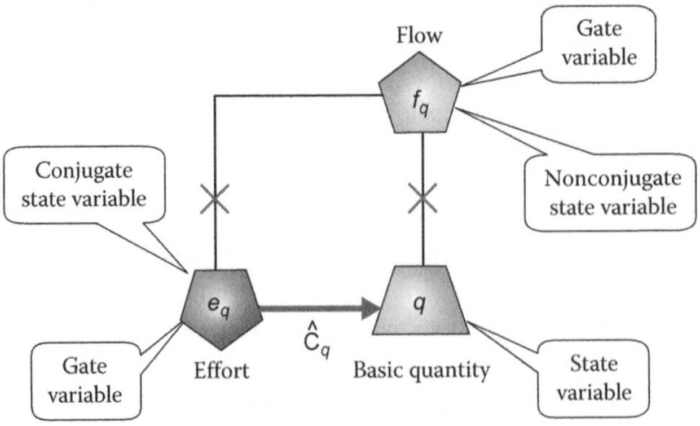

GRAPH 4.20 Formal Graph of a capacitive pole with indication of the respective roles of each variable.

4.5.2 Definition of a Pole

A pole is one of the Formal Objects in the Formal Graph theory. It has several properties:

- A pole belongs to one energy subvariety.
- A pole has no possibility of internal evolution; it has no time but it can be space distributed.
- A pole cannot exchange or convert energy alone. It needs to be in interaction with another pole.
- Five types of poles are defined, depending on the category of the subvariety they are able to store or depending on their possibility to dissipate energy.

The three fundamental poles are:

- The *inductive pole*, able to store the inductive subvariety of energy (kinetic, electromagnetic, etc.).
- The *capacitive pole*, able to store the capacitive subvariety of energy (internal, potential, etc.).
- The *conductive pole*, able only to convert energy into heat (by dissipation).

Two mixed types complete the fundamental types of poles:

- The *inductive–conductive pole*, able to store inductive energy and to dissipate it.
- The *capacitive–conductive pole*, able to store capacitive energy and to dissipate it.
    - A pole is featured by two gate variables (energies-per-entity) for allowing external connections, and, in case of storing pole, is complemented by a state variable that is an entity number.
    - A pole can be *composed* of several smaller poles, all sharing the same energy-per-entity and without interaction acting on their energies-per-entity.

An inductive or capacitive pole, fundamental or mixed, is a collection of identical energy containers, called entities, having the same amount of energy. It is able to effectively store energy only if a constitutive property exists in the system, inductance or capacitance.

A purely conductive pole is a more subtle concept. Such a pole has no entities able to contain energy in the considered subvariety; it possesses only two gate variables (effort and flow) allowing the circulation of energy that is dissipated through the conductance. Actually, a conductive pole is always associated (directly in case of a heat reservoir or indirectly in case of heat transfer) to a thermal pole which is of capacitive type, and is therefore able to store thermal energy. The entropy is the entity number that counts the containers of this peculiar energy variety (see case study A9 "Thermal Pole"). In this way, a conductive pole may also be considered as a collection of entities, but entropic ones, all at the same temperature. The association with a thermal pole is not made explicit on a Formal Graph but is implied through the conductance.

In conclusion, whatever the type of pole, storing or dissipating energy, the main feature of a pole is to consist of a *collection*, an ensemble, of identical entities sharing the *same amount of energy*, quantified by a unique energy-per-entity.

The term collection is voluntarily preferred to ensemble or set, for inducing the idea of collective behavior of all members. This important property is sometimes termed as "colligative," as said in physical chemistry (Atkins 1998).

What is behind the collective behavior of entities? A collective behavior means that all energies-per-entity in an ensemble of entities adopt the same value in any condition. When energy is exchanged between two collections, the variation of the energy amount in each collection is equally distributed over all its entities.

This being said, the question that arises is with regard to the mechanism that permits this distribution. This is not a simple question and there is no definite answer yet.

What is certain is that this collective behavior does exist, because of the adequateness of models relying on this notion with experimental facts.

In the past, the concept of field has been elaborated for modeling an action at a distance, which could explain how some information could be transmitted between entities through space. Nowadays, the concept of exchanged particle is proposed for replacing a field; for instance, the photon is the vector of the electromagnetic field. However, the question of the speed of transmission remains a problem to be solved with regard to the (apparent?) immediacy of many phenomena.

As space is a notion that is disconnected from the definition of a pole, the Formal Graph approach avoids the problem of transmission speed in having recourse to a nonlocalized theory of influence between entities that will be exposed in Chapter 7.

4.6 IN SHORT

POLE VARIABLES

1. *Gate variable:* A node bearing a gate variable is a terminal for the external links used by a pole for exchanging energy with other poles or dipoles. This role is ensured by energies-per-entity.
2. *State variable:* Variable that participates in pairs to determine the amount of energy in a given subvariety. Entity number and energy-per-entity constitute such a pair of state variables.
3. *Conjugate state variable:* Refers to the energy-per-entity corresponding to the entity number in the pair mentioned above.
4. *Global variable:* Variable, among the preceding categories, featuring the whole collection, independent from any space distribution.

FORMAL GRAPHS

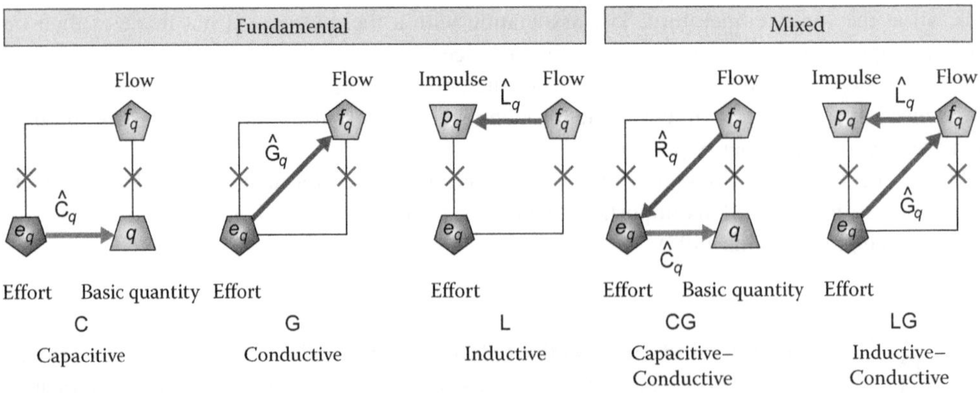

GRAPH 4.21

POLE PROPERTIES

1. A pole is a Formal Object representing a collection of entities possessing the same amount of energy in a given subvariety, for any amount of this energy.
2. Three fundamental types of poles are distinguished (inductive, capacitive, and conductive) and two mixed types (inductive–conductive and capacitive–conductive), depending on the nature of constitutive properties in the system.
3. Energy can vary in a pole only when a capacitance or an inductance exists. Dissipation of energy occurs only when a conductance exists in the system.
4. No internal evolution of energy is permitted in a pole, which is deprived from the temporal operator. (No time is considered in the concept of pole.)
5. By default, a pole is independent from space. Its variables are therefore defined at the global level of the whole collection. When energy is unequally distributed in space (its amounts or its dissipation varies from place to place), the pole is termed *space distributed pole.*
6. An elementary pole contains a single collection of entities. A composed pole comprises several collections of entities, different in nature, but sharing the same energy-per-entity. No energy exchange is permitted between these elementary poles.

5 Space Distributed Poles

CONTENTS

Some objects do not extend in space, like the geometrical or mechanical point that has no volume, no surface, and no length. Some others do occupy a piece of the universe but in such a way that there is no distinction between the various elements that comprise them. They are homogeneous in the sense that each geometrical element of the object has the same properties whatever the localization or the shape of the element. Some others extend in space too, but with a dependence of their element properties on their geometrical shape or on their location in space. They are spatially heterogeneous objects.

The poles discussed in Chapter 4 belong to the category of space-independent objects, that is, points or homogeneous bodies. Here, we examine the case of poles that have properties dependent on space.

Table 5.1 lists the case studies of poles with space distribution given in this chapter. They concern mechanics and electrodynamics (see also Figure 5.1).

Historically, two-dimensional graphs have been used first for relating state variables of a physical system, such as equivalent circuits for representing generalized Kirchhoff's laws. The two dimensions being system properties (constitutive equations) and time property (time derivation), these graphs obviously concern a limited range of physical systems. They are unable to model a physical system in which space properties play a significant role. In electrodynamics, equivalent circuits have been extended by Kron (1943) for representing electromagnetic field equations; that is, by adding the space as a third dimension. Later, several mathematicians (Tonti 1972; Deschamps 1981; Schleifer 1983) have built three-dimensional graphs in this domain of electrodynamics by taking into account the developments of the mathematical branch of geometry called Differential Forms Geometry, consisting of a topological description of the structure of our physical space (Frankel 2004). Tonti (1976, 1995) has extended this interesting approach outside electrodynamics, but restricted his graphs to generic models, without making a connection between a graph and a physical object. In short, these ancestors of Formal Graphs use the same topological approach but without thermodynamics and an organization scheme.

Having difficulties with vector analysis? Vector analysis is a branch of mathematics dealing with objects oriented in space. Among these objects are vectors (one dimension), matrices (two dimensions), and tensors (three dimensions or more) and they are related through spatial operators that can be quite complex, depending on the structure and the properties of space. A modern development of this branch bears the name of Differential Forms Geometry, generalizing vector analysis to any structure of space.

In order to keep a difficulty level as low as is reasonable, this presentation is restricted to a *continuous, and isotropic space in three dimensions*, called Euclidean space. This is the classical assumption made for modeling all physical systems that are not relevant to relativity or to string theories. However, it is possible to handle relativistic concepts without having recourse to a formalism based on a four-dimensional space (or more). One assumption that facilitates the approach is the separation between space and time made in Formal Graphs. (So we speak of *space–time* and not of *spacetime*.)

What makes vector analysis difficult is the quasisystematic use in most textbooks of space coordinates and components of vectors and tensors. This requires the acquisition of certain skills in matrix and tensor calculus before being acquainted with the physics behind it.

Here, nothing of this kind is used. Vectors and tensors are handled as mathematical beings representing a physical quantity independent of the structure of space. No components and no coordinates or spatial frames are used. This means that there are no computations or exercises for juggling with tensors, because physics must be first. A bit of mathematics, however, is useful for understanding how operators are defined. As this not a prerequisite for using Formal Graphs, this mathematical part is discussed at the end of this chapter.

TABLE 5.1
List of Case Studies of Space Poles

	Name of Case Study	Energy Variety	Subvariety	Type	Page Number
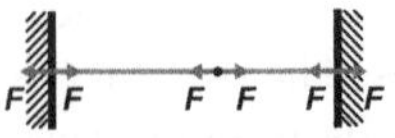	B1: Elastic Element	Translational mechanics	Capacitive	One-dimensional space	109
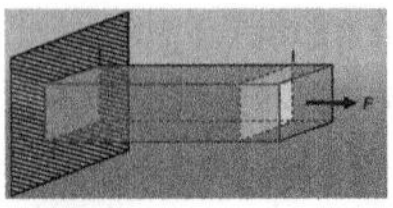	B2: Elasticity of Solids	Translational mechanics	Capacitive	Three-dimensional space	110
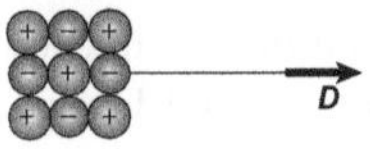	B3: Electric Space Charges	Electrodynamics	Capacitive	Three-dimensional space	113
$\nabla . \boldsymbol{E} = \frac{\rho}{\varepsilon}$	B4: Poisson Equation	Electrodynamics	Capacitive	Three-dimensional space	116
$\Phi_E = \frac{Q}{\varepsilon}$	B5: Gauss Theorem	Electrodynamics	Capacitive	Three-dimensional space	119

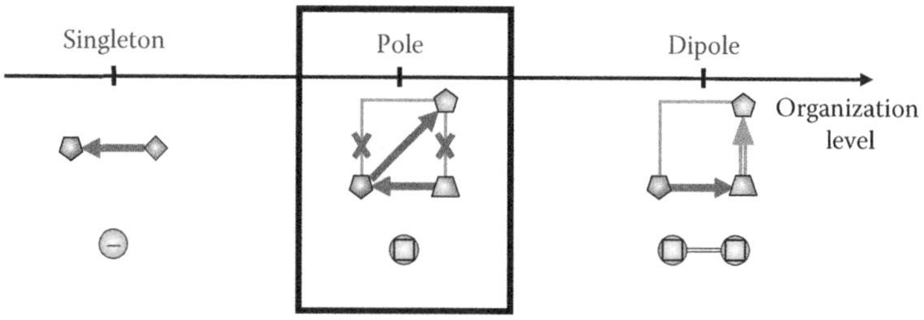

FIGURE 5.1 Limited view of the organization levels along a complexity scale around the Formal Object called a pole.

TABLE 5.2
Depths (or Levels) of Spatial Localization

Depth n_R	Localization	Variable	Generic Symbol	Generic Name
0	Whole system	*Global quantity*	u	State variable
1	Point of a curve	*Global quantity / Length*	$\mathbf{u}_{/r}$	Lineic density, field strength, gradient, etc.
2	Point of a surface	*Global quantity / Area*	$\mathbf{u}_{/A}$	Density
3	Point of a volume	*Global quantity / Volume*	$u_{/V}$	Concentration

5.1 THE ROLE OF SPACE

All the significant variables that help determine the amount of energy in a system are state variables defined globally for the system. When space plays a role, localized variables help model the system.

The description and nomenclature of localized variables have been introduced in Chapter 3, Section 3.1.6. Unless a dedicated notation exists, the generic notation used consists of adjoining a subscript ***/r*** or ***/A*** or */V* to the variable symbol for indicating on which geometric element the variable is localized—curve, surface, or volume, respectively. The main features are summarized in Table 5.2.

Global and localized variables are linked by space operators that depend on the properties of space and, in principle, not on the properties of the system. The main operators are listed with their various symbols in Table 5.3.

TABLE 5.3
Main Operators Used for Spatial Distribution of Variables

Name	Abbreviated Name	Del Notation	Differential/Partial Differential	Applies to (Operand)	Gives (Result)
Derivation	der	∇	$\frac{d}{dr} / \frac{\partial}{\partial r}$	Scalar	Scalar
Gradient	**grad**	∇	$\frac{\mathbf{d}}{\mathbf{dr}} / \frac{\partial}{\partial \mathbf{r}}$	Scalar	**Vector**
Curl	**curl**	$\nabla\times$	$\frac{\mathbf{d}}{\mathbf{dr}}\times / \frac{\partial}{\partial \mathbf{r}}\times$	**Vector**	**Vector**
Divergence	div	$\nabla.$	$\frac{\mathbf{d}}{\mathbf{dr}}\bullet / \frac{\partial}{\partial \mathbf{r}}\bullet$	**Vector**	Scalar
Surface derivation	***A***-der		$\frac{\mathbf{d}}{\mathbf{dA}} / \frac{\partial}{\partial \mathbf{A}}$	Scalar	**Vector**
Volume derivation	*V*-der		$\frac{d}{dV} / \frac{\partial}{\partial V}$	Scalar	Scalar

When the first four operators are used with a minus sign, they bear the (optional) names of contra-derivation, contra-gradient, contra-curl, and convergence.

Electrodynamics is a typical domain in which space plays an important role. An example of a curve-localized variable is the electric field (It is true that "intensity" or "strengh" is sometimes used, but since it is misleading with the electric current or a force, I prefer nothing to use). $\mathbf{E}$ resulting from spatial variation of a potential V, which is given by application of the contra-gradient operator:

$$\mathbf{E} = -\nabla V \tag{5.1}$$

The minus sign is a matter of convention for a curve-localized variable. Historically, fields have been considered as "driving forces" able to exert an action on a body and the choice of the minus sign comes from the resulting identical orientation between field and force.

The differential operator $\nabla\times$ consisting of the *cross product* of the differential vector ∇ with another vector is called the *rotational* or *curl* operator, and the notation in terms of a function *rot* or *curl* is sometimes used. An example of such a link is the relationship between the electromagnetic induction $\mathbf{B}$ and the electromagnetic potential vector $\boldsymbol{A}$, written as

$$\mathbf{B} = \nabla \times \boldsymbol{A} = -\nabla\times(-\boldsymbol{A}) \tag{5.2}$$

To adopt the same spatial operators for all variable families, the contra-curl will be systematically used, inducing the presence of a minus sign in front of the magnetic potential vector $\boldsymbol{A}$. In fact, as for the pressure P in hydrodynamics, the correct variable is always the opposite of the usual one, because historically this vector has been defined without sign rationalization.

As the sign of the curve-localized variable is a matter of convention, so too for the surface-localized variable to take minus signs (– –) or plus signs (+ +) for both operators in sequence (i.e., gradient and curl). Most Formal Graphs are built with the convention of two minus signs in sequence.

The mathematical operator consisting of the *dot product*, or *scalar product*, of the differential vector ∇ with another vector is called the *divergence operator*. For instance, the charge concentration ρ in a volume is linked to the electric displacement $\mathbf{D}$ through one of Maxwell's relationships:

$$\rho = \nabla \cdot \mathbf{D} \tag{5.3}$$

5.2 FORMAL GRAPH REPRESENTATION OF A SPACE DISTRIBUTED POLE

The graphical representation of the global and localized variables and of the spatial operators is shown in a three-dimensional Formal Graph in Graph 5.1. The preferred notation in Formal Graphs is the differential one instead of the del/nabla because it makes explicit the space variable used and allows the total/partial distinction. The use of functions with abbreviated names (*grad*, *div*, etc.) is an old notation which is not encouraged.

Which notation: *grad* V or ∇V or $dV/d\mathbf{r}$ or $(d/d\mathbf{r})V$?
These are equivalent ways of writing the gradient operator. The first one, using the function notation, is not recommended as it is obsolete. The second one, using the del or nabla symbol, is concise but, by default, is not explicit with regard to the concerned space variable. The third one uses the classical notation of a derivative, which is not very concise, but explicit. However, the recommended notation is the last one, because, in addition to being explicit, it clearly distinguishes between operator and operand (the variable upon which the operator is applied). This is in fact an algebraic translation of the Formal Graph basic principle of separation between operators and variables. Whatever the notation chosen, indication by a bold font of the vector nature of the space variable is also recommended.

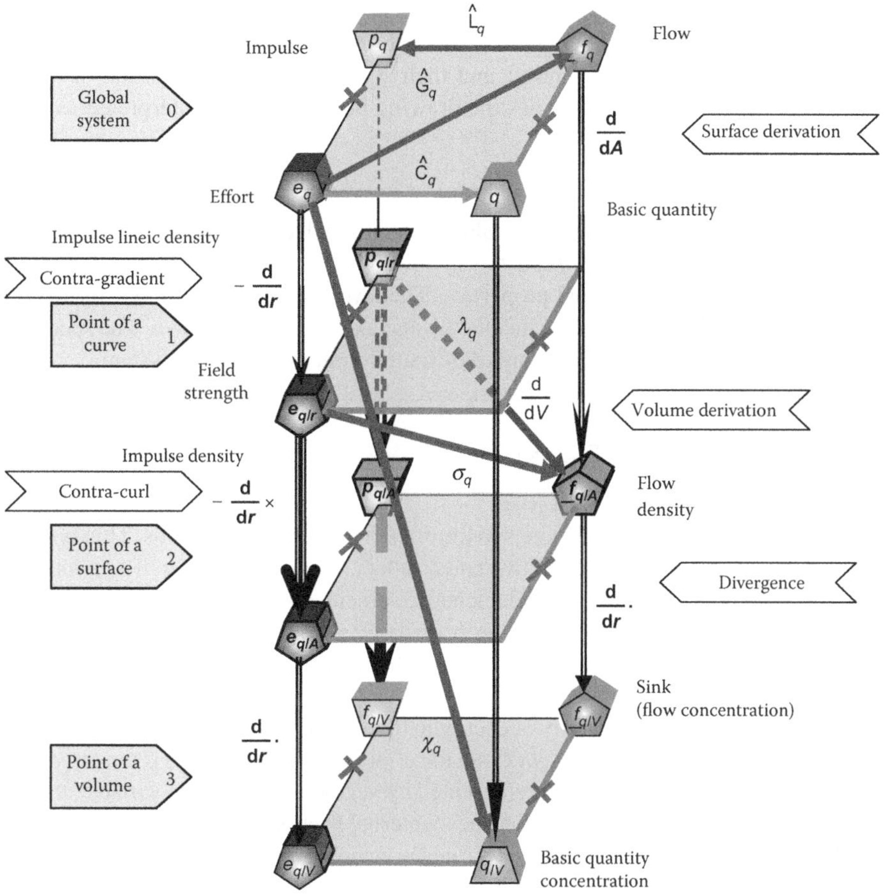

GRAPH 5.1 Projection of a generic three-dimensional Formal Graph representing the various variables and operators that may be used for modeling space distribution in a pole.

The Formal Graph shown in Graph 5.1 is a generic one; it does not represent a pole, but various variables and operators that can be met in a capacitive pole or in an inductive pole and even in a mixed pole. The same sequence of spatial operators is used for all variable families, although not represented completely. Combined operators are also represented under the form of surface and volume derivations.

5.3 CASE STUDIES OF POLES DISTRIBUTED IN SPACE

All case studies are devoted to capacitive poles distributed in space.

The first case study is in the frame of a one-dimensional space.

- B1: Elastic Element in Translational Mechanics

The other case studies are in the frame of a three-dimensional space.

- B2: Elasticity of Solids in Translational Mechanics
- B3: Electric Space Charges in Electrodynamics
- B4: Poisson* Equation in Electrodynamics
- B5: Gauss† Theorem in Electrodynamics

* Siméon Denis Poisson (1781–1840): French mathematician and physicist; Paris, France.

† Carl Friedrich Gauss (1777–1855): German mathematician and physicist; Brunswick and Göttingen, Germany.

As no examples of conductive or mixed poles are given, not both gate variables (effort and flow) are actually used by these poles, allowing the drawing of two-dimensional Formal Graphs based on the pair {entity number, energy-per-entity} and their localized variables, instead of a full three-dimensional Formal Graph. This simplifies the drawing but must not be interpreted as inexistence of the lacking variables.

> **Space reduction levels.** In tables given in each case study abstract are listed the *system constitutive properties* involved in the model that are *space-reduced*, in addition to the global properties. The levels are counted from 0 for the global level (highest) and incremented by one unit at each step down. The specificity is given relatively to the number of spatial dimensions of the system.

5.3.1 Case Study B1: Elastic Element — Translation Mechanics

The case study abstract is given in next page.

An elastic string or a rubber band belongs to the domain of elasticity of solids. This system is basically a dipole because it has intrinsically two ends. In fact, it can be considered as a multipole (i.e., made of several poles), each pole being an elementary element. As every element occupies a different location in space, the poles are said to be *spatially distributed*. When the string is stretched by applying a force $\boldsymbol{F}$ at each end, this force applies also to each element. Here we consider only one pole, as if it was possible to cut such an element.

Variables: In the capacitive subvariety of energy in translation mechanics, the basic quantity (number of entities) is the *displacement* or *length* $\boldsymbol{\ell}$ and the effort (energy-per-entity) is the *force* $\boldsymbol{F}$. Both are quantities oriented in space and therefore represented by vectors. A third state variable, the flow, here the velocity, is only used in case of dissipation due to internal friction, which is not envisaged here. The space distribution brings up spatially reduced variables, which are the *strain* or *relative elongation*

$$\varepsilon_\ell = -\frac{\mathbf{d}}{\mathbf{d}\boldsymbol{r}}\cdot \boldsymbol{\ell} = -\nabla\boldsymbol{\ell} = -\,\mathrm{div}\,\boldsymbol{\ell} = \mathrm{conv}\,\boldsymbol{\ell} \tag{B1.1}$$

and the *force lineic density* (or simply *lineic force*) that is built on the same scheme

$$F_{/r} = -\frac{\mathbf{d}}{\mathbf{d}\boldsymbol{r}}\cdot \boldsymbol{F} = -\nabla\boldsymbol{F} = -\mathrm{div}\,\boldsymbol{F} = \mathrm{conv}\,\boldsymbol{F} \tag{B1.2}$$

(These are various equivalent notations.) Both variables are scalars as they result from a convergence operation (the opposite of the divergence operation). In one dimension, along the line of the string, the space coordinate $\boldsymbol{r}$ and the length $\boldsymbol{\ell}$ are colinear, so the *relative elongation* (*strain*) is merely the ratio of the increase in length upon the initial length:

$$\varepsilon_\ell = \frac{\Delta\ell}{\ell} \tag{B1.3}$$

and the *lineic force* is equal to zero as the force is conserved along the string:

$$F_{/r} = 0 \tag{B1.4}$$

System Properties: Besides the global property of the elasticity k_e, the only useful reduced property is the reduced elasticity, called (*mechanical*) *tension*, defined as the operator vector that links the relative elongation to the force:

$$\boldsymbol{F} \overset{def\rightarrow}{=} \hat{\mathbf{T}}_\ell\, \varepsilon_\ell \tag{B1.5}$$

B1: Elastic Element **Translation Mechanics**

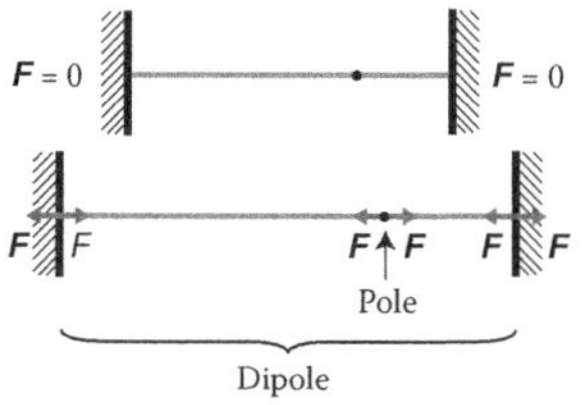

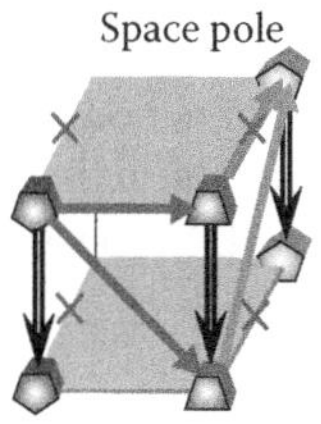

Pole
✓ Fundamental
Mixed
✓ Capacitive
Inductive
Conductive
Global
✓ Spatial

Formal Graph:

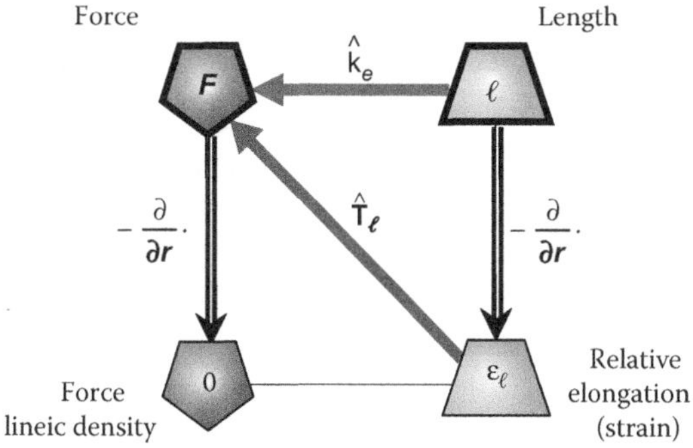

GRAPH 5.2 All vectors are taken with a uniform orientation in space, so the spatial operator leading to both curve-localized variables is the convergence operator.

Variables

Variety	Translation Mechanics		
Subvariety	Capacitive (internal)		Inductive
Category	Entity number	Energy/ entity	Energy/ entity
Family	Basic quantity	Effort	Flow
Global name	Displacement	Force	Velocity
Symbols	ℓ	F	v
Unit	[m]	[N], [J m^{-1}]	[m s^{-1}]
Local name	Strain, relative elongation	Force, lineic density	Lineic velocity
Level	1	1	1
Symbols	ε_ℓ	$F_{/r}$	$v_{/r}$
Unit	[—]	[N m^{-1}]	[m s^{-2}]

Capacitive relationship:

$$F = \hat{T}_\ell\, \varepsilon_\ell$$

Space distribution:

$$\varepsilon_\ell = -\frac{d}{dr}\cdot \ell$$

$$F_{/r} = -\frac{d}{dr}\cdot F$$

Linear model:

$$F = T_\ell\, \varepsilon_\ell$$

System Properties		
Nature	Elastance	Elastance
Level	0	0/1
Name	Elasticity	Mechanical tension
Symbol	k_e	T_ℓ
Unit	[N m^{-1}]	[N]
Specificity	—	1d

Remarks

Force versus tension. Quite often, as they have the same unit (N), force and tension are confused. This illustrates the point that an identity of unit is not sufficient for settling two quantities in the same category. What brings the Formal Graph approach is a clear distinction: here the force is a state variable, which helps determine the amount of energy (the content), whereas the tension is a specific property of the string (the container).

When to use space? It is not always necessary to use a *space pole* for modeling an elastic system. A global pole (see case study A8 "Spring End" in Chapter 4) is sufficient when the system is homogeneous in structure and properties (isotropic materials). However, the existence of inhomogeneities requires using *space poles* (see case study B2 "Elasticity of Solids").

5.3.2 Case Study B2: Elasticity of Solids — Translation Mechanics

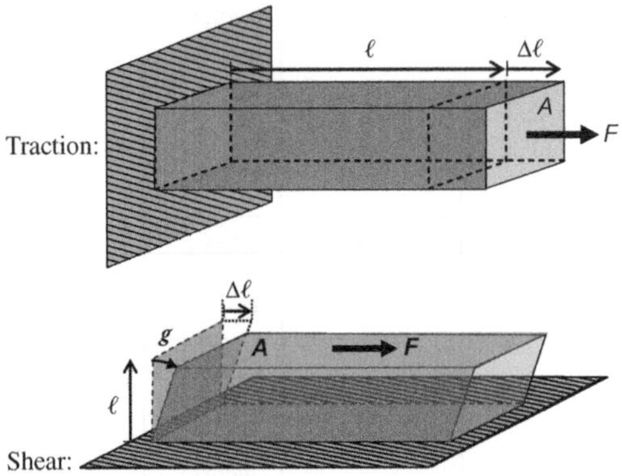

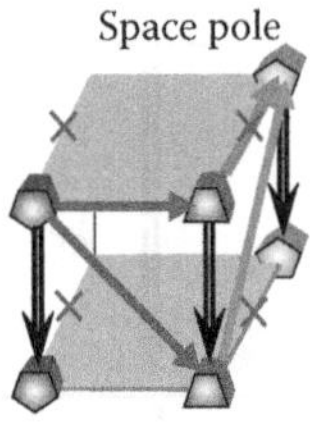

Pole
✓ Fundamental
Mixed
✓ Capacitive
Inductive
Conductive
Global
✓ Spatial

Formal Graph:

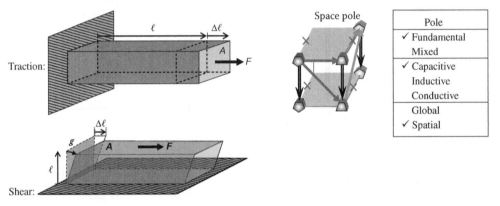

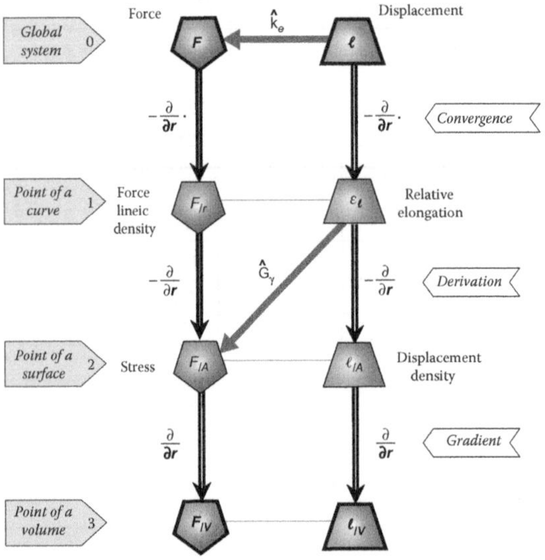

GRAPH 5.3 All vectors are taken with a uniform orientation in space, so the spatial operators leading to localized variables are the sequence convergence/contra-derivation/gradient.

Variables:

Variety	Translation Mechanics		
Subvariety	Capacitive (potential)		Inductive
Category	Entity number	Energy/entity	Energy/entity
Family	Basic quantity	Effort	Flow
Global name	Displacement	Force	Velocity
Symbols	ℓ	F	v
Unit	[m]	[N], [J m^{-1}]	[m s^{-1}]
Local name	Relative elongation	Stress (Superficial force)	Lineic velocity
Level	1	2	
Symbols	ε_ℓ	F_{IA}	v_{Ir}
Unit	[—]	[N m^{-2}]	[s^{-1}]

(continued)

Capacitive relationship:

(1) Traction: $F_{/A} = \hat{E}_Y \, \varepsilon_\ell$

(2) Shear: $F_{/A} = \hat{G}_Y \, \varepsilon_\ell$

Localized variables:

$$F_{/A} = \frac{\mathbf{d}}{\mathbf{d}A} F$$

(1) Traction: $\varepsilon_\ell = -\frac{\mathbf{d}}{\mathbf{d}r} \cdot \ell$

(2) Shear: $\varepsilon_\ell = \tan\gamma$

System Properties		
Nature	Elastance	Elastance
Level	1/2	1/2
Name	Young Modulus	Shear Modulus
Symbol	E_Y	G_Y
Unit	[N m^{-2}]	[N m^{-2}]
Specificity	3d	3d

Various elastic behaviors. The domain of elastic properties of solids is a broad subject because the number of dimensions of space allows a great variety of application modes of a force for provoking a deformation.

The spatially reduced variables are generally matrices and the operators are therefore tensors. This is not a problem for modeling with Formal Graphs, but, in a pedagogical purpose, only two modes, among the simplest ones, are studied: the *traction* and the *shearing* of an object with a simple geometric shape, a parallelepiped.

An elastic solid of parallelepipedic shape, attached on a support by one of its faces, is submitted to the action of a force on the opposite face. Naturally, this system constitutes a dipole because it has two opposed and parallel faces and the force exists on each of these two faces. As in case study A8 “Spring End” (Chapter 4), the elastic wire is divided into a chain of elementary segments, each being considered a spatial pole, and each intermediary plane between the two active faces constitutes a spatial pole for an elastic solid. Here, too, one of the ends is considered to be representative of a pole.

Two cases are considered: the force is either perpendicular or parallel to the face on which it is applied. The solid is submitted, in the first case, to traction and, in the second case, to shearing.

When a force is applied to one face, it is transmitted from element to element along the direction of application. The solid elongates (traction mode) or its parallel (to the direction of the force) internal surfaces slide past one another (shear mode). The end that represents the poles is characterized by a deformation or *strain*. In the traction mode, the rest position (absence of force) is the distance ℓ between the two faces and the deformation is the elongation $\Delta\ell$. In the case of shearing, ℓ is the solid thickness without shearing, and the deformation is the angle γ made by the face perpendicular to the force with respect to its position in the absence of force. The pole is considered to be purely capacitive, that is, without internal friction, thus being exempt from taking into account the flows (velocities of elongation).

Pressure and stress. The *stress* or *superficial force* $F_{/A}$ is a scalar analogous to a pressure, but which is not endowed with all its characteristics. It must be recalled that the pressure P is a state variable, an effort of the hydrodynamic energy variety, and that it is equal to a force divided by an area only in case of coupling with translation mechanics (see case study J5 “Piston” in Chapter 12).

This system being distributed in a space with three dimensions, three levels of spatial reduction are required for the model. Among the six reduced variables, two are therefore essential for being able to utilize specific properties, the first variable is the relative elongation ε_ℓ, which is a scalar obtained through a convergence operation on the distance in the case of the traction and by means of the tangent of the deformation angle γ in the case of shearing:

$$\begin{aligned}&\text{Traction}: \quad \varepsilon_\ell = -\frac{\mathbf{d}}{\mathbf{d}\boldsymbol{r}} \cdot \ell \\ &\text{Shear}: \quad \varepsilon_\ell = \tan\gamma\end{aligned} \tag{B2.1}$$

The second variable is also a scalar, called *stress* or *superficial force*, or *superficial density of force*, $F_{/A}$, obtained by superficial derivation of the applied force:

$$F_{/A} = \frac{\mathbf{d}}{\mathbf{d}\boldsymbol{A}} \boldsymbol{F} \tag{B2.2}$$

The relative elongation ε_ℓ results from the convergence of the distance ℓ in considering all vectors as colinear in the case of the traction. In the case of shearing, the relative elongation is obtained by dividing the length $\Delta\ell$ of the small side of the right triangle opposite to the angle γ by the remaining thickness ℓ. In both cases, the same expression is obtained:

$$\varepsilon_\ell = \frac{\Delta\ell}{\ell} \tag{B2.3}$$

The spatially reduced property of the system is the *elasticity modulus*, which is an operator in the general case. This is a specific property in a system with three dimensions of space, which links the relative elongation to the stress (superficial force) according to the definition of the capacitive relationship.

$$F_{/A} \overset{def\rightarrow}{=} \hat{G}_\gamma \, \varepsilon_\ell \tag{B2.4}$$

This property bears different names according to the mode of application of the force:

Traction: *Young's* Modulus*, notated E_Y

Shear: *Shear's Modulus*, notated G_γ

When the elongation is sufficiently small for keeping the deformation in linear dependence upon the stress, the elasticity modulus becomes a simple scalar, providing a proportionality relation analogous to Hooke's law (see case study A8 "Spring End" in Chapter 4).

$$F_{/A} \underset{lin}{=} G_\gamma \, \varepsilon_\ell \tag{B2.5}$$

Remarks

Variations. The spatial reduction scheme for a system in which all vectors are colinear markedly differs from the scheme for systems in which the vectors can take various orientations. Instead of gradients for arriving at the first reduction level, operators of divergence (or convergence) are utilized here.

* Thomas Young (1773–1829): English scientist; Cambridge and London, UK.

5.3.3 Case Study B3: Electric Charges

Electrodynamics

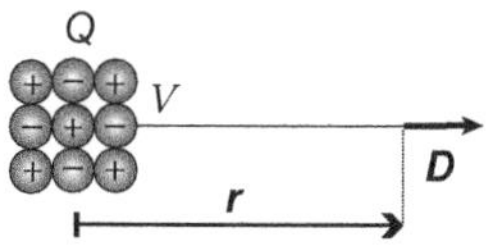

The notion of pole is extended to a set of charges, eventually with different signs, and is not restricted to a solitary electric charge.

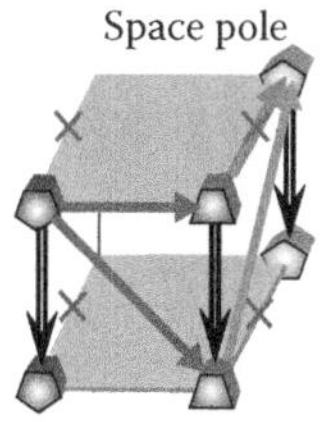

Pole	
✓	Fundamental
	Mixed
✓	Capacitive
	Inductive
	Conductive
	Global
✓	Spatial

Formal Graph:

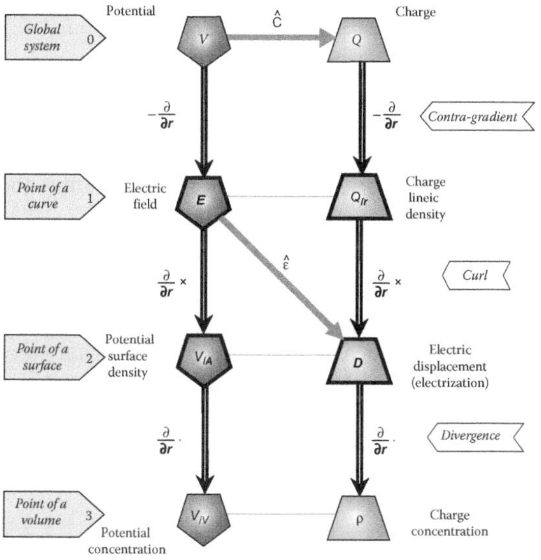

GRAPH 5.4

Variables:

Variety	Electrodynamics		
Subvariety	Capacitive (potential)		Inductive
Category	Entity number	Energy/ entity	Energy/ entity
Family	Basic quantity	Effort	Flow
Global name	Charge	Potential	Current
Symbols	Q	V, U, E, φ	I, i
Unit	[C], [A s]	[V], [J C^{-1}]	[A]
Local name	Electric displacement	Electric field	
Level	2	1	
Symbols	$\mathbf{D}$	$\mathbf{E}$	
Unit	[C m^{-2}]	[V m^{-1}]	
Local name	Charge concentration		
Level	3		
Symbols	ρ		
Unit	[C m^{-3}]		

Capacitive relationship:

$$\mathbf{D} = \hat{\varepsilon}\mathbf{E}$$

Reduced property:

$$\hat{\varepsilon} = \hat{\mathbf{A}}^{-1}\hat{C}\,\hat{\mathbf{r}}$$

Linear model:

$$\mathbf{D} = \varepsilon\mathbf{E}$$

$$C \underset{lin}{=} \varepsilon\frac{A}{\ell}$$

System Properties		
Nature	Capacitance	Capacitance
Level	0	1/2
Name	Capacitance	Permittivity
Symbol	C	ε
Unit	[F], [C V^{-1}]	[F m^{-1}]
Specificity	—	3d

A set or a cluster of electric charges *at the same potential* forms a pole of the capacitive subvariety of electric energy. The charges can be a mixing of positive and negative charges in eventually unequal number. The situation in which all the charges are of the same sign is also included in this case. Some cohesive forces (belonging to another energy variety, which must be taken into account when the total energy of such a system is evaluated) are supposed to maintain the whole cluster. The *total quantity of charge* is Q and the *electric potential* "seen" by these charges is V.

Why pole? The notion of *pole* used in Formal Graphs is directly borrowed from electrodynamics for being generalized to all energy varieties. A nuance is, therefore, brought in the electrostatic domain, precisely in including in this notion a set of charges, eventually with different signs, and not in restricting the notion to a solitary electric charge. (The case of a single charge is handled by the concept of "singleton" in the Formal Graph theory.)

According to electrostatics, each individual charge influences at a distance the other charges in the cluster to produce an electric field, which creates in every point of space an electric potential. The sum of these potentials at the point where the charges are gathered (their total spatial extension is assumed to be reduced to a point or at least to a region where V is a constant) constitutes the potential V in which the charges are bathing.

5.3.3.1 One or Several Charges?

The situation is different when a pole is made indissociable with a single electric charge. In this case, the potential V is not created by this one but must necessarily come from an external *electric field* like the field produced by another pole in the vicinity. Then, the two poles exert a *mutual influence* one upon the other. This case is not relevant for the model of a pole but of a dipole. (See case study C7 "Electric Capacitor" in Chapter 6 and case study D4 "Coulomb's Law" in Chapter 7.)

5.3.3.2 Spatial Distribution

The Formal Graph in the case study abstract features an electrostatic pole distributed in a three-dimensional space, in a *generic* representation, that is, in indicating all the possibilities of local variables (spatially reduced variables) without taking care of their actual existence, which is depending of the particularities of the system. Contrary to the representation of the global system given for a pole that is not distributed in space (see case study A5 "Electric Charges"), the node of electric current is not drawn for staying in a two-dimensional representation.

Although classical electrodynamics uses a scalar capacitance, the Formal Graph theory generalizes all the capacitive relationships by using a capacitance operator. Consequently, the relation is written at the global level:

$$Q = \hat{C}V \tag{B3.1}$$

The main local (reduced) variables of an electrostatic pole spatially distributed are the *electric field* $\boldsymbol{E}$ and the *electric displacement* (or "*electrization*") $\boldsymbol{D}$. (See the note "Anachronism" and case study B5 "Gauss Theorem.") These are vectors connected by the reduced capacitive property called *permittivity* and this is a specific property. Its unit is the farad per meter (F m^{-1}). In the Formal Graph theory, the electric capacitance is not a scalar but an operator. Consequently, the permittivity will no longer remain a scalar (even not a simple tensor taking into account the eventual anisotropy of a material), especially when the material is not homogeneous.

$$\boldsymbol{D} \overset{def}{=} \hat{\varepsilon}\boldsymbol{E} \quad \text{(B3.2)}$$

By symbolizing the spatial operators by an operator of distance **r** (with a hat ^) for the reciprocal of the contra-gradient and by a reciprocal operator of area $\hat{\mathbf{A}}$ for the integration on a surface, the relation between global property and reduced property is written as

$$\hat{\varepsilon} = \hat{\mathbf{A}}^{-1}\hat{C}\,\hat{\mathbf{r}} \quad \text{(B3.3)}$$

When the system is homogeneous, the spatial operators become divisions by a length or an area and the previous relation (inverted) becomes

$$\hat{C} \underset{hom}{=} \mathbf{A}\hat{\varepsilon}\mathbf{r}^{-1} \quad \text{(B3.4)}$$

In addition to the homogeneity case, it is only when the capacitance is a scalar, as in the case of vacuum or of "linear" materials, that the permittivity is also a scalar, a case expressed by the well-known relation (called "plate capacitor" relation):

$$C \underset{lin}{=} \varepsilon\frac{A}{\ell} \quad \text{(B3.5)}$$

Remarks

Self or mutual? It is worth mentioning that the model of a pole made of only a type of charge (all positive or all negative) does not utilize the internal capacitance (self-capacitance) but the *mutual capacitance*, which expresses the influence of another (or several) pole. (See case study D4 "Coulomb's Law" in Chapter 7.) The reason is that, in electrostatics, an electric charge does not create its own potential on the location of the charge itself but creates a potential at distance, through the electric displacement and the electric field.

Localization. The reduction scheme of variables used in space is very general and is found in other energy varieties (see for instance the Fourier equation in the thermal domain—case study G2 "Fourier's Equation of Heat Transfer" in Chapter 10). The scheme relies on the scalar nature of state variables (at the global level) and uses a sequence of spatial operators that are the contra-gradient, the curl (also called rotational), and the divergence.

Electrostatics. Two important laws of electrostatics are directly deduced from this scheme of spatial reduction: the Poisson equation and the Gauss theorem (case studies B4 and B5 in this chapter). The mutual influence between two of these poles is modeled within the framework of an electrostatic dipole using Coulomb's law.

The coupling with the electromagnetic subvariety is a basis for modeling the Maxwell equations in electrodynamics (see case study F7 "Maxwell Equations" in Chapter 9).

Anachronism. The word “displacement” is strangely borrowed from translation mechanics by analogy between the work of a force equal to its strength multiplied by the displacement (distance) $\boldsymbol{F} \cdot \mathbf{d}\ell$ and the electric “work of volumic forces” equal to $\boldsymbol{E} \cdot \mathbf{d}\boldsymbol{D}$. This is somewhat daring in our modern viewpoint but this is a historical burden. A better meaning would be ensured with *electrization*, built on the template of *magnetization*, which is also a variable localized onto a surface in the family of basic quantities.

5.3.4 Case Study B4: Poisson Equation — Electrodynamics

$$\frac{\partial^2}{\partial \boldsymbol{r}^2} V \underset{hom}{=} -\frac{\rho}{\varepsilon}$$

The Poisson equation in electrostatics relates the charge concentration contained in a system to the Laplacian* of the electric potential.

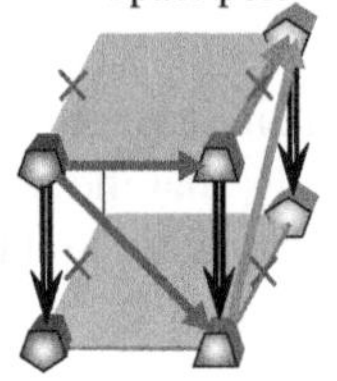

	Pole
✓	Fundamental
	Mixed
✓	Capacitive
	Inductive
	Conductive
	Global
✓	Spatial

Formal Graph:

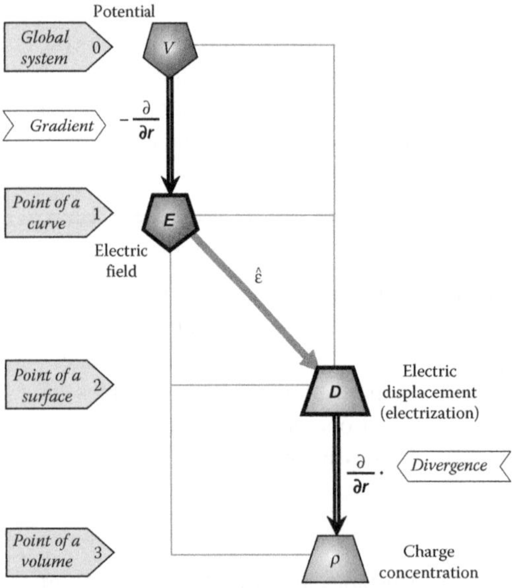

GRAPH 5.5 The Formal Graph representing the Poisson equation is a graph in two dimensions (system properties and property of space), without any role given to the time since this equation belongs to electrostatics.

Variables:

Variety	Electrodynamics		
Subvariety	Capacitive (potential)		Inductive
Category	Entity number	Energy/entity	Energy/entity
Family	Basic quantity 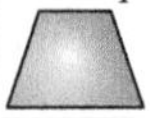	Effort 	Flow 
Global name	Charge	Potential	Current
Symbols	Q	V, U, E, φ	I, i
Unit	[C], [A s]	[V], [J C^{-1}]	[A]
Local name	Electric displacement	Electric field	
Level	2	1	
Symbols	$\boldsymbol{D}$	$\boldsymbol{E}$	
Unit	[C m^{-2}]	[V m^{-1}]	
Local name	Charge concentration		
Level	3		
Symbols	ρ		
Unit	[C m^{-3}]		

* Pierre Simon Laplace (1749–1827): French mathematician, astronomer, and physicist; Paris, France.

(continued)

Capacitive relationship:
$\boldsymbol{D} = \hat{\varepsilon}\boldsymbol{E}$
Space distribution:
$\boldsymbol{E} = -\frac{\partial}{\partial \boldsymbol{r}} V$
$\rho = \frac{\partial}{\partial \boldsymbol{r}} . \boldsymbol{D}$
Linear model:
$\boldsymbol{D} = \varepsilon \boldsymbol{E}$

System Properties	
Nature	Capacitance
Level	1/2
Name	Permittivity
Symbol	ε
Unit	[F m^{-1}]
Specificity	3d

In electrostatics, one of the important questions is the relationship between a concentration of charges in a point of space and the electric potential in the same point. The Poisson equation (published in 1813) expresses the dependence of the *electric field* **E** with the *charge concentration* ρ in a medium of *permittivity* (or dielectric constant) ε (or ε_0 in vacuum):

$$\nabla . \boldsymbol{E} = \frac{\partial}{\partial \boldsymbol{r}} . \boldsymbol{E} = \frac{\rho}{\varepsilon} \tag{B4.1}$$

This equation is very often given, not with the electric field, but with the electric potential *V*, and the utilized form assumes a permittivity independent from space (homogeneous medium) as will be demonstrated by the Formal Graph, which justifies the mention "*hom*" under the equal sign in the following equation:

$$\frac{\partial^2}{\partial \boldsymbol{r}^2} V \underset{hom}{=} -\frac{\rho}{\varepsilon} \tag{B4.2}$$

According to mathematical analysis, the spatial operator utilized is called *Laplacian* and several notations are used:

$$\Delta V = \nabla^2 V = div(\overrightarrow{grad}\, V) = \frac{\partial^2}{\partial \boldsymbol{r}^2} V \tag{B4.3}$$

The Formal Graph translating the two forms B4.1 and B4.2 is given in the case study abstract.

Each elementary path in this graph has an algebraic translation. The node, known as the electric field **E**, is defined as the spatially reduced potential at the level where the contra-gradient (i.e., the gradient with a minus sign) of electric potential is found, as shown in the following equation:

$$\boldsymbol{E} = -\frac{\partial}{\partial \boldsymbol{r}} V \tag{B4.4}$$

The electric displacement **D** is linked to the strength of the electric field **E** by a capacitive reduced relationship, which defines the electric permittivity as the operator in this relation:

$$\boldsymbol{D} \overset{def}{=} \hat{\varepsilon} \boldsymbol{E} \tag{B4.5}$$

and the *charge concentration* ρ as arising from the divergence of the electric displacement **D** in this relation:

$$\rho = \frac{\partial}{\partial \boldsymbol{r}} . \boldsymbol{D} \tag{B4.6}$$

The composition of paths provides the general Poisson equation in potential

$$\rho = \frac{\partial}{\partial r}\hat{\varepsilon}\left(-\frac{\partial}{\partial r}V\right) \tag{B4.7}$$

which is not restricted to a validity range as is the classical equation B4.2. The hypothesis of a homogeneous medium and of independence of the permittivity from space and from any state variable, added to the hypothesis of a validity domain restricted to linearity leads to the classical Poisson equation B4.2 in potential. (This is not a reference to a possible occurence but to the potential V variable!)

The Poisson equation has a double physical meaning as it is simultaneously a capacitive relationship (expressing the faculty of the system to store capacitive energy, i.e., electrostatic) and a relationship between the global and local levels for a volumic system. It allows, therefore, the recovery of an electric potential from a distribution of charges in space or inversely. It complements the Gauss theorem (see case study B5), which is also a capacitive reduced relationship that involves the charge at the global level and its relationship with the electric field.

Remarks

Localization versus operator result. It must be outlined that the variable *strength of the electric field* $\boldsymbol{E}$ is defined in the Formal Graph theory by the *location of its node* in the Formal Graph, that is, at the level of a point of a curve (level 1) and in the family of efforts. This may differ from the particular definition, given by the contra-gradient of the electric potential, when it is equal to another result, as for instance the temporal derivative of a potential vector $\boldsymbol{A}$.

Anachronism. The word "displacement" is strangely borrowed from translation mechanics by analogy between the work of a force equal to its strength multiplied by the displacement (distance) $\boldsymbol{F}.d\boldsymbol{\ell}$ and the electric "work of volumic forces" equal to $\boldsymbol{E}.d\boldsymbol{D}$. This is somewhat daring in our modern viewpoint but this is a historical burden. A better meaning would be ensured with *electrization*, built on the template of *magnetization*, which is also a variable localized onto a surface in the family of basic quantities.

Generalization. The generalization of the Poisson equation proposed here proceeds from the same approach as the one utilized for other equations of classical physics, as for instance Newton's second law in translation mechanics (see case study F1 "Accelerated Motion" in Chapter 9).

The method allowed by the Formal Graph theory consists of decomposing an algebraic equation into elementary paths in a graph, by choosing paths with *specific properties* as a criterion of choice instead of nonspecific paths. The order of paths gives prominence to the complications that may occur when the commutativity of operators, which is generally classically made implicit, is submitted to conditions of independence of system constitutive properties with respect to space.

5.3.5 Case Study B5: Gauss Theorem

Electrodynamics

$$\Phi_E = \frac{Q}{\varepsilon}$$

The Gauss theorem in electrostatics relates, in its *integral version*, the charge contained in a system to the flux of the electric field across a surface surrounding the charges.

This relation is not translated word for word in a Formal Graph, but transposed in terms of the relationship between electric field and charge. This corresponds to a *differential version* of the Gauss theorem.

Space pole

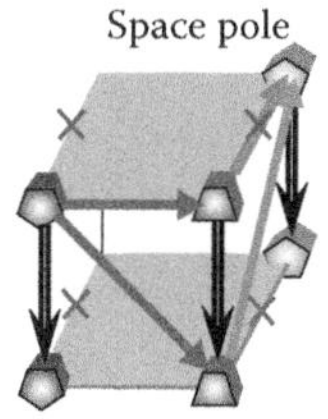

	Pole
✓	Fundamental
	Mixed
✓	Capacitive
	Inductive
	Conductive
	Global
✓	Spatial

Formal Graph:

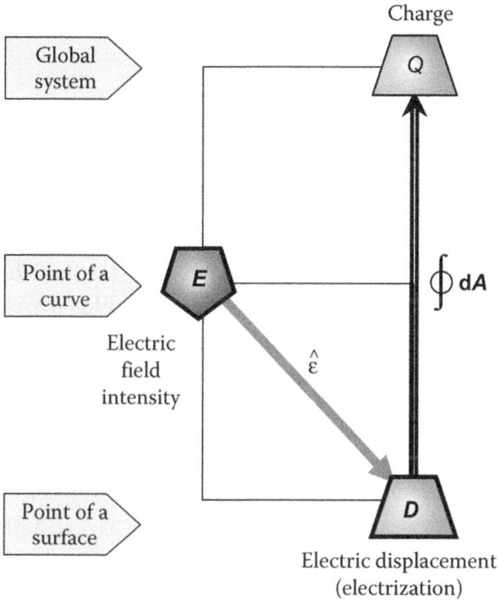

Variables:

Variety	Electrodynamics		
Subvariety	Capacitive (potential)		Inductive
Category	Entity number	Energy/ entity	Energy/ entity
Family	Basic quantity	Effort	Flow
Global name	Charge	Potential	Current
Symbols	Q	V, U, E, φ	I, i
Unit	[C], [A s]	[V], [J C^{-1}]	[A]
Local name	Electric displacemet	Electric field	
Level	2	1	
Symbols	$\boldsymbol{D}$	$\boldsymbol{E}$	
Unit	[C m^{-2}]	[V m^{-1}]	

GRAPH 5.6

Capacitive relationship:

$$\boldsymbol{D} = \hat{\varepsilon}\,\boldsymbol{E}$$

Closed-surface integration:

$$Q = \oint_S \boldsymbol{D} \,.\, \mathrm{d}\boldsymbol{A}$$

Linear model:

$$\boldsymbol{D} = \varepsilon\,\boldsymbol{E}$$

System Properties	
Nature	Capacitance
Level	1/2
Name	Permittivity
Symbol	ε
Unit	[F m^{-1}]
Specificity	3d

The Gauss theorem in electrostatics establishes a relationship between an *electric charge* Q and a *flux of electric field* Φ_E through the *permittivity* ε of the medium. It is classically presented in the following form:

$$\Phi_E = \frac{Q}{\varepsilon} \qquad \text{(B5.1)}$$

The Formal Graph as shown in Graph 5.7 represents this theorem in a different way. It is what could be a direct translation of the classical concept behind this theorem. The barred variable is the flux of electric field Φ_E defined as the integral on a closed surface of the electric field $\boldsymbol{E}$

$$\Phi_E \overset{def}{=} \oint_S \boldsymbol{E}.\,d\boldsymbol{A} \tag{B5.2}$$

Flux or flow? Here is found the correct usage of the word "flux," which refers to the scalar result of the integration on a closed surface of a vector defined on this surface. The common usage of calling a vector flow or a scalar flow as a "flux" is not recommended. This is an advantage, and a constraint, of the Formal Graph approach that is stricter with notions and nomenclature for placing the correct variables on a graph.

This flux is not drawn because a Formal Graph is built in conformation with the geometric elements of the usual space with three dimensions, and, therefore, possesses a maximum of three reduction levels beneath the global level. The result of integration on a surface must be located two levels above the integrated variable (integrand), which would place the flux of electric field above the global level. There is, therefore, a problem. Notwithstanding, if instead of the electric field $\boldsymbol{E}$ one takes the electric displacement $\boldsymbol{D}$, then the problem is solved, without using forbidden variables in a canonical Formal Graph. The only valid representation in a Formal Graph is drawn in the case study abstract. Effectively, according to the definition of an electric displacement, the integration on a closed surface is the operator that links this electric displacement to the electric charge on this surface:

$$Q \overset{def\rightarrow}{=} \oint_S \boldsymbol{D}.\,d\boldsymbol{A} \tag{B5.3}$$

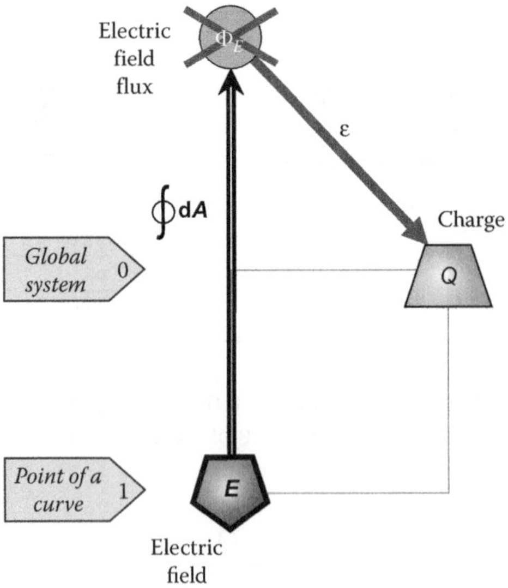

GRAPH 5.7 This graph is not a canonical Formal Graph because it involves a level above the global level.

The solution consists of replacing the electric displacement by its expression as a function of electric field, which is given by the capacitive reduced property, which in the ordinary case (linear material) utilizes a scalar ε called *electric permittivity* (or dielectric constant)

$$\boldsymbol{D} \overset{def\rightarrow}{=} \hat{\varepsilon}\,\boldsymbol{E} \tag{B5.4}$$

for obtaining a relation between electric field and charge

$$Q = \oint_S \hat{\varepsilon}\,\boldsymbol{E}\,.\,\mathbf{d}\boldsymbol{A} \tag{B5.5}$$

Remarks

Homogeneity. In the particular case where the permittivity is independent from the geometry of the system, which corresponds to a homogeneous material, and when the operator is linear, as in linear materials (vacuum), it can be removed from the integral.

$$Q \underset{hom}{\overset{lin}{=}} \varepsilon \oint_S \boldsymbol{E}\,.\,\mathbf{d}\boldsymbol{A} \tag{B5.6}$$

By using the definition (Equation B5.2) of the electric flux, the classical expression (Equation B5.1) of the Gauss theorem is found again. The indications "*hom*" and "*lin*" placed under and above the equal sign are a reminder that this equality is only valid under the conditions of homogeneity of the medium and linearity of the permittivity. The Gauss theorem in its classical writing is therefore restricted to homogeneous and linear materials. On the contrary, in its Formal Graph representation, it becomes much more general.

Inhomogeneity. What we learn from the Formal Graph theory is that the Gauss theorem is valid only for permittivities that are independent from space, which is the case in homogeneous materials. Nevertheless, it is always interesting to know that a theorem that is judged to be a fundamental one is not so fundamental. It opens new horizons for grasping some peculiar cases that are not classically tackled.

Meaningless practice. A quite widespread practice among modern physicists working in electrodynamics consists of using only one electric field, say ***E***, instead of two, ***E*** and ***D***, taking advantage of the capacitive relationship between these fields. It is true that one may consider in linear media (vacuum or free space) that both fields are equivalent, being merely proportional between them. This is less obvious in anisotropic media where their orientations differ, but this is feasible, at the expense of the physical meaning. In nonlinear media, this becomes nonsense since this medium characteristic is lost in models confusing the two fields.

Another practice consists of replacing both fields by a geometrical average which is equal to the common value:

$$\boldsymbol{D}\,\hat{\varepsilon}^{\frac{1}{2}} = \hat{\varepsilon}^{\frac{1}{2}}\,\boldsymbol{E} \tag{B5.7}$$

(*continued*)

This amounts to changing the unit system by giving to the permittivity the value 1 (and, by extension, to the celerity of light equally). This mathematical practice becomes rather counterproductive with respect to the physical role of each field, especially in nonlinear media.

The Formal Graph approach insists on the importance of keeping these notions separate as they correspond to physically different families of variables, one belonging to the entity number category and the other to the energy-per-entity category. The coherence with other domains depends on the respect of this separation.

Elimination of variables. One of the essential contributions of Formal Graphs to physics lies in the operation of sorting among all the variables that the algebraic tool can create at will. For instance, the notion of flux Φ_E of an electric field is not retained by the Formal Graph theory to the benefit of a more important role given to the electric displacement (or "*electrization*"), $\boldsymbol{D}$, whose flux through a closed surface is by definition equal to the charge on this surface.

The approach adopted here is identical to the approach utilized for representing Newton's second law in mechanics in the Formal Graph, which refutes the usage of the acceleration because it cannot be part of a canonical Formal Graph, that is, built only with the variables that directly determine the energy of the system (see case study F1 "Accelerated Motion" in Chapter 9).

5.4 SPACE OPERATORS

When a system is spatially homogeneous, none of its features depends on the localization inside the system or on its orientation in space. In that simple case, peculiar operators are not required for knowing a state variable in each geometrical subdivision of the system—curve, surface, or volume. A simple division by a length, an area, or a volume provides the pertinent quantity, a density or a concentration. On the contrary, in spatially heterogeneous systems, localized variables are dependent on their localization. In that case, a differential definition must be used around each point of the system (in a continuous space). For the first localization step, a line or a curve, the definition of the 1-localized quantity, is given by the differential equation:

$$\mathrm{d}(\textit{Global quantity}) = (\textit{Localized quantity}) \,.\, \mathrm{d}(\textit{Length})$$

This relationship can be written more concisely with the help of generic symbols for the localized variables, as indicated in Table 3.2 in Chapter 3, and by using the vector $\boldsymbol{r}$ for the length, as it is a quantity oriented in space, indicating the direction of the variation:

Curve-localized variable: $$\mathrm{d}u = -\boldsymbol{u}_{/r} . \mathbf{d}\boldsymbol{r} \tag{5.4}$$

The curve-localized variable $\boldsymbol{u}_{/r}$, also called *lineic density*, is a vector defined at a point of a line or a curve belonging to the system, oriented, through the choice of the minus sign, in the opposite direction of the greatest variation of the global variable u. This choice results from the physical status of the "driving force" of $\boldsymbol{u}_{/r}$: In all conduction or transfer processes not driven by an external cause, flows are oriented toward decreasing potentials (efforts), that is, in the same direction as $\boldsymbol{u}_{/r}$ (see Figure 5.2).

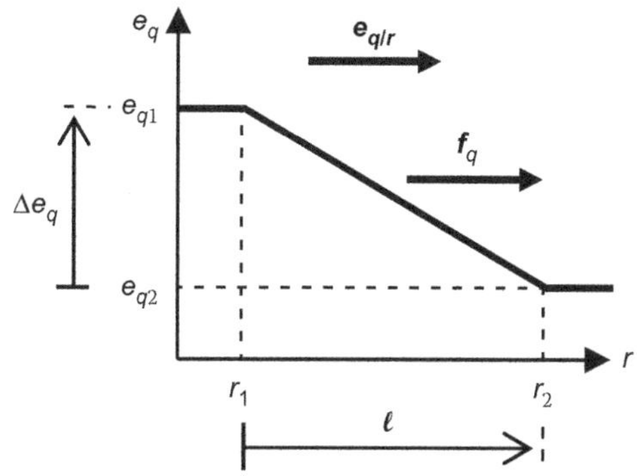

FIGURE 5.2 Scheme of a conduction between two localizations r_1 and r_2 in space with respective efforts e_{q1} and e_{q2} showing that, by convention, the flow is oriented along the field $\mathbf{e}_{q/r}$ (the "driving force") and opposes the effort difference (or gradient).

The operation of multiplication between the two vectors $\mathbf{u}_{/r}$ and $\mathbf{dr}$ is the *dot product* (or *scalar product* or *inner product*), which is zero when the two vectors are orthogonal and maximum when they are parallel.

The mathematical operator leading to $-\mathbf{u}_{/r}$ is the *gradient operator*, whose usual symbol is the nabla (or del) ∇. The geometrical representation is given in Figure 5.3a, showing a curve of the system defined by $u = C^{st}$, on which every point satisfies the condition $du = 0$. The vector $\mathbf{U}$ placed perpendicularly to an arbitrary point represents the gradient $-\mathbf{u}_{/r}$ as it is this orthogonality with the tangent vector $\mathbf{dr}$ to the curve that gives zero for the dot product. The direction of $\mathbf{U}$ is given by the direction of the maximum variation of u, represented by another curve intersecting orthogonally the previous curve, not sketched in Figure 5.3a, and defined by du = maximum. Consequently, the curve-localized variable can be written as the result of minus the gradient operator applied to the global variable u:

$$\mathbf{u}_{/r} = -\frac{d}{\mathbf{dr}} u = -\nabla u = -\text{grad } \mathbf{u} \tag{5.5}$$

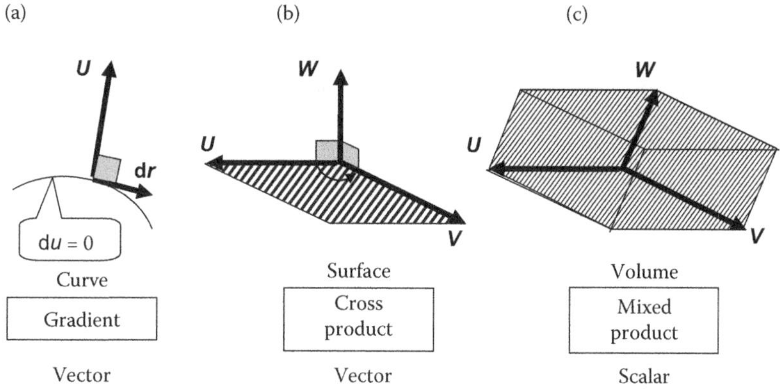

FIGURE 5.3 Geometrical illustration of representative vectors of a curve, a surface, and a volume: (a) Gradient $\mathbf{U}$ is orthogonal to a curve defined by the scalar $u = C^{st}$ and oriented toward the maximum increase in u. (b) The cross product $\mathbf{U} \times \mathbf{V}$ is defined as the vector $\mathbf{W}$ orthogonal to $\mathbf{U}$ and $\mathbf{V}$ whose modulus is equal to the area of the parallelogram $(\mathbf{U},\mathbf{V})$. (c) The mixed product of $\mathbf{U}$, $\mathbf{V}$, $\mathbf{W}$ is defined as the dot product of the cross product $\mathbf{U}.(\mathbf{V} \times \mathbf{W})$ and is equal to the volume of the parallelepiped $(\mathbf{U},\mathbf{V},\mathbf{W})$.

The localization of the global variable to the level of a surface belonging to the system is obtained by the dot product of the *surface-localized variable* $\boldsymbol{u}_{/A}$ with the vector $\mathbf{dA}$ representing the differential area:

$$\mathrm{d}\boldsymbol{u} = u_{/A}.\mathbf{d}\boldsymbol{A} \tag{5.6}$$

The differential area to be used for localization is given by using a different mathematical operator, the *vector product* (or *cross product*) of two vectors. It has the property to give a third vector whose length is equal to the area defined by the two vectors. If $\boldsymbol{W}$ is the result of the cross product of $\boldsymbol{U}$ and $\boldsymbol{V}$:

$$\boldsymbol{W} = \boldsymbol{U} \times \boldsymbol{V} \tag{5.7}$$

the area of the parallelogram defined by $\boldsymbol{U}$ and $\boldsymbol{V}$ is the modulus, or length, of $\boldsymbol{W}$:

$$\text{Area}(\boldsymbol{U},\boldsymbol{V}) = \|\boldsymbol{W}\| \tag{5.8}$$

It can be useful to recall that the vector operator is not properly defined by the expressions of its components, for they depend on the adopted space representation, Cartesian, spherical, etc., but by Equation 5.8 representing the area and by stating that $\boldsymbol{W}$ is orthogonal to both $\boldsymbol{U}$ and $\boldsymbol{V}$. Figure 5.3b shows the geometrical representation of the cross product. The ability of the cross product to express an area works also in the reciprocal way, that is, the differentiation with respect to an area is equal to the cross product of two differential vectors, supposedly oriented in different directions and distinguished by subscripts 1 and 2:

$$\frac{\mathbf{d}}{\mathbf{d}\boldsymbol{A}} = \nabla_2 \times \nabla_1 \tag{5.9}$$

By using the previous operator equation, the definition of the surface-localized variable in Equation 5.6 can be linked to the global variable

$$\boldsymbol{u}_{/A} = \frac{\mathbf{d}}{\mathbf{d}\boldsymbol{A}} u = \nabla_2 \times \nabla_1 u \tag{5.10}$$

The link between the two localized variables is easily deduced from the comparison with Equation 5.5 expressing the curve-localized variable (lineic density)

$$\boldsymbol{u}_{/A} = -\frac{\mathbf{d}}{\mathbf{d}\boldsymbol{r}} \times \boldsymbol{u}_{/r} = -\nabla_2 \times \boldsymbol{u}_{/r} = -\mathbf{curl}\ \boldsymbol{u}_{/r} \tag{5.11}$$

At last, the *volume-localized variable* is defined by an ordinary product between scalars

$$\mathrm{d}u = u_{/V}\mathrm{d}V \tag{5.12}$$

The reason is that, contrary to a line or a surface, a volume has no orientation in space and does not require a vector to represent it. The operator expressing a volume is the *mixed product* of three vectors, or *scalar triple product*, which gives a scalar equal to the volume of the parallelepiped defined by the three vectors, as shown in Figure 5.3c.

$$\text{Volume}\ (\boldsymbol{U},\boldsymbol{V},\boldsymbol{W}) = \boldsymbol{U}.(\boldsymbol{V} \times \boldsymbol{W}) \tag{5.13}$$

Following the same approach as for the surface-localized variable, the differentiation with respect to the volume is written using another differential operator corresponding to the third direction

$$\frac{\mathrm{d}}{\mathrm{d}V} = \nabla_3 . \nabla_2 \times \nabla_1 \tag{5.14}$$

leading to the relationship with the global variable

$$u_{/V} = \frac{\mathrm{d}}{\mathrm{d}V} u = \nabla_3 . \nabla_2 \times \nabla_1 u \tag{5.15}$$

The link with the surface-localized variable is directly obtained by introducing Equation 5.10 into Equation 5.15

3-localized variable: $$u_{/V} = \frac{\mathbf{d}}{\mathbf{d}\boldsymbol{r}} \boldsymbol{u}_{/A} = \nabla_3 . \boldsymbol{u}_{/A} = \mathrm{div}\ u_{/A} \tag{5.16}$$

The previous relations and variables can be represented in a Formal Graph as shown in Graph 5.8.

Polar versus Axial. A distinction is classically made between the two localized variables represented by vectors, in terms of dependence with the orientation in space. The first localized variables ($\boldsymbol{u}_{/r}$) are said to be *polar vectors*, whose orientation is independent of the choice of coordinates, whereas the second ones ($\boldsymbol{u}_{/A}$) are said to be *axial vectors* as their orientation depends on the order of the vectors implied in the cross product. In the Formal Graph theory, this classification is replaced by the notion of localization level, directly linked to the dimension of the geometrical elements of line, surface, and volume, each one having a proper orientation rule or not. In Differential Forms Geometry (Schutz 1980; Warnick et al. 1997), the classification is the same, with different names, each level being numerically identified by its reduction degree: 0-form (global variable), 1-form (curve level), 2-form (surface level), and 3-form (volume level).

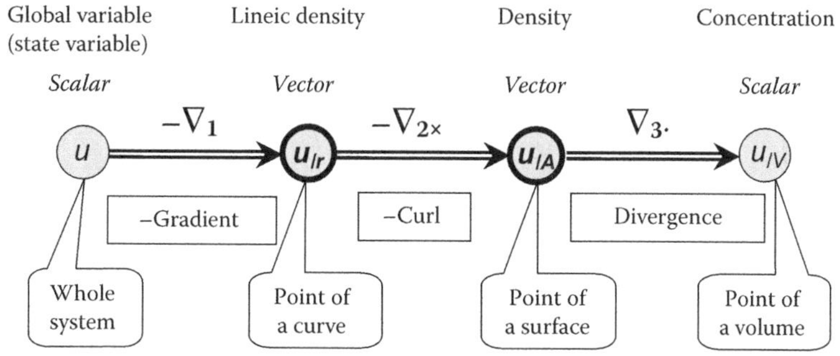

GRAPH 5.8 Generic Formal Graph of the four localization levels: global, curve, surface, and volume, with the convention of two minus signs for the two first operators.

In the previous reasoning for establishing the three spatial operators, three different operators, one for each direction in space, were used. This was only for demonstration, because in a homogeneous and isotropic space all these differential operators are identical.

$$\text{Homogeneous + isotropic space:} \qquad \nabla_3 = \nabla_2 = \nabla_1 = \nabla = \frac{\mathbf{d}}{\mathbf{d}\boldsymbol{r}} \tag{5.17}$$

Also, the previous reasoning was based on a state variable without orientation in space, represented by a scalar, at the global level. When this is not the case and the global variable is a vector, a different sequence of spatial operators must be used as the sequence of operands for each operator is different. Consequently, for representing the role of space in a Formal Graph, two ways of sequencing the operators from the global level to the deeper level (point of a volume) must be distinguished in function of the geometrical nature of the global variable, scalar or vector.

5.4.1 Scalar at the Global Level

When the variable at the global level is a scalar, the sequence of operators previously established applies. The three localization levels and the operators linking them can be represented by a linear Formal Graph that merely adapts Graph 5.8 to the case of identical operators, as shown in Graph 5.9.

The sequence of the three spatial operators (gradient, curl, and divergence) is the *only one possible* for creating, from a scalar, the sequence of variables in the right order, from global to lineic densities, then (surface) densities, and finally (volume) concentration, when differential variables need to be considered. (In homogeneously distributed systems, all operators can merely be replaced by divisions by length, area, and volume.)

Inversion of the reduction, augmentation in localization levels, is naturally made by integration along a curve Γ, on a surface S, or in a volume V:

$$U = -\int_\Gamma \mathbf{u}_{/r} \cdot \mathbf{d}\boldsymbol{r} \tag{5.18}$$

$$U = \int_S \boldsymbol{u}_{/A} \cdot \mathbf{d}\boldsymbol{A} \tag{5.19}$$

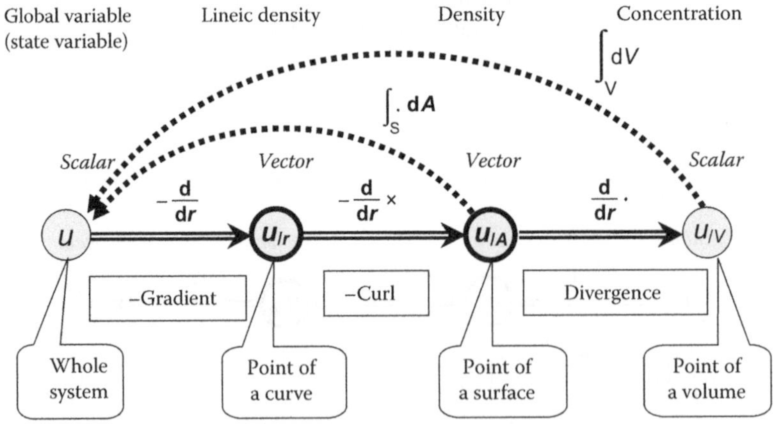

GRAPH 5.9 Generic Formal Graph of the four localization levels: global, curve, surface, and volume, with the various operators linking them: contra-gradient, contra-curl, and divergence. The two outer nodes are scalar variables, the two inner ones are vectors, symbolized with a bold circumference. Dotted connections represent inverse operators.

$$u = \int_V u_{/V} \cdot \mathrm{d}V \tag{5.20}$$

By default, the geometrical elements are supposed to be open. In some cases, the precision of the closing is necessary. A circulation along a closed curve is noted:

$$u = -\oint_\Gamma \boldsymbol{u}_{/r} \,.\, \mathbf{d}\boldsymbol{r} \tag{5.21}$$

Similarly, a collection, or flux, through a closed surface is noted

$$u = \oint_S \boldsymbol{u}_{/A} \,.\, \mathbf{d}\boldsymbol{A} \tag{5.22}$$

This question of closed geometrical element is of great importance for discriminating between physical behaviors. A closed integral equal to zero expresses independence with the path followed during a displacement, a notion related to the definition of a state of the system. Thus, a state variable is also the result of a closed integration.

Two important theorems in vector analysis involving inverse spatial operators are included in Graph 5.9. The first is the Stokes* theorem:

$$\oint_\Gamma \boldsymbol{u}_{/r} \,.\, \mathbf{d}\boldsymbol{r} = \oint_S \left(\nabla \times \boldsymbol{u}_{/r}\right) \cdot \mathbf{d}\boldsymbol{A} = -u \tag{5.23}$$

The second important relationship is the Green†–Ostrogradsky‡ theorem, or the divergence theorem (also called the Gauss theorem, but this leads to confusion with a similar theorem bearing the same name in electrostatics seen in case study B5):

$$\oint_S \boldsymbol{u}_{/A} \,.\, \mathbf{d}\boldsymbol{A} = \int_V \left(\nabla \,.\, \boldsymbol{u}_{/A}\right) \mathrm{d}V = u \tag{5.24}$$

Actually, these theorems concern only the first equality in both previous equations, the second equality being given for pedagogical purposes.

5.4.2 Vector at the Global Level

When the variable at the global level is a vector, the situation is a little more complex because it involves more sophisticated variables that are matrices and tensors. However, before addressing the whole subject, the special case of a single orientation is worth presenting because it is simpler than the general case. This occurs in systems with state variables depending on one dimension of space, which does not mean necessarily that the number of dimensions of the system is one. Case study B2 "Elasticity of Solids" about the elastic deformation of a three-dimensional solid bar illustrates this case.

When a vector state variable $\boldsymbol{u}$ is oriented in a *unique direction*, whatever its localization in space, the variation of this state variable produces a vector $\mathbf{d}\boldsymbol{u}$, which is colinear with the space

---

* George Gabriel Stokes (1819–1903): Irish mathematician and physicist; Cambridge, UK.

† George Green (1793–1841): British mathematician, UK.

‡ Mikhail Vasilievich Ostrogradsky (1801–1862): Russian mathematician, mechanician, and physicist; Paris, France and St Petersburg, Russia.

orientation **dr**. This is expressed by the proportionality between these two vectors through a simple scalar, and the same works for the relation between **du** and **dA**.

Single-orientation system:

$$\text{Curve-localized variable: } \mathbf{du} = -u_{/r}\mathbf{dr} \tag{5.25}$$

$$\text{Surface-localized variable: } \mathbf{du} = -u_{/A}\mathbf{dA} \tag{5.26}$$

$$\text{Volume-localized variable: } \mathbf{du} = \mathbf{u}_{/V}\, dV \tag{5.27}$$

For the last relation, since dV is a scalar, one needs a vector for the volume-localized variable.

Consequently, the first localized variable can be obtained by a *convergence* operator from the global variable (the minus sign is a deliberate choice on the same footing as the contra-gradient in the previous discussion):

$$-\frac{\mathbf{d}}{\mathbf{dr}}\cdot\mathbf{u} = -\nabla.\mathbf{u} = -\text{div }\mathbf{u} = \text{conv }\mathbf{u} \tag{5.28}$$

In contradistinction with the previous discussion about reducing scalar state variables, the relationship with the next localized variable cannot be obtained by a differential vector operator, but by an ordinary derivation, because here the two variables $u_{/r}$ and $u_{/A}$ are both scalars. Notwithstanding, this is possible between the two last levels by a gradient operator:

$$\mathbf{u}_{/V} = \frac{\mathbf{d}}{\mathbf{dr}}u_{/A} = \nabla u_{/A} = \text{grad }u_{/A} \tag{5.29}$$

The three localization levels and the operators linking them can be represented by a linear Formal Graph shown in Graph 5.10.

When the global variable is a vector with several possible orientations in space, as is the general case in mechanics, the sequence of operators is the same as for the scalar global variable (gradient, curl, and divergence), but dimensions of the mathematical beings used as variables are increased to

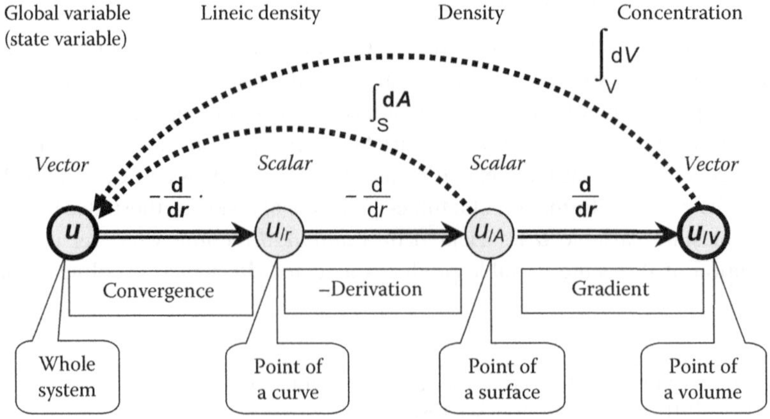

GRAPH 5.10 Localized variables and spatial operators for single-orientation systems (**du** colinear with **dr**).

reflect their dependence on multiple orientations. Instead of a vector for the 1-localized and 2-localized variables, one finds square matrices, and instead of a scalar for the 3-localized variable, one finds a vector. This will be exemplified later when dealing with the energy variety of translation mechanics.

In that case of oriented state variables, which is the most general one, the previously encountered schemes of spatial reduction must be replaced by another one, shown in Graph 5.11.

The gradient of a vector $\boldsymbol{u}$ must be understood as the square matrix $-[\boldsymbol{u}_{/r}]$ resulting from the matrix product of the 1-line matrix of the transposed differential vector ∇^T with the 1-column matrix of the vector $\boldsymbol{u}$:

$$\left[\boldsymbol{u}_{/r}\right] = -\nabla^T \boldsymbol{u} = -\frac{\mathbf{d}^T}{\mathbf{d}\boldsymbol{r}}\boldsymbol{u} \tag{5.30}$$

The same principle applies for reducing further the variables. For more details, the reader is referred to the literature on differential forms (Darling 1994; Frankel 2004). It requires the sequence of localized variables described in Graph 5.11, with the following features for a three-dimensional space:

Variable	Applied To	Symbol	Nature	Equivalent Tensor
State variable	Global	$\boldsymbol{u}$	Vector (3)	Tensor of rank 1
First localized variable	Curve point	$[\boldsymbol{u}_{/r}]$	Matrix (3×3)	Tensor of rank 2
Second localized variable	Surface point	$[\boldsymbol{u}_{/A}]$	Matrix (3×3)	Tensor of rank 2
Third localized variable	Volume point	$\boldsymbol{u}_{/V}$	Vector (3)	Tensor of rank 1

In accordance with such a scheme, the main localized variables in this domain are the following (localized variables not listed do exist, but they are not always used):

Localized basic quantity:

$$\text{Strain tensor: } [\ell_{/r}] = -\nabla^T.\ell \tag{5.31}$$

$$\text{Strain surface concentration: } [\ell_{/A}] = -\nabla \times [\ell_{/r}] \tag{5.32}$$

Localized effort:

$$\text{Force lineic density: } [F_{/r}] = -\nabla^T \cdot \boldsymbol{F} \tag{5.33}$$

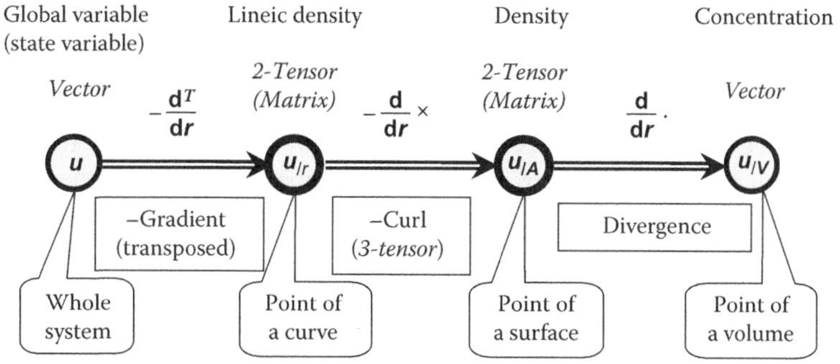

GRAPH 5.11 Generic Formal Graph of the four reduction levels (depths): global, curve, surface, and volume, when the state variable is a vector with any orientation.

$$\text{Stress tensor: } [\sigma] = -\nabla \times [F_{/r}] \tag{5.34}$$

Localized impulse:

$$\text{Momentum lineic density: } [p_{/r}] = -\boldsymbol{\nabla}^T \cdot \boldsymbol{P} \tag{5.35}$$

$$\text{Momentum surface concentration: } [p_{/A}] = -\boldsymbol{\nabla} \times [p_{/r}] \tag{5.36}$$

$$\text{Momentum concentration: } \boldsymbol{P}_{/V} = \boldsymbol{\nabla} \cdot [p_{/A}] \tag{5.37}$$

Localized flow:

$$\text{Velocity lineic density: } [v_{/r}] = -\boldsymbol{\nabla}^T \cdot \boldsymbol{v} \tag{5.38}$$

Reduced system properties are operators of different nature, depending on the linked variables. The generalized elasticity and shear modulus is a tensor of rank 3 because it links two tensors of rank 2. The same applies for the generalized viscosity.

$$\text{Elasticity tensor: } [\sigma] = \hat{\mathrm{E}}\,[\ell_{/r}] \tag{5.39}$$

$$\text{Viscosity tensor}: [\sigma] = \hat{\eta}\,[v_{/r}] \tag{5.40}$$

The generalized volumic mass is a matrix (tensor of rank 2) because the two linked variables are vectors (tensors of rank 1).

$$\text{Volumic mass: } \boldsymbol{P}_{/V} = \hat{\rho}_M \boldsymbol{v} \tag{5.41}$$

Such a generalized volumic mass is required when the mass distribution in the system is inhomogeneous in a homogeneous system; it reduces to a scalar (tensor of rank 0) identical to the classical volumic mass seen in single-orientation systems.

5.5 TRANSLATION PROBLEMS AND GENERALIZATION

The translation of the Gauss theorem into a Formal Graph cannot be carried out word for word. The Formal Graph language does not possess all the required words; in particular, the node for representing the flux of electric field cannot be used because it would be necessary to add a level above the global level, which is deprived of physical meaning in the Formal Graph theory. This apparent constraint is not a real constraint as the representation in the Formal Graph brings forth a lot to be learnt on what is hidden in this theorem.

Both the ordinary Differential Forms Geometry and the Formal Graph theory deal with ordinary physical three-dimensional space (without considering a time dimension for localization), so no deeper localization level (point of a hyper volume, gradient of the volume-localized variable) is admitted. This represents a significant constraint when translation into graphs of algebraic

TABLE 5.4
Generalized Operators for Noncontinuous Spaces (or Non-Euclidean Spaces in General)

	Continuous Space	Generalization
Curve integration	$\frac{d^{-1}}{\mathbf{d}\boldsymbol{r}^{-1}}$	$\hat{\mathbf{R}}$
Surface integration	$\frac{d^{-1}}{\mathbf{d}\boldsymbol{A}^{-1}}$	$\hat{\mathrm{A}}$
Volume integration	$\frac{d^{-1}}{dV^{-1}}$	$\hat{\mathrm{V}}$

equations containing a concentration gradient for instance is examined (see case study G3 "Fick's Law of Diffusion" in Chapter 10).

Generalized operators. Without departing from the restriction to the Euclidean space adopted for pedagogical reasons, Table 5.4 the generalization to other spaces (discrete space in particular) made possible by using different symbols. They will be used in many Formal Graphs later on as they apply to Euclidean space too.

5.6 IN SHORT

Global versus Localized Variables

1. **Global variable**: Variable, among the preceding categories, featuring the whole collection, independent from any space distribution.
2. **Localized variable**: Variable defined at a given point in space. (A state variable loses its status when localized.)

Localized Variables

Depth n_R	Localization	Variable	Generic Symbol (Scalar and Vector)		Generic Name
0	Whole system	Global quantity	u	$\mathbf{u}$	State variable
1	Point of a curve	Global quantity / Length	$\mathbf{u}_{/r}$	$u_{/r}$	Lineic density
2	Point of a surface	Global quantity / Area	$\mathbf{u}_{/A}$	$u_{/A}$	Density
3	Point of a volume	Global quantity / Volume	$u_{/V}$	$\mathbf{u}_{/V}$	Concentration

Sequence of Spatial Operators for a Scalar Global Variable

Name	Differential Operator	Applies to (Operand)	Gives (Result)
Contra-gradient	$-\frac{\mathbf{d}}{\mathbf{d}\boldsymbol{r}}$	Scalar u 0-Localized *Global (State variable)*	Vector $\boldsymbol{u}_{/r}$ 1-Localized *Point of curve*
Contra-curl	$-\frac{\mathbf{d}}{\mathbf{d}\boldsymbol{r}}\times$	Vector $\boldsymbol{u}_{/r}$ 1-Localized *Point of curve*	Vector $\boldsymbol{u}_{/A}$ 2-Localized *Point of surface*
Divergence	$\frac{\mathbf{d}}{\mathbf{d}\boldsymbol{r}}\bullet$	Vector $\boldsymbol{u}_{/A}$ 2-Localized *Point of surface*	Scalar $u_{/V}$ 3-Localized *Point of volume*

Sequence of Spatial Operators for a Vector Global Variable with Various Orientations

Name	Differential Operator	Applies to (Operand)	Gives (Result)
Transposed contra-gradient	$-\frac{\mathbf{d}^T}{\mathbf{d}\boldsymbol{r}}$	Vector $\boldsymbol{u}$ ($\mathbf{d}\boldsymbol{u} \nparallel \mathbf{d}\boldsymbol{r}$) 0-Localized *Global (State variable)*	2-Tensor (Matrix) $[u_{/r}]$ 1-Localized *Point of curve*
Contra-curl (3-Tensor)	$-\frac{\mathbf{d}}{\mathbf{d}\boldsymbol{r}}\times$	2-Tensor (Matrix) $[u_{/r}]$ 1-Localized *Point of curve*	2-Tensor (Matrix) $[u_{/A}]$ 2-Localized *Point of surface*
Divergence	$\frac{\mathbf{d}}{\mathbf{d}\boldsymbol{r}}\bullet$	2-Tensor (Matrix) $[u_{/A}]$ 2-Localized *Point of surface*	Vector $\boldsymbol{u}_{/V}$ 3-Localized *Point of volume*

Sequence of Spatial Operators for a Vector Global Variable with a Single Orientation

Name	Differential Operator	Applies to (Operand)	Gives (Result)
Convergence	$-\frac{\mathbf{d}}{\mathbf{d}\boldsymbol{r}}\bullet$	Vector $\boldsymbol{u}$ ($\mathbf{d}\boldsymbol{u} \parallel \mathbf{d}\boldsymbol{r}$) 0-Localized *Global (State variable)*	*Scalar* $u_{/r}$ 1-Localized *Point of curve*
Contra-derivation	$-\frac{\mathrm{d}}{\mathrm{d}r}$	Scalar $u_{/r}$ 1-Localized *Point of curve*	Scalar $u_{/A}$ 2-Localized *Point of surface*
Gradient	$\frac{\mathbf{d}}{\mathbf{d}\boldsymbol{r}}$	Scalar $u_{/A}$ 2-Localized *Point of surface*	Vector $\boldsymbol{u}_{/V}$ 3-Localized *Point of volume*

Node Coding

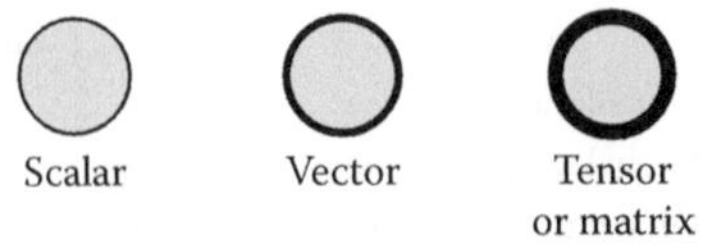

The circle (disc) is used as a generic drawing of nodes for all families of variables (basic quantities, efforts, flows, impulses).

6 Dipoles

CONTENTS

This chapter deals with the association of two poles forming a dipole. By definition, both poles must belong to the same energy subvariety or must be conductive. This new Formal Object belongs to the organization level immediately above that of poles, which is itself just below the level of the multipoles (comprising more than two poles).

Table 6.1 lists the case studies of dipoles, all working at the global level of systems (without space distribution) (see also Figure 6.1).

Systems that are dipoles:

- Two sets of electric charges
- Electric capacitor
- Magnet
- Solenoid (coil)
- Spring (elastic solid)
- Mechanical rod (two linked masses)
- Twin stars
- Apple and Earth
- Colliding objects
- Adjacent viscous layers
- Two coalescent liquid drops
- Chemical reaction
- Two phases apart from an interface
- Electrons and holes
- Conductor, pipes, etc.

TABLE 6.1
List of Case Studies of Dipoles

	Name	Energy Variety	Type	Conservation	Separable	Page Number
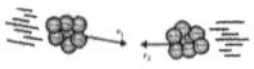	C1: Colliding Bodies	Translational mechanics	L	Conservative	Yes	145
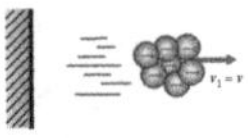	C2: Relative Motion	Translational mechanics	L	No	Yes	148
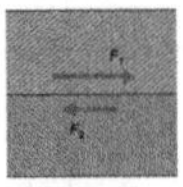	C3: Viscous Layers	Translational mechanics	LG	Conservative	Yes	151
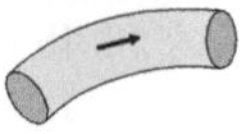	C4: Pipe	Hydrodynamics	G	No	Yes	154
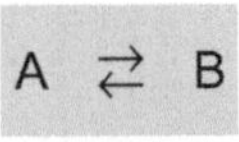	C5: First-Order Chemical Reaction	Physical chemistry	CG	Conservative	Yes	159
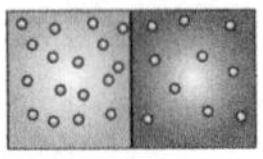	C6: Physical Chemical Interface	Physical chemistry	C	Conservative	Yes	164
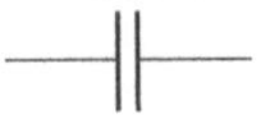	C7: Electric Capacitor	Electrodynamics	C	Conservative	Yes	168
	C8: Spring	Translational mechanics	C	Conservative	No	171
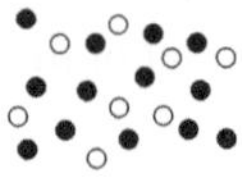	C9: Electrons and Holes	Corpuscular energy	C	Conservative	No	177

Note: C, capacitive; G, conductive; L, inductive; CG, capacitive–conductive; LG, inductive–conductive.

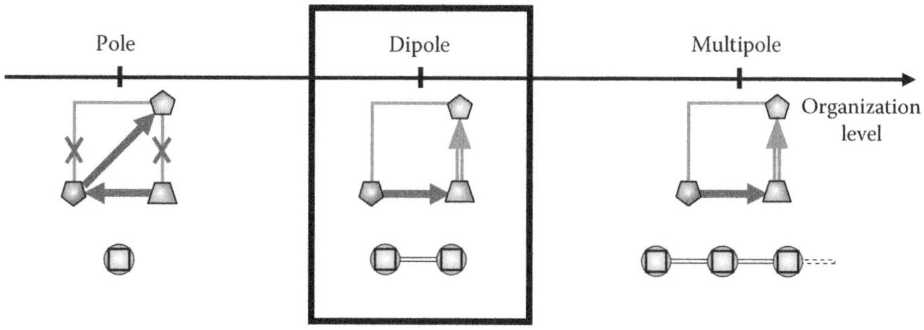

FIGURE 6.1 Position of the dipole Formal Object along the scale of organization levels.

6.1 THE DIPOLE

The progression in organization is accompanied by the appearance of new properties that could not be present in the lower level of the poles. It must be stressed that the most interesting advantage of having defined rigorously a complexity scale is precisely to allow tackling the important question of emergence of properties.

The most obvious emergent property is the possibility of *interaction* between the poles. Another new property is the *conservation* or not of the total number of entities when the energy varies in the dipole. At last, another new, although less evident, property is the separability of a dipole (i.e., whether or not a dipole is *separable*), which is related to the symmetry of the energies-per-entity between the poles and to conservation.

A pole has no possibility of exchanging energy with another system when compared with another pole in the same energy subvariety, but a dipole can be involved in many operations with other dipoles whatever the energy variety or subvariety. In particular, a dipole storing energy can convert it when associated with a dipole that does not belong to the same energy subvariety, owing to its space–time property based on the concept of time called *evolution*. Time is not yet introduced in the Formal Graph theory; it will be done later in Chapter 9 when conversion between energy subvarieties will be discussed, but it is mentioned in this chapter as a possibility for an energy storing dipole to use it. This is also an emerging feature that a pole is unable to accommodate.

The complete definition of the concept of dipole will be given after having presented several case studies of dipoles of various types and taken in a selection of domains able to give a sufficient view of this basic Formal Object.

Reductionism versus emergentism. These are two opposite conceptions of the relationship between organization levels. Reductionism is a principle stating that all properties of objects can be deduced from the properties of objects belonging to the lower level. Emergentism claims that some properties cannot be reduced to combinations of lower-level properties. This is summarized by the sentence "the whole is not the sum of all its parts."

The advantage of reductionism is to allow a true analytical approach for dividing a system into simpler elements and being able to reconstruct it in the opposite way. Emergentism requires a rigorous definition of levels and associated properties for pretending to be a valid approach.

As such a clear picture is rarely available, the most widespread and secure approach in science is reductionism. Emergentism is often considered, with reason, as pure speculation, owing to the lack of theoretical tools.

The case studies in this chapter deal with dipoles involving interaction by exchange, without involving influence and without space distribution. A short presentation of the principle of exchange will be given before discussing the case studies in detail. Chapter 7 will deal with dipoles involving interaction by influence and will also introduce space distribution in the limited case of homogeneous space and spherical geometry.

6.2 FORMAL GRAPH REPRESENTATION OF A DIPOLE

6.2.1 The Embedding Principle

At the organization level of a pole, four state variables were defined: the basic quantity q with its conjugate variable the effort e_q, and the impulse p_q with its conjugate the flow f_q. The same canonical scheme is taken again for a dipole, with the same names for the state variables, which are obviously different variables since they describe the state of a different object. The consequence is that the same set of variables and the same graph structure depict equally well a dipole and a pole. This is the *embedding principle*: The lower structure is included into the upper one, and even better, is superimposable.

There are three fundamental poles and two mixed ones, with exactly the same topological structure. The inductive or the capacitive dipole is the basic unit for storing energy in spreading it among its two poles, while being in a well-defined state of the system. The reason for this is that the first principle of thermodynamics allows the energy variation of such a dipole to be an exact differential because of the possibility of exchange between the two poles. The variations of energy are written either as a sum of pole contributions or as a single dipole contribution:

Inductive dipole: $$d\mathcal{T}_q = f_{q1}\, dp_{q1} + f_{q2}\, dp_{q2} = f_q\, dp_q \quad (6.1)$$

Capacitive dipole: $$d\mathcal{U}_q = e_{q1}\, dq_1 + e_{q2}\, dq_2 = e_q\, dq \quad (6.2)$$

When poles are to be considered in the same graph or equation, numeral subscripts are used to distinguish the pole state variables.

6.2.2 Formal Graphs

The graphical representation of a dipole without space distribution is quite simple; two groups are to be considered: a first one when only one system property is present, consisting of three fundamental poles—inductive, conductive, and capacitive, and a second one with two properties, one of them being the conductance (see Table 6.2).

TABLE 6.2
System Properties and Energetic Nature of the Five Kinds of Dipoles

		Constitutive Properties			Energetic Behavior	
Type	Nature	Inductance	Conductance	Capacitance	Storage	Dissipation
Inductive	Fundamental	☑	☐	☐	☑	☐
Inductive–conductive	Mixed	☑	☑	☐	☑	☑
Conductive	Fundamental	☐	☑	☐	☐	☑
Conductive–capacitive	Mixed	☐	☑	☑	☑	☑
Capacitive	Fundamental	☐	☐	☑	☑	☐

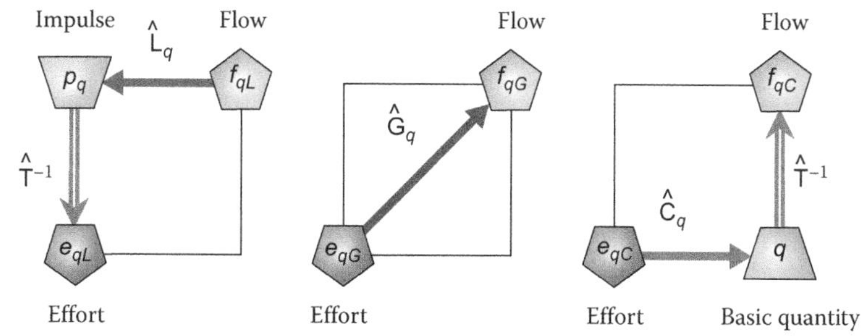

GRAPH 6.1 Inductive (left), conductive (center), capacitive (right) dipoles.

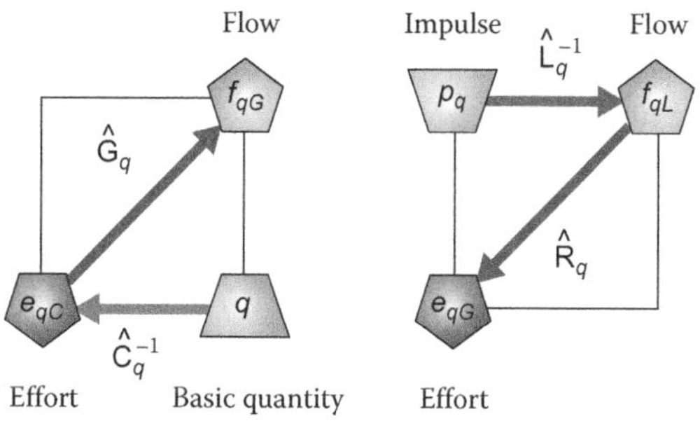

GRAPH 6.2 Inductive–conductive dipole (left). Capacitive–conductive dipole (right).

The Formal Graph representations of the three basic dipoles and of the two mixed dipoles are given in Graphs 6.1 and 6.2.

Subscripts C, L, and G have been used for the energies-per-entity symbols in the nodes according to the dipole type. This is an optional notation useful when different types of dipoles are assembled together. The three system constitutive properties already used in poles are however more elaborate properties than in the lower organization level (although notated with the same symbols, operators have different expressions). In addition to these three system constitutive properties, fundamental dipoles that can store energy are endowed with a fourth property, the *evolution*. Its reciprocal is the *inverse evolution*, represented by an operator $\hat{T}$. The reason for introducing this property is given now, although its effective use will be tackled later in Chapter 9.

6.2.3 The Circularity Principle

The Formal Graphs shown in Graph 6.3 depict the principle of connecting two dipoles, taking as examples two capacitive dipoles on the one hand and one capacitive and one inductive dipole on the other.

As dipoles are able to exchange energy, connections must exist between them. This possibility is coded in a Formal Graph by relating variables belonging to the same family. For instance, efforts are connected together, basic quantities too, and so on. The nature of these connections will be detailed later in Chapter 9 that discusses assembling dipoles; for the moment the principle is considered.

A remarkable property of graphs is used here—the notion of loop (Berge 1962). A loop is a circular path in a graph that encodes identity or equivalence or a neutral operation. Many physical meanings can be attributed to the existence of a loop; it is not possible to discuss them in detail for

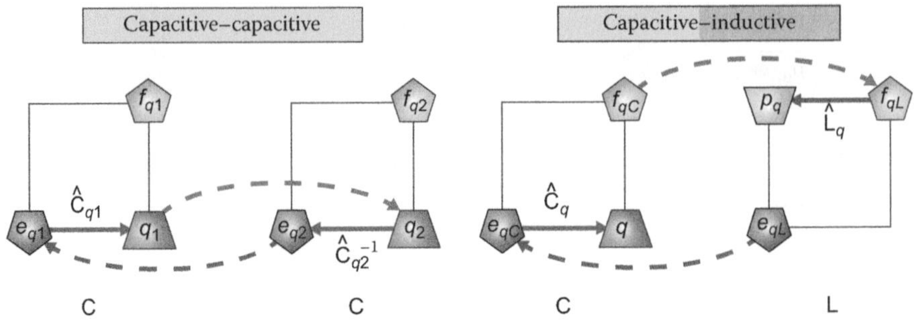

GRAPH 6.3 Associations of two capacitive dipoles (left) and one capacitive with one conductive (right). The principle of the energy exchange or conversion is to work through connections between state variables of the same family. Because the nature of the connections is not specified, dashed lines are used to represent them.

the moment, but one keeps in mind the idea that each time a circular path can be formed in a Formal Graph, it reveals something important.

In this example of association of two identical dipoles, a loop can be formed as two pairs of state variables, here two efforts and two basic quantities, can be connected and two capacitances provide the internal links. In contradistinction, no such circular path can be formed with the two different dipoles shown in Graph 6.3.

The missing link is the evolution operator that will ensure energy conversion between subvarieties or dissipation (cf. Graph 6.4). No conversion is needed between dipoles belonging to the same energy subvariety or being both conductive, which explains that this link is not required for connecting identical dipoles.

6.2.4 Gate Nodes and Variables

Gate nodes represent variables that are not necessarily state variables but that are used for connection with other Formal Objects, as explained in Section 4.5.1 in Chapter 4. The evolution property, together with a capacitance or an inductance, allows the fundamental storing dipoles to connect internally their two gate nodes that are the effort and the flow. The status of gate variable is indicated in Graph 6.4 when appropriate. These nodes are already internally connected in the conductive and mixed dipoles. It means that, for ensuring circulation of energy between associated dipoles (in following the circularity principle), both energies-per-entities are used as gate variables. They are the input and output of energy in the dipole through interconnections linking gate nodes of the same state variable family.

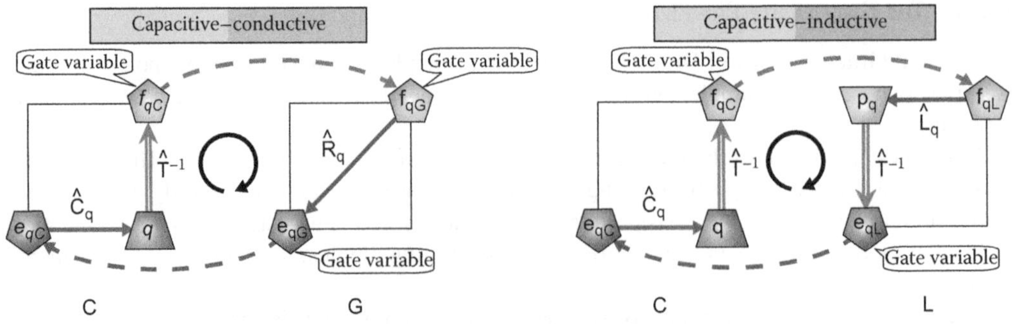

GRAPH 6.4 A loop can be formed when an evolution link is added to the properties of an energy storing dipole. Because the nature of the connections is not specified, dashed lines are used to represent them.

6.2.5 Evolution and Time

As already mentioned, the evolution link that did not exist in a pole appears. This link belongs to the category of space–time properties and not to the system constitutive properties. It is therefore invariant with the systems and their variables because space–time is a common frame for all systems.

As can be seen in Graphs 6.1 and 6.4, this space–time property is represented by an operator, since operators feature all properties in a Formal Graph. It is used for two different purposes, a first one for linking basic quantities q to flows f_q, and a second one between impulses p_q and efforts e_q:

$$f_{qC} = \hat{T}^{-1} q; \quad e_{qL} = \hat{T}^{-1} p_q \tag{6.3}$$

Symbols C and L have been added as subscripts to the resulting variable to indicate the energy subvariety of the operand variable. This convention helps to remember that the relationships shown in Equation 6.3 are not true definitions of flows and efforts, which are energies-per-entity and therefore defined as the partial derivative of energy with respect to the entity number, as explained in Chapter 2, Section 2.1.5.

Chapter 9 shows that in a *continuous* (which can be seen also as homogeneous) *space* and when energy conversion occurs, the operator of evolution is the derivation with respect to time*:

$$\hat{T}^{-1} \underset{cont}{=} \frac{d}{dt} \tag{6.4}$$

This link is graphically represented by a double line conventionally used for encoding links endowed with differential operators.

In the Formal Graph theory, time is a variable that follows the conversion of energy from one subvariety to another. It plays the role of *extent*, also called *advance*, in the progress of the conversion process. Naturally, evolution is used only when the dipole needs to convert energy, which is possible only when the dipole is connected to another one, belonging to a different energy subvariety or provoking dissipation.

Although no association or connection of dipoles is treated in this chapter, the evolution link is drawn in some Formal Graphs to remember that this space–time property may work between the connected nodes when necessary. This is obviously an optional choice, which helps to distinguish a dipole from a pole in an accentuated way.

An argued introduction and discussion of time will be made in Chapter 9, when dealing with the conversion of energy permitted by associating dipoles of different energy subvarieties, called *dipole assemblies*. In this chapter, the dynamics of dipoles, which are considered without any time involvement, will not be developed. The dynamic regimes, or kinetics out of stationary regime, will be treated in Chapter 8.

An example may help to better understand our approach concerning the question of time. Let us take an electric capacitor in various states of charge. The first state is the "Off state" in which the capacitor is loaded with $+Q$ and $-Q$ charges on its plates, which are isolated from each other. The second state is the "Intermediate state" defined by the possibility of the charges moving through an external circuit *without resistance* to combine into +/− pairs bearing no charge. (In fact, electrons

* The derivation with respect to time is a frequency operator because somewhere it divides by the time. In contradistinction, the integration with respect to time is really a time operator, so its operator notation uses the symbol T.

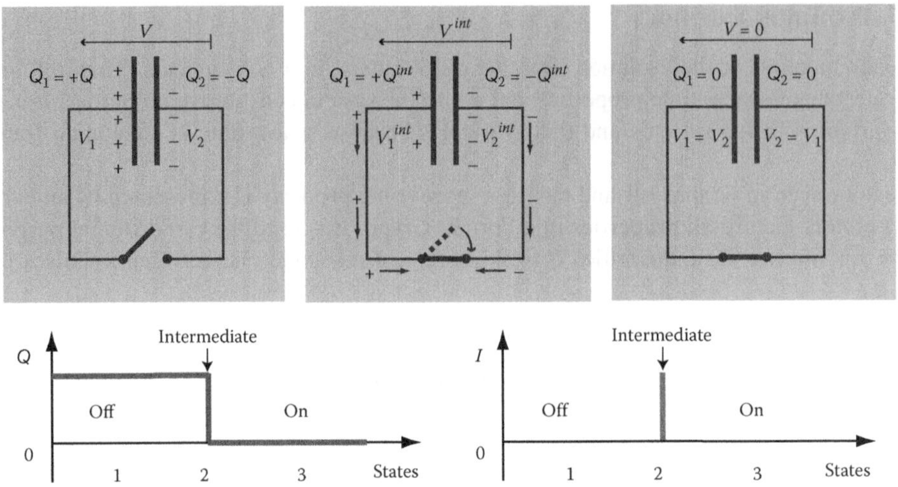

FIGURE 6.2 Three states of an electric capacitor: Off, Intermediate, On, and corresponding variations of charge and current.

flow from one side to the other to recombine with holes.) This flow is permitted by the closure of a switch in the circuit. The third state is the "On state" in which the switch is kept closed, so the charges on the capacitor plates and the potential difference are null. The total energy is conserved thanks to the formation of electron–hole pairs. These three states are sketched in Figure 6.2 with the corresponding charges and current variations.

Mathematically, the current is a Dirac* distribution, which has no thickness; physically, it means that no duration and no time can be attributed to this process. Time can only be introduced here when a resistance allows conversion into heat (dissipation) or when an inductance allows conversion into inductive energy. Naturally, this example is an ideal case study, because actually the electron–hole recombination takes time.

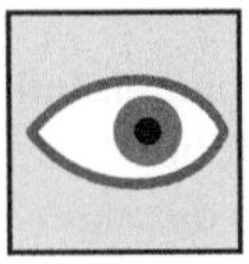

What is the absence of time? In this chapter, dipoles are treated independently of time. It means that the systems are normally considered in a static state or in a stationary regime. Notwithstanding, it does not mean that dynamics is excluded: All models developed here may work also when time allows systems to evolve by converting energy between dipoles.

The profound meaning is that all dipole properties defined in the absence of time, such as conservation, separability, interaction, exchange, influence, and so on, are obviously independent of the existence of time, which means that they remain the same with or without time. They are *time invariant* properties.

6.3 INTERACTION THROUGH EXCHANGE BETWEEN POLES

The subject of the interactions between the two poles is divided into two kinds: *influence* and *exchange*. *Influence* occurs when the interaction relies on the system constitutive properties that

* Paul Adrien Maurice Dirac (1902–1984): British physicist; Bristol, UK and Tallahassee, Florida, USA.

include capacitance, inductance, or conductance, and that allow the system to store or dissipate energy. *Exchange* is the process by which a dipole ensures a circulation of entities (basic quantities or impulses) between poles, independently of the system constitutive properties. In the case of a conductive dipole, which does not possess its own entities, they must be provided by at least one storing dipole associated to the conductive dipole. (In other words, conductive dipoles alone have zero values for their effort and flow.) Influence being a more complex phenomenon than exchange, the theory of influence will be developed in the next chapter for establishing the generalized capacitive relationship (modeling the dependence of the effort with the basic quantity).

A quantity, namely a state variable, which must be of the same family, is exchanged between the two poles. The *content* of an exchange is always an *entity number* (basic quantity or impulse) and the *vector* of this exchange is always the *energy-per-entity* (effort or flow). Flows are vectors for an exchange of basic quantities, whereas efforts are vectors for an impulse exchange. Generally, the exchange is symmetrical, each pole emitting an effort or a flow toward the other one, each pole being simultaneously an emitter and a receptor (see Graph 6.5).

6.3.1 Exchange by Flows

A classical schematic representation using arrows to symbolize the exchange by flows is shown in Figure 6.3.

A conservation law applied to each pole implies the definition of a net flow f_q allowing the sum of flows to be zero at each pole:

$$f_q + f_{q2} - f_{q1} = 0 \tag{6.5}$$

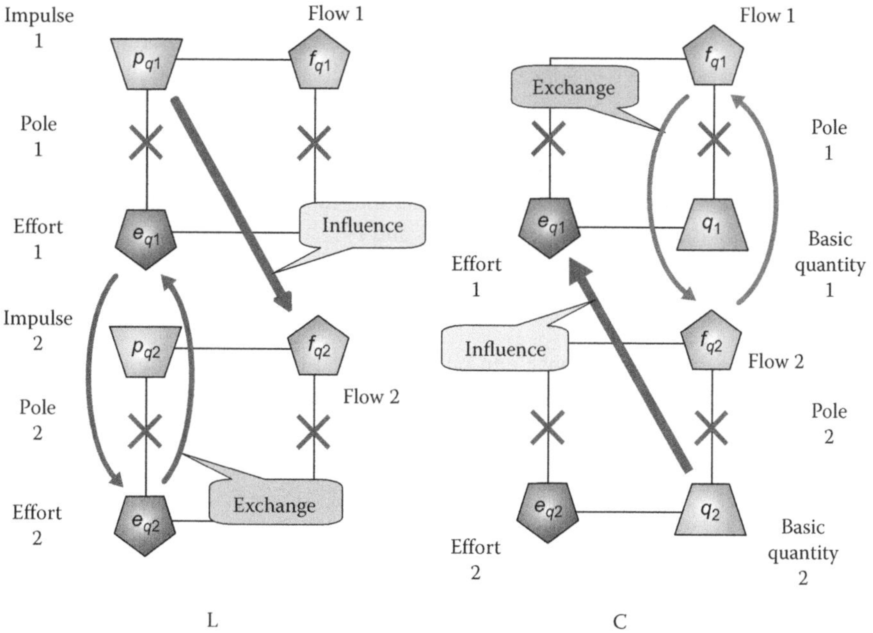

GRAPH 6.5 The two different kinds of pole–pole interactions are sketched in a pair of inductive poles (left) and in a pair of capacitive poles (right). Influence proceeds through constitutive properties: mutual reluctance (left) or mutual elastance (right). Exchange of entities proceeds by efforts (left) or by flows (right).

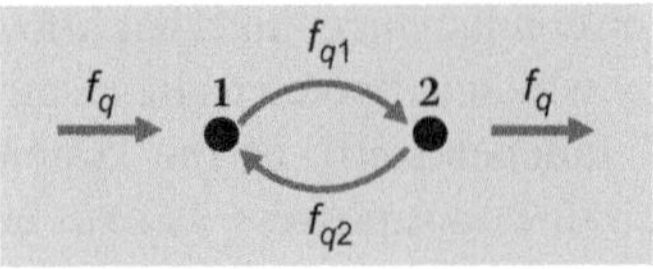

FIGURE 6.3 Symbolic representation of an exchange by flows between poles.

The net flow corresponds to the difference of polar flows, its nullity being obtained by the equality of both polar flows. In this case, it does not mean that the exchange is stopped but it still exists without net result. This notion corresponds to the *dynamic equilibrium* met in a semiconductor junction or in reaction kinetics. The reversible chemical reaction between two substances A and B is symbolically represented with the help of opposite arrows labeled with the corresponding specific rate (kinetic constant) (see Figure 6.4).

The net reaction rate $\mathfrak{I}_{AB}$ (or V_{AB}) for the transformation of A into B is defined in chemical kinetics by the difference between the two irreversible reaction rates (here taken as first-order reactions for this example):

$$\mathfrak{I}_{AB} = k_A n_A - k_B n_B \tag{6.6}$$

When the net reaction rate $\mathfrak{I}_{AB}$ is equal to zero, the two substances are said to be in *dynamic equilibrium*, each reaction in one direction being of any magnitude but equal to the other, leading to a strict compensation in terms of reacting molecules. Naturally, the correspondence with the generalized exchange scheme given previously is obtained by assimilating species A to pole 1 and species B to pole 2.

6.3.2 Exchange by Efforts

The discussion so far was concerned with the exchange by flows, but, as stated above, the exchange by efforts obeys the same scheme, although the classical use of arrows differs in this case. As an example, the interface between two adjacent layers of a viscous fluid, or solid phase, moving at different velocities $\boldsymbol{v}_1$ and $\boldsymbol{v}_2$, that is bearing different momenta $\boldsymbol{P}_1$ and $\boldsymbol{P}_2$, is the seat of contact forces. Each layer i exerts an interfacial, or contact, force $\boldsymbol{F}_i$ on the other layer, as featured in Figure 6.5.

The frictional force $\boldsymbol{F}$ appearing at the interface between the two phases results from the necessity to enforce a conservation law on each face of the interface, which is realized by equating $\boldsymbol{F}$ to the difference between the two interfacial forces $\boldsymbol{F}_1$ and $\boldsymbol{F}_2$:

$$\boldsymbol{F} = \boldsymbol{F}_1 - \boldsymbol{F}_2 \tag{6.7}$$

In this example, each layer or phase can be assimilated to a pole acting on the other, or, in other words, emitting an effort applied to the other. A *dynamic equilibrium* is reached when the two contact forces become equal. For the generalized dipole, the exchange by efforts can be written as

$$e_q + e_{q2} - e_{q1} = 0 \tag{6.8}$$

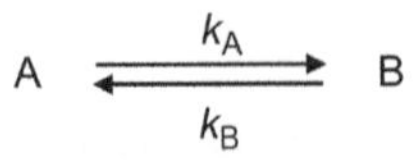

FIGURE 6.4 Reversible and simple binary chemical reaction.

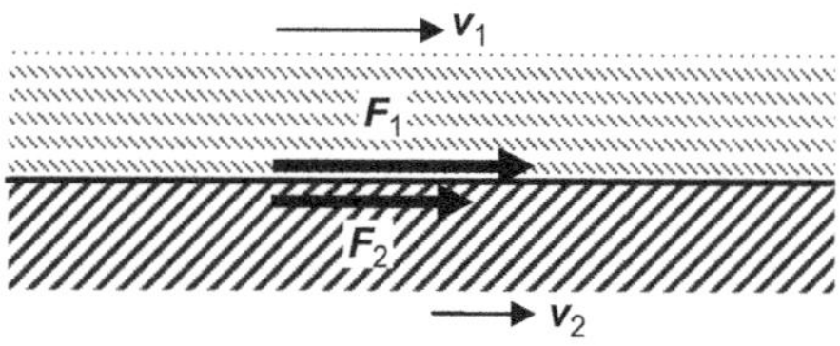

FIGURE 6.5 Symbolic representation of two phases in contact and in movement with different velocities, generating a frictional force at the interface by exchange of momenta.

To conclude on this question of exchange, the energy-per-entity (effort or flow) playing the role of the exchange vector in a dipole is formed by the difference between the corresponding *energies per entity* of each pole:

Effort as exchange vector: $$e_q = e_{q1} - e_{q2} \tag{6.9}$$

Flow as exchange vector: $$f_q = f_{q1} - f_{q2} \tag{6.10}$$

The convention is to ascribe the positive contribution to the pole labeled "1" and the negative contribution to the pole labeled "2." It is possible to view the first one as a positive pole and the second one as a negative pole, as it is commonly done in electrostatics for attracting charges, but, for the sake of generality as well as for the case of electrostatic repulsion between charges, the numbering is preferred. It can be remarked that the dipolar effort is written without any particular symbol such as delta. This is a classical abuse in many scientific domains, such as in electrodynamics, to call and to note identically a potential and a potential difference. Normally, the context should help to drop the ambiguity, so the tradition will be kept here, to facilitate the translation work between graphs and equations and to simplify the application of the embedding principle. A last remark is that content–vector couples are not belonging to the same energy subvariety but are crossed: basic quantity–flow for one and impulse–effort for the other. This outlines the fact that interpole exchanges are not energy exchanges.

6.3.3 Formal Graph Construction

The Formal Graphs of the various types of dipoles given previously in Graphs 6.1 and 6.2 are complete and self-standing models of dipoles. Dipoles use variables that are different from the variables of poles, although the same node shapes and algebraic symbols are used for both. It is not necessary to accompany such Formal Graphs with the representation of the poles from which they are built because the pole variables do not play an explicit role in the behavior of the dipole.

Nevertheless, it can be interesting to draw simultaneously all Formal Graphs of the poles and of the dipole to show graphically the connections between pole and dipole variables.

These connections represent two different algebraic operations: an equivalence between variables and a proportionality to a difference between variables. The first operation, equivalence, is merely encoded by two links starting from a dipole variable and each arriving on a pole variable.

For the second operation, there are basically two ways to encode a difference of variables in a graph: either directly, by summing algebraically weighted links on the resulting node, or by mimicking the exchange in making two summations on the pole nodes, as shown in the two Formal Graphs in Graph 6.6. The left one encodes an exchange by flows and shows the connection between dipole and pole efforts through a weight S_D that will be made explicit in Section 6.6.

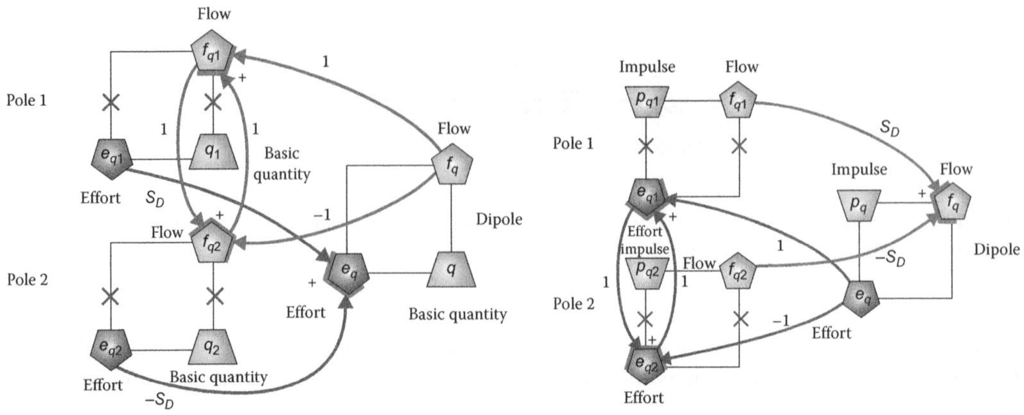

GRAPH 6.6 Connections involving an exchange by flows (left) and by efforts (right).

As will be seen in the case studies, the exchange is not always represented in the construction graphs, because exchange is only effective when the dipole is working in a dynamic regime (using time) or when conduction is ensured.

6.4 CASE STUDIES OF DIPOLE

Dipoles of various energy subvarieties that are working at the global level of the systems (not using space) are grouped in this section. They share the common feature of being able to involve an entity exchange, which is not always implemented.

All case studies are concerned by conservative and separable dipoles unless the contrary is mentioned.

Inductive
- C1: Colliding bodies in translational mechanics
- C2: Relative motion in translational mechanics (nonconservative)

Inductive–conductive
- C3: Viscous layers in translational mechanics

Conductive
- C4: Pipe in hydrodynamics (nonconservative)

Capacitive–conductive
- C5: First-order chemical reaction in physical chemistry

Capacitive
- C6: Physical chemical interface in physical chemistry
- C7: Electric capacitor in electrodynamics
- C8: Spring in translational mechanics (inseparable)
- C9: Electrons and holes in corpuscular energy (inseparable)

Independence of case studies. Although the number of covered domains is lower than for the poles, there is still little chance for the average reader to be at ease with all domains tackled in this chapter. Each case study is written as an independent topic, allowing the reader to browse according to his or her acquaintances.

6.4.1 Case Study C1: Colliding Bodies

Translational Mechanics

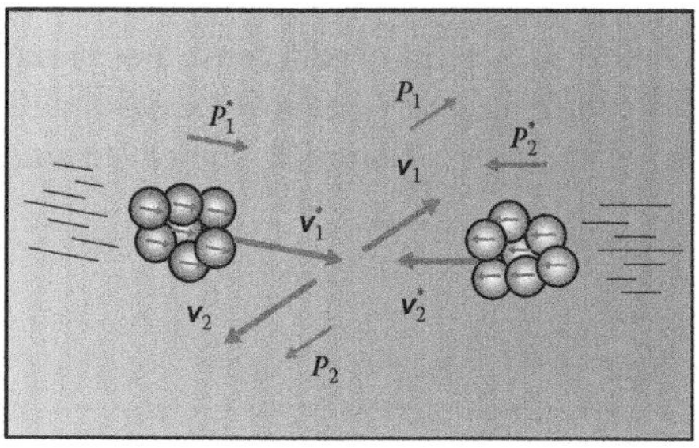

Two objects with mass running into collision make up an inductive fundamental dipole.

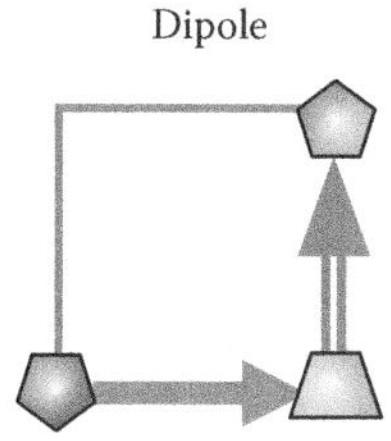

Dipole	
✓	Fundamental
	Mixed
	Capacitive
✓	Inductive
	Conductive
✓	Global
	Spatial
	Inseparable
✓	Conservative

Formal Graph (see Graph 6.7):

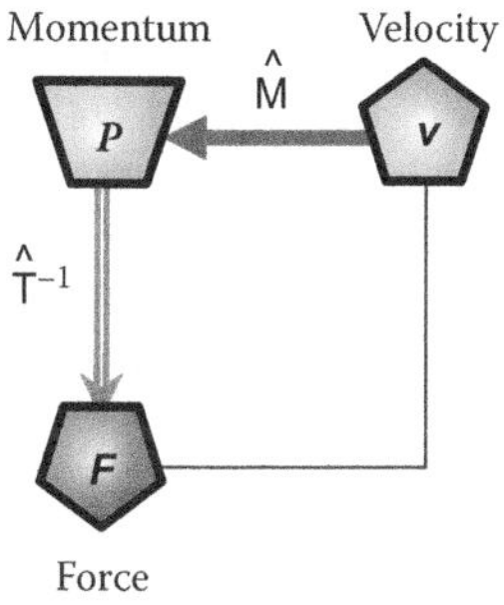

GRAPH 6.7

Variables:

Variety	Translational Mechanics		
Subvariety	Inductive (Kinetic)		Capacitive
Category	Entity number	Energy/entity	Energy/ entity
Family	Impulse	Flow	Effort
Name	Momentum (quantity of movement)	Velocity	Force
Symbols	$\boldsymbol{P}$, $(\boldsymbol{p})$	$\boldsymbol{v}$	$\boldsymbol{F}$
Unit	[N s], [J s m^{-1}]	[m s^{-1}]	[N], [J m^{-1}]

Property relationships:

$$\boldsymbol{P} = \hat{M}\boldsymbol{v}$$

Linear model:

$$\boldsymbol{P} \underset{lin}{=} M\boldsymbol{v}$$

System Property	
Nature	Inductance
Level	Global (0)
Name	Inertial mass
Symbol	M
Unit	[kg]
Specificity	—

Energy variation:

$$\mathrm{d}\mathcal{T}_{\ell} \underset{\exists M}{=} \boldsymbol{v} \cdot \mathrm{d}\boldsymbol{P}$$

The dipole belonging to translational mechanics, and more precisely to the inductive energy subvariety (i.e., kinetic energy), is a model for a large class of objects. Two balls of pool game are only an example; in fact, every couple of mechanical objects, not necessarily identical, interacting through a collision is also an example. Both objects do not need to move simultaneously, this is a question of referential; one object can be at rest and the other not. The problem of a representation by means of a unique object having a uniform aspect, especially if it is spherical, is to induce instinctively confusion between such an object and a particle. It must be recalled that each pole must be able to vary its entity number (without denaturizing the object), which is not always possible for a single particle. The scheme in the case study abstract is most respectful of this constraint in representing clusters of particles, each particle having a momentum (quantity of movement), and the sum of all momenta constitutes the impulse (momentum) of the pole. (Naturally, the clusters are supposed to be indestructible in the collision.)

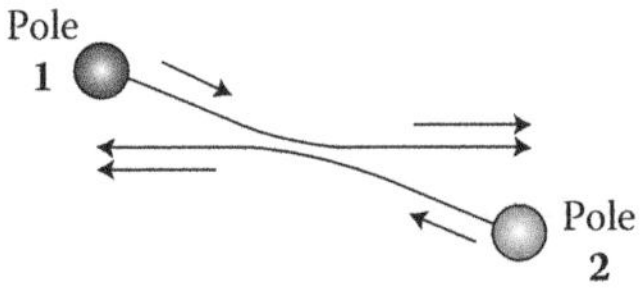

Two states of the system are considered: "before" the collision, with the quantities marked with a star, and "after" the collision. The quotes are there to help recall that one utilizes a vocabulary marked with notions related to time, even though this notion does not play any role in the conservation of impulses (momenta, quantities of movement). An inversion of time would not change anything, and even a change of its flow (acceleration of time, for instance, as relativity can predict) would be inefficient.

Energy: The variation of inductive (kinetic) energy is given in an equivalent way from individual variations of the poles and from variation of the dipole; this is the embedding principle:

$$\mathrm{d}\mathcal{T}_\ell = \mathbf{v}_1 \cdot \mathbf{d}\boldsymbol{P}_1 + \mathbf{v}_2 \cdot \mathbf{d}\boldsymbol{P}_2 = \mathbf{v} \cdot \mathbf{d}\boldsymbol{P} \tag{C1.1}$$

Conservation: The rule that applies to a dipole able to store energy is that the variations of entity numbers must be symmetric for each pole (no privileged pole with respect to the other; the choice of signs corresponding to the convention of a receptor dipole):

$$\mathbf{d}\boldsymbol{P}_1 = -\mathbf{d}\boldsymbol{P}_2 = \mathbf{d}\boldsymbol{P} \tag{C1.2}$$

It follows by integration the relation between momenta of poles and of dipole

$$\boldsymbol{P} = \boldsymbol{P}_1 - \boldsymbol{P}_1^* = -\boldsymbol{P}_2 + \boldsymbol{P}_2^* \tag{C1.3}$$

The impulse (momentum) of the dipole corresponds therefore, by integration, to the common difference of each pole. From this, it may be deduced that the conservation of momenta is expressed by the following balance equation

$$\boldsymbol{P}_1 + \boldsymbol{P}_2 = \boldsymbol{P}_1^* + \boldsymbol{P}_2^* \tag{C1.4}$$

and the poles–dipole relation for the velocities is

$$\mathbf{v} = \mathbf{v}_1 - \mathbf{v}_2 \tag{C1.5}$$

This relation comes from a combination of the energy variation and the momentum symmetry as shown in Equations C1.1 and C1.2, respectively.

System Constitutive Property: As for each pole (see case study A1 "Moving Bodies" in Chapter 4), the velocity of the dipole is related to the dipole momentum through an inductive relationship using the inertial mass M of the dipole as inductance (which is an operator in the general case):

$$\boldsymbol{P} = \hat{\mathsf{M}}\mathbf{v} \tag{C1.6}$$

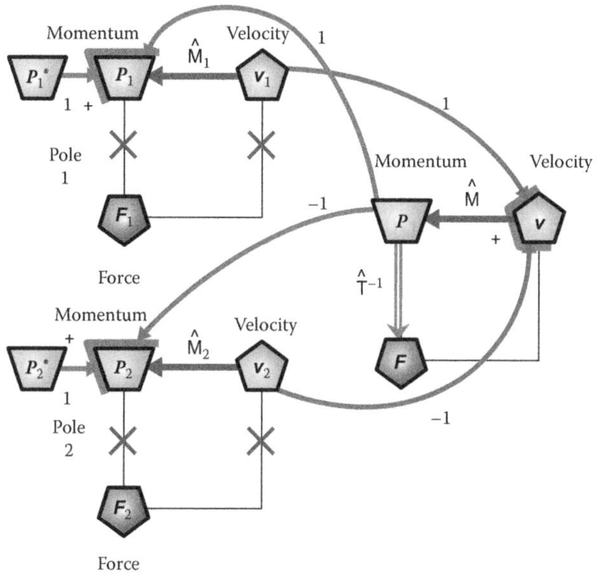

GRAPH 6.8 The two poles are located on the left part and the dipole on the right. To each momentum is associated its "initial" or "before" quantity (star superscript), which is added to the contribution of the dipole momentum (with the adequate sign) for forming each pole momentum, according to Equation C1.4. This is the way conservation of momenta is encoded by a Formal Graph. The dipole velocity is represented as resulting from a simple addition of pole velocities, weighted by positive and negative unit coefficients.

It is worth recalling that the mass, whether featuring a pole or a dipole, is an invariable scalar M (independent from another state variable) only in the frame of Newtonian mechanic, that is, for small velocities compared to the speed of light. This peculiar case is modeled by the well-known linear relation:

$$\mathbf{P} \underset{lin}{=} M\mathbf{v} \quad \text{(C1.7)}$$

Dipole Construction: Relationships C1.4 and C1.5 between pole and dipole variables can be represented by a special Formal Graph including both poles and the dipole and their connections, called *construction graph*, as shown in Graph 6.8.

Although incorporated in the poles and the dipole, the forces are not connected in this graph because they do not participate in the dynamics of the system. If friction (due to a viscous medium) were present for relating forces to velocities (see case study A10 "Motion with Friction" in Chapter 4), it would have been necessary to connect the pole forces to the dipole force. The dipole would enter the category of mixed dipoles, being of the inductive–conductive type instead of the fundamental type.

Remarks

Conservation is a property. The conservation of momentum during a collision that is elastic [i.e., with conservation of the inductive (kinetic) energy and therefore without dissipation] is classically introduced as a founding axiom of mechanics (which means without justification). The Formal Graph theory introduces it

(continued)

on the same footing as all the conservation laws of dipoles throughout all scientific domains. This conservation results from general properties (embedding principle and separability of poles) that are constitutive of the formation of a conservative dipole.

6.4.2 Case Study C2: Relative Motion — Translational Mechanics

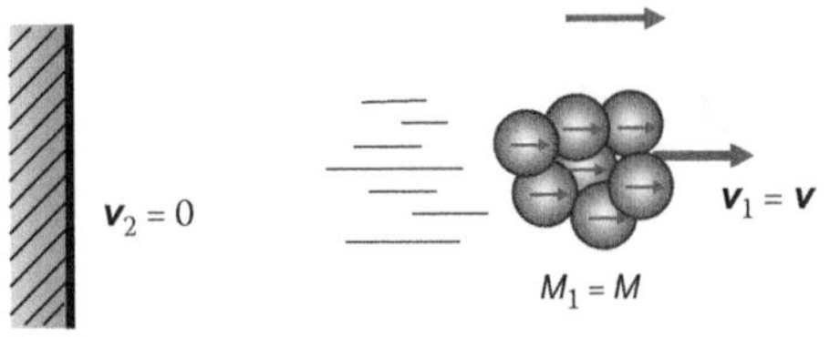

An object with mass, moving relatively to an immobile reference (with or without mass), makes up an inductive fundamental dipole which is *not conservative*.

Dipole

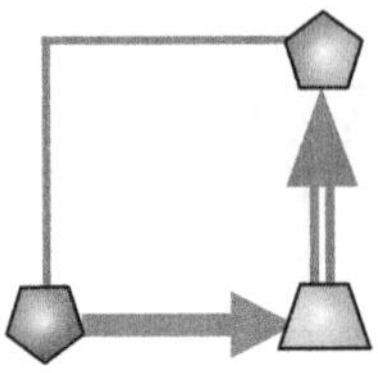

Dipole	
✓	Fundamental
	Mixed
	Capacitive
✓	Inductive
	Conductive
✓	Global
	Spatial
	Inseparable
	Conservative

Formal Graph:

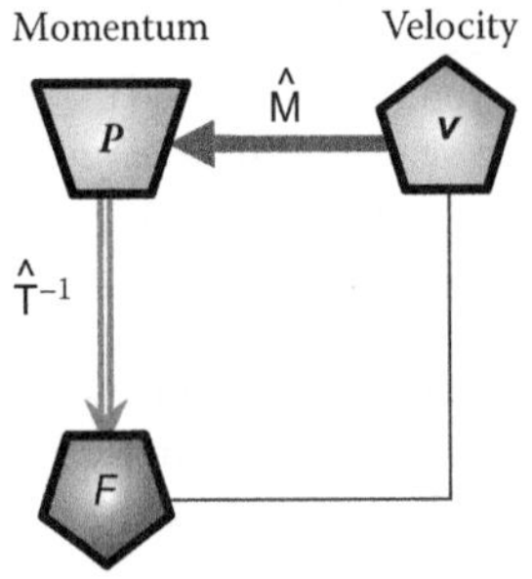

Reluctance operator trace:

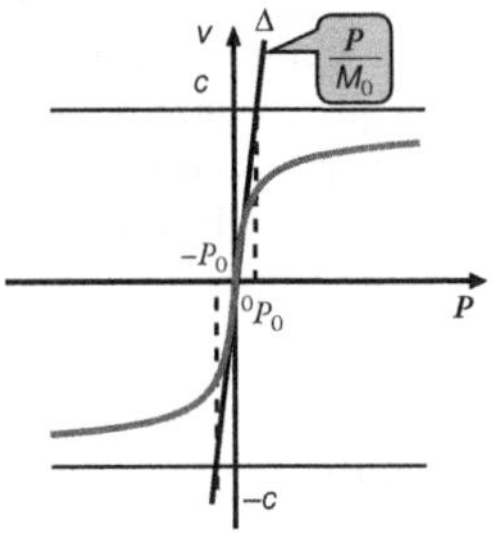

GRAPH 6.9

Variables:

Variety	Translational Mechanics		
Subvariety	Inductive (Kinetic)		Capacitive
Category	Entity number	Energy/entity	Energy/ entity
Family	Impulse 	Flow 	Effort 
Name	Momentum (quantity of movement)	Velocity	Force
Symbols	P, (p)	v	F
Unit	[N s], [J s m^{-1}]	[m s^{-1}]	[N], [J m^{-1}]

Property relationships:

$$P = \hat{M}v$$

Newton's model:

$$P \underset{lin}{=} Mv$$

Einstein's model:

$$P = M_0 c \frac{v}{\sqrt{c^2 - v^2}}$$

System Property	
Nature	Inductance
Level	Global (0)
Name	Inertial mass
Symbol	M
Unit	[kg]
Specificity	—

Energy variation:

$$d\mathcal{T}_{\ell} \underset{\exists M}{=} v \cdot dP$$

This case study concerns the same energy subvariety, inductive (kinetic), in the same energy variety, translational mechanics. The difference with the previous case study lies in the asymmetry between the two bodies, one of them is immobile now whereas the other moves with a velocity $\mathbf{v}$ relative to the immobile body.

This system is also modeled as a dipole in the Formal Graph theory, which may appear surprising as only the moving body is the subject of modeling in classical mechanics, thus suggesting that a pole would be adequate as graphical translation. The problem is that a pole cannot evolve alone as it cannot possess time, which is contrary to the physical reality.

6.4.2.1 Notion of Frame of Reference

As it has been explained in Section 6.3, the simplest method of connecting flows between poles and the dipole is to assume that the dipole velocity results from a simple addition of pole velocities, weighted by positive and negative unit coefficients.

$$\mathbf{v} = \mathbf{v}_1 - \mathbf{v}_2 \tag{C2.1}$$

This is a general rule for connecting energies-per-entity between separable poles and their dipole. In order to consider this system as a dipole, we need to consider the immobile body as the second pole, featured by zero values of its velocity and momentum:

$$\mathbf{v}_2 = 0; \quad \mathbf{P}_2 = 0 \tag{C2.2}$$

The mass of this object can be any, null or positive, as no inductive energy is attributed to this pole, which is called a *reference* or a *frame of reference*. Classically, it is viewed as a mathematical system of coordinates with an origin chosen generally at the position of the moving body at time $t = 0$.

Introducing the nullity of the velocity of this pole into Equation C2.1 leads to the attribution of the moving body velocity to the dipole

$$\mathbf{v} = \mathbf{v}_1 \tag{C2.3}$$

and similarly for the momentum and the mass

$$\mathbf{P} = \mathbf{P}_1; \quad M = M_1 \tag{C2.4}$$

As only one momentum characterizes the system, there is no possibility of conservation of the momentum, as can be achieved by other dipoles made by two mobile bodies (see case study C1 "Colliding Bodies").

Energy: The variation of inductive (kinetic) energy is given in an equivalent way from individual variations of the poles and from variation of the dipole; this is the embedding principle. However, in the peculiar case of an immobile pole, the dipole energy variation is equal to the mobile pole energy variation.

$$\mathrm{d}\mathcal{T}_\ell = \mathbf{v}_1 \cdot \mathbf{dP}_1 + \cancel{\mathbf{v}_2} \cdot \cancel{\mathbf{dP}_2} = \mathbf{v} \cdot \mathbf{dP} \tag{C2.5}$$

System Constitutive Property: As for each pole (see case study A1 "Moving Bodies" in Chapter 4), the velocity of the dipole is related to the dipole momentum through an inductive relationship using the inertial mass M of the dipole as inductance (which is an operator in the general case)

$$\mathbf{P} = \hat{\mathrm{M}}\mathbf{v} \tag{C2.6}$$

It is worth recalling that the mass, whether featuring a pole or a dipole, is an invariable scalar M (independent of another state variable) only in the frame of Newtonian mechanics, that is, for small

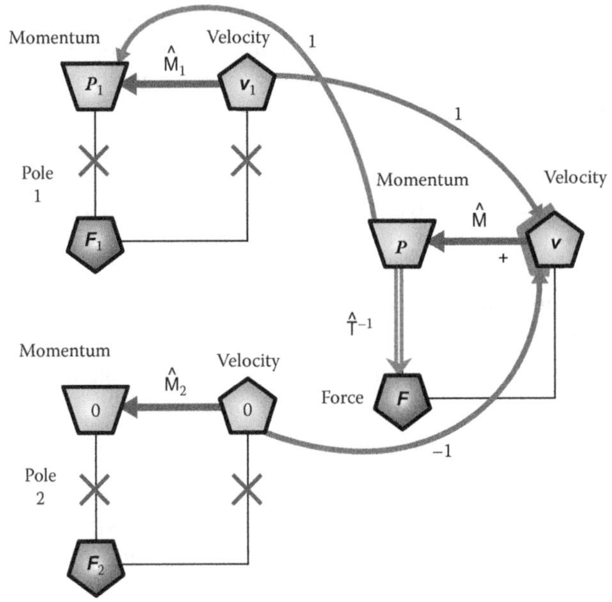

GRAPH 6.10

velocities compared to the speed of light. This peculiar case is modeled by the well-known linear relation:

$$P \underset{lin}{=} M\,v \tag{C2.7}$$

Dipole Construction: Relationships C2.1 and C2.4 between pole and dipole variables can be represented by a special Formal Graph including both poles and the dipole and their connections, called *construction graph*, as shown in Graph 6.10.

The connection between velocities is represented as an addition of pole velocities, weighted by positive and negative unit coefficients, notwithstanding the nullity of the velocity of the frame of reference. Normally, this is an assumption corresponding to the Galilean* principle of additivity of velocities, which has no effect here owing to the nullity of one of the velocities.

Although incorporated in the poles and the dipole, the forces are not connected in this graph because they do not participate in the dynamics of the system. A conversion of energy is required, into another subvariety or heat, for using forces in the model. Conversion into capacitive (potential) energy occurs for instance when a gravitational field acts on the body or when friction provokes dissipation (see case study C3 "Viscous Layers").

Remarks

Nonconservative dipole. The mechanical system made up of two bodies, one immobile and the other mobile, works without conservation of momenta, because the dipole has no way of exchanging momenta to distribute its energy among the poles. Case study C1 "Colliding Bodies" has shown the same mechanical system, but with two mobile bodies exchanging momenta in a collision. Conservation of momentum could apply in this case. This means that conservation of momentum is neither a fundamental law nor an intrinsic feature of the inductive (kinetic) energy subvariety.

* Galileo Galilei (1564–1642): Italian physicist and astronomer; Pisa and Florence, Italy.

6.4.3 Case Study C3: Viscous Layers

Translational Mechanics

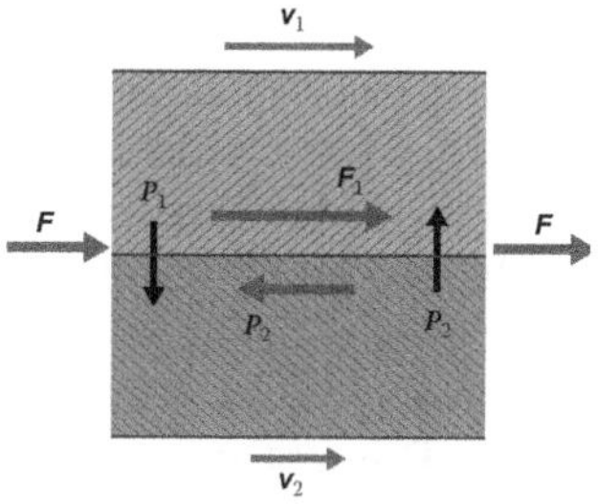

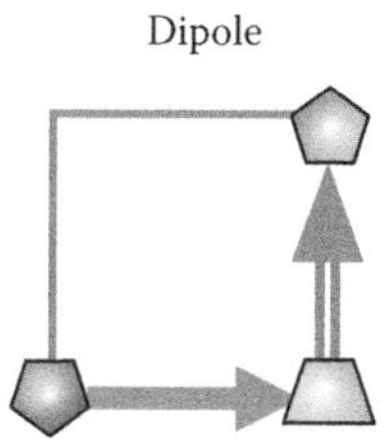

Dipole	
	Fundamental
✓	Mixed
	Capacitive
✓	Inductive
✓	Conductive
✓	Global
	Spatial
	Inseparable
✓	Conservative

Two moving objects, or two moving layers, with mass and friction between them or with their environment, make up an inductive–conductive dipole.

Formal Graph:

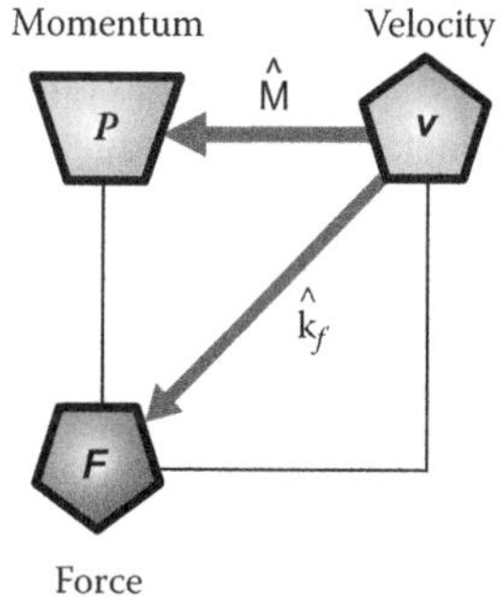

GRAPH 6.11

Variables:

Variety	Translational Mechanics		
Subvariety	Inductive (Kinetic)		Capacitive
Category	Entity number	Energy/entity	Energy/ entity
Family	Impulse	Flow	Effort
Name	Momentum (quantity of movement)	Velocity	Force
Symbols	$\boldsymbol{P}$, ($\boldsymbol{p}$)	$\boldsymbol{v}$	$\boldsymbol{F}$
Unit	[N s], [J s m^{-1}]	[m s^{-1}]	[N], [J m^{-1}]

Property relationships:

$$\boldsymbol{P} = \hat{M}\boldsymbol{v}$$

$$\boldsymbol{F} = \hat{k}_f \boldsymbol{v}$$

Linear models:

$$\boldsymbol{P} \underset{lin}{=} M\boldsymbol{v}$$

$$\boldsymbol{F} \underset{lin}{=} k_f \boldsymbol{v}$$

System Properties		
Nature	Inductance	Reistance
Levels	Global (0)	Global (0)
Name	Inertial mass	Friction resistance
Symbol	M	k_f
Unit	[kg]	[kg s^{-1}]
Specificity	—	—

Energy variation:

$$d\mathcal{T}_{\ell} \underset{\exists M}{=} \boldsymbol{v} \cdot d\boldsymbol{P}$$

The dipole belonging to translational mechanics, and more precisely to the inductive energy subvariety (i.e., kinetic energy), is a model for a large class of objects. The collision between two balls or bodies is an example (see case study C2 "Colliding Bodies"); in fact, every couple of mechanical objects, not necessarily identical, interacting through an exchange of momentum is also an example. Here is treated the case of two layers of fluid or solid, moving at two different velocities, in contact through a common interface. Both layers do not need to move simultaneously, this is a question of referential; one layer can be at rest and the other not. In addition to this ideal scheme, which is modeled by the Formal Graph in the previous case study on colliding bodies, the contact here is supposed to be accompanied by *friction*, which is a dissipative process. When the layers consist of fluid, the contact (or the fluid itself) is said to be *viscous*. This system is modeled by a mixed-type inductive–conductive dipole.

In the scheme in the case study abstract is shown the momentum exchange that occurs through the interface and the forces due to the friction, that each layer exerts on the other in a parallel direction to the interface. The variables with a subscript belong to the poles and those without (here the force $\boldsymbol{F}$) belong to the dipole.

Formal Graph: The Formal Graph of this dipole is similar in structure to the Formal Graph of a pole (see case study A8 "Spring End" in Chapter 4), except that dipole variables and properties are used instead of pole ones. Two system constitutive properties, inductance and conductance (under its reciprocal form, the resistance) are drawn.

Energy: The variation of inductive (kinetic) energy is given by the product of the dipole velocity with the variation of the dipole momentum:

$$\mathrm{d}\mathcal{T}_\ell \underset{\exists M}{=} \boldsymbol{v}\cdot\mathbf{d}\boldsymbol{P} \qquad \text{(C3.1)}$$

6.4.3.1 System Constitutive Properties

Inductive Property: As for each pole (see case study A1 "Moving Body" in Chapter 4), the velocity of the dipole is related to the dipole momentum through an inductive relationship using the inertial mass M of the dipole as inductance (which is an operator in the general case):

$$\boldsymbol{P} = \hat{\mathrm{M}}\boldsymbol{v} \qquad \text{(C3.2)}$$

It is worth recalling that the mass, whether featuring a pole or a dipole, is an invariable scalar M (independent from another state variable) only in the frame of Newtonian mechanics, that is, for small velocities compared to the speed of light. This peculiar case is modeled by the well-known linear relation:

$$\boldsymbol{P} \underset{lin}{=} M\boldsymbol{v} \qquad \text{(C3.3)}$$

Conductive Property: The friction resistance is the operator linking the velocity to the force (see the case study A10 "Motion with Friction" in Chapter 4).

$$\boldsymbol{F} = \hat{\mathrm{k}}_f\,\boldsymbol{v} \qquad \text{(C3.4)}$$

As in every energy variety, such an operator may reduce to a scalar in a restricted range of small operands, providing a simple linear relation between velocity and force. Another frequently encountered case, especially in gases, is the proportionality of the friction resistance to the velocity, leading to a quadratic dependence of the force on the velocity:

$$\boldsymbol{F} \underset{lin}{=} k_f\,\boldsymbol{v} \qquad \boldsymbol{F} \underset{quad}{\propto} \boldsymbol{v}^2 \qquad \text{(C3.5)}$$

Dipole Construction: The variation of inductive (kinetic) energy is given in an equivalent way from individual variations of the poles and from variation of the dipole; this is the embedding principle:

$$\mathrm{d}\mathcal{T}_\ell = \mathbf{v}_1 \cdot \mathbf{d}\boldsymbol{P}_1 + \mathbf{v}_2 \cdot \mathbf{d}\boldsymbol{P}_2 = \mathbf{v} \cdot \mathbf{d}\boldsymbol{P} \quad \text{(C3.6)}$$

Conservation: The rule that applies to a *conservative dipole* able to store energy is that the variations of entity numbers must be *symmetric* for each pole (i.e., no privileged pole with respect to the other; the choice of signs corresponding to the convention of a receptor dipole):

$$\mathbf{d}\boldsymbol{P}_1 = -\mathbf{d}\boldsymbol{P}_2 = \mathbf{d}\boldsymbol{P} \quad \text{(C3.7)}$$

The relation between the velocities of poles and of dipole comes by considering the two previous equations

$$\mathbf{v} = \mathbf{v}_1 - \mathbf{v}_2 \quad \text{(C3.8)}$$

The impulse (momentum) of the dipole corresponds therefore, by integration, to the common difference of each pole:

$$\boldsymbol{P} = \boldsymbol{P}_1 - \boldsymbol{P}_1^* = -\boldsymbol{P}_2 + \boldsymbol{P}_2^* \quad \text{(C3.9)}$$

Two states of the system are considered: *separated*, the layers are not in contact, with the quantities marked with a star, and in contact. From this, it may be deduced that the conservation of momenta is expressed by the following balance equation:

$$\boldsymbol{P}_1 + \boldsymbol{P}_2 = \boldsymbol{P}_1^* + \boldsymbol{P}_2^* \quad \text{(C3.10)}$$

Exchange by Effort: The two layers exchange momentum that is supported by polar forces. Entities are exchanged (they are the content) and energies-per-entity are the vectors of the exchange. The same scheme is used in capacitive dipoles, for instance in the physical chemical domain (see case study C5 "Chemical Reaction" or case study C6 "Physical Chemical Interface"), in which basic quantities (substance amounts) are exchanged, with the flows playing the role of vectors. The scheme on the right illustrates the relationships between polar forces (the vectors of the exchange) and dipolar force (the net result of the exchange), and the two following relationships express a conservation of forces situated on each pole:

$$\mathbf{F}_1 = \mathbf{F}_2 + \mathbf{F} \qquad \mathbf{F}_2 = \mathbf{F}_1 - \mathbf{F} \quad \text{(C3.11)}$$

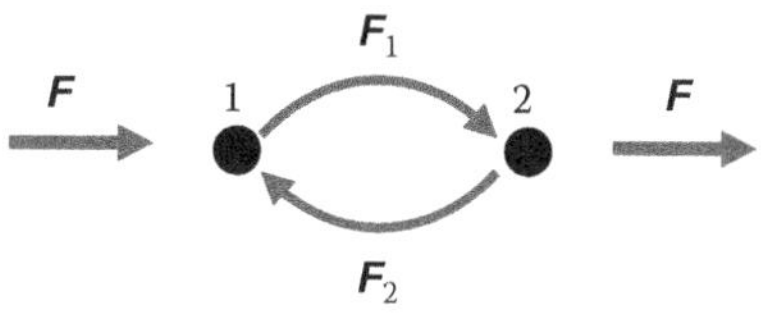

These two relations are in fact equivalent to this third one:

$$\mathbf{F} = \mathbf{F}_1 - \mathbf{F}_2 \quad \text{(C3.12)}$$

which establishes the relationship between poles and dipole in terms of forces.

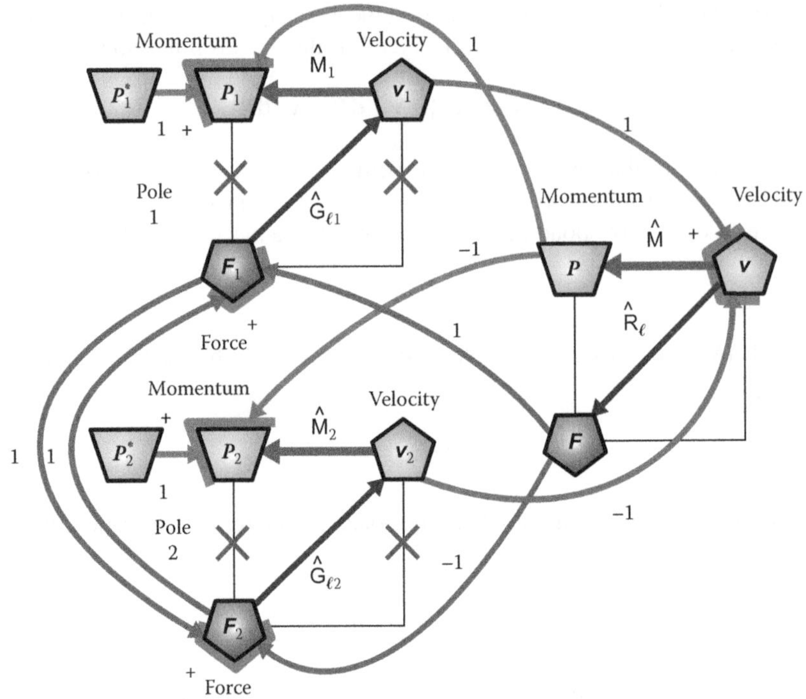

GRAPH 6.12

Construction Formal Graph: The relationships C3.8 and C3.9 between pole and dipole variables can be represented by a special Formal Graph including both poles and the dipole and their connections, called *construction graph*, as shown in Graph 6.12.

The two poles are located on the left and the dipole on the right. To each momentum is ascribed its *dissociated state* quantity (star superscript), which is added to the contribution of the dipole momentum (with the adequate sign) for forming each pole momentum, according to Equation C3.9. This is the way conservation of momenta is encoded by a Formal Graph. The dipole velocity is represented as resulting from a simple addition of pole velocities, weighted by positive and negative unit coefficients. The connections between forces are more complicated as they represent an exchange by effort. Equation C3.12 is not directly represented but, instead, the two relationships in Equation C3.11 are represented.

6.4.4 Case Study C4: Pipe — Hydrodynamics

The case study abstract is given on next page.

A pipe is a conductor of fluid. The energy variety concerned with this conduction process is hydrodynamics, in the capacitive subvariety, in which the entity is the volume of fluid. Hence, the alternate name of volume energy.

At the global level considered here, there is no consideration of the geometry of the pipe; it can be of any shape and its section may vary from one end to the other. The system is a conductive dipole, in the same manner as an electric or heat conductor. In fact, a large class of systems is modeled by a conductive dipole as any energy subvariety can be concerned. Conduction is a dissipative process, converting the energy into heat.

Variables: In this energy variety, the basic quantity is the *volume* V, the effort is the *pressure* P. The flow is the *volume flow* $\mathcal{Q}$, also called volumetric flow rate.

C4: Pipe **Hydrodynamics**

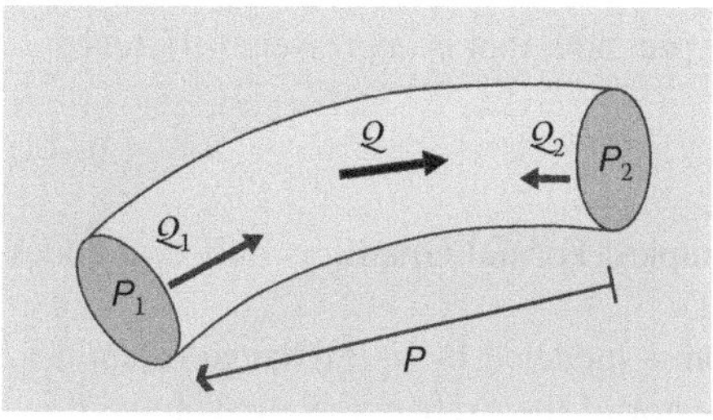

A pipe is a conductor of fluid, liquid, or gas, in the hydrodynamical energy. It has (always) two ends that are the two poles making up a conductive dipole. The pressure of the dipole is the difference between pole pressures.

Dipole

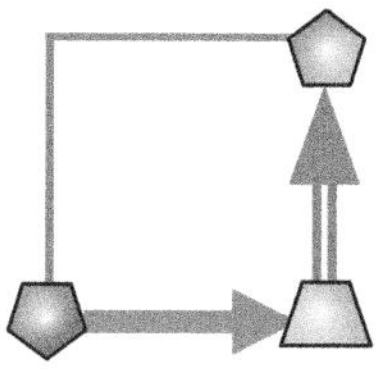

Dipole	
✓	Fundamental
	Mixed
	Capacitive
	Inductive
✓	Conductive
✓	Global
	Spatial
	Inseparable
	Conservative

Formal Graph:

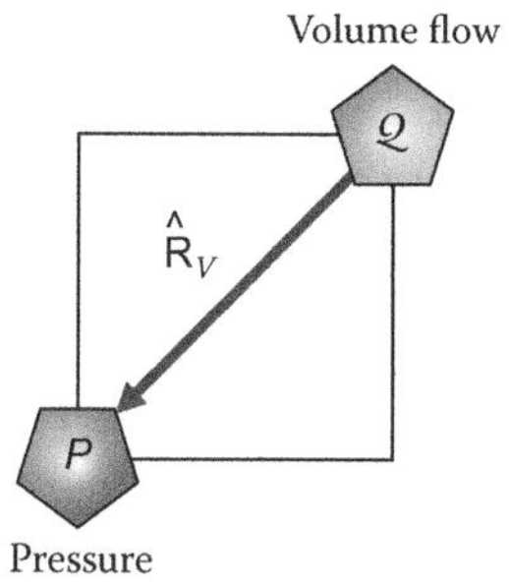

GRAPH 6.13 The system constitutive property used here is the hydrodynamical, or hydraulic, resistance, which can convert volume energy into heat by dissipation.

Variables:

Variety	Hydrodynamics	
Subvariety	Inductive	Capacitive
Category	Energy/entity	Energy/ entity
Family	Flow 	Effort 
Name	Volume flow	Pressure
Symbols	Q, d	P
Unit	[m^3 s^{-1}]	[Pa], [N m^{-2}]

Properties relationship:

$$P = \hat{R}_V\, Q$$

Linear model:

$$P \underset{lin}{=} R_V\, Q$$

System Property	
Nature	Resistance
Levels	Global (0)
Name	Hydraulic resistance
Symbol	R_V
Unit	[N s m^{-5}]
Specificity	—

Energy variation:
No energy is stored in a conductive dipole, only dissipation occurs.

System Constitutive Property: The *hydrodynamical conductance* (also called *hydraulic conductance*) G_V and its reciprocal the resistance R_V are operators linking the volume flow $\mathcal{Q}$ to the *pressure drop* P. This variable must be understood as a dipole pressure, that is, a pressure difference:

$$P = \hat{R}_V \, \mathcal{Q} \tag{C4.1}$$

Formal Graph: A conductive dipole has one of the simplest Formal Graphs as only two nodes and one link are used.

Construction: In the Formal Graph theory, conduction is modeled by an exchange of entities. Here the entity is the volume element of fluid, and the vector of the exchange is the volume flow. Thus, the poles–dipole relation between flows is written as

$$\mathcal{Q} = \mathcal{Q}_1 - \mathcal{Q}_2 \tag{C4.2}$$

This is a generalization of some systems such as a semiconductor (see case study C9 "Electrons and Holes") or a chemical reaction (see case study C5 "Chemical Reaction") in which two carriers or species participate in the exchange. The two pole flows are always counted positively whereas the dipole flow can be positive or negative depending on the net flow direction. In these cases, the two directions of the exchange are intuitively understood as related to the existence of different containers for the exchange. This is less intuitive in the case of a homogeneous fluid in which all volume elements are identical.

It would be easier to admit that one of the pole flows is equal to zero, say $\mathcal{Q}_2 = 0$, leading therefore to a simple continuity from the dipole flow $\mathcal{Q}$ entering the pipe to the inner flow $\mathcal{Q}_1$ and again from this latter to the output flow $\mathcal{Q}$. However, the choice to ascribe the zero flow to one pole and not to the other seems difficult, as it confers a dissymmetrical role to the poles. In fact, it is the role of the pole resistances to adjust each pole flow to its pole pressure. After all, it is not totally absurd to see a flow in a pipe as a dynamic exchange between a forward and a backward flow, the latter being eventually negligible.

The other poles–dipole connection involves the pressures, and a simple difference of energy-per-entity (effort) is adequate for modeling a conduction process. It corresponds to the classical expression of a "driving force" or *pressure drop.*

Receptor convention:

$$P = P_1 - P_2 \tag{C4.3}$$

Naturally, for a conductor there is no real choice other than the receptor convention for choosing the order of poles to build this difference.

Construction Graph: The construction graph is built by basing the encoding on an exchange of flow and on an effort difference.

Relationship between Resistances: In the construction graph shown in Graph 6.14, one can read the following relationship:

$$\hat{G}_V^{-1} \, \mathcal{Q} = \hat{R}_V \, \mathcal{Q} = \hat{R}_{V1} \, \mathcal{Q}_1 - \hat{R}_{V2} \, \mathcal{Q}_2 \tag{C4.4}$$

This relationship leads to an expression of the dipole resistance as a function of the pole resistances, which in the general case depends not on a linear function but on the flows:

$$\hat{R}_V = \frac{\hat{R}_{V1} \, \mathcal{Q}_1 - \hat{R}_{V2} \, \mathcal{Q}_2}{\mathcal{Q}} \tag{C4.5}$$

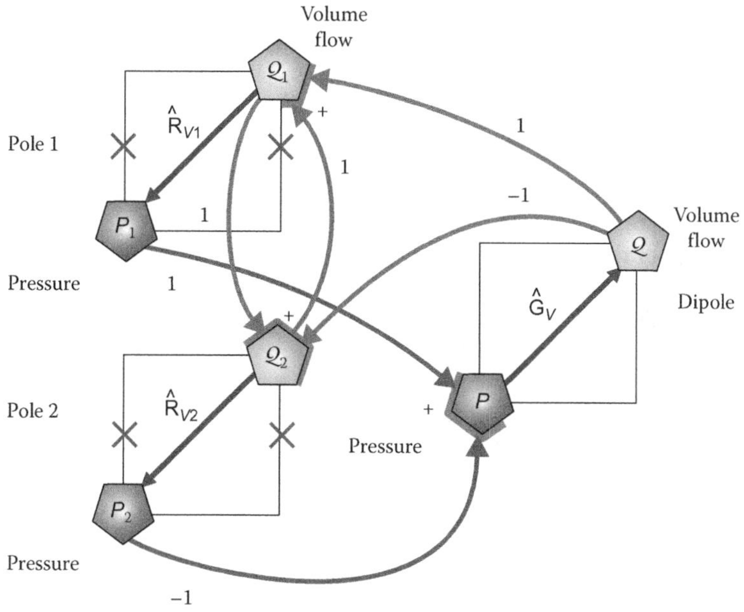

GRAPH 6.14

Even in the linear case, when resistance operators become equivalent to scalars, the dipole resistance still depends on the flows, unless one of the flows is zero or both scalar polar resistances are equal ($R_V = R_{V1} = R_{V2}$). This corresponds to the classical laminar flow, for which one has

$$P \underset{lin}{=} R_V\, \mathcal{Q} \tag{C4.6}$$

with the dipole resistance

$$R_V \underset{lin}{=} \frac{R_{V1}\, \mathcal{Q}_1 - R_{V2}\, \mathcal{Q}_2}{\mathcal{Q}} \tag{C4.7}$$

Exchange Reversibility: By defining a variable r_{V12} quantifying the reversibility of the exchange with the ratio of the backward flow on the forward one

$$r_{V12} \overset{def}{=} \frac{\mathcal{Q}_2}{\mathcal{Q}_1} \tag{C4.8}$$

three limiting regimes can be distinguished according to the values of this unique parameter

$$\text{Irreversible} \rightarrow r_{V12} \rightarrow 0 \quad \Leftrightarrow \quad \mathcal{Q}_1 \gg \mathcal{Q}_2 \quad \Leftrightarrow \quad \hat{R}_V \rightarrow \hat{R}_{V1} \tag{C4.9}$$

$$\text{Reversible} \rightleftarrows r_{V12} \approx 1 \quad \Leftrightarrow \quad \mathcal{Q}_1 \approx \mathcal{Q}_2 \quad \Leftrightarrow \quad \mathcal{Q} \approx 0 \tag{C4.10}$$

$$\text{Irreversible} \leftarrow r_{V12} \rightarrow \infty \quad \Leftrightarrow \quad \mathcal{Q}_1 \ll \mathcal{Q}_2 \quad \Leftrightarrow \quad \hat{R}_V \rightarrow \hat{R}_{V2} \tag{C4.11}$$

Also, the dipole resistance can be expressed with the help of this parameter in the linear case as

$$R_V \underset{lin}{=} \frac{R_{V1} - r_{V12} R_{V2}}{1 - r_{V12}} \tag{C4.12}$$

Naturally, this concept of quantified reversibility is not specific to hydrodynamics and works for all dipoles in which an exchange takes place. The interesting feature of this notion is that it can be easily related to the dipole effort (or to the dipole flow in case of an exchange by efforts), as it will be shown in Chapter 8.

Remarks

Sign of the pressure. For historical reasons, the pressure has been chosen with the wrong sign. Instead of positive pressure, one should consider opposite values. This is not really necessary when a system contains or dissipates only hydrodynamics energy, like here. However, when handling systems with several energy varieties, the pressure P must be taken with a minus sign for maintaining the coherence between energy varieties. In the present case of the hydrodynamical pipe, if one chooses to use $-P$, it would be necessary to take the volume flow with a negative sign.

Definition of the pressure. In the Formal Graph theory, the pressure is an effort in the hydrodynamical energy variety which has for basic quantity the volume V.

It is mathematically defined as the partial derivative of the energy with respect to the volume in maintaining constant all other entity numbers:

$$-P \overset{def}{=} \left(\frac{\partial \mathcal{E}}{\partial V}\right)_{q_i, p_j} \qquad \text{(C4.13)}$$

Classically, the pressure is seen as resulting from elastic collisions due to the motion of particles without dissipation meeting a surface (wall), as will be discussed in case study K4 "Kinetic of Gases" in Chapter 13. However, this concept is difficult to apply in all systems, notably when there are no mobile particles but some volume energy exists (elastic solids, gels, etc.) or in dissipative systems (viscous liquids, conductive channels or pipes, etc.).

Another classical approach, which also requires a surface (and a force), consists of confusing the pressure with the surface density of a force. This point will be discussed in case study J5 "Piston" in Chapter 12.

The definition here is much more general and applies to all systems that have a volume.

Generality. Exactly the same model, algebraic as well as graphic, can be used for all conductive dipoles in other energy varieties. It is just necessary to replace the pair (effort, flow) by the right pair in the wanted energy variety.

Space distribution. The system considered here is a global one. Naturally, space distribution can be added to the model, in the same way as for the poles. Correspondence between global properties and reduced ones works identically and the specificity of system properties obeys the same rules. Several case studies of conductive dipoles with space distribution are given in Chapter 10 that is devoted to transfer processes.

6.4.5 Case Study C5: First-Order Chemical Reaction

Physical Chemistry

$$A \underset{k_B}{\overset{k_A}{\rightleftarrows}} B$$

When the kinetic reaction order is one, a binary chemical reaction forms a mixed dipole (capacitive–conductive) in the physical chemical energy variety. When the order of the kinetic reaction is different from one, a binary reaction still forms a dipole, but in another variety of energy, the chemical reaction energy.

Dipole

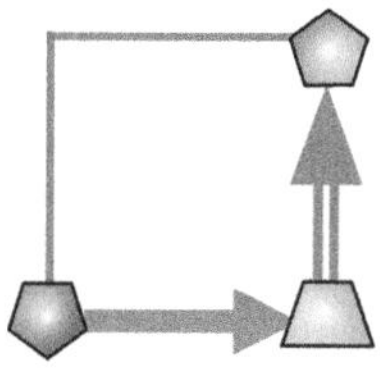

Dipole	
	Fundamental
✓	Mixed
✓	Capacitive
	Inductive
✓	Conductive
✓	Global
	Spatial
	Inseparable
✓	Conservative

Formal Graph:

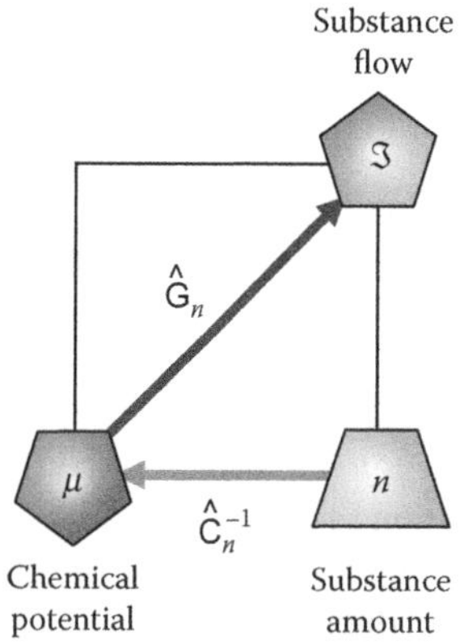

GRAPH 6.15

Capacitance operator traces:

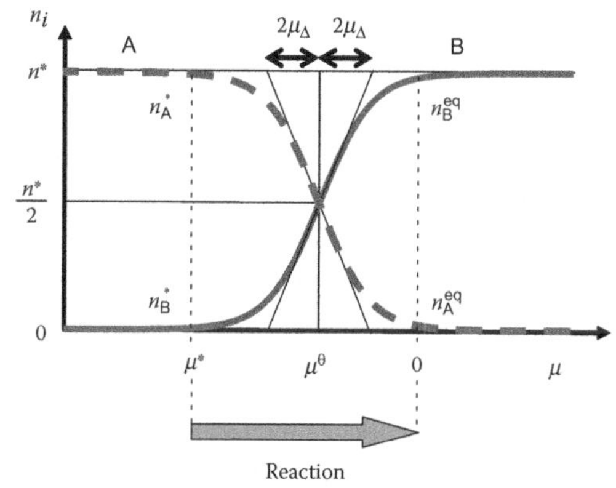

Scaling effort: $\mu_\Delta = RT$

Total substance amount: $n^* = n_A^* + n_B^*$

Variables:

Variety	Translational Mechanics		
Subvariety	Capacitive (Potential)		Inductive
Category	Entity number	Energy/entity	Entity number
Family	Basic quantity 	Effort 	Flow 
Name	Substance amount	Chemical potential	Substance amount
Symbols	n, N	μ	$\Im$, v
Unit	[mol]	[J mol^{-1}]	[mol s^{-1}]

Property relationships:

$$n = \hat{C}_n\ \mu$$

$$\Im = \hat{G}_n\ \mu$$

System Properties		
Nature	Capacitance	Conductance
Levels	Global (0)	Global (0)
Name	Physical chemical capacitance	Physical chemical conductance
Symbol	C_n	G_n
Unit	[mol^2 J^{-1}]	[mol^2 J^{-1} s^{-1}]
Specificity	—	—

Energy variation:

$$d\mathcal{U}_n \underset{\exists C_n}{=} \mu\, dn$$

A reaction between two chemical species A and B forms a dipole. The simplest case is called "elementary," meaning that only two species are involved (binary reaction) and that the number of molecules participating in the reaction is the same on each side.

$$\mathsf{A} \underset{k_\mathsf{B}}{\overset{k_\mathsf{A}}{\rightleftarrows}} \mathsf{B}$$

Transversality. The Formal Graph language offers a conceptual frame, the pole–dipole scheme, that is identical for many systems as different as a Champagne bubble, a pair of pool balls, an electric capacitor, or a chemical reaction. This means that the various sciences related to these systems are similar and that the concepts of one can be applied to another. For instance, the whole language of chemical kinetics can be used for modeling every cited system when some dissipation is involved (friction, current leakage, etc.).

The reaction is said to be reversible,* meaning that it can work in both directions, from A to B and conversely. The case of an irreversible reaction was considered under the concept of *species reactivity* (see case study A11 "Reactive Species" in Chapter 4).

Variables: As the order of the reaction is equal to one, this system belongs to the physical chemical energy, in which the basic quantity is the *substance amount* n (with unit in moles). (See Remark at the end about higher reaction orders.) The effort is the *chemical potential* μ and the flow is the *substance flow* $\mathfrak{I}$. This last state variable corresponds to the reaction rate.

Graph: The Formal Graph of a chemical reaction is the graph of a mixed capacitive–conductive dipole, endowed with all its system properties (capacitance and conductance).

System Constitutive Properties: As for a pole, the dipolar capacitive relationship is expressed through a dipolar capacitance that depends on the polar capacitances

$$n = \hat{\mathsf{C}}_n \, \mu \tag{C5.1}$$

The same scheme applies to the dipolar conductance

$$\mathfrak{I} = \hat{\mathsf{G}}_n \, \mu \tag{C5.2}$$

The capacitance and the conductance can be composed together to form a global "rate constant" k (i.e., at the level of the dipole), in the same manner as the formation of rate constants for each pole (see case study A6 "Chemical Species" in Chapter 4). However, this is a seldom used property at the level of a dipole because the operator representing this dipole "rate constant" is not a constant but is a function of substance amounts.

Graph Construction: In such a *reversible* reaction, the global reaction rate is equal to the difference between forward and backward reaction rates. In the language of Formal Graphs, it corresponds to the relationship between the flow of a dipole and those of each pole (exchange by flows)

$$\mathfrak{I} = \mathfrak{I}_\mathsf{A} - \mathfrak{I}_\mathsf{B} \tag{C5.3}$$

* "Reversible" has two meanings: one signifying the possibility for a reaction or a process to be reversed, the other featuring a kinetic behavior near equilibrium. It is the first meaning that is considered here.

The chemical reaction is considered as a *generator*, entailing a reverse order of poles in the link between the chemical potential of the dipole and the chemical potentials of each pole

$$\mu = \mu_{BA} = \mu_B - \mu_A \tag{C5.4}$$

Conservation: The chemical reaction constitutes a capacitive–dissipative dipole that is a *conservative* one, a feature that is translated by a relationship between basic quantities, imposing that their sum, whatever the state of the dipole, is equal to the sum of substance amounts of both poles when they do not form a dipole ("initial" quantities of the reaction)

$$n_A + n_B = n_A^* + n_B^* = n^* \tag{C5.5}$$

This conservation and the convention of a *generator* allow the definition of the substance amount of the dipole to be

$$n = n_A^* - n_A = n_B - n_B^* \tag{C5.6}$$

It can be checked that Equation C5.4 and the derivation of Equation C5.5 result in the same variation of energy of the dipole (respecting the embedding principle)

$$d\mathcal{U} = \mu_A dn_A + \mu_B dn_B = \mu dn \tag{C5.7}$$

The construction graph of the dipole from its poles takes into account all these equations, with a peculiar coding for Equation C5.3 between pole and dipole flows (see Graphs 6.16 and 6.17).

Convention. The coding used in the construction Formal Graph of the dipole for representing the relationship between flows is nothing other than the transposition of the representation of a dipolar component utilized in an equivalent electric circuit, which by itself is another coding of the chemical reaction A ⇄ B

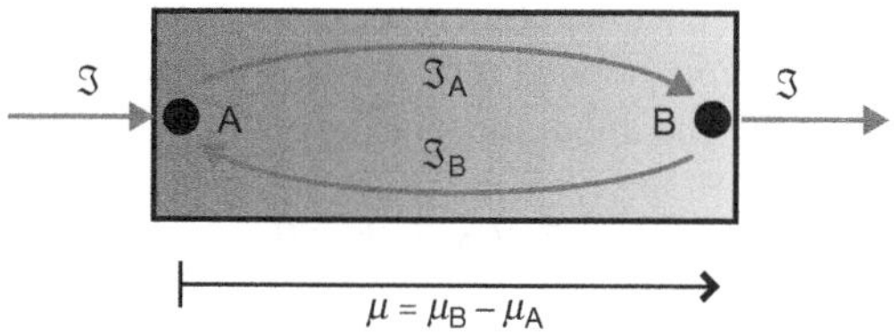

The convention used for orienting the flow and the chemical potential of the dipole is that of the generator component (i.e., same orientation). As a generator, the power (derivative of the energy with respect to time as will be seen in Chapter 11) of an exothermal reaction (which gives thermal energy to the exterior) is counted negatively.

Dynamics: Now, it is enough to consider the function of each pole for retrieving the classical description of such a reaction. When the hypothesis of a first-order kinetic reaction is made for each pole, each reaction rate is related to the quantity of reactive substance through a coefficient called "rate constant" or "specific rate of reaction" k_A and k_B for each direction of the reaction.

As no direct link is permitted between the substance amount and the substance flow in a pole, each rate constant therefore needs to be a composed path with the physical chemical elastance

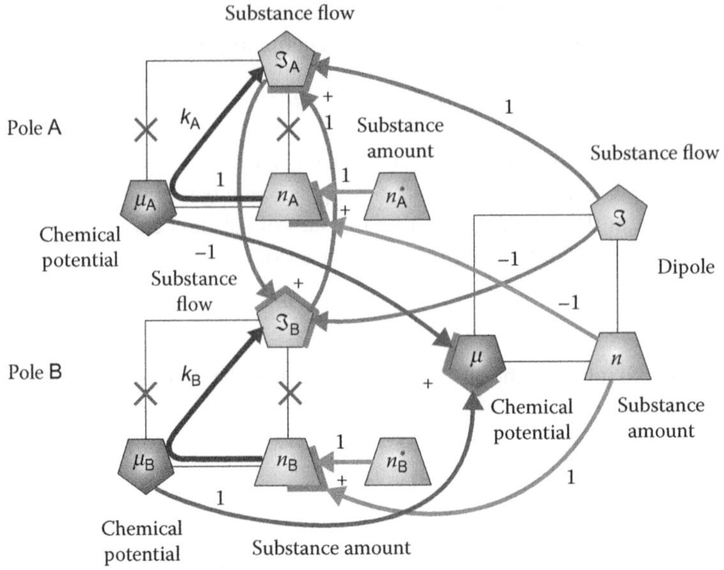

GRAPH 6.16 Formal Graph of the construction of the dipole from the two poles representing the reactants A and B of the chemical reaction.

(reciprocal of the capacitance) and with the physical chemical conductance (see case study A6 "Chemical Species" in Chapter 4).

With this hypothesis, the difference between the forward and backward rates is written as

$$\Im_A - \Im_B = k_A n_A - k_B n_B \tag{C5.8}$$

Identification of this difference with the flow of the dipole (Equation C5.3) leads to the classical expression in chemical kinetics of the algebraic model of a simple chemical reaction

$$\Im = k_A n_A - k_B n_B \tag{C5.9}$$

Concerning efforts, the classical chemical kinetics considers the chemical potential of the dipole as the *molar free enthalpy of reaction*, notated $\Delta_r G$, which amounts to Equation C5.4 as a definition of this quantity (see Remark at the end about the naming).

$$\Delta_r G = \mu_{BA} = \mu_B - \mu_A \tag{C5.10}$$

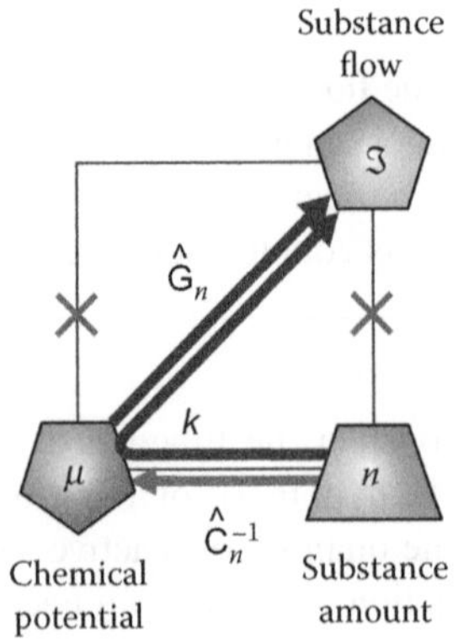

GRAPH 6.17

Dipole Capacitive Property: From two polar capacitances, or more precisely from the two expressions of *activities* (see case study A6 "Chemical Species" in Chapter 4), and from relation C5.4 between chemical potentials, the expression of the ratio between substance amounts is directly deduced

$$\frac{n_B}{n_A} = \frac{a_B}{a_A} = g_{BA} = \exp\frac{\mu - \mu^{\ominus}}{\mu_{\Delta}} \tag{C5.11}$$

with the *standard chemical potential* of the dipole given by the difference between standard chemical potentials of the poles

$$\mu^{\ominus} \overset{def}{=} \mu_B^{\ominus} - \mu_A^{\ominus} = \Delta_r G^{\ominus} \tag{C5.12}$$

This *standard chemical potential* is a characteristic of the reaction A – B; it corresponds to the standard molar free enthalpy of reaction and its definition is the value taken by the chemical potential when equality between substance amounts is realized

$$\mu = \mu^{\ominus} \Leftrightarrow n_A = n_B \tag{C5.13}$$

The ratio of activities is called the *gain* g in the Formal Graph theory and it expresses the *influence* of one pole on another. The capacitive relationship C5.1 now can be made explicit by using Equation C5.6, allowing the replacement of polar substance amounts by the dipolar one, for instance under the form of elastance

$$\mu = \mu^{\ominus} + \mu_{\Delta} \ln\frac{n_B^* + n}{n_A^* - n} \tag{C5.14}$$

This result outlines the importance of avoiding confusing a polar capacitance operator (see case study A6 "Chemical Species" in Chapter 4) from a dipolar one. The expressions of each substance amount are deduced from Equations C5.5 and C5.11

$$\frac{n_A}{n_A^* + n_B^*} = \frac{1}{1+\exp\left(\mu - \mu^{\ominus}/\mu_{\Delta}\right)} \quad \frac{n_B}{n_A^* + n_B^*} = \frac{\exp\left(\mu - \mu^{\ominus}/\mu_{\Delta}\right)}{1+\exp\left(\mu - \mu^{\ominus}/\mu_{\Delta}\right)} \tag{C5.15}$$

A plot of the two curves giving the substance amounts is given in the case study abstract table.

The starting situation (reaction out of equilibrium) is given for a chemical potential μ^* corresponding to initial quantities of A and B. The final situation corresponds to a null chemical potential (equality of chemical potentials of the poles); that is, it corresponds to the equilibrium of the dipole (i.e., energy extremum).

The scaling chemical potential μ_{Δ} is proportional to temperature, because thermal energy is responsible for the thermal agitation of molecules and comes by coupling with physical chemical energy.

$$\mu_{\Delta} = RT \tag{C5.16}$$

Remarks

Another energy variety required. The case of reaction orders higher than one cannot be handled with the sole physical chemical energy variety. It requires a supplementary energy variety, called *chemical reaction energy*, whose state variables are the reaction extent (basic quantity), the affinity (effort), and the reaction rate (flow). This subject is treated as a coupling between energy varieties in case study J1 "nth-Order Chemical Reaction" in Chapter 12.

Molar is the key point. The *molar free enthalpy of reaction* is also called *molar energy of reaction*. This is not a very important distinction because this quantity has nothing to do with enthalpy or energy! The important point is not to forget the adjective *molar* in the naming. (We recall that it is the *energy-per-entity*, an effort indeed, which is an intensive state variable, whereas energy is extensive, and is not a state variable.)

Extension to other processes. It must be noted that this model of "First-Order Chemical Reaction" is not restricted, although its title may apply to the sole chemical reaction in the sense of a process changing the chemical nature of the reactants, but works for other physical chemical processes such as an exchange of a same species across an interface between two phases (see case study C6 "Physical Chemical Interface").

Extension of this model to the case of interaction with another energy variety allows modeling, for instance, an electrochemical reaction (charge transfer). In such a reaction, both equilibrium and kinetics are dependent on the electric potential through electrochemical potentials that replace the chemical potentials. This subject is treated in case study J2 "Electrochemical Potential" in Chapter 12.

6.4.6 Case Study C6: Physical Chemical Interface — Physical Chemistry

The case study abstract is given on next page.

A chemical substance that is present on both sides of an interface constitutes a physical chemical dipole. The interface, also called separator, must be permeable to the substance for allowing exchanges between the two phases or compartments.

In each phase or compartment, there exists a different *reference* (or *standard*) *chemical potential*, expressing the physical or physical chemical properties that are specific to each phase. This difference allows each phase to contain different amounts of substance even when the chemical potentials are equal on both sides (which is a situation of dynamic equilibrium).

The case treated here is a general one, at equilibrium as well as outside equilibrium, but without consideration of the exchange kinetic, which would pertain to a mixed capacitive–conductive dipole.

Variables: The energy variety to which this system belongs is the physical chemical energy, in which the basic quantity is the *amount of substance n* (with unit in number of moles). The effort is the *chemical potential* μ and the flow is the *flow of substance* $\mathfrak{I}$. This last state variable corresponds to the rate of exchange between the two phases which is used for modeling a dynamic interface.

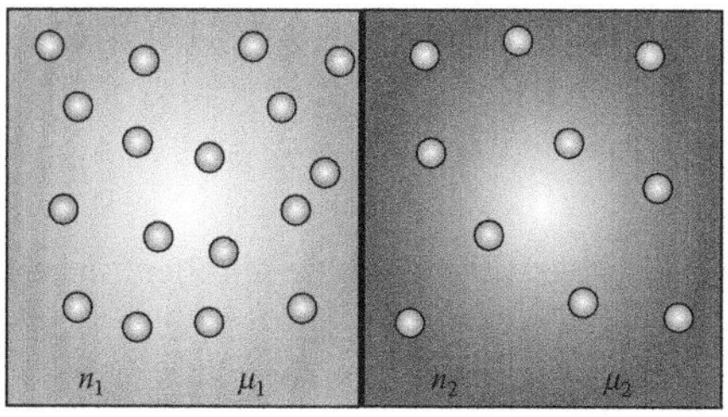

A physical chemical interface separates two populations of molecules or atoms, each one being contained in a phase of a medium. Such a system makes up a dipole (fundamental capacitive) in the physical chemical energy variety.

Dipole

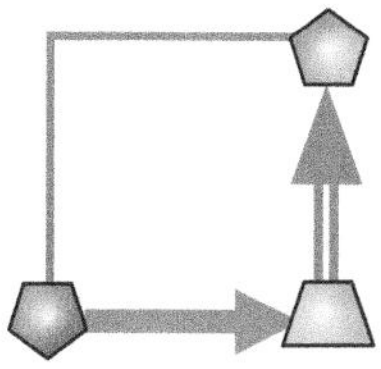

	Dipole
✓	Fundamental
	Mixed
✓	Capacitive
	Inductive
	Conductive
✓	Global
	Spatial
	Inseparable
✓	Conservative

Formal Graph:

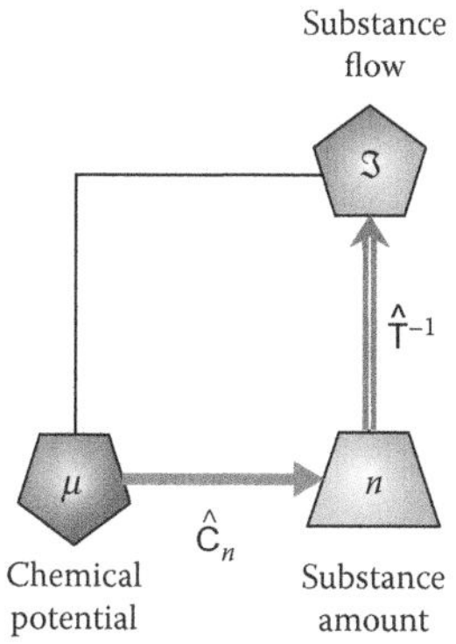

Capacitance operator traces:

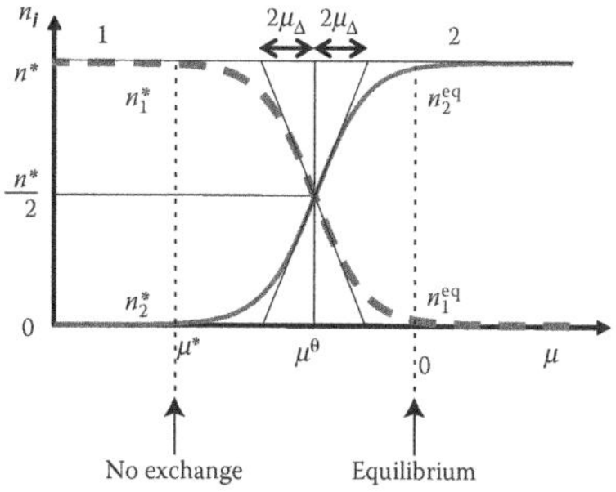

Scaling effort: $\mu_\Delta = RT$

GRAPH 6.18

Variables:

Variety	Physical Chemistry		
Subvariety	Capacitive (Potential)		Inductive
Category	Entity number	Energy/entity	Entity number
Family	Basic quantity 	Effort 	Flow 
Name	Substance amount	Chemical potential	Substance amount
Symbols	n, N	μ	$\mathfrak{I}, v$
Unit	[mol]	[J mol^{-1}]	[mol s^{-1}]

Property relationships:

$n = \hat{C}_n \, \mu$

System Property	
Nature	Capacitance
Levels	Global (0)
Name	Physical chemical capacitance
Symbol	C_n
Unit	[mol^2 J^{-1}]
Specificity	—

Energy variation:

$$\mathrm{d}\mathcal{U}_n \underset{\exists C_n}{=} \mu \, \mathrm{d}n$$

Graph: The Formal Graph represents a dipole with its three state variables (in considering only the global level). The indication of the temporal evolution between substance amount and flow is a genuine property of a dipole, which facilitates the distinction with a pole.

Energy: The *embedding principle* makes the energy variation of the system keep the same value whatever the organization level considered, with two poles or with only one dipole

$$\mathrm{d}\mathcal{U} = \mu_1 \, \mathrm{d}n_1 + \mu_2 \, \mathrm{d}n_2 = \mu \, \mathrm{d}n \qquad \text{(C6.1)}$$

Construction: The physical chemical interface constitutes a capacitive *conservative* dipole. This is mathematically translated into a relation bearing on the basic quantities of the two poles, saying that their sum for any state of the dipole is equal to their sum when the poles are not forming a dipole (dissociated state).

$$n_1 + n_2 = n_1^* + n_2^* \tag{C6.2}$$

This conservation and the arbitrary choice to attribute a positive weight to pole number 2 lead to the definition of the amount of substance of the dipole as being

$$n = n_1^* - n_1 = n_2 - n_2^* \tag{C6.3}$$

It must be stressed that this amount is an algebraic quantity for a dipole which may be negative. The link between the chemical potential of the dipole and the chemical potentials of each pole is deduced from this equality and from Equation C6.1, in choosing the generator convention as for the chemical reaction (same directions for dipole flow and effort)

$$\mu = \mu_{21} = \mu_2 - \mu_1 \tag{C6.4}$$

The construction graph, as shown in Graph 6.19, is a translation of these equations.

The same construction graph is found for other dipoles, such as the chemical reaction, the spring, the electric capacitor, and the electrons and holes (relevant in condensed matter physics). Refer to the case studies C5 and C7 through C9 in this chapter.

Capacitive Property: The expression of the activity of a substance (see case study A6 "Chemical Species" in Chapter 4) is recalled:

$$\frac{n_i}{n^{\ominus}} = a_i = \exp\left(\frac{\mu_i - \mu_i^{\ominus}}{\mu_\Delta}\right) \tag{C6.5}$$

From two expressions of the *activities* (one for each substance) and from the relation shown in Equation C6.4 between chemical potentials, the expression of the ratio of substance amounts is directly deduced as

$$\frac{n_2}{n_1} = \frac{a_2}{a_1} = g_{21} = \exp\frac{\mu - \mu^{\ominus}}{\mu_\Delta} \tag{C6.6}$$

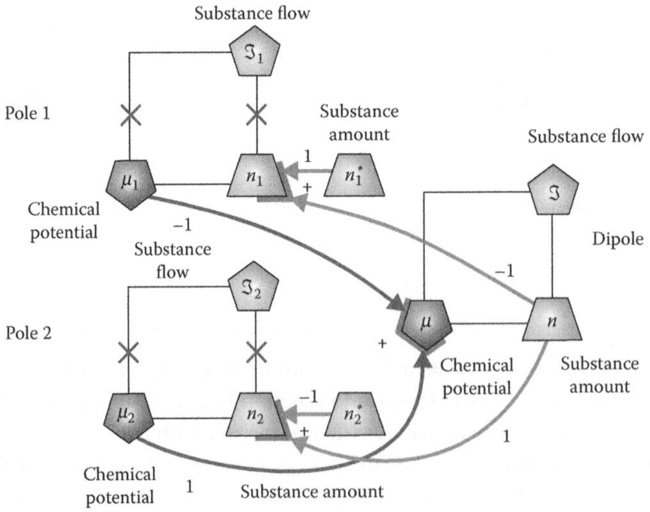

GRAPH 6.19

with the *standard chemical potential* of the dipole defined by

$$\mu^{\ominus} \overset{def}{=} \mu_2^{\ominus} - \mu_1^{\ominus} \tag{C6.7}$$

This *standard chemical potential* is a characteristic of the dipole. Its definition is the value taken by the chemical potential when equality between substance amounts is realized

$$\mu = \mu^{\ominus} \Leftrightarrow n_1 = n_2 \tag{C6.8}$$

The ratio of activities is called the *gain* g in the Formal Graph theory and it expresses the *influence* of one pole on another. The expressions of each amount of substance are obtained from Equations C6.2 and C6.6.

$$\frac{n_1}{n^*} = \frac{1}{1+\exp\left(\mu - \mu^{\ominus}/\mu_\Delta\right)} \qquad \frac{n_2}{n^*} = \frac{\exp\left(\mu - \mu^{\ominus}/\mu_\Delta\right)}{1+\exp\left(\mu - \mu^{\ominus}/\mu_\Delta\right)} \tag{C6.9}$$

The plot of the curves giving each amount of substance is given in the case study abstract.

The position of the chemical potential μ^* corresponding to the situation of an impermeable interface (no exchange is occurring so the dipole substance amount $n = 0$) is arbitrary; it depends on reference chemical potentials of each phase. The equilibrium occurs when the energy of the dipole is extremum, that is, from Equation C6.1 when $\mu = 0$,

$$\frac{d\mathcal{U}}{dn} = \mu^{eq} = 0 \tag{C6.10}$$

This means, according to Equation C6.4, that both chemical potentials are equal on each side of the interface at equilibrium. Finally, the capacitive relationship of the dipole is obtained from Equations C6.2 and C6.9

$$n = \widehat{C}_n \mu = \frac{n_1^* \exp\left(\mu - \mu^{\ominus}/\mu_\Delta\right) - n_2^*}{1+\exp\left(\mu - \mu^{\ominus}/\mu_\Delta\right)} \tag{C6.11}$$

Remarks

Transversality. The Formal Graph language affords an identical conceptual frame, the pole–dipole scheme, for systems as different as a Champagne bubble, a pair of pool balls, an electric capacitor, or a chemical reaction. This means that the various scientific theories modeling these systems are identical and that the concepts of one might be used in another. If condensed matter physics is combined with the example of the dipole formed by two populations of electrons and holes, one has an idea of the realm of the proposed unification.

Higher generality. It is worth noting that this model of interface is not limited to the case of the sole physical separation between two regions of a system, but equally applies to other physical chemical processes in the same regions, such as the chemical reaction between two species which may occur in the same phase of a medium (homogeneous reaction) or to a mixing of charged molecules or corpuscles in the same material (see the case study C9 "Electrons and Holes").

6.4.7 Case Study C7: Electric Capacitor

Electrodynamics

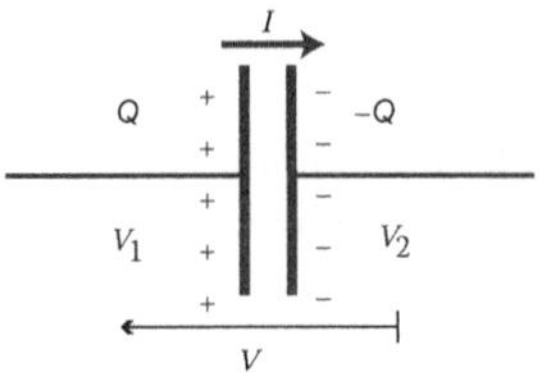

This figure shows the symbolic representation of a capacitor, schematizing two parallel plates separated by a dielectric material (which can be vacuum). Each plate bears a population of several electric charges with the same sign, but the two populations of charges have opposite signs.

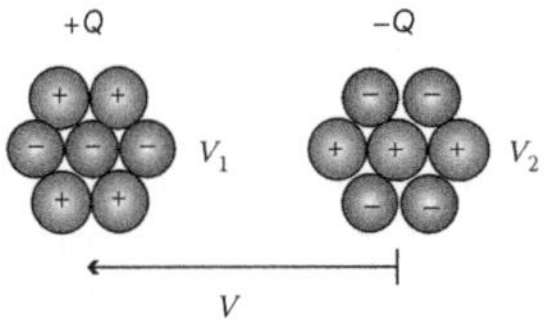

This scheme describes another well-known system made by two sets of charges separated in space. Each set forms a pole but is not necessarily composed of charges with the same sign (see case study A5 "Electric Charges" in Chapter 4 for a description of the pole).

Dipole

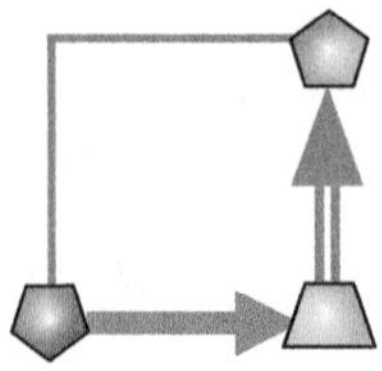

Electrostatics provides the archetype of poles and dipoles, which are very old concepts. They have been used as templates for the generalization made in the Formal Graph theory, which extends these concepts to all scientific domains.

Dipole	
✓	Fundamental
	Mixed
✓	Capacitive
	Inductive
	Conductive
✓	Global
	Spatial
	Inseparable
✓	Conservative

The electric capacitor is a fundamental capacitive dipole storing capacitive (electrostatic) energy.

Formal Graph (see Graph 6.20):

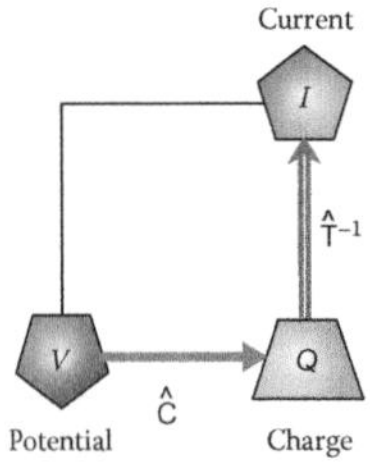

GRAPH 6.20

Variables:

Variety	Electrodynamics		
Subvariety	Capacitive (Electrostatic)		Inductive
Category	Entity number	Energy/entity	Entity number
Family	Basic quantity	Effort	Flow
Name	Charge	Potential	Current
Symbols	Q	V, U, E, φ	$I, (i)$
Unit	[C], [A s]	[V], [J C^{-1}]	[A]

Capacitance operator traces:

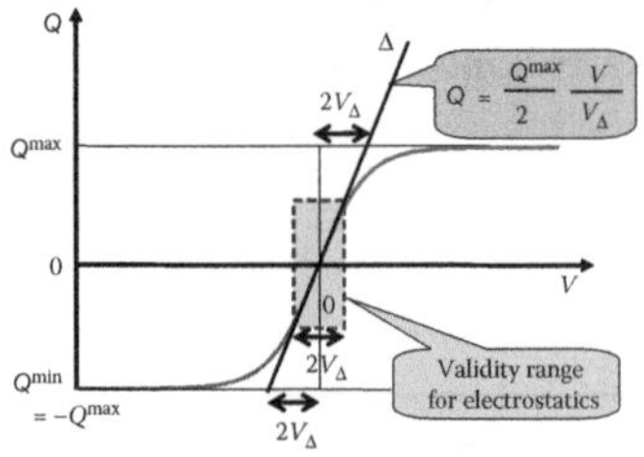

Scaling effort: $V_\Delta = \dfrac{Q^{max}}{2C}$

Property relationships:

$$Q = \hat{C}V$$

Linear model

$$Q \underset{lin}{=} CV$$

System Property	
Nature	Capacitance
Levels	Global (0)
Name	Capacitance
Symbol	C
Unit	[F], [J C^{-1}]
Specificity	—

Energy variation:

$$d\mathcal{U}_Q \underset{\exists C}{=} V\, dQ$$

The electric capacitor provides an intuitive example of dipole because it is often used as component (dipolar) of an electric circuit and also because it is often considered as made up of two electric charges (two poles) between which an electric field is established. This is a system able to store capacitive (electrostatic) energy and to see it varying when it is connected to another dipole.

Graph: Two properties are drawn in the Formal Graph of a capacitor. One is the capacitance, as for a pole, and the other is the evolution, featuring the link between charge and current. When the dipole has to convert its energy into another energy subvariety (by association with a dipole of another type, conductive or inductive, for instance, as a resistor or an inductor), the electric current (the flow) circulates from one dipole to the other and this current results from the evolution operating on the dipole charge (i.e., the temporal derivative of the charge as will be seen in Chapter 9).

$$I = \hat{\mathrm{T}}^{-1} \mathrm{Q} \underset{cont}{=} \frac{\mathrm{dQ}}{\mathrm{d}t} \quad \text{(C7.1)}$$

It must be recalled that this is not a definition of a current as a state variable (see case study A3 "Current Loop" in Chapter 4) but of a current acting as a gate variable when it is connected to another dipole.

Construction: The variation of capacitive (electrostatic) energy is given in an equivalent way from individual variations of poles and from the variation of dipole; this is the embedding principle:

$$\mathrm{d}\mathcal{U}_{\mathrm{Q}} = V_1\,\mathrm{dQ}_1 + V_2\,\mathrm{dQ}_2 = V\,\mathrm{dQ} \quad \text{(C7.2)}$$

Such a dipole is said to be *conservative* because it fulfills a balance equation between the basic quantities of the constitutive poles (see Graph 6.21).

$$\mathrm{Q}_1 + \mathrm{Q}_2 = \mathrm{Q}_1^* + \mathrm{Q}_2^* = 0 \quad \text{(C7.3)}$$

Quantities marked with a star refer to the state of dissociated poles ("before" to form a dipole). In this equation the basic quantities of each pole are opposite and their relation with the charge of dipole is (in adopting the convention of a dipole receptor):

$$\mathrm{Q} = \mathrm{Q}_1 = -\mathrm{Q}_2 \quad \text{(C7.4)}$$

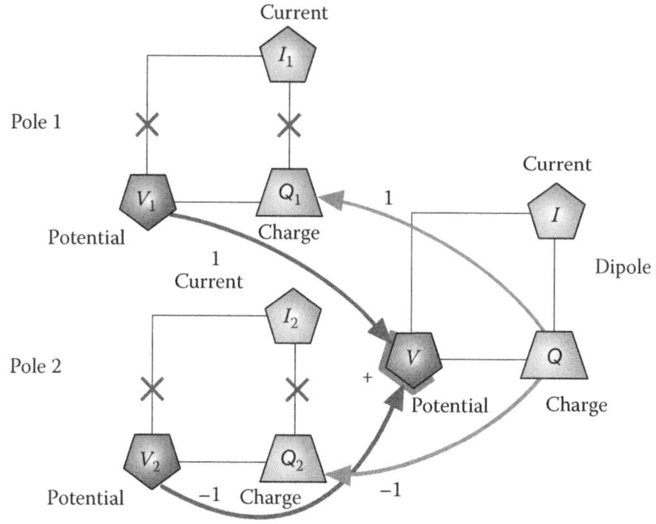

GRAPH 6.21

Introduction of this double equality C7.4 into the equation of energy variation, Equation C7.2, provides, by using the embedding principle, the relation between pole potentials and dipole potential:

$$V = V_1 - V_2 \tag{C7.5}$$

The Formal Graph of construction of the dipole from the two poles uses the preceding equations.

Constitutive Property: In the frame of the Formal Graph theory, the capacitive relation between the electric potential of the dipole and its charge is written in a very general way with the help of a capacitance operator, as for a pole (see case study A5 “Electric Charges” in Chapter 4):

$$Q = \hat{C}V \tag{C7.6}$$

In electrostatics, the capacitive relation is classically considered as a linear relation:

$$Q \underset{lin}{=} CV \tag{C7.7}$$

The Formal Graph approach establishes the capacitive relationship, for all energy varieties, on the basis of a theory of influence between entities. The result is that a capacitive relation has an exponential shape for a pole, giving a nonlinear relation for the dipole (a sigmoid shape for a conservative dipole). This means that the classical linear relation is an asymptote of the exponential function around the origin.

For each pole, its activity is (see case study A5 “Electric Charges” in Chapter 4)

$$\frac{Q_i + Q_i^m}{Q_i^m} = a_{Qi} = \exp\frac{V_i}{V_\Delta} \tag{C7.8}$$

in which Q_i^m represents the opposite of the minimum charge Q_i^{min}. The consequence of the symmetry between charges is that the maximum of one pole corresponds to the minimum of the other pole, so the relation between limiting quantities for the poles and the dipole is identical for both poles.

$$Q^{max} = Q_1^m = Q_2^m = -Q^{min} \tag{C7.9}$$

Between activities of each pole and the gain of the dipole, one has the following relation:

$$g_{Q21} = \frac{a_{Q2}}{a_{Q1}} \tag{C7.10}$$

From expression C7.8 of activities, by taking into account relation C7.5 between potentials, this gain is written as

$$\frac{Q^{max} - Q}{Q - Q^{min}} = \frac{Q^{max} - Q}{Q^{max} + Q} = g_{Q21} = \exp\frac{-V}{V_\Delta} \tag{C7.11}$$

The final expression of the dipole capacitive relationship follows by rearranging the previous equation:

$$\frac{Q}{Q^{max}} = \frac{1 - \exp(-V/V_\Delta)}{1 + \exp(-V/V_\Delta)} \tag{C7.12}$$

For small values (in absolute) of the potential V compared to the scaling potential V_Δ the previous equation reduces to the asymptotic straight line

$$Q \underset{lin}{=} \frac{Q^{max}}{2} \frac{V}{V_\Delta} \tag{C7.13}$$

which allows the identification of the scalar capacitance in the classical relation C7.7, with half the ratio of the maximum charge upon the scaling potential.

Without entering into the details of the theory of influence establishing the capacitive relationship, this theory says that the scaling potential V_Δ is dependant on the coupling with another energy variety (see Section 13.4.3 in Chapter 13). Without any coupling, the scaling potential is quasi infinite and the capacitive relation is linear over the whole potential range. It is the case in vacuum and in number of "linear" materials. In the presence of coupling, the scaling potential V_Δ becomes finite and the linearity is only observed for values of potential smaller than V_Δ (in absolute values). For instance, the coupling with the thermal variety lowers the scaling potential V_Δ down to few tens of millivolts, thus making the nonlinear behavior the dominant one when the considered potential range is wider.

6.4.8 Case Study C8: Spring — Translational Mechanics

The case study abstract is given on next page.

In translational mechanics, the *force* $\mathbf{F}$ plays the role of *energy-per-entity* in the capacitive subvariety of energy (in other words, it belongs to the family of efforts) and the *displacement* $\boldsymbol{\ell}$ plays the role of *capacitive entity number*, which is therefore a basic quantity. These are state variables oriented in space and therefore represented by vectors. The spring schematized in the case study abstract works in extension, meaning that forces are oriented from the center of the spring toward the ends, but a spring may also work in compression with forces oriented toward the center.

Graph: The modeling of this system involves three state variables and the space–time property of temporal evolution for making up a dipole.

Energy: In this energetic subvariety, the capacitive energy is called "work" in the classical language and its variation is given by the scalar product of the vector force by the variation of the vector displacement:

$$d\mathcal{U}_\ell \underset{\exists k_e}{=} \mathbf{F} \cdot \mathbf{d}\boldsymbol{\ell} \tag{C8.1}$$

Construction: Two poles are associated in the building of this dipole. The variation of capacitive energy (or *work*) of a dipole is the sum of variations of capacitive energy of the two poles:

$$d\mathcal{U}_\ell = \mathbf{F}_1 \cdot \mathbf{d}\boldsymbol{\ell}_1 + \mathbf{F}_2 \cdot \mathbf{d}\boldsymbol{\ell}_2 \tag{C8.2}$$

A spring, with no end attached to an immobile support, sees its ends behaving in a symmetric way, which is translated by a *conservation* of displacements of both poles:

$$\boldsymbol{\ell}_1 + \boldsymbol{\ell}_2 = C^{st} \tag{C8.3}$$

In other words, the variations of displacement of each end have opposite signs, and, by convention, the variation of the dipole displacement is taken with the same sign as that of the first pole. This variation is obviously equal to the double of the first end variation (the spring length is the difference of two displacements):

$$\mathbf{d}\boldsymbol{\ell} = 2\,\mathbf{d}\boldsymbol{\ell}_1 = -2\,\mathbf{d}\boldsymbol{\ell}_2 \tag{C8.4}$$

C8: Spring **Translational Mechanics**

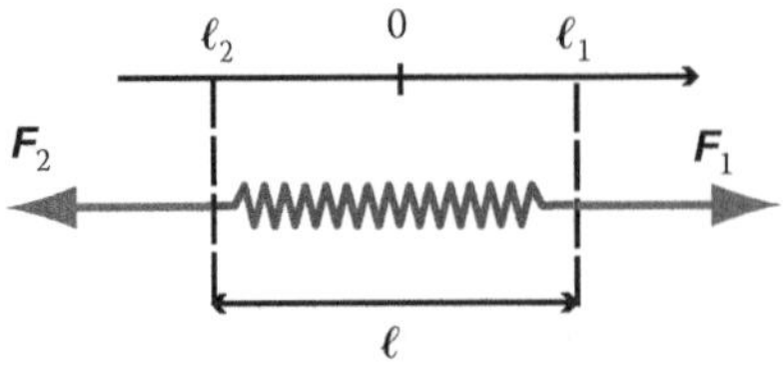

An elastic body in one dimension of space, such as a spring, a solid but deformable bar, or an elastic wire, makes up a dipole belonging to the capacitive subvariety of translational mechanics.

Dipole

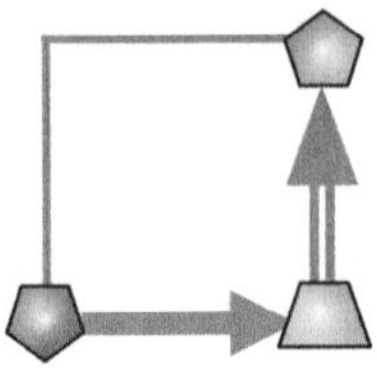

Dipole	
✓	Fundamental
	Mixed
✓	Capacitive
	Inductive
	Conductive
✓	Global
	Spatial
✓	Inseparable
✓	Conservative

Formal Graph:

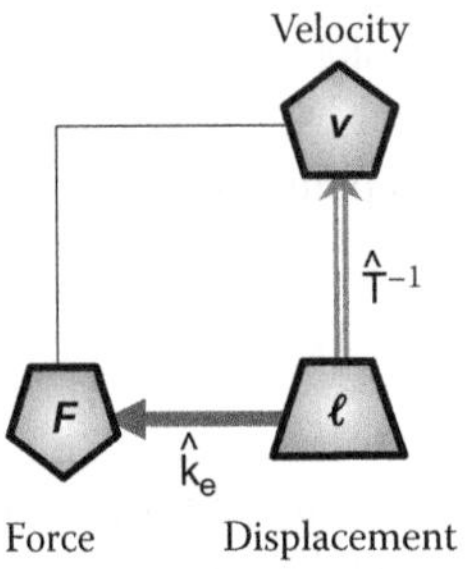

GRAPH 6.22

Capacitance operator traces:

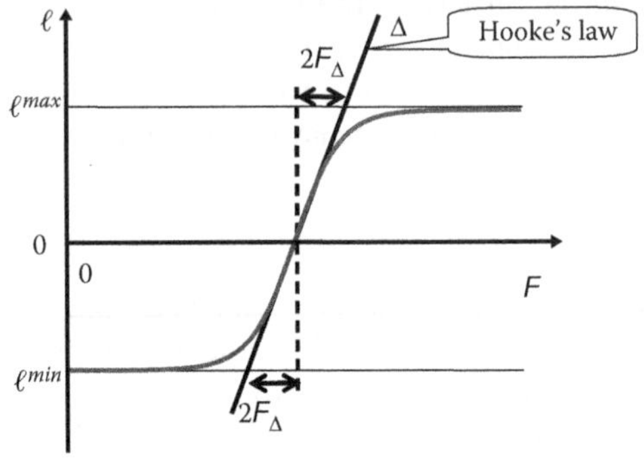

Scaling effort: $F_\Delta = k_e\left(-\ell^{min}\right)$

Variables:

Variety	Translational Mechanics		
Subvariety	Capacitive (Elastic)		Inductive
Category	Entity number	Energy/entity	Entity number
Family	Basic quantity	Effort	Flow
Name	Displacement	Force	Velocity
Symbols	ℓ	F	v
Unit	[m]	[N], [J m^{-1}]	[m s^{-1}]

Property relationships:

$F = \hat{k}_e\, \ell$

Hooke's law:

$F \underset{lin}{=} k_e\, \ell$

System Property	
Nature	Elastance
Levels	Global (0)
Name	Elasticity
Symbol	k_e
Unit	[N m^{-1}]
Specificity	—

Energy variation:

$$d\mathcal{U}_\ell \underset{\exists k_e}{=} F \cdot d\ell$$

By using as integration constants the rest positions of the poles corresponding to zero forces applied, one has

$$\ell = 2\left(\ell_1 - \ell_1^0\right) = -2\left(\ell_2 - \ell_2^0\right) \tag{C8.5}$$

Through this equation is defined the dipole displacement ℓ, which is not the actual length ℓ^{tot} of the spring but the difference between this length and the rest length ℓ^0:

$$\ell \overset{def}{=} \ell^{tot} - \ell^0 \tag{C8.6}$$

In other words, the origin of the dipole displacement ($\ell = 0$) is chosen so as to correspond to the rest state of the spring.

The dipole force is defined by the general relationship between dipole effort and pole efforts which uses a *separability factor* s_D

$$\boldsymbol{F} = s_D(\boldsymbol{F}_1 - \boldsymbol{F}_2) \tag{C8.7}$$

that may be equal to 1 (separability) or 1/2 (inseparability). Here, it can be checked by solving the system of Equations C8.5 and C8.7 for equating C8.1 and C8.2; one finds $s_D = 1/2$, with the consequence that the force dipole is half the difference of pole forces,

$$\boldsymbol{F} = \frac{\boldsymbol{F}_1 - \boldsymbol{F}_2}{2} \tag{C8.8}$$

which ensures that the variation of energy is the same for the two poles and for the dipole by application of the *embedding principle*. Another way to reach this result is to recognize that nothing entitles us to attribute a different energy to each end, so the variations of energy must be equal:

$$\boldsymbol{F}_1 \cdot \mathbf{d}\boldsymbol{\ell}_1 = \boldsymbol{F}_2 \cdot \mathbf{d}\boldsymbol{\ell}_2 \tag{C8.9}$$

By taking into account the symmetry of the displacement variations in Equation C8.4 into the previous equality in Equation C8.9, the forces at the level of the ends are opposite but equal in magnitude:

$$\boldsymbol{F}_1 = -\boldsymbol{F}_2 \tag{C8.10}$$

It is tempting to express the symmetry of pole forces in Equation C8.10 under the form of a conservation equation in writing the nullity of the sum of pole forces:

$$\boldsymbol{F}_1 + \boldsymbol{F}_2 = 0 \tag{C8.11}$$

Equilibrium or conservation? The notion of the *mechanical equilibrium law* is a particularity that is hardly transposable to other energy varieties because, in most of these, the notion of *conservation* is utilized instead of the notion of symmetry between state variables. In addition, the symmetric Equation C8.9 is perfectly valid during a movement of translation, as shown by a Formal Graph, which includes the relationships between displacement and velocity (the dipole flow). This is not compatible with the notion of static equilibrium that implies an absence of movement and therefore of time.

In mechanics, the state of rest is called state of *static equilibrium* and Equation C8.11 is viewed as a *law of equilibrium*. In fact, it is merely a *relation of symmetry* and is neither a rule nor an autonomous principle: This equation merely results from hypotheses made on the system (inseparability and indeformability) and from the application of the First Principle of Thermodynamics (Equation C8.2). It applies only in the case of two opposite forces and not in the case of several converging forces, it remains valid in dynamics (movement of the ends). In the Formal Graph theory, this classical "law" stating that a null sum of forces expresses a static equilibrium is considered unsuitable for a dipole because the notion of equilibrium is much more general than this static vision peculiar to mechanics (see Remark above). On the contrary, it applies to higher organization levels that are the assemblies of dipoles (see case study E2 "Concurring Forces" in Chapter 8).

All these relations are translated into the Formal Graph shown in Graph 6.23, made with three sub-Formal Graphs, two for the poles and one for the dipole formed by these latter.

Graph 6.23 is a *construction graph*. It represents, on the left, the two Formal Graphs of the poles, and, on the right, the Formal Graph of the dipole. The relationships between the state variables of

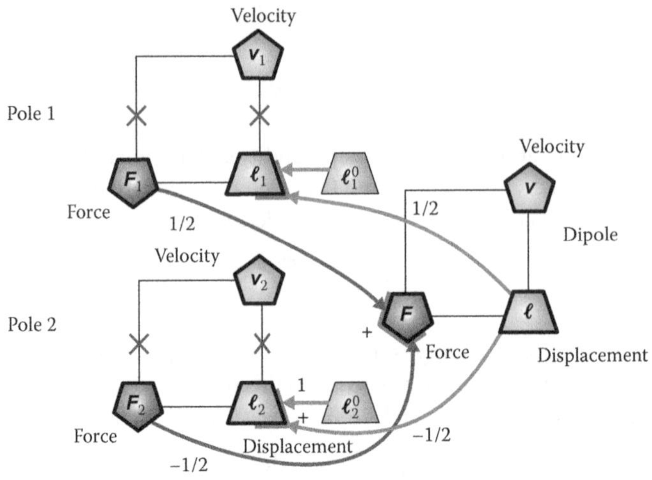

GRAPH 6.23

the poles and those of the dipole are represented by connections figured by arcs with arrows, each connection being endowed with an *operator* playing the role of a proportionality coefficient (therefore a scalar) for building the adequate linear combinations. It can be remarked that this graph is perfectly symmetric, each pole playing a symmetrical role with respect to the other.

It is worth noting that the Formal Graph above also models the static state as a dynamic function when a movement of translation elongates the spring. The absence of inertial mass for the spring (this is an ideal case) helps avoid taking into account any momentum (quantity of movement) or inductive (kinetic) energy. This is one of the reasons, it must be recalled, which justifies not linking the nullity of the sum of applied forces with the static situation of a system.

Property: Obviously, the dipole is endowed with a capacitive property, the elastance, which depends on the elastances of each pole.

$$\boldsymbol{F} = \hat{k}_e \ell \tag{C8.12}$$

The starting point for establishing the capacitive relationship is the *activity*, generalized to all energy varieties, translational mechanics being included (see case study A8 "Spring End" in Chapter 4):

$$\frac{\ell_i - \ell_i^{min}}{\ell_i^0 - \ell_i^{min}} = a_{\ell i} = \exp \frac{\boldsymbol{F}_i}{\boldsymbol{F}_\Delta} \tag{C8.13}$$

In Figure C8.1, the spring is shown in three different states depending on the application of forces: elongated state with length $\ell^0 + \ell$, rest state with length ℓ^0, and compressed state at the minimum length $\ell^0 + \ell^{min}$. In this figure are indicated the coordinates of the two ends in relation to the spring lengths. A maximum coordinate ℓ_i^{max} is defined for each end so as to locate their rest coordinate ℓ_i^0 right in the middle of their range $[\ell_i^{min}, \ell_i^{max}]$.

The characteristic displacements, maximum and minimum displacements, are in the ratio 1:2 between the poles and the dipole. The symmetry of the spring creates equality between all half-ranges, and in particular

$$\ell^{max} = \frac{\ell^{max} - \ell^{min}}{2} = 2\left(\ell_1^0 - \ell_1^{min}\right) = 2\left(\ell_2^0 - \ell_2^{min}\right) \tag{C8.14}$$

Owing to the choice of identical orientations for counting the displacements, when the spring is at its minimum length, the left end 1 is at its minimum and the right end 2 is at its maximum. The

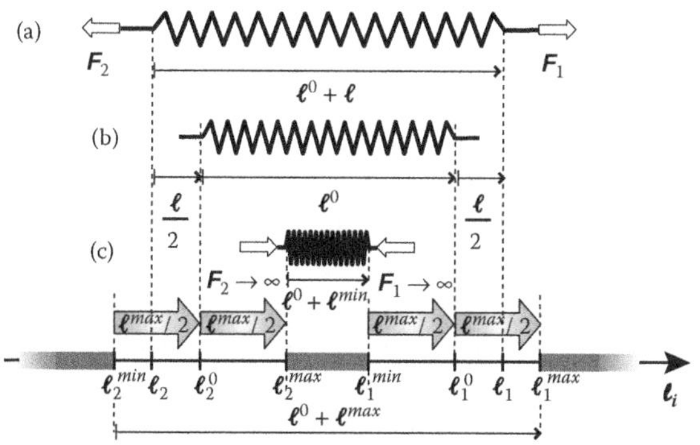

FIGURE C8.1 The three states of a spring: (a) Elongated state with divergent applied forces and length $\ell^0+\ell$. (b) Rest state with no forces and length ℓ^0. (c) Maximum compression state with convergent forces of maximum strength leading to the minimum length $\ell^0+\ell^{min}$ (ℓ^{min} is negative).

converse is observed for the maximum elongation of the spring. Using the poles–dipole relationship in Equation C8.5 between displacements, gives the following equalities:

$$\ell-\ell^{min}=2\left(\ell_1-\ell_1^{min}\right);\quad \ell^{max}-\ell=2\left(\ell_2-\ell_2^{min}\right) \tag{C8.15}$$

With these relationships, the two activities can be expressed as

$$a_{\ell 1}=\frac{\ell-\ell^{min}}{\ell^{max}};\quad a_{\ell 2}=\frac{\ell^{max}-\ell}{\ell^{max}} \tag{C8.16}$$

The algebraic expression of the capacitive relationship is based on the definition of the gain of the dipole as the ratio of pole activities and on the formation of the difference between polar efforts for constituting the dipolar effort, which gives the following ratio:

$$\frac{\ell^{max}-\ell}{\ell-\ell^{min}}=\frac{a_{\ell 2}}{a_{\ell 1}}=g_{\ell 21}=\exp\left(-\frac{F}{F_\Delta}\right) \tag{C8.17}$$

When the displacements are small, the capacitive relationship becomes linear and is known as Hooke's law:

$$F\underset{lin}{=}k_e\,\ell \tag{C8.18}$$

The relation between the various parameters of the capacitive relationship is as follows:

$$F_\Delta=k_e\left(-\ell^{min}\right) \tag{C8.19}$$

The variation of the displacement as a function of the applied force is plotted in the case study abstract, with the tangent at the inflexion point that is the asymptote for small strengths of the force compared to the scaling force F_Δ.

Separability: This dipole does not work with a single conservation relationship, concerning its entity numbers (basic quantities), but with an additional second relationship ruling the conservation of energy-per-entity (effort), which must be better viewed as a *symmetry rule* (antisymmetry in fact). The two relationships in Equations C8.5 and C8.10 are recalled and rewritten hereafter:

$$\ell_1+\ell_2=\ell_1^0+\ell_2^0=0 \tag{C8.20}$$

$$F_1=-F_2 \tag{C8.21}$$

The existence of one conservation rule (which is also a symmetry) and one symmetry rule features an *inseparable dipole*, which is a characteristic of this system as mentioned in the study of the spring end (which is a virtual concept; see case study A8 "Spring End" in Chapter 4). It is indeed impossible to cut a spring for isolating one end, except maybe in going under the size of an elementary segment (a few grains of matter), which would make the spring lose its essential properties. Another example of an inseparable dipole is a semiconductor material populated by electrons and holes where the Fermi levels play the role of efforts and the numbers of charge carriers are the basic quantities (the physical reason being that a hole is defined by the absence of electron; see case study C9 "Electrons and Holes"). In contradistinction, the other case studies of dipole—chemical reaction, electric capacitor, colliding bodies, physical chemical interface—are all *separable* as they possess only one conservation rule and no symmetry rule (see case studies C5 through C7).

Remarks

Applicability of Newton's third law. This law, mostly known under the abridged version "action equals reaction," has been proposed by its inventor for the case of attraction between two gravitational masses, a case that corresponds in the Formal Graph theory to a capacitive dipole in the energy variety called gravitation. It equally applies in translational mechanics. It is this law that is utilized in the Formal Graph theory for describing an elastic spring or wire by means of a capacitive dipole in translational mechanics.

Is action really equal to reaction? The third law of Newton in mechanics is very often misinterpreted when it is shortened to the formulation "action is equal to reaction" without specifying on what these forces are acting. In fact, this law applies either to the elastic forces, as in this case study, or to the action of a body on another one, when symmetric forces are acting on both bodies (by equating the magnitudes). It is often utilized abusively as a model for the convergence of two forces at the same point, which is not the case described here (see case study E2 "Concurring Forces" in Chapter 8).

Beyond the concept of equilibrium. Using concepts that are common to all energy varieties leads to abandoning the notion of static equilibrium as a unique basis of the rule that the sum of forces is zero. It is replaced, in the case of two forces, by the much more powerful notion of symmetry relationship between forces, deriving itself from the inseparability of poles.

Ideality versus reality. The simple model developed in this case study corresponds to an ideal spring made of a homogeneous and purely elastic material. In a real solid, one must take into account discrepancies between spring elements that may lead to a distribution of dipoles having different characteristics. This helps compose a capacitive relationship with several exponential functions. A *distributed dipole* constitutes a Formal Object having a higher complexity than those studied in this book.

6.4.9 Case Study C9: Electrons and Holes

Corpuscular Energy

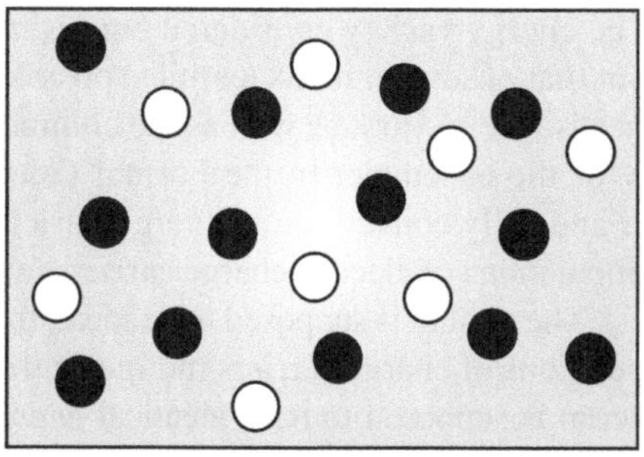

An ensemble of electrons and holes, assumed to be ideal with only two populations of charge carriers with opposite sign, is an example of inseparable dipole, that is to say conservative for the basic quantities and symmetrical for the energies-per-entity. In this respect, it is analogous to the elastic dipole in mechanics (see case study C8 "Spring").

Dipole

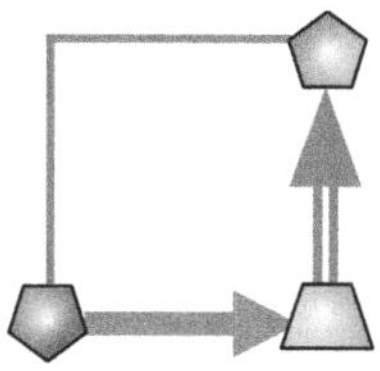

Dipole	
✓	Fundamental
	Mixed
✓	Capacitive
	Inductive
	Conductive
✓	Global
	Spatial
✓	Inseparable
✓	Conservative

The remarkable feature of this case study is to show that recourse to statistic considerations is not mandatory in this domain (as in others).

Formal Graph:

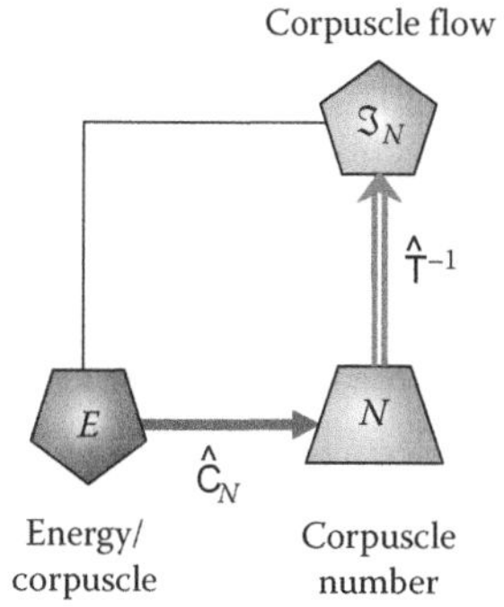

GRAPH 6.24

Capacitance operator trace:

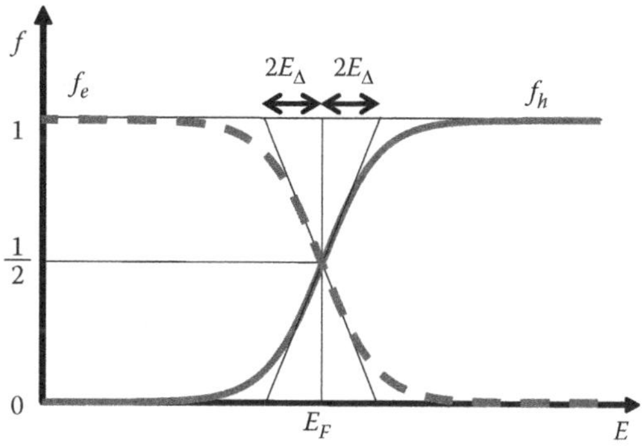

Scaling effort: $E_\Delta = k_B T$

Variables:

Variety	Corpuscular Energy		
Subvariety	Capacitive (Internal)		Inductive
Category	Entity number	Energy/ entity	Entity number
Family	Basic quantity 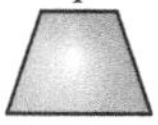	Effort 	Flow 
Name	Corpuscle number	Corpuscular energy	Corpuscular flow
Symbols	N, n	E	$\Im_N$
Unit	[corp]	[J corp^{-1}]	[corp s^{-1}]

Property relationships:

$$N = \hat{C}_N E$$

System Property	
Nature	Capacitance
Levels	Global (0)
Name	Corpuscular capacitance
Symbol	C_N
Unit	[J^{-1}]
Specificity	—

Energy variation:

$$d\mathcal{U}_N \underset{\exists C_N}{=} E\,dN$$

Under the name of corpuscles one finds a large diversity of objects such as particles (electrons, photons, quarks, etc.), molecules, atoms, nuclei, clusters, proteins, and microorganisms, in fact all that is not counted by moles but by numbers of items. The energy variety considered here is not that of a single item with its own *energy-per-corpuscle* but that of several items (entities) possessing the same *energy-per-corpuscle*. The number of entities is able to vary, as well as the common *energy-per-corpuscle* may vary, for varying the energy of the ensemble. In the Formal Graph language, this ensemble forms a *collection of corpuscles* and fully constitutes an energy variety.

The system studied here is a material containing two populations of electric charge carriers, one with a negative sign (electrons) and the other positive (holes). The system is supposed to be ideal, that is to say homogeneous and containing only these two populations of charge carriers (no impurities, no defaults, no doping agents, etc.). Furthermore, the material is supposed to have identical *energy states*, which is a notion in physics corresponding to the notion in chemistry of *sites* in a material that can be *occupied* by molecules, and which must not be confused with the thermodynamic notion of *energetic state* of the system. The pole corresponding to a single type of corpuscle is described in case study A7 "Group of Corpuscles" in Chapter 4. A semiconductor is a material more complex than the system studied here because energy states are not identical, meaning that there exists a *distribution of energy states* (see "Extension to Semiconductors" at the end of this case study).

Variables: In the capacitive subvariety of this variety, the *energy-per-corpuscle* E plays the role of an *energy-per-entity* (*effort*) and the *number of corpuscles* N plays the role of *number of entities* (*basic quantity*). Each population of charge carriers represents a pole, the negative one possessing an *energy-per-corpuscle* E_e and a number of electrons N_e, and the positive one possessing an *energy-per-corpuscle* E_h and a number of holes N_h. This representation of a collection of objects by a pole is also used for a chemical substance (see case study A6 "Chemical Species" in Chapter 4), which is merely an energy variety related to the one of corpuscles as the basic quantities are proportional between them via the Avogadro constant.

Graph: The modeling by the Formal Graph of this system uses three state variables, namely the *flow of corpuscles* f_N in supplement to the two capacitive state variables and to space–time property of temporal evolution, to form a dipole.

Energy: In this energetic variety (said to be "corpuscular" for making short), the variation of capacitive energy is given by the product of the *energy-per-corpuscle* by the variation of the *number of corpuscles*:

$$\mathrm{d}\mathcal{U}_N \underset{C_N \exists}{=} E\,\mathrm{d}N \tag{C9.1}$$

Construction: The building of this Formal Graph makes use of two poles associated with forming a dipole. The variation of capacitive energy of the dipole is the sum of the variations of capacitive energy of each pole:

$$\mathrm{d}\mathcal{U} = E_e \mathrm{d}N_e + E_h \mathrm{d}N_h \tag{C9.2}$$

The particularity of this system is to be a symmetric dipole because the definition of a hole is the exact opposite of the definition of the electron: A hole is created in a material by the vacancy of an electron. The sum of the numbers of electrons and holes is therefore a constant.

$$N_e + N_h = N_e^* + N_h^* \tag{C9.3}$$

The generalization afforded by the Formal Graph theory makes the relationship between pole efforts and dipole effort depending on the separability of the dipole. The general relation is written with a *separability factor* s_D which may be equal to 1 (separability) or to 1/2 (inseparability):

$$E = \mathsf{s}_D (E_e - E_h) \tag{C9.4}$$

In the present case, the inseparability between the poles, as it is equally the case for a spring (see previous case study C8), gives to the separability factor the value one-half:

$$s_D = \frac{1}{2} \tag{C9.5}$$

The inseparability imposes symmetry between the *energies-per-entities* (efforts) of the two poles. This is expressed by the null sum of *energies-per-corpuscle* or by equating one of them to the opposite of the other one:

$$E_e = -E_h \tag{C9.6}$$

These last three equations ensure that all *energies-per-corpuscle* are equal in absolute value:

$$E_e = -E_h = E \tag{C9.7}$$

Insertion of this equality into Equation C9.2 allows defining, in application of the embedding principle, the variation of the number of corpuscles of the dipole as the variation of the number of electrons minus the variation of the number of holes, which can be alternatively written by taking into account the separability:

$$\mathrm{d}N = 2\mathrm{d}N_e = -2\,\mathrm{d}N_h \tag{C9.8}$$

By integration, the relation between numbers of corpuscles is written with the help of suitable constants

$$\frac{N}{2} = N_e - N_e^* = N_h^* - N_h \tag{C9.9}$$

allowing one to recall conservation C9.3 of basic quantities. A dipole endowed with a double conservation is said to be *inseparable* because the poles cannot work alone when they are contained in a material (in contradistinction with free behavior in which the absence of an electron does not create a hole, in other words when the notion of energy state or of site is irrelevant). Another example of symmetry inducing inseparability is the spring (see case study C8 "Spring").

All these relations are translated into a special Formal Graph called *construction* Formal Graph, made of three sub-Formal Graphs, two for the poles and one for the dipole. Flows are connected on the model of the *exchange by flows*, used for the chemical reaction notably (see case study C5 "First-Order Chemical Reaction").

The Formal Graph in Graph 6.25 is a *construction graph* by opposition to the simplest graph modeling directly an object (as shown in Graph 6.24): On the left are shown the two Formal Graphs of poles, and on the right the Formal Graph of the dipole. It can be remarked that this graph is fully symmetric, each pole playing a symmetric role with respect to the other. The relations between the state variables of the pole and those of the dipole are represented by connections made with bent arrows, each connection being endowed with an *operator* acting as a proportionality coefficient (therefore of scalar nature) for building the suitable linear combinations.

The connections between *energies-per-corpuscle* are coded with the *separability factor* s_D according to the relation shown in Equation C9.5, as is the case for forces in mechanics of elasticity (see case study C8 "Spring").

Capacitive Property: The dipole is endowed with a capacitance that links the *energy-per-corpuscle E* of the dipole to the corpuscle number *N* of the dipole:

$$N = \hat{\mathrm{C}}_N E \tag{C9.10}$$

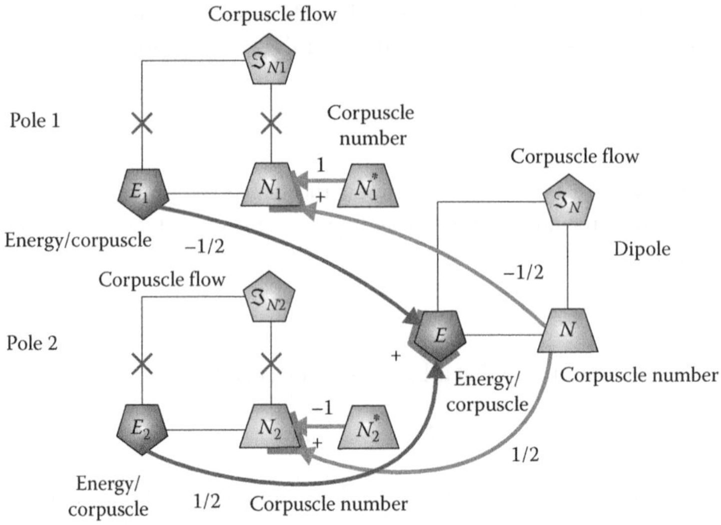

GRAPH 6.25 Formal Graph depicting the construction of the dipole from two poles.

The capacitive relations at the level of each pole are established in the Formal Graph theory by using the notion of *activity*, generalized to this domain as well as to others (by using a general theory of influence between poles). There is no threshold quantity for the number of corpuscles below which the corpuscles are inactive in the mutual influence (in contradistinction with a solid, such as a spring—see case study C8—which cannot be compressed below a certain limit without losing its essential properties). This lack of threshold renders the activity directly proportional to the number of carriers:

$$\frac{N_e}{N_i^0} = a_{Ne} = \exp\left(\frac{E_e - E_e^0}{2E_\Delta}\right) \quad \frac{N_h}{N_i^0} = a_{Nh} = \exp\left(\frac{E_h - E_h^0}{2E_\Delta}\right) \tag{C9.11}$$

The basic quantity of reference N_i^0, which is defined as the common value that may take the two numbers of electrons and holes, is called the *intrinsic number of charge carriers*. Here, as well as in the simple case of linear distribution of energy states, this number is indicated with a superscript zero to avoid confusion with the number N_i which can be determined in the more complex case of a material with an irregular distribution of energy states. The denominator of the argument of the exponential is the ratio of the *scaling energy-per-corpuscle* E_Δ on the *separability factor* s_D and the numerator contains a *reference energy-per-corpuscle* corresponding to the unit activity

$$E_e = E_e^0 \iff a_{Ne} = 1 \iff N_e = N_i^0 \tag{C9.12}$$

$$E_h = E_h^0 \iff a_{Nh} = 1 \iff N_h = N_i^0 \tag{C9.13}$$

By taking into account equality C9.7 (with plus and with minus) of energies-per-corpuscle between them due to the inseparability of the dipole, it is possible to write the same equality between the *reference energies-per-corpuscle* and to define a common reference. This common *energy-per-corpuscle* is known as *Fermi* level* in the domain of condensed matter physics (see Remark "Fermi level or energy?") and it is notated E_F:

$$E_F = E_e^0 = -E_h^0 = s_D\left(E_e^0 - E_h^0\right) \tag{C9.14}$$

* Enrico Fermi (1901–1954): Italian physicist; Florence and Rome, Italy and Chicago, USA.

Equalities C9.7 and C9.14 lead to a unique and same writing of the capacitive relations at the level of the poles:

$$a_{Ne} = \frac{1}{a_{Nh}} = \frac{N_e}{N_i^0} = \frac{N_i^0}{N_h} = \exp\left(\frac{E - E_F}{2E_\Delta}\right) \quad \text{(C9.15)}$$

Two interesting relations can be deduced that feature the inseparability of the dipole:

$$a_{Ne}\, a_{Nh} = 1; \qquad N_e N_h = \left(N_i^0\right)^2 \quad \text{(C9.16)}$$

Relation C9.16 on the right between numbers of charge carriers is called in physics the "law of mass action" (see Remark "Action of which mass?") because of its resemblance with the equation of a dismutation (disproportionation) equilibrium.

The theory of influence between poles, which underlies this generalization of the activity to all energy varieties, utilizes the notion of *gain of the dipole* g defined as the ratio of the activities, which corresponds here to the ratio of the numbers of charge carriers:

$$g_{Nhe} = \frac{a_{Nh}}{a_{Ne}} = \frac{N_h}{N_e} = \exp\left(-\frac{E - E_F}{E_\Delta}\right) \quad \text{(C9.17)}$$

In contradistinction with the domain of physical chemical energy in which the modeling is made in terms of ratios of substance amounts (or their concentrations), it is usual in condensed matter physics to utilize the notion of *function of distribution* f_k, defined as the derivative of the number of corpuscles of a given type k with respect to the maximum number of corpuscles.

$$f_k \overset{def}{=} \frac{dN_k}{dN^{max}} \underset{\text{equiv. states}}{=} \frac{N_k}{N^{max}} \quad \text{(C9.18)}$$

When energy states are all equivalent, that is to say *not distributed**, the *function of distribution* f_k, is independent from the maximum number of corpuscles N^{max} in the system, which here is equal to the sum of both corpuscle numbers. In this simplified case, one finds

$$f_e = \frac{N_e}{N_e + N_h} \qquad f_h = \frac{N_h}{N_e + N_h} \qquad f_e + f_h = 1 \quad \text{(C9.19)}$$

These are ratios without dimension ranging from 0 to 1 and their sum over all types of corpuscles is equal to 1. Their expressions, here in function of the energy-per-corpuscle E deduced from Equation C9.17, are therefore

$$f_e = \frac{1}{1 + \exp\left(E - E_F / E_\Delta\right)} \qquad f_h = \frac{\exp\left(E - E_F / E_\Delta\right)}{1 + \exp\left(E - E_F / E_\Delta\right)} \quad \text{(C9.20)}$$

A comparison with the classical models of condensed matter physics provides for the scaling energy-per-corpuscle E_Δ the equality with the product of the Boltzmann constant k_B by the temperature T:

$$E_\Delta = k_B T \quad \text{(C9.21)}$$

From the Formal Graph viewpoint, this relation between two energies-per-entity of different varieties means a coupling with the thermal variety, exactly as a scaling chemical potential μ_Δ is related to the temperature through the gas constant R. The expressions in Equation C9.20 with $k_B T$ are called in physics *distributions of Fermi–Dirac*† and they are classically established from statistics that bear the same name.

* Two distributions are considered here: distribution of corpuscles and distribution of energy states.

† Paul Adrien Maurice Dirac (1902–1984): British physicist; Bristol, UK and Tallahassee, Florida, USA.

The curves of variation of these *functions of distribution* in function of the energy-per-corpuscle and with the Fermi level as a parameter are plotted in the case study abstract.

This model proposed by the Formal Graph theory is much more general than the Fermi–Dirac model as it applies to any dependence of E_Δ on an energy variety other than the one responsible for the thermal agitation, and notably applies to every system for which statistics are irrelevant.

The physical meaning of the Fermi level is "energy" (understood as "per corpuscle") below which the population of electrons is a majority. More precisely, in an ideal system, only coupled to the thermal energy variety, the Fermi level constitutes the upper limit at 0 K (in fact, at $E_\Delta = 0$, which is a more general condition from the Formal Graph viewpoint) that can be reached by the energy-per-corpuscle in this fundamental state.

Extension to semiconductors: A semiconductor is featured by a distribution of energy states that are not equivalent as in the simple material studied above. The determination of the capacitive relationships requires integration of the previous *functions of distribution* f_e and f_h over all energy states knowing their distribution as a function of the energy-per-corpuscle E. In the Formal Graph hierarchy of systems, such a semiconductor belongs to the level of *distributed dipoles*. Here are schematized the principles for modeling this higher complexity level.

For an *intrinsic semiconductor*, that is, not doped by impurities, the capacitive relations are calculated according to the *energy band theory* that separates electrons and holes in two bands (levels), the *conduction band* for electrons and the *valence band* for holes. Each band is featured by energy-per-corpuscle E_C and E_V, respectively, with the Fermi level located between them. When the separation $E_C - E_V$ is several times higher than the scaling energy-per-corpuscle E_Δ, the semiconductor is said to be *nondegenerate* (which corresponds, in the language of condensed matter physics, to a change from Fermi–Dirac statistics to Boltzmann statistics), computation can be made analytically, providing the following capacitive relations:

$$N_e = N_C \exp\left(-\frac{E_C - E_F}{E_\Delta}\right) \quad N_h = N_V \exp\left(\frac{E_V - E_F}{E_\Delta}\right) \qquad \text{(C9.22)}$$

The two reference amounts of electrons and holes N_C and N_V are constant quantities depending on the masses of corpuscles. They are called *effective number of energy states* (or *effective densities of energy states* when divided by the volume as often done).

Remarks

Fermi level or energy? The Fermi level is also classically called Fermi "energy." The Formal Graph theory insists, for the sake of coherence from one domain to the other, on the importance of not confusing an effort with energy. The first one is an *energy-per-entity* which is an *intensive state variable* whereas the second one is *not a state variable* and is by contrast an *extensive* variable (it can be added in any conditions and independently from the varieties). It is true that in particle physics, the basic object is unique and is not a collection like here, and therefore that the distinction made has no importance, as long as a single particle is considered.

In condensed matter physics, the assertion that the Fermi level is a *chemical potential* is currently met. It is partly true, but misleading, as this chemical potential is not the effort μ of the physical chemical energy variety, but rather the product of this latter with the Avogadro constant N_A. However, the interest of this incomplete assertion is at least to signify the nature of the Fermi level as being an *energy-per-entity*.

Action of which mass? The naming "law of mass action" is incorrect for several reasons. First, from the viewpoint of chemistry, which in modern literature (although not always) prefers the naming "law of reaction equilibrium," because the reference to the mass of a chemical substance into kinetic equations or into equilibrium equations is abandoned for long to the benefit of the *number of moles* (which is a measure of the quantity of substance based on a counting of discretized entities).

From the viewpoint of science history, the *law of mass action*, proposed in 1864 by Guldberg* and Waage,† is not a law of equilibrium but of kinetics instead. It sets up the dependence of a reaction rate with the mass of reactant (or nowadays with the quantity of substance). The name of their law must be understood as the action of the mass of reactant (proportional to the number of moles through the molar mass) on the reaction rate.

Next, and this is the most important, from the viewpoint of physics, in copying the chemical concept of dismutation reaction to which this law corresponds, the classical approach gives credence to the existence of quantities of electron–hole pairs predicted by this reaction. However, their amount must be immediately neglected for keeping the system of equations amenable to a solution (i.e., ideal system)! In addition, to consider a third "species" in equilibrium with the two others is contrary to the definition of a hole, which is the exact opposite of an electron, and therefore contrary to the fact that both annihilate. This is the kind of long-lasting paradox in physics! (However, it must be acknowledged that the concept of separability is still not very well mastered in physics, even in quantum physics.)

Concentration or density? In the domain of condensed matter physics (alternatively called solid state physics), it is customary not to use as a variable the number of charge carriers but their volumic concentration (also called density). This habit is not compatible with the standardization brought by Formal Graphs. It prevents, for instance, the application of the First Principle of Thermodynamics as volumes are relevant to another energy variety (basic quantity of hydrodynamics) and in addition are not always constant.

Condensed matter approach. It must be recalled that the studied system is a simple case of a material with equivalent energy states, in which the number of electrons is merely proportional to the total number of electrons and holes through the distribution function f_e (cf. Equation C9.19).

The treatment of a material with a distribution of energy states, such as a semiconductor, requires the knowledge of the number of energy states in function of the energy-per-corpuscle.

The number of charge carriers in each energy state, or energetic density of carriers, is the product of three terms:

1. The *energy state degeneracy* N_g (generally notated g but it is confusing with the gain and the gravity field modulus) giving the maximum occupancy of an energy state, defined as the derivative of N^{max} with respect to the number of energy states $\mathcal{N}$:

$$N_g \overset{def}{=} \frac{dN^{max}}{d\mathcal{N}} \tag{C9.23}$$

* Cato Maximilian Guldberg (1836–1902): Norwegian physical chemist; Oslo, Norway.
† Peder Waage (Wåge in Norwegian) (1833–1900): Norwegian chemist; Oslo, Norway.

2. The *density of energetic states* $\mathcal{D}$ in the conduction band ($E > E_C$), determined by quantum considerations. The notion of energetic density is utilized here (and in physics in general) as the derivative of a variable with respect to the energy-per-corpuscle and $\mathcal{D}$ is therefore defined by

$$\mathcal{D} \overset{def}{=} \frac{d\mathcal{N}}{dE} \tag{C9.24}$$

3. The *function of distribution* $f_e(E - E_F)$ giving the proportion of electrons with respect to the maximum number of corpuscles in each energetic state, defined in Equation C9.18.

The product of the three terms is algebraically written for the electrons as follows:

$$dN_e = N_g\, \mathcal{D}(E - E_c)\, f_e(E - E_F)\, dE \tag{C9.25}$$

The total number of charge carriers is then the integral over all allowed energies-per-corpuscle of this energetic density of carriers (i.e., for $E > E_C$). The analytic calculus is not always possible (it depends on the mathematical form of the energy states density $\mathcal{D}$, which in turn depends on the number of space dimensions).

Physicists have recourse then to an approximation that consists of replacing Fermi–Dirac statistics by Boltzmann's statistics, in assuming that all energies-per-corpuscle are higher than the Fermi level by several times the quantity $k_B T$. In the Formal Graph language, it amounts to approximate the function of distribution f_e by the gain of the dipole g, which corresponds to the assumption of a constant number of holes equal to N^* (cf. Equation C9.17).

Statistics versus influence. In condensed matter physics, the capacitive relationships are classically considered under a statistical angle of view. Electrons (and their counterpart, the holes) are particles called *fermions*, that is, they obey the exclusion principle of Pauli* (not two identical fermions in the same energy state) and the probability for them to occupy a site, or an energy state, is relevant to Fermi–Dirac statistics.

The Formal Graph theory proceeds by adopting an approach based on the notion of *influence*, which does not require the notion of time or the notion of event. The unity aimed at implies that the same model applies also to purely static phenomena (electrostatics) or concerning immobile objects (elasticity of solids).

Not to have recourse to statistics does not mean that the Formal Graphs is a deterministic theory. Determined operators may apply to fluctuating variables or may obey a statistic distribution.

These important points are discussed in Section 6.5.

6.5 DIPOLE PROPERTIES

6.5.1 Poles–Dipole Relationship

Variables of a dipole are dependent on the variables of the two poles forming the dipole. An important rule governing these relationships is the restriction to relations within the same family of variables:

Construction Rule: Only variables belonging to the same canonical family are linked in a poles–dipole relation.

* Wolfgang Ernst Pauli (1900–1958): Austrian physicist; Vienna, Austria, Zürich, Switzerland, and Princeton, USA.

For instance, the only dipole variable that a pole effort may be related to is the dipole effort. This rule concerns only the relationship between the dipole and its poles, and does not apply to pole–pole relationships. Chapter 7 shows that interaction between poles (through mutual influence) consists of relating variables that are not in the same family.

Relations between poles and dipole variables are not identical for all families; they depend on the category of the variables, entity number or energy-per-entity, the existence of a conservation law, the separability, and the existence of an exchange.

A frequently encountered relation is the dependence of a dipole energy-per-entity (effort or flow) on the difference of the two corresponding energies-per-entity in the poles. These differences are written as follows:

$$e_{q12} \overset{def}{=} e_{q1} - e_{q2} \tag{6.11}$$

$$f_{q12} \overset{def}{=} f_{q1} - f_{q2} \tag{6.12}$$

Let us take the case of the relationship between efforts (the case of flows is similar and can be deduced from the present case). This relationship between dipole effort and effort difference is not as simple as one could expect. One must consider several aspects. The first one is the nature of the physical phenomenon requiring this link. Two phenomena are to be considered: exchange by efforts and conservation of basic quantities. The first one has been seen in the beginning of this chapter, and the second one will be tackled soon after, thus necessitating a brief introduction.

The second aspect concerns the conservation of basic quantities and, more precisely, the way a conservative dipole stores energy, which can be done by using symmetrically both poles. This behavior is quantified through a *separability factor* s_D playing the role of a proportionality factor in the relationship between efforts:

$$e_q \overset{def\rightarrow}{=} s_D e_{q12} = s_D (e_{q1} - e_{q2}) \tag{6.13}$$

This relation defines the *separability factor* s_D and not the dipole effort, which will be defined later in Section 6.6 where the signification of this separability factor will be detailed. A progressive understanding of this notion will come with some case studies in Section 6.4. For the moment it is sufficient to limit this discussion to the general case, which is the separable dipole, for which the separability factor is equal to ±1. Note that, by extension, the previous relation also applies to purely conductive dipoles (that do not store energy) in taking the same absolute value 1 for the separability factor.

$$\text{Separable dipole or pure conduction:} \qquad |s_D| = 1 \tag{6.14}$$

The sign of this factor depends on the third aspect, which is a matter of convention for expressing an effort difference in conjunction with a flow difference.

Two conventions are classically used in electric circuits for orienting voltage differences and currents (that can be sometimes thought of as differences between forward and backward currents, as in a diode), depending on the status of the dipole component: generator or receptor of energy. If e_q and f_q are the dipole energies-per-entity, in the generator convention, they correspond to the following differences

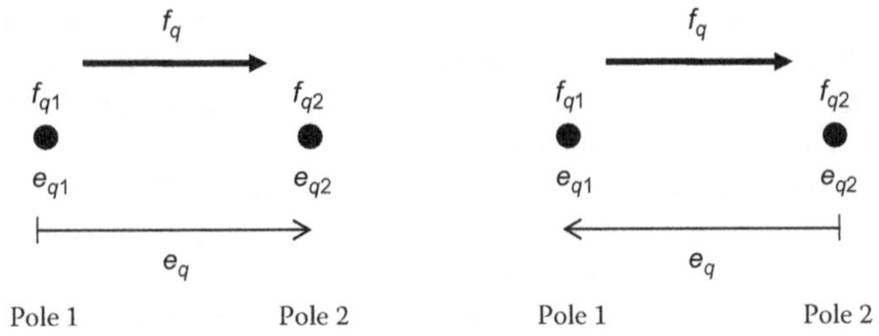

FIGURE 6.6 Symbolic representation of the two possible conventions for orienting dipole differences between pole variables. In the generator convention (left), dipole effort and flow have the same orientation, whereas in the receptor convention (right), they have opposite directions.

Generator (separable):

$$e_q \underset{gen}{=} e_{q21} = e_{q2} - e_{q1} \tag{6.15}$$

$$f_q = f_{q12} = f_{q1} - f_{q2} \tag{6.16}$$

in having assumed a separable dipole or pure conduction, that means $s_D = -1$. In the receptor convention, it is naturally the opposite ($s_D = +1$).

Receptor (separable):

$$e_q \underset{rec}{=} e_{q12} = e_{q1} - e_{q2} \tag{6.17}$$

$$f_q = f_{q12} = f_{q1} - f_{q2} \tag{6.18}$$

In fact, we have kept unchanged the dipole flow orientation and allowed the dipole effort to adapt to the chosen convention. These conventions are sketched in Figure 6.6, where it is clear that the traditional rule used for orienting a dipole flow (a net current is positive when it goes "down" from a higher value to a lower one) does not follow the traditional rule for an effort (positive when it "climbs" from the lower to the higher value). This discrepancy is a heritage that we follow due to its widespread use.

By default, the adopted convention is the *receptor* one; nevertheless, to respect the tradition in physical chemistry, the generator convention will be considered in this domain.

6.5.2 Exchange Reversibility

The unique case study on conduction, case study C4 "Pipe" in hydrodynamics, has introduced the notion of *exchange reversibility* in an exchange by flows. This notion is generalized through the definition of a new variable r_{q12} quantifying the reversibility of an exchange as the ratio of the backward flow f_{q2} on the forward one f_{q1}:

$$r_{q12} \overset{def}{=} \frac{f_{q2}}{f_{q1}} \tag{6.19}$$

A symmetrical definition using efforts instead of flows naturally works for the case of an exchange by efforts. As flows or efforts participating in an exchange (and attributed to the poles) are defined as positive quantities, the exchange reversibility is also a positive quantity. The interest of

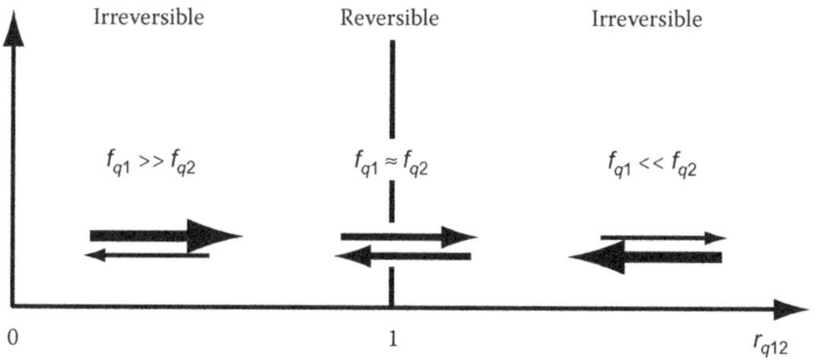

FIGURE 6.7 Graphical representation of the three limiting regimes of an exchange by flows as a function of the exchange reversibility variable.

this parameter is to handle a continuum of regimes in which the reversible or irreversible behavior of the exchange evolves gradually from one limiting regime to the next one when the exchange reversibility varies from zero to infinite. Three limiting regimes can be distinguished according to the values of this parameter:

$$\text{Irreversible} \quad \rightarrow \quad r_{q12} \rightarrow 0 \quad \Leftrightarrow \quad f_{q1} \gg f_{q2} \tag{6.20}$$

$$\text{Reversible} \quad \rightleftarrows \quad r_{q12} \approx 1 \quad \Leftrightarrow \quad f_{q1} \approx f_{q2} \quad \Leftrightarrow \quad f_q \approx 0 \tag{6.21}$$

$$\text{Irreversible} \quad \leftarrow \quad r_{q12} \rightarrow \infty \quad \Leftrightarrow \quad f_{q1} \ll f_{q2} \tag{6.22}$$

In Figure 6.7 is shown the sequence of regimes that can be met from irreversible in the forward direction to irreversible in the backward direction in going through the quasireversible regime around $r_{q12} = 1$. The dynamic equilibrium, defined by the equality between forward and backward flows, is reached exactly for $r_{q12} = 1$.

6.5.3 Standard Energy-per-Entity

In some case studies has been met a notion not yet discussed, the *standard energy-per-entity*, which is in fact characteristic of every dipole. Energy varieties in which this notion is used are mainly physical chemical and corpuscular energies.

It features a storing dipole by quantifying the energy-per-entity required for having equal amounts of entities in both poles. The dipole does not need to be conservative to use this notion. For a capacitive dipole, it is defined as the value taken by the effort when equality between basic quantities is realized (cf. Figure 6.8):

$$e_q = e_q^{\ominus} \Leftrightarrow q_1 = q_2 \tag{6.23}$$

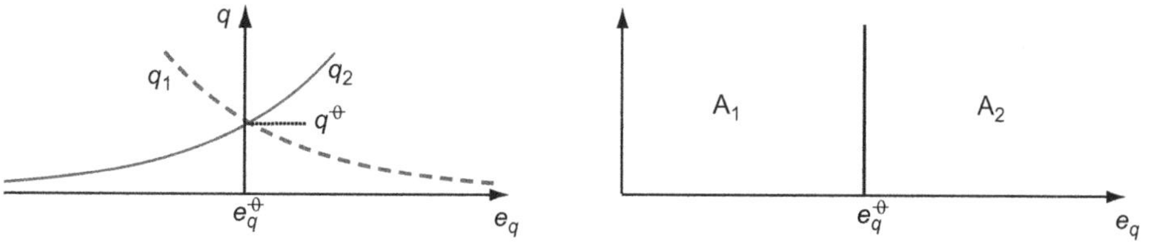

FIGURE 6.8 Plot of the two capacitances of a dipole with their intersection defining the standard effort (left) and preponderance or stability diagram (right).

Figure 6.8 shows a schematized interpretation of the plot called *preponderance diagram* or *stability diagram*. It indicates domains in which one pole is dominant over its partner, with the border placed at the standard effort. This kind of representation is mainly used in physical chemistry for delimiting areas in which a chemical species is present in majority.

Following the template of Equation 6.13 relating dipole effort to pole efforts, the relation between the standard efforts is written as

$$e_q^{\theta} \overset{def\rightarrow}{=} s_D e_{q12}^{\theta} = s_D\left(e_{q1}^{\theta} - e_{q2}^{\theta}\right) \tag{6.24}$$

Although not exemplified before, this notion is not reserved to capacitive dipoles; it also works for an inductive dipole. It is analogously defined as the value taken by the flow when equality between impulse amounts in each pole is realized:

$$f_q = f_q^{\theta} \Leftrightarrow p_{q1} = p_{q2} \tag{6.25}$$

The relation between standard flows is written as

$$f_q^{\theta} \overset{def\rightarrow}{=} s_D f_{q12}^{\theta} = s_D\left(f_{q1}^{\theta} - f_{q2}^{\theta}\right) \tag{6.26}$$

with a separability factor s_D, as for the capacitive case above.

In physical chemistry, the standard flow is called the standard chemical potential of the dipole, which can be an interface separating the same species in two media or a chemical reaction between two species. In this latter case, the name *standard molar free enthalpy of reaction* is also given. Also, in the case of coupling between physical chemistry and electrodynamics, the name *half-wave potential* is given to the value of the electric potential for which equal amounts of redox species are present on the charge transfer site. This naming comes from the sigmoid shape of the electrochemical current in response to a potential sweep in stationary regime, which follows the capacitive relationship plotted in the mentioned case studies.

In solid state physics, although the same sigmoid shape is met, under the form of the Fermi–Dirac distribution of electrons, the name *Fermi level* is given to the standard effort.

Naturally, this characteristic parameter works for any dipole, conservative or not, for all energy varieties and subvarieties. It physically represents a threshold parameter, meaning the predominance of one pole over the other above its value.

Combinations of several dipoles in multipoles (see Chapter 7) or in dipole distributions can be analyzed in terms of successive values of the standard efforts featuring each dipole.

6.5.4 Conservation and Separability

The *embedding principle*, enunciated in Section 6.2 of this chapter, requires that a dipole behaves in the same manner as a single pole in terms of energy variation described by a couple of state variables. First, its capacitive energy must be borne by a *dipole entity number* q, and second, deriving from the first, the *dipole effort* must be defined by the partial derivative of the total capacitive energy $\mathcal{U}$ of the system with respect to this *dipole basic quantity* q:

$$\text{Dipole effort: } e_{qC} \underset{def}{=} \left(\frac{\partial \mathcal{U}}{\partial q}\right)_{other\ q} \tag{6.27}$$

(The subscript "other q" being indicated for the sake of generality, when the system is composed of more than one dipole, their basic quantities must be held constant.) The capacitive energy variation can therefore be written under the concise form which is similar to the one used for a single pole:

$$\mathrm{d}\mathcal{U}_q = e_{qC}\,\mathrm{d}q \tag{6.28}$$

Alternatively, the variation of the capacitive energy of the dipole can also be expressed as the sum of the energy variations of the two poles:

$$\mathrm{d}\mathcal{U}_q = e_{q1}\,\mathrm{d}q_1 + e_{q2}\,\mathrm{d}q_2 \tag{6.29}$$

A comparison between these equations gives

$$e_{q1}\,\mathrm{d}q_1 + e_{q2}\,\mathrm{d}q_2 = e_{qC}\,\mathrm{d}q \tag{6.30}$$

In assuming that none of the pole efforts are equal to zero, the existence of the conservation law in a dipole means that the *entity numbers* of the two poles considered together remains invariable

Conservative dipole: $$q_1 + q_2 = q_1^* + q_2^* \tag{6.31}$$

This conservation is written with the help of constant quantities q_1^* referring to the dissociated situation of poles (separated, off state). (They are constant quantities because they are unable to vary in a dissociated pole.) The variations of these entity numbers are consequently opposite

$$\mathrm{d}q_1 = -\mathrm{d}q_2 \tag{6.32}$$

allowing the relationship between efforts (Equation 6.30) to be written differently:

$$e_{qC}\,\mathrm{d}q = (e_{q1} - e_{q2})\,\mathrm{d}q_1 = -(e_{q1} - e_{q2})\,\mathrm{d}q_2 \tag{6.33}$$

This equality between products can be recast into an equality between ratios, allowing the definition of a *dipole separability factor* s_D as their common value:

$$s_D \overset{def}{=} \frac{e_{qC}}{e_{q1} - e_{q2}} = \frac{\mathrm{d}q_1}{\mathrm{d}q} = -\frac{\mathrm{d}q_2}{\mathrm{d}q} \tag{6.34}$$

This separability factor stands as a quantifier of the existence or not of the separability described above with the following values.

Separable: $$|s_D| = 1 \tag{6.35}$$

Inseparable: $$|s_D| = \frac{1}{2} \tag{6.36}$$

These values are somewhat arbitrary and they figure the simplest set of possible values for expressing the relationship between poles and their dipole. The fact that other sets can be used means that the present theory is not closed and restricted to the actual cases. With this separability factor, the relationship between variations of the basic quantities is written as

$$s_D\,\mathrm{d}q = \mathrm{d}q_1 = -\mathrm{d}q_2 \tag{6.37}$$

and, by integration, the *dipolar basic quantity*, multiplied by the separability factor, appears as the common *entity number difference*, taken in opposite order for the other pole, and from the equality 6.34 between ratios, the *dipolar effort* is proportional to the effort difference through the separability factor:

$$s_D q = q_1 - q_1^* = q_1^* - q_2 \tag{6.38}$$

$$e_{qC} = s_D (e_{q1} - e_{q2}) \tag{6.39}$$

The incorporation of the generator/receptor convention into the separability factor through its sign was introduced in Section 6.2.

Receptor: $$s_D > 0 \tag{6.40}$$

Generator: $$s_D < 0 \tag{6.41}$$

As the only domain using the generator convention is physical chemistry, in which no inseparability is found, negative values of the separability factors are seldom utilized. Unless specified, separability factors are used with positive values.

A conservative capacitive dipole is therefore featured by the preceding relationships between state variables that satisfy the embedding principle (Section 6.2) and the conservation law (Equation 6.31) in all cases of separability. When inseparability makes the two efforts opposed ($s_D = 1/2$), one has equality between absolute values of the efforts of the poles and of the dipole, as was the case for the forces in the previous case study of the mechanical spring.

Inseparable dipole: $$e_{qC} = e_{q1} = -e_{q2} \tag{6.42}$$

On the other hand, when separability prevails, $s_D = 1$ and the same relationship as in Equation 6.9 for the conductive dipole is obtained for the efforts.

Separable dipole: $$e_{qC} = e_{q1} - e_{q2} \tag{6.43}$$

6.5.5 Comparison between Dipoles

Through the various case studies given of dipoles, the question of the separability has been met in several cases. It is worthwhile to compare the main features of two categories of dipoles, separable and inseparable ones, as done in Table 6.3, which lists only capacitive dipoles

All these dipoles are conservative ones with respect to the entity number, the basic quantity in this case of capacitive dipoles. The interesting feature is that the separability is linked to the symmetry between energies-per-entity (here efforts): When the dipole is inseparable, both efforts are equal in magnitude but opposed in direction (for vectors) or value (for scalars). The converse is not true; such symmetry may be found in peculiar cases for separable dipoles. For instance, the two potential values V_1 and V_2 of a capacitor may be equal in magnitude and opposite; however, this happens only in case of equal pole capacitances. This is a frequent case in electrodynamics when the two capacitor plates are strictly identical, as in the case of planar capacitor, but this is not general as nonidentical shapes or geometries can also be found. Note that, in physical chemistry, this never happens, because it would correspond to identical partners in a chemical reaction of to identical phases in an interface!

TABLE 6.3
Comparison between Four Dipoles in Terms of Conservation, Symmetry, and Separability

	Physical Chemical Interface (or Reaction)	Electric Capacitor	Mechanical Spring	Electrons and Holes
Conservation of basic quantities	Yes	Yes	Yes	Yes
	$n_1 + n_2 = n_1^* + n_2^*$	$Q_1 + Q_2 = 0$	$\ell_1 + \ell_2 = 0$	$N_e + N_h = N_e + N_h^*$
Symmetry of basic quantities	No	Yes	Yes	No
	$n_1 \neq -n_2$	$Q_1 = -Q_2$	$\ell_1 = -\ell_2$	$N_e \neq -N_h$
Symmetry of efforts	No	No	Yes	Yes
	$\mu_1 \neq -\mu_2$	$V_1 \neq -V_2$	$F_1 = -F_2$	$E_e = -E_h$
Separability	Yes	Yes	No	No
	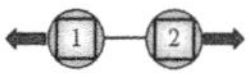	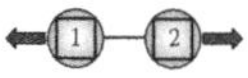		

The rule establishing the effort symmetry can be summarized as follows:

$$\textit{Inseparability} \Rightarrow \{\textit{Energy-per-entity } 1\} = -\{\textit{Energy-per-entity } 2\}$$

Physically speaking, it means that the poles are behaving in an exactly symmetrical way in storing energy; they share equally the total energy, each pole bearing the same amount, as demonstrated by the following sequence of equations.

Dipole energy variation: $$d\mathcal{U}_q = e_{q1}\, dq_1 + e_{q2}\, dq_2 = e_{qC}\, dq \quad (6.44)$$

Inseparable dipole: $$e_{qC} = e_{q1} = -e_{q2} \quad (6.45)$$

Inseparable dipole: $$dq = 2\, dq_1 = -2\, dq_2 \quad (6.46)$$

Inseparable dipole: $$d\mathcal{U}_{q1} = d\mathcal{U}_{q2} = \frac{1}{2} d\mathcal{U}_q \quad (6.47)$$

The impossibility to separate the poles results from the need for the poles to communicate through the capacitances with each other in order to equilibrate energy between them. Breaking this communication entails violation of the symmetry between efforts, thus making the dipole lose its properties.

Table 6.4 gives the correspondence between basic quantities featuring the two poles of a dipole for some capacitive systems belonging to various energy varieties.

Table 6.5 gives the correspondence between basic quantities and efforts for the dipoles.

6.5.6 Spontaneous Direction

In the short case study of the spontaneous evolution of a capacitor discussed in Section 6.2.5 it was admitted without argumentation that a dipole charged with energy could lose it freely in an external circuit. This point merits a more thorough examination now that we have a deeper knowledge of dipole properties.

To provide energy to an external circuit or another system, the dipole must behave as a generator, which is expressed by the condition imposed on the variation of energy to be negative during the process as shown:

$$d\mathcal{T}_q < 0 \quad \text{or} \quad d\mathcal{U}_q < 0 \quad (6.48)$$

For varying the energy content, let us take a process consisting of transferring entities (through external circulation or by exchange in a mixed dipole) from pole 1 to pole 2 in a conservative dipole.

TABLE 6.4
Correspondence between Basic Quantities of the Poles of a Dipole for Various Systems

Pole Basic Quantities	Chemical Reaction (i = A, B) Interface (i = 1, 2)	Electric Capacitor (i = 1, 2)	Mechanical Spring (i = 1, 2)	Electrons and Holes (i = e, h)
q_i	n_i	Q_i	ℓ_i	N_i
q_i^0	n_i	0	ℓ_i^0	N_i^*
q_i^{min}	0	Q_i^{min}	ℓ_i^{min}	0
$q_i^{\ominus}$	$n^{\ominus}$	0	ℓ_i^0	N_i^0 (*intrinsic*)

TABLE 6.5
Correspondence between Basic Quantities and Efforts of a Dipole for Various Systems

Dipole Variables	Chemical Reaction (or Interface)	Electric Capacitor	Mechanical Spring	Electrons and Holes
s_D	1	1	1/2	1/2
Q	$-n$	Q	ℓ	N
q^0	0 ($\neq n^{mid}$)	0	0	0 ($\neq N^{mid}$)
q^{min}	$-n_A^*$	$Q^{min} = Q_1^{min}$	$\ell^{min} = -2(\ell^0 - \ell_1^{min})$	$-2\,N_e^*$
q^{max}	n_B^*	$Q^{max} = -Q_2^{min}$	$\ell^{max} = 2(\ell_2^0 - \ell_2^{min})$	$2\,N_h^*$
q^{mid}	$n^{mid} = (n_B^* - n_A^*)/2$	$0 = (Q^{max} + Q^{min})/2$	$0 = (\ell^{max} + \ell^{min})/2$	$N^{mid} = N_h^* - N_e^*$
e_{qC}	μ	V	**F**	E
$e_{qC}^{\ominus}$	$\mu^{\ominus}$	0	0	E_F

Note: Subscripted variables (by 1 and 2 or A and B or *e* and *h*) correspond to poles and nonsubscripted ones to dipoles.

The variation of the number of entities must therefore be positive in pole 2 and negative in pole 1. The conservation of entities ensures that the variations of the numbers of entities have a sum equal to zero. Consequently, from Equations 6.1and 6.2 giving the variations of energy, this condition becomes:

$$\text{Condition for } 1 \rightarrow 2: \qquad d\mathcal{U}_q = (e_{q1} - e_{q2})\, dq_1 < 0 \Leftrightarrow e_{q2} < e_{q1} \tag{6.49}$$

$$\text{Condition for } 1 \rightarrow 2: \qquad d\mathcal{T}_q = (f_{q1} - f_{q2})\, dp_{q1} < 0 \Leftrightarrow f_{q2} < f_{q1} \tag{6.50}$$

For instance, a chemical reaction A → B requires a positive variation of the substance amount of species B for occurring in the direction of its formation, which implies that the chemical potential of the product of the reaction must be less than that of the initial species.

$$\text{Condition for A} \rightarrow \text{B}: \qquad d\mathcal{U}_n = (\mu_B - \mu_A)\, dn_B < 0 \Leftrightarrow \mu_B < \mu_A \tag{6.51}$$

(This example shows that the chosen convention between receptor and generator has no consequence on this result.)

Figure 6.9 graphically represents, under the form of energy-per-entity levels, the condition that the pole efforts must fulfill for allowing the process 1 → 2 to occur spontaneously in a capacitive dipole. (The same scheme works for an inductive dipole.)

This energy-per-entity level representation *versus* the pole numbers i, called *system coordinate* or *reaction coordinate*, is a common one in many domains but often presented in terms of "energy levels."

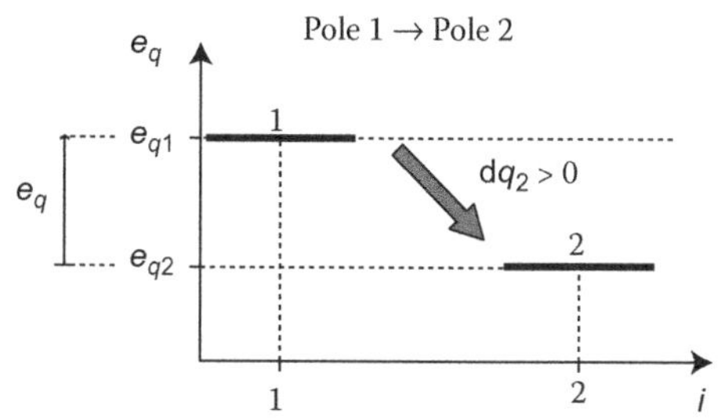

FIGURE 6.9 A capacitive dipole can undertake freely an entity transfer from pole 1 to pole 2 only if the effort of pole 2 is lower than the effort of pole 1. The horizontal axis uses a "pole coordinate" numbering the poles.

6.5.7 Definition of a Dipole

A dipole is one of the Formal Objects in the Formal Graph theory. It has several properties:

- A dipole belongs to one energy subvariety (inductive or capacitive).
- A dipole can be space distributed.
- A dipole cannot exchange or convert energy alone. It needs to be connected to another dipole.
- Five types of dipoles are defined depending on the category of the subvariety they are able to store or depending on their possibility to dissipate energy.
  The three fundamental dipoles are:
  - The *inductive dipole*, able to store the inductive subvariety of energy (kinetic, electromagnetic, etc.). It has the possibility of internal evolution.
  - The *capacitive dipole*, able to store the capacitive subvariety of energy (internal, potential, etc.). It has the possibility of internal evolution.
  - The *conductive dipole*, able only to convert energy into heat (by dissipation).
  - The two mixed types complete the fundamental types:
  - The *inductive–conductive dipole*, able to store inductive energy and to dissipate it.
  - The *capacitive–conductive dipole*, able to store capacitive energy and to dissipate it.
- A dipole is featured by two gate variables (energies-per-entity) for allowing external connections, and, in case of storing dipole, is complemented by a state variable that is an entity number.
- In a storing dipole, *conservation* of entity numbers may apply. Such a dipole is said to be conservative.
- Poles can be *separable* or not. Inseparability means that the two poles allow themselves to take opposite values of energy-per-entity, sharing exactly the same energy amount in case of storing dipole.

6.6 COMMON FEATURES RESULT IN NEW IDEAS

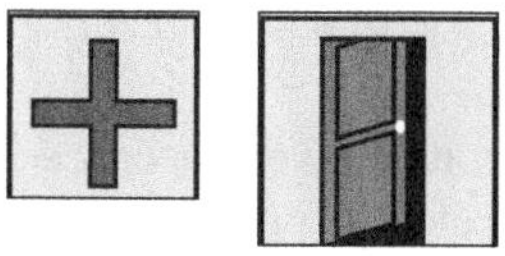

This section develops some interesting consequences of the existence of the same model for capacitive dipoles throughout all energy varieties. This is clearly a new insight into fundamental subjects that livens up the traditional way of thinking. It can be easily skipped by nonphysicists.

The various case studies given cover a wide span of scientific domains, from mechanics and electrodynamics to semiconductors and physical chemistry.

Exactly the same structures of Formal Graphs and the same shapes of operators have been used. This identity of operator shapes must be understood as the same mathematical functions but having eventually different parameters (scaling and origin variables).

Consequences are enumerated and discussed as follows.

6.6.1 Linearity is Limited to a Validity Domain

The general shape of the capacitance operator is the exponential function. However, in a narrow region limited approximately by twice the scaling effort $e_{q\Delta}$ used for the argument, this function can be assimilated to a straight line around the origin of its argument. Roughly, two domains can be distinguished:

$$\left|e_q\right| < 2\ e_{q\Delta} \Leftrightarrow \text{Linear capacitance} \tag{6.52}$$

$$\left|e_q\right| \geq 2\ e_{q\Delta} \Leftrightarrow \text{Nonlinear capacitance} \tag{6.53}$$

Using a linear operator presents many advantages, summarized by its definition:

$$\hat{O}(\lambda x) = \lambda\ \hat{O}(x) \tag{6.54}$$

$$\hat{O}(x + y) = \hat{O}(x) + \hat{O}(y) \tag{6.55}$$

The last relation expresses the *principle of superposition*, due to Duhamel,* stating that the response of a system to a sum of perturbations $x + y$ is equal to the sum of the responses to individual and separated perturbations x and y. The summation being a commutative operation, it also states that, despite the order of imposition of the perturbations to the system, the response will be the same.

The problem is that this superposition principle is the basis of many theories, for instance electromagnetic and gravitational field theories. The string theory has a large part of its elemental assumptions based on it too. One of the most used properties of the wave function in quantum mechanics is precisely its linearity.

This means that all these theories are only valid in a restricted region, in the same manner as the Newtonian gravitation is valid in a region limited by velocities lower than, roughly, one-tenth of the maximum velocity c.

Einstein's theory of spacetime (relativity) is a nonlinear theory (cf. case study A1 "Moving Body," on the mechanical inductive pole, in Chapter 4). One of the reasons, among others, of the difficulty for physicists to find unification between quantum mechanics and relativity lies certainly in this problem of linearity.

6.6.2 Constancy of Constitutive Properties is Not the Rule

In some domains, electrodynamics and gravitation mainly, the system constitutive properties, are assumed to be constant scalars. The capacitance C of a capacitor and the electric permittivity ε are basically constants in vacuum, the same for the gravitational constant G. Note that this latter is also

* Jean Marie Constant Duhamel (1797–1872): French mathematician, France.

a constant in Einstein's theory of gravitation (general relativity) (Einstein 1961). They are considered as foundations of their respective domain.

In some other domains, no constancy is assumed, as in mechanics with Hooke's law where the elasticity or rigidity strongly depends on the material. In addition, one of the main reasons for this variability is that the general form of a capacitance is a nonlinear operator, as is the case in hydrodynamics with the compressibility relating a relative volume to the pressure, and evidently in physical chemistry with the dependence of the chemical potential with the substance amount.

This means that the general rule is that capacitances are not constants, the exception being for a restricted domain and a limited class of materials. Such exceptions are met in electrodynamics where nonlinear permittivities are recognized for some materials (Jonscher 1996), but not yet in gravitation theories. Note that this last assertion must be tempered by the attempts of some astrophysicist to remove the constancy of the gravitational constant in order to decrease or suppress the recourse to dark matter for modeling the composition of the universe (Magueijo 2003; Moffat 2008).

6.6.3 Conservation, Statistics, and Fermions

Physicists give the name *fermion* to every particle that behaves according to Fermi–Dirac statistics. This theory applies to systems possessing sites (also called energy states) that can be occupied by objects or left void (*site* does not necessarily mean spatial localization). The Fermi–Dirac statistics models the distribution of objects as a function of their "energy" (read *energy-per-entity*) based on the following assumptions:

- All objects are identical.
- Objects are present in large numbers (as for Boltzmann's statistics).
- Each object occupies one *energy state* or *site*.
- No two objects can occupy the same *energy state* or *site*.

The last assumption is known as the *Pauli exclusion principle*. It features the behavior of all fermions, including electrons, protons, neutrons, and neutrinos. In quantum physics, these particles are endowed with a half-integer spin, while particles that do not obey the Pauli exclusion principle, and therefore Fermi–Dirac statistics, have an integer spin. They are called bosons because they follow Bose*–Einstein statistics instead. Examples of bosons are photons and alpha particles.

The Formal Graph theory models the dependence of a population of electrons in a material as a function of their energy-per-corpuscle in a completely different way, and the result is exactly the same as with Fermi–Dirac statistics.

Coincidence? Artifact somewhere? This merits analysis. The assumptions made in the Formal Graph approach is that electrons are not really alone, in the sense that they have for partners the holes, defined as vacancies created in the material by absence of electrons. Holes and electrons are considered as poles, because they form two collections of corpuscles, each one featured by its own energy-per-corpuscle. This forms a dipole by the fact that both populations coexist in the same material (without annihilation). Then, it is assumed that a *conservation* of the total number of electrons and holes prevails.

The key point is the conservation of the total number of entities for a pair of objects. It translates Pauli's exclusion principle in the following manner: The constancy of the total number of objects can be viewed as a fixed number of available sites or allowed energy states. A site (or energy state) can contain one electron or one hole but not both (no electron–hole pairs and no annihilation), because, if we could put more than one electron or hole on one site, the conservation would not be respected. This relies on the assumption that the conservation must

* Satyendra Nath Bose (1894–1974): Indian physicist; Calcutta, India.

be enforced site by site and not as an average between several sites containing more than one electron or hole.

Pair conservation: 1 site = 1 energy state = 1 electron or 1 hole

This is good news for physics. The peculiarity that represents the Pauli exclusion principle working only in a restricted (although quite wide) area of physics is removed by substituting a more general rule of conservation for a pair. It means, and the large number of examples given confirms it, that the same rule applies throughout physics and chemistry without frontiers and particularisms. It is not exaggerated to see in the notion of conservation one of the most important guidelines in science, with the notion of symmetry.

Taking advantage of the generalization made, it is worth making an interesting comparison with a purely macroscopic case, such as the spring, as shown in one of the case studies of inseparable dipoles. What we learn from bringing together this system with electrons and holes is that each spring end obeys the Pauli exclusion principle too: The conservation rule applies here to elongations ℓ_1 and ℓ_2 of the spring end nr 1 and nr 2, respectively. For comparison, let us suppose first that lengths are discretized. The Pauli exclusion principle in this case means that one end can only be located at one elongation value and cannot spread over two values! Or, alternatively:

Pair conservation: 1 elongation value = 1 energy state = 1 end nr 1 or 1 end nr 2

It is worth remarking that discretization is not mandatory; this equivalence also works in the continuous case, which is especially enlightening when returning to electrons that do not need to be discretized particles to obey the Pauli exclusion principle. This may help to understand better the notion of quantum object that may appear under several forms.

To use the equivalence between Pauli's exclusion principle and the conservation rule, a partner, a counter-object, is associated with the considered object, for making up a dipole. The electron cannot be thought to be independent of the hole in a hosting material. This is a feature of the association material–object to enforce conservation or not.

The complete picture about equivalences is achieved by asserting that objects that do not follow the Pauli exclusion principle do not form conservative dipoles. The Second Principle of Thermodynamics, in which entropy can only stay constant or increase during a transformation, can be translated to state that a thermal dipole is not conservative or that entropy behaves as bosons.

Beyond the wide extent in generalization to many domains, the Formal Graph approach presents the non-negligible advantage of removing the requirement of a large number of objects for using statistical models, as poles and dipoles are not subject to this limitation. It explains why a "Fermi–Dirac behavior" (and not statistics) can be observed with a few electrons.

Nothing has been said about the method by which the Formal Graph approach retrieves the same results as those produced by the statistical approach. This will be tackled in Chapter 7 with the subject of influence between poles.

6.7 IN SHORT

DIPOLE DEFINITION

1. A dipole is the association of *two poles* of the same type in the same energy variety:
    - Inductive association: two inductive poles of the same energy subvariety.
    - Capacitive association: two capacitive poles of the same energy subvariety.
    - Dissipative association: two conductive poles of the same energy variety.
    - Mixed associations: two inductive–conductive or capacitive–conductive poles.

(*continued*)

2. A dipole can be connected to other dipoles, whatever the energy subvariety or variety. A dipole, or multipole, association belongs to the following higher organization level of *dipole assemblies* (above multipoles).
3. A dipole is able to *store* and/or to *dissipate* energy. To allow energy to vary or circulate, a dipole must be connected to another dipole.
4. *Energy conversion* is possible between two poles of different energy subvarieties or varieties.

DIPOLE VARIABLES

Dipole variables are state variables at the global level (no space distribution). Nature and number of state variables depend on the ability to store and/or to dissipate energy.

1. Capacitive subvariety *storage*:
   - Basic quantity q as *entity number*
   - Effort e_q as *energy-per-entity*
   - Flow f_q as *gate variable* (inductive *energy-per-entity)*
2. Inductive subvariety *storage*
   - Impulse p_q as *entity number*
   - Flow f_q as *energy-per-entity*
   - Effort e_q as *gate variable* (capacitive *energy-per-entity)*
3. Both subvarieties *dissipation*
   - Flow f_q as *gate variable* (inductive *energy-per-entity)*
   - Effort p_q as *gate variable* (capacitive *energy-per-entity)*

FORMAL GRAPHS

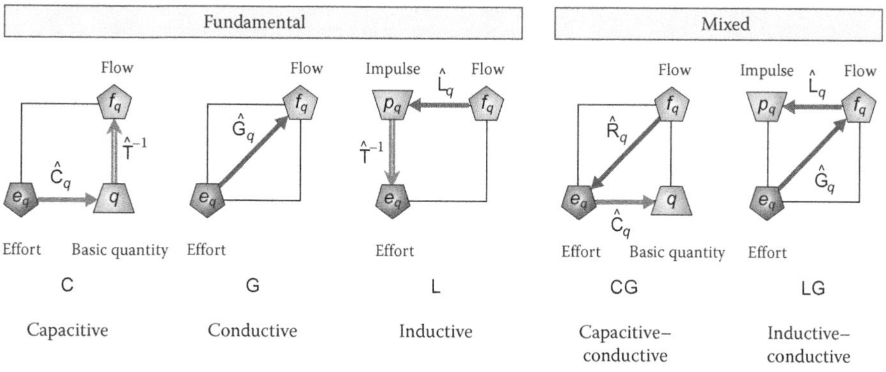

GRAPH 6.26

DIPOLE PROPERTIES

1. *Nature*
   - Three fundamental dipoles: Inductive (L), Capacitive (C), Conductive (G)
   - Two mixed dipoles: Inductive–Conductive (L–G), Capacitive–Conductive (C–G)
2. *Interaction*
   - *Exchange of entities*
   - *Influence*: Self-influence and/or mutual influence
3. *Separability*: Quantified through a *separability factor* s_D
   - $s_D = \pm 1 \Leftrightarrow$ Separable poles
   - $s_D = \pm 1/2 \Leftrightarrow$ Inseparable poles

   The sign depends on the generator (–)/receptor (+) convention chosen
4. *Conservation*
   - *Inductive dipole*: Conservation of impulses p_q
   - *Capacitive dipole*: Conservation of basic quantities q
5. *Evolution*: Existence in capacitive and inductive fundamental dipoles of a space–time property using time for energy conversion
6. *Space*: A dipole can be space distributed
7. *Reversibility* of exchange in a conductive or mixed dipole
8. *Standard effort/flow*: Corresponds to equal number of entities in both poles (inductive, capacitive, or mixed dipole)

STANDARD ENERGY PER ENTITY

Inductive dipole	$f_q = f_q^{\ominus} \Leftrightarrow p_{q1} = p_{q2}$
Capacitive dipole	$e_q = e_q^{\ominus} \Leftrightarrow q_1 = q_2$

CONNECTIONS BETWEEN ENERGIES-PER-ENTITY IN CONSTRUCTION FORMAL GRAPHS

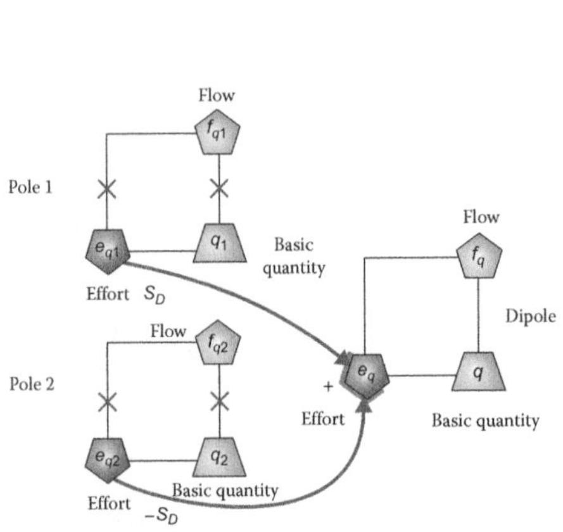

$$e_q = s_D e_{q12} = s_D(e_{q1} - e_{q2})$$

GRAPH 6.27 Connections between efforts.

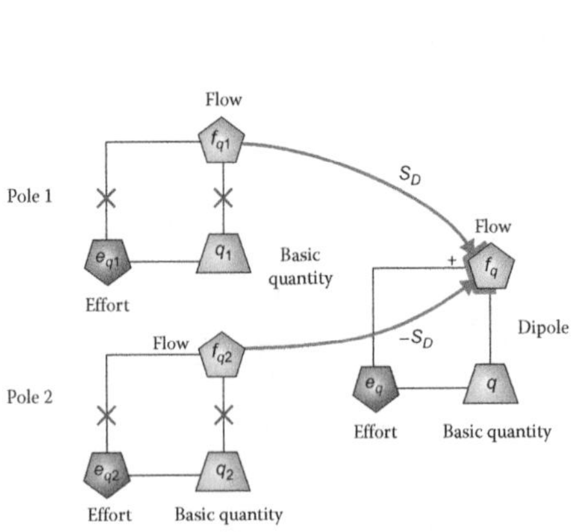

$$f_q = s_D f_{q12} = s_D(f_{q1} - f_{q2})$$

GRAPH 6.28 Connections between flows.

EXCHANGE OF ENTITIES

1. Exchange of *impulses* p_q (content) with efforts e_q (exchange vectors):

$$e_q = e_{q1} - e_{q2} \tag{6.9}$$

Exchange reversibility: $$r_{q12} \overset{def}{=} \frac{e_{q2}}{e_{q1}}$$

2. Exchange of *basic quantities* q (content) with flows f_q (exchange vectors):

$$f_q = f_{q1} - f_{q2} \tag{6.10}$$

Exchange reversibility: $$r_{q12} \overset{def}{=} \frac{f_{q2}}{f_{q1}} \tag{6.19}$$

EXCHANGE REPRESENTATIONS IN CONSTRUCTION FORMAL GRAPHS

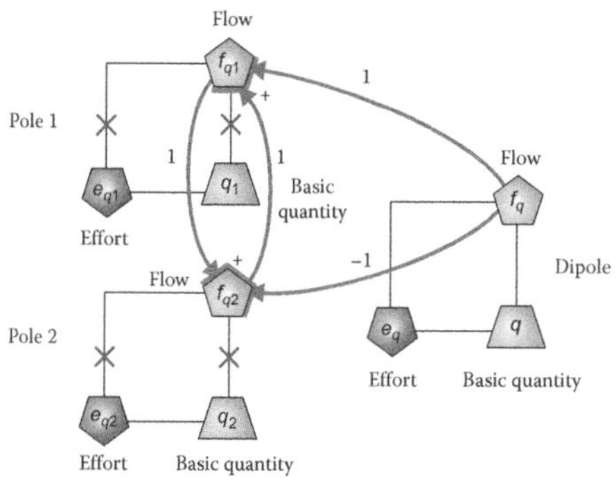

$f_q = f_{q1} - f_{q2}$

GRAPH 6.29 Exchange of basic quantities by flows.

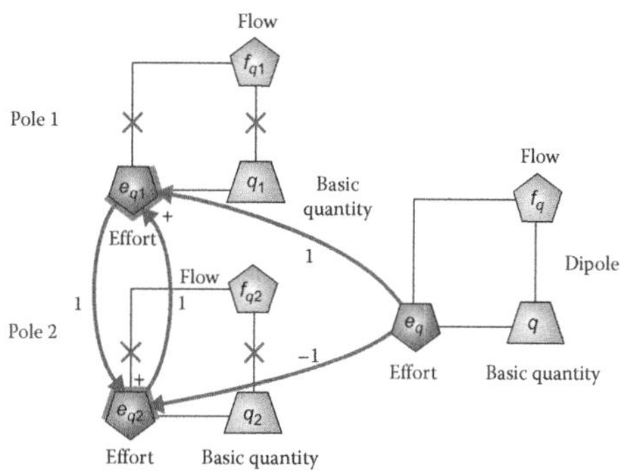

$e_q = e_{q1} - e_{q2}$

GRAPH 6.30 Exchange of impulses by efforts.

CONNECTIONS BETWEEN ENTITY NUMBERS IN CONSTRUCTION FORMAL GRAPHS (CONSERVATIVE DIPOLES)

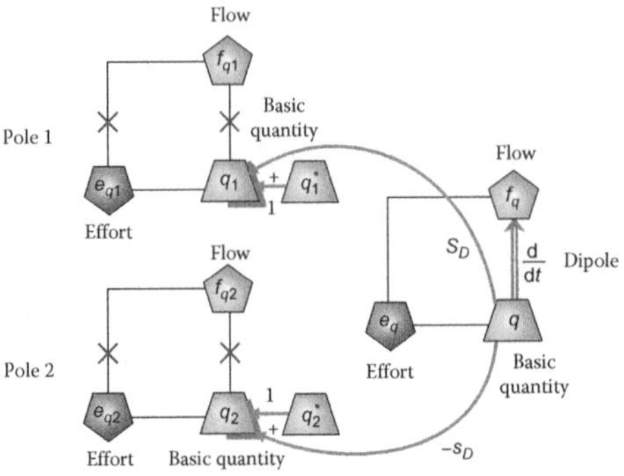

$$s_D\, q = q_1 - q_1^* = q_2^* - q_2$$

GRAPH 6.31 Connections between conserved basic quantities.

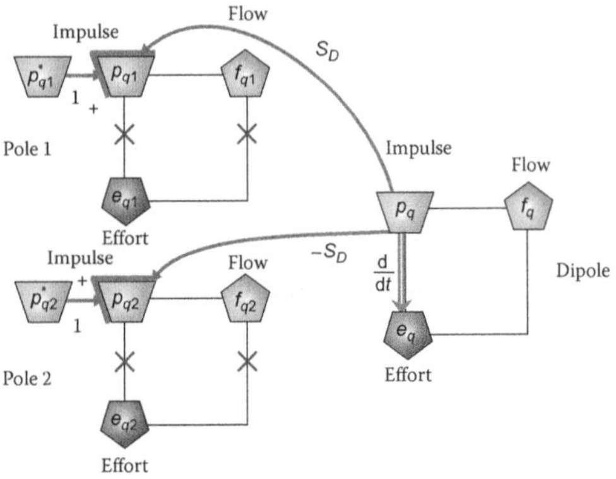

$$s_D\, p_q = p_{q1} - p_{q1}^* = p_{q2}^* - p_{q2}$$

GRAPH 6.32 Connections between conserved impulses.

7 Influence between Poles

CONTENTS

The phenomenon that makes the two poles of a dipole dependent on each other without exchanging entities but using system constitutive properties is called *influence*. This chapter first presents the main features of this phenomenon and then gives a series of case studies of dipoles. Most of these case studies illustrate the role of space in the influence through space-reduced properties. Finally, it will be demonstrated that the strong link between system constitutive properties and influence leads to the formulation of the mathematical operators representing the capacitance and the conductance.

A fundamental topic. With this subject, one enters into a very important question of physics dealing with the interdependency of elementary physical beings. The general concept widely in use has several names: communication or interaction or influence, without a clear definition of the physical processes involved. In the present approach, several aspects are distinguished.

- A first aspect is the *immaterial communication* between objects, defined as an exchange of information, through vectors that are not entities, such as waves (or fields without particle exchange). No particles having mass are involved in such a process.
- A second aspect is the *exchange* of entities, which may include particles, that has been studied in Chapter 6.
- A third aspect is the *influence*, which ensures adjustment between state variables of the objects for allowing the system to distribute its energy between them. Some links (here capacitance, inductance, or conductance) must exist to allow the state variables to communicate.

These two last aspects are grouped under the term of *interaction*, which is therefore a more general concept than influence.

Domains tackled in this chapter are: electromagnetism, electrostatics, magnetism, gravitation, and physical chemistry. This last domain is not classically considered as relevant to the concept of influence. However, it is a sound way to put on solid basis the concept of *activity* used for substances.

Table 7.1 lists the case studies of dipoles involving an influence phenomenon given in this chapter. (See also Figure 7.1 for the position of the dipoles along the complexity scale of Formal Objects.)

7.1 INTERACTION BETWEEN POLES

Interactions between two poles is of two kinds: *influence* and *exchange*. *Influence* occurs when the interaction relies on the system constitutive properties (capacitance, inductance, or conductance) that allow the system to store or dissipate energy. *Exchange* is the means for a dipole to ensure a circulation of entities (basic quantities or impulses) between poles, independent of the system constitutive properties. In the case of a conductive dipole, which does not possess its own entities, they must be provided by at least one storing dipole associated to the conductive dipole. (In other words, conductive dipoles alone have zero values for their effort and flow.) Influence being a more complex phenomenon than exchange, it requires a longer discussion that will be developed at the end of this chapter in order to establish the generalized capacitive relationship (modeling the dependence of the basic quantity with the effort).

TABLE 7.1
List of Case Studies of Dipoles That Implement Influence

	Name	Energy Variety	Type	Localization	Page Number
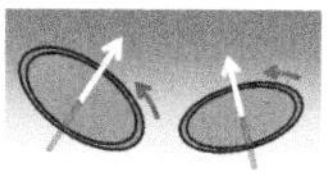	D1: Electromagnetic Influence	Electrodynamics	Inductive	Global	208
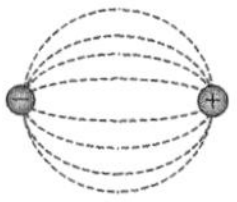	D2: Electrostatic Influence	Electrodynamics	Capacitive	Global	211
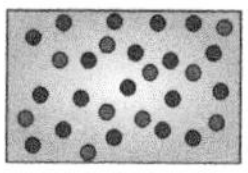	D3: Physical Chemical Influence	Physical chemistry	Capacitive	Global	215
	D4: Coulomb's Law	Electrodynamics	Capacitive	Three-dimensional space	218
	D5: Newtonian Gravitation	Gravitation	Capacitive	Three-dimensional space	224
	D6: Magnetic Interaction	Magnetism	Inductive	Three-dimensional space	230

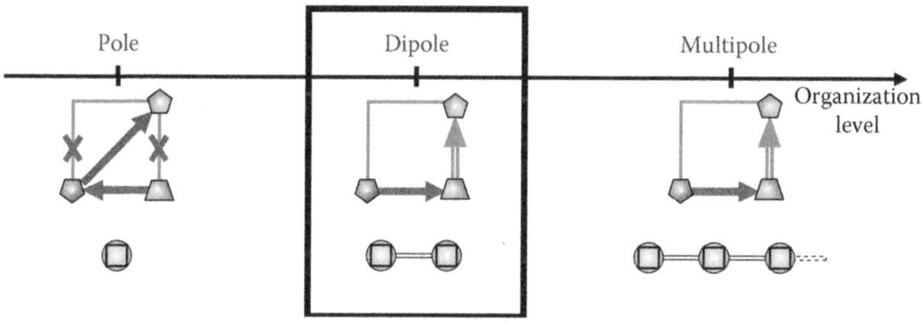

FIGURE 7.1 Position of the dipole along the complexity scale of Formal Objects.

7.1.1 Interaction through Influence

Influence describes a phenomenon in which a first system exerts an action on a second system. One of the most well-known examples is the influence of the Moon on the sea on Earth, which is mediated by the gravitational field existing between the two bodies. Other examples are a magnet able to attract magnetizable materials, the electromagnetic induction occurring between two wires in which current flows, or the electrostatic attraction or repulsion between two charged bodies. All these cases will be treated in detail in Section 7.5 devoted to dipoles using influence.

The Formal Graph theory helps to define more precisely this phenomenon by enunciating the following properties:

- Influence occurs only between poles (in any number, but not between dipoles).
- Poles involved in influence must belong to the same type (same energy subvariety, inductive or capacitive, or both conductive).
- Influence provides means for the system to store or dissipate energy. There is, consequently, an inductive, a capacitive, and a conductive influence.

- Influence requires the existence of system constitutive properties (inductances or capacitances or conductances) to be effective.
- *Self-influence* is supported by system constitutive properties that are internal to the poles.
- *Mutual influence* is supported by system constitutive properties that are external to the poles (featuring the medium in which they are bathing).
- In the general case, both self-influence and mutual influence may coexist in the same system. However, the most fundamental case corresponds to mutual influence.

Inductive influence is algebraically modeled by a system of two equations relating impulses p_{qi} to flows f_{qi}, which are written with the help of two self-reluctances (reluctance is the reciprocal of inductance) and two mutual reluctances (that are in fact identical, as will be shown later).

$$f_{q1} = \hat{L}^{-1}_{q11} p_{q1} + \hat{L}^{-1}_{q12} p_{q2} \tag{7.1}$$

$$f_{q2} = \hat{L}^{-1}_{q21} p_{q1} + \hat{L}^{-1}_{q22} p_{q2} \tag{7.2}$$

Capacitive influence is modeled on the same template, with elastances relating basic quantities q_i to efforts e_{qi}, which are reciprocal of capacitances.

$$e_{q1} = \hat{C}^{-1}_{q11} q_1 + \hat{C}^{-1}_{q12} q_2 \tag{7.3}$$

$$e_{q2} = \hat{C}^{-1}_{q21} q_1 + \hat{C}^{-1}_{q22} q_2 \tag{7.4}$$

It would have been possible to choose another way, by using inductances and capacitances instead of reluctances and elastances, to express entity numbers (impulses p_{qi} or basic quantities q_i) in function of energies-per-entity (flows f_{qi} or efforts e_{qi}) for the inductive and capacitive influences, but models are found to be more complex in that case, except when operators reduce to scalars, in which case both ways are equivalent and equally convenient. The Formal Graphs in Graph 7.1 translate these two systems of equations.

The common feature in both inductive and capacitive influences is that energies-per-entity are expressed in function of entity numbers. It should not be deduced that this means causality or irreversibility. Influence is a reversible phenomenon, based on a mutual adaptation of variables, without dedicated emitters and receivers, at least at the global level, without space–time consideration. When space–time plays a role, this adaptation may require propagation of a signal, but this is a subject that requires a better understanding of space–time.

Conductive influence follows the same scheme that is written as a system of two equations. A first writing is used when an exchange by flows ensures the conduction

$$f_{q1} = \hat{G}_{q11} e_{q1} + \hat{G}_{q12} e_{q2} \tag{7.5}$$

$$f_{q2} = \hat{G}_{q21} e_{q1} + \hat{G}_{q22} e_{q2} \tag{7.6}$$

whereas an exchange by efforts corresponds to the second writing, which is not equivalent to the first one unless linearity of operators is assumed.

$$e_{q1} = \hat{R}_{q11} f_{q1} + \hat{R}_{q12} f_{q2} \tag{7.7}$$

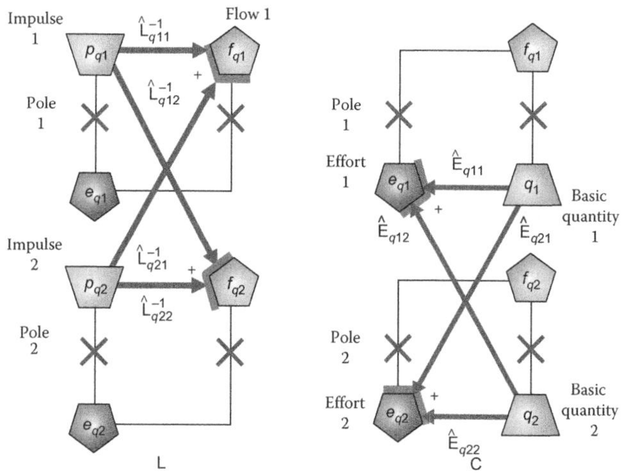

GRAPH 7.1 Formal Graph representation of the inductive influence between two poles, mixing self-influence and mutual influence (left), and of the similar cases of capacitive influence (right).

$$e_{q2} = \hat{R}_{q21} f_{q1} + \hat{R}_{q22} f_{q2} \tag{7.8}$$

The Formal Graphs in Graph 7.2 translate the two last systems of equations.

The above presentation of influence is voluntarily succinct, intended as an introduction providing a better grasp of the case studies that follow. A more detailed discussion will be undertaken in the example of *physical chemical influence*, which is not a classical concept but is introduced by the Formal Graph theory as a general frame encompassing all capacitive influence phenomena. After the case studies, the discussion will be deepened (Section 7.4) by demonstrating the exponential shape of the capacitive relationship that is proposed as a universal model.

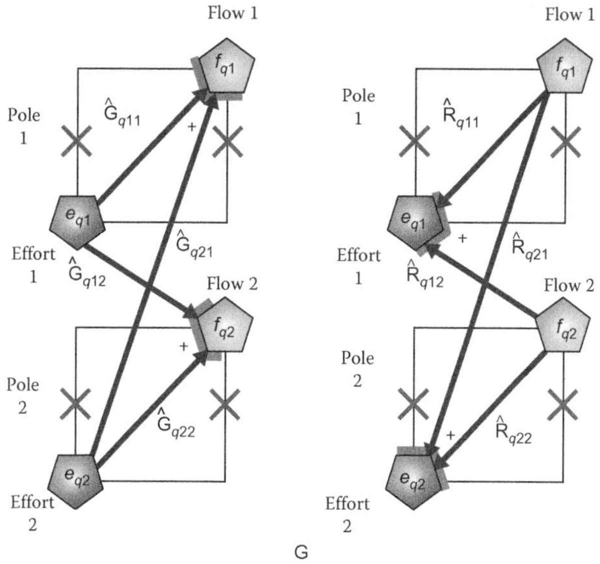

GRAPH 7.2 Formal Graph representation of the conductive influence between two poles, mixing self-influence and mutual influence in the case of exchange by flows (left) and in the case of exchange by efforts (right).

Fields and particles. In classical physics, electromagnetic or gravitational fields are vectors of mutual influence between charges or masses. The notion of force expresses the action of a field on these quantities.

In the modern conception of fields, influence is seen to result from an exchange of particle (*photon* for the electromagnetic field and *graviton* for the gravitational field, although this latter has never been observed) (Hecht 1996). This point of view is compatible with the Formal Graph representation, in the sense that influence can be represented as a combination of paths that includes, under certain conditions, the paths of an exchange of *entities* (not necessarily *particles*).

7.1.2 Space Distribution

Dipoles with space distribution are represented according to the same structure as space-distributed poles given as case studies in Chapter 6. However, the case studies using space distribution in this chapter have been chosen in a peculiar case corresponding to the simplified frame in which Coulomb's law and Newton's gravitation are established. In this frame, variables are assumed to be regularly distributed (homogeneous case) and surfaces are spherical ones, thereby leading to peculiar expressions for the space operators used, contragradient and surface integration:

$$-\frac{\partial}{\partial \mathbf{r}} e_q \underset{hom}{=} e_q \frac{\mathbf{r}}{r^2} \tag{7.9}$$

$$\oint_{sphere} \mathbf{q}_{/\mathbf{A}} . d\mathbf{A} \underset{hom}{=} \pm 4\pi |r| \mathbf{q}_{/\mathbf{A}} . \mathbf{r} \tag{7.10}$$

The sign for the surface integration depends on the orientation of the vector $\mathbf{q}_{/\mathbf{A}}$ with respect to the radius vector $\mathbf{r}$ (the subscript $/\mathbf{A}$ denotes a variable localized onto a surface). Note that this simplification is made only to retrieve from Formal Graphs the classical models mentioned. By restoring the differential operators used previously in keeping the same Formal Graph, one gets a much more general model than the ones issued from classical theories exposed in the case studies.

7.1.3 General Interaction

The two kinds of interaction, through exchange and through influence, are not exclusive of each other. They may coexist in superimposition, as, for instance, a capacitor or an inductor in electrodynamics, having a finite internal resistance, which means that a mutual influence and an interaction through an exchange by flows exist simultaneously. Such a dipole corresponds to the category of mixed dipoles, called capacitive–conductive or inductive–conductive.

Complete case studies with explicit inclusion of exchange and influence have not been given as the discussion would require longer explanations. In fact, a combination of Formal Graphs modeling these two aspects is straightforward and can be easily done without further explanation.

7.2 CASE STUDIES OF DIPOLES BASED ON INFLUENCE

The following case studies are representatives of the phenomenon of influence through the main domains in which it is acknowledged by the classical theory: electrodynamics and gravitation. The

Formal Graph theory extends it to many other domains by generalizing the classical model. The most interesting extension is to the physical chemical energy variety as it allows the concept of activity to be founded on a rigorous basis. The demonstration of the exponential shape of the capacitive relationship will rely on this extension.

Inductive global
- D1: Electromagnetic Influence in Electrodynamics

Capacitive global
- D2: Electrostatic Influence in Electrodynamics
- D3: Physical Chemical Influence in Physical Chemistry

Capacitive three-dimensional space
- D4: Coulomb's Law in Electrodynamics
- D5: Newtonian Gravitation in Gravitation

Inductive three-dimensional space
- D6: Magnetic Interaction in Magnetism

7.2.1 Case Study D1: Electromagnetic Influence — Electrodynamics

The case study abstract is given in next page.

In electromagnetism, a circuit with a current I flowing in it develops an electromagnetic induction field $\boldsymbol{B}$ around, which can influence another circuit in the vicinity. The current in this latter circuit can in turn influence the first circuit, meaning that the influence is mutual.

Both circuits contain some inductive (electromagnetic) energy, which means that they each possess a property of inductance called *self-inductance*, notated here L_{11} and L_{22}. The influence between the two circuits is translated into two *mutual inductances* L_{12} and L_{21} and one usually writes each induction quantity (flux) as the sum of each contribution, in assuming that all inductances are scalars:

$$\begin{aligned} \Phi_{B1} &\underset{lin}{=} L_{11}\, I_1 + L_{12}\, I_2 \\ \Phi_{B2} &\underset{lin}{=} L_{21}\, I_1 + L_{22}\, I_2 \end{aligned} \tag{D1.1}$$

Due to the assumed linearity, this system of two equations can be inverted to express each current as resulting from the contribution of two induction quantities (fluxes):

$$\begin{aligned} I_1 &\underset{lin}{=} \mathfrak{R}_{11}\, \Phi_{B1} + \mathfrak{R}_{12}\, \Phi_{B2} \\ I_2 &\underset{lin}{=} \mathfrak{R}_{21}\, \Phi_{B1} + \mathfrak{R}_{22}\, \Phi_{B2} \end{aligned} \tag{D1.2}$$

The inductive properties of the system used here are the *reluctances*, which are reciprocal of the inductances. Inverse scheme D1.2 is mathematically and physically (see Remark [1]) equivalent to scheme D1.1, but inverse scheme D1.2 is the only one retained for the representation in Formal Graph of two inductive poles influencing each other.

This choice is based on the rule used for a dipole assembly. In a mounting in series, the efforts are added to form the total effort (and the flow is common to both dipoles), whereas in parallel mounting it is the converse, flows are added as the effort is shared by both dipoles. In any case, the basic quantities or the impulses are involved in such an operation, so it is not very easy to justify physically an addition of induction quantities for modeling influence. The energies-per-entity are entitled in the Formal Graph theory to be formed from several contributions contrary to entity numbers. This rule applies equally to the poles.

D1: Electromagnetic Influence **Electrodynamics**

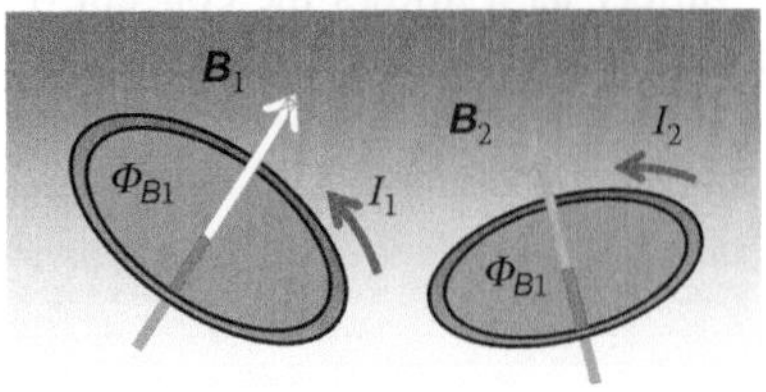

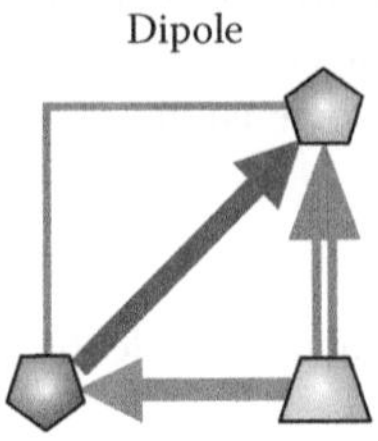

Dipole	
✓	Fundamental
	Mixed
	Capacitive
✓	Inductive
	Conductive
✓	Global
	Spatial
	Inseparable
✓	Conservative

Here, two loops having some current flowing influence each other through electromagnetic fields. Influence is a phenomenon that is extremely important, not only in electrodynamics, because it ensures the "communication" between poles. The presentation made in this case study is limited to the global level of system, which is a first step in the approach of the phenomenon of influence.

Formal Graph:

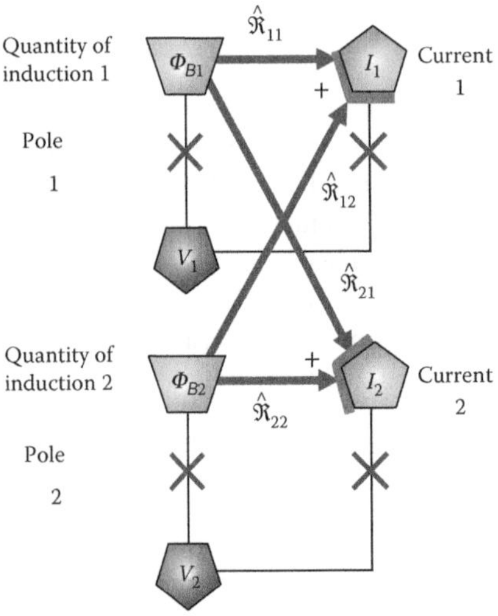

GRAPH 7.3

Differential Formal Graph:

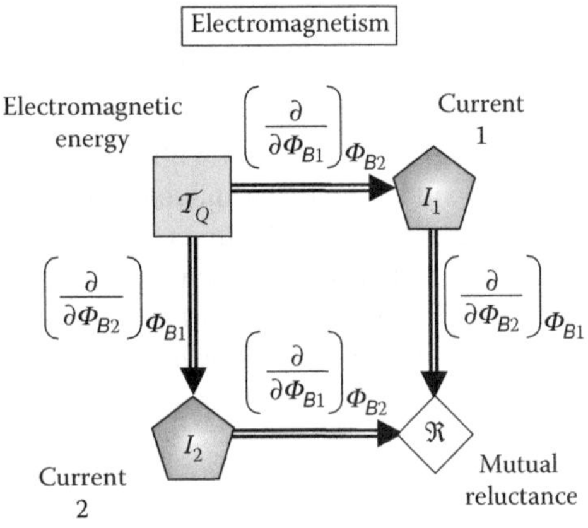

GRAPH 7.4

Variables:

Variety	Electrodynamics		
Subvariety	Inductive (Electromagnetic)		Capacitive
Category	Entity number	Energy/entity	Entity number
Family	Impulse	Flow	Effort
Name	Quantity of induction (flux)	Current	Potential
Symbols	Φ_B	I, i	V, U, E, φ
Unit	[Wb], [J A^{-1}]	[A]	[V], [J C^{-1}]

Property relationships:

$$I_1 = \hat{\mathfrak{R}}_{11}\,\Phi_{B1} + \hat{\mathfrak{R}}_{12}\,\Phi_{B2}$$

$$I_2 = \hat{\mathfrak{R}}_{21}\,\Phi_{B1} + \hat{\mathfrak{R}}_{22}\,\Phi_{B2}$$

$$\hat{\mathfrak{R}}_{12} = \hat{\mathfrak{R}}_{21} = \hat{\mathfrak{R}}$$

Classical electromagnetism

$$\mathfrak{R} = \frac{1}{L} \underset{lin}{\equiv} scalar$$

System Property	
Nature	Reluctance
Levels	0
Name	Reluctance
Symbol	$\mathfrak{R}$, L^{-1}
Unit	[H^{-1}]
Specificity	—

Energy variation:

$$d\mathcal{T}_Q \underset{\exists L}{=} I\,d\Phi_B$$

7.2.1.1 Equality of Mutual Elastances

The energy of a dipole is a state function; otherwise, the system could not store energy. The variation of this inductive (electromagnetic) energy is necessarily an exact differential

$$\mathrm{d}\mathcal{T}_Q = I_1\,\mathrm{d}\Phi_{B1} + I_2\,\mathrm{d}\Phi_{B2} \tag{D1.3}$$

the condition for having an exact differential being

$$\left(\frac{\partial I_1}{\partial \Phi_{B2}}\right)_{\Phi_{B1}} = \left(\frac{\partial I_2}{\partial \Phi_{B1}}\right)_{\Phi_{B2}} \tag{D1.4}$$

and, as the definition of mutual reluctances comes from the set of equations in scheme D1.2,

$$\Re_{12} \overset{def}{=} \left(\frac{\partial I_1}{\partial \Phi_{B2}}\right)_{\Phi_{B1}} \;;\; \Re_{21} \overset{def}{=} \left(\frac{\partial I_2}{\partial \Phi_{B1}}\right)_{\Phi_{B2}} \tag{D1.5}$$

equality of these mutual reluctances directly follows:

$$\Re_{12} = \Re_{21} = \Re \tag{D1.6}$$

This can be demonstrated in an entirely graphic way as shown by the *Differential Formal Graph* shown in the case study abstract.

There is effectively a "demonstration" by the fact that the graph is *closed*, that the paths converge at the same node, which is thereby the node of the common mutual reluctance.

Remarks

[1] Cause and effect are pointless. Influence phenomena are reversible (i.e., can be reverted): a current can create an induction flux (impulse) while going through another closed circuit and an induction flux (impulse) can provoke (induce) a current circulation in another closed circuit. No notions of cause and effect are involved in these phenomena, which do not imply any notion of time because the energy needs "simultaneously" the system properties, the entities, and their energy-per-entity in order to exist in the dipole. However, the question of the physical transmission from one pole to another may imply a propagation through space and time, which intrinsically depends on properties of space–time and not on properties of the system itself.

Interest of Differential Formal Graphs. A Differential Formal Graph is a Formal Graph of a different kind, as no system constitutive properties are used as links and nodes are used for various kinds of variables besides state variables. (A canonical Formal Graph is built only with state variables, or assimilated variables.)

The graph closure expresses the property of *commutativity of operators* that are involved. What is expressed by the *symmetry of graph* (same operator horizontally and vertically) is the independence of followed path (indifferent to the order of operators) for any variation of the state function representing the energy. There are many fundamental things in this simple Differential Formal Graph.

Transversality of the notion of influence. This phenomenon of influence is not limited to systems storing inductive (here electromagnetic) energy but exists equally on the capacitive side. In other Formal Graphs, the same crossed structures are found again for representing the influence links between basic quantities and efforts (see case study D2 "Electric Influence").

Electrodynamics is the domain by excellence for notions of poles and dipoles and their assemblies under various forms. The concepts of influence and action at a distance by means of a distribution in space of fields (electric displacements, induction, magnetization) are essential in this domain and are a basis for a large part of the physics of interactions.

The case study given here works at the global level of the system, without reference to space. This latter is included in Formal Graphs in two dimensions (system properties + properties of space) or in three dimensions (adding time). The system properties involved in the mutual influence are then reduced properties by space and they are specific to the supporting medium of the influence. As an example, one may refer to Coulomb's law in electrostatics and to Newtonian gravitation. These case studies do not involve inductive energy as here, but capacitive energy. With regard to the inductive subvariety, case study D6 of the interaction of magnetization (between poles of a magnet) helps to understand the difference between electromagnetic energy and energy of magnetization (or, if no confusion occurs, magnetic energy).

Interest of Differential Formal Graphs. The mutual influence is a phenomenon particularly interesting to model with Formal Graphs because the theory has a very interesting tool, the differential graph. It is a Formal Graph, but in contradistinction with a "canonical'" graph, where nodes are not all filled with state variables; one finds for instance energy or even some system properties. Notwithstanding, all the links are only differential operators with respect to two variables, generally, state variables.

The interest of a Differential Formal Graph is to translate in a particularly simple way the notion of state function and to demonstrate in an elegant way the necessary equality between crossed coefficients, here mutual elastances, or in other words, nondiagonal elements of matrices.

Some other very fundamental notions are equally handled by this tool, such as the notion of commutativity of operators and its link with the notions of graph symmetry and of graph closure or of circularity (paths in a loop).

7.2.2 Case Study D2: Electrostatic Influence — Electrodynamics

The case study abstract is given in next page.

In electrostatics, two electric charges can influence each other by participating in the electric potential of the other charge.

Normally, an isolated elementary charge does not influence itself, that is, it cannot create its own potential, but here we consider a pole composed of several elementary charges (a collection with an entity number that may vary), with eventually a mixing of signs, and in this case each pole influences itself and is influenced by another pole at a distance. As both poles contain some capacitive (electrostatic) energy, they each possess a property of capacitance, or, equivalently, of elastance (reciprocal of the capacitance) notated here E_{11} and E_{22}. The influence between the two poles is

D2: Electrostatic Influence **Electrodynamics**

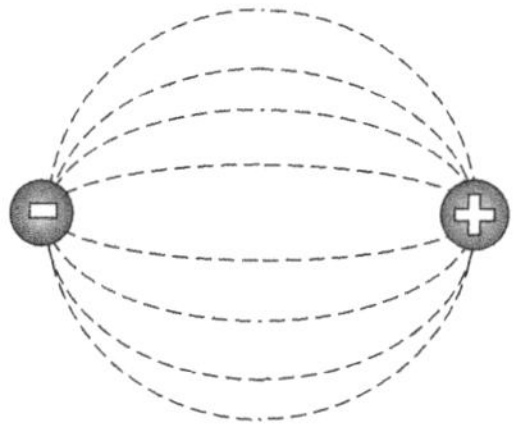

Here, two electrostatic poles influence the potential of the opposite pole through electric fields. Influence is a phenomenon that is extremely important, not only in electrodynamics, because it ensures the "communication" between poles. The presentation made in this case study is limited to the global level of system, which is a first step in the approach of the phenomenon of influence.

Dipole

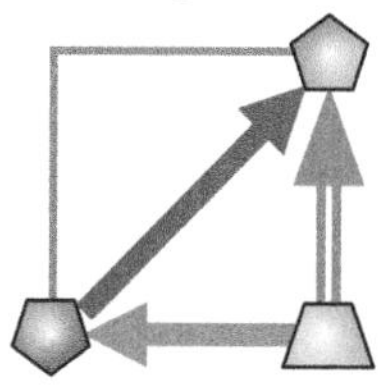

Dipole	
✓	Fundamental
	Mixed
✓	Capacitive
	Inductive
	Conductive
✓	Global
	Spatial
	Inseparable
✓	Conservative

Formal Graph:

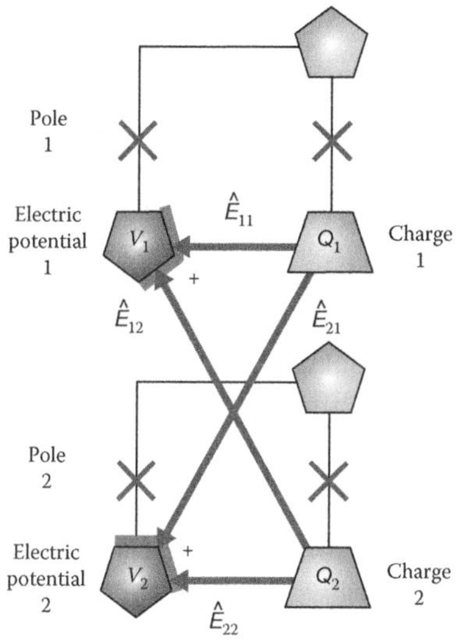

GRAPH 7.5

Differential Formal:

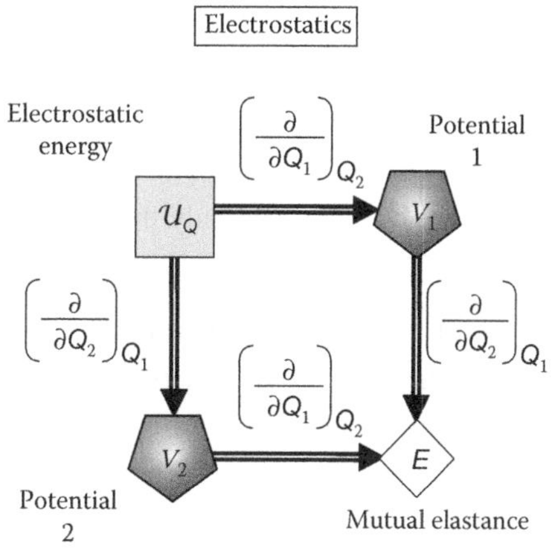

GRAPH 7.6

Variables:

Variety	Electrodynamics		
Subvariety	Capacitive (Electrostatic)		Inductive
Category	Entity number	Energy/ entity	Entity number
Family	Basic quantity	Effort	Flow
Name	Charge	Potential	Current
Symbols	Q	V, U, E, φ	I, i
Unit	[C], [A s]	[V], [J C^{-1}]	[A]

Property relationships:

$$V_1 = \hat{E}_{11}\, Q_1 + \hat{E}_{12}\, Q_2$$

$$V_2 = \hat{E}_{21}\, Q_1 + \hat{E}_{22}\, Q_2$$

$$\hat{E}_{12} = \hat{E}_{21} = \hat{E} = \hat{C}^{-1}$$

System Property	
Nature	Elastance
Levels	0
Name	Elastance
Symbol	E, C^{-1}
Unit	[F^{-1}], [V C^{-1}]
Specificity	—

Energy variation:

$$d\mathcal{U}_Q \underset{3C}{=} V\, dQ$$

translated into the existence of two *mutual elastances* E_{12} and E_{21} that are necessarily equal, as shown later. One writes each electric potential as the sum of each contribution, in assuming classically linear relationships:

$$\begin{aligned} V_1 &\underset{lin}{=} E_{11}\, Q_1 + E_{12}\, Q_2 \\ V_2 &\underset{lin}{=} E_{21}\, Q_1 + E_{22}\, Q_2 \end{aligned} \tag{D2.1}$$

Naturally, this system of two equations can be inverted to express each charge as resulting from the contribution of two electric potentials:

$$\begin{aligned} Q_1 &\underset{lin}{=} C_{11}\, V_1 + C_{12}\, V_2 \\ Q_2 &\underset{lin}{=} C_{21}\, V_1 + C_{22}\, V_2 \end{aligned} \tag{D2.2}$$

These two schemes D2.1 and D2.2, direct and inverse, are mathematically and physically (see Remark [1]) equivalent due to the linearity, but only scheme D2.1 is retained for the representation in Formal Graph of two capacitive poles which are influencing each other.

This choice is based on the rule used for assembling two dipoles. When the serial mounting is chosen, the efforts are added, and when flow circulates, it goes through the two dipoles (common flow). Conversely, for the parallel mounting, the effort is common to both dipoles and the flows are added for building up the total current. The basic quantities or the impulses are never combined in such an operation. It means that only the energies-per-entity (effort and flow) are entitled to result from several contributions contrary to entity numbers. This rule applies equally to the poles.

7.2.2.1 Equality of Mutual Elastances

The energy of a dipole is a state function; otherwise, the system could not store energy. The variation of this capacitive (electrostatic) energy is necessarily an exact differential

$$d\mathcal{U}_Q = V_1\, dQ_1 + V_2\, dQ_2 \tag{D2.3}$$

the condition here for having an exact differential being

$$\left(\frac{\partial V_1}{\partial Q_2}\right)_{Q_1} = \left(\frac{\partial V_2}{\partial Q_1}\right)_{Q_2} \tag{D2.4}$$

and, as the definition of mutual elastances comes from equations in scheme D2.1,

$$E_{12} \overset{def}{=} \left(\frac{\partial V_1}{\partial Q_2}\right)_{Q_1} \qquad E_{21} \overset{def}{=} \left(\frac{\partial V_2}{\partial Q_1}\right)_{Q_2} \tag{D2.5}$$

the equality of these mutual elastances directly follows:

$$E_{12} = E_{21} = E = C^{-1} \tag{D2.6}$$

This can be demonstrated in an entirely graphic way as shown by the *Differential Formal Graph* (which is not a canonical Formal Graph as explained in the remark below).

The demonstration is brought by the fact that the graph is *closed*, that is, the paths converge at the same node. This common node is the common mutual elastance.

Remarks

Cause and effect. The phenomenon of influence is reversible (in the sense of process direction): a charge can be at the origin of an electric potential in another pole and, conversely, an electric potential may make a charge appear in another pole. What has to be retained is that this is the conjunction of a charge and of an associated potential (sometimes this is enunciated as "seen" by the charge) which stores electrostatic energy. No notions of cause and effect are involved in these phenomena, which do not imply any notion of time because the energy needs "simultaneously" the system property, the entities, and their energy-per-entity to exist in the dipole. However, the question of the physical transmission from one pole to another may imply a propagation through space and time, which intrinsically depends on properties of space–time and not on properties of the system itself.

Closure and symmetry. A Differential Formal Graph is a Formal Graph of a different kind, as no system constitutive properties are used as links and nodes are used for various kinds of variables besides state variables. (A canonical Formal Graph is built only with state variables or assimilated variables.)

The graph closure expresses the property of *commutativity of operators* that are involved. What is expressed by the *symmetry of graph* (same operator horizontally and vertically) is the independence of followed path (indifferent to the order of operators) for any variation of the state function representing the energy. There are many fundamental things in this simple Differential Formal Graph.

Capacitive versus inductive. This phenomenon of influence is not limited to systems storing capacitive (here electrostatic) energy but exists equally on the inductive side. In other Formal Graphs, the same crossed structures are found again for representing the influence links between impulses and flows (see case study D1 "Electromagnetic Influence").

Electrodynamics is the domain by excellence for notions of poles and dipoles and of their assemblies under various forms. The concepts of influence and of action at a distance by means of a distribution in space of fields (electric displacements, induction, magnetization) are essential in this domain and are a basis for a large part of the physics of interactions.

This phenomenon will be generalized and used for modeling capacitive relationships in many other domains, as will be seen at the end of this chapter.

Extension to space. The case study given here works at the global level of the system, without reference to space. This latter is included in Formal Graphs in two dimensions (system properties + properties of space) or in three dimensions (adding time). The system properties involved in the mutual influence are then reduced properties by space and they are specific to the supporting medium of the influence. As an example, one may refer to Coulomb's law in electrostatics (case study D4) and to Newtonian gravitation (case study D5).

Properties of Differential Formal Graphs. The mutual influence is a phenomenon particularly interesting to model with Formal Graphs because this theory possesses a very interesting tool, the differential graph. It is a Formal Graph, but in contradistinction with a "canonical" graph, where nodes are not all filled with state variables; one finds for instance energy or even some system properties. Notwithstanding, all the links are only differential operators with respect to two variables, generally, state variables.

The interest of a differential Formal Graph is to translate in a particularly simple way the notion of *state function* and to demonstrate in an elegant way the necessary equality between crossed coefficients, here mutual reluctances, or nondiagonal elements of matrices.

Some other very fundamental notions are equally handled by this tool, such as the notion of *commutativity* of operators and its link with the notions of graph symmetry and of graph closure or of *circularity* (paths in a loop).

7.2.3 Case Study D3: Physical Chemical Influence — Physical Chemistry

The case study abstract is given in next page.

Influence is a phenomenon in which an action or a change is operated by an object on another one. Self-influence refers to an action from one part of an object on another part and mutual influence means a crossed action between two objects. This phenomenon is classically and mainly encountered in electrodynamics, in magnetism and in gravitation as illustrated by the case studies D1 on electromagnetic influence and D2 on electrostatic influence involving global poles (without space distribution). Mutual influence through space will be exemplified with Coulomb's law, magnetic interaction, and Newtonian gravitation (case studies D4 through D6).

An originality of the Formal Graph theory is to postulate the existence of this phenomenon in all other domains, both for mutual influence and for self-influence. One of the most striking domains is the physical chemical energy variety because it opens the way for establishing the capacitive relationship between a *substance amount* and its *chemical potential*. An interesting peculiarity of this domain is the use of the notion of *activity*, classically related to the *chemical potential* through an exponential function (see case study A6 "Chemical Species" in Chapter 4), which will be generalized to all other energy varieties.

The system studied here is of a dipole formed by mixing two chemical species in the same medium phase, each species i being featured by a *chemical potential* μ_i and a *substance amount* n_i. To simplify the demonstration, both species are assumed to be inert (no reaction between them), thus allowing one to model only the relationship between two poles without considering the whole dipole (that is treated in case study C5 "First-Order Chemical Reaction" in Chapter 6). Naturally, all that is developed in this frame will be also valid in the case of a reaction (dynamic regime).

Graph: The Formal Graph represents two poles, each one with its three state variables, plus a new variable called *activity*, inserted into the capacitance path in a separate node. This variable is in charge of the interaction between the poles in being connected to the *activity* of the other pole.

In this decomposed representation of a capacitance, this latter is equivalent to the combination of an "activation" operator and a "self-influence" operator that is algebraically written as:

$$\widehat{C}_{ni} = \hat{I}_{ii}\ \widehat{N}_i \quad \text{(D3.1)}$$

The double subscripts for the self-influence operator allow differentiating from mutual influence (that will distinguish two subscripts).

This decomposition means that it is not necessarily the entity number that is the pertinent variable for modeling an influence, because entities need to be activated for interacting with another

D3: Physical Chemical Influence

Physical Chemistry

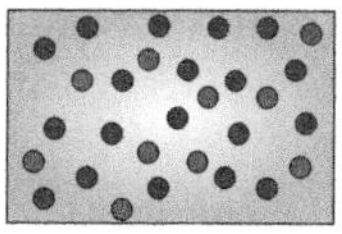

The system studied here is a dipole formed by mixing two chemical species in the same medium phase.

Dipole

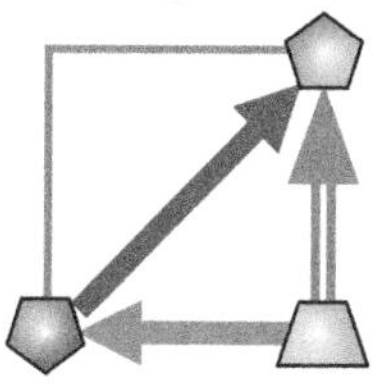

Dipole	
✓	Fundamental
	Mixed
✓	Capacitive
	Inductive
	Conductive
✓	Global
	Spatial
	Inseparable
✓	Conservative

Formal Graph: Self-influence only

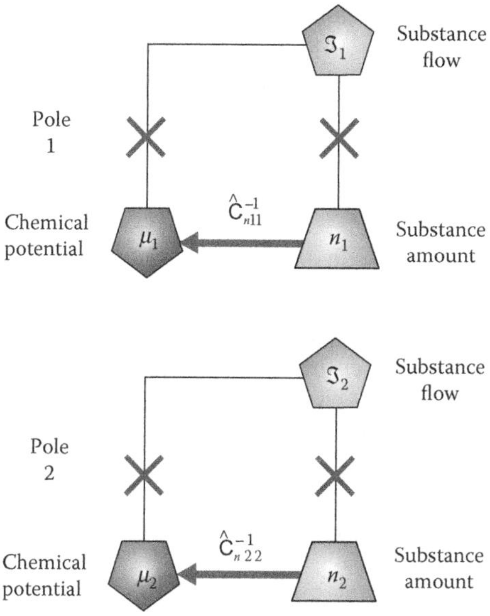

GRAPH 7.7

Formal Graph: Self-influence + Interaction

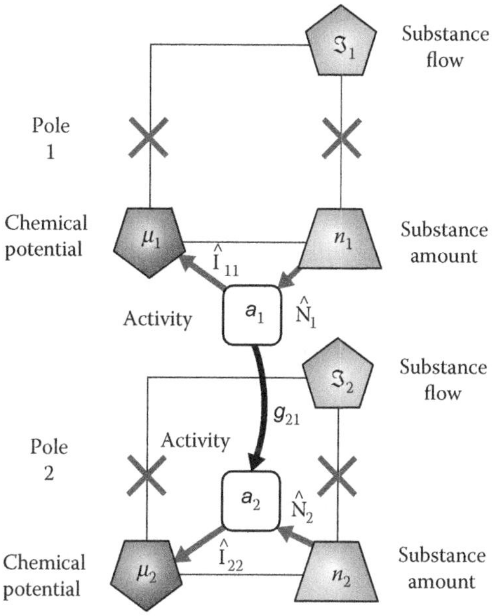

GRAPH 7.8

Variables:

Variety	Physical Chemistry		
Subvariety	Capacitive (Internal)		Inductive
Category	Entity number	Energy/ entity	Entity number
Family	Basic quantity	Effort	Flow
Name	Substance amount	Chemical potential	Substance flow
Symbols	n, (N)	μ	$\mathfrak{I}$, v
Unit	[mol]	[J mol^{-1}]	[mol s^{-1}]

Property relationships:

Self-influence:

$$\mu_i = \hat{I}_{ii}\, a_i$$

$$a_i = \hat{N}_i\, n_i$$

Self-elastance:

$$\hat{C}_{ni}^{-1} = \hat{I}_{ii}\, \hat{N}_i$$

Interaction:

$$g_{21} = \frac{a_2}{a_1}$$

System Property	
Nature	Capacitance
Levels	0
Name	Physical chemical capacitance
Symbol	C_n
Unit	[mol^2 J^{-1}]
Specificity	—

Energy variation:

$$d\mathcal{U}_n \underset{\exists C_n}{=} \mu\, dn$$

pole and for acting (influencing) on the energy-per-entity (here the chemical potential). This is the basic assumption that allows generalizing the capacitance model to every energy variety.

$$a_i = \widehat{N}_i \, n_i \tag{D3.2}$$

In a totally active population of substance, which is the general case in physical chemistry, one has $n^m = 0$ for the quantity of inactive substance (see Section 7.4.2), so the activity is proportional to the substance amount:

$$a_i = \frac{n_i}{n^\ominus} \tag{D3.3}$$

Most of the time, the substance reference amount is taken to be equal to 1 mol, so activity and substance amount have equal values, which may introduce confusion and some difficulties. This reference amount is made explicit in all formulas to avoid problems with units.

Self-influence: The physical chemical self-influence operator is defined as relating the activity to the chemical potential for the same pole:

$$\mu_i \overset{def\rightarrow}{=} \hat{I}_{ii} \, a_i \tag{D3.4}$$

This operator will be made explicit in the following expressions. It can be checked that combining Equations D3.4 and D3.2 leads to the capacitance expression in Equation D3.1.

Interaction: An analogy with signal processing theories is made in defining a *gain* g as quantifying the amplification of activities in the same manner as signal intensities are treated.

$$g_{21} \overset{def}{=} \frac{a_2}{a_1} \tag{D3.5}$$

Another relationship between the two poles is ensured by the relationship between chemical potentials of the poles and of the dipole:

$$\mu_{21} = \mu_2 - \mu_1 \tag{D3.6}$$

The relationship between the dipole chemical potential and the gain is dependent on the interaction between the two poles. This question is addressed later (see Section 7.4.2), as it requires a theoretical development that is not specific to the physical chemical domain but is a general one. The result is an exponential function linking the two variables

$$g_{21} = \exp \frac{\mu_{21} - \mu^\ominus_{21}}{\mu_\Delta} \tag{D3.7}$$

where $\mu^\ominus{}_{21}$ is a constant specific to the couple of species called the *standard chemical potential* and μ_Δ is a *scaling chemical potential.*

In an ideal substance, as molecules are submitted to thermal agitation only, without any other energy variety (able to influence the molecules) than the thermal one, this *scaling chemical potential* is merely proportional to the temperature through the gas constant R:

$$\mu_\Delta = RT \tag{D3.8}$$

When additional energy varieties participate in the energy of the system (surface energy for instance), this relationship may become more complicated. This is an advantage of the Formal Graph approach to individualize this variable instead of writing directly the product RT as the classical approach does.

From the dependence of the gain on the dipole chemical potential (Equation D3.7), the capacitive relationship for a physical chemical pole is deduced (see Section 7.4.2):

$$\frac{n_i}{n^{\vartheta}} = a_i = \exp\left(\frac{\mu_i - \mu_i^{\vartheta}}{\mu_{\Delta}}\right) \tag{D3.9}$$

The relationship between *standard chemical potentials* follows the relationship between chemical potentials as shown in Equation D3.6:

$$\mu_{21}^{\vartheta} = \mu_2^{\vartheta} - \mu_1^{\vartheta} \tag{D3.10}$$

Finally, the expression of the influence operator corresponds to the alternate way of writing the *influence relationship* between activity and chemical potential:

$$\mu_i = \hat{\mathrm{I}}_{ii}\; a_i = \mu_i^{\vartheta} + \mu_{\Delta} \ln a_i \tag{D3.11}$$

Remarks

Statistics is not always applicable. The influence relationship in physical chemistry is classically introduced (Atkins 1998) as a definition of the activity in arguing that it obeys Boltzmann statistics. This is generally a perfectly acceptable model as, in a gas or a liquid, thermal agitation induces fluctuations in the system that can be treated in terms of probability of events. However, there are cases in which molecules are immobilized in a lattice or on a surface and which still obey the capacitive relationship shown in Equation D3.9. This clearly means that the powerful tool and concept of probability cannot be used in all cases.

Where does *RT* come from? The product *RT* is not deduced from an interaction model but from the thermodynamical analysis of a system that includes thermal energy and not merely physical chemical energy. This subject is treated in Chapter 13 where the identification of the *scaling chemical potential* μ_{Δ} with the product *RT* is demonstrated.

7.2.4 Case Study D4: Coulomb's Law — Electrodynamics

The case study abstract is given in next page.

The attraction force between two electric charges Q_1 and Q_2 separated by a distance $\boldsymbol{r}$ is given in electrostatics by the double equation called Coulomb's law.

$$\boldsymbol{F}_1 = -\boldsymbol{F}_2 = -\frac{1}{4\pi\varepsilon}\frac{Q_1\,Q_2}{\boldsymbol{r}^2}\frac{\boldsymbol{r}}{|\boldsymbol{r}|} \tag{D4.1}$$

The scaling factor is the *electric permittivity (or dielectric constant)* ε that may vary depending on the milieu in which the charges are placed. In vacuum, this permittivity is notated ε_0 and is approximately equal to 8.85×10^{-12} A s V^{-1} m^{-1} (a more precise value is given in the Appendix 2).

D4: Coulomb's Law **Electrodynamics**

$$\boldsymbol{F}_1 = -\boldsymbol{F}_2 = -\frac{1}{4\pi\varepsilon}\frac{Q_1 Q_2}{\boldsymbol{r}^2}\frac{\boldsymbol{r}}{|\boldsymbol{r}|}$$

Space dipole

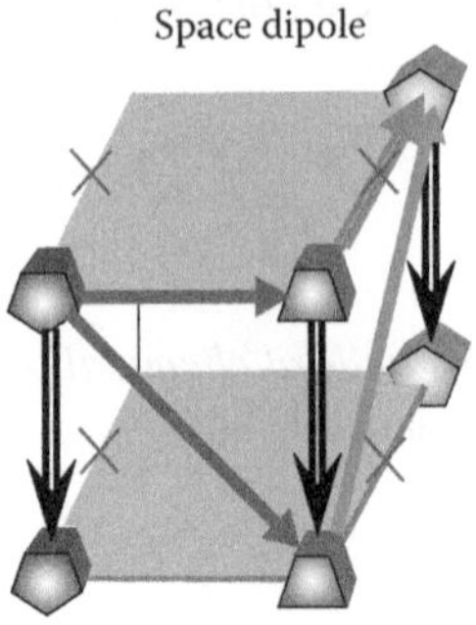

Dipole	
✓	Fundamental
	Mixed
✓	Capacitive
	Inductive
	Conductive
	Global
✓	Spatial
	Inseparable
✓	Conservative

Formal Graph: (in three dimensions)

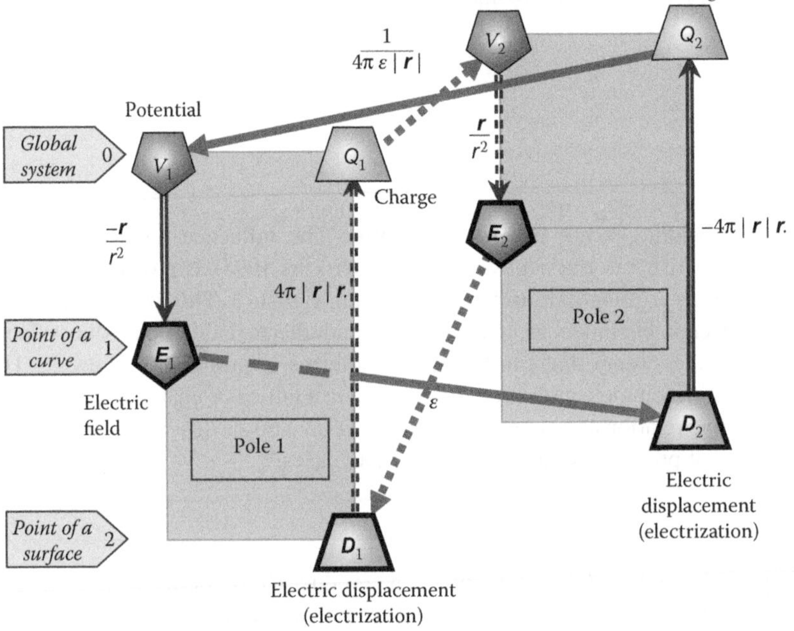

GRAPH 7.9

Variables:

Variety	Electrodynamics		
Subvariety	Capacitive (Electrostatic)		Inductive
Category	Entity number	Energy/ entity	Entity number
Family	Basic quantity 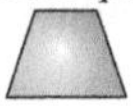	Effort 	Flow 
Name	Charge	Potential	Current
Symbols	Q	V, U, E, φ	I, i
Unit	[C], [A s]	[V], [J C^{-1}]	[A]

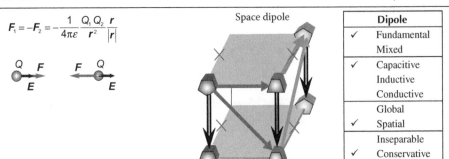

Property relationships:

Mutual influence:

$V_1 = \cancel{\hat{E}}_{11}\, Q_1 + \hat{E}_{12}\, Q_2$

$V_2 = \hat{E}_{21}\, Q_1 + \cancel{\hat{E}}_{22}\, Q_2$

$\hat{E}_{12} = \hat{E}_{21} = \hat{E} = \hat{C}^{-1}$

$\boldsymbol{D}_i = \hat{\varepsilon}_{ij}\, \boldsymbol{E}_j$

$\hat{\varepsilon}_{ij} = \hat{\varepsilon}_{ji} = \hat{\varepsilon}$

Coulomb:

$C \underset{lin}{=} 4\pi\varepsilon|r|$

System Property	
Nature	Capacitance
Levels	1 / 2
Name	Permittivity
Symbol	ε
Unit	[F m^{-1}]
Specificity	Three-dimensional

Energy variation:

$d\mathcal{U}_Q \underset{\exists C}{=} V\, dQ$

The ratio of the vector $\boldsymbol{r}$ on its modulus determines the orientation of the vector force $\boldsymbol{F}_1$ (colinear to the vector distance), oriented from charge 1 toward charge 2. This means that the two charges attract themselves when they have opposite signs and mutually repel in the inverse case.

The left scheme in Figure D4.1 shows the four possibilities of placing two charges along an oriented horizontal axis. The theory of Coulombian influence involves vector quantities that are shown in this scheme and that are:

1. The forces of attraction $\boldsymbol{F}_1$ and $\boldsymbol{F}_2$.
2. The electric fields $\boldsymbol{E}_1$ and $\boldsymbol{E}_2$.

The right scheme in Figure D4.1 illustrates the orientations taken by the two vectors $\boldsymbol{D}_1$ and $\boldsymbol{D}_2$ that are the electric displacements (or "electrizations"; see Remarks [1] and [2]) determined on spheres of radius $\boldsymbol{r}$.

In fact, the double equation D4.1 results from the composition of four relations, two for each charge:

$$\boldsymbol{F}_1 = Q_1\,\boldsymbol{E}_1 \qquad \boldsymbol{F}_2 = Q_2\,\boldsymbol{E}_2 \tag{D4.2}$$

$$\boldsymbol{E}_1 = -\frac{1}{4\pi\varepsilon}\frac{Q_2}{\boldsymbol{r}^2}\frac{\boldsymbol{r}}{|\boldsymbol{r}|} \qquad \boldsymbol{E}_2 = \frac{1}{4\pi\varepsilon}\frac{Q_1}{\boldsymbol{r}^2} \tag{D4.3}$$

The vectors $\boldsymbol{E}_1$ and $\boldsymbol{E}_2$ represent the electric fields exerted by a charge at the point where the other charge is located ($\boldsymbol{E}_1$ is created by Q_2 and $\boldsymbol{E}_2$ by Q_1). The localizations of the various variables belonging to the model are (cf. Figure D4.1):

$\boldsymbol{F}_1$, Q_1, V_1, $\boldsymbol{E}_1$, and $\boldsymbol{D}_2$ (this is subscript two) at the place of the first charge Q_1.
$\boldsymbol{F}_2$, Q_2, V_2, $\boldsymbol{E}_2$, and $\boldsymbol{D}_1$ (this is subscript one) at the place of the second charge Q_2.

Force and field. The two relationships D4.2 between electric field and mechanical force are relevant to another subject, that is, the coupling between translational mechanics and electrodynamics (see case study J3 "Electric Force" in Chapter 12), and are not demonstrated here.

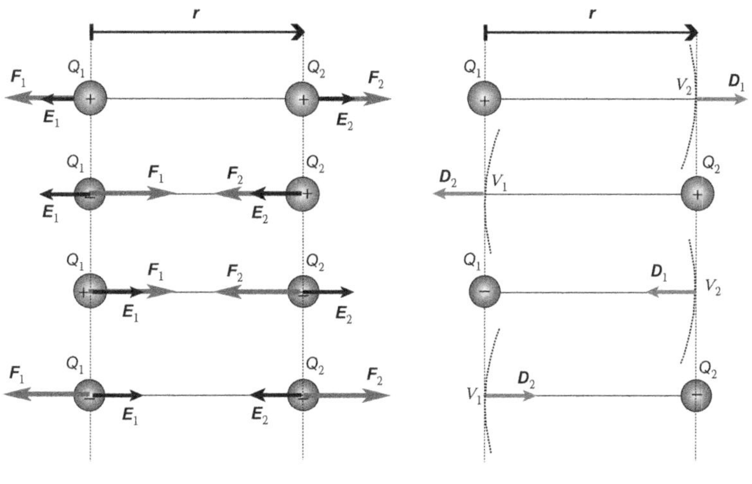

FIGURE D4.1 Schemes of the various orientations taken by the vectors $\boldsymbol{F}$ and $\boldsymbol{E}$ (left) and $\boldsymbol{D}$ (right) according to the four ways of placing two opposite charges.

The Formal Graph in the case study abstract shows the relationships D4.3 and the two circular paths that determine them. It reproduces in part (by limiting to two-dimensional space to simplify the graph, which does not mean that the model cannot work in three-dimensional space) for each pole the graph given in case study B3 "Electric Space Charges" in Chapter 5.

Each loop contains the relationship of mutual influence between the two spatially distributed poles that are made up of both charges. The dashed-line loop is the influence of charge 2 on charge 1 and the solid-line loop is the opposite influence. The upper horizontal paths reflect the influences between state variables (at global level) and the lower paths reflect reciprocal influences between spatially reduced variables that are the electric fields and the electric displacements (or "electrizations"; see Remark [1]). Spatial operators (usually a gradient and an integral on a surface) are written in their peculiar expression, which corresponds to the assumptions made to find the Coulombian model represented in scheme D4.3, that is, a homogeneous distribution in space and an electric displacement (*electrization*) determined on a sphere. The demonstration of this graph is as follows. It proceeds in several steps each corresponding to a physical assumption.

7.2.4.1 Mutual Influence

The model is based on the principle of influence of each charge on the other but not on itself (non-zero mutual influence and zero self-influence), resulting at the global level in each electrical potential equal to a cross-elastance applied to the charge of the other pole.

$$\begin{aligned} V_1 &= \cancel{\hat{C}_{11}^{-1}}\, Q_1 + \hat{C}_{12}^{-1}\, Q_2 \\ V_2 &= \hat{C}_{21}^{-1}\, Q_1 + \cancel{\hat{C}_{22}^{-1}}\, Q_2 \end{aligned} \qquad \text{(D4.4)}$$

In the absence of self-influence, this mutual influence means that each pole has its electric potential determined by the other pole. The electric energy of the dipole being a *state function*, its variation is an exact differential

$$d\mathcal{U}_Q = V_1\, dQ_1 + V_2\, dQ_2 \qquad \text{(D4.5)}$$

which implies that the derivatives of electrical potential with respect to the other charge must be equal. By taking into account the relationships in scheme D4.4, this results in equal cross-elastances:

$$\hat{C}_{12}^{-1} = \hat{C}_{21}^{-1} \qquad \text{(D4.6)}$$

7.2.4.2 Spherical Surfaces

The surface element of the system is considered to be spherical, which assumes isotropic space (equivalent in all directions). The relationship between the charge contained within a sphere and the electric displacement (*electrization*) $\boldsymbol{D}$ on the surface is therefore an integral over the total surface of the sphere (see case study B5 "Gauss Theorem" in Chapter 5). Each electric displacement (*electrization*) produced by a positive charge is a vector directed outward from the sphere (and vice versa toward the center in case of a negative charge). The sphere radius is equal to the distance between the charges, which, in view of the orientation (cf. Figure D4.1, right) of the distance vector from charge 1 to charge 2, gives

$$Q_1 = \oint_{sphere} \boldsymbol{D}_1 . d\boldsymbol{A} = 4\pi |r| \boldsymbol{D}_1 . \boldsymbol{r} \qquad Q_2 = \oint_{sphere} \boldsymbol{D}_2 . d\boldsymbol{A} = -4\pi |r| \boldsymbol{D}_2 . \boldsymbol{r} \qquad \text{(D4.7)}$$

One should consider the concept of electric displacement $\boldsymbol{D}$ not as a "charge density" representing a real distribution of charge on the surface, which would be contrary to one's perception of the structural unity of electric charges, but as a *field* of "electrization" analogous to the *magnetization* $\boldsymbol{M}$ and to the *gravitization* $\boldsymbol{G}$ (or *gravitational displacement*). This plays the same role as the electric field (to which it is proportional through the electric permittivity) that exists at a distance from the body producing it and that can propagate through space. Equations D4.7 are none other than the expression of the Gauss theorem (case study B5 in Chapter 5) in electrostatics that expresses the total charge contained in a sphere as being equal to the integral of the electric displacement (*electrization*) on the surface thereof (or as a function of the electric field through the permittivity, as in Equations D4.10 with the electric field).

7.2.4.3 Spatial Homogeneity (Linearity)

The electric potential is assumed to be regularly distributed through space, which amounts to assuming a *homogeneous space*. In these circumstances, the contragradient is equivalent to a division of the electrical potential by the distance and to a multiplication by plus or minus the unit vector (ratio of the vector $\boldsymbol{r}$ on its modulus) to meet the direction of each electric field (refer to the schemes in Figure D4.1):

$$\boldsymbol{E}_1 = -\frac{\partial}{\partial \boldsymbol{r}} V_1 \underset{hom}{=} -\frac{1}{\boldsymbol{r}} V_1 = -V_1 \frac{\boldsymbol{r}}{r^2} \quad \boldsymbol{E}_2 = -\frac{\partial}{\partial \boldsymbol{r}} V_2 \underset{hom}{=} \frac{1}{\boldsymbol{r}} V_2 = V_2 \frac{\boldsymbol{r}}{r^2} \tag{D4.8}$$

7.2.4.4 Electric Permittivity

The spatially reduced capacitances that are *specific* of the medium are electric permittivities in this energy variety. To each capacitance at the global level corresponds an electric permittivity ε_{ij} and, therefore, the equality D4.6 corresponds to an equality between mutual permittivities. Their common value is a scalar constant, meaning that the assumption of a *homogeneous space* constituting the system is made.

$$\hat{\varepsilon}_{12} = \hat{\varepsilon}_{21} = \hat{\varepsilon} \underset{hom}{=} \varepsilon \tag{D4.9}$$

Consequently, the two reduced capacitive relations of influence are written:

$$\boldsymbol{D}_1 \underset{hom}{=} \varepsilon \boldsymbol{E}_2; \quad \boldsymbol{D}_2 \underset{hom}{=} \varepsilon \boldsymbol{E}_1 \tag{D4.10}$$

Synthesis: In the Formal Graph in the case study abstract, elastances (reciprocal of capacitances) are given directly with their expression particular to the considered system as a result of the treatment carried out below. Before the solution is given, the reasoning leading to it is to consider the two loops that are formed by each elastance and its reduced property and to apply the *circularity principle*:

$$\begin{aligned} &\hat{C}_{12}^{-1}\left(-4\pi|r|\boldsymbol{r}.\right)\varepsilon\left(-\frac{\boldsymbol{r}}{r^2}\right) = 1 \\ &\hat{C}_{21}^{-1}\left(4\pi|r|\boldsymbol{r}.\right)\varepsilon\left(\frac{\boldsymbol{r}}{r^2}\right) = 1 \end{aligned} \tag{D4.11}$$

From these identities is deduced the common solution

$$\hat{C}_{12}^{-1} = \hat{C}_{21}^{-1} = \frac{1}{4\pi\varepsilon|r|} \tag{D4.12}$$

that establishes the proportionality of the mutual elastance with the inverse of the distance. The two capacitive relations are, consequently,

$$V_1 = \frac{1}{4\pi\varepsilon}\frac{Q_2}{|r|}$$
$$V_2 = \frac{1}{4\pi\varepsilon}\frac{Q_1}{|r|} \qquad \text{(D4.13)}$$

By injecting the terms of the two electric potentials in relations D4.8, we obtain the sought relations D4.3, which achieves the demonstration.

Remarks

Is there really some displacement? The term "displacement," normally meaning a change of position in space, is strangely borrowed from translational mechanics by analogy between the work of a force equal to its strength multiplied by the displacement (distance) $\mathbf{F.d\ell}$ and the "work of electric volume forces" equal to $\mathbf{E.dD}$.

Electrization? For naming $\mathbf{D}$, the term "electric induction" is more respectful of the physical reality, but "induced" is classically related to the electromagnetic induction and refers to the inductive energy that can be confusing. We propose the word *electrization*, on the template of *magnetization* $\mathbf{M}$, which is used for the equivalent variable in magnetism (see case study D6 "Magnetic Interaction"). Such a termination in "-ization" has the advantage of meaning an *action*, which is appropriate. These two variables $\mathbf{D}$ and $\mathbf{M}$ share the common feature of corresponding to the reduction depth on a surface of a density of entity number.

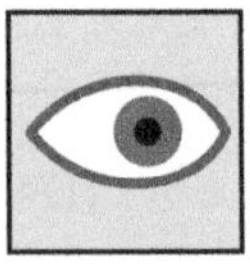

Charles-Augustin de Coulomb. He gave his name to this electrostatic model of interaction between electric charges. He also established the same form of expression for the interaction between "magnetic charges" (see case study D6 "Magnetic Interaction"). It is important to consider these models not as "fundamental laws" (despite their historical name) but as approximate models (nevertheless very well tested in their validity domain).

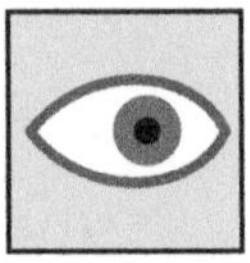

Analogy with gravitation. This model of influence between electric charges is based on identical assumptions with Newton's gravity. The Newtonian model expressing the two forces acting on gravitational masses is strictly of the same form as the dual Equation D4.1 and of course is exactly represented by the same Formal Graph. Just two details vary between these two models: to find the full Newtonian model from Coulomb's model, the electric permittivity ε should be replaced by the gravitational permittivity $1/4\pi G$ and the electrical charges Q_1 and Q_2 should be replaced by the gravitational masses with a minus sign, that is, $-M_{G1}$ and $-M_{G2}$. This is due to a fundamental difference in behavior; in gravity, two masses of the same sign (positive) attract, whereas two electric charges of the same sign repel each other.

Simplifications may be removed. The model is based on several simplifying assumptions that the Formal Graphs highlight. These are based, firstly, on the homogeneity of the medium to use a reduced capacitance (permittivity) independent from distance and orientation (scalar electric permittivity) and, secondly, on the homogeneity and isotropy of space.

These assumptions are classic and Formal Graphs can shed some light on what improvements can be made to this model by expanding each simplification.

7.2.5 Case Study D5: Newtonian Gravitation — Gravitation

The case study abstract is given in next page.

The force of attraction between two gravitational masses M_{G1} and M_{G2} separated by a distance $\boldsymbol{r}$ is given in Newtonian gravitation by the double equation

$$\boldsymbol{F}_1 = -\boldsymbol{F}_2 = G\frac{M_{G1}\,M_{G2}}{\boldsymbol{r}^2}\frac{\boldsymbol{r}}{|\boldsymbol{r}|} \qquad \text{(D5.1)}$$

The scaling factor is the *constant of gravitation* G approximately equal to 6.7×10^{-11} N m^2 kg^{-2} (or m^3 kg^{-1} s^{-2}) and the ratio of the vector distance $\boldsymbol{r}$ on its modulus determines the orientation of the vector force $\boldsymbol{F}_1$ colinear to the vector distance, oriented from mass 1 toward mass 2, which means that the two masses attract themselves. In fact, this double equation results from the composition of four relations, two for each mass,

$$\boldsymbol{F}_1 = M_{G1}\,\boldsymbol{g}_1 \qquad\qquad \boldsymbol{F}_2 = M_{G2}\,\boldsymbol{g}_2 \qquad \text{(D5.2)}$$

$$\boldsymbol{g}_1 = G\frac{M_{G2}}{\boldsymbol{r}^2}\frac{\boldsymbol{r}}{|\boldsymbol{r}|} \qquad\qquad \boldsymbol{g}_2 = -G\frac{M_{G1}}{\boldsymbol{r}^2}\frac{\boldsymbol{r}}{|\boldsymbol{r}|} \qquad \text{(D5.3)}$$

in which the vectors $\boldsymbol{g}_1$ and $\boldsymbol{g}_2$ represent the gravitational fields exerted by one mass at the level of the other mass. The relations D5.2 between gravitational field and force are relevant to a subject different from the present one, which is the coupling between gravitation and translational mechanics, treated in case study J4 "Gravitational Force" in Chapter 12, analogous to the coupling between electrodynamics and mechanics (see case study J3 "Electric Force" in Chapter 12).

The theory of the Newtonian gravitation involves a certain number of vector quantities represented in the scheme in Figure D5.1:

1. The forces of attraction $\boldsymbol{F}_1$ and $\boldsymbol{F}_2$.
2. The fields of gravity $\boldsymbol{g}_1$ and $\boldsymbol{g}_2$.
3. The gravitational displacements (or "gravitizations") $\boldsymbol{G}_1$ and $\boldsymbol{G}_2$ determined on spheres of radius $\boldsymbol{r}$ (see Remark [2]).

The Formal Graph in the case study abstract represents the relations D5.3 together with the two circular paths that are their translation.

Each loop contains the relations of mutual influence between the two poles distributed spatially that constitute the two masses. The loop in dotted lines features the influence of mass 2 on mass 1 and the loop in solid line features the other influence. The horizontal upper paths express the influences between state variables (global level) and the slanted lower paths express the reciprocal influences between spatially reduced variables that are the *gravitational field* $\boldsymbol{g}$ and the *gravitational displacement* $\boldsymbol{G}$ (for which the name "gravitization" is more appropriate; see Remarks [1] and [2]). (For building a loop, these influences have been reversed: This is a gravitational displacement

D5: Newtonian Gravitation **Gravitation**

$$\boldsymbol{F}_1 = -\boldsymbol{F}_2 = G\frac{M_{G1}\,M_{G2}}{\boldsymbol{r}^2}\frac{\boldsymbol{r}}{|\boldsymbol{r}|}$$

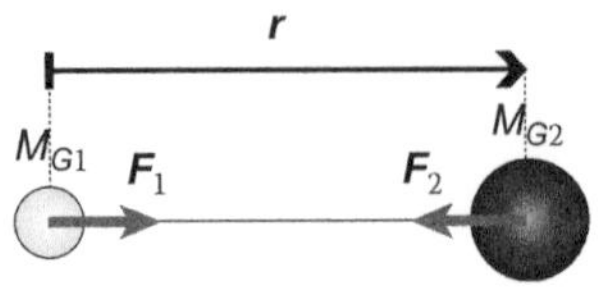

Space dipole

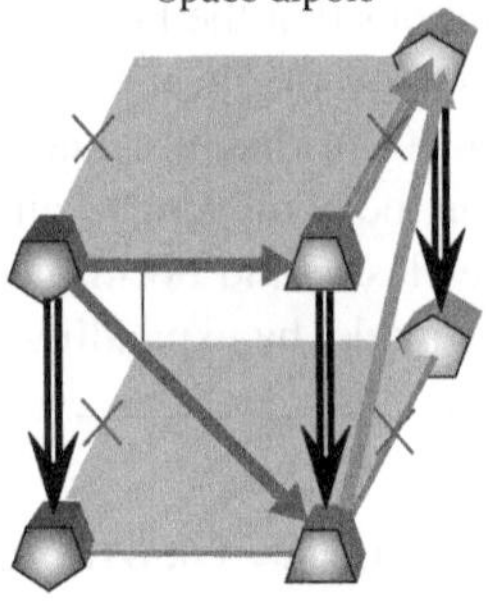

Dipole	
✓	Fundamental
	Mixed
✓	Capacitive
	Inductive
	Conductive
	Global
✓	Spatial
	Inseparable
✓	Conservative

Formal Graph (see Graph 7.10): (in three dimensions)

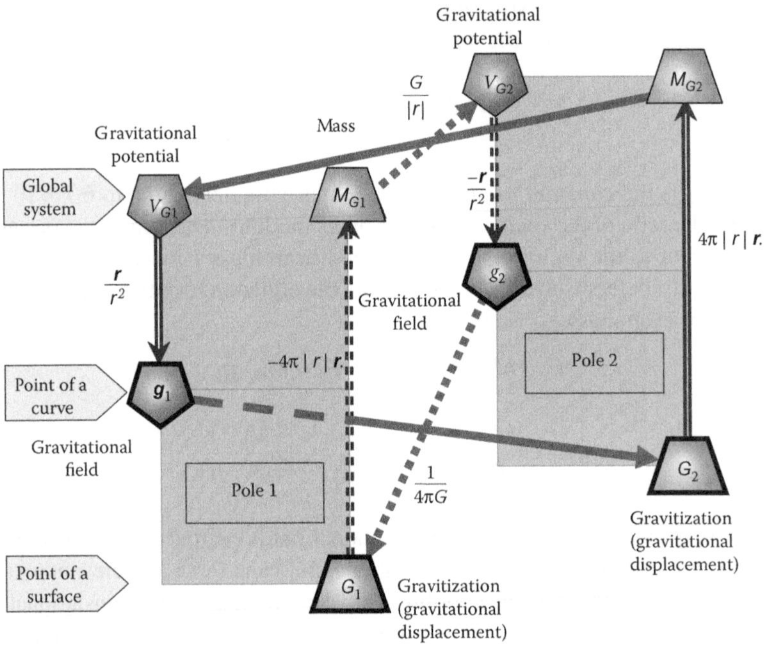

GRAPH 7.10

Variables:

Variety	Gravitation		
Subvariety	Capacitive (Potential)		Inductive
Category	Entity number	Energy/entity	Entity number
Family	Basic quantity	Effort	Flow
Name	Gravitational mass	Gravitational potential	Mass flow rate
Symbols	M_G	V_G	f_M
Unit	[kg]	[m² s⁻²], [J kg⁻¹]	[kg s⁻¹]

Property relationships:

Mutual influence:

$$V_{G1} = \hat{C}_{M11}^{-1}\,M_{G1} + \hat{C}_{M12}^{-1}\,M_{G2}$$

$$V_{G2} = \hat{C}_{M21}^{-1}\,M_{G1} + \hat{C}_{M22}^{-1}\,M_{G2}$$

$$\hat{C}_{M12}^{-1} = \hat{C}_{M21}^{-1} = \hat{C}_{M}^{-1}$$

$$\boldsymbol{G}_i = \hat{\varepsilon}_{Mij}\,\boldsymbol{g}_j$$

$$\hat{\varepsilon}_{M12} = \hat{\varepsilon}_{M21} = \hat{\varepsilon}_{M}$$

Newton:

$$C_M \underset{lin}{=} \frac{|r|}{G}$$

System Property	
Nature	Elastance
Levels	2/1
Name	Gravitation constant (× 4π)
Symbol	$4\pi G$
Unit	[m³ kg⁻¹ s⁻²]
Specificity	Three-dimensional

Energy variation:

$$d\mathcal{U}_M \underset{\exists C_M}{=} V_G\,dM_G$$

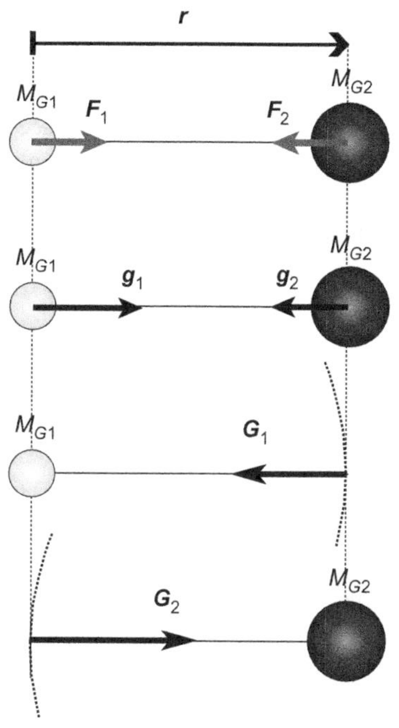

FIGURE D5.1 Scheme showing the orientations taken by the vectors ***F***, ***g***, and ***G***.

that creates the gravitational field on the other pole, but every link in a Formal Graph can be reversed.) The spatial operators are written with their particular expression corresponding to the hypothesis made for retrieving the Newtonian model shown in scheme D5.3, which is a homogeneous distribution in space and a gravitational displacement determined on a sphere.

The demonstration of this graph is as follows. It proceeds in several steps, each step corresponding to a physical hypothesis.

7.2.5.1 Mutual Influence

The model relies on the principle of each mass influencing the other but not itself (nonzero mutual influence and zero self-influence), which is translated at the global level by each gravitational potential being equal to a cross-elastance applied to the mass of the other pole.

$$\begin{aligned} V_{G1} &= \cancel{\hat{\mathrm{C}}^{-1}_{M11}}\, M_{G1} + \hat{\mathrm{C}}^{-1}_{M12}\, M_{G2} \\ V_{G2} &= \hat{\mathrm{C}}^{-1}_{M21}\, M_{G1} + \cancel{\hat{\mathrm{C}}^{-1}_{M22}}\, M_{G2} \end{aligned} \tag{D5.4}$$

The gravitational energy of the dipole being a state function, its variation is an exact differential

$$\mathrm{d}\mathcal{U}_M = V_{G1}\,\mathrm{d}M_{G1} + V_{G2}\,\mathrm{d}M_{G2} \tag{D5.5}$$

which implies that the derivatives of the gravitational potentials with respect to the other mass are equal, and which is therefore translated by the equality of the cross-elastances taking into account the relations D5.4.

$$\hat{\mathrm{C}}^{-1}_{M12} = \hat{\mathrm{C}}^{-1}_{M21} \tag{D5.6}$$

7.2.5.2 Spherical Surfaces

The surface element of the system is supposed to have a spherical shape, which amounts to assuming an *isotropic* space, equivalent in all directions. The relation between the mass contained inside a sphere and the *gravitational displacement (gravitization)* $\boldsymbol{G}$ on the surface of this sphere is therefore an integral on the total area of the sphere. Each gravitational displacement (*gravitization*) is a vector oriented toward the center of the sphere. The sphere radius is taken to be equal to the distance between the masses, which, taking into account the orientation of pole 1 toward pole 2 (cf. Figure D5.1), gives

$$
\begin{aligned}
M_{G1} &= \oint_{sphere} \boldsymbol{G}_1.\mathbf{d}\boldsymbol{A} = -4\pi|r|\boldsymbol{G}_1.\boldsymbol{r} \\
M_{G2} &= \oint_{sphere} \boldsymbol{G}_2.\mathbf{d}\boldsymbol{A} = 4\pi|r|\boldsymbol{G}_2.\boldsymbol{r}
\end{aligned}
\tag{D5.7}
$$

This notion of vector *gravitational displacement* (*gravitization*) $\boldsymbol{G}$ must not be considered as a "density of mass" representative of a real distribution of mass on the surface, which should be contrary to one's perception of a structural unity of massive bodies, but as a *field* of "gravitization" analogous to the *electric displacement* $\boldsymbol{D}$ for which the term "electrization" is equivalently proposed. This latter plays exactly the same role as the electric field (to which it is proportional through the electric permittivity) which exists at a distance from the charged body that creates it and can propagate through space. Equations D5.7 are nothing other than the analogue of the Gauss theorem in electrostatics (case study B5 in Chapter 5), which states that the totality of the charge contained in a sphere is equal to the integral of the electric displacement (*electrization*) on the surface of this sphere (or as a function of the electric field $\boldsymbol{E}$ through the permittivity, as in Equations D5.10, by analogy with the gravitational field $\boldsymbol{g}$).

7.2.5.3 Space Homogeneity (Linearity)

The spatial distribution of gravitational potential is assumed to be regular, which amounts to assuming a *homogeneous space*. In these conditions, the contragradient is equivalent to a division of the gravitational potential by the distance and to a multiplication by plus or minus the unit vector (ratio of the vector $\boldsymbol{r}$ on its modulus) in order to respect the orientation of each gravitational field (refer to the scheme in Figure D5.1, and consider as negative scalars the gravitational potentials shown later):

$$
\boldsymbol{g}_1 = -\frac{\partial}{\partial \boldsymbol{r}} V_{G1} \underset{hom}{=} -V_{G1}\frac{\boldsymbol{r}}{r^2} \qquad \boldsymbol{g}_2 = -\frac{\partial}{\partial \boldsymbol{r}} V_{G2} \underset{hom}{=} V_{G2}\frac{\boldsymbol{r}}{r^2}
\tag{D5.8}
$$

7.2.5.4 Gravitational Permittivity

The spatially reduced capacitances are in fact "gravitational permittivities" in this domain of energy, analogous to electric permittivities. To each capacitance at the global level corresponds a gravitational permittivity $\varepsilon_{M\,ij}$ and, therefore, the equality D5.6 corresponds to an equality between cross-permittivities. According to Newton, their common value is a scalar supposed to be a constant, that is, the hypothesis of a *homogeneous medium* constituting the system is made. To find the shape of Equation D5.1, this scalar must be the inverse of the constant of gravitation divided by 4π

$$
\hat{\varepsilon}_{M12} = \hat{\varepsilon}_{M21} = \hat{\varepsilon}_M \underset{hom}{=} \frac{1}{4\pi G}
\tag{D5.9}
$$

Consequently, the two reduced capacitive relationships of influence are written as:

$$
\boldsymbol{G}_1 \underset{hom}{=} \frac{\boldsymbol{g}_2}{4\pi G} \qquad \boldsymbol{G}_2 \underset{hom}{=} \frac{\boldsymbol{g}_1}{4\pi G}
\tag{D5.10}
$$

Synthesis: In the Formal Graph above, the elastances (reciprocals of capacitances) are given directly with their expression particular to the considered system, which results from the treatment below. Before knowing the solution, the reasoning consists of examining the two loops formed by each elastance and its reduced property and of applying the *circularity principle*:

$$\begin{aligned} \hat{C}_{M12}^{-1}\left(4\pi|r|\boldsymbol{r}.\right)\frac{1}{4\pi G}\left(\frac{\boldsymbol{r}}{r^2}\right) &= 1 \\ \hat{C}_{M21}^{-1}\left(-4\pi|r|\boldsymbol{r}.\right)\frac{1}{4\pi G}\left(-\frac{\boldsymbol{r}}{r^2}\right) &= 1 \end{aligned} \tag{D5.11}$$

From this is deduced the common solution

$$\hat{C}_{M12}^{-1} = \hat{C}_{M21}^{-1} = \frac{G}{|r|} \tag{D5.12}$$

which establishes the proportionality of the mutual elastance with the inverse of the distance. The two capacitive relationships are therefore

$$\begin{aligned} V_{G1} &= \frac{GM_{G2}}{|r|} \\ V_{G2} &= \frac{GM_{G1}}{|r|} \end{aligned} \tag{D5.13}$$

By injecting the expressions of the two gravitational potentials into the relations D5.8, one obtains the relations D5.3, which achieves the demonstration.

It can be remarked that in replacing the right members of the previous expressions by the relations D5.3, multiplied (through a scalar product) by the distance, one obtains two markedly simple equations:

$$\begin{aligned} V_{G1} &\underset{lin}{=} \boldsymbol{g}_1 \cdot \boldsymbol{r} \\ V_{G2} &\underset{lin}{=} -\boldsymbol{g}_2 \cdot \boldsymbol{r} \end{aligned} \tag{D5.14}$$

Contrary to what could be deduced from the relationship between potentials and fields through a gradient (and which is sometimes asserted), the gravitational fields $\boldsymbol{g}$ do not need to be constant for validating these equations. In contradistinction, they rely on hypotheses of homogeneity and constancy of gravitational permittivities (linearity) made in the relations in Equation D5.11, implying that the gravitational potential V_G varies in $1/r$ (Equation D5.13) and the gravitational field $\boldsymbol{g}$ in $1/r^2$ (Equation D5.3). These equations are used directly when one is looking for expressions of the gravitational potentials in terms of the localized gravitational field $\boldsymbol{g}$, as in the case of gravity (one of the two masses is assumed to be much more important than the other, so its action is formulated in terms of the gravitational field; see Figure D5.2).

Gravity: This notion corresponds to the action of a gravitational field on a mass, especially in the vicinity of an important mass attracting a smaller one, as for instance in the case of the Earth and all bodies submitted to its attraction.

Figure D5.2 illustrates the case of gravity, in which a mass M_{G2} is attracted downward by a bigger mass M_{G1}. By notating r the modulus of the distance $\boldsymbol{r}$ and g the modulus of the gravitational field $\boldsymbol{g}_2$ at the level of the mass M_{G2}, the gravitational potential at this level V_{G2}, given by the second Equation D5.8, is therefore equal to the product $g\,r$ (which is a positive quantity equal to the scalar

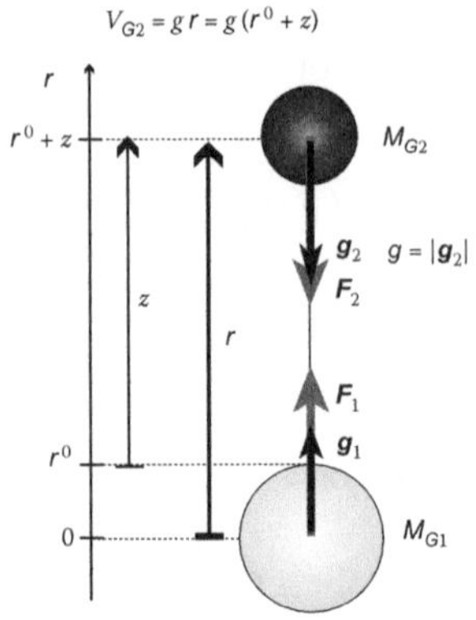

FIGURE D5.2 Coordinate scheme for gravity exerted on a probe body in the upper position.

product $-\boldsymbol{g}_2 . \boldsymbol{r}$ according to Equation D5.14). Furthermore, the distance r can be replaced by the sum of the radius r^0 of the attracting mass (Earth) and of the height z (altitude).

$$V_{G2} \underset{lin}{=} g\,r = g\left(r^0 + z\right) \tag{D5.15}$$

This expression is used in the example of a gas column submitted to gravity (see case study K3 "Barometric Equation" in Chapter 13).

Remarks

Behind Newton's model. The subject of gravitation has been central in physics for millennia, and, without any doubt, will remain so for many years. The merit of Newton has been to model the phenomenon of attraction between bodies endowed with a mass, celestial or on Earth, thus unifying sky and Earth. His model relies on several simplifying hypotheses that the Formal Graphs bring to light. These hypotheses are based, on the one hand, on the medium homogeneity for being able to utilize a capacitive reduced property that is independent of distance and of orientation (scalar gravitational permittivity) and, on the other hand, on the *homogeneity* and the *isotropy* of space (as for electrostatic influence, the medium may be different from vacuum).

[1] Which displacement? The term "displacement," normally meaning a change of position in space, is classically used for the vector $\boldsymbol{G}$ due to the resemblance of the expression $\boldsymbol{g.dG}$ to the expression $\boldsymbol{F.d\ell}$, giving in translational mechanics the work of a force. The same analogy is invoked for naming the "electric displacement" $\boldsymbol{D}$ with the expression of the "work of electric volume forces" equal to $\boldsymbol{E.dD}$.

[2] Gravitization. For naming the *gravitational displacement* $\boldsymbol{G}$, we propose the word "gravitization," on the template of *magnetization*, which is used for the equivalent variable in magnetism (see case study D6 "Magnetic Interaction"). Such a termination in "-ization" has the advantage of meaning an *action*, which is appropriate. On the same template, we propose for the *electric displacement* $\boldsymbol{D}$ the word "electrization." All these three variables share the common feature of corresponding to the reduction depth on a surface of a density of entity number.

Analogy with electrodynamics. This model of influence between massive bodies relies on identical elements with electrodynamics. Coulomb's law expressing the forces that appear between two electric charges has strictly the same shape as the double Equation D5.1 and is obviously represented with exactly the same Formal Graph.

Improvement. These hypotheses are classical and feature, by including the invariance of the inertial mass, the whole Newtonian mechanics. The Formal Graph allows us to put a finger on some improvements that can be brought to this model in widening or removing each simplification. The improvements brought by Einstein in his theory of general relativity notably concern the introduction of a curvature of space instead of homogeneity (linearity).

Is the gravitational constant a constant? According to Newton and Einstein, the "gravitational permittivity" $1/4\pi G$ is a constant. If one has confidence in the Formal Graph approach, comparison with electrodynamics, where the electric permittivity ε may vary depending on the medium, leads to removal of this constancy for some media. This is still a subject of research within the framework of a theory that would unify all forces in physics.

7.2.6 Case Study D6: Magnetic Interaction — Magnetism

The case study abstract is given in next page.

Magnetism must not be confused with electromagnetism, the inductive subvariety of electrodynamics. This is a peculiar energy variety contained in magnetized materials (magnets or diamagnetic matter) under the form of the inductive subvariety. To avoid confusion, the term "magnetization energy" may be used. The entity number is the magnetic (or magnetization) impulse (and not a "magnetic charge") p_m that has for unit the ampere meter (A m), and the energy-per-entity is the *current (or flow) of magnetization* or the *magnetic current* f_m.

The force of attraction between two magnetic poles separated by a distance $\boldsymbol{r}$ is given in magnetism by a double equation similar to Coulomb's law in electrostatics (see case study D4 "Coulomb's Law"), in noting that the formal equivalents (both are entity numbers) to the electrostatic charges are the magnetic impulses p_{m1} and p_{m2} (see case study A4 "Magnetization" in Chapter 4):

$$\boldsymbol{F}_1 = -\boldsymbol{F}_2 = -\frac{\mu}{4\pi}\frac{p_{m1}\,p_{m2}}{r^2}\frac{\boldsymbol{r}}{|\boldsymbol{r}|} \qquad \text{(D6.1)}$$

The scaling factor is the *magnetic permeability* μ, which may vary according to the medium in which the magnetic poles are placed. In vacuum, this permeability is equal to $4\pi \times 10^{-7}$ V s A^{-1} m^{-1} and the ratio of the vector distance $\boldsymbol{r}$ on its modulus determines the orientation of the vector force $\boldsymbol{F}_1$ colinear to the vector distance, oriented from pole 1 toward pole 2, which means that the two poles attract themselves if they have opposite signs (north and south) and repel each other in the inverse case.

In fact, the double Equation D6.1 results from the composition of four relations, two for each pole,

$$\boldsymbol{F}_1 = p_{m1}\,\boldsymbol{f}_{m/r1} \qquad \boldsymbol{F}_2 = p_{m2}\,\boldsymbol{f}_{m/r2} \qquad \text{(D6.2)}$$

D6: Magnetic Interaction **Magnetism**

$$\boldsymbol{F}_1 = -\boldsymbol{F}_2 = -\frac{\mu}{4\pi}\frac{p_{m1}\,p_{m2}}{\boldsymbol{r}^2}\frac{\boldsymbol{r}}{|\boldsymbol{r}|}$$

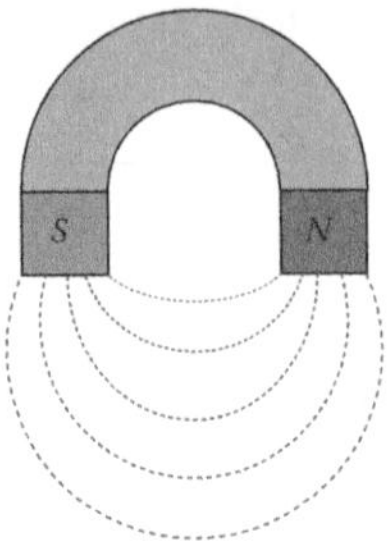

Space dipole

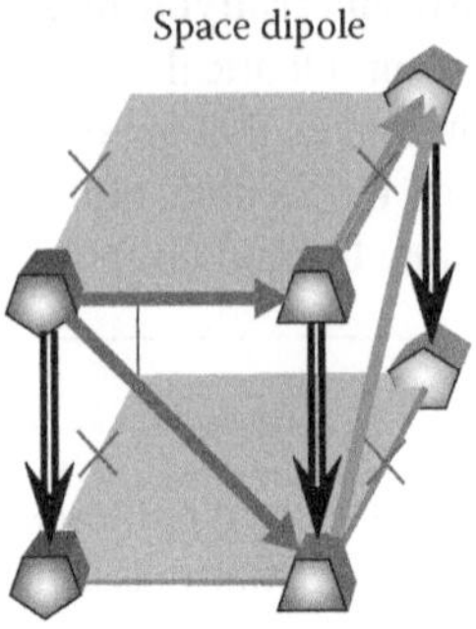

Dipole	
✓	Fundamental
	Mixed
	Capacitive
✓	Inductive
	Conductive
	Global
✓	Spatial
	Inseparable
✓	Conservative

Formal Graph (see Graph 7.11): (in three dimensions)

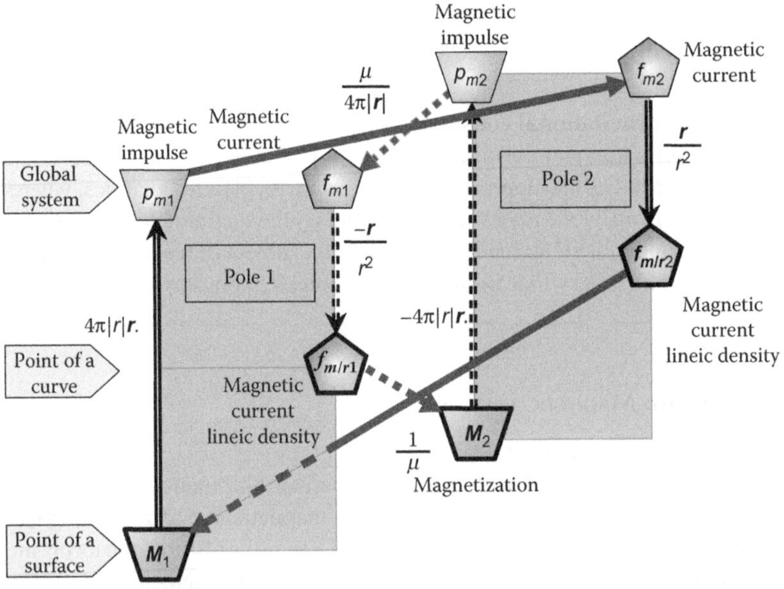

GRAPH 7.11

Variables:

Variety	**Magnetism**		
Subvariety	Inductive (Magnetic)		Capacitive
Category	Entity number	Energy/ entity	Entity number
Family	Impulse 	Flow 	Effort 
Name	Magnetic impulse	Magnetic current	Magnetic potential
Symbols	p_m	f_m	e_m
Unit	[A s]	[J A^{-1} s^{-1}]	[A]

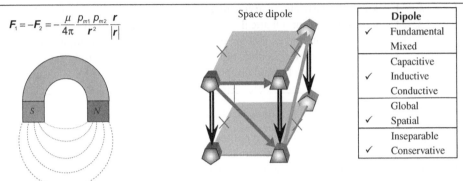

Property relationships:

Mutual influence:

$$f_{m1} = \cancel{\hat{L}}^{-1}_{m11}\,p_{m1} + \hat{L}^{-1}_{m12}\,p_{m2}$$

$$f_{m2} = \hat{L}^{-1}_{m21}\,p_{m1} + \cancel{\hat{L}}^{-1}_{m22}\,p_{m2}$$

$$\hat{L}^{-1}_{m12} = \hat{L}^{-1}_{m21} = \hat{L}^{-1}_{m}$$

$$\hat{\mu}_{12} = \hat{\mu}_{21} = \hat{\mu}$$

Linear model:

$$L_m = \frac{4\pi|r|}{\mu}$$

System Property	
Nature	Reluctance
Levels	2/1
Name	Magnetic permeability
Symbol	μ
Unit	[H m^{-1}]
Specificity	Three-dimensional

Energy variation:

$$\mathrm{d}\mathcal{T}_m \underset{\exists L_m}{=} e_m\,\mathrm{d}p_m$$

$$\mathbf{f}_{m/r1} = -\frac{\mu}{4\pi}\frac{p_{m2}}{\mathbf{r}^2}\frac{\mathbf{r}}{|\mathbf{r}|} \qquad \mathbf{f}_{m/r2} = \frac{\mu}{4\pi}\frac{p_{m1}}{\mathbf{r}^2}\frac{\mathbf{r}}{|\mathbf{r}|} \tag{D6.3}$$

in which the vectors $\mathbf{f}_{m/r1}$ and $\mathbf{f}_{m/r2}$ represent the lineic densities (contragradients) of magnetic currents (not to be confused with the electromagnetic fields $\mathbf{H}$ or $\mathbf{B}$) exerted by one of the poles at the level of the other pole ($\mathbf{f}_{m/r1}$ is created by p_{m2} and $\mathbf{f}_{m/r2}$ by p_{m1}). The localizations of the various variables involved in the model are (cf. Figures D6.1 and D6.2):

$\mathbf{F}_1$, p_{m1}, f_{m1}, $\mathbf{f}_{m/r1}$, and $\mathbf{M}_2$ (this is the subscript two) at the level of the first pole p_{m1}.
$\mathbf{F}_2$, p_{m2}, f_{m2}, $\mathbf{f}_{m/r2}$, and $\mathbf{M}_1$ (this is the subscript one) at the level of the second pole p_{m2}.

The relations D6.2 between contragradient of magnetic current and force are relevant to a subject different from the one exposed here, which is the coupling between magnetism and translational mechanics. This subject is completely analogous to the force resulting from a coupling between translational mechanics and other energy varieties such as gravity and electrostatics (see case studies J4 "Gravitational Force" and J3 "Electric Force" in Chapter 12).

In the Formal Graph in the case study abstract, two circular paths formed by the system constitutive and space properties can be seen. They represent the relations D6.2 and D6.3 and each loop contains the relation of mutual influence between the two poles distributed spatially that constitute the two magnetic impulses. The loop in dotted lines features the influence of pole 2 on pole 1 and the loop in solid line features the other influence. The horizontal upper paths express the influences between state variables (global level) and the slanted lower paths express the reciprocal influences

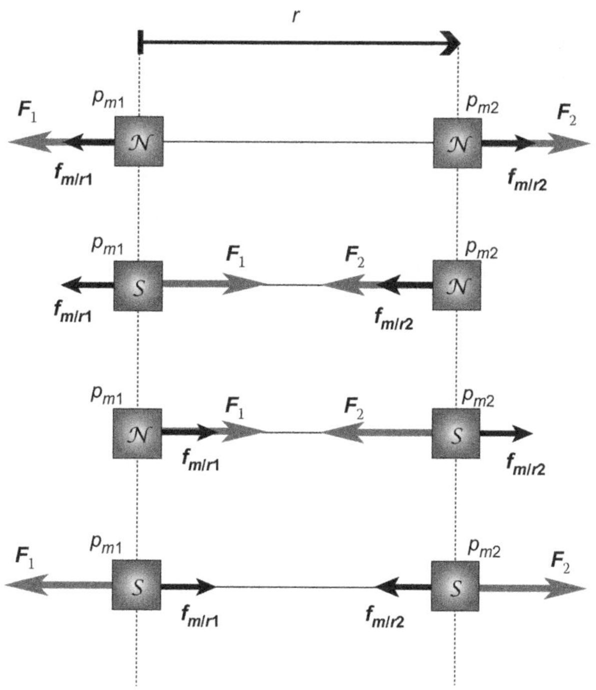

FIGURE D6.1 Scheme showing the four possibilities of placing two magnetic poles along an oriented horizontal axis. The theory of the magnetic influence involves the vector quantities shown in this scheme, which are: (1) The forces of attraction $\mathbf{F}_1$ and $\mathbf{F}_2$. (2) The lineic densities of magnetic currents $\mathbf{f}_{m/r1}$ and $\mathbf{f}_{m/r2}$.

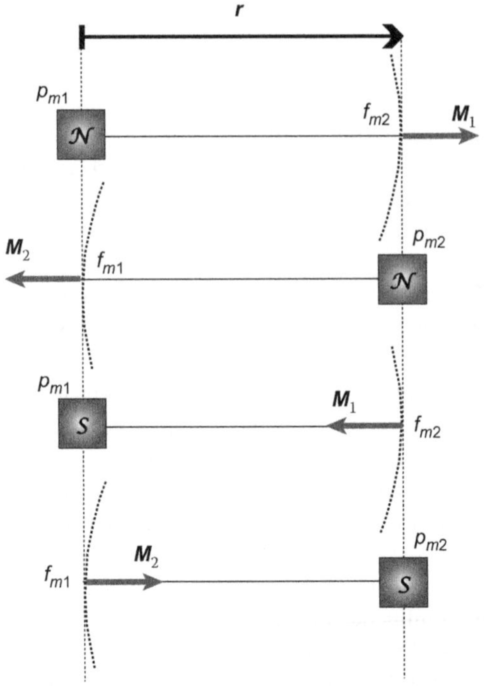

FIGURE D6.2 Scheme illustrating the orientation taken by the two vectors that are the magnetizations $\boldsymbol{M}_1$ and $\boldsymbol{M}_2$ determined on spheres of radius $\boldsymbol{r}$.

between spatially reduced variables that are the *lineic densities of magnetic current* $\mathbf{f}_{m/r}$ and *magnetization* $\boldsymbol{M}$. (For building a loop, these influences have been reversed: This is a *magnetization* $\boldsymbol{M}$ that creates the *magnetic current density* $\mathbf{f}_{m/r}$ on the other pole, but every link in a Formal Graph can be reversed without inducing a real physical meaning.)

The spatial operators are written with their particular expression corresponding to the hypothesis for retrieving the magnetization model D6.3, which is a homogeneous distribution in space and a *magnetization* $\boldsymbol{M}$ determined on a sphere.

The demonstration of this graph is as follows. It proceeds in several steps, each step corresponding to a physical hypothesis.

7.2.6.1 Mutual Influence

The model relies on the principle of each pole influencing the other but not itself (nonzero mutual influence and zero self-influence), which is translated at the global level by each *magnetic current* being equal to a magnetic cross-reluctance applied to the *magnetic impulse* of the other pole.

$$\begin{aligned} f_{m1} &= \hat{\cancel{L}}^{-1}_{m11}\, p_{m1} + \hat{L}^{-1}_{m12}\, p_{m2} \\ f_{m2} &= \hat{L}^{-1}_{m21}\, p_{m1} + \hat{\cancel{L}}^{-1}_{m22}\, p_{m2} \end{aligned} \tag{D6.4}$$

This mutual influence in absence of auto-influence means that the magnetic current of each pole is determined by the other pole. The magnetization energy of the dipole being a state function, its variation is an exact differential

$$\mathrm{d}\mathcal{T}_m = f_{m1}\, \mathrm{d}p_{m1} + f_{m2}\, \mathrm{d}p_{m2} \tag{D6.5}$$

which implies that the derivatives of magnetic currents with respect to the other magnetic impulse are equal, which is therefore translated by the equality between cross-reluctances taking into account relations D6.4.

$$\hat{L}_{m12}^{-1} = \hat{L}_{m21}^{-1} \tag{D6.6}$$

7.2.6.2 Spherical Surfaces

The surface element of the system is supposed to have a spherical shape, which amounts to assuming an *isotropic* space, equivalent in all directions. The relation between the magnetic impulse contained inside a sphere and the *magnetization* $\boldsymbol{M}$ on the surface of the sphere is therefore an integral on the total area of the sphere (see case study B5 "Gauss Theorem" in Chapter 5, which expresses it in electrostatics). Each magnetization created by a positive (north pole) magnetic impulse is a vector oriented toward the exterior of the sphere (and conversely toward the center in the case of a negative—south pole—magnetic impulse). The sphere radius is taken to be equal to the distance between the magnetic impulses, which, taking into account the orientation of pole 1 toward pole 2 (cf. scheme in Figure D6.1), gives

$$p_{m1} = \oint_{sphere} \boldsymbol{M}_1.\mathbf{d}\boldsymbol{A} = 4\pi|r|\boldsymbol{M}_1.\boldsymbol{r} \qquad p_{m2} = \oint_{sphere} \boldsymbol{M}_2.\mathbf{d}\boldsymbol{A} = -4\pi|r|\boldsymbol{M}_2.\boldsymbol{r} \tag{D6.7}$$

This notion of vector *magnetization* $\boldsymbol{M}$ must not be considered as a "density of magnetic impulse" representative of a real distribution of entities on the surface, which should be contrary to one's perception of a structural unity of an entity (or, more exactly here, of an impulse), but as a *magnetization field* $\boldsymbol{M}$ analogous to the *gravitational displacement* (*gravitization*) $\boldsymbol{G}$ or to the *electric displacement* (*electrization*) $\boldsymbol{D}$. This latter plays exactly the same role as the electric field (to which it is proportional through the electric permittivity), which exists at a distance from the charged body that creates it and can propagate through space. The relations in Equation D6.7 are nothing other than the analogue of the Gauss theorem in electrostatics, which states that the totality of the charge contained in a sphere is equal to the integral of the electric displacement on the surface of this sphere (or as a function of the electric field through the permittivity, as in Equations D6.10, by analogy with the gravitational field but with the lineic density of magnetic current).

7.2.6.3 Space Homogeneity (Linearity)

The spatial distribution of magnetic current density is assumed to be regular, which amounts to assuming a *homogeneous space*. In these conditions, the contragradient is equivalent to a division of the magnetic current density by the distance and to a multiplication by plus or minus the unit vector (ratio of the vector $\boldsymbol{r}$ on its modulus) in order to respect the orientation of each magnetic current density (refer to the scheme in Figure D6.1, right, and consider as negative scalars the magnetic current densities shown later):

$$\begin{aligned} \boldsymbol{f}_{m/r1} &= -\frac{\partial}{\partial \boldsymbol{r}} f_{m1} \underset{hom}{=} -\frac{1}{\boldsymbol{r}} f_{m1} = -f_{m1}\frac{\boldsymbol{r}}{r^2} \\ \boldsymbol{f}_{m/r2} &= -\frac{\partial}{\partial \boldsymbol{r}} f_{m2} \underset{hom}{=} -\frac{1}{\boldsymbol{r}} f_{m2} = -f_{m2}\frac{\boldsymbol{r}}{r^2} \end{aligned} \tag{D6.8}$$

7.2.6.4 Magnetic Permeability

The spatially reduced reluctances that are specific to the medium are magnetic permeabilities in this domain of energy. To each reluctance at the global level corresponds a magnetic permeability μ_{ij} and, therefore, the equality D6.6 corresponds to an equality between mutual permeabilities. Their common value is a scalar assumed to be a constant, that is, the hypothesis of a *homogeneous medium* constituting the system is made.

$$\hat{\mu}_{12} = \hat{\mu}_{21} = \hat{\mu} \underset{hom}{=} \mu \tag{D6.9}$$

Consequently, the two reduced inductive relationships of influence are written as:

$$\boldsymbol{M}_1 \underset{hom}{=} \frac{1}{\mu}\boldsymbol{f}_{m/r2} \qquad \boldsymbol{M}_2 \underset{hom}{=} \frac{1}{\mu}\boldsymbol{f}_{m/r1} \tag{D6.10}$$

Synthesis: In the Formal Graph above, the reluctances (reciprocals of inductances) are given directly with their expression particular to the considered system, which results from the treatment below. Before knowing the solution, the reasoning consists of examining the two loops formed by both reluctances and their reduced property and of applying the *circularity principle*:

$$\begin{aligned} &\hat{L}^{-1}_{m12}\left(-4\pi|r|\boldsymbol{r}.\right)\frac{1}{\mu}\left(-\frac{\boldsymbol{r}}{r^2}\right)=1 \\ &\hat{L}^{-1}_{m21}\left(4\pi|r|\boldsymbol{r}.\right)\frac{1}{\mu}\left(\frac{\boldsymbol{r}}{r^2}\right)=1 \end{aligned} \tag{D6.11}$$

From this is deduced the common solution

$$\hat{L}^{-1}_{m12} = \hat{L}^{-1}_{m21} = \frac{\mu}{4\pi|r|} \tag{D6.12}$$

which establishes the proportionality of the mutual reluctance with the inverse of the distance. The two inductive relationships are therefore

$$\begin{aligned} f_{m1} &= \frac{\mu}{4\pi}\frac{p_{m2}}{|r|} \\ f_{m2} &= \frac{\mu}{4\pi}\frac{p_{m1}}{|r|} \end{aligned} \tag{D6.13}$$

By injecting the expressions of two magnetic currents into the relations D6.8, one obtains the relations D6.3, which achieves the demonstration.

Remarks

The inventor. Although no particular name of inventor is attached to it, the formula giving the force exerted between two magnetic poles was proposed (with the unsuitable concept of "magnetic charge") by Charles-Augustin de Coulomb in 1785 based on the model of his law (Coulomb's law) for the force appearing between two electric charges (Hecht 1996).

Analogies. This model of influence between magnetization impulses is based on identical elements with the electrostatic influence (cf. case study D4 "Coulomb's Law") and with the attractive influence in the domain of gravitation (cf. case study D5 "Newtonian Gravitation"). The Newtonian model expressing the forces acting on two gravitational masses has strictly the same shape as the double Equation D6.1 and is obviously represented with exactly the same Formal Graph. Just two details are different between the two models: for retrieving the Newtonian model entirely, the magnetic permeability μ must be replaced by the gravitational permittivity $1/4\pi G$ and the magnetic impulses p_{m1} and p_{m2} by the gravitational masses with a minus sign, that is, $-M_{G1}$ and $-M_{G2}$. This is due to an intrinsic difference in behavior; in gravity, two masses with the same sign (positive) attract themselves whereas two magnetization impulses with the same sign repel each other.

This model can be improved. The model relies on several simplifying hypotheses that the Formal Graphs bring to light. These hypotheses are based, on the one hand, on the medium homogeneity for being able to utilize an inductive reduced property that is independent of the distance and the orientation (scalar magnetic permeability) and, on the other hand, on the homogeneity and the isotropy of space. (As for electrostatic influence, the medium may be different from vacuum.) These hypotheses are classical and the Formal Graphs allows one to put a finger on some improvements that can be brought to this model in widening or removing each simplification.

7.3 POLES–DIPOLE CONSTITUTIVE PROPERTIES

This section develops the role played by the link between state variables of the two poles realized by the influence phenomenon. The first point deals with the demonstration of the identity of mutual constitutive properties, capacitances, and inductances in the generalized case, and the second point establishes the models of dipole properties in function of the pole properties.

7.3.1 EQUALITY BETWEEN MUTUAL SYSTEM PROPERTIES

Through the various case studies given in this chapter, the equality between mutual elastances or reluctances has been discussed and demonstrated in each specific domain. Now is presented the generalization of this essential identity between mutual constitutive properties, which comes from the fundamental property of energy as a state function. The capacitive energy subvariety is treated in detail by the algebraic approach and the result for the inductive subvariety is merely given without demonstration, as it follows by simple substitution of relevant variables. The reasoning for both subvarieties is then transposed into the Formal Graph.

The generalized capacitive relationships have been given in Equations 7.3 and 7.4 and are recalled here:

$$e_{q1} = \hat{C}^{-1}_{q11} q_1 + \hat{C}^{-1}_{q12} q_2$$
$$e_{q2} = \hat{C}^{-1}_{q21} q_1 + \hat{C}^{-1}_{q22} q_2$$

The two mutual capacitances (or their reciprocal, the mutual elastances) are equal by virtue of the properties of energy as demonstrated in the following expressions.

The energy of a dipole is a state function; otherwise, the system could not store energy. The variation of this capacitive energy is necessarily an exact differential

$$d\mathcal{U}_q = e_{q1}\, dq_1 + e_{q2}\, dq_2 \tag{7.11}$$

The condition here for having an exact differential being

$$\left(\frac{\partial e_{q1}}{\partial q_2}\right)_{q_1} = \left(\frac{\partial e_{q2}}{\partial q_1}\right)_{q_2} \tag{7.12}$$

and as from Equations 7.3 and 7.4 the definition of mutual elastances is

$$\hat{C}_{q12}^{-1} = \hat{E}_{q12} \overset{def}{=} \left(\frac{\partial e_{q1}}{\partial q_2}\right)_{q_1} \qquad \hat{C}_{q21}^{-1} = \hat{E}_{q21} \overset{def}{=} \left(\frac{\partial e_{q2}}{\partial q_1}\right)_{q_2} \tag{7.13}$$

the equality of these mutual elastances directly follows by comparing the last equations

$$\hat{C}_{q12}^{-1} = \hat{C}_{q21}^{-1} = \hat{E}_{q12} = \hat{E}_{q21} = \hat{C}_{q}^{-1} = \hat{E}_{q} \tag{7.14}$$

The same reasoning is followed for the mutual reluctances in starting from Equations 7.1 and 7.2, leading to similar identities:

$$\hat{L}_{q12}^{-1} = \hat{L}_{q21}^{-1} = \widehat{\mathfrak{R}}_{q12} = \widehat{\mathfrak{R}}_{q21} = \hat{L}_{q}^{-1} = \widehat{\mathfrak{R}}_{q} \tag{7.15}$$

This development is a purely algebraic one for establishing the rigorously founded basis of the demonstration which the reader may recognize. However, another approach, as rigorous as algebra, can be used with the Formal Graph with the concepts of *Differential Formal Graph* and of *graph closure* introduced in Chapter 2, Section 2.1.

The *Differential Formal Graphs* in Graph 7.12 are built for each energy subvariety in using only differential operators that are arranged in pairs. Horizontal links are for the same partial

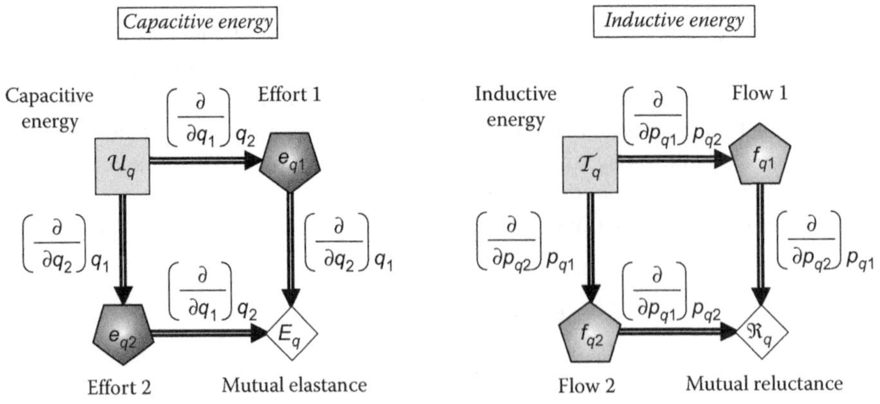

GRAPH 7.12 Two differential Formal Graphs establishing the identity of the mutual system properties by virtue of the property of energy to be a state function.

differentiation with respect to the first entity number and vertical ones with respect to the second entity number.

The closure of the graphs expresses the uniqueness of the node at the end of the two parallel composed paths issued from the capacitive or inductive energy. This amounts to saying that the order of the differential operator in a composed path is meaningless, which is the mathematical translation of the property of energy to define system states that are independent of the followed path for reaching them.

7.3.2 Dipole Capacitance from Pole Capacitances

The question that is tackled now is the relationship between the pole constitutive properties and the corresponding dipole property. To derive this dipole constitutive relationship, the influence relationships introduced in Section 7.1 at the level of the poles need to be combined with the poles–dipole connections between variables. We take the case of the capacitive relationships for detailing the model; the other cases of inductive relationships being similar will be merely given without justification.

The poles–dipole relationships between state variables have been established in Chapter 6. They work both for a separable dipole and for an inseparable one by adjusting the separability factor between the values ±1 (separable) and ± 1/2 (inseparable), the sign depending on the orientation convention chosen (+ for a receptor, – for a generator).

$$s_D\, q = q_1 - q_1^* = q_2^* - q_2 \tag{7.16}$$

$$e_{qC} = s_D e_{q12} = s_D\left(e_{q1} - e_{q2}\right) \tag{7.17}$$

The stars used as superscripts indicate constant quantities q_i* referring to the dissociated situation of poles (dissociated, "off state"). By replacing the pole efforts by their expressions in Equations 7.3 and 7.4, the dipole effort is written as:

$$e_{qC} = s_D\left(\left\{\hat{C}_{q11}^{-1} - \hat{C}_{q21}^{-1}\right\}q_1 + \left\{\hat{C}_{q12}^{-1} - \hat{C}_{q22}^{-1}\right\}q_2\right) \tag{7.18}$$

Now, by defining two apparent elastances for the poles,

$$\hat{C}_{q1}^{-1}\, q \overset{def}{=} s_D\left\{\hat{C}_{q11}^{-1} - \hat{C}_{q21}^{-1}\right\}\left(q_1^* + s_D\, q\right) \tag{7.19}$$

$$\hat{C}_{q2}^{-1}\, q \overset{def}{=} -s_D\left\{\hat{C}_{q22}^{-1} - \hat{C}_{q12}^{-1}\right\}\left(q_2^* - s_D\, q\right) \tag{7.20}$$

and one dipole elastance,

$$\hat{C}_{q}^{-1} \overset{def}{=} \hat{C}_{q1}^{-1} + \hat{C}_{q2}^{-1} \tag{7.21}$$

the capacitive relationship for the dipole can be written concisely as:

$$q = \hat{C}_q\, e_{qC} \tag{7.22}$$

This is the general relationship for any shape of capacitive relationship in which the dipole capacitance depends on the state variables. Naturally, this dependence vanishes when both pole

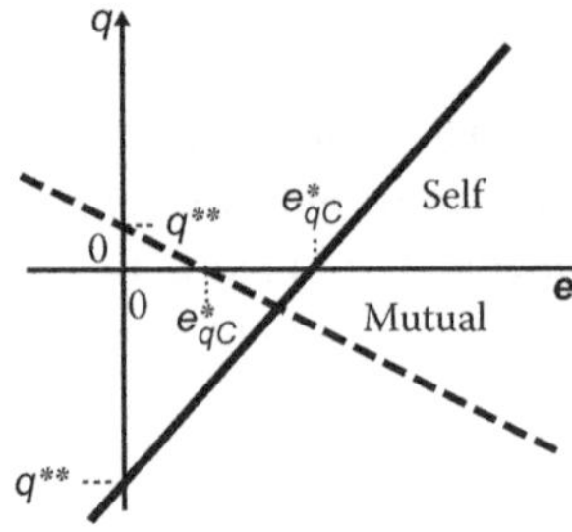

FIGURE 7.2 Plot of the dipole capacitance in the linear case and in the two extreme situations of pure self-influence (solid line) and pure mutual influence (dashed line). The slopes (in absolute values) and the values of the effort origins are not the same, unless all capacitances are equal.

capacitances are scalars. This corresponds to the linear case in which a dipole elastance and an effort origin can be defined independent of the basic quantities as follows:

$$\textit{Dipole elastance}: \quad C_q^{-1} \underset{lin}{\overset{def}{=}} s_D^2\left(C_{q11}^{-1} + C_{q22}^{-1} - C_{q12}^{-1} - C_{q21}^{-1}\right) \tag{7.23}$$

$$e_{qC}^* \underset{lin}{\overset{def}{=}} s_D\left(C_{q11}^{-1} - C_{q21}^{-1}\right)q_1^* + s_D\left(C_{q12}^{-1} - C_{q22}^{-1}\right)q_2^* \quad \text{and} \quad q^{**} \underset{lin}{=} C_q e_{qC}^* \tag{7.24}$$

With these definitions, the linear capacitive relationship is written as:

$$q \underset{lin}{=} C_q\left(e_{qC} - e_{qC}^*\right) \quad \text{or} \quad e_{qC} \underset{lin}{=} C_q^{-1}\left(q - q^{**}\right) \tag{7.25}$$

The special, but not so uncommon, case of equality between the self-elastances (mutual elastances are equal by definition) leads to simpler relations for the dipole elastance and the effort origin:

$$C_{q11}^{-1} = C_{q22}^{-1} \Leftrightarrow C_q^{-1} \underset{lin}{=} 2s_D^2\left(C_{q11}^{-1} - C_{q21}^{-1}\right); \quad q^{**} \underset{lin}{=} \frac{q_1^* - q_2^*}{2s_D} \tag{7.26}$$

The interesting point in this derivation is the equivalent and symmetrical role played by the self-capacitances and the mutual capacitances of the poles in the overall capacitance, as shown in Figure 7.2 for the simplified case of linear capacitances.

7.3.3 Dipole Inductance from Pole Inductances

With the following definitions of apparent reluctances for the poles

$$\hat{L}_{q1}^{-1} p_q \overset{def}{=} s_D\left\{\hat{L}_{q11}^{-1} - \hat{L}_{q21}^{-1}\right\}\left(p_{q1}^* + s_D p_q\right) \tag{7.27}$$

$$\hat{L}_{q2}^{-1} p_q \overset{def}{=} -s_D\left\{\hat{L}_{q22}^{-1} - \hat{L}_{q12}^{-1}\right\}\left(p_{q2}^* - s_D p_q\right) \tag{7.28}$$

and with the definition of the dipole reluctance as

$$\hat{L}_q^{-1} \overset{def}{=} \hat{L}_{q1}^{-1} + \hat{L}_{q2}^{-1} \tag{7.29}$$

the inductive relationship for the dipole is written as:

$$f_{qL} = \hat{L}_q^{-1}\, p_q \tag{7.30}$$

The linear case is described by the following definitions of dipole reluctance and origins for flow and impulse:

$$L_q^{-1} \underset{lin}{\overset{def}{=}} s_D^2\left(L_{q11}^{-1} + L_{q22}^{-1} - L_{q12}^{-1} - L_{q21}^{-1}\right) \tag{7.31}$$

$$f_{qL}^{*} \underset{lin}{\overset{def}{=}} s_D\left(L_{q11}^{-1} - L_{q21}^{-1}\right)p_{q1}^{*} + s_D\left(L_{q12}^{-1} - L_{q22}^{-1}\right)p_{q2}^{*} \quad \text{and} \quad p_q^{**} \underset{lin}{=} L_q f_q^{*} \tag{7.32}$$

With these definitions, the linear inductive relationship is written as:

$$p_q \underset{lin}{=} L_q\left(f_{qL} - f_{qL}^{*}\right) \quad \text{or} \quad f_{qL} \underset{lin}{=} L_q^{-1}\left(p_q - p_q^{**}\right) \tag{7.33}$$

The same shapes as in Figure 7.2 are observed for this simplified case.

7.3.4 Dipole Conductance from Pole Conductances

The conductive influence relationships have been introduced in Section 7.1 in Equations 7.5 and 7.6; they are recalled hereafter for the case of a conduction ensured by exchange by flows.

$$f_{q1} = \hat{G}_{q11}\, e_{q1} + \hat{G}_{q12}\, e_{q2}$$
$$f_{q2} = \hat{G}_{q21}\, e_{q1} + \hat{G}_{q22}\, e_{q2}$$

The poles–dipole relationships between state variables have been established in Chapter 6. For an exchange by flows, one has

$$f_q = f_{q1} - f_{q2} \tag{7.34}$$

The replacement of the pole flows by their expressions 7.5 and 7.6 provides a mixed poles–dipole conductive relationship

$$f_q = \hat{G}_{q1}\, e_{q1} - \hat{G}_{q2}\, e_{q2} \tag{7.35}$$

in having defined two apparent conductances

$$\hat{G}_{q1} \overset{def}{=} \hat{G}_{q11} - \hat{G}_{q21}; \quad \hat{G}_{q2} \overset{def}{=} \hat{G}_{q22} - \hat{G}_{q12} \tag{7.5}$$

This relationship is mixed because it relates the pole efforts to the dipole flow, which is useful when these efforts are known. When it is not the case, the dipole conductive relationship can be written as a dipole property by using a dipole conductance

$$f_q = \hat{G}_q\, e_q \tag{7.36}$$

which is defined as (apart from the special case of equal pole efforts)

$$\hat{G}_q = \frac{\hat{G}_{q1}\, e_{q1} - \hat{G}_{q2}\, e_{q2}}{e_{q1} - e_{q2}} \qquad \left(e_{q1} \neq e_{q2}\right) \tag{7.37}$$

in having used the following pole–dipole connection for the efforts

$$e_q = e_{q1} - e_{q2} \tag{7.38}$$

This is not a simple model because the dipole conductance is dependent on the pole efforts, even when the pole conductances are scalar, at least as long as the pole conductances are different. In the linear case, the conductance expression keeps the same shape:

$$G_q \underset{lin}{=} \frac{G_{q1}\, e_{q1} - G_{q2}\, e_{q2}}{e_{q1} - e_{q2}} \qquad \left(e_{q1} \neq e_{q2}\right) \tag{7.39}$$

When the pole conductances are scalar and equal, the dipole conductance becomes equal to their common value (notated with an equal sign between the subscripts to indicate that their order does not matter):

$$G_{q1} = G_{q2} = G_{q1=2} \Rightarrow G_q = G_{q1=2} \tag{7.40}$$

Dynamic equilibrium and anisotropy. Contrary to a widespread belief, the condition for having a null flow in an exchange (i.e., dynamic equilibrium featured by equality between forward and backward pole flows) is not always the equality of efforts between the poles. This works only in case of equal conductances, meaning that the medium in which conduction occurs must be isotropic and homogeneous.

7.4 INFLUENCE THEORY

The fact that the same mathematical operator may work for all capacitances throughout the various domains encountered is not fortuitous. However, according to classical theories, the exponential shape that has been met in some case studies is the result of Boltzmann statistics and the linear shape met in other case studies comes from fundamental assumptions, as in electrostatics or in gravitation.

The proposition of a unique exponential shape, which may tend toward a linear asymptote in a restricted range, is a first step in evidencing the homogeneity among domains that is induced by the observed similarity between graph structures. The problem is that this approach is purely heuristic and not based on firm roots.

The second step that is necessary is to propose rigorous foundations for this exponential shape that are compatible with the characteristics of the different systems encountered. The main difficulty is that the statistical approach cannot be used in all cases because it requires special conditions, such as thermal agitation, and that another approach, valid for immobile systems as well, must be found.

To pave the way toward the proposed solution, an example of a system possessing a behavior classically modeled by an exponential law is discussed. The optical transmission of light through a material being not subjected to thermal agitation, the development that leads to an exponential function is worth analyzing.

This discussion will be followed by applying the same reasoning to the influence phenomenon in the general case. As influence proceeds by using system constitutive properties, the outcome will be the demonstration of the exponential function featuring the generalized capacitance operator.

7.4.1 The Lambert–Beer–Bouguer Law for Radiation Transmission

In optics, the Lambert*–Beer†–Bouguer‡ law models the transmission of radiation through an absorbing medium. This physical phenomenon follows an exponential law that can be established without involving any statistical arguments, as will be proved in case study H8 "Attenuated Propagation" in Chapter 11. It may lead to a new comprehension of the role of influence between objects.

A radiation beam of wavelength λ crossing an absorbing medium sees its intensity, called *irradiance*, $\mathcal{I}_{\lambda q}$ (the subscript q refers to the energy variety of the radiation, electrodynamics for instance) decaying with an exponential law as a function of the medium thickness Δl, according to the Lambert–Beer–Bouguer law (Figure 7.3):

$$T_{\lambda q} \overset{def}{=} \frac{\mathcal{I}_{\lambda q}}{\mathcal{I}^{\circ}_{\lambda q}} = \exp\left(-\frac{\Delta l}{l_\Delta}\right) \tag{7.41}$$

The first scaling factor is the beam irradiance (intensity) $\mathcal{I}^{\circ}_{\lambda q}$ at the entrance of the absorbing medium (incident radiation). Division of the beam irradiance by this factor gives the *transmittance* $T_{\lambda q}$ (also called *attenuation* in the language of signal processing) of the device. The subscript λ indicates dependence on the wavelength (and avoids confusion with the temperature).

The second one is the scaling length l_Δ, which depends on the absorbing property of the medium. The physical model of the medium assumes that total absorption is caused by N objects each having a cross-section (projected area) κ_N, also called *specific absorbance*. So, the ratio of lengths in the Lambert–Beer–Bouguer law must be understood as the ratio of the absorbing area upon the total area, which is also equal to a ratio of volumes:

$$\frac{\Delta l}{l_\Delta} = \frac{N\kappa_N}{A} = \frac{N\kappa_N \Delta l}{V} = c_N \kappa_N \Delta l = c_n \kappa_n \Delta l \tag{7.42}$$

It is customary to express these ratios as proportional to the volumic concentration c_N (in corp m^{-3}) or c_n (in mol m^{-3}) of absorbing obstacles met along the beam path. When the choice is made to count in moles, the specific absorbance becomes a *molar specific absorbance* κ_n (in m^2 mol^{-1}).§

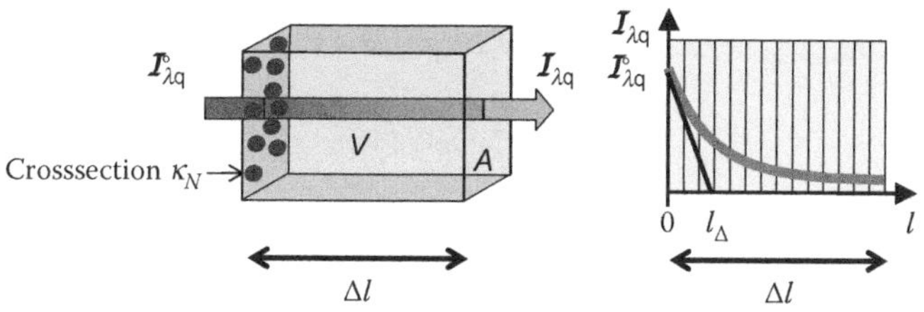

FIGURE 7.3 Three-dimensional view of a beam crossing a medium of thickness Δl and area A filled with N absorbing objects of cross-section κ_N (left) and plot of the intensity decay as a function of the penetration length (right). The theory of the decay relies on slicing the medium into successive layers.

* Johann Heinrich Lambert (1728–1777): Swiss mathematician, physicist, and astronomer; Berlin, Germany.

† August Beer (1825–1863): German physicist and mathematician; Bonn, Germany.

‡ Pierre Bouguer (1698–1758): French mathematician, geophysicist, and astronomer; Le Havre and Paris, France.

§ In physical chemistry, it is customary to write the Lambert–Beer–Bouguer law in decadic mode (as a power of ten) (Atkins 2010). The decadic *molar specific absorbance* ε_n is then equal to $\kappa_n / \ln 10$.

Intensity, radiation intensity, and irradiance. A comparison between the models of absorption (Lambert–Beer–Bouguer law) and conduction (see case study A11 in Chapter 4) may induce the conclusion that irradiance and flow are similar because both models use an exponential function. In addition, the notion of radiation *intensity*, often used in lieu of irradiance, induces the same confusion because it is also commonly used in electrodynamics (one speaks about *current intensity*, or even *intensity* alone, which is not recommended). This similarity is not physically true: in a (static) *conduction* process the flow or current is *constant* throughout the conductor whereas in an absorbing medium the irradiance decreases and is therefore lower at the output than at the input. The irradiance is not a state variable but a *power density* (in W m^{-2}), which is the spatially reduced product of the two energies-per-entity (i.e., effort times flow), as seen later in Chapters 9 and 11. It means that radiant energy is not an energy variety per se and that analogies must be handled with care. To conclude, *irradiance* is a better word for avoiding confusion.

Demonstration of this important Lambert–Beer–Bouguer law (often presented as empirical) is very simple and instructive for the purpose of this study. It relies on the basic idea that each layer of the medium contributes to a modification of the transmittance (attenuation) in a multiplicative manner. Everything is said with this sentence because it is implicit that the layer thicknesses are additive. Nevertheless, a mathematical derivation is worth giving. The scheme in Figure 7.4 shows the effect on the beam of a first layer with thickness Δl_{12}, spanning from distance l_1 to l_2, which is mathematically expressed as

$$l_2 = \Delta l_{12} + l_1; \quad I_{\lambda q2} = T_{\lambda q12} I_{\lambda q1} \tag{7.43}$$

and for a second layer

$$l_3 = \Delta l_{23} + l_2; \quad I_{\lambda q3} = T_{\lambda q23} I_{\lambda q2} \tag{7.44}$$

When this second layer is adjoined to the first one, the two layers behave as a single one, but each with its own features, as illustrated in Figure 7.4.

$$l_3 = \Delta l_{13} + l_1; \quad I_{\lambda q3} = T_{\lambda q13} I_{\lambda q1} \tag{7.45}$$

The important result that comes from a comparison of the two last cases is

$$\Delta l_{13} = \Delta l_{12} + \Delta l_{23} \tag{7.46}$$

$$T_{\lambda q13} = T_{\lambda q12}\, T_{\lambda q23} \tag{7.47}$$

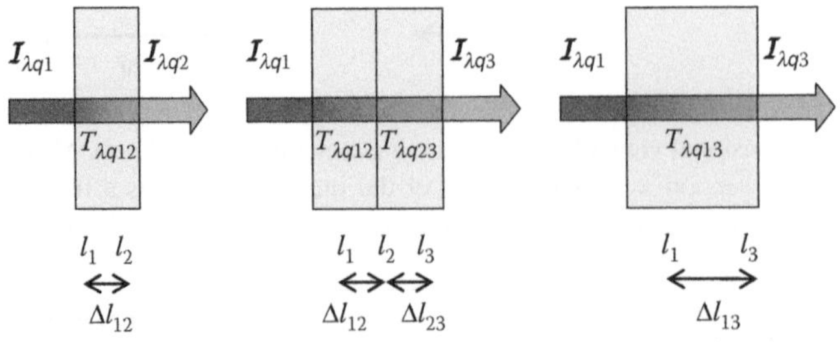

FIGURE 7.4 Geometric view of a beam crossing one layer (left), two layers (middle) and one equivalent of two layers (right) of the medium.

That is, one has an addition on one side and a multiplication on the other side. This is precisely the remarkable property of the exponential function to obey this duality. This function can be introduced by deriving the previous relationships:

$$\mathrm{d}\Delta l_{13} = \mathrm{d}\Delta l_{12} + \mathrm{d}\Delta l_{23} \tag{7.48}$$

$$\frac{\mathrm{d}T_{\lambda q13}}{T_{\lambda q13}} = \frac{\mathrm{d}T_{\lambda q12}}{T_{\lambda q12}} + \frac{\mathrm{d}T_{\lambda q23}}{T_{\lambda q23}} \tag{7.49}$$

The crucial assumption to be made now is that there should be independence between each layer, so that each term in these two equations corresponds one to one. This means that each member of the first equation is proportional to its matching member through the same factor, noted l_Δ and taken with a negative sign for counterbalancing the negative variation of transmittance. For instance, the first members of the right-hand side of these equations are related with this factor:

$$\mathbf{d}\Delta l_{12} = -l_\Delta \frac{\mathbf{d}T_{\lambda q12}}{T_{\lambda q12}} \tag{7.50}$$

By integration, one gets an exponential function, and in choosing integration constants such as a transmittance equal to 1 (no attenuation) for a null thickness

$$T_{\lambda q12} = \exp\left(-\frac{\Delta l_{12}}{l_\Delta}\right) \tag{7.51}$$

which is identical to Equation 7.41 in replacing the layer thickness Δl_{12} by the total thickness Δl and the transmittance $T_{\lambda q12}$ by the total transmittance $T_{\lambda q}$.

7.4.2 Interaction and Influence

This section is devoted to the demonstration of the capacitive relationship as formed with an exponential function. It can be skipped by readers who can admit it without proof.

The following development establishes the model proposed by the Formal Graph theory of the phenomenon of pole–pole interaction and of influence of activities on efforts. The exponential function used in capacitive relationships is introduced by following a rigorous reasoning based on the assumption of independency between variables that depends on the type of influence, self or mutual.

This new theory is not restricted to the physical chemical domain but valid for all energy varieties. For this reason, generic state variables, efforts e_q and basic quantities q, will be used (the subscript q refers to the energy variety that has q for basic quantity).

The key idea in this new theory is to assume that not all entities are directly involved in the influence between poles but only a part of them, called *active entities*. A good example is the dipole formed by a spring or an elastic solid, which cannot be compressed below a certain limit without losing its capacitive (elastic) properties. It means that only an active length of the spring in participating in the influence on the forces acting on the spring. To take this into account, a new variable, borrowed from physical chemistry and generalized to all domains, called the *activity* is introduced and inserted into the capacitance path in a Formal Graph.

7.4.2.1 Activation

The activity is a variable that reflects the availability of an entity for participating in an influence or interaction process. The activity is related to the entity number through an activation operator defined as follows:

$$\text{Activity}: \quad a_{qi} \overset{def}{=} \hat{N}_{qi}\, q_i \overset{def}{=} \frac{q_i - q_i^{min}}{q_i^{\theta} - q_i^{min}} \tag{7.52}$$

Indeed, not every entity may be active in the influence, depending on their ability to vary when capacitive energy varies. If q_i^{min} is the number of inactive entities, the amount of active entities is equal to the difference $q_i - q_i^{min}$, so q_i^{min} represents a minimum value. At the denominator, the reference number of entities q_i^{θ} is chosen to correspond to a value 1 for the activity. The plot in Figure 7.5 illustrates this notion of availability threshold.

Chapter 6 discussed a case study of the mechanics of solids (case study C8 "Spring") in which a threshold was associated with the impossibility of compressing a solid below a certain limit. The thicknesses (which are entity numbers in translational mechanics) of the elastic solid elements cannot be reduced to zero owing to the existence of incompressible grains (atoms or clusters) in the material. In contradistinction, all molecules in a collection of physical chemical entities are available for interaction and therefore no threshold exists in such a case.

The two types of mutual influence are shown in the Formal Graphs in Graphs 7.13 and 7.14, giving two different interpretations of the phenomenon. Graph 7.13 shows the direct translations of the classical theory and Graph 7.14 shows the new models involving activities that are inserted into the capacitive paths.

7.4.2.1.1 Mutual Influence

Mutual Capacitance Operator: Mutual capacitance is composed of the activation and the mutual influence operators:

$$\hat{C}_{qij} = \hat{I}_{qij}\, \hat{N}_{qj} \tag{7.53}$$

Mutual Influence Operator: The mutual influence operator is defined as relating the activity to the effort of the opposite pole:

$$e_{qi} \overset{def\rightarrow}{=} \hat{I}_{qij}\, a_{qj} \tag{7.54}$$

This operator will be made explicit as follows. It can be checked that combining Equations 7.52 and 7.54 leads to the mutual capacitance expression 7.53.

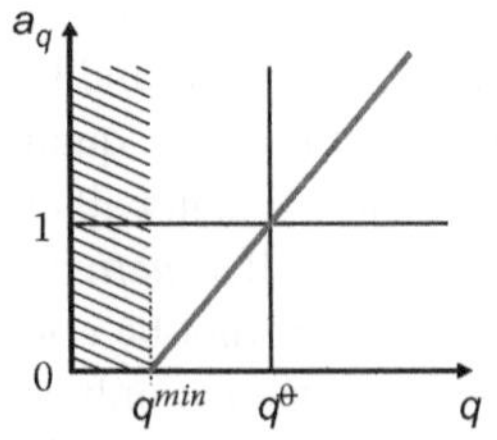

FIGURE 7.5 Plot of the activity of a collection of entities as a function of the number of entities. No activity is present below the threshold q^{min}, which may not exist depending on the energy variety.

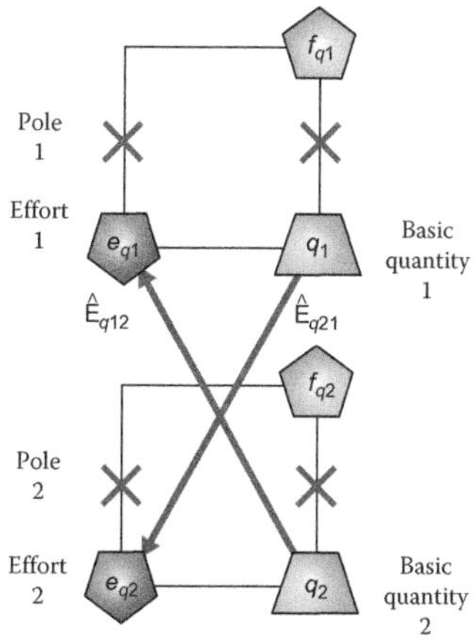

GRAPH 7.13 Formal Graph of *mutual influence* between two poles translated from the classical approach, relating directly basic quantities to efforts.

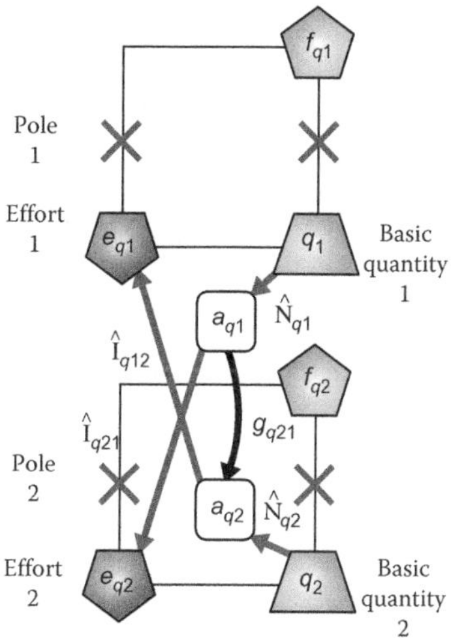

GRAPH 7.14 Formal Graph of *mutual influence* between two poles according to the decomposition operated by the Formal Graph approach involving activities.

7.4.2.1.2 Self-Influence (see Graphs 7.15 and 7.16)

Self-Capacitance Operator: Self-capacitance is composed of the activation and the self-influence operators:

$$\hat{C}_i = \hat{I}_{ii}\ \hat{N}_i \tag{7.55}$$

Self-influence operator: The physical chemical self-influence operator is defined as relating the activity to the chemical potential for the same pole:

$$e_{qi} \overset{def\rightarrow}{=} \hat{I}_{qii}\ a_{qi} \tag{7.56}$$

This operator will also be made explicit as follows. It can be checked that combining Equations 7.52 and 7.56 leads to the self-capacitance expression 7.55.

7.4.2.2 Interaction

An analogy with signal processing theories is made in defining a *gain* g as quantifying the amplification of activities in the same manner as signal intensities are treated.

Gain 1 → 2:
$$g_{q21} \overset{def}{=} \frac{a_{q2}}{a_{q1}} \tag{7.57}$$

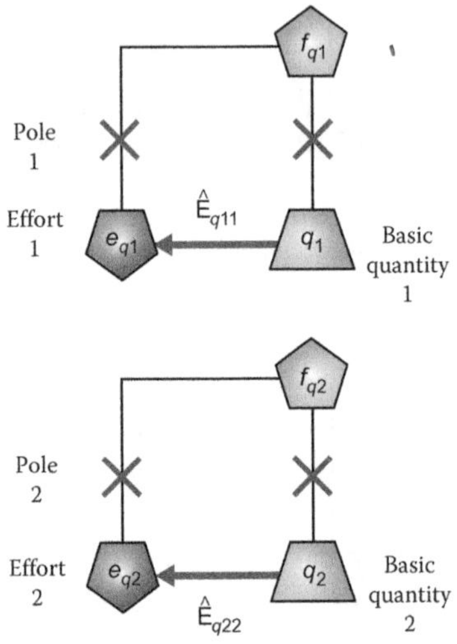

GRAPH 7.15 Formal Graph of *self-influence* between two poles translated from the classical approach, relating directly basic quantities to efforts.

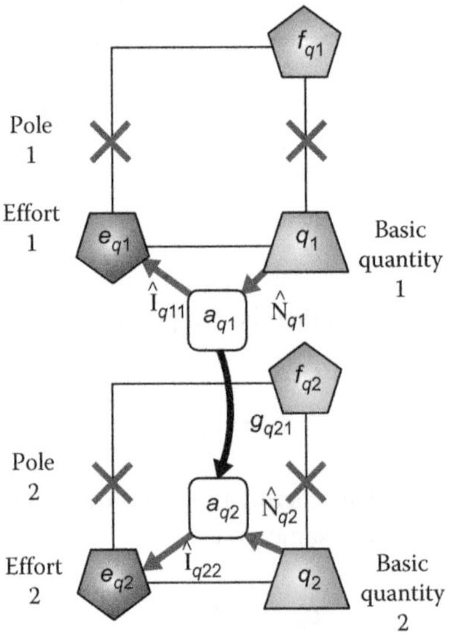

GRAPH 7.16 Two Formal Graphs of *self-influence* between two poles with the decomposition operated by the Formal Graph approach involving activities.

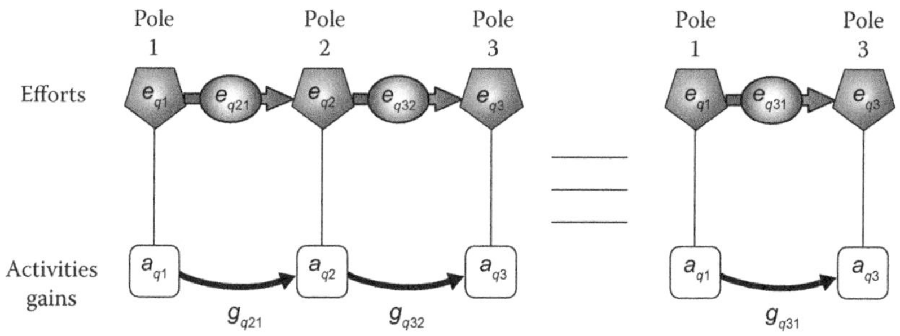

GRAPH 7.17 Equivalence between a pole chain and a dipole.

Another relationship between the two poles is obviously ensured by the relation between efforts of the poles and their differences:

$$e_{q21} = -e_{q12} = e_{q2} - e_{q1} \tag{7.58}$$

To understand the way gain and effort interact, it is necessary to extend the system by considering a chain of poles (multipole) and by examining how the association of several poles becomes equivalent to a single dipole, whatever the number of poles.

The Formal Graph (which is only partly canonical) in Graph 7.17 shows the equivalence between a chain of three poles and the dipole formed by the two ends of the chain. This equivalence is algebraically expressed by two relations:

$$e_{q31} = e_{q21} + e_{q32} \qquad\qquad g_{q31} = g_{q21}\, g_{q32} \tag{7.59}$$

The remarkable point is that one relation is an *addition* while the other is a *multiplication*. This is the main property of the exponential function, to establish a correspondence between a sum of its arguments and the product of the results. It is worth demonstrating this mathematical property because it brings to light the underlying assumptions. It begins with the derivation of the two previous relations for expressing variations:

$$\mathrm{d}e_{q31} = \mathrm{d}e_{q21} + \mathrm{d}e_{q32} \qquad\qquad \frac{\mathrm{d}g_{q31}}{g_{q31}} = \frac{\mathrm{d}g_{q21}}{g_{q21}} + \frac{\mathrm{d}g_{q32}}{g_{q32}} \tag{7.60}$$

The three terms in each equation can be made to correspond in pairs if the independence between dipoles is assumed.

Hypothesis. The gain (between its two poles) of a dipole is independent of the other dipole efforts.

This means that the gain of the interaction between two poles depends only on the effort of the considered dipole. A common coefficient $e_{q\Delta}$ is introduced as a proportionality factor for all pairs:

$$\mathrm{d}e_{q31} = \pm e_{q\Delta}\frac{\mathrm{d}g_{q31}}{g_{q31}} \qquad \mathrm{d}e_{q21} = \pm e_{q\Delta}\frac{\mathrm{d}g_{q21}}{g_{q21}}$$
$$\mathrm{d}e_{q32} = \pm e_{q\Delta}\frac{\mathrm{d}g_{q32}}{g_{q32}} \tag{7.61}$$

The choice of the sign depends on the type of influence: *positive for self-influence, negative for mutual influence*. Integrating the second equation relative to the two first poles, the exponential function is found, by introducing an integration constant e_q^{θ} corresponding to the value of the effort giving a unit gain. This constant is specific to each couple of species and is called the *standard effort*.

$$\text{Interaction: } \quad g_{q21} = \exp\left(\pm\frac{e_{q21} - e_{q21}^{\theta}}{e_{q\Delta}}\right) \tag{7.62}$$

The splitting between the poles can be made by using the same method, by replacing in the previous equation the dipole variables with their expressions as function of pole variables 7.57 and 7.58, then deriving

$$\mathrm{d}e_{q2} - \mathrm{d}e_{q1} = \pm e_{q\Delta}\frac{\mathrm{d}a_{q2}}{a_{q2}} \mp e_{q\Delta}\frac{\mathrm{d}a_{q1}}{a_{q1}} \tag{7.63}$$

The correct separation of these variables can only be made by specifying the kind of influence.

7.4.2.2.1 Mutual Influence

By assuming complete independency inside each pole between effort and activity required by the absence of self-influence (i.e., an effort depends only on the activity of the other pole), one is led to write an analogous equation to the three relations in Equation 7.61, but involving pole variables instead of dipole ones:

$$\mathrm{d}e_{qi} = e_{q\Delta}\frac{\mathrm{d}a_{qj}}{a_{qj}} \tag{7.64}$$

Integration of this differential equation provides the final result that was sought and that expresses the mutual influence relationship for a physical chemical dipole.

$$\text{Mutual influence: } \quad a_{qj} = \exp\left(\frac{e_{qi} - e_{qi}^{\theta}}{e_{q\Delta}}\right) \tag{7.65}$$

The expression of the mutual influence operator corresponds to an alternate way of writing the relationship between activity and effort:

$$e_{qi} = \hat{\mathrm{I}}_{qij}\, a_{qj} = e_{qi}^{\theta} + e_{q\Delta} \ln a_{qj} \tag{7.66}$$

7.4.2.2.2 Self-Influence

By assuming complete independency between poles, required by the absence of mutual influence (i.e., an effort depends only on the activity of its own pole), one is led to write an analogous equation to the three relations in Equation 7.61, but involving pole variables instead of dipole ones:

$$de_{qi} = e_{q\Delta} \frac{da_{qi}}{a_{qi}} \tag{7.67}$$

Integration of this differential equation provides the final result that was sought and that expresses the self-influence relationship for a physical chemical pole.

$$\text{Self-influence}: \quad a_{qi} = \exp\left(\frac{e_{qi} - e_{qi}^{\ominus}}{e_{q\Delta}}\right) \tag{7.68}$$

The expression of the self-influence operator corresponds to an alternate way of writing the relationship between activity and effort:

$$e_{qi} = \hat{\mathrm{I}}_{qii}\, a_{qi} = e_{qi}^{\ominus} + e_{q\Delta} \ln a_{qi} \tag{7.69}$$

Hypothesis. The effort of a pole is independent of the activity, whether of its own pole, as in case of mutual influence, or of the other pole, as in case of self-influence.

Standard Efforts: Naturally, the relationship between *standard efforts* follows the relationship between efforts as shown in Equation 7.58:

$$e_{q21}^{\ominus} = -e_{q12}^{\ominus} = e_{q2}^{\ominus} - e_{q1}^{\ominus} \tag{7.70}$$

Scaling Effort: It remains to discuss the constant $e_{q\Delta}$ introduced in the relations in Equation 7.61 for establishing the proportionality between effort variation and gain variation. On examination of the arguments of the exponential functions, it appears that this constant plays the role of a scaling variable for the efforts. This feature gives to it the name of *scaling effort*.

Nothing permits one to give a value to this *scaling effort* with the only elements provided by the present model. A more detailed modeling is required in terms of possible other energy varieties that can be involved in the system. In particular, being a system composed of molecules that may be submitted to thermal agitation, inclusion of the thermal energy variety in the model is often compulsory. The complete model is based on the same assumptions as for an ideal gas, and a thermodynamical development (too long to include here but discussed in Chapter 13) leads to a simple proportionality with the temperature through the gas constant R or the Boltzmann constant k_B, depending on the energy variety considered, physical chemical or corpuscular. In other energy varieties, one comes across various combinations of constants, such as in electrodynamics for instance, where the ratio of the Boltzmann constant k_B and the elementary charge e is found.

$$\mu_\Delta = RT; \quad E_\Delta = k_B T; \quad V_\Delta = \frac{k_B T}{e} = \frac{RT}{F} \tag{7.71}$$

When additional energy varieties participate in the energy of the system (surface energy for instance), this relationship becomes more complicated. This is an advantage of the Formal Graph approach to individualize this variable instead of keeping the product RT (or k_BT) as the classical approach does.

Dipole Effort: As sometimes met in the case studies and already explained, the difference between e_{qlj} is not always equal to the dipole effort e_{qC}. It depends on the separability of the poles and on the generator/receptor convention chosen. In Chapter 6 (Section 6.5), the dipole effort has been related to the difference of pole efforts through the following relationships

$$e_{qC} = s_D e_{q12} = s_D\left(e_{q1} - e_{q2}\right) = -s_D e_{q21} \tag{7.72}$$

and between standard efforts one has the same relation shape

$$e^{\theta}_{qC} = s_D e^{\theta}_{q12} = s_D\left(e^{\theta}_{q1} - e^{\theta}_{q2}\right) = -s_D e^{\theta}_{q21} \tag{7.73}$$

The only impact of this subtlety is the presence of the *separability factor* s_D into expression 7.62 of the dipole gain when expressed as a function of the dipole effort

$$\text{Interaction}: \quad g_{q21} = \frac{a_{q2}}{a_{q1}} = \exp\left(\mp\frac{e_{qC} - e^{\theta}_{qC}}{s_D e_{q\Delta}}\right) \tag{7.74}$$

where the minus sign stands for self-influence and the plus sign for mutual influence.

Remarks

Analogy with a chain reaction. The equivalence between a chain of several poles, making a multipole, and the dipole formed by the two ends of the chain is a well-known property of chains composed of linear relationships. A representative example in physical chemistry is the chain formed by a sequence of first-order chemical reactions. The poles between the ends are the intermediate species and the poles at the ends are the substrate A and product species Z. In such a chain, the overall reaction rates and equilibrium constants correspond to those of the equivalent dipole A – Z.

$$A \rightleftarrows B \rightleftarrows C \rightleftarrows \cdots \rightleftarrows Z$$

See Chapter 8 where multipoles are studied.

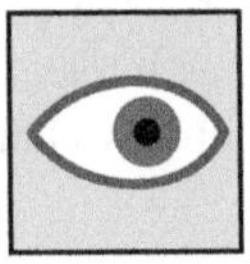

Activity. The notion of *activity* has been introduced by Gilbert Newton Lewis in 1907 and described in the following way: "Activity in chemistry is a measure of how different molecules in a non-ideal gas or solution *interact* with each other." Its definition proceeds from the relationship between effort and activity through the exponential or logarithmic function, as given in Equation 7.15.

Where does k_BT comes from? The classical theory explains the exponential function and the presence of the product k_BT (or RT) at the denominator of its argument in terms of Boltzmann statistics. The Formal Graph theory does not use any statistics for establishing models that are irrelevant from the notion of probability. This approach applies both to physical chemistry and to electrodynamics and gravitation, which are domains known to be far away from statistical treatment. The exponential function is retrieved in merely assuming that molecules put together influence each other according to their activity. The activities of various collections (poles) behave similarly to signal intensities in a transmission line, being amplified or attenuated, from pole to pole.

7.4.3 Statistics Is Not the Only Explanation

The treatment of the influence phenomenon has led to the formulation of an exponential law that is identical to the one known as Boltzmann's statistics (also known as Maxwell–Boltzmann's statistics). This classical approach relies on the concept of probability, which predicts the number of times an event may occur when a large number of triggering actions are performed. (It must be noted that the notions of event and of sequences of events require the existence of time). In the derivation of Boltzmann's statistics, two assumptions are required (Landau and Lifchitz 1958):

- Independence between events.
- Large number of events [for assimilating a factorial to a logarithm using Stirling's* approximation (Riley et al. 2006)].

The most important consequence of the proposed approach is to offer an alternate foundation for establishing many laws that rely on an exponential function. Until this proposal, the only foundation for the notion of activity was Boltzmann's statistics, the only one able to explain the exponential behavior, and correlatively the only way to introduce the product RT in models (Atkins and De Paula 2010).

It must be outlined that the proposed approach is not a restoration of the old determinism; it is not because a link between some variables is determined that fluctuations of these variables are not permitted. It must be stressed also that one speaks of an alternative, not a substitution. The proposed theory of influence does not eliminate Boltzmann's statistics, which certainly remains valid in numerous cases and remains one of the most beautiful achievements for relating microscopic objects to macroscopic ones. However, there are cases where the number of objects is so small that it forbids the approximations allowed by very large numbers for finding an exponential function. There are cases where the notion of probability is irrelevant or cannot be applied, either because no time is pertinent (static conditions) or because no freedom is left to objects for fluctuating or moving (immobilized objects), thus forbidding any statistical distribution.

It can be objected however that as statistics is able to provide an explanation for many of the systems for which one proposes the influence theory, it dismisses the validity of the proposed approach, which in the best case becomes unnecessary. This is not a correct reasoning. In many cases, a statistical approach may propose a model for a physical phenomenon that can be modeled by another theoretical approach not using statistics. For instance, the area of a target included in a larger and known area may be determined by the measure of the number of hits of a dart launched by a player, assuming a random process in the launch. This is the basis of stochastic methods such

* James Stirling (1692–1770): Scottish mathematician, UK.

as Monte Carlo for simulating or modeling numerous systems. The case of absorption of a radiation beam (see Section 7.4.1) can also be treated in terms of probability for a photon to be absorbed by an obstacle and a statistical interpretation of the Lambert–Beer–Bouguer law can perfectly be put forward. However, it is obvious that numerous other methods (geometrical, weighting, time of coverage, electrochemical, etc.) exist for measuring an area. The conclusion is that it is not because statistics may apply that it forbids all other explanations. When these latter do exist.

7.4.4 The Fundamental Capacitive Relationship

After leading to the formulation of dependence 7.68 of the activity on the effort of the considered pole (in the self-influence case), the dependence of the basic quantity, related to the activity by its definition in Equation 7.52, on the pole effort remains to be expressed directly. This expression constitutes the capacitive relationship for a pole that was postulated until now and which is now demonstrated.

$$\text{Pole capacitance:} \quad q_i = q_i^{min} + \left(q_i^{\theta} - q_i^{min}\right) \exp\left(\frac{e_{qi} - e_{qi}^{\theta}}{e_{q\Delta}}\right) \tag{7.75}$$

In Figure 7.6 is plotted the trace of the capacitance operator given by the preceding equation, indicating the tangent Δ around the reference point which has for equation

$$q_i \underset{lin}{=} q^{\theta} + \frac{q_i^{\theta} - q_i^{min}}{e_{q\Delta}}\left(e_{qi} - e_{qi}^{\theta}\right) \tag{7.76}$$

This tangent provides the linear pole capacitance:

$$C_{qi} \underset{lin}{=} \frac{q_i^{\theta} - q_i^{min}}{e_{q\Delta}} \tag{7.77}$$

Inverting the capacitance expression gives the "elastive" relationship or the pole elastance:

$$e_{qi} = e_{qi}^{\theta} + e_{q\Delta} \ln\frac{q_i - q_i^{min}}{q_i^{\theta} - q_i^{min}} \tag{7.78}$$

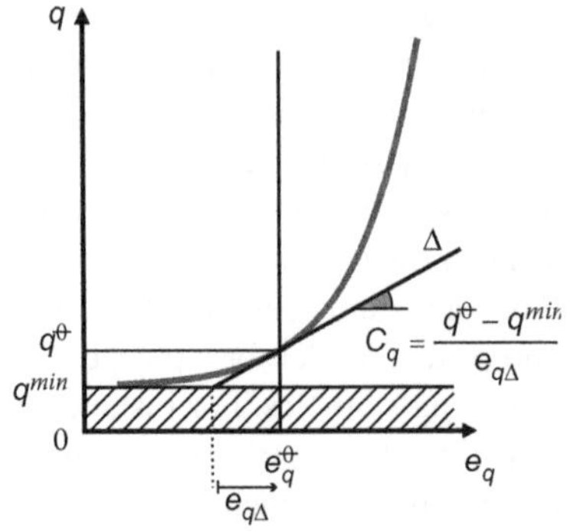

FIGURE 7.6 Plot of the generalized capacitance of a pole.

Why is self-influence demonstrated by pole–pole interaction? It may appear surprising to have to consider another pole for establishing an internal relationship to a pole. In fact this results from the definition of a pole as a collection of entities. Whether the interaction concerns entities belonging to two collections (of identical entities) or a single one it is not very different; only the effort differs between two collections. The self-influence amounts to an interaction between two collections sharing the same effort. In other words, the mechanism by which an entity influences another one is the same whether or not they belong to the same collection.

7.4.5 The Dipole Capacitive Relationship

The fundamental capacitive relationship works at the level of a pole and constitutes a fundamental law on which many models can be built. The dipole capacitive relationship is one of them. Several case studies in the previous chapter have illustrated this important relationship which is established now in the general case of conservative dipoles.

The general expression of the gain is as follows:

$$\frac{q^{max}-q}{q-q^{min}} = g_{q21} = \frac{a_{q2}}{a_{q1}} \tag{7.79}$$

which uses two limits q^{min} and q^{max} for the dipole basic quantity. In mathematical terms, this expression amounts to a simple technical "normalization," but that one has to establish on physical grounds.

According to definition 7.52 of the activity, the two activities of the poles are written as

$$a_{q1} \overset{def}{=} \frac{q_1 - q_1^{min}}{q_1^{\theta} - q_1^{min}}; \quad a_{q2} \overset{def}{=} \frac{q_2 - q_2^{min}}{q_2^{\theta} - q_2^{min}} \tag{7.80}$$

with an availability threshold corresponding to the minimum number of entities q_i^{min} as explained earlier and a reference basic quantity q_i^{θ} for each pole.

In Chapter 6, the relationships between the poles and the dipole have been established, recalled here briefly. The starting point is always the Gibbs equation giving the capacitive energy variation in the system:

$$d\mathcal{U}_q = e_{q1}\,dq_1 + e_{q2}\,dq_2 = e_{qC}\,dq \tag{7.81}$$

By inserting the general relationship 7.72 between efforts into the Gibbs equation, the relationship between basic quantities variations is deduced as

$$s_D\,dq = dq_1 = -dq_2 \tag{7.82}$$

In the general case, one needs to take into account some integration constants for writing the integrated relationship

$$❸ \quad s_D\left(q - q^0\right) = q_1 - q_1^0 = -\left(q_2 - q_2^0\right) \tag{7.83}$$

(Circled numbers indicate equations that are graphically represented in Figure 7.7 and that will be combined with others.) Three integration constants have been used, which exceeds the normal

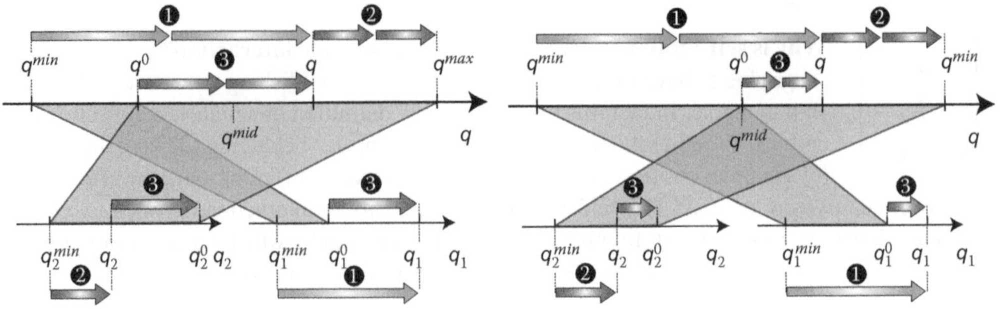

FIGURE 7.7 Schemes of the relationships between basic quantities of the poles (bottom axes) and the dipole (top axis) in the cases of an asymmetric dipole (left) and of a symmetric one (right). The filled polygons indicate the two definitions of the dipole extrema (Equation 7.85) and the numbered arrows indicate the correspondences set by relationships in Equations 7.83 and 7.86. (Numbers by the arrows refer to these relationships.) The separability factor is set to 1/2 so the dipole scale is twice the pole scales.

two constants required for two equations. This leaves some latitude in the choice of these constants for modeling various cases, as will be seen. Two of these constants are related to the poles and correspond to the *conservation equation*.

$$\text{Conservative dipole}: \quad q_1 + q_2 = q_1^0 + q_2^0 \tag{7.84}$$

The choice is to identify the integration constants q_i^0 with the reference basic quantities q_i^θ and to relate the differences of pole basic quantities appearing in the expressions of the activities as functions of the dipole corresponding differences.

The first step consists of setting the extrema of the dipole basic quantity. This is done by setting all differences with respect to the references as proportional through the separability factor s_D. The definition of the dipole limits q^{max} and q^{min} (more exactly, their difference with the dipole reference q^0) is determined by relating the dipole minimum to the first pole and the dipole maximum to the second one:

$$❷+❸\ s_D(q^{max} - q^0) \overset{def}{=} q_2^0 - q_2^{min}; \quad ❶-❸\ \text{-}s_D(q^{min} - q^0) \overset{def}{=} q_1^0 - q_1^{min} \tag{7.85}$$

The principle of the conservation of the basic quantities between the poles is that, when the basic quantity of the first pole is minimum, the basic quantity of the dipole is equal to q^{min}. It is the converse for the second pole; when its basic quantity is minimum, the dipole basic quantity is equal to q^{max}.

Now, the second step consists of combining by subtraction or by addition the corresponding terms in Equations 7.83 and 7.86 to reach the expressions of the two numerators:

$$❷\ s_D\left(q^{max} - q\right) = q_2 - q_2^{min}; \quad ❶\ s_D\left(q - q^{min}\right) = q_1 - q_1^{min} \tag{7.86}$$

These relationships are sketched in Figure 7.7 in which the differences established in Equations 7.83 and 7.86 are represented by arrows with numbers allowing their identification. All arrows with the same number have identical lengths.

In Figure 7.7, is placed a convenient parameter q^{mid}, which is the middle of the allowed range for the dipole basic quantity, defined by

$$q^{mid} \overset{def}{=} \frac{q^{max} + q^{min}}{2} \tag{7.87}$$

After the numerators, the denominators remain to be determined. Two cases are considered, depending on the symmetry of the dipole, as illustrated in Figure 7.7.

(i) *Asymmetric Dipole*: Asymmetry means that the sum of the basic quantities of both poles, which is a constant as stated in the conservation Equation 7.84, is not null:

$$q_1 + q_2 = q_1^0 + q_2^0 \neq 0 \quad (7.88)$$

The consequence is that the maximum and the minimum of the basic quantity are not symmetrical with respect to the dipole reference q^0, which corresponds to the nonequality

$$q^0 \neq q^{mid} \quad (7.89)$$

This situation occurs, for instance, in physical chemical dipoles (see case studies C5 and C6 in Chapter 6) and in corpuscular dipoles (see case study C9 in Chapter 6). In these dipoles, no threshold exists for the basic quantities of their poles:

$$q_i^{min} = 0 \quad (7.90)$$

Also, a common reference q_i^{min} could be chosen, which simplifies the modeling:

$$q^{\theta} = q_1^{\theta} = q_2^{\theta} \quad (7.91)$$

With these settings, one finds for the activities

$$a_{q1} = \frac{q_1}{q^{\theta}} = s_D \frac{q - q^{min}}{q^{\theta}}; \quad a_{q2} = \frac{q_2}{q^{\theta}} = s_D \frac{q^{max} - q}{q^{\theta}} \quad (7.92)$$

The ratio of these two expressions gives the result that was sought.

(ii) *Symmetric dipole*: Symmetry for a dipole means that the sum of the pole basic quantities, which is a constant as stated in the conservation Equation 7.84, is null:

$$q_1 + q_2 = q_1^0 + q_2^0 = 0 \quad (7.93)$$

Symmetric dipoles are met each time the basic quantity of one pole is the opposite of that of the other pole, such as in a capacitor (see case study C7 in Chapter 6) or a spring (see case study C8 in Chapter 6). The consequence is that the maximum and the minimum of the basic quantity are symmetrical with respect to the dipole reference q^0, which corresponds to the equality

$$q^{max} - q^0 = q^0 - q^{min} \quad (7.94)$$

Consequently, the final expressions for the two activities are

$$a_{q1} = \frac{q - q^{min}}{q^{max} - q^0}; \quad a_{q2} = \frac{q^{max} - q}{q^{max} - q^0} \quad (7.95)$$

Again, the ratio of these two activities provides the right expression 7.79 for the gain of the dipole. As mentioned above, the exceeding number of integration constants in Equation 7.84 offers the possibility to choose arbitrarily one of them, done by placing the origin of the dipole basic quantity right in the middle of its range, which also corresponds to the reference basic quantity (according to Equations 7.87 and 7.94)

$$q^0 = q^{mid} = 0 \tag{7.96}$$

(iii) *Both Cases*: The general case of asymmetric and symmetric dipoles is modeled by a single algebraic expression of the capacitive relationship, which stems directly from the previous developments and by using the gain dependence 7.74 with the effort for a self-capacitance:

$$\frac{q^{max} - q}{q - q^{min}} = \frac{a_{q2}}{a_{q1}} = g_{q21} = \exp\left(-\frac{e_{qC} - e_{qC}^{\theta}}{e_{q\Delta}}\right) \tag{7.97}$$

A plot of the function representing the capacitance operator for a dipole is shown in Figures 7.8 and 7.9 for both cases of symmetry.

The plots in these figures indicate the range of variation of the basic quantity of the dipole, defined as

$$\Delta q^m \overset{def}{=} 2\Delta q^m_{1/2} = q^{max} - q^{min} \tag{7.98}$$

The mid-range basic quantity 7.87, which can be null in case of a symmetric dipole, corresponds to the dipole reference effort

$$e_{qC} = e_{qC}^{\theta} \quad \Leftrightarrow \quad q = q^{mid} \tag{7.99}$$

The mid-range basic quantity is also used for expressing the equation of the asymptote around the inflexion point that corresponds to the linear case

$$q \underset{lin}{=} q^{mid} + C_q\left(e_{qC} - e_{qC}^{\theta}\right) \tag{7.100}$$

with the linear dipole capacitance

$$C_q \underset{lin}{=} \frac{\Delta q^m}{4e_{q\Delta}} \tag{7.101}$$

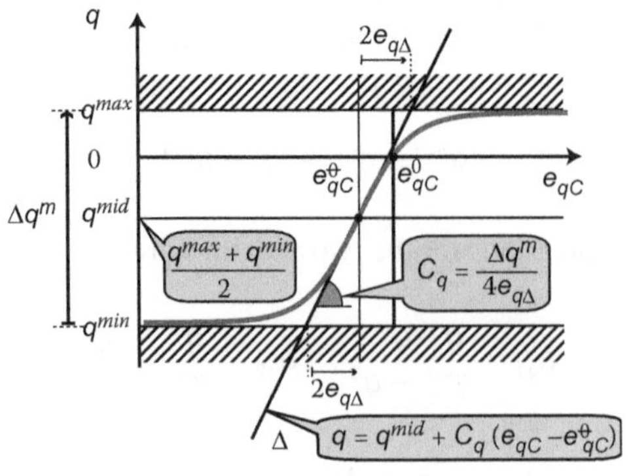

FIGURE 7.8 Plot of the dipole capacitance in the general case of an *asymmetric* dipole. The asymptote around the inflexion point is the linear approximation.

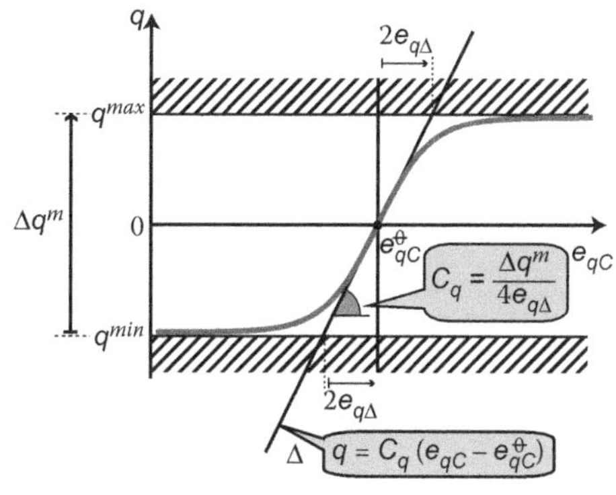

FIGURE 7.9 Plot of the dipole capacitance in the general case of a *symmetric* dipole. The asymptote around the inflexion point is the linear approximation.

Earlier in Section 7.3.2 the capacitive linear relationship of a dipole has been derived from the pole capacitances, which was written under two equivalent short expressions in Equation 7.25 recalled here

$$q \underset{lin}{=} C_q\left(e_{qC} - e^{*}_{qC}\right); \quad e_{qC} \underset{lin}{=} C_q^{-1}\left(q - q^{**}\right)$$

The correspondence between the various expressions is as follows:

$$e^{*}_{qC} = e^{\theta}_{qC} + C_q^{-1}\, q^{mid}; \quad q^{**} = q^{mid} - C_q\, e^{\theta}_{qC} \tag{7.102}$$

7.4.6 Pole and Dipole Parameters for Some Energy Varieties

Table 7.2 gives the correspondence between basic quantities featuring the two poles of a dipole for some capacitive systems belonging to several energy varieties and Table 7.3 gives the correspondence between basic quantities and efforts for the dipoles.

In Table 7.4, the four case studies are sorted according to their characteritics.

TABLE 7.2
Correspondence between Basic Quantities of a Pole for Various Systems

Pole Basic Quantities	Chemical Reaction (i = A, B) or Interface (i = 1, 2)	Electric Capacitor (i = 1, 2)	Mechanical Spring (i = 1, 2)	Electrons and Holes ($i = e, h$)
q_i	n_i	Q_i	ℓ_i	N_i
q_i^0	n_i	0	ℓ_i^0	N_i^*
q_i^{min}	0	Q_i^{min}	ℓ_i^{min}	0
q_i^{θ}	n^{θ}	0	ℓ_i^0	N_i^0 (*intrinsic*)

TABLE 7.3
Correspondence between Basic Quantities and Efforts of a Dipole for Various Systems

Dipole Variables	Chemical Reaction (or Interface)	Electric Capacitor	Mechanical Spring	Electrons and Holes
s_D	1 Separable	1 Separable	1/2 Inseparable	1/2 Inseparable
q	$-n$	Q	$\boldsymbol{\ell}$	N
q^0	$0\ (\neq n^{mid})$	0	0	$0\ (\neq N^{mid})$
q^{min}	$-n^*_A$	$Q^{min} = Q_1^{min}$	$\ell^{min} = -2(\ell_1^0 - \ell_1^{min})$	$-2\,N^*_e$
q^{max}	n^*_B	$Q^{max} = -Q_2^{min}$	$\ell^{max} = 2(\ell_2^0 - \ell_2^{min})$	$2\,N^*_h$
q^{mid}	$n^{mid} = (n^*_B - n^*_A)/2$	$0 = (Q^{max} + Q^{min})/2$	$0 = (\ell^{max} + \ell^{min})/2$	$N^{mid} = N^*_h - N^*_e$
	Asymmetric	Symmetric	Symmetric	Asymmetric
e_{qC}	μ	V	$\boldsymbol{F}$	E
e^{θ}_{qC}	μ^{θ}	0	0	E_F

Subscripted variables (by 1 and 2 or A and B or e and h) correspond to poles and nonsubscripted ones to dipoles.

TABLE 7.4
Classification of the Dipole Case Studies Given in Chapter 6 According to Their Separability and Symmetry Characteristics

Characteristic	Asymmetric	Symmetric
Separable	Case studies C5 "First-Order Chemical Reaction" and C6 "Physical Chemical Interface" A⇄B 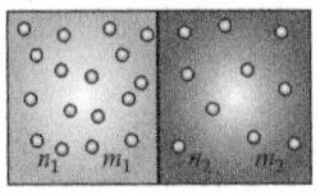	Case study C7 "Electric Capacitor" 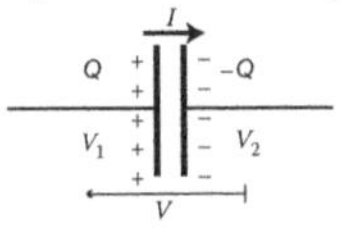
Inseparable	Case study C9 "Electrons and Holes" 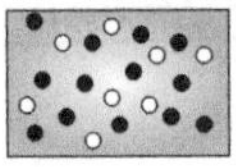	Case study C8 "Spring" 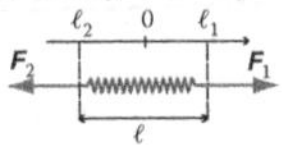

7.5 IN SHORT

<table>
<tr><th colspan="2">DIPOLE SEPARABILITY AND POLE ASSOCIATION</th></tr>
<tr><td>Separable dipole: $s_D = \pm 1$
(+1: Receptor, −1: Generator convention)
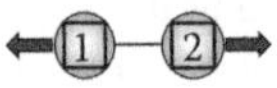</td><td>Inseparable dipole: $s_D = \pm 1/2$
(+1/2: Receptor, −1/2: Generator convention)
</td></tr>
<tr><td>q_1^*  q_2^*
No interaction</td><td>q_1  q_2
Interaction</td></tr>
<tr><td colspan="2">q_1* and q_2* refer to the dissociated state or "off state" of the poles, that is, without interaction, whatever the separability of the dipole.</td></tr>
</table>

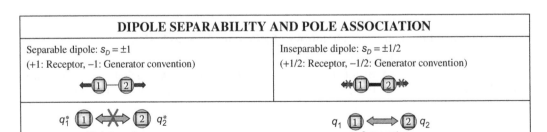

DIPOLE EFFORT FROM POLES

Dipole effort: $e_{qC} = s_D e_{q12} = s_D(e_{q1} - e_{q2}) = -s_D e_{q21}$ (7.72)

Standard dipole effort: $e_{qC}^{\ominus} = s_D e_{q12}^{\ominus} = s_D(e_{q1}^{\ominus} - e_{q2}^{\ominus}) = -s_D e_{q21}^{\ominus}$ (7.73)

Separable dipole: $s_D = \pm 1$, Inseparable dipole: $s_D = \pm 1/2$
(+: Receptor, –: Generator convention)

INFLUENCE THROUGH CONSTITUTIVE PROPERTIES

Influence is interdependence between variables through system properties, featuring poles or the medium between poles.

1. Two (nonexclusive) possibilities:
   - *Self-influence* (within a pole): Supported by system constitutive properties (featuring poles), (self-) inductance, (self-) capacitance, and (self-) conductance.
   - *Mutual influence* (between poles): Supported by system properties featuring the interpole medium; mutual inductance, mutual capacitance, and mutual conductance.
2. Three kinds: inductive, capacitive, and conductive

THREE KINDS OF INFLUENCE

Inductive

$$f_{q1} = \hat{L}^{-1}_{q11}\, p_{q1} + \hat{L}^{-1}_{q12}\, p_{q2} \quad (7.1)$$

$$f_{q2} = \hat{L}^{-1}_{q21}\, p_{q1} + \hat{L}^{-1}_{q22}\, p_{q2} \quad (7.2)$$

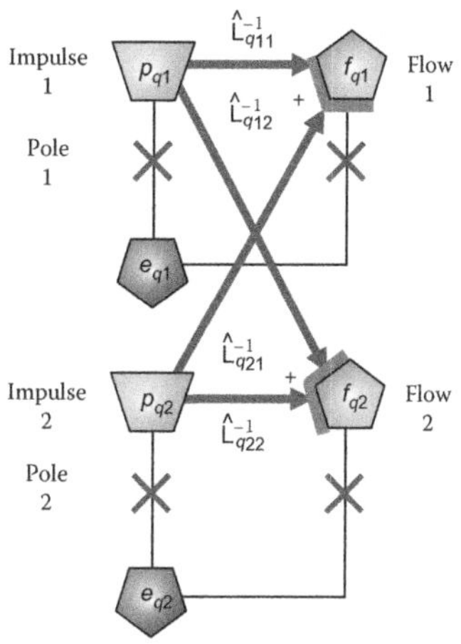

GRAPH 7.18

Capacitive

$$e_{q1} = \hat{C}^{-1}_{q11}\, q_1 + \hat{C}^{-1}_{q12}\, q_2 \quad (7.3)$$

$$e_{q2} = \hat{C}^{-1}_{q21}\, q_1 + \hat{C}^{-1}_{q22}\, q_2 \quad (7.4)$$

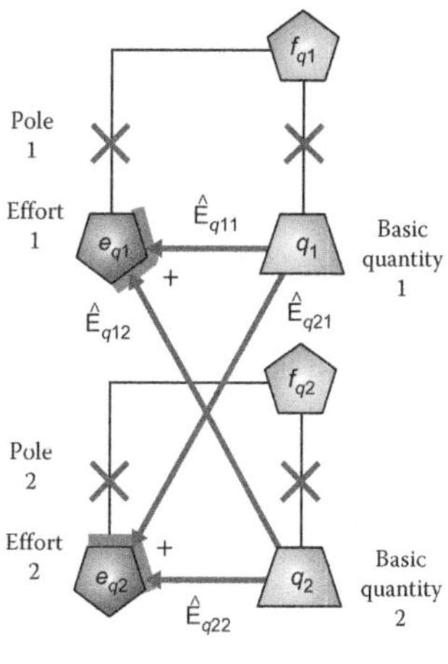

GRAPH 7.19

Conductive
(in case of exchange by flows)

$$f_{q1} = \hat{G}_{q11}\, e_{q1} + \hat{G}_{q12}\, e_{q2} \quad (7.5)$$

$$f_{q2} = \hat{G}_{q21}\, e_{q1} + \hat{G}_{q22}\, e_{q2} \quad (7.6)$$

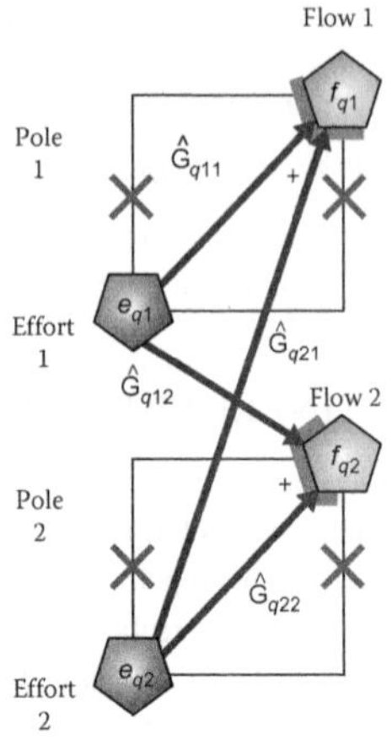

GRAPH 7.20

Conductive
(in case of exchange by efforts)

$$e_{q1} = \hat{R}_{q11}\, f_{q1} + \hat{R}_{q12}\, f_{q2} \quad (7.7)$$

$$e_{q2} = \hat{R}_{q21}\, f_{q1} + \hat{R}_{q22}\, f_{q2} \quad (7.8)$$

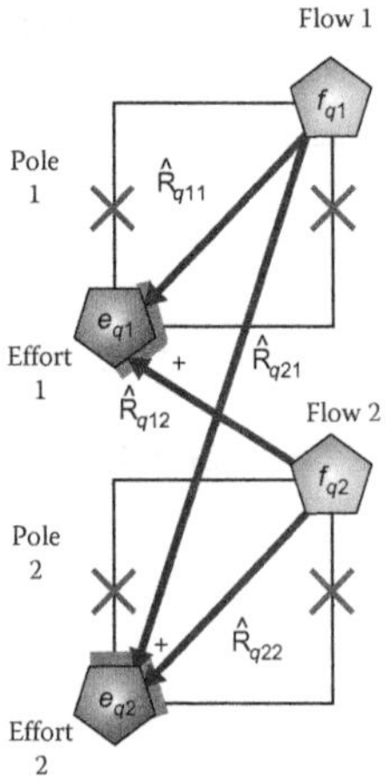

GRAPH 7.21

DIPOLE INDUCTIVE RELATIONSHIP FROM POLES		
Dipole inductive relationship:	$f_q = \hat{L}_q^{-1} p_q$	(7.30)
Dipole reluctance:	$\hat{L}_q^{-1} \stackrel{def}{=} \hat{L}_{q1}^{-1} + \hat{L}_{q2}^{-1}$	(7.29)
Apparent reluctance pole 1:	$\hat{L}_{q1}^{-1} p_q \stackrel{def}{=} s_D\left\{\hat{L}_{q11}^{-1} - \hat{L}_{q21}^{-1}\right\}\left(p_{q1}^* + s_D p_q\right)$	(7.27)
Apparent reluctance pole 2:	$\hat{L}_{q2}^{-1} p_q \stackrel{def}{=} -s_D\left\{\hat{L}_{q22}^{-1} - \hat{L}_{q12}^{-1}\right\}\left(p_{q2}^* - s_D p_q\right)$	(7.28)
Linear case		
Inductive relationships:	$p_q \underset{lin}{=} L_q\left(f_{qL} - f_q^*\right)$; $f_{qL} \underset{lin}{=} L_q^{-1}\left(p_q - p_q^{**}\right)$	(7.33)
Dipole reluctance:	$L_q^{-1} \overset{def}{\underset{lin}{=}} s_D^2\left(L_{q11}^{-1} + L_{q22}^{-1} - L_{q12}^{-1} - L_{q21}^{-1}\right)$	(7.31)
Origins:	$p_q^{**} \underset{lin}{=} L_q f_{qL}^*$; $f_{qL}^* \overset{def}{\underset{lin}{=}} s_D\left(L_{q11}^{-1} - L_{q21}^{-1}\right)p_{q1}^* + s_D\left(L_{q12}^{-1} - L_{q22}^{-1}\right)p_{q2}^*$	(7.32)

DIPOLE CAPACITIVE RELATIONSHIP FROM POLES

Dipole capacitive relationship: $q = \hat{C}_q\, e_q$ (7.22)

Dipole elastance: $\hat{C}_q^{-1} \overset{def}{=} \hat{C}_{q1}^{-1} + \hat{C}_{q2}^{-1}$ (7.21)

Apparent elastance pole 1: $\hat{C}_{q1}^{-1} q \overset{def}{=} s_D\left\{\hat{C}_{q11}^{-1} - \hat{C}_{q21}^{-1}\right\}\left(q_1^* + s_D\, q\right)$ (7.19)

Apparent elastance pole 2: $\hat{C}_{q2}^{-1} q \overset{def}{=} -s_D\left\{\hat{C}_{q22}^{-1} - \hat{C}_{q12}^{-1}\right\}\left(q_2^* - s_D\, q\right)$ (7.20)

Linear case

Capacitive relationships: $q \underset{lin}{=} C_q\left(e_{qC} - e_{qC}^*\right);\quad e_{qC} \underset{lin}{=} C_q^{-1}\left(q - q^{**}\right)$ (7.25)

Dipole elastance: $C_q^{-1} \underset{lin}{\overset{def}{=}} s_D^2\left(C_{q11}^{-1} + C_{q22}^{-1} - C_{q12}^{-1} - C_{q21}^{-1}\right)$ (7.23)

Origins: $q^{**} \underset{lin}{=} C_q e_{qC}^* \quad e_{qC}^* \underset{lin}{\overset{def}{=}} s_D\left(C_{q11}^{-1} - C_{q21}^{-1}\right)q_1^* + s_D\left(C_{q12}^{-1} - C_{q22}^{-1}\right)q_2^*$ (7.24)

DIPOLE CONDUCTIVE RELATIONSHIP FROM POLES

Dipole conductive relationship: $f_q = \hat{G}_q\, e_q$ (7.36)

Dipole conductance: $\hat{G}_q = \dfrac{\hat{G}_{q1}\, e_{q1} - \hat{G}_{q2}\, e_{q2}}{e_{q1} - e_{q2}} \quad \left(e_{q1} \neq e_{q2}\right)$ (7.37)

Linear case

Dipole conductance: $G_q \underset{lin}{=} \dfrac{G_{q1}\, e_{q1} - G_{q2}\, e_{q2}}{e_{q1} - e_{q2}} \quad \left(e_{q1} \neq e_{q2}\right)$ (7.39)

POLE CAPACITIVE RELATIONSHIP

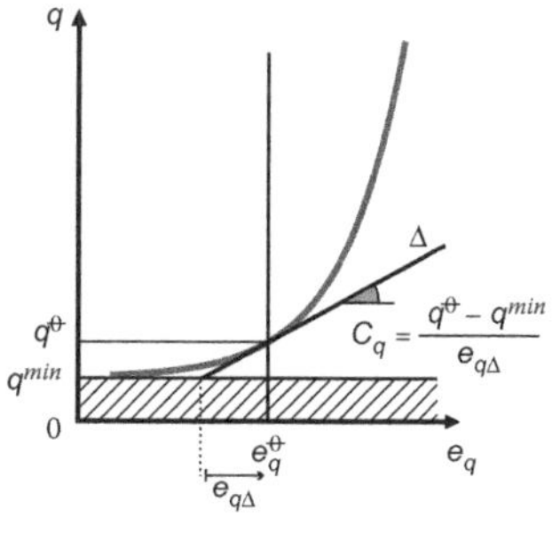

FIGURE 7.6

Pole capacitance: $q_i = q_i^m + \left(q^\theta - q_i^m\right)\exp\left(\dfrac{e_{qi} - e_{qi}^\theta}{e_{q\Delta}}\right)$ (7.75)

Pole elastance: $e_{qi} = e_{qi}^\theta + e_{q\Delta}\ln\dfrac{q_i - q_i^{min}}{q^\theta - q_i^{min}}$ (7.78)

q^θ, Reference basic quantity; q_i^{min}, Availability threshold (or *minimum*) basic quantity; e_{qi}^0, Reference effort; $e_{q\Delta}$, Scaling effort.

Examples of scaling efforts: $\mu_\Delta = RT;\quad E_\Delta = k_B T;\quad V_\Delta = \dfrac{k_B T}{e} = \dfrac{RT}{F}$

Linear case

Pole capacitance: $C_q \underset{lin}{=} \dfrac{q^\theta - q_i^{min}}{e_{q\Delta}}$ (7.77)

INFLUENCE THROUGH ACTIVITIES

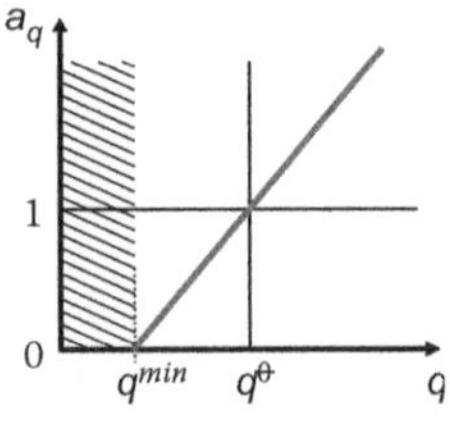

FIGURE 7.5

Activity a_q is a variable defined at the level of a capacitive pole and quantifying the availability of its entity for participating to an interaction with another pole by influence. It has no dimension and is normalized for being equal to 1 in the reference state ($q = q^{\theta}$).

$$\text{Activity (pole } i\text{):} \quad a_{qi} \stackrel{def}{=} \hat{N}_{qi}\, q_i \stackrel{def}{=} \frac{q_i - q_i^{min}}{q^{\theta} - q_i^{min}} \tag{7.52}$$

q^{θ}, Reference basic quantity; q_i^{min}, Availability threshold (or *minimum*) basic quantity.

The *gain of activity* g_q measures the variation (amplification) of activity from one pole to another.

$$\text{Gain } 1 \to 2: \quad g_{q21} \stackrel{def}{=} \frac{a_{q2}}{a_{q1}} \tag{7.57}$$

Interaction by mutual influence

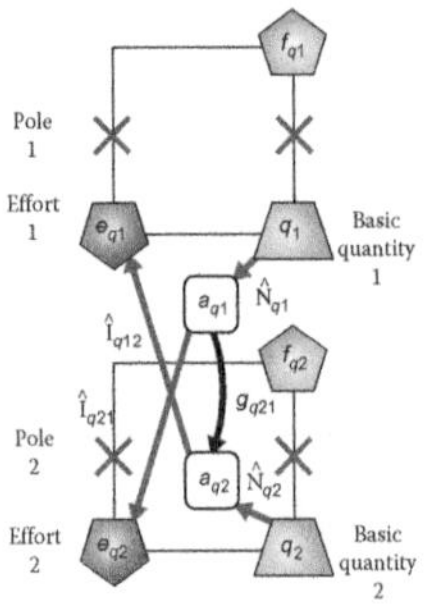

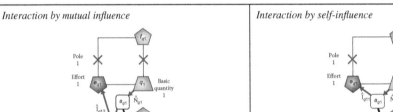

GRAPH 7.14

$$a_{qj} = \hat{I}_{qij}^{-1}\, e_{qi} = \exp\left(\frac{e_{qi} - e_{qi}^{\theta}}{e_{q\Delta}}\right) \tag{7.65}$$

$$g_{q21} = \exp\left(+\frac{e_{qC} - e_{qC}^{\theta}}{s_D e_{q\Delta}}\right) \tag{7.74}$$

Interaction by self-influence

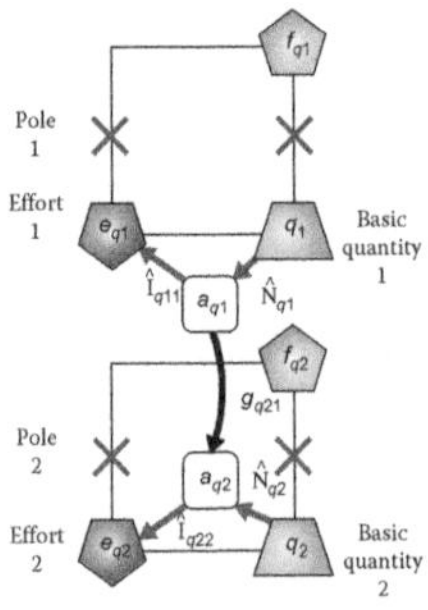

GRAPH 7.16

$$a_{qi} = \hat{I}_{qii}^{-1}\, e_{qi} = \exp\left(\frac{e_{qi} - e_{qi}^{\theta}}{e_{q\Delta}}\right) \tag{7.68}$$

$$g_{q21} = \exp\left(-\frac{e_{qC} - e_{qC}^{\theta}}{s_D e_{q\Delta}}\right) \tag{7.74}$$

DIPOLE CAPACITIVE RELATIONSHIP (CONSERVATIVE DIPOLE)

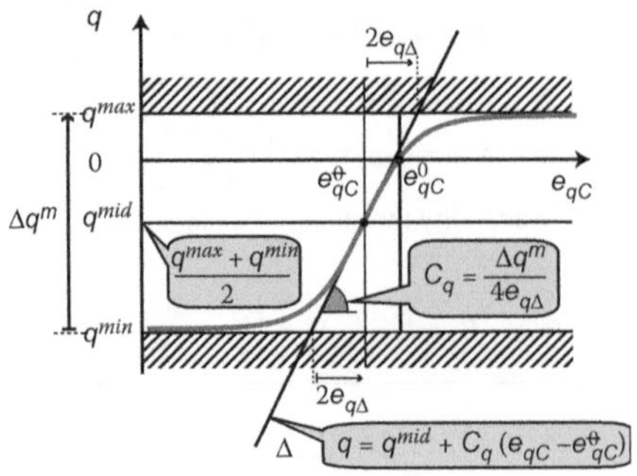

FIGURE 7.8

Dipole capacitive relation:	$\dfrac{q^{max}-q}{q-q^{min}}=\dfrac{a_{q2}}{a_{q1}}=g_{q21}=\exp\left(-\dfrac{e_{qC}-e_{qC}^{\ominus}}{e_{qD}}\right)$	(7.97)
Mid-range basic quantity:	$q^{mid}\overset{def}{=}\dfrac{q^{max}+q^{min}}{2}$	(7.87)
Standard effort $\Leftrightarrow$ Mid range:	$e_{qC}=e_{qC}^{\ominus}\quad\Leftrightarrow\quad q=q^{mid}$	(7.99)
Reference effort $\Leftrightarrow$ Null basic quantity:	$e_{qC}=e_{qC}^{0}\quad\Leftrightarrow\quad q=0$	(7.99)
Asymmetric dipole: $q^{mid}\neq 0$; *Symmetric* dipole: $q^{mid}=0$		
$q_1+q_2\neq 0$	$q_1+q_2=0$	
Linear case		
Capacitive linear relation:	$q\underset{lin}{=}q^{mid}+C_q\left(e_{qC}-e_{qC}^{\ominus}\right)$	(7.100)
Capacitive short expression:	$q\underset{lin}{=}C_q\left(e_{qC}-e_{qC}^{*}\right)$ (7.25); $e_{qC}^{*}=e_{qC}^{\ominus}+C_q^{-1}\,q^{mid}$	(7.102)
Elastive short expression:	$e_{qC}\underset{lin}{=}C_q^{-1}\left(q-q^{**}\right)$ (7.25); $q^{**}=q^{mid}-C_q\,e_{qC}^{\ominus}$	(7.102)
Linear dipole capacitance:	$C_q\underset{lin}{=}\dfrac{\Delta q^{m}}{4e_{q\Delta}}=\dfrac{q^{max}-q^{min}}{4e_{q\Delta}}$	(7.101)

8 Multipoles

CONTENTS

What happens when, instead of two interacting objects, more than two objects are put together?

The answer is new things, which were not taken into account in a single dipole. With this numerical step from two to three or more, one enters a vast world of composed systems that occupies actually all the upper levels in the complexity scale.

A multipole is a Formal Object occupying just the next level above the dipole because it is composed of identical poles, belonging to the same energy variety in the same subvariety and with the same energetic nature (i.e., all inductive, capacitive, or conductive for fundamental multipoles, and for mixed ones, all inductive–conductive or capacitive–conductive). This identity of energetic nature does not mean identity of characteristics: each variable in a pole may have a value that differs from the other poles, in particular the energies-per-entity.

Table 8.1 lists the case studies of global multipoles (without space distribution) given in this chapter. (See also Figure 8.1 for the position of the multipoles.)

TABLE 8.1
List of Case Studies of Global Multipoles

	Name	Energy Variety/ Subvariety	Structure	Regularity	Page Number
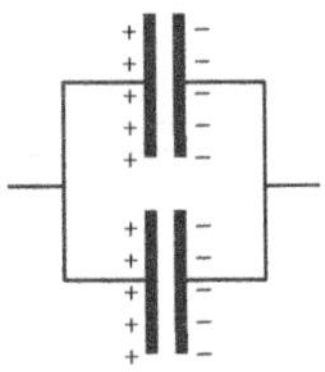	E1: Parallel Capacitors	Electrodynamics/capacitive	Ladder	Irregular	274
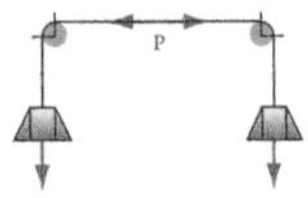	E2: Concurring Forces	Translational mechanics/ capacitive	Chain	Irregular	277
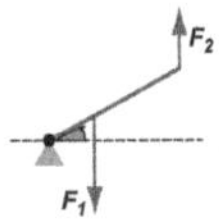	E3: Levers and Balance	Rotational mechanics/ capacitive	Chain	Irregular	281
$A_1 \rightleftarrows A_2 \rightleftarrows A_3$	E4: First-order Chemical Reactions in Series	Physical chemistry/ capacitive–conductive	Chain	Irregular	284
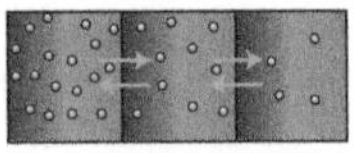	E5: Diffusion through Layers	Corpuscular energy/ capacitive–conductive	Chain	Regular	290
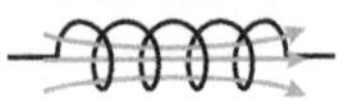	E6: Solenoid	Electrodynamics/inductive	Chain	Regular	296

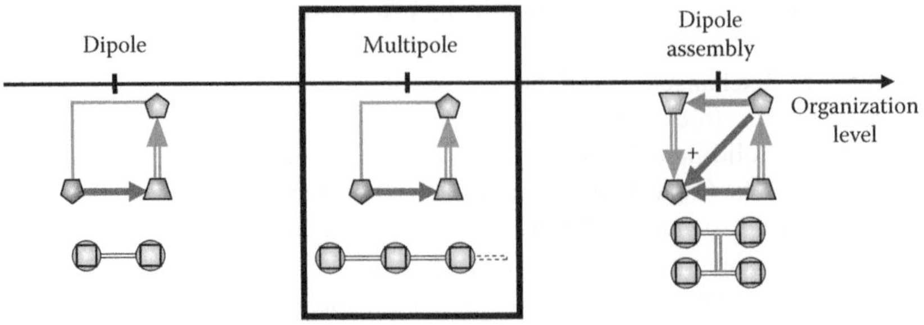

FIGURE 8.1 Position of the multipole along the complexity scale of Formal Objects.

8.1 THE MULTIPOLE

A multipole is merely an extension of the concept of dipole by increasing the number of poles. As for the dipole, by definition, poles must be of the same energetic nature, that is, all belonging to the same energy subvariety or all being purely conductive in the same energy variety. The case of mixed natures (for instance, capacitive–inductive) is relevant to the upper Formal Object called *dipole assembly*.

Which difference with a composition of poles? Both are made up with poles. But energy exchanges between poles are forbidden in a composed pole whereas they are permitted in a multipole. This allows a multipole to equilibrate energy between its poles and to behave as a dipole. On the other hand, a composed pole remains a pole.

In a multipole, all poles are interconnected, but not necessarily with all other poles. Some poles are connected to the outside, playing the role of *ports* (or *gates*, but this word is reserved to variables and nodes in a Formal Graph). A first characteristic of a multipole is quantified by the number of poles and ports. These numbers are necessarily higher than one, and even two for the number of poles, if one considers that two poles make a dipole, which belongs to a lower level of Formal Objects. Figure 8.2 shows a five-port multipole made of eight poles.

8.1.1 Formal Graph Representation of a Multipole

The graphical representation of a multipole without space distribution obeys the same classification as for dipoles; two kinds are to be considered: a first one when only one system property is present, consisting of three fundamental multipoles, inductive, conductive, and capacitive, and a second one with two properties, one being the conductance.

The Formal Graph representation of the five basic multipoles is identical to the representation of dipoles, which is shown in Graphs 8.1 and 8.2.

Time unnecessary. By definition as a multipole is made up of poles belonging to the same energetic nature (same energy subvariety), no energy conversion may occur in a multipole. Consequently, evolution and time are not required in a multipole, unless the multipole is connected to other multipoles (or dipoles) having different natures. This potentiality, which will be applied in Chapter 9, is indicated in the Formal Graphs of multipoles for generality as an evolution operator for not making explicit use of time.

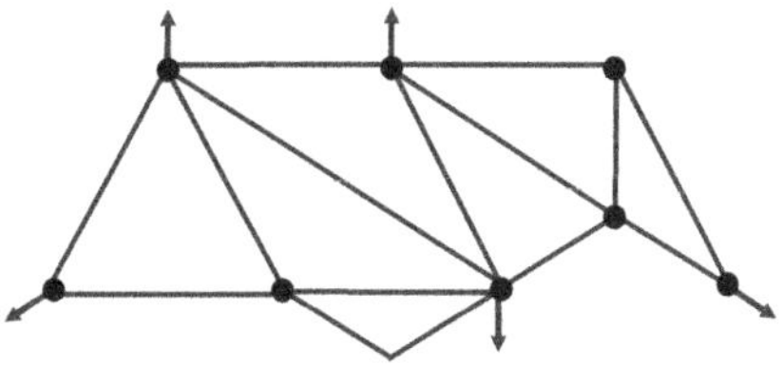

FIGURE 8.2 A multipole can be actually represented by an ordinary graph with poles as nodes. Some of them act as ports (marked with an arrow), in connection with the exterior, the others are internal nodes. Links can be of various kinds, expressing that poles are exchanging energy-per-entity or are merely connected as in a dipole.

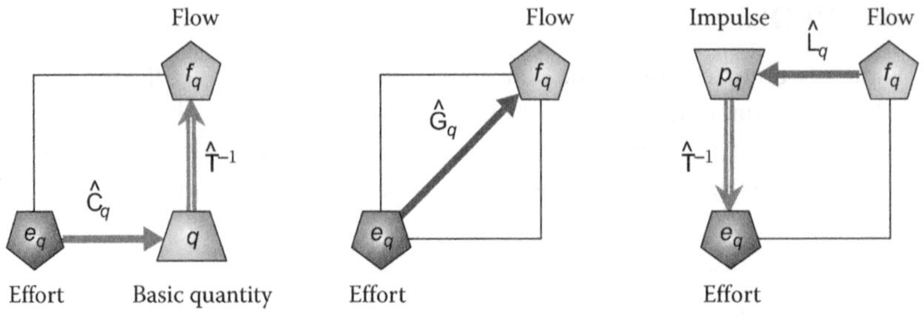

GRAPH 8.1 Capacitive (left), conductive (center), and inductive (right) multipoles.

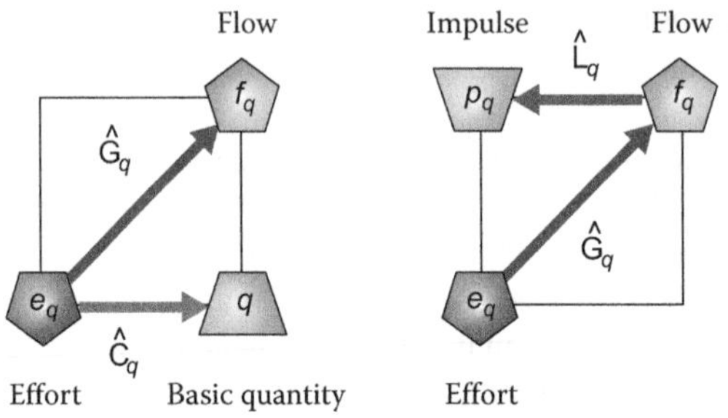

GRAPH 8.2 Capacitive–conductive multipole (left). Inductive–conductive multipole (right).

8.1.2 Decomposing a Multipole

A multipole may appear not very different from a dipole; actually this simple progression in the complexity scale reveals new features linked to the way poles are arranged.

There are basically two ways to decompose a multipole in lower Formal Objects, either as a set of poles or as a set of dipoles. For analyzing structures, the appropriate way is to consider Formal Objects at the immediately lower level, that is, dipoles. Each dipole is connected to one or several dipoles through its two gate variables, which are both energies-per-entity.

The rule for connecting two dipoles is that they must share the same energy-per-entity, either a flow or an effort. In the language of electric circuits, it corresponds to either a serial or a parallel mounting. This naming comes from the usual representation of dipoles as longitudinally oriented in space or as components of a circuit with two ends. In the Formal Graph approach, serial mounting means a common flow (same flow/current through all dipoles) and parallel mounting means a common effort (same effort/potential across all dipoles). These two configurations are sketched in Figure 8.3 for two dipoles according to the usual electric circuit representation.

When increasing the number of poles, the possibility of combining serial mounting for some dipoles and parallel mounting for some others makes the picture extremely diverse. In Figure 8.4 are shown the various possible combinations for only three dipoles, comparing the two levels of

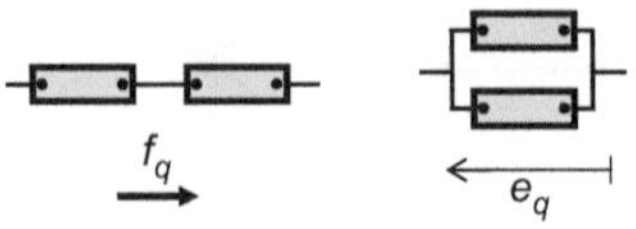

FIGURE 8.3 Serial mounting (left) and parallel mounting (right) of two dipoles in the language of electric circuits.

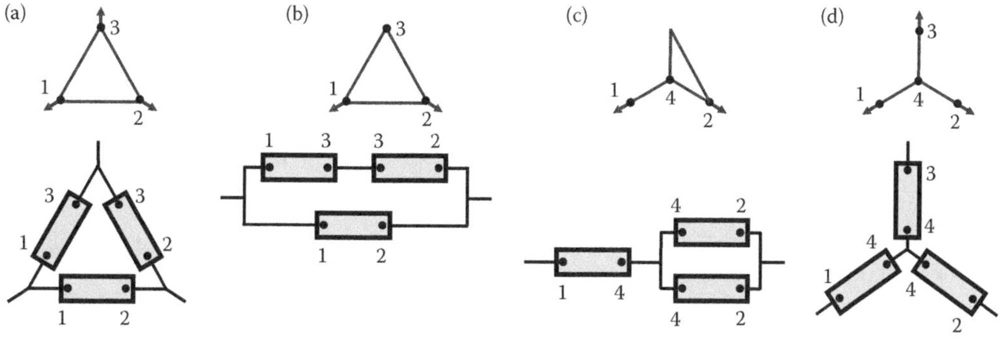

FIGURE 8.4 The four different ways to arrange three poles (upper row) and dipoles (lower row): (a) three-ports circular or tangential, (b) two-ports chain ladder, (c) two-ports ladder chain, and (d) three-ports radial. The first three structures have three poles whereas the fourth one has four poles.

decomposition, the level of poles with ordinary graphs and the level of dipoles with circuit components.

However, the fact that all dipoles are of the same energetic nature exempts one from having to treat all cases, because simple rules for combining identically mounted dipoles can be used. In other words, the category of multipole is self-similar in the sense that several multipoles assembled form again a multipole. Moreover, the number of configurations to study can also be reduced by considering only multipoles having two ports.

Only two-port multipoles. Multipoles with more than two ports can be decomposed as arrangements of two-port multipoles. This makes normally a Formal Object of higher complexity that is not especially interesting to discuss from the methodological viewpoint, because no new property appears. Consequently, only multipoles with two ports are treated in this chapter.

In practice, it is enough to consider as multipole, a set of identical poles in which all dipoles share a same energy-per-entity. This makes two categories, called *chain* and *ladder*, according to the images formed respectively by serial and parallel dipolar components in a circuit.

In the pole chain, all dipoles share their flows (in a circuit, they are mounted in series).

In the pole ladder, all dipoles share their efforts (in a circuit, they are mounted in parallel).

In Figure 8.5 are sketched in the language of electric circuits these two opposed ways of mounting several dipoles that form a multipole.

Chain or ladder, which one is the most important? The chain, without any doubt. A number of processes in physics and chemistry are serial processes that require special concepts, such as barrier, limiting step, reversibility, and so on, whereas parallel processes are less important in the sense that involved concepts are less fundamental (although important in percolation processes for instance). The case studies devoted to chain processes will be many in comparison with those devoted to ladder processes.

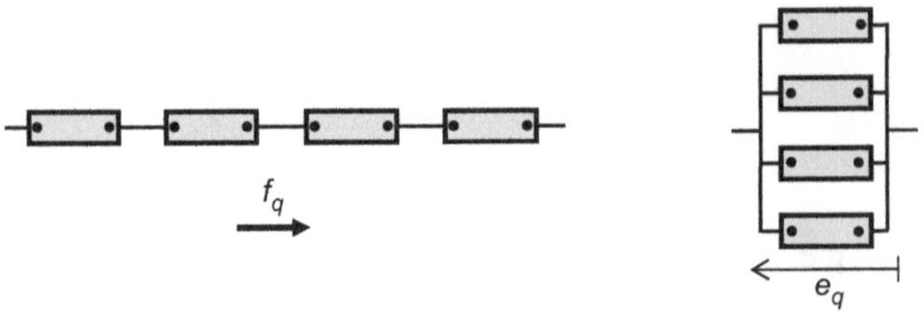

FIGURE 8.5 Equivalent electric circuits showing the two structures of a multipole composed of four dipoles: chain or serial (left) and ladder or parallel (right).

The chain multipole allows one to establish a simple theory of conduction, leading to formulation of the conductive relationship as an exponential function, as was done for the capacitive relationship in Chapter 7. The well-known Arrhenius* law of conduction will be demonstrated in part (without explicit dependence on the temperature for the moment).

$$k = A \exp\left(-\frac{E_a}{RT}\right) \quad (8.1)$$

Equally well-known dynamic equations such as the charge transfer kinetic equation in electrochemistry and the diode characteristic in semiconductor materials will be established.

Upward and downward emergence. The clear distinction between organization levels allows putting into evidence that emergence of new properties is not uniquely an upward process from lower to upper levels but is a more complex process. Some properties of a lower level can only be justified by considering the behavior of the upper level. For instance, the capacitive or conductive relationships are system constitutive properties that are attributed to a pole, which can be established only by analyzing dipole and multipole properties. One must go up to the level of energy-coupled Formal Objects for completely establishing the temperature dependence. It does not mean that properties at low levels cannot be stated and utilized in models, but that they have some indeterminations (about scaling parameters, for instance) that will be removed in ascending the organization levels. In other words, lower levels have more freedom when considered alone than when included into bigger ensembles. This is a subject of investigation pertaining to "multiscale" physics.

8.1.3 Multipole Regularity

Another characteristic of a dipole is the diversity of the dipoles or their similarity. A multipole made up of identical dipoles (in the sense of same values of system constitutive properties) is said to be regular. An example of a regular multipole is the solenoid, also called coil, which is made of an electrically conducting wire enrolled around a tube. Each turn is a dipole and all turns are strictly identical, having the same quantity of induction (induction flux), the same inductance, and the same current, so the multipole is a regular chain.

* Svante August Arrhenius (1859–1927): Swedish chemist, Stockholm, Sweden.

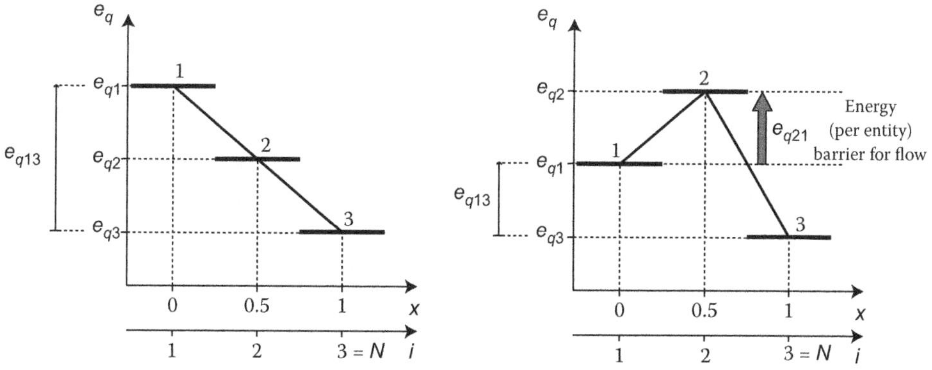

FIGURE 8.6 Two graphic representations of the notion of regularity for a chain multipole. In the left scheme of a regular multipole, all dipole efforts are equal, whereas in the right scheme of an irregular multipole, inequality is effective. The existence of an energy-per-entity barrier is shown in this last case.

An example of an irregular multipole is a rigid rod with a fixed end and submitted to various forces applied on points at different distances along the rod. It forms a set of different levers that are interdependent and constitute an irregular chain. This arrangement results from the fact that all dipoles share the same angle and therefore the same angular velocity (flow) on rotation of the rod, but with not necessarily equal torques (couples) applied. These two case studies are described in detail in Section 8.2.

This regularity characteristic is illustrated in Figure 8.6 for a chain multipole composed of two dipoles in series. The position along the chain is quantified by a chain coordinate x related to the pole number i through the relation

$$x = \frac{i-1}{N-1} \tag{8.2}$$

in which N is the total number of poles.

A regular multipole is shown in Figure 8.6 (left) with a regular decrease in the pole efforts along the chain, which means equal dipole efforts (and negatively counted). It can be as well a regular increase, the question of the sign being irrelevant.

An irregular multipole is shown in Figure 8.6 (right) in the peculiar case of an intermediate effort higher than the two efforts of the extreme poles. This case is interesting to outline because it introduces the very important concept of *barrier* in a process, which occurs when one of the energies-per-entity, here an effort, creates a jump in the opposite direction of the general variation of the energies-per-entity in the chain.

Energy or energy-per-entity? It is customary in physics to speak of *energy* about what is actually an *energy-per-entity* and more precisely an effort. In the present case, physicists speak of "energy barrier" and not of "energy-per-entity barrier." This usage comes from particles physics where the energy container is unique, a single particle or corpuscle. Notwithstanding, when dealing with collections of containers, energy and energy-per-entity are not the same concept. It is not superfluous to recall that rigorous naming of concepts is always beneficial for everybody.

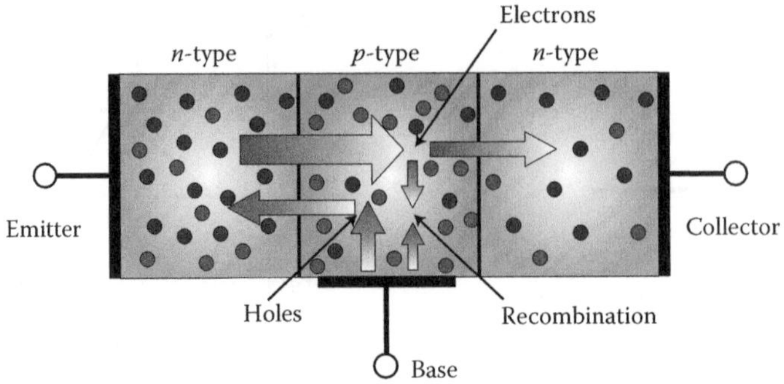

FIGURE 8.7 Bipolar junction transistor made with a sequence of *n–p–n*-type semiconductor layers. This system is a three-port irregular multipole with the height of the barrier controlled by an external voltage applied between the *base* and the *emitter*.

An example of an irregular multipole in which the existence of a barrier is essential is the transistor, made up of two semiconductor junctions (*n–p* and *p–n*, for instance) in series, which allows one to control the flow of electrons through the multipole by adjusting the height of the barrier (see Figure 8.7). This example is not detailed in this chapter because it is a three-port multipole, but also mainly because a transistor involves more than one energy variety for modeling its behavior (at least electrodynamics and corpuscular energy).

 Regularity is not so common. A look at Figure 8.8 suggests that regularity is not the most frequent configuration because to obtain the regularity of three poles in a chain multipole, the intermediate effort must match exactly the middle between the two extreme efforts. The condition is still more stringent with an increase in the number of poles. However, when the multipole is free to adjust its intermediate efforts, regularity is the most stable configuration.

$$\text{Irregular} \quad \text{Regular} \quad \text{Irregular}$$

$$e_{q2} \ll e_{q2}^{reg} \quad e_{q2} \approx e_{q2}^{reg} \quad e_{q2} \gg e_{q2}^{reg}$$

$$e_{q2}^{reg} = \frac{e_{q1} + e_{q3}}{2} \quad e_{q2}$$

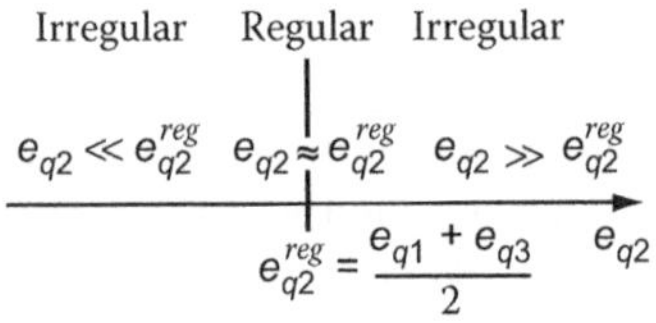

FIGURE 8.8 Regularity of a three-pole chain is quantified by the value of the intermediate effort.

8.2 CASE STUDIES OF MULTIPOLES

The case studies presented in this chapter illustrate the great variety of domains in which the concept of multipole, with its variations of structure, is pertinent for modeling different systems. The choice to take irregular multipoles in some domains and regular ones in other domains is arbitrary, or merely dictated by the exemplarity of the system. Every multipole presented as irregular could have been studied in the regular configuration and conversely.

Irregular Ladder
- E1: Capacitors in Parallel in Electrodynamics (Capacitive)

Irregular Chain
- E2: Concurring Forces in Translational Mechanics (Capacitive)
- E3: Levers and Balance in Rotational Mechanics (Capacitive)
- E4: First-Order Chemical Reactions in Series in Physical Chemistry (Capacitive–Conductive)

Regular Chain
- E5: Diffusion through Layers in Corpuscular Energy (Capacitive–Conductive)
- E6: Solenoid in Electrodynamics (Inductive)

The simple fact that these so different systems obey the same model is enlightening. To be able to think of a set of rotating levers and a chemical reaction in series in identical terms, for instance, without forgetting the mentioned example in the text of the transistor, is quite instructive.

8.2.1 Case Study E1: Parallel Capacitors — Electrodynamics

The case study abstract is given on next page.

Two capacitors associated by joining their two ends in pairs form a new system that is used for its increased storage capacity of capacitive energy (electrostatic energy). The multipole formed by these two components mounted in parallel is in the general case, an irregular ladder, unless the two capacitances are equal, entailing regularity.

The correspondence pole–dipole is the following for a dipole i:

$$Q_i = Q_{1:i} = -Q_{2:i} \quad V_i = V_{1:i} - V_{2:i} \tag{E1.1}$$

(according to the notation in the scheme in the case study abstract) and for a pole j

$$V_j = V_{j:1} = V_{j:2} \tag{E1.2}$$

as the poles labeled 1:1 and 1:2 are merged into a unique pole 1 and poles 2:1 and 2:2 into 2.

8.2.1.1 Construction Formal Graph from Dipoles

In electrodynamics, it is not customary to consider systems under the thermodynamical viewpoint. The Formal Graph approach being based on thermodynamics, the starting point for analyzing a system—and an electric circuit does not escape from this—is the Gibbs equation ruling the variation of capacitive energy. This energy variation is written according to the various ways to decompose a system, here by considering two dipoles and one multipole.

$$\mathrm{d}\mathcal{U}_Q \underset{\exists C}{=} V_1\,\mathrm{d}Q_1 + V_2\,\mathrm{d}Q_2 = V\,\mathrm{d}Q \tag{E1.3}$$

The key feature of a system of two components in parallel is that the dipoles share the same effort, here the same potential

$$V = V_1 = V_2 \tag{E1.4}$$

E1: Parallel Capacitors **Electrodynamics**

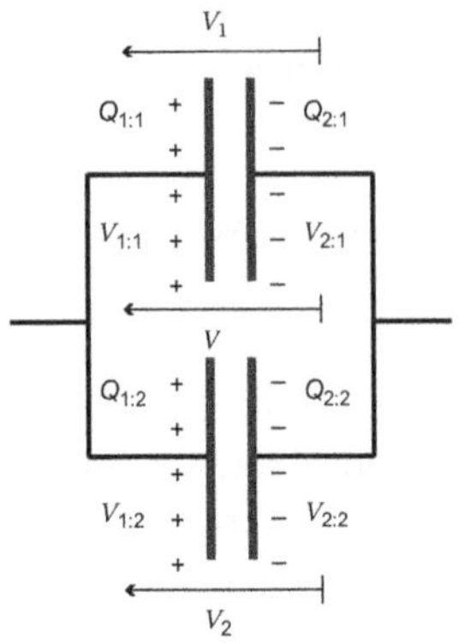

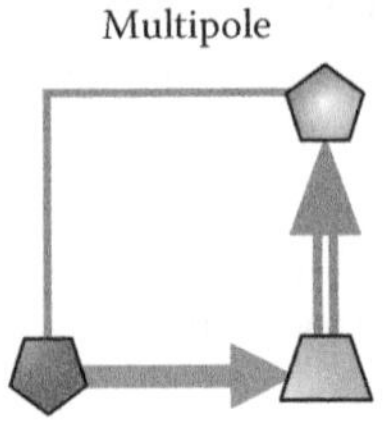

Multipole	
✓	Fundamental
	Mixed
✓	Capacitive
	Inductive
	Conductive
	Chain/serial
✓	Ladder/parallel
	Regular
✓	Global
	Spatial
✓	Conservative

Formal Graph:

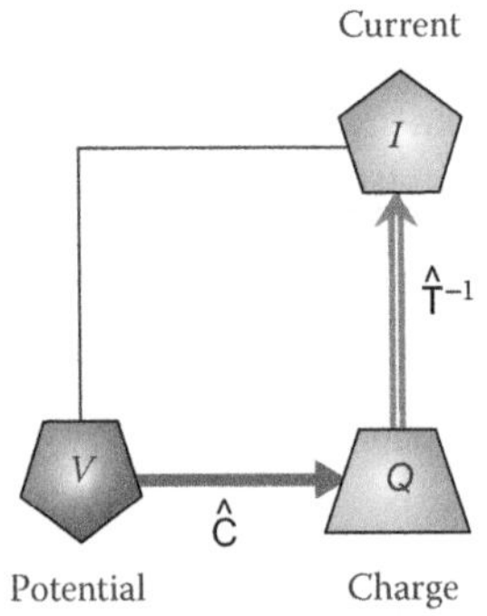

GRAPH 8.3

Construction from two dipoles:

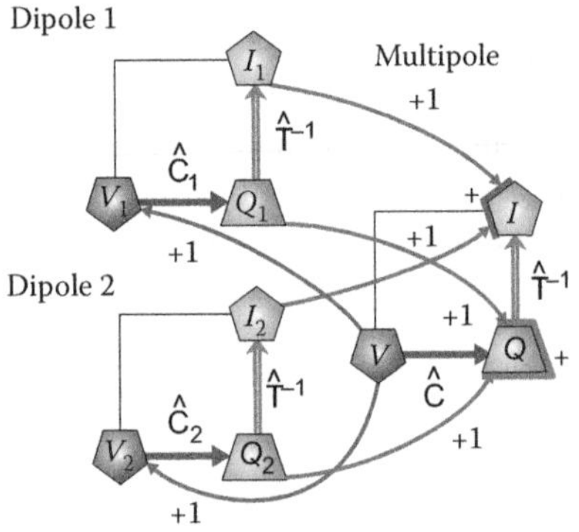

GRAPH 8.4

Variables:

Variety	Electrodynamics		
Subvariety	Capacitive (Electrostatic)		Inductive
Category	Entity number	Energy/ entity	Entity number
Family	Basic quantity	Effort	Flow
Name	Charge	Potential	Current
Symbols	Q	V, U, E, φ	I, i
Unit	[C], [A s]	[V], [J C^{-1}]	[A]

Property relationships:

$$Q = \hat{C}\,V$$

Linear model:

$$Q \underset{lin}{=} C\,V$$

System Property	
Nature	Capacitance
Levels	0
Name	Capacitance
Symbol	C
Unit	[F], [J C^{-1}]
Specificity	—

Energy variation:

$$d\mathcal{U}_Q \underset{\exists C}{=} V\,dQ$$

From the Gibbs equation, the variation of the multipole charge variations appears as the sum of the dipole charge variations

$$dQ = dQ_1 + dQ_2 \tag{E1.5}$$

This means that the association in parallel of two capacitors is effectively the way for increasing the storage capacity, in contradistinction with the serial mounting (see Section 8.3).

By assuming proportionality between flows and basic quantity variations through the evolution operator (see Chapter 9), the multipole current follows the same dependence with the dipole currents

$$I = I_1 + I_2 \tag{E1.6}$$

Equations E1.4 and E1.6 form Kirchhoff* circuit laws for a parallel mounting, which are the classical starting point in electrodynamics for modeling circuits. By integration of the charge variations, one obtains the multipole charge as equal to the total charge:

$$Q = Q_1 + Q_2 \tag{E1.7}$$

Integration constants are not required for electric dipoles as the conservation of the charges in a capacitor is ruled by a null sum of its pole charges (see case study C7 "Electric Capacitor" in Chapter 6).

These relationships are translated in the construction Formal Graph given in the case study abstract.

8.2.1.2 Constitutive System Properties

A capacitance is defined for each element, dipoles and multipole, thus providing three capacitive relationships:

$$Q = \hat{C}V; \quad Q_1 = \hat{C}_1 V_1; \quad Q_2 = \hat{C}_2 V_2 \tag{E1.8}$$

By substituting the charges in the summation in Equation E1.7, in remarking that the grouping operators in a sum is not possible in the general case of nonlinear capacitances one writes

$$\hat{C}V = \hat{C}_1 V + \hat{C}_2 V \tag{E1.9}$$

The overall operator can be identified with the multipole capacitance

$$\hat{C} = \frac{\hat{C}_1 V + \hat{C}_2 V}{V} \tag{E1.10}$$

* Gustav Robert Kirchhoff (1824–1887): German physicist; Heidelberg and Berlin, Germany.

Physically speaking, it means that the multipole behaves exactly as a new capacitor with a capacitance depending on the potential and on the individual capacitances.

Naturally, in the linear case, this relationship simplifies

$$C \underset{lin}{=} C_1 + C_2 \tag{E1.11}$$

which is the classical expression for modeling the parallelism of two capacitors.

Extension to several capacitors is trivial with the summation extended to N dipoles

$$\hat{C} V = \sum_{i=1}^{N} \hat{C}_i V \underset{lin}{=} V \sum_{i=1}^{N} C_i \tag{E1.12}$$

8.2.2 Case Study E2: Concurring Forces — Translational Mechanics

The case study abstract is given on next page.

In translational mechanics, the *force* $\mathbf{F}$ plays the role of *energy-per-entity* in the capacitive subvariety of energy; in other words, it belongs to the family of efforts. The *displacement* $\boldsymbol{\ell}$ plays the role of *capacitive entity number* which is therefore a basic quantity. These are state variables oriented in space and therefore they are represented by vectors. In this energetic subvariety, the capacitive energy is called "work" and its variation is given by the scalar product of vector force by the variation of vector displacement:

$$\mathrm{d}\mathcal{U}_\ell = \mathbf{F} \cdot \mathbf{d}\boldsymbol{\ell} \tag{E2.1}$$

The case studied here is the simultaneous and concurrent action of two forces on the same point P in the situation called "state of equilibrium of forces," a situation that is classically qualified of *static*, this is to say in the absence of movement and thereby of time. One shall see that this limitation to the static regime is not an absolute requirement for applying the model proposed by the Formal Graph approach.

For experimentally realizing this case study, numerous systems are possible but the most used in textbooks is that of the wire tightened between two identical pulleys placed horizontally, with two equal gravitational masses suspended to the vertical ends of the wire. This system is not as simple as for instance two springs, each one fixed to an immobile support and exerting their traction on a same point. The complexity arises from the usage of masses in a gravitational field that involve a second energy variety for creating a force by virtue of the proportionality between weight (gravity force) and gravitational mass through the gravitational field (assumed to be uniform) (see case study J4 "Gravitational Force" in Chapter 12). In the scheme in the case study abstract, the studied part is outlined by a rectangle that delimitates between the two pulleys the subsystem that is relevant to the sole translational mechanics. With the aim of generalizing more easily to the case of non-colinear forces, the case of three unequal forces created by three different masses is shown after the case study abstract.

E2: Concurring Forces **Translational Mechanics**

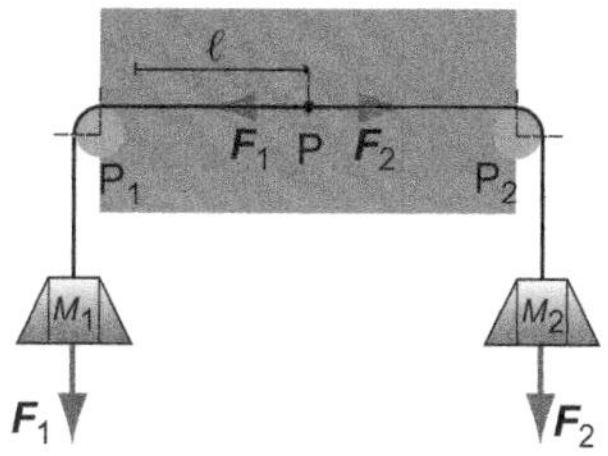

Multipole

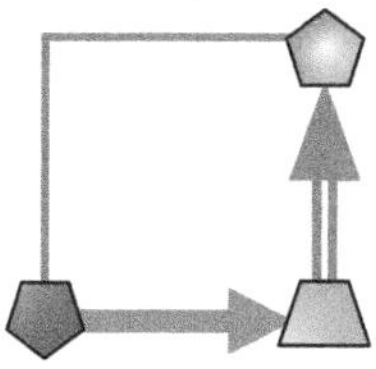

Multipole	
✓	Fundamental
	Mixed
✓	Capacitive
	Inductive
	Conductive
✓	Chain/serial
	Ladder/parallel
	Regular
✓	Global
	Spatial
✓	Conservative

Formal Graph:

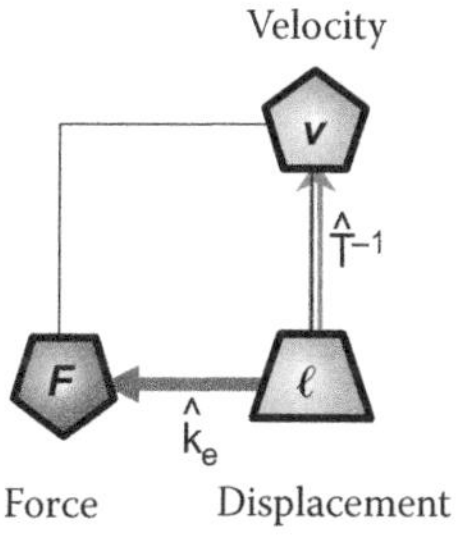

GRAPH 8.5

Construction from two dipoles:

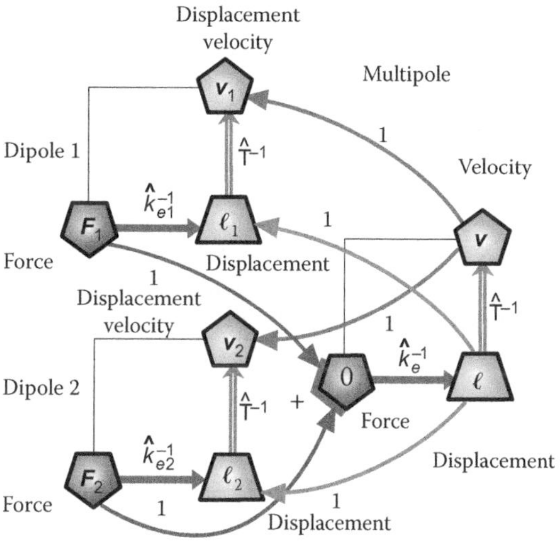

GRAPH 8.6

Variables:

Variety	Translational Mechanics		
Subvariety	Capacitive (Potential)		Inductive
Category	Entity number	Energy/ entity	Energy/ entity
Family	Basic quantity 	Effort 	Flow 
Name	Displacement	Force	Velocity
Symbols	ℓ	$\boldsymbol{F}$	$\boldsymbol{v}$
Unit	[m]	[N], [J m^{-1}]	[m s^{-1}]

System Property	
Nature	Elastance
Levels	0
Name	Elasticity
Symbol	k_e
Unit	[N m^{-1}]
Specificity	—

Which capacitance? Normally, the validity of the differential Equation E2.1 giving the variation of capacitive energy is dependent on the existence of a capacitance in the system, according to the rule saying that capacitance or inductance is the necessary support of energy storage. Classically, this is expressed by the statement that the "force is working" (i.e., produces work, that is, stores or consumes stored energy). Here, the system does not possess a mechanical capacitance but a gravitational one.

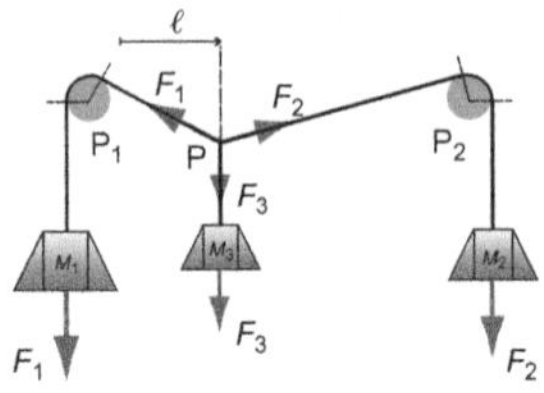

The modeling with a Formal Graph of a system of two concurrent forces does not utilize one dipole (as it could be inferred from the fact that only two forces are involved) but *two dipoles* sharing the same displacement ℓ and therefore in series (common flow). The justification of this level of organization in multipole comes from the requirement that the same model should apply to the action of two forces as well as more than two forces on the same point, as the physical principles used here must not depend on the number of forces. Each dipole used is made of one segment of wire between the converging point P and the contact point on the pulley, where the same but opposite force acts (see case study C8 "Spring" in Chapter 6).

The variation of capacitive energy (or *work* in the classical language) is the sum of variations of capacitive energy of each dipole

$$d\mathcal{U}_\ell = \mathbf{F}_1 \cdot \mathbf{d}\ell_1 + \mathbf{F}_2 \cdot \mathbf{d}\ell_2 \quad \text{(E2.2)}$$

The constraint of uniqueness of the point of application, or more generally when the physical link between the dipoles is ensured by a body having a non-negligible extension, the constraint of *indeformability* of the link, is expressed by equality between the variations of displacement of each dipole

Coupling. Precisely in this system, the two energy varieties, translational mechanics and gravitation, cannot convert energy in order to vary their respective part in the total energy. This situation corresponds to a *coupling* between the two varieties (see Chapter 12).

$$d\ell = d\ell_1 = d\ell_2 \quad \text{(E2.3)}$$

The *isolation of the system* with respect to the exterior (no external force applied) and the *absence of energy exchange* with the gravitational variety (see remark on "Coupling") entail the nullity of the variation of mechanical energy in the capacitive subvariety:

$$d\mathcal{U} = 0 \quad \text{(E2.4)}$$

These three hypotheses are expressed by the nullity of the sum of the dipole forces:

$$\mathbf{F}_1 + \mathbf{F}_2 = 0 \quad \text{(E2.5)}$$

The generalization to more than two forces, colinear or not, is written as

$$\sum_i \boldsymbol{F}_i = 0 \tag{E2.6}$$

In mechanics, the state of rest is called the state of *static equilibrium* and Equation E2.6 is viewed as the *law of equilibrium*. In fact, it is a *mounting rule* (similar to the Kirchhoff rule in electricity for components in series) based on a *law of conservation* and not on an autonomous principle: This equation results merely from hypotheses made on the system [indeformability (E2.3), isolation, and the absence of exchange between varieties] and from the application of the First Principle of Thermodynamics (Equation E2.2). In the Formal Graph theory, this "law" saying that a null sum of forces expresses a static equilibrium is considered as unsuitable because the notion of equilibrium is much more general than this mechanical conception.

A too peculiar notion. The notion of *static equilibrium* is a particularity that cannot be easily transposed to other energy varieties, in which, for most of them, a mounting rule of the Kirchhoff type is utilized. Moreover, the conservation equation E2.6 is perfectly valid during a movement of rotation as shown by the Formal Graph, which includes the relationships between angular velocities (flows). This is not compatible with the definition of a static equilibrium, which implies the absence of movement and therefore of evolution (and of time).

Nevertheless, the pertinence of the notion of static equilibrium is found when an external constraint, an external constant force (action of a static electric field on an electric charge, for instance) is imposed on this system, because the notion of conservation is lost in an open system.

All these relations are translated in the Formal Graph in the case study abstract (Graph 8.6), which utilizes two dipoles for combining them in only one assembly in series of dipoles. (The mounting is effectively in series as, the displacements being equal, the displacement velocities are also equal, and therefore the flow is the same for the two dipoles.)

The Formal Graph in Graph 8.6 is a *graph of construction*: It represents on one hand on the left the two Formal Graphs of dipoles independently, and on the other hand on the right the resulting Formal Graph of the assembly in series of the two dipoles. The relations between the state variables of dipoles and those of the assembly are represented by connections featured by arrowed arcs, each connection being endowed with an operator *identity*. The sum of dipole forces is expressed by an operation of addition on the node of the effort in the assembly, equal to zero because the system is isolated from the exterior.

It is worth remarking that the Formal Graph in the case study abstract (Graph 8.6) models equally well a static regime and a dynamic regime when a translation moves the point P. The absence of inertial mass in the wire (ideal case) results in the absence of impulse (quantity of movement) and of kinetic energy that do not need to be taken into account (for the purely translational part of the system, which could alternatively be realized with springs, as aforementioned). It has to be recalled once more that this is one of the reasons that prevents one from making a logical connection between the nullity of the sum of applied forces and the static situation of a system.

Remarks

Dipole mounting. To determine the type of dipole mounting, one needs to determine which one is the common state variable between the dipoles: Here it is the displacement, which is an entity number (basic quantity). Consequently, the variables that are added together are the energies-per-entity (efforts), here the forces. In electricity, this is a mounting in series of components.

Concurrence. This case study tackles the core of essential notions of static mechanics and the Formal Graphs help to better understand the notion of concurrent forces in generalizing the fundamental concepts to the whole physics and to physical chemistry.

Mounting rule versus static equilibrium. What the Formal Graph theory shows is that the action of two forces, equal in magnitude but opposite in direction, is analogous to two capacitors in series or to an elementary chemical reaction.

However, the necessity of utilizing a maximum of concepts common to all energy varieties leads to abandoning the notion of static equilibrium as a basis of the rule saying that the sum of forces is zero. This rule is replaced by the much more logical rule of dipole mounting, which relies on a law of conservation.

Newton's laws in mechanics. This "law of equilibrium" establishing a link with the rule of zeroing the sum of concurrent forces must not be confused with Newton's laws of mechanical movement that are recalled below for insisting on the differences.

1. The first law stipulates that the velocity of a body can only change under the action of a force (or most rigorously that the absence of force able to change the velocity means that the body is moving in an inertial referential). This law applies only to a body endowed with an inertial mass.
2. The second law says that when the velocity changes, the applied force is proportional to the temporal derivative of the velocity. As for the first statement, this law applies only to a body endowed with an inertial mass.
3. The third law is generally the most misinterpreted one; because it is often reduced to the formulation "action is equal to reaction" without specifying on what these forces act. Actually, it applies either to an elastic body (translational mechanics) or to the action of a body on another one (gravitation) in symmetrizing the forces involved (by equating the magnitudes). It is often utilized abusively as a model for the convergence of two forces on the same body, case which is described here.

Equilibrium of forces. In fact, the Formal Graph theory tells us that Newton's third law, although thought of by its inventor in the restricted case of gravitation, applies only to a mechanical dipole or multipole. This is this law that is utilized in the Formal Graph theory for describing a capacitive dipole in translational mechanics (see case study C8 "Spring" in Chapter 6).

The true founder of the rule of "forces in equilibrium," although involuntary (this is the analysis permitted by the Formal Graph approach that comes to this conclusion), is Gustav Kirchhoff and not Isaac Newton!

8.2.3 Case Study E3: Levers and Balance

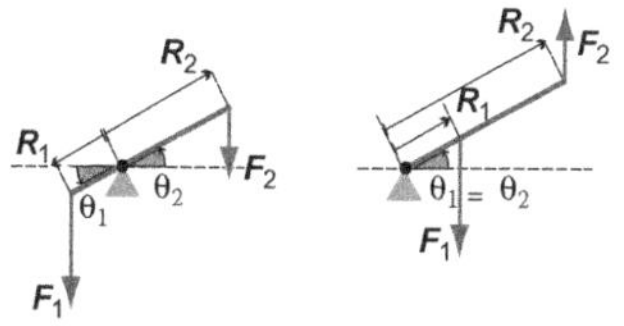

A lever (right) or a balance (left) is modeled with the help of an assembly of two dipoles in series, one for each arm of a lever, both belonging to the capacitive subvariety of mechanics of rotation.

Multipole

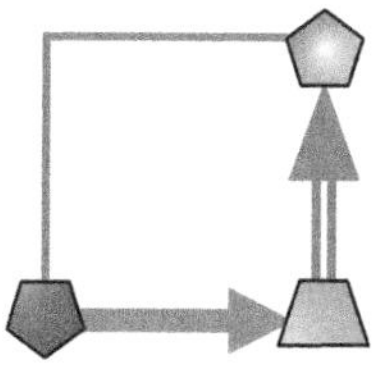

Multipole	
✓	Fundamental
	Mixed
✓	Capacitive
	Inductive
	Conductive
✓	Chain/serial
	Ladder/parallel
	Regular
✓	Global
	Spatial
✓	Conservative

Formal Graph:

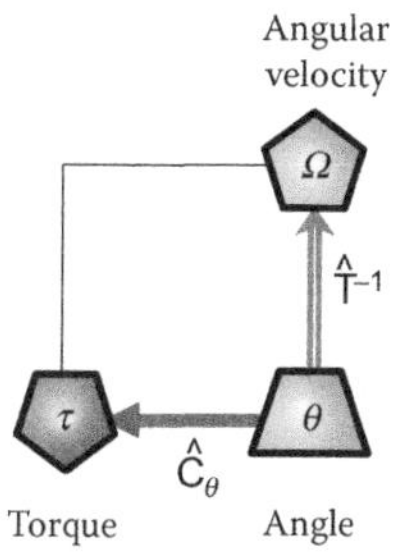

GRAPH 8.7

Construction from two dipoles:

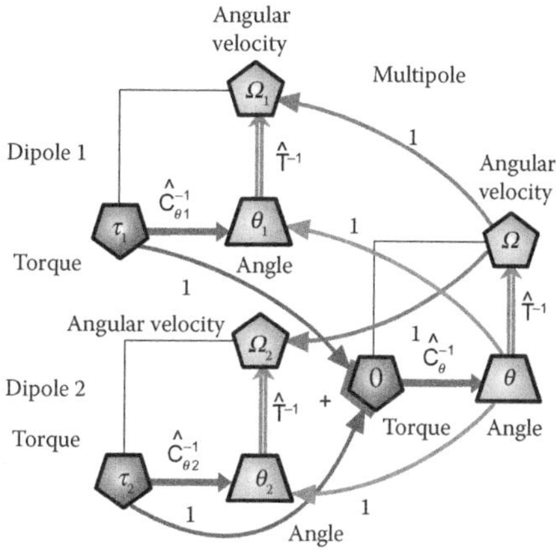

GRAPH 8.8

Variables:

Variety	Rotational Mechanics		
Subvariety	Capacitive (Potential)		Inductive
Category	Entity number	Energy/entity	Energy/entity
Family	Basic quantity 	Effort 	Flow 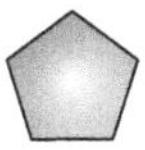
Name	Angle	Torque (couple)	Angular velocity
Symbols	$\boldsymbol{\theta}, \boldsymbol{\alpha}$	$\boldsymbol{\tau}$	$\boldsymbol{\Omega}$
Unit	[rad]	[N m rad^{-1}], [J rad^{-1}]	[rad s^{-1}]

Property relationships:

$$\boldsymbol{\tau} = \hat{C}_\theta\, \boldsymbol{\theta}$$

Linear model:

$$\boldsymbol{\tau} = C_\theta\, \boldsymbol{\theta}$$

System Property	
Nature	Elastance
Levels	0
Name	Constant of torsion
Symbol	E_θ, C_α
Unit	[N m rad^{-2}]
Specificity	—

Energy variation:

$$\mathrm{d}\mathcal{U}_\theta \underset{\exists C_\theta}{=} \boldsymbol{\tau} \cdot \mathbf{d}\boldsymbol{\theta}$$

In mechanics of rotation, the force $\boldsymbol{F}$ acting at a distance $\boldsymbol{R}$ from a center around which a body can rotate corresponds to a state variable $\boldsymbol{\tau}$ from the family of efforts called *torque* (or *couple*). All these variables are oriented in space and therefore represented by vectors. The relationship between force and torque is expressed by a vector product of the distance (radius of gyration) by the force

$$\boldsymbol{\tau} = \boldsymbol{R} \times \boldsymbol{F} \tag{E3.1}$$

A lever is made of two interdependent and indeformable bodies able to rotate around the same point; that is, the angles of rotation of both bodies remain identical whatever the forces applied for producing the rotation (in the limit of deformability of a solid body). Each body constitutes what is called a "lever arm" and, in practice, they are part of the same bar, rod, or board. The figure in the case study abstract shows two relative dispositions of two bodies with respect to the center of rotation, either on opposite sides (balance with unequal arms, on the left) or on the same side (lever, on the right).

The variation of capacitive energy (or *internal energy* in the classical language) is the sum of variations of capacitive energy of each body, each variation being given by the scalar product of two vector quantities, the torque (the effort) and the angle variation (the basic quantity in rotation mechanics) $\boldsymbol{\theta}$

$$\mathrm{d}\mathcal{U}_\theta = \boldsymbol{\tau}_1 \cdot \mathrm{d}\boldsymbol{\theta}_1 + \boldsymbol{\tau}_2 \cdot \mathrm{d}\boldsymbol{\theta}_2 \tag{E3.2}$$

The constraint of indeformability of bodies is expressed by equating the angles of rotation

$$\boldsymbol{\theta} = \boldsymbol{\theta}_1 = \boldsymbol{\theta}_2 \tag{E3.3}$$

The isolation of the system with respect to the exterior makes the variation of capacitive energy equal to zero, which, by taking into account the equality of angles and therefore of their variations, is translated by zeroing the sum of dipolar torques

$$\boldsymbol{\tau}_1 + \boldsymbol{\tau}_2 = 0 \tag{E3.4}$$

In mechanics, the state of rest is called state of *static equilibrium* and Equation E3.4 is viewed as the *law of equilibrium*. Actually, it is rather an *equation of conservation* than a law or an autonomous principle: This equation merely results from hypotheses made on the system (isolation and indeformability) and from the application of the First Principle of Thermodynamics (Equation E3.2). In the Formal Graph theory, this classical notion of static equilibrium, expressed by a zero sum of forces or of torques, is considered as unsuitable because it cannot be generalized outside mechanics (see remark on "Notion hard to generalize").

Notion hard to generalize. The notion of *static equilibrium* is a particularity that cannot be easily transposed to other energy varieties, in which, for most of them, a mounting rule of the Kirchhoff type is utilized. Moreover, the conservation equation E3.4 is perfectly valid during a movement of rotation as shown by the Formal (Graph 8.8), which includes the relationships between angular velocities (flows). This is not compatible with the definition of a static equilibrium, which implies the absence of movement and therefore of evolution (and time).

All these relations are translated into the Formal Graph in the case study abstract (Graph 8.8), which utilizes two dipoles for combining them in an assembly in a series of dipoles. (The mounting is effectively in series because the angles being equal; the angular velocities are also equal, therefore the flow is the same for the two dipoles.)

One of the Formal Graphs (Graph 8.8) in the case study abstract is a *graph of construction*: It represents on the one hand, on the left, the two Formal Graphs of dipoles, and on the other hand, on the right, the Formal Graph resulting from the assembly in series of two dipoles. The relationships between the dipolar state variables and those of the assembly are represented by connections featured by arrowed arcs, each connection being endowed with an *identity* operator. The sum of dipolar torques is translated into an operation of addition on the effort node of the assembly, which is equal to zero because the system is isolated from the exterior.

It is worth remarking that the Formal Graph models equally well a static state as a dynamic regime, when a movement of rotation drives the lever (stationary or not).

Balance: A balance is an instrument based on the principle of the lever made for comparing two weights. The *Roman balance* works with unequal arms (adjusting the distance of one of the suspended weights for obtaining an horizontal position of the of arms) whereas the *equal-arm balance*, or the *Roberval* balance*, is featured by the equality of lengths of opposite arms (which form the beam), and therefore the conservation of torques is transposed into an equality of forces (and therefore of weights when masses are hung at the lever arms)

$$\boldsymbol{R} = \boldsymbol{R}_1 = -\boldsymbol{R}_2 \quad \Rightarrow \quad \tau_1 + \tau_2 = \boldsymbol{R} \times (\boldsymbol{F}_1 - \boldsymbol{F}_2) = 0 \quad \Rightarrow \quad \boldsymbol{F}_1 = \boldsymbol{F}_2 \qquad \text{(E3.55)}$$

[*libra* in Latin means the pound (unit of weight), from which the word *equilibrium*, equi-libra, is issued, meaning "balance," equality of weight].

Remarks

Generality. What the Formal Graph theory may say on these elementary notions of mechanics is that a lever is analogous to a couple of capacitors in series or to a chemical reaction. In fact, it corresponds to two capacitive dipoles forming a multipole.

However, the necessity of using common concepts for all energy varieties leads to abandoning the notion of static equilibrium as the basis of the rule saying that the sum of torques must be zero. It is replaced by a much more powerful rule of conservation.

Equilibrium versus conservation. The notion of *static equilibrium* is to proscribe, if one is interested in better understanding the notion of reversibility and related concepts, such as the role of time, the distinction between dynamic and static, the separation thermodynamics—thermodynamics of irreversible processes, etc. In replacing this notion of *static equilibrium* by a *rule of conservation*, one has access to a more interesting conceptual level.

* Gilles Personne de Roberval (1602–1675): French mathematician and physicist; Paris, France.

8.2.4 Case Study E4: First-Order Chemical Reactions in Series — Physical Chemistry

$$A_1 \rightleftarrows A_2 \rightleftarrows A_3$$

Multipole

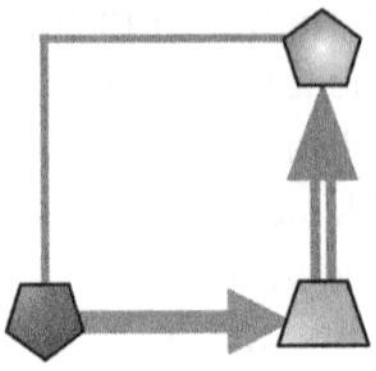

Multipole	
	Fundamental
✓	Mixed
✓	Capacitive
	Inductive
✓	Conductive
✓	Chain/serial
	Ladder/parallel
	Regular
✓	Global
	Spatial
✓	Conservative

Formal Graph:

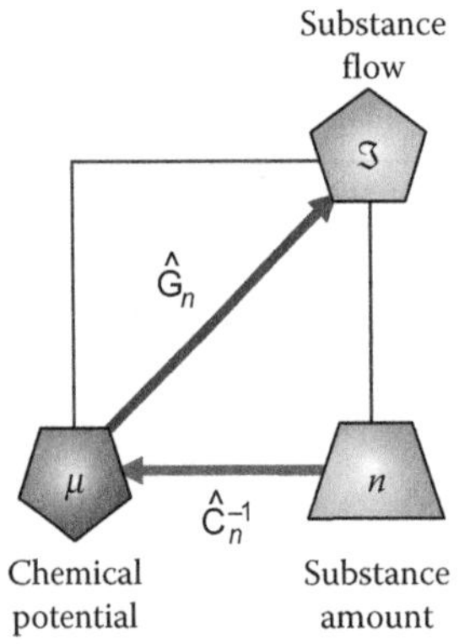

GRAPH 8.9

Construction from two dipoles:

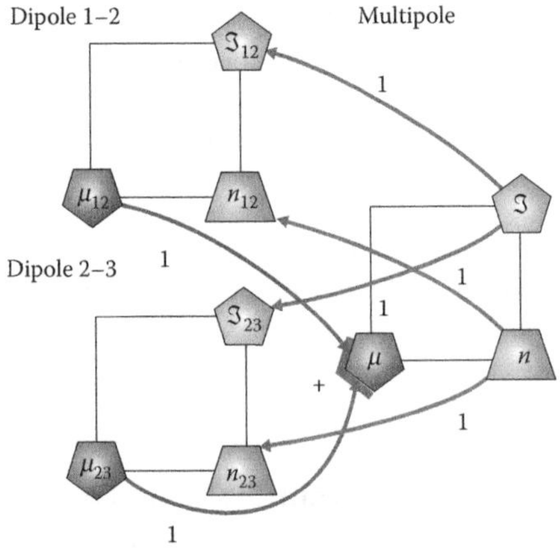

GRAPH 8.10

Variables:

Variety	Physical Chemical Energy		
Subvariety	Capacitive (Internal)		Inductive
Category	Entity number	Energy/ entity	Energy/ entity
Family	Basic quantity	Effort	Flow
Name	Substance amount	Chemical potential	Substance flow
Symbols	n	μ	$\mathfrak{I}$
Unit	[mol]	[J mol^{-1}]	[mol s^{-1}]

Property relationships:

$$n = \hat{C}_n \, \mu$$

$$\mathfrak{I} = \hat{G}_n \, \mu$$

Linear model:

$$n \underset{lin}{=} C_n \, \mu$$

System Property	
Nature	Capacitance
Levels	0
Name	Physical chemical capacitance
Symbol	C_n
Unit	[mol^2 J^{-1}]
Specificity	—

Energy variation:

$$d\mathcal{U}_n \underset{\ni C_n}{=} \mu \, dn$$

A sequence of chemical reactions with a first-order kinetics is the archetype of chain processes in which the concepts of intermediate species, transition state, and activation energy barrier find a thorough application. This is for saying that the study of such a system with this case study of two reactions in series is of paramount importance.

As stated in case study A11 of a reacting species in Chapter 4, a chemical reaction is actually a process that combines energy storage through a capacitance (or its reciprocal an elastance) and dissipation through a conductance (or resistance). In Chapter 7, the theory of influence allowed the demonstration of the exponential shape of the capacitance operator. In this chapter (Section 8.5), the same theoretical reasoning will be adapted to establish the conductive relationship in showing that an exponential function also models a simple conduction process. This concomitant presence of the exponential function leads to the demonstration of the chemical kinetic law stating that the substance flow (reaction rate) of a first-order reaction is proportional to the substance amount.

8.2.4.1 Formal Graph

The Formal Graph shown in the case study abstract represents a capacitive–conductive multipole that is identical to a dipole Formal Graph. Two properties are present in this graph: the capacitance and the conductance.

$$n = \widehat{C}_n \; \mu \tag{E4.1}$$

$$\mathfrak{I} = \widehat{G}_n \; \mu \tag{E4.2}$$

The combination of both operators, elastance and conductance, is not indicated on the Formal Graph as it was done for a pole (see case study A11 in Chapter 4) because it does not correspond to the usual practice for a set of several reactions as is the case here. On the contrary, at the level of a single pole, it is customary in chemistry to relate directly the substance amount of a species i to the substance flow (or reaction rate) through a proportionality factor called "rate constant" or "specific rate of reaction" k_i.

First-order kinetics:

$$\mathfrak{I}_i = k_i \, n_i \tag{E4.3}$$

This coefficient corresponds to the combination of pole conductance and pole elastance as already stated in Chapter 4 and this fundamental statement in chemical kinetics will be demonstrated in Chapter 11 dealing with the subject of dissipation.

8.2.4.2 Construction Formal Graph

1. *Dipoles–multipole connections*: The sequence of reactions between two ports (meaning that for the intermediate species no side reaction or transport of substance occur) is featured by a conservative transmission of the dipole flows along the reaction chain:

$$\mathfrak{I} = \mathfrak{I}_{12} = \mathfrak{I}_{23} \tag{E4.4}$$

On another hand, the variation of capacitive energy is the same in the system whatever the considered level, dipole or multipole,

$$\mathrm{d}\mathcal{U}_n \underset{\exists C_n}{=} \mu \, \mathrm{d}n = \mu_{12} \, \mathrm{d}n_{12} + \mu_{23} \, \mathrm{d}n_{23} \tag{E4.5}$$

The conservative transmission of the dipole flows has the effect of producing equal variations of the dipole substance amounts (as the evolution operator is a derivation):

$$\mathrm{d}n = \mathrm{d}n_{12} = \mathrm{d}n_{23} \tag{E4.6}$$

This equality introduced into the capacitive energy variation leads to the rule of adding dipole efforts for obtaining the multipole chemical potential

$$\mu = \mu_{12} + \mu_{23} \tag{E4.7}$$

Integration of the variations of dipole substance amounts (without having to consider integration constants as dipole substance amounts are already counted as differences between pole quantities) gives

$$n = n_{12} = n_{23} \tag{E4.8}$$

All these relationships are represented in the construction Formal Graph given in the case study abstract.

2. *Poles–dipoles connections*: The multipole being conservative, the sum of substance amounts is a constant

$$n_1 + n_2 + n_3 = n_1^* + n_2^* + n_3^* \tag{E4.9}$$

and the general decomposition into pole substance amounts, respecting this conservation, is written as

$$n_1 = n_1^* + n_{12} \tag{E4.10}$$

$$n_2 = n_2^* - n_{12} + n_{23} \tag{E4.11}$$

$$n_3 = n_3^* - n_{23} \tag{E4.12}$$

The two dipole chemical potentials are related to pole chemical potentials through simple differences as an exchange by flows exists in the dipole

$$\mu_{12} = \mu_1 - \mu_2; \quad \mu_{23} = \mu_2 - \mu_3 \tag{E4.13}$$

In addition, each reaction is modeled by an exchange by flows written as a pole flow difference

$$\mathfrak{I}_{12} = \mathfrak{I}_{1:2} - \mathfrak{I}_{2:1}; \quad \mathfrak{I}_{23} = \mathfrak{I}_{2:3} - \mathfrak{I}_{3:2} \tag{E4.14}$$

3. *All connections*: The construction Formal Graph given in Graph 8.11 shows the various connections between nodes for all organization levels, poles, dipoles, and multipole.

Among the various properties, only evolutions are drawn for readability. Refer to case study E5 "Diffusion through Layers," in which a construction Formal Graph bears elastances and conductances.

It can be remarked that the equality E4.8 between dipole basic quantities amounts to writing the invariance of the substance amount of the second pole

$$n_2 = n_2^* \quad \text{(E4.15)}$$

This invariance is sometimes described as a *stationary state* for the intermediate pole.

Simplification? It would have been graphically simpler to represent directly this invariance by omitting on the second pole the two connections coming from the dipoles, but this would have diminished the generality of the model. Keeping these connections outlines the fact that the invariance of the intermediate pole substance amount results from the equality E4.4 expressing the conservative transmission of the dipole flows. If this constraint is removed, by introducing a third port branched on the intermediate pole, this conservation does not hold anymore; the connections between dipoles and multipole are modified but not those between poles and dipoles.

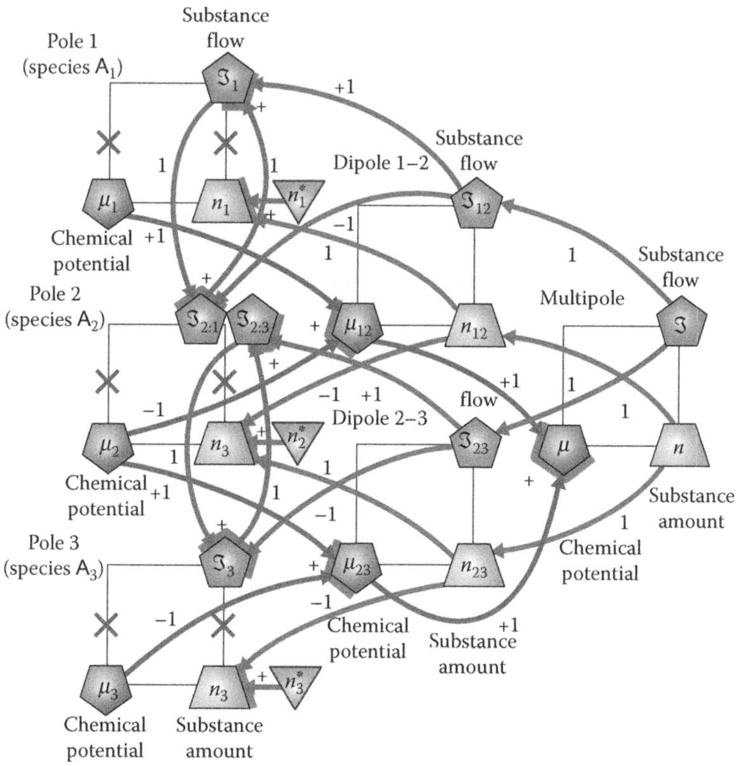

GRAPH 8.11 Construction Formal Graph of a three-multipole from two dipoles and three poles representing a chain reaction composed of two steps in series.

8.2.4.3 System Constitutive Properties

As demonstrated in Chapter 7, the capacitive relationship for a pole is featured by an exponential function and as will be demonstrated in Section 8.4, the conductive relationship for a pole obeys an Arrhenius-type equation, which has the same shape.

$$n_1 = \hat{C}_{n1}\,\mu_1 = n^{\theta}\exp\left(\frac{\mu_1 - \mu_1^{\theta}}{\mu_{\Delta}}\right) \tag{E4.16}$$

$$\Im_{1:2} = \hat{G}_{n1:2}\,\mu_1 = \Im^{0}_{1=2}\exp\left(\frac{\mu_1 - \mu_{1=2}^{eq}}{\mu_{\Delta}}\right) \tag{E4.17}$$

Both relations use scaling variables and reference chemical potentials that are specific to each system property, except the scaling chemical potential μ_{Δ} which is common and depends only on the couplings with other energy varieties. (Thermal agitation leads to proportionality with the temperature T through the gas constant R.) The presence of an equal sign between two subscripts indicates that their order does not matter, whereas a colon is used for subscripts indicating a reaction direction (and that the order matters).

By combining the two relations and by comparing with the first-order kinetics law as in Equation E4.3 which is rewritten as

$$\Im_{1:2} = k_{1:2}\,n_1 = \hat{G}_{n1:2}\,\hat{C}_{n1}^{-1}n_1 \tag{E4.18}$$

the expression of the apparent specific rate is obtained

$$k_{1:2} = \frac{\Im^{0}_{1=2}}{n^{\theta}}\exp\left(\frac{\mu_1^{\theta} - \mu_{1=2}^{eq}}{\mu_{\Delta}}\right) \tag{E4.19}$$

This was for the first pole and for the forward reaction from species A_1 to A_2; now, the same approach is used for the second pole and the backward reaction from species A_2 to A_1

$$n_2 = \hat{C}_{n2}\,\mu_2 = n^{\theta}\exp\left(\frac{\mu_2 - \mu_2^{\theta}}{\mu_{\Delta}}\right) \tag{E4.20}$$

$$\Im_{2:1} = \hat{G}_{n2:1}\,\mu_2 = \Im^{0}_{1=2}\exp\left(\frac{\mu_2 - \mu_{1=2}^{eq}}{\mu_{\Delta}}\right) \tag{E4.21}$$

$$\Im_{2:1} = k_{2:1}\,n_2 = \hat{G}_{n2:1}\,\hat{C}_{n2}^{-1}\,n_2 \tag{E4.22}$$

$$k_{2:1} = \frac{\Im^{0}_{1=2}}{n^{\theta}}\exp\left(\frac{\mu_2^{\theta} - \mu_{1=2}^{eq}}{\mu_{\Delta}}\right) \tag{E4.23}$$

It remains to insert the expressions of the pole flows into the dipole flow equation E4.14 in order to determine the classical kinetic equation of a first-order reversible reaction

$$\mathfrak{I}_{12} = k_{1:2}\, n_1 - k_{2:1}\, n_2 \tag{E4.24}$$

The ratio of specific rates obeys the *chemical equilibrium law* (see case study C5 "First-Order Chemical Reaction" in Chapter 6).

$$\frac{k_{2:1}}{k_{1:2}} = \exp\left(-\frac{\mu_{12}^{\theta}}{\mu_{\Delta}}\right) = \exp\left(-\frac{\Delta_r^{\theta} G_{12}}{\mu_{\Delta}}\right) \tag{E4.25}$$

In chemical thermodynamics, the dipole reference chemical potential is called *molar free energy of reaction*.

In this case study, no demonstration is really brought of the fundamental postulate E4.3 of chemical kinetics for a first-order reaction because the expression of the conductive relationship is merely postulated here without justification. It will be demonstrated at the end of this chapter in Section 8.5.7.

The following case study, dealing with transport of corpuscles through sandwiched medium, is similar to the present one. Details that have not been given here, such as inclusion of capacitance and conductance properties in the construction Formal Graph, will be shown. Moreover, comparison between chemical reaction and substance transport will be discussed, demonstrating the perfect equivalence between them (provided the reaction is of a first order).

8.2.5 Case Study E5: Diffusion through Layers — Corpuscular Energy

The case study abstract is given on next page.

In this system, made up of adjacent layers of a material able to contain some corpuscles that are free to move, transport of corpuscles is ensured by diffusion from one layer to another.

All layers are supposed to be identical in terms of geometrical and physical properties, same thickness, same transport properties, ensuring homogeneity throughout the system. The geometrical configuration of this system allows modeling it with a regular chain multipole. To simplify, the number of layers is taken equal to three. Each layer is considered as a pole and the transport of corpuscles is ensured by exchanges by flows that occur across the interfaces between layers.

8.2.5.1 Formal Graph

The Formal Graph shown in the case study abstract (Graph 8.12) represents a capacitive–conductive multipole which is identical to a dipole Formal Graph. Two properties are present in this graph: the capacitance and the conductance.

$$N = \hat{\mathsf{C}}_N\, E \tag{E5.1}$$

$$\mathfrak{I}_N = \hat{\mathsf{G}}_N\, E \tag{E5.2}$$

Naturally, conductance and elastance can be combined to form a single operator relating the corpuscle number to the flow. This possibility will be discussed later.

8.2.5.2 Construction Formal Graph

The serial mounting is featured by a conservative transmission of the dipole flows across all layers

$$\mathfrak{I}_N = \mathfrak{I}_{N12} = \mathfrak{I}_{N23} \tag{E5.3}$$

E5: Diffusion through Layers **Corpuscular Energy**

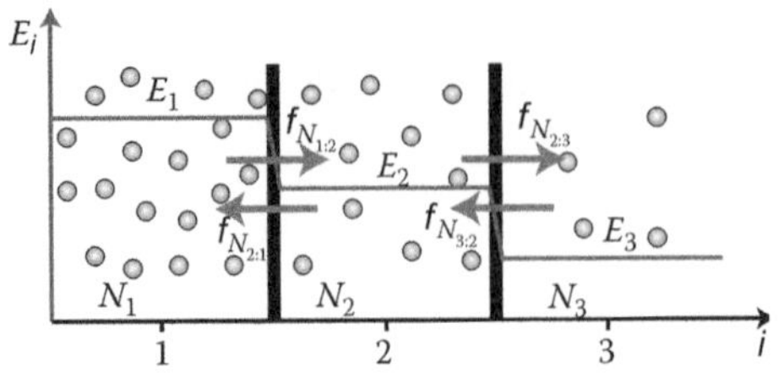

Multipole

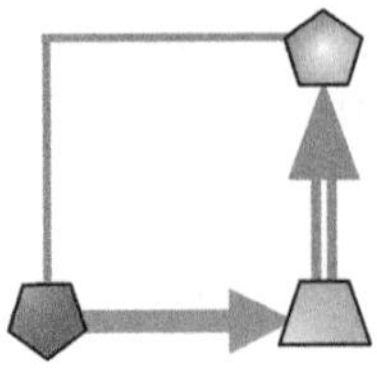

Multipole	
	Fundamental
✓	Mixed
✓	Capacitive
	Inductive
✓	Conductive
✓	Chain/serial
	Ladder/parallel
✓	Regular
✓	Global
	Spatial
✓	Conservative

Formal Graph:

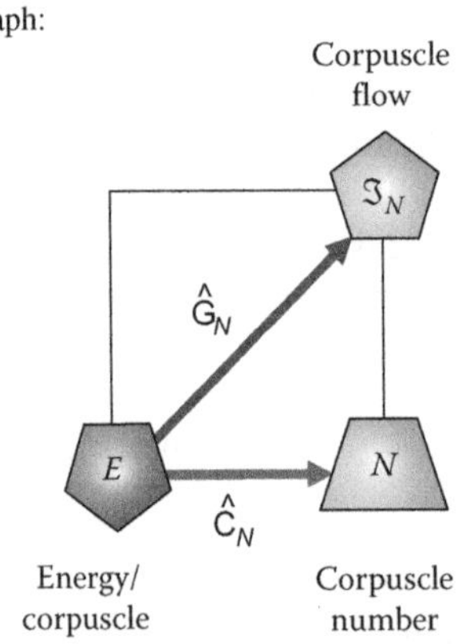

GRAPH 8.12

Construction from three dipoles:

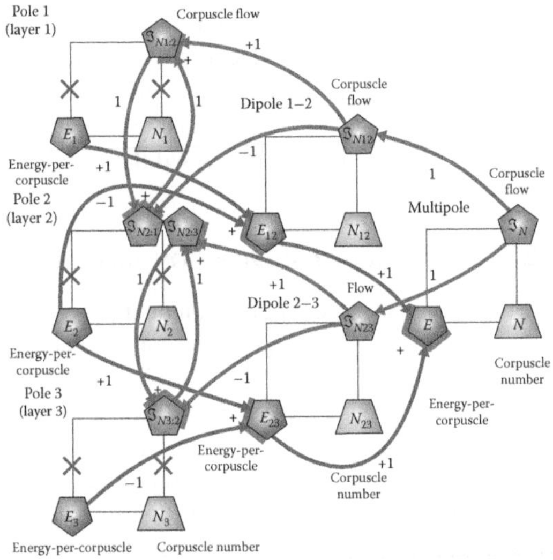

GRAPH 8.13

Variables:

Variety	**Corpuscular Energy**		
Subvariety	Capacitive (Potential)		Inductive
Category	Entity number	Energy/entity	Energy/ entity
Family	Basic quantity	Effort	Flow
Name	Corpuscle number	Corpuscular energy	Corpuscular flow
Symbols	N, n	E	$\mathfrak{I}_N$
Unit	[corp]	[J corp^{-1}]	[corp s^{-1}]

Property relationships:

$$N = \hat{C}_N\, E$$

$$\mathfrak{I}_N = \hat{G}_N\, E$$

System Property	
Nature	Capacitance
Levels	0
Name	Corpuscular capacitance
Symbol	C_N
Unit	[J^{-1}]
Specificity	—

Linear model:

$$N \underset{lin}{=} C_N\, E$$

Energy variation:

$$\mathrm{d}\mathcal{U}_N \underset{\exists C_N}{=} E\,\mathrm{d}N$$

Each dipole is the siege of an exchange by flows across its interface

$$\mathfrak{I}_{N12} = \mathfrak{I}_{N1:2} - \mathfrak{I}_{N2:1}; \quad \mathfrak{I}_{N23} = \mathfrak{I}_{N2:3} - \mathfrak{I}_{N3:2} \tag{E5.4}$$

Whatever the considered level, dipole or multipole, the variation of capacitive energy is the same in the system.

$$\mathrm{d}\mathcal{U}_N \underset{\exists C_N}{=} E\,\mathrm{d}N = E_{12}\,\mathrm{d}N_{12} + E_{23}\,\mathrm{d}N_{23} \tag{E5.5}$$

The two dipole energies-per-corpuscle are related to pole energies-per-corpuscle as usual when an exchange by flows exists in the dipole

$$E_{12} = E_1 - E_2; \quad E_{23} = E_2 - E_3 \tag{E5.6}$$

The conservative transmission of the dipole flows has the effect of producing equal variations of the dipole corpuscle numbers (by virtue of the differential nature of the evolution operator)

$$\mathrm{d}N = \mathrm{d}N_{12} = \mathrm{d}N_{23} \tag{E5.7}$$

This equality introduced into the capacitive energy variation leads to the rule of adding dipole efforts to obtain the multipole effort

$$E = E_{12} + E_{23} \tag{E5.8}$$

All these relationships concerning flows and efforts are represented in the construction Formal Graph given in the case study abstract. The relationships among basic quantities are not represented for not drawing over with too many connections. They are drawn in a separate Formal Graph that is discussed now.

Integration of the variations of dipole corpuscle numbers (without having to consider integration constants, as dipole basic quantities are already counted as differences between pole quantities), gives

$$N = N_{12} = N_{23} \tag{E5.9}$$

The multipole being conservative, the sum of corpuscle numbers is a constant

$$N_1 + N_2 + N_3 = N_1^* + N_2^* + N_3^* \tag{E5.10}$$

and the general decomposition into pole corpuscle numbers, respecting this conservation, is

$$N_1 = N_1^* + N_{12} \tag{E5.11}$$

$$N_2 = N_2^* - N_{12} + N_{23} \tag{E5.12}$$

$$N_3 = N_3^* - N_{23} \tag{E5.13}$$

The construction Formal Graph given in Graph 8.14 shows the relationships between corpuscle numbers and the various properties, elastances, conductances, and evolutions, for every organization levels, poles, dipoles, and multipole.

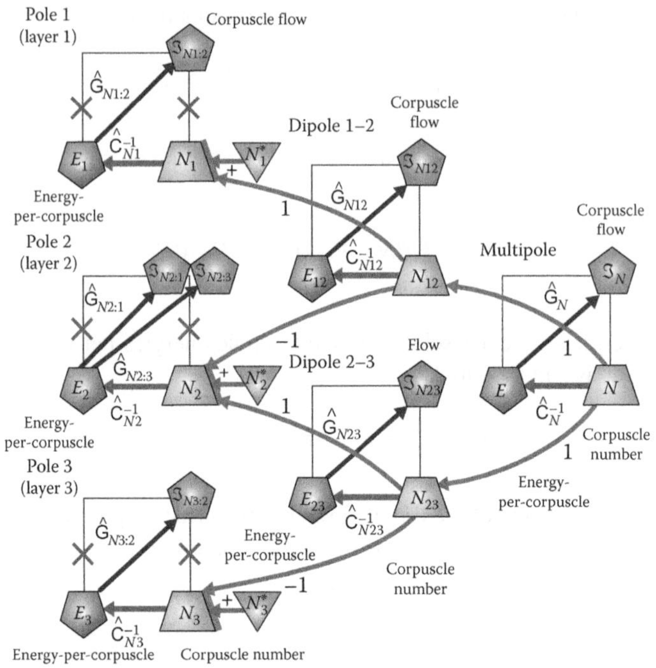

GRAPH 8.14 Construction Formal Graph of a three-multipole from two dipoles and three poles representing a transport chain composed of two interfaces in series (three layers).

It can be remarked that the equality E5.9 between dipole basic quantities amounts to writing the invariance of the corpuscle number of the second pole

$$N_2 = N_2^* \tag{E5.14}$$

This invariance is sometimes described as a *stationary state* for the intermediate pole.

Simplification? It would have been graphically simpler to represent directly this invariance by omitting on the second pole the two connections coming from the dipoles, but this would have diminished the generality of the model. Keeping these connections outlines the fact that the invariance of the intermediate pole corpuscle number results from the equality E5.3 expressing the conservative transmission of the dipole flows. If this constraint is removed, by introducing a third port branched on the intermediate pole, this conservation does not hold anymore; the connections between dipoles and multipole are modified but not those between poles and dipoles.

8.2.5.3 System Constitutive Properties

Diffusion of an ensemble of corpuscles is governed by Fick's* law that states that the flow density of corpuscles is proportional to the contragradient of the corpuscles concentration at a given location of the diffusing medium.

$$\boldsymbol{J}_{Ni} = -D_i \frac{\mathrm{d}}{\mathbf{d}\boldsymbol{\ell}} c_{Ni} \tag{E5.15}$$

* Adolf Eugen Fick (1829–1901): German physiologist; Zürich, Switzerland and Würzburg, Germany.

The proportionality factor D_i is called the *diffusivity* (or diffusion coefficient); it is a characteristic of the corpuscle and of the medium. In the Formal Graph theory, this law applies to a space pole in the corpuscular energy variety. Naturally, this equation can be rewritten with global variables instead of localized ones as

$$\frac{\mathrm{d}}{\mathrm{d}\boldsymbol{A}} \mathfrak{I}_{Ni} = \boldsymbol{J}_{Ni} = -D_i \frac{\mathrm{d}}{\mathrm{d}\boldsymbol{\ell}} \frac{\mathrm{d}}{\mathrm{d}V} N_i \tag{E5.16}$$

and when the volume is sufficiently small for allowing one to consider a homogeneous medium, the following simplified equation can be used

$$\mathfrak{I}_{Ni} \underset{hom}{=} \frac{D_i \boldsymbol{A}}{\boldsymbol{\ell} V} N_i \tag{E5.17}$$

By analogy with a first-order chemical reaction, a kinetic rate k_{Ni} is defined as

$$k_{Ni} \overset{def}{=} \frac{D_i \boldsymbol{A}}{\boldsymbol{\ell} V} \tag{E5.18}$$

so as to write the homogeneous Fick's law under the concise formula

$$\mathfrak{I}_{Ni} \underset{hom}{=} k_{Ni} N_i \tag{E5.19}$$

Diffusion and reaction. Diffusion and chemical reaction are, from a physical chemical point of view, different processes. The difference lies in the change of chemical nature that the reaction provokes on the reacting species whereas diffusion is a purely physical change of space location. However, from a strictly formal point of view, diffusion and first-order chemical reaction appear as similar processes as shown above. It is therefore not surprising to see some situations that can be interpreted in either way (see Figure 8.9).

For instance a sequence of elementary irreversible first-order reactions, all identical, between species exchanging the same reactant, can be viewed as a sequence of transport steps between regularly spaced locations, separated by a distance $\boldsymbol{\ell}$ and characterized by a diffusivity D. It can be added that the reasoning can obviously be extended to reversible reactions and forward–backward diffusion.

The condition for the physical chemical equivalence is the same as for the formal one, that is to say that distances should be small enough in order to consider the reactant exchange as a jump (called hopping) between spatially distributed and close species.

In short, it can be said that the Fickian diffusion is the localized process of certain first-order chemical reactions, which are the global process, in the same way as conductivity features the localized process of conduction.

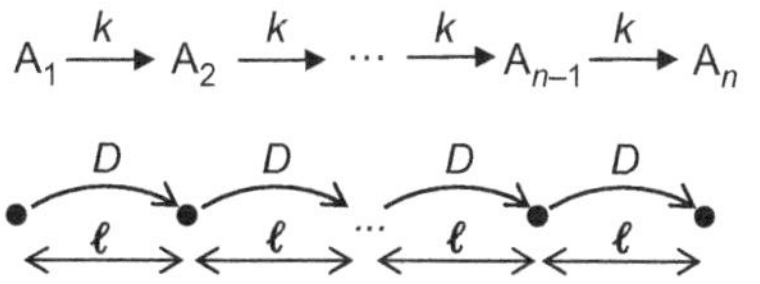

FIGURE 8.9 Equivalence between diffusion and reaction processes.

8.2.5.4 Conductive and Capacitive Relationships

As demonstrated in Chapter 7, the capacitive relationship for a pole is featured by an exponential function and as will be demonstrated in Section 8.4, the conductive relationship for a pole obeys an Arrhenius-type equation, which has the same shape.

$$N_1 = \hat{C}_{N1}\, E_1 = N^{\ominus} \exp\left(\frac{E_1 - E_1^{\ominus}}{E_\Delta}\right) \tag{E5.20}$$

$$\mathfrak{I}_{N1:2} = \hat{G}_{N1:2}\, E_1 = \mathfrak{I}^0_{N1=2} \exp\left(\frac{E_1 - E^{eq}_{1=2}}{E_\Delta}\right) \tag{E5.21}$$

Both relations use scaling variables and reference efforts that are specific to each system property, except the scaling energy-per-entity E_Δ which is common and depends only on the couplings with other energy varieties. (Thermal agitation leads to proportionality with the temperature T through the Boltzmann constant k_B.) The presence of an equal sign between two subscripts indicates that their order does not matter, whereas a colon is used for subscripts indicating a transport direction (and that the order matters).

By combining the two relations and by comparing with the homogeneous Fick's law E5.19 which is rewritten as

$$\mathfrak{I}_{N1:2} = k_{N1:2}\, N_1 = \hat{G}_{N1:2}\, \hat{C}^{-1}_{N1}\, N_1 \tag{E5.22}$$

the expression of the apparent specific rate is obtained

$$k_{N1:2} = \frac{\mathfrak{I}^0_{N1=2}}{N^{\ominus}} \exp\left(\frac{E_1^{\ominus} - E^{eq}_{1=2}}{E_\Delta}\right) \tag{E5.23}$$

This was for the first pole and for the forward transport; now, the same approach is used for the second pole and the backward transport from pole 2 to pole 1.

$$N_2 = \hat{C}_{N2}\, E_2 = N^{\ominus} \exp\left(\frac{E_2 - E_2^{\ominus}}{E_\Delta}\right) \tag{E5.24}$$

$$\mathfrak{I}_{N2:1} = \hat{G}_{N2:1}\, E_2 = \mathfrak{I}^0_{N1=2} \exp\left(\frac{E_2 - E^{eq}_{1=2}}{E_\Delta}\right) \tag{E5.25}$$

$$\mathfrak{I}_{N2:1} = k_{N2:1}\, N_2 = \hat{G}_{N2:1}\, \hat{C}^{-1}_{N2}\, N_2 \tag{E5.26}$$

$$k_{N2:1} = \frac{\mathfrak{I}^0_{N1=2}}{N^{\ominus}} \exp\left(\frac{E_2^{\ominus} - E^{eq}_{1=2}}{E_\Delta}\right) \tag{E5.27}$$

It remains to insert the expressions of the pole flows into the dipole flow equation E5.4 to determine the classical kinetic equation of a first-order reversible reaction

$$\mathfrak{I}_{N12} = k_{N1:2}\, N_1 - k_{N2:1}\, N_2 \tag{E5.28}$$

As for a chemical reaction, the ratio of apparent specific rates obeys an equilibrium law (see case studies C5 "First-Order Chemical Reaction" and C6 "Physical Chemical Interface" in Chapter 6).

$$\frac{k_{N2:1}}{k_{N1:2}} = \exp\left(-\frac{E_{12}^{\ominus}}{E_{\Delta}}\right) = \exp\left(-\frac{\Delta_r^{\ominus} G_{N12}}{E_{\Delta}}\right) \tag{E5.29}$$

The second writing is for making a bridge with chemical thermodynamics in which the dipole reference energy-per-corpuscle is called *free energy (per-corpuscle) of reaction*.

8.2.6 Case Study E6: Solenoid — Electrodynamics

The case study abstract is given on next page.

A solenoid, or coil, is made up of a series of loops in which an electric current flows, creating an electromagnetic induction $\boldsymbol{B}$ and containing a quantity of induction (induction flux) Φ_B. The system contains electromagnetic energy, which is the inductive subvariety of electric energy.

Every loop, or turn, is a dipole and all dipoles are identical, so the multipole is a regular one. Since the current I is the same all along the circuit and since all individual potentials V_i across a turn are added for constituting the total potential V, the solenoid is a *multipole chain*. Each turn contains an individual quantity of induction (induction flux) Φ_{Bi} (see Figure 8.10).

If N is the number of turns of the solenoid, the relationships between dipole–multipole are as follows:

$$\Phi_B = N\,\Phi_{Bi} \tag{E6.1}$$

$$V = N\,V_i \tag{E6.2}$$

In electromagnetism, the total quantity of induction Φ_B is classically "linkage flux" of induction. The electromagnetic energy contained in one turn is supported by an inductance L_i linking the current to the quantity of induction

$$\Phi_{Bi} = \hat{\mathrm{L}}_i\, I \tag{E6.3}$$

The multipole inductance L is therefore a multiple of the individual inductance

$$\hat{\mathrm{L}} = N\,\hat{\mathrm{L}}_i \tag{E6.4}$$

In a fundamental inductive dipole, the induction quantity is related to the potential across the solenoid through the evolution operator

$$V = \hat{\mathrm{T}}^{-1}\,\Phi_B \tag{E6.5}$$

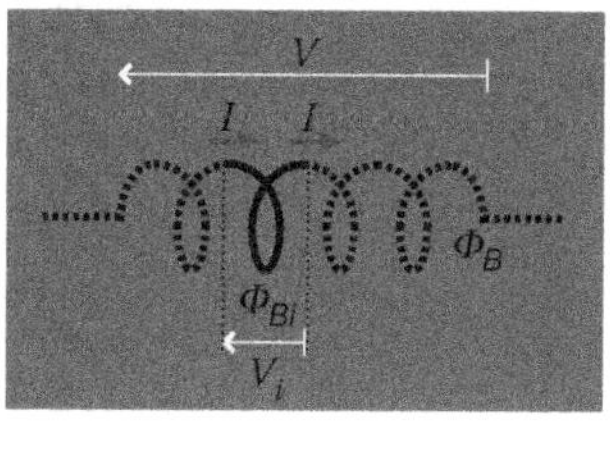

FIGURE 8.10 Scheme positioning the variables for one turn and for the whole solenoid.

E6: Solenoid **Electrodynamics**

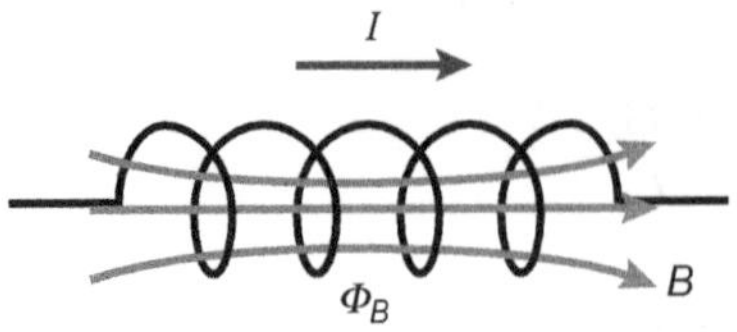

A solenoid is made up with N identical turns of a conducting material (wire).

Multipole

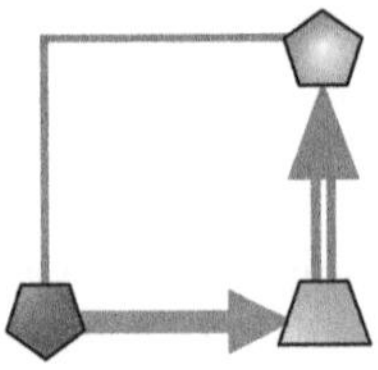

Multipole	
✓	Fundamental
	Mixed
	Capacitive
✓	Inductive
	Conductive
✓	Chain/serial
	Ladder/parallel
✓	Regular
✓	Global
	Spatial
✓	Conservative

Formal Graph:

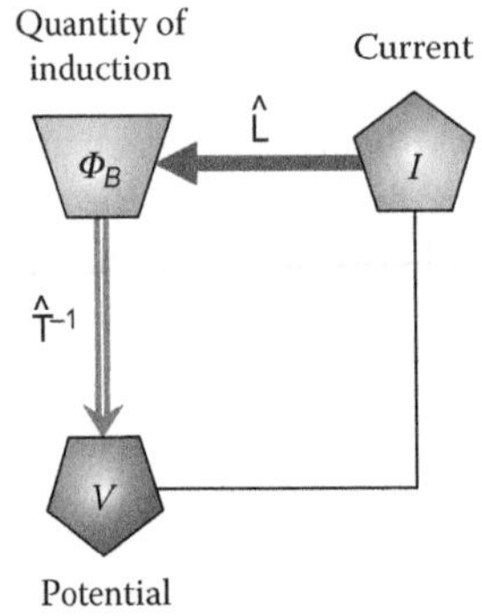

GRAPH 8.15

Construction from N identical poles:

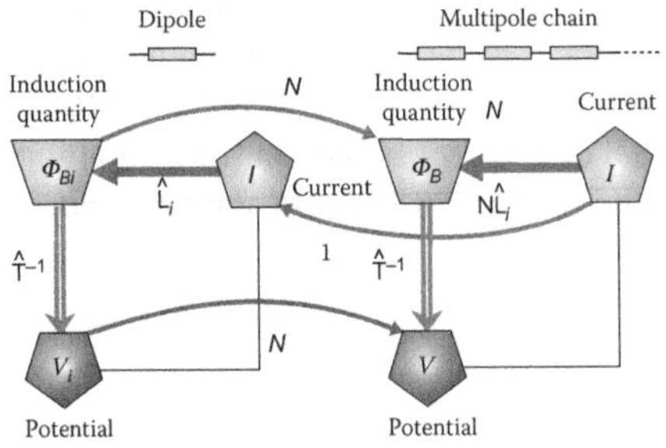

GRAPH 8.16

Variables:

Variety	Electrodynamics		
Subvariety	Inductive (kinetic)		Capacitive
Category	Entity number	Energy/ entity	Energy/ entity
Family	Impulse 	Flow 	Effort 
Name	Quantity (flux) of induction	Current	Potential
Symbols	Φ_B	I, i	V, U, E, φ
Unit	[Wb], [J A^{-1}]	[A]	[V], [J C^{-1}]

Inductance:

$$\Phi_B = \hat{L}\, I$$

System Property	
Nature	Inductance
Levels	0
Name	Inductance
Symbol	L
Unit	[H]
Specificity	—

Construction

$$\Phi_B = N\Phi_{Bi}$$
$$I = I$$
$$V = NV_i$$

Energy variation:

$$d\mathcal{T}_Q \underset{\exists L}{=} I\, d\Phi_B$$

Classically, this law is written with a time derivation and corresponds in that case to Faraday's law of induction. Here, one has written a generalized law valid independently of any definition of time. This law is also customarily written with a minus sign because of the choice of the generator convention for orienting current and potential difference. For harmonization, it is preferable to use the receptor convention used in most domains. (This point will be made clearer when a solenoid will be associated to other different dipoles as in Chapter 9.)

From the expression of the energy variation, two ways of grouping variables can be chosen, corresponding in fact to the chain mounting or to the ladder mounting.

$$d\mathcal{T}_Q \underset{\exists L}{=} I\,d\Phi_B = I\,d(N\,\Phi_{Bi}) = (NI)\,d\Phi_{Bi} \tag{E6.6}$$

It is true that, from the strict formal point of view, an equivalence with a multipole ladder is possible, corresponding to the following relations:

$$\Phi_B = N\,\Phi_{Bi} \tag{E6.7}$$

$$V = N\,V_i \tag{E6.8}$$

$$N\,I = N\,I \tag{E6.9}$$

From this viewpoint, as the evolution operator can be assumed to be a linear operator, the generalized Faraday's law of induction must be written as

$$V = N\,\hat{T}^{-1}\,\Phi_{Bi} = \hat{T}^{-1}\,\Phi_B \tag{E6.10}$$

The construction Formal Graph illustrating the two possibilities, chain and ladder, is given in Graph 8.17.

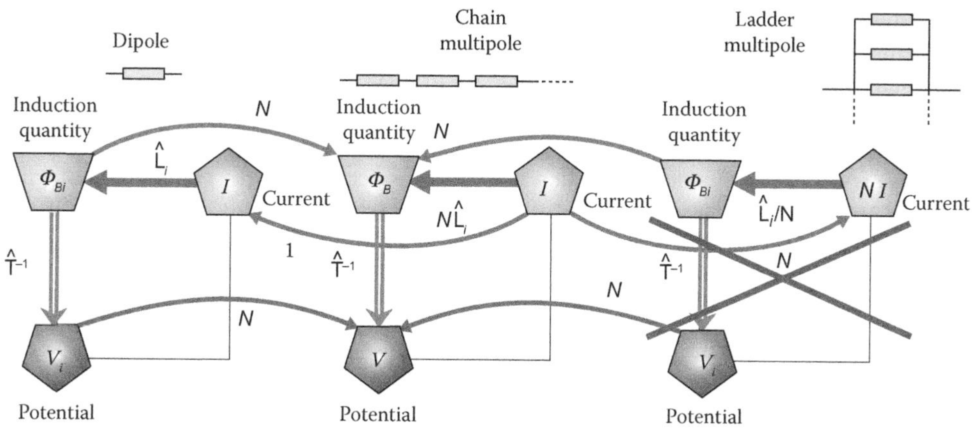

GRAPH 8.17 Construction Formal Graph showing the relationships between the individual turn represented by a dipole and the solenoid represented by a multipole chain. The relationships with a multipole ladder, which is not a physically suitable model, are represented on the right.

In other words, the solenoid can be viewed as an apparent inductive dipole with the individual quantity of induction Φ_{Bi} as impulse (entity number) and the product NI of the number of turns N by the current I as flow (energy-per-entity).

The problem is that the ladder mounting is a purely virtual equivalence that has no physical ground. In no way is there a physical reasoning that allows one to consider a flow NI of current in the solenoid. That is the reason why the Formal Graph of the ladder is barred in Graph 8.17. Unhappily, this way of thinking is classically used in electromagnetism, which can be understood as originating from the lack of rules for identifying physical quantities with only the algebraic tool.

8.3 DECOMPOSITION INTO DIPOLES

By definition of a multipole, dipoles that compose it belong to the same category, capacitive, inductive, or conductive (but may have different properties in case of an irregular multipole). This uniqueness of energy subvariety forbids any energy conversion between dipoles, but storing dipoles are able to exchange energy between them.

8.3.1 External Exchange of Energy

An *external exchange* implies an equilibration process working without time considerations, what is expressed by "instantaneous" process. As long as no dissipative or conversion process is involved, two capacitors, for instance, having their ends connected in pairs, will match their *energy-per-entities* (potentials) instantaneously by exchanging some charges. This follows the same process as for the discharge of a single capacitor discussed in Chapter 6, Section 6.2, except that the switch is replaced by a capacitor.

Exchange of entities and external exchange. Between the two poles of a dipole may exist an *exchange by flows* or an *exchange by efforts* that are vectors of the exchange. In a reversible exchange, these vectors work in pairs, in opposite directions (forward–backward). Otherwise, in an irreversible exchange, only one direction is effective. In both cases, the exchange content is a set of entities that circulate between the poles. This is an *internal exchange* of energy, with respect to the dipole.

The *external exchange* is an almost similar process allowing two dipoles to exchange entities and energy. The difference is that the connected poles, belonging to different dipoles, do not form a new dipole but a single pole by merging their energies-per-entity. In Figure 8.11 are sketched the two kinds of exchange in the case of flows playing the role of vectors. In the case of an external exchange (between dipoles), no decomposition in forward–backward vectors is possible because no difference in energies-per-entities exists. For this reason, such an exchange is not named in reference to flows or efforts as vectors, but the preferred naming is *external exchange* in order to avoid confusion with the exchange inside a dipole.

Both exchanges, internal and external, do not modify the nature of energy; it remains within the same subvariety, but energy amounts vary within each partner.

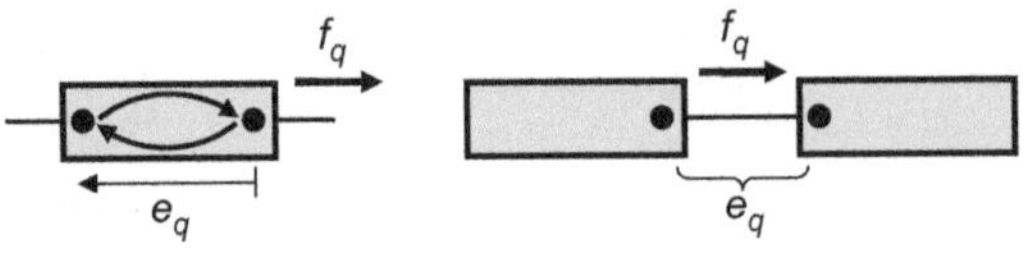

FIGURE 8.11 Exchange by flows inside a dipole (left) and external exchange between two dipoles sharing a common pole (right).

Now the question to tackle is the *external exchange* of a dipole with another one, which is only possible when two dipoles share the same *energy-per-entity.*

8.3.2 Forming a Multipole by Combination of Dipoles

As mentioned in Section 8.1, there are basically two ways of assembling two dipoles in order to allow sharing of efforts or flows. The chain structure or serial mounting corresponds to a common flow whereas the ladder structure or parallel mounting corresponds to a common effort, as shown in Figure 8.12.

The Formal Graph representations of these two ways of mounting are drawn in Graph 8.18.

The connections drawn in these Formal Graphs correspond to mounting rules known in electrodynamics as Kirchhoff laws. These rules are encoded in electric circuits through the literal symbols of potentials and currents written along the components and the wires, as illustrated in Figure 8.12. These laws are merely derived from the First Principle of Thermodynamics through the Gibbs equation as shown in the following section.

8.3.3 Kirchhoff Laws

Kirchhoff laws and the properties of a multipole can be directly established from its dipole components when three ingredients are gathered:

(i) The First Principle of Thermodynamics in its variational form, that is, the Gibbs equation written under the form of a sum of dipole energy variations that equates the multipole energy variation, meaning that the energy in the system does not depend on the subdivision chosen.
(ii) The choice between common effort and common flow for all dipoles (i.e., chain or ladder structures but not mixed arrangements).
(iii) The linearity of the link between entity number and energy-per-entity belonging to different subvarieties (through the evolution operator).

This last requirement deserves some explanation. As stated in Chapter 7, a fundamental storing dipole possesses a space–time property called evolution that allows its energy-per-entity to be linked to the entity number of the other energy subvariety. This is mathematically expressed by two relationships involving the same evolution operator, that is written here under the form of one pair

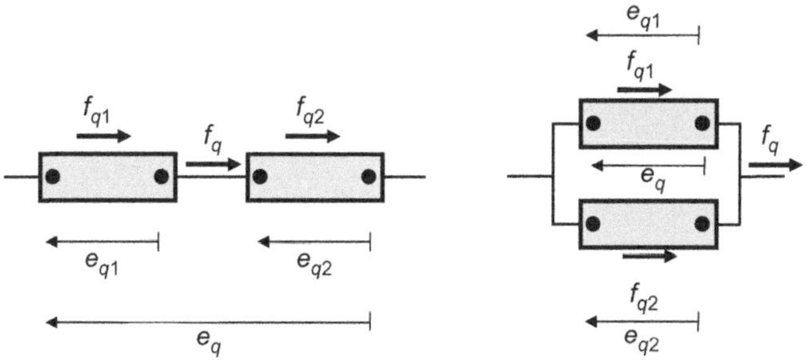

FIGURE 8.12 In the symbolic language of equivalent electric circuits, two dipoles mounted in series share a common dipole flow and mounted in parallel, a common dipole effort. Conjugated energies-per-entity are added to form the multipole energy-per-entity, according to Kirchhoff mounting laws.

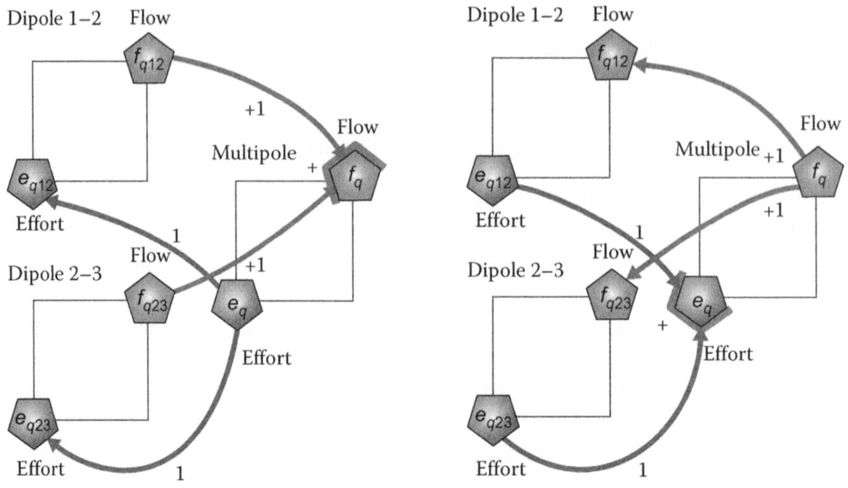

GRAPH 8.18 Formal Graphs of two multipoles composed of two dipoles, one multipole mounted as a ladder (left) and the other as a chain (right).

of equation (the operator $\hat{T}$ is used for the inverse evolution and the reciprocal operator is used for the evolution) for each evolution link:

$$p_{qi} = \hat{T}\, e_{qi} \quad \text{or} \quad e_{qi} = \hat{T}^{-1} p_{qi} \tag{8.3}$$

$$q_i = \hat{T}\, f_{qi} \quad \text{or} \quad f_{qi} = \hat{T}^{-1} q_i \tag{8.4}$$

The translation into Formal Graph of these relationships is given in Graph 8.19.

These relations apply equally well to the dipoles and the multipole as the Formal Graph of a multipole is similar to the one of a dipole and possesses an evolution link too (in case of energy storage in fundamental dipoles).

The linearity of this operator means that one must have, for instance, in the case of a common effort leading to a sum of two dipole flows for forming the multipole flow,

$$f_q = f_{q1} + f_{q2} \Rightarrow q = \hat{T}\, f_q = \hat{T}\left(f_{q1} + f_{q2}\right) = \hat{T}\, f_{q1} + \hat{T}\, f_{q2} = q_1 + q_2 \tag{8.5}$$

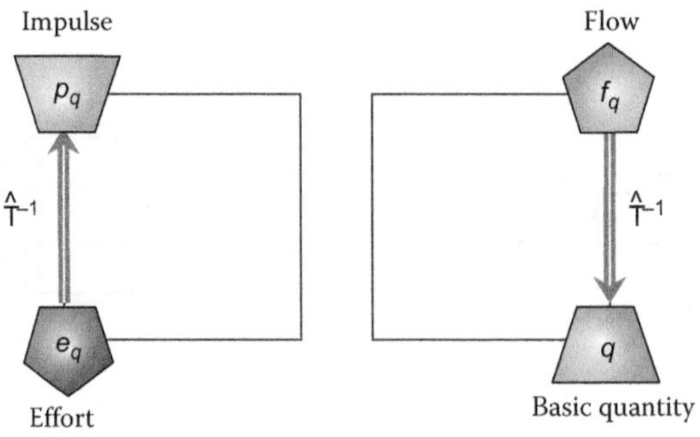

GRAPH 8.19 Formal Graphs showing the space–time property linking energies-per-entity to entity numbers called inverse evolution.

The converse is also true, so this linearity condition is written for both evolution links:

$$\hat{\mathrm{T}}\ \text{linear}:\quad f_q = f_{q1} + f_{q2} \Leftrightarrow q = q_1 + q_2 \tag{8.6}$$

$$\hat{\mathrm{T}}\ \text{linear}:\quad e_q = e_{q1} + e_{q2} \Leftrightarrow p_q = p_{q1} + p_{q2} \tag{8.7}$$

This linearity of evolution is a standard feature of our classical space–time, as in the continuous and homogenous case, the operator is the time derivation, as will be seen in Chapter 9,

$$\hat{\mathrm{T}}^{-1} \underset{cont}{=} \frac{\mathrm{d}}{\mathrm{d}t} \tag{8.8}$$

which is indeed a linear operator.

In Tables 8.2 and 8.3 are established (without comments) the mounting relationships (Kirchhoff laws) in the case of a multipole composed of two dipoles. Extension to a greater number of dipoles is straightforward as all sums of two terms are easily generalized to any number.

The system constitutive properties of the multipole are also established as a function of the corresponding dipole properties, in the peculiar case of linear operators as well as in the general case of nonlinear ones. This modeling is independent of the shapes of the dipole constitutive properties and is therefore completely general.

The main results from the preceding developments are the mounting rules, or Kirchhoff laws, that are recalled hereafter, but adapted to the case of N dipoles (Table 8.4).

These rules have been demonstrated by using the storage property of dipoles (Gibbs equation) and by assuming the linearity of the space–time property that composes the links between flow and effort. Such a link exists in purely inductive and purely capacitive dipoles that are able to store energy, but also in mixed dipoles, storing and dissipating energy. The question to tackle now is the applicability of these rules to the case of purely conductive dipoles that do not store energy and do not possess this link.

8.3.4 Combining Two Conductive Dipoles

As a conductive dipole has no energy container, and therefore no entity number, it possesses neither the evolution property nor the ability to store energy, properties that were used in the reasoning leading to the establishment of the mounting rules. Another approach must be followed, which relies on an inductive reasoning.

By comparing the three fundamental dipoles, inductive, conductive, and capacitive, one can notice that all of them link the two energies-per-entity, flow and effort, as shown in Graph 8.24.

This possibility of going through several paths in linking two nodes corresponds to a concept called *apparent path*. It means that various operators (or combinations of operators) applied to the same variable provide the same result. This concept exists in electrodynamics under the name of *impedance*, which is an operator linking a current to a potential. The interest of an apparent operator like the impedance is to exempt one from knowing the actual operators that are effective when making a model. In other words, it plays the role of the unknown x in algebra whose value does not need to be determined before writing an equation.

TABLE 8.2

Ladder Multipole Composed of Two Dipoles, Either Inductive (Left) or Capacitive (Right)

Inductive LL | **Capacitive CC**

$$\mathrm{d}\mathcal{T}_q \underset{\exists L_q}{=} f_{q1}\,\mathrm{d}p_{q1} + f_{q2}\,\mathrm{d}p_{q2} = f_q\,\mathrm{d}p_q \quad (8.9)$$

$$\mathrm{d}\mathcal{U}_q \underset{\exists C_q}{=} e_{q1}\,\mathrm{d}q_1 + e_{q2}\,\mathrm{d}q_2 = e_q\,\mathrm{d}q \quad (8.10)$$

Common effort: $e_q = e_{q1} = e_{q2}$ = **Ladder** (Parallel) (8.11)

Inverse evolution: $p_{qi} = \hat{\mathrm{T}}\,e_{qi}$ (8.12)

(+8.11) $\Rightarrow$ $p_q = p_{q1} = p_{q2}$ (8.13)

$\Rightarrow$ $\mathrm{d}p_q = \mathrm{d}p_{q1} = \mathrm{d}p_{q2}$ (8.14)

(8.10 + 8.11) $\Rightarrow$ $\mathrm{d}q = \mathrm{d}q_1 + \mathrm{d}q_2$ (8.15)

$\Rightarrow$ $q = q_1 + q_2$ (8.16)

$\hat{\mathrm{T}}$ *Linear* $f_{qi} = \hat{\mathrm{T}}^{-1} q_i$ (8.17)

(8.9 + 8.14) or (8.16 + 8.17) $\Rightarrow$ $f_q = f_{q1} + f_{q2}$ (8.18)

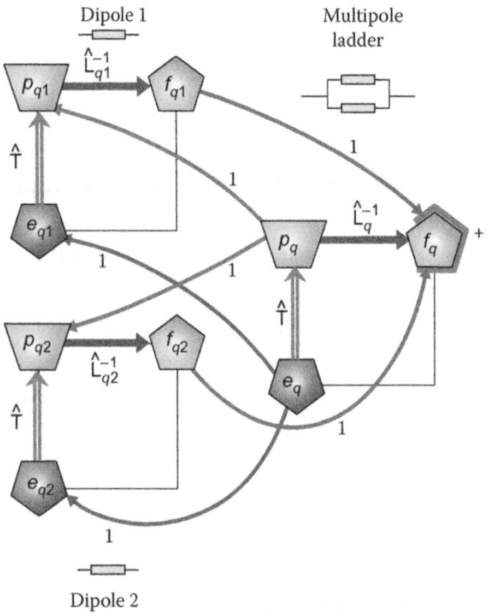

GRAPH 8.20

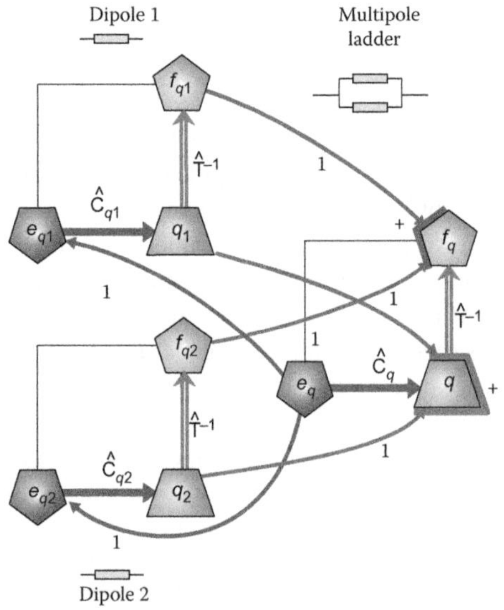

GRAPH 8.21

$$f_{qi} = \hat{\mathrm{L}}_{qi}^{-1}\, p_{qi} \quad (8.19)$$

$$(+8.19) \Rightarrow f_q = \hat{\mathrm{L}}_{q1}^{-1}\, p_q + \hat{\mathrm{L}}_{q2}^{-1}\, p_q \quad (8.20)$$

$$\Rightarrow \boxed{\hat{\mathrm{L}}_q^{-1} = \hat{\mathrm{L}}_{q1}^{-1} + \hat{\mathrm{L}}_{q2}^{-1}} \quad (8.21)$$

Linear case:

$$f_{qi} \underset{lin}{=} L_{qi}^{-1}\left(p_{qi} - p_{qi}^{**}\right) \quad (8.25)$$

$$p_q^{**} \underset{lin}{\overset{def}{=}} \frac{L_{q1}^{-1}\, p_{q1}^{**} + L_{q2}^{-1}\, p_{q2}^{**}}{L_{q1}^{-1} + L_{q2}^{-1}} \quad (8.26)$$

$$\Rightarrow L_q^{-1} \underset{lin}{=} L_{q1}^{-1} + L_{q2}^{-1} \quad (8.27)$$

$$\Rightarrow f_q \underset{lin}{=} L_q^{-1}\left(p_q - p_q^{**}\right) \quad (8.28)$$

$$q_i = \hat{\mathrm{C}}_{qi}\, e_{qi} \quad (8.22)$$

$$(+8.19) \Rightarrow q = \hat{\mathrm{C}}_{q1}\, e_q + \hat{\mathrm{C}}_{q2}\, e_q \quad (8.23)$$

$$\Rightarrow \boxed{\hat{\mathrm{C}}_q = \hat{\mathrm{C}}_{q1} + \hat{\mathrm{C}}_{q2}} \quad (8.24)$$

Linear case:

$$q_i \underset{lin}{=} C_{qi}\left(e_{qi} - e_{qi}^{*}\right) \quad (8.29)$$

$$e_q^{*} \underset{lin}{\overset{def}{=}} \frac{C_{q1}\, e_{q1}^{*} + C_{q2}\, e_{q2}^{*}}{C_{q1} + C_{q2}} \quad (8.30)$$

$$\Rightarrow C_q \underset{lin}{=} C_{q1} + C_{q2} \quad (8.31)$$

$$\Rightarrow q \underset{lin}{=} C_q\left(e_q - e_q^{*}\right) \quad (8.32)$$

TABLE 8.3
Chain Multipole Composed of Two Dipoles, Either Inductive (Left) or Capacitive (Right)

Inductive LL | **Capacitive CC**

$$d\mathcal{T}_q \underset{\exists L_q}{=} f_{q1}\,dp_{q1} + f_{q2}\,dp_{q2} = f_q\,dp_q \quad (8.9)$$

$$d\mathcal{U}_q \underset{\exists C_q}{=} e_{q1}\,dq_1 + e_{q2}\,dq_2 = e_q\,dq \quad (8.10)$$

Common flow: $f_q = f_{q1} = f_{q2}$ = **Chain** (Serial) (8.33)

$(8.9 + 8.33) \Rightarrow dp_q = dp_{q1} + dp_{q2}$ (8.34)

Inverse evolution: $q_i = \hat{T} f_{qi}$ (8.37)

$\Rightarrow \quad p_q = p_{q1} + p_{q2}$ (8.35)

$(+8.34) \Rightarrow q = q_1 = q_2$ (8.38)

$\hat{T}$ *Linear* $\quad e_{qi} = \hat{T}^{-1} p_{qi}$ (8.36)

$\Rightarrow dq = dq_1 = dq_2$ (8.39)

(8.35 + 8.36) or (8.10 + 8.39) $\Rightarrow \quad e_q = e_{q1} + e_{q2}$ (8.40)

See Graph 8.22

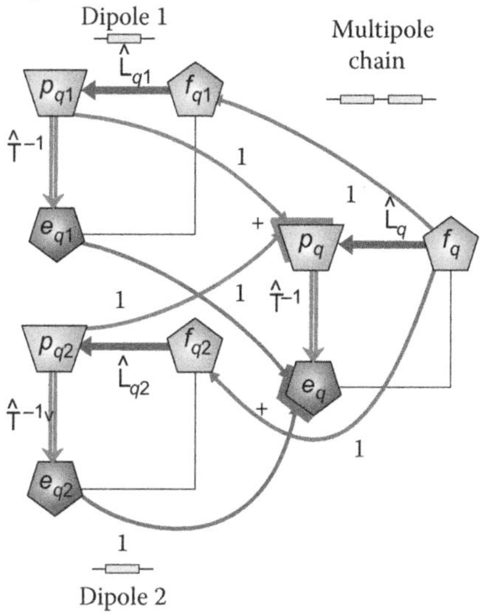

GRAPH 8.22

See Graph 8.23

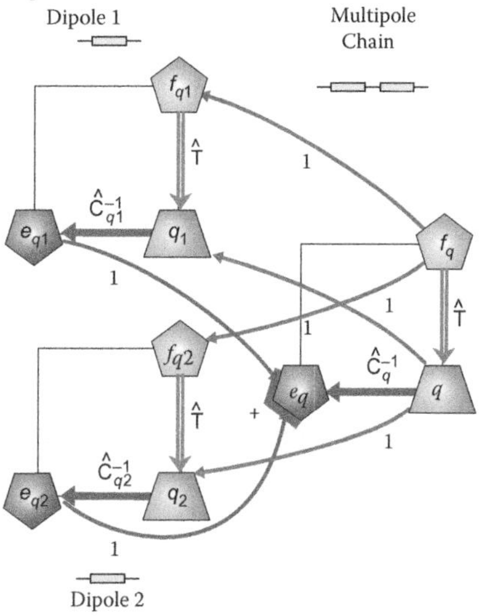

GRAPH 8.23

$$p_{qi} = \hat{L}_{qi}\, f_{qi} \quad (8.41)$$

$$(+8.41) \Rightarrow p_q = \hat{L}_{q1} f_q + \hat{L}_{q2} f_q \quad (8.42)$$

$$\Rightarrow \boxed{\hat{L}_q = \hat{L}_{q1} + \hat{L}_{q2}} \quad (8.43)$$

Linear case:

$$p_{qi} \underset{lin}{=} L_{qi}\left(f_{qi} - f_{qi}^{*}\right) \quad (8.47)$$

$$f_q^{*} \underset{lin}{\overset{def}{=}} \frac{L_{q1} f_{q1}^{*} + L_{q2} f_{q2}^{*}}{L_{q1} + L_{q2}} \quad (8.48)$$

$$\Rightarrow L_q \underset{lin}{=} L_{q1} + L_{q2} \quad (8.49)$$

$$\Rightarrow p_q \underset{lin}{=} L_q\left(f_q - f_q^{*}\right) \quad (8.50)$$

$$e_{qi} = \hat{C}_{qi}^{-1}\, q_i \quad (8.44)$$

$$(+8.41) \Rightarrow e_q = \hat{C}_{q1}^{-1} q + \hat{C}_{q2}^{-1} q \quad (8.45)$$

$$\Rightarrow \boxed{\hat{C}_q^{-1} = \hat{C}_{q1}^{-1} + \hat{C}_{q2}^{-1}} \quad (8.46)$$

Linear case:

$$e_{qi} \underset{lin}{=} C_{qi}^{-1}\left(q_i - q_i^{**}\right) \quad (8.51)$$

$$q^{**} \underset{lin}{\overset{def}{=}} \frac{C_{q1}^{-1} q_1^{**} + C_{q2}^{-1} q_2^{**}}{C_{q1}^{-1} + C_{q2}^{-1}} \quad (8.52)$$

$$\Rightarrow C_q^{-1} \underset{lin}{=} C_{q1}^{-1} + C_{q2}^{-1} \quad (8.53)$$

$$\Rightarrow e_q \underset{lin}{=} C_q^{-1}\left(q - q^{**}\right) \quad (8.54)$$

TABLE 8.4
Mounting Rules, also Called Kirchhoff Laws in Electrodynamics, for Assembling N Dipoles

Common Effort = Ladder (Parallel)		Common Flow = Chain (Serial)	
$e_q = e_{qk},\ k = 1, N$	(8.55)	$f_q = f_{qk},\ k = 1, N$	(8.57)
$f_q = \sum_{k=1}^{N} f_{qk}$	(8.56)	$e_q = \sum_{k=1}^{N} e_{qk}$	(8.58)

The reciprocal of the impedance is known as *admittance*:

$$\hat{Y}_q = \hat{Z}_q^{-1} \tag{8.59}$$

The transcription of the three paths used in the fundamental dipoles shown in Graph 8.24 in operator combinations is as follows:

Inductive impedance: $$\hat{Z}_q = \hat{T}^{-1}\,\hat{L}_q \underset{hom\,time}{=} \frac{d}{dt}\hat{L}_q \tag{8.60}$$

Conductive impedance: $$\hat{Z}_q = \hat{R}_q \tag{8.61}$$

Capacitive impedance: $$\hat{Z}_q = \hat{C}_q^{-1}\,\hat{T} \underset{hom\,time}{=} \hat{C}_q^{-1}\frac{d^{-1}}{dt^{-1}} \tag{8.62}$$

"Alternate" usage. Beware of not restricting, even in electrodynamics, the notions of impedance/admittance to the sole domain of alternative current circuits! Too often these notions are defined in the frame of periodically varying currents or potentials when in fact they are much more general operators working with any signal shape.

According to the approach used in the Formal Graph theory, the concept of impedance is generalized in these equations for all energy varieties, as in the single Formal Graph given in Graph 8.25.

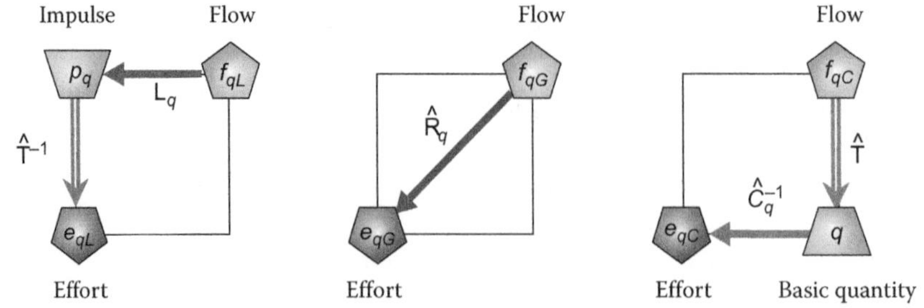

GRAPH 8.24 Three paths that link the same starting and arriving nodes, but going through different elementary paths.

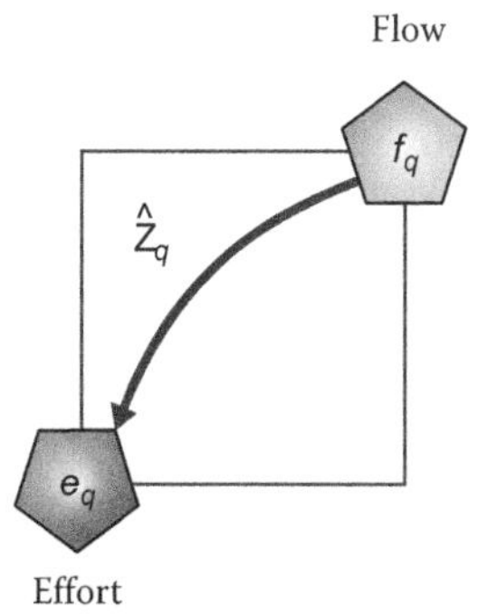

GRAPH 8.25 The concept of impedance as an apparent path equivalent to a conductance, or an inductance + evolution, or an inverse evolution + elastance, or any path that links the same nodes.

In case of mixed constitutive properties, in inductive–conductive or in capacitive–conductive multipoles, the impedance is a combination of two paths, one per nature of system constitutive properties. This fusion of purely inductive or capacitive paths with a purely conductive one justifies the adoption of common mounting rules for all natures of dipoles.

It must be remarked that, for the moment, all studied Formal Objects, from pole to multipole, cannot contain by definition more than one energy subvariety. This means that in such systems the impedance cannot be a mix or composition of inductive and capacitive paths. In Chapter 9, this cohabitation will be considered, and the impedance can then be of any path composition.

Common Effort = *Ladder* (Parallel)		Common Flow = *Chain* (Serial)	
$f_{qi} = \hat{Y}_{qi}\ e_{qi} = \hat{Z}_{qi}^{-1}\ e_{qi}$	(8.63)	$e_{qi} = \hat{Z}_{qi}\ f_{qi} = \hat{Y}_{qi}^{-1}\ f_{qi}$	(8.66)
$\hat{Y}_q\ e_q = \hat{Y}_{q1}\ e_q + \hat{Y}_{q2}\ e_q$	(8.64)	$\hat{Z}_q\ f_q = \hat{Z}_{q1}\ f_q + \hat{Z}_{q2}\ f_q$	(8.67)
$\hat{Y}_q = \hat{Y}_{q1} + \hat{Y}_{q2}$	(8.65)	$\hat{Z}_q = \hat{Z}_{q1} + \hat{Z}_{q2}$	(8.68)

By adapting the impedance to the case of a purely conductive dipole through its identification with a resistance as given by Equation 8.61, the previous general results are written in terms of conductances and resistances (Table 8.5).

8.3.5 Regular Multipoles

When a dipole is reproduced in N identical copies that form a regular chain or a regular ladder, the energy variations can be written according to three ways of grouping the number of dipoles with the state variables of the dipole:

$$d\mathcal{T}_q \underset{\exists L_q}{=} N\left(f_q\ dp_q\right) = \left(N\ f_q\right) dp_q = f_q d\left(N\ p_q\right) \tag{8.73}$$

$$d\mathcal{U}_q \underset{\exists C_q}{=} N\left(e_q\ dq\right) = \left(N\ e_q\right) dq = e_q d\left(N\ q\right) \tag{8.74}$$

Consequently, the Formal Graph of the regular multipole can be equivalently represented according to either configuration, whether the entity number of the dipole or the energy-per-entity is multiplied by N, as shown in Graph 8.28.

TABLE 8.5
Formation of the Multipole Conductive Properties as a Function of Dipole Properties

Common Effort = Ladder (Parallel)

$$\hat{G}_q = \hat{G}_{q1} + \hat{G}_{q2} \quad (8.69)$$

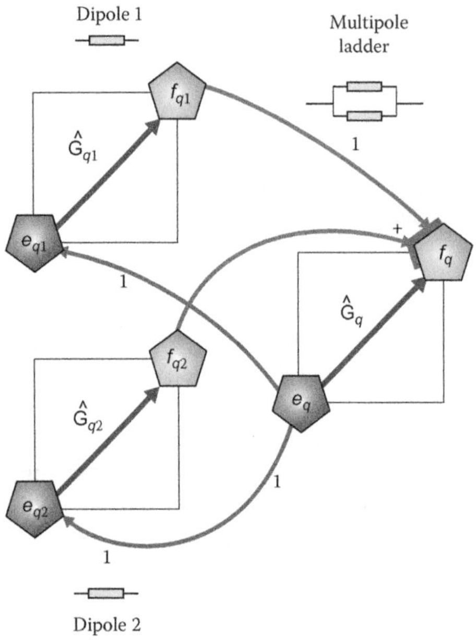

GRAPH 8.26

Linear case:

$$G_q \underset{lin}{=} G_{q1} + G_{q2} \quad (8.71)$$

Common Flow = Chain (Serial)

$$\hat{R}_q = \hat{R}_{q1} + \hat{R}_{q2} \quad (8.70)$$

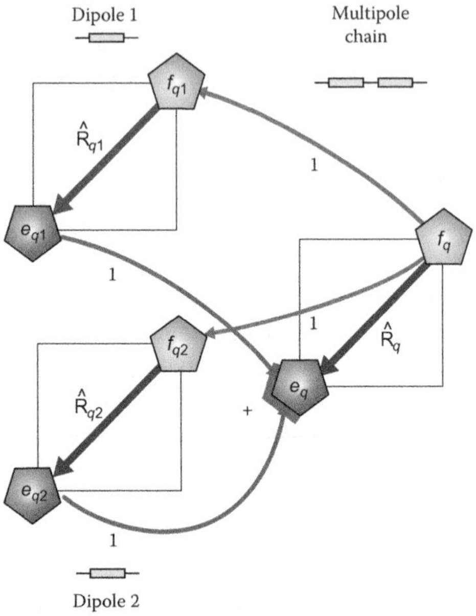

GRAPH 8.27

Linear case:

$$R_q \underset{lin}{=} R_{q1} + R_{q2} \quad (8.72)$$

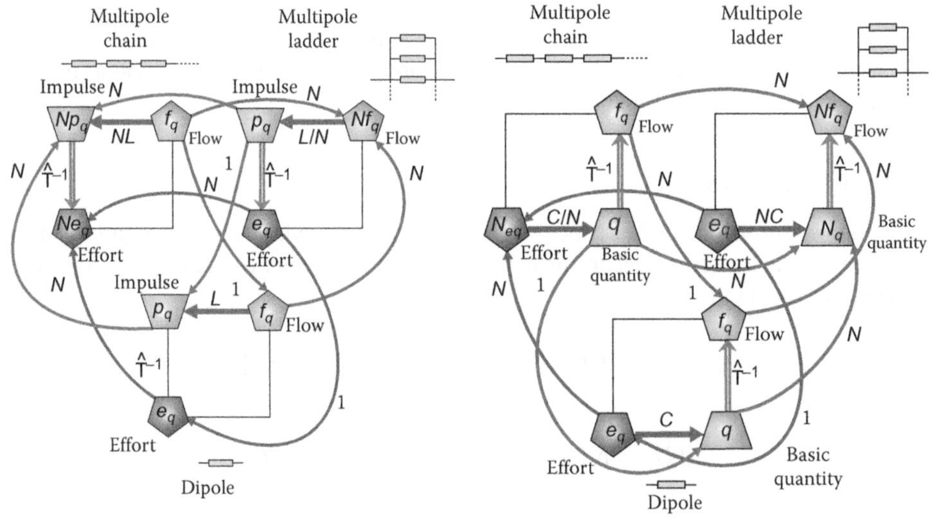

GRAPH 8.28 Equivalences among regular multipoles between chain multipole, ladder multipole, and their dipole unit. On the left are the inductive multipoles and on the right the capacitive multipoles.

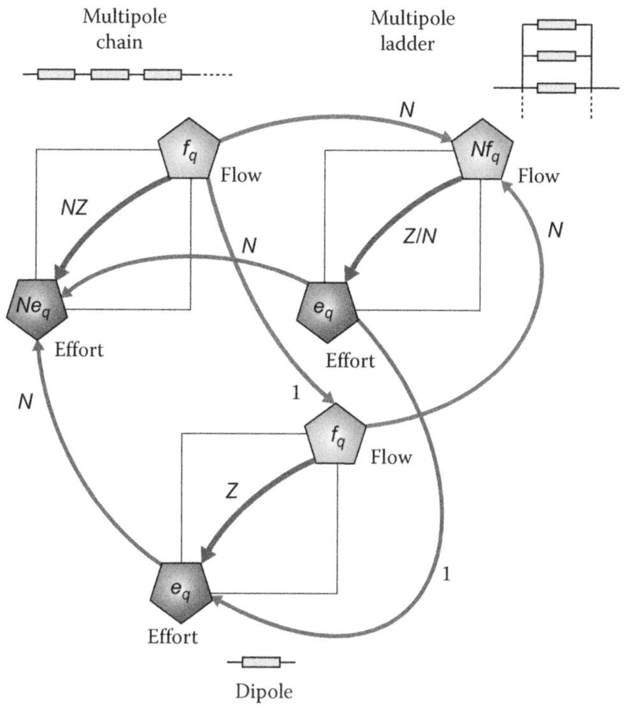

GRAPH 8.29 Equivalence between chain and ladder representations for a regular multipole of purely conductive type.

When the system constitutive properties of the dipole are scalars, the corresponding properties of the different configurations are related by proportionalities using the number N of dipoles as a factor. This was the case for the inductances and capacitances in the previous graphs in Graph 8.28 of energy storing multipoles, and this is also the case for impedances used in Graph 8.29 that represent resistances as well as combinations of inductances and evolutions or inverse evolutions and elastances.

Mathematics versus physics. Care must be exercised with regard to the meaning of this chain–ladder equivalence, which is purely mathematical and may not correspond to physical systems. In case study E6 of the solenoid, it has been shown that the ladder configuration was not acceptable because the same current was flowing across each turn of the coil and could never appear physically in the system as multiplied by N.

It should be clear that if one were to mention this possibility of chain–ladder equivalence, it is only to allow the reader to make connections with usual practices and not to encourage their development. The actual interest lies in the equivalence between dipole and one configuration, chain or ladder, but not in the nonphysical equivalence between chain and ladder.

Table 8.6 gives the algebraic relationships between energies-per-entity according to the various possible configurations chain/ladder and irregular/regular multipoles.

TABLE 8.6
Various Expressions Used for Determining the Multipole Energies-per-Entities According to the Chain or Ladder Structure and to Its Regularity

Multipole	Chain		Ladder	
Irregular	$e_q = \sum_{k=1}^{N} e_{qk}$	(8.75)	$f_q = \sum_{k=1}^{N} f_{qk}$	(8.77)
	$f_q = f_{qk};\ k = 1, N$	(8.76)	$e_q = e_{qk};\ k = 1, N$	(8.78)
Regular	$e_q = N e_{qk};\ k = 1, N$	(8.79)	$f_q = N f_{qk};\ k = 1, N$	(8.81)
	$f_q = f_{qk};\ k = 1, N$	(8.80)	$e_q = e_{qk};\ k = 1, N$	(8.82)

8.4 DECOMPOSITION INTO POLES

It is always possible to decompose a Formal Object into other ones belonging to every lower level. This increases the complexity of the Formal Graphs detailing the decomposition and this is not always useful. For instance, to decompose a set of capacitors in parallel (case study E1) into electrostatic poles is not interesting as experimental evidence or measurement at the level of the poles is not of easy access.

Notwithstanding, there are some cases where poles can be accessed, at least some of them, and decomposing a multipole down to this level may present an interest. This is the case in physical chemistry with a chemical reaction (case study E4) in which at least initial and final reactants are known or in corpuscular energy with a diffusion process through layers in series (case study E5) where measurement of the amount of corpuscles is sometimes feasible inside each layer. In some other cases, such as levers and balance (case study E3), it may be pertinent to go down the level of poles in order to take into account individual properties.

As a multipole can be considered simultaneously as a set of poles and a set of dipoles, the algebraic notation must distinguish three organization levels (no problem with Formal Graphs). This is done in numbering poles with a single number and dipoles with pairs of subsequent numbers.

Pole k	Dipole $k, k+1$	Multipole
One subscript	Two subscripts	No subscript
x_{qk}	$x_{qk,\,k+1}$	x_q

8.4.1 Merging Poles

In a multipole, poles can be connected to more than one pole and consequently more than one state variable per family can be necessary for handling several interconnections.

A typical configuration is shown in Figure 8.13 with four poles interconnected through exchanges by flows and arranged in a radial layout around a central pole.

The central pole, labeled 2, has to support three exchanges, which means that three flow variables must be borne by this pole.

There are two ways of modeling the central pole with Formal Graphs: either by using three distinct canonical graphs, each one built with one node per state variable family, or by using a single graph built with multiple nodes where necessary. In Graph 8.30 the two ways are drawn with the interconnections between them, in restricting the graphs to the gate variables only, efforts and flows. In this multipole, the effort of pole 2 is unique and is common to all graphs, whereas flows are in the number of exchanges as each exchange works with two opposite flows, one for each related pole.

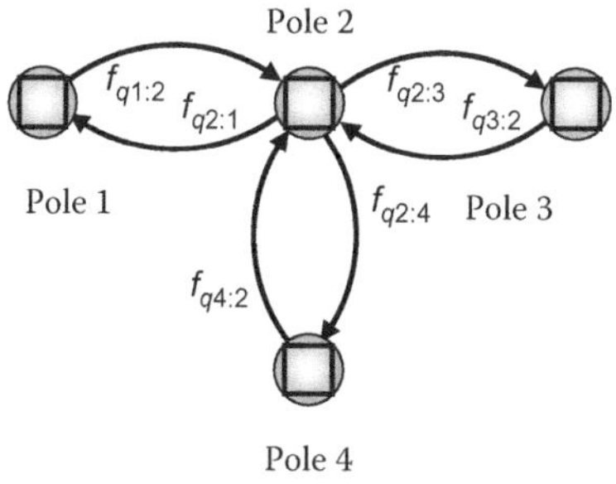

FIGURE 8.13 Multipole made up of a central pole interconnected with three other poles through exchanges by flows.

Therefore, the merged Formal Graph on the right of Graph 8.30 contains one effort and three flows, one for each exchange. When conduction processes work in the system, three conductances are used in the merged pole, in order to link the unique effort to the three flows.

The previous equivalence between three distinct Formal Graphs and a merged one is used for simplifying models and is not compulsory. The case study E4 "First-Order Chemical Reactions in Series" uses this merging principle to simplify the Formal Graph of this chain multipole.

The Formal Graph shown in Graph 8.31 models a similar system, but generalized to any energy variety (the unique Formal Graph is split for clarity). The multipole is composed of three capacitive–conductive poles and two exchanges by flows work between them. The intermediate pole is represented as a merged pole with two nodes for the flows and two conductances. Only one capacitance is used because a single effort features this pole.

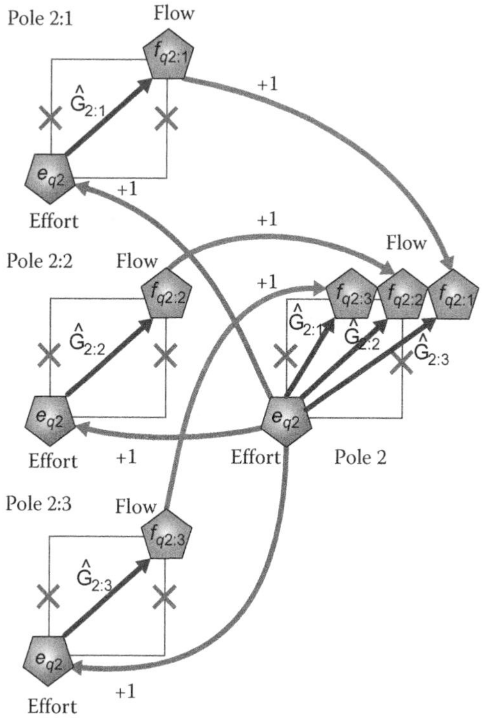

GRAPH 8.30 Construction of the merging of the three Formal Graphs modeling pole 2 into a single Formal Graph.

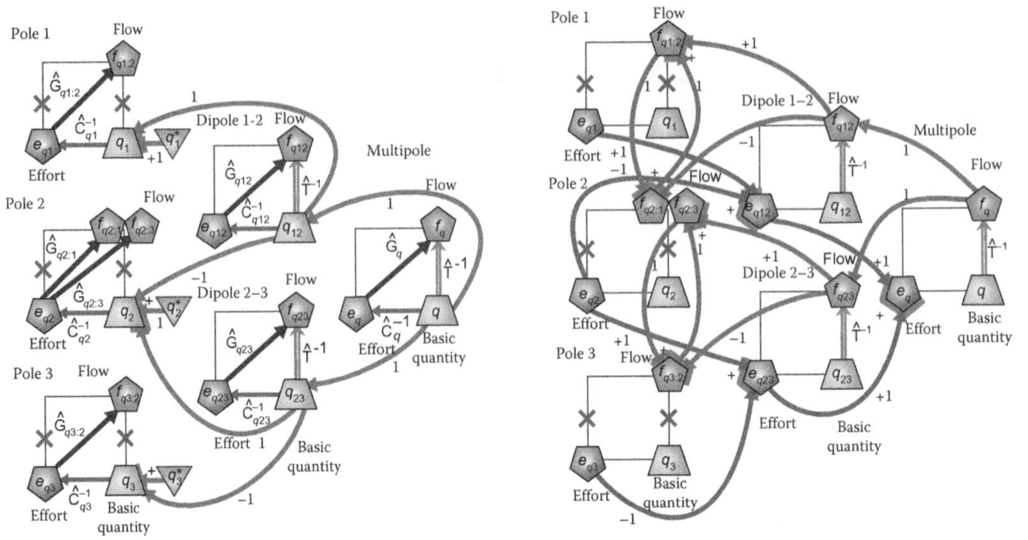

GRAPH 8.31 Two Formal Graphs modeling the same chain multipole made up of three capacitive–conductive poles. The left one shows only the system and space–time properties together with the connections between basic quantities; the right one shows only the connections between flows and between efforts, which are characteristic of a serial mounting (common flow).

The same technique of pole merging can be used for other multipole configurations, chain or ladder, and for the other nodes, even basic quantities and impulses, when it is necessary to distinguish them in the same pole.

8.5 THEORY OF CONDUCTION

Classically, the notion of conduction refers to a process of exchange of entities, particles (electrons, molecules, etc.) or energy (heat), between two or several locations in space. It has close connections with the notion of transport or transfer and quite often both notions are seen as equivalent. Here, by conduction is meant any physical process that relies on the system constitutive property of conductance and therefore dissipates energy. This is in part an extension of the classical notion, for instance in including the chemical reaction or the mechanical friction in the category of conduction processes and for another part a slight restriction in linking necessarily the dissipation to the conduction.

Since the first case studies from Chapter 4 to now, the model of conductance utilized was an exponential function similar to the one used for the capacitance expression. This model could not be justified at the levels of poles and dipoles, because not enough elements were gathered for establishing it until the study of multipoles. It was the same situation for the capacitance but it has been possible to anticipate in Chapter 7 on the multipole properties by extrapolating dipole behaviors to a sequence of three poles, thus evidencing the role of the exponential function in the model of capacitance.

The peculiarity of the capacitance, compared to conductance, is to be a support for one energy subvariety and to allow systems to store capacitive energy, either internally in a pole (self-capacitance) or externally to a pole in the phenomenon of influence (mutual capacitance). This pole–pole interaction is the expression of a physical phenomenon specific to the capacitive energy, which is not retrieved either in inductive energy (the inductance obeys completely different principles) or in energy dissipation.

FIGURE 8.14 The hopping of particles from site to site is a conduction process modeled by assimilating sites to poles arranged in a chain.

The theory of influence could be developed on the ground of the notion of activity that was known in physical chemistry and that has been generalized to all other energy varieties. Nothing similar may be found for dissipation in any domain, so it is somewhat adventurous to invent from scratch a similar notion for conductive processes, although the same theoretical tool can be perfectly used to bring up an exponential function in a conductance model.

Several case studies given in this chapter have introduced the key role of a chain multipole made up of conductive dipoles, especially in case studies E4 "Chemical Reaction" and E5 "Diffusion through Layers." Other systems involving conductive dipoles could have been taken also by extending the examples of single dipoles to several dipoles, given in Chapter 6, case studies C3 "Viscous Layers" and C4 "Pipe," for instance.

The chain multipole is the support for modeling the conduction process for several reasons. The first one is that it mimicks a well-known process called hopping conduction in which particles jump from site to site, as illustrated in Figure 8.14.

In this case, the process is discretized because it is decomposed in as many steps as there are sites in the chain.

The second reason lies in the possibility of a multipole containing a sufficiently large number of poles for mimicking a continuous process (the number of poles is not limited). In this way, a multipole may model a transmission line in which dissipation occurs.

The third reason is that a multipole behaves according to the embedding principle, which states that an ensemble of dipoles may be equivalent to a single apparent dipole, when they are all of the same nature. When applied to an exchange of entities, as is the case for the conduction process, it means that a sequence of exchanges is equivalent to a single exchange. The converse is true, meaning that an exchange viewed as working between two poles can be always decomposed into a sequence of exchanges.

At last, the fourth reason comes from the property of a chain to implement naturally the notion of composed properties in going along the chain from pole to pole. It means that when a transmission of entities occurs, each link depends on the behavior of the preceding one. This dependence does not work in only one direction but is truly bidirectional as an exchange is always based on forward and backward processes, even when one is dominating the other. The key notion here is the one of reversibility which is a property of an exchange that can be easily understood as a dipole property that must be composed with other dipole reversibilities for featuring the reversibility of the whole chain. In other words, the reversibility of a sequence of processes depends on the reversibility of each element.

For all these reasons, the chain multipole endowed with exchanges of entities is an ideal support for modeling conduction.

8.5.1 Formal Graph of a Conductive Chain

Let us take the case of the simplest chain multipole made up of three poles and two exchanges, as sketched in Figure 8.15.

For ensuring conduction along the chain, each pole must possess conductance properties, one for each exchange supported by the pole. The inner pole must have two conductances while the two outer poles have only one. The reason is that conductances are linking pole effort to flows and these latter ones are individualized when they participate in exchanges, as discussed previously about the merging of poles.

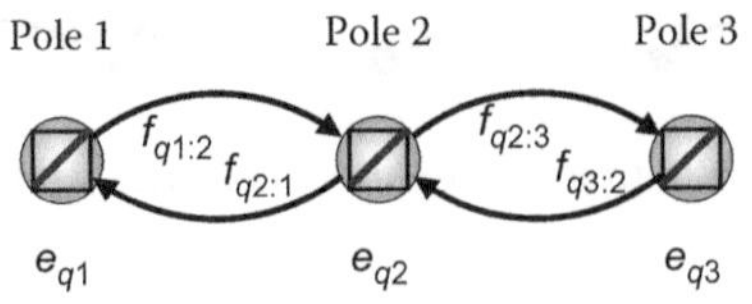

FIGURE 8.15 Three poles constituting a chain multipole endowed with conduction through the two exchanges between poles and conductance properties within poles.

The complete Formal Graph of the three levels, poles, dipoles, and multipole, and their connections, is drawn in Graph 8.32.

This construction is based on a common flow through the two dipoles and the summation of dipole efforts for giving the multipole effort. Between dipoles and poles, the flows are coded according to the requirement of an exchange and the dipole efforts are formed as a difference of pole efforts. The intermediate pole is represented as a merged pole, in accordance with the principles used in Graph 8.30.

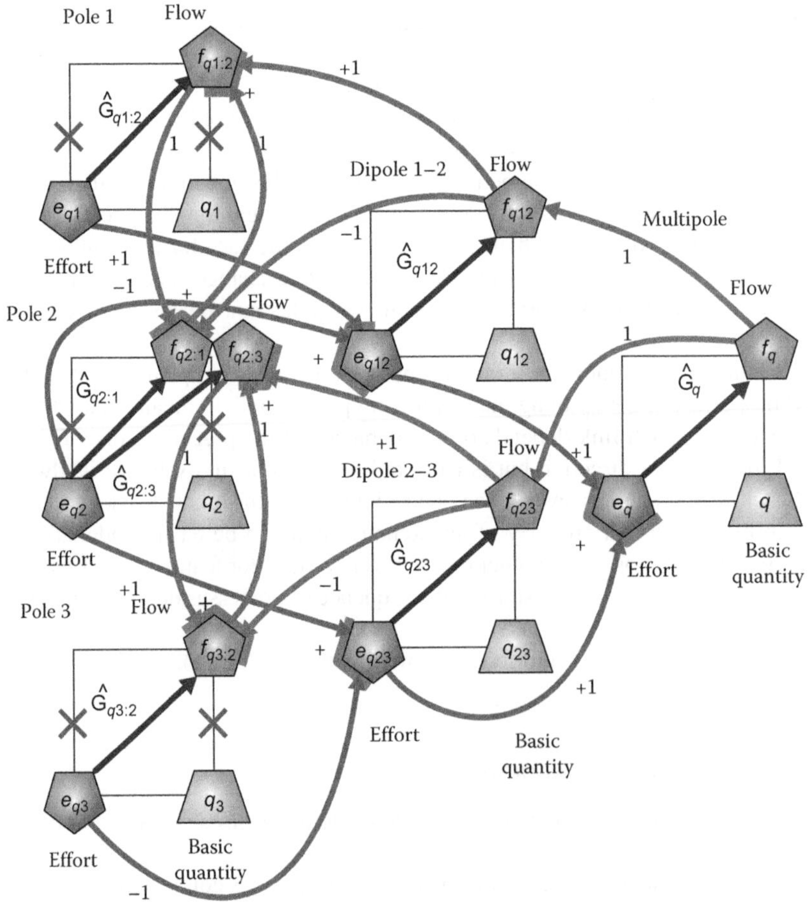

GRAPH 8.32 Construction Formal Graph of a chain multipole composed of two dipoles (with common flow) and three poles, able to ensure conduction through exchanges by flows and conductances. Pole 2 in the middle is a merged pole bearing two pole flows and two pole conductances.

8.5.2 Exchange Reversibility

Chapter 6, devoted to dipoles, defines the *exchange reversibility* $r_{qi,j}$ which quantifies the behavior of the exchange by flows between pole i and pole j:

$$r_{qi,j} \stackrel{def}{=} \frac{f_{qj:i}}{f_{qi:j}} \tag{8.83}$$

By definition, the pole flows are positive quantities, so the exchange reversibility is also always positive. It ranges from zero, meaning total irreversibility in the $i \rightarrow j$ direction, to infinite, meaning total irreversibility in the opposite direction. The peculiar value one corresponds to an exchange in dynamic equilibrium (equal flows in both directions) and a tolerance around this value delimitates a reversible regime (or *quasireversible* when the tolerance is large).

However, the most interesting feature of this variable, in addition to its role in determining the exchange regime in terms of irreversibility or reversibility, is to be a multiplicative variable when considering a sequence of exchanges. This means that for a chain multipole made up of several dipoles, the overall reversibility of the equivalent dipole is expressed as the product of all individual reversibilities. For any conductive chain of N poles we have

$$r_{q1,N} = \prod_{k=1}^{N} r_{qk,k+1} \tag{8.84}$$

This important result will be demonstrated soon. Its physical meaning is that the overall reversibility of a chain is the product of the reversibilities of all links. An interesting principle is deduced from this property, which is the *rule of the weakest link* that determines the regime of the chain. Alternatively, the corollary is that all individual reversibilities must be equal to one for having a fully reversible chain. Let us take as example (see Figure 8.16) a chain made up with five poles in which one of the exchange is irreversible, say the third one, meaning that $r_{q3:4} = 0$. The overall reversibility is therefore also equal to zero, conferring to the chain the irreversibility as resulting behavior ($r_{q1:5} = 0$).

A logical reasoning leads to another interesting consequence that it is the first irreversible exchange in the sequence that determines the behavior of the chain. In other words, all preceding exchanges must be reversible (otherwise the first irreversible exchange would not be the one considered) and all subsequent exchanges may have any reversibility; they cannot influence the overall irreversibility. In the language of chemical kinetics, this is called the "rule of the first rate-determining step."

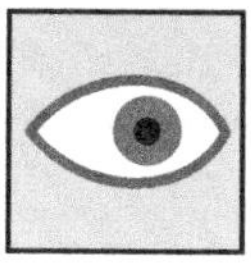

Logical order of reasoning. It must be clear that the presentation of the physical meaning of the *exchange reversibility* as resulting from the mathematical properties is not perfectly true. Actually, the converse is the true reasoning. These physical properties are logical ones; they can be stated without having recourse to the quantitative notion of reversibility. Thus, the mathematical properties are consequences of the physical and logical ones, which legitimate the multiplicative property of the exchange reversibility.

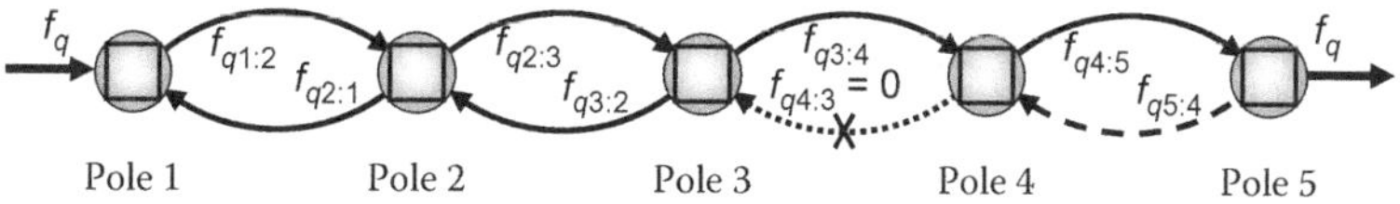

FIGURE 8.16 The rule of the weakest link illustrated on a five-pole chain. The third exchange (3–4) is totally irreversible in this example, constraining the whole chain to be irreversible whatever the reversibility of the following exchanges.

8.5.3 Equivalence of a Conductive Chain with a Single Dipole

The subject of the equivalence of a conductive chain with a single conductive dipole is of paramount importance in conduction theories and in chemical reactions modeling. In the Formal Graph theory, its fundamental importance comes from constituting the basis for establishing from scratch the conductive relationship, allowing demonstration of many empirical or semiempirical conduction models such as the Arrhenius law in physical chemistry or transition state theories in chemical kinetics.

The simplest chain is made up of three poles arranged in two dipoles in series, labeled 12 and 23, establishing two exchanges by flows between the poles. Figure 8.17 shows the principle of the equivalence with the dipole formed by the first and the last poles.

With regard to the equivalent dipole, as every exchange, the relationships between pole flows and dipole flow and reversibility are written under the form of two equations:

$$f_{q13} = f_{q1:3} - f_{q3:1} \tag{8.85}$$

$$r_{q13} = \frac{f_{q3:1}}{f_{q1:3}} \tag{8.86}$$

They constitute a system of two equations with two unknown variables, which can be solved for expressing both pole flows, outside the special case of dynamic equilibrium:

$$f_{q1:3} = \frac{1}{1-r_{q13}} f_{q13}; \quad f_{q3:1} = \frac{r_{q13}}{1-r_{q13}} f_{q13}; \; (r_{q13} \neq 1) \tag{8.87}$$

It means that the knowledge of the two dipole variables, flow and reversibility, leads to determining the exchange flows.

The logical consequence is that if one is able to determine the chain flow and the chain reversibility, the knowledge of the equivalent dipole characteristics ensues.

To establish rigorously this equivalence, the reasoning will be made in two steps. A first step consists of relating dipole flows and a second step establishes relationships between reversibilities at the deeper level of pole flows.

The first step is simple and straightforward. The multipole flow of the sequence of two dipoles in series is equal to the dipole flows, by definition of the serial mounting

$$f_q = f_{q12} = f_{q23} \tag{8.88}$$

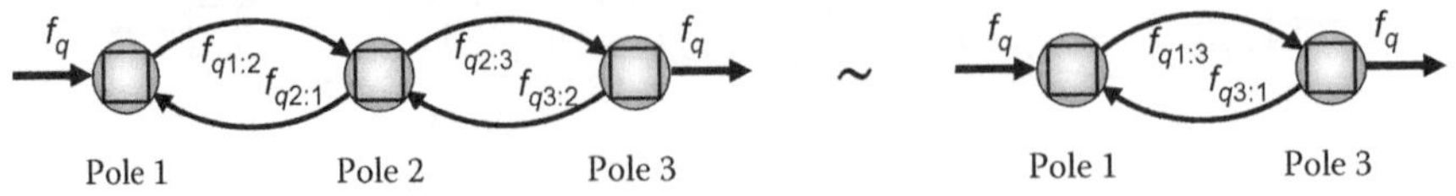

FIGURE 8.17 Scheme showing a sequence of two exchanges by flows arranged in series which is equivalent to one dipole between the most extreme poles.

These simple equalities express in an implicit way the assumption that there is no extra flow on the central pole other than within the two dipoles in series. No side flow through a lateral branch is permitted by construction. To ensure the equivalence chain–dipole, the necessary condition is to have

$$f_q = f_{q13} \tag{8.89}$$

The second step is less straightforward and demands a few lines for establishing it. The reasoning relies on the multiplicative property of the exchange reversibility which is written here as

$$r_{q13} = r_{q12}\, r_{q23} \tag{8.90}$$

The demonstration of this relation is done hereafter.

8.5.4 Demonstration of the Multiplicative Property of the Reversibility

Two flow conservation equations are written for the two extreme poles 1 and 3, which are each multiplied by the flow belonging to the central pole 2 to the concerned exchange

$$f_{q1:2} = f_q + f_{q2:1} \quad \times f_{q2:3} \Rightarrow f_{q1:2}\, f_{q2:3} = f_q\, f_{q2:3} + f_{q2:1}\, f_{q2:3} \tag{8.91}$$

$$f_{q2:3} = f_q + f_{q3:2} \quad \times f_{q2:1} \Rightarrow f_{q2:1}\, f_{q2:3} = f_{q2:1}\, f_q + f_{q2:1}\, f_{q3:2} \tag{8.92}$$

Substituting the product of the central pole flows in the first equation by the expression given by the second equation gives

$$f_{q1:2}\, f_{q2:3} = f_q \left(f_{q2:3} + f_{q2:1}\right) + f_{q2:1}\, f_{q3:2} \tag{8.93}$$

By rearrangement, one gets the expression of the chain flow as a difference between two terms:

$$f_q = \frac{f_{q1:2}\, f_{q2:3}}{f_{q2:3} + f_{q2:1}} - \frac{f_{q2:1}\, f_{q3:2}}{f_{q2:3} + f_{q2:1}} = f_{q1:3} - f_{q3:1} \tag{8.94}$$

As the chain flow f_q is also equal to the difference (Equation 8.85) between the flows of the equivalent exchange between extreme poles, each pole flow is written according to a term to term identification:

$$f_{q1:3} = \frac{f_{q1:2}\, f_{q2:3}}{f_{q2:3} + f_{q2:1}} \qquad f_{q3:1} = \frac{f_{q2:1}\, f_{q3:2}}{f_{q2:3} + f_{q2:1}} \tag{8.95}$$

No common additive constant is required in this separation, because when one flow is null, the equivalent flow in the same direction must be equal to zero. There is no possibility for the equivalent exchange to be reversible when one of the elementary exchanges is irreversible (see Figure 8.18).

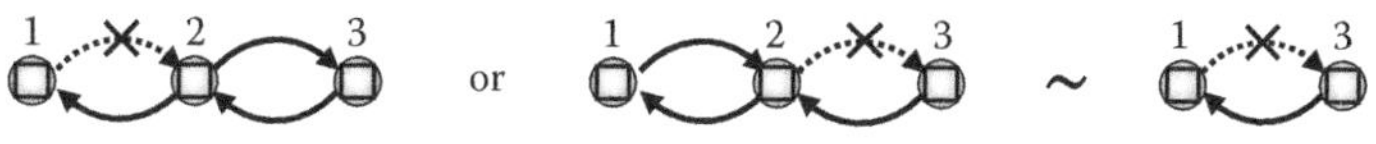

FIGURE 8.18 The nullity of only one polar flow in one direction entails the nullity of the overall flow in the same direction.

The ratio of the two equivalent flows provides the final result:

$$\frac{f_{q3:1}}{f_{q1:3}} = \frac{f_{q2:1}}{f_{q1:2}} \frac{f_{q3:2}}{f_{q2:3}} \tag{8.96}$$

Using the definition in Equation 8.82 of the exchange reversibility leads to result sought.

8.5.5 Dependence of the Reversibility on the Effort Difference

The relationships between pole efforts and effort differences are written for each dipole:

$$e_{q12} = e_{q1} - e_{q2}; \quad e_{q23} = e_{q2} - e_{q3}; \quad e_{q13} = e_{q1} - e_{q3} \tag{8.97}$$

From these three equations the relationship between dipole effort differences is deduced and Equation 8.90 is recalled:

$$e_{q13} = e_{q12} + e_{q23} \tag{8.98}$$

$$r_{q13} = r_{q12}\, r_{q23} \tag{8.90}$$

Exactly the same reasoning as the one employed in Chapter 7 for demonstrating the relation between gain and effort difference of a dipole leading to a model of influence is made. The procedure for going from one dipole to the other consists of two simultaneous operations, one addition and one multiplication.

Hypothesis. The exchange reversibility (between its two poles) of a dipole is independent of the other dipole efforts.

In assuming complete independence between dipoles, that is, in stating that the exchange reversibility of one dipole does not depend on the effort of another dipole, one is led to formulate the relationship between these two variables with the help of an exponential function

$$r_{qij} = \exp\left(-\frac{e_{qij}}{e_{q\Delta}}\right) \tag{8.99}$$

There is no need of a reference effort because the *dynamic equilibrium* defined by the equality of forward and backward flows corresponds also in a conductive dipole to the equality between polar efforts.

$$e_{qij} = 0 \quad \Leftrightarrow \quad r_{qij} = 1 \quad \Leftrightarrow \quad f_{qi:j} = f_{qj:i} = f^{eq}_{q\,j=i} \tag{8.100}$$

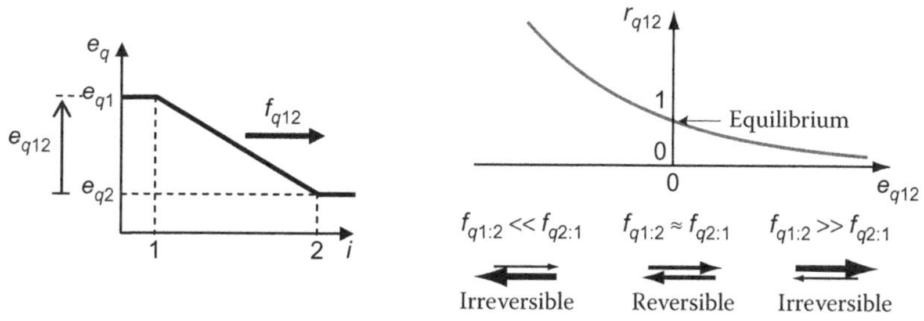

FIGURE 8.19 Sign convention between efforts and flow for a receptor dipole (left) and dependence of the exchange reversibility on the effort difference (right).

By convention, the common value at equilibrium of the flows is notated with an equal sign separating the subscripts i and j for indicating that the order of the subscripts can be any. A constant is introduced, which is the *scaling effort* $e_{q\Delta}$, a positive quantity already introduced for modeling the capacitive relationship. The same quantity is chosen to allow the composition of a conductance with a capacitance to be independent of the effort, which is the case in most situations. Formally, a different scaling effort can be chosen at this stage, but this remains a very hypothetical case as the scaling effort is determined by the existence of energy couplings. And there is no reason to have different couplings for the same system depending on the system property to model.

The choice of the negative sign in the argument of the exponential function in Equation 8.99 comes from the receptor convention to attribute a positive flow to a process governed by a decrease between efforts. This is illustrated in Figure 8.19 (left) for a dipole 1–2 where the second effort e_{q2} is lower than the first one e_{q1}, meaning a positive difference e_{q12}.

Applied to the first dipole 1–2, its exchange reversibility is plotted against the effort difference in Figure 8.19 (right).

From the definition in Equation 8.20 of the exchange reversibility the relationship between pole flows and effort difference is deduced:

$$\frac{f_{q2:1}}{f_{q1:2}} = \exp\left(-\frac{e_{q12}}{e_{q\Delta}}\right) \tag{8.101}$$

8.5.6 Conductive Relationships

The preceding discussion was held at the dipole level and now one has to go down at the pole level by separating the two flows in Equation 8.101. One adopts exactly the same procedure, which is not repeated, as for separating activities in the influence theory developed in Chapter 7, based on the assumption of independence between poles.

> ? **Hypothesis.** The pole flow of a dipole is independent of the effort of the other pole.

Each pole flow is assumed to depend only on the effort of its own pole and not on the other pole of the dipole. This leads to the two expressions

$$f_{q1:2} = f_{q1=2}^{eq} \exp\left(\frac{e_{q1} - e_{q1=2}^{eq}}{e_{q\Delta}}\right); \quad f_{q2:1} = f_{q1=2}^{eq} \exp\left(\frac{e_{q2} - e_{q1=2}^{eq}}{e_{q\Delta}}\right) \tag{8.102}$$

in which the scaling flow ("preexponential factor") is the common value of the pole flows at equilibrium, as stated in Equation 8.100, and the equilibrium effort, superscripted with "*eq*," is the same for both flows. As these constants are specific to the dipole, they bear the corresponding subscript.

8.5.6.1 Exchange Coefficients

The point in the preceding separation of efforts is that pole efforts are not always known individually but only their difference, the dipole effort, is known. A supplementary assumption is required for relating these efforts; this is done through *exchange coefficients*, also called *transfer coefficients* when the exchange models a transfer process between sites or corpuscles. In the Formal Graph theory, these coefficients express a distribution of dipole energy among the poles, which implies associating a storing dipole with the conductive dipole discussed here. This is not a real constraint as a conductive dipole needs to be supplied with energy for dissipating it.

These coefficients are defined as the proportion in which the supplied energy $\mathcal{T}_{qij}$ or $\mathcal{U}_{qij}$ to the conductive dipole is shared between its poles.

Exchange coefficients (inductive):

$$\alpha_{qi:j} \overset{def}{=} \frac{\partial \mathcal{T}_{qi}}{\partial \mathcal{T}_{qij}} \tag{8.103}$$

Exchange coefficients (capacitive):

$$\alpha_{qi:j} \overset{def}{=} \frac{\partial \mathcal{U}_{qi}}{\partial \mathcal{U}_{qij}} \tag{8.104}$$

Both definitions respect a sum equal to 1

$$\alpha_{q1:2} + \alpha_{q2:1} = 1 \tag{8.105}$$

These definitions help to understand that exchange coefficients express a sharing of energy between poles that can be eventually asymmetric (when the coefficients differ from 0.5), and that they are not necessarily constant but may adopt any dependence from other variables.

The correspondence between efforts in the storing dipole and in the conductive dipole (see Chapter 11 devoted to dissipation in dipole assemblies) makes these coefficients play the role of a proportionality factor between the conductive dipole effort and one of the pole efforts (see Graph 8.33).

Exchange coefficients:

$$\alpha_{q1:2} = \frac{e_{q1} - e_{q1=2}^{eq}}{e_{q12}}; \quad \alpha_{q2:1} = \frac{e_{q1=2}^{eq} - e_{q2}}{e_{q12}} \tag{8.106}$$

With these relationships between efforts, the pole flows are now written as

$$f_{q1:2} = f_{q1=2}^{eq} \exp\left(\alpha_{q1:2} \frac{e_{q12}}{e_{q\Delta}}\right); \quad f_{q2:1} = f_{q1=2}^{eq} \exp\left(-\alpha_{q2:1} \frac{e_{q12}}{e_{q\Delta}}\right) \tag{8.107}$$

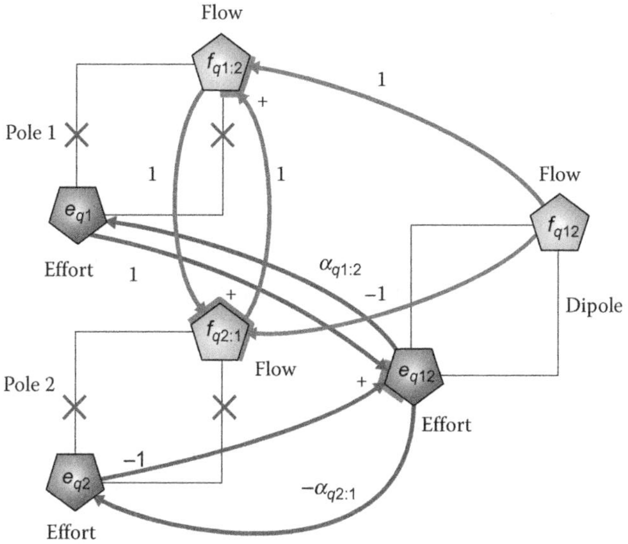

GRAPH 8.33 Principle of the modulation of the sharing of energy between poles through exchange coefficients. (Only one of these coefficients must be given because the construction of the Formal Graph implements the equality of their sum to 1, and the equilibrium potential has been set to 0.)

Their variations against the dipole effort are plotted in Figure 8.20 for two values of the first exchange coefficient $\alpha_{q1:2}$, 0.5 and 0.33.

As one has by definition of an exchange by flows a dipole flow that is equal to the pole flow difference,

$$f_{q12} = f_{q1:2} - f_{q2:1} \tag{8.108}$$

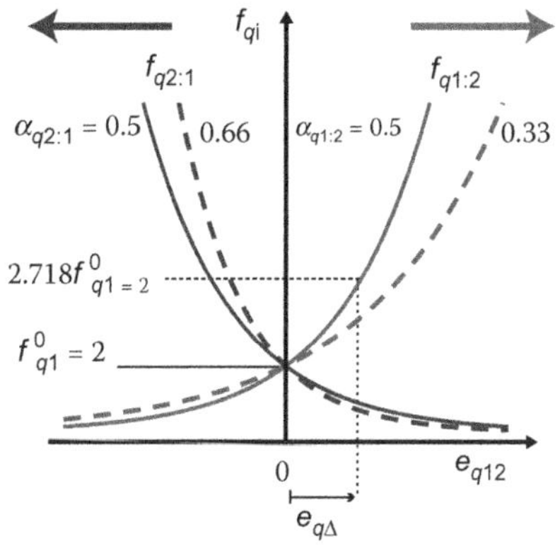

FIGURE 8.20 Plot of the pole flows as a function of the dipole effort for two sets of exchange coefficients. The first set (solid curves) is the fully symmetrical exchange with both coefficients equal to one half. The second set (dashed curves) is a nonsymmetrical exchange in favor of the backward flow that varies steeper than the forward flow with the dipole effort.

it allows one to write the dipole flow in following classical form:

$$f_{q12} = f^{eq}_{q1=2}\left[\exp\left(\alpha_{q1:2}\frac{e_{q12}}{e_{q\Delta}}\right) - \exp\left(-\alpha_{q2:1}\frac{e_{q12}}{e_{q\Delta}}\right)\right] \tag{8.109}$$

This expression is the conductive relationship for a dipole. Many systems based on an exchange by flows are modeled by this expression; mentioned here are two significant examples, in electrochemistry and in semiconducting materials. Both examples are models of capacitive–conductive dipoles.

8.5.6.2 Electrochemistry

The Butler*–Volmer† equation is a well-known model of charge transfer between two electroactive species (Bard and Faulkner 2001). It relates the overpotential η, defined as the difference between the electrode potential and its value at dynamic equilibrium, to the electrode current I, in near equilibrium conditions (see Figure 8.21):

$$I \underset{\approx eq}{=} I^0\left(\exp\frac{\alpha_A \eta}{V_\Delta} - \exp\frac{-\alpha_B \eta}{V_\Delta}\right);\; V_\Delta = \frac{RT}{\nu_e F} \tag{8.110}$$

Identification of the variables in this domain with the generalized ones does not require any comment. Except for the *scaling potential* V_Δ which corresponds to a transfer of ν_e electrons per electroactive molecule and which is determined by energy coupling with the thermal energy. (I^0 is the *exchange current*, R is the *gas constant*, T the *temperature*, and F the *Faraday constant*.)

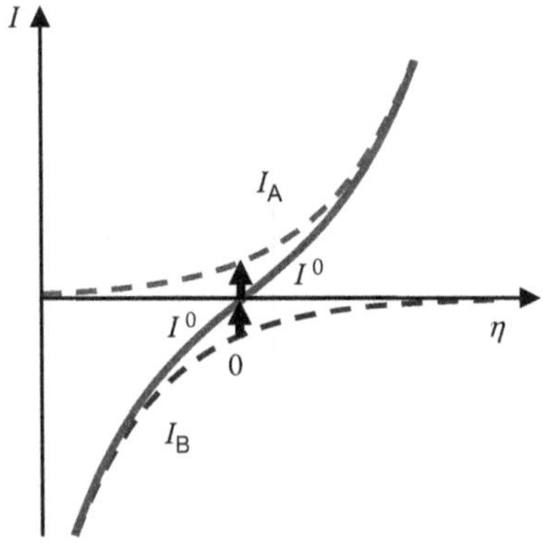

FIGURE 8.21 Plot of the electrochemical cell current given by the Butler–Volmer model as a function of the difference of electrode potential with the equilibrium potential (electrode overpotential). I_A and I_B refer to the forward and backward currents, respectively.

---

* John Alfred Valentine Butler (1899–1977): British electrochemist; Swansea and Edinburgh, UK.

† Max Volmer (1885–1965): German physical chemist; Berlin and Hamburg, Germany.

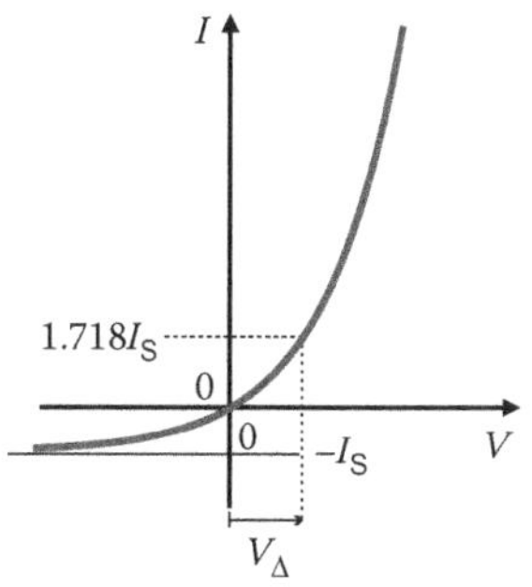

FIGURE 8.22 Plot of the current given by the Shockley equation as a function of the difference of potential across the junction.

8.5.6.3 Solid-State Physics

The Shockley* equation models the behavior of an ideal diode or *p-n* junction made by joining two semiconductors differently charged by doping, one positively (*p*-type semiconductor) and the other negatively (*n*-type) (see Figure 8.22):

$$I \underset{\approx eq}{=} I_S\left(\exp\frac{V}{V_\Delta}-1\right); \quad V_\Delta = \frac{k_B\,T}{e} \tag{8.111}$$

This model is also valid in near-equilibrium conditions and the correspondence with the general model is ensured by setting one of the exchange coefficients equal to one and the other to zero. It means that the potential across the junction acts only on the *p*-type material and not on the *n*-type. This features an *ideal diode* and discrepancies with these extreme values are observed for real diodes (Smith 1959). Again, the *scaling potential* V_Δ is determined by energy coupling with the thermal energy (agitation). (I_S is the *saturation current*, k_B the *Boltzmann constant*, T the *temperature*, and e is the *elementary charge*.)

Oversimplification. It must be mentioned that the systems modeled by these expressions are more complex systems than simple dipoles, as can be inferred from the simplicity of the equations. They involve more than one energy variety for being correctly modeled and it is an apparent dipole that is considered here. This subject is tackled in Chapter 12 in case study J2 "Electrochemical Potential."

Nonideal electron transfers and junctions. In these examples, the *exchange coefficients* are assumed to be constants; however, this is not a constraint and the general model proposed does not require such a condition. This means that more elaborated theories, such as the Marcus† theory of electron transfer or models of real diodes, are potentially included in the development in this study.

* William Bradford Shockley (1910–1989): American physicist; Stanford, California, USA.
† Rudolph Arthur Marcus (1923–): Canadian chemist; Ottawa, Canada and Pasadena, USA.

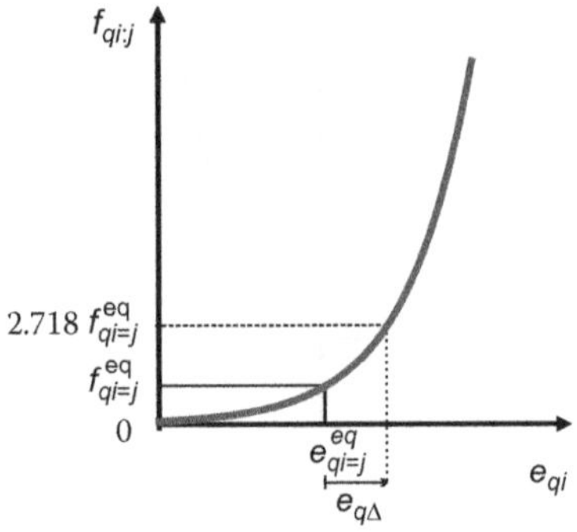

FIGURE 8.23 Plot of the pole flow given by the conductive equation as a function of the potential of the pole.

The two expressions gathered in Equation 8.102 gave the respective flows of the two poles in the same dipole. This presentation may hide the important result that with these expressions one has established the fundamental conductive relationship for a single pole. The key point is that the conductance operator of a single pole is written under the form of the exponential function

$$f_{qi} = f_{qi:j} = f^{eq}_{qi=j} \exp\left(\frac{e_{qi} - e^{eq}_{qi=j}}{e_{q\Delta}}\right) = f^0_{qi} \exp\left(\frac{e_{qi}}{e_{q\Delta}}\right) \quad \text{with} \quad f^0_{qi} = f^{eq}_{qi=j} \exp\left(-\frac{e^{eq}_{qi=j}}{e_{q\Delta}}\right) \tag{8.112}$$

Normally, the notation used for a pole *i* alone does not refer to another pole *j*, as was the notation in Chapter 4 devoted to poles. However, for coherence with this chapter, the notation referring to a pole partner is maintained.

In Figure 8.23 the variation of the pole flow versus the pole effort is plotted.

As long as the pole is isolated, the reference flow can be chosen arbitrarily as, in an exponential function, the couple formed by the preexponential factor and the argument shift can be taken at any point of the function. It is the formation of a dipole that determines the reference flow when dynamic equilibrium is reached.

8.5.7 Kinetics: The Capacitive–Conductive Relationship

Chapter 7 demonstrated the capacitive relationship between effort and basic quantity in a pole through the development of a theory of influence. The important results were expressed in the form of two relationships, the first one expressing the self-influence and relating the effort to the activity

$$a_{qi} = \exp\left(\frac{e_{qi} - e^{\ominus}_{qi}}{e_{q\Delta}}\right) \tag{8.113}$$

and the second one expressing the availability of the entities through the dependence of the activity on the basic quantity

$$a_{qi} \overset{def}{=} \frac{q_i - q_i^m}{q^{\ominus} - q_i^m} \tag{8.114}$$

Two parameters are used in this definition, a reference or standard basic quantity $q^{\ominus}$ corresponding to a unit activity and a standard effort for the pole, and a threshold or minimum basic quantity q_i^m quantifying the amount of inactive entities.

From the first relation 8.113 one may express the pole effort and then it can be introduced in the pole conductance expression 8.112:

$$f_{qi:j} = f_{qi=j}^{eq} \exp\left(\frac{e_{qi}^{\ominus} - e_{qi=j}^{eq}}{e_{q\Delta}} \right) a_{qi} \tag{8.115}$$

By defining a *kinetic rate* (or rate "constant") $k_{qi:j}$

$$k_{qi:j} \overset{def}{=} \frac{f_{qi=j}^{eq}}{q^{\ominus}} \exp\left(\frac{e_{qi}^{\ominus} - e_{qi=j}^{eq}}{e_{q\Delta}} \right) \tag{8.116}$$

the kinetic equation (first-order kinetics) is written in the general case

$$f_{qi:j} = k_{qi:j}\, q^{\ominus}\, a_{qi} \tag{8.117}$$

When all entities are active in the self-influence phenomenon, the well-known relationship between flow [often called in this case reaction rate or transport ("mass") flux] and basic quantity for a first-order kinetic process is found from the activity definition (Equation 8.114).

$$f_{qi:j} \underset{q_i^m=0}{=} k_{qi:j}\, q_i \tag{8.118}$$

This kinetic rate corresponds to the combination of elastance and capacitance operators existing in the pole

$$k_{qi:j} = \hat{G}_{qi:j}\, \hat{C}_{qi}^{-1} \tag{8.119}$$

A classical postulate demonstrated. This may appear a negligible side effect of the present conduction theory, but here is merely demonstrated one of the classical postulates of kinetics, which is that for a first-order process, the flow is proportional to the basic quantity through a rate "constant." This is a direct consequence of the similar dependence of the conductance and the capacitance on the effort (see Graph 8.34).

8.5.8 Generalized Arrhenius Model

The definition in Equation 8.53 of the kinetic rate can be put in the form of the Arrhenius equation 8.1

$$k_{qi} = A_{qi} \exp\left(-\frac{E_a}{RT} \right) \tag{8.120}$$

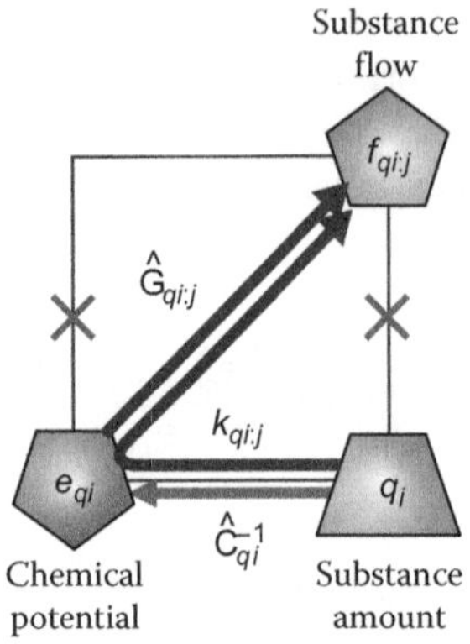

GRAPH 8.34 Formal Graph of a capacitive–conductive pole *i* showing the combination of elastance and conductance into a kinetic rate that links the substance amount to the substance flow according to a first-order kinetics.

in making the following correspondences:

$$k_{qi} = k_{qi:j};\quad A_{qi} = \frac{f^{eq}_{qi=j}}{q^{\theta}};\quad E_a = e^{eq}_{qi=j} - e^{\theta}_{qi};\quad RT = e_{q\Delta} \tag{8.58}$$

The preexponential factor is identified with the ratio of reference parameters of the conductive and of the capacitive relationships; the scaling effort is written as proportional to the temperature through the gas constant *R*; and the *activation energy-per-entity* is identified with the difference between the dynamic equilibrium effort and the standard effort.

The fact that reference parameters are used does not mean that they are absolutely constant parameters. They are constant parameters with respect to the state variables in the considered energy variety, but certainly not with respect to other variables, in particular when other energy varieties are present in the system (they are often considered as temperature dependent). Their modeling is therefore relevant of higher levels of organization.

This is the same for the scaling effort. The proportionality to the temperature means that thermal energy must be accounted in the system. Physically speaking, heat is stored in the system under the form of thermal agitation and the coupling with the energy variety under consideration determines the scale of the energy-per-entity in the system constitutive properties. This subject will be tackled in the Chapter 12 devoted to energy coupling.

In this correspondence between models, there is no notion of overcoming an *activation barrier* for the process to work. The reason is that the notion of barrier does not exist in an elementary process as the one working in a dipole. To be able to use the notion of barrier, a more complex process must be envisaged, such as a process made up of serial steps in a chain multipole, as explained in Section 8.1.2.

8.5.9 Activation Barrier

Let us return to the sequence of two steps of the 1–3 chain multipole for putting into evidence the effect of an intermediate dipole efforts on the whole kinetics. The interesting product of reversibilities in Equation 8.90 is rewritten in making explicit the ratio of multipole flows:

$$\frac{f_{q3:1}}{f_{q1:3}} = r_{q12}\, r_{q23} = \frac{r_{q23}}{r_{q21}} \tag{8.122}$$

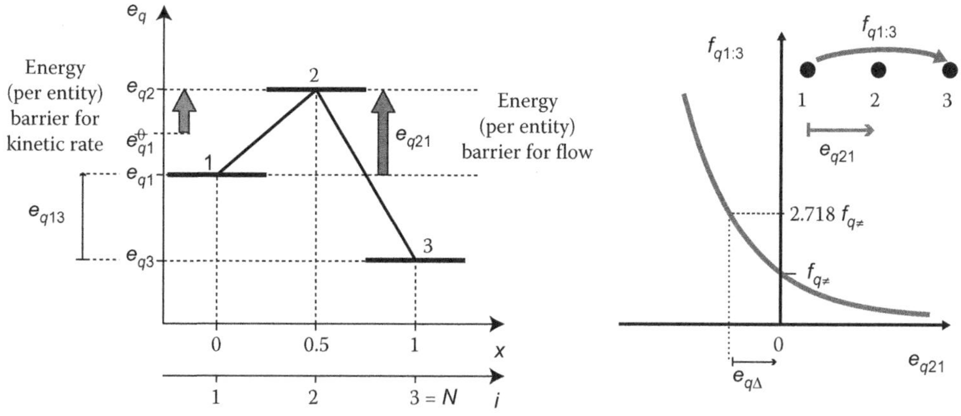

FIGURE 8.24 Energy-per-entity levels for an irregular three-poles chain in the case of a barrier (left) and plot of the pole flow as a function of the barrier height (right).

As for any equality between two ratios, it is interesting to make the cross multiplication and to use the common value of the resulting products or ratios as a new variable. Thus, an *intermediate flow* $f_{q\neq}$ is defined as the common value of the new ratios obtained by swapping one numerator with one denominator:

$$f_{q\neq} \overset{def}{=} \frac{f_{q1:3}}{r_{q21}} = \frac{f_{q3:1}}{r_{q23}} \tag{8.123}$$

Then, by using the dependence of the reversibilities in Equation 8.99 on the dipole efforts, one gets

$$f_{q1:3} = f_{q\neq} \exp\left(-\frac{e_{q21}}{e_{q\Delta}}\right); \quad f_{q3:1} = f_{q\neq} \exp\left(-\frac{e_{q23}}{e_{q\Delta}}\right) \tag{8.124}$$

This makes explicit the dependence of the overall flows on the effort difference between the considered pole 1 or 3 and the intermediate pole 2. The plot of the forward flow is given in Figure 8.24 and puts in evidence the decrease of the flow with increase of the barrier height.

Now, by associating the conductive relationship of the first pole, as done above in the case of a single dipole for developing the kinetic model, one can replace the pole effort e_{q1} by its dependence with the activity. This gives the forward overall kinetic rate

$$f_{q1:3} = f_{q\neq} \exp\left(-\frac{e_{q2} - e^{\theta}_{q1}}{e_{q\Delta}}\right) a_{q1} \tag{8.125}$$

Finally, assuming a complete availability of entities for the influence ($q^m_i = 0$), the expression of the overall kinetic rate is

$$k_{q1:3} = \frac{f_{q\neq}}{q^{\theta}} \exp\left(-\frac{e_{q2} - e^{\theta}_{q1}}{e_{q\Delta}}\right) \tag{8.126}$$

This has the same shape as the Arrhenius equation and the correspondence with Arrhenius parameters is

$$A_{q\neq} = \frac{f_{q\neq}}{q^{\theta}}; \quad e_{q\Delta} = RT; \quad E_a = e_{q2} - e_{q1}^{\theta} \tag{8.127}$$

This time, the effect of a barrier is explicit, by giving to the effort of the intermediate pole an effective role in increasing (by low values) the rate of the process from pole 1 to pole 3 and in slowing down with high values.

Barrier for a flow and barrier for a rate. According to the genuine Arrhenius equation, established for a kinetic rate k, the activation energy-per-entity E_a is not exactly the effort difference between pole 2 and pole 1 because for a kinetic rate one must take into account the dependence of the basic quantity on the effort of pole 1. On the contrary, in the overall flow equation, the determining variable is the right difference.

Arrhenius for flow:

$$f_{q1:3} = f_{q\neq} \exp\left(-\frac{E_a}{e_{q\Delta}}\right) \quad \text{with } E_a = e_{q2} - e_{q1} \tag{8.128}$$

Arrhenius for rate:

$$k_{q1:3} = \frac{f_{q\neq}}{q^{\theta}} \exp\left(-\frac{E_a}{e_{q\Delta}}\right) \quad \text{with } E_a = e_{q2} - e_{q1}^{\theta} \quad (8.126) + (8.127)$$

A standard graphic representation of the Arrhenius model is shown in Figure 8.25.

8.5.10 Comparison with Statistics

An interesting development is worth noting for expressing in an alternate way the dependence of the pole flows on the effort difference. When one substitutes in expression 8.87 the exchange reversibility by its dependence on the effort difference given by Equation 8.99, one gets

$$\frac{f_{q1:3}}{f_{q13}} = \frac{1}{1 - \exp\left(-\frac{e_{q13}}{e_{q\Delta}}\right)}; \quad \frac{f_{q3:1}}{f_{q13}} = \frac{1}{\exp\left(+\frac{e_{q13}}{e_{q\Delta}}\right) - 1} \quad e_{q13} \neq 0 \tag{8.129}$$

These expressions have the same shape as those given by Bose*–Einstein† statistics in assimilating each ratio of the pole flow on the "total" flow (dipole flow) to a distribution function.

* Satyendra Nath Bose (1894–1974): Indian physicist; Calcutta, India.

† Albert Einstein (1879–1955): German, Swiss, American physicist; Bern, Switzerland and Princeton, USA.

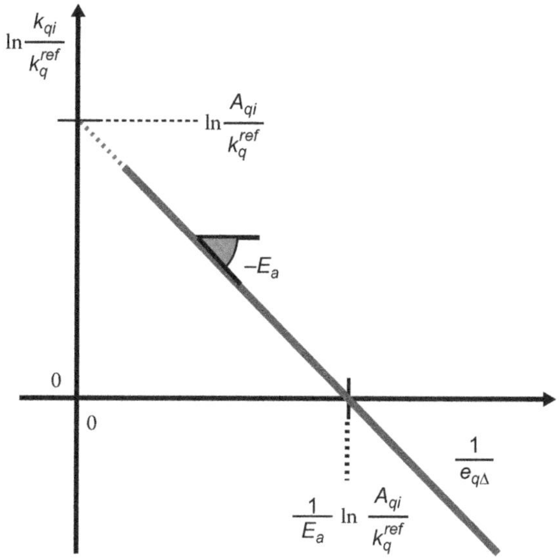

FIGURE 8.25 This plot is called an "Arrhenius plot"; it allows interpretation of experimental data through the linearization obtained by plotting the logarithm of the observed kinetic rate (normalized by an arbitrary reference kinetic rate) versus the inverse of the temperature. As this is a generalized model, this temperature is embedded into the scaling effort $e_{q\Delta}$ supposed to be equal to *RT*. Whatever the reference chosen, the negative of the slope of the straight line gives the activation energy-per-entity E_a.

Bose–Einstein statistics differs from Fermi*–Dirac† statistics by the nonapplicability of the Pauli‡ exclusion principle, meaning that objects may occupy energy levels in any number. In Chapter 7, this principle has been shown as being equivalent to a conservation law in a dipole. Translated in Formal Graph language, it means that Bose–Einstein statistics is a model of nonconservative dipole. The conductive dipole is by excellence an example of system in which no possibility of conservation exists as there are no entities to conserve. Table 8.7 gives the main differences between these statistical models and the correspondence with the Formal Graph approach.

TABLE 8.7
Comparison between the Classical Approach and the Formal Graph Approach for Various Statistical Models and Corresponding Formal Objects

Statistics	Maxwell–Boltzmann	Fermi–Dirac	Bose–Einstein
Classical view	Identical distinguishable particles	Indistinguishable particles subject to Pauli exclusion principle	Indistinguishable particles that may occupy the same energy level
Formal Object Relationships	Pole (i.e., collection)	Conservative dipole $q_1 + q_2 = q_1^* + q_2^*$ $e_{q2} = e_{q1} + e_{q21}$	Nonconservative dipole $f_{q2} = f_{q1} - f_{q12}$ $e_{q2} = e_{q1} - e_{q12}$

* Enrico Fermi (1901–1954): Italian physicist; Florence and Rome, Italy and Chicago, USA.
† Paul Adrien Maurice Dirac (1902–1984): British physicist; Bristol, UK and Tallahassee, Florida, USA.
‡ Wolfgang Ernst Pauli (1900–1958): Austrian physicist; Vienna, Austria, Zürich, Switzerland, and Princeton, USA.

8.6 IN SHORT

FORMAL GRAPHS

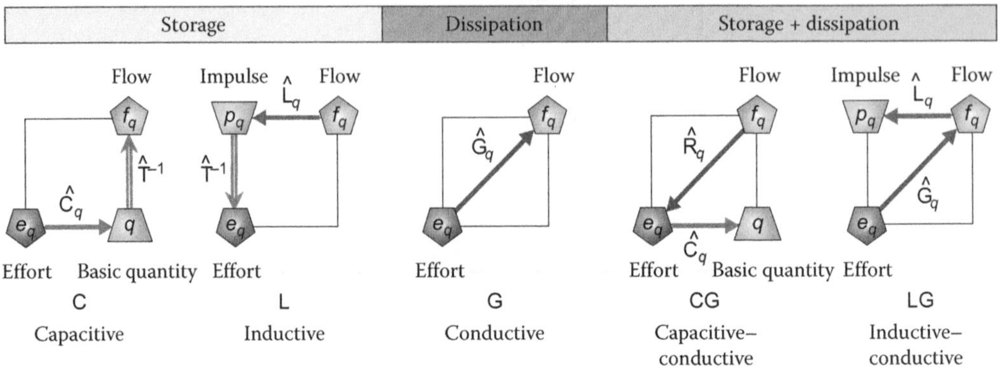

GRAPH 8.35

MULTIPOLE STRUCTURES

1. Chain
   Common flow = serial mounting of dipoles

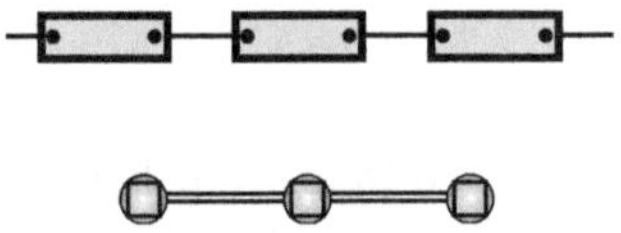

2. Ladder
   Common effort = parallel mounting of dipoles

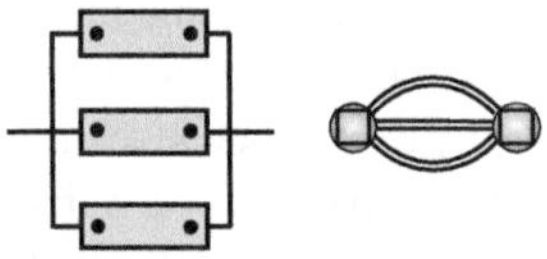

3. Other
   No common energy-per-entity

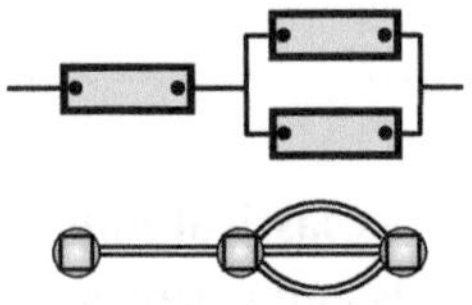

CHAIN MULTIPOLES

Chain coordinate $x = \frac{i-1}{N-1};\quad i = 1, N$ (N = number of poles) (8.2)

Regularity
y_{qi}: *Pole energy-per-entity (effort or flow)*

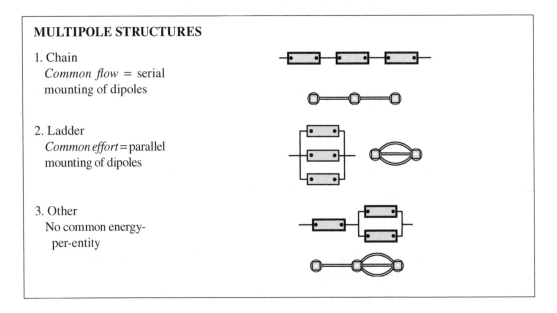

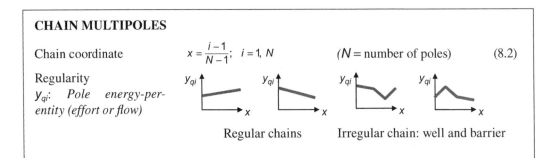

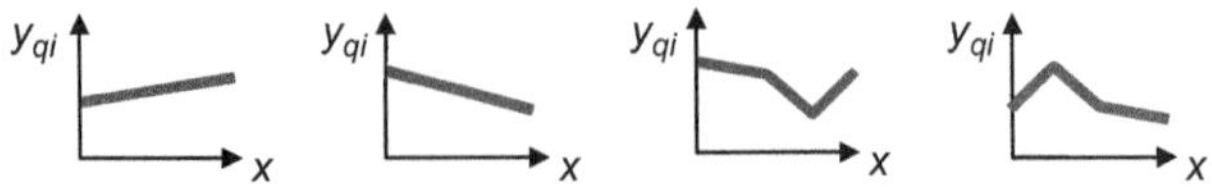

Regular chains Irregular chain: well and barrier

DIPOLE MOUNTING RULES (GENERALIZED KIRCHHOFF LAWS)

Common effort = *ladder* (parallel)

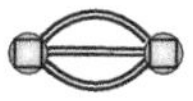

$$e_q = e_{qk}, \quad k = 1, N \qquad (8.55)$$

$$f_q = \sum_{k=1}^{N} f_{qk} \qquad (8.56)$$

Common flow = *chain* (serial)

$$f_q = f_{qk}, \quad k = 1, N \qquad (8.57)$$

$$e_q = \sum_{k=1}^{N} e_{qk} \qquad (8.58)$$

DIPOLES–MULTIPOLE RELATIONSHIPS BETWEEN ENERGIES PER ENTITIES

Multipole	Chain		Ladder	
Irregular	$e_q = \sum_{k=1}^{N} e_{qk,k+1}$	(8.130)	$f_q = \sum_{k=1}^{N} f_{qk,k+1}$	(8.132)
	$f_q = f_{qk,k+1}; \quad k = 1, N$	(8.131)	$e_q = e_{qk,k+1}; \quad k = 1, N$	(8.133)
Regular	$e_q = N e_{qk,k+1}; \quad k = 1, N$	(8.134)	$f_q = N f_{qk,k+1}; \quad k = 1, N$	(8.136)
	$f_q = f_{qk,k+1}; \quad k = 1, N$	(8.135)	$e_q = e_{qi}; \quad k = 1, N$	(8.137)

DIPOLE WITH EXCHANGE BY FLOWS

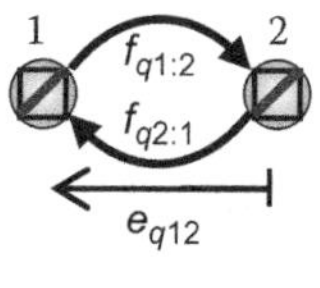

Exchange reversibility:

$$r_{q12} \overset{def}{=} \frac{f_{q2:1}}{f_{q1:2}} = \exp\left(-\frac{e_{q12}}{e_{q\Delta}}\right) \qquad (8.101)$$

POLE CONDUCTIVE RELATIONSHIP

Pole–dipole notation:

$$f_{qi:j} = f_{qi=j}^{eq} \exp\left(\frac{e_{qi} - e_{qi=j}^{eq}}{e_{q\Delta}}\right) \qquad (8.112)$$

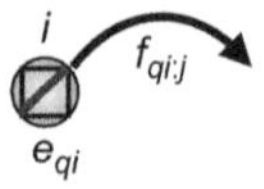

$f^0_{qi=j}$: Equilibrium value of the flows between poles *i* and *j*. $e^{eq}_{qi=j}$: Equilibrium effort

Single pole notation:

$$f_{qi} = f_{qi:j} = f^0_{qi} \exp\left(\frac{e_{qi}}{e_{q\Delta}}\right) \text{ with } f^0_{qi} = f^{eq}_{qi=j} \exp\left(-\frac{e^{eq}_{qi=j}}{e_{q\Delta}}\right) \quad (8.112)$$

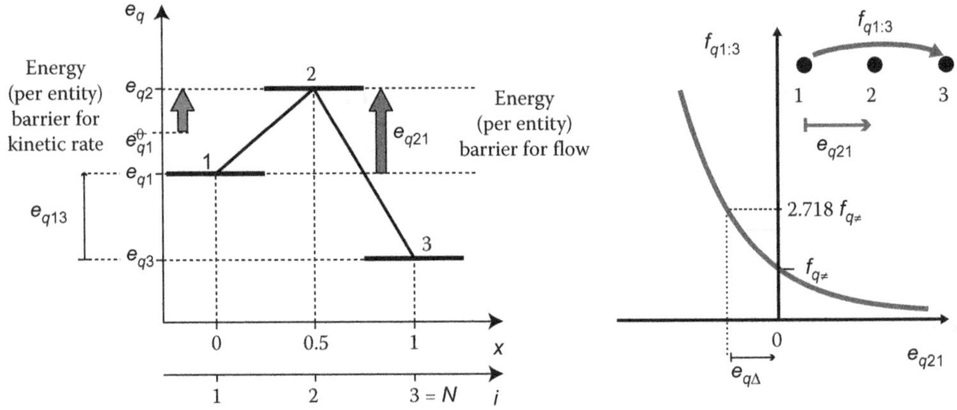

FIGURE 8.24

DIPOLE EXCHANGE RELATIONSHIP

Exchange coefficients: $\alpha_{q1:2} + \alpha_{q2:1} = 1$ (8.105)

Inductive–conductive LG

$$\alpha_{qi:j} \overset{def}{=} \frac{\partial \mathcal{T}_{qi}}{\partial \mathcal{T}_{qij}} \quad (8.103)$$

Capacitive–conductive CG

$$\alpha_{qi:j} \overset{def}{=} \frac{\partial \mathcal{U}_{qi}}{\partial \mathcal{U}_{qij}} \quad (8.104)$$

Conductive G

$$\alpha_{q1:2} \overset{def}{=} \frac{e_{q1} - e^{eq}_{q1=2}}{e_{q12}}$$

$$\alpha_{q2:1} \overset{def}{=} \frac{e^{eq}_{q1=2} - e_{q2}}{e_{q12}} \quad (8.106)$$

Dipole flow

$$f_{q12} = f^{eq}_{q1=2}\left[\exp\left(\alpha_{q1:2}\frac{e_{q12}}{e_{q\Delta}}\right) - \exp\left(-\alpha_{q2:1}\frac{e_{q12}}{e_{q\Delta}}\right)\right] \quad (8.109)$$

$f^0_{q1=2}$: equilibrium value of the pole flows ($e_{q12} = 0$)

POLE KINETIC RELATIONSHIP

First-order kinetic rate:

$$k_{qi:j} \overset{def}{=} \frac{f^{eq}_{qi=j}}{q^{\ominus}} \exp\left(\frac{e^{\ominus}_{qi} - e^{eq}_{qi=j}}{e_{q\Delta}}\right) \tag{8.116}$$

General first-order kinetic equation:

$$f_{qi:j} = k_{qi:j}\, q^{\ominus}\, a_{qi} \tag{8.117}$$

Full activity case:

$$f_{qi:j} \underset{q^m_i=0}{=} k_{qi:j}\, q_i \tag{8.118}$$

Capacitance and conductance:

$$k_{qi:j} = \hat{G}_{qi:j} \hat{C}^{-1}_{qi} \tag{8.119}$$

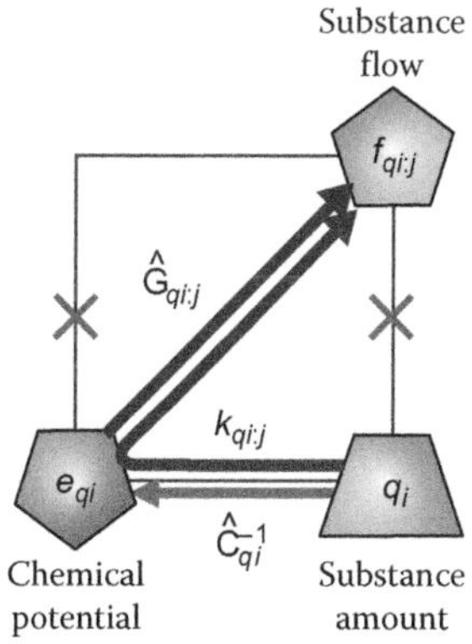

GRAPH 8.34

9 Dipole Assemblies

CONTENTS

When dipoles belonging to different energy subvarieties are assembled together, one enters a completely new area, the domain of energy conversions. The novelty comes from the introduction of time in the Formal Graph theory.

The role of time will be discussed in two steps. This chapter deals with reversible behaviors of systems, that is, conversions without dissipation, in studying oscillators and the role of space–time in various systems (propagating waves, Maxwell equations in electrodynamics, Hall effect). Chapters 10 and 11, subsequently, will be devoted to irreversible behaviors, when conduction and dissipation play a non-negligible role in conversion processes.

Table 9.1 lists the case studies of dipole assemblies given in this chapter. (See also Figure 9.1 for the position of the dipole assemblies.)

9.1 THE DIPOLE ASSEMBLY

The association of several dipoles of identical nature (in terms of system constitutive properties) has been studied in Chapter 8. Such an association was described as forming a multipole in which energy is stored and/or dissipated but in keeping the same form, that is, without conversion from one subvariety to another or into heat. The name *dipole assembly* refers to a more general concept mixing several dipoles or multipoles having different natures but belonging to the same energy variety. For instance, two dipoles (or multipoles), one capacitive and one inductive, form a dipole assembly, provided that both subvarieties are in the same energy variety. The case of putting together dipoles of different energy varieties is relevant from a category of Formal Objects which is not precisely defined because it is very wide. This case will be examined among others in Chapter 12, which tackles the coupling of Formal Objects (poles, dipoles, dipole assemblies) belonging to different energy varieties. In Table 9.2, the various Formal Objects mentioned are sorted according to the energy subvarieties and varieties involved.

In the present theory, the term *energy conversion* is restricted to a change from one subvariety to another. Whether or not the two subvarieties belong to the same energy variety is included in the notion. A conversion process involves two pairs of *energies-per-entity* and requires some time.

When energy goes between identical objects, the process is called an *energy exchange*, which is an equilibration process. It involves the state variables of only one energy subvariety (two *entity numbers* and two *energies-per-entity*) and works without time considerations (what is expressed by "instantaneous" process). As long as no dissipative process is involved, two capacitors, for instance, having their ends connected in pairs, will match their *energy-per-entities* instantaneously.

Energy exchange or conversion? Between two identical dipoles containing energy (i.e., either inductive dipoles or capacitive dipoles), energy can be merely exchanged. Energy remains in the same subvariety, so there is no need of conversion into another subvariety or into heat. This is different when the two dipoles do no belong to the same subvariety. Energy conversion is then required. The important difference is that an energy exchange occurs without involving the notion of time (it is instantaneous), whereas an energy conversion takes time.

TABLE 9.1
List of Case Studies of Dipole Assemblies (Reversible)

	Name	Energy Varieties	Subvariety	Type	Page Number
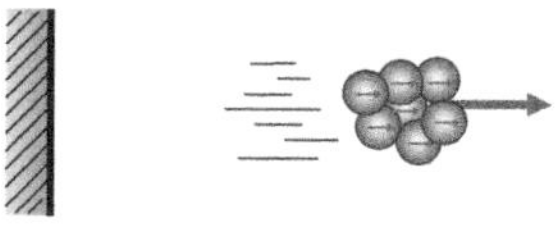	F1: Accelerated Motion	Translational mechanics	Inductive	Powered temporal	355
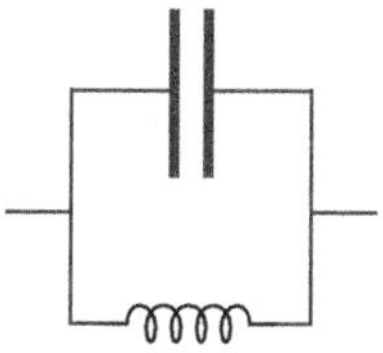	F2: LC Circuit	Electrodynamics	Inductive + capacitive	Isolated temporal	362
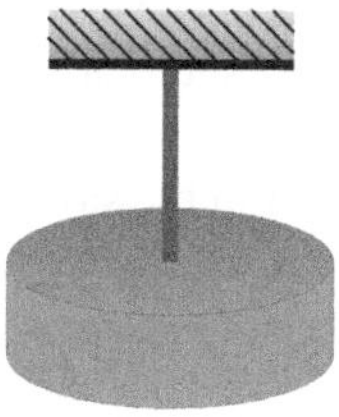	F3: Torsion Pendulum	Rotational mechanics	Inductive + capacitive	Isolated temporal	365
	F4: Vibrating String	Translational mechanics	Inductive + capacitive	Isolated one-dimensional spatiotemporal	369
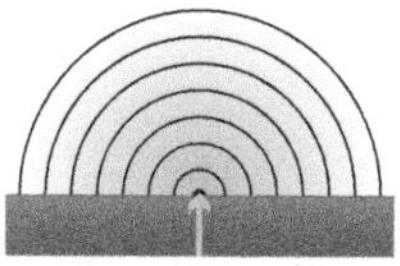	F5: Sound Propagation	Hydrodynamics	Inductive + capacitive	Isolated three-dimensional spatiotemporal	373
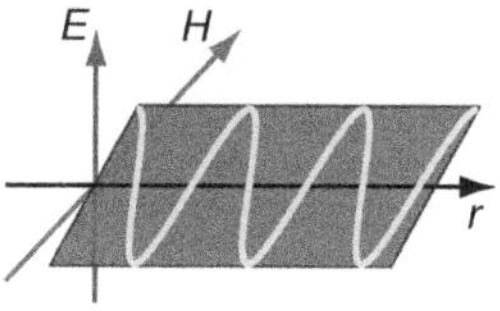	F6: Light Propagation	Electrodynamics	Inductive + capacitive	Isolated three-dimensional spatiotemporal	377
$\nabla . D = \rho$ $\nabla . B = 0$ $\nabla \times E = -\frac{\partial}{\partial t} B$ $\nabla \times H = j + \frac{\partial}{\partial t} D$	F7: Maxwell Equations	Electrodynamics	Inductive + capacitive	Three-dimensional spatiotemporal	382
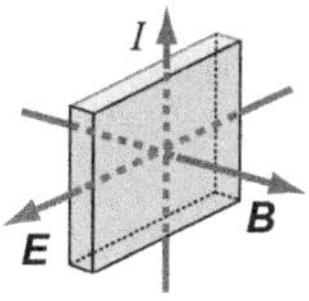	F8: Hall Effect	Electrodynamics	Inductive + capacitive	Powered three-dimensional spatiotemporal	386

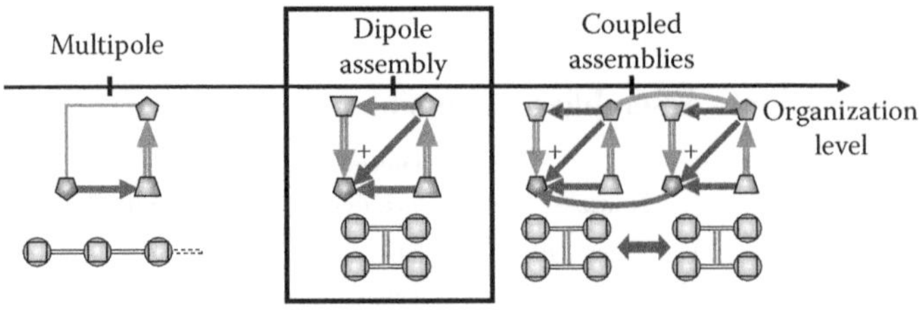

FIGURE 9.1 Position of the dipole assembly along the complexity scale of Formal Objects.

In Figure 9.2 are sketched the two different ways—through energy exchange and conversion—of combining the three different kinds of fundamental dipoles (inductive, capacitive, and conductive) according to their energy subvariety.

In Figure 9.2 are laid down the relationships between energy subvarieties in the frame of one energy variety only. The two passages for energy implying dipoles are *conversion*, when subvarieties differ and with evolution involved, and *exchange* between identical objects (same subvariety) and without evolution.

The *energy exchange* process does not modify the nature of energy; it remains within the same subvariety, but energy amounts vary within each partner.

The *energy conversion* process makes energy change in nature and in amounts from one subvariety to another.

When energy moves from a storing dipole to a conductive dipole, the process is merely an *energy circulation* as will be seen in Chapter 11 about dissipation.

This scheme is drawn again in Graph 9.1 with the individual Formal Graphs of the dipoles and the connections that model the energy exchange and conversion.

9.2 EVOLUTION AND TIME

In Chapter 6, the general rule for energy variations in a dipole according to each energy subvariety has been established. For a single capacitive dipole and a single inductive one, these rules were expressed on the same model as simpler systems by application of the embedding principle.

TABLE 9.2
Energetic Features of Formal Objects in Term of Exchange, Conversion, Circulation, or Coupling Inside Each Object

Energy varieties	Only one				Two or more
Energy subvarieties or dissipation	One subvariety or dissipation			Two Subvarieties or one subvariety and dissipation	Any
Internal energetic process	No exchange	Exchange		Conversion or circulation	Conversion and/or coupling
Formal Object	Pole	Dipole	Multipole	Dipole assembly	Coupling

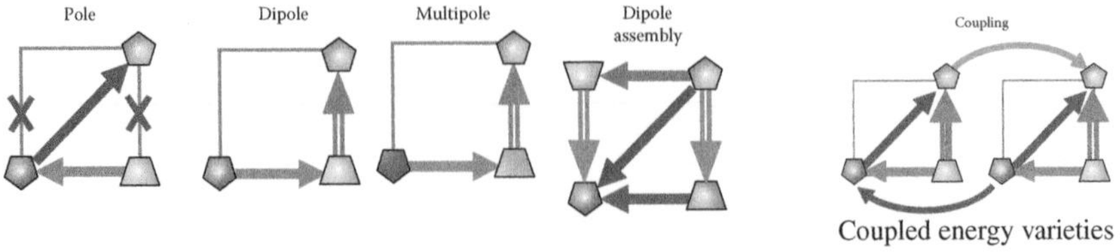

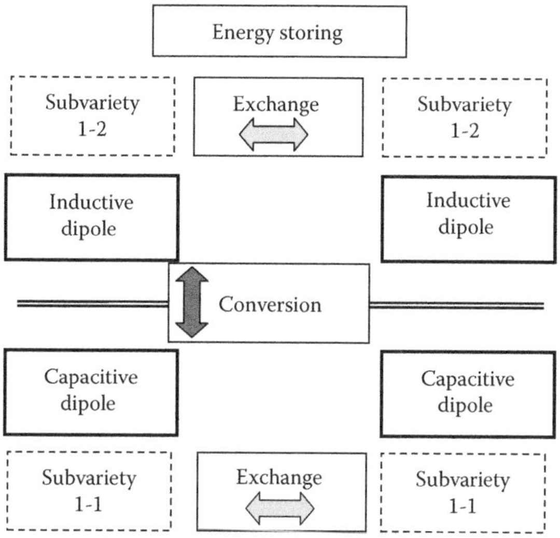

FIGURE 9.2 Scheme of the energy conversion and exchange between dipoles of various natures (without dissipation).

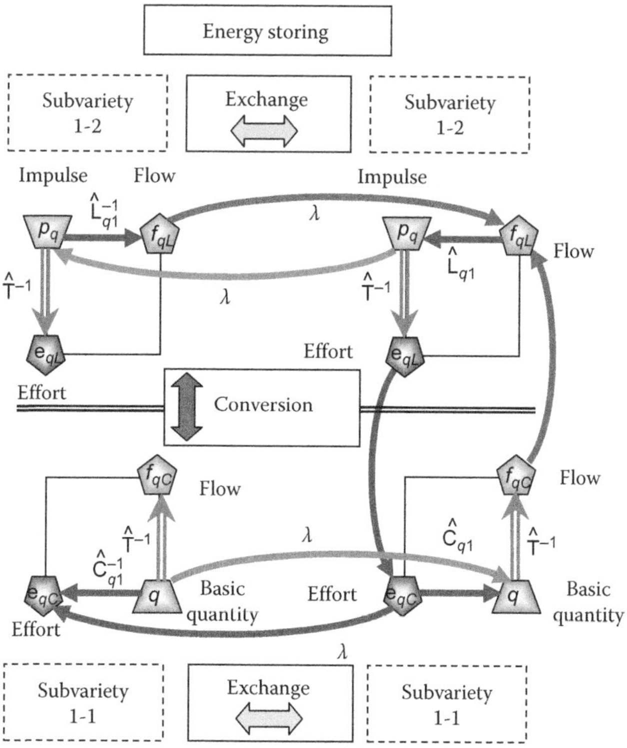

GRAPH 9.1 Energy exchange and conversion (without dissipation) between different storing dipoles represented by interconnected Formal Graphs.

Capacitive dipole:
$$d\mathcal{U}_q \underset{\exists C_q}{=} e_{qC}\, dq \tag{9.1}$$

Inductive dipole:
$$d\mathcal{T}_q \underset{\exists L_q}{=} f_{qL}\, dp_q \tag{9.2}$$

Each type of dipole possesses a dipole property, capacitance of inductance, relating its state variables, which are recalled here:

$$e_{qC} = \hat{C}_q^{-1}\, q \tag{9.3}$$

$$f_{qL} = \hat{L}_q^{-1}\, p_q \tag{9.4}$$

Until now, the notion of time has not been necessary in all the developments made. In particular, the definition of flows, classically defined as the time derivative of a basic quantity, has been based on the concept of *energy-per-entity*, existing independent of any notion of time. Their definition stems from the First Principle of Thermodynamics as written above in Equations 9.1 and 9.2.

$$e_q \overset{def}{=} \left(\frac{\partial \mathcal{U}_q}{\partial q}\right)_{other\ q} \tag{9.5}$$

$$f_q \overset{def}{=} \left(\frac{\partial \mathcal{T}_q}{\partial p_q}\right)_{other\ p_q} \tag{9.6}$$

Moreover, the concept of exchange between the two poles has been developed without reference to the time eventually needed for carrying it out. This has allowed conservation laws to be developed without reference to time-related notions such as stability, stationary, permanence, static, and so on.

The two poles of a dipole cannot convert energy between them as they belong to the same energy subvariety. Two dipoles of the same energy subvariety can only equilibrate their common *energies-per-entity*, an exchange process that can be ideally done without needing time, that is to say instantaneously (cf. discussion in Chapter 6, Section 6.2.5). In contradistinction, two dipoles with different energy subvarieties can convert energy. This takes time.

In Chapter 6 and followings, an unknown *inverse evolution operator* $\hat{T}$ had been introduced on the model of the system constitutive properties, for linking the *energy-per-entity* of one energy subvariety to the *entity number* of the other subvariety, as recalled in Graph 9.2. This was required to allow circulation of energy inside the inductive and capacitive dipoles, which was only possible in other types of dipoles involving a conductance, because the two energies-per-entities were directly connected through this constitutive property.

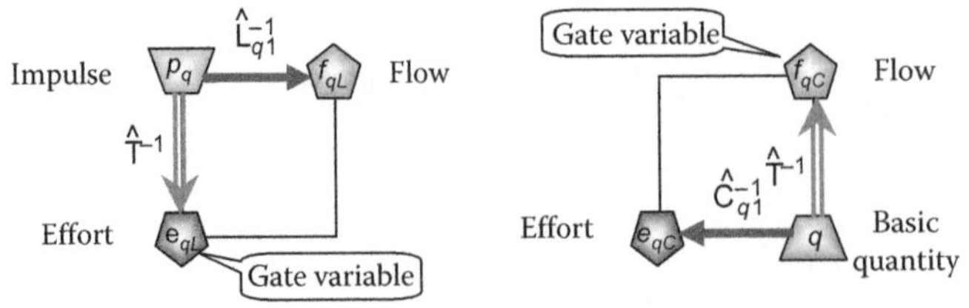

GRAPH 9.2 Formal Graphs of an inductive dipole (left) and of a capacitive dipole (right) with their evolution link allowing their two energies-per-entity to be connected.

The symmetry of the situation had led to adopting the same operator for both subvarieties, but this remains to be demonstrated. Now, this evolution operator can be defined through the following reasoning.

A postulate of the Formal Graph theory is that the orientation of the connection between the two state variables must be reversible, like all system properties already defined, meaning that the *inverse evolution* operator $\hat{T}$ must have a reciprocal $\hat{T}^{-1}$ merely called *evolution* operator.

Evolution:

$$(energy\text{-}per\text{-}entity)_{\text{subvariety 2}} = \hat{T}^{-1}(entity\ number)_{\text{subvariety 1}} \tag{9.7}$$

Inverse evolution:

$$(entity\ number)_{\text{subvariety 1}} = \hat{T}\left(energy\text{-}per\text{-}entity\right)_{\text{subvariety 2}} \tag{9.8}$$

9.2.1 Definition of Time

This operator $\hat{T}$ acts to transform a certain amount of *energy-per-entity* y_2 into an amount of entities x_1 in the other subvariety. The Formal Graph theory states that such a conversion is gradual, in the sense that the advance of the transformation can be physically evaluated and quantified. The difference between two degrees of advance corresponding to two states of the conversion is called the *elementary duration* of the transformation (see Figure 9.3). A first definition of the *elementary duration* is to identify it with the interval of advance degrees, notated Δt, required for changing the *entity number* between x_1 and $x_1 + \Delta x_1$ when the source y_2 is held constant:

$$\Delta t = \frac{\Delta x_1}{y_2} \tag{9.9}$$

Here, this constancy is thought of only as a pedagogical step; it is obvious that the second step consists of a more precise definition eliminating the need of a constant y_2:

$$\Delta t \overset{def}{=} \int_{x_1}^{x_1+\Delta x_1} \frac{dx_1}{y_2} \tag{9.10}$$

From this integral definition, one may introduce a *time* variable, the variation of which is defined by

$$dt \overset{def}{=} \frac{dx_1}{y_2} \tag{9.11}$$

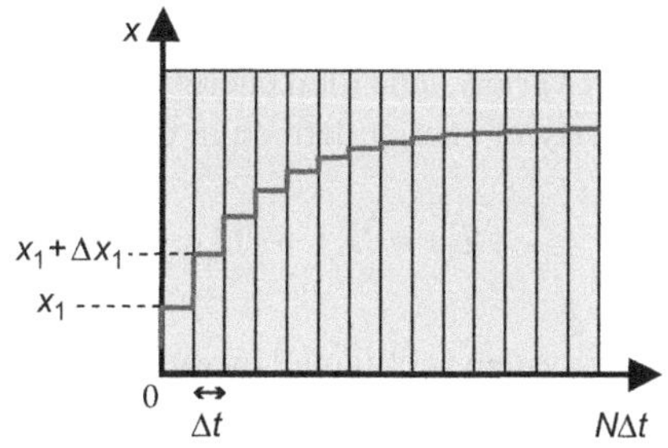

FIGURE 9.3 Evolution of an entity number x resulting from a step-by-step energy conversion. (Here, N is the number of elementary durations.)

Formal algebra

$$B = \frac{d}{dt} A$$

Formal graph

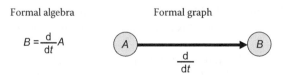

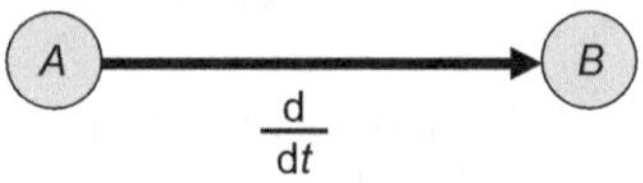

GRAPH 9.3 The evolution operator in algebraic language and in Formal Graph language.

allowing one to write the duration as an integration of this variable:

$$\Delta t = \int_{t_0}^{t_0 + \Delta x_1 / y_2} dt \tag{9.12}$$

It should be remarked that the definition in Equation 9.11 of the time is relative, only the time variation is defined. No time origin is involved here (the boundary t_0 in the integration being arbitrary). In addition, this time is not absolute as it is associated with a conversion processes that may occur only in some regions of space. This differs notably from the Newtonian concept of absolute and universal time, identical in all space. This new definition can be rewritten in the form of the evolution operator relationship (see Equation 9.7)

$$\text{Continuous conversion:} \quad (\textit{energy-per-entity})_{\text{subvariety 2}} = \frac{d}{dt}(\textit{entity number})_{\text{subvariety 1}} \tag{9.13}$$

This temporal derivation can be seen as a peculiar form of the *evolution operator* $\hat{T}^{-1}$, as it conveys the continuous change of state of the system caused by the energy conversion. A precision must be made about the reference to a *continuous* process, as this is not a compulsory requirement for the Formal Graph theory. This justifies the distinct use of the concept of *inverse evolution operator*, which can be represented by a temporal integration when the conversion process is continuous or by another operator otherwise. As it is simpler to represent derivations with algebraic symbols than integrations, it will be made mainly using the reciprocal $\hat{T}^{-1}$ instead of the $\hat{T}$ operator.

$$\text{Continuous conversion :} \quad \hat{T}^{-1} \underset{cont}{=} \frac{d}{dt} \tag{9.14}$$

Its Formal Graph representation obeys the same convention as for the system properties (see Graph 9.3).

9.2.2 Uniqueness of Time

Once this mathematical form for the evolution operator is established, one has to show that the same operator works for all energy subvarieties. Let us take a system (a dipole assembly) containing the two subvarieties of the same energy variety. The total energy is the sum of energies of the subvarieties, according to the additive property of energy defined in Chapter 2:

$$\mathcal{E}_q = \mathcal{T}_q + \mathcal{U}_q \tag{9.15}$$

The total energy variation of the system is obtained by replacing each energy subvariety given by Equations 9.1 and 9.2

$$d\mathcal{E}_q = e_{qC} dq + f_{qL} dp_q \tag{9.16}$$

Let us suppose that the advances of each energy conversion are quantified by different times t_C and t_L:

Hypothesis 1: $$dt_C = \frac{dq}{f_{qC}} \tag{9.17}$$

Hypothesis 2: $$dt_L = \frac{dp_q}{e_{qL}} \tag{9.18}$$

The energy variation (Equation 9.3) becomes

$$d\mathcal{E}_q = e_{qC}\, f_{qC}\, dt_C + f_{qL}\, e_{qL}\, dt_L \tag{9.19}$$

Depending on the assembling mode, one writes

Common dipole flow: $$d\mathcal{E}_q = f_q \left(e_{qC} dt_C + e_{qL} dt_L\right) \tag{9.20}$$

Common dipole effort: $$d\mathcal{E}_q = e_q \left(f_{qC} dt_C + f_{qL} dt_L\right) \tag{9.21}$$

and assembling rules (generalized Kirchhoff's rules) are expressed by

Common dipole flow: $$e_q = e_{qC} + e_{qL} \tag{9.22}$$

Common dipole effort: $$f_q = f_{qC} + f_{qL} \tag{9.23}$$

The embedding principle, saying that one should express the total energy variation in a concise form by using state variables of the global system and therefore a time t specific to the global system too, is written for both assembling modes:

$$d\mathcal{E}_q = e_q\, f_q\, dt \tag{9.24}$$

This unique writing expresses the principle that the energy variation in the two dipoles must be independent of the way they are assembled. Comparison with energy variations (Equations 9.20 and 9.21) according to each assembling mode, leads to the following equalities:

Common dipole flow: $$e_q dt = e_{qC} dt_C + e_{qL} dt_L \tag{9.25}$$

Common dipole effort: $$f_q \mathrm{d}t = f_{qC} \mathrm{d}t_C + f_{qL} \mathrm{d}t_L \quad (9.26)$$

Whatever the assembling mode, the only possibility, besides an absence of time variation, for satisfying both assembling rules and the embedding principle is to equate the two time variables to the global one

$$\mathrm{d}t = \mathrm{d}t_C = \mathrm{d}t_L \quad (9.27)$$

which proves that time is independent of the nature of the conversion it accompanies; that is, it is fundamentally a transversal concept through all energy subvarieties. This is a characteristic borne by space–time properties, eventually locally defined, but independently from any system.

This demonstration has been carried out in the frame of a single energy variety, but extension to any mixing of varieties is straightforward once the concept of *energy coupling* is explained, as no time is involved in this notion (see Chapter 12).

9.2.3 Gate Variables

Applied to both energy subvarieties, the unique evolution operator leads to the definition of two *energies-per-entity* acting as *gate variables* (see Chapter 6, Section 6.2.3) referenced by their origin (C or L) as follows:

Capacitive gate variable: $$f_{qC} \overset{def}{=} \hat{T}^{-1} q \underset{cont}{=} \frac{\mathrm{d}}{\mathrm{d}t} q \quad (9.28)$$

Inductive gate variables: $$e_{qL} \overset{def}{=} \hat{T}^{-1} p_q \underset{cont}{=} \frac{\mathrm{d}}{\mathrm{d}t} p_q \quad (9.29)$$

Examples of couples of state variables linked by the first space–time property are numerous, as such relationships are found in every energy variety:

Electric current: $$I = \frac{\mathrm{d}Q}{\mathrm{d}t} \quad (Q\text{: Electric charge}) \quad (9.30)$$

Displacement velocity: $$\boldsymbol{v} = \frac{\mathrm{d}\boldsymbol{\ell}}{\mathrm{d}t} \quad (\boldsymbol{\ell}\text{: Displacement}) \quad (9.31)$$

Angular velocity: $$\boldsymbol{\Omega} = \frac{\mathrm{d}\boldsymbol{\theta}}{\mathrm{d}t} \quad (\boldsymbol{\theta}\text{: Angle}) \quad (9.32)$$

Substance ("mass") flow: $$\Im = \frac{\mathrm{d}n}{\mathrm{d}t} \quad (n\text{: Substance amount}) \quad (9.33)$$

Reaction rate (or flow): $$v = \frac{\mathrm{d}\xi}{\mathrm{d}t} \quad (\xi\text{: Advance}) \tag{9.34}$$

Entropy production rate: $$f_S = \frac{\mathrm{d}S}{\mathrm{d}t} \quad (S\text{: Entropy}) \tag{9.35}$$

Examples of couples of state variables linked by the second space–time property are found less frequently than for the previous one, as impulses are not used in every energy variety:

Electric potential: $$V = \frac{\mathrm{d}\Phi_B}{\mathrm{d}t} \quad (\text{Faraday's law of induction}) \tag{9.36}$$

Mechanical force: $$\boldsymbol{F} = \frac{\mathrm{d}\boldsymbol{P}}{\mathrm{d}t} \quad (\text{Newton's second law in mechanics}) \tag{9.37}$$

Mechanical torque (couple): $$\boldsymbol{\tau} = \frac{\mathrm{d}\boldsymbol{L}}{\mathrm{d}t} \quad (\boldsymbol{L}\text{: Angular momentum}) \tag{9.38}$$

Hydrodynamical pressure: $$P = \frac{\mathrm{d}F}{\mathrm{d}t} \quad (F\text{: Percussion}) \tag{9.39}$$

Generally, some of these laws are given with a minus sign in arguing a simple matter of convention. This is in fact a more subtle question and the main reason comes from the classical confusion between the different origins of efforts or flows (i.e., whether they participate in the energy storage or they are issued from an energy conversion). This point will be discussed later in the case of an oscillator circuit.

Which notation? $\frac{dQ}{dt}$ or $\frac{\mathrm{d}}{dt}Q$? These are equivalent ways of writing the evolution operator and its operand (the variable on which the operator applies). The first one is the classical way of expressing the derivative of a variable, by making the ratio of two differential variations. This notation has been used above for pedagogical reasons, in trying not to change the habit of the reader. However, the recommended notation is the second one, because it clearly puts into evidence the distinction between operator and operand, which is one of the basic principles on which Formal Graphs are built.

9.2.4 Power as Energy Conversion Rate

Power is defined as the rate of evolution of energy in a system in a given energy variety. Its unit is the Watt which corresponds to the Joule per second.

Received power: $$\mathcal{P}_q \overset{def}{=} \frac{\mathrm{d}}{\mathrm{d}t}\mathcal{E}_q \tag{9.40}$$

The positive implicit sign in this definition means that power is counted as it is received by the system, because it quantifies an increase of energy in the system, which is therefore a receptor. As a generator give energy to the outside, power is counted negatively in this case.

No distinction is made between energy subvarieties for defining power. It refers to the total energy, encompassing systematically inductive and capacitive subvarieties. This confers to this notion a status of global variable describing the conversion rate between systems independently from their inductive, capacitive, or conductive nature. However, this does not forbid individualizing it by subscripting the symbol when distinction is requested.

As time does not depend on the system features, the power variable receives all attributes of energy. It is an extensive and additive variable, the total power being the sum of partial powers and of individual powers quantifying the energy variations according to all energy varieties contained in the system. For a set of dipoles of various kinds, the total received power is therefore

$$\mathcal{P} = \sum_{q,i} \mathcal{P}_{qi} \tag{9.41}$$

the summation being done over all energy varieties q and natures of dipoles indexed by i.

For a capacitive dipole, definitions of the capacitive energy variation in Equation 9.1 and the capacitive gate variable in Equation 9.28 lead to the following expression for the received power:

$$\mathcal{P}_{qC} = \frac{\mathrm{d}}{\mathrm{d}t}\mathcal{U}_q = f_{qC}\, e_{qC} \tag{9.42}$$

For an inductive dipole, from the definition of the inductive energy variation in Equation 9.2 and the inductive gate variable in Equation 9.29, the following expression for the received power is obtained:

$$\mathcal{P}_{qL} = \frac{\mathrm{d}}{\mathrm{d}t}\mathcal{T}_q = f_{qL}\, e_{qL} \tag{9.43}$$

For a conductive dipole, energy of the considered variety $\mathcal{E}_q$ is not stored but only received. The power of a conductive dipole is therefore always positive. To find the relationship between power and the energies-per-entity of the conductive dipole, one considers a system composed of three dipoles, one of each nature. The two energy storing dipoles (seen as equivalent to capacitor and inductor components) globally behave as a generator giving energy to the conductive dipole. The three previous equations are combined to express the total power

$$\mathcal{P}_q = e_{qL} f_{qL} + e_{qC} f_{qC} + \mathcal{P}_{qG} \tag{9.44}$$

Now, let us suppose that the three dipoles are assembled with a common flow, which corresponds to a serial mounting as shown in Figure 9.4 in the language of equivalent circuits.

The two algebraic equations describing the mounting are written as

$$f_q = f_{qL} = f_{qC} = f_{qG} \tag{9.45}$$

$$e_q = e_{qL} + e_{qC} + e_{qG} \underset{isol}{=} 0 \tag{9.46}$$

having encoded the isolation of the circuit from the external world by setting to zero the total effort. Then, in setting the total power also to zero and by replacing in Equation 9.44 all variables

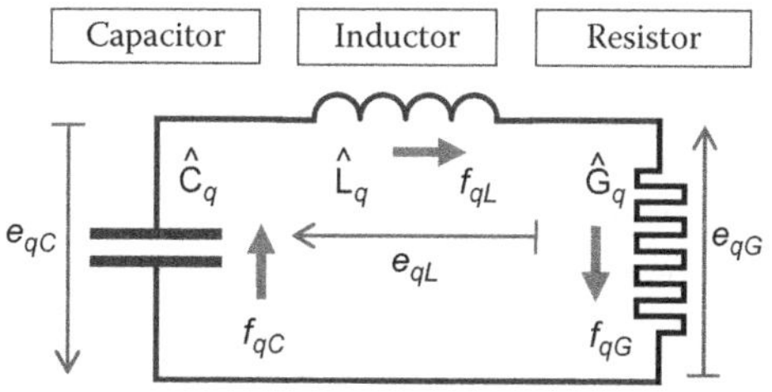

FIGURE 9.4 Equivalent circuit made up of three dipoles, capacitive, inductive, and conductive, sharing their flow.

that are relevant from the capacitive and inductive dipoles by using the two previous equations, one finds

$$(0 - e_{qG})f_q + \mathcal{P}_{qG} = 0 \tag{9.47}$$

This intermediate equation provides the expression for the power received by a conductive dipole

$$\mathcal{P}_{qG} = f_{qG} e_{qG} \tag{9.48}$$

This result has been obtained with a serial mounting; it could have been obtained as well with a parallel mounting and with any number of capacitive or inductive dipoles. This is an important relationship, which, by comparison with the expressions 9.42 and 9.43, establishes the property of the power to be given by the product flow × effort whatever the nature of the dipole.

For a set of dipoles of various kinds, the total received power is therefore

$$\mathcal{P} = \sum_{q,i} f_{qi} e_{qi} \tag{9.49}$$

Note that when the system is isolated, the received power is naturally zero, as it has been considered in the previous dipole assembly. In such a situation, Equation 9.49 set to zero is known as Tellegen's theorem (Tellegen 1952).

Time and equilibrium. A frequent misunderstanding about the relationship between equilibrium and evolution is met in science. A noticeable example is the conception of thermodynamics as relevant from static conditions in opposition to kinetics which corresponds to dynamic conditions.

With the late introduction of time in the Formal Graph approach, well after having modeled influence and exchange phenomena, this distinction is not pertinent. The important point to stress is that it is not because poles, dipoles, and multipoles have been defined and modeled without any reference to time that all their properties are valid only in static conditions. They work exactly in the same way when time becomes a significant parameter in models. In other words, but not in restricting this assertion to this single domain, all concepts of thermodynamics are valid in dynamic conditions. It is too early to discuss this point, but the traditional approach of the concept of *reversibility* consisting of linking this concept to infinitely slow kinetics is totally misleading in view of the present approach. No separation between equilibrium and far from equilibrium is made in the Formal Graph theory. This is an important conceptual step forward.

9.3 FORMAL GRAPH REPRESENTATION OF A DIPOLE ASSEMBLY

As the dipole assembly is made up of dipoles (or multipoles) all belonging to the same energy variety, no supplementary variables and properties are required for modeling it, other than the already considered state variables. This would normally mean that the embedding principle can be used to represent a dipole assembly by the same structure of a Formal Graph as for the dipole or multipole. However, this simple principle is insufficient for modeling all possible ways to assemble several dipoles. What a single square unit (up to four nodes) may model is an assembly, called *elementary dipole assembly*, in which only one common node is shared by the dipoles. This means that all dipoles must have either one common effort or one common flow to be represented by the Formal Graph square unit and when several common nodes are required, several units need to be associated for modeling the assembly. The basic rule is:

1 common node = 1 elementary dipole assembly = 1 square unit Formal Graph

Naturally, when several units are used, they need to be connected together and this is ensured by supplementary square units, on the same template as for the dipoles constituting a multipole. This case of more complex structures will be tackled in subsequent chapters. In this chapter, only elementary dipole assemblies will be treated, so elementary Formal Graphs (square unit) will be sufficient.

To discuss the way to represent a dipole assembly having in common only one energy-per-entity (i.e., elementary), we take again the case of three dipoles, one capacitive, one inductive, and one conductive.

9.3.1 Common Flow

The first case, a common flow, corresponds to a *serial mounting* of components, and the mounting equations are written by notating f_q the common flow and by taking into account an external source of effort notated e_q

Common flow:
$$f_q = f_{qL} = f_{qC} = f_{qG} \tag{9.50}$$

Powered effort:
$$e_q = e_{qL} + e_{qC} + e_{qG} \tag{9.51}$$

This allows the received power (Equation 9.44) to be written in the concise form

$$\mathcal{P}_q = e_q f_q \tag{9.52}$$

When the RLC circuit is isolated, the received power is null, so the second mounting equation is written as

$$e_{qL} + e_{qC} + e_{qG} \underset{isol}{=} 0 \tag{9.53}$$

We recall that this situation is referred to as a *static equilibrium* in many domains (Newton's third law of equilibrium of forces in mechanics, for instance), but not in all domains, where it is seen as a *conservation law* of signed quantities (Kirchhoff's law of addition of tensions (voltages) in electricity, for instance).

9.3.2 Common Effort

In the second case, a common effort, or *parallel mounting* of components in the circuit language, the mounting equations are

Common effort:
$$e_q = e_{qL} = e_{qC} = e_{qG} \tag{9.54}$$

Powered flow: $$f_q = f_{qL} + f_{qC} + f_{qG} \tag{9.55}$$

and the received power (Equation 9.44) is written naturally in the same concise form as before

$$\mathcal{P}_q = e_q f_q \tag{9.56}$$

Again, when the RLC parallel circuit is isolated, the received power is null, and the only solution for an effective conversion of energy (non-null effort) is a resulting flow equal to zero

$$f_{qL} + f_{qC} + f_{qG} \underset{isol}{=} 0 \tag{9.57}$$

9.3.3 Formal Graphs

Formal graphs of the two mounting ways are coded as usual by indication of the resulting state variable with a polygonal thick line grouping all summed contributions and by writing the symbol of the performed algebraic operation in the vicinity.

The two Formal Graphs in Graph 9.4 model the most general dipole assemblies using only one common node which are composed of the three natures of dipoles simultaneously. Simpler Formal Graphs using fewer properties are drawn when a lower number of dipole natures form the assembly, as will be shown in this chapter.

To illustrate how *elementary dipole assemblies* are modeled by a Formal Graph, two cases are worth presenting before discussing several case studies. The first case deals with the subject of the isolation or the connection of a dipole assembly with an external source of power; the second case deals with the representation of the most important system treated in this chapter, which is the oscillator.

9.3.4 Power Supply

An effort or a flow, which is a gate for communicating with the exterior, can be imposed (or supplied) by an external system (another dipole or dipole assembly, for instance). An example of external effort is when a force is imposed on a mass placed in a gravitational field. An external flow may correspond to the convection phenomenon, when a fluid transporting an object imposes its own velocity.

When nothing is imposed on efforts or flows, one of them must be equal to zero. The other may also be equal to zero, but this depends on the ability of the system to store energy. If an inductance or a capacitance is present in the system, the other energy-per-entity is different.

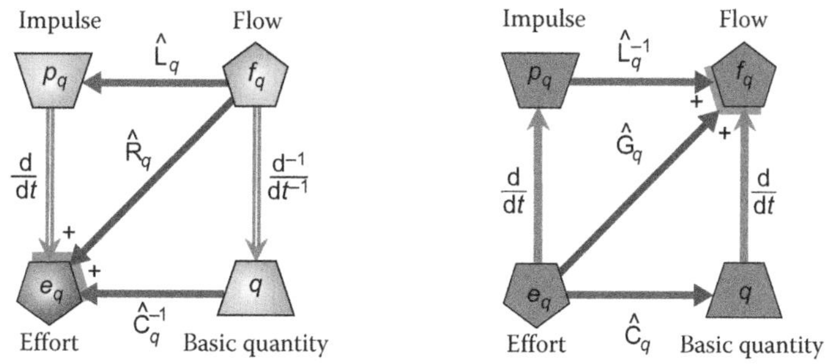

GRAPH 9.4 Serial RLC (common flow and summed effort) dipole assembly (left) and parallel RLC (common effort and summed flow) dipole assembly (right). No indication is made about the isolation or the connection with a power supply in these models.

To impose an effort or a flow means that power is transferred from the exterior to the system. Only one energy-per-entity can be imposed, the other being free to adapt to the conditions. The imposition of both energies-per-entity simultaneously would be a violation of thermodynamical rules and would entail the destruction of the system, unable to cope with this constraint.

This subject could have been tackled before, when dipoles or multipoles were treated, because the possibility for these systems to work whether alone or in conjunction with other systems also existed. However, this was not really necessary because the subject could be treated in an implicit manner by giving literal symbols to all nodes without specifying whether or not they corresponded to variables with zero values.

When modeling a dipole, its two energies-per-entity, effort e_q and flow f_q, are used in a general way, without taking care of the possibility of one of them being imposed by an external source or being equal to zero due to the isolation of the dipole from any source.

When modeling a dipole assembly, the question needs to be taken into account because such a system may work whether in an autonomous manner or in relationship with another system, considered as external in this case. When power is supplied to the dipole assembly, it influences the way energy is exchanged or converted between dipoles. The consequence is that the energetic behavior is different according to the isolation or the connection of the assembly to exterior.

The external source, also called generator, is considered as a dipole belonging to a different energy subvariety with respect to the other dipoles in the assembly, which is considered as a receptor. The energies-per-entity of the generator are connected to the corresponding energies-per-entity of the receptor. The connectors are identity operators as both efforts are equal as is the case for both flows. The node of the receptor on which the connection from the generator arrives is the powered energy-per-entity. The other connection links the energy-per-entities that are common to the generator and the receptor and defines the mounting mode. Its orientation is opposite to the previous link between powered energies-per-entity.

Normally two square units are necessary for representing these connections in a Formal Graph but a simplification can be brought by omitting the square unit of the external source. As values are identical between the generator and the receptor, it is enough to represent the receptor only, but in coding it so as to take into account the constraint on the powered energy-per-entity. For this, a circle is added around the node, as shown in Graph 9.5 for the case of a generator of effort.

This indication is necessary when using a Formal Graph for simulation owing to its similarity with neural networks used in computation. For theoretical studies—for just modeling the physical law governing a process—the fact that a dipole assembly is powered or not may be less important.

Graphs 9.6 and 9.7 give in the three languages—equivalent circuit, algebraic equation, and Formal Graph—the models of a powered dipole, or dipole assembly, featured by an admittance Y_q.

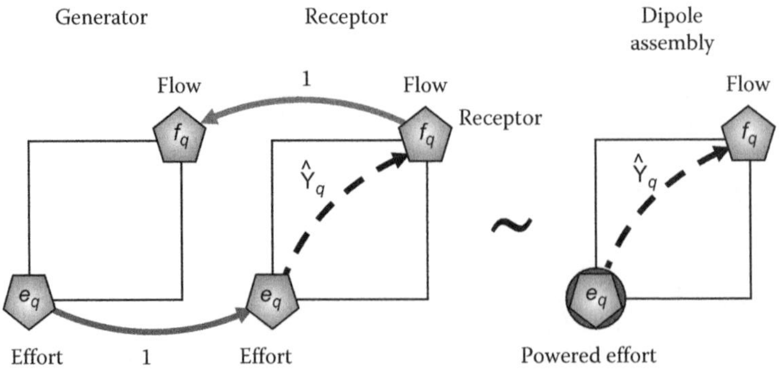

GRAPH 9.5 Equivalence between the two Formal Graphs of a dipole assembly with an external generator imposing an effort. The Formal Graph on the right is the compact version in which the circle around the effort indicates that it is imposed.

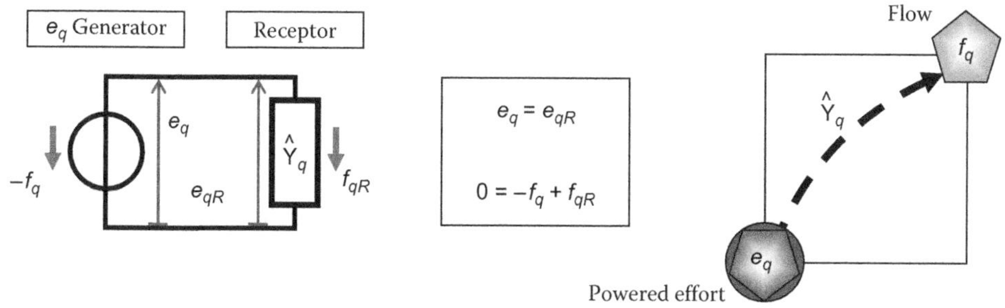

GRAPH 9.6 External (powered) effort (common effort).

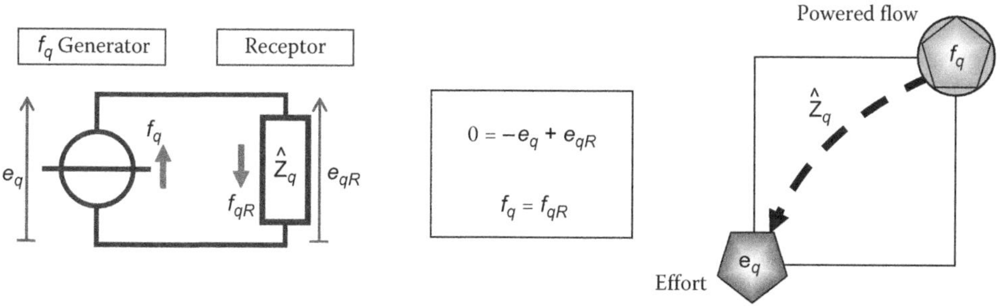

GRAPH 9.7 External (powered) flow (common flow).

The two possible cases of imposition of an energy-per-entity, whether effort or flow, are distinguished. Graph 9.6 features the case of a common flow that allows the effort to be supplied, whereas Graph 9.7 corresponds to the other case of common effort and powered flow.

In both equivalent circuit and algebraic equations, the subscript R refers to the receptor, because distinction between the generator and the receptor is necessary in those languages. This is not necessary in the Formal Graph due to the identity of energies-per-entities (by pair) between generator and receptor.

Note that in the electrical circuit language, the power source is represented by two different names and symbols, according to international conventions. The *tension generator* is symbolized with a circle and a diameter linking the input and the output, whereas in the *current generator* the diameter (slightly protruding the circle) is orthogonal.

9.3.5 Oscillators

A system worth studying is the oscillator. The temporal oscillator uses only time and no space; the converse is true for the spatial oscillator. The spatiotemporal oscillator combines both space–time properties for modeling the propagation of waves: sound, light, or mechanical vibrations.

Truly essential notions. The interest of the detailed study of oscillators is to establish from scratch the *wave function*. This is one of the most important tools in quantum physics and to be able to know its foundations is of paramount interest. All the more, the wave function is laid down axiomatically in most textbooks as a specificity of quantum physics, which contributes to set apart this branch of physics.

(continued)

In the Formal Graph approach, the wave function is "naturally" introduced as a pertinent variable (and not exactly as a "function") endowed with the role of containing the information about the state of an oscillator at any time or for any space location.

To avoid obscure steps in this introduction, even the *imaginary number* is brought logically (i.e., demonstrated) as a useful factor for combining variables. For the same reason, the *exponential function* is entirely defined from its basic properties and harmonic (circular) functions are deduced. No prerequisite about these elementary mathematical notions is therefore required for understanding what a wave function is really.

The purpose, as can be guessed, is to introduce quantum physics on solid bases without any trick. Quantum operators representing the action of space–time on a wave function are merely defined as tools for introducing important notions such as the pulsation ω, the space vector $\boldsymbol{k}$, and the phase velocity $\boldsymbol{u}$. Contrary to the classical approach, they are defined as space–time properties, independently from the system they are connected with. This clear separation between system and space–time is a condition for understanding their interaction.

A temporal oscillator is made up of two energy storing dipoles of different energy subvarieties. One of the dipole must be inductive and the other capacitive. This is a compulsory requirement for allowing the system to oscillate as time elapses. The physical process provoking oscillations is the impossibility for a system to store simultaneously the two energy subvarieties without conversion from one subvariety to another. As shown in the Formal Graphs (Graph 9.8 and Graph 9.9) of an elementary dipole assembly, the two evolution operators work in opposite directions, meaning that when conversion occurs from basic quantity into flow, capacitive energy is converted into inductive energy. The opposite conversion from impulse into effort occurs concomitantly, converting inductive energy into capacitive energy. In an isolated oscillator, the total energy remains constant, so the amount in one energy subvariety is always the complement of the other (see Figure 9.5).

A spatial oscillator works on the same principle: Energy is distributed in space either among localized entities or among localized energies-per-entity, but not both on the same location. The model of a spatial oscillator is completely similar to that of the temporal oscillator, only the names of the variables are changed. For this reason, all reasoning will be made for the temporal oscillator.

As will be demonstrated in this chapter, the rate at which oscillations occur in a temporal oscillator depends on both inductive and capacitive properties of the oscillator, indicating that there exists a correlation between time and the ability of a system to store more or less energy. It will be

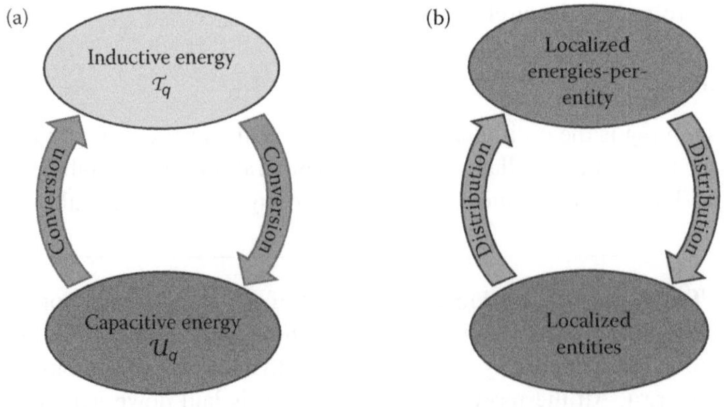

FIGURE 9.5 Schemes of the principle of an oscillator. A temporal oscillator converts energy from one subvariety to another and vice versa (a). A spatial oscillator distributes alternatively energy among localized entities and localized energies-per-entity (b).

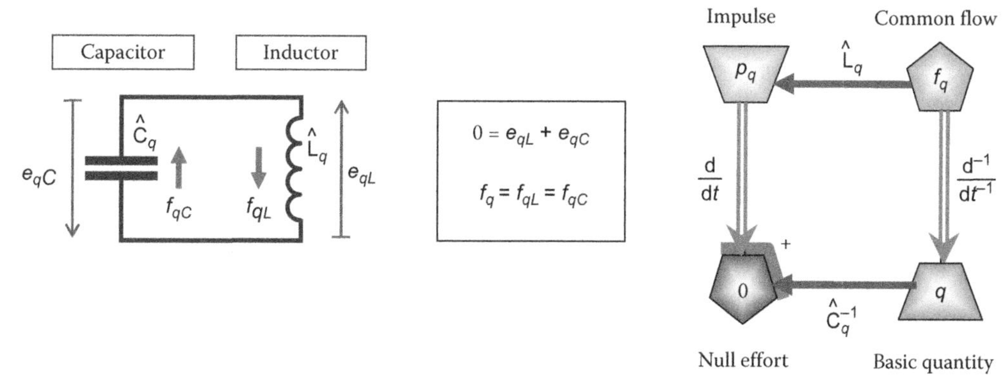

GRAPH 9.8 Isolated LC oscillator in serial assembly (common flow).

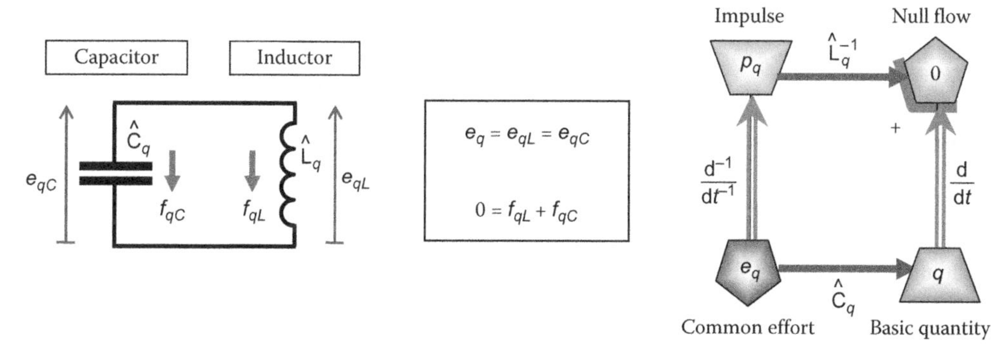

GRAPH 9.9 Isolated LC oscillator in parallel assembly (common effort).

seen that the more energy can be stored for a given effort and a given flow, the lower is the frequency of the oscillations, meaning that it takes more time to convert more energy.

As can be expected, there exist two methods of mounting the inductive dipole with a capacitive dipole, whether the common node is the flow or the effort. However, for an isolated oscillator, this does not make a difference, as shown in Graphs 9.8 and 9.9 which give in the three usual languages the representations of the serial mounting and of the parallel mounting.

It can be remarked that the difference between Graphs 9.8 and 9.9 is clearly visible in the Formal Graphs but is only perceptible in the equivalent circuit through the (optional) indications of the effort and flow directions.

Analogies between temporal oscillators are frequently discussed topics in physics. A certain number of simple examples are given in Table 9.3. The characterization of each system by its generalized capacitance and inductance (eventually space reduced) is the proof that the respective energy subvarieties are supported by the system. In this table, all these system constitutive properties are assumed to be scalars or vectors, but not operators. The notion of analogy is not to be understood here as a fortuitous coincidence but as recognition of the same fundamental mechanism of energy conversion.

9.4 CASE STUDIES OF DIPOLE ASSEMBLIES

The first case study is devoted to an important subject in mechanics which is the acceleration of a body under the effect of an external force. It offers the opportunity to precisely define several important concepts that are linked to Newton's second law of movement, such as the *acceleration* and the *distance* covered by the object. As neither of these variable are state variables, the Formal Graph cannot be a word-for-word translation of the classical model. This treatment evidences the requirement of a

TABLE 9.3
A Few Examples of Various Temporal Oscillators

System	Capacitance C_q	Inductance L_q
Mass and spring	$1/k_e$ k_e: Spring constant	M Inertial mass
Torsion pendulum	$1/C_\theta$ C_θ: Torsion constant	J Inertia
Vibrating string	$1/\boldsymbol{T}_\ell$ $\boldsymbol{T}_\ell$: Mechanical tension	M/ℓ Lineic mass density
Capacitor and inductor	C Electric capacitance	L Self-inductance
Expanding gas in an elastic envelop	$1/\chi$ χ: Envelop compressibility	ρ_M $\rho_M = M/V$ volumic mass
Pendulum under gravity	$\ell/M_G\,\boldsymbol{g}$ M_G = Gravitational mass $\boldsymbol{g}$: Gravitational acceleration ℓ: Pendulum length	M Inertial mass (= Gravitational mass)
Plasma sphere in a potential well	$2\mathcal{U}_{well}/r^2$ r: Curvature of potential well	M_{ion} Ionic mass

constant inertial mass for applying Newton's theory. Incidentally, thermodynamics is used for analyzing the acceleration occurring in circular motions. The First Principle of Thermodynamics leads directly to the relationship between the *centripetal force* and the velocity in a circular uniform motion.

The other case studies are all based on the exchange of energy between the two subvarieties, according to the three ways of considering the space–time: purely temporal, purely spatial, and both in one- and three-dimensional space. The systems studied are temporal and spatial oscillators and wave propagations in the spatiotemporal case.

The two last case studies belong to electrodynamics (as for case study F6 "Light Propagation") and are concerned with Maxwell equations in free space and to the Hall effect in a conducting material. The latter is a process working through the synchronization of two space–time velocities, which is an example of internal coupling. (This subject will be tackled in Chapter 13.)

Unless otherwise mentioned, all systems are isolated.

Temporal Inductive

- F1: Accelerated Motion in Translational Mechanics (Powered)

Temporal Inductive + Capacitive

- F2: LC Circuit in Electrodynamics
- F3: Torsion Pendulum in Rotational Mechanics

One-dimensional Spatiotemporal Inductive + Capacitive

- F4: Vibrating String in Translational Mechanics

Three-dimensional Spatiotemporal Inductive + Capacitive

- F5: Sound Propagation in Hydrodynamics
- F6: Light Propagation in Electrodynamics
- F7: Maxwell Equations in Electrodynamics
- F8: Hall Effect in Electrodynamics (Powered)

9.4.1 Case Study F1: Accelerated Motion — Translational Mechanics

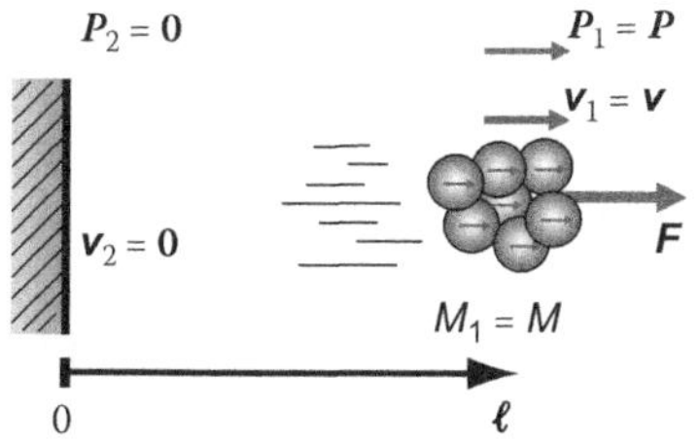

Formal Object:

Dipole assembly

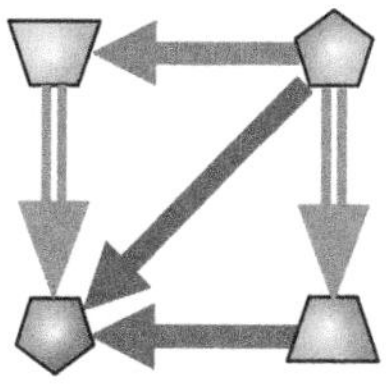

Dipole Assembly	
✓	Fundamental
	Mixed
	Capacitive
✓	Inductive
	Conductive
✓	Common effort
	Common flow
✓	Global
	Spatial

Formal Graph:

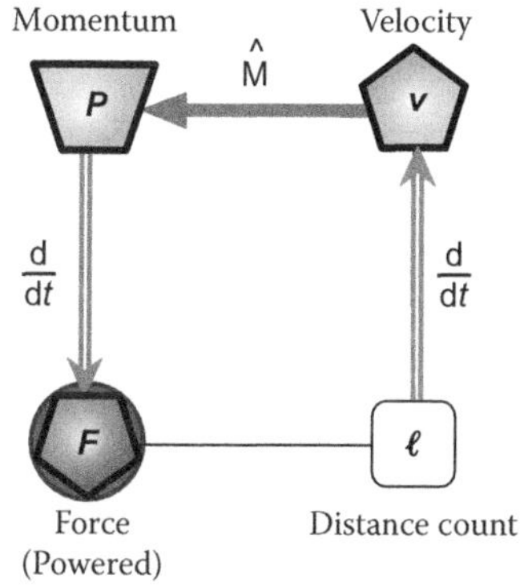

GRAPH 9.10

The circle around the node of the force is a convention for an external force applied ("powered"). The drawing convention for the distance node means that this variable is neither a state variable nor a gate variable but a counted variable (or merely a count).

Generator + receptor dipoles:

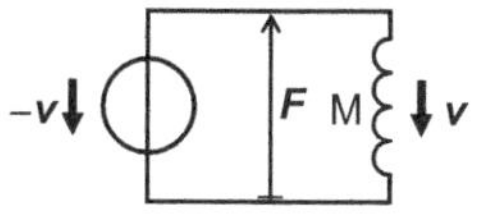

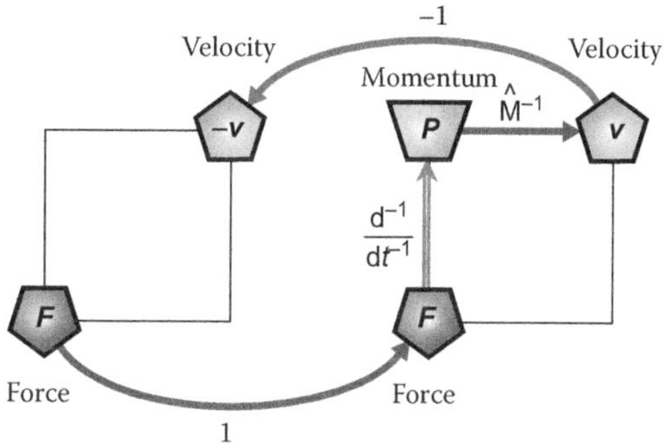

GRAPH 9.11

Variables:

Variety	Translation Mechanics		
Subvariety	Inductive (kinetic)		Capacitive
Category	Entity number	Energy/ entity	Energy/ entity
Family	Impulse 	Flow 	Effort 
Name	Momentum (quantity of movement)	Velocity	Force
Symbols	***P***, (***p***)	v	F
Unit	[N s], [J s m^{-1}]	[m s^{-1}]	[N], [J m^{-1}]

Property relationships:

$$P = \hat{M} v$$

$$F = \frac{dP}{dt}$$

Distance count

$$\ell = \frac{d^{-1}}{dt^{-1}} v = \int_0^t v \, dt$$

System Property	
Nature	Inductance
Levels	0
Name	Inertial mass
Symbol	M
Unit	[kg]
Specificity	—

Energy variation:

$$d\mathcal{T}_\ell \underset{\exists M}{=} v . dP$$

When an external force is applied to a moving body, it accelerates the body. This external force can be for instance a gravitational field, an electric field, or an electromagnetic field applied on a charged body.

Why is this case study relevant for a dipole assembly? One may wonder about the place of this case study in this chapter, instead of in Chapter 6 devoted to dipoles. This is due to the external force that is seen to be generated by a specific dipole (a generator of effort, similar to a generator of tension in electrodynamics). As the moving body is modeled as an inductive dipole, the whole system is considered as an assembly of two dipoles exchanging energy. The equivalent electric circuit in the case study abstract is a model of such an assembly, built with a common effort (force), that is, as a parallel mounting. To simplify the Formal Graph, the generator dipole can be omitted by surrounding the node of the force by a circle, which is a convention recalling the symbol of a generator in the electric circuit language.

In translational mechanics, Newton's second law of motion establishes proportionality between the acceleration $\boldsymbol{a}$ endured by a moving body and the force $\boldsymbol{F}$ acting on this body. The proportionality factor is the inertial mass M.

$$\boldsymbol{F} \underset{lin}{=} M \boldsymbol{a} \qquad \text{(F1.1)}$$

The acceleration $\boldsymbol{a}$ is defined as the derivative of the displacement velocity $\boldsymbol{v}$ with respect to time or, alternatively, as the temporal second derivative of the displacement $\boldsymbol{\ell}$ itself:

$$\boldsymbol{a} \overset{def}{=} \frac{\mathrm{d}^2}{\mathrm{d}t^2} \boldsymbol{\ell} \qquad \text{(F1.2)}$$

A translation of this law into a Formal Graph could give Graph 9.12.

In fact this is not an acceptable representation because a Formal Graph accepts in its nodes only state variables or gate variables (cf. Chapter 4) and the acceleration $\boldsymbol{a}$ is neither a state variable (it does not participate in the determination of the quantity of energy in the system) nor a gate variable (it has no correspondent variable in other energy varieties). Moreover, it would be necessary to have a node at our disposal in the prolongation of the evolution path linking the displacement $\boldsymbol{\ell}$ to the displacement velocity $\boldsymbol{v}$. This would make an exception with respect to the other Formal Graphs in different energy varieties, and, at last, it would be a break in the interesting symmetry of the square pattern adopted for all *canonical Formal Graphs* (meaning that they follow the rule of placing only state variables, gate variables, and localized variables, in the nodes).

Respecting this canonical rule demands therefore building only graphs based on a square pattern with the four families of state variable from Helmholtz. For translational mechanics, they are represented by the displacement $\boldsymbol{\ell}$ (basic quantity), the force $\boldsymbol{F}$ (effort), the momentum $\boldsymbol{P}$ (impulse or quantity of movement), and the velocity $\boldsymbol{v}$ (flow), as shown in the Formal Graph in the case study abstract. The inertial mass M plays the role of an inductive property of the system in linking the velocity $\boldsymbol{v}$ to the momentum (impulse) $\boldsymbol{P}$. The composed path starting from the displacement $\boldsymbol{\ell}$ and

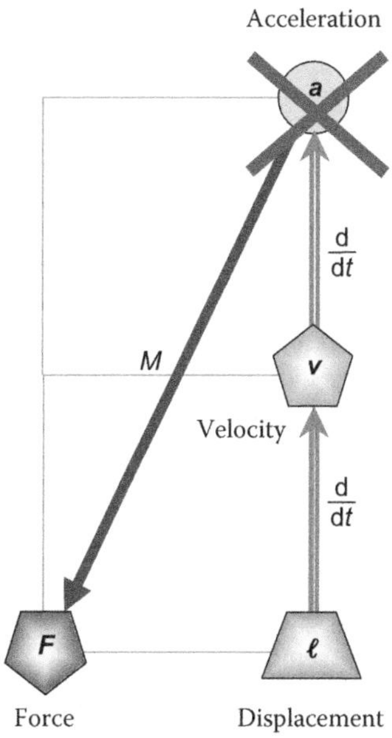

GRAPH 9.12

arriving onto the force $\boldsymbol{F}$ takes a first evolution (temporal derivation) path, then the inertial mass M, and finally a second evolution path. This is written algebraically as

$$\boldsymbol{F} = \frac{\mathrm{d}}{\mathrm{d}t}\hat{\mathrm{M}}\frac{\mathrm{d}}{\mathrm{d}t}\ell \tag{F1.3}$$

This relation F1.3 constitutes the generalized version of Newton's second law of motion, having a validity range that allows a dependence of the inertial mass on one of the state variables, which is compatible with the theory of space–time (special relativity) from Einstein* in which the mass depends on the displacement velocity.

Restriction to Newtonian frame. It is only in assuming constant the inertial mass M that Newton's second law as written in Equations F1.1 and F1.2 is retrieved from the expression F1.3. In the Formal Graph theory, one is therefore invited to indicate this restriction of validity range in writing a "*lin*" under the equal sign of Newton's second law (when classically composed with Equations F1.1 and F1.2), to recall the linearity condition of the operator that is an inertial mass.

$$\boldsymbol{F} \underset{lin}{=} M\frac{\mathrm{d}^2}{\mathrm{d}t^2}\ell \tag{F1.4}$$

* Albert Einstein (1879–1955): German, Swiss, American physicist; Bern, Switzerland and Princeton, USA.

9.4.1.1 Notion of Counted Distance

By definition, a dipole belongs to a single energy subvariety, capacitive or inductive (except the purely conductive dipole, which does not actually belong to an energy subvariety as it only dissipates energy). It means that a dipole cannot possess two kinds of entities; only one, basic quantity or impulse, is allowed.

Here, we have a momentum, which is the inductive entity (impulse), and a distance, which is a capacitive entity (basic quantity). As only one system constitutive property (the inertial mass) exists in this system, this dipole is an inductive dipole and the momentum $\boldsymbol{P}$ is clearly a state variable for this system. On the contrary, the distance $\boldsymbol{\ell}$ cannot be a state variable because no capacitance allows storing capacitive energy in the system.

Also, it cannot play the role of a gate variable because this role is played by the two energies-per-entity, the velocity $\boldsymbol{v}$ and the force $\boldsymbol{F}$. Recall that a gate variable is used as an input/output port for establishing connections with other dipoles and for exchanging or converting energy. No such processes are involved in Newton's second law.

If the distance $\boldsymbol{\ell}$ is neither a state variable nor a gate variable, it has normally no place in a canonical Formal Graph. However, by adopting a different drawing, a rounded square, one may depart from the canonical rule by representing even so a node in the Formal Graph.

Apart from this representation problem, it remains to discuss the physical meaning of this variable. This is typically the kind of question raised by the Formal Graph approach that is not met in the classical use of algebra, which does not bother so much about statuses of variables.

What we may observe is that the distance results from the knowledge of the velocity by integration along time.

$$\ell = \frac{\mathrm{d}^{-1}}{\mathrm{d}t^{-1}} \boldsymbol{v} = \int_0^t \boldsymbol{v}\, \mathrm{d}t \qquad \text{(F1.5)}$$

This result corresponds to the covered distance from an origin of time, in an analog manner to the passed charge in a current loop, measured by a counter located at a point of the circuit (see case study A3 "Current Loop" in Chapter 4). The plot given beside shows the case of a constant velocity, producing a distance increasing monotonically and linearly.

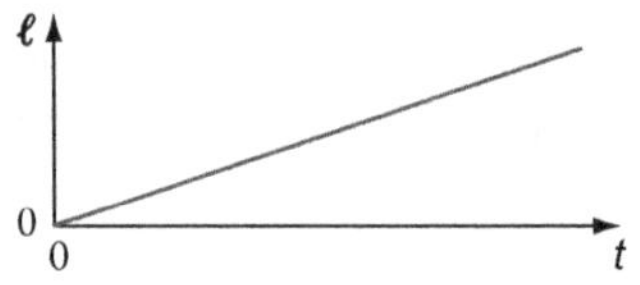

This indefinite growth is a proof that this variable is not a state variable, because with a constant velocity (and mass), the inductive (kinetic) and total energy of the moving body remains constant.

The general name given to this category of variable is "counted variable."

System constitutive property: As for each pole (see case study A1 "Moving Bodies" in Chapter 4), the velocity of the dipole is related to the dipole momentum through an inductive relationship using the inertial mass M of the dipole as inductance (which is an operator in the general case):

$$\boldsymbol{P} = \hat{M} \boldsymbol{v} \qquad \text{(F1.6)}$$

It is worth recalling that the mass, whether featuring a pole or a dipole, is an invariable scalar M (independent of another state variable) only in the frame of Newtonian mechanic, that is, for small velocities compared to the speed of light. This peculiar case is modeled by the well-known linear relation

$$\boldsymbol{P} \underset{lin}{=} M \boldsymbol{v} \qquad \text{(F1.7)}$$

Energy: The variation of inductive (kinetic) energy is given in an equivalent way from individual variations of the poles and from variation of the dipole, this is the embedding principle. However, in the peculiar case of an immobile pole, the dipole energy variation is equal to the mobile pole energy variation.

$$\mathrm{d}\mathcal{T}_\ell \underset{\exists M}{=} \boldsymbol{v}.\mathbf{d}\boldsymbol{P} \tag{F1.8}$$

It is recalled that the necessary condition for using this formula is the existence of mass in the body.

Evolution: The exchange of energy with the exterior is done by conversion into another subvariety through the force, equal to the temporal derivative of the momentum. This is indeed the correct way of writing Newton's second law in mechanics.

$$\boldsymbol{F} = \frac{\mathbf{d}\boldsymbol{P}}{\mathrm{d}t} \tag{F1.9}$$

Inversion of this relationship provides the evolution of the quantity of movement as a function of the applied force

$$\boldsymbol{P} = \frac{\mathrm{d}^{-1}}{\mathrm{d}t^{-1}}\boldsymbol{F} = \int_0^t \boldsymbol{F}\,\mathrm{d}t \tag{F1.10}$$

When this force is positive (same direction as the movement), the quantity of movement increases, and so does the velocity. This is the case depicted by the title of this case study. However, a negative force, which would decelerate the body, can also be handled with the same model without modification (in fact any shape of external perturbation).

9.4.1.2 Centripetal Force

The case of circular uniform motion is worth studying in view of the traditional role of acceleration in the derivation of the expression of the force that acts on the moving object.

The uniform circular motion is classically said to be accelerated owing to the regular change in direction of the velocity, thus creating a force tending to curb the object trajectory. This force is said to be *centripetal*,* signifying that the force is directed inward toward the center of curvature of the trajectory. The classical approach uses an acceleration, established by a geometrical reasoning involving distance covered during a time interval (Hecht 1996), and the force is deduced by applying Newton's second law (Equation F1.1):

$$\boldsymbol{a}_c = -\frac{\boldsymbol{v}^2}{\boldsymbol{r}}; \quad \boldsymbol{F} \underset{lin}{=} M\,\boldsymbol{a}_c \Rightarrow \boldsymbol{F} \underset{lin}{=} -\frac{M\boldsymbol{v}^2}{\boldsymbol{r}} \tag{F1.11}$$

The negative sign indicates the opposed directions of the force and the radius vectors, as can be seen in Figure 9.6.

This may appear surprising, but the Formal Graph in Graph 9.13 exactly models the system represented in Figure 9.6. Indeed, the presence of the four state variables of translational mechanics allows one to deduce how the total energy of this system may vary. A purely thermodynamical reasoning using the First Principle and its expression under the form of a Gibbs equation gives

* Centripetal comes from the Latin words centrum (center) and petere (to tend toward).

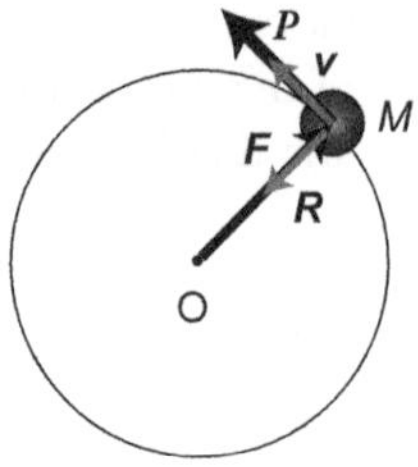

FIGURE 9.6 Circular motion of a mass M.

$$d\mathcal{E}_\ell = \boldsymbol{v} \cdot \mathbf{d}\boldsymbol{P} + \boldsymbol{F} \cdot \mathbf{d}\boldsymbol{R} \underset{isol}{=} 0 \tag{F1.12}$$

The variation of total energy is zero because the system, in contradistinction with the linearly accelerated motion seen previously, is isolated. Now, in a uniform motion the velocity and the radius have constant modules and the angle they form is a constant right angle, so the integration of the Gibbs equation provides the following equality between dot products

$$\boldsymbol{v} \cdot \boldsymbol{P} \underset{isol}{=} -\boldsymbol{F} \cdot \boldsymbol{R} \tag{F1.13}$$

which leads to the general expression

$$\boldsymbol{F} \underset{isol}{=} -\left(\boldsymbol{v} \cdot \widehat{M}\boldsymbol{v}\right)\frac{\boldsymbol{R}}{R^2} \tag{F1.14}$$

When linearity of the inertial mass operator is assumed, the classical expression is retrieved.

This simple reasoning based on thermodynamics is not usual but it shows that the notion of acceleration is not essential and can be put aside. It also highlights the nonrequirement of any notion of time for establishing this centripetal force.

However, it may be objected that the way we use the Gibbs equation is incorrect. Normally, for taking into account any energy subvariety in a Gibbs equation, the supporting system constitutive property must exist. This is the case for the inductive energy (kinetic) with the inertial mass but not for the capacitive energy in this case study, in which no capacitance or elastance is involved. The reason is that in fact the model is not complete. What is lacking is the other force counterbalancing

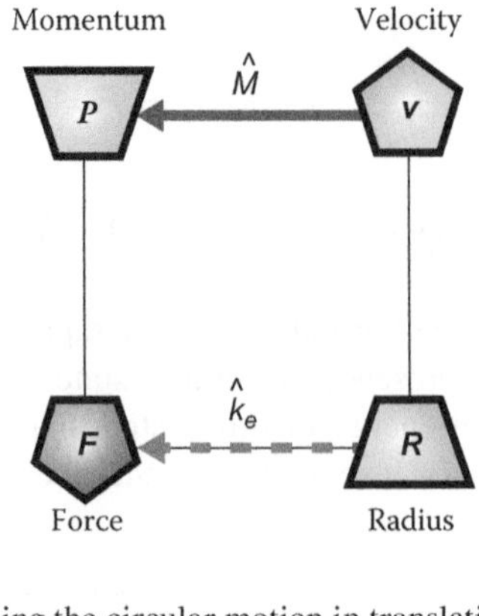

GRAPH 9.13 Formal Graph modeling the circular motion in translational mechanics.

the centripetal force in order to allow the system to be stable. For instance, a wire with elasticity or an electromagnetic field or a gravitational field are means of providing the lacking constitutive property to the system.

Remarks

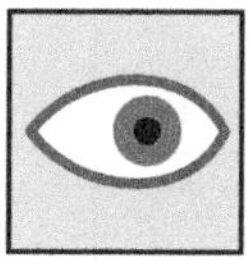

Methodology. This case study is instructive for many reasons.

First, it gives prominence to a fundamental difference between the two languages, algebraic and graphic, which is that a Formal Graph cannot represent all the variables handled by algebra. Not because this is technically impossible (one may always create a node for a variable and a link with another node), but because the rules of construction forbid it. Among the imperative rules, one is to build a Formal Graph called canonical only with state or gate variables of the system or their reduced variables (division by a geometric element—length, area, or volume). The acceleration of a movement in translational mechanics is neither a state variable nor a gate variable; it cannot therefore be placed in a canonical graph.

Second, the constraint of limiting the Formal Graph to a reduced set of variables (this a free choice of not representing all the variables that algebra can propose in profusion) is not a default, on the contrary. In effect, such a constraint does not forbid modeling a phenomenon, which is algebraically modeled with an "illegal" variable, but obliges one to privilege a different "legal" solution. Quite often, the solution respecting the rules of Formal Graphs brings an interesting, or even, new light on the behavior of energy or on the conditions of validity of the algebraic model. This is the case here, as the necessity of having a constant inertial mass for being able to utilize Newton's second law of movement is evidenced.

Unavoidable thermodynamics. One may ask for the reason that justifies these canonical rules and the constraints they induce. After all, it may appear presumptuous to pretend, one century after Einstein, that Newtonian mechanics is only valid in a restricted domain, with so little reasoning. The true reason comes from the prominence given in the Formal Graph theory to energy and to its specific properties, in other words to the strict respect of thermodynamics.

Banned variables. As other examples of variables that cannot be represented canonically, one may cite the "electric field flux" (see case study B5 "Gauss Theorem" in Chapter 5) or the "gradient of concentration" (see case study G3 "Fick's Law of Diffusion" in Chapter 10).

Acceleration is not pointless! It is not because this variable is issued from Newton's theory that it cannot be defined beyond. The following definition works normally in all cases:

$$\boldsymbol{a} \overset{def}{=} \widehat{\mathrm{M}}^{-1} \frac{\mathrm{d}}{\mathrm{d}t} \widehat{\mathrm{M}} \boldsymbol{v} \qquad \text{(F1.15)}$$

9.4.2 Case Study F2: LC Circuit

Electrodynamics

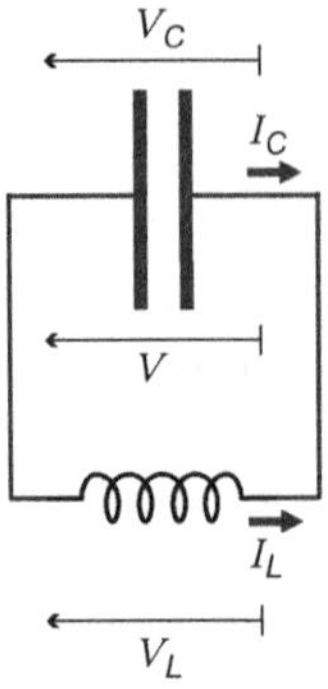

Formal Object:

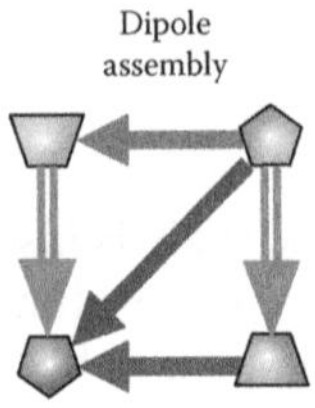

Dipole Assembly	
✓	Fundamental
	Mixed
✓	Capacitive
✓	Inductive
	Conductive
✓	Common effort
	Common flow
✓	Global
	Spatial

Formal Graph:

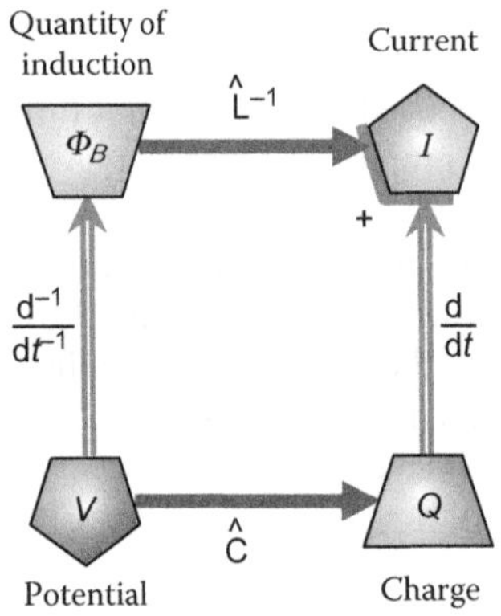

GRAPH 9.14

Variations with time:

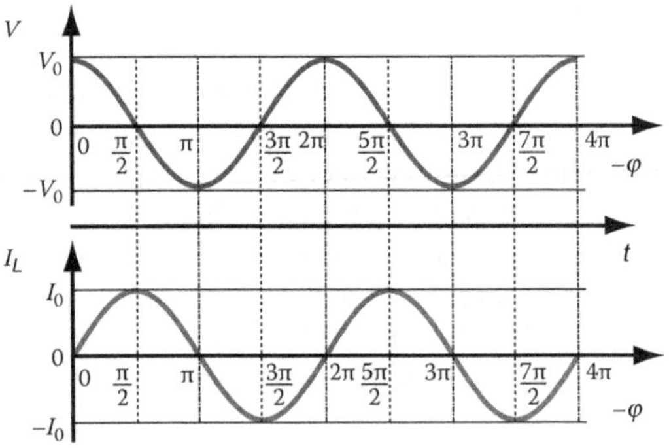

$$V = V_0 \cos \omega_Q t; \quad I_L = -I_C = I_0 \sin \omega_Q t$$

$$V = V_L = V_C; \quad \omega = \frac{-\varphi}{t} = \omega_Q \overset{def}{=} \frac{1}{\sqrt{LC}}$$

Isolated: $I = I_L + I_C \underset{isol}{=} 0$

Variables:

Variety	Electrodynamics			
Subvariety	Inductive (kinetic)		Capacitive	
Category	Entity number	Energy/ entity	Entity number	Energy/ entity
Family	Impulse	Flow	Basic quantity	Effort
Name	Quantity (flux) of induction	Current	Charge	Potential
Symbols	Φ_B	I	Q	V
Unit	[Wb]	[A]	[C]	[V]

System Properties		
Nature	Inductance	Capacitance
Levels	0	0
Name	Inductance	Capacitance
Symbol	L	C
Unit	[H]	[F]
Specificity	—	—

$$\Phi_B \underset{lin}{=} L\,I; \quad Q \underset{lin}{=} CV$$

Energy variation:

$$d\mathcal{E}_Q \underset{\exists L,C}{=} I\,d\Phi_B + V dQ$$

The association of an inductor with a capacitor allows the two subvarieties of electrodynamical energy, inductive (electromagnetic) and capacitive (electric or electrostatic), to be stored in the same system. The two system constitutive properties, inductance and capacitance, are the supports for the storage of energy and they link the state variables according to the following relations:

$$\Phi_B = \hat{L}\, I_L \underset{lin}{=} L\, I_L \tag{F2.1}$$

$$Q = \hat{C} V_C \underset{lin}{=} C V_C \tag{F2.2}$$

As the classical model of the electrodynamical oscillator is built in the frame of linear constitutive properties, the present case study is also developed in the linear case.

When conversion of energy occurs between the two subvarieties, the converted energies-per-entities are given by derivation of the entity numbers as follows:

$$I_C = \frac{d}{dt} Q \tag{F2.3}$$

$$V_L = \frac{d}{dt} \Phi_B \tag{F2.4}$$

Now, in this case study the choice is made to assemble the two dipoles according to a parallel mounting, or in the language of Formal Graphs, with a common potential. In addition, the assembly is isolated from external power supply, which results in writing the mounting equations

$$I = I_L + I_C \underset{isol}{=} 0 \tag{F2.5}$$

$$V = V_L = V_C \tag{F2.6}$$

Note that when an assembly of two dipoles is isolated, the parallel or the serial mountings are equivalent. (The only difference lies in the choice of the energy-per-entity for being the common one.) The Formal Graph in the case study abstract contains all the previous equations and represents therefore a general model of electric oscillator.

9.4.2.1 Algebraic Solution

The combination of the six Equations F2.1 through F2.6 gives the global equation

$$\hat{L}^{-1} \frac{d^{-1}}{dt^{-1}} V + \frac{d}{dt} \hat{C} V \underset{isol}{=} 0 \tag{F2.7}$$

By defining an *electric natural pulsation* ω_Q, also called natural angular frequency, encompassing in a single variable the system constitutive properties

$$\omega_Q \overset{def}{=} \frac{1}{\sqrt{LC}} \tag{F2.8}$$

and in restricting the model to the linear case and constant properties, the oscillator behavior is found as governed by this second-order differential equation

$$\frac{d^2}{dt^2}V + \omega_Q^2 V \underset{lin}{=} 0 \tag{F2.9}$$

Solving this equation consists of taking recourse to a known mathematical recipe that provides the following results:

$$V \underset{lin}{=} V_0 \cos \omega t \tag{F2.10}$$

$$I_L = -I_C \underset{lin}{=} I_0 \sin \omega t \tag{F2.11}$$

The reference to the electric pulsation has been dropped for generalizing the solution to any oscillator, electric or not. Physically speaking, this is justified by the distinction made in this generalizing approach between the oscillation generated by the oscillator, called *temporal wave* and featured by a *wave pulsation* ω, and the oscillator itself. When the wave generation is not accompanied by dissipation (i.e., the oscillator conserves its energy), both pulsations are equal:

$$\omega \underset{cst\,\mathcal{E}_Q}{=} \omega_Q \tag{F2.12}$$

The plots of the potential and current variations as a function of the time are given in the case study abstract. However, a more convenient variable proportional to time, the opposite of the phase angle φ, is used for the horizontal axis, allowing graduating in radians. Its relation with time is

$$\varphi = -\omega t \tag{F2.13}$$

The advantage of using the phase angle is to make the horizontal scale independent of the characteristics of the oscillator, featured by the system pulsation ω_Q. This may appear as a technical normalization but it relies on physical grounds explained in Section 9.5. The same phase angle is used for any energy variety and will be used again with the other oscillators, spatial and spatiotemporal. After the case studies, in Section 9.5, the detailed demonstration of the above solution will be given in the general case (for any energy variety).

Remarks

Oscillator natural pulsation and wave pulsation. The Formal Graph theory, by construction, makes a clear distinction between variables featuring the system, such as the constitutive properties, and variables belonging to energy, such as state variables. This distinction is extended to the pulsation, which is classically considered as a unique variable, but is separated in the Formal Graph approach between a *natural pulsation* ω_q also called oscillator pulsation, notated with a subscript indicating the energy variety of the oscillator and the *wave pulsation* ω, or merely *pulsation*, without subscript, also called *angular frequency*.

> For brevity, this naming scheme follows the usual distinction between the emitter of a wave, the oscillator, and what is emitted, the wave. The *system natural pulsation* features the oscillator, whereas the *wave pulsation* is one of the variables of a wave. Without entering in too many details, it will be just said that the latter is a state variable, more precisely an effort of an energy variety present in waves called *phase* or *wave energy*. It becomes equal to the system natural pulsation ω_q only in peculiar cases of matching between emitter and emitted signal, as in the conservative harmonic oscillator treated here, but in a damped oscillator (see case study H7 in Chapter 11) or in case of forced oscillations for instance, the two variables will not coincide.
>
> (See case study F3 about the difference with the angular velocity Ω.)

9.4.3 Case Study F3: Torsion Pendulum — Rotational Mechanics

The case study abstract is given in next page.

In rotational mechanics, the *rotation angle* θ is the basic quantity, the *torque* (or couple) τ is the effort, the *angular momentum* (or *kinetic momentum*) $\boldsymbol{L}$ (or $\boldsymbol{\sigma}$) is the impulse, and the *angular velocity* $\boldsymbol{\Omega}$ is the flow. All these state variables are vectors. (The angle $\boldsymbol{\theta}$ is a vector perpendicular to the plane of rotation, colinear to the angular velocity.)

The torsion pendulum is made up of a vertical torsion wire (or rod) having one end fixed onto a hanging device and having hung at the other end an inertial (or massive) body free to rotate around the wire axis.

The presence in the system of inductive (rotational kinetic) energy and of capacitive (torsion) energy is made effective by the existence of the two system constitutive properties, *rotational inertia* (or "*moment*" *of inertia*, also notated I) and *torsion elastance* (or more classically torsion "constant"):

Inertia:
$$\boldsymbol{L} = \hat{\mathrm{J}}\,\boldsymbol{\Omega}_L \underset{lin}{=} J\,\boldsymbol{\Omega}_L \tag{F3.1}$$

Torsion constant:
$$\boldsymbol{\tau}_C = \hat{\mathrm{C}}_\theta\,\boldsymbol{\Theta} \underset{lin}{=} C_\theta\,\boldsymbol{\Theta} \tag{F3.2}$$

Two possibilities of evolution are linking the entity numbers to the gate variables, torque and angular velocity:

Gate torque:
$$\boldsymbol{\tau}_L = \frac{\mathrm{d}}{\mathrm{d}t}\boldsymbol{L} \tag{F3.3}$$

Gate angular velocity:
$$\boldsymbol{\Omega}_C = \frac{\mathrm{d}}{\mathrm{d}t}\boldsymbol{\Theta} \tag{F3.4}$$

The mechanical mounting of this system results in the angular velocity of the inertial body coinciding with the angular velocity of the wire at the end at which the body is fixed. This means that the flow is common to both dipoles, wire and inertial body, or in the language of equivalent circuit, that they are mounted in series. The consequence is that the dipole torques are added for

F3: Torsion Pendulum **Rotational Mechanics**

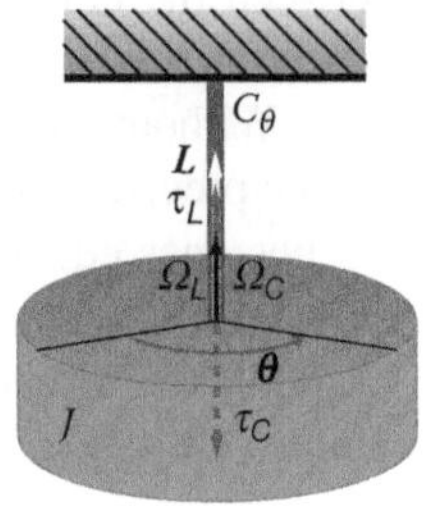

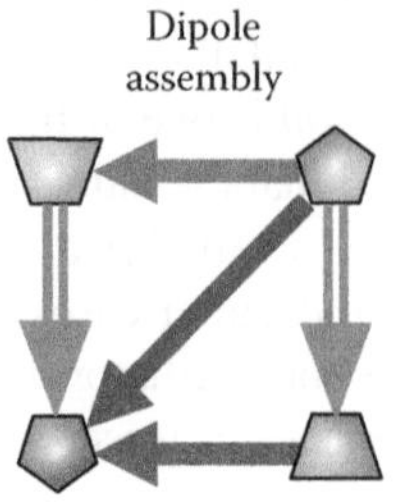

Dipole Assembly	
✓	Fundamental
	Mixed
✓	Capacitive
✓	Inductive
	Conductive
	Common effort
✓	Common flow
✓	Global
	Spatial

Formal Graph:

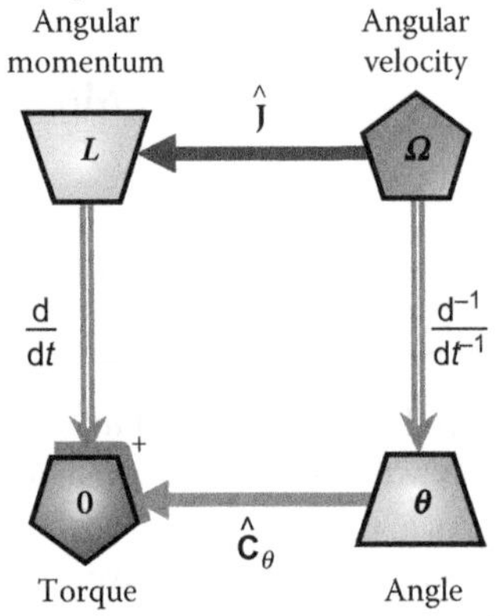

Variations with time:

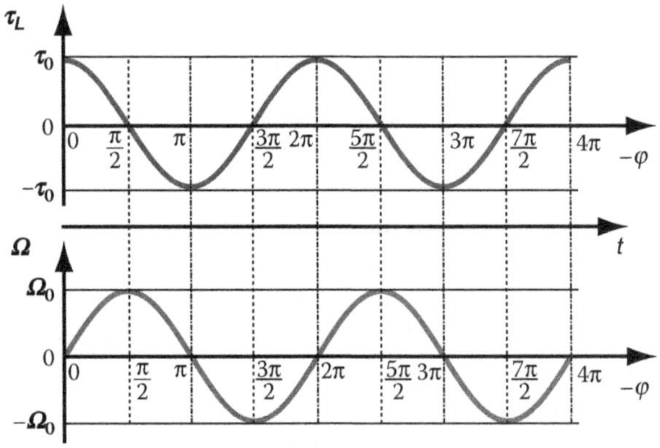

GRAPH 9.15

$$\tau_L = -\tau_C = \tau_0 \cos \omega t; \quad \Omega = \Omega_0 \sin \omega t$$

Isolated: $\tau = \tau_L + \tau_C \underset{isol}{=} 0$

$$\Omega = \Omega_L = \Omega_C; \quad \omega = \frac{-\varphi}{t} = \omega_\theta \overset{def}{=} \sqrt{\frac{C_\theta}{J}}$$

Variables:

Variety	Rotational Mechanics			
Subvariety	Inductive (kinetic)		Capacitive	
Category	Entity number	Energy/ entity	Entity number	Energy/ entity
Family	Impulse	Flow	Basic quantity	Effort
Name	Angular momentum (kinetic moment)	Angular velocity	Angle	Torque (couple)
Symbols	L, (σ)	Ω	θ, α	τ
Unit	[N m s rad^{-1}], [J s rad^{-1}]	[rad s^{-1}]	[rad]	[N m rad^{-1}], [J rad^{-1}]

Properties

System Properties		
Nature	Inductance	Elastance
Levels	0	0
Name	Rotational inertia	Constant of torsion
Symbol	J	E_θ, C_θ
Unit	[kg m^2 rad^{-2}]	[N m rad^{-2}]
Specificity	One-dimensional	One-dimensional

$L \underset{lin}{=} J\,\Omega \quad \tau_C \underset{lin}{=} C_\theta\,\Theta$

Energy variation:

$$d\mathcal{E}_\theta \underset{\exists J, C_\theta}{=} \Omega . dL + \tau \cdot d\Theta$$

forming the total torque, which is null owing to the isolation of the pendulum from external perturbation.

Common flow: $$\boldsymbol{\Omega} = \boldsymbol{\Omega}_L = \boldsymbol{\Omega}_C \quad \text{(F3.5)}$$

Isolated: $$\boldsymbol{\tau} = \boldsymbol{\tau}_L + \boldsymbol{\tau}_C \underset{isol}{=} \mathbf{0} \quad \text{(F3.6)}$$

Once the elementary relations modeling this system are set, they can be combined to establish the global algebraic equation (by substituting the variables in the previous equation)

$$\frac{\mathrm{d}}{\mathrm{d}t}\hat{\mathrm{J}}\boldsymbol{\Omega} + \hat{\mathrm{C}}_\theta \frac{\mathrm{d}^{-1}}{\mathrm{d}t^{-1}}\boldsymbol{\Omega} \underset{isol}{=} \mathbf{0} \quad \text{(F3.7)}$$

This algebraic model has been obtained by making a classical reasoning in which the Formal Graph given in the case study abstract is not really used. In fact, the common flow representation of this Formal Graph is not suited for a direct exploitation, in contradistinction with the loop representation shown in Graph 9.16.

By walking along the circular path represented in this graph, a translation into a composition of operators can be found by starting from the angular velocity for instance:

$$\frac{\mathrm{d}}{\mathrm{d}t}\hat{\mathrm{C}}_\theta^{-1}\left(-\frac{\mathrm{d}}{\mathrm{d}t}\right)\hat{\mathrm{J}}\,\boldsymbol{\Omega} = \boldsymbol{\Omega} \quad \text{(F3.8)}$$

The interest of this approach is clearly demonstrated as the algebraic model F3.7 is retrieved by rearranging this equation. However, it should be mentioned that this powerful method is not universal and works mainly when the circularity principle can be applied.

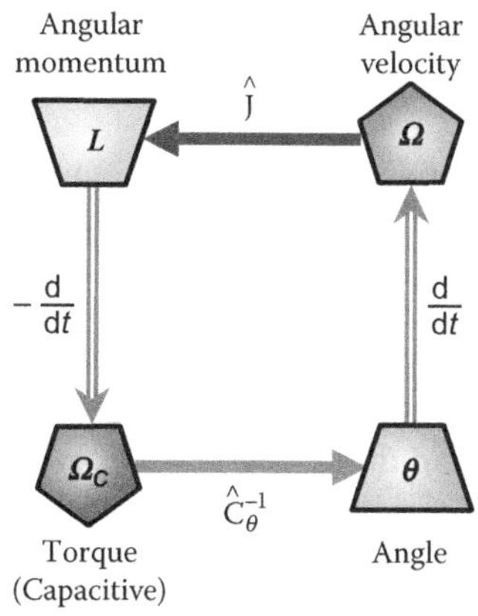

GRAPH 9.16

Now, by assuming first that the constitutive properties are not operators but simple scalars (linear case) and second that they are independent of time (constant properties), a *pendulum pulsation* featuring the system can be defined:

$$\omega_\theta \overset{def}{=} \sqrt{\frac{C_\theta}{J}} \tag{F3.9}$$

These hypotheses and definition allow one to write the global equation F3.7 or F3.8 as a second-order differential equation

$$\frac{\mathrm{d}^2}{\mathrm{d}t^2}\boldsymbol{\Omega} + \omega_\theta^2\,\boldsymbol{\Omega} \underset{lin}{\overset{isol}{=}} \mathbf{0} \tag{F3.10}$$

As all temporal oscillators have the same Formal Graph, it is logical that the same algebraic equation is also found. This means that this system is relevant to a more general concept called *temporal wave*, which can be generated by any oscillator. Here the wave production is made without loss, because the pendulum is an ideal system, without friction or dissipation (the oscillator is said to be conservative). This is translated in the statement that the *wave pulsation* ω is the same as the *pendulum pulsation* ω_θ

$$\omega \underset{cst\ \mathcal{E}_\theta}{=} \omega_\theta \tag{F3.11}$$

The solution of the second-order differential equation, which is a classical and simple one, is well known, using harmonic functions *cosine* or *sine* according to the initial conditions

$$\boldsymbol{\Omega} \underset{lin}{=} \boldsymbol{\Omega}_0 \cos \omega t \tag{F3.12}$$

$$\boldsymbol{\tau}_L = -\boldsymbol{\tau}_C \underset{lin}{=} \boldsymbol{\tau}_0 \sin \omega t \tag{F3.13}$$

The plots of the torque and angular velocity variations as a function of time are given in the case study abstract. However, a more convenient variable proportional to time, the opposite of the *phase angle* φ, is used for the horizontal axis, allowing one to graduate it in radians. Its relation with time is

$$\varphi = -\omega t \tag{F3.14}$$

In fact, the *phase angle* is not merely a convenient variable but is a state variable featuring the wave, as explained in Section 9.5. This section also demonstrates the above solution.

Remarks

The swinging pendulum. The classical pendulum consisting of hanging a massive body to a rigid wire fixed to a support and free to oscillate in a vertical plane is another example of a mechanical oscillator involving rotational energy. However, this system is not detailed here because it is a more complex case than the torsion pendulum. Indeed, gravitational energy participates under its capacitive subvariety (potential) in complement to the rotational (inertial) energy. This simultaneous existence of two different energy varieties comes out of the frame of a dipole assembly that works with only one energy variety. It is relevant to the category of coupled energy varieties tackled in Chapter 12.

Angular velocity and pulsations. A frequent confusion arises in mechanics in assimilating the *angular velocity* $\boldsymbol{\Omega}$ to a *pulsation* ω (also called "*angular frequency*" which aggravates the confusion) introduced in case study F2 "LC Circuits." It is clear that both variables having the same unit in rad s^{-1} does not facilitate the distinction. Moreover, the existence of several pulsations with different status complicates the subject. The Formal Graph theory makes the following distinctions:

- The *angular velocity* $\boldsymbol{\Omega}$ is a state variable (vector, in rad s^{-1}), a *flow*, in rotational mechanics.
- The *wave pulsation* ω, or *angular frequency*, is a state variable (scalar, in rad s^{-1}), an *effort*, of the wave energy.
- The *natural pulsation* ω_q, or *natural angular frequency*, is a system *property* (scalar), in the considered energy variety.

In order to comply with the tradition in the domain of signal processing, the term angular frequency is preferred for the pulsation ω when it is considered as the Fourier-transformed variable of the time derivation.

The relationship with the *frequency* ν (scalar, in Hz) is $\omega = 2\pi\nu$.

9.4.4 Case Study F4: Vibrating String — Translational Mechanics

The case study abstract is given in next page.

A string tightened between two fixed points, able to vibrate by moving in a plane passing by its two ends, is an oscillator in one dimension. The vibration plane may be fixed or in rotation around the axis joining the ends. The mode of vibration n of the string is defined as the number of crests (antinodes) present on one side of the axis (in the figure below $n = 2$). The coordinates of a point are the distance $\boldsymbol{\ell}$ from the axis and the position $\boldsymbol{r}$ along the axis.

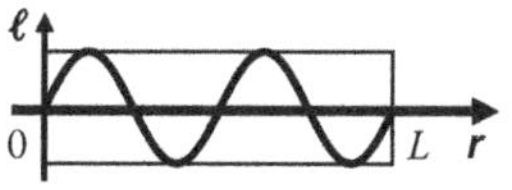

A vibrating string possesses two *system properties* that are vectors, a capacitive one, the *mechanical tension* $\boldsymbol{T_\ell}$ which is a spatially reduced elastance, and an inductive one, also spatially reduced, the *lineic mass* or more precisely *lineic density of mass* (or mass by unit of length) $\boldsymbol{\mu_M}$. They are specific properties in a system with one dimension of space.

$$\boldsymbol{F_C} \underset{lin}{=} \boldsymbol{T_\ell}\, \varepsilon_\ell \qquad \text{(F4.1)}$$

$$P_{lr} \underset{lin}{=} \boldsymbol{\mu_M} . \boldsymbol{v} \qquad \text{(F4.2)}$$

F4: Vibrating String **Translational Mechanics**

Formal Object:

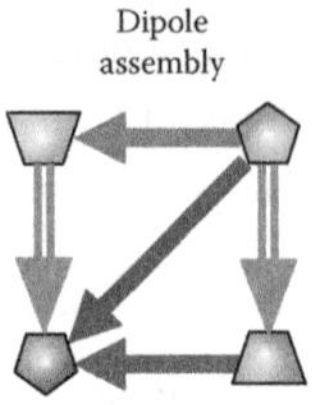

Dipole Assembly	
✓	Fundamental
	Mixed
	Capacitive
✓	Inductive
	Conductive
✓	Common effort
	Common flow
✓	Global
	Spatial

Formal Graph:

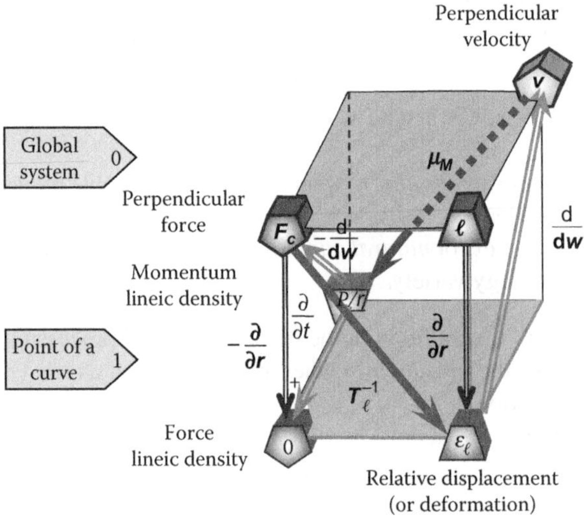

GRAPH 9.17

Variables:

Variety	Translational Mechanics			
Subvariety	Inductive (kinetic)		Capacitive	
Category	Entity number	Energy/ entity	Entity number	Energy/ entity
Family	Impulse	Flow	Basic quantity	Effort
Name	Momentum (quantity of movement)	Velocity	Displacement	Force
Symbols	***P***, (***p***)	***v***	ℓ	***F***
Unit	[N s], [J s m^{-1}]	[m s^{-1}]	[m]	[N], [J m^{-1}]

Properties

System Properties		
Nature	Inductance	Capacitance
Levels	0/1	1/0
Name	Lineic mass	Mechanical tension
Symbol	μ_M	T_ℓ
Unit	[kg m^{-1}]	[N]
Specificity	One-dimensional	One-dimensional

The Formal Graph is built according to a mounting in series of the dipoles (common flow), with the displacement velocity as common flow and with a summation of lineic densities of force, equal to zero because the system is isolated (no external influence which would have "forced" the oscillations with a nonzero value)

$$-\frac{\partial}{\partial r}\boldsymbol{F_C} + \frac{\partial}{\partial t}P_{lr} \underset{isol}{=} 0 \quad \text{(F4.3)}$$

The substitution of the capacitive force and of the lineic density of impulse by their expression in function of the distance $\boldsymbol{\ell}$ gives

$$-\frac{\partial}{\partial r}\boldsymbol{T_\ell}\frac{\partial}{\partial r}\ell + \frac{\partial}{\partial t}\mu_M\frac{\partial}{\partial t}\ell = 0 \quad \text{(F4.4)}$$

The hypothesis of constant system properties (restriction to the *linear case*) allows one to group the terms in the following way:

$$\frac{\boldsymbol{T_\ell}}{\mu_M}\frac{\partial^2}{\partial \boldsymbol{r}^2}\ell \underset{lin}{=} \frac{\partial^2}{\partial t^2}\ell \quad \text{(F4.5)}$$

By defining an *oscillator velocity* as a scalar combination of system constitutive properties

$$u_\ell \overset{def}{=} \sqrt{\frac{\boldsymbol{T_\ell}}{\mu_M}} \quad \text{(F4.6)}$$

and in writing that the modulus of the propagation velocity of the wave is equal to the oscillator velocity owing to the conservation of energy in the system

$$\|\boldsymbol{u}\| \underset{cst\ \mathcal{E}_\ell}{=} u_\ell \quad \text{(F4.7)}$$

the classical equation of propagation of a wave is obtained

$$\boldsymbol{u}^2\frac{\partial^2}{\partial \boldsymbol{r}^2}\ell = \frac{\partial^2}{\partial t^2}\ell \quad \text{(F4.8)}$$

9.4.4.1 Formal Graph Approach

A more straightforward approach is permitted by using the Formal Graph in the case study abstract, which is based on a diagonal link on the space–time sides of the graph. The space–time operator in this diagonal is defined as the composition of the space integration with the time derivation:

$$\frac{\mathrm{d}}{\mathbf{dw}} \overset{def}{=} \frac{\partial}{\partial t}\frac{\partial^{-1}}{\partial \boldsymbol{r}^{-1}} = \frac{\partial^{-1}}{\partial \boldsymbol{r}^{-1}}\frac{\partial}{\partial t} \quad \text{(F4.9)}$$

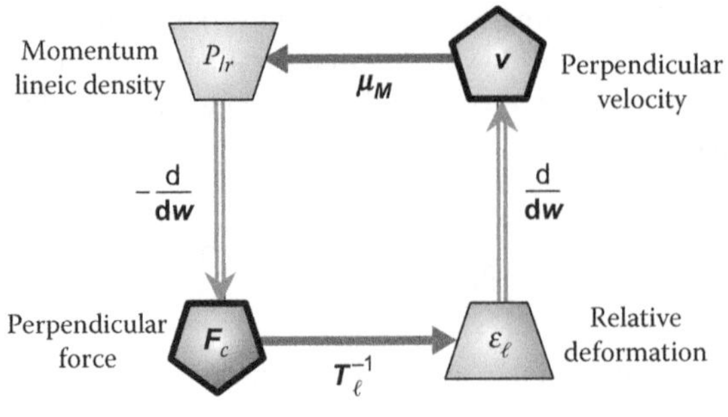

GRAPH 9.18

A projection in two dimensions of the three-dimensional Formal Graph retaining only the useful nodes clearly evidences the circular path modeling the oscillation of the string (see Graph 9.18).

Application of the principle of circularity allows one to write directly the relationship

$$\left(-\frac{d}{dw}\right)\mu_M\,\frac{d}{dw}\,T_\ell^{-1}\,F_C = F_C \tag{F4.10}$$

from which the second-order differential equation is derived

$$\frac{d^2}{dw^2}F_C + u^2\,F_C = 0 \tag{F4.11}$$

The wave equation is then readily found by using Equations F4.9 and F4.7 and rearranging them.

Case study F5 "Sound Propagation" allows the similarity between the models to be observed, despite the different number of space dimensions.

Remarks

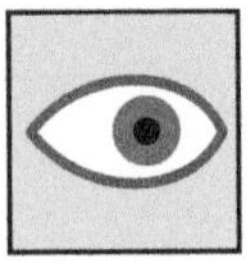

A simple oscillator. The Formal Graph of a vibrating string is one of the simplest in terms of spatial levels because the system has only one dimension. The constitutive properties of this system, the tension of the string and its lineic mass (mass by unit of length), are effectively one dimension-specific. In this graph, the loop of paths is characteristic of an oscillator because it includes two evolution paths (temporal derivation) and characteristic of a wave propagation because of the two paths of spatial distribution (spatial derivation).

String theory. The one-dimensional oscillator (with one space dimension) is an object of the highest importance in physics. The whole area of wave mechanics and quantum mechanics is based on its properties and particularly in the manner space–time ensures the conversions between the capacitive and inductive subvarieties. This system is at the basis of the string theory, whose name comes from the vibrating string described here.

Principle of circularity. The principle of circularity, applied to the loop of paths in this Formal Graph, immediately provides a general wave equation, which has therefore the same shape whatever the energy variety. (The classical wave equation is found at the condition of being within the restricted framework of the linearity of system properties.) The link between the property of space time and the system properties stems from this circularity. In this system, this link is created by the relationship between the propagation velocity (velocity of phase) and the ratio of the mechanical tension on the lineic mass.

9.4.5 Case Study F5: Sound Propagation — Translational Mechanics

The case study abstract is given in next page.

The sound is a hydrodynamical vibration that propagates with a velocity depending on the properties of elasticity and inertia of the medium. The representation of this phenomenon by a Formal Graph consists of a three-dimensional graph in the hydrodynamical energy variety with a circular path going successively through four variables, a first state variable which is the pressure P, and three spatially reduced variables: the volume relative expansion $V_{/V}$ (reduced from the volume V), the fluid velocity $\boldsymbol{v}$ (reduced from the flow rate $\mathcal{Q}$), and the lineic density of percussion $\boldsymbol{F}_{/r}$ (reduced from the percussion F).

The spatially reduced basic quantity $V_{/V}$ called *volume relative expansion* is without dimension, it is defined as the volumic derivative of the volume expansion, and, in the case of a homogeneous distribution, it is equal to the ratio of the volume expansion on the volume. Besides, it is related to the pressure by the reduced capacitance χ, which bears the name of *compressibility*

$$V_{/V} \overset{def}{=} -\frac{\partial}{\partial V}\Delta V \underset{hom}{=} \frac{\Delta V}{V} \underset{lin}{=} -\chi P \tag{F5.1}$$

The spatially reduced impulse is the *lineic density of percussion*, defined as the contragradient of the percussion F [or the volumic derivative of the impulse (quantity of movement) $\boldsymbol{P}$]; it is related to the fluid velocity by the reduced inductance ρ_M, which is the *volumic mass.*

$$\boldsymbol{F}_{/r} \overset{def}{=} -\frac{\partial}{\partial \boldsymbol{r}}F = -\nabla F = \frac{\partial}{\partial V}\boldsymbol{P} \underset{lin}{=} \rho_M \boldsymbol{v} \tag{F5.2}$$

On the side of energies-per-entity, one equally finds reduced variables that play the same role with respect to space–time as the sink or source term in a continuity equation. The reduced effort is the *pressure field*, which may also come from the evolution of the lineic density of percussion

$$\boldsymbol{P}_{/r} \overset{def}{=} \frac{\partial}{\partial \boldsymbol{r}}P = -\nabla(-P) = \frac{\partial}{\partial t}\boldsymbol{F}_{/r} \tag{F5.3}$$

The reduced flow is the *lineic density of velocity*, or divergence of the fluid velocity, which may equally come from an evolution of the volume relative expansion

$$v_{/r} \overset{def}{=} -\nabla.\boldsymbol{v} = -\frac{\partial}{\partial t}V_{/V} \tag{F5.4}$$

F5: Sound Propagation **Hydrodynamics**

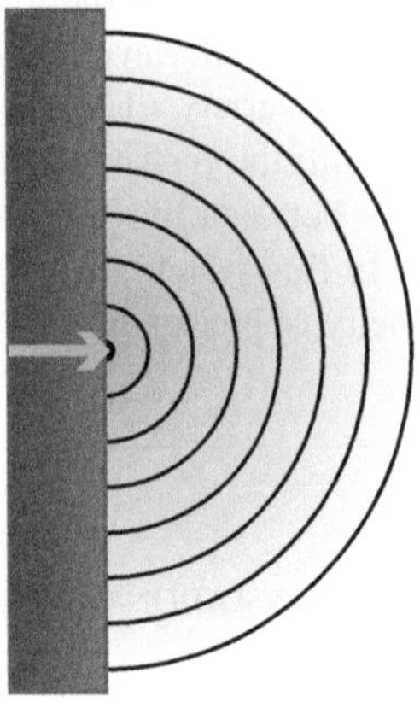

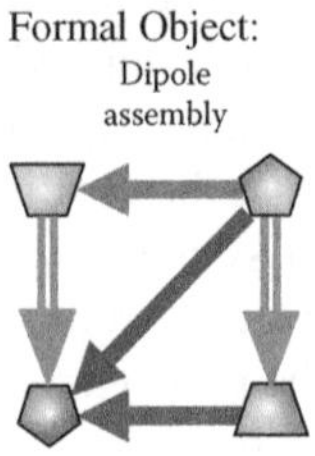

Dipole Assembly	
✓	Fundamental
	Mixed
	Capacitive
✓	Inductive
	Conductive
✓	Common effort
	Common flow
✓	Global
	Spatial

Formal Graph:

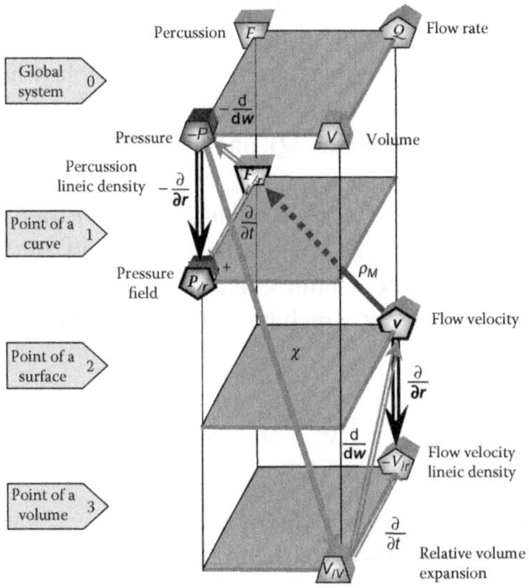

GRAPH 9.19

Variety	Hydrodynamics			
Subvariety	Inductive (kinetic)		Capacitive	
Category	Entity number	Energy/ entity	Entity number	Energy/ entity
Family	Impulse	Flow	Basic quantity	Effort
Name	Percussion	Flow rate	Volume	Pressure
Symbols	F	Q	V	P
Unit	[Pa s]	[m^3 s^{-1}]	[m^3]	[Pa], [N m^{-2}]

System Properties		
Nature	Inductance	Elastance
Levels	2/1	0/3
Name	Volumic mass	Compressibility
Symbol	ρ_M	χ
Unit	[kg m^{-3}]	[Pa^{-1}], [m^2 N^{-1}]
Specificity	Three-dimensional	Three-dimensional

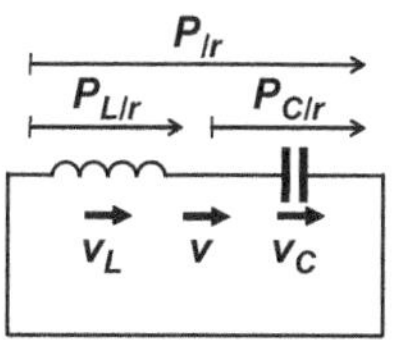

The modeling of an oscillatory movement of fluid is done by mounting in series an inductive dipole and a capacitive dipole that can be represented by an equivalent circuit in series, despite the fact that both are distributed in space.

This serial mounting is algebraically translated into equality of velocities and by a sum of pressure fields of the dipoles:

$$\boldsymbol{P}_{/r} = -\frac{\partial}{\partial \boldsymbol{r}} \chi^{-1} V_{/V} + \frac{\partial}{\partial t} \rho_M \frac{\partial^{-1}}{\partial \boldsymbol{r}^{-1}} \frac{\partial}{\partial t} V_{/V} \tag{F5.5}$$

To ensure the propagation, an exchange of capacitive and inductive energies is necessary; this requires closing circuit for forming a loop, in other words, an oscillator. This is expressed by the nullity of the sum of pressure fields:

$$\boldsymbol{P}_{/r} = 0 \tag{F5.6}$$

Equation F5.5 is therefore written as a classical wave equation

$$\frac{\partial^2}{\partial \boldsymbol{r}^2} V_{/V} = \boldsymbol{u}_V^2 \frac{\partial^2}{\partial t^2} V_{/V} \tag{F5.7}$$

having defined an *oscillator velocity* $\boldsymbol{u}_V$ featuring the medium in which the sound wave propagates. In the absence of dissipation in this medium, it will be seen that the propagation velocity of the sound (more exactly its modulus) is equal to this system property:

$$\left\|\boldsymbol{u}_V\right\| = u_V \overset{def}{=} \left(\chi\,\rho_M\right)^{-\frac{1}{2}} \tag{F5.8}$$

Another approach of this model is worth presenting. It relies on the selection of relevant variables in the functioning of the loop, which can be drawn as a vertical two-dimensional projection (seen from above) of the Formal Graph given in the case study abstract (see also Graph 9.20).

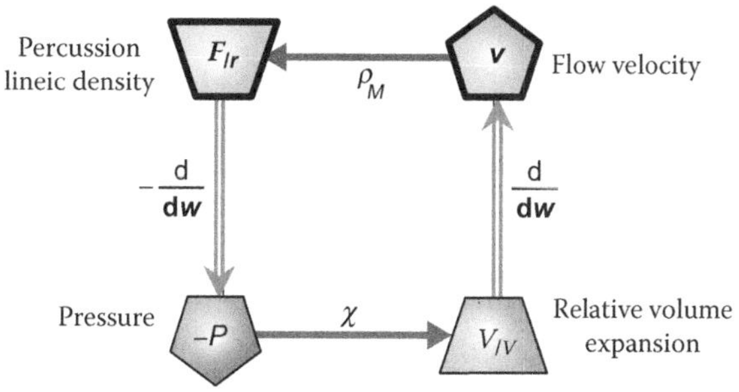

GRAPH 9.20

Instead of representing individually the role of space by a spatial derivative and the role of time by a temporal derivative, their joint action is represented by a space–time operator defined as

$$\frac{\mathrm{d}}{\mathbf{d}\boldsymbol{w}} \overset{def}{=} \frac{\partial^{-1}}{\partial \boldsymbol{r}^{-1}} \frac{\partial}{\partial t} \tag{F5.9}$$

Graph 9.20 utilizes the loop representation, consisting of placing in the effort node the pressure given by the capacitive relationship. The whole path of the loop, from the volume relative expansion, is algebraically translated by the equivalence

$$\chi\left(-\frac{\mathrm{d}}{\mathbf{d}\boldsymbol{w}}\right)\rho_M \frac{\mathrm{d}}{\mathbf{d}\boldsymbol{w}} V_{IV} = V_{IV} \tag{F5.10}$$

As here the system constitutive properties are assumed to be constants, the previous equation can be rearranged into equality between a second-order differential operator and its eigen-value:

$$\frac{\mathrm{d}^2}{\mathbf{d}\boldsymbol{w}^2} V_{IV} = -u^2\, V_{IV} \tag{F5.11}$$

This is another viewpoint on the wave equation that is generally presented as an ordinary second-order differential equation with respect to two variables, time and space vector. The eigen-value is a function of the *wave velocity* $\boldsymbol{u}$ which represents the propagation speed of the wave and which is a concept that differs from the *oscillator velocity* u_V introduced in definition F5.8. However, although concepts are different, the modulus of the *wave velocity* becomes equal to the *oscillator velocity* in a conservative system (without dissipation):

$$\|\boldsymbol{u}\| \underset{cst\ \mathcal{E}_V}{=} u_V \tag{F5.12}$$

This distinction originates from the necessity in the Formal Graph approach to sort among variables featuring the energy (that can be placed in nodes), variables that describe the properties of the system and variables describing the space–time (that constitute links in a Formal Graph). Here, the *wave velocity* $\boldsymbol{u}$ is a variable featuring the space–time whereas the *oscillator velocity* u_V is a specific property of the propagating medium.

By returning to the individual operators embedded in the space–time operator defined in Equation F5.9, the wave equation F5.7 can be rewritten under the classical form of the *equation of wave propagation* with the usual wave velocity:

$$\frac{\partial^2}{\partial \boldsymbol{r}^2} V_{IV} = \boldsymbol{u}^2\, \frac{\partial^2}{\partial t^2} V_{IV} \tag{F5.13}$$

The solution of this differential equation is expressed with the help of a harmonic function

$$V_{IV} = V_{IV0} \cos(\omega t - \boldsymbol{k}.\boldsymbol{r}) \tag{F5.14}$$

with the following relation between pulsation ω and wave vector $\boldsymbol{k}$:

$$\boldsymbol{u} = \frac{\omega}{\boldsymbol{k}} \tag{F5.15}$$

The details of the resolution are given in Section 9.5.

It is worth comparing with a one-dimensional oscillator such as the vibrating string in referring to case study F4.

Remarks

Nonclassical variables. The Formal Graph utilizes some localized variables that are not defined in classical hydrodynamics, because in algebra it is always possible to ignore intermediate variables by combining several operators. The reason for using these variables, and therefore having to make them explicit, comes from the interest to use reduced properties of a system which are specific, such as the compressibility and the volumic mass.

A general model. An oscillating circuit distributed in space models the propagation of a wave, independently from its energy variety (hydrodynamics, mechanics, electrodynamics, and so on). This is therefore a key concept in physics and the same equation of wave is found for modeling the propagation in every domain. What may vary from one domain to another is the choice of the specific properties of systems, between the global level and the volumic level as for the compressibility, or between the lineic level (of a curve point) and the superficial level (of a surface point), as for the volumic mass in the present case. In electrodynamics for instance, only two levels are used, the lineic level and the superficial level, for modeling the light propagation.

Principle of circularity. The Formal Graph gives prominence to the functioning of the propagation by means of the existence of a loop of paths, going through the properties of space, of time, and of system. The principle of circularity allows one to introduce the propagation velocity $\boldsymbol{u}$ as an eigen-value of a space–time operator. This wave velocity should not be confused with the fluid velocity $\boldsymbol{v}$.

9.4.6 Case Study F6: Light Propagation — Electrodynamics

The case study abstract is given in next page.

Light is an electrodynamical wave that propagates in medium featured by a *permittivity* ε and a *permeability* μ. (In vacuum these properties are notated ε_0 and μ_0.) They are specific properties in a three-dimensional space. The permittivity is the capacitive reduced property that links the *electric field* $\boldsymbol{E}$ to the *electric displacement* $\boldsymbol{D}$ (or "*electrization*") and the permeability is the inductive reduced property that links the *electromagnetic field* $\boldsymbol{H}$ to the *electromagnetic induction* $\boldsymbol{B}$. The fields $\boldsymbol{E}$ and $\boldsymbol{H}$ are defined as the one-dimensional localized potential and current, respectively. All these relationships are written, assuming linear system constitutive properties as it is usual in electrodynamics:

$$\boldsymbol{E} \overset{def}{=} -\frac{\partial}{\partial \boldsymbol{r}} V = -\nabla V = \hat{\varepsilon}^{-1} \boldsymbol{D} \underset{lin}{=} \frac{1}{\varepsilon} \boldsymbol{D} \tag{F6.1}$$

F6: Light Propagation **Electrodynamics**

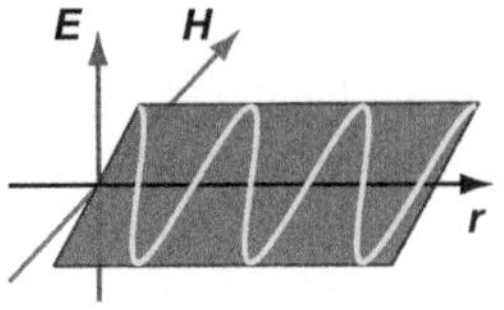

Formal Object:

Dipole assembly

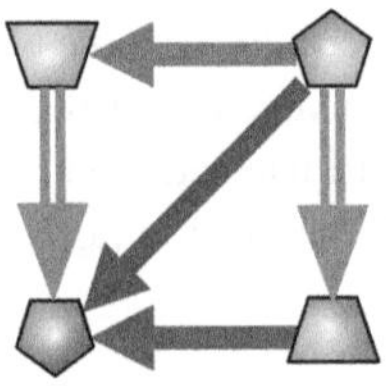

Dipole Assembly	
✓	Fundamental
	Mixed
✓	Capacitive
✓	Inductive
	Conductive
✓	Common effort
	Common flow
	Global
✓	Spatial

Formal Graph (see Graph 9.21):

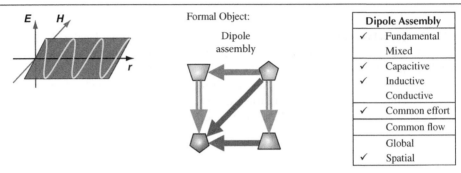

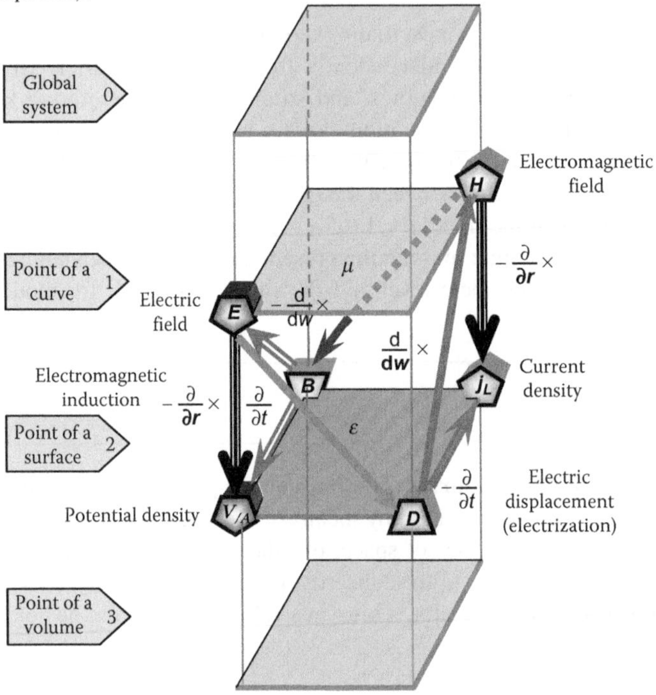

GRAPH 9.21

Variety	Electrodynamics			
Subvariety	Inductive (kinetic)		Capacitive	
Category	Entity number	Energy/entity	Entity number	Energy/entity
Family	Impulse	Flow	Basic quantity	Effort
Name	Induction quantity (flux)	Current	Charge	Potential
Symbols	Φ_B	I	Q	V
Unit	[Wb], [J A^{-1}]	[A]	[C]	[V], [J C^{-1}]

System Properties		
Nature	Inductance	Capacitance
Levels	1/2	1/2
Name	Permeability	Permittivity
Symbol	μ	ε
Unit	[Wb m^{-1}]	[F m^{-1}]
Specificity	Three-dimensional	Three-dimensional

$$\boldsymbol{H} \overset{def}{=} -\frac{\partial}{\partial \boldsymbol{r}} I = -\nabla I = \hat{\mu}^{-1} \; \boldsymbol{B} \underset{lin}{=} \frac{1}{\mu} \boldsymbol{B} \tag{F6.2}$$

Normally the two fields should be subscripted with *L* or *C* to indicate their capacitive or inductive nature, but in order to be faithful to the classical use, they are omitted.

At the level of a surface point, a localized capacitive potential is defined as the contrarotational of the electric field and a localized inductive potential is defined as the time derivative of the electromagnetic induction:

$$\boldsymbol{V}_{IA_L} \overset{def}{=} \frac{\partial}{\partial t}\boldsymbol{B}; \quad V_{IAc} \overset{def}{=} -\frac{\partial}{\partial \boldsymbol{r}} \times \boldsymbol{E} \tag{F6.3}$$

Note that these variables are not used in classical electrodynamics in order to spare the number of variables. In fact they are implicit in the Maxwell–Faraday equation that states their equality (see Maxwell's equations in case study F7).

Still at the level of a surface point, but in the opposite corner of a Formal Graph, an inductive current density is defined as the contrarotational of the electromagnetic field and a capacitive current density is defined as the time derivative of the electric displacement

$$\boldsymbol{j}_L \overset{def}{=} -\frac{\partial}{\partial \boldsymbol{r}} \times \boldsymbol{H}; \quad \boldsymbol{j}_C \overset{def}{=} \frac{\partial}{\partial t}\boldsymbol{D} \tag{F6.4}$$

The preceding equations depict the general model of electrodynamics in a three-dimensional space. Now, the assumption of a parallel mounting of two dipoles, that is, sharing a common localized potential, is made together with the assumption of an isolated dipole assembly, which is algebraically translated by

$$V_{IA} = V_{IA_L} = V_{IAc} \tag{F6.5}$$

$$\boldsymbol{j}_L + \boldsymbol{j}_C = 0 \tag{F6.6}$$

All these relationships are represented in the three-dimensional Formal Graph given in the case study abstract.

In this Formal Graph is also drawn a space–time operator combining a space integration with a time derivation. It is notated as a derivative with respect to a variable $\boldsymbol{w}$ which is a vector that combines space and time:

$$\frac{\mathrm{d}}{\mathrm{d}\boldsymbol{w}} \overset{def}{=} \frac{\partial^{-1}}{\partial \boldsymbol{r}^{-1}} \frac{\partial}{\partial t} \tag{F6.7}$$

The interest of this operator is to link directly the *electric displacement* $\boldsymbol{D}$ to the *electromagnetic field* $\boldsymbol{H}$ on the one hand and the *electromagnetic induction* $\boldsymbol{B}$ to the *electric field* $\boldsymbol{E}$ on the other hand:

$$\boldsymbol{H} = \frac{\mathrm{d}}{\mathrm{d}\boldsymbol{w}} \times \boldsymbol{D}; \quad \boldsymbol{E} = -\frac{\mathrm{d}}{\mathrm{d}\boldsymbol{w}} \times \boldsymbol{B} \tag{F6.8}$$

This simplifies the Formal Graph representation because a projection from above, in keeping only the pertinent variables, can be drawn as shown in Graph 9.22.

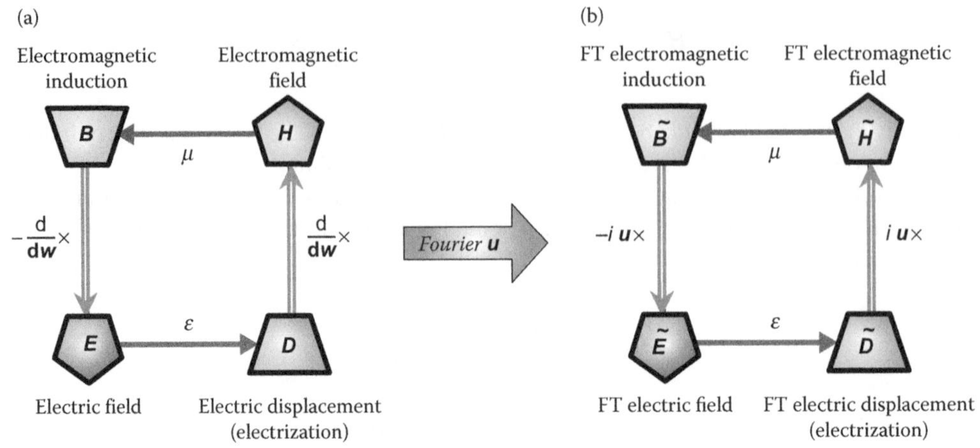

GRAPH 9.22 Normal Formal Graph of an electrodynamical oscillator (a) and its Fourier-transformed Formal Graph (b).

The algebraic translation of this Formal Graph provides a sequence of operator (illustrating the principle of circularity)

$$-\frac{\mathrm{d}}{\mathrm{d}\mathbf{w}}\times\,\mu\,\frac{\mathrm{d}}{\mathrm{d}\mathbf{w}}\times\varepsilon\,\mathbf{E}\;=\mathbf{E} \tag{F6.9}$$

On the left of Graph 9.22 is drawn another Formal Graph obtained by Fourier transforming the normal Formal Graph using as parameter a *wave velocity* $\mathbf{u}$ (instead of the usual pulsation ω or wave vector $\mathbf{k}$). In Section 9.5.8 is explained the interest of the Fourier transform in case of linear constitutive properties and its role in replacing operators by their eigen-values (the wave velocity is defined as the eigen-value of the derivation with respect to the space–time vector $\mathbf{w}$). In this Fourier Formal Graph the following products of variables can be read by application of the circularity principle:

$$-i\,\mathbf{u}\times\mu\;i\,\mathbf{u}\times\varepsilon\;\tilde{\mathbf{E}}=\tilde{\mathbf{E}} \tag{F6.10}$$

The introduction of the imaginary number i is made here in anticipation of the step-by-step demonstration made in Section 9.5 of the mathematical solution of second-order differential equations. Here, it is sufficient to admit that $-i^2 = 1$ for using this mathematical tool.

As the permittivity and the permeability are scalars, which means that the oscillator is harmonic, an electrodynamical *oscillator velocity* featuring the oscillator is defined as a function of these system constitutive properties:

$$u_Q \overset{def}{=} (\varepsilon\,\mu)^{-\frac{1}{2}} \tag{F6.11}$$

Inserting this scalar variable into the circular relation F6.10 allows one to identify the modulus of wave velocity $\mathbf{u}$ (which is a space–time property) with the oscillator velocity u_Q (which is a system property) as the electric energy is conserved:

$$\|\mathbf{u}\| \underset{cst\,\mathcal{E}_Q}{=} u_Q = (\varepsilon\,\mu)^{-\frac{1}{2}} \tag{F6.12}$$

This equivalence results in fact from the hypotheses of a conservative oscillator and linear system constitutive properties. (This equivalence would not be observed in an absorbing medium provoking dissipation for instance.)

In a homogeneous space, the system constitutive properties are independent of space–time, so Equation F6.9 can be compacted into a simple second-order differential equation

$$\frac{d^2}{d\boldsymbol{w}^2}\boldsymbol{E} = \frac{\partial^{-2}}{\partial \boldsymbol{r}^{-2}}\frac{\partial^2}{\partial t^2}\boldsymbol{E} = -\boldsymbol{u}^2\,\boldsymbol{E} \tag{F6.13}$$

By rearranging this equation, the classical wave equation is obtained:

$$\frac{\partial^2}{\partial \boldsymbol{r}^2}\boldsymbol{E} = \boldsymbol{u}^2\,\frac{\partial^2}{\partial t^2}\boldsymbol{E} \tag{F6.14}$$

The classical approach links the oscillator velocity to the ratio of the natural pulsation ω_Q to the wave vector $\boldsymbol{k}$ assuming that the wave can be decomposed into a temporal wave and a spatial one

$$\boldsymbol{u} = \frac{\omega}{k} \tag{F6.15}$$

This allows splitting the wave equation into symmetrical terms

$$\frac{1}{k^2}\frac{\partial^2}{\partial \boldsymbol{r}^2}\boldsymbol{E} = \frac{1}{\omega^2}\frac{\partial^2}{\partial t^2}\boldsymbol{E} \tag{F6.16}$$

and the solutions are expressed as a harmonic function *cosine* and *sine* as follows:

$$\boldsymbol{B} = -\frac{\boldsymbol{u}\times\boldsymbol{E}_0}{\boldsymbol{u}^2}\sin(\omega t - k.\boldsymbol{r});\quad \boldsymbol{H} = -\varepsilon\boldsymbol{u}\times\boldsymbol{E}_0\,\sin(\omega t - k.\boldsymbol{r}) \tag{F6.17}$$

$$\boldsymbol{E} = \boldsymbol{E}_0\,\cos(\omega t - k.\boldsymbol{r});\quad \boldsymbol{D} = \varepsilon\boldsymbol{E}_0\,\cos(\omega t - k.\boldsymbol{r}) \tag{F6.18}$$

It can be observed that the solution for $\boldsymbol{E}$ is also a solution of two distinct differential equations, which once combined give back the wave equation F6.16.

$$\frac{\partial^2}{\partial \boldsymbol{r}^2}\boldsymbol{E} = -k^2\boldsymbol{E};\quad \frac{\partial^2}{\partial t^2}\boldsymbol{E} = -\omega^2\boldsymbol{E} \tag{F6.19}$$

These solutions are demonstrated later in Section 9.7. From these variables, the inductive and capacitive energies can be deduced, or more precisely their concentrations, because when space is homogeneous, the product of a lineic density ($\boldsymbol{E}$, $\boldsymbol{H}$) and a surface density ($\boldsymbol{D}$, $\boldsymbol{B}$) gives a volumic concentration. In addition, the restriction to linear constitutive properties allows one to obtain simple expressions that are classically used

$$\mathcal{T}_{Q/V} \underset{hom}{=} \int \boldsymbol{H}.\mathrm{d}\boldsymbol{B} \underset{lin}{=} \frac{1}{2}\mu\boldsymbol{H}^2 \tag{F6.20}$$

$$\mathcal{U}_{Q/V} \underset{hom}{=} \int \boldsymbol{E}.\mathrm{d}\boldsymbol{D} \underset{lin}{=} \frac{1}{2}\varepsilon\boldsymbol{E}^2 \tag{F6.21}$$

$$\mathcal{E}_{Q/V} = \mathcal{T}_{Q/V} + \mathcal{U}_{Q/V} \underset{lin}{=} \frac{1}{2}\left(\mu \boldsymbol{H}^2 + \varepsilon \boldsymbol{E}^2\right) \tag{F6.22}$$

Then, the total energy concentration is found as invariant with time as space:

$$\mathcal{E}_{Q/V} \underset{lin}{=} \frac{1}{2}\varepsilon \boldsymbol{E}_0^2 \tag{F6.23}$$

Another important variable featuring the propagation of a wave is the *irradiance*, often called *radiation intensity*, introduced in Chapter 7 when the Lambert–Beer–Bouguer law was invoked in support of the theory of influence. In Section 9.7.6 it will be shown that in a three-dimensional space the irradiance is given by the product of the energy concentration times the wave velocity $\boldsymbol{u}$, which make this variable dependent on the wave frequency or wavelength:

$$I_\lambda \underset{lin}{=} \frac{1}{2}\varepsilon \boldsymbol{u} \boldsymbol{E}_0^2 \tag{F6.24}$$

Remarks

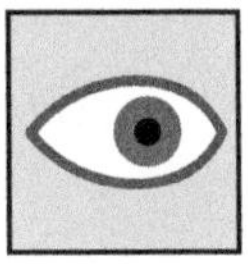

Speed of light in free space. As mentioned in the introduction of this case study, in vacuum, the system constitutive properties are notated ε_0 and μ_0. In this medium, the oscillator velocity given by Equation F6.11 must therefore be adapted

Vacuum: $$c = \left(\varepsilon_0\, \mu_0\right)^{-\frac{1}{2}} \tag{F6.25}$$

This velocity is called the *speed of light* and is notated c. However, the vacuum constitutive properties are known with less precision than c which is therefore not deduced from the previous relationship. According to the 17th Conférence Générale des Poids et Mesures (CGPM), the speed of light has been set by convention to an exact value, from which the definition of the meter is issued (CGPM 1983).

Speed of light in free space: $$c = 299792458 \text{ m s}^{-1} \tag{F6.26}$$

The special relativity sets this constant as the maximum velocity for any material object. It means in particular that all oscillator and wave velocities are lower than or equal to this value.

By convention the vacuum permeability is taken as a fixed value also, so the vacuum permittivity follows.

Vacuum permeability: $$\mu_0 = 4\pi \times 10^{-7}\,\text{H m}^{-1} \cong 12.566370614 \times 10^{-7}\,\text{H m}^{-1} \tag{F6.27}$$

Vacuum permittivity: $$\varepsilon_0 \cong 8.854187817 \times 10^{-12}\,\text{F m}^{-1} \tag{F6.28}$$

Oscillator and wave variables. The link between an oscillator and a wave is not always tight as in the case of propagation in a nonabsorbing medium. There are some cases where the oscillator and wave must be considered as two independent systems. The Formal Graph distinguishes two velocities, one for each system, attributing the symbol $\boldsymbol{u}$ to the wave velocity. In the present case of a harmonic and conservative oscillator, both velocities coincide (in considering the modulus of the wave velocity vector):

$$\|\boldsymbol{u}\| \underset{cst\ \mathcal{E}_Q}{=} u_Q \qquad \text{(F6.29)}$$

The wave velocity and the oscillator velocity differ in their physical nature: the wave velocity is a vector and space–time property whereas the oscillator velocity is a scalar and a system property, which is a function of the constitutive properties.

9.4.7 Case Study F7: Maxwell Equations — Electrodynamics

The case study abstract is given in next page.

In electrodynamics, the Maxwell* equations (formulated in 1864) are classically written as

Gauss (electrostatics):
$$\rho = \nabla \cdot \boldsymbol{D} \qquad \text{(F7.1)}$$

Gauss (electromagnetism):
$$0 = \nabla . \boldsymbol{B} \qquad \text{(F7.2)}$$

Maxwell–Faraday:
$$\boldsymbol{0} = \nabla \times \boldsymbol{E} + \frac{\partial}{\partial t}\boldsymbol{B} \qquad \text{(F7.3)}$$

Maxwell–Ampere:
$$\boldsymbol{j} = \nabla \times \boldsymbol{H} - \frac{\partial}{\partial t}\boldsymbol{D} \qquad \text{(F7.4)}$$

These four Maxwell equations are represented in Graph 9.24 in two Formal Graphs in two dimensions, with, in "pedagogical supplement," the algebraic writing of the operation represented placed inside a balloon in the vicinity of the node it concerns. (It must be recalled that all the Formal Graphs are totally explicit and do not need to be accompanied by algebra.) These graphs obviously describe links between dipoles, as only dipoles can include a temporal derivation, and they are distributed in space.

The four Maxwell equations describe the behavior of four fields that are the following vectors:

$\boldsymbol{E}$: electric field
$\boldsymbol{D}$: electric displacement
$\boldsymbol{H}$: electromagnetic field
$\boldsymbol{B}$: electromagnetic induction

* James Clerk Maxwell (1831–1879): Scottish physicist; Edinburgh and Cambridge, UK.

F7: Maxwell Equations **Electrodynamics**

$$\nabla . \boldsymbol{D} = \rho$$
$$\nabla . \boldsymbol{B} = 0$$
$$\nabla \times \boldsymbol{E} = -\frac{\partial}{\partial t}\boldsymbol{B}$$
$$\nabla \times \boldsymbol{H} = \boldsymbol{j} + \frac{\partial}{\partial t}\boldsymbol{D}$$

Formal Object:

Dipole assembly

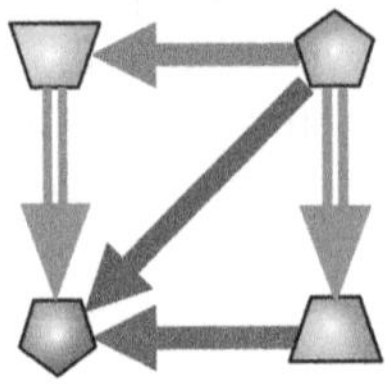

Dipole Assembly	
✓	Fundamental
	Mixed
	Capacitive
✓	Inductive
	Conductive
✓	Common effort
	Common flow
✓	Global
	Spatial

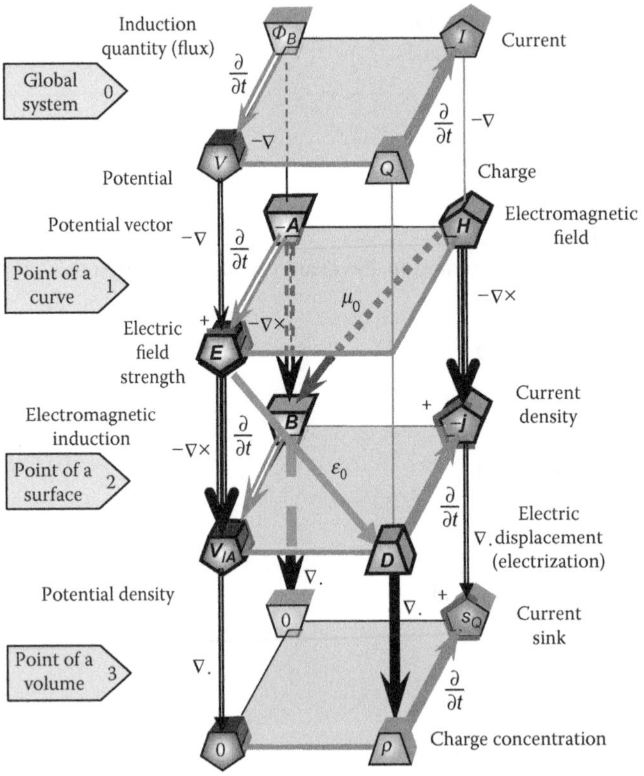

GRAPH 9.23

Variety	Electrodynamics			
Subvariety	Inductive (kinetic)		Capacitive	
Category	Entity number	Energy/ entity	Entity number	Energy/ entity
Family	Impulse 	Flow 	Basic quantity 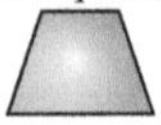	Effort 
Name	Induction quantity (flux)	Current	Charge	Potential
Symbols	Φ_B	I	Q	V
Unit	[Wb]	[A]	[C]	[V]

System Properties		
Nature	Inductance	Capacitance
Level	1/2	1/2
Name	Permeability	Permittivity
Symbol	μ	ε
Unit	[Wb m^{-1}]	[F m^{-1}]
Specificity	Three-dimensional	Three-dimensional

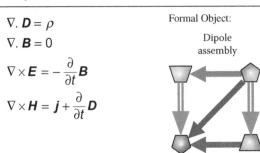

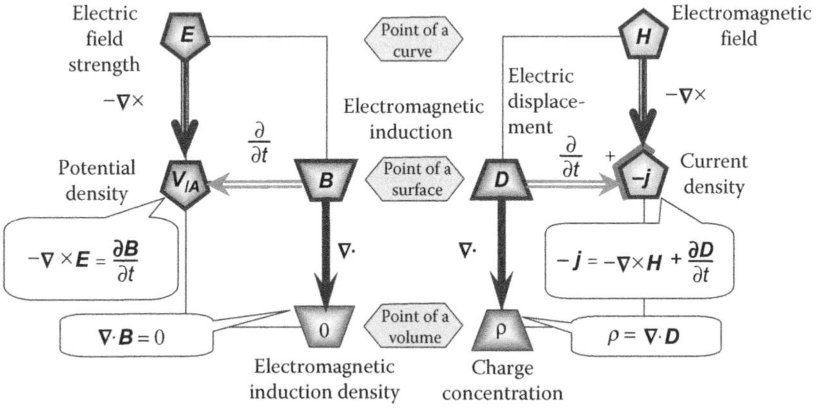

GRAPH 9.24

It must be remarked that these equations do not include the system constitutive properties (i.e., permittivity and permeability, which are specific properties in a three-dimensional space). They are considered as implicit in the above formulation of Maxwell equations because they link the previous fields in pairs.

The Formal Graph in three dimensions containing the Maxwell equations and the specific properties of the system (here the vacuum) is shown in the case study abstract.

In translating Maxwell equations, the following points must be noted:

- The representation in Formal Graph requires one to make explicit a reduced variable at the level of a surface, which is the surface density of potential $\boldsymbol{V}_{/A}$ and that works as the common value of the temporal derivative of the electromagnetic induction $\boldsymbol{B}$ and of the (contra-) curl of electric field $\boldsymbol{E}$.
- The choice of identical signs for all operators of spatial derivation whatever the state variable on which they apply, leads to represent the current density $\boldsymbol{j}$ with a minus sign. (It is the same for the potential vector $\boldsymbol{A}$.)
- The addition of converging paths on the node of the current density $-\boldsymbol{j}$ expresses the mounting in parallel of two dipoles distributed in space, a capacitive and an inductive dipole.
- The fact that the result of the divergence of the electromagnetic induction $\boldsymbol{B}$ (Gauss equation in electromagnetism) is equal to zero means that the (volumic) concentration of electromagnetic entities, improperly seen as "magnetic charges" (by abusive assimilation of the entity "quantity (flux) of induction" to a capacitive basic quantity) is nonexistent, in contradistinction with the concentration of electric charge.

Remarks

Several representations. These equations may be represented in Formal Graphs as well by two two-dimensional graphs as by a single three-dimensional graph. The four fields, $\boldsymbol{E}$ and $\boldsymbol{D}$ for the capacitive (electrostatic) energy and $\boldsymbol{B}$ and $\boldsymbol{H}$ for the inductive (electromagnetic) energy, are the reduced variables of the four families of state variables, effort for $\boldsymbol{E}$, basic quantity for $\boldsymbol{D}$, impulse for $\boldsymbol{B}$, and flow for $\boldsymbol{H}$. The Formal Graph in three dimensions shows that the two dipoles, capacitive and inductive, are mounted in parallel (addition of spatially reduced flows) in a "circuit" distributed in space.

The core of physics. Maxwell's equations are the heart of classical electrodynamics because they completely model the electrostatic and electromagnetostatic duality together with dynamics, which works between the two domains. They rely on the fundamental properties of space–time and they are considered as the foundation of every physical theory of phenomena occurring in this space–time. The theory of the relativity from Einstein, for instance is based on the invariance of these equations during a regular motion (constant velocity).

Not so fundamental. Notwithstanding the widespread belief in the most fundamental status of these laws, this assertion must be nuanced in adding that they apply only to *linear media* such as vacuum (free space).

As predicted by the Formal Graph theory, the linear behavior of system properties, at least capacitive ones, is valid as long as a coupling with another energy variety is not significant (see case study J6 "Ion Distribution" in Chapter 12, for instance). The question that the Formal Graph theory allows to be asked is, how far is the validity limit in vacuum?

9.4.8 Case Study F8: Hall Effect — Electrodynamics

The case study abstract is given in next page.

When an electromagnetic induction flux and a current are simultaneously imposed on a material containing electric charges, an electric field may appear inside the material that creates a potential difference between its edges. This is the Hall* effect, which is mainly used for measuring electromagnetic induction fields (see Figure 9.7).

According to the Formal Graph viewpoint, the Hall effect is a special conversion process (without involving time) between two subvarieties belonging to the same energy variety, that is, between electromagnetic and electrostatic energies.

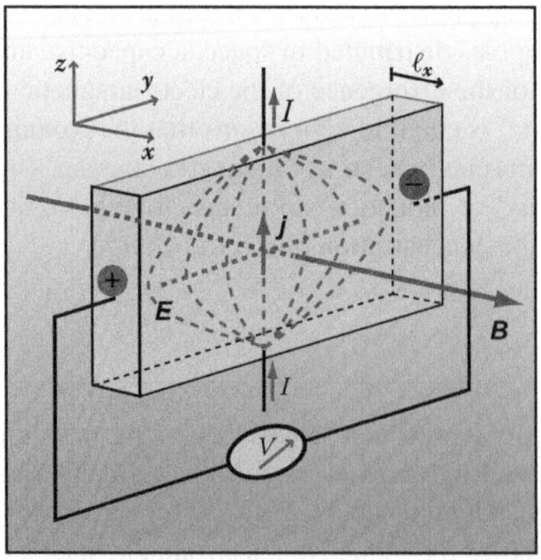

FIGURE 9.7 Scheme illustrating the principle of the Hall effect in a parallelepiped placed in an electromagnetic induction field $\boldsymbol{B}$. When a current I flows perpendicularly to the field $\boldsymbol{B}$, a potential difference V appears along the third perpendicular axis.

* Edwin Herbert Hall (1855–1928): North American physicist; Baltimore, Maryland, USA.

F8: Hall Effect **Electrodynamics**

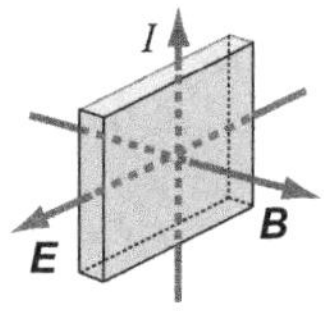

Formal Object:

Dipole assembly

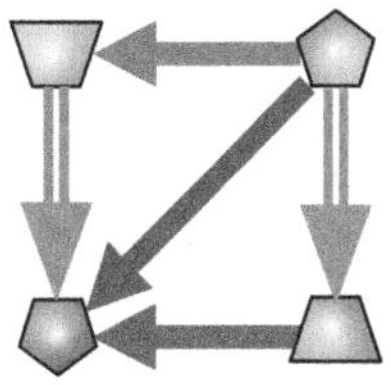

Dipole Assembly	
✓	Fundamental
	Mixed
	Capacitive
✓	Inductive
	Conductive
✓	Common effort
	Common flow
✓	Global
	Spatial

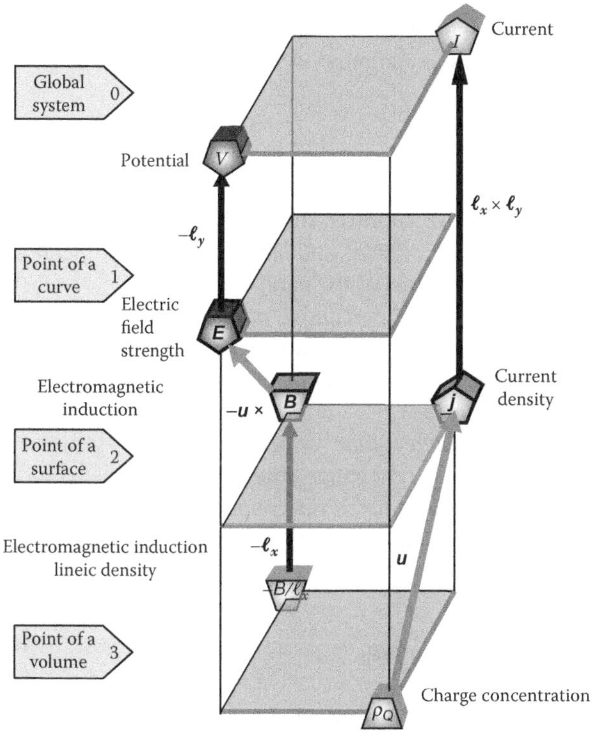

GRAPH 9.25

Variety	Electrodynamics			
Subvariety	Inductive (kinetic)		Capacitive	
Category	Entity number	Energy/entity	Entity number	Energy/entity
Family	Impulse	Flow	Basic quantity	Effort
Name	Induction quantity (flux)	Current	Charge	Potential
Symbols	Φ_B	I	Q	V
Unit	[Wb], [J A^{-1}]	[A]	[C]	[V], [J C^{-1}]

System Properties		
Nature	Inductance	Capacitance
Level	0	0
Name	Inductance	Capacitance
Symbol	L	C
Unit	[H]	[F]
Specificity	—	—

For simplification, the probe material is chosen with a parallelepiped shape. The electromagnetic induction field $\boldsymbol{B}$ is applied along the x-axis, corresponding to the smallest length of the material (thickness), which is supposed to be horizontal, in order to maximize the lineic density of electromagnetic induction field B/ℓ_x inside the material. The current I is imposed in the direction of another z-axis perpendicular to the first x-axis, and that we assume to be vertically oriented. The electric field $\boldsymbol{E}$ will then develop along the third y-axis supposed to be horizontal and perpendicular to the two first axes. The potential difference is expressed as a function of the modulus of the electromagnetic induction field B, of the current I, of the thickness ℓ_x, and of the volumic concentration n_e (in corp m^{-3}) of charge carriers (here electrons with a charge $-e$), by the formula

$$V = -\frac{I}{n_e e}\frac{B}{\ell_x} \tag{F8.1}$$

This formula is given in one of the most used shapes, with the concentration of charges expressed as the product $-n_e\, e$ instead of the dedicated variable ρ_Q (another well-used shape will be given at the end of this case study).

$$\rho_Q = -n_e e \tag{F8.2}$$

The formula in Equation F8.1 is valid in the case of a uniform distribution of fields and current lines in the material volume. The Formal Graph utilizes a three-dimensional representation for a homogeneous material and with a uniform distribution of fields and current lines. The operators of spatial integration are consequently replaced by the lengths of the parallelepiped.

What this Formal Graph shows is that not any system constitutive property links the variables, only some space–time properties are implicit, and that there is no explicit connection between variables of the left side and of the right side of the graph. For the formula in Equation F8.1, that models the phenomenon, is stating that such a relation does exist!

The explanation is called *internal coupling of space–time velocities*. The two subvarieties, inductive (electromagnetic) and capacitive (electrostatic), exchange energy through the space–time links that connect the entity numbers to the energies-per-entity. It is a coupling, or a kind of dynamic exchange, but between the two subvarieties of the same energy variety, which is justified from the First Principle of Thermodynamics. The Gibbs equation (generalized to any subvariety) applied to the two subvarieties that are present in the system is

$$\mathrm{d}\mathcal{E} = V\,\mathrm{d}Q + I\,\mathrm{d}\Phi_B \tag{F8.3}$$

For a stable system with as much electromagnetic energy entering the material as energy leaving it (and without any current used for measuring the potential) and with a distribution of entity numbers in a volume V, one obtains a null variation of energy, leading to

$$V\frac{\mathrm{d}Q}{\mathrm{d}V} = -I\frac{\mathrm{d}\Phi_B}{\mathrm{d}V} \tag{F8.4}$$

By replacing the volumic derivatives by the concentrations ρ_Q and $-B/\ell_x$ (the latter being the *lineic density of electromagnetic induction field*) and by rearranging the ratios for evidencing a *space–time volume rate* U (in m^3 s^{-1}), the previous equality becomes analogous to a coupling relationship:

$$U = \frac{V\ell_x}{B} = \frac{I}{\rho_Q} \tag{F8.5}$$

The formula in Equation F8.1 of the Hall effect immediately follows by replacing the charge concentration by its expression in Equation F8.2. The Formal Graph representing this velocity matching is given in Graph 9.26 (it may be considered as a part of the graph in the case study abstract by filtering only pertinent nodes).

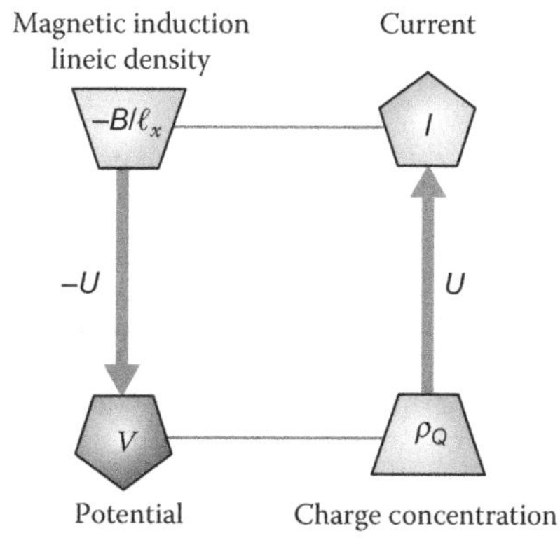

GRAPH 9.26

The role of space–time is to match energy exchanges in two directions, from capacitive toward inductive and vice versa, in such a way that the total energy, that is, the sum of inductive and capacitive energy, is a constant. However, in contradistinction with an oscillator or with the propagation of a wave, the role of space–time is not to proceed by derivation of the entity numbers but to make a change of scale to convert them into energies-per-entity. Therefore, there are no oscillations; the system is stable, even when velocities vary with time. This energy exchange without conversion is analogous to a dynamic equilibrium in the sense that forward and backward energies compensate each other (see Graph 9.26).

The previous relations can be demonstrated from classical equations of electrodynamics, allowing in passing the decomposition of the volume velocity in terms of length and space–time velocity $\boldsymbol{u}$. The demonstration brings to light the necessary precisions about the signs and orientations of vectors, which is detailed below.

A first starting point is the fact that the electric field is not defined in the Formal Graph theory by an operator producing a value, but by its position as a reduced variable at the level of a point of a curve or of a line in the family of electric potentials. The total, or apparent, electric field (noted with an over bar) is linked to (at least) two variables, the electric potential V from which it comes through a contragradient (here a derived field merely notated $\boldsymbol{E}$) and the electromagnetic induction field from which it comes through a vector product with the space–time velocity $\boldsymbol{u}$. Here the latter field is distinguished with a subscript indicating the field $\boldsymbol{B}$. The two expressions are

$$\boldsymbol{E} = -\frac{\partial}{\partial \boldsymbol{r}} V \qquad \text{(F8.6)}$$

$$\boldsymbol{E}_B = \boldsymbol{u} \times \boldsymbol{B} \qquad \text{(F8.7)}$$

The (apparent) electric field that exerts on a moving charge in the presence of a potential distribution is the sum of these two fields, as indicated by the Formal Graph in Graph 9.27. It must be remarked that the Lorentz force originates from this apparent potential. (See remark on *Lorentz force* at the end of the case study.)

$$\bar{\boldsymbol{E}} = \boldsymbol{E} + \boldsymbol{E}_{\boldsymbol{B}} \qquad \text{(F8.8)}$$

This sum of fields corresponds to a mounting of dipoles in series (in the same way as summing efforts). It is equal to zero because of the absence of other electric fields, notably external fields. Another way to say this is to assert that the electric field $\boldsymbol{E}_{\boldsymbol{B}}$ created by the induction field provokes a distribution of electric potential V, which in turn helps create a derived field $\boldsymbol{E}$ compensating the first field. From this compensation is deduced a "convective" relationship between the two vectors $\boldsymbol{E}$ and $\boldsymbol{B}$, which determines the orientation of the derived electric field (cf. scheme on next page).

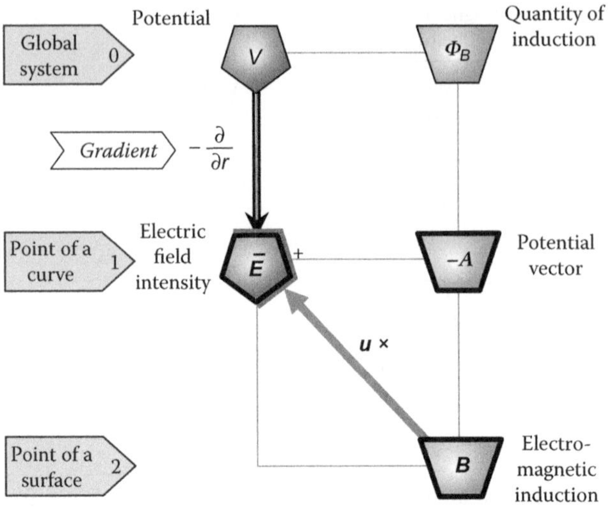

GRAPH 9.27

$$\boldsymbol{E} = -\boldsymbol{u} \times \boldsymbol{B} \tag{F8.9}$$

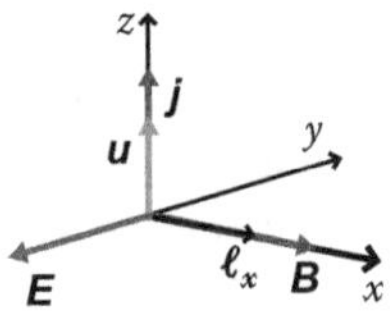

Assuming a uniform distribution of the electric field, the difference of electric potential on the parallelepiped edges is proportional to the electric field (cf. scheme on the below).

$$V \underset{hom}{=} -\boldsymbol{E} \cdot \ell_y \tag{F8.10}$$

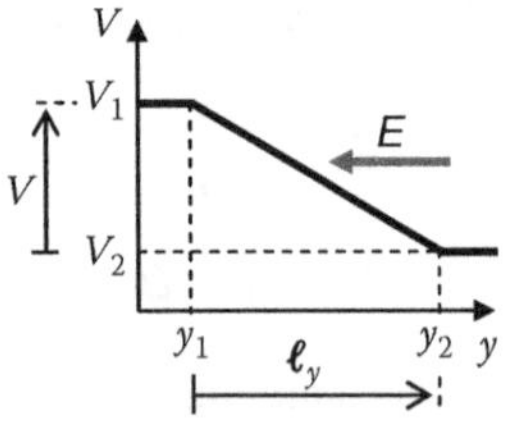

A second starting point is the convective relationship between charge concentration ρ_Q and current density $\boldsymbol{j}$ that links these two variables through the space–time velocity $\boldsymbol{u}$:

$$\boldsymbol{j} = \boldsymbol{u}\, \rho_Q \tag{F8.11}$$

Combining these three last relations (F8.9 through F8.11) gives

$$V \underset{hom}{=} \frac{(\boldsymbol{j} \times \boldsymbol{B}) \cdot \ell_y}{\rho_Q} = \frac{(\boldsymbol{B} \times \ell_y) \cdot \boldsymbol{j}}{\rho_Q} \tag{F8.12}$$

The vector $\boldsymbol{B}$ of the electromagnetic induction field being colinear to the vector thickness $\boldsymbol{\ell}_x$, it can be expressed with the modules of these vectors B and ℓ_x:

$$\boldsymbol{B} = \frac{B}{\ell_x}\boldsymbol{\ell}_x \tag{F8.13}$$

Then comes the expression of the potential

$$V \underset{hom}{=} \frac{B}{\ell_x}\frac{\left(\boldsymbol{\ell}_x \times \boldsymbol{\ell}_y\right)\cdot \boldsymbol{j}}{\rho_Q} \tag{F8.14}$$

In the case of a homogeneous distribution of current lines, the total current I is the scalar product of the surface vector by the current density:

$$I \underset{hom}{=} \left(\boldsymbol{\ell}_x \times \boldsymbol{\ell}_y\right)\cdot \boldsymbol{j} \tag{F8.15}$$

Finally, the sought-after expression is

$$V \underset{hom}{=} \frac{B}{\ell_x}\frac{I}{\rho_Q} \tag{F8.16}$$

Note that specialists in this domain do not express this relation as a function of the inverse of the charge concentration ρ_Q as above, but as a function of a *Hall coefficient* equal to this inverse:

$$\mathcal{R}_H = \frac{1}{\rho_Q} \tag{F8.17}$$

Remarks

Lorentz force. The Lorentz[*] force is the product of the charge by the apparent potential, which amounts to saying that this force is the sum of forces exerted by the simultaneous action of an electric field $\boldsymbol{E}$ and of an electromagnetic induction field $\boldsymbol{B}$ on a charge Q moving with a velocity $\boldsymbol{u}$.

$$\boldsymbol{F} \underset{lin}{=} Q\left(\boldsymbol{E} + \boldsymbol{u} \times \boldsymbol{B}\right) \tag{F8.18}$$

[The indication "*lin*" under the equality sign is there to remind one that the validity domain of this expression is restricted to the linearity domain of electrical capacitances (see case study J3 "Electrical Force" in Chapter 12)]. In classical mechanics, the nullity of the Lorentz force, according to Newton's second law (see case study F1 "Accelerated Motion"), means that the movement of charges, endowed with a mass, is uniform and this implies the stability of current lines. In the Formal Graph generalization, it corresponds to an isolated (nonpowered) system.

Charge accumulation? In a large number of textbooks, the proposed explanation of the phenomenon is that the movement of charges is deviated by the electromagnetic induction field and that it is the accumulation of charges on one side of the probe that creates the potential difference.

(continued)

[*] Hendrik Antoon Lorentz (1853–1928): Dutch physicist; Leiden, Netherlands.

The treatment allowed by the Formal Graphs theory shows that such is not the case. The charges are not deviated, because the potential difference establishes itself in order to compensate the effect of the electromagnetic field on the moving charges. On the contrary, the classic formula can only be established in the case of a uniform distribution of current lines and of charges.

To corroborate this remark, it is worth making a comparison with the Righi*–Leduc† effect, called thermal Hall effect, which is the equivalent of the Hall effect, but here it is a temperature gradient that appears under the effect of a electromagnetic field (instead of a gradient of electric potential) (Nolas et al. 1962). If one makes the analogy with the explanation found in textbooks, one should have an accumulation of entropy (heat) on one side. However, entropy is not known to be sensitive to a thermal field for producing a force!

(The Righi–Leduc effect is not treated in this book but other thermoelectric effects are tackled in Chapter 12.)

Concept of internal coupling. The internal coupling of space–time velocities illustrated by this case study is a fundamental concept in the Formal Graph theory. Its feature is to associate closely a capacitive subvariety with an inductive subvariety of energy without implying any derivation operator with respect to time. An internal coupling of frequencies is equally a concept that naturally stems from the previous one in case of the absence of distribution in the space. Each energy subvariety can therefore be associated with the other one without involving time. This is very important in *singletons* when entities are not in variable quantity (which would mean that they could be derivated with respect to the time) but are indissociable, as are numbers of quantum objects.

Neither mechanics nor particles. This case study shows in addition that it is possible to model, and then to think of, such a system in remaining in the frame of a unique energy variety, without involving either the mechanical variety or the variety of particles collection: neither the notion of force nor the notion of displacement nor the one of mass has been necessary. Not anymore the notion of number of charge carriers, except for retrieving at the end the classic formula linking the charge concentration to this "particular" notion. This means that the notion of particle is not a requirement for thinking of such a system. This is a considerable advantage for tackling in an intelligible way the notion of wave–corpuscle duality.

9.5 TEMPORAL OSCILLATOR

When the system is isolated from the exterior, the energy in the variety is conserved, and the only possible change is the energy conversion between the two subvarieties. As energy variation of an isolated system is null, so is the power received by the system. According to its definition in Equation 9.40, the latter is obtained by dividing the total energy variation equation 9.16 of a system composed of two energy subvarieties by $\mathrm{d}t$:

$$f_{qL}\frac{\mathrm{d}}{\mathrm{d}t}p_q + e_{qC}\frac{\mathrm{d}}{\mathrm{d}t}q \underset{isol}{=} 0 \tag{9.58}$$

* Augusto Righi (1850–1920): Italian physicist; Bologna, Italy.

† Sylvestre Anatole Leduc (1856–1937): French physicist; Paris, France.

If only one space–time property is assumed, say the one linking the basic quantity to the capacitive flow (Equation 9.28),

$$f_{qC} \overset{def}{=} \frac{\mathrm{d}}{\mathrm{d}t} q \tag{9.59}$$

the derivative of the basic quantity can be replaced by the capacitive gate variable f_{qC}, allowing the capacitive effort to be expressed as

$$e_{qc} \underset{isol}{=} -\frac{f_{qL}}{f_{qc}} \frac{\mathrm{d}}{\mathrm{d}t} p_q \tag{9.60}$$

Now, when the flows are common,

$$f_q = f_{qL} = f_{qc} \tag{9.61}$$

the flow ratio becomes equal to unity, leading to

$$e_{qc} \underset{isol}{=} -\frac{\mathrm{d}}{\mathrm{d}t} p_q \tag{9.62}$$

which is the classical expression when the reference to the subvariety is dropped. Comparison with the potential gate variable definition in Equation 9.29 given earlier shows that coherence with the previous presentation is verified; that is, the resulting effort is null for a common flow mounting and for an isolated system.

Dipole assembly effort: $$e_{qL} + e_{qc} \underset{isol}{=} 0 \tag{9.63}$$

This development is also a justification of the identity between the two operators for the space–time properties (i.e., for conversion from capacitive to inductive subvarieties on the one hand and in the opposite direction on the other hand). Without this identity, the conservation of the energy in the variety could not be ensured. The Formal Graph representing the set of Equations 9.3, 9.4, 9.59, 9.61, and 9.62 is drawn in Graph 9.28a and b by using the null potential in Equation 9.63.

A symmetrical reasoning is made in case of common effort, by assuming the link between impulse and inductive potential given in Equation 9.29:

$$e_q = e_{qc} = e_{qL} \overset{def}{=} \frac{\mathrm{d}}{\mathrm{d}t} p_q \tag{9.64}$$

The isolation of the system leads to the symmetrical relation of Equation 9.28

$$f_{qL} \underset{isol}{=} -\frac{\mathrm{d}}{\mathrm{d}t} q \tag{9.65}$$

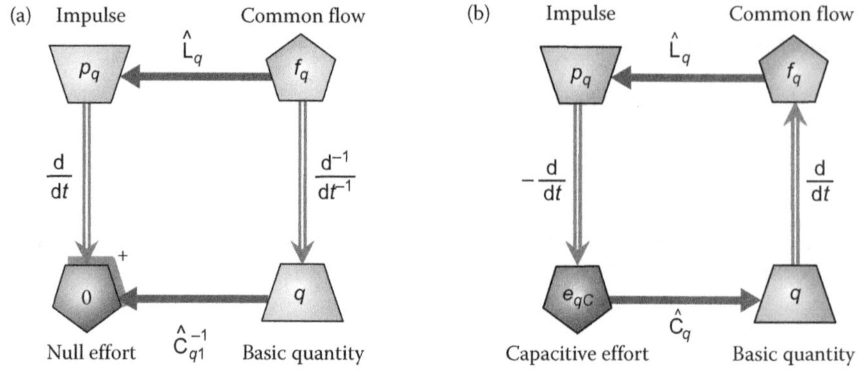

GRAPH 9.28 Common flow (serial mounting) representation (a) and loop representation (b) of an isolated conservative oscillator.

and isolation of the dipole assembly flow is written in these conditions

$$f_{qL} + f_{qC} \underset{isol}{=} 0 \tag{9.66}$$

All these relationships are represented in Graph 9.29, either in the common effort representation or in the loop representation.

Closed-loop graphs express a rotation. The energy conversion between the two subvarieties, after having proceeded from one variety toward the other, reverses the conversion direction, and so on. There is no stationary solution for the four equations shown in Graph 9.28b or in Graph 9.29b other than all state variables being equal to zero, which would mean that the system is deprived from energy. Following the direction given by the two space–time properties and the two system properties, a rotation with a positive sign stems from Graphs 9.28b and 9.29b, suggesting the cycling conversion energy and the periodicity of the oscillator. In taking for instance Graph 9.28b, the following relationship can be read in following the paths represented by the properties, starting from the basic quantity:

$$\hat{C}_q\left(-\frac{d}{dt}\right)\hat{L}_q\frac{d}{dt}q = q \tag{9.67}$$

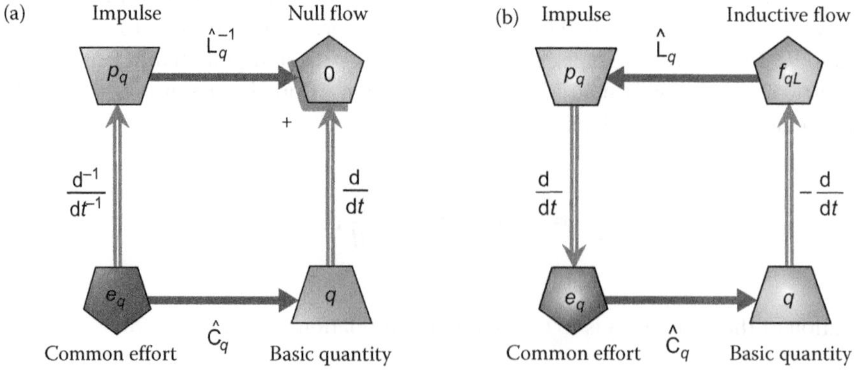

GRAPH 9.29 Common effort (parallel mounting) representation (a) and loop representation (b) of an isolated conservative oscillator.

This circular relationship constitutes the algebraic model issued from the Formal Graph of a conservative oscillator.

A mathematical trick? Solving the previous differential equation in the general case of operators is a difficult problem. Notwithstanding, simplifications can be envisaged that make the search for a solution a trivial problem for many students trained in very basic analytical techniques.

The *classical approach* assumes that system properties are time-independent *scalars*, meaning that both capacitive and inductive relationships are *linear*. The constitutive properties can therefore be grouped into a pulsation variable called *pulsation* (without more precision in the classical approach)

$$\omega \underset{lin}{=} L_q^{-1/2}\, C_q^{-1/2} \tag{9.68}$$

which is suitable for compacting the differential equation into

$$\frac{d^2}{dt^2} q \underset{lin}{=} -\omega^2 q \tag{9.69}$$

The algebraic solution of this equation is proposed in every textbook and its expression as a function of time uses harmonic functions, *sine* or *cosine*. In the case of a nonzero variable at time $t = 0$, the solution is written as

$$q \underset{lin}{=} q(0) \cos \omega t \tag{9.70}$$

This is never demonstrated in basic textbooks but justified *a posteriori* by the derivation rules given with the definition of the harmonic functions

$$\frac{d}{dt} \cos\omega t = -\omega \sin\omega t \tag{9.71}$$

$$\frac{d}{dt} \sin\omega t = \omega \cos\omega t \tag{9.72}$$

This is a real pity, because many interesting and fundamental properties of the physical phenomenon are missed. In the following sections, the demonstration will allow one to define a crucial concept, the *wave function*, by going through the definition of the imaginary number and of the exponential function. All are fundamental tools that are not just mathematical beings but important concepts helping to understand physics as much as possible.

9.5.1 Reduction of the Formal Graph

The Formal Graph tool can be used to find the time dependence of the state variables of an oscillator. It is worthwhile to detail the procedure because of the importance of oscillating systems in the whole of physics and especially in wave mechanics. This allows one to show how a wave function can be built graphically and to pave the way for introducing Fourier-transformed Formal Graphs.

The fact that the time derivation occurs twice in following the sequence of operators in Graph 9.28 or 9.29, or, algebraically speaking, the fact that Equation 9.67 is a second-order differential equation, lends to searching for reducing the degree from 2 to 1, as is conventionally done in solving algebraically such equations. Graphically speaking, the technique consists of finding two Formal Graphs that are solutions of the "second-order" Formal Graph, as shown in Graph 9.30.

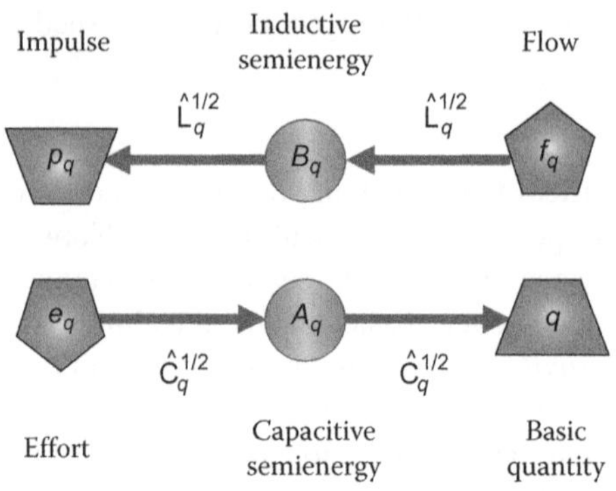

GRAPH 9.30 Insertion of halfway intermediate variables called "semienergies" into each system property to allow the separation into equivalent pieces.

At halfway between energies-per-entity and entity numbers are inserted new variables that will allow the separation of the Formal Graph in two parts. These halfway intermediate variables are defined as:

$$A_q \overset{def}{=} \hat{C}_q^{1/2} e_q \overset{def}{=} \hat{C}_q^{-1/2} q \tag{9.73}$$

$$B_q \overset{def}{=} \hat{L}_q^{1/2} f_q \overset{def}{=} \hat{L}_q^{-1/2} p_q \tag{9.74}$$

In doing this partition, the constitutive properties of the system are split because their operators are raised to the power one-half. They are abbreviated *semiproperties* in the following development.

The halfway intermediate variables A_q and B_q are in fact energetic variables having for unit the square root of the Joule, as proven by the following equations that are deduced from the previous definitions:

$$\begin{aligned} A_q^2 &= \hat{C}_q^{1/2} e_q \ \hat{C}_q^{-1/2} q \underset{lin}{=} e_q\, q \underset{lin}{=} 2\mathcal{U}_q \\ B_q^2 &= \hat{L}_q^{1/2} f_q \ \hat{L}_q^{-1/2} p_q \underset{lin}{=} f_q\, p_q \underset{lin}{=} 2\mathcal{T}_q \end{aligned} \tag{9.75}$$

In the linear case, when the system constitutive properties are scalars, the squares of these variables become equal to twice the energies. In the nonlinear case, they will have different values from twice the energy, but the unit $J^{-1/2}$ remains in all cases. Hence, the term *semienergies* for A_q and B_q (in the multiplicative meaning of the term).

In an isolated system, the sum of inductive and capacitive energies being equal to total energy, the sum of squared semienergies can be shown, in the linear case, as proportional to this total energy:

$$A_q^2 + B_q^2 \underset{lin}{=} 2\mathcal{E}_q \tag{9.76}$$

With these semienergies and semiproperties, the Formal Graph of a conservative oscillating system (Graph 9.28b) can be split into two "first-order" Formal Graphs, that is, using only one time derivation each, as is shown in Graph 9.31.

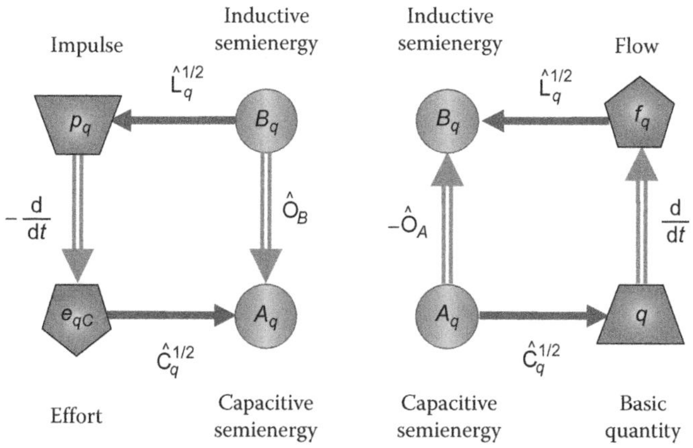

GRAPH 9.31 Splitting of the Formal Graph of a conservative oscillator into two "first-order" graphs. Each graph is a solution of the whole Formal Graph given in Graph 9.28b.

In these graphs, the relationships between semienergies are featured by two operators $\hat{O}_A$ (when applying to semienergy A_q) and $\hat{O}_B$ (when applying to semienergy B_q), which are to be determined. In fact, if commutativity exists, they are identical because they are involved in the same operator loop, as can be algebraically proven. The algebraic sequences of operators forming a loop are, for both graphs, identical to the identity operator, so the unknown operators can be expressed as functions of the known operators

$$\hat{O}_B = \hat{C}_q^{1/2}\left(-\frac{d}{dt}\right)\hat{L}_q^{1/2} \tag{9.77}$$

$$\hat{O}_A = -\hat{L}_q^{1/2}\left(\frac{d}{dt}\right)\hat{C}_q^{1/2} \tag{9.78}$$

Their identity can be proved when system properties are time-independent operators. (This may not be the unique possibility.) The condition to fulfill, independency of system properties with time, is expressed by the requirement of commutativity of these operators with the time derivation, which can be written using Poisson's brackets:

Restriction.

Commutativity: $$\left[\frac{d}{dt}, \hat{C}_q^{1/2}\right] = \frac{d}{dt}\hat{C}_q^{1/2} - \hat{C}_q^{1/2}\frac{d}{dt} = 0 \tag{9.79}$$

Commutativity: $$\left[\frac{d}{dt}, \hat{L}_q^{1/2}\right] = \frac{d}{dt}\hat{L}_q^{1/2} - \hat{L}_q^{1/2}\frac{d}{dt} = 0 \tag{9.80}$$

By virtue of this commutativity, the unknown operators become

$$\hat{O}_B = \hat{C}_q^{1/2}\hat{L}_q^{1/2}\left(-\frac{d}{dt}\right) \tag{9.81}$$

$$\hat{O}_A = -\hat{L}_q^{1/2}\hat{C}_q^{1/2}\left(\frac{d}{dt}\right) \tag{9.82}$$

When in addition one assumes the commutativity of the constitutive semiproperties,

Restriction.

Commutativity: $$\left[\hat{L}_q^{1/2}, \hat{C}_q^{1/2}\right] = \hat{L}_q^{1/2}\hat{C}_q^{1/2} - \hat{C}_q^{1/2}\hat{L}_q^{1/2} = 0 \tag{9.83}$$

it appears that both operators are clearly the same operator

$$\hat{O}_A = \hat{O}_B = \hat{O} \tag{9.84}$$

As this operator is composed of a derivation, and as constitutive semiproperties are assumed to be independent of time, it can be described by a derivation with respect to a new variable φ, which will be seen shortly after as corresponding to the *phase angle* that was sought:

$$\frac{d}{d\varphi(t)} \stackrel{def}{=} -\hat{C}_q^{1/2}\hat{L}_q^{1/2}\frac{d}{dt} \tag{9.85}$$

Once this variable is explicitly defined as being dependent on time, the reference to time can be momentarily dropped in order to substitute the unknown operator by the derivation with respect to the phase angle φ:

$$\hat{O} = \frac{d}{d\varphi} \tag{9.86}$$

Toward generality. The reasoning was to define semienergies as exactly halfway (in the multiplicative meaning) between energies-per-entity and entity numbers. Then, an attempt was made to find the conditions for having equal but opposite operators linking these semienergies, which were to impose commutativity between operators. Another approach is possible in reversing the reasoning; keeping the noncommutativity of operators and finding other positions, different from halfway, for the "semi"-energies. That would allow a more general treatment (but too cumbersome for to discuss here). The idea to retain is that the factor 1/2 used almost everywhere corresponds only to the peculiar case of linearity of system constitutive properties.

Finally, taking into account the equality in Equation 9.84 between operators in both disassembled "first-order" Formal Graphs in Graph 9.31 and the common expression in terms of angular derivation, leads to these two relationships between semienergies:

$$B_q = -\frac{d}{d\varphi}A_q \tag{9.87}$$

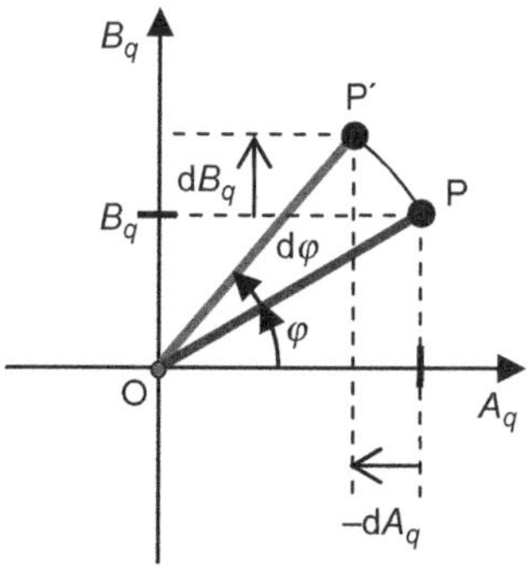

FIGURE 9.8 Differential rotation of a vector OP around its origin O with an angle $d\varphi$ giving a vector OP′.

$$A_q = \frac{d}{d\varphi} B_q \tag{9.88}$$

It is here that the meaning of a rotation angle can be attributed to the variable φ when the two half variables are seen as components $[A_q, B_q]$ of a vector OP whose representative point P rotates around its origin O in a plane. A variation $d\varphi$ of the angle provides component variations

$$dA_q = -B_q d\varphi \tag{9.89}$$

$$dB_q = A_q d\varphi \tag{9.90}$$

so the new rotated vector is related to the former one

$$\begin{bmatrix} A_q + dA_q \\ B_q + dB_q \end{bmatrix} = \begin{bmatrix} 1 & -d\varphi \\ d\varphi & 1 \end{bmatrix} \begin{bmatrix} A_q \\ B_q \end{bmatrix} \tag{9.91}$$

through a matrix that is a rotation operator. The geometrical picture of this rotation is given in Figure 9.8.

The rotation not only applies to the loop in a Formal Graph but also features the cycling conversion of energy in an oscillator. The energy of the oscillator alternately changes from one subvariety to another, each turn replacing the system in the same energy state, that is, with the same proportion of capacitive and inductive energy. In this process, the role of the phase angle φ is to quantify the conversion advance.

Returning to the representation of an oscillator with a Formal Graph, the two disassembled graphs can be associated again as shown in Graph 9.32.

Once this intrinsic feature of rotation is established, it is possible to pursue the reduction of the number of variables representing the oscillator, which has led from the quadruplet of state variables (q, e_q, p_q, f_q) to the construction of the couple (A_q, B_q) of semienergies, in searching for a single variable, depending on the phase angle, for quantifying the oscillator.

9.5.2 Definition of the Wave Function

Graph 9.33 is built with a linear combination of semienergies arbitrarily chosen to model a conservative (and linear) oscillator in the simplest possible way.

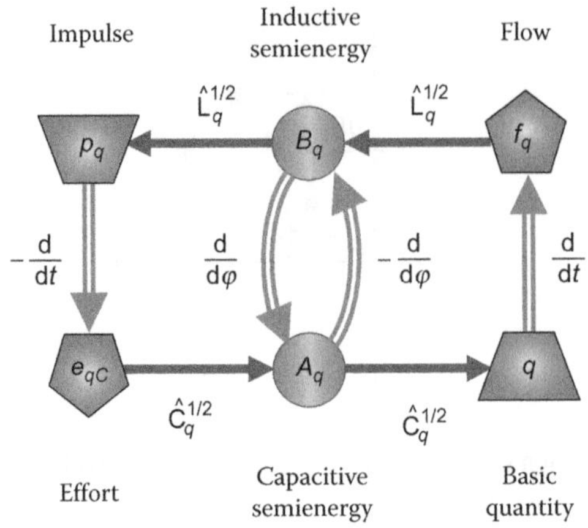

GRAPH 9.32 Insertion into the Formal Graph of an oscillator of halfway intermediate variables ("semienergies") leads to a reversible direct relationship between these intermediate variables expressing a rotation.

Graph 9.33 represents the linear combination using an unknown algebraic coefficient notated i, which can be of any value, given that one shall define as a *temporal wave function* $\psi_q(t)$, that is, depending on time (and not on space). Later, when dealing with the role of space, one shall define a *spatial wave function* $\psi_q(\mathbf{r})$; the product of these two functions will be shown as corresponding to the *wave function* of a propagating wave.

Linear combination:
$$\psi_q \overset{def}{=} A_q + i\, B_q \tag{9.92}$$

For being exhaustive, this definition relies on the definition of the coefficient i that follows.

9.5.3 Definition of the Imaginary Number

In Graph 9.33 is also represented another linear combination

$$i\,\psi_q = i\,A_q + \left(-i^2\right)\left(-B_q\right) \tag{9.93}$$

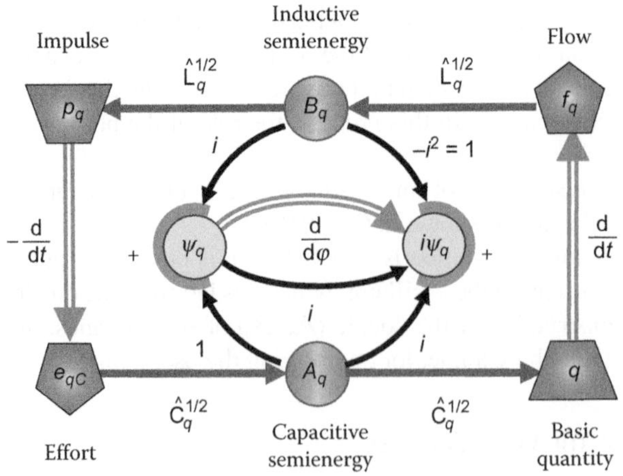

GRAPH 9.33 The connections between the various ensembles of variables able to quantify an oscillator.

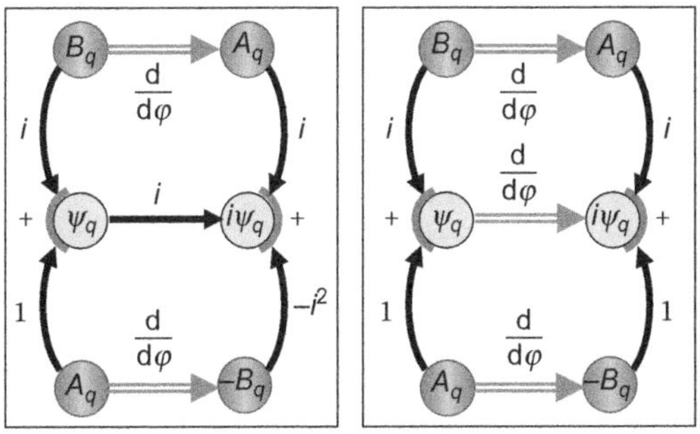

GRAPH 9.34 Two similar graphs representing the same relationships between semienergies and their linear combination. They demonstrate the relationship $-i^2 = 1$ by purely graph rules.

obtained when the rotation operator is applied to the semienergies. In order to decompose in a clearer way the reasoning that has led to the previous graph, two other graphs are proposed in Graph 9.34. On the left are shown the two previous combinations and on the right is represented the result of the application of the rotation operator to the wave function, which is also a linear combination:

$$\frac{d}{d\varphi}\psi_q = i\,A_q - B_q \tag{9.94}$$

These two graphs are similar but not identical. Unless a special value is chosen for the coefficient i in such a manner as to have the identity

Deliberate choice: $$-i^2 \overset{def}{=} 1 \tag{9.95}$$

Obviously, the resemblance with the imaginary number is difficult to hide anymore to advanced readers, but the purpose is to clearly evidence that the use of complex numbers is far from being a computational trick as generally presented.

Besides this rather elegant justification of the necessity to employ imaginary numbers, the real interest of this development lies much more in Graph 9.35, which is the central part of Graph 9.33, constituting the conclusion of the identity of the two previous graphs.

This Formal Graph encodes, through an equivalence of two paths, the link between the eigen-function and the eigen-value of the rotation operator, which is expressed algebraically as

$$\frac{d}{d\varphi}\psi_q = i\,\psi_q \tag{9.96}$$

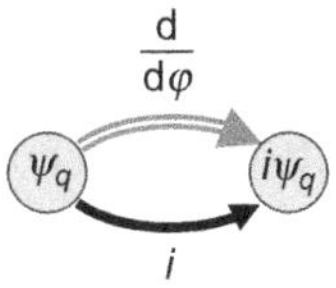

GRAPH 9.35 The eigen-function and eigen-value of a derivation operator viewed by the Formal Graph language.

Thus, the imaginary number is the eigen-value of the derivation with respect to the phase angle when the eigen-function is the wave function.

9.5.4 Definition of the Exponential Function

By replacing $i\,\varphi$ with x and by dropping the reference to the q variety for the sake of mathematical generalization, a trivial recast of this equation under the form

$$\frac{d\psi(x)}{\psi(x)} = dx \tag{9.97}$$

allows one to put into evidence a fundamental property of a wave function, as shown in the following. When the wave function exists for two independent values of its argument, the addition of two previous equations is written as

$$\frac{d\psi(x_1)}{\psi(x_1)} + \frac{d\psi(x_2)}{\psi(x_2)} = dx_1 + dx_2 \tag{9.98}$$

It can be observed that the left-hand term of this equation may be viewed as the result of the derivation of the product of the two wave functions and by setting

$$x = x_1 + x_2 \tag{9.99}$$

the right-hand term can be identified with Equation 9.97, leading to the following differential equation between wave functions:

$$\frac{d\left(\psi(x_1)\,\psi(x_2)\right)}{\psi(x_1)\,\psi(x_2)} = \frac{d\psi(x)}{\psi(x)} \tag{9.100}$$

This differential equation can be integrated by introducing as integration constant the wave function for the null argument

$$\frac{\psi(x)}{\psi(0)} = \frac{\psi(x_1)}{\psi(0)}\frac{\psi(x_2)}{\psi(0)} \tag{9.101}$$

This result demonstrates the multiplicative property of wave functions, which is an extremely important property because it provides the foundation for establishing the wave function of oscillators simultaneously based on time and space that are the support of propagating waves. This result also provides the definition of the exponential function: This universally known function, notated exp, is defined as obeying the rule of derivation given by the differential equation 9.97, leading to the above multiplication property and having the value 1 when its argument equates zero. The wave function normalized by its value for the null argument corresponds to such a case:

Exponential Function:
$$\frac{\psi(x)}{\psi(0)} \overset{def\rightarrow}{=} \exp(x) \tag{9.102}$$

Universal function. Each time multiplication and addition are simultaneously required for linking two pairs of variables, the exponential function is there. It is not astonishing to find it in a lot of domains because this combination of simple operations is quite frequent. In the Formal Graph theory, it has already been demonstrated that influence between poles and conduction were ruled by this property, entailing the expression of the capacitive and conductive relationships to use the exponential function (Chapters 7 and 8). The physical meaning corresponds to the amplification of a variable on incremental increase of a command variable.

The preceding development and definition is summarized by the following correspondence between equations involving the wave function:

$$\frac{\mathrm{d}}{\mathrm{d}\varphi}\psi_q = i\,\psi_q \quad \Leftrightarrow \quad \psi_q = \psi_q(0)\exp(i\varphi) \tag{9.103}$$

9.5.5 Definition of the Harmonic Functions

This definition of the exponential function leads to a closed-form expression of the solution of the differential equation 9.96, in recalling also the linear combination in Equation 9.92 of the semienergies

Wave function:
$$\psi_q \stackrel{def}{=} A_q + i\,B_q = \psi_q(0)\exp(i\,\varphi) \tag{9.104}$$

When one defines the *conjugate complex* of the wave function as

$$\psi_q^* \stackrel{def}{=} A_q - i\,B_q = \psi_q(0)\exp(-i\,\varphi) \tag{9.105}$$

one finds by linear combinations the expression for each semienergy:

$$A_q = \psi_q(0)\,\frac{\exp(i\varphi)+\exp(-i\varphi)}{2} \tag{9.106}$$

$$B_q = \psi_q(0)\,\frac{\exp(i\varphi)-\exp(-i\varphi)}{2\,i} \tag{9.107}$$

It is just a matter of convention to shorten these mathematical expressions in defining two new functions

$$\cos\varphi \stackrel{def}{=} \frac{\exp(i\varphi)+\exp(-i\varphi)}{2} \tag{9.108}$$

$$\sin\varphi \stackrel{def}{=} \frac{\exp(i\varphi)-\exp(-i\varphi)}{2\,i} \tag{9.109}$$

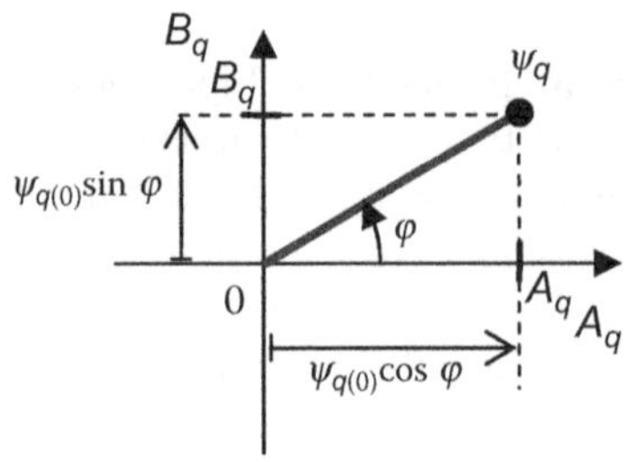

FIGURE 9.9 Polar representation of the wave function as a vector with components A_q and B_q, according to Euler's formula.

that are called *harmonic functions*. The short notation for the semienergies is therefore

$$A_q = \psi_q(0)\cos\varphi \tag{9.110}$$

$$B_q = \psi_q(0)\sin\varphi \tag{9.111}$$

Note that the main properties of the harmonic functions in derivation operations can be retrieved from the relationships between semienergies.

It is worth deriving a last interesting relationship. From Equations 9.108 and 9.109 the Euler's* formula can be directly deduced:

$$\exp(i\varphi) = \cos\varphi + i\sin\varphi \tag{9.112}$$

This is one of the most important formulas in physics. One of the reasons is that it gives support to the graphical representation of complex numbers in polar coordinates, as illustrated in Figure 9.9 with the wave function represented by a vector having the semienergies as components.

Not so basic. It may appear superfluous in such a book presenting a new physical theory, to reformulate such primary and well-known mathematics as done here. The reason is that behind these apparently purely mathematical objects is hidden a physical meaning that constitutes foundation of many physical theories. The role of the multiplicative property of wave functions is crucial in quantum mechanics and the duality between multiplicative/additive properties is the seed for proposing exponential functions in many domains. In support of this insistence, one may invoke Feynman's judgment, who does not hesitate to present Euler's formula as the "most beautiful formula ever written" (Feynman 1964).

9.5.6 Wave Function Modulus

According to its definition in Equation 9.92 as a linear combination of semienergies, the wave function has the same unit, the $J^{1/2}$. This is naturally the same for the *conjugate* of the wave function and the product of these two variables gives the squared modulus of the wave function.

* Leonhardt Euler (1707–1783): Swiss mathematician; Basel, Switzerland and St. Petersburg, Russia.

$$\left\|\psi_q\right\|^2 = \psi_q\,\psi_q^* = A_q^2 + B_q^2 \underset{lin}{=} 2\mathcal{E}_q \tag{9.113}$$

This is peculiarly interesting as it establishes the proportionality between the modulus of the wave function and the square root of the total energy of the oscillator (by recalling that it is only in the case of linear system constitutive properties that coefficient 2 is found). It must be outlined that this modulus is a constant only in the case of a conservative oscillator.

Is normalization compulsory? This is more or less a matter of taste because the scale of the wave function is indifferent when using its differential or exponential expression 9.103. Many definitions of the wave function are made by dividing it by the square root of $A_q^2 + B_q^2$, thus normalizing it in the interval [−1, +1] and depriving the wave function from unit. This requires holding two variables for characterizing the state of an oscillator, the normalized wave function and the quantity by which it has been normalized. What may become different from a question of taste is when the second variable is abandoned. In particular, as a result of this, the normalized wave function is unable to model a nonconservative oscillator (as in the case of a damped oscillator, for instance). This is one of the difficulties met in quantum physics for taking into account dissipation and irreversibility.

9.5.7 Oscillator Pulsation and Wave Pulsation

Putting together the definition of the phase angle (Equation 9.85) and the differential equation of the wave function (Equation 9.96) provides a differential equation

$$\frac{d}{d\varphi}\psi_q = -\hat{C}_q^{1/2}\,\hat{L}_q^{1/2}\,\frac{d}{dt}\psi_q = i\,\psi_q \tag{9.114}$$

that leads to the expression of the time derivative of the wave function

$$\frac{d}{dt}\psi_q = \hat{L}_q^{-1/2}\,\hat{C}_q^{-1/2}\,(-i)\,\psi_q \tag{9.115}$$

This differential equation can be made more compact by defining a *pulsation operator*

$$\hat{\omega} \overset{def}{=} i\frac{d}{dt} \tag{9.116}$$

and by introducing a *wave pulsation* variable ω identified with the eigen-value of the pulsation operator when its eigen-function is the wave function.

$$\hat{\omega}\,\psi_q = \omega\,\psi_q \tag{9.117}$$

This *wave pulsation*, also called *angular frequency*, ω, is a real scalar and it has the status of a state variable for the energy of a wave. Some precisions will be given soon.

Now, the interesting constraint to impose for using this *wave pulsation* is to assume that the last two equations constitute the same model as Equation 9.115, which may occur only when the composition of the two constitutive properties is equivalent to a scalar for eliminating the imaginary number. For handling this, a *natural pulsation* ω_q is defined as the result of this composition when it is a scalar (i.e., when the operator can be substituted by its eigen-value).

Restriction. Harmonicity

Natural pulsation: $$\omega_q \overset{def}{=} \hat{L}_q^{-1/2} \hat{C}_q^{-1/2} \sim scalar \quad (9.118)$$

Note that this restriction is less stringent than the classical assumption in Equation 9.68 requiring that both constitutive properties should be individually scalars. Notably, it allows nonlinear reciprocal functions that combine in a commutative way to produce a scalar. This is the hypothesis that makes the system a harmonic oscillator because its state variables behave "in just proportions" (which is the Greek etymology of *harmony*). The above definition of the *natural pulsation* ω_q corresponds exactly to the classical definition of the pulsation ω, made for modeling an oscillator, without distinguishing the two pulsations.

The consideration of harmonicity, energy conservation, and of the last four equations helps find a match between the pulsation variable ω and the natural pulsation ω_q:

$$\omega \underset{cst\,\mathcal{E}_q}{=} \omega_q \quad (9.119)$$

Nonlinearity and commutativity. These are operator properties that are not incompatible. For instance, the two operators defined by the following functions

$$y = a^{\frac{1}{p_2-1}} x^{p_1}; \quad y = a^{\frac{1}{p_1-1}} x^{p_2}$$

are nonlinear operator but they commute.

Energy, system, and space–time. Attention must be drawn to the different physical natures of the pulsation. The *pulsation operator* $\hat{\omega}$ belongs to space–time, the *natural pulsation* ω_q belongs to the system, and the *wave pulsation* ω belongs to energy. If this distinction is generally made in quantum physics for the pulsation operator which is clearly defined by Equation 9.116, this is rarely true in classical physics for the pulsation variable ω which is generally confused with the natural pulsation ω_q and therefore directly defined from the system constitutive properties through Equation 9.119.

The Formal Graph approach introduces the wave pulsation ω as an eigen-value of the pulsation operator, before establishing the link with the natural pulsation ω_q that depends on the system constitutive properties of the oscillator. This separation between system, space–time, and energy is of paramount importance for understanding their mutual relationship.

An analogous situation can be found with the time constant τ for a relaxation process. It is also a variable pertaining to space–time, defined as an eigen-value of the time derivation (without imaginary number as here), which matches a *system time constant* τ_q defined as a composition of system constitutive properties in a given system (and relatively to an energy variety q).

Furthermore, translation of quantum physics relations into Formal Graphs requires the wave pulsation ω to be defined (this is indeed the true definition) as a *state variable* of a peculiar energy variety, called *phase energy*, contained in waves* (borne by photons, for instance). This is incompatible with the classical definition of the pulsation as a function of an inductance times a capacitance (which are hard to define for a photon).

9.5.8 Fourier Transform

Another method of establishing the relationship between the pulsation operator and its eigen-value is to resort to an integral transform called Fourier transform. With this mathematical operation, which uses a parameter that can be notated here ω, the derivation operator is transformed into the product of this parameter with the imaginary number:

$$\frac{\mathrm{d}}{\mathrm{d}t} \xrightarrow{\text{Fourier } \omega} i\omega \tag{9.120}$$

Several definitions of the Fourier transform are possible for obtaining this result, either by directly using ω or by replacing it with the product $2\pi\,\nu$, which amounts to using the frequency as a parameter instead of the pulsation. Here the first definition is used and the transform of a derivable function $f(t)$ is written in this case (Woan 2003):

Fourier ω:
$$\tilde{f}(\omega) = \frac{1}{\sqrt{2\pi}} \int_{-\infty}^{+\infty} f(t) \exp(-i\omega t)\,\mathrm{d}t \tag{9.121}$$

The demonstration of the expected result (Equation 9.120) is rather straightforward using an integration by part:

$$\int_{-\infty}^{+\infty} \frac{\mathrm{d}}{\mathrm{d}t} f(t) \exp(-i\omega t)\,\mathrm{d}t = \left[f(t)\exp(-i\omega t)\right]_{-\infty}^{+\infty} + i\omega \int_{-\infty}^{+\infty} f(t)\exp(-i\omega t)\,\mathrm{d}t \tag{9.122}$$

If the function $f(t)$ is derivable, the term between brackets vanishes and the relationship in Equation 9.120 is demonstrated. It can be remarked that the coefficient in front of the integral operator in the Fourier definition is indifferent in this demonstration; it plays only the role of a normalizing factor for having a unitary transform, that is, which gives the same result when using the inverse transform (which is written in a symmetrical expression by changing just the sign of the parameter ω).

There is no mystery in this "happy coincidence" as the Fourier transform is built on the properties of the exponential function and on the classical rules of derivation of ordinary functions with the aim to replace derivations with multiplications. In other words, the Fourier transform is not a purely technical tool as often said, but a true link between derivation operator and eigen-value. Naturally, variables other than the pair (ω, t) can be used for defining other Fourier transforms, as

* In quantum physics, Planck's formula ($E = \hbar\omega$) relates the wave pulsation to the energy-per-corpuscle.

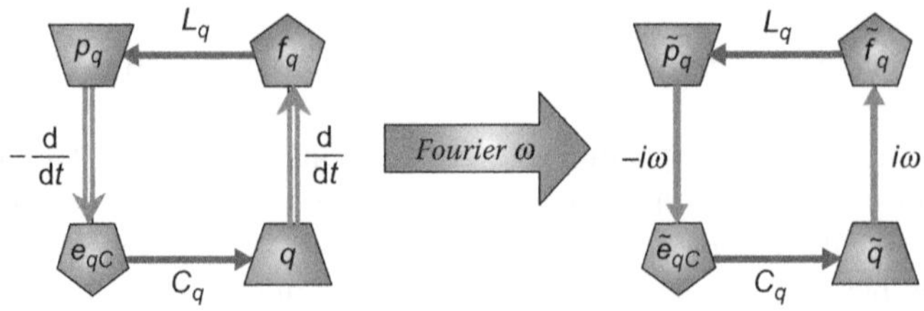

GRAPH 9.36 In the case of linear system constitutive properties, the Fourier transform of a Formal Graph is another Formal Graph using the same constitutive properties but transformed state variables and space–time properties.

for instance in the case of the spatial oscillator. (That is the explanation of the indication of the parameter in the calling of the transformation.)

A property of the Fourier transform that may not appear necessary to evidence is the linearity of this operator. It means on the one hand that the transform of a sum of two functions is the sum of the two transforms, and on the other hand that the multiplication with a constant coefficient is indifferent in the transformation. Because it has nonnegligible consequence in the present development, this last property is made explicit:

$$a = C^{st} \Rightarrow af(t) \xrightarrow{\text{Fourier } \omega} a\tilde{f}(\omega) \tag{9.123}$$

The purpose of this introduction of the Fourier transform in this study is to provide another insight in this important and fundamental relationship between operator and eigen-value. An interesting use of the Fourier transform is its application to Formal Graphs in order to obtain transformed ones. This is feasible in the general case of nonlinearity of operators, but mainly useful in the linear case because linear operators independent of time are conserved through the transformation.

In restricting the constitutive properties of the oscillator to the classical case of linearity, Graph 9.36 gives the result of the transformation of the Formal Graph of a conservative temporal oscillator.

The interest of this Fourier-transformed Formal Graph is to provide directly a transposition of the differential Equation 9.67 by applying the circularity principle and the definition of the natural pulsation (Equation 9.118)

Linear case: $$C_q(-i\omega)L_q(i\omega)\tilde{q} = \omega^2\, C_q L_q\, \tilde{q} = \omega^2\, \omega_q^{-2}\, \tilde{q} = \tilde{q} \tag{9.124}$$

from which is deduced the equivalence between wave pulsation and natural pulsation. It can be remarked that, if a conductance existed in the system (case of a nonconservative oscillator), no loop could be formed in such a simple manner and the circularity principle could not be applied to establish this equivalence between pulsations.

A powerful but limited tool. The Fourier transform is extremely interesting from the theoretical viewpoint for tackling the relationship between derivation operators and eigen-values, and therefore as a technical tool for simplifying differential equations. The perception of the symmetry between time and frequency (or pulsation) or between space and wavelength (or wave vector) facilitated by the Fourier approach is widely recognized as an essential advance of physics.

(continued)

Notwithstanding this importance, caution must be exercised about the limitation of its practical applicability to linear relationships between functions and variables. From a more philosophical viewpoint, the beautiful simplicity of the Fourier transform must not hide the fact that physics occupies and depicts a wider world than the linear domain.

9.5.9 Dependence on Time

Before ending the discussion on the temporal oscillator, it remains to formulate the mathematical solution in terms of the dependence of the state variables of the oscillator with time.

With the restriction of harmonicity (Equation 9.118) and the equivalence (Equation 9.119) between pulsations, the phase angle definition in Equation 9.85 can now be written as

$$\frac{\mathrm{d}}{\mathrm{d}\varphi} = -\frac{1}{\omega}\frac{\mathrm{d}}{\mathrm{d}t} \tag{9.125}$$

or, more practically, as

$$\frac{\mathrm{d}\varphi}{\mathrm{d}t} = -\omega \tag{9.126}$$

which establishes by simple integration, as the pulsation is time independent, proportionality between time and phase angle

$$\varphi(t) = -\omega\, t \tag{9.127}$$

From Equations 9.110 and 9.111 the semienergies are expressed as functions of time:

$$A_q = \psi_q(0)\cos\omega\, t \tag{9.128}$$

$$B_q = -\psi_q(0)\sin\omega\, t \tag{9.129}$$

The last step consists of recalling the link between these semienergies and the state variables of the system in order to obtain the expressions of these latter as functions of time:

$$q = \hat{C}_q^{1/2}\left(\psi_q(0)\cos\omega\, t\right) \tag{9.130}$$

$$p_q = \hat{L}_q^{1/2}\left(-\psi_q(0)\sin\omega\, t\right) \tag{9.131}$$

The initial value of the wave function is therefore

$$\psi_q(0) = \hat{C}_q^{-1/2}\, q(0) \tag{9.132}$$

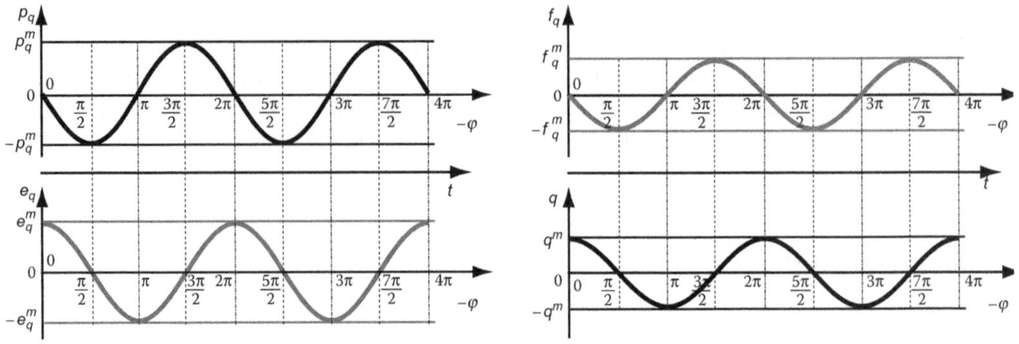

FIGURE 9.10 Plots of the variations of the four state variables as a function of minus the phase angle.

Now, when the system constitutive properties are both linear operators, the classical solution is retrieved:

$$q \underset{lin}{=} q^m \cos \omega t \tag{9.133}$$

$$p_q \underset{lin}{=} -L_q^{1/2} C_q^{-1/2} q^m \sin \omega t = -p_q^m \sin \omega t \tag{9.134}$$

The harmonic functions are symmetrical functions with respect to time, confirming the insensitivity of the oscillator behavior on a time reversal $t \rightarrow -t$. Naturally, all the state variables also follow a harmonic function, *cosine* (an even function) for capacitive state variables, effort e_q, and basic quantity q, and *sine* (an odd function) for inductive state variables (impulse p_q and flow f_q). In Figure 9.10 are plotted the variations of these four state variables assuming scalar constitutive properties (linear case).

It is noteworthy to outline that both basic quantity and impulse are in phase opposition; when one is at its maximum or minimum, the other is zero. It is the same for the energy-per-entities; effort and flow are also in phase opposition. This conveys very well the way the system converts energy, by giving an alternate role to each energy subvariety. In Figure 9.11 are plotted the variations of inductive energy and capacitive energy in the case of a linear system (i.e., having scalar constitutive properties). The plot is built on the following algebraic equations:

$$d\mathcal{U}_q = e_q \, dq \underset{lin}{=} -\psi_q(0)^2 \sin\varphi \cos\varphi \, d\varphi; \quad \mathcal{U}_q \underset{lin}{=} \frac{\psi_q(0)^2}{2} \cos^2 \varphi \tag{9.135}$$

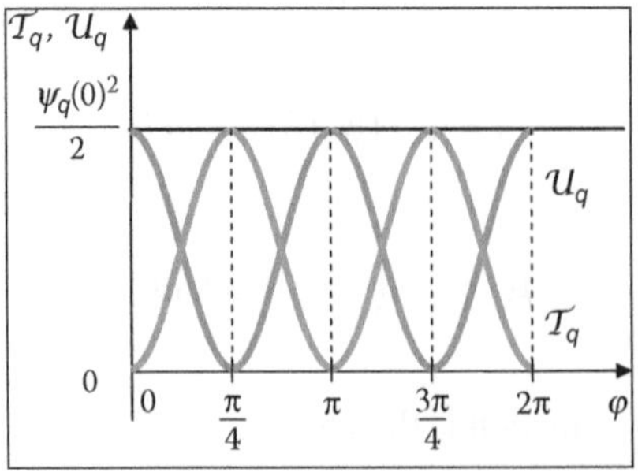

FIGURE 9.11 Variations of inductive energy $\mathcal{T}_q$ and capacitive energy $\mathcal{U}_q$ in a conservative temporal oscillator as a function of the phase angle φ.

$$d\mathcal{T}_q = f_q dp_q \underset{lin}{=} \psi_q(0)^2 \sin\varphi \cos\varphi \, d\varphi; \quad \mathcal{T}_q \underset{lin}{=} \frac{\psi_q(0)^2}{2} \sin^2\varphi \tag{9.136}$$

Nevertheless, this is not at all compatible with a rigid link (in the sense of time independent) between the state variables of these two energy subvarieties as it happens when a conductance property exists in the system. This strong constraint is solved by the system in converting a part of the energy into the thermal variety at each rotation, oscillations are damped and the oscillator loses little by little its original energy. This degradation process will be described in Chapter 11.

Frequency and period. Two other variables are useful for featuring a temporal wave. They are the *frequency* ν or f and the *period* T. The period is defined as the duration between two points of the same amplitude and same slope in a temporal wave as illustrated in the following figure. More rigorously, it corresponds to the angle 2π between two values of the phase angle φ giving the same result for the wave function.

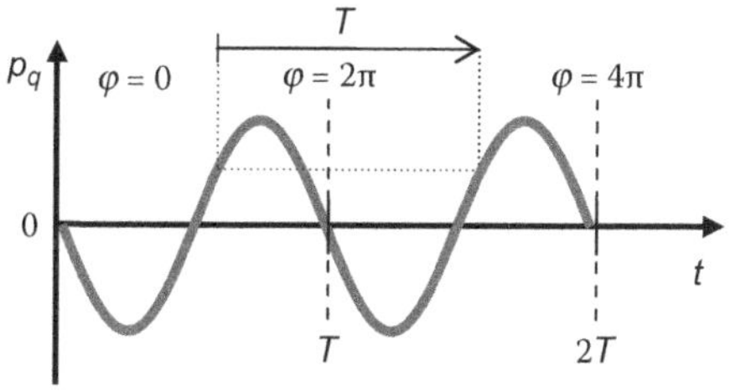

The *frequency* ν quantifies the number of periods contained in a time unit. It is therefore the inverse of the period T. Their relationship with the pulsation is written as

$$\omega = \frac{2\pi}{T} = 2\pi\,\nu \tag{9.137}$$

The period is the formal analogue of the *wavelength* λ of a spatial wave.

9.6 SPATIAL OSCILLATOR

A spatial oscillator works in a purely stationary condition, using only space dimensions for developing a wave-shaped spatial structure presenting regular periodicity. A spatial wave may exist in one, two, or three dimensions, linearly or in a circular way for the simplest cases. A good example is the interface between a soil and a fluid, which can be shaped by the wind (sand in the desert) or by sea flows (sand undulation on the sea shore). By analogy with the temporal oscillator, the Formal Graph of a spatial oscillator is given in Graph 9.37.

The system constitutive properties featured in this graph are space-reduced properties from the global level (0) to the first depth (level 1) and vice versa. To distinguish them, they are subscripted with the reduction levels. One property is the 0–1-reduced capacitance (subscripted /01) and the other is the 0–1-reduced elastance, which is equal to the reciprocal of the 1–0-reduced capacitance (subscripted /10, which is the one drawn in Graph 9.37).

$$q = \hat{\mathbf{C}}_{q/10}\, \mathbf{e}_{q/rC} \tag{9.138}$$

$$e_q = \hat{\mathbf{C}}_{q/01}^{-1}\, \mathbf{q}_{/r} \tag{9.139}$$

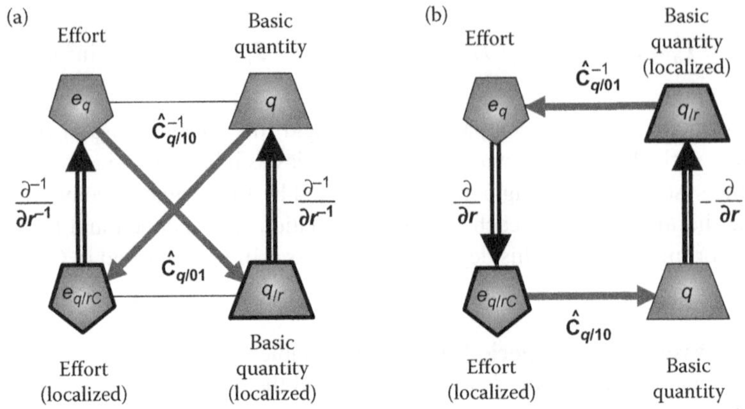

GRAPH 9.37 Formal Graph of the spatial oscillator in serial mounting in the loop representation, with parallel orientation of space derivations (a) and with antiparallel orientation (b).

9.6.1 Wave Function of a Spatial Oscillator

The difference with the Formal Graph of the temporal oscillator previously given in Graph 9.28b lying just in the replacement of the derivation variable t with $-r$, it ensues that the same treatment can be carried out to translate this graph into a wave function graph. By defining semienergies as follows,

$$A_q = \hat{C}_{q/10}^{1/2}\, \mathbf{e}_{q/rC} = \hat{C}_{q/10}^{-1/2}\, q \tag{9.140}$$

$$B_q = \hat{C}_{q/01}^{-1/2}\, \mathbf{q}_{/r} = \hat{C}_{q/01}^{1/2}\, e_q \tag{9.141}$$

and by defining a *phase angle* as dependent on the space vector $\boldsymbol{r}$

$$\frac{\mathrm{d}}{\mathrm{d}\varphi(\boldsymbol{r})} \overset{def}{=} \hat{C}_{q/10}^{1/2}\hat{C}_{q/01}^{-1/2}\frac{\mathrm{d}}{\mathbf{dr}} \tag{9.142}$$

one naturally assumes the commutativity between constitutive properties and space derivation as was done for the temporal oscillator. Note the positive sign in this definition in contradistinction with the equivalent expression 9.85 for the temporal oscillator. However, this definition allows the two semienergies to be related in the same way (see Graph 9.38):

$$B_q = -\frac{\mathrm{d}}{\mathrm{d}\varphi}A_q \tag{9.143}$$

$$A_q = \frac{\mathrm{d}}{\mathrm{d}\varphi}B_q \tag{9.144}$$

The definition of the wave function $\psi_q(\boldsymbol{r})$ is rigorously identical to the time-dependent wave function

$$\psi_q(\boldsymbol{r}) \overset{def}{=} A_q + i\,B_q \tag{9.145}$$

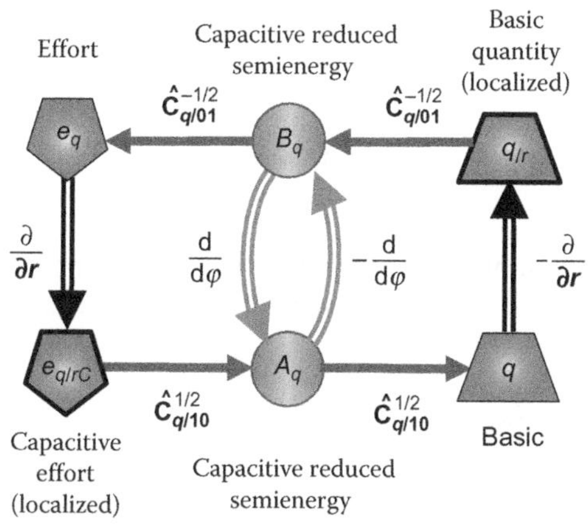

GRAPH 9.38 Insertion into the Formal Graph of a spatial oscillator of halfway intermediate variables ("semienergies") allows the definition of the derivation with respect to the phase angle φ.

as a linear combination of the semienergies using the imaginary number i previously defined by $-i^2 = 1$.

$$\frac{d}{d\varphi}\psi_q(\mathbf{r}) = i\,\psi_q(\mathbf{r}) \tag{9.146}$$

The differential definition of the exponential function can be recognized in Equation 9.146, leading to the simple expression

$$\psi_q(\mathbf{r}) = \psi_q(0)\,\exp(i\varphi) \tag{9.147}$$

allowing the solution of the differential equation to be expressed as a harmonic function

$$q = q^{max}\cos\varphi \tag{9.148}$$

9.6.2 Dependence on Space

To find the dependence of the wave function and of the phase angle on the space vector $\mathbf{r}$, the technique is the same as for the temporal oscillator. It consists of the definition of the *wave vector operator* in a similar manner as for the wave pulsation operator, that is, by taking a derivation operator, with a proportionality factor chosen for obtaining a real solution:

$$\hat{\mathbf{k}} \overset{def}{=} -i\frac{d}{d\mathbf{r}} \tag{9.149}$$

To this operator is associated an eigen-vector of the eigen-function ψ_q, corresponding to the wave vector variable $\boldsymbol{k}$

$$\hat{\mathbf{k}}\,\psi_q(\mathbf{r}) = k\,\psi_q(\mathbf{r}) \tag{9.150}$$

From the definition in Equation 9.142 of the phase angle and from the differential equation 9.146 of the wave function, one may write

$$-\frac{\mathrm{d}}{\mathrm{d}\boldsymbol{r}}\psi_q(\boldsymbol{r}) = \hat{\mathbf{C}}_{q/10}^{1/2}\,\hat{\mathbf{C}}_{q/01}^{-1/2}\; i\;\psi_q(\boldsymbol{r}) \tag{9.151}$$

The assumption that the combination of reduced constitutive properties does not result in an operator but in a vector is made in an identical manner as for the temporal oscillator.

Restriction: *Harmonicity*

Natural space vector: $\boldsymbol{k}_q \overset{def}{=} \hat{\mathbf{C}}_{q/01}^{1/2}\hat{\mathbf{C}}_{q/10}^{-1/2} \sim \textit{vector}$ (9.152)

This assumption, with the conservation of energy confers the property of harmonicity on the oscillator. It leads to the simple expression for the *wave vector*

$$\boldsymbol{k} \underset{cst\,\mathcal{E}_q}{=} \boldsymbol{k}_q \tag{9.153}$$

Substitution of the combination of reduced constitutive properties by the wave vector into the definition (Equation 9.142) of the phase angle gives

$$\frac{\mathrm{d}\varphi}{\mathrm{d}\boldsymbol{r}} = \boldsymbol{k} \tag{9.154}$$

By integrating in assuming independency of the wave vector operator with the space vector and in taking both origins equal to zero, the final result is got

Phase angle: $\varphi(\boldsymbol{r}) = \boldsymbol{k}\cdot\boldsymbol{r}$ (9.155)

Wavelength and wave vector. Another variable is frequently used in lieu of the wave vector which is the wavelength $\boldsymbol{\lambda}$. It is a vector too, but working in a reciprocal way and related to the wave vector in such a way that the dot product of the two vectors equates the full angle 2π. Another way to express their relationship is to write

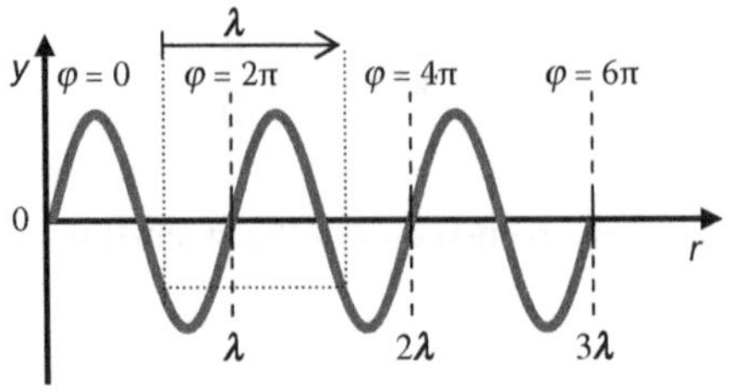

(continued)

$$k = \frac{2\pi}{\lambda} \tag{9.156}$$

The wavelength has the advantage of being a less abstract notion because it corresponds to the distance between two points of the same amplitude and same slope in a spatial wave as illustrated in the above figure. The wavelength is the formal analogue of the *period* T of a temporal wave.

9.7 SPATIOTEMPORAL OSCILLATOR

The typical Formal Graph of a space–time oscillator is shown in Graph 9.39 in three dimensions. Such a three-dimensional Formal Graph implying two space levels has been given for the vibrating string and for the electrodynamical wave in case studies F4 and F6. The three-dimensional Formal Graph for the hydrodynamic acoustic wave in case study F5 used four space levels but its structure can be exactly compacted to a lower number of levels, as done in Graph 9.39, without losing the basic structure of a space–time oscillator. The nonscalar system properties in this energy variety are not an issue because the requirement of scalar system properties, as found in the classical examples of vibrating string and electrodynamical wave, is not compulsory (only their product should be).

The diagonal link on each side wall is a *space–time operator* that is represented by a derivation versus a space–time vector $\boldsymbol{w}$, with a negative sign borne by the right-side link. This *space–time operator* is defined as the commutative combination of the time derivation with the space integration

$$\frac{\partial}{\partial \boldsymbol{w}} \overset{def}{=} \frac{\partial}{\partial t}\frac{\partial^{-1}}{\partial \boldsymbol{r}^{-1}} = \frac{\partial^{-1}}{\partial \boldsymbol{r}^{-1}}\frac{\partial}{\partial t} \tag{9.157}$$

This vector operator (it applies to a scalar for giving a vector or the reverse) represents the composition of space and time that is used for simplifying the writing.

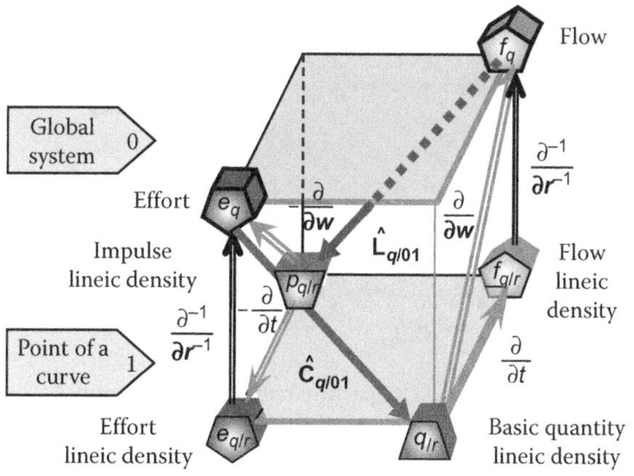

GRAPH 9.39 Typical three-dimensional Formal Graph of a space–time oscillator.

As before, two semienergies are defined at a half-operating distance of each system property:

Capacitive semienergy: $$A_q \overset{def}{=} \hat{\mathbf{C}}_{q/01}^{1/2}\, \mathbf{e}_q \overset{def}{=} \hat{\mathrm{C}}_{q/01}^{-1/2}\, q_{/r} \tag{9.158}$$

Inductive semienergy: $$B_q \overset{def}{=} \hat{\mathrm{L}}_{q/01}^{1/2}\, \mathbf{f}_q \overset{def}{=} \hat{\mathrm{L}}_{q/01}^{-1/2}\, p_{q/r} \tag{9.159}$$

Projection from above of the previous three-dimensional Formal Graph without the unnecessary localized energy-per-entities $e_{q/rC}$ and $f_{q/r}$ gives an analogue of Graph 9.32, and when semienergies are inserted, the following Graph 9.40 is obtained. By applying the circularity principle to this graph, one finds that the sequence of space–time and system operators is equivalent to an operator identified with the derivation with respect to a variable φ that is a *phase angle*.

$$\frac{\mathrm{d}}{\mathrm{d}\varphi} \overset{def}{=} -\hat{\mathbf{C}}_{q/01}^{1/2}\, \hat{\mathrm{L}}_{q/01}^{1/2} \frac{\partial}{\partial \mathbf{w}} \tag{9.160}$$

The assumption of commutativity and independence with space and time of the system properties is necessary in order to have a single phase angle. The parallelism of the paths leads to the expression of the reversibility of the link between the two semienergies in terms of two symmetrical differential equations

$$B_q = -\frac{\mathrm{d}}{\mathrm{d}\varphi} A_q \tag{9.161}$$

$$A_q = \frac{\mathrm{d}}{\mathrm{d}\varphi} B_q \tag{9.162}$$

Then, a wave function is defined by linearly combining the semienergies with the help of the imaginary number *i* previously defined

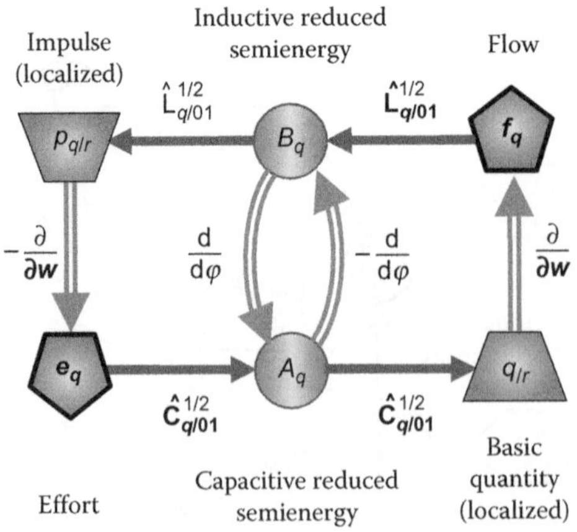

GRAPH 9.40 Insertion into the Formal Graph of a space–time oscillator of halfway intermediate variables ("semienergies").

$$\psi_q \overset{def}{=} A_q + i\,B_q \tag{9.163}$$

obeying the differential equation because of the choice $-i^2 = 1$

$$\frac{d}{d\varphi}\psi_q = i\,\psi_q \tag{9.164}$$

This equation is the differential definition of the exponential function as already seen, thus modeling the wave function of the spatiotemporal oscillator with exactly the same expression as for the temporal and spatial oscillators

$$\psi_q = \psi_q(0)\,\exp(i\,\varphi) \tag{9.165}$$

The consequence is that the solutions in terms of harmonic functions for the semienergies and state variables are exactly identical to the previous solutions given.

9.7.1 Relationship with Temporal Oscillator and Spatial Oscillator

It seems feasible that the spatiotemporal oscillator, which is the generator of wave propagation, can be described as a combination of two oscillators, a temporal one and a spatial one. However, this has to be proven.

Each oscillator possesses a pair of constitutive properties, an inductive one and a capacitive one; they determine the pulsation ω (hence the frequency) and the wave vector $\boldsymbol{k}$ (hence the wavelength) of the propagating wave. To establish this relationship between the spatiotemporal oscillator and its constituents, a relationship between these pairs of constitutive properties has to be found.

In Graph 9.41 are drawn some of the space–time and system constitutive properties that are involved in the temporal oscillator and in the spatial one. (All properties are not drawn for the sake of clarity.) The pair of global constitutive properties is indicated in the upper level whereas only one member of the other pairs is shown, because it will be sufficient to establish the sought-after relationship.

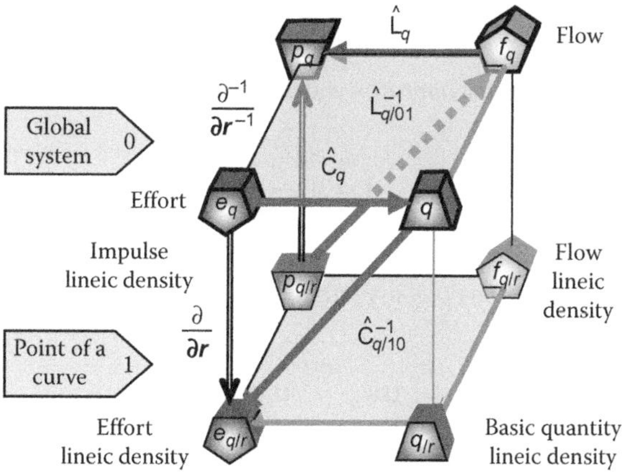

GRAPH 9.41 Correspondence between spatial, global, and reduced constitutive properties of a system.

In the rear plane of the graph are drawn the global inductance of the temporal oscillator and the reduced reluctance (reciprocal of the inductance) used in the spatiotemporal oscillator. The combination of these two inductive properties amounts to space integration between the localized and global impulse.

In the front plane, the spatiotemporal oscillator is featured by the global capacitance whereas the spatial oscillator is represented by its reduced elastance (reciprocal of the capacitance). As for the previous properties, their combination gives a spatial operator, but this time, it is a space derivation. These equivalences are expressed algebraically by two equations:

$$\frac{\partial^{-1}}{\partial \boldsymbol{r}^{-1}} = \hat{\mathsf{L}}_q \, \hat{\mathsf{L}}_{q/01}^{-1} \tag{9.166}$$

$$\frac{\partial}{\partial \boldsymbol{r}} = \hat{\mathsf{C}}_{q/10}^{-1} \, \hat{\mathsf{C}}_q \tag{9.167}$$

In order to relate the combination of spatiotemporal constitutive properties to those of the other oscillators, the following sequence is built that evidences each pair:

$$\hat{\mathsf{C}}_{q/01}^{-1} \, \hat{\mathsf{L}}_{q/01}^{-1} = \hat{\mathsf{C}}_{q/01}^{-1} \, \hat{\mathsf{C}}_{q/10}^{-1} \, \hat{\mathsf{C}}_{q/10}^{-1} \, \hat{\mathsf{C}}_q \, \hat{\mathsf{C}}_q^{-1} \, \hat{\mathsf{L}}_q^{-1} \, \hat{\mathsf{L}}_q \, \hat{\mathsf{L}}_{q/01}^{-1} \tag{9.168}$$

By replacing the property combinations by their equivalent space operator or by their corresponding eigen-value (pulsation or wave vector), according to relations 9.119 and 9.153 expressing the harmonicity of the oscillators, one finds the following relationship:

$$\hat{\mathsf{C}}_{q/01}^{-1} \, \hat{\mathsf{L}}_{q/01}^{-1} = \boldsymbol{k}_q^{-2} \frac{\partial}{\partial \boldsymbol{r}} \omega_q^2 \frac{\partial^{-1}}{\partial \boldsymbol{r}^{-1}} = \frac{\omega_q^2}{\boldsymbol{k}_q^2} \tag{9.169}$$

which has been simplified by virtue of the independency of the system pulsation with space. Now, by replacing this combination of constitutive properties in Equation 9.160 defining the derivation with respect to the phase angle, and in rearranging it, a relationship between eigen-function and eigen-vector is obtained

$$-i \frac{\omega_q}{\boldsymbol{k}_q} \frac{\mathrm{d}}{\mathrm{d}\varphi} \psi_q = \frac{\omega_q}{\boldsymbol{k}_q} \psi_q = i \frac{\partial}{\partial \mathbf{w}} \psi_q \tag{9.170}$$

A new operator, the *wave velocity operator* is defined

$$\hat{\mathbf{u}} \stackrel{def}{=} i \frac{\partial}{\partial \mathbf{w}} \tag{9.171}$$

allowing one to define a velocity $\boldsymbol{u}$ as eigen-vector

$$\hat{\mathbf{u}} \psi_q \stackrel{def \rightarrow}{=} \boldsymbol{u} \, \psi_q \tag{9.172}$$

This eigen-value $\boldsymbol{u}$ is called *space–time velocity*, or more specifically, *propagation velocity* or *phase velocity* when applied to a wave.

Two different velocities. The *space–time velocity* $\boldsymbol{u}$ is a vector that should not be confused with the *velocity* $\boldsymbol{v}$ of an object in translational mechanics, which is a state variable contrary to the phase or propagation velocity $\boldsymbol{u}$. The latter is a purely space–time property having no energetic content.

$$\boldsymbol{u} \neq \boldsymbol{v}$$

Notwithstanding this distinction, it happens in some mechanical systems that both velocities have the same value when the displacements of an object and of a wave coincide, but this is not always the case.

By comparing the last three equations, the propagation velocity appears to be equivalent to the ratio of the pulsation on the wave vector, by recalling that the two oscillators are supposed to be harmonic and consequently by equating wave variables to system properties.

$$\boldsymbol{u} = \frac{\omega}{k} \underset{cst\,\mathcal{E}_q}{=} \frac{\omega_q}{k_q} = \boldsymbol{u}_q \tag{9.173}$$

This is a result and not a definition, as this equivalence may not be true when harmonicity is not observed in both oscillators. Replacing this ratio in Equation 9.170 leads to the following equivalence between derivations:

$$\frac{\partial}{\partial \boldsymbol{w}} = -\boldsymbol{u}\frac{\mathrm{d}}{\mathrm{d}\varphi} \tag{9.174}$$

9.7.2 The Fourier Transform Approach

In Section 9.5.8, the Fourier transform has been introduced as a convenient way to relate an operator to its eigen-value, and in the case of a temporal oscillator it relates the time derivation to the pulsation ω. In the case of a spatial oscillator, the space derivation is related to the space vector $\boldsymbol{k}$. In the present case of a spatiotemporal oscillator, the same transform, but in using the *space–time velocity* $\boldsymbol{u}$ in lieu of the pulsation or the space vector, can be used

$$\frac{\mathrm{d}}{\mathrm{d}\boldsymbol{w}} \xrightarrow{\text{Fourier } \boldsymbol{u}} i\,\boldsymbol{u} \tag{9.175}$$

Applied to the Formal Graph given in Graph 9.39, or more exactly to its two-dimensional projection from above (keeping only pertinent nodes), and by assuming linear operators for the system constitutive properties for simplification, the transformed Formal Graph in Graph 9.42 (right) is obtained.

One of the algebraic translations of the sequence of transformed operators pertaining to the loop in the Fourier-transformed Formal Graph is

$$-i\,\boldsymbol{u}\,\boldsymbol{L}_{q/01}\,i\,\boldsymbol{u}\,\boldsymbol{C}_{q/01}\,\tilde{\boldsymbol{e}}_q \underset{lin}{=} \tilde{\boldsymbol{e}}_q \tag{9.176}$$

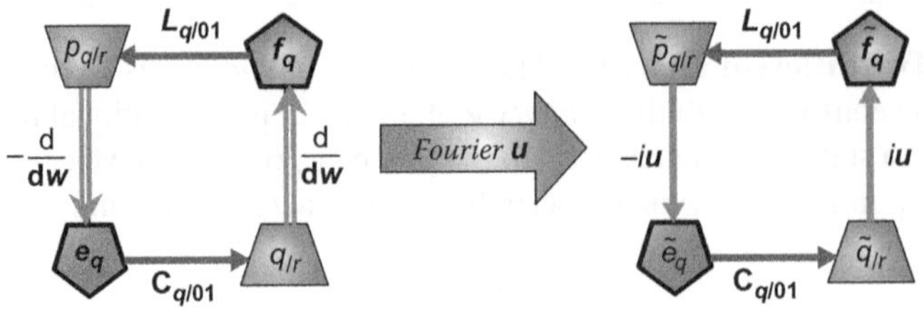

GRAPH 9.42 Transformation of the Formal Graph of a linear spatiotemporal oscillator (left) into a Fourier-transformed Formal Graph (right) by using as transform variable the *space–time velocity* **u**.

As commutativity works between all these scalars and vectors, by replacing the composition of constitutive properties by their expression in Equation 9.169, the equality in Equation 9.173 between the space–time velocity and the system constitutive velocity $\boldsymbol{u}_q$ is directly retrieved:

$$\boldsymbol{u}^2\,\tilde{\boldsymbol{e}}_q \underset{lin}{=} \boldsymbol{u}_q^2\,\tilde{\boldsymbol{e}}_q \tag{9.177}$$

It must be outlined that the Fourier transform approach is an alternate approach that works well when all system constitutive properties are linear operators (in fact, scalars or vectors) and which is less straightforward in the case of nonlinearity.

9.7.3 Relationship between Wave Functions

It remains now to find the dependence of the phase angle with the space–time properties. This is not as simple a task as before when treating oscillators working with only time or space because of the entanglement of the two space–time components in a propagating wave. The presence of a single derivation with respect to one space–time variable directly allowed establishing a proportional link between this variable and the phase angle, through the eigen-values ω or $\boldsymbol{k}$. Here, one needs to assess a double dependence of the phase angle on time and space, expressed by the decomposition of its variation in a sum of two independent variations:

$$\mathrm{d}\varphi = \left(\frac{\partial \varphi}{\partial t}\right)_{\boldsymbol{r}} \mathrm{d}t + \left(\frac{\partial \varphi}{\partial \boldsymbol{r}}\right)_{t} \cdot \mathbf{dr} \tag{9.178}$$

Now, one needs to consider two separate wave functions $\psi_q(t)$ and $\psi_q(\boldsymbol{r})$, that is, one for each component oscillator. Through the differential equations 9.96 and 9.146 featuring the wave functions, the individual variations of the phase angle can be expressed:

$$\mathrm{d}\varphi(t) = \left(\frac{\partial \varphi}{\partial t}\right)_{\boldsymbol{r}} \mathrm{d}t = \frac{\mathrm{d}\psi_q(t)}{i\,\psi_q(t)} \tag{9.179}$$

$$\mathrm{d}\varphi(\boldsymbol{r}) = \left(\frac{\partial \varphi}{\partial \boldsymbol{r}}\right)_{t} \cdot \mathbf{dr} = \frac{\mathrm{d}\psi_q(\boldsymbol{r})}{i\,\psi_q(\boldsymbol{r})} \tag{9.180}$$

By using the corresponding differential equation 9.164 for the spatiotemporal wave function and by replacing the individual variations of the phase angle in Equation 9.178 giving its total variation, one gets

$$\frac{\mathrm{d}\psi_q}{\psi_q} = \frac{\mathrm{d}\psi_q(t)}{\psi_q(t)} + \frac{\mathrm{d}\psi_q(\boldsymbol{r})}{\psi_q(\boldsymbol{r})} \tag{9.181}$$

which is the expression of the derivative of a product that is written as

$$\psi_q = \psi_q(t)\ \psi_q(\boldsymbol{r}) \tag{9.182}$$

This is an important result. It proves that the total wave function of the spatiotemporal oscillator is equivalent to the product of the wave functions of the temporal and spatial oscillators. This result is very general, as it has not been necessary to detail the exact dependence of the phase angle on time and space.

9.7.4 The Propagation Equation

Now that the wave function is known to be equivalent to the product of individual wave functions, one can find easily all derivatives with respect to time or space and construct the propagation equation of the wave.

This is done by applying separately to the wave function the operators that were defined previously for each oscillator. By applying first the *pulsation operator* $\hat{\omega}$ defined by Equation 9.116 that provided, through Equation 9.117, the pulsation eigen-value ω in the case of the temporal oscillator, one concludes that the pulsation is also an eigen-function of the same operator applied to the wave function of the spatiotemporal oscillator. The same conclusion is reached, by using Equation 9.150, for the wave vector $\boldsymbol{k}$ that appears also as an eigen-function of the wave vector operator $\hat{\mathbf{k}}$ defined by Equation 9.149.

$$i\frac{\partial}{\partial t}\psi_q = i\frac{\partial}{\partial t}\psi_q(t)\,\psi_q(\boldsymbol{r}) = \omega\,\psi_q \tag{9.183}$$

$$-i\frac{\partial}{\partial \boldsymbol{r}}\psi_q = -i\frac{\partial}{\partial \boldsymbol{r}}\ \psi_q(\boldsymbol{r})\psi_q(t) = \boldsymbol{k}\,\psi_q \tag{9.184}$$

As the wave function is the same in both equations, its extraction provides the following equalities:

$$\psi_q = \frac{i}{\omega}\frac{\partial}{\partial t}\psi_q = -\frac{i}{\boldsymbol{k}}\frac{\partial}{\partial \boldsymbol{r}}\psi_q \tag{9.185}$$

This means that once correctly scaled, both operators are equivalent. The imaginary number can be eliminated by applying again each scaled operator, which leads to the wave equation by replacing the pulsation and the wave vector with their ratio $\boldsymbol{u}$:

$$\frac{\partial^2}{\partial t^2}\psi_q = \boldsymbol{u}^2\frac{\partial^2}{\partial \boldsymbol{r}^2}\psi_q \tag{9.186}$$

The propagation equation of the wave is expressed with one of the state variables $\boldsymbol{e}_q$ or $\boldsymbol{f}_q$, or one of the localized variables $q_{q/r}$ or $p_{q/r}$, as argument. The usual propagation equation is written by using a generic variable y standing for any of the four variables instead of the wave function in the previous equation:

$$\frac{\partial^2}{\partial t^2} y = \boldsymbol{u}^2 \frac{\partial^2}{\partial \boldsymbol{r}^2} y \tag{9.187}$$

9.7.5 Dependence on Space–Time

The double dependence of the phase angle φ on time and space, previously stated in Equation 9.178, can be developed a little more by replacing the partial derivatives with the pulsation and the wave vector. In the treatment of the temporal oscillator, the derivative of the phase angle was found to be equal to the opposite of the pulsation (Equation 9.126) and in the spatial oscillator, it was found to be equal to the wave vector (Equation 9.154). This gives

$$\left(\frac{\partial \varphi}{\partial t}\right)_r = \frac{\mathrm{d}\varphi(t)}{\mathrm{d}t} = -\omega \tag{9.188}$$

$$\left(\frac{\partial \varphi}{\partial \boldsymbol{r}}\right)_t = \frac{\mathrm{d}\varphi(\boldsymbol{r})}{\mathbf{d}\boldsymbol{r}} = \boldsymbol{k} \tag{9.189}$$

The substitution of the partial derivatives into Equation 9.178 provides the simple equation for the variation of the phase angle:

$$\mathrm{d}\varphi = -\omega \, \mathrm{d}t + \boldsymbol{k} \cdot \mathbf{d}\boldsymbol{r} \tag{9.190}$$

As the pulsation is time invariant and the wave vector is space invariant too, integration is straightforward and gives the linear relationship

$$\varphi = -\omega \, t + \boldsymbol{k} \cdot \boldsymbol{r} \tag{9.191}$$

The model of the wave function as a function of time and space is finally obtained by introducing this expression into the argument of the exponential function (Equation 9.165)

$$\psi_q = \psi_q(0) \, \exp(-i\omega \, t + i\boldsymbol{k} \cdot \boldsymbol{r}) \tag{9.192}$$

It leads to the solution of the wave propagation Equation 9.187 in terms of the harmonic function *cosine* when the initial value is not zero (or *sine* when this is not the case):

$$y = y_0 \cos(\omega t - \boldsymbol{k} \cdot \boldsymbol{r}) \tag{9.193}$$

The four state variables and localized variables are expressed as follows:

$$\boldsymbol{p}_{q/r} = -\frac{1}{\boldsymbol{u}} e_q(0) \sin(\omega t - \boldsymbol{k} \cdot \mathbf{r}); \quad f_q = -\boldsymbol{u} \, \boldsymbol{C}_{q/01} \, e_q(0) \sin(\omega t - \boldsymbol{k} \cdot \boldsymbol{r}) \tag{9.194}$$

$$e_q = e_q(0) \cos(\omega t - \boldsymbol{k} \cdot \boldsymbol{r}); \quad \boldsymbol{q}_{/r} = \boldsymbol{C}_{q/01} \, e_q(0) \cos(\omega t - \boldsymbol{k} \cdot \boldsymbol{r}) \tag{9.195}$$

9.7.6 Irradiance or Radiation Intensity

A wave that propagates in space contains energy and is called a radiation. In order to describe the transported energy, the notion of *radiated* (or *radiant*) *power* is defined as the time derivative of the energy $\mathcal{E}$ of the wave at a given wavelength λ (or frequency $\nu = \omega/2\pi$):

$$\mathcal{P}_{\lambda q} \overset{def}{=} \frac{d}{dt}\mathcal{E}_{\lambda q} \tag{9.196}$$

A wave may propagate in a one-dimensional space (as the vibrating string studied in case study F4), or onto a surface, or through a three-dimensional space (as the sound or the light in case studies F5 and F6). Localized variables are therefore more useful than global variables for quantifying energy.

In a two- or three-dimensional space, a variable called *irradiance* or *radiation intensity* $\mathcal{I}_{\lambda q}$ is defined for a given wavelength λ as the radiated power surface density of the wave (with unit in W m^{-2}):

$$\mathcal{I}_{\lambda q} \overset{def}{=} \frac{d}{d\mathbf{A}}\mathcal{P}_{\lambda q} \tag{9.197}$$

Rigorously speaking, the notion of *radiation intensity* refers to spherical conditions of propagation and corresponds to a power per unit of solid angle (watt per steradian). But when the wave propagates along a line, irradiance and intensity are often confused.

Classically, this notion is used as the average over time of this vector, but the Formal Graph approach requires keeping its instantaneous vector nature to be able to place it at the right level of space reduction, as is done in Graph 9.43. This Formal Graph is a noncanonical one because it does not use state variables but energetic variables. It uses time and space operators in the same manner as an ordinary two- or three-dimensional canonical Formal Graph.

At the bottom level (points of a volume) of Graph 9.43, are located the volumic concentrations of the power and of the energy that allow modeling a three-dimensional propagation. The space–time velocity $\boldsymbol{u}$ of the wave relates the energy concentration to the irradiance (radiated intensity):

$$\mathcal{I}_{\lambda q} = \boldsymbol{u}\,\mathcal{E}_{\lambda q/V} \tag{9.198}$$

To pursue this study, we take the case of the propagation in a three-dimensional space of a wave based on fields and surface densities (like an electromagnetic wave), as shown in Graph 9.44. On this three-dimensional Formal Graph are represented the two half-energies that are exactly located in the middle of the capacitive and inductive face of the graph. This localization expresses their square as an energy concentration, as can be verified with the following relationships deduced from Graph 9.44 and adapted from Equation 9.75:

$$A_q^2 = \hat{C}_{q12}^{1/2}\, e_{q/r}\; \hat{C}_{q12}^{-1/2}\, q_{/V} \underset{lin}{=} e_{q/r}\, q_{/V} \underset{lin}{=} 2\mathcal{U}_{q/V} \tag{9.199}$$

$$B_q^2 = \hat{L}_{q12}^{1/2}\, f_{q/r}\; \hat{L}_{q12}^{-1/2}\, p_{q/V} \underset{lin}{=} f_{q/r}\, p_{q/V} \underset{lin}{=} 2\mathcal{T}_{q/V} \tag{9.200}$$

From the definition of the wave function, recalled here,

$$\psi_q \overset{def}{=} A_q + i\,B_q \tag{9.201}$$

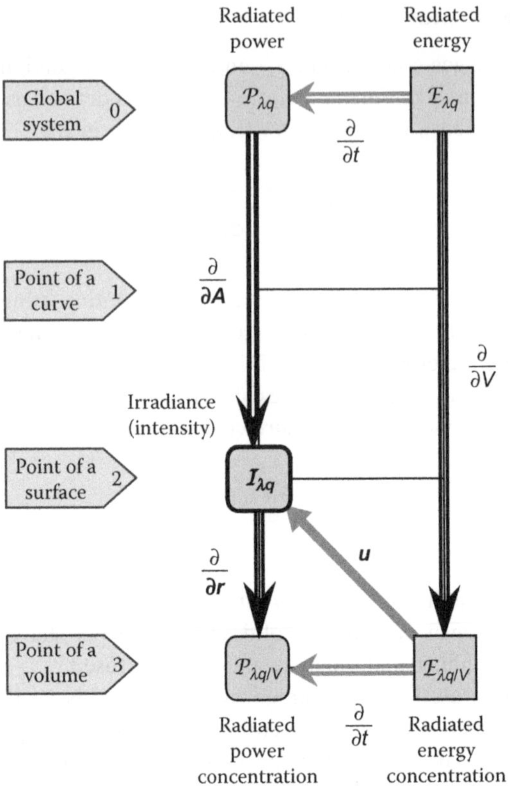

GRAPH 9.43 Noncanonical Formal Graph modeling the relationships between variables used for describing radiation.

the total energy concentration is given by half the modulus of the wave function

$$\mathcal{E}_{q/V} = \mathcal{T}_{q/V} + \mathcal{U}_{q/V} \underset{lin}{=} \frac{A_q^2 + B_q^2}{2} = \frac{\psi_q \psi_q^*}{2} = \frac{1}{2}\left\|\psi_q\right\|^2 \tag{9.202}$$

so the irradiance is

$$I_{\lambda q} \underset{lin}{=} \frac{\boldsymbol{u}}{2}\left\|\psi_q\right\|^2 \tag{9.203}$$

This relationship is quite general, although restricted to the linear case, and works for any wave function. In the case of a conservative spatiotemporal oscillator, or wave, the wave function modulus is a constant:

$$I_{\lambda q} \underset{lin}{=} \frac{\boldsymbol{u}}{2}\, \psi_q(0)^{\,2} = \frac{\boldsymbol{u}}{2}\, \boldsymbol{C}_{q/01}\, e_q(0)^2 \tag{9.204}$$

The dependence on the wavelength is ensured through the initial value of the wave function and through the space–time velocity $\boldsymbol{u}$ of the wave which is directly proportional to the wave pulsation ω through Equation 9.173.

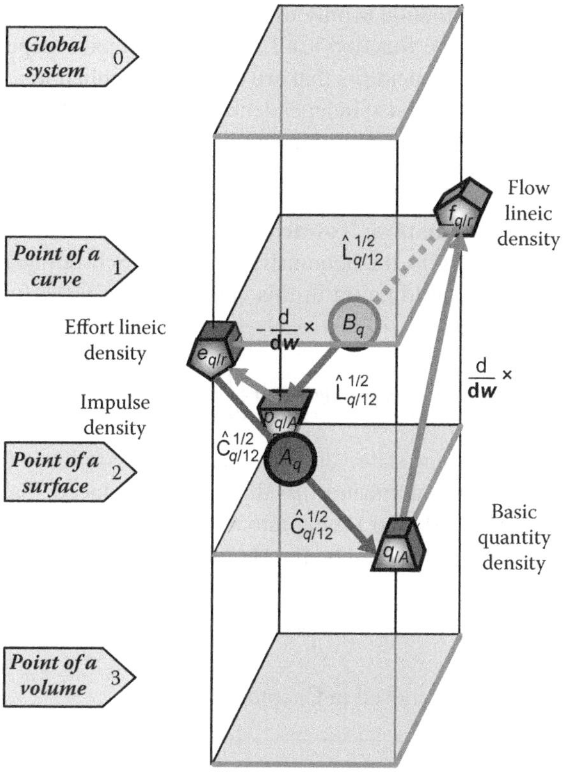

GRAPH 9.44 Formal Graph of a propagation wave in a three-dimensional space with the four significant nodes plus the two half-energies between the curve level and the surface level.

9.7.7 Probalistic Viewpoint

In quantum physics, the wave function is viewed as a support of the probabilistic approach that is an essential feature of this branch of physics. Classically, probability is defined as the chance of finding an object at a certain time and position. This can be generalized by replacing these space–time-related notions with the phase angle φ, which avoids focusing on the ambiguous subject of what is really a "position" for nonmechanical objects. Thus, here one speaks of the chance of finding an object at a certain phase angle φ in a given interval $[\varphi_1, \varphi_2]$. This probability is mathematically expressed as the finite integral of the squared modulus of the normalized wave function featuring the object.

$$\mathcal{p}(\varphi_1, \varphi_2) \overset{def}{=} \frac{1}{\wp_q(\varphi_1, \varphi_2)} \int_{\varphi_1}^{\varphi_2} \left\| \psi_q(\varphi) \right\|^2 \mathrm{d}\varphi \tag{9.205}$$

The normalization function $\wp_q(\varphi_1, \varphi_2)$ of the wave function is determined by the constraint of a probability comprised between 0 and 1.

For a conservative oscillator, the modulus of the wave function is a constant and, according to Equation 9.113, the square of this modulus in the linear case is equal to twice the total energy of the system. As the probability of existence of oscillations in any range of phase angles is equal to 1, one must have for the normalization function in this case

$$\wp_q(\varphi_1, \varphi_2) \underset{lin}{=} 2\mathcal{E}_q(\varphi_2 - \varphi_1) \tag{9.206}$$

However, this normalization function is only used for transposing the wave function into probability and has no effect on the wave function itself, which is defined independent of any notion of probability. In particular, the main operators that are defined as applied to a wave function, such as the various space–time operators, are also independent. By extension, it is also the case for all models based on these operators, such as the Schrödinger equation.

Schrödinger equation: Toward demonstration. The most difficult part of the route leading to the demonstration of the Schrödinger equation has been made with the building up in this chapter of the wave function. By using its eigen-values

$$i\frac{\partial}{\partial t}\psi = \omega\,\psi; \quad -i\frac{\partial}{\partial \boldsymbol{r}}\psi = \boldsymbol{k}\,\psi \qquad (9.207)$$

and by relating them to the other state variables featuring a corpuscle (i.e., the energy-per-corpuscle E and the corpuscle momentum $\boldsymbol{p}$) through the reduced Planck constant $\hbar$, the Schrödinger equation is deduced after taking into account the adequate energy varieties that are present in the corpuscle and handled by the Hamiltonian operator $\hat{\mathrm{H}}$

$$i\hbar\frac{\partial}{\partial t}\psi = \hat{\mathrm{H}}\psi \qquad (9.208)$$

This demonstration will be achieved in Chapter 14.

9.8 IN SHORT

TIME AND EVOLUTION

In a dipole assembly, the conversion of energy between two subvarieties is expressed in the general case by an *inverse evolution* operator inside each storing dipole

$$x_1 = \hat{\mathrm{T}}\,y_2 \qquad (9.8)$$

where x_i is the *entity number* of subvariety i and y_i is the *energy-per-entity*. (See Graph 9.2.) The *duration* of the conversion is defined as

$$\Delta t \overset{def}{=} \int_{x_1}^{x_1+\Delta x_1} \frac{dx_1}{y_2} \underset{cont}{=} \int_{x_1}^{x_1+\Delta x_1} dt \qquad (9.10)$$

providing the definition of the time variable in case of continuous conversion

$$\hat{\mathrm{T}}^{-1} \underset{cont}{=} \frac{d}{dt} \qquad (9.14)$$

A capacitive and an inductive *gate variables* are defined as

Capacitive gate variable: $$f_{qC} \overset{def}{=} \hat{\mathrm{T}}^{-1} q \underset{cont}{=} \frac{d}{dt} q \qquad (9.28)$$

Inductive gate variables: $$e_{qL} \overset{def}{=} \hat{\mathrm{T}}^{-1} p_q \underset{cont}{=} \frac{d}{dt} p_q \qquad (9.29)$$

POWER

The power received by a system is the temporal derivative of the total energy (in a given variety) and for a dipole it is equal to the product flow × effort:

$$\mathcal{P}_q \overset{def}{=} \frac{d}{dt}\mathcal{E}_q = f_q\, e_q \tag{9.40}$$

In a dipole assembly, the total power is the sum of all powers received:

$$\mathcal{P} = \sum_{q,i} \mathcal{P}_{qi} \tag{9.41}$$

ISOLATED VERSUS POWERED SYSTEM

Powered (isolated) system	Isolated (Nonpowered)
$d\mathcal{E}_q = 0$	$d\mathcal{E}_q > 0$

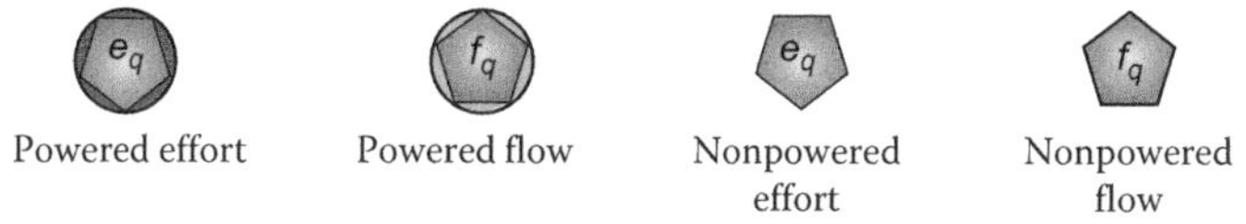

GRAPH 9.45

TEMPORAL OSCILLATOR

An assembly of an inductive dipole with a capacitive one makes up a temporal oscillator owing to the regular conversion of inductive energy into capacitive energy and vice versa. When the system is isolated, the total energy of the oscillator is conserved.

Common effort oscillator is represented in Graph 9.29b and common flow oscillator in Graph 9.28b, in representation loop.

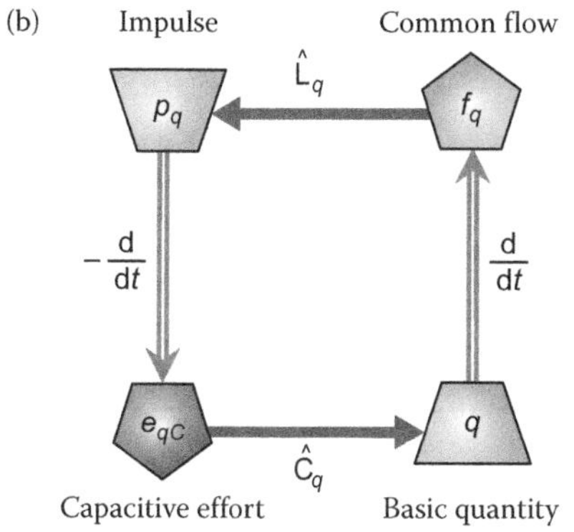

GRAPH 9.28b

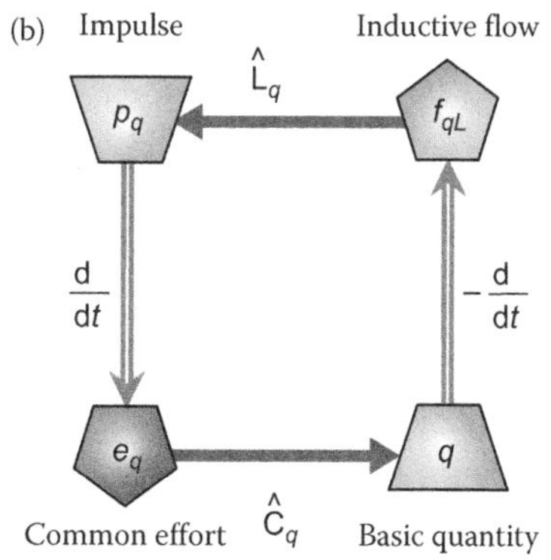

GRAPH 9.29b

TEMPORAL OSCILLATOR: PULSATIONS (UNIT: rad s^{-1})			
Frame	Energy	Space–time	System (oscillator) (energy variety q)
Status	State variable	Operator	Property (apparent)
Name	Wave pulsation	Wave pulsation operator	Natural pulsation
Symbol	ω	$\hat{\omega}$	ω_q
Definition	Effort of phase energy	$i\frac{d}{dt}$	$\hat{L}_q^{-1/2}\hat{C}_q^{-1/2}$

SPATIAL OSCILLATOR

A spatial oscillator appears in a space dipole when the space distribution of *energy-per-entity* and *entity number* are able to exchange energy in an analogous way to the conversion between two subvariety, except that time is replaced by space.

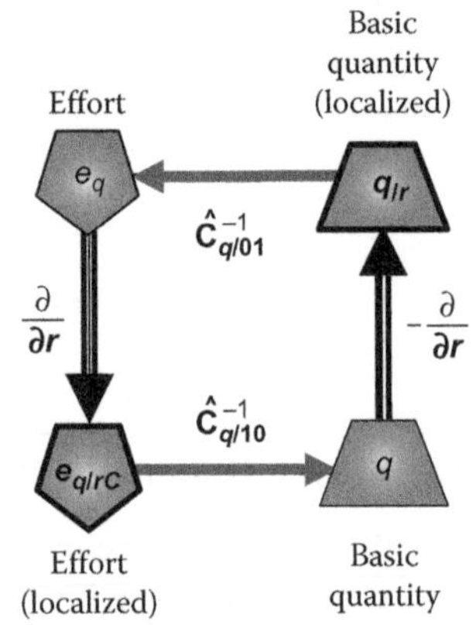

GRAPH 9.46

SPATIAL OSCILLATOR: WAVE VECTORS (UNIT: rad m^{-1})			
Frame	Energy	Space–time	System (oscillator) (energy variety q)
Status	State variable	Operator	Property (apparent)
Name	Wave vector	Wave-vector operator	Natural wave vector
Symbol	k	$\hat{\mathbf{k}}$	k_q
Definition	Impulse of space wave energy	$-i\frac{d}{dr}$	$k_q = \hat{C}_{q/01}^{1/2}\hat{C}_{q/10}^{-1/2}$

SPATIAL OSCILLATOR

The assembly of two space dipoles, one inductive and one capacitive, forms a spatial oscillator.

The three-dimensional Formal Graph of this system can be simplified by defining a space–time operator, as a composition of space integration with time derivation:

$$\frac{\partial}{\partial \mathbf{w}} \overset{def}{=} \frac{\partial}{\partial t}\frac{\partial^{-1}}{\partial \mathbf{r}^{-1}} = \frac{\partial^{-1}}{\partial \mathbf{r}^{-1}}\frac{\partial}{\partial t} \tag{9.157}$$

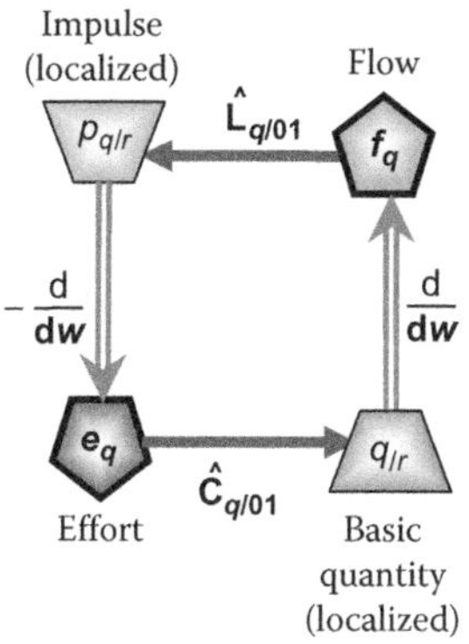

GRAPH 9.47

COMPARISON BETWEEN OSCILLATORS

Operator or Variable	Temporal	Spatial	Spatiotemporal
Capacitive semienergy A_q	$\hat{C}_q^{1/2} e_q = \hat{C}_q^{-1/2} q$	$\hat{\mathbf{C}}_{q/10}^{1/2}\, \mathbf{e}_{q/rC} = \hat{\mathbf{C}}_{q/10}^{-1/2}\, q$	$\hat{\mathbf{C}}_{q/01}^{1/2}\, \mathbf{e}_q = \hat{\mathbf{C}}_{q/01}^{-1/2}\, q_{/r}$
Capacitive semienergy B_q	$\hat{L}_q^{1/2} f_q = \hat{L}_q^{-1/2} p_q$	$\hat{\mathbf{C}}_{q/01}^{-1/2}\, \mathbf{q}_{/r} = \hat{\mathbf{C}}_{q/01}^{1/2}\, e_q$	$\hat{\mathbf{L}}_{q/01}^{1/2}\, \mathbf{f}_q = \hat{L}_{q/01}^{-1/2}\, p_{q/r}$
Natural property (harmonicity)	$\omega_q = \hat{L}_q^{-1/2}\hat{C}_q^{-1/2}$	$k_q = \hat{\mathbf{C}}_{q/01}^{1/2}\hat{\mathbf{C}}_{q/10}^{-1/2}$	$\frac{\omega_q}{k_q} = \hat{\mathbf{C}}_{q/01}^{-1/2}\, \hat{\mathbf{L}}_{q/01}^{-1/2}$
Derivation/Phase angle $\frac{d}{d\varphi}$	$-\hat{C}_q^{1/2}\hat{L}_q^{1/2}\frac{d}{dt}$	$\hat{\mathbf{C}}_{q/10}^{1/2}\hat{\mathbf{C}}_{q/01}^{-1/2}\frac{d}{d\mathbf{r}}$	$-\hat{\mathbf{C}}_{q/01}^{1/2}\, \hat{L}_{q/01}^{1/2}\frac{d}{d\mathbf{w}}$
Space–time operator	$\hat{\omega} \overset{def}{=} i\frac{d}{dt}$	$\hat{\mathbf{k}} \overset{def}{=} -i\frac{d}{d\mathbf{r}}$	$\hat{\mathbf{u}} \overset{def}{=} i\frac{d}{d\mathbf{w}}$
Eigen-value or eigen-vector (wave variables)	$\hat{\omega}\,\psi_q = \omega\,\psi_q$	$\hat{\mathbf{k}}\,\psi_q = k\,\psi_q$	$\hat{\mathbf{u}}\,\psi_q = \mathbf{u}\,\psi_q$
Energy conservation (wave = oscillator)	$\omega = \omega_q$	$k = \mathbf{k}_q$	$\mathbf{u} = \frac{\omega_q}{\mathbf{k}_q}$
Phase angle φ (wave variable)	$-\omega\, t$	$k \cdot \mathbf{r}$	$-\omega\, t + \mathbf{k}\cdot\mathbf{r}$

WAVE FUNCTION

	Equations	Graphics
Imaginary number	$-i^2 \overset{def}{=} 1$	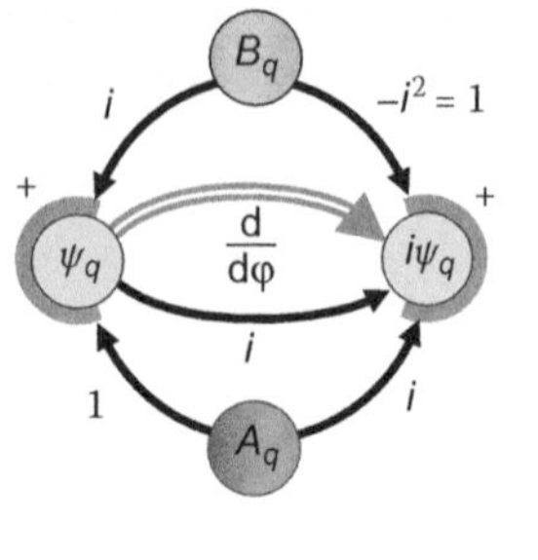
Wave function	$\psi_q \overset{def}{=} A_q + i\, B_q$	
Differential equation	$\frac{d}{d\varphi}\psi_q = i\,\psi_q$	GRAPH 9.48
Exponential function	$\psi_q = \psi_q(0)\exp(i\varphi)$	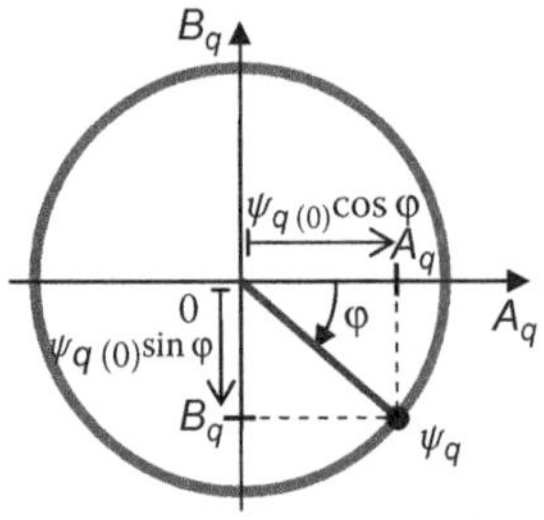
Semienergies	$A_q = \psi_q(0)\cos\varphi$ $B_q = \psi_q(0)\sin\varphi$	

RADIATION (SEE GRAPH 9.43)

Wave or radiation energy (/J): $$\mathcal{E}_{\lambda q} \underset{lin}{=} \frac{1}{2}\left\|\psi_q\right\|^2 \quad (9.196)$$

Radiated power (/W): $$\mathcal{P}_{\lambda q} \overset{def}{=} \frac{d}{dt}\mathcal{E}_{\lambda q} \quad (9.196)$$

In a three-dimensional space:

Radiated energy concentration (/J m^{-3}):

$$\mathcal{E}_{\lambda q/V} \overset{def}{=} \frac{d}{dV}\mathcal{E}_{\lambda q} \quad (9.209)$$

Irradiance (radiated intensity) (/W m^{-2}):

$$\mathcal{I}_{\lambda q} \overset{def}{=} \frac{d}{d\boldsymbol{A}}\mathcal{P}_{\lambda q} = \boldsymbol{u}\,\mathcal{E}_{\lambda q/V} \quad (9.210)$$

General (linear) case: $$\mathcal{I}_{\lambda q} \underset{lin}{=} \frac{\boldsymbol{u}}{2}\left\|\psi_q\right\|^2 \quad (9.203)$$

Conservative wave: $$\mathcal{I}_{\lambda q} \underset{lin}{=} \frac{\boldsymbol{u}}{2}\psi_q(0)^2 \quad (9.204)$$

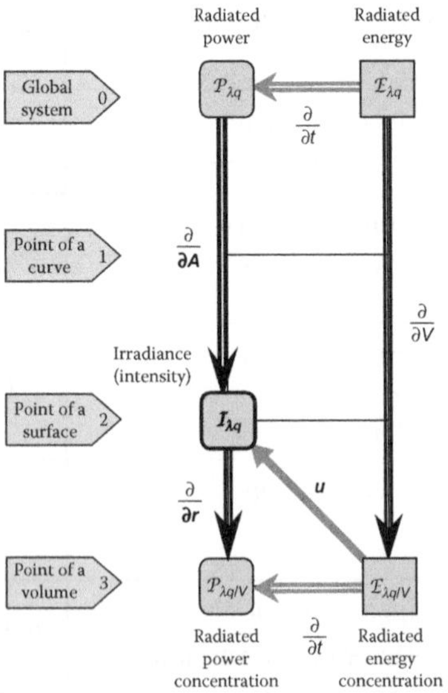

GRAPH 9.43

10 Transfers

CONTENTS

Charge, heat, mass, and momenta transfers (and the list is not limited to these four quantities) are processes having the same physical background which is different from propagation and from relaxation phenomena occurring in space. This specificity justifies a whole chapter devoted to the many aspects of transfers. It requires the introduction of new concepts, although the Formal Objects used for modeling transfers are much simpler than the last ones studied, as poles and dipoles are sufficient for hosting transfer processes.

Spatially reduced Ohm's law in electrodynamics, Fourier's equation of heat transfer, Fick's law for diffusion, and Newton's law in hydrodynamics are among the subjects treated and their comparison is enlightening. The various transfers tackled in this chapter are *stationary diffusion*, *transient diffusion*, *thin-layer evolution*, *migration*, and *convection*, in addition to less classical cases such as *anomalous diffusion* and the model of *constant phase element* (CPE).

Besides the usual standardization brought by Formal Graphs, two main contributions are outlined: the *higher level of modeling* allowed by using operators instead of the classical solving of differential equations (which is traditional in this area) and the supplement of physical meaning allowed by the representation of processes by means of *paths in a graph*. Thus, a transfer can be interpreted in terms of *energy behavior* and the proportion of conserved or dissipated energy can be quantified, experimentally analyzed, and directly measured. This feature has direct consequences for engineers having to design practical devices. It also sheds some physical signification upon the mysterious fractional derivation introduced for mathematical reasons.

A side effect of the higher level of modeling is the proposal of a new concept, the *mass transfer operator*, presented here in the frame of mass transfers but that may be generalized to any transfer. With the help of a single operator taking the various forms corresponding to all kinds of mass transfers (in the same way as an impedance may take on several forms depending on the considered electrical component), it becomes possible to build models without knowing in advance the real process (and even without knowing how to model it). The latter can then be determined directly from experimental data.

Table 10.1 lists the case studies of poles or dipoles implementing a transfer given in this chapter. (See also Figure 10.1 for the position of the poles and dipoles.)

TABLE 10.1
List of Case Studies Involving Transfer

	Name	Energy Variety	Constitutive Properties	Formal Object	Page Number
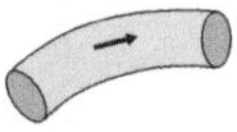	G1: Reduced Ohm's law	Electrodynamics	Conductive	Pole	441
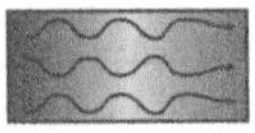	G2: Fourier's equation of heat transfer	Thermics	Conductive	Pole	444
$J_{isotropic} = -D\frac{d}{dr}c_r$	G3: Fick's law of diffusion	Physical chemistry	Capacitive + conductive	Pole	447
	G4: Newton's law of viscosity	Translational mechanics	Inductive + conductive	Pole	453
	G5: Capacitive transmission line	Electrodynamics	Capacitive + conductive	Dipole	457
	G6: Constant phase element	Electrodynamics	Capacitive or inductive + conductive	Dipole	465
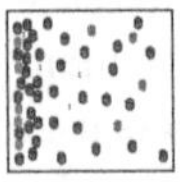	G7: Transient diffusion	Physical chemistry	Capacitive + conductive	Dipole	469
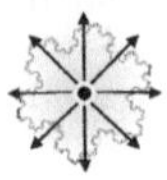	G8: Anomalous diffusion	Physical chemistry	Capacitive + conductive	Dipole	473
$\hat{m}$	G9: Generalized mass transfer	Physical chemistry	Capacitive + conductive	Dipole	427

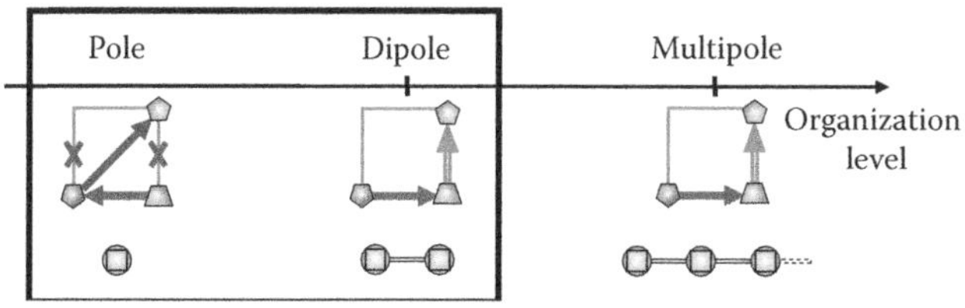

FIGURE 10.1 Position of the pole and dipole Formal Objects along the complexity scale.

10.1 DEFINITION OF TRANSFER

In physics, transfer means a process in which an object or a quantity of material, energy, heat, etc. is moved from one location or site to another one. Contrary to relaxation that is modeled by a dipole assembly, a transfer is modeled by a simpler Formal Object, a dipole or even a pole, but it involves more complex relationships.

10.1.1 TRANSFER, TRANSPORT, AND EVOLUTION

In fact, two different processes can be distinguished under the notion of transfer (see Figure 10.2). When the two sites are separated in space, one speaks of transport and when a single site sees its occupancy changed owing to reaction or transformation, the notion of evolution or *temporal transfer* is used. An example is an immobilized molecule or enzyme changing its oxidation state by exchange of an electron from the surroundings. From the viewpoint of the electron, there is transport, while for the molecule there is no displacement but change in quantity as another molecule (or redox state), quantified by another number of entities, is produced from it. Transport implies necessarily space, whereas time is necessarily involved when no transport is effective in an on-site transformation.

Transport can work without time, in steady state, as, for instance, a river flowing between two reservoirs or two lakes with different but constant water levels. It may also imply time, as in the case of transient diffusion of molecules occurring after perturbation of their space distribution. There is simultaneously transport due to the movement of molecules from one site to another, and temporal transfer corresponding to the change of the number of molecules at every location (each site sees its

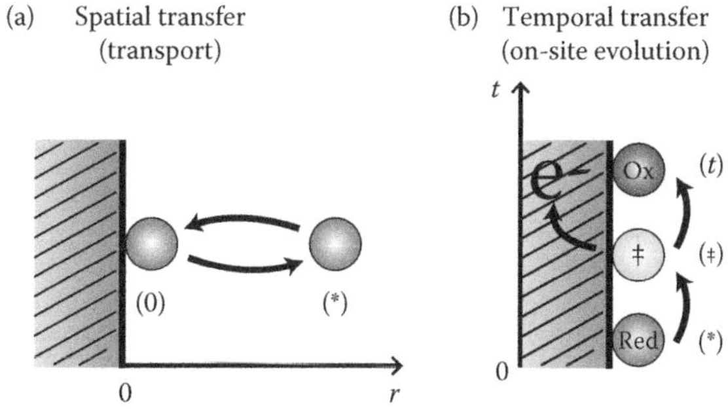

FIGURE 10.2 Two transfer processes: one of transport (displacement) of objects (molecules) through space (a) and one of temporal transfer (evolution) through time (b). In the case of evolution, there is no displacement of molecules but they see their nature changed. In this example, they change from reduced to oxidized state by exchange of electrons with their support (which is well a displacement, but not of molecules).

number of molecules changed). One shall see that in normal (ideal) systems both processes participate equally in the transient transfer.

If one quantifies empirically by a number p the proportion of transport playing a role in each category of transfer, one has:

- $p = 1$ for the *stationary transfer* (pure transport).
- $p = 1/2$ for the *transient transfer* (half transport–half evolution).
- $p = 0$ for the *temporal transfer* (pure evolution).

However, the frontier between transport and temporal transfer is not always clearly defined. In some systems where geometry is not simple, being described by fractal geometry for instance, the proportion of transport and temporal transfer fluctuates according to the geometric dimensions of the system (Le Méhauté and Crépy 1983). A value for p between 0 and 1 but different from 1/2 characterizes the transfer.

Nontraditional viewpoint. To consider *evolution* on a site as a *transfer* is not common. The traditional conception of space and time separates these two notions and transfer is most often seen as a synonym of transport. However, there are situations in which a continuous increase of the distance allowed for the transfer makes the transition between temporal transfer (very small distance) and spatial transfer (larger distance) progressive. For instance, when a thin layer of electroactive molecules deposited on an electrode undertakes an exchange of electrons with the latter, the modification of the nature of the molecules is quantified by a flow (corresponding to the current of electrons) and by a variation of the number of reacting molecules on the electrode. No displacement of these molecules occurs during this electrochemical process. The interesting case is when the molecules produced are soluble in the medium surrounding the electrode (the electrolyte). In the early stages of the process, they also form a thin layer on the electrode, but they progressively diffuse in the electrolyte, forming a layer (called diffusion layer) becoming thicker and thicker. Soon, a transport by diffusion takes the place of the purely temporal transfer (see Figure 10.3).

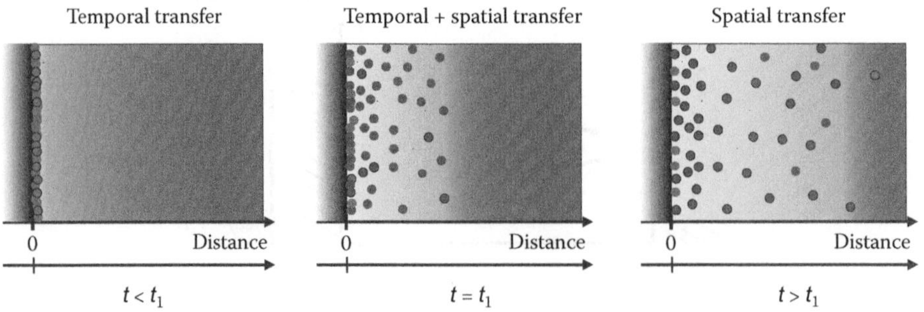

FIGURE 10.3 Transition from temporal transfer to spatial transfer during the electrochemical dissolution of a substance deposited on an electrode.

This interplay between space and time is the key for understanding the notion of transfer and its role in many essential processes. In particular, the knowledge of the number p, which we call *evolution mode* (or *transfer regime*), is one of the important issues in the experimental characterization of transfer processes.

10.1.2 Formal Graph Representations

A transfer is a localized process, even in the case of pure temporal transfer. The Formal Object hosting a transfer is therefore necessarily a spatial object. However, as the variables that can be experimentally determined belong to the global level, a first approach consists of discussing what can be globally represented without detailing the role of space.

There are two ways to model a transfer with a Formal Graph at the global level: either by relating two energies-per-entity or by relating one entity number to one energy-per-entity. Depending on the considered subvariety, this makes three possible representations, as depicted in Graph 10.1.

As no evolution link is made explicit nor reference to time is made in these Formal Graphs, they represent indifferently poles or dipoles. The differences in the three Formal Graphs in Graph 10.1 lie in making explicit or not the involved subvariety in the transfer. Algebraic models also follow this classification; one may cite on the one hand Ohm's law relating a potential (effort) to a current (flow) without making explicit what is transported, thus modeled by an admittance or impedance, and on the other hand Fick's law of diffusion (mass transfer) relating a substance concentration (localized basic quantity) to a flow density, which belongs to the category of models using explicitly the capacitive subvariety.

The latter case of explicit subvariety corresponds to the classical notion of kinematic or kinetic process, that is, processes in which the number of entities plays a significant role and the energies-per-entity a secondary role, or even no role at all. To make explicit the subvariety, the global operator modeling the transfer in a Formal Graph is decomposed into an inductive (or capacitive) operator and a *kinematic operator* $\hat{N}_q$ (respectively, a *kinetic operator* $\hat{M}_q$), generally used in its spatially reduced form $\hat{\mathbf{n}}_q$ (respectively, $\hat{\mathbf{m}}_q$, allowing modeling Fick's law, for instance). This is not a classical approach, so it will be detailed later when discussing the case of mass transfer.

10.1.3 Complemented Dipoles

To model the interplay between space and time with a Formal Graph, a new way of composing paths, called path complementing, is necessary. In Chapter 6 introducing the concept of dipole two

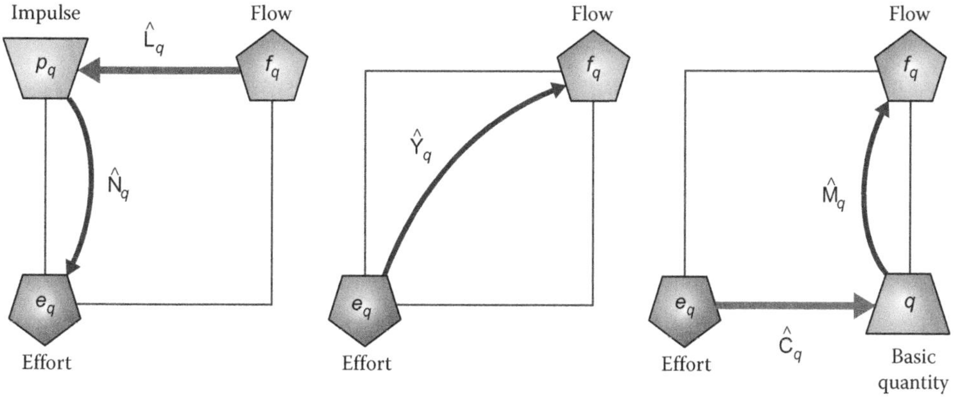

GRAPH 10.1 Three Formal Graph representations of a transfer at the global level. On the left is the representation used when the inductive subvariety is explicitly implied, on the right when it is capacitive, and in the middle when no subvariety is specified.

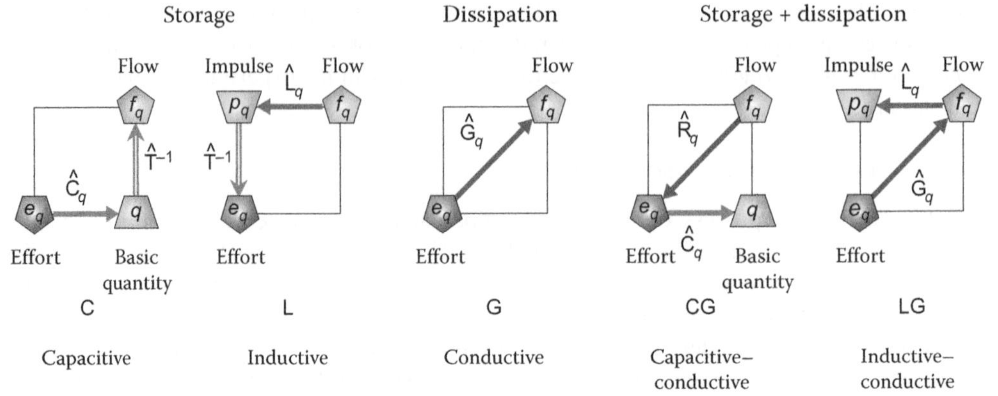

GRAPH 10.2 The three fundamental dipoles (two storing and one dissipating) and the two mixed dipoles (one conductive–capacitive and one conductive–inductive).

categories of dipoles, fundamental dipoles and mixed ones, were distinguished. The fundamental dipoles were defined as built with only one system constitutive property whereas the mixed ones were made up of two of these properties, one of them being necessarily the conductance. The peculiarity of the inductive and the capacitive dipoles (which belong to the fundamental category) being the lack of bond between the two energies-per-entity (the two gate nodes), the evolution link (space–time property) was required in order to achieve this bond. In the mixed dipoles, this was not necessary because the two gate nodes were linked by two system constitutive properties. These categories of dipole are recalled in Graph 10.2.

It appears that this nonrequirement of evolution link in a mixed dipole is not an intrinsic characteristic and that an evolution link may perfectly be added in a mixed dipole when necessary. This is the case for modeling transfers involving evolution such as the *transient transfer*.

However, the inclusion of an evolution link in a mixed dipole obeys a peculiar rule that consists of making two parallel paths complementary to each other, which differs from the previously seen rule of path addition on a node (see Graph 10.3).

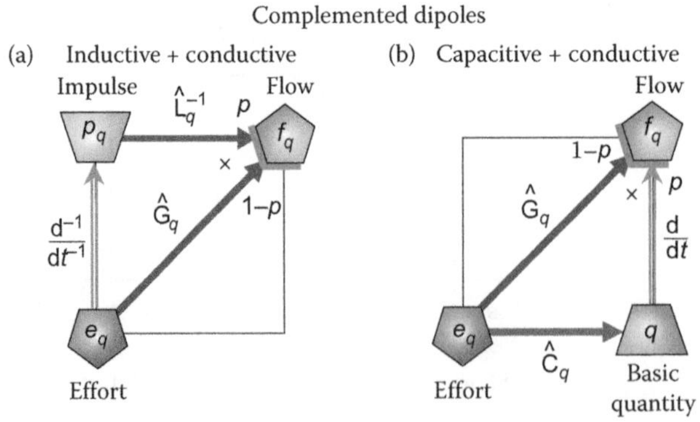

GRAPH 10.3 Two mixed dipoles implementing path complement: one inductive–conductive dipole (a) and one capacitive–conductive (b). The complement is made by attributing a weight to each path, with a sum of weights equal to 1. These weights are indicated near the arrow heads and the complement operation is symbolized by a cross.

The method of translating into algebraic terms this path complementing will be detailed through the case studies treating transient diffusion and will be recalled in Section 10.7. For the moment let us give the algebraic expression of the operator resulting from the complement of two paths, each one featured by an operator $\hat{O}_i$ and a weight p_i:

Path complement: $$\hat{O} = \hat{O}_1^{p_1} \times \hat{O}_2^{p_2} = \hat{O}_2^{p_2} \times \hat{O}_2^{p_1} \quad (10.1)$$

The cross (×) means an ordinary multiplication between the results of the two operators (and not a composition, which would express that paths are in series). Ordinary multiplication is commutative, meaning here that the order of the paths is indifferent. This path complement works only for parallel paths, that is, when the operand of both operators is the same.

The noticeable feature of the implementation of this operation in the case of transfers is that one of the two paths is necessarily a conductance.

10.1.4 Conductance, Admittance, and Transport

Conduction and transport are two intimately related processes. Conduction is generally seen as a flow of entities (charges, molecules, etc.) through a piece of material having an extension through space, so the movement of entities is considered as a transport under the influence of a difference of efforts (although it can be the converse). Conductance is the system constitutive property that models the behavior of conduction at the global level. As transport requires space, it is not the conductance that is really able to model transport but the spatially reduced property called *conductivity*.

However, there are processes featured by a conductance that are not considered as involving transport, such as mechanical friction or chemical reaction. They are seen as nonlocalized processes and therefore no influence of space is awaited in their modeling. Although some systems show a close similarity between reaction and transport, as in the case of a chain reaction behaving as a transport from site to site or vice versa. This point has been discussed in case study E5 "Diffusion through Layers" in Chapter 8 devoted to multipoles.

In fact, the question of the existence of transport underlain within the conductance depends on the considered system and on the role of space in its behavior, and there is no general rule for linking the two notions. In other words, a conductance does not imply necessarily a transport whereas the converse is true.

The previous discussion about the inclusion of temporal processes without transport into the category of transfer means that one must be able to model transfer without conductance. Inductive or capacitive dipoles are therefore used for this case, together with purely conductive and mixed dipoles for the other cases.

As will be seen in the various case studies, there are two classical approaches of transfer modeling, a global one using the concept of component of an equivalent electrical circuit and a local one based on the writing of differential or discretized equations. The latter approach is the most general one, allowing a wide variety of phenomena and of geometries to be handled, but it is the most laborious one as no general solutions come out owing to the wide variety of boundary conditions.

The technique of modeling by equivalent electrical circuit is widely used, mainly in conjunction with AC signal techniques, but also implicitly with large perturbation techniques using steps or ramp signals (or any signal shape).

Basically, four electrical components are utilized in this approach, listed in Table 10.2 with their algebraic models in two versions, the integral admittance for use with large perturbation and the complex admittance for small AC perturbation (differential admittance measurement).

TABLE 10.2
Correspondence between Components of an Equivalent Electric Circuit and Algebraic Models of (Integral) Admittances and Complex Admittances in the Linear Approximation

Components	Integral Admittance	Complex Admittance
Inductor ($p = -1$)	$\underset{lin}{\hat{Y}} = L^{-1}\frac{d^{-1}}{dt^{-1}}$	$\underset{lin}{\tilde{Y}} = (i\omega L)^{-1}$
Conductor ($p = 0$)	$\underset{lin}{\hat{Y}} = G = R^{-1}$	$\underset{lin}{\tilde{Y}} = G$
Capacitor ($p = 1$)	$\underset{lin}{\hat{Y}} = \frac{d}{dt}C$	$\underset{lin}{\tilde{Y}} = i\omega C$
CPE ($-1 < p < 1$)($p \neq 0$)	$\underset{lin}{\hat{Y}} = G\,\tau^{p}\frac{d^{p}}{dt^{p}}$	$\underset{lin}{\tilde{Y}} = G(i\omega\tau)^{p}$

Note: For each component the degree p of derivation is indicated with respect to time. A negative value means integration.

In addition to the three fundamental components, the Constant Phase Element (CPE), which is defined as being modeled by an admittance proportional to the time derivation raised to a noninteger exponent p, is indicated. In the expressions of this admittance, a characteristic time τ that plays the role of scaling factor for the time is used. A few words about the fractional derivation or integration are also given shortly after. The parameter p associated with the name of each component is the degree of derivation of the temporal derivation used in the integral model. It will be shown through the last case studies that it corresponds to a physical being quantifying the behavior of energy along the transfer process.

Fractional derivation/integration. It may be useful to recall how a fractional integral can be evaluated with the classical tools of mathematical analysis. The Riemann[*]–Liouville[†] p integral, originally established to determine how to reduce a multiple integration of order p to a single one, provides the tool by extending the order of integration to noninteger values (Samko and Ross 1993). The p integration is a convolution with a Green[‡] function, defined as the integration variable raised to the power p–1 and divided by the Gamma, or Euler,[§] function of p (Oldham and Spanier 1974)

$$\frac{d^{-p}}{dt^{-p}} f(t) = \frac{t^{p-1}}{\Gamma(p)} * f(t) = \frac{1}{\Gamma(p)} \int_0^t \frac{f(\lambda)}{(t-\lambda)^{1-p}}\, d\lambda \quad (p > 0) \tag{10.2}$$

(continued)

[*] Georg Friedrich Bernhardt Riemann (1826–1866): German mathematician; Göttingen, Germany.
[†] Joseph Liouville (1809–1882): French mathematician; Paris, France.
[‡] George Green (1793–1841): British mathematician; Nottingham, UK.
[§] Leonhardt Euler (1707–1783): Swiss mathematician; Basel, Switzerland and St. Petersburg, Russia.

This formula works only for integer or noninteger integrations and for working in the opposite way, the rule of composition of operators raised to various powers allows evaluation of a noninteger derivation from the previous formula

$$\frac{d^p}{dt^p} = \frac{d}{dt}\frac{d^{-(1-p)}}{dt^{-(1-p)}} \tag{10.3}$$

Numerical computations are naturally feasible. They are based on various techniques of integration (one of the simplest, the trapezoidal discretization, is well suited for that). Analog computations are also feasible by using electrical transmission lines as the one shown in case study G5 "Capacitive Transmission Line."

The Green function in the Riemann–Liouville integration formula is a partial derivative of the Heaviside* function (step function):

$$\frac{d^{1-p}}{dt^{1-p}}\mathcal{H}(t) = \frac{t^{p-1}}{\Gamma(p)} \tag{10.4}$$

This is a peculiarly interesting formula because it provides the response of systems modeled by a fractional derivation when they are perturbated by a Heaviside or step function, defined as

$$\mathcal{H}(t) \overset{def}{=} \begin{vmatrix} 0 & (t<0) \\ 1/2 & (t=0) \\ 1 & (t>0) \end{matrix} \tag{10.5}$$

In case study G9 "Generalized Mass Transfer," plots for some characteristic values of the power p are given. Here are some noticeable values of the Euler function:

$$\Gamma(1/4) = 3.625610\ldots \quad \Gamma(1/2) = \sqrt{\pi} = 1.772454\ldots \quad \Gamma(3/4) = 1.225417\ldots \quad \Gamma(1) = 1$$

For the value zero of its argument, the Euler function is infinite, but the derivative of the Heavidside function is perfectly defined as the Dirac† function, or peak function

$$\frac{d}{dt}\mathcal{H}(t) = \delta(0) \tag{10.6}$$

This function is equal to zero for all values of its argument, except for zero where the value of the function is infinite.

* Oliver Heaviside (1850–1925): British mathematician and physicist; Newcastle upon Tyne, UK.

† Paul Adrien Maurice Dirac (1902–1984): British physicist; Bristol, UK and Tallahassee, Florida, USA.

10.2 CASE STUDIES OF TRANSFERS

Two rather different transfer processes are discussed in this series of case studies. A first category comprises the simplest processes that are modeled by a pole, in distinguishing those that are relevant for a pure conductive pole, because they work as a pure conduction process, and those involving an energy storage in series with conduction.

Conductive poles

- G1: Reduced Ohm's law in electrodynamics
- G2: Fourier's equation of heat transfer in thermics

Capacitive–conductive pole

- G3: Fick's law of diffusion in physical chemistry

Inductive–conductive pole

- G4: Newton's law of viscosity in translational mechanics

The second category comprises more elaborate processes that require a dipole for being modeled because energy conversion (in addition to dissipation by conduction) occurs in parallel.

Capacitive + conductive dipoles

- G5: Capacitive transmission line in electrodynamics
- G6: Constant phase element in electrodynamics (or inductive + conductive)
- G7: Transient diffusion in physical chemistry
- G8: Anomalous diffusion in physical chemistry
- G9: Generalized mass transfer in physical chemistry

The last case study is not a study of a known subject but a proposal of a new tool for modeling mass transfers in particular and all others in general.

10.2.1 CASE STUDY G1: OHM'S LAW (REDUCED) ELECTRODYNAMICS

The case study abstract is given in next page.

Conduction is a fundamental process that is mathematically translated by an operator of conductance at the global level and by an operator of conductivity between spatially reduced levels. This scheme is found again in all energy varieties, whether as elementary path, as in the thermal conduction (Fourier equation), or as participating in a composed path, as in the diffusion of matter (Fick's law). A symmetric scheme (inverting the role of two energies-per-entity) works equally for the viscous friction in a fluid (Newton's law).

Ohm's[*] law, in its original form, models the behavior of a homogeneous electric conductor by linking the electric current I to the potential difference (electric tension) V through a scalar R called electric resistance:

$$V \underset{lin}{=} RI \qquad \text{(G1.1)}$$

[*] Georg Simon Ohm (1789–1854): German physicist; Cologne and Munich, Germany.

G1: Ohm's Law (Reduced) **Electrodynamics**

$$j = \hat{\sigma} E \quad \text{or} \quad j = -\hat{\sigma}\frac{d}{dr}V$$

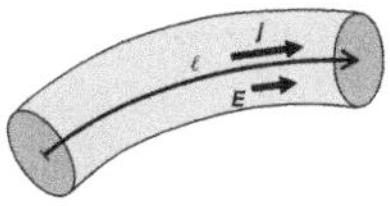

Formal Object

Space pole

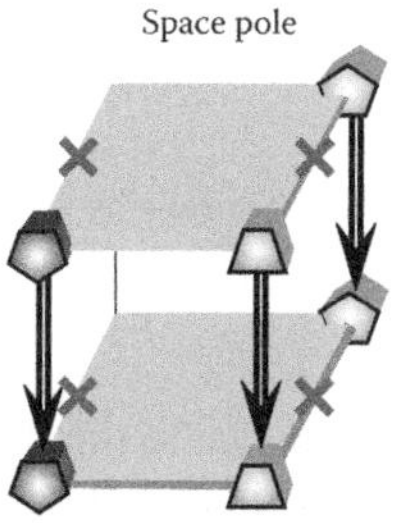

Pole	
✓	Fundamental
	Mixed
	Capacitive
	Inductive
✓	Conductive
	Global
✓	Spatial

Formal Graph:

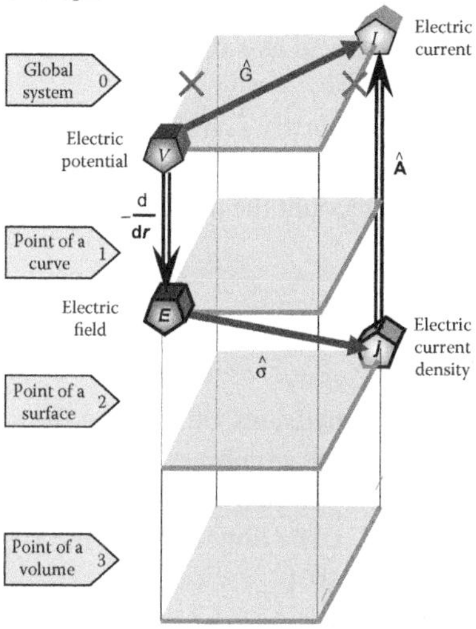

Variation with time:

Current density response to a potential step from 0 to V^∞

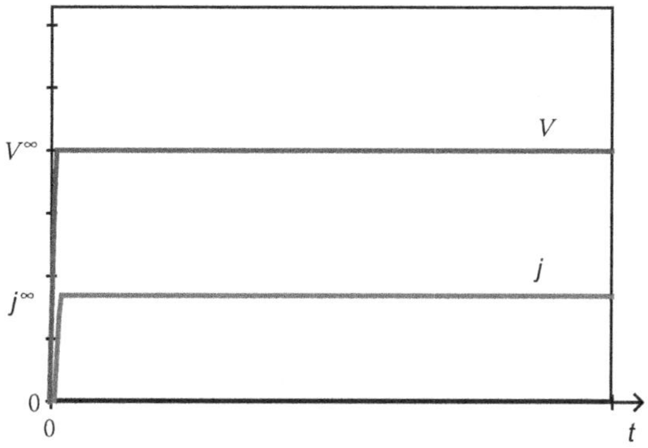

GRAPH 10.4

Variables:

Variety	Electrodynamics	
Subvariety	Capacitive	Inductive
Category	Energy/entity	Entity number
Family	Effort	Flow
Name	Potential	Current
Symbols	V, U, E, φ	$I, (i)$
Unit	[V], [J C^{-1}]	[A]

Properties:

System Properties		
Nature	Conductance	Conductivity
Levels	0	1/2
Name	Electrical conductance	Electrical conductivity
Symbol	G	σ
Unit	[S]	[S m^{-1}]
Specificity	–	Three-dimensional

(This scalar corresponds to the linear case.) The electric conductance G is the inverse of the resistance and allows expression of the current as a function of the electric tension:

$$I \underset{lin}{=} GV \tag{G1.2}$$

The Formal Graph (Graph 10.4) of the electrodynamical energy variety generalizes Ohm's law to any heterogeneous conductor by using not a scalar but an operator for the conductance and establishes the link with the spatially reduced conductance which is the conductivity.

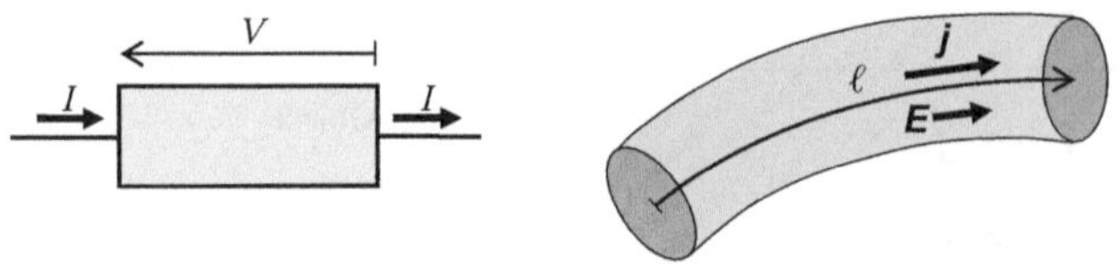

FIGURE 10.4 Representation of a conductor as a circuit component (dipole) with the receptor convention (left) and as a tube of conducting material (right).

The Formal Graph gives prominence to the dissipative nature of the resistance and conductance by the undertaken diagonal path, which results in the whole electric energy passing through a conductor being converted into heat. A conductor is therefore represented in the Formal Graph theory by a *dissipative dipole* (as a "slice" of conductor is represented by a dissipative pole, the Formal Graph in the case study abstract represents a pole but can also be viewed as a dipole in the absence of temporal evolution).

The *reduced* Ohm's law expresses the dependence of the *current density* $\boldsymbol{j}$ on the *electric field* $\mathbf{E}$ with the *electric conductivity* σ as a scalar or tensor coefficient. (In effect, this relation is generalized by means of an operator or tensor of conductivity to take into account the anisotropy of certain materials.)

$$\boldsymbol{j} = \hat{\sigma}\,\mathbf{E} \tag{G1.3}$$

This conductivity is a *specific property* of a material in three dimensions (it remains the same whatever the considered volume of material), which can be read in the graph because the sum of linked levels (a level is quantified by the number of dimensions of the geometrical element, 1 for the line of current and 2 for the section of a tube of current) is equal to 3, the dimension of the system considered here. The conventional representation is shown in Figure 10.4.

The nodes (not the variables) bearing the localized variables in Equation G1.3 are defined as the result of derivation with respect to the pertinent geometric element, curve and surface:

$$\mathbf{E} \underset{\text{(node)}}{\overset{def}{=}} -\frac{d}{d\mathbf{r}}V \tag{G1.4}$$

$$\boldsymbol{j} \underset{\text{(node)}}{\overset{def}{=}} \frac{d}{d\mathbf{A}}I = \hat{\mathbf{A}}^{-1} I \tag{G1.5}$$

In Equation G1.5, the integration on a surface (section of a tube of current) is represented algebraically by an operator (that can be called operator of surface collect) $\hat{\mathbf{A}}$.

$$I = \hat{\mathbf{A}}\boldsymbol{j} = \int_S \boldsymbol{j}.d\mathbf{A} \tag{G1.6}$$

In substituting in this last relation the current density by its expression in Equation G1.3 and then by expressing the electric field as a function of the electric tension in Equation G1.4, the conductance operator is written as

$$\hat{G} = \hat{\mathbf{A}}\,\hat{\sigma}\left\{-\frac{d}{d\mathbf{r}}\right\} \tag{G1.7}$$

This algebraic expression in the Formal Graph is read as an equivalence of paths between the elementary path of the global property and the composed path of the spatial derivation (contragradient), of the conductivity and of collect on the considered surface. In reasoning on the reciprocal operator of resistance, it can be observed that this equivalence corresponds to a loop making use of the *circularity principle*.

Remarks

Homogeneous medium. A frequently encountered case is a conducting material that is homogeneous (and isotropic, linear, etc), for which a simplified relation can be deduced from the Formal Graph. The integration along a line of electric field over a length ℓ provides in this case the electric tension

$$V \underset{hom}{=} \ell \cdot \boldsymbol{E} \tag{G1.8}$$

and the current density is merely obtained by division of the current by the area (section) of the considered tube of current

$$\boldsymbol{j} \underset{hom}{=} \frac{I}{\boldsymbol{A}} \tag{G1.9}$$

The reduced Ohm's law as in Equation G1.3 leads therefore to a well-known expression linking the conductivity to the scalar conductance used in Equation G1.2 or maybe still better known under its reciprocal form using the resistivity ρ:

$$G \underset{hom}{=} \sigma \frac{\mathbf{A}}{\ell} \quad \text{or} \quad R \underset{hom}{=} \rho \frac{\ell}{\boldsymbol{A}} \tag{G1.10}$$

However, for symbol and variable economy, it is not recommended to use the resistivity ρ as the conductivity can be perfectly substituted.

Higher generality. The relationship between global property and reduced property (here conductance and conductivity) is traditionally presented in the particular case of a homogeneous material, without influence of local levels (geometric elements). On the contrary, the Formal Graphs in three dimensions work in the general case of nonlinearity and handle right away any kind of geometry and of inhomogeneity. The reasoning used here to demonstrate this global/local link helps to understand the role of the *specificity* of a system constitutive property (which depends on the number of dimensions of the system) in the choice of different paths that may compose a global property. This reasoning is not particular to the dissipative pole or dipole but is valid for any system distributed in space, whatever its geometry.

10.2.2 Case Study G2: Fourier's Equation (Heat Transfer) — Thermics

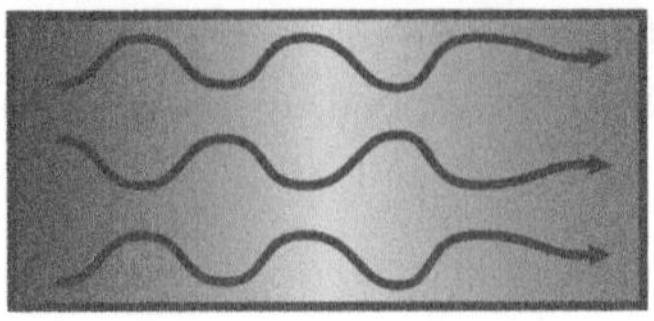

$$\frac{\mathrm{d}}{\mathrm{d}t}\mathcal{Q} = -A\kappa\frac{\partial}{\partial \boldsymbol{r}}T$$

Formal Object

Space pole

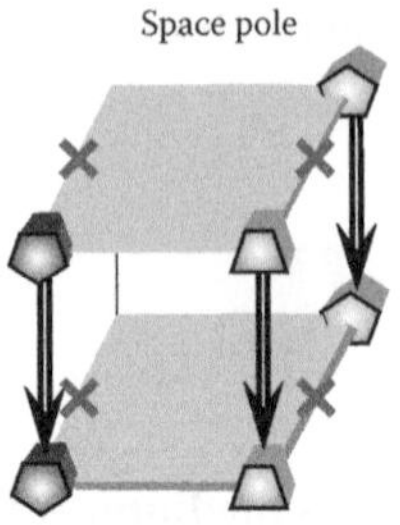

Pole	
✓	Fundamental
	Mixed
	Capacitive
	Inductive
✓	Conductive
	Global
✓	Spatial

Formal Graph:

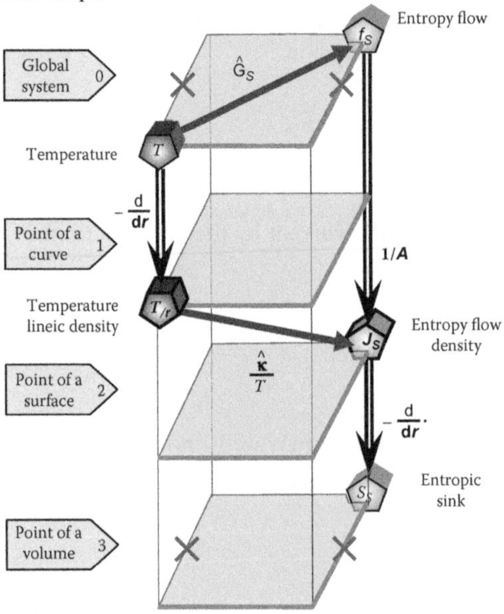

Variation with time:
Entropy flow response to a temperature (difference) step from 0 to T^∞

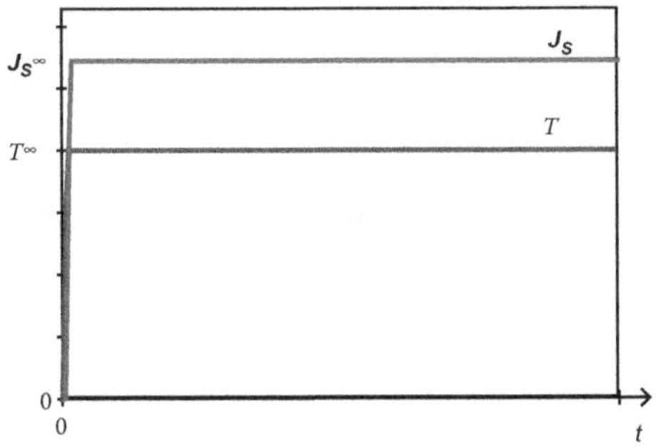

GRAPH 10.5

Variables:

Variety	Thermics (Heat)		
Subvariety	Capacitive (potential)		Inductive
Category	Entity number	Energy/entity	Energy/entity
Family	Basic quantity 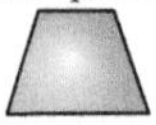	Effort 	Flow 
Name	Entropy	Temperature	Entropic flow
Symbols	S	T	f_S
Unit	[J K^{-1}]	[K]	[J K^{-1} s^{-1}]

Properties:

System Properties		
Nature	Conductance	Conductivity
Levels	0	1/2
Name	Entropic conductance	Entropic conductivity
Symbol	G_S	$\sigma_S = \kappa/T$
Unit	[J K^{-2} s^{-1}]	[J K^{-2} m^{-1} s^{-1}]
Specificity	–	Three-dimensional

Thermics. The thermal energy variety, or heat, or thermics, is a special domain because of the particular role played by the *entropy* and the *Second Principle of Thermodynamics*, which results in systematically coupling the thermal domain to all the other domains as soon as dissipation occurs through a conductance. Notwithstanding, the relationships between state variables and localized variables in this domain are of the same nature as in the other varieties of energy. Thus, the heat transfer follows a law that is identical to the reduced conduction in physical chemistry. In both cases, the transferred energy is capacitive energy, but inductive energy (kinetic) can also be transferred as in the case of viscous friction (Newton's law for fluids). The same law can also be found in electrodynamics with the reduced Ohm's law in which capacitive as well as inductive energy can be transferred.

The heat transfer obeys the Fourier* equation which expresses the proportionality of the thermal energy flow, classically called "heat flux," with the temperature gradient between two points in space, the proportionality coefficient being the product of the *thermal conductivity* κ by the area $\boldsymbol{A}$ on which the heat flow is counted.

$$\frac{\mathrm{d}}{\mathrm{d}t}\mathcal{Q} \underset{lin}{=} -\boldsymbol{A}\kappa \frac{\partial}{\partial \boldsymbol{r}} T \tag{G2.1}$$

The direct representation of the Fourier equation is not possible in a Formal Graph because heat is energy (of the thermal variety) and not a state variable, and also because the "heat flux" does not correspond to a variable that can be represented in a Formal Graph. The solution consists of taking recourse to state variables of the thermal variety that are the *entropy* S (basic quantity), the *temperature* T (effort), and the *entropic flow* f_S (also called entropy production rate). The variation of entropy is linked to the variation of heat according to the energy-per-entity times the variation of entity number

$$\mathrm{d}\mathcal{Q} = T\,\mathrm{d}S \tag{G2.2}$$

and the *entropic flow* f_S is linked, when a conversion of energy occurs, to the temporal derivative of the entropy

$$f_S = \frac{\mathrm{d}}{\mathrm{d}t} S \tag{G2.3}$$

The correspondent at the localized level of a surface of this flow, which is a gate variable at the global level, is the *density of entropic flow*, defined by

$$\boldsymbol{J}_S \underset{(node)}{\overset{def}{=}} \frac{\mathrm{d}}{\mathrm{d}\boldsymbol{A}} f_S \tag{G2.4}$$

but which, in the case of a *homogeneous surface*, is merely proportional to the entropic flow

$$\boldsymbol{J}_S \underset{hom}{=} \frac{f_S}{\boldsymbol{A}} \tag{G2.5}$$

* Jean-Baptiste Joseph Fourier (1768–1830): French mathematician and physicist; Grenoble, Lyon, and Paris, France.

It remains to gather the different relationships for rewriting the Fourier equation under a form that can be graphically represented

$$\boldsymbol{J_S} = \frac{\hat{\boldsymbol{\kappa}}}{T}\left\{-\frac{\mathrm{d}}{\mathbf{dr}}\right\}T \tag{G2.6}$$

This form is identical to the spatially reduced Ohm's law (see case study G1) in which the current density $\boldsymbol{j}$ results from an *electrical conductivity* tensor applied to the electric field $\boldsymbol{E}$

$$\boldsymbol{j} = \hat{\boldsymbol{\sigma}}\left\{-\frac{\mathrm{d}}{\mathbf{dr}}\right\}V = \hat{\boldsymbol{\sigma}}\ \boldsymbol{E} \tag{G2.7}$$

From this similarity an *entropic conductivity tensor* $\hat{\boldsymbol{\sigma}}_S$ can be defined according to

$$\hat{\boldsymbol{\sigma}}_S = \frac{\hat{\boldsymbol{\kappa}}}{T} \tag{G2.8}$$

and a *field of temperature* $\boldsymbol{T_{/r}}$ can be utilized defined on the same model as the electric field

$$\boldsymbol{T_{/r}} \underset{(node)}{\overset{def}{=}} -\frac{\mathrm{d}}{\mathbf{dr}}T \tag{G2.9}$$

to obtain an equation of transfer of entropy (and therefore of heat) analogous to Ohm's law

$$\boldsymbol{J_S} = \hat{\boldsymbol{\sigma}}_S\ \boldsymbol{T_{/r}} \underset{isotropic}{\overset{lin}{=}} \sigma_S\ \boldsymbol{T_{/r}} \tag{G2.10}$$

but with an *entropic conductivity* that depends explicitly (and strongly) on the temperature, even though in electrodynamics the dependence of the electric conductivity is less pronounced, because it is generally given by an exponential law of the Arrhenius type (see Chapter 8, Section 8.5.6). This algebraic model closer to the traditional equation can be written as follows:

$$\boldsymbol{J_S} \underset{isotropic}{\overset{lin}{=}} -\frac{\kappa}{T}\frac{\mathrm{d}}{\mathbf{dr}}T \tag{G2.11}$$

From this equation, the original Fourier's equation is retrieved by substituting the entropic flow density $\boldsymbol{J_S}$ times the temperature T by the heat flow density.

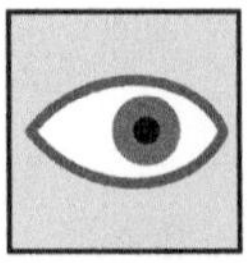

What is transported? One of the peculiarities of the thermal energy variety is to think of the thermal conduction in terms of energy transported, when other domains consider *entities* as the transported quantity (charges, momenta, etc.). According to their definition, as entities bear energy, there is no physical consequence due to this disparity. There are naturally historical reasons for this, but also a conceptual difficulty in our modern minds to view the entropy as a quantity able to be transported. This is certainly due to the influence of the statistical definition of entropy as a measure of order/disorder in the system, considered as a whole with implicitly a uniform entropy distribution.

10.2.3 Case Study G3: Fick's Law of Diffusion — Physical Chemistry

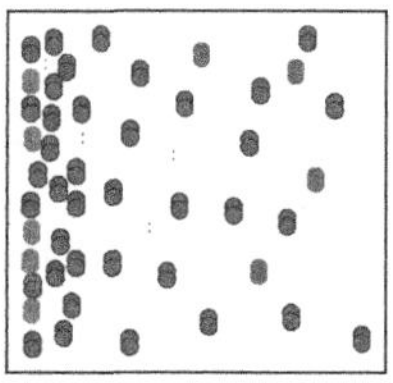

$$J \underset{\text{isotropic}}{=} -D \frac{\mathrm{d}}{\mathrm{d}r} c_n$$

Formal Object:

Space pole

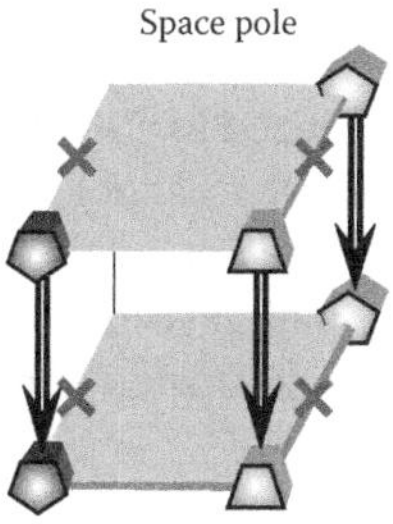

	Pole
	Fundamental
✓	Mixed
✓	Capacitive
	Inductive
✓	Conductive
	Global
✓	Spatial

Formal Graph:

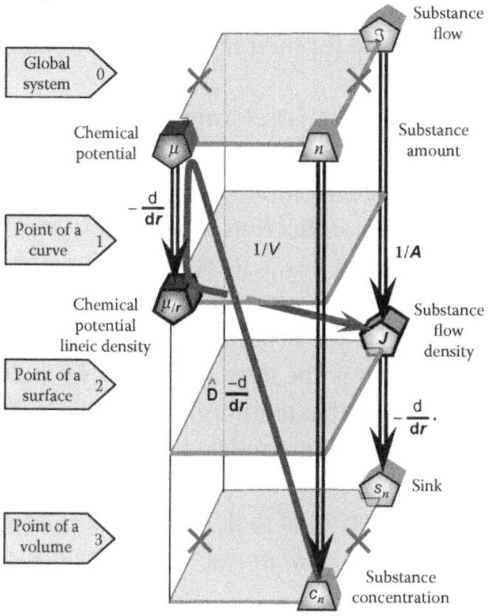

Variation with time:
Substance flow response to a concentration step from 0 to c^*

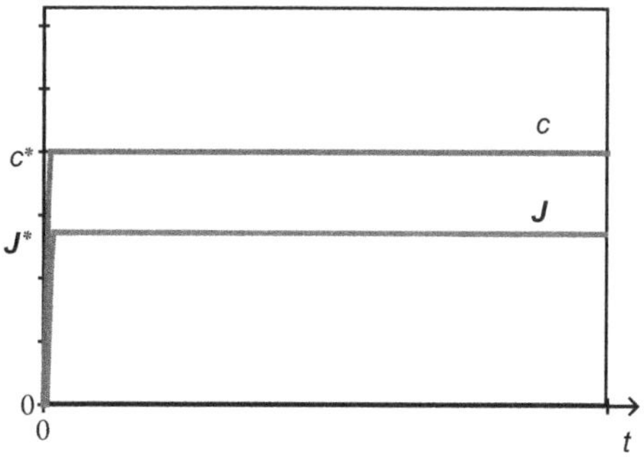

GRAPH 10.6

Variables:

Variety	Physical Chemistry		
Subvariety	Capacitive (potential)		Inductive
Category	Entity number	Energy/entity	Energy/entity
Family	Basic quantity	Effort	Flow
Name	Substance amount	Chemical potential	Substance flow
Symbols	n, N	μ	$\mathfrak{J}, v$
Unit	[mol]	[J mol^{-1}]	[mol s^{-1}]

Properties:

System Properties		
Nature	Conductance	Conductivity
Levels	0/3	1/2
Name	Physical chemical capacitivity	Physical chemical conductivity
Symbol	χ_n	σ_n
Unit	[mol^2 J^{-1} m^{-3}]	[mol^2 J^{-1} m^{-1} s^{-1}]
Specificity	Three-dimensional	Three-dimensional

Diffusion. An uneven distribution of entities in space tends to be leveled by a movement proportional to the gradient of effort or concentration.

Fick's law states that the flow density is proportional, through an operator called *diffusivity tensor*, to the gradient of concentration. In isotropic media, this tensor is reduced to a scalar D (called merely *diffusivity* or *diffusion coefficient*).

Here is a typical example of supplement of physical meaning brought by Formal Graphs. As the diffusivity is a constant (with respect to time), it cannot be represented by a path involving time; the only path possible takes the physicochemical conductivity path, meaning that the energy is entirely dissipated during a stationary diffusion process.

Diffusion is a transport of objects, particles, molecules, corpuscles, etc. under the influence of their uneven distribution in space. No external field is responsible for their movement tending toward a leveling of their distribution.

Depending on the nature of the diffusing objects, this process is relevant of several domains. Fick's law describes the phenomenon of diffusion of molecules or corpuscles. It belongs to the physical chemical energy variety when molecules are counted in moles, and to the corpuscular energy variety when the diffusing objects are considered as a collection of particles, molecules, atoms, and so on. As the difference between the two energy varieties is merely a matter of proportionality between their basic quantities (see the discussion on the energetic equivalence in Chapter 12), it is enough for modeling to consider one variety, for instance the physical chemical one. Taken alone, this law works in a stationary regime, which is the case modeled by a pole (or a stationary dipole) in the Formal Graph theory. In conjunction with a temporal evolution (by using a continuity equation), the process is an evolving or transient diffusion, modeled by a dipole (see case study G6 "Transient Diffusion"; Graph 10.7).

Algebraically, Fick's law relates the contragradient of concentration to the *substance flow density* $\boldsymbol{J}$ [also called, but improperly, "mass flux" (see remark on *old naming*) or what would be better

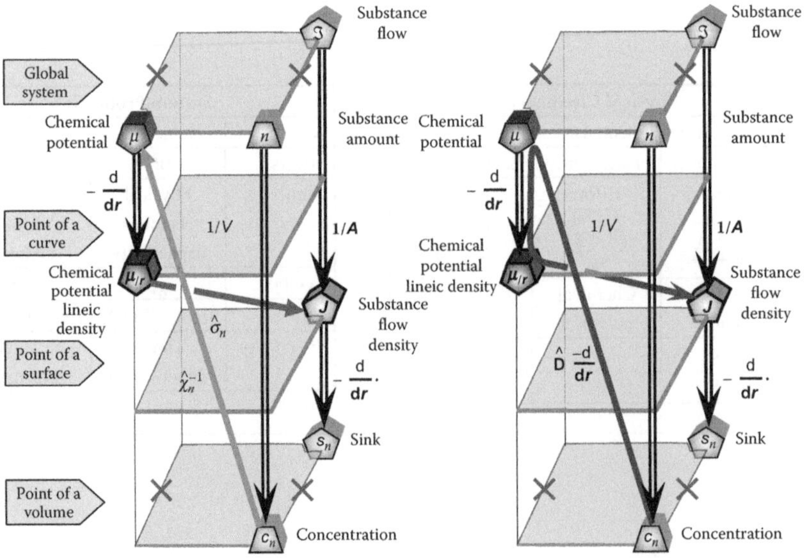

GRAPH 10.7 Two Formal Graphs representing Fick's law in a pole: the left one with elementary properties and the right one with the composed path corresponding to the operator relating concentration and flow density.

"mass flux density"] through a tensor $\hat{\mathbf{D}}$ called *diffusivity tensor*, as can be suspected by the fact that two vectors nonnecessarily colinear have to be related.

Fick's law:

$$\boldsymbol{J} = \hat{\mathbf{D}}\left\{-\frac{\mathrm{d}}{\mathbf{dr}}\right\}c_n \tag{G3.1}$$

When the diffusing medium is isotropic, the tensor reduces to a scalar D called *diffusivity*. (This naming is preferable to "diffusion coefficient," because such a qualifier is not in use for most other system properties; for instance, one does not use a "conduction coefficient" when speaking about conductivity.)

Isotropic medium:

$$\boldsymbol{J} \underset{isotropic}{=} -D\frac{\mathrm{d}}{\mathbf{dr}}c_n \tag{G3.2}$$

This restriction of Fick's law is mentioned here to comply with a more basic approach, but the requirement of an isotropic medium is not compulsory in the Formal Graph approach.

The simplicity of Fick's law contrasts strongly with the complexity of the Formal Graph that models the same phenomenon. It would have been simpler to build a graph on the same model as the other laws of transfer, Fourier equation or Ohm's law, for instance (see case studies G1 and G2). These graphs utilize a single elementary property (constitutive of the system) together with space properties, as is shown in Graph 10.8 (but in two dimensions instead of three dimensions). Fick's law is translated "word for word" by taking recourse to a variable called the "contragradient of concentration," which is then transformed by application of the diffusivity $\hat{\mathbf{D}}$ in order to arrive at the substance flow density ("mass flux") $\boldsymbol{J}$.

This graph is not a canonical one, in the sense of using variables that are not state variables (although they come from state variables by spatial reduction) but that are defined at a level where no geometric element exists in the usual space with three dimensions. Below the level of points of a volume, it would be necessary to involve a hypervolume with four dimensions to host these variables, which is not an approach having a physical meaning (but mathematical).

It appears that a simple solution exists with a canonical graph consisting of looking for a composed path going through a contragradient and through specific properties of the system. The passing through a temporal property being forbidden because a diffusivity is independent from

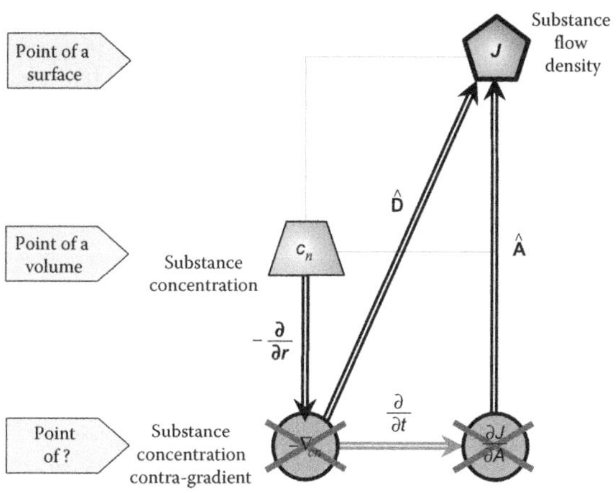

GRAPH 10.8 This is not a canonical Formal Graph! Modeling Fick's law with such a graph would require a fourth space dimension, which is not the simplest thing to do.

time, the simplest path using only one space property, the contragradient, is the composed path starting from the concentration c_n going through the elementary path of the reciprocal of the capacitive specific property (physical chemical elasticity in fact) which goes up to the chemical potential μ. From there, the contragradient path goes down to the lineic density of chemical potential $\boldsymbol{\mu}_{/r}$ and then the physical chemical conductivity reaches diagonally and downward the substance flow density ("mass flux") $\boldsymbol{J}$. All these elementary links have for algebraic equations:

$$\boldsymbol{J} = \hat{\sigma}_n\, \boldsymbol{\mu}_{/r} \tag{G3.3}$$

$$\boldsymbol{\mu}_{/r} \underset{(node)}{\overset{def}{=}} -\frac{\mathrm{d}}{\mathrm{d}\boldsymbol{r}}\mu \tag{G3.4}$$

$$\mu = \hat{\chi}_n^{-1}\, c_n \tag{G3.5}$$

with the reduced physical chemical capacitance ("*capacitivity*") defined as (this property is specific for three-dimensional systems)

$$\hat{\chi}_n = \frac{\mathrm{d}}{\mathrm{d}V}\hat{C}_n \tag{G3.6}$$

Consequently, the comparison of the equation corresponding to this composed path with Fick's law provides an equivalent for the diffusivity tensor

$$\hat{\mathbf{D}} = -\hat{\sigma}_n \left\{-\frac{\mathrm{d}}{\mathrm{d}\boldsymbol{r}}\right\} \hat{\chi}_n^{-1} \frac{\mathrm{d}^{-1}}{\mathrm{d}\boldsymbol{r}^{-1}} \tag{G3.7}$$

This expression is a much more general model of diffusion than suspected through a superficial analysis. It appears clearly that the diffusivity normally depends on space and that the case of invariance is a peculiar one.

Ideality, linearity, isotropy, homogeneity, dilution, etc. When the diffusing substance is an ideal one (i.e., sufficiently diluted in order to avoid interactions), the capacitive relationship is written, as demonstrated in Chapter 7, with the help of an exponential (or here a logarithmic) function

$$\mu = \hat{C}_n^{-1}\, n = \mu^{\ominus} + \mu_\Delta \ln\frac{n}{n^{\ominus}} \tag{G3.8}$$

using standard (reference) variables superscripted with a zero bar and a scaling chemical potential

$$\mu_\Delta \underset{ideal}{=} RT \tag{G3.9}$$

The lineic density of the chemical potential follows from its definition in Equation G3.4:

$$\boldsymbol{\mu}_{/r} = -\frac{\mu_\Delta}{n}\frac{\mathrm{d}}{\mathrm{d}\boldsymbol{r}}n \underset{hom}{=} -\frac{\mu_\Delta}{c_n}\frac{\mathrm{d}}{\mathrm{d}\boldsymbol{r}}c_n \tag{G3.10}$$

The second equality stems from the hypothesis of a homogeneous and isotropic medium making the volume independent of the localization. Now, if the physical chemical conductivity is assumed to be a scalar,

$$\hat{\sigma}_n \underset{isotropic}{\overset{lin}{=}} \sigma_n \tag{G3.11}$$

the diffusivity can be expressed merely as a scalar proportional to the temperature, which corresponds to the Nernst*–Einstein† relationship

$$D \underset{ideal+lin}{\overset{isotropic}{=}} \frac{\sigma_n}{c_n} RT \tag{G3.12}$$

The ratio of the conductivity on the concentration is called *physical chemical mobility* with the symbol u_n, which is considered to be specific to the substance in a given medium and in a limited temperature range. (There are as many mobilities as energy varieties, so it is important not to omit the indication of the domain.) It is clear from the previous development that this relationship is approximate and should not be considered as fundamental.

Adolf Fick‡ established his law in 1855 on an empirical basis and it is still often presented as such. When the diffusivity is considered as a space invariant scalar, it can be said that it is indeed an approximation. This is a true fundamental relationship when the diffusivity is taken as a tensor depending on space and on the concentration.

10.2.3.1 Physical Chemical Migration

Ions are charged molecules that may move under the action of an electric field. Their movement is called migration and is generally quantified by the physical chemical flow density $\boldsymbol{J}$ and the classical model of this process is written as its proportionality to the electric field, or equivalently to the contragradient of the potential.

Electrical+physical chemical migration:

$$\boldsymbol{J} \underset{isotropic}{\overset{lin}{=}} -u_Q c_n \frac{\mathrm{d}}{\mathrm{d}\boldsymbol{r}} V \tag{G3.13}$$

The coefficient here is the product of the *electric mobility* u_Q times the substance concentration. This relationship is not easy to translate in a Formal Graph because it involves two energy varieties, which complicates the model. A simpler model not mixing two energy varieties would have been to use a charge concentration ρ instead of the substance concentration c_n, which would have given a current density $\boldsymbol{j}$, but this is not the custom in physical chemistry.

Electrical migration:

$$\boldsymbol{j} \underset{isotropic}{\overset{lin}{=}} -u_Q \rho \frac{\mathrm{d}}{\mathrm{d}\boldsymbol{r}} V \tag{G3.14}$$

In fact, this equation is nothing other than Ohm's law (cf. case study G1) by remarking that the conductivity is the product of the *electrical mobility* u_Q by the charge concentration:

$$\sigma = u_Q \rho \tag{G3.15}$$

Another alternative to avoid mixing two energy varieties is to shift entirely to physical chemistry, which is written as

$$\boldsymbol{J} \underset{isotropic}{\overset{lin}{=}} -\sigma_n \frac{\mathrm{d}}{\mathrm{d}\boldsymbol{r}} \mu \tag{G3.16}$$

---

* Walther Hermann Nernst (1864–1941): German physical chemist; Göttingen and Berlin, Germany.

† Albert Einstein (1879–1955): German, Swiss, American physicist; Bern, Switzerland and Princeton, USA.

‡ Adolf Eugen Fick (1829–1901): German physiologist; Zürich, Switzerland and Würzburg, Germany.

This is done first by remarking again that the physical chemical conductivity is the product of the *physical chemical mobility* u_n by the charge concentration and second by remarking that one has the following correspondence between the two mobilities and the two efforts:

$$u_Q = zFu_n \tag{G3.17}$$

$$\mu = zFV \tag{G3.18}$$

The coefficient utilized in the product of the charge number z of the ion times the Faraday F ($\cong 9.6 \times 10^4$ C mol^{-1}; see the values of constants in Appendix 2). The subject of systems containing more than one energy variety will be tackled in Chapter 12 about energetic coupling, where it will appear that this correspondence is not a fundamental law but depends on a certain number of conditions.

The interest of this development is to demonstrate that migration is in fact a conduction process and that it may be seen from two viewpoints, depending on the considered energy variety.

Remarks

"Mass flux" is old naming. The term "flux" is used classically with two meanings: first as equivalent to "flow" and second as the result of integration on a closed surface of a vector crossing it. In the present case, the two meanings are pertinent. Nevertheless, if "flux density" is the sole remaining possibility for naming a vector on a surface, one comes upon a circular reasoning (in fact, an identity operation resulting from the combination of an operator with its reciprocal, for instance the son of the father of X). We prefer to keep the term "flux" for the second meaning for which there is no practical equivalent.

With regard to the term "mass," it presents an ambiguity as it creates confusion with another energy variety, which is the gravitation. This naming has ancient roots, going back to when the concept of mole was not clear enough and the mass of a substance was the practical means to quantify it. Nowadays, it is preferable to abandon this old naming in order to explicitly place the notion in the right energy variety.

Energy behavior. The interest of this composition with paths playing different roles concerning the behavior of the energy is to outline the total dissipation of energy during a stationary diffusion process because of the conductive path undertaken, despite passing through the capacitive path. The latter is in series and cannot therefore divert a part by storing energy. It would be necessary to have a conservative path in parallel to allow this diversion, which may happen when time is involved (see case study G6 "Transient Diffusion"), allowing (due to the conservation expressed by the continuity equation) a part of the energy to be conserved (transient diffusion regime).

Fruitful constraint. The method, utilized for finding the right path corresponding to Fick's law, uses the remarkable constraint provided by the obligation to conform to a physical space with three dimensions and not four as it would be tempting to do from a purely mathematical viewpoint. The constraint of remaining in a canonical Formal Graph, using only state variables or localized variables in a space with a maximum of three dimensions, is also a method utilized for translating other laws of physics (see case study G2 "Fourier Equation" about heat transfer, case study F1 "Accelerated Motion" about Newton's second law in Chapter 9, and case study B5 "Gauss Theorem" about the relationship between field and charge in electrodynamics in Chapter 5, for instance).

10.2.4 Case Study G4: Newton's Law for a Fluid — Translational Mechanics

$$P \underset{isotropic}{=} -\eta \frac{d}{\mathbf{dr}}.\mathbf{v} \quad \text{and} \quad P \underset{isotropic}{=} -\nu \frac{d}{\mathbf{dr}} \boldsymbol{P}_{/v}$$

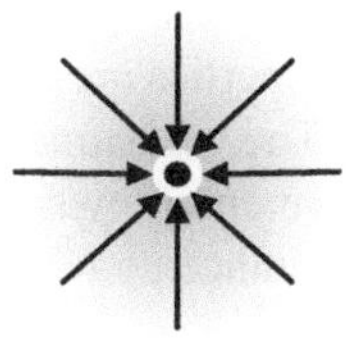

Formal Object

Space pole

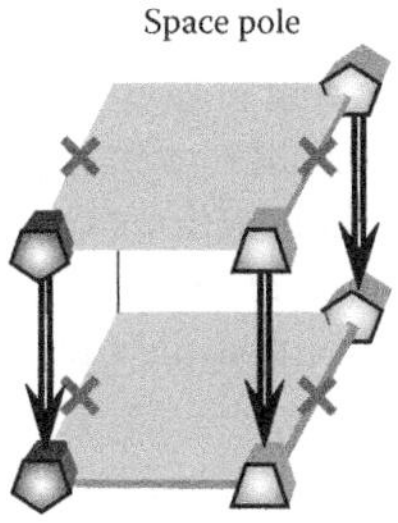

Pole	
	Fundamental
✓	Mixed
	Capacitive
✓	Inductive
✓	Conductive
	Global
✓	Spatial

Formal Graph:

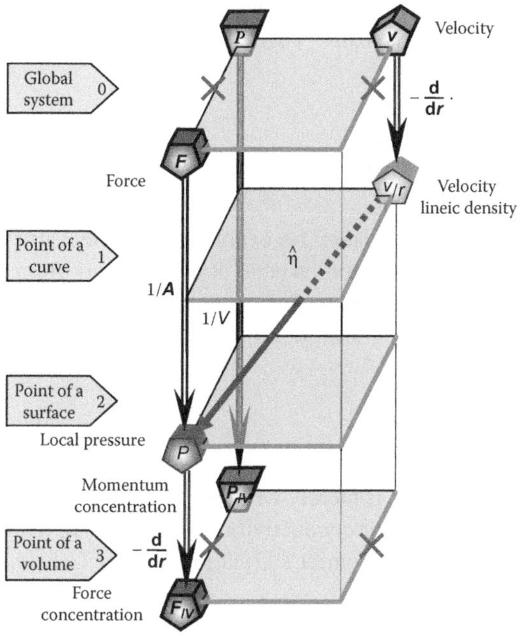

GRAPH 10.9

Variation with time:

Pressure response to a velocity (difference) step from 0 to $\boldsymbol{v}^{\infty}$

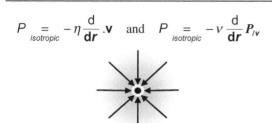

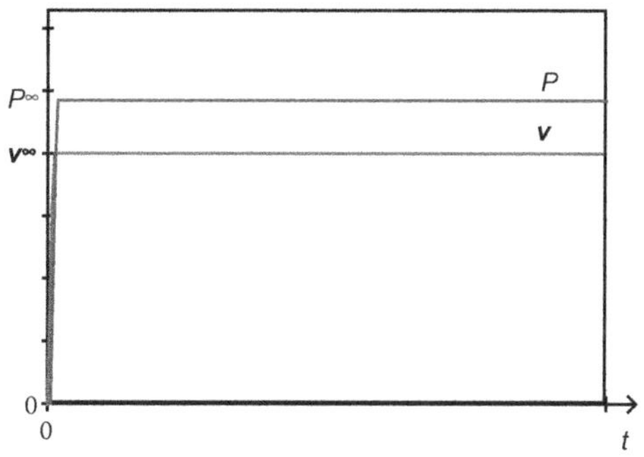

Variables:

Variety	Translation Mechanics		
Subvariety	Inductive (kinetic)		Capacitive
Category	Entity number	Energy/entity	Energy/entity
Family	Impulse	Flow	Force
Name	Momentum (quantity of movement)	Velocity	Force
Symbols	***P***, (***p***)	***v***	***F***
Unit	[N s], [J s m^{-1}]	[m s^{-1}]	[N], [J m^{-1}]

Properties:

System Properties		
Nature	Resistance	Resistance
Levels	0	1/2
Name	Friction resistance	Dynamic viscosity
Symbol	k_f	η
Unit	[N s m^{-1}]	[N s m^{-2}]
Specificity	–	Three-dimensional

Viscosity is a notion somewhat confused in fluid mechanics because it encompasses two very different properties: The *dynamic* viscosity η and the *kinematic* viscosity ν. These archaic adjectives "dynamic" and "kinematic" are misleading because there is no real notion of time or movement involved in these system properties.

The Formal Graph approach shows that both properties are reduced in space but that they differ by their actions on the energy in a system. The "dynamic" viscosity is a pure (reduced) inductance allowing the system to store inductive energy whereas the "kinematic" viscosity is a more complex property composed of a resistance converting the energy into heat (dissipation).

Newton's law in hydrodynamics, or in fluid mechanics, models the viscous friction between fluid elements. It links what is classically called a "gradient" of velocity (in fact a *lineic density*) to the *local pressure* P through a coefficient η called *dynamic viscosity*. This coefficient is a scalar in isotropic media and a tensor otherwise (Phan-Tien 2002).

$$P = \hat{\eta}\left\{-\frac{\mathrm{d}}{\mathbf{d}\boldsymbol{r}}\right\}.\boldsymbol{v} \underset{isotropic}{\overset{lin}{=}} -\eta \frac{\mathrm{d}}{\mathbf{d}\boldsymbol{r}}.\boldsymbol{v} \tag{G4.1}$$

The name "gradient" is improper (although widely used) because, for expressing a lineic density, a gradient normally applies to a scalar variable and not to a vector variable. It is also a pity from a pedagogical viewpoint because the physical meaning is more obvious when one refers to the effect of a convergence in a point of a fluid element, as shown in the illustration in the case study abstract.

$$v_{/r} \underset{(node)}{\overset{def}{=}} -\frac{\mathrm{d}}{\mathbf{d}\boldsymbol{r}}.\boldsymbol{v} \tag{G4.2}$$

With this lineic density, Newton's law as in Equation G4.1 is merely written as a dissipative relationship in a spatially reduced form, similar to the resistivity or conductivity relationships in many domains (see case studies G1 "Reduced Ohm's Law" and G2 "Fourier's Equation of Heat Transfer" for instance)

$$P = \hat{\eta}\, v_{/r} = \hat{\eta}\left\{-\frac{\mathrm{d}}{\mathbf{d}\boldsymbol{r}}\right\}.\boldsymbol{v} \tag{G4.3}$$

The physical phenomenon behind this law is explained as the effect of the convergence of velocities in fluid elements in a point, that is mathematically expressed by the operator "minus divergence = convergence," which creates a local pressure and dissipates the energy into heat. The inverse effect, a negative convergence, or positive divergence, creates a local depression and still dissipates.

Note that the adequate energy variety for modeling this law by a Formal Graph is not hydrodynamics but translational mechanics, because the pressure here is not a global variable (i.e., state variable) but a *local one* (i.e., localized variable) in the viscous phenomenon.

As for Ohm's law, which works equally well at the global level and in spatial reduction, for which one can link the conductance to the conductivity, it is possible to link the coefficient of friction (or of viscous friction) k_f defined at the global level to the dynamic viscosity. The relation between these two system properties derives from the *principle of circularity*:

$$\hat{k}_f = \frac{\mathrm{d}^{-1}}{\mathbf{d}\boldsymbol{A}^{-1}}\hat{\eta}\left\{-\frac{\mathrm{d}}{\mathbf{d}\boldsymbol{r}}\right\}. \tag{G4.4}$$

In making the hypothesis of a homogeneous, isotropic fluid and of volume elements having the form of cylindrical tubes, the integration on a tube section provides the *Stokes** *relation*

$$k_f \underset{hom}{=} 6\pi\, r^2 \eta \frac{1}{r} \tag{G4.5}$$

The path circularity leading to this model is shown in Graph 10.10 (left), together with the reduced inductive property which is the volumic mass ρ_M. This property can be combined with the dynamic viscosity in an analog manner to what has been done for modeling the diffusivity in case study G3 "Fick's Law of Diffusion," as shown in Graph 10.10 (right).

The Formal Graphs in Graph 10.10 are identical and merely illustrate two ways of considering the paths, elementary paths on one side, composed path on the other side. Each graph describes a *pole in translational mechanics*, but when the influence of time must be taken into account, it can be a *dipole* by adding temporal evolution links. The Formal Graph on the left shows the two spatially reduced system properties that are the *volumic mass* ρ_M (inductive property) and the *dynamic viscosity* η (conductive property). Both are specific for the fluid (i.e., invariant with the size of the considered fluid element) and represented by elementary paths.

The volumic mass links the velocity $\boldsymbol{v}$ of a fluid element to the volumic concentration of momentum $\boldsymbol{P}_{/V}$

$$\boldsymbol{P}_{/V} = \hat{\rho}_M \boldsymbol{v} \tag{G4.6}$$

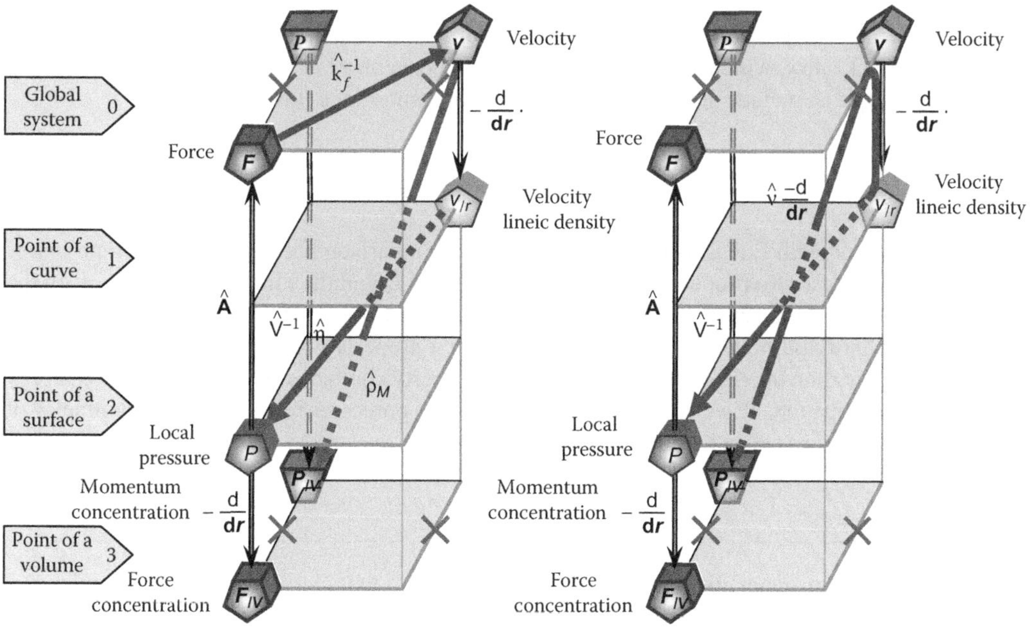

GRAPH 10.10 Two Formal Graphs representing Newton's law in a pole: the left one with elementary properties (system constitutive and space–time) and the right one with the composed path corresponding to the operator relating momentum concentration and local pressure.

* George Gabrielle Stokes (1819–1903): Irish mathematician; Cambridge, UK.

The presence of this property in the system allows one to compose with the dynamic viscosity η a path linking the concentration of momentum $\boldsymbol{P}_{/V}$ to the local pressure P (shear stress), which constitutes a definition for the operator ambiguously named *kinematic viscosity*

$$P \overset{def\to}{=} \hat{\nu}\left\{-\frac{\mathrm{d}}{\mathbf{d}r}\right\}.\boldsymbol{P}_{/V} \quad \text{(G4.7)}$$

As already mentioned, the Formal Graph on the right shows the composed path representing this *kinematic viscosity* ν. The relationship between the two viscosities is written in the general case as the composition of operators

$$\hat{\nu} = -\frac{\mathrm{d}^{-1}}{\mathbf{d}r^{-1}}\hat{\eta}\left\{-\frac{\mathrm{d}}{\mathbf{d}r}\right\}.\hat{\rho}_M^{-1} \quad \text{(G4.8)}$$

which, in the case of homogeneity (and isotropy) of the medium, simplifies in the classical relation

$$\nu \underset{hom}{=} \frac{\eta}{\rho_M} \quad \text{(G4.9)}$$

Improper definition. It is worth remarking that this relation is systematically presented as the definition of the kinematic viscosity, which has the disadvantage of preventing one from properly modeling a heterogeneous fluid having a dynamic viscosity depending on the distance or on the geometry.

An interesting comparison is frequently made with the phenomenon of diffusion of matter, described by Fick's law, which links in a similar way a concentration of entity number (substance concentration c_n) to a surface density of energy-per-entity (substance flow density $\boldsymbol{J}$)

$$\boldsymbol{J} = \hat{\mathbf{D}}\left\{-\frac{\mathrm{d}}{\mathbf{d}r}\right\}c_n \quad \text{(G4.10)}$$

The comparison with Equation G4.3 and mainly the comparison between Formal Graphs show the symmetry of these two properties that are the diffusivity D and the kinematic viscosity ν. In the case of diffusion, this is a substance amount that is transported, and in the case of kinematic viscosity, this is a quantity of movement (momentum). The first property combines storage of capacitive energy with dissipation into heat, whereas the second property combines storage of inductive energy (kinetic) with dissipation. The symmetry between these two processes results from the symmetry of these two subenergy varieties in a Formal Graph.

Remarks

Higher generality. The Formal Graphs bring a first clarification of these notions due to the need to make explicit all the variables that algebra exempts one from considering. A second clarification comes from the comparison between the structures and properties of the different domains. For instance, an evident symmetry between the electric resistivity ruled by *Ohm's law* and Newton's law for dynamic viscosity and another symmetry between this latter and the diffusion of a substance ruled by *Fick's law* for kinematic viscosity enlighten one about the similarity of physical phenomena involved.

Viscoelasticity. When in addition, the fluid or the material stores some capacitive energy, an elastic property comes in supplement to the dynamic and kinematic viscosities for featuring the system. The storage and the transport of capacitive energy is then supported by a combined property of viscoelasticity, analogous to a kinetic constant in physical chemistry (see case study A11 "Reactive Chemical Species" in Chapter 4 and case study H6 "Viscoelastic Relaxation" in Chapter 11).

10.2.5 CASE STUDY G5: CAPACITIVE TRANSMISSION LINE — ELECTRODYNAMICS

Coaxial cable with one end short-circuited

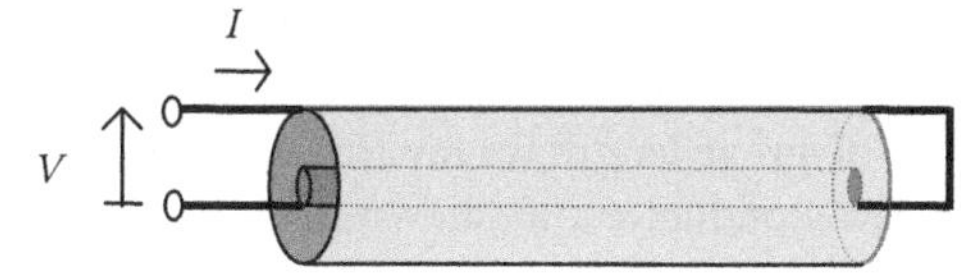

Dipole

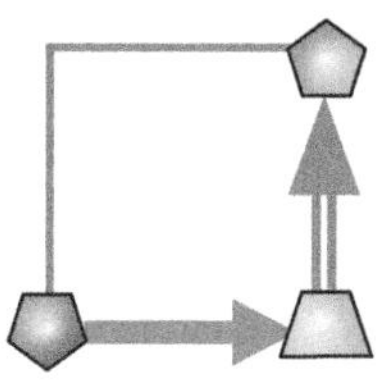

Dipole	
✓	Fundamental
	Mixed
✓	Capacitive
	Inductive
✓	Conductive
	Global
✓	Spatial
	Inseparable
✓	Conservative

Formal Graph:

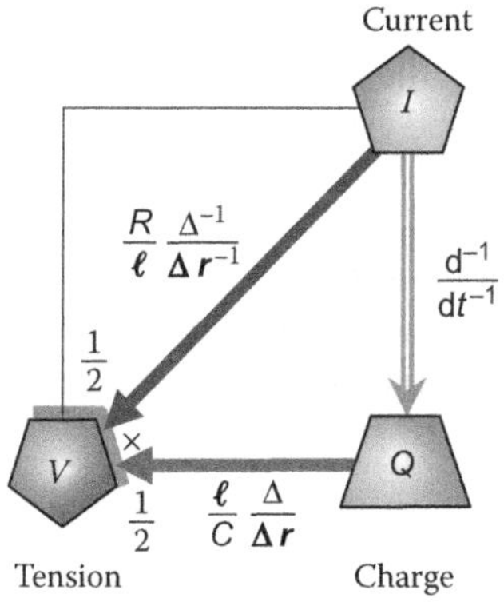

GRAPH 10.11

Variations with time:

Response to a potential step from 0 to V^{∞}

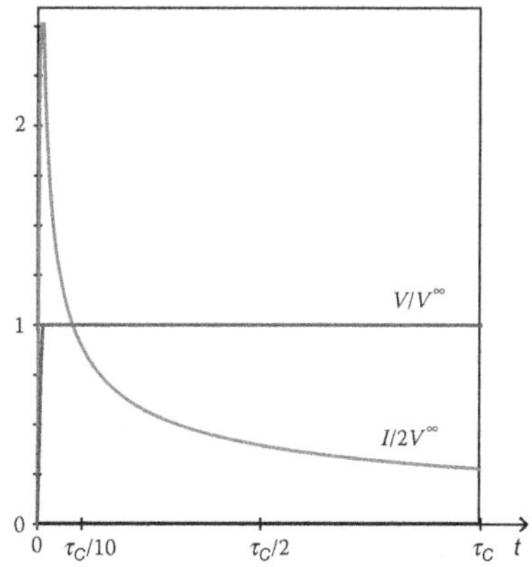

Variables:

Variety	Electrodynamics		
Subvariety	Capacitive (electrostatic)		Inductive
Category	Entity number	Energy/entity	Entity number
Family	Basic quantity	Effort	Flow
Name	Charge	Potential, tension	Current
Symbols	Q	V, U, E, φ	$I, (i)$
Unit	[C], [A s]	[V], [J C^{-1}]	[A]

Properties:

System Properties		
Nature	Capacitance	Resistance
Levels	0	0
Name	Electrical capacitance	Electrical resistance
Symbol	C	R
Unit	[F]	[Ω]
Specificity	—	—

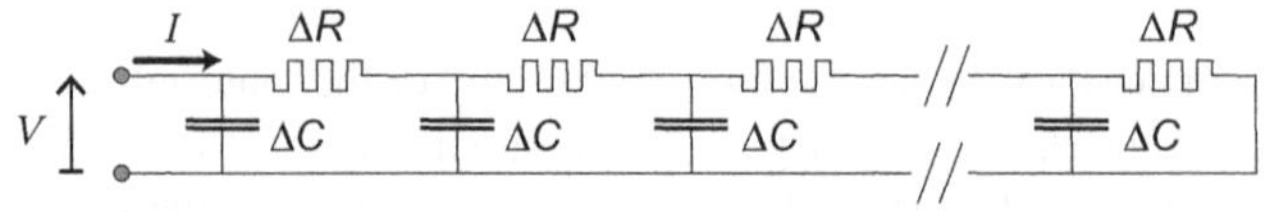

FIGURE 10.5 Discrete electric circuit modeling a capacitive coaxial cable. Each finite element possesses a serial resistance ΔR and a parallel capacitance ΔC.

This case study in electrokinetics illustrates algebraically how the concept of *path complement* may apply and how fractional exponents may appear in impedance expressions. Although the algebraic modeling is well known, the calculation of the impedance of this case study is detailed step by step in order to demonstrate the validity of the proposed concept and for comparing with the much simpler treatment permitted by the Formal Graph approach. Consider a coaxial cable made with two electronic conductors of great length, compared to its diameter, terminated at one end by a short circuit, linking the two conductors, as shown in the figure in the case study abstract.

An electric current I enters the cable at the other end and goes out by the other conductor and a tension V is present between the two conductors at this end. Such a device is also named a semi-infinite line.

The insulating material between the two conductors is a dielectric behaving as a capacitor with a capacitance C, and the material used for the conductors has a non-null longitudinal resistance R. The whole model is developed in the classical frame of linear properties. There is no direct translation into known finite components of electric circuits for such a device, as the capacitance and the resistance develop all along the length ℓ of the cable, making the tension and the current regularly decrease until they both reach a zero value at the short-circuited extremity. The conventional representation uses discrete elements each one representing a cable slice of finite length Δr, which are put in series (Figure 10.5).

A detailed circuit representing such a slice is shown in Figure 10.6.

The algebraic coding of this circuit element is made as follows: an electric charge ΔQ is present in the element capacitor of capacitance ΔC, related to the potential difference V between the capacitor poles by the capacitive relationship

$$\Delta Q = \Delta C\, V \tag{G5.1}$$

From the relationship between this charge and the element current ΔI flowing through the capacitor

$$\Delta I = \frac{d}{dt}\Delta Q \tag{G5.2}$$

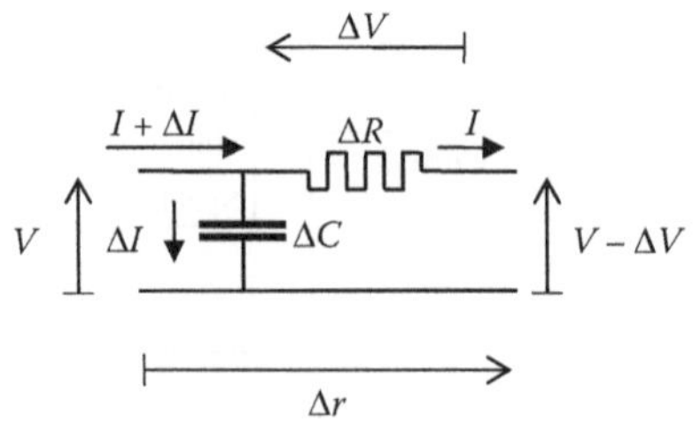

FIGURE 10.6 Electric circuit of a slice of a semi-infinite line.

one obtains by derivation the tension rate of change, a first equation of a system of two equations given below. On the other hand, Ohm's law applied to the discrete resistor with a resistance ΔR, having a potential difference ΔV corresponding to a crossing current I, provides a second equation for this system of equations:

$$\frac{\mathrm{d}}{\mathrm{d}t}V = \frac{\Delta I}{\Delta C}; \quad I = \frac{\Delta V}{\Delta R} \tag{G5.3}$$

It is assumed that the materials and the geometry of the cable are homogenous, so each property of a slice is proportional to the corresponding total cable property and to the slice thickness $\Delta \boldsymbol{r}$:

$$\Delta C = \frac{\Delta \boldsymbol{r}}{\ell}C; \quad \Delta R = \frac{\Delta \boldsymbol{r}}{\ell}R \tag{G5.4}$$

The previous system of equations is then

$$\frac{\mathrm{d}}{\mathrm{d}t}V = \frac{\ell}{C}\frac{\Delta I}{\Delta \boldsymbol{r}}; \quad I = \frac{\ell}{R}\frac{\Delta V}{\Delta \boldsymbol{r}} \tag{G5.5}$$

Making the thickness tend toward zero, one replaces discrete variations by differential ones, expressing the relationships between potentials (tensions) and currents at every point of the cable:

$$\frac{\partial}{\partial t}V = \frac{\ell}{C}\frac{\partial}{\partial \boldsymbol{r}}I; \quad I = \frac{\ell}{R}\frac{\partial}{\partial \boldsymbol{r}}V \tag{G5.6}$$

Here emerges the concept of parallel paths. Both previous equations relate the current to the tension, with different operators, and this is translated in graph language as two parallel connections (cf. Formal Graph in the case study abstract). This point will be reexamined a little further, for the usual algebraic treatment ignores this fact and proceeds by "eliminating" one variable to mix both equations into a single one (here, for instance, by eliminating the current)

$$\frac{\partial}{\partial t}V = \frac{\ell^2}{RC}\frac{\partial^2}{\partial \boldsymbol{r}^2}V \tag{G5.7}$$

This second-order partial differential equation constitutes the compact model of the semi-infinite cable.

Transient transport. This second-order partial differential equation is well known in physics as the *equation of diffusion*, modeling many processes such as neutron diffusion, heat transport, transient diffusion, or viscous processes in physical chemistry; in fact, every *transient transport* process. This remark is not fortuitous; all what is developed here may apply to these phenomena and justifies the length of the present development. However, care must be exercised about algebraic analogies, which do not imply identical physics, as commented in case study G7 "Transient Diffusion."

Instead of the conventional treatment for solving this equation using the Laplace* transformation, the Formal Graph introduction of the concept of time (Chapter 9) provides a more elegant

* Pierre Simon Laplace (1749–1827): French mathematician, astronomer, and physicist; Paris, France.

way of finding the solution algebraically. It consists of using the evolution operator $\hat{\mathrm{T}}^{-1}$ identified with the time derivation in the continuous case. Extension to the space derivation is ensured by identifying this latter with the operator $\hat{\mathbf{R}}^{-1}$, viewed as a *space distribution* operator (cf. Chapter 5). With these equivalences, the previous equation can be written as an ordinary parabolic equation between two operators $\hat{\mathrm{T}}^{-1}$ and $\hat{\mathbf{R}}^{-1}$ (the common operand is omitted as required by the method):

$$\hat{\mathrm{T}}^{-1} = \frac{\ell^2}{RC}\hat{\mathbf{R}}^{-2} \tag{G5.8}$$

This method has originally been developed by Heaviside and is known as *operational calculus* (Kullstam 1991). It consists of treating each operator in the same way as a scalar variable. The solution in terms of $\hat{\mathbf{R}}^{-1}$ of the second-degree algebraic equation is obtained by taking the square root of all terms:

$$\hat{\mathrm{T}}^{-1/2} = \pm\frac{\ell}{\sqrt{RC}}\hat{\mathbf{R}}^{-1} \tag{G5.9}$$

The return to the original "normal" differential equation is not straightforward. Care must be exercised about integration constants arising when the power of a differential operator is decreased (an increase would mean use of derivation). However, the boundary conditions $V = 0$ at one end, together with a null current at the same end (so the derivative of V versus $\boldsymbol{r}$ is also zero at this point, by virtue of Ohm's law expressed by Equation G5.6), wipe off any integration constant and eliminate the negative solution, so the resulting solution is written as

$$\frac{\partial^{1/2}}{\partial t^{1/2}}V = \frac{\ell}{\sqrt{RC}}\frac{\partial V}{\partial \boldsymbol{r}} \tag{G5.10}$$

The power one half of the time derivation means that when this operator is applied twice, a normal derivation is obtained, giving the reason to name it "semiderivation." The space derivative of the tension V at any point in the cable can be replaced by the current with the help of Equation G5.6, leading to an admittance relationship

$$\frac{\mathrm{d}^{1/2}}{\mathrm{d}t^{1/2}}V = \sqrt{\frac{R}{C}}\,I \tag{G5.11}$$

which can be inverted into impedance by using the classical rules between reciprocals of differential operators

$$V = \sqrt{\frac{R}{C}}\frac{\mathrm{d}^{-1/2}}{\mathrm{d}t^{-1/2}}I \tag{G5.12}$$

and written alternatively as

$$V = \frac{R}{\sqrt{\tau_c}}\frac{\mathrm{d}^{-1/2}}{\mathrm{d}t^{-1/2}}I \quad \text{with } \tau_c = RC \tag{G5.13}$$

The reciprocal of the semiderivation operator is the semi-integration. It can be calculated rigorously by a Riemann–Liouville integration, which is in fact a temporal convolution with a Green function (Oldham and Spanier 1974):

$$\frac{d^{-1/2}}{dt^{-1/2}} f(t) = \frac{1}{\sqrt{\pi t}} * f(t) = \frac{1}{\sqrt{\pi}} \int_0^t \frac{f(\lambda)}{\sqrt{t-\lambda}} d\lambda \tag{G5.14}$$

The final result is the expression of the impedance operator of a semi-infinite line (short-circuited coaxial cable) in terms of the power one half on the time integration

$$\hat{Z} = \frac{R}{\sqrt{\tau_c}} \frac{d^{-1/2}}{dt^{-1/2}} \tag{G5.15}$$

The Fourier transformation of the semi-integral operator is the Green function given above, but as a function of $i\omega$ instead of πt (with i the imaginary number and ω the *angular frequency*). The Fourier transformation of the impedance operator is consequently

$$\tilde{Z} = R\sqrt{\frac{1}{i\omega\tau_c}} = \frac{R}{2}\sqrt{\frac{2}{\omega\tau_c}}(1-i) \tag{G5.16}$$

which corresponds to what is measured by using an AC technique (impedancemetry or conductimetry at frequency $\omega/2\pi$) to characterize electrical components. The classical representation of this complex impedance in the Nyquist* plane (also called Cole†–Cole‡ representation), imaginary part (opposite) of the impedance versus the real one, is a 45° straight line when the frequency is varied, as Figure 10.7 illustrates.

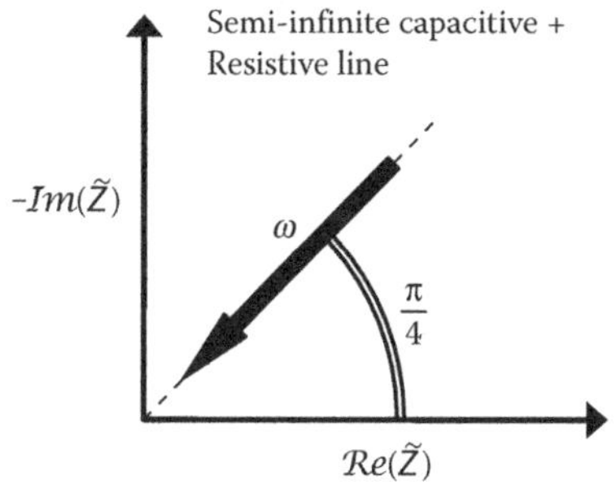

FIGURE 10.7 Nyquist (Cole–Cole) representation of the electric complex impedance of a semi-infinite coaxial cable. The direction of the arrow corresponds to increasing angular frequency.

---

* Harry Nyquist (1889–1976): Swedish and American electronic engineer; Murray Hill, New Jersey, USA.

† Kenneth Stewart Cole (1900–1984): American biophysicist; New York, USA.

‡ Robert H. Cole: Kenneth Stewart Cole's brother.

Formal Graph approach. What precedes is the classical treatment of the semi-infinite electric line, somewhat enhanced by the more elegant solving allowed by operational calculus compared to Laplace transformations. Instead of using an electric circuit, the Formal Graph approach is still simpler, as is shown here in modeling again the same device. Taking, in a first approach, the same starting point, a slice with a thickness $\boldsymbol{\Delta r}$ storing a charge ΔQ having a potential difference ΔV and crossed by current I, characterized by two constitutive relations, one capacitive and one conductive, one writes

$$V = \frac{\Delta Q}{\Delta C}; \quad I = \frac{\Delta V}{\Delta R} \tag{G5.17}$$

The same proportionalities as in Equation G5.4 between total cable and slice properties are assumed, leading to

$$V = \frac{\ell}{C}\frac{\Delta Q}{\boldsymbol{\Delta r}}; \quad I = \frac{\ell}{R}\frac{\Delta V}{\boldsymbol{\Delta r}} \tag{G5.18}$$

The last ingredient is to invoke the space–time property linking charge and current, that will be used in the reverse way, that is, by using a time integration (between 0 and t and symbolically represented by the reciprocal of the time derivative)

$$V = \frac{\mathrm{d}^{-1}}{\mathrm{d}t^{-1}} I \tag{G5.19}$$

The three relationships are represented in the Formal Graph in the case study abstract, by choosing the resistive orientation instead of the conductive one for the link between current and potential, as one is interested in modeling impedance rather than admittance.

From the Formal Graph in the case study abstract, recognition of the parallel feature is obvious for the two paths starting from the current and converging toward the electric potential when one considers the concomitance of both capacitive and resistive processes. In the present case of strongly correlated parallel processes, attribution of an equal weight to each one and application of the *paths complement* rule, introduced previously in Section 10.1.3, are immediately translated into operator multiplication

$$V = \left\{\frac{R}{\ell}\frac{\Delta^{-1}}{\boldsymbol{\Delta r}^{-1}}\right\}^{1/2} \times \left\{\frac{\ell}{C}\frac{\Delta}{\boldsymbol{\Delta r}}\frac{\mathrm{d}^{-1}}{\mathrm{d}t^{-1}}\right\}^{1/2} I \tag{G5.20}$$

leading to the result sought

$$V = \sqrt{\frac{R}{C}}\frac{\mathrm{d}^{-1/2}}{\mathrm{d}t^{-1/2}} I \tag{G5.21}$$

which can be expressed under the final form of impedance (Equation G5.15) as

$$\hat{Z} = \frac{R}{\sqrt{\tau_C}}\frac{\mathrm{d}^{-1/2}}{\mathrm{d}t^{-1/2}} \tag{G5.22}$$

A still more straightforward approach. A criticism, which can apply to both the classic and the last approaches, is that in view of the final result, involving only global system properties and no space-dependent parameters, the detour through space considerations is superfluous and in fact expresses an ill-conditioned

(continued)

problem. It is indeed simpler to start directly with global variables instead of variables at the level of a slice, such as the following two equalities:

$$V = \frac{Q}{C}; \quad V = RI \tag{G5.23}$$

and after adjoining the relationship in Equation G5.19 between the global charge and the current, to apply immediately the *path complement* with equal weights

$$V = \{R\}^{1/2} \times \left\{ \frac{1}{C} \frac{\mathrm{d}^{-1}}{\mathrm{d}t^{-1}} \right\}^{1/2} I \tag{G5.24}$$

to obtain directly Equation G5.21 and then the impedance G5.

$$\hat{Z} = \frac{R}{\sqrt{\tau_c}} \frac{\mathrm{d}^{-1/2}}{\mathrm{d}t^{-1/2}} \tag{G5.25}$$

The reason for not having directly modeled the semi-infinite line with this Formal Graph approach is that the demonstrative virtue of the comparison would have been lost.

An example of response of the capacitive transmission line is given in the case study abstract. The plot of the current resulting from a tension step (Heaviside function) shows the current increasing extremely fast to a high value then decreasing more slowly as the inverse square root of the time. (This function is the Green function used in the convolution Equation G5.14.) Mathematically, the current should climb to infinite value, but instrumental and physical limitations prevent it.

Equivalence between the two processes is translated by the exponent one half of time derivative operators. The physical meaning is that half of the energy entering a slice is stored in the slice capacitor, that is to say conserved, and the other half is lost as electric energy, that is to say converted into thermal energy by the Joule* effect in the conduction process.

Remarks

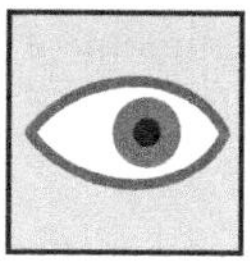

Warburg's impedance. In physical chemistry, the impedance found for the semi-infinite electric line is called Warburg† impedance.

Most of the time, this impedance is measured by electric (electrochemical) techniques and is thought of as an electric property of diffusing substances.

$$\tilde{Z} = R\sqrt{\frac{1}{i\omega\tau_c}} \tag{G5.26}$$

where τ_C is a time constant and R is an electric resistance equivalent to the physical chemical resistance (i.e., multiplied by the Faraday F). This notion is treated in case study G7 "Transient Diffusion."

* James Prescott Joule (1818–1889): English physicist and brewer; Manchester, UK.
† Emil Gabriel Warburg (1846–1931): German physical chemist; Berlin, Germany.

Energy behavior. The equivalence between the two processes, conductive and capacitive, is expressed by the value one half of the exponents of time derivation operators. The physical meaning is that half of the energy entering a slice is stored in the slice capacitor, that is to say conserved, and the other half is lost as electric energy, that is to say converted into thermal energy by the Joule effect in the conduction process.

Elegance. In a scale designed for sorting analytical solving techniques based on the elegance of the process (in the sense of using minimum steps), the following order is observed:

Classical algebra < Laplace transformation < Heaviside operational calculus < Formal Graph

10.2.6 Case Study G6: Constant Phase Element — Electrodynamics

The case study abstract is given in next page.

The CPE is a dipolar component of electrical circuit that has an intermediate behavior between a storage component (inductor or capacitor) and a dissipative one (resistor), but not exactly half-way of these two behaviors.

A component having exactly one-half of its energy stored and the other half dissipated is called differently, depending on the domain using such a component: semi-infinite line or transmission cable impedance in electrodynamics, Warburg element or diffusional impedance in physical chemistry, are among the most known callings. These are not unique because there is no uniform definition across domains. Case study G5 "Capacitive Transmission Line" was devoted to the transmission cable and G7 "Transient Diffusion" will deal with the diffusional impedance. In fact, this half-way behavior can be viewed as a peculiar case of the constant phase element.

10.2.6.1 The Classical Approach

The basic approach consists of defining on a purely mathematical basis a CPE by an impedance or an admittance expressed as a complex number proportional to a power p of the product of the imaginary number times the angular frequency ω, which is equivalent to say that the argument of an exponential function is a multiple of the same product

$$\tilde{Y} \underset{lin}{=} G(i\omega\tau)^p = G(\omega\tau)^p \exp\left(ip\frac{\pi}{2}\right) \quad (-1 < p < 1) \tag{G6.1}$$

(The Euler formula demonstrated in Chapter 9 is used here with $\pi/2$ as argument for expressing the imaginary number in polar coordinates.) This definition is formulated in the linear case with G the scalar conductance of the element and τ is a time constant featuring the element. In basic and frequent definitions, the product $G\tau^p$ is merged into a single parameter called preexponential factor, often notated $Q°$, without physical meaning. According to this model, the phase angle is independent of the frequency, being a constant, hence the name of this element.

$$\varphi = p\frac{\pi}{2} \tag{G6.2}$$

G6: Constant Phase Element **Electrodynamics**

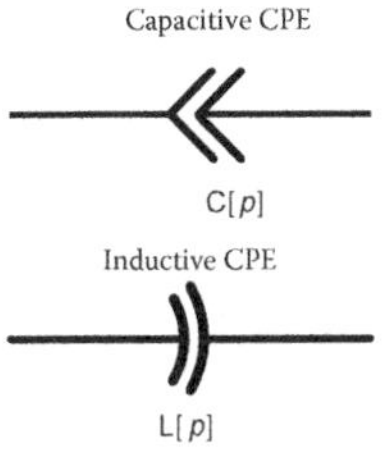

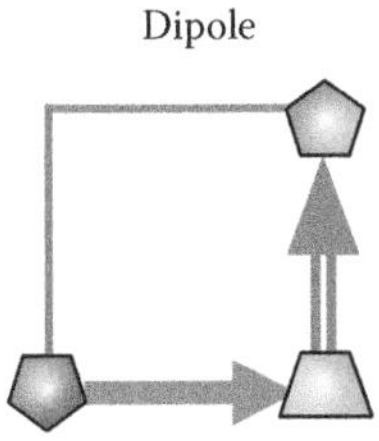

Dipole	
✓	Fundamental
	Mixed
✓	Capacitive
✓	Inductive
✓	Conductive
	Global
✓	Spatial
	Inseparable
✓	Conservative

Formal Graphs:

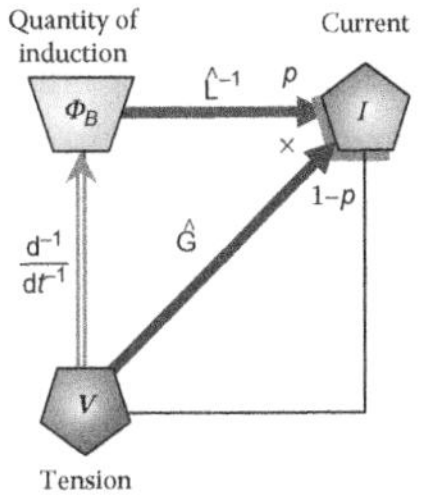

GRAPH 10.12 Inductive constant phase element.

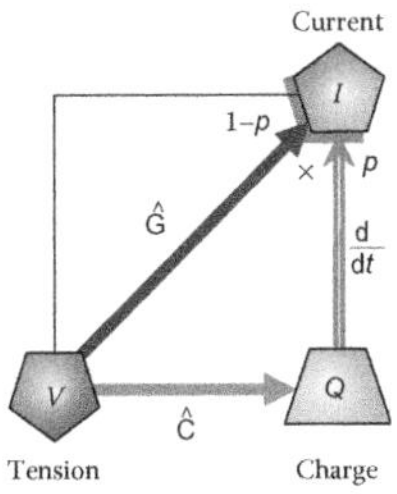

GRAPH 10.13 Capacitive constant phase element.

Variation with time:

Response to a potential step from 0 to V

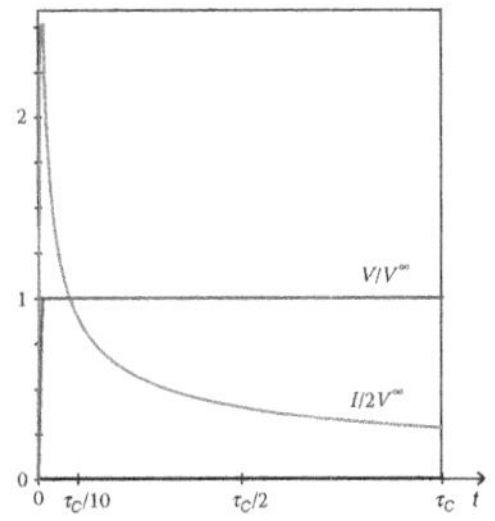

Variables:

Variety	Electrodynamics			
Subvariety	Inductive (kinetic)		Capacitive	
Category	Entity number	Energy/ entity	Entity number	Energy/ entity
Family	Impulse	Flow	Basic quantity	Effort
Name	Induction quantity (flux)	Current	Charge	Potential
Symbols	Φ_B	I	Q	V
Unit	[Wb], [J A^{-1}]	[A]	[C]	[V], [J C^{-1}]

Properties:

System Properties			
Nature	Capacitance	Resistance	Inductance
Levels	0	0	0
Name	Electrical capacitance	Electrical resistance	Electrical inductance
Symbol	C	R	L
Unit	[F]	[Ω]	[H]
Specificity	—	—	—

This concept of CPE is mainly in use among analysts using AC techniques for translating experimental data into equivalent electrical circuit. The physical meaning is not understood but the circuit representation facilitates comparisons better than complex number representations. (The most used are the Nyquist or Cole–Cole plot: imaginary part versus real part of the impedance or admittance as a function of the imposed signal frequency $\omega/2\pi$ and Bode* plots: logarithm of the modulus versus logarithm of the frequency for one plot, phase angle versus logarithm of the frequency for the other.)

There are no real standard symbols for use in electric circuit drawings; in the case study abstract the most general symbols are given consisting of deformations of the capacitor symbol. One needs to distinguish between inductive and capacitive CPEs, so in this book the curved deformation will be used for inductive CPE and the linear one for capacitive CPE.

* Hendrik Wade Bode (1905–1982): American electronic engineer; New York, USA.

Translated in terms of integral admittance (through an inverse Fourier transformation; cf. Chapter 9), the expression of the CPE is

$$\underset{lin}{\hat{Y}} = G\,\tau^p \frac{d^p}{dt^p} \quad (-1 < p < 1) \tag{G6.3}$$

The partial derivation or integration with respect to time is a shorthand notation for the convolution with a function that is a power $1-p$ of the time, as detailed earlier in Section 10.1.4. This model allows one to predict or identify the large signal response of a CPE.

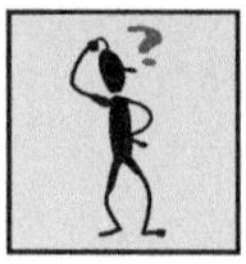

A mysterious object. The purely mathematical definition of the CPE naturally hides the physical nature of this element, which remains a mystery among physicists, although several experiments have given some indications on the phenomenon. Most observations of a noninteger (and nonhalf) exponent are made when at least one the following conditions is met:

- Interface roughness
- Fractal geometry
- Varying composition of materials
- Nonuniform distribution of flows or efforts

Depending on the scientific domain, some specific reasons are invoked, as for instance

- Inhomogeneous activity of electrode surfaces (electrochemistry)
- Distribution of reaction rates (catalysis)
- Distribution of relaxation constants (physical chemistry)
- Distribution of space charges (electrodynamics, condensed matter)
- Gradients of porosity in the material (condensed matter)
- Fringing fields near the electrode (electrodynamics)
- Imperfect joining of material grains (mechanics)

The most frequently invoked cause being the presence of fractal interface in the system, it has fostered the imagination of many and added mystery to the phenomenon, which is still a profitable source of numerous scientific publications.

10.2.6.2 The Formal Graph Approach

The way a CPE is represented in a Formal Graph differs significantly from the algebraic approach as can be imagined. It presents the advantage of injecting more physical meaning in this issue, although it does not answer completely the question of the physical origin of the phenomenon.

The key concept allowing a Formal Graph handling of this subject is the concept of *complementary paths*. When during a process the energy is susceptible to undertaking simultaneously several paths in parallel (starting and arriving on identical nodes) in a Formal Graph, each path may convey a different part of the energy. This is quantified by attributing to each path i a weight p_i between 0 and 1, and by imposing that the sum of all weights equates 1.

Path complement:
$$\sum p_i = 1 \quad (0 < p_i < 1) \tag{G6.4}$$

The second rule is that the operator of the overall path must be equivalent to the multiplication of all individual operators raised to the weight of the path. Applied to admittance operators, this is written as

Complemented admittance: $$\hat{Y} = \hat{Y}_1^{p1} \times \hat{Y}_2^{p2} \ldots \qquad \text{(G6.5)}$$

The third rule is optional and features a system in which there is no preference among the possible paths, implying that the order of multiplication must be indifferent, which is expressed in terms of requirement of commutative operators. These properties were introduced earlier in Section 10.1.3 and are detailed in Section 10.7.

This concept of *path complement* is applied to the modeling of the CPE by considering that the admittance of the system results from the contribution of two paths, one conductance and the other an energy storing path that can be either an inductive path or a capacitive one. These two alternatives are drawn in Graph 10.14 by using a weight p for the energy storing path and its complement to one for the conductive path.

Naturally, the Formal Graph modeling is not restricted to the admittance and may also represent an impedance, as shown in Graph 10.15 for the inductive and capacitive CPEs.

The demonstration of the correctness of these graphical models follows; by taking the Formal Graphs of the admittances given in Graphs 10.14 and 10.15, the multiplication of the two paths starting from the potential node and converging onto the current node of the inductive CPE is algebraically written as

$$\hat{Y} \underset{lin}{=} \left\{L^{-1}\frac{d^{-1}}{dt^{-1}}\right\}^{p} \times \left\{G\right\}^{1-p} \qquad \text{(G6.6)}$$

Note that the Formal Graph uses general properties represented by operators and that demonstration is restricted to the linear case in which the classical concept of CPE is defined. It remains to assemble the product of the conductance with the inductance into a time constant, referenced as inductive distinguishing with the capacitive case,

$$\tau_L \underset{lin}{=} G\,L \qquad \text{(G6.7)}$$

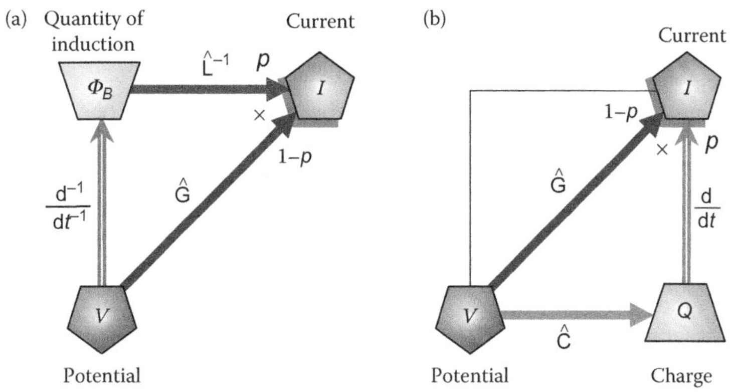

GRAPH 10.14 Formal Graph representations of the admittance of an inductive constant phase element (a) and of a capacitive one (b). The cross near the flow node (current) symbolizes a multiplication of operators (and not an addition) with the respective weights indicated near the arrow head.

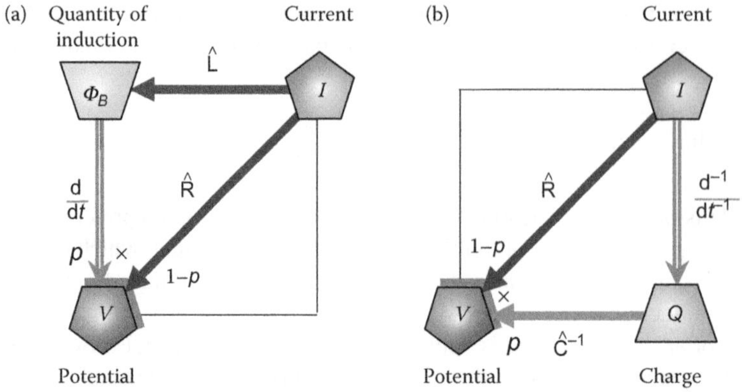

GRAPH 10.15 Formal Graph representations of the impedance of an inductive constant phase element (a) and of a capacitive one (b). The cross near the effort (potential) node symbolizes a multiplication of operators with the respective weights indicated near the arrow heads.

to obtain the algebraic model of the inductive CPE

$$\hat{Y} \underset{lin}{=} G\,\tau_L^{-p}\frac{d^{-p}}{dt^{-p}} \quad (0 < p < 1) \tag{G6.8}$$

Turning one's attention to the other CPE, the multiplication of the two paths is written as

$$\hat{Y} \underset{lin}{=} \left\{\frac{d}{dt}C\right\}^{p} \times \left\{G\right\}^{1-p} \tag{G6.9}$$

and the product of the resistance and the capacitance is used as (capacitive) time constant

$$\tau_C \underset{lin}{=} R\,C = \frac{C}{G} \tag{G6.10}$$

so as to obtain the following expression:

$$\hat{Y} \underset{lin}{=} G\,\tau_C^{p}\frac{d^{p}}{dt^{p}} \quad (0 < p < 1) \tag{G6.11}$$

Both expressions G6.8 and G6.11 are identical by exchange $-p \leftrightarrow p$ of one of the weights, which corresponds to the classical formulation in Equation G6.3, which uses a signed exponent p instead of two positive ones in the translations from Formal Graphs.

It can be remarked that a supplement of information is brought by the Formal Graph approach by giving to the time constants a more physical significance than does the preexponential factor in basic models, by relating them to the system constitutive properties.

The interest of modeling with Formal Graphs these mathematical beings that are the CPEs does not end there. The conceptualization in two parallel processes, each one contributing in a different manner to the overall process, offers new horizons for approaching the physical reality behind them.

As will be discussed later in Section 10.4, a specific role in the behavior of energy can be attributed to each path. In this case, the fact that one of the paths dissipates the energy while the other stores it provides an extremely interesting viewpoint of the physics underlying a CPE. This viewpoint will be thoroughly detailed later in the mentioned section.

10.2.7 Case Study G7: Transient Diffusion — Physical Chemistry

$$\frac{\partial}{\partial t} c_n = D \frac{\partial^2}{\partial r^2} c_n$$

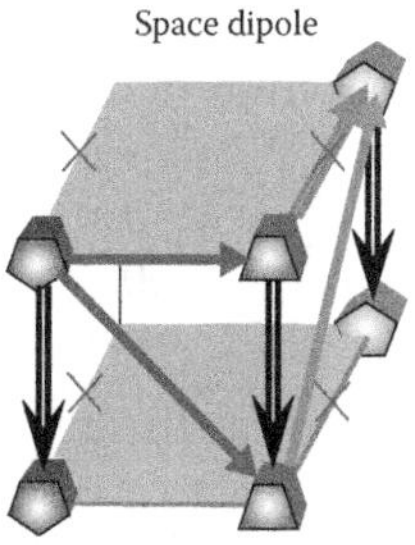

Dipole
✓ Fundamental
Mixed
✓ Capacitive
Inductive
✓ Conductive
Global
✓ Spatial
Inseparable
✓ Conservative

Formal Graph:

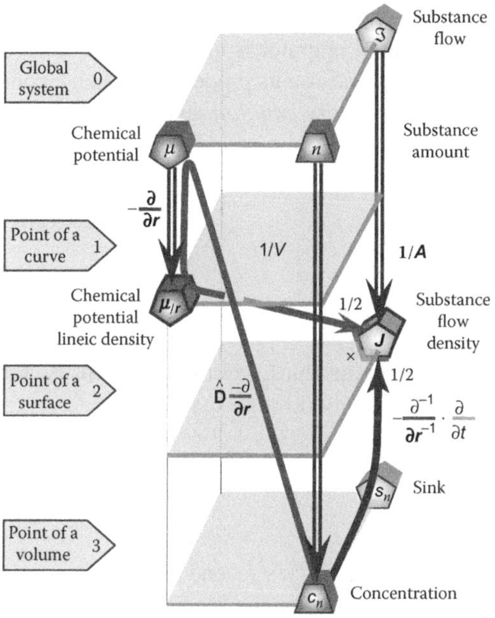

Variation with time: Response to a concentration step from c to c^*

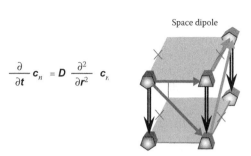

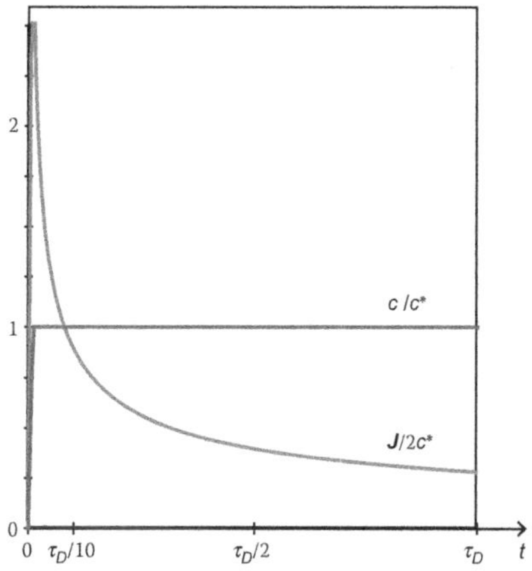

GRAPH 10.16

Variables:

Variety	Physical Chemistry		
Subvariety	Capacitive (potential)		Inductive
Category	Entity number	Energy/entity	Energy/entity
Family	Basic quantity	Effort	Flow
Name	Substance amount	Chemical potential	Substance flow
Symbols	n, N	μ	$\mathfrak{I}, v$
Unit	[mol]	[J mol^{-1}]	[mol s^{-1}]

Properties:

System Properties		
Nature	Capacitance	Conductance
Levels	0/3	1/2
Name	Physical chemical capacitivity	Physical chemical conductivity
Symbol	χ_n	σ_n
Unit	[mol^2 J^{-1} m^{-3}]	[mol^2 J^{-1} m^{-1} s^{-1}]
Specificity	Three-dimensional	Three-dimensional

Transient diffusion is a process mixing diffusion and evolution, that is, involving Fick's law and time. In case study G3 Fick's law working in stationary regime has been presented, featured by a constant substance flow density, which requires a constant difference of efforts in the physical chemical dipole. In homogeneous media the diffusivity tensor is replaced by a scalar D, so two cases are to be distinguished according to the anisotropy or isotropy of the diffusing medium

Fick's law: $$\mathbf{J} = -\hat{D}\,\frac{\partial}{\partial \mathbf{r}} c_n \quad \text{or} \quad \mathbf{J} \underset{isotropic}{=} -D\frac{\partial}{\partial \mathbf{r}} c_n \tag{G7.1}$$

When the effort difference (or concentration difference as these variables are not independent due to the capacitive relationship) is perturbated or is left free to evolve, Fick's law is still valid but is not sufficient for modeling the system and a supplementary relationship linking the concentration and the flow density and involving an evolution operator must be added to the model.

As concentration and flow density are variables not belonging to the same spatial level (volume for the concentration and surface for the flow density), a spatial operator is also required. Such a relationship characterizes the space–time and consequently depends on its properties. When a continuous space–time is assumed, this relationship is called *continuity equation* and is written as a balance equation:

$$\frac{\partial}{\partial t} c_n + \frac{\partial}{\partial \mathbf{r}} \cdot \mathbf{J} = 0 \tag{G7.2}$$

10.2.7.1 The Classical Approach

The classical approach consists of eliminating one of the two variables, say the flow density, for writing only one equation, by assuming a constant scalar diffusivity through space:

$$\frac{\partial}{\partial t} c_n \overset{hom}{\underset{isotropic}{=}} D\frac{\partial^2}{\partial \mathbf{r}^2} c_n \tag{G7.3}$$

This second-order partial differential equation, abusively called Fick's second law (but it is convenient to do so) represents the basic algebraic model of transient diffusion.

Mathematical analogy versus physical resemblance. In case study G5 "Transmission Line" the algebraic model of a semi-infinite electric line has been established, having an equation

$$\frac{\partial}{\partial t} V = \frac{\ell^2}{RC}\frac{\partial^2}{\partial \mathbf{r}^2} V \tag{G7.4}$$

This is also a second-order partial differential equation, but the algebraic similarity is misleading. An electric potential is not in the same family as a substance concentration, so the Formal Graphs are not identical, for it is only when Formal Graphs are identical that the analogy may have a physical meaning. This is an interesting example of true algebraic analogy but false physically speaking. However, the mathematical similarity is interesting for solving identically both equations. The solution in terms of semi-integration found for expressing the tension as a function of the current can be used here too (except that the short-circuit at the end of the transmission line imposes zero for the integration constant).

By using the technique of operational calculus (see case study G5 "Transmission Line"), the search for a solution goes through the division by 2 of the degree of operators, which amounts to integration with respect to space. An integration constant c_n^* representing the unperturbed concentration far from the localization of the diffusional process is therefore introduced in the expression of the solution. A sufficiently long distance is required to maintain the constancy of this concentration all along the duration of the transfer, which develops in space as time elapses. This configuration of space is called infinite (or semi-infinite, depending on the space symmetry).

Infinite transient diffusion: $$c_n - c_n^* \underset{isotropic}{\overset{hom}{=}} \frac{1}{\sqrt{D}} \frac{d^{-1/2}}{dt^{-1/2}} J \qquad \text{(G7.5)}$$

The semi-integration is an operator that is equivalent to an ordinary integration when applied twice. It is computed through the Riemann–Liouville formula introduced previously in case study G5 "Transmission Line."

10.2.7.2 The Formal Graph Approach

The Formal Graph representation of Fick's law has been given in case study G3 "Fick's Law" and is reproduced in the case study abstract. It remains to tackle the continuity equation.

This equation featuring the space–time is in fact already included in every three-dimensional Formal Graph by construction. The deepest variable in the flow family is the *sink* variable s_n, defined as the convergence of the flow density (cf. Chapter 5).

Convergence on sink: $$s_n \underset{(node)}{\overset{def}{=}} -\frac{\partial}{\partial \boldsymbol{r}} \cdot \boldsymbol{J} \qquad \text{(G7.6)}$$

The convergence operator, as signified by its name, accumulates substance in a point of the volume when flow vectors are oriented in majority toward this point (they are converging). Consequently, the physical meaning of the reverse operation, expressed by the reciprocal of the convergence, is to spread the accumulated amount in the sink, hence the proposed name of *spreading* for this operator. (The creation of nontraditional names must not be encouraged but is unavoidable when one attempts to give more physical meaning to anonymous mathematical concepts.)

Sink spreading: $$\boldsymbol{J} = -\frac{\partial^{-1}}{\partial \boldsymbol{r}^{-1}} s_n \qquad \text{(G7.7)}$$

In the other family of basic quantities, the deepest variable is the concentration. The two variables are linked by an evolution path, in the same way the substance flow $\mathfrak{I}$ is related to the substance amount n

Evolution: $$s_n = \frac{\partial}{\partial t} c_n \qquad \text{(G7.8)}$$

The path from the concentration to the flow density is therefore the composition of operators, which is a space–time velocity operator independent of the system, leading to the writing of the continuity equation in the form of a convection equation (when the composition of operators is replaced by a velocity vector; cf. the treatment of convection in Section 10.5).

Space–time convection:
$$\mathbf{J} = -\frac{\partial^{-1}}{\partial \mathbf{r}^{-1}}\frac{\partial}{\partial t}c_n \tag{G7.9}$$

The path featured by this relationship is the Formal Graph translation of the continuity equation G7.2.

The true parallelism of the path representing Fick's law and the one representing the continuity equation allows one to use the concept of path complement (see Section 10.1.3). The previous algebraic model in Equation G7.3 is retrieved when equal weights are attributed to the paths, which is algebraically translated by

Transient diffusion:
$$\mathbf{J} = \left\{-\frac{\partial^{-1}}{\partial \mathbf{r}^{-1}}\frac{\partial}{\partial t}\right\}^{1/2} \times \left\{\hat{\mathbf{D}}\left\{-\frac{\partial}{\partial \mathbf{r}}\right\}\right\}^{1/2} c_n \tag{G7.10}$$

This relationship is the general model of transient diffusion as proposed by the Formal Graph theory. It can be simplified by choosing peculiar conditions. In particular, the multiplication annihilates the space operators for leaving only the time derivation, in assuming as before that the diffusivity is space invariant (homogeneous medium) and that the medium is isotropic:

Infinite transient diffusion:
$$J \underset{isotropic}{\overset{hom}{=}} \sqrt{D}\,\frac{d^{1/2}}{dt^{1/2}}\left(c_n - c_n^*\right) \tag{G7.11}$$

An integration constant has been voluntarily added for allowing this relation to be inverted, as explained earlier in the classical approach, in order to retrieve Equation G7.5. This integration constant is the unperturbed concentration far from the place where the diffusional process occurs, meaning that this model applies to infinite (or semi-infinite) transient diffusion.

Warburg's impedance: When infinite transient diffusion is characterized by AC techniques, such as impedance spectroscopy in electrochemistry, the electrical complex impedance measured by imposing a periodic signal with a constant angular frequency is called Warburg impedance.

$$\tilde{Z} = R\sqrt{\frac{1}{i\omega\tau_C}} \tag{G7.12}$$

where τ_C is a time constant and R is an electrical resistance proportional to the physical chemical resistance R_n.

Transposed in the physical chemical energy variety, an impedance relates a substance flow to a chemical potential, so the physical chemical Warburg impedance is written in an analogous form:

$$\tilde{Z}_n = R_n\sqrt{\frac{1}{i\omega\tau_C}} \tag{G7.13}$$

The electrical impedance, that is, the genuine Warburg impedance, is proportional to this physical chemical impedance (see Figure 10.8).

10.2.8 Case Study G8: Anomalous Diffusion — Physical Chemistry

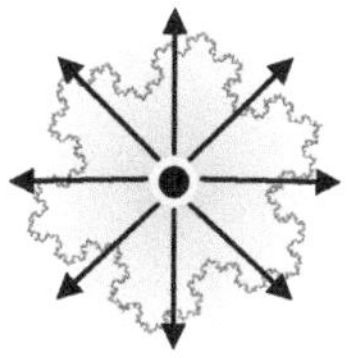

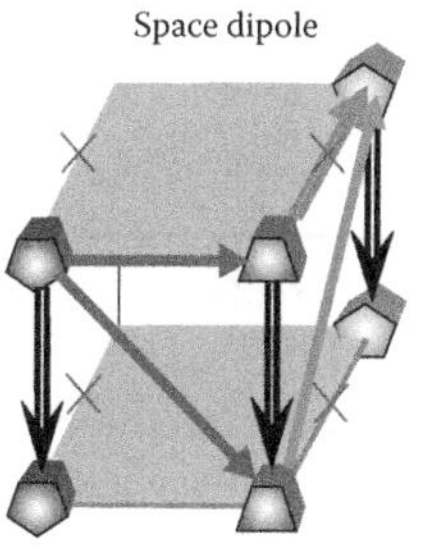

Dipole	
✓	Fundamental
	Mixed
✓	Capacitive
	Inductive
✓	Conductive
	Global
✓	Spatial
	Inseparable
✓	Conservative

Formal Graph:

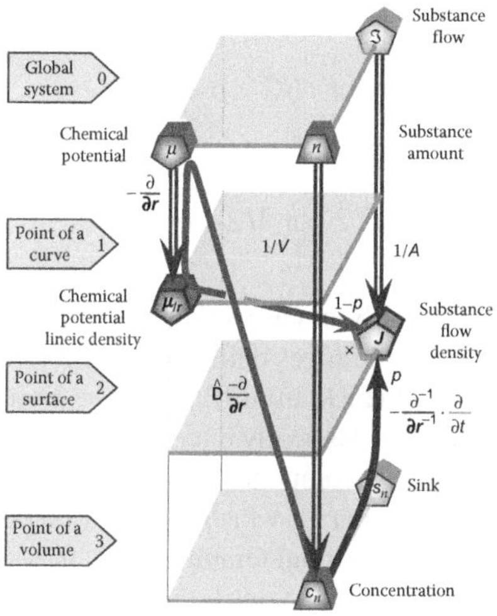

GRAPH 10.17

Variation with time: Response to a concentration step from c to c^*

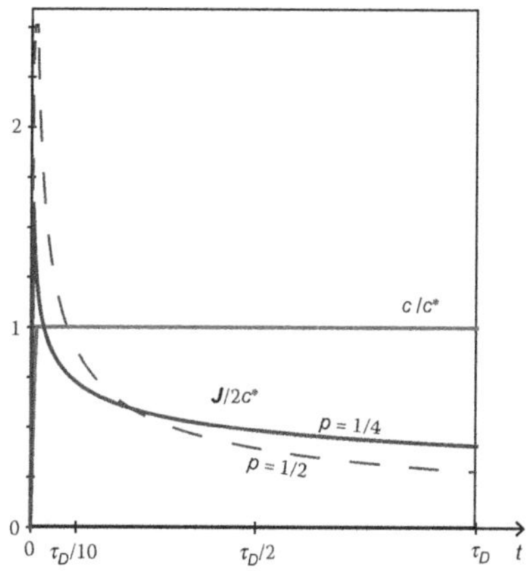

Variables:

Variety	Physical Chemistry		
Subvariety	Capacitive (potential)		Inductive
Category	Entity number	Energy/entity	Energy/entity
Family	Basic quantity	Effort	Flow
Name	Substance amount	Chemical potential	Substance flow
Symbols	n, N	μ	$\mathfrak{I}, v$
Unit	[mol]	[J mol^{-1}]	[mol s^{-1}]

Properties:

System Properties		
Nature	Capacitance	Conductance
Levels	0/3	1/2
Name	Physical chemical capacitivity	Physical chemical conductivity
Symbol	χ_n	σ_n
Unit	[mol^2 J^{-1} m^{-3}]	[mol^2 J^{-1} m^{-1} s^{-1}]
Specificity	Three-dimensional	Three-dimensional

Anomalous diffusion describes a transport process analogous to transient diffusion which does not follow the classical model based on Fick's law. The dependence with the frequency of the electrical impedance featuring this transfer, which is normally 1/2 for the Fickian diffusion in infinite space, is different from this value. The same is true for the degree

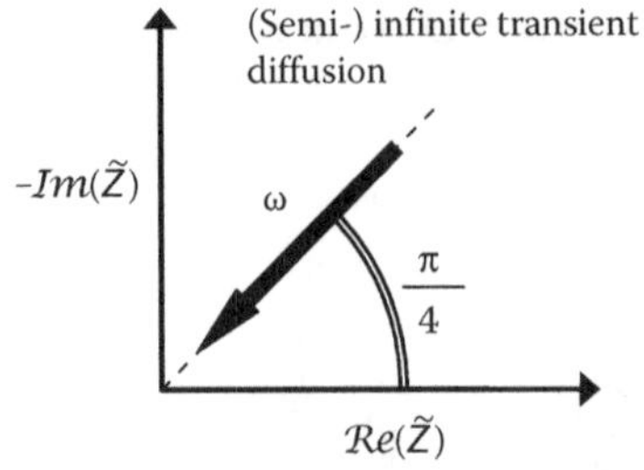

FIGURE 10.8 Nyquist (Cole–Cole) representation of the electric complex impedance of infinite transient diffusion. The direction of the arrow corresponds to increasing angular frequency.

of temporal derivation or integration or the integral impedance. The models usually adopted in this case are

Infinite anomalous diffusion: $$\tilde{Z} = R(i\omega\tau)^{-p} \quad (0 < p < 1/2 \text{ or } 1/2 < p < 1) \tag{G8.1}$$

Infinite anomalous diffusion: $$\hat{Z} = R\tau^{-p}\frac{d^{-p}}{dt^{-p}} \quad (0 < p < 1/2 \text{ or } 1/2 < p < 1) \tag{G8.2}$$

In fact, these models are identical for the CPE seen previously in case study G6, restricting the possible range for the exponent to positive values. Two separated ranges delimitate the possible values for the exponent p as the value 1/2 corresponds to normal (Fickian) diffusion. The Nyquist (or Cole–Cole) plots in Figure 10.9 depict these two cases, which are classically interpreted as attributing to the imaginary component a capacitive meaning and to the real component a resistive meaning. If this analysis holds for a CPE, this is not true for anomalous diffusion for which the criterion is indeed in terms of conservation/dissipation, as will be evidenced by the Formal Graph approach.

The physics underlying anomalous diffusion is not well understood until now, as is the case for the CPE, but the Formal Graph approach allows a deeper insight by adopting the same decomposition on the process in two complementing paths. However, there is a marked difference with the CPE in the nodes used and consequently in the physical role of the paths.

The Formal Graph models every diffusion process as a path between concentration and flow density; path that may be a single path (as for Fick's law in steady state) or may result from complementing two paths as in the case of transient diffusion (see case study G7). The Formal Graph in the case study abstract uses a path complement with the continuity equation and Fick's law, as for the transient diffusion, but with unequal weights, p being the weight of the path representing the continuity equation.

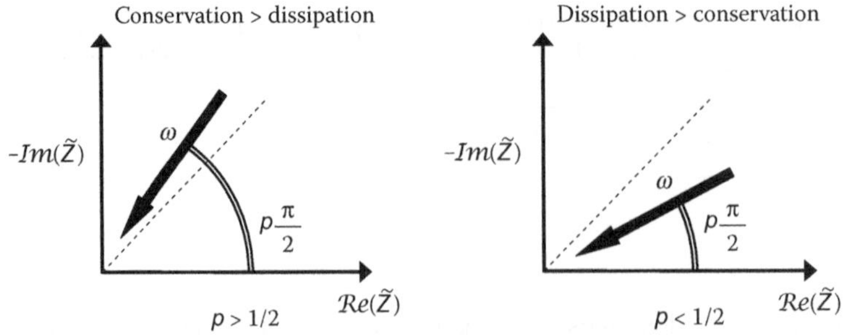

FIGURE 10.9 Two cases of infinite anomalous diffusion. When $p > 1/2$, the transport is more conservative than dissipative (left), when $p < 1/2$, the converse is true. The classical analysis in terms of capacitive/resistive does not hold in anomalous diffusion.

As the continuity equation has already been discussed in case study G7 "Transient Diffusion," only the essential result is recalled here, which is that the operator describing its path is the *space–time velocity* algebraically written as a *convection equation*:

Space–time convection: $$\boldsymbol{J} = \hat{\mathbf{U}}\, c_n \tag{G8.3}$$

with an operator defined as

Space–time velocity $$\hat{\mathbf{U}} \overset{def}{=} -\frac{\partial^{-1}}{\partial \boldsymbol{r}^{-1}}\frac{\partial}{\partial t} \tag{G8.4}$$

It is important to recall that this path features a conservation of energy, whereas the other path representing Fick's law, as explained in case study G3, features a dissipation of energy

Fick's law: $$\boldsymbol{J} = \hat{\mathbf{D}}\left\{-\frac{\partial}{\partial \boldsymbol{r}}\right\} c_n \tag{G8.5}$$

Complementing the paths with unequal weights is algebraically written as

$$\boldsymbol{J} = \left\{-\frac{\partial^{-1}}{\partial \boldsymbol{r}^{-1}}\frac{\partial}{\partial t}\right\}^{p} \times \left\{\hat{\mathbf{D}}\left\{-\frac{\partial}{\partial \boldsymbol{r}}\right\}\right\}^{1-p} c_n \tag{G8.6}$$

This relationship is the general model of anomalous diffusion as proposed by the Formal Graph theory. It can be simplified by choosing conditions more peculiar than those for transient diffusion.

The first simplification relies on the assumption of a homogeneous space, so the spatial integration amounts to a simple multiplication by a vector $\boldsymbol{\ell}$ featuring a *characteristic distance* of the transfer (thickness of a thin layer for instance):

$$-\frac{\partial^{-1}}{\partial \boldsymbol{r}^{-1}} \xrightarrow[hom]{} \boldsymbol{\ell} \tag{G8.7}$$

In this condition, the *space–time convection* becomes a purely temporal transfer

$$\boldsymbol{J} = \boldsymbol{\ell}\,\frac{\partial}{\partial t}\left(c_n - c_n^{*}\right) \tag{G8.8}$$

An integration constant has been added, which is the initial concentration and simultaneously the unperturbed concentration far from the place where the diffusional process occurs. This means that the model applies to (semi-) infinite space, as for the normal diffusion treated earlier.

The second simplification comes from assuming as before that the diffusivity is space invariant (homogeneous medium) and that the medium is isotropic:

Infinite anomalous diffusion: $$J \underset{isotropic}{\overset{hom}{=}} \left\{\ell\,\frac{\mathrm{d}}{\mathrm{d}t}\right\}^{p}\left\{\frac{D}{\ell}\right\}^{1-p}\left(c_n - c_n^{*}\right) \tag{G8.9}$$

It is interesting to define a *diffusion characteristic time* for making the model more meaningful, as this variable stands for a scaling time in all diffusional processes:

$$\tau_D \overset{def}{=} \frac{\ell^2}{D} \tag{G8.10}$$

thus allowing one to write the algebraic model of anomalous diffusion under the simplified form

$$\text{Infinite anomalous diffusion:} \quad J \underset{isotropic}{\overset{hom}{=}} \frac{D}{\ell}\tau_D^p \frac{\mathrm{d}^p}{\mathrm{d}t^p}\left(c_n - c_n^*\right) \quad \text{(G8.11)}$$

The interpretation in terms of energy behavior, that is, conservation versus dissipation, will be discussed later in Section 10.4. The link with the electrical admittance or impedance will be explained in Chapter 12 when the subject of energy variety coupling will be tackled.

Remarks

Restricted diffusion. The model presented in this case study works in semi-infinite or infinite space, that is, when a sufficiently wide space is left for the transfer which never meets a limit, at least in one direction of expansion of the diffusion layer. When a limit is encountered, the fixed concentration cannot stay constant and the model does not apply anymore. In Figure 10.10, the fixed concentration at distance ℓ is zero as long as the diffusion layer does not reach the other side.

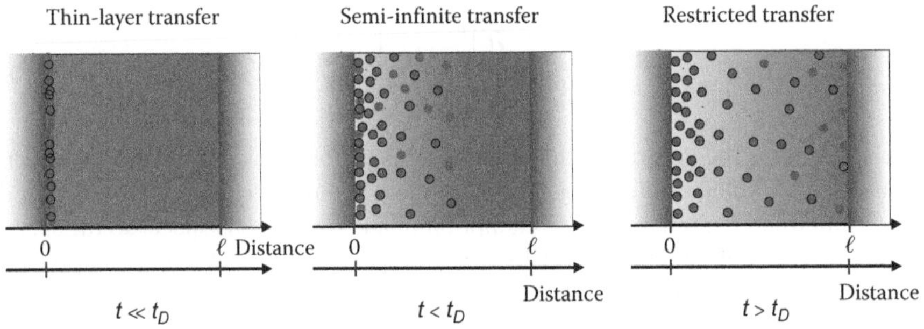

FIGURE 10.10 Evolution of a transient mass transfer in a restricted space from the early stage (thin-layer behavior due to the small diffusion layer) to the restricted diffusion due to the finite space.

Depending on the permeability of the other wall, the semi-infinite transient transfer will evolve toward a stationary transfer (permeable wall) or a thin-layer transfer (impermeable wall).

This means that the evolution mode given by the weight p evolves from the value 1/2 toward 1 or 0. Confusion can arise, mainly in l ow diffusion velocities as is the case in solids, between restricted diffusion and anomalous diffusion, because at a given time the evolution mode p is different from the standard values. The diagnosis of anomalous diffusion cannot be made without checking the constancy of the exponent over a sufficiently wide window of time.

10.2.9 Case Study G9: Generalized Mass Transfer — Physical Chemistry

The case study abstract is given in next page.

Mass transfer is a generic name encompassing different phenomena having in common a transport of molecules and, by extension of all entities, representing pieces of matter. The reference to the mass is a historical heritage from the time when molecules were not quantified as such but by measures of variations of masses appearing or disappearing in a reaction or a transfer. A better term would be *substance transfer* but the habit is well rooted. The cause, or "driving force," of a mass transfer can be diverse, ranging from the uneven distribution of entities (diffusion, osmosis) to the effect of an

$\hat{\mathrm{m}}$

$J = \hat{\mathrm{m}}\,(c_{n1} - c_{n2})$

Space dipole

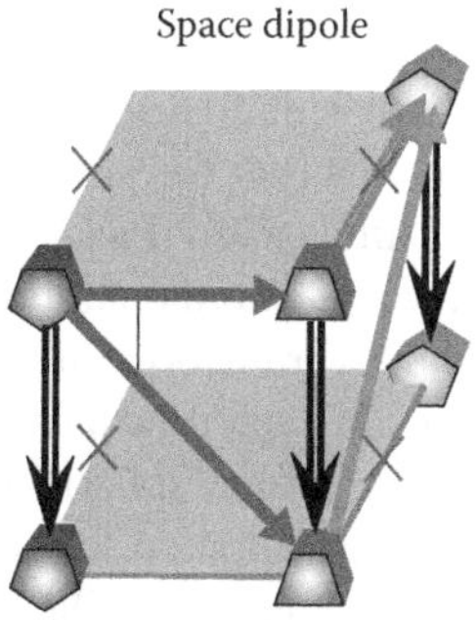

	Dipole
	Fundamental
✓	Mixed
	No
	Definite
	Nature
	Global
✓	Spatial
	Inseparable
✓	Conservative

Formal Graph:

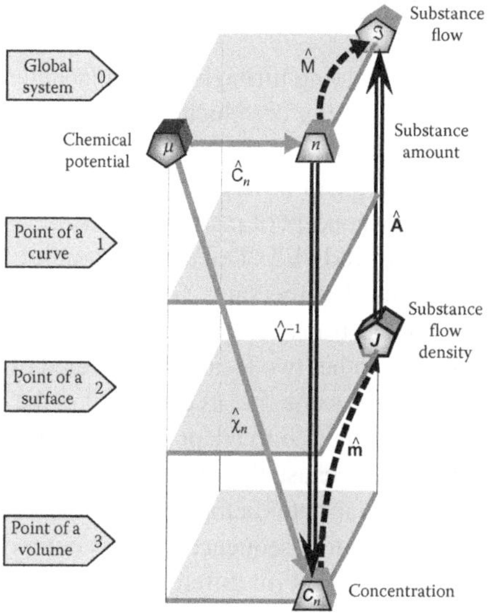

Variations with time:

Response to a concentration step from c to c^* for various values of the evolution mode p

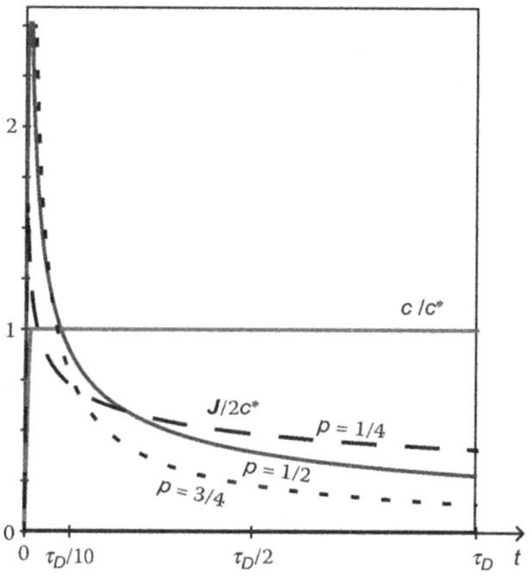

GRAPH 10.18

Variables:

Variety	Physical Chemistry		
Subvariety	Capacitive (potential)		Inductive
Category	Entity number	Energy/entity	Energy/entity
Family	Basic quantity 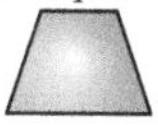	Effort 	Flow 
Name	Substance amount	Chemical potential	Substance flow
Symbols	n, N	μ	$\mathfrak{I}, v$
Unit	[mol]	[J mol^{-1}]	[mol s^{-1}]

Properties:

System Properties		
Nature	Capacitance	Conductance
Levels	0/3	1/2
Name	Physical chemical capacitivity	Physical chemical conductivity
Symbol	χ_n	σ_n
Unit	[mol^2 J^{-1} m^{-3}]	[mol^2 J^{-1} m^{-1} s^{-1}]
Specificity	Three-dimensional	Three-dimensional

Unsatisfying situation. Usually, mass or heat transfer is modeled at the level of differential equations and not at the higher level of operators. A good example of this widespread practice is transient diffusion which is treated by combining Fick's law with the continuity equation for arriving at a second-order partial differential equation (traditionally called Fick's second law*). By choosing boundary conditions adapted to the geometry and initial conditions of the system, this equation is solved (generally using Laplace or Fourier transformation).

It is clear that this method must be used again each time a change of boundary conditions occurs. Thus, there exist different solutions for planar, spherical, and cylindrical geometry, and, within each geometry, solutions have been computed for constant initial flow, for constant effort imposed, for linearly varying efforts, and so on. The landscape is as diversified as there are different signal shapes and geometries.

external field (electrical migration, gravitational convection) in going through the displacement by a supporting medium (hydrodynamic convection, transport by a vector or vehicle, etc.).

The Formal Graph theory making an extensive use of operators is well adapted for proposing a more intelligent approach avoiding the drawbacks of the classical solving of differential equations. This new modeling is based on two ideas: analogy with a reaction and the concept of path complement.

The first idea is the analogy already pointed out (see case study E5 "Diffusion through Layers" in Chapter 8) between reaction and transfer, which suggests that the same formalism used for a dipole modeling a chemical reaction can be adopted for a mass transfer.

After all, a transfer works between two states of a substance, either two locations separated by space (two sites) or two amounts separated by the time course but on the same site, as explained in the beginning of this chapter. In the simplest case, one of the states/sites is free to adapt its content under the effect of the transfer, while the other is maintained fixed, either because the site is far from the region of space perturbed by the transfer (bulk) or because it is the initial state (before the transfer is active). A less simple case is met when the mass transfer is embedded into a sequence of steps composing a reactional chain in a heterogeneous mechanism. Other mass transfers (in different phases), chemical reactions or adsorption steps onto a surface, electron transfer steps, etc. may work in series (or for some of them in parallel) with the mass transfer, making both concentrations vary on the two states/sites.

These two states can be viewed as the analogs of the initial (substrate) and final (product) reaction, having for "kinetic constants" two mass transfer operators, one for each direction.

$$A_1 \underset{\widehat{\mathbf{m}}_2}{\overset{\widehat{\mathbf{m}}_1}{\rightleftarrows}} A_2$$

This dipole works in the same way as a reacting dipole, exchanging flows according to the usual relationship between pole and dipole flows (cf. Chapter 6):

$$\boldsymbol{J} = \boldsymbol{J}_1 - \boldsymbol{J}_2 \tag{G9.1}$$

Each pole flow density is expressed as the result of the mass transfer operator applied to the concentration on the site

Pole i:
$$\boldsymbol{J}_i \overset{def\rightarrow}{=} \widehat{\mathbf{m}}_i \, c_{n\,i} \tag{G9.2}$$

* Fick has proposed only one law of diffusion, empirically based on steady-state observations.

This operator is indeed a *velocity*, with unit in m s^{-1}, analogous to a *heterogeneous rate constant* used for reactions occurring in nonhomogeneous media or at the interface between two phases. The general name of *kinetic operator* for generalizing this operator to all energy varieties comes from this analogy and also from the fact that it links concentrations of basic quantities to flow densities, which is precisely the frame of *kinetics* in the general sense.

As a reacting dipole has been considered exactly in the same manner as any dipole exchanging flows, the symbolism used for electric circuit components can also be utilized, by taking care that concentrations are used instead of efforts for quantifying the "state" of each pole (properly speaking it is not a state but an occupancy) (see Figure 10.11).

The second idea founding the proposed approach of mass transfer is the path complement between pure diffusion and pure space–time convection, as already discussed in the previous case studies. This operation is made at the level of the poles.

Pure diffusion is modeled by Fick's law using a diffusivity operator featuring each pole. As the same procedure applies to both poles, one presents here only the modeling of the free pole. This law is written for this pole as

$$J_i = \hat{\mathbf{D}}_i \left\{ -\frac{\partial}{\partial r} \right\} c_{n\,i} \tag{G9.3}$$

and the space–time convection is written as

$$J_i = \hat{\mathbf{U}}\, c_{n\,i} \tag{G9.4}$$

The path complement with a weight p for the space–time convection provides the model of mass transfer operator for this pole

$$\hat{\mathbf{m}}_i = \left\{ \hat{\mathbf{U}} \right\}^p \times \left\{ \hat{\mathbf{D}}_i \left\{ -\frac{\partial}{\partial r} \right\} \right\}^{1-p} \quad (0 \le p \le 1) \tag{G9.5}$$

and the same expression holds for the initial/bulk pole.

At the level of the dipole formed by the two poles, the dipole concentration is defined as the difference between pole concentrations, and a dipole mass transfer operator relates this difference to the dipole flow density:

$$J = \hat{\mathbf{m}}\,(c_{n\,1} - c_{n\,2}) \tag{G9.6}$$

From the expression G9.2 defined for each pole, the following relation holds:

Dipolar mass transfer velocity:
$$\hat{\mathbf{m}} \overset{def}{=} \frac{\hat{\mathbf{m}}_1\, c_{n\,1} - \hat{\mathbf{m}}_2\, c_{n\,2}}{c_{n\,1} - c_{n\,2}} \tag{G9.7}$$

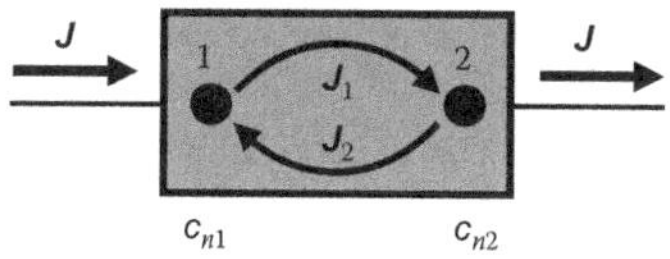

FIGURE 10.11 Symbolic representation of generalized mass transfer in the manner of a circuit component.

which is based on the same template as the conductance of a dipole expressed as a function of individual conductances of the poles (see case study C4 "Pipe" in Chapter 6).

The last three equations, adding one equation for the fixed pole, constitute the general model of mass transfer. The Formal Graph modeling it is the same as the one established for the anomalous diffusion, based on the path complement, so it is not repeated here. Instead, a more synthetic graph is drawn in the case study abstract using merely the dipole mass transfer velocity. If the graph in case study G8 "Anomalous Diffusion" is more operational as no extra information is required for using it in a neural network for instance, the graph given here is more suitable for higher-level modeling.

10.2.9.1 Simplification

The general model can be simplified in the same way as the previous models, by assuming isotropy and homogeneity of the system.

The first assumption of isotropy makes the two mass transfer velocities equal at the level of the poles:

$$\hat{\mathbf{D}} \underset{isotropic}{=} \hat{\mathbf{D}}_1 = \hat{\mathbf{D}}_2 \Leftrightarrow \widehat{\mathbf{m}}_1 = \widehat{\mathbf{m}}_2 \tag{G9.8}$$

which does not imply that the dipole transfer velocity is equal to the common pole velocity in the general case. For obtaining this result, a second assumption about the space homogeneity is required:

$$-\frac{\partial^{-1}}{\partial \boldsymbol{r}^{-1}} \xrightarrow[hom]{} \ell \tag{G9.9}$$

In this condition, the *space–time velocity* becomes proportional to a temporal derivation

$$\hat{\mathbf{U}} \underset{hom}{=} \ell \frac{\mathrm{d}}{\mathrm{d}t} \tag{G9.10}$$

and the diffusivity tensor reduces to a scalar in homogeneous and isotropic medium

$$\hat{\mathbf{D}} \overset{hom}{\underset{isotropic}{=}} D \tag{G9.11}$$

Finally, all these hypotheses allow one to write a simpler model

$$\widehat{\mathbf{m}} \overset{hom}{\underset{isotropic}{=}} \left\{\ell \frac{\mathrm{d}}{\mathrm{d}t}\right\}^{p} \left\{\frac{D}{\ell}\right\}^{1-p} \quad (0 \le p \le 1) \tag{G9.12}$$

and the definition of a *diffusion characteristic time* playing the role of scaling time

$$\tau_D \overset{def}{=} \frac{\ell^2}{D} \tag{G9.13}$$

leads to the simplified expression of mass transfer velocity

Generalized transfer: $$\widehat{\mathbf{m}} \overset{hom}{\underset{isotropic}{=}} \frac{D}{\ell} \tau_D^p \frac{\mathrm{d}^p}{\mathrm{d}t^p} \quad (0 \le p \le 1) \tag{G9.14}$$

(a)

$$\hat{m} \overset{hom}{\underset{isotropic}{=}} \frac{D}{\ell}\tau_D^p \frac{d^p}{dt^p}$$

$$\tau_D \overset{def}{=} \frac{\ell^2}{D}$$

$$(0 \le p \le 1)$$

(b)

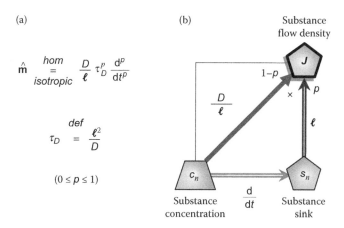

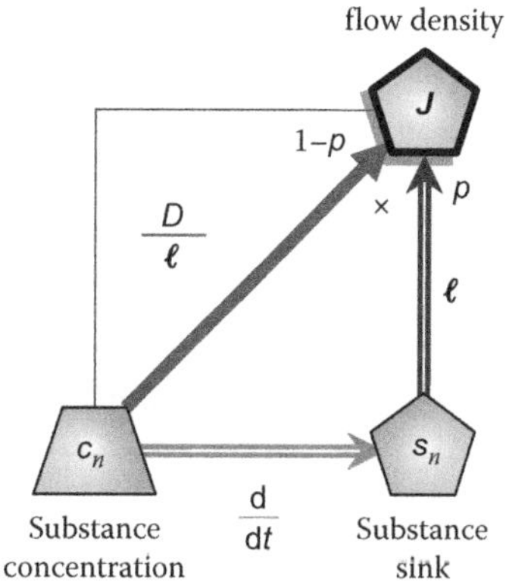

GRAPH 10.19 Homogeneous mass transfer viewed according to two representations: algebraic formulas (a) and Formal Graph with elementary links (b). The two paths are composed of different weights $1-p$ and p through multiplication on the flow node (p is positive for a mass transfer). This two-dimensional graph is an extract of the three-dimensional Formal Graph modeling a physical chemical dipole. (As the variables in the nodes are dipole variables, the concentration is the difference of pole concentrations.)

The difference with the model established for the anomalous diffusion in case study G8 is that the range of possible values for p includes the three peculiar values 0, 1/2, and 1.

This formula has been proposed by the author of this book before the development of the Formal Graph theory in making a heuristic reasoning based on the various known expressions that could be attributed to this operator (Vieil 1991a,b).

Graph 10.19 shows the lower part of the right side of the three-dimensional Formal Graph given in the case study abstract, but in decomposing the mass transfer path in its two complements in the case of a homogeneous medium.

Beyond the mentioned interest of this general mass transfer operator for modeling systems at a level higher than that of differential equations, the great interest of this approach lies in the physical meaning that emerges from the path complement used. Case Study G3 "Fick's Law of Diffusion" has shown that a pure diffusional path was composed with a conductivity path which results in dissipation of the energy flowing through it. On the contrary, the continuity path is composed only of space–time properties that conserve the energy.

Thus, the weight p is endowed with the physical meaning of a ratio of conserved energy on total energy, that is to say corresponds to a *conservation yield* featuring the transfer:

$$p = \frac{\text{conserved energy}}{\text{conserved} + \text{dissipated energy}} \quad (0 \le p \le 1) \qquad \text{(G9.15)}$$

In fact, the weight p defines an *evolution mode* or *mass transfer regime*, which in turn corresponds to an *energy conservation yield*, as shown in Figure 10.12 for various transfer regimes.

This important aspect of the theory is more general than the strict mass transfer and will be discussed in Section 10.4.

The inclusion in the model of the three peculiar values 0, 1/2, and 1 means that the three fundamental regimes of mass transfer can be deduced from the generalized mass transfer operator as shown in Table 10.3.

In Table 10.3 the significant distances delimiting the space region in which the mass transfer occurs have been mentioned, according to each regime. The reasoning is based on a fixed time

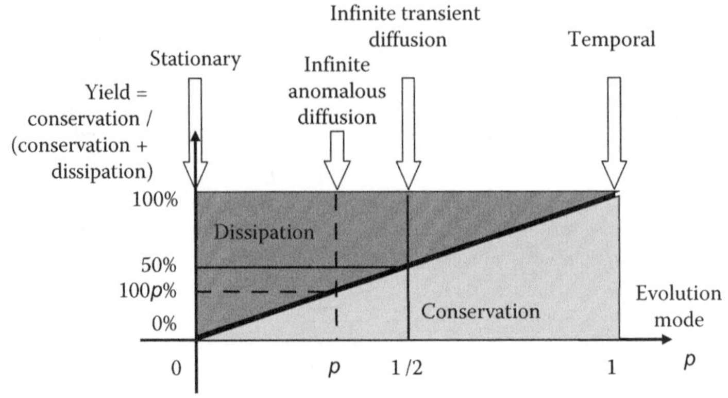

FIGURE 10.12 Dependence of the energy conservation yield on the weight p ("evolution mode") for the main regimes of mass transfer.

window τ for observing the transfer, which corresponds to a distance $\sqrt{D\tau}$ when the transfer characteristic time matches this time window.

In the stationary regime, the space region concerned with the transfer is much higher than the region perturbated by the transient diffusion. In the temporal regime, it is the opposite; the effective distance in which the transfer occurs is significantly smaller than in the transient diffusion regime.

This observation can be utilized in another way, when one tries to impose a fixed space region to the mass transfer.

Remarks

Practical application for system designers. If the allowed space for mass transfer in a physical chemical system is relatively thick, the stationary regime may develop easily; but if a very small layer is left to the transfer, the temporal regime will be favored. The consequence on the proportion of dissipated energy is clear: *smaller is better* for avoiding loss of energy. This kind of consideration is useful for engineers in charge of designing systems such as energy converters, generators, storage devices, and so on.

TABLE 10.3
Three Fundamental Regimes of Mass Transfer with Their Algebraic Models and Their Physical Features

Regime	Stationary	Semi-Infinite Transient Diffusion	Temporal
Mass transfer operator	$\widehat{m} = \frac{D}{\ell}$ (G9.16)	$\widehat{m} = \sqrt{D}\,\frac{d^{1/2}}{dt^{1/2}}$ (G9.17)	$\widehat{m} = \ell\,\frac{d}{dt}$ (G9.18)
Energetic behavior	Dissipation $p = 0$	50% Conservation 50% Dissipation $p = 1/2$	Conservation $p = 1$
Distance	$\ell \gg \sqrt{D\tau}$	$\ell = \sqrt{D\tau}$	$\ell \ll \sqrt{D\tau}$
Occupancy	Thick layer	Moving boundary layer	Ultra-thin layer

Kinetics for mass transfer. The classical method based on modeling with differential equations is not suitable in case of unknown and varying boundaries as is the case in mass transfers embedded in a chain (cf. Figure 10.13) (unless one takes into account all the steps in a unique system of differential equations, which complicates the solving).

The use of a mass transfer operator offers a remarkable tool for modeling complex mechanisms by using the techniques developed for electrical circuits.

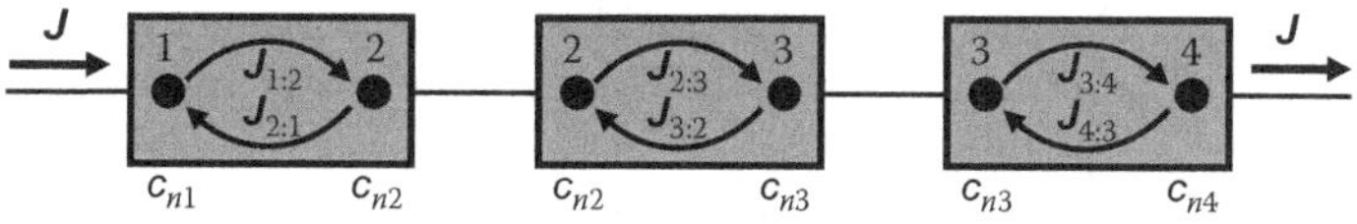

FIGURE 10.13 When a mass transfer (or any process) is in the middle of other processes, its boundaries variables are not known since they depend on the functioning of the whole mechanism.

Moreover, it offers a still more remarkable tool for modeling complex mechanisms in the same manner as a set of reactions in chemical kinetics.

$$A_1 \rightleftarrows A_2 \rightleftarrows A_3$$

All the concepts (equilibrium, reversibility, rate determining step, global equivalent steps, etc.) used in this domain and the corresponding techniques can be used for modeling mechanisms including mass transfer steps.

Unknown mass transfer. The invention of the unknown variable x in algebra has been a decisive progress for handling quantities that are to be determined by solving equations. In electrodynamics, the concept of impedance Z, which is not necessarily known prior to modeling, is another significant advance. It is a pity that the same tool is not imported into physical chemistry for modeling mass transfer without being obliged to know its nature (and the boundary conditions) in advance (see Figure 10.14).

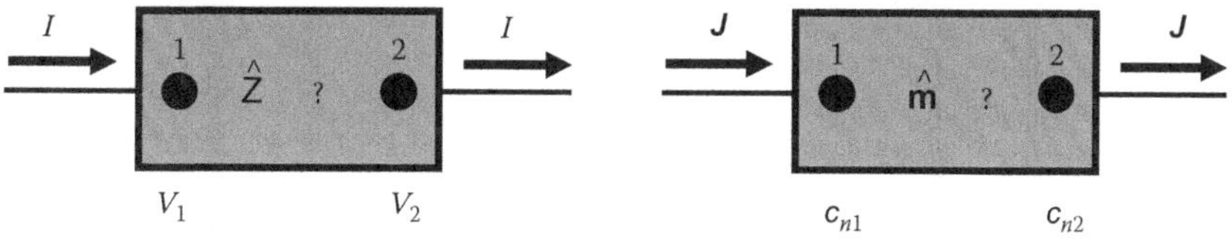

FIGURE 10.14 Two usages of operators for modeling unknown processes: the well-known concept of impedance in electrodynamics (left) and the proposed concept of mass transfer operator in physical chemistry (right).

Using a mass transfer operator to build models with the aim of identifying an unknown mass transfer allows one to interpret experimental data and isolate the transfer properties (see Vieil 2011). It is even possible to use methods providing direct identification without having recourse to the classical method of best fit between theoretical model and experimental data (see Miomandre and Vieil 1994; Vieil and Miomandre 1995). Thus anomalous or nonclassical transfers can be found or discovered.

10.3 COMPARISON BETWEEN ENERGY VARIETIES

The comparison between energy varieties is made in Graph 10.20 for the transfer by pure conduction and in Graph 10.21 for diffusivity and kinematic viscosity models.

What comes out from this comparison is that not all transfers are identically modeled and that two categories of transfer processes must be distinguished:

- *Purely conductive transfers*. In all the four energy varieties in Graph 10.20, transfers are modeled by a pure conduction through space, represented by a single reduced property, the conductivity. Ohm's law (case study G1) and Fourier's equation (case study G2) are modeled exactly in the same way, provided the change from heat Q to entropy S is made in the model of Fourier's equation. Fick's law is clearly not modeled in the same manner as a reduced capacitance must be included. What is comparable to the other transfers is the physical chemical migration using a physical chemical conductivity σ_n (see case study G3). Newton's law, although appearing with a different symmetry, also works identically with a mechanical resistivity, which is the dynamic viscosity η (see case study G4). The algebraic models of these transfers in the linear and isotropic case are:

Ohm: $$\boldsymbol{j} \overset{lin}{\underset{isotropic}{=}} -\sigma \frac{\mathrm{d}}{\mathbf{dr}} V \tag{10.7}$$

Fourier: $$\boldsymbol{J}_\mathsf{S} \overset{lin}{\underset{isotropic}{=}} -\frac{\kappa}{\mathrm{T}} \frac{\mathrm{d}}{\mathbf{dr}} T \tag{10.8}$$

Physical chemical migration: $$\boldsymbol{J} \overset{lin}{\underset{isotropic}{=}} -\sigma_n \frac{\mathrm{d}}{\mathbf{dr}} \mu \tag{10.9}$$

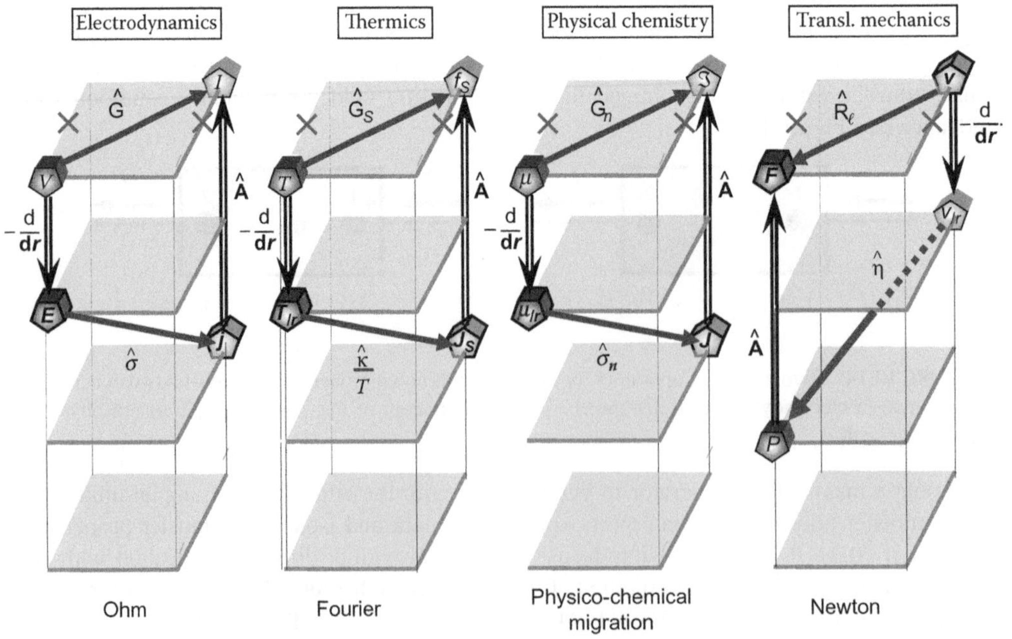

GRAPH 10.20 Four energy varieties and their conductivities.

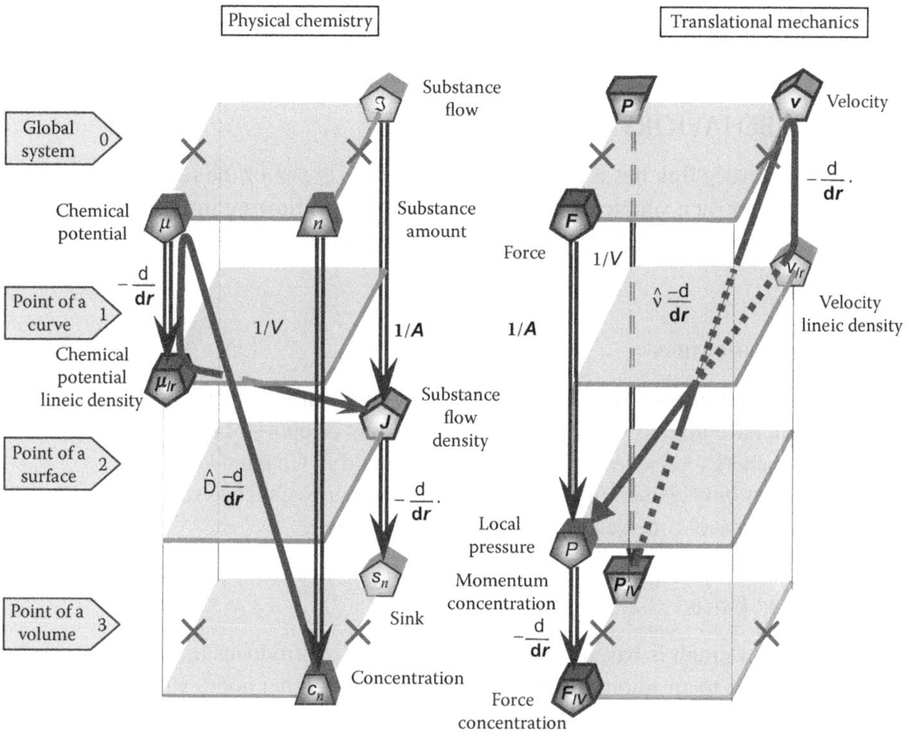

GRAPH 10.21 Diffusivity and kinematic viscosity compared.

Newton: $$P \underset{isotropic}{\overset{lin}{=}} -\eta \frac{\mathrm{d}}{\mathbf{dr}}.\boldsymbol{v} \tag{10.10}$$

Their common feature is to relate two energies-per-entity through a space derivation and a reduced system constitutive property. These models belong to the category of representations without explicit use of an energy subvariety, as discussed in Section 10.1.2.

- *Mixed transfers.* Contrary to the previous category, these transfers do not use a single system constitutive property but two. In addition to the conductivity, a storage property, inductance or capacitance (in their reduced form), is used. Their models make explicit use of an energy subvariety through the concentrations of entities. Graph 10.21 shows the two cases of capacitive storage with diffusivity (Fick's law) and of inductive storage with kinematic viscosity. The corresponding algebraic models in the linear and isotropic case are:

Fick's law: $$\boldsymbol{J} \underset{isotropic}{\overset{lin}{=}} -D\frac{\mathrm{d}}{\mathbf{dr}}c_n \tag{10.11}$$

Kinematic viscosity: $$P \underset{isotropic}{\overset{lin}{=}} -\nu\frac{\mathrm{d}}{\mathbf{dr}}.\boldsymbol{P}_{lV} \tag{10.12}$$

The algebraic models are so similar between purely conductive and mixed transfers that many conclusions are drawn about the identity of physical processes involved. The Formal Graph representation being more rigorous than the algebraic writing shows that the sole analogy between equations is misleading. With the proposed approach it is not possible to assimilate a concentration of entities to an effort and to jump to erroneous conclusions.

10.4 ENERGY BEHAVIORS

In a Formal Graph, every link has a physical meaning. This is one of the remarkable features that places this tool well above a simple computational network by allowing interpretations in terms of physical processes. Each link plays a specific role in the interplay between the system and its contained energy.

10.4.1 Links Signification

The link categories are limited to six, including three system constitutive properties and two space–time properties that have already been presented, plus a new category of *energetic coefficients* used for connecting two energy varieties, which will be presented in Chapter 11.

Table 10.4 contains only elementary links; composed properties, such as relaxation time constants (or reaction rates) or space–time velocities, have not been listed.

10.4.2 Energetic Paths

Contrary to algebra, a graph is based on topology. It not only introduces the concept of path when one goes from one node to an adjoining one but it also connects other nodes somewhere in the graph by taking a sequence of links (properties of graphs are described in Berge 1962). This possibility of composing several links into a path is an interesting feature that will be used in conjunction with the physical meaning attributed to each link for finding out physical interpretations.

When a path is not composed with conductance, the path means that conservation of energy prevails in the process, and, as soon as a conductance is encountered along the path, dissipation of energy occurs, even when storage of energy participates in the path.

$$\text{Conductance in the path} \Leftrightarrow \text{Dissipation}$$

$$\text{No conductance in the path} \Leftrightarrow \text{Conservation}$$

TABLE 10.4
Various Links Used in a Formal Graph with Their Physical Meaning

Link Category	Elementary Path	Operator	Physical Meaning
System constitutive properties	Inductance	$\hat{L}_q$	Energy storage
	Capacitance	$\hat{C}_q$	Energy storage
	Conductance	$\hat{G}_q$	Energy dissipation
Space–time properties	Evolution	$\hat{T}^{-1}, \frac{d}{dt}$	Energy conversion
	Space derivation	$\hat{\mathbf{R}}^{-1} \cdot \frac{d}{d\mathbf{r}}$	Energy distribution/spreading (through space)
Energetic coefficients	Energy coupling	λ	Cooperation of several energy varieties

Note: The space–time operators have been given in two versions, general and continuous.

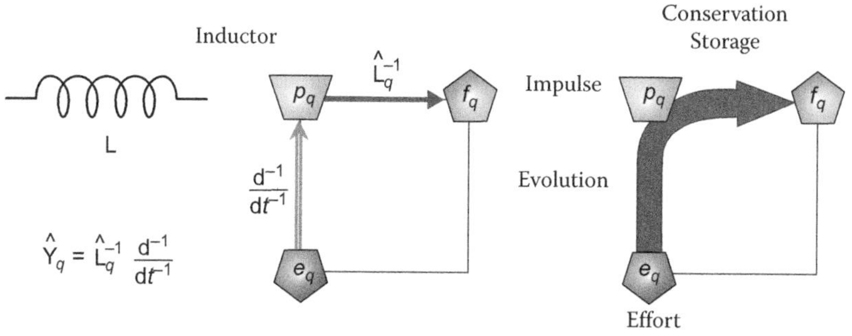

GRAPH 10.22 Generalized inductor according to four representations: circuit symbol and algebraic formula (left), Formal Graphs with elementary links (center) and with the path describing the energetic behavior (right).

In the formulas and graphs given in Graphs 10.22 through 10.26, the various electric components listed in Table 10.2 are generalized to any energy variety. They show the three different models that describe the same component: first the algebraic formula of the admittance, then the elementary Formal Graph, and at last the Formal Graph representing paths composed of elemental links.

Translation of the general algebraic model of the CPE cannot be done with a single Formal Graph. According to the sign of the exponent p, one has to distinguish two CPEs, one inductive ($-1 < p < 0$) and the other capacitive ($0 < p < 1$). These two translations are shown in Graphs 10.25 and 10.26.

10.4.3 Energetic Behavior

The interesting specificity of this new representation lies in the difference of energetic behavior between the two paths. One path conserves energy because it is composed of evolution and storage but no conductance whereas the other dissipates energy because of the presence of conductance. As the exponent p represents the weight of the conservation path and $1-p$ the weight of the dissipation, one has access to a physical meaning of this quantity that is called *evolution mode*. The scheme in Figure 10.15 gives the proportion of energy that is conserved in a generalized CPE (which corresponds to a yield) as a function of this evolution mode p.

This scheme may induce the false perception that a continuous evolution of the exponent p (evolution mode) is possible in a CPE. The strict definition of a CPE is contained in its name, meaning that the angle must be a constant. In this respect, another more customary scheme adopting the Fresnel[*] representation of a vector is given in Figure 10.16. It shows the connection between the angle made by a vector of constant magnitude, called director, with a horizontal axis, and the energetic behavior of a CPE.

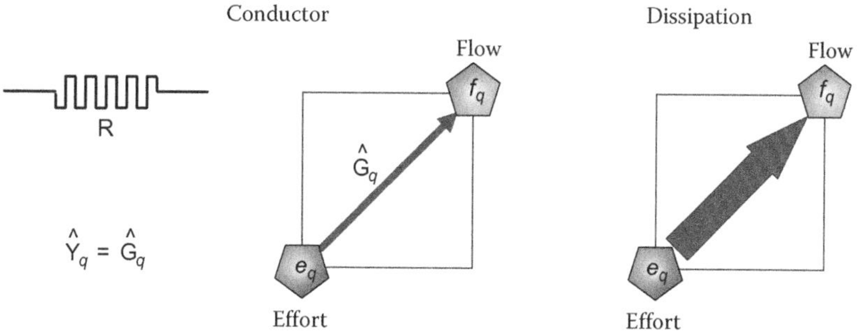

GRAPH 10.23 Generalized conductor according to four representations: circuit symbol and algebraic formula (left), Formal Graphs with elementary links (center) and with the path describing the energetic behavior (right).

[24] Augustin Fresnel (1788–1827): French physicist; Paris, France.

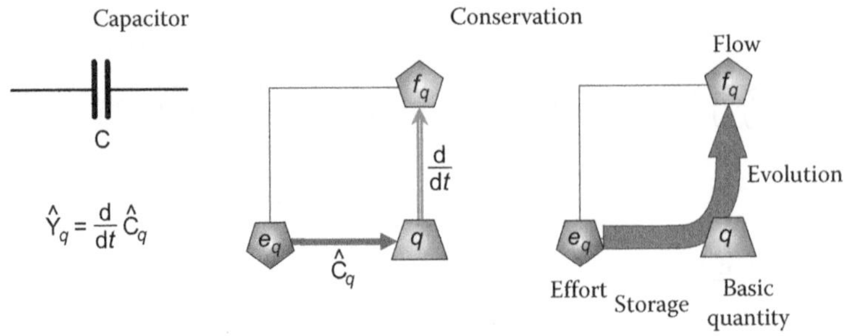

GRAPH 10.24 Generalized capacitor according to four representations: circuit symbol and algebraic formula (left), Formal Graphs with elementary links (center) and with the path describing the energetic behavior (right).

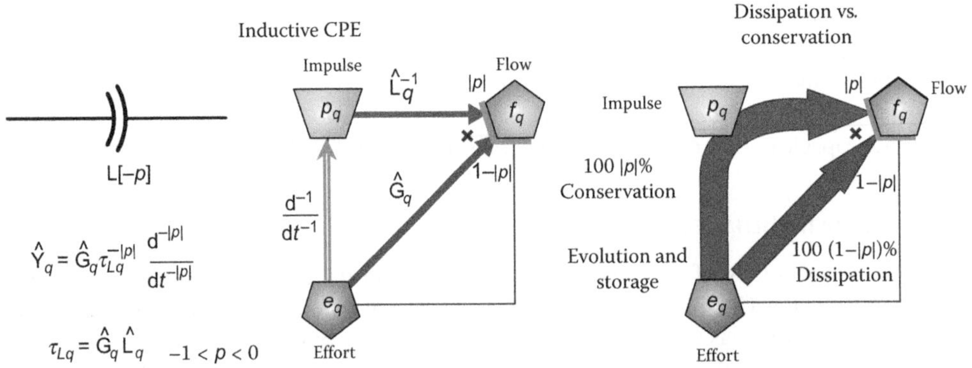

GRAPH 10.25 Generalized inductive constant phase element viewed according to four representations: circuit symbol and algebraic formulas (left), Formal Graph with elementary links (center), and Formal Graph with composed paths describing the energetic behavior (right). The two paths are composed of different weights $|p|$ and $1-|p|$ when arriving on the flow node (p is negative for an inductive CPE).

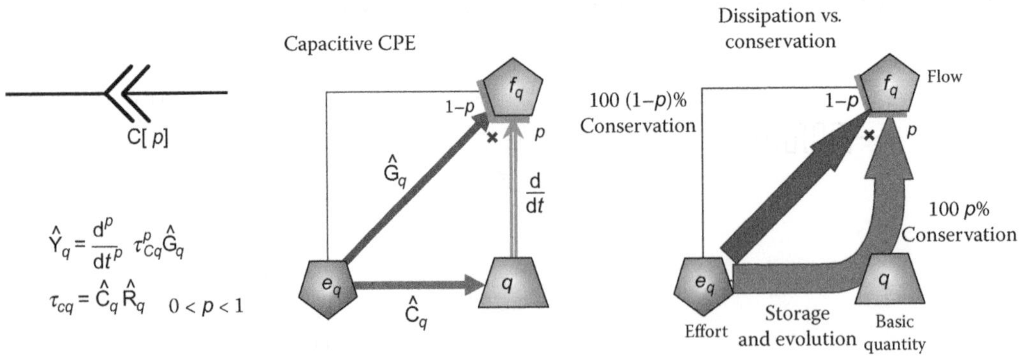

GRAPH 10.26 Generalized capacitive constant phase element viewed according to four representations: circuit symbol and algebraic formulas (left), Formal Graph with elementary links (center), and Formal Graph with composed paths describing the energetic behavior (right). The two paths are composed of different weights $1-p$ and p when arriving on the flow node (p is positive for a capacitive CPE).

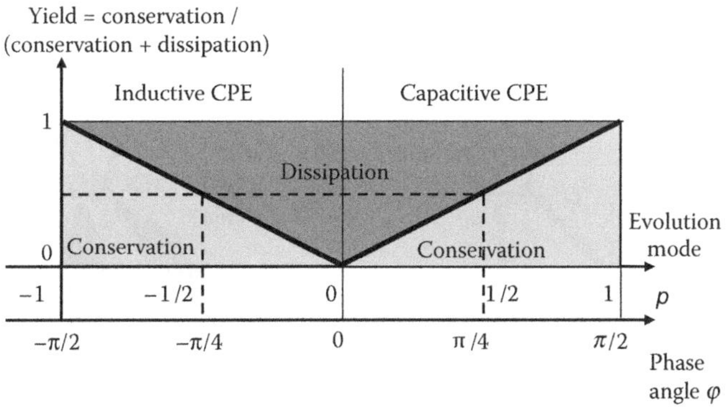

FIGURE 10.15 Dependence of the yield of the mass transfer (proportion of conserved energy) on the exponent p ("evolution mode") or the phase angle in a constant phase element.

In Figure 10.16 are compared the Fresnel representations of a generalized CPE and of a mass transfer. The only difference is the range of possible values for the evolution mode p which can be negative in the case of a generalized CPE when inductive energy exists, and which is always positive in a mass transfer, owing to the absence of inductive energy in physical chemistry.

Intuitive interpretation of experiences. One of the direct applications of the relation between the value of the evolution mode and the behavior of energy is that an immediate diagnosis can be made without a computer, "by hand," of what happens in the system under study. The phase angle measured by the slope of a Nyquist (Cole–Cole) plot or a Bode plot gives directly the proportion of conserved energy during the process for a given frequency.

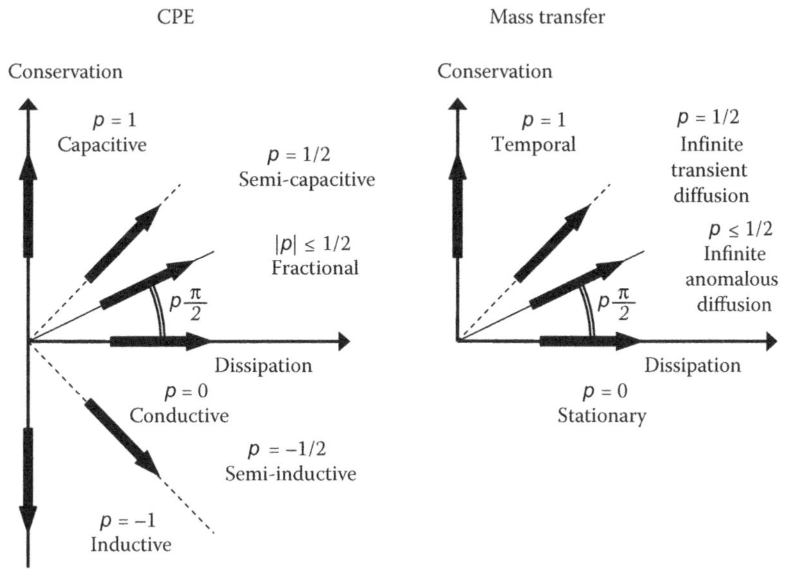

FIGURE 10.16 Fresnel representation of the director of a constant phase element (left) and of a mass transfer (right). The angle of the director is the phase angle and its two projections on the orthogonal axes give the amount of conservation (vertically) and the amount of dissipation (horizontally).

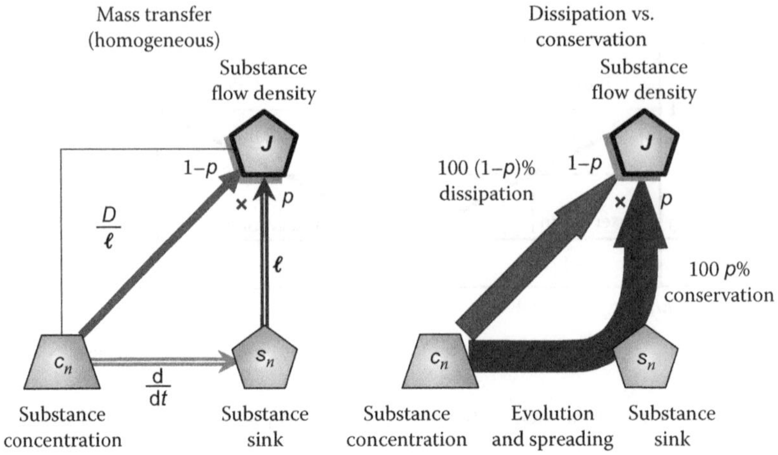

GRAPH 10.27 Mass transfer modeled by a Formal Graph with elementary links (a) and a Formal Graph with composed paths describing the energetic behavior (b). The two paths are composed of different weights $1-p$ and p when arriving on the flow node (p is positive for a mass transfer).

The two Formal Graphs in Graph 10.27 mimic those drawn earlier for the generalized components of an equivalent electrical circuit. They model and describe a mass transfer in physical chemistry with the same principle of path complement, but, instead of defining an admittance, this path complement defines a mass transfer operator (or velocity). (The distinction between these two notions has been explained in Section 10.1.2.)

It can be remarked that the structure of the Formal Graphs is exactly identical to those previously drawn for a generalized CPE; only the nature of the nodes and variables is different. The Formal Graphs here are extracted from the lower right side of a three-dimensional graph, so the diagonal link represents a stationary velocity instead of a conductance. Both dissipate energy because the stationary velocity also involves a conductance in its composition.

10.5 CONVECTION

Convection is a transfer process occurring when entities are carried by a supporting medium moving by itself. This process is well known at least for charges, molecules, and corpuscles, that is to say in the energy varieties of electrodynamics and physical chemistry or corpuscular energy. The algebraic equations describing the process are written in a simple form in which the flow density is given by the product of the medium velocity $\boldsymbol{v}$ by the volumic concentration of entities.

Electrodynamical convection: $$\boldsymbol{j} = \boldsymbol{v}\,\rho \tag{10.13}$$

Physical chemical convection: $$\boldsymbol{J} = \boldsymbol{v}\,c_n \tag{10.14}$$

Corpuscular convection: $$\boldsymbol{J}_N = \boldsymbol{v}\,c_N \tag{10.15}$$

The Formal Graph representation is shown in Graph 10.28 together with the representation of a similar process which is the propagation of a phenomenon (wave, perturbation, etc.) through space

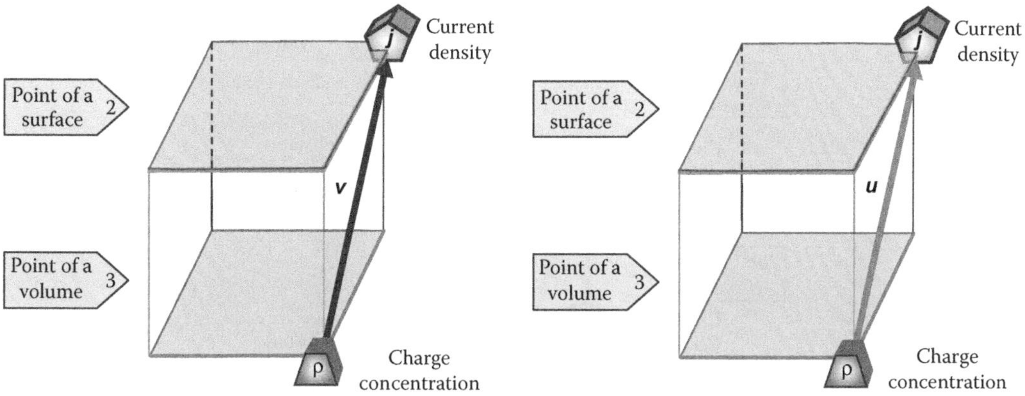

GRAPH 10.28 Lower parts of three-dimensional Formal Graphs showing two processes represented by the same structure: convection with a medium velocity $\boldsymbol{v}$ (left) and space–time propagation with a velocity $\boldsymbol{u}$ (right).

with a space–time velocity $\boldsymbol{u}$. The energy variety chosen in this graph is electrodynamics, although it may apply to any variety.

The propagation of a wave has been studied in Chapter 9 (case studies F5 "Sound Propagation" and F6 "Light Propagation") and will be studied again in Chapter 11 (case study H8 "Attenuated Propagation") in the case of absorbing medium. Wave propagation in fact involves two velocities as two energy subvarieties are exchanged, but they are symmetrically set in a spatial Formal Graph so the representation of only one velocity is sufficient for the present discussion.

This juxtaposition brings to light an interesting insight about the nature of propagation which can be seen as a kind of convection by space–time, playing the role of supporting medium. It also sheds some light on the nature of convection in terms of energy behavior. The propagation of a wave in a nonabsorbing medium is a conservative process because no dissipation occurs. The same conclusion may be drawn for convection. This is by itself a nondissipative process although it could be assimilated with a stationary transfer in view of a similar algebraic model as seen above. For instance, the stationary mass transfer by diffusion in a homogeneous medium obeys an equation that looks like a convection equation with a velocity given by the ratio of the diffusivity on the characteristic length of the transfer.

Stationary diffusion: $$\boldsymbol{J} \underset{isotropic}{\overset{hom}{=}} \frac{D}{\ell} c_n \tag{10.16}$$

As no time operator is involved in both stationary diffusion and convection, one may conclude from the discussion in the previous section about path complementing that convection is also a stationary process accompanied by dissipation. This reasoning forgets the analysis made in case study G3 describing Fick's law as a composed path involving three processes: capacitive storage, space distribution, and conduction. This last step inevitably includes dissipation as the final outcome for energy. Again, the analogy between algebraic models is a false criterion for finding the physical meaning behind the formula.

In Graph 10.29 the simultaneous occurrence of stationary diffusion and of convection, expressed by the sum of the two respective paths on the flow density is modeled. This example illustrates the way to take into account by an addition of flows several transfers working in parallel. This parallelism must not be confused with the path complement seen previously, working also in parallel but when the two paths are interdependent (complementary). Here, the two paths are completely independent.

The equation modeling this mass transfer is

Stationary diffusion + convection: $$\boldsymbol{J} \underset{isotropic}{\overset{hom}{=}} \frac{D}{\ell} c_n + \boldsymbol{v}\, c_n \tag{10.17}$$

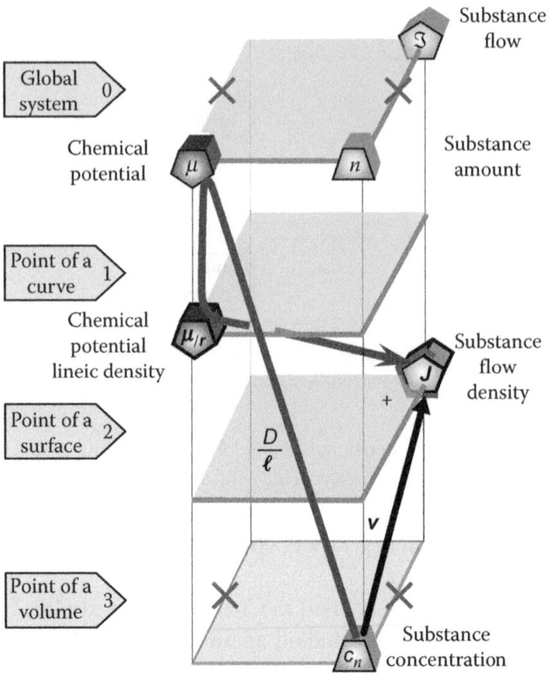

GRAPH 10.29 Formal Graph of a double mass transfer, the first one ensured by stationary diffusion in a homogeneous and isotropic medium, the second one ensured by convection. The diffusion path goes through a conduction step, which dissipates energy, whereas the convection one does not undertake any dissipative path.

When we say that convection does not dissipate energy, it must be added that the movement of the supporting medium may be dissipative on its own when the motion is achieved by conduction. However, this does not contradict the assertion that convection is a nondissipative phenomenon in itself for the transported entities.

10.6 ASSEMBLIES AND CIRCUITS

A transfer seldom works alone. An energy source must be supplied for driving the transfer or a perturbation signal needs to be sent for allowing relaxation. Characterizing a transfer often requires adjoining other dipoles with state variables that can be measured. And most of the time, a transfer participates in a multistep mechanism, which may also include other transfer steps.

10.6.1 Formal Graph Coding

Poles are grouped and connected in pairs for forming a dipole as explained in Chapter 6. The formation of multipoles, when poles are in greater number, has been described in Chapter 8. The method of assembling dipoles (or multipoles) of the same energy variety to form an assembly has been discussed in Chapter 9. What remains to be discussed is the inclusion of nonstandard dipoles such as CPE components and the way to build assemblies with some identical dipoles.

This study focuses on the connections between global state variables, not because connections between localized variables are not possible, but because of the traditional importance of equivalent circuit modeling in this area, which works only at the global level. Connections between localized levels are much more used for coupling energy varieties and will be studied in Chapter 13.

In an equivalent electrical circuit all components are supposed to belong to the same energy variety, as is the case in a dipole assembly in the Formal Graph classification of objects. The question raised when representing a dipole assembly by a Formal Graph is with regard to the number of units

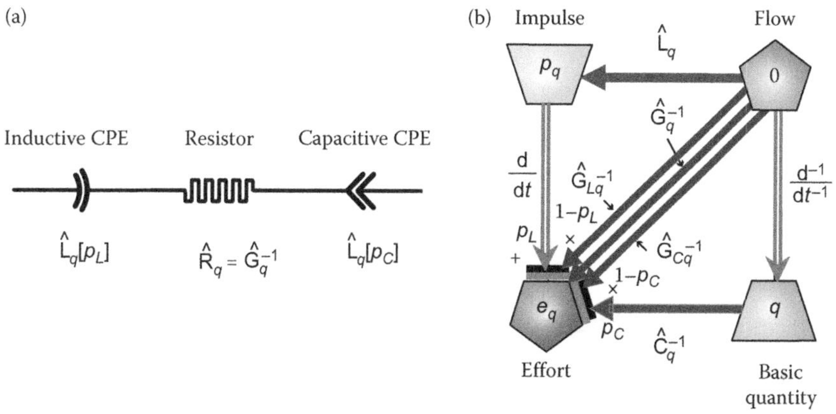

GRAPH 10.30 Three dipoles with common flow: inductive CPE, resistor, and capacitive CPE and their representation as an equivalent electrical circuit in series (a) and as a single unit of Formal Graph (b). Each CPE is built by complementing its own paths with its own evolution mode before being added to the resistance. As the linear circuit is not powered, the common flow is zero.

(square scheme) required. A single unit cannot contain more than one node for each energy subvariety, but it may contain as many links as wanted. Consequently, when two capacitors or two inductors are to be modeled in the same energy variety, two Formal Graph units will be necessary.

An example of representation with a single unit is shown in Graph 10.30. It models a system containing three dipoles sharing a common flow, one purely conductive and two generalized CPE of different subvarieties.

Such a compact coding is possible because only one basic quantity and only one impulse feature in the system and also because only one energy-per-entity works as common gate variable. In other words, a mix of serial and parallel mountings would require as many units as the number of common variables.

An example of representation with two units is shown in Graph 10.31.

This system combines two capacitive dipoles: a pure capacitor and a semicapacitor (or Warburg dipole) and two common variables, a common flow for the conductor and the Warburg dipole for one branch, and a common effort to both branches of the circuit. In Graph 10.31, an external effort (powered effort) is imposed on the system for exemplifying a practical situation.

Randles circuit. In electrochemistry, the system chosen as example for Graph 10.31 is well known under the name of Randles* circuit when a resistor is added in series to the two parallel branches. This resistor corresponds to the resistance of the electrolyte while the resistor in one of the parallel branches models the charge transfer step. The associated mass transfer step is represented by a semicapacitor (Warburg "impedance") when semi-infinite transient diffusion prevails. The capacitor in the other branch models the *double layer*, which is the charged interface between the electrode and the electrolyte (Bard and Faulkner 2001). This circuit is a classical model of electrochemical half-cell (the full cell has two electrodes for allowing the current to flow). Note that all these components are nonlinear dipoles and that operators must be used instead of scalars. In particular, the dipole featuring the charge transfer is modeled by an exchange by flows and each conductance (forward and backward) is given by an exponential function of the potential (see Section 8.5.6 in Chapter 8).

* John Edward Brough Randles (1912–1998): British electrochemist; Birmingham, UK.

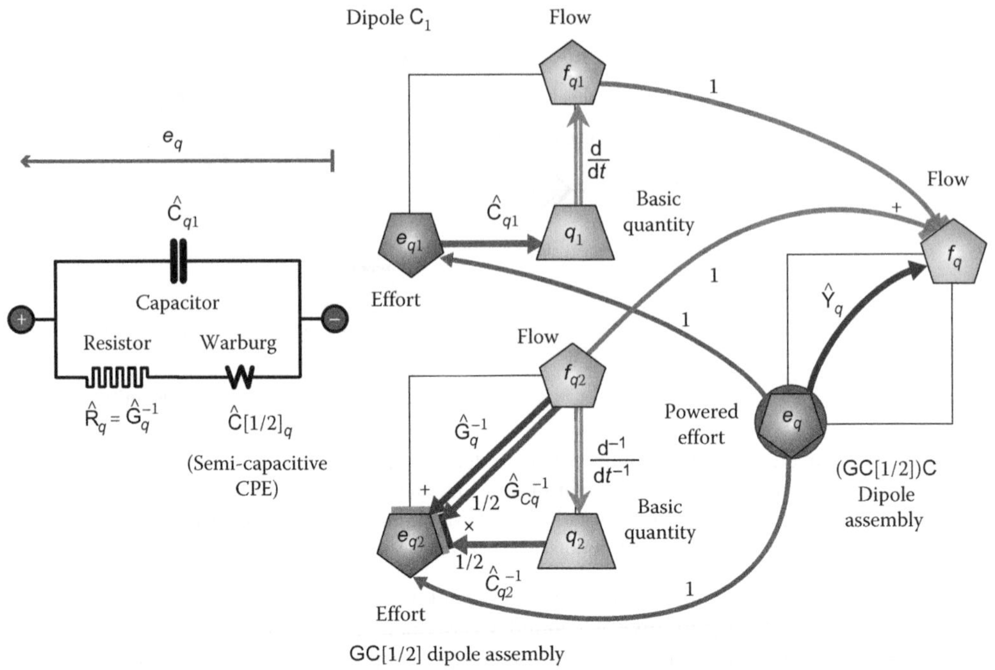

GRAPH 10.31 Equivalent circuit (left) and Formal Graph (right) of a circuit combining in parallel a capacitor with a resistor and a semicapacitor (called also Warburg dipole) in series. The circuit is powered with an imposed effort.

In order to highlight the computing ability of Formal Graphs, Figure 10.17 shows two responses of the previous circuit (Randles circuit) by different techniques. The left plot represents the variation of the current resulting from a linear variation with time of the imposed potential (potential ramp)* and the right plot represents the complex impedance in the Cole–Cole (Nyquist) representation. The parallel capacitance C_{q1} is supposed to be sufficiently important for dominating the

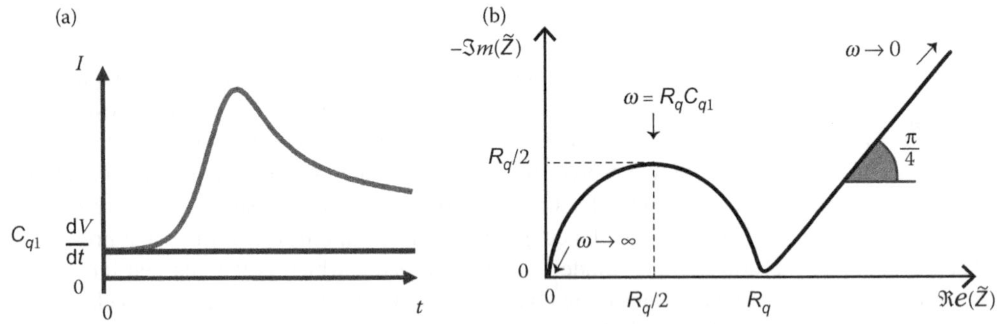

FIGURE 10.17 Characterization by two techniques of the Randles circuit: plot of the current measured on imposition of a potential ramp (a) and Cole–Cole (Nyquist) plot of the complex impedance (b). The Warburg impedance is supposed to be small in comparison with the parallel capacitance.

* This analytical technique bears the name of linear sweep voltammetry and of cyclic voltammetry when the ramp is followed by a reversed ramp.

Warburg impedance, thus allowing the semicircle to appear fully developed. These responses are computed with the Formal Graph given in Graph 10.31 by giving for the large signal perturbation nonlinear shapes (exponential) to the conductances and to the Warburg capacitance C_{q2} as is the rule in physical chemistry.

Naturally, these examples are just a brief outline of the subject of representing assemblies with Formal Graphs, which are not limited in number of elementary units. The embedding principle saves a lot of room by allowing one to choose a degree of resolution (depth in the Formal Objects scale) adapted to the problem. The modularity of the tool is an advantage for modeling multiscale processes and complex systems.

10.6.2 Translation from Electrical Circuits with Software

In order to help the reader to encode Formal Graphs, a small software has been designed for translating electrical circuits into Formal Graphs.

The reverse translation is not implemented, for the main reason that it is not always possible to find an equivalent electrical circuit for all Formal Graphs, which are able to model far more complex systems than the electrical circuit can.

The user has at disposal a storehouse with dipolar components (capacitors, resistors, CPEs, etc.) and wires, that s/he may place at will on a bench (mounting board) between predesigned connections. The circuit is build by successive addition or removal of components and wires and the corresponding Formal Graph is displayed at each step. The choice is left to the user to draw the Formal Graph in the time domain or in the frequency domain (Fourier transformed). In addition, the algebraic equation modeling the system is displayed simultaneously in the frame of large signal perturbations (i.e., with differential operators) or in the Fourier space (i.e., with imaginary angular frequency).

This small program is limited to circuits in a single energy variety with no more than two identical components and no more than two common gate variables. This corresponds to an encoding using one or two square schemes only.

The Circuit-to-Graph translator is provided on the accompanying CD-ROM (together with the color bitmap files of the Formal Graphs) and explanations are given in Appendix 5.

10.7 IN SHORT

GLOBAL TRANSFERS

A transfer normally occurs between localized variables as space and time necessarily play a significant role. However, the characterization of transfers is generally made in terms of state variables and operators at the global level of a system. Hence the importance of relating global operators to spatially reduced ones.

Three different ways of representing a transfer at the global level are possible:

- By a single impedance or admittance operator, avoiding specifying if a subvariety is involved.
- By a capacitance $\hat{C}_q$ and a basic quantity transfer (kinetic) operator $\hat{M}_q$.
- By an inductance $\hat{L}_q$ and an impulse transfer (kinematic) operator $\hat{N}_q$.

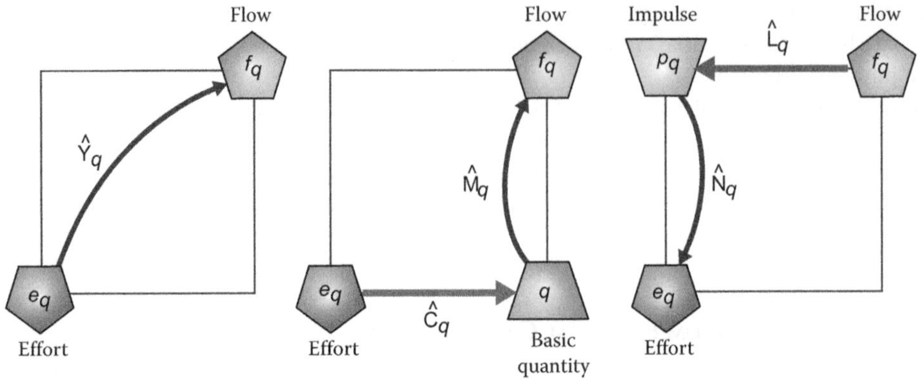

GRAPH 10.32

On the left of Graph 10.32 is drawn a transfer modeled by an admittance, that is, without indication of the energy subvariety. (If any, when no subvariety is involved, the admittance becomes a conductance.) In the middle and on the right, the energy subvariety is specified by decomposing the admittance or the impedance into an energy storage operator and an entity transfer operator. The admittance is a more general model as it includes the other representations, which are mainly used in the frame of kinetics or kinematics.

IMPEDANCES FOR CONDUCTION AND INDUCTIVE TRANSFERS

Four Formal Graphs in Graph 10.33 depict the characteristic modes of conduction and inductive transfers according to the value of the *evolution mode* p.

The stationary mode, or steady state, is a transfer ensured only by conduction. The temporal mode corresponds to a pure storage of inductive energy (ultra-thin layer case, i.e., without transport). The other modes mix the two pure modes in various proportions (by complementing paths).

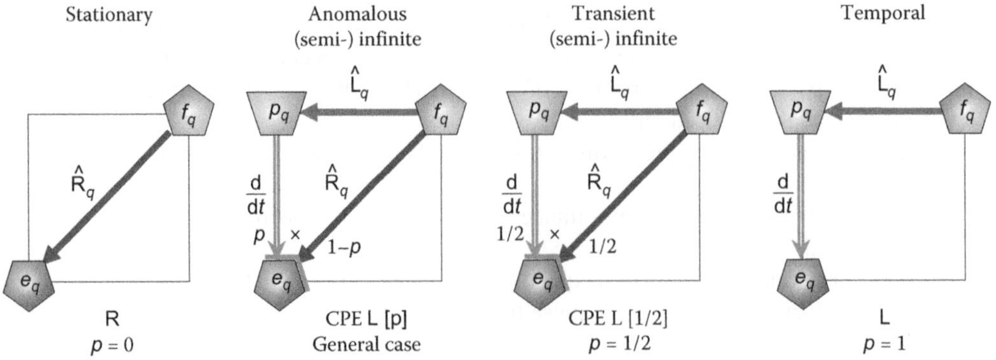

GRAPH 10.33

ADMITTANCES FOR CONDUCTION AND CAPACITIVE TRANSFERS

Four Formal Graphs in Graph 10.34 depict the characteristic modes of conduction and capacitive transfers according to the value of the *evolution mode* p.

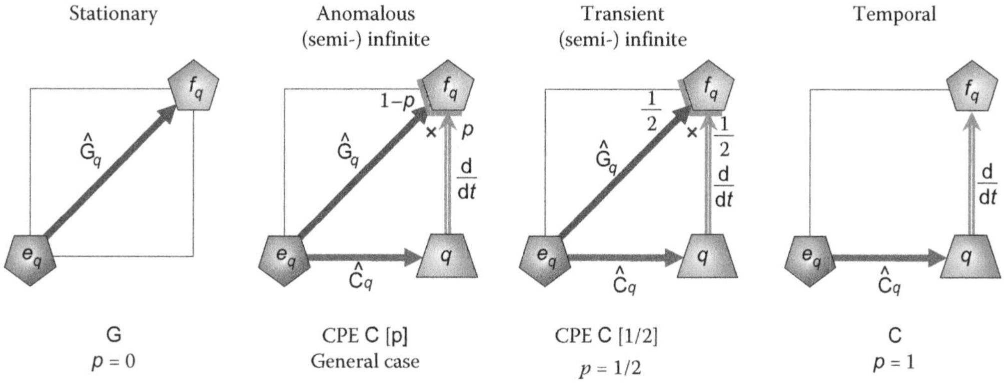

GRAPH 10.34

The stationary mode, or steady state, is a transfer ensured only by conduction. The temporal mode corresponds to a pure storage of capacitive energy (ultra-thin layer case, i.e., without transport). The other modes mix the two pure modes in various proportions (by complementing paths).

SPATIALLY REDUCED TRANSFERS: THE FOUR MODELS

Depending on the explicit reference to a subvariety or not and on the capacitive or inductive nature of the transferred entities, four models are distinguished. The transfer of basic quantities is modeled either by a reduced conductance (i.e., conductivity), linking an effort field (or lineic density) to a flow density, or by a composition of reduced capacitance and transfer (kinetic) operator (Graph 10.35a and c). The transfer of impulses uses either a reduced resistance (i.e., resistivity), linking the flow lineic density to the effort density, or a composition of reduced inductance and transfer (kinematic) operator (Graph 10.35b and d).

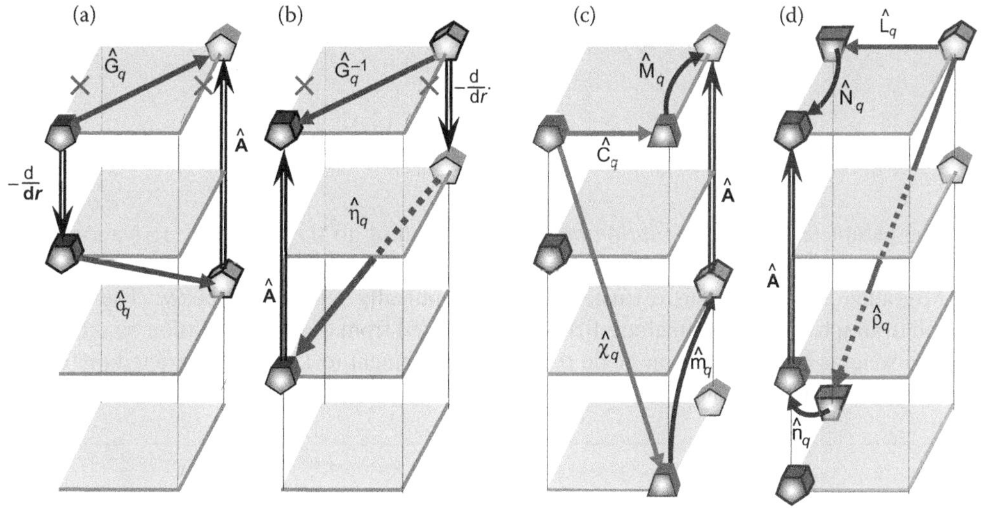

GRAPH 10.35

SPATIALLY REDUCED TRANSFERS: MASS TRANSFERS

Mass transfer can be ensured by various fundamental processes:

- *Migration* (generally for charged substance): This is a conduction process (cf. previous frame) and not a kinetic process.

$$p = 0 \qquad \mathbf{J} \underset{isotropic}{\overset{lin}{=}} -\sigma_n \frac{\mathrm{d}}{\mathrm{d}\mathbf{r}} \mu \qquad \text{(G3.16)}$$

- *Stationary diffusion*: Transport modeled by generalized Fick's law.

$$p = 0 \qquad \mathbf{J} = \hat{\mathbf{D}}\left\{-\frac{\mathrm{d}}{\mathrm{d}\mathbf{r}}\right\} c_n \qquad \text{(G3.1)}$$

- *Transient diffusion* in (semi-) infinite space: Mixing diffusion and continuity in equal proportions (by complementing paths).

$$p = 1/2 \qquad \mathbf{J} = \left\{-\frac{\partial^{-1}}{\partial \mathbf{r}^{-1}}\frac{\partial}{\partial t}\right\}^{1/2} \times \left\{\hat{\mathbf{D}}\left\{-\frac{\partial}{\partial \mathbf{r}}\right\}\right\}^{1/2} c_n \qquad \text{(G7.10)}$$

- *Temporal transfer*: Modeled by a continuity equation in the absence of transport.

$$p = 1 \qquad \mathbf{J} = -\frac{\partial^{-1}}{\partial \mathbf{r}^{-1}}\frac{\partial}{\partial t} c_n \qquad \text{(G7.9)}$$

- *Convection*: Transport ensured by the displacement of the supporting medium.

$$p = 0 \qquad \mathbf{J} = \mathbf{v}\, c_n \qquad \text{(10.14)}$$

- *Anomalous diffusion* in (semi-) infinite space: Mixing diffusion and continuity in unequal proportions (by complementing paths).

$$0 < p < 1/2 \text{ or } 1/2 < p < 1 \qquad \mathbf{J} = \left\{-\frac{\partial^{-1}}{\partial \mathbf{r}^{-1}}\frac{\partial}{\partial t}\right\}^{p} \times \left\{\hat{\mathbf{D}}\left\{-\frac{\partial}{\partial \mathbf{r}}\right\}\right\}^{1-p} c_n \qquad \text{(G8.6)}$$

The parameter p is the *evolution mode*, corresponding to the degree of derivation with respect to time.

Migration and anomalous diffusion are not graphically represented below. The Formal Graph in Graph 10.36 of anomalous diffusion is deduced from transient diffusion by substituting the weights by the evolution mode p and its complement to 1. (Space restricted diffusion is not listed because it is not fundamental.)

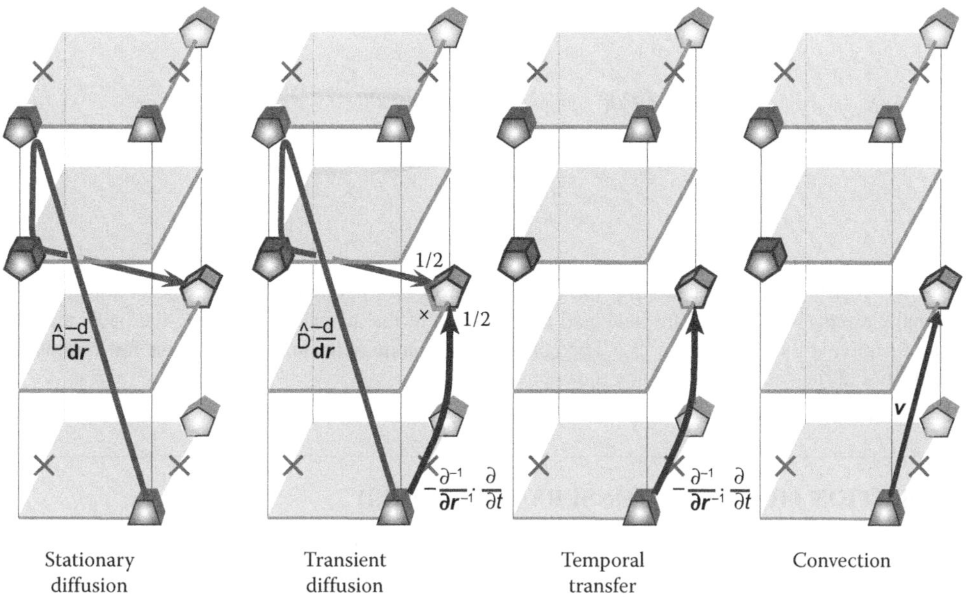

GRAPH 10.36

COMPLEMENTING PATHS

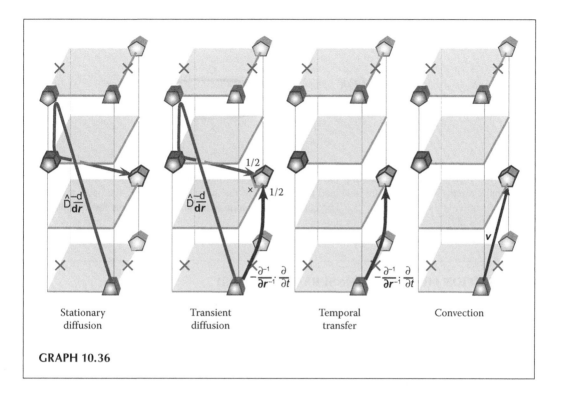

In the Formal Graph theory, the transient transfer is seen as a double process, working with two parallel paths, one inductive or capacitive, and the other purely conductive. However, in some cases, the two paths are not equivalent (anomalous transfer or evolving transfer in the case of restricted space). The importance of each path, or weight, is featured by a coefficient which is the exponent of their operator when the two paths are combined on their arrival node.

The operation for combining the two paths is a multiplication between the results of the two operators, each one raised to the power corresponding to the weight of the path. This is illustrated in the scheme in Figure 10.18, in which the operators are notated $\hat{O}_1$ and $\hat{O}_2$. The weights p_1 and p_2 must comply with the requirement of positive numbers and with a sum equal to 1. As the order of the paths is indifferent (true parallelism), the operators must commute with respect to the multiplication.

Naturally, this operation can be extended to any number of paths, provided the sum of the weights is still equal to 1.

(continued)

Formal algebra

$$B = \hat{O}_1^{p1} \times \hat{O}_1^{p2} A$$

$$\text{and } B = \hat{O}_2^{p2} \times \hat{O}_1^{p1} A$$

Formal Graph

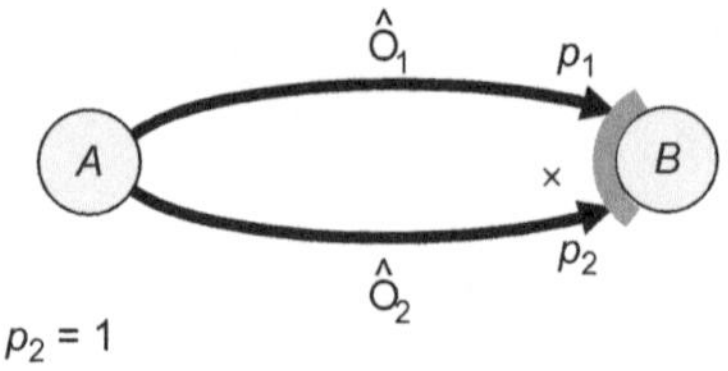

$$p_1 + p_2 = 1$$

FIGURE 10.18 Scheme illustrating the complement of two parallel paths and the coding used in Formal Graphs. The weights are indicated near the end of the arrows symbolizing the paths and the multiplication is indicated by an "x." The contour of the node is made locally thicker for joining the arrow ends of the paths concerned by the complement.

EVOLUTION MODE AND CONSERVATION YIELD

The *evolution mode* and the *conservation yield* are two closely related notions that differ only by the allowed range. Both are notated p as the context is generally sufficient for distinguishing them.

The *evolution mode* is a signed parameter expressing the importance of evolution in the transfer and it corresponds to the degree of derivation with respect to time of the transfer operator: it is a negative parameter when antievolution, that is, temporal integration, is effective, and positive when it is the normal evolution, that is, temporal derivation. By convention, the direction of the transfer is chosen as corresponding to an admittance for appreciating the degree of derivation. Consequently, a negative value means that the storage is made under the inductive form of energy whereas a positive value corresponds to capacitive energy.

The *conservation yield* is the ratio of conserved energy on the total energy involved in the transfer. It is therefore always positive or null and less than or equal to one:

$$p = \frac{\text{conserved energy}}{\text{conserved} + \text{dissipated energy}} \quad (0 \le p \le 1)$$

As the *conservation yield* is used as a weight for combining (complementing) two paths in a Formal Graph, it appears to correspond to the absolute value of the *evolution mode*.

Evolution mode p: $-1 \le p \le 1$ *Conservation yield* = | *Evolution mode* |

In the Formal Graphs in Graphs 10.37 and 10.38, the parameter p represents the *conservation yield*.

Inductive transfer:

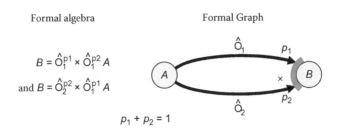

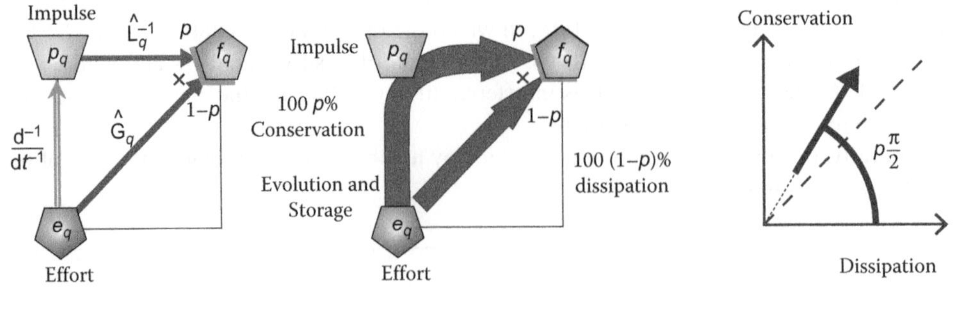

GRAPH 10.37

Capacitive transfer:

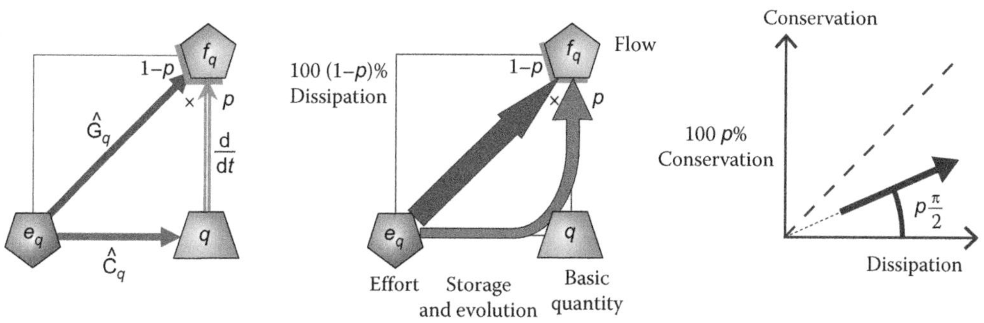

GRAPH 10.38

11 Assemblies and Dissipation

CONTENTS

With this chapter we enter the peculiar world of instability. A world considered by many as imperfect because the symmetry of time (meaning that an inversion of sign of the time variable is without effect on processes) is broken, conferring an arrow to the time.* This is the domain of *out-of-equilibrium* systems and of *irreversible processes*, as opposed to equilibrium resulting from reversible processes.

Many negative notions are associated with it: friction, losses, leaks, damping, dissipation, degradation, disorder, and so on, contributing to the poor image of this field among many physicists, from quantum physics to relativistic physics. The irreversible conversion into heat (called dissipation) is well handled in physical chemistry (chemical reaction) and in electrodynamics (Joule effect), for instance, but this is not the case in all domains. In many modern theories, in particular in mechanics, the part devoted to dissipation in their fundamental laws and concepts is still very small keeping in mind the importance of this field. And yet, our universe is far from equilibrium and our ecosystem is driven out of equilibrium through the flow of energy coming from the sun owing to the instability of the universe.

The key notion in this domain is the *Second Principle of Thermodynamics* that governs the behavior of entropy, which confers to the thermal energy variety some unique features.

Table 11.1 lists the case studies of dipole assemblies given in this chapter. (See also Figure 11.1 for the position of the dipoles.)

11.1 DISSIPATION AND CONVERSION

Conversion of energy into heat may occur in two ways: through the process of conduction or by coupling of two storage systems, one of them being the thermal capacitor. The first process is known as dissipation and it is an irreversible conversion. The second process is a reversible conversion, for instance, as in the thermoelectric conversion from electrical energy into heat, known as the Seebeck effect, and in the reverse process known as the Peltier effect (see Chapter 12).

Dissipation is the name of the irreversible conversion into heat of any other energy. It occurs specifically in a conductor that belongs to the converted energy variety and is not a heat conductor. Considered globally, this process requires a thermal container for receiving the produced heat, a conductor that converts energy, and an energy source. In the Formal Graph theory, this system is modeled by a *dipole assembly* made of an inductor or capacitor and a conductor, associated to a storing dipole in the thermal energy variety.

It must be recalled that, in the Formal Graph approach, the notion of conductor is generalized to any device or material possessing the constitutive property of conductance. This includes the classical concept of conductor of particles (charges, molecules, etc.) but also conductors of entities that are momenta, impulses, volumes, lengths, and so on. Friction, for instance, is a conduction process of mechanical momentum in a viscous fluid or of geometric entities (surfaces, etc.) in a solid (in that case one speaks of internal friction between solid elements).

Many processes involving dissipation are not purely conduction processes but combine energy storage processes with conduction. This is the case of the chemical reaction or of the viscoelastic relaxation, for instance.

* This term has been coined in 1927 by Arthur Eddington (1882–1944), a British astronomer; Cambridge, UK.

TABLE 11.1
List of Case Studies of Global Dipole Assemblies

	Name	Energy Variety	Nature	Type	Page Number
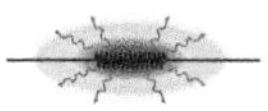	H1: Electrical Joule Effect	Electrodynamics	C + G	Global	522
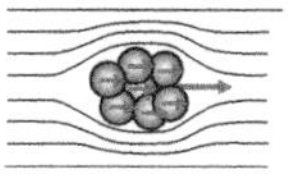	H2: Slowed Down Motion	Translational mechanics	L + G	Global	524
$A \xrightarrow{k} B$	H3: Occurring Reaction	Physical chemistry	C + G	Global	528
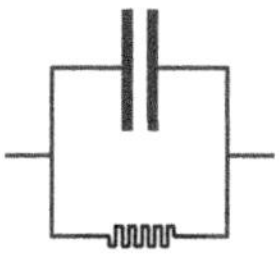	H4: RC Circuit	Electrodynamics	C + G	Global	534
	H5: Dielectric Relaxation	Electrodynamics	C + G	Spatial three-dimensional	540
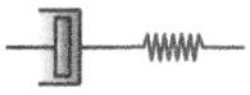	H6: Viscoelastic Relaxation	Translational mechanics	C + G	Spatial three-dimensional	544
	H7: Damped Mechanical Oscillator	Translational mechanics	L + C + G	Global	551
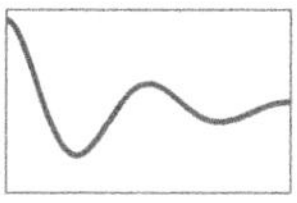	H8: Attenuated Propagation	Electrodynamics	L + C + G	Spatial three-dimensional	555

Note: C, capacitive; G, conductive; L, inductive.

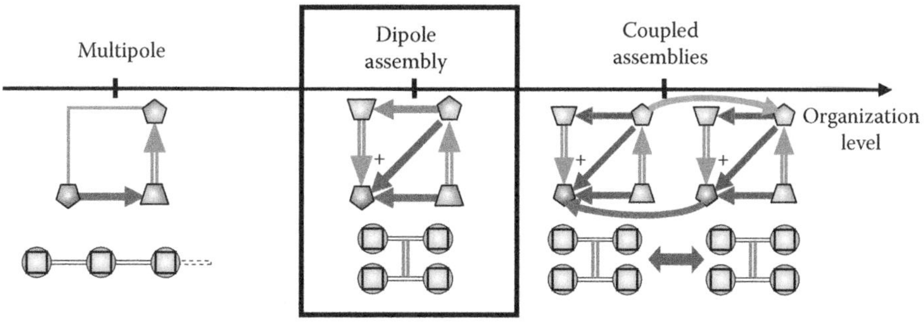

FIGURE 11.1 Position of the dipole assembly along the complexity scale of formal objects.

11.1.1 Passages of Energy between Objects

A short description of the various passages energy can undertake to move between objects is useful for clarifying this fundamental subject.

The *energy exchange* is a two-way process occurring between two identical storing objects. It does not modify the nature of energy and remains within the same subvariety, but energy amounts vary within each partner according to their storage capability. Entity numbers are necessarily involved in the exchange, which is not the case for other passages.

The *energy conversion* process makes energy change in nature and in amounts from one subvariety to the other or from one variety to the other. In this chapter, it applies to the change in the nature of energy when passing from a conductive dipole to a thermal one (i.e., dissipation).

The *energy circulation* is merely a process that makes the energy move from one object to the other through the gate variables without conversion. Contrary to the exchange, the two partners need not be identical, and the process is not necessarily a two-way process. Here, it applies to the one-way transfer of energy from a storing dipole to a dissipative one.

These three passages are illustrated in Graph 11.1 showing how gate variables are connected in pairs for ensuring these various processes.

Note that two passages implying dipoles with different varieties or subvarieties are not indicated in Graph 11.1: *conversion* when subvarieties or varieties differ and when time is involved, and *coupling* between identical subvarieties but different varieties and without time (this is an extension of the exchange to the realm of systems containing multiple varieties). They will be treated in subsequent chapters. They are mentioned here to clarify the whole picture by outlining their specificities.

11.1.2 Circularity Principle and Reversibility

In anticipation of a fundamental result that will be discussed in this chapter, it is useful to grasp the properties of these various processes to decode the essential information contained in this Formal Graph in terms of reversibility.

The notion of reversibility has been encountered in Chapter 6 (devoted to dipoles) under the form of a variable used for describing the behavior of an exchange of energy-per-entity between two poles. In the Formal Graph theory, this variable is called the *exchange reversibility* and is quantitatively defined as the ratio of the backward on the forward energies-per-entity involved in the exchange. Indeed, this quantitative definition applies only to exchange processes and the notion of reversibility is classically extended to any process in a qualitative way to describe the possibility of reversing a process. However, one shall see that a more precise definition can be given with the help of the Formal Graph approach.

In Graph 11.1, each passage is represented by two opposite connections between pairs of state variables or gate variables, adopting an identical scheme for all of them. Figure 11.2 and Graph 11.2 show the similarity between an energy-per-entity exchange and a dipole interconnection, taking the case of connections between gate variables. (An exchange of energy uses only one pair of gate variables for one direction and the other direction is ensured by a pair of entity numbers, as shown in Graph 11.1, but the principle of forward–backward connections is identical.)

Note that in a reaction or an exchange, the quantification of the reversibility by a variable (*exchange reversibility*) is possible because the exchanged quantities belong to only one family of state variables (exchange of basic quantities or exchange of impulses). In a circulation or conversion process, several state variables are involved, which complicates the task.

As for the exchange, the same distinction applies concerning the *possibility* of reversibility and the *effectiveness* of reversibility. The first notion applies to the nature of the routes that a process may undertake whereas the second one applies to the behavior adopted by the process. Draw two opposite and parallel arrows for representing a process demonstrates merely the existence of two routes, one for each direction; it does not imply systematically a reversible behavior. The existence of the two routes leaves open the possibility for the process to use one of them or both, but does not

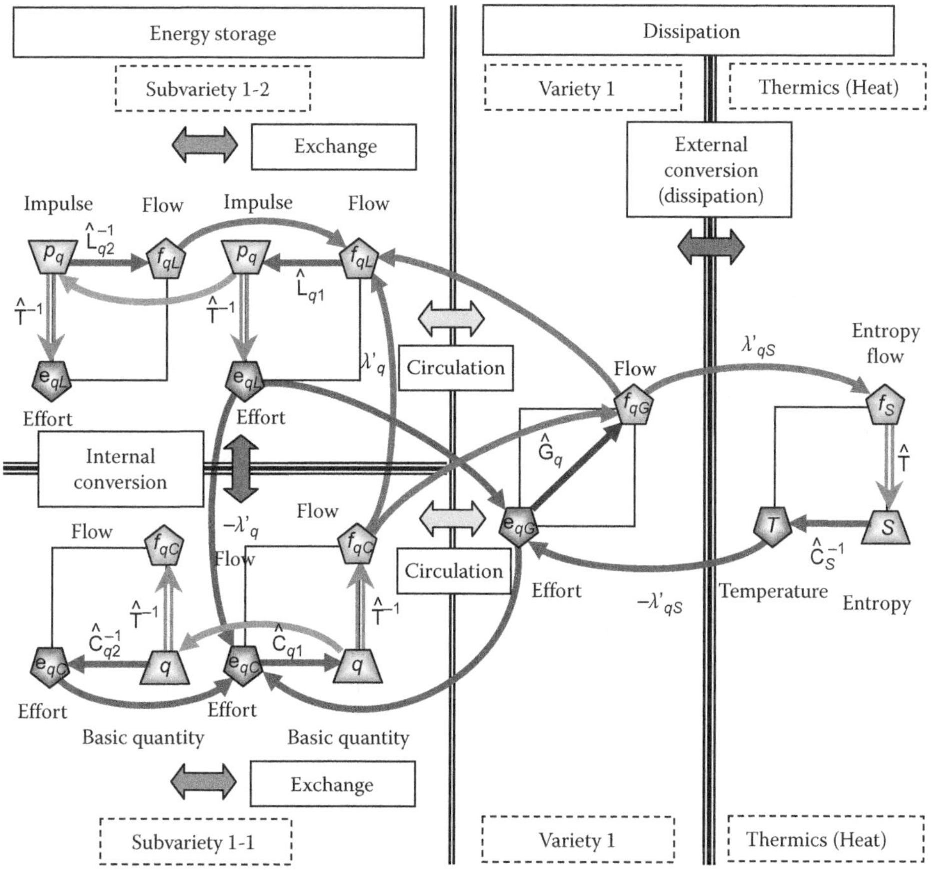

GRAPH 11.1 The three passages for the energy between a dipole and other dipoles: *exchange* between identical variety and subvariety, *conversion* between different subvarieties (capacitive–inductive) or different varieties, and *circulation* between storing and conductive dipoles in the same variety. A fourth passage is not shown here, which is the *coupling* between different energy varieties with the same subvariety.

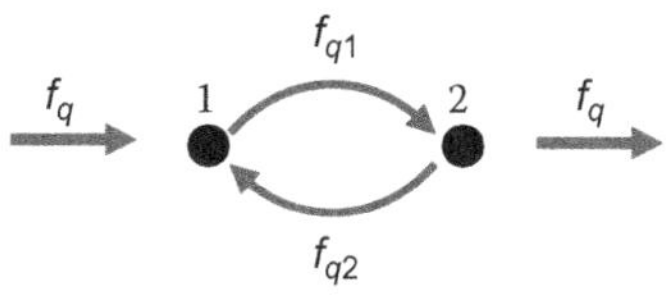

FIGURE 11.2 Scheme of an exchange by flows between two poles.

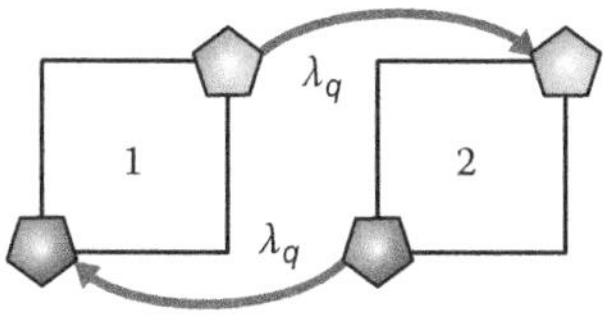

GRAPH 11.2 Formal Graph of an interconnection between two dipoles through their gate variables.

oblige the process to behave reversibly. This distinction is borrowed from chemical kinetics where the following conventions apply:

Reaction that always behaves irreversibly: $\mathrm{A} \longrightarrow \mathrm{B}$

Reaction that may behave reversibly or irreversibly: $\mathrm{A} \rightleftarrows \mathrm{B}$

If the existence of two parallel routes in opposite directions is a necessary condition for allowing a process to behave reversibly, it is not sufficient. A second condition is the similarity of the two routes in terms of operations. It is here that the Formal Graph becomes appreciable owing to its intrinsic use of operators associated to paths.

A close examination of Graph 11.1 reveals the existence of many opposite paths that can be compared in terms of the nature of operators they use. The idea is to consider the opposition in the direction of two parallel paths as forming a loop, without attaching importance to the nodes. Taking into account the links inside the dipoles, one may observe that closed loops are formed in all connections between dipoles. (The direction of internal links, even evolution operators, is not important.) This formation of loops obeys the *circularity principle* that says that any passage of energy must be ensured by a loop in order to respect the First Principle of Thermodynamics.

Table 11.2 gives the number of evolution operators met in the various loops encountered in Graph 11.1, omitting those involving an inductive dipole.

The interesting feature is that according to the parity of the number of evolution operators, the loop behaves differently in terms of reversibility. An even number confers a possibility of reversibility to the process whereas for an uneven number it is necessarily the opposite behavior (i.e., irreversibility) that prevails.

Even number of evolution operators in a loop ⇔ **Reversible** process

Uneven number of evolution operators in a loop ⇔ **Irreversible** process

The explanation is that if one of the two parallel paths composing a loop converts energy and not the other path, the dissymmetry in the treatment of energy does not allow the routes to fulfill the condition of similarity for a reversible behavior.

Reversibility corresponds to temporal symmetry, that is, time plays the same role for all paths (irrespective of the direction of evolution), whereas irreversibility is associated with asymmetry of time. Consequently, the circulation of energy between a capacitor or an inductor and a conductor is always irreversible, from the storing dipole toward the conductor. The fact that the reverse is not allowed means that the conductor cannot provide energy to the storing dipole.

The dissipation obeys the same rule, but in a reverse way, the conductor providing energy to the thermal storage. This may appear contradictory to the previous behavior, but energy does not change its nature in circulating between dipoles belonging to the same energy variety, which is not the case for dissipation. This asymmetry between circulation and dissipation is a key feature for understanding the behavior of energy and the specificity of thermics.

Why a capacitive storage for heat? It can be reasonably questioned why the thermal dipole is a capacitive one and not another type. A conductive thermal dipole would destroy the above analysis in modifying the number of evolution operators in the loop for instance. The reason will be given soon by applying fundamental thermodynamical arguments as discussed in the next section. Regarding the use of an inductive dipole, the nonexistence of the inductive subvariety in the thermal energy variety prohibits this possibility.

TABLE 11.2
Association of the Passage Reversibility with the Number of Evolution Operators in the Loop Featuring the Passage

Passage	Connected Objects	Number of Evolution Operators in the Loop	Reversibility of the Passage
Exchange	Storage ↔ Storage 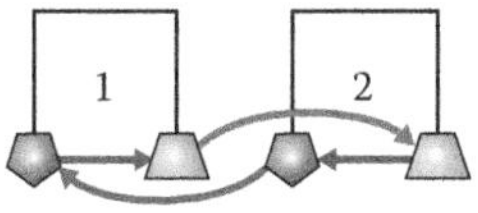 **GRAPH 11.3**	0	Potentially reversible
Circulation	Storage → Conductor 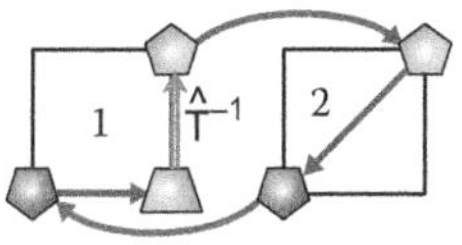 **GRAPH 11.4**	1	Irreversible
Dissipation	Conductor → Thermal storage 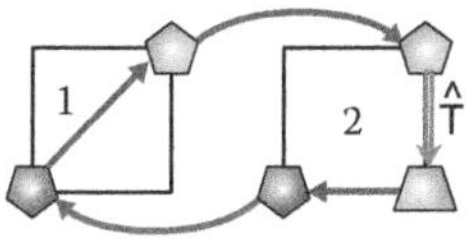 **GRAPH 11.5**	1	Irreversible
Conversion	Storage ↔ Storage 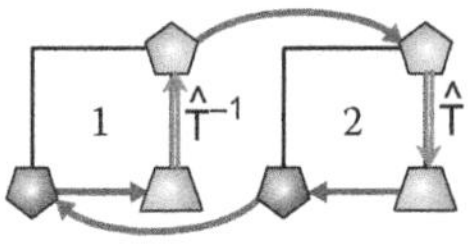 **GRAPH 11.6**	2	Potentially reversible

Note: An even number means a potentially reversible passage whereas an uneven number means an irreversible one. The direction of the evolution operators is irrelevant for the count.

The paradox of a reversible reaction that dissipates energy. A chemical reaction may produce heat (exothermicity) or may absorb heat from its surrounding (endothermicity). This depends on the relative proportion of dissipation and conversion (without dissipation) of stored energy involved in the process. The interesting point to outline is the possibility of a reaction behaving reversibly to dissipate energy. This statement contradicts the usual paradigm for which reversibility means necessarily an absence of dissipation. Another formulation is that the systematic association of dissipation with a purely irreversible process is nuanced by stating that dissipation may occur also in a reversible process.

(continued)

The explanation is that the same notion of reversibility is used for two different processes, exchange and dissipation. Conduction, which is a process inherent to the chemical reaction as seen in Chapter 4, works as an exchange of substance amounts, as explained in Chapter 6. This exchange behaves reversibly when the forward flow (reaction rate) is close to the backward flow. In the meantime, the exchange produces heat owing to the nature of the conduction process. This production is an irreversible process that is different from the exchange.

A detailed model built with Formal Graphs helps distinguish the two processes in evidencing the exchange working between the two poles as graphically separated from the conversion (dissipation) between the two dipoles.

11.1.3 Dissipation and Entropy

Until now, the conductance property has been presented as one of the three fundamental properties constituting a system, with the physical role of having to convert energy into heat. This role has never been detailed, not only for methodological or pedagogical reasons but because the related concepts of time and of energy conversion are required in order to discuss the subject.

As already stated, the conversion process underlying the phenomenon of dissipation occurs between the thermal energy variety (thermics) and any other energy variety, which differs slightly from the conversion between two subvarieties belonging to the same energy variety studied in Chapter 10. The principle is the same, connecting gate variables in pairs, but the energy varieties are not the same. Graph 11.7 shows a Formal Graph built by associating a conductive dipole with a thermal dipole and joining the nodes by a *conversion coefficient* λ_{qS} and by completing the graph with an impedance inside the thermal dipole, which will be detailed soon.

The choice of drawing an impedance instead of its reciprocal, an admittance, is dictated by the possibility of forming a loop going through the conductance in the q variety. It is always interesting to put into evidence such loops that enforce the *circularity principle* useful for understanding the behavior of energy in systems.

The association of two energy varieties brings one to invoke thermodynamics for modeling this system. A consequence of the additive property of energy is that the variation of total energy is the sum of the individual variations of each energy variety.

$$\mathrm{d}\mathcal{E} = \delta\mathcal{E}_q + \delta\mathcal{U}_S \tag{11.1}$$

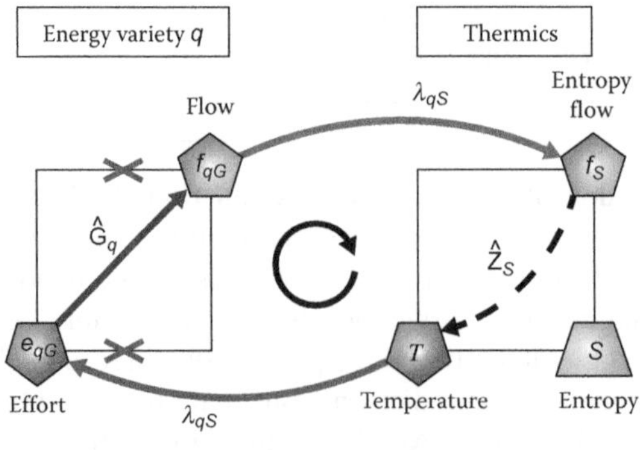

GRAPH 11.7 Formal Graph of the association of a thermal dipole to the conductive dipole in an energy variety q. A loop can be formed by using the thermal impedance (which avoids specifying the nature of the thermal dipole) for implementing the circularity principle, symbolized by a circular arrow.

As no inductive subvariety is recognized in the thermal energy, only the capacitive subvariety, called heat, is involved in this equation. It is recalled that an individual variation of energy is not an exact differential, which is indicated by a delta (δ) instead of an upright d. The First Principle of Thermodynamics, in its variational version, states that each individual variation of energy is the product of the energy-per-entity times the entity number variation, which obliges one to choose the subvariety involved in the process. One begins first by choosing the inductive subvariety; so the First Principle of Thermodynamics can be written as:

$$\mathrm{d}\mathcal{E} = f_{qL}\mathrm{d}p_q + T\mathrm{d}S \underset{isol}{=} 0 \tag{11.2}$$

The total variation of energy is equal to zero because the system is assumed to be isolated from outside, being able to store only the thermal energy produced by dissipation. This means that the thermal dipole must be of capacitive nature for playing this role, thus requiring the participation of the entropy S as written in the previous equation.

Another consequence of the involvement of thermodynamics in the modeling is that some inductive energy must be present in the system. As a conductive dipole does not contain energy, a supplementary dipole must be considered for storing energy. The complete model able to comply with thermodynamics is shown in Graph 11.8.

The association of a conductor with an inductor forms a dipole assembly, which is normally modeled by a single square unit using three nodes, as explained in Chapter 9, Section 9.3.3, but for the sake of pedagogy the decomposition into two dipoles is maintained.

Once the existence of two storing dipoles in the system has been determined, their connections need to be established. By dividing Equation 11.2 that expresses the First Principle of Thermodynamics by the time variation and by using the definitions of the gate variables,

$$e_{qL} \overset{def}{=} \frac{\mathrm{d}p_q}{\mathrm{d}t} \tag{11.3}$$

$$f_S \overset{def}{=} \frac{\mathrm{d}S}{\mathrm{d}t} \tag{11.4}$$

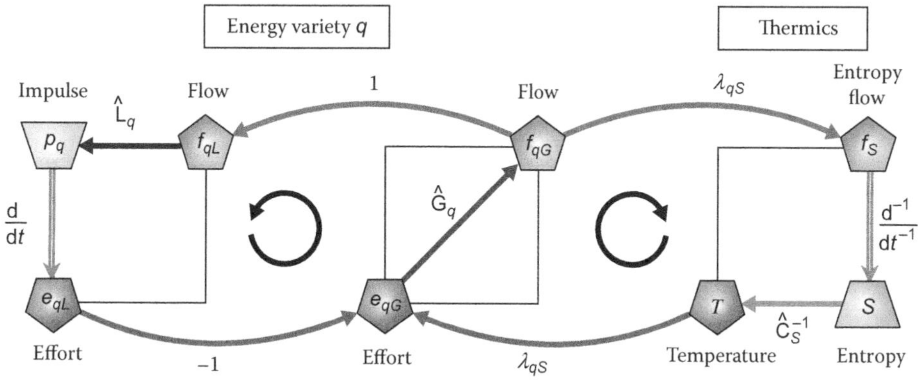

GRAPH 11.8 Complete Formal Graph of the conversion of inductive energy stored in an inductor into heat stored in a thermal capacitive dipole, going through a conductor.

the expression of the total power ensues

$$\mathcal{P} \overset{def}{=} \frac{d\mathcal{E}}{dt} = f_{qL} e_{qL} + T f_S \underset{isol}{=} 0 \tag{11.5}$$

This total power is naturally also equal to zero owing to the isolation of the system. This allows defining a *conversion coefficient* λ_{qS} as the common value of two ratios:

$$\lambda_{qs} \overset{def}{=} \frac{f_S}{f_{qL}} = -\frac{e_{qL}}{T} \tag{11.6}$$

Although this thermodynamical reasoning incites one to connect directly the two storing dipoles, in linking the entropy flow to the inductive flow and the temperature to the inductive effort, the requirement of implying the conductive dipole entails doing otherwise. The correct configuration consists of the association of the conductive dipole with the thermal one as explained previously, and then to a second association of the two dipoles belonging to the same subvariety, the inductive and the conductive dipoles. For assembling two such dipoles, the choice is between a common flow and a common effort, which is expressed by the following relationships:

Common flow (serial): $$e_{qL} + e_{qG} = 0; \qquad f_{qL} = f_{qG} \tag{11.7}$$

Common effort (parallel): $$f_{qL} + f_{qG} = 0; \qquad e_{qL} = e_{qG} \tag{11.8}$$

The Formal Graph in Graph 11.8 models a common flow (i.e., serial mounting) as can be recognized by the negative sign on the connecting coefficient between efforts. (The alternate choice of a common effort can be represented by displacing the minus sign from the effort connection to the flow connection.) This allows substituting the inductive state variables with the conductive ones in the expression of the conversion coefficient:

Conversion coefficient: $$\lambda_{qs} = \frac{f_S}{f_{qG}} = \frac{e_{qG}}{T} \tag{11.9}$$

Equation 11.1 of the First Principle of Thermodynamics has been previously adapted to the case of an inductive energy storage for providing Equation 11.9. This equation is also valid in the case of a capacitive storage as is demonstrated now by adapting Equation 11.1 to the capacitive sub-variety

$$d\mathcal{U} = e_{qC}\, dq + T\, dS \underset{isol}{=} 0 \tag{11.10}$$

and the need to store capacitive energy is ensured by the presence of a capacitive dipole in the model, as shown in Graph 11.9.

By using definition 11.4 of the entropy flow and by defining the capacitive flow (as a gate variable),

$$f_{qC} \overset{def}{=} \frac{dq}{dt} \tag{11.11}$$

and then by dividing by the time variation, one gets, from the previous writing of the First Principle, the expression of the total power

$$\mathcal{P} \overset{def}{=} \frac{d\mathcal{E}}{dt} = f_{qC} e_{qC} + T\, f_S \underset{isol}{=} 0 \tag{11.12}$$

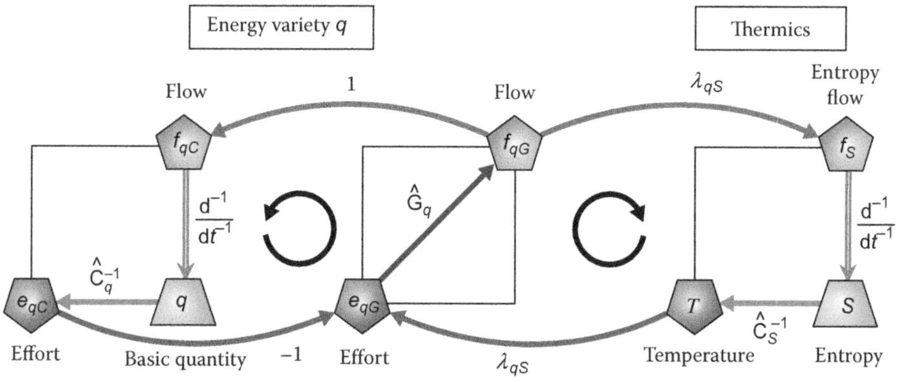

GRAPH 11.9 Complete Formal Graph of the conversion of capacitive energy stored in a capacitor into heat stored in a thermal capacitive dipole, going through a conductor.

which is also equal to zero. The same method consisting of recasting this equation in two ratios and of defining their common value as the conversion coefficient provides an expression similar to definition 11.6:

$$\lambda_{qS} \overset{def}{=} \frac{f_S}{f_{qC}} = -\frac{e_{qC}}{T} \tag{11.13}$$

By choosing to assemble the two dipoles in the q variety as sharing a common flow as previously done, one retrieves Equation 11.9 using the conductive dipole variables.

Dissipation needs conductance. If the conductor is removed between the storing dipole and the thermal capacitor, the new loop contains two evolution operators. The consequence is that dissipation is replaced by a conversion process between the considered energy variety and thermics, which is a potentially reversible process.

It appears clearly that the association of the conductive and thermal dipoles is not affected by the nature of the energy subvariety which is supplied to the conductive dipole. The right parts of Graphs 11.8 and 11.9 are identical, which means that a simplification of these Formal Graphs is possible by omitting the storing dipole, as is done in Graph 11.10.

However, the Formal Graph would not be complete without indicating that the conductor is powered by an external source, which is done by drawing a disk around one of the conductor's nodes. (Here the flow has been chosen for receiving the power.) The absence of this indication would change completely the meaning of the Formal Graph as it would model a thermal source dissipating in a nonthermal conductor, which is a violation of the laws of thermodynamics.

Another violation of thermodynamics can be encountered when one tries to provoke dissipation into a thermal conductor instead of a thermal capacitor, as shown in Graph 11.11.

This dipole assembly is not correct because the system cannot store the heat produced by dissipation. In fact, neither dissipation nor conversion occurs because a thermal conductor is only able to achieve heat transfer without storage. This is an exception because all conductors in other energy varieties not only ensure conduction or transfer but also ensure dissipation simultaneously. One of the specificities of the thermal energy variety is pointed out here.

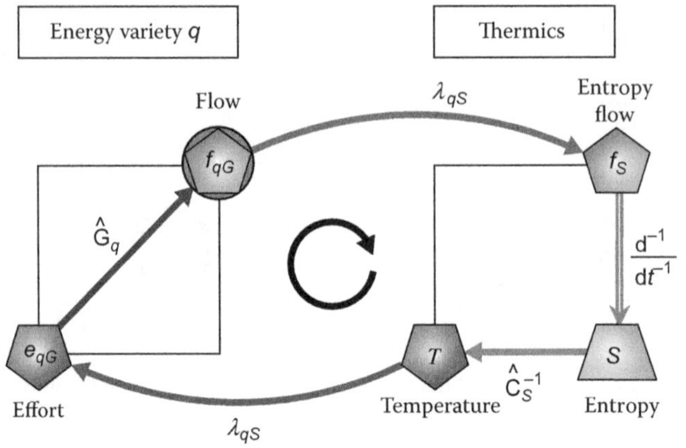

GRAPH 11.10 The simplified Formal Graph of the dissipation by a conductance receiving energy from an external source, which can be as well inductive as capacitive energy. The choice of the powered energy-per-entity (symbolized with a surrounding circle) has no influence on the process.

From the two expressions 11.5 and 11.12 of the total power and from the mounting relationships, the thermal power received by the thermal capacitor can be given by a unique expression as a function of the nonthermal conductor variables:

$$\mathcal{P}_S \overset{def}{=} T\, f_S \underset{isol}{=} -f_{qG}e_{qG} > 0 \tag{11.14}$$

The power $f_{qG}\, e_{qG}$ in the q variety is negative as it is generated by the nonthermal conductor, whereas the thermal power is a positive quantity, confirming the role of receptor played by the thermal capacitor.

It must be recalled that one is dealing with dipoles and that their efforts or flows are differences between state variables of the poles that constitute them. It is therefore impossible to know in advance, without supplemental information, the sign of the temperature or the entropy flow of a dipole. One needs an external rule to help decide the right choices for these signs; it is the role of the Second Principle of Thermodynamics.

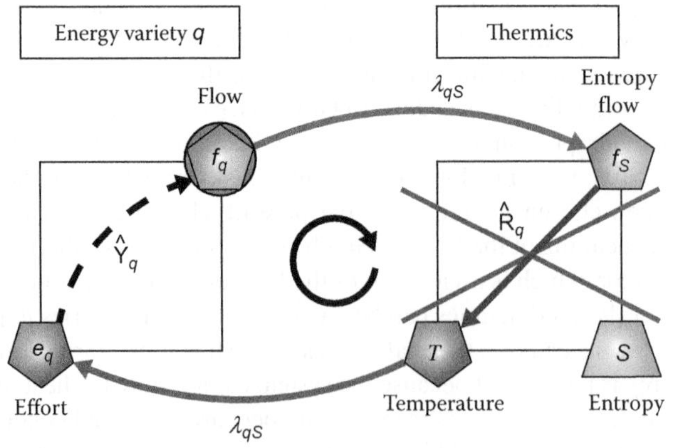

GRAPH 11.11 Unsuitable representation with a Formal Graph of the dissipation process. A conductive thermal dipole cannot be the receiver of the produced heat, only a storing dipole can be a receiver.

11.1.4 The Second Principle of Thermodynamics

This well-known principle ruling the evolution of entropy can be expressed in several ways and this is enunciated first in a nonconventional way, by saying abruptly that the entropy flow is always positive in a dissipative process occurring in an isolated system.

Second Principle of Thermodynamics $$f_S \underset{isol}{>} 0 \quad \text{for dissipation} \tag{11.15}$$

The conventional way states that during a transformation occurring in an isolated system containing thermal energy, the variation of entropy is always positive or null (Planck 1922).

Second Principle of Thermodynamics: $$dS \underset{isol}{\geq} 0 \tag{11.16}$$

This statement is apparently more general than the first one since it encompasses all cases of transformation, that is, exchange or conversion, that may occur. In fact, they are equivalent when one admits that the case of constant entropy corresponds to processes in which no dissipation occurs at all. In other words, each time a process involves dissipation, the entropy increases in the isolated system.

Second Principle of Thermodynamics: $$dS \underset{isol}{>} 0 \quad \text{for dissipation} \tag{11.17}$$

When time can be invoked, which is the case in any conversion by definition, division by the time variation (which is positive and must not be infinitely small) provides the nonconventional statement of the Second Principle given above.

One of the first consequences, but not the most important, of this principle is that the dipole temperature (which is a difference) of the thermal conductor must be always positive as can be deduced from the thermal power expression 11.14.

Thermal capacitor: $$T \underset{isol}{>} 0 \tag{11.18}$$

The most interesting consequence is that the direction of all dissipation processes is unique and cannot be reversed because the entropy cannot decrease. This is called *irreversibility*.

Another, but in fact not less interesting, consequence is that *time always increases* because entropy flow and entropy are related by the evolution operator. In other words, the arrow of time is determined and, as it is the same variable that quantifies the advance of every conversion, this orientation is universal, for oscillations as well as for dissipation.

Integral or differential entropy? In classical thermodynamics, the concept of temperature is introduced first, then the concept of entropy, which is defined as the difference between two states of a system (Planck 1922). Most often, a sequence of transformations of an ideal system (perfect gas), called

(*continued*)

Carnot* cycle, is taken as support for defining the variation of entropy between two stages of the cycle. Definitions based on statistical mechanics provide also a definition related to the state (or degree of complexity) of the system (Landau and Lifchitz 1958). Mathematical approaches without statistics are less frequent, especially when the temperature is not supposed to be defined before (see, for instance, Giles 1964). However, all these approaches define entropy as a state function, which is an integral quantity defining the state of a system. The problem is that the passage from an integral notion to a differential one is not straightforward, as seen by the paradox of the infinitely slow transformation required for avoiding an increase of entropy in a reversible transformation.

A more general definition. By defining the entropy as an entity number in the thermal energy variety, the Formal Graph theory provides a greater generality and a better accuracy in the understanding of this concept. Several levels of complexity are evidenced by distinguishing between forms of the variable quantifying the entropy:

- Discrete values at the quantum level in considering one entity (singleton).
- Continuous values (differential) with collections of entities (poles).
- Integral values (differences) with the storage of thermal energy (dipoles).

A consequence of this approach is that the temperature is defined as the energy-per-entity once the entropy is defined. This is the opposite of the classical approach.

Discrete or continuous time. The link between the entropy flow f_S and the entropy S is the evolution operator. As already discussed in Chapter 10, this operator is not necessarily a differential one and discretization of time is easily admitted as an alternative to the classical theory. This opens some interesting perspectives on thermodynamics in discretized space–time.

11.2 BASIC PROCESSES INVOLVING DISSIPATION

Each time a conductance in a nonthermal energy variety appears in a system, dissipation occurs. Among all mechanisms able to provoke dissipation, two basic processes deserve attention because of their fundamental importance. They are the relaxation process and the damping of oscillations.

11.2.1 Relaxation

Relaxation, in the rigorous sense of the term, is the process that makes a system driven out of equilibrium return progressively to its stable state of equilibrium. This strict definition applies to systems that are isolated from outside after having been placed out of equilibrium, and then left free to relax by dissipating the excess of energy with respect to the equilibrium state (see Figure 11.3).

* Nicolas Léonard Sadi Carnot (1796–1832): French physicist; Paris, France.

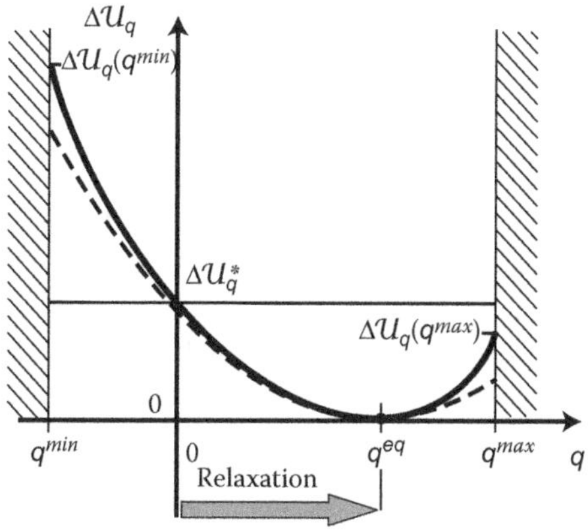

FIGURE 11.3 Example of energy profile of a capacitive dipole as a function of the dipole basic quantity. The initial basic quantity is zero by definition and the equilibrium value corresponds to the minimum of energy. The dashed curve is the classical parabolic profile in the case of linear capacitance.

In contradistinction with a simple energy exchange between two energy containers of the same energy subvariety (for instance, two reservoirs or two capacitors) which is instantaneous, relaxation implies an energy conversion (into heat), thus necessitating time.

The Formal Graphs in Graph 11.12 show the simplest model of relaxation that uses two paths in parallel, with a temporal operator in one of them and a nontemporal operator in the other. The two representations are equivalent; they differ in terms of encoding the role of the two paths. In the additive encoding (left), the two paths contribute to the destination node through an addition, while in the loop representation (right), they work in opposite directions.

The required irreversibility of the dissipation results from the presence of a temporal derivation in only one path, the other being featured by a time-independent operator. This latter is called a *kinetic operator*, which in most systems reduces to a scalar, called *kinetic constant* or *relaxation frequency*. Its inverse is the *relaxation time* or *time constant.*

An extension of the notion of relaxation is used when the system is not isolated and continuously perturbed by imposition of an external source of energy. In that case one speaks of *forced relaxation* in contrast to the *free relaxation* described above. The model is the same; but instead of establishing the dependence upon time of the state variables during their evolution toward equilibrium, the preferred modeling consists of finding transfer functions, that is, impedance or admittance, featuring the behavior of the system independently from the shape of the perturbation signal.

For this reason, the two aspects of free relaxation of an isolated system and forced relaxation of an open system are developed in detail. This latter case will be treated in the frame of

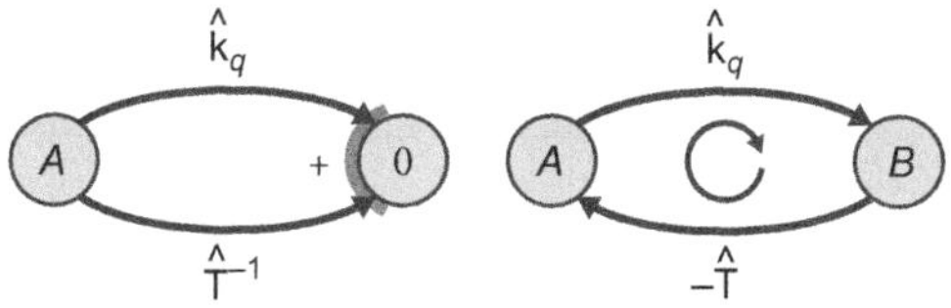

GRAPH 11.12 Two equivalent ways of representing the same phenomenon of relaxation in an isolated system by Formal Graphs. The left one encodes the relaxation as an additive contribution of two processes whereas the right one uses the loop representation. In both cases, the operators are different with regard to the role of time, making the relaxation an irreversible process.

small AC perturbations, allowing one to discuss the model in terms of complex admittance or impedance.

11.2.2 Analogy between Relaxation and Influence

In Chapter 7 dealing with the theory of influence, an important law governing the transmission of a radiation through an absorbing medium, the Lambert–Beer–Bouguer law, has been taken as a template for demonstrating the exponential shape of the decay as a function of the medium thickness (see Section 7.4.1 in Chapter 7). A similar reasoning establishes the shape of the time dependence of the conversion occurring when an initial energy amount is dissipated into a conductor (free relaxation).

Let us take some entities quantified by a state variable q that are initially present with the amount $q(0)$ in a storing dipole. This initial amount is constant until the dipole is connected to a conductor, provoking the circulation of energy which is then entirely dissipated. The advance variable t of the conversion (dissipation) is set by convention to zero (time origin of the process).

The analogy with the absorbing medium works in comparing the quantity q to the radiation intensity and the advance variable (elapsed time) to the thickness (run distance). As for the absorbing medium that was cut into successive layers in which the decay of radiation was followed by the series of intensities $I_1, I_2, \ldots, I_n$, the conversion is divided into sequences of durations $\Delta t_1, \Delta t_2, \ldots, \Delta t_n$, corresponding to evolving quantities $q_1, q_2, \ldots, q_n$. Each sequence features an amplification or transmission factor $T_{qi,i+1}$ defined as the ratio of the output on the input quantities.

The reasoning establishing the equivalence between two sequences and their sum is sketched in Figure 11.4.

The details of the reasoning are not repeated here, only the last steps are recalled. The equivalence between the product of transmissions and the sum of durations is expressed by the following two equations:

$$\frac{dT_{q13}}{T_{q13}} = \frac{dT_{q12}}{T_{q12}} + \frac{dT_{q23}}{T_{q23}} \tag{11.19}$$

$$\Delta t_{13} = \Delta t_{12} + \Delta t_{23} \tag{11.20}$$

Identical assumptions concerning the independence between sequences are made, allowing one to propose for each sequence the same proportionality coefficient between each term of the two preceding equations. For the first sequence, this is written as

$$\frac{dT_{q12}}{T_{q12}} = -k\Delta t_{12} \tag{11.21}$$

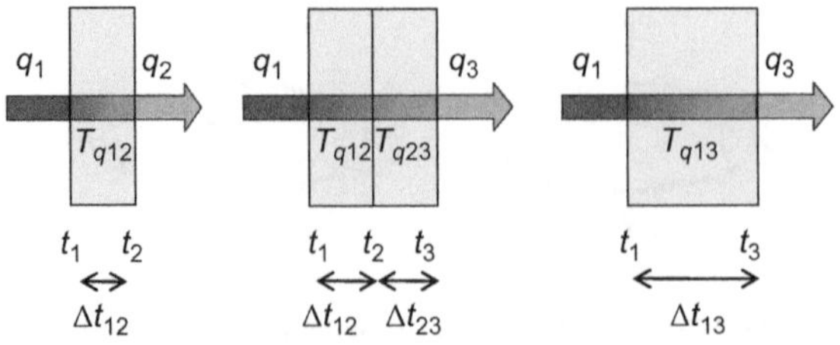

FIGURE 11.4 Three steps of the reasoning leading to the equivalence of the whole duration with the sequence of two shorter durations.

having explicitly used a negative value $-k$ for the common proportionality factor (assuming a positive value for k) because the variation of the transmission factor is negative. By integration the expected exponential function is retrieved

$$T_{q12} = \frac{q_2}{q_1} = \exp(-k\Delta t_{12}) \tag{11.22}$$

and by extension to all sequences one gets the model of relaxation

$$q = q(0)\exp(-k\,t) \tag{11.23}$$

in which t stands for the sum of all durations.

The interest of this development is to justify again the introduction of the exponential function on a physical basis and not as a mathematical solution of a differential equation. It is because at each step of the conversion a certain amount of entities disappears independently from the other steps that the exponential model is found. Another interest lies in the nonrequirement of a continuous time for justifying this model. The analogy with the layers of the absorbing medium makes clear the possibility to consider discrete values for the time without introducing too much complexity.

Figure 11.5 illustrates a discontinuous conversion, which can be understood as an effect of discretization of time or as a process occurring by sudden steps even with a continuous time.

The noticeable result is that the exponential function is obtained without specifying the shape of the evolution operator. However, for facilitating the discussion, one shall take in the case studies the classical differential operator as the standard shape for the evolution operator, which allows retrieving the exponential function as the outcome of a differential calculus.

As the Lambert–Beer–Bouguer law is often introduced as an empirical law (as it was originally), it is important for the solidity of the presented reasoning to say that this is not true nowadays. The Lambert–Beer–Bouguer law is demonstrated in case study H8 "Attenuated Propagation" and in Section 11.7 devoted to the attenuated propagation of a wave.

11.2.3 Difference between Transfer and Relaxation

Chapter 10 tackled the subject of transfer, which is a general process having multiple aspects as it may involve, among other elementary processes, various proportions of conduction and evolution. These two ingredients are also used in a relaxation process, but in a different way.

According to the Formal Graph theory, the key concept in a transfer is the path complement between two parallel paths that have different contributions to the overall process. One of these paths involves an evolution operator that can be combined with another space–time property or with a system constitutive property with the exception of the conductance. The other path is composed in

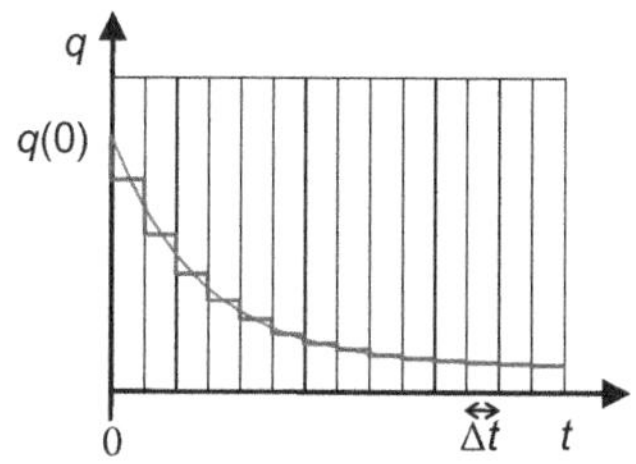

FIGURE 11.5 The model of relaxation in case of discontinuous time.

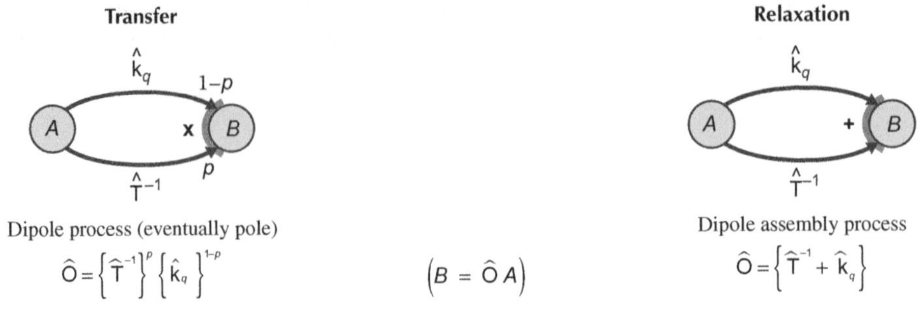

GRAPH 11.13 Difference between the processes of transfer (left) and relaxation (right) in Formal Graph language and in algebraic language. The operators of the individual paths are identical in both cases (evolution and system constitutive properties), but their combination is different: Weighted composition (product) for the transfer and addition for the relaxation. (The case of *free relaxation* corresponds to $B = 0$ and the case of *forced relaxation* to $B \neq 0$.)

the opposite way, involving conduction without evolution. For comparison, this conductive path is represented here by a kinetic operator $\hat{k}_q$ alone, although the operator representing the process may be a simple conductance or may involve space operators and other system constitutive properties (capacitance, inductance). On the same principle, the evolution operator is used alone for the other path, although other properties may participate (but nonconductive ones).

This is in fact the same path specialization that applies to relaxation, which relies also on the simultaneous contribution of two paths, one with conduction without evolution and vice versa for the other. The basic relaxation studied here works with the simplest possible compositions of paths, but the phenomenon can be obviously generalized to more complex compositions. In a relaxation process—and here is found the main difference with a transfer—the two parallel paths always contribute in the same proportion through a cumulative participation, mathematically expressed by an addition of the two paths.

Graph 11.13 depicts these differences in the Formal Graph and algebraic languages.

Another difference is that a transfer occurs within a single dipole (or even a pole when a pure conduction ensures the transfer) and that relaxation requires at least two dipoles (in a dipole assembly). This is naturally a Formal Graph concept hardly transposable in classical terms, but it may help grasp the physical specificities of these processes.

It should be clear that this is a basic scheme and that more complex mechanisms can be built with these bricks. A transfer may be incorporated within a relaxation process reciprocally, leading to a variety of systems that are not always easy to describe but that the analytical decomposition into Formal Objects allows modeling nevertheless.

Transfer:	Conductive and evolutive paths complemented in various proportions.
Relaxation:	Conductive and evolutive paths added in equal proportions.

The damping process studied hereafter can be added to this scheme, which is a double relaxation process, but with opposite directions.

11.2.4 Damping

Damping is the name given to the phenomenon of decrease of the amplitude of temporal oscillations under the effect of dissipation of the energy. In case of propagating oscillations the phenomenon is also referred to as the absorption or attenuation of a wave by a medium. For a spatial oscillator, the notion of damping is also used, although there is no dissipation as will be seen in Section 11.6.

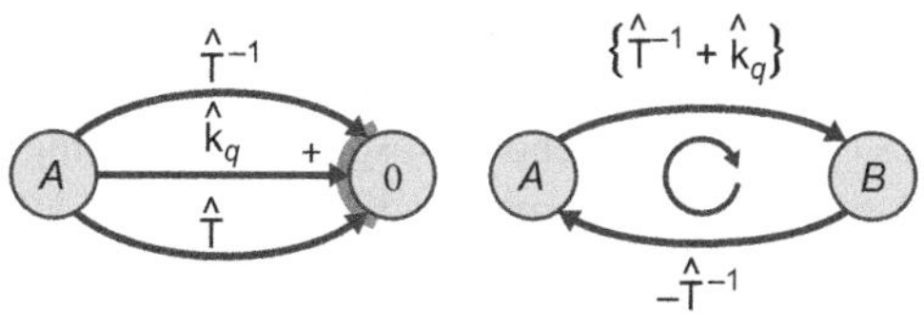

GRAPH 11.14 Two equivalent ways of representing the same phenomenon of damped oscillations in an isolated system by Formal Graphs. The left one encodes the overall process as an additive contribution of three elementary processes whereas the right one uses the loop representation, made possible by combining two parallel paths in a single one. In both cases, the operators are different with regard to the role of time, making the damping an irreversible process.

In all cases, the phenomenon is due to the continuous diversion of a part of the oscillator energy by a conduction process, entailing a progressive decrease of this energy and, consequently, of the amplitude of the oscillations.

In Graph 11.14 are shown two Formal Graphs schematizing the minimum ingredients for modeling the phenomenon in an isolated temporal oscillator (these are noncanonical ones). The first one simply adds a parallel path representing conduction or relaxation, which comes as a diverting route in the symmetrical network made by the two evolution paths. The second one modifies one of the evolution operators by including the relaxation, destroying the required temporal symmetry for a conservative (i.e., nondamped) oscillator. Both representations are equivalent but each one conveys a different meaning—energy diversion or temporal asymmetry.

The relaxation and the damping processes are the main subjects of the following case studies (with the exception of the electrical Joule effect). They will be analyzed in more detail after the series of case studies.

11.3 CASE STUDIES OF DISSIPATIVE DIPOLE ASSEMBLIES

The six first case studies deal with the phenomenon of relaxation caused by the dissipation associated with conduction, mainly in the capacitive subvariety as only one works in the inductive subvariety with the kinetic relaxation (decelerated motion). Both global and space distributed systems are studied and also free and forced relaxation.

The RC circuit allows one to develop the notions of real and imaginary responses of the system when the relaxation is forced by a periodic signal (i.e., in characterizing systems by impedancemetry).

Relaxation (global)

- H1: Electrical Joule effect in electrodynamics (C + G)
- H2: Slowed down motion in translational mechanics (L + G)
- H3: Occurring reaction in physical chemistry (C + G)
- H4: RC circuit in electrodynamics (C + G)

The relaxations of dielectric materials and of viscous fluids are very similar phenomena, both experimentally studied by impedancemetry in general. Two different but classical models are compared, the equivalent electric circuit (using resistors and capacitors) and the rheological circuit (using dashpots and springs). Their fundamental difference is evidenced with the RC circuit in parallel and the Maxwell model in series. Both model the same physical process, as attested by the Formal Graphs, but in using opposite conventions with regard to the notions of across and through variables for a dipolar component.

Relaxation (three-dimensional)

- H5: Dielectric relaxation in electrodynamics (C + G)
- H6: Viscoelastic relaxation in translational mechanics (C + G)

The two last case studies enter the realm of damped oscillators and attenuated propagation, also called wave absorption. Several examples other than the two tackled could have been chosen, but they are good representatives of the phenomenon of oscillation damping.

The last case study deserves a special comment as it deals with the attenuation provoked by spatial constraints and not by a conduction process as was the case for all previous case studies in this chapter (and as usually done for this case). The Maxwell equations need to be modified to take into account this peculiarity, thus allowing demonstration of the Lambert–Beer–Bouguer law of light absorption in materials that can be electrically nonconducting and noncharged (which is the most frequently encountered situation).

Damping (global)

- H7: Damped mechanical oscillator in translational mechanics (L + C + G)

Attenuation (three-dimensional)

- H8: Attenuated propagation in electrodynamics (L + C + G)

From this last case study one learns the role of *space distribution* as a possible cause of *irreversibility* of processes.

11.3.1 Case Study H1: Joule Effect — Electrodynamics

The case study abstract is given on next page.

The Joule* effect in electrodynamics, also known as Joule's first law,† refers to the production of heat by an electrical conductor due to the circulation of charges. Joule studied the phenomenon in the 1840s and expressed his first law as the relationship between the current and the heat production rate:

Joule's first law:
$$\mathcal{P}_S = \frac{d\mathcal{Q}}{dt} = R\, I^2 \quad \text{(H1.1)}$$

where R is the scalar resistance of the conductor and $\mathcal{Q}$ is the thermal energy (heat) that appears by dissipation.

The Formal Graph in the case study abstract models a thermal capacitor receiving energy from an electrical conductor, which is powered by an electrical current source. The fact that it is the current that is imposed and not the potential is not important for the demonstration.

Joule's first law is easily demonstrated in using the general definition of the power entering a dipole:

Received power:
$$\mathcal{P}_q \overset{def}{=} \frac{d\mathcal{U}_q}{dt} = e_q f_q \quad \text{(H1.2)}$$

Applied to the electrical power supply connected to the conductor, it gives

Electrical power:
$$\mathcal{P}_Q = \frac{d\mathcal{U}_Q}{dt} = -V\, I \quad \text{(H1.3)}$$

* James Prescott Joule (1818–1889): English physicist and brewer; Manchester, UK.

† Joule's second law describes the energy dependence of an ideal gas on its state variables.

H1: Joule Effect **Electrodynamics**

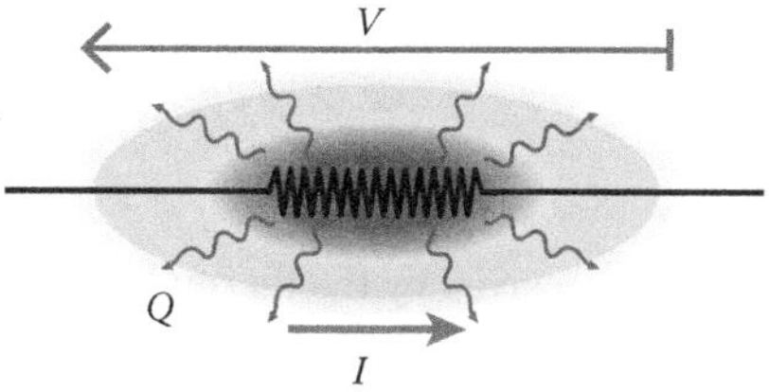

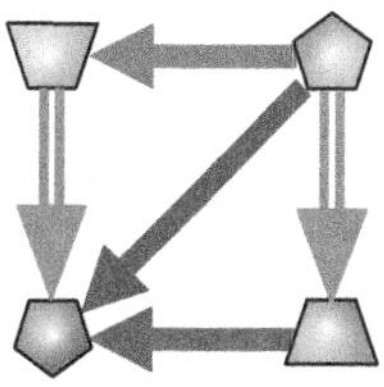

Dipole Assembly	
	Fundamental
✓	Mixed
✓	Capacitive
	Inductive
✓	Conductive
	Common effort
✓	Common flow
	Global
✓	Spatial

Formal Graph (see Graph 11.5):

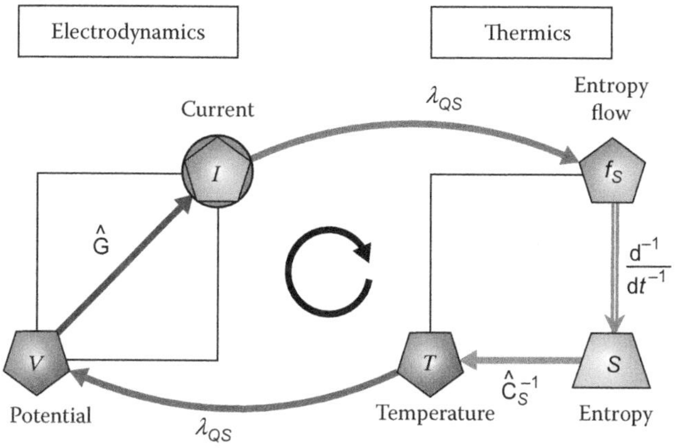

GRAPH 11.15

Variations with time:

$$\mathcal{P}_S = \frac{dQ}{dt} = R\, I^2$$

Variables:

Variety	Electrodynamics		
Subvariety	Capacitive (electrostatic)		Inductive
Category	Entity number	Energy/ entity	Entity number
Family	Basic Quantity	Effort	Flow
Name	Charge	Potential	Current
Symbols	Q	V, U, E, φ	$I, (i)$
Unit	[C], [A s]	[V], [J C^{-1}]	[A]

System Property		
Nature	Capacitance	Resistance
Levels	0	0
Name	Capacitance	Resistance
Symbol	C	R
Unit	[F]	[Ω]
Specificity	—	—

This power is negative because of the decrease in the energy in the generator. The generator potential is linked to the current according to Ohm's law taken in the classical linear case of scalar resistance (note that in the Formal Graph the general operator is used) as shown:

Ohm's law: $$V \underset{lin}{=} R\, I \qquad \text{(H1.4)}$$

The total power, electrical plus thermal, is naturally equal to zero owing to the isolation of the whole system:

$$\mathcal{P}_Q + \mathcal{P}_S \underset{isol}{=} 0 \tag{H1.5}$$

Then, by combining the last three equations, the thermal power received by the thermal dipole is written as

$$\mathcal{P}_S \underset{lin}{=} R\, I^2 \tag{H1.6}$$

which is the equation expressing Joule's first law.

Now, it is interesting to go beyond this model to determine what happens in the thermal dipole. As the variation of thermal energy (heat) is given by

$$\mathrm{d}\mathcal{U}_S = T\,\mathrm{d}S \tag{H1.7}$$

the received thermal power is

$$\mathcal{P}_S = \frac{\mathrm{d}\mathcal{U}_S}{\mathrm{d}t} = T\frac{\mathrm{d}S}{\mathrm{d}t} = T\, f_S \tag{H1.8}$$

By integration of this equation one finds the change in entropy to be

$$\Delta S = R\int_0^t \frac{I^2}{T}\,\mathrm{d}t \tag{H1.9}$$

As the Second Principle of Thermodynamics states that the entropy cannot decrease during dissipation, the dipole temperature, which is a difference between two pole temperatures, must be positive.

The previous result is a general one that can be adapted to peculiar cases when the capacitive relationship between entropy and temperature is known. For instance, as when the material of the conductor is made with a substance possessing thermal capacitance. This property can be defined at constant pressure or volume and for instance in this latter case its definition is the partial derivative of the capacitive energy of one mole of substance at constant volume.

Molar heat capacity at $V = C^{st}$:

$$C_V = \left(\frac{\partial \mathcal{U}(n=1)}{\partial T}\right)_V \tag{H1.10}$$

11.3.2 Case Study H2: Slowed Down Motion — Translational Mechanics

The case study abstract is given on next page.

The slowing down provoked by a nonideal medium opposing friction against motion is modeled by a conductive relationship:

$$\mathbf{F_G} = \hat{\mathrm{k}}_f\,\mathbf{v} \underset{lin}{=} k_f\,\mathbf{v} \tag{H2.1}$$

The operator in this relation is the mechanical resistance $\hat{\mathrm{k}}_f$ also called friction coefficient when it is a scalar.

H2: Slowed Down Motion **Translational Mechanics**

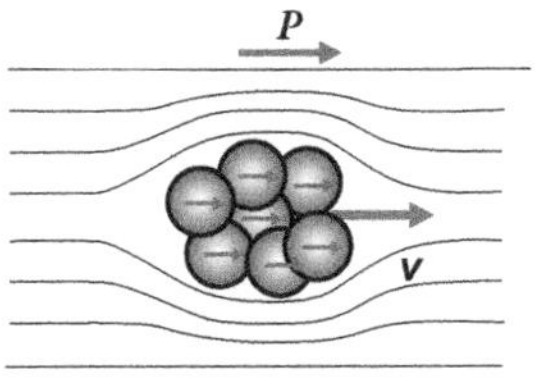

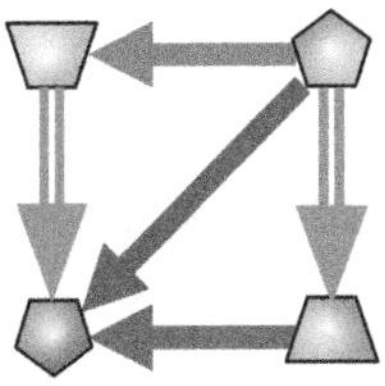

Dipole Assembly	
✓	Fundamental
	Mixed
	Capacitive
✓	Inductive
✓	Conductive
	Common effort
✓	Common flow
✓	Global
	Spatial

Formal Graph:

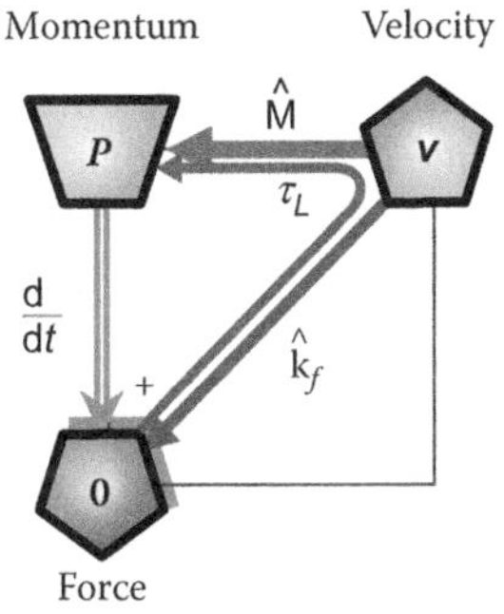

Variations with time:

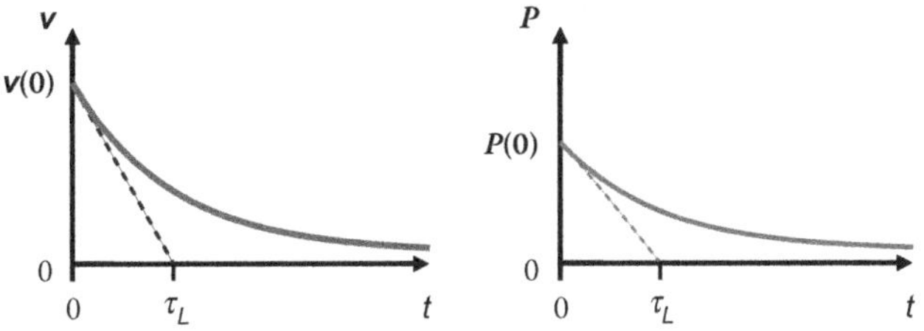

$$\mathbf{v} = \mathbf{v}(0)\exp\left(-\frac{t}{\tau_{RL}}\right);\quad \tau_{RL} = \hat{M}\,\hat{k}_f^{-1} \underset{lin}{=} \frac{M}{k_f}$$

$$\mathbf{v} = \mathbf{v}_L = \mathbf{v}_G$$

GRAPH 11.16

Isolated: $\mathbf{F} = \mathbf{F}_L + \mathbf{F}_G \underset{isol}{=} \mathbf{0}$

Variables:

Variety	Translation Mechanics		
Subvariety	Inductive (kinetic)		Capacitive
Category	Entity number	Energy/entity	Energy/entity
Family	Impulse	Flow	Effort
Name	Momentum (quantity of movement)	Velocity	Force
Symbols	$\mathbf{P}$, $(\mathbf{p})$	$\mathbf{v}$	$\mathbf{F}$
Unit	[N s], [J s m^{-1}]	[m s^{-1}]	[N], [J m^{-1}]

System Property		
Nature	Inductance	Resistance
Levels	0	0
Name	Inertial Mass	Friction coefficient
Symbol	M	k_f
Unit	[kg]	[N s m^{-1}]
Specificity	—	—

$$\mathbf{P} \underset{lin}{=} M\mathbf{v};\quad \mathbf{F}_G \underset{lin}{=} k_f\mathbf{v}$$

Energy variation:

$$\mathrm{d}\mathcal{T}_{\ell} \underset{\exists M}{=} \mathbf{v}.\mathrm{d}\mathbf{P}$$

The other relationships modeling this system are the same as in the case of a moving body through an ideal medium without friction (see case study C1 "Colliding Bodies" in Chapter 6). Newton's second law of motion is decomposed in the definition of the inductive force (see case study F1 "Accelerated Motion" in Chapter 9).

$$\mathbf{F}_L = \frac{\mathrm{d}}{\mathrm{d}t}\mathbf{P} \tag{H2.2}$$

and the inductive relationship

$$\boldsymbol{P} = \hat{\mathrm{M}}\,\boldsymbol{v} \underset{lin}{=} M\,\boldsymbol{v} \qquad \text{(H2.3)}$$

The variation of inductive (kinetic) energy of the inductive dipole is

$$\mathrm{d}\mathcal{T}_\ell \underset{\exists M}{=} \boldsymbol{v}.\mathrm{d}\boldsymbol{P} \qquad \text{(H2.4)}$$

The friction provokes a conversion of inductive (kinetic) energy into heat, which is not made explicit in the Formal Graph given in the case study abstract. The following Formal Graph (Graph 11.17) associates thermics (the short name for thermal energy) with the previous one, without detailing how the dissipation process provides energy to the thermal dipole (which is of a capacitive type). Such details can be given only after introducing the concept of energy coupling, which will be done in Chapter 12.

The system is initially in a dynamic state with a mass moving at a relative velocity $\boldsymbol{v}(0)$, which is the initial dipole velocity. The isolation of the dipole beyond the initial time $t = 0$ means that no external force is applied, implying that the sum of internal forces is equal to zero. The node of the dipole bearing the effort is built as the result of this sum (coded as a synapse with a plus sign near the converging paths) and consequently has zero for value:

$$\boldsymbol{F} = \boldsymbol{F}_L + \boldsymbol{F}_G \underset{isol}{=} 0 \qquad \text{(H2.5)}$$

The vertical path at this node is the inductive force $\boldsymbol{F_L}$ given by Equation H2.3 whereas the oblique path is the friction force $\boldsymbol{F_G}$ expressed by Equation H2.1. By relating with Equation H2.2 this latter to the velocity and by combining the conductive and inductive operators into a time "constant" $\hat{\tau}_{RL}$ (which can be a scalar, especially in the case of linearity of the property operators), the resulting differential equation is obtained:

$$\frac{\mathrm{d}}{\mathrm{d}t}\boldsymbol{P} + \hat{\tau}_L^{-1}\boldsymbol{P} = 0; \qquad \hat{\tau}_L = \hat{\mathrm{M}}\,\hat{\mathrm{k}}_f^{-1} \qquad \text{(H2.6)}$$

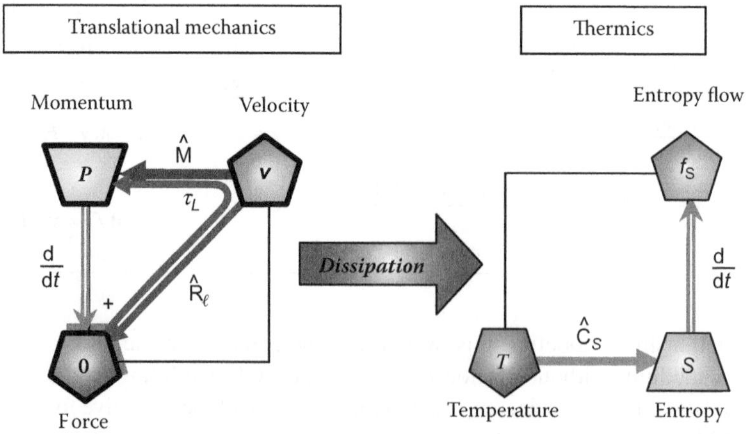

GRAPH 11.17 The principle of dissipation illustrated by a transfer and conversion of energy from the mechanical dipole toward the thermal dipole.

To solve this differential equation, we need to know the shape of the time "constant" operator or to restrict the generality to the linear case. Indeed, this equation can easily be integrated in assuming a time constant independent from any varying state variable and equal to a scalar:

$$\hat{\tau}_L \underset{lin}{=} \tau_L \tag{H2.7}$$

The linearity of both system properties is not required for fulfilling this condition, but their combination must be needed. The integration gives an exponential decrease in the dipole momentum with time:

$$\boldsymbol{P} \underset{lin}{=} \boldsymbol{P}(0) \exp\frac{-t}{\tau_L} \tag{H2.8}$$

By using this equation together with Equation H2.2 that gives the velocity (linear case) and Equation H2.4 that gives the mechanical dipole energy variation, the kinetic (inductive) energy is obtained by simple integration:

$$\mathcal{T}_\ell \underset{lin}{=} \mathcal{T}_\ell(0) \exp\frac{-2t}{\tau_L} \tag{H2.9}$$

The initial amount of inductive energy is equal to the total energy of the system, which is the sum of the kinetic (inductive) energy and of thermal (capacitive) energy (heat):

$$\mathcal{T}_\ell(0) = \mathcal{U}_S + \mathcal{T}_\ell \underset{lin}{=} \frac{\boldsymbol{P}(0)^2}{M} \underset{lin}{=} M\,\boldsymbol{v}(0)^2 \tag{H2.10}$$

The variations with time of the momentum and of the two inductive and capacitive energies are plotted in the following figure.

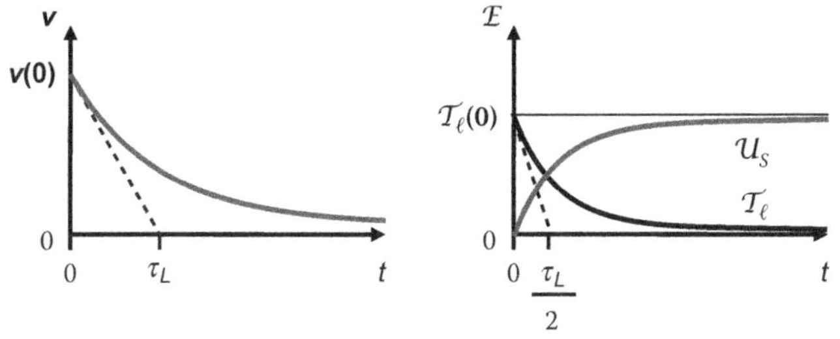

This model of assembly between inductive and conductive dipoles that relax when isolated from external constraints is not limited to translational mechanics, but is general for all energy varieties possessing the inductive subvariety. For instance, a rotating cylinder is a viscous fluid (rotational mechanics) slowing down or a nonremanent material (magnetic energy) relaxing from an imposed magnetization, both modeled in exactly the same way. Still more general is the transposition to capacitive and conductive dipoles where symmetrical Formal Graphs and hence identical exponential functions model the variations of the state variables.

11.3.3 Case Study H3: Occurring Reaction

Physical Chemistry

$$\mathrm{A} \underset{k_B}{\overset{k_A}{\rightleftarrows}} \mathrm{B}$$

Dipole assembly

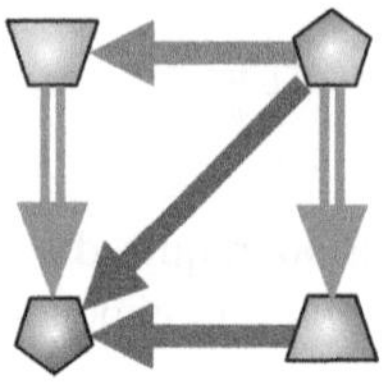

Dipole Assembly	
✓	Fundamental
	Mixed
✓	Capacitive
	Inductive
✓	Conductive
✓	Common effort
	Common flow
✓	Global
	Spatial

Formal Graph:

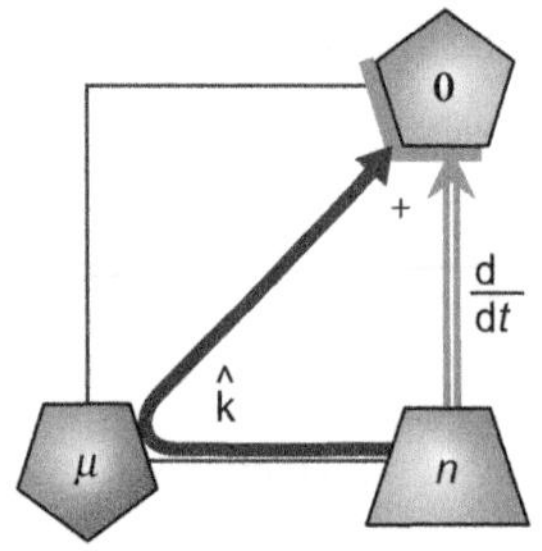

GRAPH 11.18

Isolated: $\Im = \Im_C + (-\Im_G) \underset{isol}{=} 0$

Variations with time:

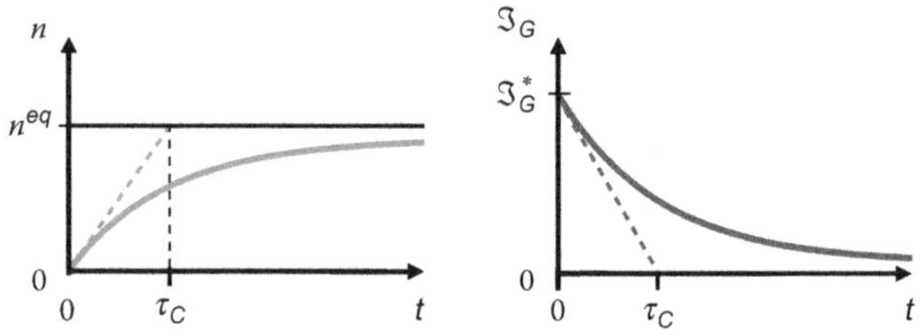

$$n = n_A^* - n_A = n_B - n_B^* = n^{eq}\left(1 - \exp(-kt)\right)$$

$$\Im_G = \Im_G^* \exp(-kt); \quad \Im_G^* \overset{def}{=} k_A n_A^* - k_B n_B^*$$

$$k \overset{def}{=} k_A + k_B; \quad n^{eq} = \frac{\Im_G^*}{k}; \tau_C = \frac{1}{k}$$

Variables:

Variety	Physical Chemistry		
Subvariety	Capacitive (potential)		Inductive
Category	Entity number	Energy/entity	Energy/entity
Family	Basic quantity 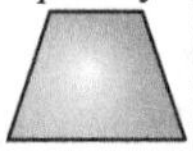	Effort 	Flow 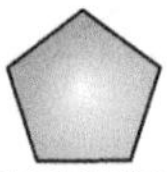
Name	Substance amount	Chemical potential	Substance flow
Symbols	n, N	μ	$\Im, v$
Unit	[mol]	[J mol^{-1}]	[mol s^{-1}]

Evolution (capacitive)

$$\Im_C = \frac{d}{dt} n$$

Exchange (conductive):

$$\Im_G = k_A n_A - k_B n_B$$

$$\Im_G = -\hat{k}\, n$$

$$\hat{k} = \frac{n - n^{eq}}{n} k$$

System Properties		
Nature	Capacitance	Conductance
Levels	0	0
Name	Physical chemical capacitance	Physical chemical conductance
Symbol	C_n	G_n
Unit	[mol^2 J^{-1}]	[mol^2 J^{-1} s^{-1}]
Specificity	—	—

The chemical reaction considered here is a simple binary reaction (i.e., two reacting species A and B) with first-order kinetics.

$$\mathrm{A} \underset{k_\mathrm{B}}{\overset{k_\mathrm{A}}{\rightleftarrows}} \mathrm{B}$$

This subject has been already studied in Chapter 6 (about dipoles) in case study C5 "First-Order Chemical Reaction" in steady-state conditions. The *embedding principle* allows one to relate the poles (representing the species) to the dipole (representing the reaction) by stating that the variation of (capacitive) energy can be written indifferently in terms of poles or of dipole variations:

$$\mathrm{d}\mathcal{U}_n = \mu_\mathrm{A}\mathrm{d}n_\mathrm{A} + \mu_\mathrm{B}\mathrm{d}n_\mathrm{B} = \mu\,\mathrm{d}n \quad \text{(H3.1)}$$

One recalls the relationships between the two poles and the dipole that are deduced (in adopting the generator convention for the chemical potentials difference):

$$n = n_\mathrm{A}^* - n_\mathrm{A} = n_\mathrm{B} - n_\mathrm{B}^* \quad \text{(H3.2)}$$

$$\mathfrak{I}_G = \mathfrak{I}_\mathrm{A} - \mathfrak{I}_\mathrm{B} \quad \text{(H3.3)}$$

$$\mu = \mu_\mathrm{B} - \mu_\mathrm{A} \quad \text{(H3.4)}$$

A slight change in the notation has been introduced with the "G" subscript for the dipole flow owing to the necessity (for comparison with other relaxation processes) to distinguish the flow involved in the exchange between the poles (which is a conductive process) from the flow resulting from evolution of the capacitive energy stored (subscript "C").

To facilitate comparisons with the other case studies in this chapter, it is also important to recall the way the kinetic constants, or specific rates, were defined as resulting from the composition of system constitutive properties at the level of each pole. Graph 11.19 reproduces the Formal Graph of one pole i that was given in case study A6 "Chemical Species" in Chapter 4.

The algebraic translations of the conductance and capacitance properties are

$$\mathfrak{I}_i = \hat{\mathrm{G}}_{ni}\,\mu_i \quad \text{(H3.5)}$$

$$n_i = \hat{\mathrm{C}}_{ni}\,\mu_i \quad \text{(H3.6)}$$

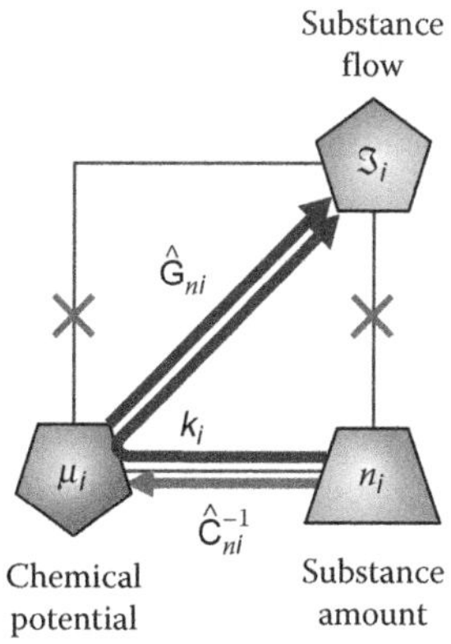

GRAPH 11.19 Formal Graph of the pole representing one of the reactants, A or B, of a chemical reaction.

and their composition gives the *specific rate* or *kinetic constant*

$$k_i \underset{homot}{\overset{def}{=}} \hat{G}_{ni}\,\hat{C}_{ni}^{-1} \tag{H3.7}$$

which is a scalar when the two constitutive properties are homothetic, that is, proportional between them,

$$\mathfrak{I}_i \underset{homot}{=} k_i n_i \tag{H3.8}$$

This is the standard case in the physical chemical energy variety, and the *specific rate* is a constant for a first-order reaction (i.e., independent from the substance amount and from time).

Turning one's attention now to the dipole modeling the reaction, the chemical process can be viewed as two simultaneous processes, a capacitive storage shared by the poles and an exchange by flows of substance amounts between the two poles.

Normally, one should express the conductive flow as a function of the chemical potentials, as for any conductive relationship, but in the physical chemical energy variety there are no direct means for measuring chemical potentials. On the other hand, substance amounts are measurable quantities. Consequently, models are built without chemical potentials, which can be deduced from the other state variables through the knowledge of the constitutive properties. The dipole capacitive relationship has been given in case study C5 "First-Order Chemical Reaction" and is recalled here.

$$\mu = \mu^{\theta} + \mu_{\Delta} \ln \frac{\overset{*}{n}_B + n}{\overset{*}{n}_A - n} \quad \text{with } \mu_{\Delta} = RT \tag{H3.9}$$

The pole substance amounts n_A and n_B have been replaced by their expressions in function of the dipole substance amount according to Equation H3.2. The *standard chemical potential* μ^{θ} is a characteristic of the dipole and is equal to the difference of the species standard chemical potentials.

As chemical potentials are not useful for modeling, relationships between flows and substance amounts are sought. The conductive flow, expressed by the exchange relationship H3.3 between flows, is rewritten by using the kinetic constants defined by Equation H3.7

$$\mathfrak{I}_G = k_A n_A - k_B n_B \tag{H3.10}$$

By replacing the species amounts using Equation H3.2, this flow is expressed as a function of the dipole substance amount

$$\mathfrak{I}_G = -(k_A + k_B)n + k_A \overset{*}{n}_A - k_B \overset{*}{n}_B \tag{H3.11}$$

One defines now an overall kinetic constant as the sum of the two individual kinetic constants

$$k \overset{def}{=} k_A + k_B \tag{H3.12}$$

and a dedicated variable is used for representing the initial value of the conductive flow

$$\overset{*}{\mathfrak{I}}_G \overset{def}{=} k_A \overset{*}{n}_A - k_B \overset{*}{n}_B \tag{H3.13}$$

in order to express the conductive flow in a simpler form:

$$\mathfrak{I}_G = -k\,n + \overset{*}{\mathfrak{I}}_G \tag{H3.14}$$

As already stated, the second process constitutive of the reaction is the capacitive storage. The dipole capacitive flow, that is, the flow supported by the evolution path preceded by the capacitive storage, is defined as the time derivative of the dipole substance amount

$$\mathfrak{I}_C \overset{def}{=} \frac{d}{dt} n \tag{H3.15}$$

To relate the conductive and the capacitive flows, one writes the classical kinetic equations expressing the evolution rates of the two substances as functions of the substance amounts

$$-\frac{d}{dt} n_A = \frac{d}{dt} n_B = k_A n_A - k_B n_B \tag{H3.16}$$

By using the previous relationships, the two flows appear to be identical

$$-\frac{d}{dt} n_A = \frac{d}{dt} n_B = \mathfrak{I}_G = -\frac{d}{dt}\left(n_A^* - n\right) = \frac{d}{dt}\left(n_B^* + n\right) = \frac{d}{dt} n = \mathfrak{I}_C \tag{H3.17}$$

which corresponds to the isolation of the system when the sum of these two flows is identified to the flow of the dipole assembly

$$\mathfrak{I} = \mathfrak{I}_C + \left(-\mathfrak{I}_G\right) \underset{isol}{=} 0 \tag{H3.18}$$

The isolation of the system is implicit in the classical formulation of the kinetic Equation H3.16, leading to the equality between capacitive and conductive flows. All the previous relationships are represented in the Formal Graph shown in Graph 11.20.

Despite the fact that the *overall kinetic constant* k is not explicit in this graph, this Formal Graph is a complete model of the reaction studied in this case study. All the dipole variables can be deduced from the pole variables through the connections and properties that are drawn. The overall kinetic constant k is not required because the conductive relationship between the dipole substance amount n and the conductive flow $\mathfrak{I}_G$ is composed with the pole–dipole connections and the individual kinetic constants of the poles. However, a more explicit derivation needs to be developed by algebra to demonstrate the validity of the graphic model and to describe the relaxation behavior.

11.3.3.1 Algebraic Model of Relaxation

From the two Equations H3.14 and H3.15 giving the conductive and capacitive flows, the algebraic model is directly deduced

$$\frac{d}{dt} n + k n = \mathfrak{I}_G^* \tag{H3.19}$$

This first-order differential equation has for solution the well-known exponential function

$$n = \frac{\mathfrak{I}_G^*}{k}\left(1 - \exp(-k t)\right) \tag{H3.20}$$

assuming that the dipole substance amount n is equal to zero initially. The expression of the conductive flow follows immediately:

$$\mathfrak{I}_G = \mathfrak{I}_G^* \exp(-k t) \tag{H3.21}$$

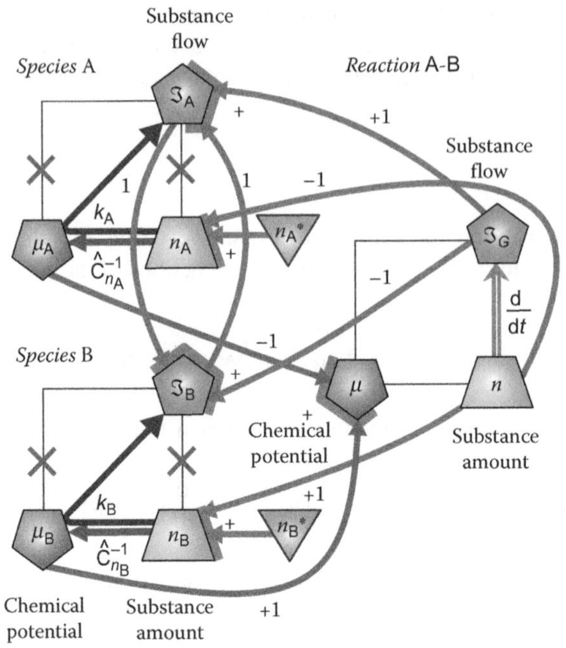

GRAPH 11.20 Formal Graph of the construction of the dipole assembly from the two poles representing the reactants A and B of the chemical reaction.

In the case study abstract are plotted the variations of the dipole substance amount and of the conductive flow as functions of time.

As can be observed in these curves, for infinitely long time, the equilibrium of the reaction is reached (i.e., dynamic equilibrium of the exchange), the conductive flow is equal to zero, and the substance amount reaches its equilibrium value, given by

Dynamic equilibrium: $$t \to \infty \quad \Leftrightarrow \quad \Im_G^{eq} = 0 \quad \Leftrightarrow \quad n^{eq} = \frac{\Im_G^*}{k} \qquad \text{(H3.22)}$$

This equilibrium value n^{eq} allows one to express the initial conductive flow as proportional to the shift of substance amount with respect to its equilibrium value, or as the result of a kinetic operator applied to the substance amount itself.

$$\Im_G = -k\,(n - n^{eq}) = -\hat{k}\,n \qquad \text{(H3.23)}$$

This operator is defined by the expression

$$\hat{k} = \frac{n - n^{eq}}{n}k \qquad \text{(H3.24)}$$

leading to a simpler form of the algebraic differential equation H3.19

$$\frac{d}{dt}n + \hat{k}\,n = 0 \qquad \text{(H3.25)}$$

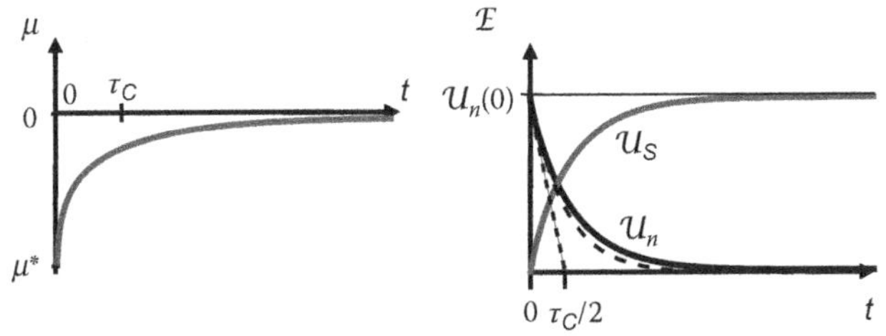

FIGURE 11.6 Plots of the variations of the dipole chemical potential (left) and of the physical chemical and thermal energies (right) as a function of time during a chemical reaction. The dashed curve in the energy plot is the exponential function.

Such an operator allows one to build a Formal Graph of the sole dipole (without its poles) as shown in the case study abstract.

11.3.3.2 Relaxation Regimes

The fact that chemical potentials are not experimentally handled, being neither directly measurable nor subject to control, prevents one from using impedance techniques to characterize chemical reactions occurring alone, that is, not participating in a mechanism with controllable steps. (For instance, in an electrochemical mechanism the electron transfer provides means for controlling the flow of chemical reactions in series.)

Consequently, it is useless to model the chemical reaction in terms of transfer functions (i.e., impedance or admittance) as will be done in the other case studies. The notion of forced regime is therefore not as applicable as it is for other systems in different energy varieties, in which both efforts and flows are experimentally accessible.

The variation of the chemical potential, given by Equations H3.9 and H3.20 and plotted in Figure 11.6 (left), shows clearly that the end of the reaction occurs when the condition of equilibrium is reached, that is, when the difference of chemical potentials between the two species becomes equal to zero. Initially this difference was negative, because the chemical potential of the starting species A was higher than that of the product B.

The variations of the physical chemical energy, decreasing with time, and of the thermal energy (heat), increasing with time, are plotted in Figure 11.6 (right). The dashed curve is drawn for comparison with a normal exponential decrease, which is not observed here due to the nonlinearity of the capacitive relationship H3.9. It appears that the decrease in physical chemical energy is slightly slower than for another energy variety in which the capacitive relationship is linear.

The production of heat due to the dissipation occurring during the course of the reaction is due to the conductive property of the mixed dipole representing the chemical reaction. Graph 11.21 represents the transfer of energy and simultaneous conversion from the physical chemical dipole to the thermal dipole. This scheme is identical to the one given in the previous case study H2 "Slowed Down Motion," except that the source of the dissipation was a mixed inductive–conductive dipole.

Remarks

Exothermic or endothermic? The example studied in this case study is a bit oversimplified in not detailing the connections between the two energy varieties. The reaction here is supposed to deliver the totality of its available energy to the physical chemical conductance and hence to the thermal dipole. This simple scheme, corresponding to an *exothermic* reaction, may not be always

(continued)

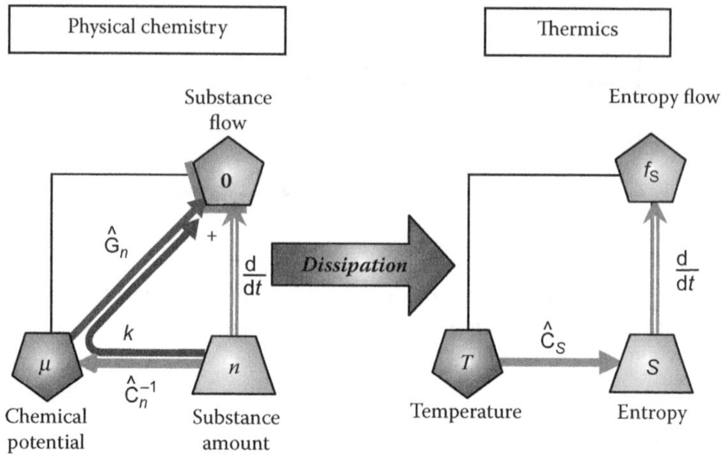

GRAPH 11.21 Illustration by two Formal Graphs of the transfer and conversion of physical chemical energy toward the thermal energy occurring by dissipation during a chemical reaction.

the case and a reverse exchange of energy may sometimes occur in addition, leading eventually to a decrease in thermal energy in the system. The reaction is said to be *endothermic* in that case. According to the Formal Graph viewpoint, this is modeled by a connection between all the pairs of state variables (i.e., including basic quantities) allowing a reversible exchange of part of the energy. However, this is a more complex process than the pure relaxation described here.

Dissipation made explicit. The fact that most of chemical reactions provoke a production of heat is well known, but the rationale explaining why and how it works is not easy to grasp with the conventional formalism. The Formal Graph language helps to visualize the reason of the dissipation in evidencing the nature of the notion of kinetic constant, which involves clearly a conductance, responsible for the conversion into heat.

11.3.4 Case Study H4: Parallel RC Circuit — Electrodynamics

The case study abstract is given on next page.

The electric circuit made up with a capacitor and a resistor in parallel is of great importance throughout physics because it models many relaxation phenomena and imperfections of energy storage. Leaking capacitors, viscoelastic behaviors, permeable barriers, or membranes, in fact all bad (nonideal) energy containers, are modeled by the association of these two components when using an equivalent electrical circuit.

This circuit is treated here in two situations: free relaxation using large signal models and forced regime using small amplitude AC signals.

The circuit, with the indication of the various currents and of the common potential, is drawn in Figure 11.7. The two components are modeled with operators, which is not the classical approach but which is justified by the frequent use in other domains of this model as an equivalent circuit for representing similar phenomena (for which linearity is not always the case).

H4: Parallel RC Circuit **Electrodynamics**

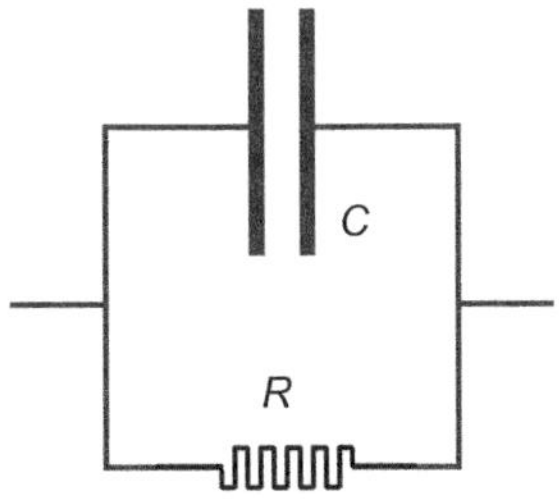

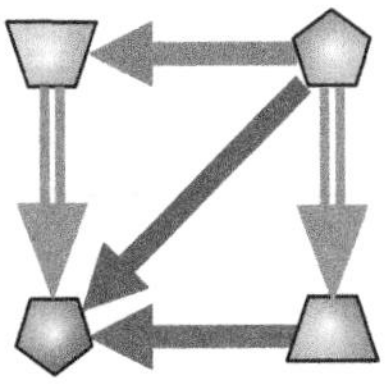

Dipole Assembly	
✓	Fundamental
	Mixed
✓	Capacitive
	Inductive
✓	Conductive
✓	Common effort
✓	Common flow
✓	Global
	Spatial

Formal Graph:

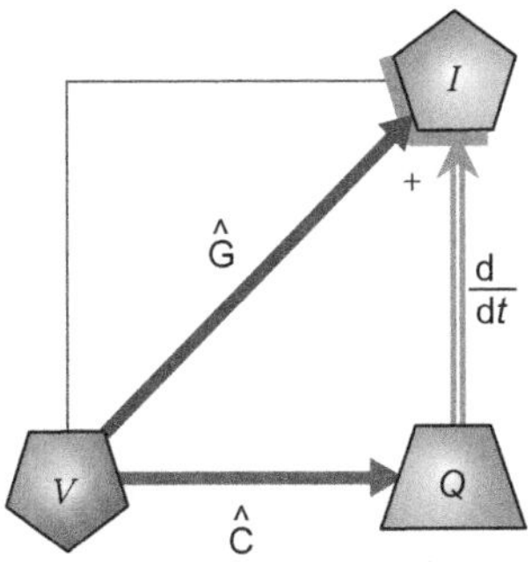

Variations with time:

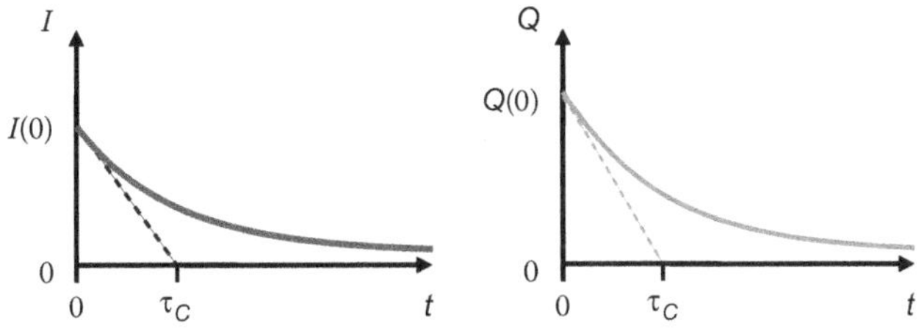

$$I_G \underset{isol}{=} I(0)\exp\left(-\frac{t}{\tau_C}\right); \quad Q \underset{lin}{=} Q(0)\exp\left(-\frac{t}{\tau_C}\right)$$

$$\tau_C \overset{def}{=} RC; \quad I(0) = \frac{Q(0)}{\tau_C}$$

GRAPH 11.22

Variables:

Variety	Electrodynamics		
Subvariety	Capacitive (electrostatic)		Inductive
Category	Entity number	Energy/entity	Entity number
Family	Basic quantity 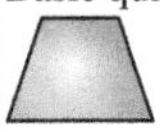	Effort 	Flow 
Name	Charge	Potential	Current
Symbols	Q	V, U, E, φ	$I, (i)$
Unit	[C], [A s]	[V], [J C^{-1}]	[A]

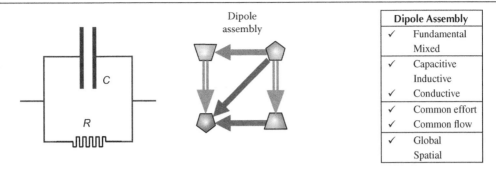

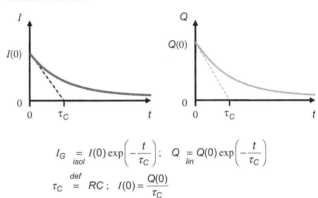

$$I_C = \frac{d}{dt}Q$$

$$I_G = \hat{G}\,V$$

$$Q = \hat{C}\,V$$

System Properties		
Nature	Capacitance	Conductance
Levels	0	0
Name	Electrical capacitance	Electrical conductance
Symbol	C	G
Unit	[F]	[S]
Specificity	—	—

Energy variation:

$$\delta\mathcal{U}_Q \underset{\exists\hat{C}}{=} V\,dQ$$

The three properties featuring the relaxation are two constitutive properties, capacitance and conductance, and one space–time property, evolution, as shown in the Formal Graph in the case study abstract and algebraically translated as follows:

$$Q = \hat{C}\,V \tag{H4.1}$$

$$I_G = \hat{G}\,V \tag{H4.2}$$

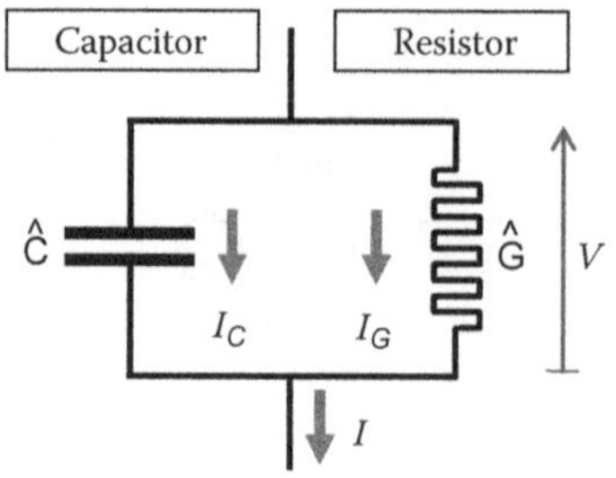

FIGURE 11.7 Electrical circuit of the parallel mounting of a capacitor and a resistor.

$$I_C = \frac{\mathrm{d}}{\mathrm{d}t} Q \tag{H4.3}$$

The parallel mounting of the two components corresponds to a common potential and to a total current given by the sum of individual currents

$$V = V_C = V_G \tag{H4.4}$$

$$I = I_C + I_G \tag{H4.5}$$

The algebraic equation modeling this circuit is written as a first-order differential equation

$$\frac{\mathrm{d}}{\mathrm{d}t} \hat{C}\, V + \hat{G}\, V = I \tag{H4.6}$$

By assuming homothetic constitutive properties, making the composition of operators a scalar variable

Homothetic properties: $$\hat{G}\,\hat{C}^{-1} \underset{homot}{=} k_C \quad (= scalar) \tag{H4.7}$$

one gets a simpler equation which is in fact identical to the classical model but which is generally established in the linear frame of both constitutive properties taken as scalars.

$$\frac{\mathrm{d}}{\mathrm{d}t} Q + k_C Q \underset{homot}{=} I \tag{H4.8}$$

The Formal Graph representing this model is shown in Graph 11.23 (right). This model is quite general (within the frame of homothetic properties); it can be used as well in free relaxation as in forced conditions.

11.3.4.1 Free Relaxation

One begins by discussing the free relaxation permitted by the isolation of the circuit, setting the current to zero, which corresponds to a modified model using one of the individual currents, the capacitive one for instance:

$$I_C = \frac{\mathrm{d}}{\mathrm{d}t} Q \underset{isol}{=} -k_C\, Q \tag{H4.9}$$

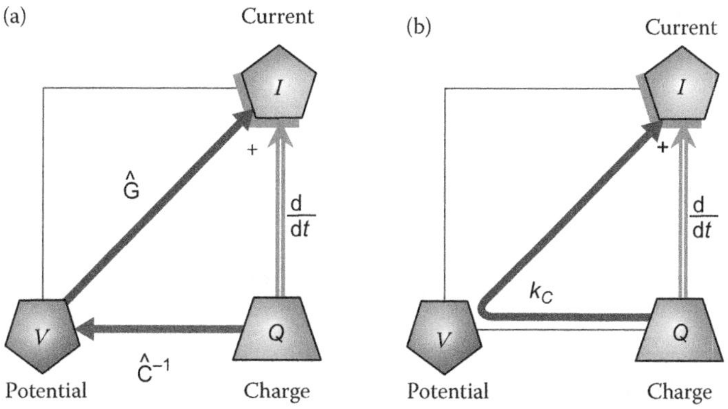

GRAPH 11.23 Formal Graph of the parallel RC circuit in the general case with elementary properties (a) and when the conductance is proportional to the capacitance (homothetic conditions), leading to the existence of a scalar kinetic constant representing their composition (b).

The Formal Graph of the free relaxation is given in Graph 11.23 by setting the total current equal to zero. The opposite of the kinetic constant appears as the eigen-value of the time derivation, which means that the eigen-function is the exponential function as demonstrated in Chapter 7 when establishing the influence theory. The solutions for the three state variables of this system are therefore:

$$Q \underset{isol}{=} Q(0)\exp(-k_C\, t) \tag{H4.10}$$

$$I_G \underset{isol}{=} -I_C \underset{isol}{=} I(0)\exp(-k_C\, t) \tag{H4.11}$$

$$V \underset{isol}{=} \hat{R}\, I(0)\exp(-k_C\, t) \tag{H4.12}$$

The relation between initial parameters is

$$I(0) = k_C\, Q(0) \tag{H4.13}$$

As the electrical power (received by the circuit) is equal to minus the product of the current times the potential, one has the opposite for the thermal power generated by dissipation:

$$\mathcal{P}_S \underset{isol}{=} I(0)\exp(-k_C\, t)\, \hat{R}\, I(0)\exp(-k_C\, t) \tag{H4.14}$$

In the linear and classical case, when both conductance and capacitance are scalars, the well-known capacitive relationship is retrieved

$$Q \underset{lin}{=} C\, V \tag{H4.15}$$

and it is common to use a (capacitive) *relaxation time* defined as the product of the resistance times the capacitance, equal in this frame to the reciprocal of the *kinetic constant*

$$\tau_C \overset{def}{=} RC \underset{lin}{=} \frac{1}{k_C} \tag{H4.16}$$

The expression of the potential variation as a function of time becomes

$$V \underset{lin}{=} RI(0)\exp\left(-\frac{t}{\tau_C}\right) \tag{H4.17}$$

and for the electrical capacitive energy one has

$$\mathcal{U}_Q \underset{lin}{=} \mathcal{U}_Q(0)\exp\left(-\frac{2t}{\tau_C}\right) \tag{H4.18}$$

The heat generated during dissipation is naturally the complement of this electrical capacitive energy, as illustrated in Figure 11.8. The plots of the current and charge variations are given in the case study abstract.

11.3.4.2 Forced Regime

From the general Equation H4.6, the relationship between the potential and the current is derived from the form of an admittance relationship

$$I = \left\{\frac{\mathrm{d}}{\mathrm{d}t}\hat{\mathrm{C}} + \hat{\mathrm{G}}\right\} V = \hat{\mathrm{Y}} V \tag{H4.19}$$

This is a useful method of modeling this system when large signals are imposed, and the expression between the braces is the transfer function that can be directly used for computing the response of the circuit. When a periodic signal is imposed, whatever its amplitude, this useful method goes through the Fourier transformation based on the angular frequency ω of the imposed signal

$$\hat{\mathrm{Y}} = \left\{\frac{\mathrm{d}}{\mathrm{d}t}\hat{\mathrm{C}} + \hat{\mathrm{G}}\right\} \xrightarrow{\text{Fourier } \omega} \tilde{\mathrm{Y}} = i\omega\,\tilde{\mathrm{C}} + \tilde{\mathrm{G}} \tag{H4.20}$$

This is also a general model valid for any shape of system constitutive property operators, but which is only useful when their Fourier transformation can be analytically expressed. It is interesting to proceed in a similar way to the free relaxation case by combining the Fourier transformations of the constitutive properties into a transformed relaxation time because it provides a scaling of the imaginary term

$$\tilde{\mathrm{Y}} = \left\{i\omega\,\tilde{\tau}_C\right\}\tilde{\mathrm{G}} + \tilde{\mathrm{G}}; \quad \tilde{\tau}_C = \tilde{\mathrm{G}}\tilde{\mathrm{G}}^{-1} \tag{H4.21}$$

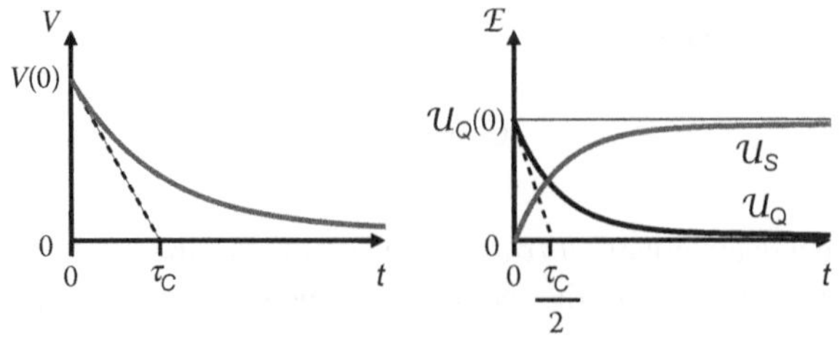

FIGURE 11.8 Plots of the potential variation (left) and of the electrical and thermal energies variations (right) during free relaxation of an RC circuit.

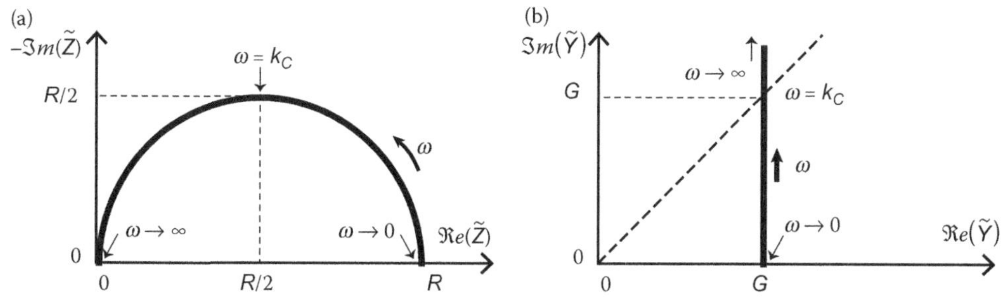

FIGURE 11.9 Cole–Cole (Nyquist) plots of the impedance (a) and of the admittance (b) of a parallel RC circuit in the linear case. The apex of the semicircle occurs when the angular frequency ω equates the kinetic constant k_C (=1/τ_C), which corresponds to the point intersecting the bisector in the admittance plot.

When the Fourier-transformed conductance can be put in factor (distributive property), the expression of the impedance can be written as

Distributive transformed conductance: $$\tilde{Z} = \frac{\tilde{R}}{1 + i\omega\,\tilde{\tau}_C} \qquad \text{(H4.22)}$$

To simplify the model, the classical assumption of scalar constitutive properties can be made, leading to the impedance defined in the linear frame and known as the Debye* model (or equation)

Debye's model: $$\tilde{Z} \underset{lin}{=} \frac{R}{1 + i\omega\,\tau_C} \qquad \text{(H4.23)}$$

The variations of the impedance and of the admittance as functions of the angular frequency are given in Figure 11.9 in the form of Cole†–Cole‡ (or Nyquist§) plots, that is, with the imaginary part (with a minus sign for the impedance) plotted against the real part. In these plots, reference is made to the *kinetic constant* k_C, which plays the role of scaling frequency, and not to its reciprocal, the *relaxation time.*

Unnecessary abstraction. A strong habit among users of AC techniques is to speak in terms of apparent capacitance represented by a complex number and defined by the relation

$$\tilde{Y} \underset{lin}{=} i\omega C + G \overset{def\to}{=} i\omega\,\overline{C} \qquad \text{(H4.24)}$$

This complex number is then decomposed into real and imaginary parts, but always with positive components in using a minus sign for the imaginary part (contrary to the usual mathematical decomposition)

$$\overline{C} = C' - iC'' \qquad \text{(H4.25)}$$

(*continued*)

* Petrus Josephus Wilhelmus Debye (1884–1966): American physicist; Maastricht (Netherland) and Ithaca, New York, USA.

† Kenneth Stewart Cole (1900–1984): American biophysicist; New York, USA.

‡ Robert H. Cole, brother of Kenneth Stewart Cole.

§ Harry Nyquist (1889–1976): Swedish American electronic engineer; Murray Hill, New Jersey, USA.

Consequently, the real part represents the true capacitance, normally independent of the angular frequency, whereas the imaginary part is frequency-dependent and is associated with the dissipation (or "dielectric loss" as often said in material sciences)

$$C' = C; \; C'' = \frac{G}{\omega} \tag{H4.26}$$

This usage is not recommended for two reasons:

First, the definition depends on the mounting of the two components. In a serial mounting the apparent capacitance has a different expression and this important detail is implicit (or often forgotten) when models are given (in fact, it must be understood that the model with a minus sign implies necessarily the parallelism of the two components).

Second, the physical meaning is more difficult to grasp and (together with the lack of physical justification of the imaginary number) contributes to rendering this domain more abstract than it is (especially when constant phase elements are introduced).

11.3.5 Case Study H5: Dielectric Relaxation — Electrodynamics

The case study abstract is given on next page.

Dielectric relaxation is a standard name for conductive–capacitive processes occurring in materials submitted to electric field and current under a dynamic regime. The two system constitutive properties involved are the *permittivity* ε, also called *dielectric constant*, which is the spatially reduced capacitance, and the *conductivity* σ, which is the spatially reduced conductance. The permittivity has been studied in Chapter 5 (dealing with space-distributed poles) in case studies B3 "Electric Space Charges," B4 "Poisson Equation," and B5 "Gauss Equation." It relates the *electric field* $\boldsymbol{E}$ to the *electric displacement* ("*electrization*") $\boldsymbol{D}$

$$\boldsymbol{D} \overset{def\rightarrow}{=} \hat{\varepsilon}\,\boldsymbol{E} \tag{H5.1}$$

The two localized variables are defined by their position in the Formal Graph:

$$\boldsymbol{E} \underset{(node)}{\overset{def}{=}} -\frac{\mathrm{d}}{\mathbf{dr}}V; \quad \boldsymbol{D} \underset{(node)}{\overset{def}{=}} \frac{\mathrm{d}}{\mathbf{dA}}Q \tag{H5.2}$$

The conductivity relates the electric field $\boldsymbol{E}$ to the *current density* $\boldsymbol{j}$

$$\boldsymbol{j} \overset{def\rightarrow}{=} \hat{\sigma}\,\boldsymbol{E} \tag{H5.3}$$

recalling that the current density is also defined by its position in the Formal Graph

$$\boldsymbol{j} \underset{(node)}{\overset{def}{=}} \frac{\mathrm{d}}{\mathbf{dA}}I \tag{H5.4}$$

The simplest model of relaxation using these reduced properties is shown in the Formal Graph in the case study abstract. Two projections from above in a two-dimensional horizontal plane are given in Graph 11.25, one with the elementary properties and one with the composed path formed by a combination of the reciprocal of the permittivity with the conductivity, assuming that the two properties are homothetic (i.e., proportional).

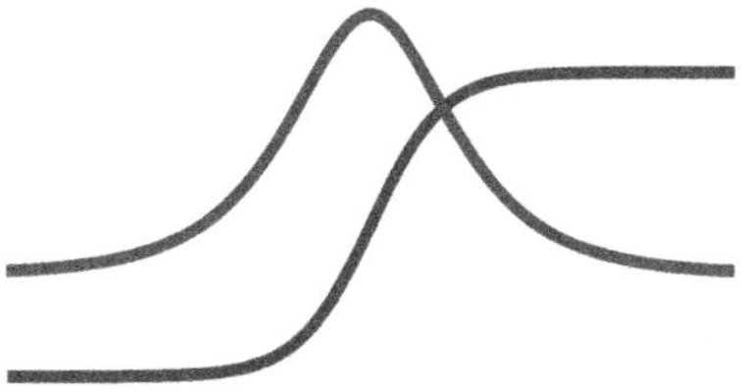

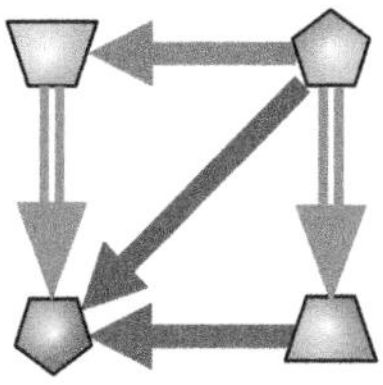

Dipole Assembly	
	Fundamental
✓	Mixed
✓	Capacitive
	Inductive
✓	Conductive
	Common effort
✓	Common flow
	Global
✓	Spatial

Formal Graph (see Graph 11.24):

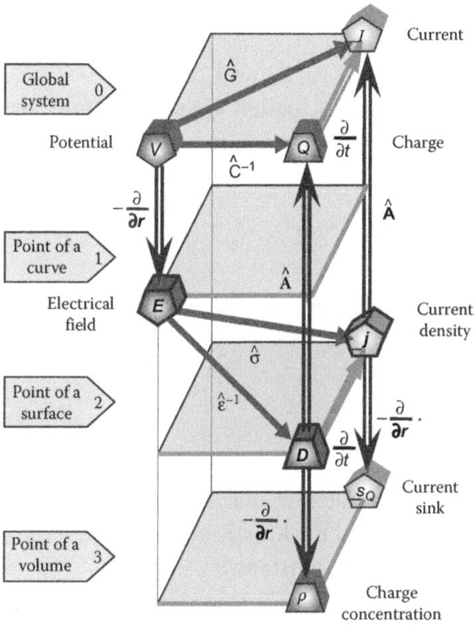

GRAPH 11.24

Impedancemetry:

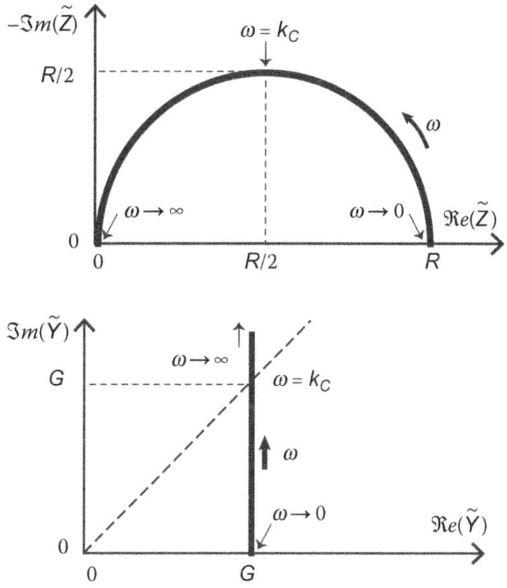

Cole–Cole plots of the impedance (top) and of the admittance (bottom) ($k_C = \sigma/\varepsilon$)

Variables:

Variety	**Electrodynamics**		
Subvariety	Capacitive (electrostatic)		Inductive
Category	Entity number	Energy/entity	Entity number
Family	Basic quantity	Effort	Flow
Name	Charge	Potential	Current
Symbols	Q	V, U, E, φ	$I, (i)$
Unit	[C], [A s]	[V], [J C^{-1}]	[A]

System Property		
Nature	Capacitance	Conductance
Levels	1/2	1/2
Name	Permittivity	Conductivity
Symbol	ε	σ
Unit	[F m^{-1}]	[S m^{-1}]
Specificity	3-d	3-d

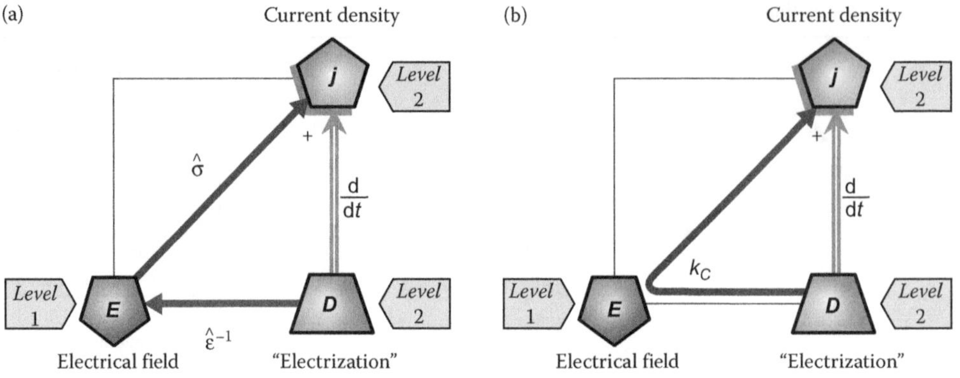

GRAPH 11.25 Two-dimensional projections from the three-dimensional Formal Graph in the case study abstract for case study H5, with decomposed kinetic path (a) and with composed kinetic path in the case of homothetic properties (b).

This proportionality makes their composition result in a scalar k_C, called *kinetic constant*, by analogy with chemical kinetics, where subscript C specifies its capacitive nature.

$$k_C \underset{homot}{=} \hat{\sigma}\,\hat{\varepsilon}^{-1} \tag{H5.5}$$

This variable is well adapted to the case of relaxation provoked by large signal perturbations, which is modeled by the first-order differential equation

$$\frac{\mathrm{d}}{\mathrm{d}t}\boldsymbol{D} + k_C\boldsymbol{D} \underset{homot}{=} \boldsymbol{j} \tag{H5.6}$$

Naturally, the localized variables are not measured directly but through the intermediate of global state variables, so the real model allowing the determination of the kinetic constant is (assuming that the charge can be dynamically measured, by spectroscopy for instance)

$$\frac{\mathrm{d}}{\mathrm{d}t}Q + k_C Q \underset{homot}{=} I \tag{H5.7}$$

However, the main method for studying dielectric relaxation is not based on large signals but on small amplitude AC signals in the frame of forced relaxation. The classical and basic approach makes three restrictions:

- Isotropy of space (invariance of system constitutive properties with space directions).
- Homogeneous distribution of variables through space (spatial operators replaced by scalars).
- Linearity of system constitutive properties (i.e., scalars owing to the isotropy).

Although the Formal Graph (and also some less basic classical) approach allows one to remove these restrictions, we adopt this simplified frame, in which the following relationships relate local variables to global ones:

$$R \underset{hom}{=} \frac{\ell}{\boldsymbol{A}}\frac{1}{\sigma};\; C \underset{hom}{=} \frac{\boldsymbol{A}}{\ell}\varepsilon \tag{H5.8}$$

The general relationships can be read in the Formal Graph in the case study abstract. From there, the development of the model is identical to what has been done with global state variables in case study H4 "Parallel RC Circuit."

One merely recalls the expression of the complex impedance (i.e., Fourier transformed) that was derived as the Debye model

Debye's model: $$\tilde{Z} \underset{lin}{=} \frac{R}{1+i\omega\,\tau_C}; \quad \tau_C = \frac{\varepsilon}{\sigma} \qquad \text{(H5.9)}$$

The relaxation time corresponds here to the inverse of the kinetic constant. The plots of this impedance and of the corresponding admittance were given in case study H4 "Parallel RC Circuit" (Figure 11.9) and are reproduced in the case study abstract.

However, in most materials, the ideal behavior that the Debye model depicts is rarely observed and some more sophisticated models are currently proposed for taking into account deviations from ideality. One of the widely used ingredients is the constant phase element behavior, explained in Chapter 10 in case study G6 "Constant Phase Element" which can be adapted to spatially reduced properties.

According to the Formal Graph viewpoint, the lost ideality is due to the capacitive storage, supported by the permittivity and distributed through space, which is not able to ensure the totality of the energy transfer and is therefore complemented by a parallel conductive process.

Section 10.6 in Chapter 10 exposed the Formal Graph coding of a complemented transfer in serial with other processes. The same principle of embedding the complement into a sum of contributions is used here, but on the flow node because one deals with a parallel mounting (i.e., common flow), as shown in Graph 11.26.

In the general frame of large amplitude signals and inhomogeneous, anisotropic, and nonlinear properties, the integral admittance is built from the Formal Graph according to the rules given in Chapter 10 for modeling a path complement

$$\hat{Y} = \hat{\mathbf{A}}\left(\hat{\sigma} + \hat{\sigma}_C^{\,1-p}\left\{\frac{d}{dt}\hat{\varepsilon}\right\}^p\right)\hat{\mathbf{R}}^{-1} \qquad \text{(H5.10)}$$

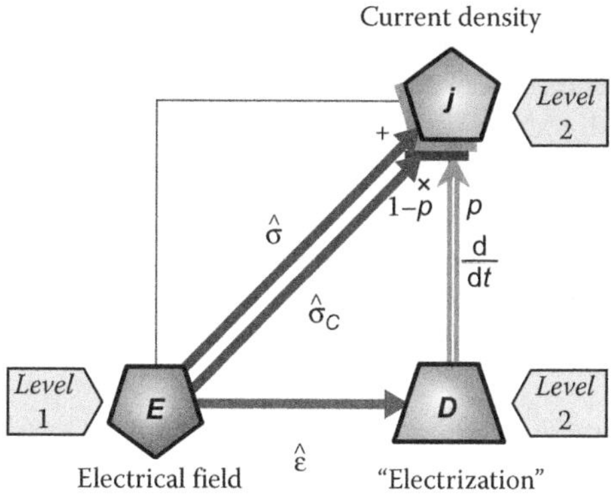

GRAPH 11.26 Formal Graph of a pure conduction in parallel with a complemented transfer between the capacitive path and another conductive path (two-dimensional projections from the three-dimensional Formal Graph in the case study abstract for case study H5).

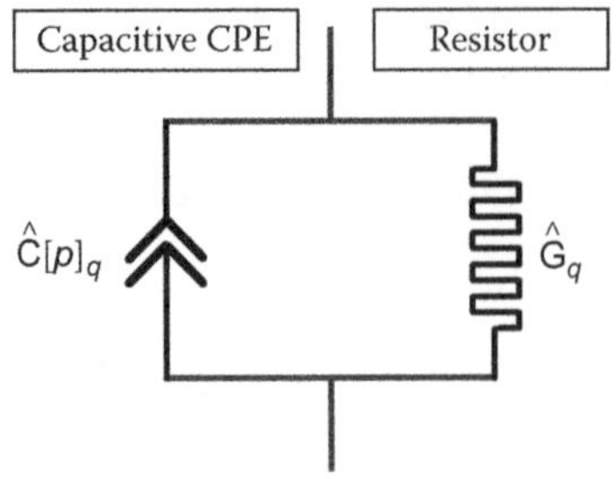

FIGURE 11.10 Electrical circuit of the parallel mounting of a capacitive common phase element and a resistor representing the generalized Cole–Cole model of nonideal dielectric relaxation.

In the restricted frame of small amplitude signals and homogeneous, isotropic, and linear properties, the complex admittance is deduced from the previous equation by Fourier transformation

$$\tilde{Y} \underset{hom}{\overset{lin}{=}} \frac{\mathbf{A}}{\ell}\left(\sigma + \sigma_C^{1-p} \varepsilon^p \left(i\omega\right)^p\right) \qquad \text{(H5.11)}$$

By rearranging this equation, the expression of the complex impedance is obtained

$$\text{Cole–Cole model:} \qquad \tilde{Z} \underset{lin}{=} \frac{R}{1+\left(i\omega\,\tau_C\right)^p}; \quad \tau_C = \frac{\varepsilon}{\bar{\sigma}}; \quad \bar{\sigma} = \sigma^{\frac{1}{p}}\,\sigma_C^{1-\frac{1}{p}} \qquad \text{(H5.12)}$$

having defined differently from the ideal case (Debye) the relaxation time with the help of an apparent conductivity combining the two elementary conductivities. This equation is well known as the Cole–Cole model and is introduced as an empirical model (Jonscher 1996) (as for the common phase element component). Figure 11.10 translates in the language of equivalent circuits the Formal Graph given in Graph 11.26.

11.3.6 Case Study H6: Viscoelastic Relaxation — Translational Mechanics

The case study abstract is given on next page.

Viscoelastic properties of materials are mechanical properties describing the behavior of a viscous system submitted to stress and deformation under a dynamic regime. Contrary to pure elastic behavior that can be observed in steady state, viscoelastic behavior is dependent on time, and more precisely constitutes a relaxation phenomenon. As indicated by the name, both capacitive energy storage and dissipation are constituents of relaxation processes in these materials. A profound similarity exists with dielectric relaxation phenomena seen in case study H5. Electrical permittivity and mechanical elasticity are both reduced constitutive properties supporting capacitive energy storage on the one hand, and conductivity and viscosity are the reduced properties provoking dissipation through a conduction process on the other hand.

Under static conditions, the elastic properties of solids have been studied in case study B2 "Elasticity of Solids" in Chapter 5, in which the capacitive relationship was modeled by two reduced elastances (elasticities), depending on the mode of deformation, by traction or by shear

$$\text{Traction}: F_{/A} = \hat{\mathrm{E}}_Y\,\varepsilon_\ell; \quad \text{Shear}: F_{/A} = \hat{\mathrm{G}}_\gamma\,\varepsilon_\ell \qquad \text{(H6.1)}$$

These two elasticities are *Young's** *modulus*, notated E_Y (traction) and *Shear modulus*, notated G_γ (shear).

* Thomas Young (1773–1829): English scientist; Cambridge and London, UK.

H6: Viscoelastic Relaxation

Translational Mechanics

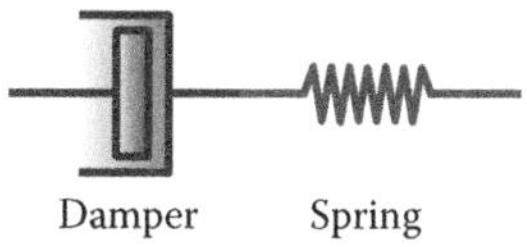

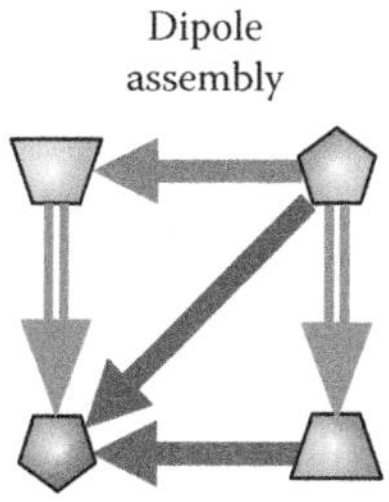

Dipole Assembly	
	Fundamental
✓	Mixed
✓	Capacitive
	Inductive
✓	Conductive
	Common effort
✓	Common flow
	Global
✓	Spatial

Formal Graph:

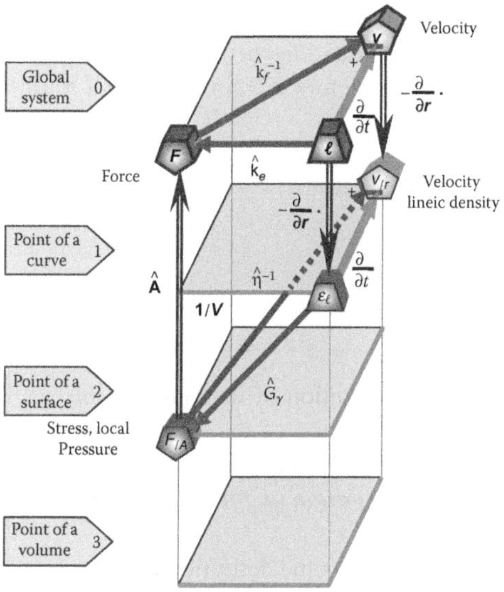

GRAPH 11.27

Impedancemetry:

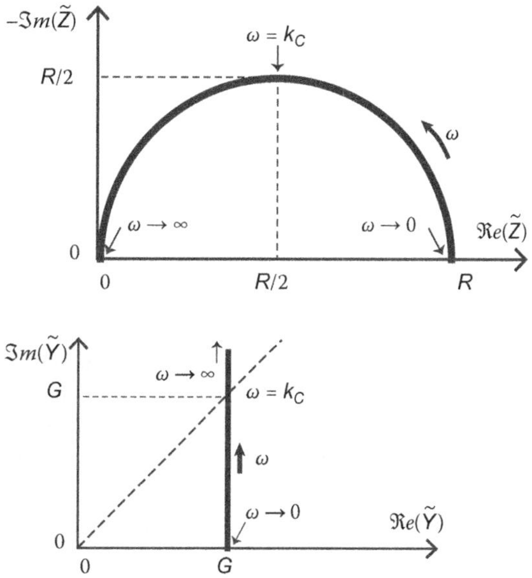

Cole–Cole plots of the impedance (top) and of the admittance (bottom) ($k_C = G_\gamma/\eta$).

Variables:

Variety	Translational Mechanics		
Subvariety	Capacitive (electrostatic)		Inductive
Category	Entity number	Energy/entity	Entity number
Family	Basic quantity 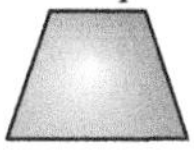	Effort 	Flow 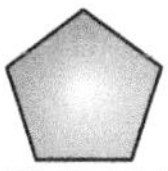
Name	Displacement	Force	Velocity
Symbols	ℓ	$\boldsymbol{F}$	$\boldsymbol{v}$
Unit	[m]	[N], [J m^{-1}]	[m s^{-1}]

System Property		
Nature	Capacitance	Conductance
Levels	1/2	1/2
Name	Shear modulus	Dynamic viscosity
Symbol	G_γ	η
Unit	[N m^{-2}]	[N s m^{-2}]
Specificity	3-d	3-d

The two localized variables in these relationships are the *relative elongation* (or *strain*), notated ε_ℓ or γ (but this symbol is preferred for superficial tension) also defined according to the deformation mode (cf. case study B2 "Elasticity of Solids"),

$$\text{Relative elongation:} \qquad \text{Traction}: \varepsilon_\ell \overset{def}{\underset{(node)}{=}} -\frac{\mathbf{d}}{\mathbf{d}\boldsymbol{r}}\cdot \ell \qquad \text{Shear}: \varepsilon_\ell = \tan\gamma \tag{H6.2}$$

and the *stress*, or *local pressure*, $F_{/A}$, classically notated σ (but confusion with all conductivities leads to the adoption of the universal subscript for a surface density $_{/A}$), defined by

$$\text{Stress (local pressure):} \qquad F_{/A} \overset{def}{\underset{(node)}{=}} \frac{\mathbf{d}}{\mathbf{d}\boldsymbol{A}}\boldsymbol{F} \tag{H6.3}$$

By examining the Formal Graphs in the case study abstract, it can be observed that the "gradient" of velocity (also called *shear rate*) is nothing other than the *lineic density of velocity* along a line of flow in the fluid. It is defined as the convergence of velocity vectors at a point, which is mathematically expressed as the negative of the operator of divergence.

$$\text{Velocity lineic density:} \qquad v_{/r} \overset{def}{\underset{(node)}{=}} -\frac{\mathbf{d}}{\mathbf{d}\boldsymbol{r}}\cdot \boldsymbol{v} \tag{H6.4}$$

Unsuitable notation and naming. A classical notation of the *velocity lineic density* is the symbol of the *relative elongation* ε (or of the *strain* γ) overhead by a point: $\dot{\varepsilon}$ or $\dot{\gamma}$, meaning that this variable is the time derivative of ε (or of γ). This is in accordance with the term *elongation* (*deformation*) *rate* (or *shear rate*) also widely in use.

The Formal Graph theory forbids such a practice that contradicts the definition of flows as *energies-per-entity* without needing any time concept for that. The concept of velocity must exist, as for a current in electrodynamics, in steady state. (see case studies A1 "Moving Body" and A3 "Current Loop" in Chapter 4).

The second system constitutive property involved in viscoelastic relaxation is the dynamic viscosity defined as the operator linking the lineic density of velocity to the stress

$$F_{/A} \overset{def\rightarrow}{=} \hat{\eta}\, v_{/r} = \hat{\eta}\left\{-\frac{\mathrm{d}}{\mathrm{d}\boldsymbol{r}}\right\}\cdot \boldsymbol{v} \tag{H6.5}$$

This relationship is a generalization of Newton's law of viscosity already studied in case study G4 in Chapter 10. The dynamic viscosity is the spatially reduced form of the friction coefficient k_f defined as a mechanical resistance at the global level. It consequently dissipates energy in the same manner as conductivity.

This mechanical property illustrates the ambivalence of a conductance or a resistance which are system properties supporting conduction either of entities of inductive energy or of capacitive energy. If the conduction or transport of momenta can be relatively well understood, as stated for the *kinematic viscosity* ν discussed in case study G4 (by comparison with the diffusion of molecules), this is less obvious for the "conduction of lengths," which are the entities in the capacitive subvariety in translational mechanics (static mechanics). Notwithstanding this conceptual difficulty, one is forced to envisage this aspect of the viscous process (in absence of inductive energy) brought by the comparison with all other energy varieties.

Rheological concepts. In rheology, it is customary to use, instead of the local pressure, the notion of *shear stress S* or σ_ℓ applied to one side of a layer or plate defined as the ratio of a tangential force on the surface of contact. Also, the velocity "gradient" is replaced by the notion of *shear rate* defined as the time derivative of the relative elongation $\ell_{/r}$ (classically notated ε or γ). The viscosity is then defined "in the steady state" by the relation

Shear stress:
$$\sigma_\ell = \frac{F}{A} = -\eta \frac{d}{dt}\ell_{/r} \qquad \text{(H6.6)}$$

The Formal Graph theory cannot follow this approach which contradicts itself in using time derivatives in steady state (even if time is carefully hidden under symbols such as $\dot{\varepsilon}$ or $\dot{\gamma}$). This is because of the tight link in this science between flow and time, in which it is customary to distinguish between solids and fluids according to the time scale of relaxations or deformations (Phan-Tien 2002).

The Formal Graph in the case study abstract represents the simplest model of relaxation using these two reduced properties. Two projections from above in a two-dimensional horizontal plane are given in Graph 11.28, the first one on the left with the elementary properties and the second one

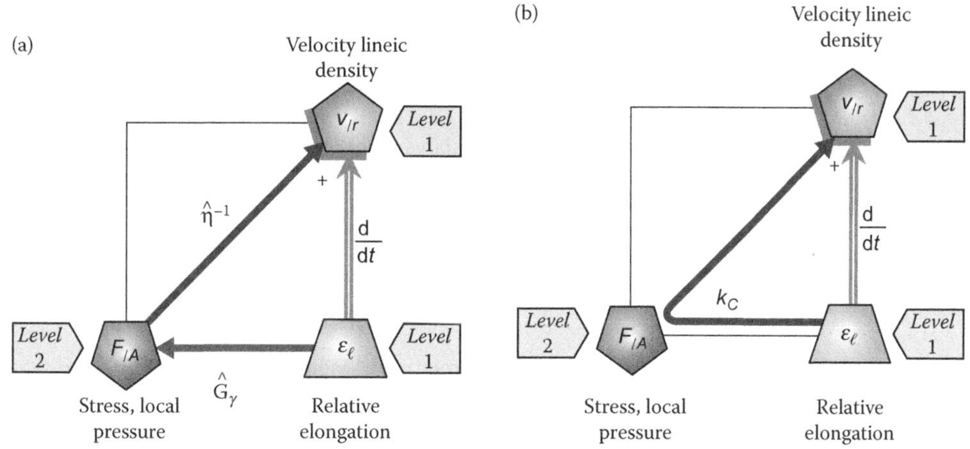

GRAPH 11.28 Two-dimensional projections from the three-dimensional Formal Graph in the case study abstract for case study H6, with decomposed kinetic path (a) and with composed kinetic path (b). This latter is featured by a scalar when the composed operators are proportional (homothetic system constitutive properties). The Formal Graphs are built with two dipoles assembled by sharing a common flow (parallel RC mounting in the electric circuit language).

on the right with the composed path formed by composition of the elasticity (here the shear modulus) with the reciprocal of the dynamic viscosity.

In the Formal Graph in Graph 11.28 (left), the result of the property composition is a scalar because the two properties are assumed to be homothetic (i.e., proportional), which is a quite general case.

$$k_C \underset{homot}{=} \hat{\eta}^{-1} \hat{G}_\gamma \quad \text{(H6.7)}$$

This result is a *kinetic constant* k_c by analogy with a chemical reaction (see case study H3 "Occurring Reaction"). Its unit is s^{-1}. From the Formal Graph is deduced the kinetic equation modeling the relaxation process

$$\frac{d}{dt}\varepsilon_\ell + k_C\,\varepsilon_\ell \underset{homot}{=} v_{lr} \quad \text{(H6.8)}$$

which can be rewritten as operation on the relative elongation

$$v_{lr} \underset{homot}{=} \left\{\frac{d}{dt} + k_C\right\}\varepsilon_\ell \quad \text{(H6.9)}$$

In case of free relaxation, the lineic density of the velocity is equal to zero (owing to the system isolation), so the variation of the relative elongation as a function of time is given by an exponential function

$$\varepsilon_\ell \overset{isol}{\underset{homot}{=}} \varepsilon_\ell(0)\exp(-k_C\,t) \quad \text{(H6.10)}$$

This is the standard relaxation model for any kind of signal amplitude. The small signal amplitude case is treated through the Fourier transformation. Due to the scalar nature of the kinetic constant, the Fourier transformation of Equation H6.9 is very simple

$$\tilde{v}_{lr} \underset{homot}{=} (k_C + i\omega)\,\tilde{\varepsilon}_\ell \quad \text{(H6.11)}$$

To put this relationship under the form of an impedance, one needs to link with the stress, which can be done easily by assuming the linearity of the elasticity (around a working point) and the isotropy of space, so its Fourier transformation being a scalar, one has

$$\tilde{F}_{lA} \underset{lin}{=} G_\gamma\,\tilde{\varepsilon}_\ell \quad \text{(H6.12)}$$

By replacing the Fourier transformation of the relative elongation with the help of Equation H6.11, one gets the reduced complex impedance relationship

$$\tilde{F}_{lA} \underset{lin}{=} \frac{\eta}{1+i\omega\tau_C}\tilde{v}_{lr};\quad \tau_C = \frac{\eta}{G_\gamma} \quad \text{(H6.13)}$$

Here, in this ideal material, that is, modeled by an assembly of two dipoles in the same energy variety (without distributed dipoles for instance), the relaxation time is the inverse of the kinetic constant. The reduced complex admittance relationship is deduced from the previous relationship and can be put in a compacted form

$$\tilde{v}_{/r} \underset{lin}{=} \left(\frac{1}{\eta} + i\frac{\omega}{G_\gamma} \right) \tilde{F}_{/A} = i\frac{\omega}{\overline{G}_\gamma}\tilde{F}_{/A} \quad \text{(H6.14)}$$

The principle of the equivalence is also shown in Graph 11.29 with two Formal Graphs that are the Fourier transformations of the normal ones (cf. Chapter 9, Section 9.5.8 for explanations).

The apparent complex elasticity modulus can obviously be decomposed into components according to

$$\overline{G}_\gamma = G'_\gamma + iG''_\gamma \quad \text{(H6.15)}$$

and the expressions of these components are

$$G'_\gamma \underset{lin}{=} G_\gamma \frac{\omega^2 \tau_C^2}{1+\omega^2 \tau_C^2} \quad \text{(H6.16)}$$

$$G''_\gamma \underset{lin}{=} G_\gamma \frac{\omega \tau_C}{1+\omega^2 \tau_C^2} \quad \text{(H6.17)}$$

A plot in the Cole–Cole (Nyquist) representation is given in Figure 11.11.

As for the dielectric relaxation, the localized variables and reduced properties are not directly accessible by measurement contrary to state variables at the global level. To establish a link between them, a space–time model must be provided. The simplest, and classically chosen, one is the isotropic and homogeneous space and material, in which spatial operators are replaced by scalars or vectors. The correspondences between variables are

Homogeneous material: $$v_{/r} \underset{hom}{=} \frac{\mathbf{v}}{\ell}; \quad F_{/A} \underset{hom}{=} \frac{\mathbf{F}}{\mathbf{A}} \quad \text{(H6.18)}$$

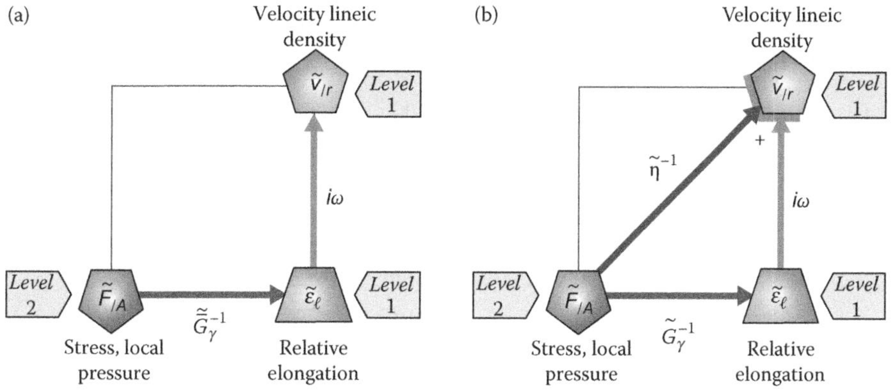

GRAPH 11.29 Two Fourier-transformed Formal Graphs depicting the equivalence between an apparent elasticity modulus (a) and the sum of individual paths modeling the relaxation (b). (In the linear case, the Fourier transformations of the system constitutive properties are replaced by scalars.)

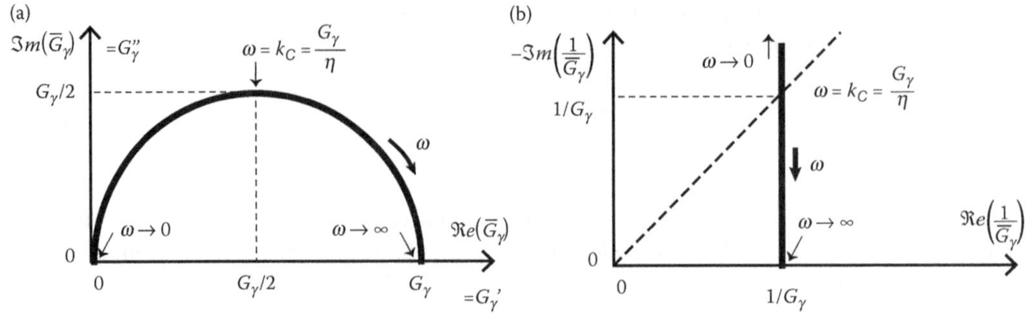

FIGURE 11.11 Cole–Cole (Nyquist) plot of the complex apparent elasticity modulus (admittance, a) and of its inverse (impedance, b). The apex of the semicircle occurs when the angular frequency ω equates the kinetic constant k_C (=$1/\tau_C$), which corresponds to the point intersecting the bisector in the impedance plot.

and between system constitutive properties

$$k_f \underset{hom}{=} \frac{\ell}{\mathbf{A}}\eta; \quad k_e \underset{hom}{=} \frac{\mathbf{A}}{\ell}G_\gamma \tag{H6.19}$$

The expression of the global complex impedance is derived from the reduced one (Equation H6.13), taking into account these correspondences

Debye's model: $$\tilde{Z} \underset{lin}{=} \frac{k_f}{1+i\omega\,\tau_C}; \quad \tau_C = \frac{\eta}{G_\gamma} = \frac{k_f}{k_e} \tag{H6.20}$$

The Cole–Cole (Nyquist) plots of the impedance and of the admittance are shown in the case study abstract. They are strictly identical to the ones in the dielectric relaxation case study (shapes and axes) and the shapes (only) are identical to the previous plots of the complex apparent elasticity modulus.

The paradox of Maxwell's model. A popular representation of models in rheology mimics the equivalent electrical circuits with dipolar components. The elastic component is naturally symbolized by a spring and the viscous component by a damper or dashpot (a piston filled with a viscous fluid able to circulate). The viscoelastic relaxation is thus represented with these two components mounted in series, as shown in Figure 11.12a and is known as Maxwell's model (Oswald 2005). (In this representation, the customary notation is used for facilitating comparison with the literature.)

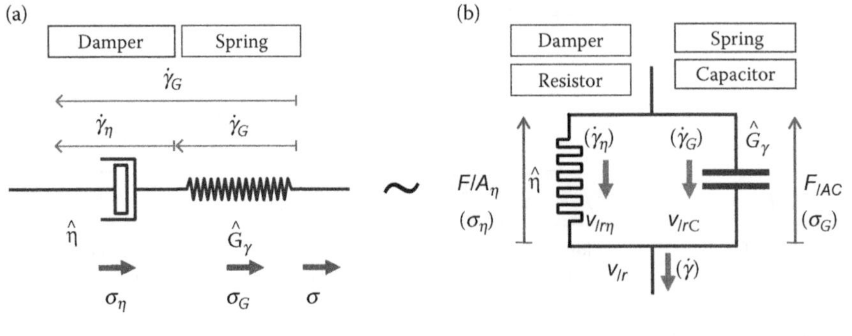

FIGURE 11.12 Maxwell's rheological model (a) and equivalent circuit model (b).

(continued)

The equivalent electric circuit modeling the same phenomenon of viscoelastic relaxation is also shown in Figure 11.12b. Their algebraic translation is identical:

Shear rates: $\dot{\gamma}_\eta + \dot{\gamma}_G = \dot{\gamma}; \quad V_{/r\eta} + V_{/rC} = V_{/r}$
Stresses: $\sigma_\eta = \sigma_G = \sigma; \quad F_{/A\eta} = F_{/AC} = F_{/A}$

The striking difference lies in the mounting, in series for the Maxwell model and in parallel for the equivalent electric circuit. However, this is purely a question of convention. In electricity, and therefore in an equivalent circuit, the current flows through the components and the potential difference develops between their ends. In rheology, it is the opposite, stress circulates through the components and the shear rate develops across them.

The Formal Graph representation is unique, like the algebraic one, but the difference is that the question of the mounting is irrelevant in algebra as long as no physical meaning in terms of across/through variables is attributed. With a Formal Graph, it matters but not exactly in the same way: the key concept is not the mounting, parallel or serial, but which gate variable is common. In viscoelastic relaxation, there is no ambiguity; the effort (stress) is common to both the conduction and the storage.

This paradox highlights the hidden bias of circuit representations with dipolar devices. They need to be accompanied by the convention chosen for stating which is the across variable and which one is the through variable.

One is so impregnated by the scheme flow or current = through variable, effort or potential = across variable, that difficulties arise in thinking otherwise. Difficulties have been met in Chapter 6 (about dipoles) for explaining how an exchange by effort could work.

In conclusion, it is preferable (although one uses them for pedagogical reasons owing to their widespread usage), not to use circuits, equivalent or not, for modeling purposes.

11.3.7 Case Study H7: Damped Mechanical Oscillator — Translational Mechanics

The case study abstract is given on next page.

The system depicted in this case study consists of a mobile mass M attached to a spring and laid on a horizontal support (for avoiding influence of gravity). The contact between the mass and the support is not ideal and creates friction when the mass moves, represented by an operator $\hat{k}_f$. The spring is featured by an elastance (spring "constant") $\hat{k}_e$ and has its other end fixed to the support.

Three constitutive properties work in this system, each one modeled by a dipole and the three dipoles are associated in a dipole assembly. The key factor deciding the mounting is the fact that the velocity is common to all dipoles (the physical explanation is that the mobile end of the spring and the mass are moving at the same speed).

The Formal Graph in the case study abstract provides the model of this system in the general case of nonlinear operators, which is a frequent case, at least for the friction and the elastance operators. The velocity is common to all dipoles and the three forces corresponding to the three constitutive properties are summed onto a null node because the system is isolated from any external influence.

Unsuitable notion of equilibrium. In classical mechanics, a null sum of forces is viewed as the expression of an "equilibrium law" (see case study E2 "Concurring Forces" in Chapter 8). This is a notion peculiar to mechanics and not applicable to other energy varieties, in which the notion of conservation is rather in use. In the Formal Graph theory it is considered as a simple conservation rule of efforts resulting from a common flow and in the language of equivalent circuits as one of "Kirchhoff's laws" expressing a serial mounting.

H7: Damped Mechanical Oscillator **Translational Mechanics**

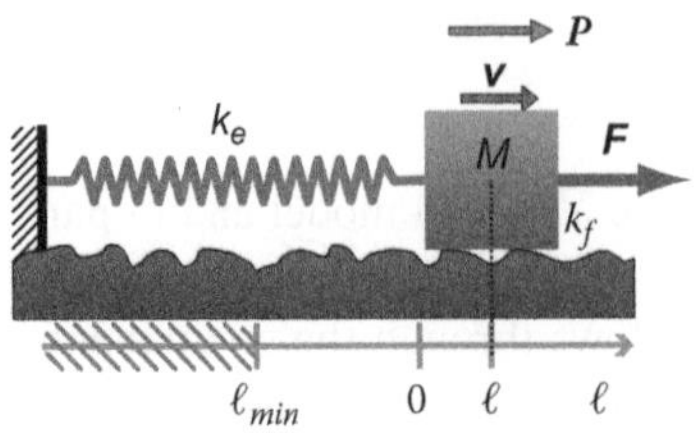

Dipole assembly

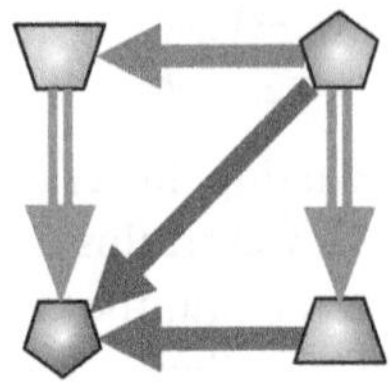

Dipole Assembly	
✓	Fundamental
	Mixed
✓	Capacitive
✓	Inductive
✓	Conductive
	Common effort
✓	Common flow
✓	Global
	Spatial

Formal Graph (see Graph 11.30):

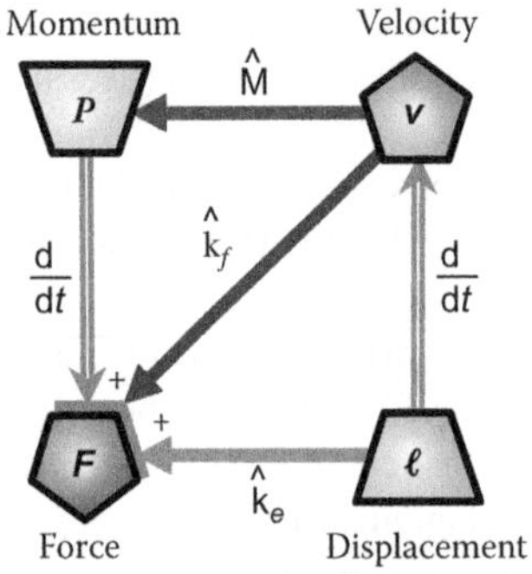

Variations with time:

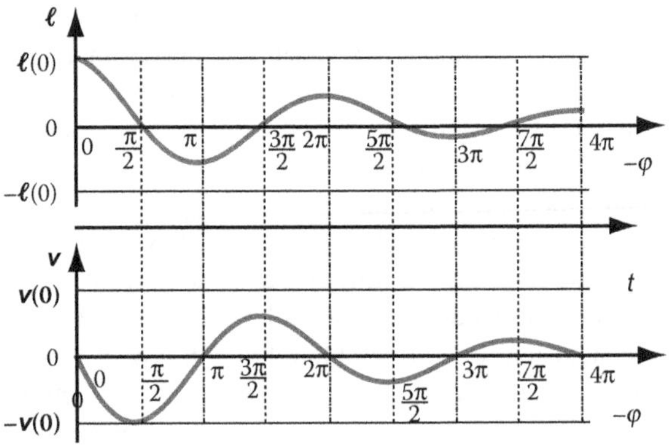

GRAPH 11.30

Isolated: $\boldsymbol{F} = \boldsymbol{F}_L + \boldsymbol{F}_G + \boldsymbol{F}_C \underset{isol}{=} 0$

$$\ell \underset{lin}{=} \ell(0)\exp\left(-\gamma_{\ell L}\, t\right)\cos\omega t,\quad \omega \underset{lin}{=} \sqrt{\frac{k_e}{M} - \gamma_{\ell L}^2};\quad \gamma_{\ell L} \underset{lin}{=} \frac{k_f}{2M}$$

$$\boldsymbol{v} = \boldsymbol{v}_L = \boldsymbol{v}_G = \boldsymbol{v}_C \underset{lin}{=} -\omega\, \ell(0)\exp\left(-\gamma_{\ell L}\, t\right)\sin\omega t;\quad \varphi = -\omega\, t$$

Variables:

Variety	Translational Mechanics			
Subvariety	Inductive (kinetic)		Capacitive (potential)	
Category	Entity number	Energy/entity	Entity number	Energy/entity
Family	Impulse	Flow	Basic quantity	Effort
Name	Momentum	Velocity	Displacement	Force
Symbols	***P***	***v***	$\boldsymbol{\ell}$	***F***
Unit	[N s]	[m s^{-1}]	[m]	[N]

System Properties			
Nature	Inductance	Elastance	Resistance
Levels	0	0	0
Name	Inertial mass	Elasticity constant	Friction resistance
Symbol	M	k_e	k_f
Unit	[kg]	[N m^{-1}]	[N s m^{-1}]
Specificity	–	–	–

The algebraic translation of the Formal Graph in the case study abstract is as follows:

$$\boldsymbol{F}_L \overset{def}{=} \frac{\mathrm{d}}{\mathrm{d}t}\boldsymbol{P} = \frac{\mathrm{d}}{\mathrm{d}t}\hat{M}\,\boldsymbol{v} \tag{H7.1}$$

$$F_G \overset{def}{=} \hat{k}_f \boldsymbol{v} \tag{H7.2}$$

$$F_C \overset{def}{=} \hat{k}_e \ell = \hat{k}_e \frac{d^{-1}}{dt^{-1}} \boldsymbol{v} \tag{H7.3}$$

$$\boldsymbol{F} = \boldsymbol{F}_L + \boldsymbol{F}_G + \boldsymbol{F}_C \underset{isol}{=} 0 \tag{H7.4}$$

Both the Formal Graph and the six algebraic equations can be supplied to a neural network or a digital solver to obtain the solution in terms of variation with time of the state variables. This is the most practical way in case of nonlinear operators.

However, a simpler graph can be proposed which helps to analyze the behavior of the oscillator. In a conservative oscillator, without dissipation, it is easy to set the directions of all paths in the same orientation, as done in Section 9.5 (devoted to the temporal oscillator) in Chapter 9, for evidencing the circularity and then to build a wave function for the oscillator.

This simple choice of orientations is not possible when all three constitutive properties exist and a more subtle transformation of the graph must be made. The full demonstration will be made in Section 11.5 for any energy variety, so just the essential steps are given here.

The inductance (inertial) and conductance (friction) are combined in an *inductive damping operator*, with a coefficient one-half (because each half will be shared by the two evolution paths for forming a new evolution operator)

Inductive damping operator: $$\hat{\gamma}_{\ell_L} \overset{def}{=} \frac{1}{2} \hat{k}_f \hat{M}^{-1} \tag{H7.5}$$

A new evolution operator is defined by adding this *inductive damping operator*, which amounts to define an apparent time $\bar{t}$

Apparent evolution operator: $$\bar{T}^{-1} = \frac{d}{d\bar{t}} \overset{def}{=} \left\{ \frac{d}{dt} + \hat{\gamma}_{\ell L} \right\} \tag{H7.6}$$

This new operator allows one to draw a simpler Formal Graph without the diagonal conductance, which is hidden because it is incorporated into the new paths, as shown in Graph 11.31.

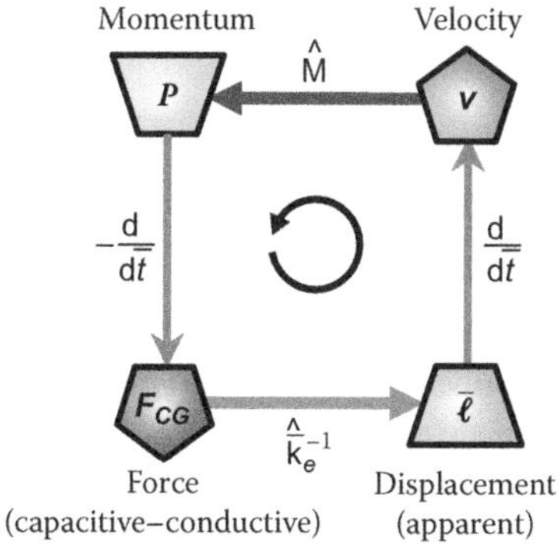

GRAPH 11.31 This circular Formal Graph models a damped temporal oscillator by using an apparent time, an apparent capacitance, and an apparent displacement.

However, to draw this graph, an apparent displacement and a partial force need to be defined with the help of this new evolution operator

$$\bar{\ell} \overset{def}{=} \frac{\mathrm{d}^{-1}}{\bar{\mathrm{d}}t^{-1}} \mathbf{v}; \quad -\boldsymbol{F}_{CG} = \boldsymbol{F}_{LG} \overset{def}{=} \frac{\mathrm{d}^{-1}}{\bar{\mathrm{d}}t^{-1}} \boldsymbol{P} \tag{H7.7}$$

and to relate the capacitive–conductive partial force $\boldsymbol{F}_{CG}$ to the apparent displacement, an apparent capacitance operator needs to be created

$$\bar{\ell} \overset{def\rightarrow}{=} \hat{\mathrm{k}}_e^{-1} \boldsymbol{F}_{CG} \tag{H7.8}$$

Graph 11.31 is now amenable to analysis in terms of circularity, as was done for all the conservative oscillators treated in Chapter 9. The algebraic model deduced from this Formal Graph is particularly simple

$$-\frac{\mathrm{d}}{\bar{\mathrm{d}}t} \hat{\mathrm{M}} \frac{\mathrm{d}}{\bar{\mathrm{d}}t} \hat{\mathrm{k}}_e^{-1} \boldsymbol{F}_{CG} = \boldsymbol{F}_{CG} \tag{H7.9}$$

but solving it is not always easy in all cases. Section 11.5 discusses in detail a peculiar solution relying on the restrictive assumptions of commutative operators and homothetic constitutive properties, but for the moment the discussion is specialized to the simplest case of scalar constitutive properties for the system.

This still more limited case corresponds to the classical treatment of the harmonic oscillator, given here in using the improvement brought by the convolution tool described in Section 11.4.3.

Restriction

$$\hat{\mathrm{M}} \underset{lin}{=} M \quad (= scalar); \quad \hat{\mathrm{k}}_e \underset{lin}{=} k_e \quad (= scalar); \quad \hat{\mathrm{k}}_f \underset{lin}{=} k_f \quad (= scalar) \tag{H7.10}$$

With the above restrictions, the algebraic model is written as a second-order differential equation

$$\frac{\mathrm{d}^2}{\bar{\mathrm{d}}t^{2}} \boldsymbol{F}_{CG} + \omega^2 \boldsymbol{F}_{CG} = 0 \tag{H7.11}$$

with a wave pulsation defined as

Wave pulsation: $$\omega \underset{lin}{\overset{def}{=}} \sqrt{\frac{\bar{k}_e}{M}}$$

which is different from the natural pulsation of the oscillator, defined as

Oscillator natural pulsation: $$\omega_\ell \underset{lin}{\overset{def}{=}} \sqrt{\frac{k_e}{M}}$$

The distinction between these two pulsations is made in the Formal Graph theory to avoid the classical confusion of the pulsation ω of the wave associated with the oscillator (which is in fact an energy variety) and the pulsation ω_ℓ featuring the mechanical oscillator. The relationship between them, explained in Section 11.5, is expressed in function of the *oscillator natural pulsation* and of the *inductive damping factor*

$$\omega \underset{lin}{=} \sqrt{\omega_\ell^2 - \gamma_{\ell L}^2} \quad \text{with } \gamma_{\ell L} \underset{lin}{=} \frac{k_f}{2M} \tag{H7.12}$$

It is only when the oscillator is conservative that the pulsation of the associated wave corresponds exactly to the natural pulsation of the oscillator. In the present case of a damped oscillator, the wave pulsation is lower than the natural pulsation, with the rule that the higher the damping factor the lower the wave pulsation. With an upper limit that is the natural pulsation, because over this threshold the diversion of energy into the conductance does not leave enough energy for maintaining the oscillations, which are disappearing. In this case, the system undertakes a simple relaxation.

$$\gamma_{\ell L} < \omega_\ell \Leftrightarrow \omega \sim \text{real} \Leftrightarrow \text{oscillations} \tag{H7.13}$$

$$\gamma_{\ell L} > \omega_\ell \Leftrightarrow \omega \sim \text{imaginary} \Leftrightarrow \text{relaxation} \tag{H7.14}$$

This study is limited to the regime of oscillations assuming a damping factor lower than the natural pulsation. In the regime of oscillations, the solution of the second-order differential equation is

$$\boldsymbol{F}_{CG} \underset{lin}{=} \bar{k}_e\, \ell(0) \exp(-\gamma_{\ell L} t) \cos \omega t \tag{H7.15}$$

This result shows that the solution in presence of damping consists of multiplying the response without damping by a decaying exponential. From this result, the other state variables are deduced:

$$\ell \underset{lin}{=} \ell(0) \exp(-\gamma_{\ell L} t) \cos \omega t \tag{H7.16}$$

$$\boldsymbol{v} = \boldsymbol{v}_L = \boldsymbol{v}_G = \boldsymbol{v}_C \underset{lin}{=} -\omega\, \ell\, \mathbf{(0)} \exp(-\gamma_{\ell L} t) \sin \omega t \tag{H7.17}$$

In the case study abstract are plotted their variations as a function of the time.

11.3.8 Case Study H8: Attenuated Propagation — Electrodynamics

The case study abstract is given on next page.

The subject of this case study is the propagation of an electromagnetic wave in a medium that absorbs energy owing to spatial damping. The propagation in a conducting medium may lead also to attenuation but this process is relevant to temporal damping, which is not the subject treated here.

In the medium, the two electrical constitutive properties that are specific to a three-dimensional material are the electrical permittivity and the permeability, defined by the following relationships:

$$D = \varepsilon\, \boldsymbol{E}; \quad \boldsymbol{B} = \mu\, \boldsymbol{H} \tag{H8.1}$$

Throughout this case study these reduced properties will be assumed to be scalars for complying with the classical treatment of the propagation of an electromagnetic wave. In other words, the medium supporting the propagation is assumed to be a linear material. In using these properties without referring to vacuum properties (subscripted with a zero), provision is made for the case of polarizable or magnetizable materials.

Dipole assembly

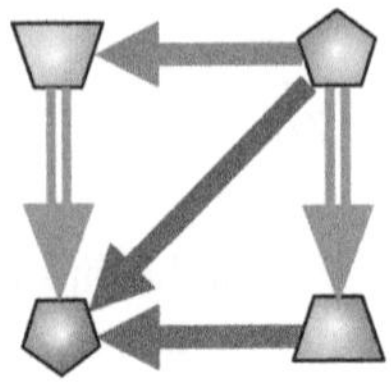

Dipole Assembly	
✓	Fundamental
	Mixed
✓	Capacitive
✓	Inductive
	Conductive
✓	Common effort
	Common flow
	Global
✓	Spatial

Formal Graph:

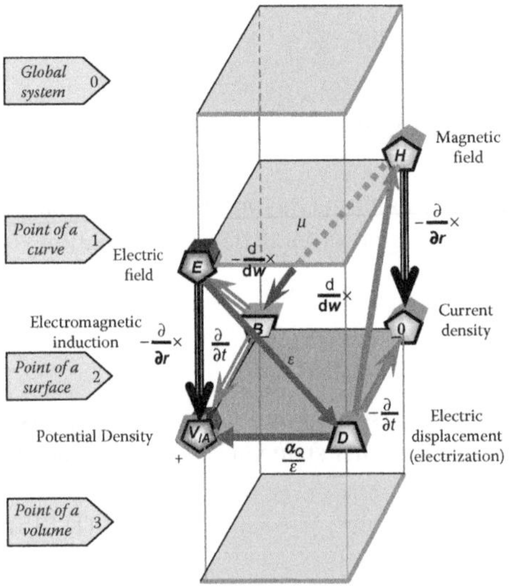

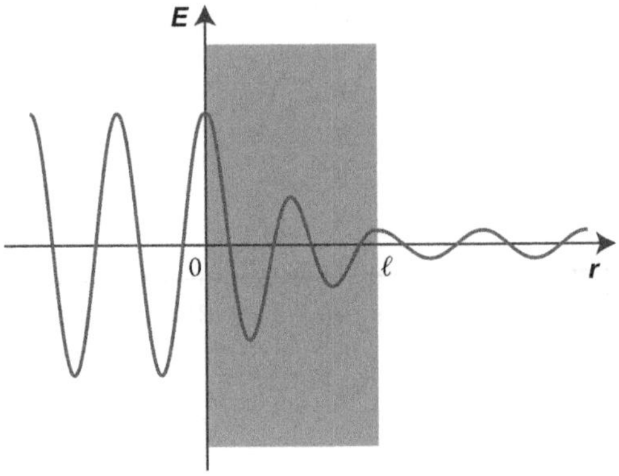

Plot of the electric field crossing a layer (thickness ℓ) of absorbing medium as a function of the distance.

GRAPH 11.32

Variables:

Variety	Electrodynamics			
Subvariety	Inductive (kinetic)		Capacitive	
Category	Entity number	Energy/ entity	Entity number	Energy/ entity
Family	Impulse	Flow	Basic Quantity	Effort
Name	Induction quantity (flux)	Current	Charge	Potential
Symbols	Φ_B	I	Q	V
Unit	[Wb], [J A^{-1}]	[A]	[C]	[V], [J C^{-1}]

System Properties			
Nature	Inductance	Capacitance	Elastance
Levels	1/2	1/2	2/2
Name	Permeability	Permittivity	Superficial elastance
Symbol	μ	ε	α_Q/ε
Unit	[H m^{-1}]	[F m^{-1}]	[F^{-1}]
Specificity	3-d	3-d	–

To establish the algebraic model of an attenuated propagating wave, one follows the classical scheme starting from Maxwell equations and then building the differential equation leading to the expression of the electric field. Section 11.7 gives another approach starting from the Formal Graph, then building the wave function, from which is deduced the expression of the electric field (in the general frame of any energy variety). This should facilitate the grasping of this rather cumbersome subject by readers not accustomed to differential calculus.

The four Maxwell equations were given in case study F7 in Chapter 9 for a nonabsorbing medium but, as it is classically done, in making provision for the existence of currents flowing in the medium (via the Maxwell–Ampère equation H8.5). To model a spatial attenuation one does not need to involve currents, but one needs to take into account an additional reduced capacitance which is able to divert some energy from the oscillation energy. This is done through a modification of the Maxwell–Faraday equation replacing the zero by an additional term proportional to the electric field, so the modified Maxwell equations become:

Gauss (electrostatics): $$\rho = \frac{\partial}{\partial \boldsymbol{r}} \cdot \boldsymbol{D} \tag{H8.2}$$

Gauss (electromagnetism): $$0 = \frac{\partial}{\partial \boldsymbol{r}} \cdot \boldsymbol{B} \tag{H8.3}$$

Maxwell–Faraday (modified): $$\boldsymbol{\alpha}_Q \times \boldsymbol{E} = \frac{\partial}{\partial \boldsymbol{r}} \times \boldsymbol{E} + \frac{\partial}{\partial t} \boldsymbol{B} \tag{H8.4}$$

Maxwell–Ampère: $$\boldsymbol{j} = \frac{\partial}{\partial \boldsymbol{r}} \times \boldsymbol{H} - \frac{\partial}{\partial t} \boldsymbol{D} \tag{H8.5}$$

A departure from the classical theory for a while explains how this new term is introduced. In the Formal Graph theory, there exists a node beneath the *electric field* $\boldsymbol{E}$ (at the level of a surface) bearing a vector called *potential density*, notated in a generic form $\boldsymbol{V}_{/A}$ because there is no such variable in classical electrostatics. This node is linked on the one hand to the *electromagnetic induction* $\boldsymbol{B}$ through the evolution operator

$$\boldsymbol{V}_{/A} = \frac{\partial}{\partial t} \boldsymbol{B} \tag{H8.6}$$

and on the other hand to a double contribution of the electric field $\boldsymbol{E}$ through the contracurl operator and of the electric displacement $\boldsymbol{D}$ (*electrization*) through a reduced capacitance

$$\boldsymbol{V}_{/A} = -\frac{\partial}{\partial \boldsymbol{r}} \times \boldsymbol{E} + \frac{\boldsymbol{\alpha}_Q}{\varepsilon} \times \boldsymbol{D} \tag{H8.7}$$

This reduced capacitance is notated by a ratio of two constitutive properties for sparing the number of symbols. The denominator is the permittivity ε and the numerator $\boldsymbol{\alpha}_Q$ is the *energetic spatial damping factor* also called *energetic absorption factor*. This last parameter is a vector used for quantifying the absorbing property of a medium owing to spatial damping, and it can be experimentally determined through the analysis of the decay of the wave energy or intensity. (The subscript Q, which refers to the electrodynamical energy variety, is not classically used but is required in the Formal Graph theory because the phenomenon of absorption of waves exists in other energy varieties; in case of sound propagation for instance.)

The previous relationship can be written with the help of relation H8.1 as a unique operator acting on the electric field

$$\boldsymbol{V}_{/A} = \left\{ -\frac{\partial}{\partial r} + \alpha_Q \right\} \times \boldsymbol{E} \tag{H8.8}$$

The shape of this operation mimics Equation H6.9 written in case study H6 about the viscoelastic relaxation that is rewritten here

$$V_{/r} \underset{homot}{=} \left\{ \frac{\mathrm{d}}{\mathrm{d}t} + k_C \right\} \varepsilon_\ell \tag{H6.9}$$

This analogy means that the *energetic absorption factor* $\boldsymbol{\alpha_Q}$ plays the same role with respect to the space derivation as the kinetic constant with respect to the temporal derivation. Such a role is illustrated by the two Formal Graphs in Graph 11.33 that explains how the *energetic absorption factor* $\boldsymbol{\alpha_Q}$ is formed by a combination of two links paralleling the spatial operator.

One returns now to the classical approach that was left for introducing the *energetic absorption factor* $\boldsymbol{\alpha_Q}$. The conditions governing the simplest phenomenon of spatial attenuation of a wave are as follows:

$$\boldsymbol{j} = \boldsymbol{0}; \quad \rho = C^{st}; \quad \alpha_Q = \boldsymbol{C}^{st} \tag{H8.9}$$

No current flows in the medium (no conduction), no charges or more precisely no charge gradient exists in the medium, and the medium is assumed to be homogeneous for having a constant *energetic absorption factor* $\boldsymbol{\alpha_Q}$ all along the wave path.

The following text box deals with the calculus leading to the differential equation for the electric field and can be skipped by those uninterested in such details.

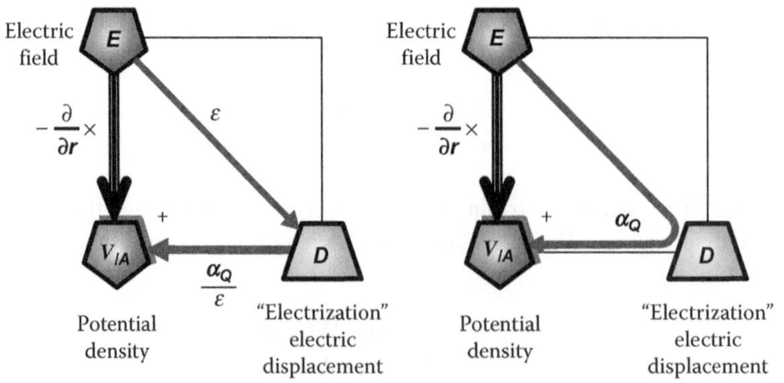

GRAPH 11.33 Two-dimensional Formal Graphs of the constraint that models the spatial damping as a three-dimensional oscillator in electrodynamics. The constraint comes from the existence of an extra capacitance between the electric displacement ("electrization") $\boldsymbol{D}$ and the potential density $\boldsymbol{V}_{/A}$.

Getting a differential equation for *E*. Solving the set of Maxwell equations in order to find the electric field (for instance) requires eliminating the unnecessary variables. The first step consists of looking for the curl of the potential density from Equation H8.6 and of using the Maxwell–Ampère equation H8.5 for replacing the curl of the electromagnetic field

$$\frac{\partial}{\partial \mathbf{r}} \times \mathbf{V}_{IA} = \frac{\partial}{\partial t}\frac{\partial}{\partial \mathbf{r}} \times \mathbf{B} = \mu\frac{\partial}{\partial t}\left(\frac{\partial}{\partial \mathbf{r}} \times \mathbf{H}\right) = \varepsilon\mu\frac{\partial^2}{\partial t^2}\mathbf{E} \tag{H8.10}$$

The second step consists of finding the curl of the potential density from its expression H8.7 as a function of the electric field. The curl of the curl of the electric field and the curl of a vector product are developed according to known formulas of vector calculus (Woan 2003):

$$\frac{\partial}{\partial \mathbf{r}} \times \left(\frac{\partial}{\partial \mathbf{r}} \times \mathbf{E}\right) = \frac{\partial}{\partial \mathbf{r}}\left(\frac{\partial}{\partial \mathbf{r}} \cdot \mathbf{E}\right) - \frac{\partial^2}{\partial \mathbf{r}^2}\mathbf{E} \underset{\rho=C^{st}}{=} -\frac{\partial^2}{\partial \mathbf{r}^2}\mathbf{E} \tag{H8.11}$$

$$\frac{\partial}{\partial \mathbf{r}} \times (\boldsymbol{\alpha}_Q \times \mathbf{E}) \underset{\alpha_Q=C^{st}}{=} \boldsymbol{\alpha}_Q\left(\frac{\partial}{\partial \mathbf{r}} \cdot \mathbf{E}\right) \tag{H8.12}$$

In developing the first item, the fact that, according to Gauss equation H8.2, the divergence of the electric field is proportional to the charge concentration has been taken into account, so its gradient is zero because the charge concentration is supposed to be uniform or equal to zero. For the second item, the absorption factor has been taken as a constant. With these developments, the curl of the potential density becomes

$$\frac{\partial}{\partial \mathbf{r}} \times \mathbf{V}_{IA} \underset{\alpha_Q=C^{st}}{\overset{\rho=C^{st}}{=}} \frac{\partial^2}{\partial \mathbf{r}^2}\mathbf{E} + \boldsymbol{\alpha}_Q\left(\frac{\partial}{\partial \mathbf{r}} \cdot \mathbf{E}\right) \tag{H8.13}$$

It remains to eliminate the curl of the potential density between Equations H8.10 and H8.13 in order to obtain the desired result

$$\frac{\partial^2}{\partial \mathbf{r}^2}\mathbf{E} + \boldsymbol{\alpha}_Q\left(\frac{\partial}{\partial \mathbf{r}} \cdot \mathbf{E}\right) = \varepsilon\mu\frac{\partial^2}{\partial t^2}\mathbf{E} \tag{H8.14}$$

which is a second-order partial differential equation with the electric field as unknown variable.

The differential equation H8.14 is the propagation equation of an attenuated wave due to absorption by the medium. It can be remarked that when the absorption factor is null, the undamped wave equation is retrieved

$$\boldsymbol{\alpha}_Q = 0 \quad \Rightarrow \quad \frac{\partial^2}{\partial \mathbf{r}^2}\mathbf{E} = \varepsilon\mu\frac{\partial^2}{\partial t^2}\mathbf{E} \tag{H8.15}$$

This means that the product of constitutive properties $\varepsilon\mu$ determines the *natural velocity* $\mathbf{u}_Q$ of the wave, that is, without absorption, as seen in Chapter 9. The adjective *natural* is used here to indicate a wave propagating in a nonabsorbing medium ("intrinsic" can be used as well). It has been seen in Chapter 9 when discussing the model of wave propagation that this velocity is equal to the ratio of the *natural pulsation* ω_Q to the *natural wave-vector* $\mathbf{k}_Q$ of the oscillator generating the wave. The two equalities in the following equations recall these relationships:

$$\varepsilon\mu = \frac{1}{\mathbf{u}_Q^2} = \frac{\mathbf{k}_Q^2}{\omega_Q^2} \tag{H8.16}$$

The replacement of the product of constitutive properties $\varepsilon\mu$ by the ratio of natural properties of the oscillator allows separating Equation H8.14 into two terms, the left one exclusively relevant to space and the right one exclusively relevant to time:

$$\frac{1}{k_Q^2}\left(\frac{\partial^2}{\partial r^2}\boldsymbol{E}+\alpha_Q\left(\frac{\partial}{\partial r}\cdot\boldsymbol{E}\right)\right) = \frac{1}{\omega_Q^2}\frac{\partial^2}{\partial t^2}\boldsymbol{E} \tag{H8.17}$$

If the left member is new and specific to the spatially attenuated wave, the right member has been already encountered in Chapter 9 in case study F6 of a nonattenuated wave. It is not astonishing to find again a second derivative with respect to time, alone without other temporal derivatives, as the role of time is unaltered in both cases of wave propagation. The spatial damping only affects space operators and the temporal behavior is identical because no temporal damping is active. Consequently, the wave pulsation is equal to the natural pulsation of the oscillator and the temporal behavior of the electric field is ruled by an independent second-order differential equation identical to the free propagation.

No temporal damping: $$\omega = \omega_Q; \quad \frac{\partial^2}{\partial t^2}\boldsymbol{E} = -\omega^2\boldsymbol{E} \tag{H8.18}$$

By replacing this temporal second derivative in Equation H8.17, a purely spatial differential equation is obtained

$$\frac{1}{k_Q^2}\left(\frac{\partial^2}{\partial r^2}\boldsymbol{E}+\alpha_Q\left(\frac{\partial}{\partial r}\cdot\boldsymbol{E}\right)\right) = -\boldsymbol{E} \tag{H8.19}$$

which can be recast into the expansion of a squared binomial

$$\frac{\partial^2}{\partial r^2}\boldsymbol{E}+2\zeta_Q\left(\frac{\partial}{\partial r}\cdot\boldsymbol{E}\right)+\zeta_Q^2\boldsymbol{E} = -\left(k_Q^2 - \zeta_Q^2\right)\boldsymbol{E} \tag{H8.20}$$

The *spatial damping factor* ζ_Q that appears in this expansion is nothing other than half the *energetic absorption factor*

$$\alpha_Q = 2\zeta_Q \tag{H8.21}$$

It may appear superfluous to use two symbols for what seems to be the same notion, but this is not totally true. The *energetic absorption factor* α_Q refers to the decay of energy in going through an absorbing medium, whereas the *spatial damping factor* ζ_Q refers to the damping of state variables (or reduced ones) in the same medium. The coefficient 1/2 relating these two quantities works only in the linear case of invariant constitutive properties, which justifies the distinction made between these two notions.

Now, the difference of squared variables in the right member of the previous differential equation is identified with the square of the wave-vector $\boldsymbol{k}$ of the wave

$$k^2 = k_Q^2 - \zeta_Q^2 \tag{H8.22}$$

The distinction made in the Formal Graph theory between the oscillator and the wave that is associated to it, finds a justification in the present case of attenuated propagation with the previous relation H8.22 showing that the wave-vector $\boldsymbol{k}$ of the wave differs from the wave-vector $\boldsymbol{k_Q}$ of the oscillator owing to the attenuation. It is assumed that $\boldsymbol{k}$ is real for keeping the oscillation of the wave, which ensures the damping factor is not higher than the natural wave-vector (in term of modulus)

$$\|\zeta_Q\| < \|k_Q\| \tag{H8.23}$$

If this condition is not fulfilled, the wave is completely extinguished by the medium.

Equation H8.17 is now rewritten in two equalities with these new variables

$$\frac{1}{k^2}\left\{\frac{\partial}{\partial \boldsymbol{r}} + \zeta_Q\right\}^2 \cdot \boldsymbol{E} = \frac{1}{\omega^2}\frac{\partial^2}{\partial t^2}\boldsymbol{E} = -\boldsymbol{E} \tag{H8.24}$$

Degree 2 of these differential equations can be reduced by taking the square root of the operators, which leads to two symmetrical solutions according to the sign of the imaginary number

$$\frac{1}{k}\left\{\frac{\partial}{\partial \boldsymbol{r}} + \zeta_Q\right\} \cdot \boldsymbol{E} = \frac{1}{\omega}\frac{\partial}{\partial t}\boldsymbol{E} = \pm i\,\boldsymbol{E} \tag{H8.25}$$

The two individual solutions are

$$\boldsymbol{E}_+ = \boldsymbol{E}_0 \exp(-\zeta_Q \cdot \boldsymbol{r})\ \exp(+i(-\omega\, t + \boldsymbol{k} \cdot \boldsymbol{r})) \tag{H8.26}$$

$$\boldsymbol{E}_- = E_0 \exp\left(-\zeta_Q \cdot \mathbf{r}\right)\ \exp(-i(-\omega\, t + \boldsymbol{k} \cdot \boldsymbol{r})) \tag{H8.27}$$

The common solution of the differential equation H8.24 is half the sum of the individual solutions

$$\boldsymbol{E} \underset{lin}{=} \boldsymbol{E}_0 \exp(-\zeta_Q \cdot \mathbf{r})\ \cos(-\omega\, t + \boldsymbol{k} \cdot \boldsymbol{r}) \tag{H8.28}$$

The end of the calculus is reached with this expression of the electric field as a function of time and space. The mention "*lin*" under the equal sign is a reminder of the restriction brought in this case study to work in linear media exclusively.

Now, a certain number of considerations can be drawn from this result.

11.3.8.1 Wave Velocity and Wave Equation

The *wave velocity*, or *phase velocity*, $\boldsymbol{u}$ is defined as the ratio of the wave pulsation upon the wave-vector of the wave:

$$\boldsymbol{u} \overset{def}{=} \frac{\omega}{k} \tag{H8.29}$$

With this variable, the differential Equation H8.24 can be rewritten as a wave equation adapted to the case of spatial damping:

Spatially damped wave equation:
$$\frac{\partial^2}{\partial t^2}\boldsymbol{E} = \boldsymbol{u}^2 \left\{\frac{\partial}{\partial \boldsymbol{r}} + \zeta_Q\right\}^2 \boldsymbol{E} \tag{H8.30}$$

This velocity $\boldsymbol{u}$ is a space–time property and must not be confused with the velocity $\boldsymbol{v}$ of an object, which is a state variable (in the family of flows). In particular, for sufficiently absorbing material (high ζ_Q meaning low $\boldsymbol{k}$) or sufficiently high frequency (high ω), for instance for x-rays, the space–time velocity can be higher than the speed of light. This does not contradict the theory of relativity because a wave is not an inertial body (no mass).

11.3.8.2 Radiation Intensity or Irradiance

From the expression obtained for $\boldsymbol{E}$ and by using Maxwell equations and relation H8.16, the electromagnetic field is derived:

$$\boldsymbol{H} \underset{lin}{=} \varepsilon\, \boldsymbol{u}_{\mathbf{Q}} \times E_0 \exp(-\zeta_Q \cdot \boldsymbol{r}) \sin(-\omega\, t + \boldsymbol{k}.\boldsymbol{r}) \quad \text{(H8.31)}$$

In case study F6 in Chapter 9, the expression for the volumic concentration of energy as a function of $\boldsymbol{H}$ and $\boldsymbol{E}$ was established, which gives here an exponential decay

$$\mathcal{E}_{Q/V} = \mathcal{T}_{Q/V} + \mathcal{U}_{Q/V} \underset{lin}{=} \frac{1}{2}(\mu\, \boldsymbol{H}^2 + \varepsilon\, \boldsymbol{E}^2) = \frac{1}{2}\varepsilon\, \boldsymbol{E}_0^2 \exp(-2\zeta_Q \cdot \boldsymbol{r}) \quad \text{(H8.32)}$$

The irradiance, or radiation intensity, was established as being proportional to this energy concentration that can be written with the energetic absorption coefficient in using relation H8.21:

$$I_\lambda = \boldsymbol{u}\, \mathcal{E}_{Q/V} \underset{lin}{=} \frac{1}{2}\varepsilon\, \boldsymbol{u}\, \boldsymbol{E}_0^2 \exp(-\boldsymbol{\alpha}_Q \cdot \boldsymbol{r}) \quad \text{(H8.33)}$$

The value of the irradiance before the wave enters the absorbing medium was given in case study F6 (Chapter 9) as the irradiance of a free propagation or undamped irradiance:

Undamped irradiance:

$$I_\lambda^0 \underset{lin}{=} \frac{1}{2}\varepsilon\, \boldsymbol{u}\, \boldsymbol{E}_0^2 \quad \text{(H8.34)}$$

so the irradiance within the absorbing medium or damped irradiance is written as

Damped irradiance:

$$I_\lambda \underset{lin}{=} I_\lambda^0 \exp(-\boldsymbol{\alpha}_Q \cdot \mathbf{r}) \quad \text{(H8.35)}$$

This is a more general expression of the Lambert–Beer–Bouguer law studied in Chapter 7, Section 7.4.1 which is retrieved when vectors are colinear. The correspondence between their modules and the parameters of the law are as follows:

$$\left\|\boldsymbol{\alpha}_Q\right\| = c_N \kappa_N = c_n \kappa_n; \quad \left\|\boldsymbol{r}\right\| = \Delta l \quad \text{(H8.36)}$$

establishing the energetic absorption coefficient as the product of the concentration of obstacles (molecules) times the cross-section of these obstacles. This development is a demonstration of the Lambert–Beer–Bouguer law that we used in Chapter 7 for building the theory of influence without applying statistical arguments.

11.3.8.3 Index of Refraction

The *index of refraction*, or *refractive index*, is an adimensional variable characterizing the velocity of a wave in an absorbing medium, which is defined as the ratio of the speed of light in vacuum on the modulus of the wave velocity

Index of refraction:

$$n_\lambda \overset{def}{=} \frac{c}{\left\|\boldsymbol{u}\right\|} \quad \text{(H8.37)}$$

The wavelength λ used as a subscript indicates the dependence on the wave pulsation (or frequency or wavelength) of this variable. From Equation H8.29 the index of refraction appears proportional to the wave-vector modulus of the wave:

$$n_\lambda = \frac{c}{\omega}\|\boldsymbol{k}\| \tag{H8.38}$$

11.3.8.4 Complex Variables

Expression H8.28 of the electric field contains a harmonic function cosine that is defined as a sum of exponential function having imaginary arguments, as explained in Section 9.5.5 in Chapter 9. Using this definition gives the following expression:

$$\boldsymbol{E} \underset{lin}{=} \frac{\boldsymbol{E}_0}{2}(\exp(-\boldsymbol{\zeta}_Q \cdot \boldsymbol{r} - i\omega t + i\boldsymbol{k} \cdot \boldsymbol{r}) + \exp(-\boldsymbol{\zeta}_Q \cdot \boldsymbol{r} + i\omega t - i\boldsymbol{k} \cdot \boldsymbol{r})) \tag{H8.39}$$

which can be shortened by using a *complex wave-vector* and its conjugate:

$$\boldsymbol{E} \underset{lin}{=} \frac{\boldsymbol{E}_0}{2}(\exp(i\tilde{\boldsymbol{k}} \cdot \boldsymbol{r} - i\omega t +) + \exp(-i\tilde{\boldsymbol{k}}^* \cdot \boldsymbol{r} + i\omega\, t)) \tag{H8.40}$$

From such a complex wave-vector, a *complex index of refraction* can be defined based on expression H8.38

$$\tilde{\boldsymbol{k}} \overset{def}{=} \boldsymbol{k} + i\boldsymbol{\zeta}_Q \Leftrightarrow \tilde{n}_\lambda = n_\lambda + i\kappa_\lambda \overset{def}{=} \frac{c}{\omega}\left(\|\boldsymbol{k}\| + i\|\boldsymbol{\zeta}_Q\|\right) \tag{H8.41}$$

so the imaginary part κ_λ of the complex index is

$$\kappa_\lambda = \frac{c}{\omega}\|\boldsymbol{\zeta}_Q\| \tag{H8.42}$$

Remarks

A more classical approach. The model developed in this case study has been built with a semiclassical method, starting from Maxwell equations and then solving the differential equations to determine the solution for the electric field. The Formal Graph approach has been used only in the modification of the Maxwell–Faraday equation for taking into account the reduced capacitance creating the spatial constraint.

An example of a completely classical approach can be found in Feynman et al. (2006) and Jackson (1998). Their approach can be summarized as follows. It relies on the electric polarization of the medium, which absorbs part of the energy owing to the separation of charges induced by the electric field acting on the **charged** material. To take into account this effect, called dielectric loss, the *free space (vacuum) permittivity* ε_0 is replaced by a *material permittivity* ε equal to the *free space permittivity* multiplied by the *relative permittivity* ε_r. The Maxwell equations are written with this *material permittivity* ε but without spatial constraint or conduction current. In doing so, the wave pulsation ω and the wave-vector $\boldsymbol{k}$ are unchanged, equal to the natural variables of the oscillator ω_Q. and $\boldsymbol{k}_Q$. The wave velocity $\boldsymbol{u}$ is therefore equal to the natural velocity $\boldsymbol{u}_Q$ which directly depends on the *material permittivity* ε, thus

(continued)

bearing the influence of the polarization. The choice is then to transfer this influence on the wave-vector to keep the pulsation unaltered.

$$\|\boldsymbol{k}\| = \frac{\omega}{\|\boldsymbol{u}\|} = \frac{\omega}{c}\sqrt{\frac{\varepsilon_0 \mu_0}{\varepsilon \mu}} = \frac{\omega}{c}\sqrt{\varepsilon_r \mu_r} \tag{H8.43}$$

A second step assumes that the *relative permittivity* ε_r is a complex number, on the same pattern as for the *dielectric relaxation* (see case study H5), consequently conferring the complex nature to the wave-vector too

$$\tilde{k} = k' + ik'' \tag{H8.44}$$

Introduction of this expression into the argument of a cosine function makes explicit an exponential function with a negative real argument proportional to the space coordinate, thus modeling the wave decay with the distance.

The **paradox** with such a reasoning is that the dielectric relaxation-like model invoked relies on the influence of time (and on the existence of conduction), which modifies the wave pulsation and contradicts the form of Maxwell equations used (no current involved).

The reasoning based on Formal Graphs avoids this contradiction and works also for noncharged and nonpolarizable materials, being applicable in addition to other energy varieties (in which the phenomenon of polarization is unknown).

Generality. The model of spatial attenuation of a wave developed here in the case of an electromagnetic wave is general and also applies to any energy variety. With the exception of a simpler treatment in case of scalar localized variables, the same treatment can be worked out for mechanical waves (vibrating string, phonon propagation) or for hydrodynamical waves (sound or seismic propagation) for instance. Strictly identical concepts are derived, such as absorption coefficient, wave intensity decay, index of refraction (real or complex) and so on.

Formal Graph approach. The Formal Graph approach has been used voluntarily in a limited but essential aspect, which is the justification of the incorporation of a spatial constraint into the Maxwell–Faraday equation. The reason comes from the necessity to prove in the most classical way the validity of the proposed modification. Nevertheless, the Formal Graph given in the case study abstract provides a complete model (in the linear frame) that can be analytically solved or numerically simulated for getting the variations of the different variables of the system.

As already illustrated with nondamped propagations in Chapter 9, a more general approach based on the construction of a *wave function* with the help of Formal Graphs is also possible. This original approach is given in detail in Sections 11.5 through 11.7.

11.4 RELAXATION MODELS

The subject of relaxation has been illustrated by several case studies in the previous section: slowed down motion, occurring reaction, RC circuit, dielectric relaxation, and viscoelastic relaxation. Depending on the peculiarity and tradition of each domain (linearity, measurability, use of forced

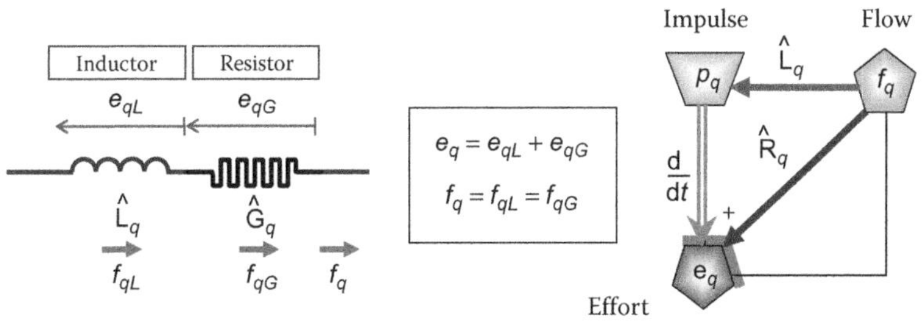

GRAPH 11.34 RL serial assembly (common flow).

conditions, large or small amplitude techniques) the treatment and the algebraic models could slightly differ but the Formal Graphs models were identical (within the same energy subvariety). The following discussion is a restatement in generalizing the modeling to any energy variety and in developing more details.

11.4.1 Relaxing Assemblies

Graphs 11.34 through 11.37 present the various ways of assembling a storing dipole with a conductor. Three models are presented each time, the equivalent electrical circuit, the algebraic model (Kirchhoff's* equations) used for the mounting, and the Formal Graph. Depending on the energy-per-entity chosen to play the role of a common gate variable and on the subvariety for the storage, it makes four different cases.

In these systems, the precision of the isolation or of an energy supply is not given for the sake of generality. It is clear that in case of isolation, the common energy-per-entity is equal to zero and the choice of the mounting is indifferent.

11.4.2 Free Relaxation

We take the capacitive case for discussing this subject, mentioning that the inductive case is handled in the same way and easily deduced from the capacitive one. The two mountings described above, with a common flow and with a common effort, are reproduced in Graph 11.38, but in setting to zero the energy-per-entity which is not shared owing to the isolation of the system. In this configuration, the two mountings are equivalent, which is shown by the third equivalent Formal Graph (on the

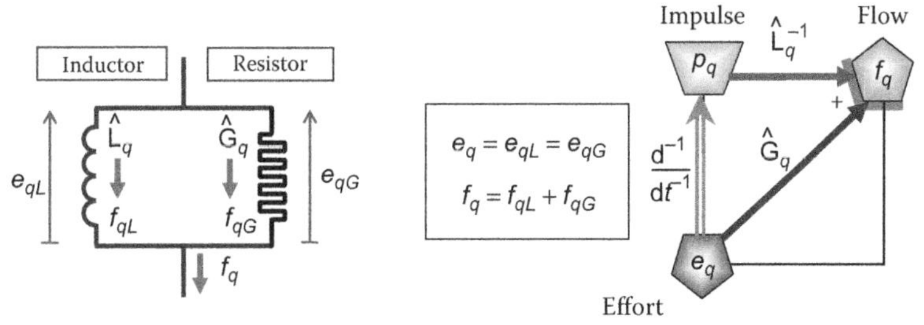

GRAPH 11.35 RL parallel assembly (common effort).

* Gustav Robert Kirchhoff (1824–1887): German physicist; Heidelberg and Berlin, Germany.

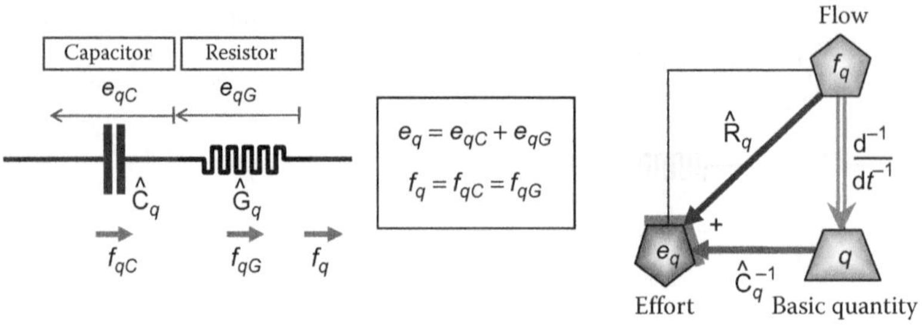

GRAPH 11.36 RC serial assembly (common flow).

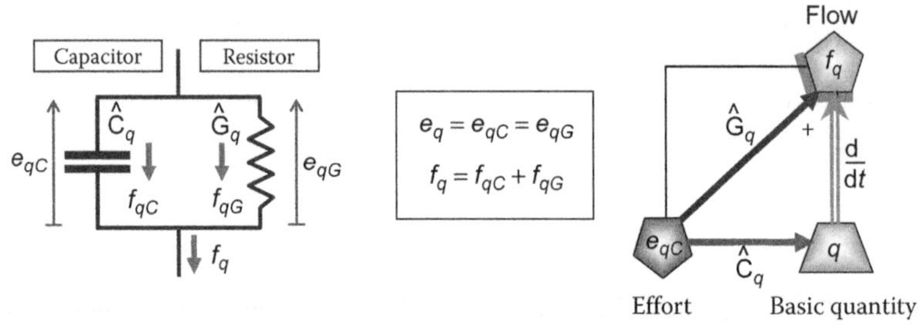

GRAPH 11.37 RC parallel assembly (common effort).

right) using a partial flow (the capacitive one) and a kinetic operator formed by a combination of the capacitance and resistance.

The generalized capacitor is assumed to contain some capacitive energy before the connection between the two dipoles in the assembly is closed at time t equal to zero. From this instant, the discharge of the capacitor will begin. The composition rules on the flow node (summation) and on the basic quantity node (transmission) results in the following algebraic equation with the effort as variable, valid after the link is closed

Common flow:
$$\hat{R}_q f_q + \hat{C}_q^{-1} \frac{d^{-1}}{dt^{-1}} f_q \underset{isol}{=} 0 \tag{11.24}$$

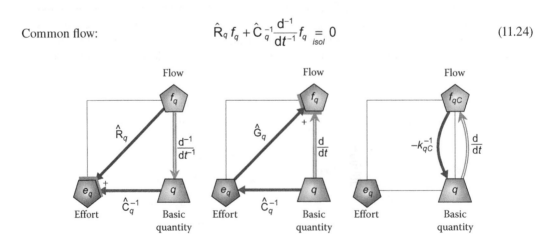

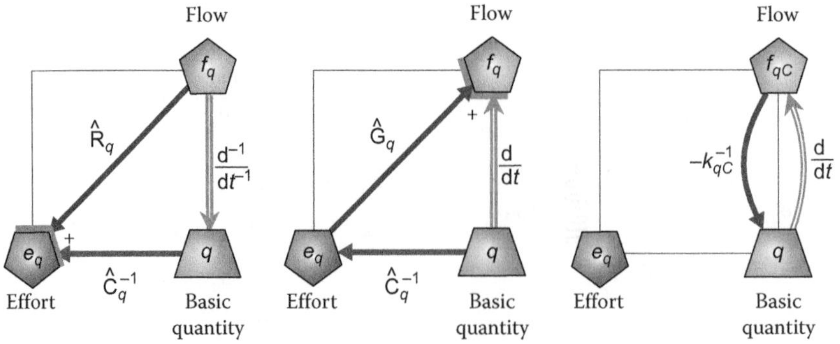

GRAPH 11.38 Three Formal Graphs of free relaxation in an isolated dipole assembly: with a common flow (left), with a common effort (center), and with a combination of paths into a kinetic operator (right). The left and center ones use the normal representation of two additive contributions, and the right one implements a loop evidencing the irreversibility of the relaxation.

Common effort:
$$\frac{\mathrm{d}}{\mathrm{d}t}\hat{C}_q\, e_q + \hat{G}_q\, e_q \underset{isol}{=} 0 \tag{11.25}$$

Alternatively, instead of using the energies-per-entity, flow or effort, the basic quantity can be used, leading to another algebraic equation having the basic quantity as variable:

$$\frac{\mathrm{d}}{\mathrm{d}t}q + \hat{G}_q\,\hat{C}_q^{-1}\, q \underset{isol}{=} 0 \tag{11.26}$$

These three equations are equivalent models of the free relaxation but the last translation has the advantage of allowing the definition of an apparent operator resulting from the combination of the elastance with the conductance:

$$\hat{k}_{qc} \overset{def}{=} \hat{G}_q\,\hat{C}_q^{-1} \tag{11.27}$$

This combination of system properties has already been viewed in Chapter 4 (case study A11 "Reactive Chemical Species") when describing an irreversible chemical reaction in terms of a pole. In that peculiar case, the combination of operators was giving a scalar k_A corresponding to the specific rate (rate constant) of the reaction $A \rightarrow B$. This operator $\hat{k}_{qc}$ known as the *kinetic operator* (in its conductive–capacitive version) is similar to a natural pulsation or frequency operator of an oscillating system but working in nonperiodic conditions (see Section 9.5 in Chapter 9).

In operational calculus, Equation 11.26 describing the behavior of the dipole assembly amounts to an equality between two operators

$$-\frac{\mathrm{d}}{\mathrm{d}t} = \hat{k}_{qC} \tag{11.28}$$

which is not especially useful in the general case when the constitutive properties are unknown operators. However, a general turnaround consists of searching for peculiar values of the state variable called eigen-vectors so as to find proportionality between the result of the operation and the state variable, which works therefore as an eigen-value of the operator:

Eigen-value definition:
$$\hat{k}_{qC}q \overset{def\rightarrow}{=} k_{qC}q \tag{11.29}$$

This is the same reasoning previously made for the wave function of an oscillator. A convenient way to express this proportionality is to assert that the composition of the constitutive properties results in a scalar, but this is in fact an assumption that needs to be put into evidence. This peculiar condition amounts to the assumption of homothetic constitutive properties.

Hypothesis. Proportionality between conductance and capacitance (homothecy).

Homothetic constitutive properties: $\hat{G}_q \propto \hat{C}_q$

$$\hat{G}_q\,\hat{C}_q^{-1} \underset{homot}{=} k_{qC} \quad (= scalar) \tag{11.30}$$

Even with this assumption that transforms the equation between operators into an ordinary differential equation, this latter cannot be solved without knowing the time dependence of the eigen-value. The simplest case is the invariance with time.

Hypothesis. Time invariance of the *kinetic constant.*

Time independence: $$\frac{dk_{qC}}{dt} = 0 \Leftrightarrow k_{qC} = \text{Relaxation Constant} \qquad (11.31)$$

This time invariant eigen-value k_{qC} is called *kinetic constant* or *natural relaxation constant.* The adjective *natural* indicates that this operator is specific to the considered system as it depends on its constitutive properties (on the template of "natural frequency of resonance") but term *intrinsic* can be used as well. In this system, the *time constant* is defined as its reciprocal (it can be defined differently in other systems):

$$\tau_{qC} \overset{def}{=} \frac{1}{k_{qC}} = \hat{C}_q \hat{R}_q \qquad (11.32)$$

With this assumption, the eigen-function of the derivation is the exponential function, allowing the solution of Equation 11.26 to be written as

$$q = q(0)\exp(-k_{qC}t) \qquad (11.33)$$

This model of relaxation is strictly identical to the model of decaying radiation through an absorbing medium (see Section 11.7) or to the model of relaxation (see Equation 11.23) developed earlier in Section 11.2.2 which was based on a same reasoning. What one has now in addition is a dependence of the *relaxation constant* on physical characteristics of the system. What one has lost compared to the previous reasoning is the possible inclusion in the model of time discretization because the present development is based on the temporal derivation as evolution operator for being able to find a solution in analytically solving the differential equation.

In contrast with the previously discussed conservative oscillator (Chapter 9), the symmetry of the time is not respected here. An inversion $t \to -t$ produces an exponential increase of both the basic quantity and the effort that would make the energy in the subvariety increase exponentially. This is never verified experimentally; moreover, this would contradict the Second Principle of Thermodynamics, never allowing the entropy to decrease in an isolated system.

11.4.3 Forced Relaxation

We take the case of a common effort and an imposed flow by an external power source. The algebraic model 11.25 is modified by replacing the zero flow by the imposed one:

$$\frac{d}{dt}\hat{C}_q\, e_q + \hat{G}_q\, e_q = f_q \qquad (11.34)$$

The same hypotheses of homothetic constitutive properties shown in Equation 11.30 and of invariance with time of the kinetic constant shown in Equation 11.31 lead to the simplified differential equation

$$\frac{\mathrm{d}}{\mathrm{d}t} q + k_{qC}\, q \underset{homot}{=} f_q(t) \tag{11.35}$$

The solution is again an exponential function, but with a time-dependent scaling factor

$$q = q'(t)\exp(-k_{qC}\, t) \tag{11.36}$$

which must satisfy the following condition:

$$\frac{\mathrm{d}}{\mathrm{d}t} q'(t) \underset{homot}{=} f_q(t)\exp(k_{qC}\, t) \tag{11.37}$$

The solution by integration of this differential equation is formulated in a general manner with the help of an integration variable t' varying in the interval $[0, t]$:

$$q'(t) \underset{homot}{=} \int_0^t f_q(t')\exp(k_{qC} t')\,\mathrm{d}t' \tag{11.38}$$

Then, by injecting this scaling factor into expression 11.36 of the basic quantity, one obtains the general solution under the form of a convolution equation, also called *convolution product* and notated as a multiplication by means of a star operator (dyadic operator):

$$\text{Convolution:} \quad q \underset{homot}{=} \int_0^t \exp(-k_{qC}(t-t'))\, f_q(t')\,\mathrm{d}t'; \quad q \underset{homot}{=} \exp(-k_{qC}\, t) * f_q(t) \tag{11.39}$$

The convolution operation has been met in the previous chapter for expressing a partial integration in an amenable way to calculus (Section 11.1.4).

Convolution, signal processing, and Formal Graphs. The great advantage of this mathematical way of viewing relationships between variables linked by an operator is the possibility to adopt the symbolism of the signal processing theories using a box with input–output terminals in specifying the transfer function in it.

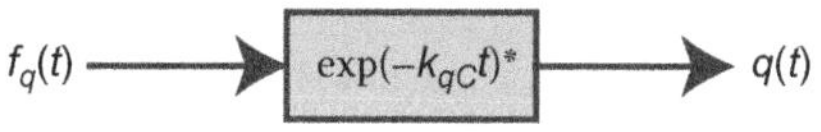

This representation is akin to the circuit symbolism, but in generalizing to any relationship using operators, in being more explicit, and without being limited to impedances. It may even be extended further than the homothetic or linear domain in replacing directly the specification of the transfer function by the operator or the differential equation.

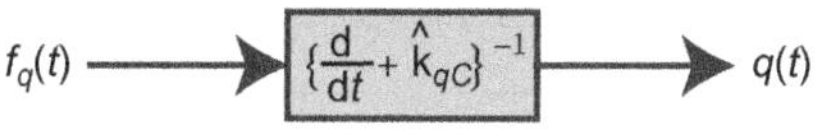

(*continued*)

Indeed, this is exactly what is done (in the reverse way here) with a Formal Graph.

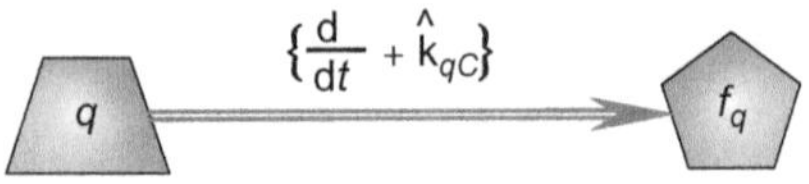

GRAPH 11.39

A rapid conclusion could be that nothing new is brought by Formal Graphs, but this is forgetting the topology, the rules based on thermodynamics and the classification in Formal Objects, which are not negligible complements.

The comparison becomes interesting when one borrows some concepts from one side, such as the concept of *black box* used for signifying that the transfer function is unknown and needs to be experimentally determined.

From the preceding considerations, it is clear that the Formal Graphs shown in Graph 11.40 model the same relaxation phenomenon (free or forced according to the flow value).

Before continuing the discussion, one gives some more details about the powerful tool that is the convolution.

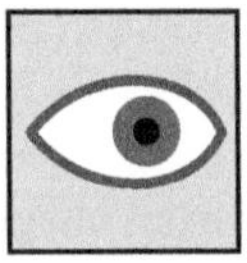

Convolution and Laplace transform. For mathematicians, the convolution integral or convolution product is a Volterra* integral equation using a *kernel* function $K(t)$, also called Green† function, as partner of the product with the function of interest $f(t)$. The characteristic of a convolution integral, compared to ordinary integration, is the presence under the summation of one (or two) of the integration boundaries, here the time.

(continued)

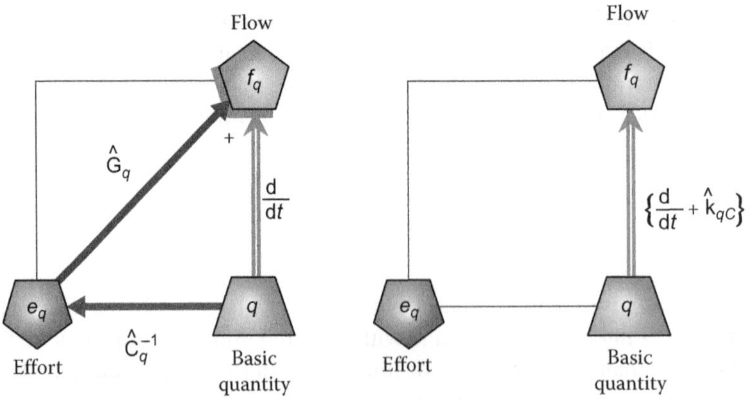

GRAPH 11.40 Two equivalent Formal Graphs of the relaxation process: free ($f_q = 0$) or forced ($f_q \neq 0$).

* Vito Volterra (1860–1940): Italian mathematician and physicist; Rome, Italy.
† George Green (1793–1841): British mathematician; Nottingham, UK.

Convolution: $$g(t)=\int_0^t K(t-t')\,f(t')\,\mathrm{d}t';\quad g(t)=K(t)*f(t) \tag{11.40}$$

For physicists, it models the behavior of a signal processing device with an input $f(t)$ and an output $g(t)$, related by the transfer function $K(t)$ featuring the device. Note that the convolution product is interesting for physicists because it helps distinguish the transfer function from the input signal.

An interesting property of the convolution product is that it can be transformed into an ordinary product of transformed functions. The transformation adapted to the time range of the convolution is not the Fourier* transform, which works on a full range $-\infty$ to $+\infty$ (two-sided transform), but the Laplace† transform, which is analogous but working on a half range from 0 to infinite (one-sided transform).

Fourier ω: $$\tilde{f}(\omega)=\frac{1}{\sqrt{2\pi}}\int_{-\infty}^{+\infty} f(t)\exp(-i\omega t)\,\mathrm{d}t \tag{11.41}$$

Laplace $i\omega$: $$\tilde{f}(i\omega)=\int_0^{+\infty} f(t)\exp(-i\omega t)\,\mathrm{d}t \tag{11.42}$$

(Usually, the Laplace transform is written with a real or complex variable s that we have replaced by the product of the imaginary times the angular frequency, as frequently done.)

Some characteristic transformed functions are worth citing.

Dirac: $$\delta(t)\xrightarrow{\text{Laplace } i\omega} 1 \tag{11.43}$$

Heaviside: $$\mathcal{H}(t)\xrightarrow{\text{Laplace } i\omega}\frac{1}{i\omega} \tag{11.44}$$

$$\frac{t^{p-1}}{\Gamma(p)}\xrightarrow{\text{Laplace } i\omega}(i\omega)^{-p} \tag{11.45}$$

$$\exp(-at)\xrightarrow{\text{Laplace } i\omega}\frac{1}{a+i\omega} \tag{11.46}$$

Applied to the convolution product, this transformation gives

$$K(t)*f(t)\xrightarrow{\text{Laplace } i\omega}\tilde{K}(i\omega)\,\tilde{f}(i\omega) \tag{11.47}$$

(continued)

[13] Jean-Baptiste Joseph Fourier (1768–1830): French mathematician and physicist; Grenoble, Lyon, and Paris, France.
[14] Pierre Simon Laplace (1749–1827): French mathematician, astronomer, and physicist; Paris, France.

and when the kernel is the exponential function, as is the case for relaxation, one has

$$\exp(-at) * f(t) \xrightarrow{\text{Laplace } i\omega} \frac{\tilde{f}(i\omega)}{a + i\omega} \tag{11.48}$$

Finally, the transformation of the time derivation gives

$$\frac{\mathrm{d}}{\mathrm{d}t} f(t) \xrightarrow{\text{Laplace } i\omega} i\omega\, \tilde{f}(i\omega) - f(0) \tag{11.49}$$

Three peculiar cases are worth discussing.

1. The free relaxation case is retrieved by giving to the imposed flow the shape of a pure impulse, mathematically expressed as the Dirac* distribution $\delta(t)$. In that case the convolution amounts to the identity operator and the outcome is the exponential function itself giving Equation 11.33. (This case is just for checking one of the possibilities of the model and does not justify the usage of the powerful tool of convolution.)
2. The constant perturbation case is modeled by choosing for the imposed flow a step signal, in mathematical terms, a function proportional to the Heaviside function, or unit step, $\mathcal{H}(t)$, previously defined in Chapter 10, Section 10.1.4. The response is a signal proportional to the integral of the kernel function according to the following scheme:

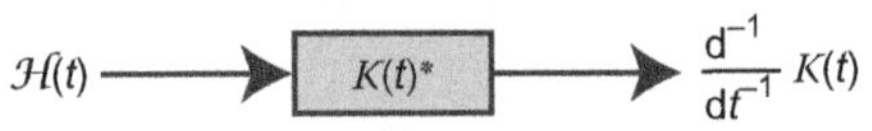

Note that this constitutes a particularly interesting method for characterizing unknown transfer functions, easier to practice than a pure impulse that is physically difficult to realize. (A perfect step jumping infinitely fast is not really easier but the first instants are not always significant.) Here in the case of the relaxation and in assuming linear constitutive properties, the response is therefore

$$q \underset{lin}{=} \frac{f_q(0)}{k_{qC}} (1 - \exp(-k_{qC}\, t)) \tag{11.50}$$

3. The small amplitude perturbation provoked by imposing an AC current is handled by using the mentioned property of the convolution to be transformed into an ordinary product

$$\exp(-k_{qc}t) * f_q(t) \xrightarrow{\text{Laplace } i\omega} \frac{\tilde{f}_q(i\omega)}{k_{qc} + i\omega} \tag{11.51}$$

* Paul Adrien Maurice Dirac (1902–1984): British physicist; Bristol, UK and Tallahassee, Florida, USA.

Mathematical transforms may be superfluous. It can be remarked that the shape of the denominator of this transformed function follows (11.51) the shape of the first-order differential equation leading to the exponential function.

$$\left\{\frac{\mathrm{d}}{\mathrm{d}t}+k_{qc}\right\}f_q(t) \xrightarrow{\text{Laplace } i\omega} \frac{\tilde{f}_q(i\omega)}{k_{qc}+i\omega} \tag{11.52}$$

This means that a differential equation can be directly transformed without having to solve the equation (this is the main interest of Laplace transform). The corollary is also that a careful use of operational calculus mimics what can be done with transformed functions, thus exempting one from making the transformations. This is the reason why we do not favor such techniques that introduce superfluous abstraction.

Nonclassical space–time. More generally, the convolution models any relationship between two variables linked by a time-dependent operator. (Naturally, the convolution is not limited to the time dependence and can be used for space dependence as well.) For instance, the usage of the convolution in Chapter 10 corresponded to the replacement of the temporal integration at a power p by a convolution with a power p–1 of the time variable

$$\frac{\mathrm{d}^{-p}}{\mathrm{d}t^{-p}} f(t) = \frac{t^{p-1}}{\Gamma(p)} * f(t); \quad (p>0) \tag{11.53}$$

with $\Gamma(p)$ the Euler or Gamma function (cf. Section 10.1.4 in Chapter 10). This correspondence allows one to generalize the evolution operator as the convolution product with a specific $T(t)$ kernel:

Evolution operator: $$\hat{\mathrm{T}}\, f(t) = \int_0^t T(t-t')\; f(t')\mathrm{d}t'; \quad \hat{\mathrm{T}}\, f(t) = T(t) * f(t) \tag{11.54}$$

The classical and continuous case of the time integration corresponds to a kernel equal to unity (or more exactly to a Heaviside function $\mathcal{H}(t)$). The choice of a different kernel is a way for attributing nonclassical properties to space–time.

11.4.4 Complex Transfer Functions

In the restricted case of linear constitutive properties, by multiplying the previous transformed function (shown in Equation 11.51) by the capacitance, one is able to express the transformed effort, and then deduce the well-known expressions of the complex transfer functions that are the admittance and the impedance:

Parallel RC: $$\tilde{Y}_q \underset{lin}{=} G_q(1+i\omega\,\tau_{qC}); \quad \tilde{Z}_q \underset{lin}{=} \frac{R_q}{1+i\omega\,\tau_{qC}} \quad \text{with } \tau_{qC} = \frac{1}{k_{qC}} = \frac{1}{R_q C_q} \tag{11.55}$$

In case studies H4 through H6 were given the plots of these two complex transfer functions in one of the most used representation, the Cole–Cole plot, also called a Nyquist plot in physical chemistry. For a capacitive impedance (having a negative imaginary part), it consists of plotting the opposite of the imaginary part $-\Im m(Z_q)$ versus the real part $\Re e(Z_q)$ for various angular frequencies. If the representation of a capacitive admittance is extremely simple (a vertical straight line positioned at the value of the conductance on the real axis) and does not need to be plotted, the representation of the capacitive impedance under the form of a semicircle is less straightforward and needs some explanations.

The complex expression of the impedance can be separated into a real part and an imaginary part:

$$\text{Parallel RC:} \qquad \Re e\left(\tilde{Z}\right) \underset{lin}{=} R\frac{1}{1+\omega^2\tau_C^2}; \quad \Im m\left(\tilde{Z}\right) \underset{lin}{=} -R\frac{\omega\,\tau}{1+\omega^2\tau_C^2} \tag{11.56}$$

These two functions are plotted in Figure 11.13 (right) as *dispersion** *curves*, representing each one as a function of the logarithm of the angular frequency, $-\Im m(Z_q)$ versus $\ln(\omega/k_{cq})$ and the same figure shows the Cole–Cole (Nyquist) plot, placing under it another dispersion curve of the real part, but symmetrically with respect to the bisector, $\Re e(Z_q)$ versus $\ln(\omega/k_{cq})$. This is to explain how a semicircle is obtained in the Cole–Cole (Nyquist) representation.

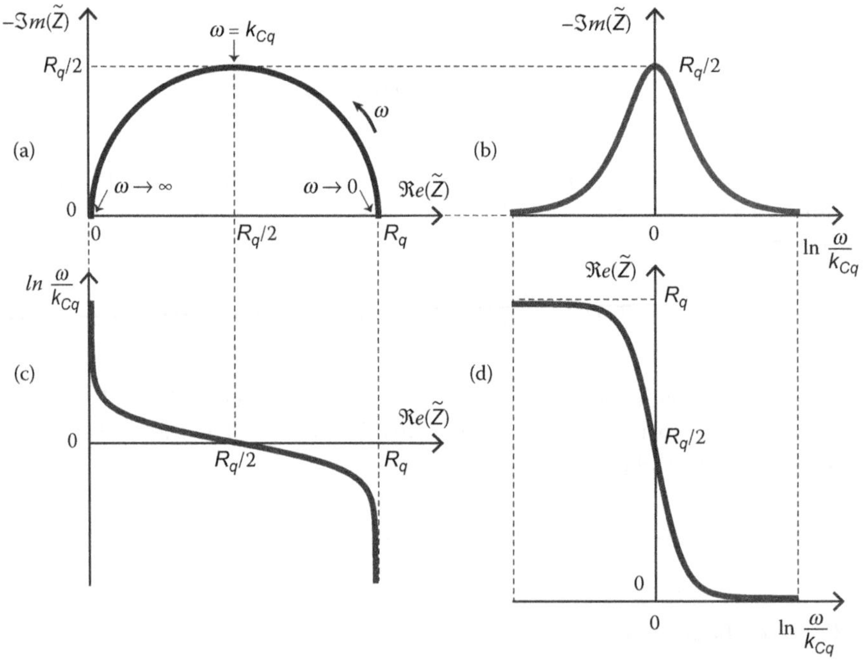

FIGURE 11.13 Plots of the complex impedance of a parallel RC circuit according to the Cole–Cole (Nyquist) representation (a) and to the dispersion representation (b, d). A 45° mirror copy (c) of the dispersion curve of the real part is placed under the Cole–Cole plot (a) for evidencing the construction of the semicircle through its projections (b, c) in the dispersion representation.

* Dispersion describes a frequency-dependent response of a medium in forced relaxation or wave propagation.

This semicircle shape can be algebraically evidenced in showing that the real and imaginary parts satisfy the equation of a circle centered at a point on the real axis at the distance corresponding to the circle radius $R/2$:

$$\left(\Re\mathrm{e}\left(\tilde{Z}\right)-\frac{R}{2}\right)^2+\Im\mathrm{m}\left(\tilde{Z}\right)^2 \underset{lin}{=} \frac{R^2}{4} \tag{11.57}$$

The kinetic constant k_{cq} is used preferentially in these plots instead of the relaxation time because it plays the role of a scaling variable.

11.5 DAMPED OSCILLATOR (TEMPORAL)

The series of case studies gives only one example of damped oscillator taking the case of the translational mechanical energy variety. It would have been possible to give many more examples in mechanics and in other energy varieties but it would not have added much to the discussion, because exactly the same Formal Graphs are used for modeling. This would have merely strengthened the demonstration of the transversality of the approach of this study, which is perhaps superfluous at present.

What remains to be developed is the generalization of the model to any energy variety and to any dipole mounting. Beyond these objectives, one of the purposes of this section is to establish the wave function of a damped oscillator and of attenuated wave propagation. It is sufficient in this aim to restrain this study to the case of a free damped oscillator.

Given here are the two basic mountings of the three dipoles that are conceivable for an isolated system, providing the mounting algebraic equations and the Formal Graph for each equivalent circuit.

The isolation in the common effort assembly shown in Graph 11.41 requires a null flow. However, in the common flow, it requires a null effort (see Graph 11.42).

These two Formal Graphs appear truly symmetric by inversion of the two energies-per-entity, which allows treating only one case, the other being easily deduced by simple permutation, so this study is devoted to the common flow assembly.

It is not completely fortuitous that the topology of the Formal Graph allows one to consider the way the energy circulates in the system. Although the shape looks more like a square than like a circle, the alternate conversions between inductive energy and capacitive energy are ensured by the two evolution paths embedded in a circular path with the inductance and the capacitance. The presence of a conductance diagonally set perturbs the endless rotation of energy that featured a conservative oscillator.

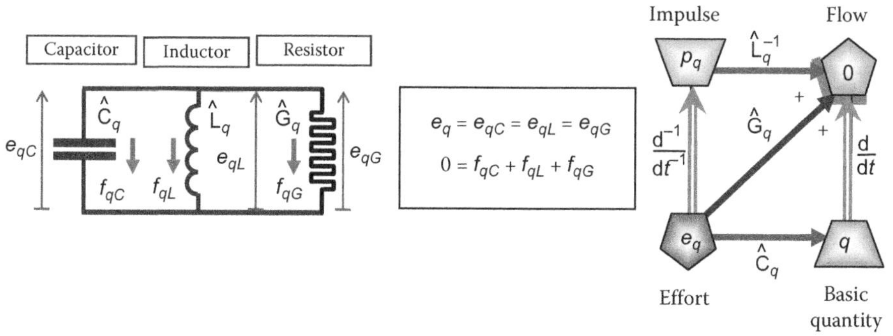

GRAPH 11.41 Damped oscillator in parallel mounting, or common effort assembly.

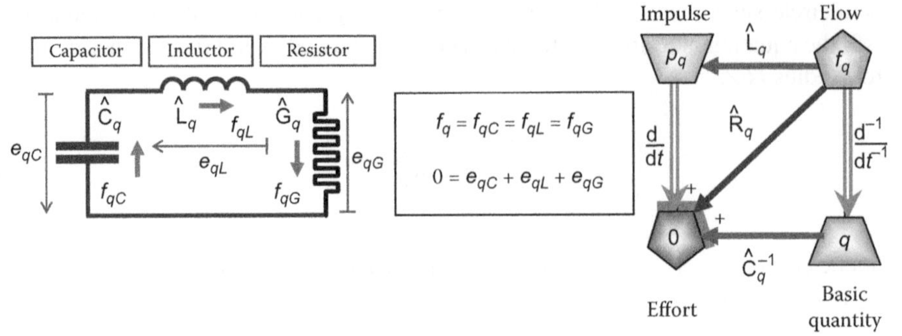

GRAPH 11.42 Damped oscillator in serial mounting, or common flow assembly.

Figure 11.14 schematizes the behavior of the energy in a damped oscillator. The circle represents the conversions of the energy between the two subvarieties, inductive and capacitive, which is a cyclic process. At each turn, part of the energy is diverted into the resistor which converts it into heat. Progressively (depending on the values of the constitutive properties as will be seen), the energy of the oscillator is lost by dissipation and the oscillations vanish.

11.5.1 Algebraic Model

The algebraic translation of the Formal Graph of the damped oscillator is given by the following set of equations:

$$e_{qL} \stackrel{def}{=} \frac{d}{dt} p_q = \frac{d}{dt} \hat{L}_q f_q \tag{11.58}$$

$$e_{qG} \stackrel{def}{=} \hat{R}_q f_q \tag{11.59}$$

$$e_{qC} \stackrel{def}{=} \hat{C}_q^{-1} q = \hat{C}_q^{-1} \frac{d^{-1}}{dt^{-1}} f_q \tag{11.60}$$

Dipole assembly effort:

$$e_{qL} + e_{qG} + e_{qC} \underset{isol}{=} 0 \tag{11.61}$$

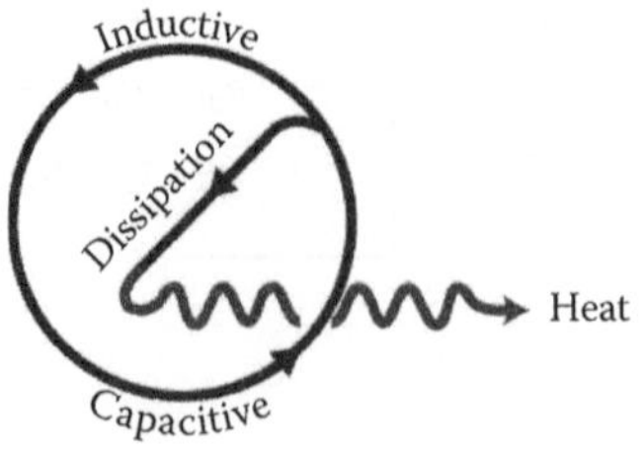

FIGURE 11.14 Schematic behavior of energy illustrating its diversion occurring at each cycle due to the effect of dissipation.

The last equation is the translation of the serial mounting of the oscillator taken as isolated from outside. Assembling this set of equations provides an algebraic equation

$$\left\{\frac{d}{dt}\hat{L}_q + \hat{R}_q + \hat{C}_q^{-1}\frac{d^{-1}}{dt^{-1}}\right\} f_q = 0 \tag{11.62}$$

which can be recast by factorizing chosen terms and by replacing the flow f_q by its expression as a function of the capacitive effort e_{qC} in order to obtain

$$\left\{\frac{d^2}{dt^2} + \hat{R}_q \hat{L}_q^{-1}\frac{d}{dt} + \hat{C}_q^{-1}\frac{d^{-1}}{dt^{-1}}\hat{L}_q^{-1}\frac{d}{dt}\right\}\frac{d^{-1}}{dt^{-1}}\hat{L}_q \frac{d}{dt}\hat{C}_q\, e_{qC} = 0 \tag{11.63}$$

By defining an *inductive damping operator*, which is a system combined property, as

Inductive damping operator: $$\hat{\gamma}_{qL} \overset{def}{=} \frac{1}{2}\hat{R}_q \hat{L}_q^{-1} \tag{11.64}$$

and by defining the square of an *natural pulsation operator* as

Natural pulsation operator (squared): $$\hat{\omega}_q^2 \overset{def}{=} \hat{C}_q^{-1}\frac{d^{-1}}{dt^{-1}}\hat{L}_q^{-1}\frac{d}{dt} \tag{11.65}$$

the general equation modeling the damped oscillator is written as

$$\left\{\frac{d^2}{dt^2} + 2\hat{\gamma}_{qL}\frac{d}{dt} + \hat{\omega}_q^2\right\}\hat{\omega}_q^{-2} e_{qC} = 0 \tag{11.66}$$

This is a more general model than the classical one that assumes linear system constitutive properties instead of operators, allowing one to find solutions for all variables as the model becomes a classical second-order differential equation. However, a higher degree of generality can be kept by merely assuming commutativity between the *inductive damping operator* and the temporal derivation.

Restriction.

Commutativity: $$\left[\frac{d}{dt}, \hat{\gamma}_{qL}\right] = \frac{d}{dt}\hat{\gamma}_{qL} - \frac{d}{dt}\hat{\gamma}_{qL} = 0 \tag{11.67}$$

This commutativity allows one to rearrange the algebraic equation into

$$\left\{\left\{\frac{d}{dt} + \hat{\gamma}_{qL}\right\}^2 + \hat{\omega}^2\right\}\hat{\omega}_q^{-2} e_{qC} = 0 \tag{11.68}$$

having used the square of the *wave pulsation operator*, which will be defined later once the wave function is established, but identified for the moment as

Wave pulsation operator (squared): $$\hat{\omega}^2 = \hat{\omega}_q^2 - \hat{\gamma}_{qL}^2 \tag{11.69}$$

This new equation is easier to solve as the solutions are given by the equality between squared operators applied to the same variable y (representing the term outside the brackets in the previous equations)

$$\left\{\frac{\mathrm{d}}{\mathrm{d}t}+\hat{\gamma}_{qL}\right\}^2 y=-\hat{\omega}^2 y \tag{11.70}$$

When both the natural pulsation and damping factor are scalars, which corresponds to the hypotheses of *harmonicity* as already seen for the conservative oscillator and of *linear damping*

Restriction.

Harmonicity: $\hat{\omega}_q=\omega_q \sim \mathit{scalar}$ (11.71)

Restriction.

Linear damping: $\hat{\gamma}_{qL}=\gamma_{qL} \sim \mathit{scalar}$ (11.72)

the wave pulsation becomes also a scalar

Wave pulsation: $$\omega=\sqrt{\omega_q^2-\gamma_{qL}^2} \tag{11.73}$$

which can be a real or complex number depending on the sign of the difference under the radical as will be discussed soon.

With these restrictive conditions, the solution of the second-order Equation 11.70 is

$$y=\exp(-\gamma_{qL}t)\left(y_1(0)\exp(i\omega t)+y_2(0)\exp(-i\omega t)\right) \tag{11.74}$$

The first exponential, with an always real argument, models a relaxation, which is a decaying process corresponding to the dissipation. The other exponentials model either another relaxation or an oscillating process, depending on whether their argument is imaginary or real. Indeed, the presence of oscillations depends on the sign of the difference of operators in Equation 11.69

$$\gamma_{qL}<\omega_q \quad \Leftrightarrow \quad \omega \sim \mathit{real} \quad \Leftrightarrow \quad \mathit{oscillations} \tag{11.75}$$

$$\gamma_{qL}>\omega_q \quad \Leftrightarrow \quad \omega \sim \mathit{imaginary} \quad \Leftrightarrow \quad \mathit{relaxation} \tag{11.76}$$

(Oscillation can be evidenced by using the Euler formula demonstrated earlier in Chapter 9.) The validity domain of the regimes of the oscillator is sketched in Figure 11.15.

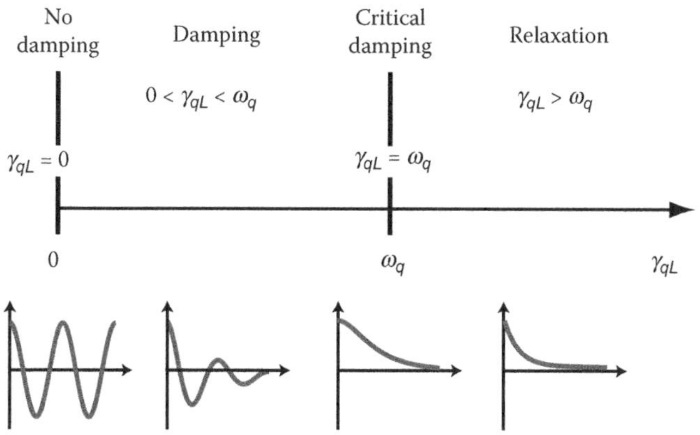

FIGURE 11.15 Influence of the *inductive damping factor* on the behavior of the oscillator.

Alternate solving. It would have been shorter to proceed by Fourier transforming the algebraic model of this system as discussed in Chapter 9. However, some generality would have been lost as the return from Fourier space to normal space requires linear operators for being tractable. Here, only some combinations of operators are requested to be scalar, which is a bit more general.

Now stepping into the Formal Graph approach leads more easily to the building up of a wave function, from which the solutions in terms of the time dependence of the state variables of the system will be derived.

11.5.2 Making the Graph Circular

By splitting the conductive link into halves, as shown in Graph 11.43a, an *inductive–conductive effort* and a *capacitive–conductive effort* can be defined as

$$e_{qLG} \overset{def}{=} \left\{ \frac{d}{dt}\hat{L}_q + \frac{1}{2}\hat{R}_q \right\} f_q \tag{11.77}$$

$$e_{qCG} \overset{def}{=} \left\{ \hat{C}_q^{-1} \frac{d^{-1}}{dt^{-1}} + \frac{1}{2}\hat{R}_q \right\} f_q \tag{11.78}$$

Naturally, their sum is also equal to zero by virtue of the isolation of the dipole assembly

Dipole assembly effort: $$e_{q_{LG}} + e_{q_{CG}} \underset{isol}{=} 0 \tag{11.79}$$

The previously given definition (see Equation 11.64) of the inductive damping operator is used to define an apparent evolution operator as the sum of this inductive damping operator with the temporal derivation

Apparent evolution operator: $$\bar{T}^{-1} = \frac{\bar{d}}{dt} \overset{def}{=} \left\{ \frac{d}{dt} + \hat{\gamma}_{qL} \right\} \tag{11.80}$$

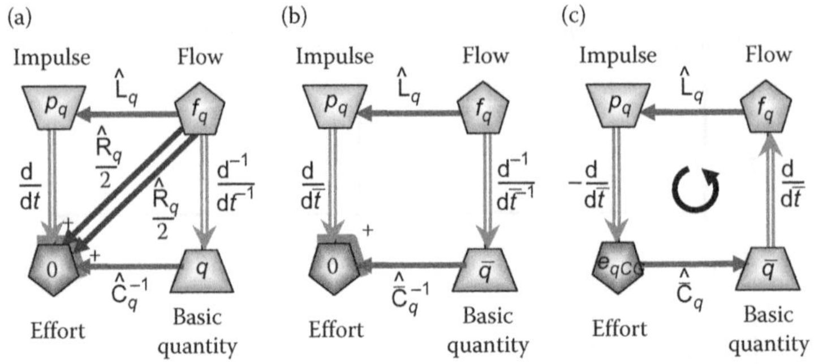

GRAPH 11.43 Steps leading to the transformation of a three-component Formal Graph into a circular graph. (a) The Formal Graph of three components in series with a split conduction. (b) Formal Graph of two components in series with an apparent capacitor. (c) The Formal Graph here is the circular version of Graph (b). The two last Formal Graphs use an *apparent temporal operator* instead of the normal time derivation/integration.

This allows the relationship between the impulse and the *inductive–conductive effort* to be written in the simple form

$$e_{qLG} = \frac{d}{d\bar{t}}\, p_q \tag{11.81}$$

Now, the key idea for obtaining an equivalent Formal Graph built up with two components is to impose the same operator on the right side of the graph, which amounts to defining an *apparent basic quantity* as

$$\bar{q} \overset{def}{=} \frac{d^{-1}}{d\bar{t}^{-1}} f_q \tag{11.82}$$

and an apparent capacitance as

$$\bar{q} \overset{def\rightarrow}{=} \hat{\hat{C}}_q\, e_{qCG} \tag{11.83}$$

These definitions are represented in Graph 11.43b, which models a simpler oscillator made by assembling in series an inductor with an apparent capacitor, but using an apparent temporal operator in lieu of the normal one.

In Graph 11.43b, the relationship between apparent and normal operators for time integration can be read as corresponding to

$$\frac{d^{-1}}{d\bar{t}^{-1}} = \hat{\hat{C}}_q \hat{C}_q^{-1} \left\{ \frac{d^{-1}}{dt^{-1}} + \hat{\gamma}_{qC}^{-1} \right\} \tag{11.84}$$

having defined a capacitive damping (through its reciprocal) as the combination of capacitive and resistive operators

$$\hat{\gamma}_{qC}^{-1} \overset{def}{=} \hat{C}_q \frac{1}{2} \hat{R}_q \tag{11.85}$$

Transformation of this Formal Graph of two components in series into a circular one is trivial; it relies on placing in the effort node the *capacitive–conductive effort*, which by virtue of the isolation relationship 11.79 is linked to the impulse by merely reverting the apparent temporal derivation in using a minus sign

$$e_{qCG} = -\frac{d}{d\bar{t}} p_q \tag{11.86}$$

The circular Formal Graph thus obtained is represented in Graph 11.43c. As this model exactly mimics the behavior of a conservative oscillator, the treatment follows the same procedure used in Chapter 9, which will not be detailed as much.

Energy conservation and time. It is worth giving a word of caution about the differences between the present oscillator and the conservative oscillator seen in Chapter 9. At first glance, the identity of Formal Graphs, both being circular, between these two systems could induce the feeling that a conservation of oscillator energy is enforced also in the damped oscillator. This is obviously a false deduction as the damping is due to dissipation of the oscillator energy, which decreases owing to conversion into heat. The explanation is that time is not the same in both models. In the apparent space–time of the damped oscillator, its energy appears to be a constant, but not in the normal space–time.

This works also for the reversibility of the oscillator. With an apparent time, an oscillator that is considered irreversible, owing to an uneven number of derivation operators with respect to the ordinary time, becomes apparently reversible.

This illustrates the link between time, symmetry, and energy conservation, which is a great principle in classical physics. It also outlines the importance of defining the validity domain of any invariance rule or principle and that the notion of reversibility depends on the definition of time.

11.5.3 Splitting the Circular Graph

The Formal Graph in Graph 11.44 is merely an adaptation of the one drawn in Chapter 9 to the apparent variables used in this model. It results from the symmetrical splitting of Graph 11.43c made for solving the model (by reducing the degree of temporal operators on the template of a second-order polynomial equation).

In Graph 11.44, the intermediate variables placed halfway along the system properties, which are called *semienergies*, are defined as:

$$\bar{A}_q \overset{def}{=} \hat{C}_q^{1/2} e_{qCG} \overset{def}{=} \hat{C}_q^{-1/2} \bar{q} \tag{11.87}$$

$$B_q \overset{def}{=} \hat{L}_q^{1/2} f_q \overset{def}{=} \hat{L}_q^{-1/2} p_q \tag{11.88}$$

They are linked by the angular derivation defined with an *apparent phase angle*

$$\frac{d}{d\bar{\varphi}(\bar{t})} \overset{def}{=} \hat{C}_q^{1/2} \left\{ -\frac{d}{d\bar{t}} \right\} \hat{L}_q^{1/2} = -\hat{L}_q^{1/2} \frac{d}{d\bar{t}} \hat{C}_q^{1/2} \tag{11.89}$$

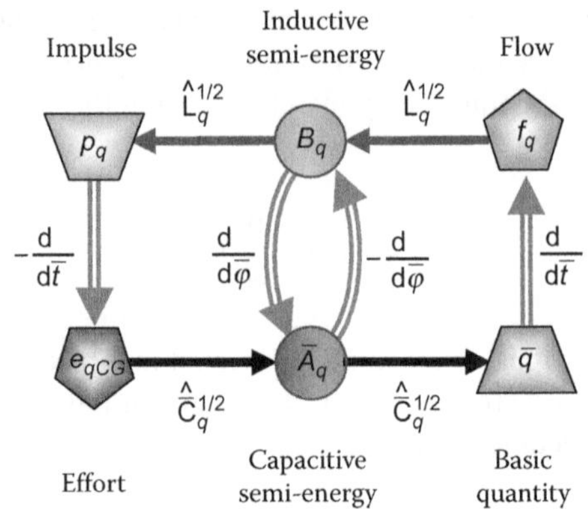

GRAPH 11.44 Insertion of halfway intermediate variables called *semienergies*, leading to a reversible direct relationship between them.

The second equality in the above expression results from the assumption of commutativity rules between the systems properties and the apparent time derivation.

Hypothesis.

Commutativity: $$\left[\frac{d}{d\bar{t}}, \widehat{C}_q^{1/2}\right] = \frac{d}{d\bar{t}}\widehat{C}_q^{1/2} - \widehat{C}_q^{1/2}\frac{d}{d\bar{t}} = 0 \tag{11.90}$$

Commutativity: $$\left[\frac{d}{d\bar{t}}, \hat{L}_q^{1/2}\right] = \frac{d}{d\bar{t}}\hat{L}_q^{1/2} - \hat{L}_q^{1/2}\frac{d}{d\bar{t}} = 0 \tag{11.91}$$

Commutativity: $$\left[\hat{L}_q^{1/2}, \widehat{C}_q^{1/2}\right] = \hat{L}_q^{1/2}\widehat{C}_q^{1/2} - \widehat{C}_q^{1/2}\hat{L}_q^{1/2} = 0$$

These rules allow one to write the angular derivation in a simpler form:

$$\frac{d}{d\varphi(\bar{t})} = -\widehat{C}_q^{1/2}\hat{L}_q^{1/2}\frac{d}{d\bar{t}} \tag{11.92}$$

11.5.4 Wave Function

From Graph 11.44 are deduced the relationships between semienergies

$$B_q = -\frac{d}{d\varphi}\bar{A}_q \tag{11.93}$$

$$\bar{A}_q = \frac{d}{d\varphi}B_q \tag{11.94}$$

These two relationships express the rotation of the wave function, which is defined as the linear combination of semienergies as was done for the wave function of a conservative oscillator in Chapter 9

Wave function: $$\psi_q \overset{def}{=} \overline{A}_q + i\, B_q \tag{11.95}$$

From these rotation relationships and wave function definition are deduced the two models of the wave function, the differential one and the integral one resulting from the definition of the exponential function seen in Section 10.5.4 in Chapter 10:

$$\boxed{\frac{d}{d\overline{\varphi}}\psi_q = i\,\psi_q} \quad \Leftrightarrow \quad \boxed{\psi_q = \psi_q(0)\ \exp\left(i\overline{\varphi}\right)} \tag{11.96}$$

Reproducing the reasoning and using the definitions of harmonic functions made in Section 9.5.5 in Chapter 9 leads to an identical writing of the solutions:

$$\overline{A}_q = \psi_q(0)\cos\overline{\varphi} \tag{11.97}$$

$$B_q = \psi_q(0)\sin\overline{\varphi} \tag{11.98}$$

It can be anticipated from the following treatment in saying that the simplicity of these solutions is only apparent, for it is clear that the apparent phase angle $\overline{\varphi}$ hides a complex dependence with time.

11.5.5 Dependence on Time

By combining the two differential equations 11.92 and 11.96 one gets an interesting equality between operators:

$$i\frac{d}{dt}\psi_q = \hat{L}_q^{-1/2}\,\hat{\overline{C}}_q^{-1/2}\,\psi_q \tag{11.99}$$

As for the conservative oscillator, a wave is associated with this damped oscillator, featured by an operator defined as

Wave pulsation operator: $$\hat{\omega} \overset{def}{=} i\frac{d}{dt} \tag{11.100}$$

and by a *wave pulsation* defined as the eigen-value of this operator when the eigen-function is the wave function

Wave pulsation eigen-value: $$\hat{\omega}\,\psi_q = \omega\,\psi_q \tag{11.101}$$

These definitions lead to equality between the wave pulsation operator and the combination of system properties used in Equation 11.99:

$$\hat{\omega} = \hat{L}_q^{-1/2}\,\hat{\overline{C}}_q^{-1/2} \tag{11.102}$$

It is difficult to go further without detailing the exact form of this operator combination, so the same restrictions as the ones in the algebraic development are made. The eigen-value of the pulsation operator is easily determined when the natural pulsation and the damping factor are both scalars, as explained earlier. The following hypotheses were made for this development.

Hypothesis.

Harmonicity: $\hat{\omega}_q = \omega_q \sim scalar$ (11.103)

Hypothesis.

Linear damping: $\hat{\gamma}_{qL} = \gamma_{qL} \sim scalar$ (11.104)

With these conditions, the wave pulsation also becomes a scalar and the relationship shown in Equation 11.73 still holds

Wave pulsation:
$$\omega = \sqrt{\omega_q^2 - \gamma_{qL}^2} \tag{11.105}$$

and Equation 11.99 now can be written as

$$i\frac{\mathrm{d}}{\mathrm{d}t}\psi_q = \omega\,\psi_q \tag{11.106}$$

The assumption of *linear damping* allows decomposition of the apparent temporal derivation, and, by redistributing the terms, the following first-order differential equation is obtained

$$\frac{\mathrm{d}}{\mathrm{d}t}\psi_q = -\left(\gamma_{qL} + i\,\omega\right)\psi_q \tag{11.107}$$

which has for integral solution

$$\psi_q = \psi_q(0)\exp\left(-\gamma_{qL}t\right)\exp\left(-i\,\omega t\right) \tag{11.108}$$

The apparent phase angle $\bar{\varphi}$ is a complex variable expressed as

$$\bar{\varphi} = -\left(\omega - i\gamma_{qL}\right)t \tag{11.109}$$

and the expressions of the semienergies are:

$$\bar{A}_q = \psi_q(0)\exp\left(-\gamma_{qL}t\right)\cos\omega t \tag{11.110}$$

$$B_q = \psi_q(0)\exp\left(-\gamma_{qL}t\right)\sin\omega t \tag{11.111}$$

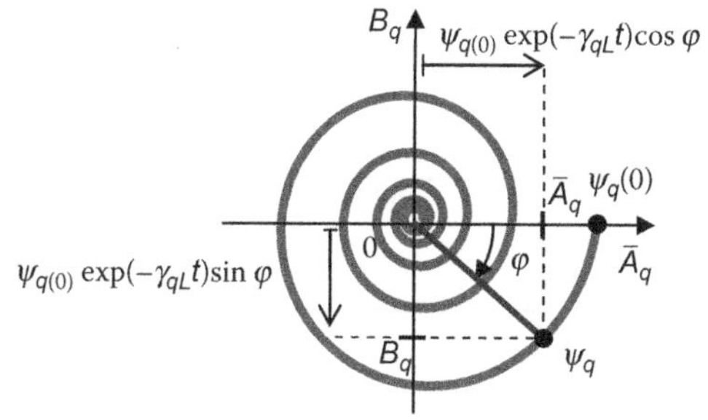

FIGURE 11.16 Plot of the wave function of a damped oscillator in polar coordinates. The vector ψ_q rotates clockwise as time elapses and its modulus decreases exponentially.

The representative vector of the wave function rotates with a phase angle

$$\varphi = -\omega\, t \tag{11.112}$$

proportional to the time and its modulus decays exponentially with time, as illustrated by the plot in Figure 11.16.

Now, the energies-per-entity of the oscillator can be given by using definitions 11.87 and 11.88 of the half-energies:

$$f_q = \hat{\mathsf{L}}_q^{-1/2}\,\psi_q(0)\exp\left(-\gamma_{qL}t\right)\sin\varphi \tag{11.113}$$

$$e_{qCG} = \hat{\hat{\mathsf{C}}}_q^{-1/2}\,\psi_q(0)\exp\left(-\gamma_{qL}t\right)\cos\varphi \tag{11.114}$$

The apparent capacitance is related to the true capacitance through

$$\hat{\hat{\mathsf{C}}}_q^{-1/2} = \hat{\mathsf{C}}_q^{-1/2}\cos\phi \tag{11.115}$$

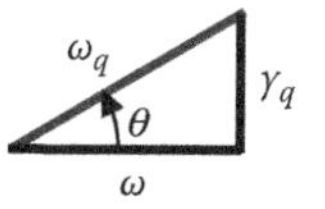

which uses a *shift angle* defined as (cf. scheme above)

$$\sin\phi \overset{def}{=} \frac{\gamma_q}{\omega_q} \tag{11.116}$$

This *shift angle* expresses the relative amount of damping and corresponds to the phase shift in the oscillations of the various state variables of the oscillator, as can be seen in the expressions of the individual efforts:

$$e_{qC} = \hat{\mathsf{C}}_q^{-1/2}\,\psi_q(0)\exp\left(-\gamma_{qL}t\right)\cos\left(\varphi+\phi\right) \tag{11.117}$$

$$e_{qL} = -\hat{\mathsf{C}}_q^{-1/2}\,\psi_q(0)\exp\left(-\gamma_{qL}t\right)\cos\left(\varphi-\phi\right) \tag{11.118}$$

$$e_{qG} = \hat{\mathsf{R}}_q\hat{\mathsf{L}}_q^{-1/2}\,\psi_q(0)\exp\left(-\gamma_{qL}t\right)\sin\varphi \tag{11.119}$$

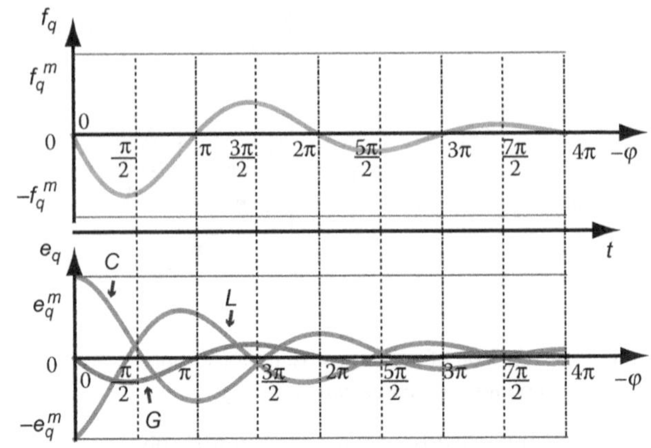

FIGURE 11.17 Example of variations of the energies-per-entity of a damped oscillator as a function of the opposite of the phase angle, which is proportional to time. The curves of the capacitive, inductive, and conductive components of the effort are labeled C, L, and G, respectively.

These variables are plotted in Figure 11.17 against the opposite $-\varphi$ of the phase angle which is proportional to the time.

11.6 SPATIALLY DAMPED OSCILLATOR

The Formal Graphs of a spatial oscillator that is subject to damping are given in Graph 11.45, in two versions according to the chosen orientation for the spatial derivations.

The damping is not caused by a conductance link as for the temporal oscillator but by another capacitance link between the localized effort and basic quantity (termed *fully reduced capacitance* and notated C_{q11}). Therefore, dissipation is not implied and another mechanism is involved here. A purely spatial oscillator must be viewed as the siege of a conflict between two distributions of energy into space, as is the case between inductive and capacitive energies that are in conflict through the time operator working in opposite directions. The energy is distributed between the global level and a localized level, but in two different ways. The two capacitances supporting the distributed energies work in opposite directions, one linking a localized effort to a global basic quantity (notated C_{q10}) and the other linking a global effort to a localized basic quantity (notated

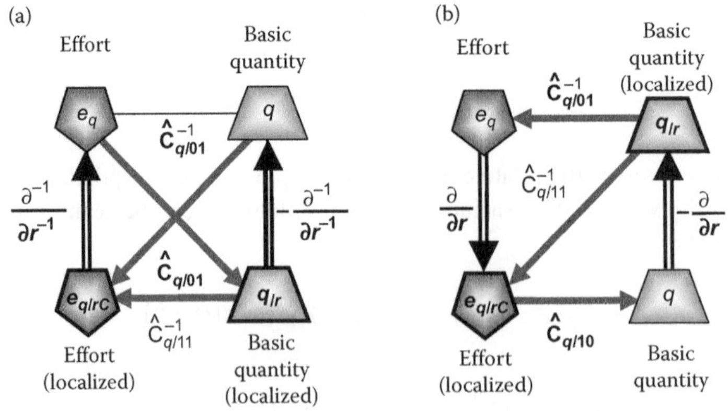

GRAPH 11.45 Two equivalent Formal Graphs of a spatially damped oscillator, with parallel orientation of space derivations (a) and with antiparallel orientation (b). The two graphs correspond by vertical permutation of the basic quantities.

C_{q01}). Physically, energy is distributed whether among energies-per-entity or among entities, which cannot be ensured at the same location, thus leading to conflict.

The supplementary link comes as a constraint preventing the phase opposition between localized variables from working normally. Without the fully reduced capacitance these variables may follow opposite variations, one being at its maximum while the other is at its minimum and the converse at another moment of the cycle. This is not compatible with a "rigid" link that can only cope with these contradictory behaviors in diverting a part of the oscillating energy toward another form of distribution (fully localized) which is not in conflict with the two others.

Once this physical insight is given and the analogy with the temporal damped oscillator is established, the modeling follows an identical route. The diagonal link in Graph 11.45b is split, as shown in Graph 11.46a for allowing the creation of two paths starting from the localized basic quantity and converging at the effort node, instead of the three initial paths. The suppression of the diagonal link is realized by creating apparent variables and operator, namely effort, localized basic quantity, reduced capacitance, and spatial derivation. Then one of the two localized efforts (combined on the same null node), the one issued from the global effort, is placed in the localized effort node for building up a loop as shown in Graph 11.46b.

Algebraically, this procedure corresponds to the creation of two localized efforts as follows:

$$\mathbf{e}_{q/rq} = \left\{ \hat{\mathbf{C}}_{q10}^{-1} \left\{ -\frac{\mathrm{d}^{-1}}{\mathbf{d}\mathbf{r}^{-1}} \right\} + \frac{\hat{\mathrm{C}}_{q11}^{-1}}{2} \right\} q_{/r} \tag{11.120}$$

$$\mathbf{e}_{q/re} = \left\{ \frac{\mathrm{d}}{\mathbf{d}\mathbf{r}} \hat{\mathbf{C}}_{q01}^{-1} + \frac{\hat{\mathrm{C}}_{q11}^{-1}}{2} \right\} q_{/r} \tag{11.121}$$

respecting the isolation of the oscillator that is expressed by the null summing of these two variables

$$\mathbf{e}_{q/re} + \mathbf{e}_{q/rq} \underset{isol}{=} \mathbf{0} \tag{11.122}$$

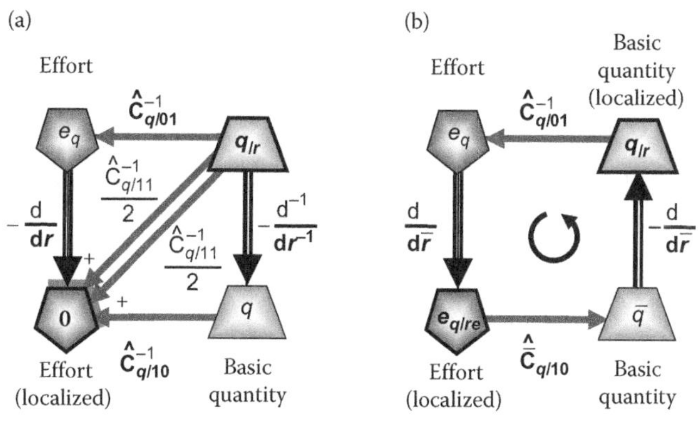

GRAPH 11.46 Two steps in the transformation of the Formal Graph of a spatially damped oscillator. (a) Split of the damping capacitance (diagonal link) and (b) apparent variables and operators are chosen for obtaining a circular graph.

A *spatial damping operator* (symbolized by ζ) is defined in the same manner as a temporal damping operator

$$\hat{\zeta}_q \overset{def}{=} \frac{\hat{C}_{q11}^{-1}}{2} \hat{C}_{q01} \tag{11.123}$$

and an apparent spatial derivation is defined accordingly

$$\frac{d}{d\bar{r}} \overset{def}{=} \left\{ \frac{d}{dr} + \hat{\zeta}_q \right\} \tag{11.124}$$

The apparent capacitance and the apparent basic quantity are defined as

$$\bar{q} \overset{def}{=} -\frac{d^{-1}}{d\bar{r}^{-1}} q_{/r} \overset{def\rightarrow}{=} \hat{C}_{q/10}\, e_{q/re} \tag{11.125}$$

These definition being established, the splitting of the Formal Graph into halves is made by defining half-energies as

$$\bar{A}_q \overset{def}{=} \hat{C}_{q/10}^{1/2} . e_{q/re} \overset{def}{=} \hat{C}_{q/10}^{-1/2} \bar{q} \tag{11.126}$$

$$B_q \overset{def}{=} \hat{C}_{q/01}^{-1/2} . q_{/r} \overset{def}{=} \hat{C}_{q/01}^{1/2} e_q \tag{11.127}$$

As for the temporal damped oscillator, these half-energies are related by two derivations (with + and – signs) with respect to an apparent phase angle $\bar{\varphi}$ defined by

Apparent phase angle:

$$\frac{d}{d\bar{\varphi}(\bar{r})} \overset{def}{=} \hat{C}_{q/10}^{1/2} \hat{C}_{q/01}^{-1/2} \frac{d}{d\bar{r}} \tag{11.128}$$

assuming that commutativity existed among these operators.

Definition 11.95 of the wave function as a linear combination of half-energies still holds, leading to the same exponential solution 11.96, which is not repeated here.

11.6.1 Dependence on Space

As for the temporal oscillators, a spatial wave is associated with this damped oscillator, featured by a *wave-vector operator* defined as

Wave-vector operator:

$$\hat{\mathbf{k}} \overset{def}{=} -i \frac{d}{d\bar{r}} \tag{11.129}$$

From definition 11.128 of the phase angle this operator can be identified with the composition of reduced capacitances

$$\hat{\mathbf{k}} = \hat{C}_{q/01}^{1/2} \hat{C}_{q/10}^{-1/2} \tag{11.130}$$

As the general solution cannot be found without specifying the shape of these capacitances, we need to restrict the generality to the case where an eigen-vector can be determined according to the equality

Wave-vector eigen-value:
$$\hat{\mathbf{k}}\,\psi_q(\boldsymbol{r}) = \boldsymbol{k}\,\psi_q(\boldsymbol{r}) \tag{11.131}$$

This eigen-vector is the *wave-vector* of the spatial wave and it is related to the *natural wave-vector* $\boldsymbol{k}_q$ and to the *damping vector* $\boldsymbol{\zeta}_q$, which are the eigen-vectors of their operators. By assuming that these operators reduce to vectors (which is similar to temporal operators reducing to scalars),

Hypothesis.

Harmonicity:
$$\hat{\mathbf{k}}_q = \boldsymbol{k}_q = \hat{\mathbf{C}}_{q/01}^{1/2}\hat{\mathbf{C}}_{q/10}^{-1/2} \quad \sim \textit{vector} \tag{11.132}$$

Hypothesis.

Linear damping:
$$\hat{\boldsymbol{\zeta}}_q = \boldsymbol{\zeta}_q = \frac{\hat{C}_{q11}^{-1}}{2}\hat{\mathbf{C}}_{q01} \quad \sim \textit{vector} \tag{11.133}$$

the wave-vector becomes

Wave-vector:
$$k = \sqrt{k_q^2 - \zeta_q^2} \tag{11.134}$$

With these restrictions, the apparent space derivation defined in Equation 11.124 can be rewritten in terms of the damping vector $\boldsymbol{\zeta}_q$ in lieu of its operator and can be identified with the product of the imaginary number times the wave-vector, owing to definition 11.129 of the wave-vector operator:

$$\frac{\mathrm{d}}{\mathbf{d}\boldsymbol{r}}\psi_q(\boldsymbol{r}) = \left\{\frac{\mathrm{d}}{\mathbf{d}\boldsymbol{r}} + \boldsymbol{\zeta}_q\right\}\psi_q(\boldsymbol{r}) = i\,\boldsymbol{k}\,\psi_q(\boldsymbol{r}) \tag{11.135}$$

The second equality in this equation can be recast into

$$\frac{\mathrm{d}}{\mathbf{d}\boldsymbol{r}}\psi_q(\boldsymbol{r}) = \left(-\boldsymbol{\zeta}_q + i\,\boldsymbol{k}\right)\psi_q(\boldsymbol{r}) \tag{11.136}$$

which is an ordinary differential equation having for integral solution

$$\psi_q(\boldsymbol{r}) = \psi_q(\mathbf{0})\exp\left(-\boldsymbol{\zeta}_q \cdot \boldsymbol{r}\right)\exp\left(i\,\boldsymbol{k}\cdot\boldsymbol{r}\right) \tag{11.137}$$

The apparent phase angle is a complex variable whose expression is deduced from Equation 11.128

$$\bar{\varphi}(\boldsymbol{r}) = (\boldsymbol{k} + i\,\zeta_q)\cdot\boldsymbol{r} \tag{11.138}$$

and the expressions of the semienergies follow:

$$\bar{A}_q = \psi_q(0)\exp(-\zeta_q \cdot \boldsymbol{r})\cos \boldsymbol{k}\cdot\boldsymbol{r} \tag{11.139}$$

$$B_q = \psi_q(0)\exp(-\zeta_q \cdot \boldsymbol{r})\sin \boldsymbol{k}\cdot\boldsymbol{r} \tag{11.140}$$

The representative vector of the wave function rotates with a phase angle

$$\varphi(\boldsymbol{r}) = \boldsymbol{k}\cdot\boldsymbol{r} \tag{11.141}$$

The different variables of the oscillator are deduced from the half-energies by using Equations 11.126 and 11.127, leading to similar solutions to the previously studied temporal damped oscillator, which are not detailed here.

"Spatial irreversibility." By analogy with time, the concept of reversibility as defined in Section 11.1.2 can perfectly be generalized for being also used with space distribution properties. It was shown that the reversibility corresponded to temporal symmetry, whereas irreversibility was associated with asymmetry of time (the arrow of time). A quantitative criterion was defined consisting of counting the number of temporal operators along the paths forming a loop in a Formal Graph. The same association with the symmetry/asymmetry of space and the same criterion can be used for defining the spatial irreversibility, occurring when a loop counts an uneven number of space operators.

Even number of spatial operators in a loop ⇔ Spatially **Reversible** distribution

Uneven number of spatial operators in a loop ⇔ Spatially **Irreversible** distribution

(Note: The spatial reversibility depends on the definition of space. Different conclusions are drawn according to the use of ordinary space or apparent space, as using this latter modifies the number of space derivations.)

"Arrow of space"? By considering the notion of spatial irreversibility, one introduces the correlated notion of "arrow of space" which is not a concept easy to grasp owing to one's everyday experience of the impossibility of going back to the past while one has absolutely no doubt of one's ability to move back to a former place.

This is forgetting the fact that it depends on the reference chosen. If on our planet the spatial reversibility seems obvious, when one refers to the movement of the earth around the sun and more generally to its trajectory across the universe, one cannot assess that spatial reversibility is the general rule.

Inverse problem. The association of the concept of spatial irreversibility with the spatial distribution can be illustrated with the phenomenon of deposition of elastic layers onto an irregularly shaped surface. A good example is the deposition of snow over a ground parsed with objects (such as round rocks and rectangular containers). The more the thickness of snow, the less the shape of the upper surface of snow conforms to the ground profile, smoothing the upper surface layer by layer until the perception of the objects almost disappears when a lot of snow has fallen.

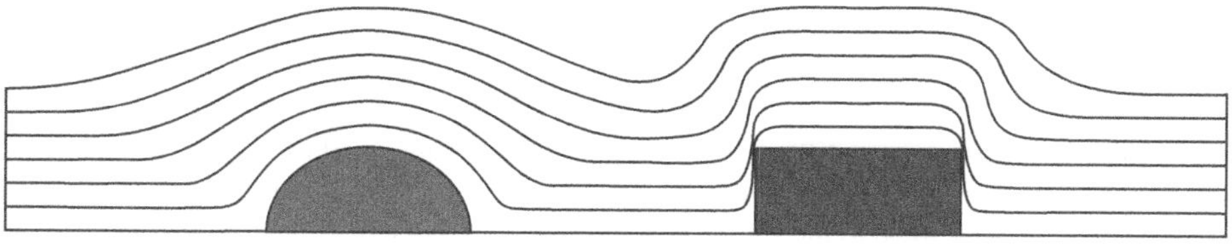

To try to reconstitute the ground profile from the observation of the free surface of the snow is recognized as a difficult problem and even as an impossible task when the thickness of deposit is much higher than the size of rocks. This is known as "inverse problem" and the determination of underlying surfaces is one of the hard challenges in imaging techniques.

11.7 ATTENUATED PROPAGATION

The propagation of a wave in an absorbing medium corresponds to a spatiotemporal oscillator in which a constraint prevents the system from oscillating freely. This constraint can be a conductance, or more precisely a conductivity, in which part of the energy is dissipated, as well as a reduced capacitance ("capacitivity") or inductance ("inductivity") that creates a deviation of the spatial distribution of energy, as for a purely spatial oscillator. These constraints are not exclusive of each other, so one may find a spatiotemporal oscillator damped twice, spatially and temporally. Consequently, three kinds of damping of a spatiotemporal oscillator are distinguished:

- The purely **temporal damping** due to dissipation.
- The purely **spatial damping** due to deviated spatial distribution.
- The **spatial and temporal damping** due to simultaneous dissipation and deviated distribution.

In Graph 11.47 are shown the Formal Graphs in three dimensions of three typical examples of a spatiotemporal oscillator according to the above classification: The temporal damping on the left is caused by a conductivity linking the global effort to the flow lineic density; the spatial damping in the center results from the presence of a reduced capacitance ("capacitivity") at the level of a point of a curve (level 1); and the spatial and temporal damping on the right ensues from the superimposition of the two previous reduced constitutive properties. As for the conservative oscillator studied in Chapter 9, these Formal Graphs model an oscillator working between two adjacent levels of spatial reduction. This corresponds exactly to the vibrating string that uses the global level and the level of curve points. It equally models an electromagnetic wave by shifting down one level to work between curve points and surface points. To model the sound propagation, one needs to expand the Formal Graph by extending it on four levels, but the principle of the damping constraints remains the same.

For a temporal damping, the only possibility is the existence of the diagonal link of the conductivity linking the effort to the flow lineic density. But for a spatial damping, there are two possibilities: either with a reduced capacitance (capacitive damping) or with a reduced inductance (inductive damping), both at level 1 (points of a curve). This makes in fact five different cases to study, but, as

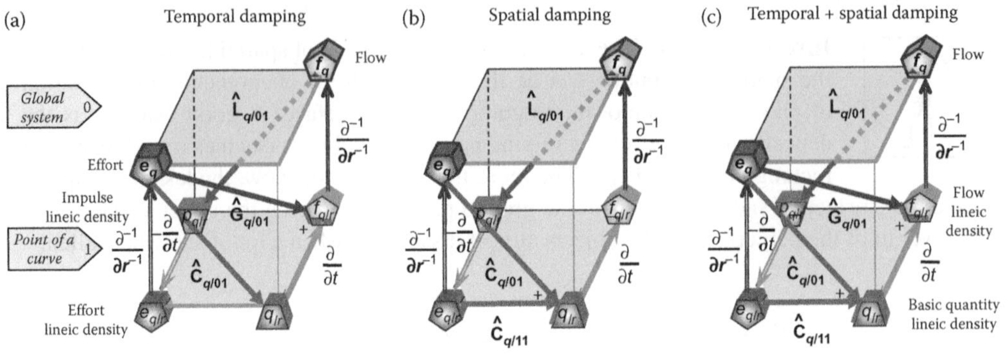

GRAPH 11.47 Three-dimensional Formal Graphs of three spatiotemporal oscillators differently damped. The conductivity makes a purely temporal damping (a), the spatially reduced capacitance (at level 1) makes a purely spatial damping (b), whereas the presence of both properties creates a simultaneous temporal and spatial damping (c). A spatially reduced inductance at level 1 would provoke the same effect as the reduced capacitance.

was discussed for the spatial conservative oscillator, there is no difference at the end in the models between the capacitive and the inductive damping. To facilitate the comparison, the capacitive damping is chosen for this study.

The previous developments in Sections 11.5 and 11.6 about the purely temporal oscillator and the purely spatial oscillator provide the main elements required to establish the model of attenuated propagation of a wave. It must be recalled how the two parameters featuring the double damping are related to the system constitutive properties. The temporal damping is characterized by the inductive damping operator, defined in the same way as done previously:

Temporal inductive damping operator: $$\hat{\gamma}_{qL} \overset{def}{=} \frac{1}{2}\hat{G}_{q/01}^{-1}\hat{L}_{q/01}^{-1} \tag{11.142}$$

The spatial damping is characterized by an operator also defined as for the purely spatial oscillator:

Spatial damping operator: $$\hat{\zeta}_q \overset{def}{=} \frac{1}{2}\hat{C}_{q11}^{-1}\hat{\mathbf{C}}_{q01} \tag{11.143}$$

The general definitions are given in terms of operators, but to simplify the discussion, linear operators are considered from now and the two damping factors are assumed to be lower than their respective thresholds (natural pulsation and space vector) for allowing oscillations to exist. As the modeling of this system proceeds according to the previously detailed models, only the final steps are developed here.

In Chapter 9, the wave function of a spatiotemporal oscillator has been demonstrated as being the product of two individual wave functions, one for the temporal oscillator and the other for the spatial one. From this property, the phase angle (here an apparent one for taking into account the damping) was expressed as a double contribution of the individual phase angles, according to the relationship

$$d\bar{\varphi} = \left(\frac{\partial \bar{\varphi}}{\partial t}\right)_r dt + \left(\frac{\partial \bar{\varphi}}{\partial \mathbf{r}}\right)_t \cdot \mathbf{dr} \tag{11.144}$$

The dependences with respect to time and to space are taken in the respective models, from Equations 11.109 and 11.138

$$\left(\frac{\partial \bar{\varphi}}{\partial t}\right)_r = \frac{d\bar{\varphi}(t)}{dt} = -\omega + i\,\gamma_{qL} \tag{11.145}$$

$$\left(\frac{\partial \bar{\varphi}}{\boldsymbol{r}}\right)_t = \frac{d\bar{\varphi}(\boldsymbol{r})}{\mathbf{dr}} = \boldsymbol{k} + i\,\boldsymbol{\zeta}_q \tag{11.146}$$

The substitution of the partial derivatives into Equation 11.144 provides the following equation for the variation of the apparent phase angle:

$$d\bar{\varphi} = \left(-\omega + i\,\gamma_{qL}\right)dt + \left(\boldsymbol{k} + i\,\boldsymbol{\zeta}_q\right)\cdot\mathbf{dr} \tag{11.147}$$

As the wave characteristics are time invariant and space invariant, integration is straightforward and gives the complex number

$$\bar{\varphi} = \left(-\omega t + \boldsymbol{k}\cdot\boldsymbol{r}\right) + i\left(\gamma_{qL}\,t + \boldsymbol{\zeta}_q\cdot\boldsymbol{r}\right) \tag{11.148}$$

Insertion of this complex number into the argument of the exponential function featuring the wave function provides the expression of this latter:

$$\psi_q = \psi_q(0)\exp\left(-\gamma_{qL}\,t - \boldsymbol{\zeta}_q\cdot\boldsymbol{r}\right)\;\exp\left(-i\omega\,t + i\boldsymbol{k}\cdot\boldsymbol{r}\right) \tag{11.149}$$

This function is the product of the two individual wave functions as mentioned.

The different state variables involved in this model can be deduced from their relationship to the half-energies (that are the real and imaginary parts of the wave function) in the same manner as for the various models previously developed.

The energy amount contained in a wave was stated in Chapter 9, Section 9.7.6, as being equal to one half the modulus of the wave function (assuming linear constitutive properties for getting the coefficient 1/2). For the wave going through the absorbing medium studied here, one gets

$$\mathcal{E}_\lambda \underset{lin}{=} \frac{1}{2}\left\|\psi_q\right\|^2 = \psi_q(0)^2\exp\left(-2\gamma_{qL}\,t - 2\boldsymbol{\zeta}_q\cdot\boldsymbol{r}\right) \tag{11.150}$$

which establishes the exponential decay in time and/or through space through the medium. To retrieve known expressions, the following alternative coefficients may be used:

Inductive relaxation constant: $$k_{qL} \overset{def}{=} 2\gamma_{qL} \underset{homot}{=} \hat{G}_{q/01}^{-1}\hat{L}_{q/01}^{-1} \tag{11.151}$$

Spatial energy-damping factor: $$\boldsymbol{\alpha}_q \overset{def}{=} 2\boldsymbol{\zeta}_q \underset{homot}{=} \hat{C}_{q11}^{-1}\hat{\boldsymbol{C}}_{q01} \tag{11.152}$$

and with these coefficients the irradiance, or radiation intensity, defined in Chapter 9, Section 9.7.6, becomes

$$I_{\lambda q} \underset{lin}{=} \frac{\boldsymbol{u}}{2}\left\|\psi_q\right\|^2 = \frac{\boldsymbol{u}}{2}\psi_q(0)^2\exp\left(-k_{qL}\,t - \boldsymbol{\alpha}_q\cdot\boldsymbol{r}\right) \tag{11.153}$$

These equations model the *spatial and temporal damping* due to simultaneous dissipation and deviated distribution, which is one of the kinds of damping. The two other kinds are:

- The purely temporal damping (due to dissipation):
  The equations modeling this case are obtained by setting $\alpha_q = 0$, giving

$$\psi_q = \psi_q(0)\exp\left(-\gamma_{qL}\,t\right)\ \exp\left(-i\omega\,t + ik\cdot\boldsymbol{r}\right) \tag{11.154}$$

$$I_{\lambda q} \underset{lin}{=} I^0_{\lambda q}\exp\left(-\,k_{qL}\,t\right) \tag{11.155}$$

  The wave function and the irradiance (radiation intensity) obey classical relaxation models, as described earlier for various systems. Damped oscillations are similar to those described in case study H7 "Damped Mechanical Oscillator."

- The purely spatial damping (due to deviated spatial distribution):
  By setting $k_{qL} = 0$ in the spatiotemporal equations, one has

$$\psi_q = \psi_q(0)\exp\left(-\,\boldsymbol{\zeta}_q\cdot\boldsymbol{r}\right)\ \exp\left(-i\omega\,t + ik\cdot\boldsymbol{r}\right) \tag{11.156}$$

$$I_{\lambda q} \underset{lin}{=} I^0_{\lambda q}\exp\left(-\,\boldsymbol{\alpha}_q\cdot\boldsymbol{r}\right) \tag{11.157}$$

The irradiance (radiation intensity) follows a simple exponential law depending only on distance, which is a demonstration of the Lambert–Beer–Bouguer law seen previously in Sections 7.2.2 and 7.4.1 in Chapter 7, bringing the proof that no statistical arguments are used for establishing this exponential dependence. Recall that this law was taken as a template for modeling the influence phenomenon and finding the capacitive relationship in a pole.

Spatial irreversibility and inverse problem. As the purely temporal damping is associated with temporal irreversibility and the purely spatial damping is associated with spatial irreversibility, the combination of both phenomena in the attenuation of wave propagation can be attributed to both irreversibilities (and not to only one). As for the spatial damping, the task of retrieving the original configuration (spatial and temporal) of the wave source from distant and delayed observations is a difficult problem, known as the "inverse problem." Inverse problems are currently studied in many fields of fundamental and applied science in which the propagation (and reflection) of waves is used for analyzing the geometry of hidden objects through the absorbing medium surrounding them (electromagnetic scattering, acoustics, seismic soundings, mechanics, quantum physics).

11.8 IN SHORT

MOVEMENTS OF ENERGY

- *Energy* **conversion** is a change of subvarieties or varieties, involving two pairs of *energies-per-entity* and which requires some time.
- *Energy* **exchange** is an equilibration process between identical objects working without time considerations, expressed by "instantaneous" process.
- *Energy* **circulation** is a generic name describing any process in which energy moves from one source (or generator) to a sink (or receiver). In this chapter, this calling is used for the shifting of energy from a storing dipole (inductor or capacitor) into a conductor.
- **Dissipation** is the energy conversion when the source is a nonthermal conductor and the target is the thermal capacitive dipole.

CONVERSION INTO HEAT, REVERSIBILITY, AND DISSIPATION

Graphs 11.48 through 11.50 show various processes for the conversion of energy into heat, from three dipoles in the energy variety q, inductive (top), capacitive (middle), and conductive (bottom). The two first conversions are reversible processes because they involve an even number of evolution operators, whereas the last one is an irreversible one owing to an odd number of evolution operators.

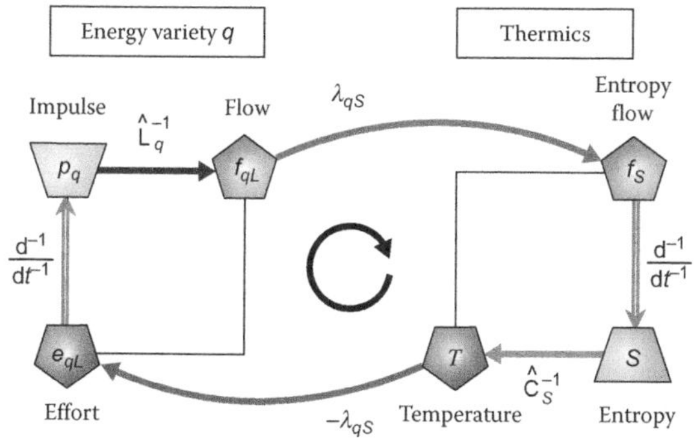

GRAPH 11.48

REVERSIBLE CONVERSION
Two evolution operators
(DIFFERENT SUBVARIETIES)

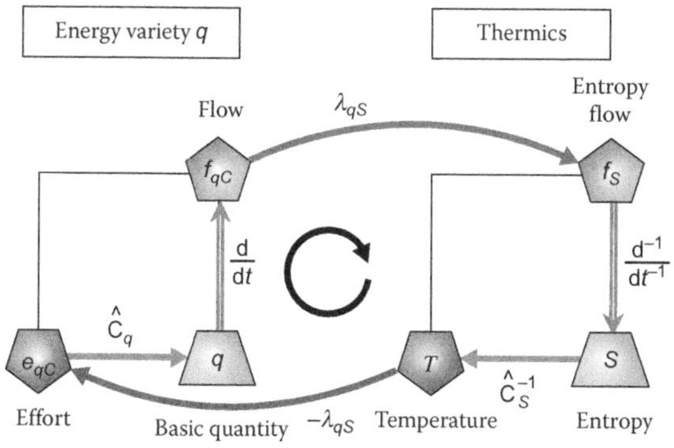

GRAPH 11.49

REVERSIBLE CONVERSION
Two evolution operators
(SAME SUBVARIETY)

(continued)

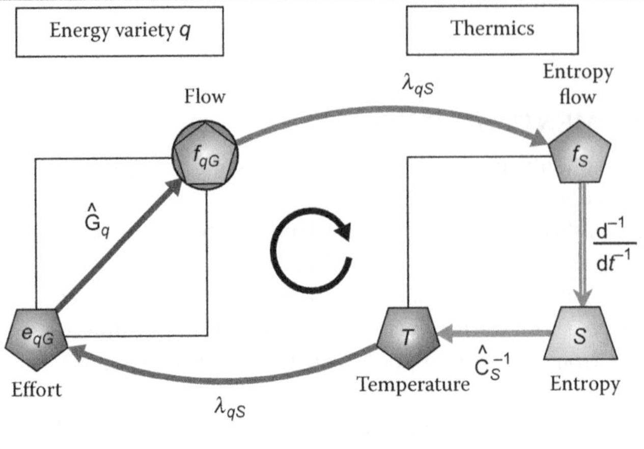

GRAPH 11.50

IRREVERSIBLE CONVERSION
One evolution operator (DISSIPATION)

TRANSFER–RELAXATION DIFFERENCES		
	Transfer	Relaxation
Formal Graph	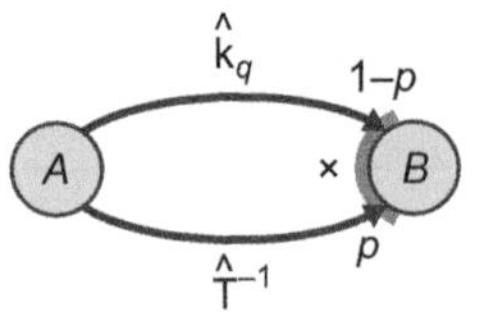 **GRAPH 11.51**	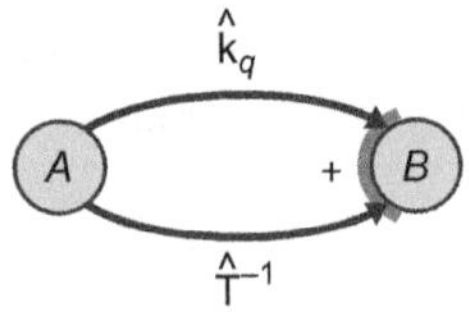 **GRAPH 11.52**
Graph operation	Path complement	Path addition
Formal Object	Dipole (eventually pole)	Dipole assembly
Equivalent operator $\left(B = \hat{O}\,A\right)$	$\hat{O} = \left\{\hat{T}^{-1}\right\}^{p}\left\{\hat{k}_q\right\}^{1-p}$	$\hat{O} = \left\{\hat{T}^{-1} + \hat{k}_q\right\}$

RELAXATION		
Two equivalent representations		
	Added contributions	Opposed contributions
Formal Graph	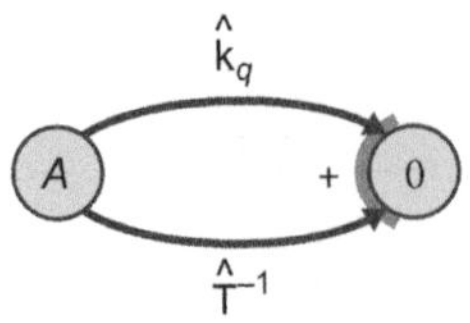 **GRAPH 11.53**	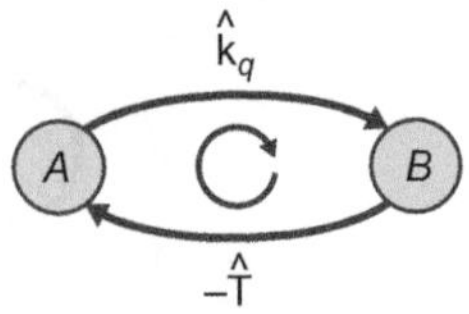 **GRAPH 11.54**
Graph operation	Path addition	Path nesting

DAMPING OF OSCILLATIONS		
Two equivalent representations		
	Added contributions	Opposed contributions
Formal Graph	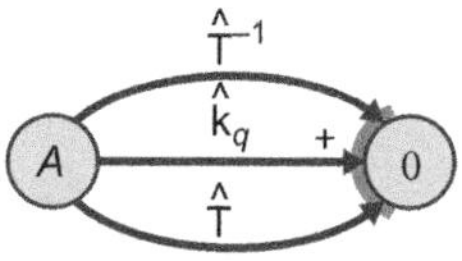 GRAPH 11.55	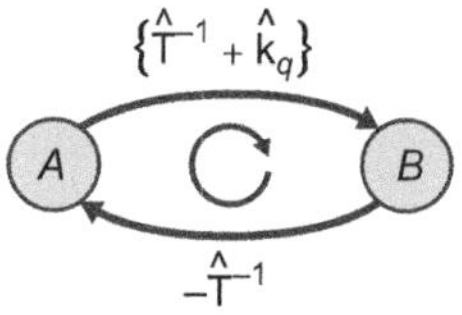 GRAPH 11.56
Graph operation	Path addition	Operator embedding
Equivalent operator		$\hat{\hat{T}}^{-1}=\left\{\hat{T}^{-1}+\hat{k}_q\right\}$

WAVE FUNCTION OF A DAMPED OSCILLATOR		
Equations		Graphics
Imaginary number	$-i^2 \overset{def}{=} 1$	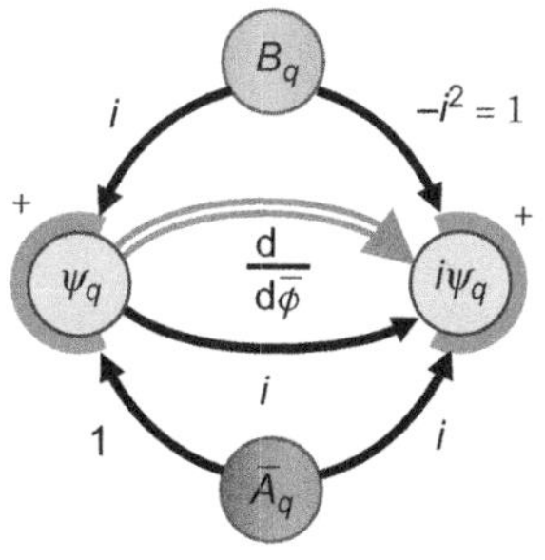 GRAPH 11.57
Wave function	$\psi_q \overset{def}{=} \bar{A}_q + i\,B_q$	
Differential equation	$\frac{d}{d\varphi}\psi_q = i\,\psi_q$	
Integral function	$\psi_q = \psi_q(0)\exp(-\gamma_{qL}t)\exp(i\varphi)$	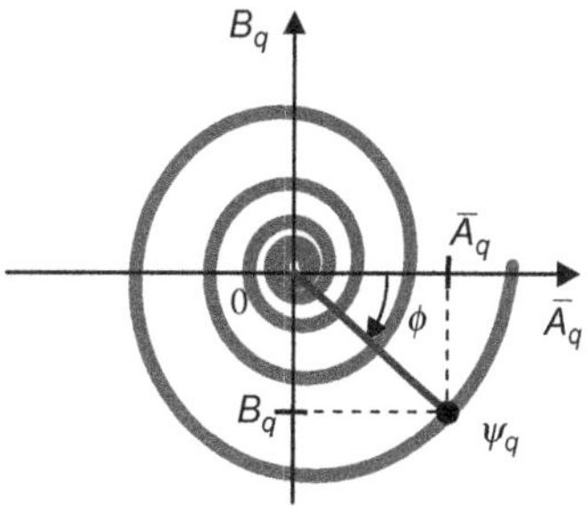
Semienergies	$\bar{A}_q = \psi_q(0)\exp(-\gamma_{qL}t)\cos\varphi$ $B_q = \psi_q(0)\exp(-\gamma_{qL}t)\sin\varphi$	

12 Coupling between Energy Varieties

CONTENTS

Until now, systems containing only one energy variety have been considered; the unique case where two varieties were taken into account was dissipation. The interaction between two energy varieties is examined here.

By virtue of the property of energy to be additive (cf. Chapter 2, Section 2.1.2), the energy of a system is the sum of all energies taken in the different energy varieties present in the system.

It is not because such a summation is stated that the energy varieties interact with each other: they may be completely independent when no mechanism allows exchanges between them. When exchanges are possible, the question is to know what machinery is behind the exchange. This is a great subject going through all scientific domains which is still imperfectly mastered.

Two classical approaches are used to tackle this question:

- Lagrangian* mechanics, based on the concept of minimization of energy in a system.
- Thermodynamics, based on its First Principle.

The problem is that the two approaches are independent: Lagrangian mechanics works only with energy varieties possessing the two subvarieties (mechanics and electrodynamics) and needs the time factor (Longair 2003), whereas thermodynamics treats only the capacitive subvariety and works in static or equilibrium conditions (in its usual acceptance, although a branch called thermodynamics of irreversible processes deals with out-of-equilibrium processes and time).

The Formal Graph approach helps clarify the question (without pretending to solve all problematic aspects) with its standardization of concepts throughout the scientific domains. The essential concept unifying the classical approaches is that of **energetic coupling**.

This is a fundamental concept in the Formal Graph theory because it provides an efficient tool for understanding the behavior of physical systems in terms of interdependence and interplay between energy varieties.

Table 12.1 lists the case studies of coupling given in this chapter. (See also Figure 12.1 for the position of the coupled assemblies.)

12.1 PASSAGES OF ENERGY

The previous chapters have dealt with some of the possibilities for energy to move from one Formal Object to another, by exchange, circulation, or internal conversion, and the passage of energy between energy varieties has only been tackled with the subject of dissipation, which is a special case of external conversion. In Graph 12.1 are sketched some of the new possibilities offered by the interactions between energy varieties, equivalence, coupling, and external conversion (between any varieties and any subvarieties).

To the list of energy passages that are recalled below added now are the ways for energy to interact or to move between varieties.

Energy exchange is a process occurring between two storing dipoles belonging to the same subvariety (C–C or L–L).

Energy circulation is a process occurring between a conductive dipole and a storing dipole belonging to the same subvariety (G–L or G–C).

The **energy equivalence** does not imply some change in the energy amounts among objects but is merely a correspondence between two scales for counting the number of entities. It expresses the obvious statement that the amount of energy does not depend on the units used for the entity numbers.

Energy coupling associates two subvarieties in different energy varieties by making interdependent their amounts. It is an extension of the exchange to several energy varieties. It can

* Joseph-Louis Lagrange (1736–1813): French mathematician and physicist; Paris, France.

TABLE 12.1
List of Case Studies of Coupling

	Name	Energy Variety	Energy Variety	Coupling	Page Number
$\nu A \longrightarrow$	J1: *n*th-Order Chemical Reaction	Physical chemistry	Chemical reaction	Equivalence	616
$\bar{\mu}$	J2: Electrochemical Potential	Electrodynamics	Physical chemistry	Equivalence	623
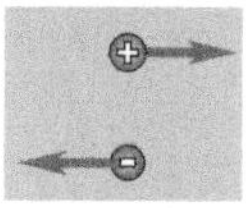	J3: Electrical Force	Electrodynamics	Translational mechanics	Equivalence	629
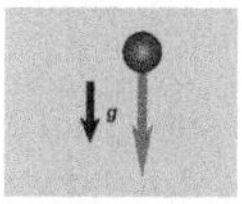	J4: Gravitational Force	Gravitation	Translational mechanics	Equivalence	633
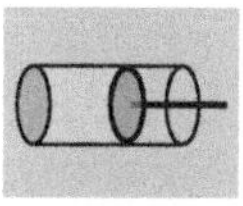	J5: Piston	Translational mechanics	Hydrodynamics	Invariable	637
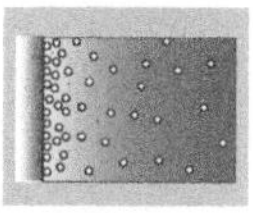	J6: Ionic Distribution	Electrodynamics	Corpuscular energy	Invariable	640
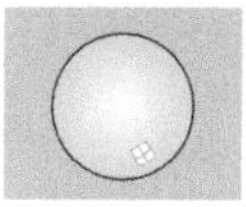	J7: Bubble	Hydrodynamics	Surface energy	Variable	643
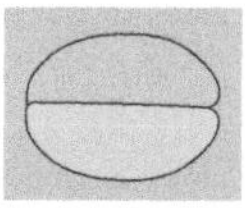	J8: Thermal Junction	Electrodynamics	Thermics	Invariable	646
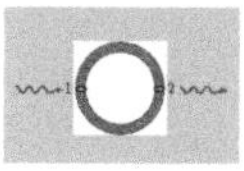	J9: Seebeck Effect	Thermics	Electrodynamics	Invariable conversion	650
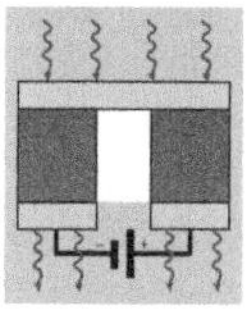	J10: Peltier Effect	Electrodynamics	Thermics	Invariable conversion	655
	J11: Thomson Effect	Electrodynamics	Thermics	Variable conversion	658
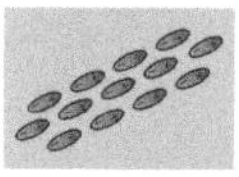	J12: Electric Polarization	Electrodynamics	Polarization	Variable three-dimensional space	664
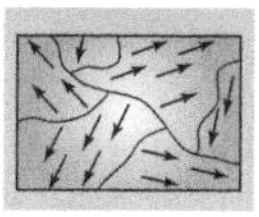	J13: Magnetism	Magnetism	Electrodynamics	Variable three-dimensional space	669

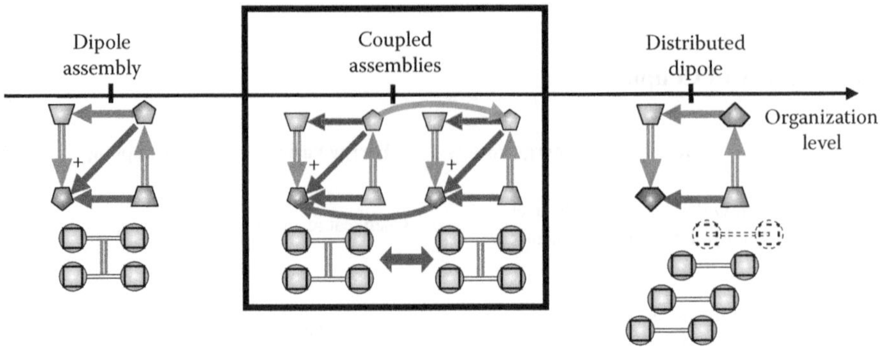

FIGURE 12.1 Position of the coupled assemblies along the complexity scale of Formal Objects.

be static or dynamic according to the role of time: an external conversion is based on a dynamic coupling.

The **energy conversion** process changes the nature and amounts of energy from one subvariety to another (internal conversion) or from one variety to another (external conversion). In this latter case, energetic coupling is involved.

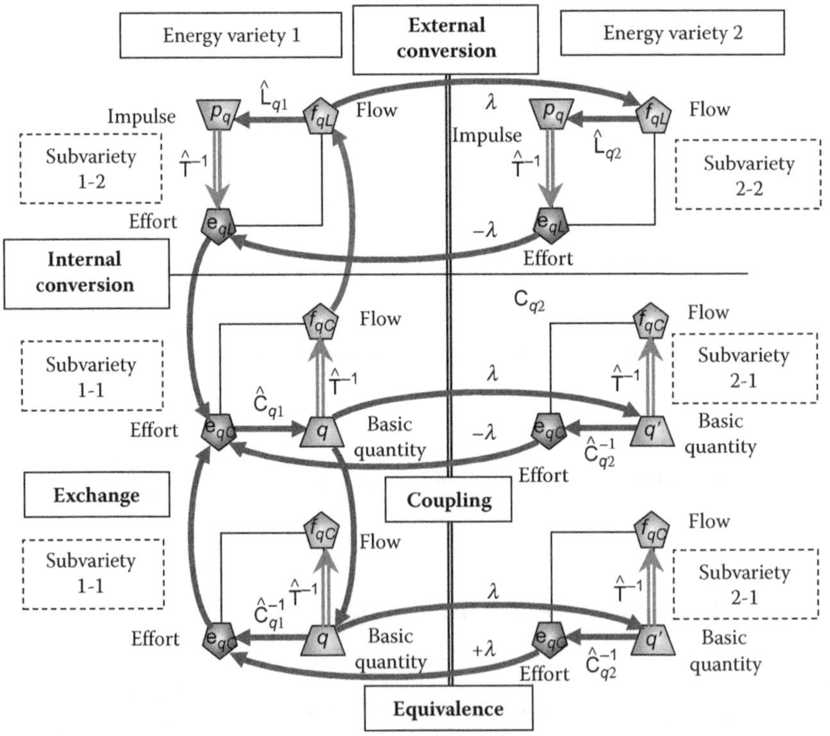

GRAPH 12.1 Examples of energetic *equivalence* and of the three passages for the energy between storing dipoles: *Exchange* between identical variety and subvariety, *conversion* between different subvarieties (here capacitive–inductive = internal) or different varieties (here inductive–inductive = external), *coupling* between different varieties but same subvariety (here capacitive–capacitive).

12.2 ENERGETIC EQUIVALENCE

An energy subvariety is defined in two steps. First, with the identification of a container, as explained in Chapter 1, through its qualitative nature (e.g., volume, charge, molecules, etc.). Second, with the choice of a basic unit, the entity, and of a scale for counting the entity numbers in a collection. In this second step, several choices may sometimes be offered, but whatever the unit chosen, it does not change the nature of the energy nor the amount of energy. It is the role of the system of units to ensure coherence and to avoid multiplication of varieties that are in fact identical in nature. The *Système International* (SI) is important in the landscape of energies for limiting the number of varieties to a minimum. However, there are cases where the existence of two varieties for the same nature of energy is justified, which require examination.

Here the case of the capacitive subvariety is examined, the case of the inductive subvariety being strictly similar. For the a and b varieties, the variations of energy are given by

$$\delta \mathcal{U}_a \underset{\exists C_a}{=} e_a \mathrm{d}q_a; \quad \delta \mathcal{U}_b \underset{\exists C_b}{=} e_b \mathrm{d}q_b \tag{12.1}$$

Owing to the embedding principle (cf. Chapter 6, Section 6.2.1), such writings of the energy variation can be used indifferently for poles or dipoles (and even dipole assemblies). This means that the Formal Objects considered in such relationships may belong to various complexity levels. Nevertheless, the convention of taking objects in the same complexity levels (two poles or two dipoles but not one pole and one dipole) is adopted. It is recalled (cf. Chapter 2, Section 2.1.2) that the notation with an upright δ signifies a nonexact differential, expressing the lack of a partner for allowing energy to vary in the considered object.

To connect or compare two energy varieties, the key parameter is the operator linking their entity numbers, which is in fact a scalar or a vector.

The *differential coupling factor* linking the first basic quantity to the second one is defined as the derivative of the first entity number with respect to the second one:

Differential coupling factor:

$$\lambda'_{ba} \overset{def}{=} \frac{\mathrm{d}q_a}{\mathrm{d}q_b} \tag{12.2}$$

The term *coupling* is used here as relative to the state variables and not to the energies.

Simplicity of the coupling factor? Nothing forbids using operators for relating energy varieties. However, it appears that simple scalars or vectors are able to model most couplings met in classical physics and physical chemistry, and this study assumes that it is so. However, there is no apparent reason for this limitation.

It is convenient, mainly for use in Formal Graphs, to define also an integral coupling factor relating directly the basic quantities taken as relative to their minimum values:

Integral coupling factor:

$$\lambda_{ba} \overset{def}{=} \frac{q_a - q_a^{min}}{q_b - q_b^{min}} \qquad \left(q_b \neq q_b^{min}\right) \tag{12.3}$$

In the general case, these two factors are distinct, they correspond through integration/derivation, but there are some cases where they are equal and constant. This point will be discussed later.

The energetic equivalence is said to occur when both energy variations are equal, which lends to proportionality between both energies-per-entity through the differential coupling factor:

Equivalence: $$\delta \mathcal{U}_a = \delta \mathcal{U}_b \Leftrightarrow \frac{e_b}{e_a} \underset{equivalence}{=} \lambda'_{ba} \qquad (e_a \neq 0) \tag{12.4}$$

An example is worth giving because of its importance in physical chemistry and in corpuscular physics. These two energy varieties have been met at different occasions, and, for instance, they were considered separately in Chapter 6 when examples of dipoles were given. In fact, they are equivalent varieties because their basic quantities, the mole for the physical chemical variety and the number of corpuscles for the corpuscular one, are proportional through the Avogadro* constant N_A.

The general reasoning made above can be applied to the present case, starting from the writing of the two energy variations:

$$\delta \mathcal{U}_n = \mu \, \mathrm{d}n; \quad \delta \mathcal{U}_N = E \, \mathrm{d}N \tag{12.5}$$

Then the proportionality between basic quantities defines the Avogadro constant:

Avogadro constant: $$N_A \overset{def}{=} \frac{N}{n} \tag{12.6}$$

The accepted value for this constant is $6.02214179(30) \times 10^{23}$ mol^{-1}. In fact, this definition is dated from 1971 and is subject to evolution.

> **Mole definition.** According to the Conférence Générale des Poids et Mesures (CGPM), the mole is defined as (IUPAC 2007): ". . . the amount of substance of a system which contains as many elementary entities as there are (unbound and at rest) atoms in 0.012 kilogram of carbon 12." Abandoning this definition based on the kilogram for a newer one based on a conventional and fixed value of the Avogadro constant is being questioned (CGPM 2007).

Anyway, this factor is a constant, which therefore can be equated to the derivative of the corpuscle number versus the substance amount. Consequently, by equating the two variations of energy, the well-known proportionality between chemical potential and energy-per-corpuscle is found

Equivalence: $$\frac{\mu}{E} = N_A = \frac{\mathrm{d}N}{\mathrm{d}n} \tag{12.7}$$

* Avogadro Lorenzo Romano Amedeo Carlo (1776–1856): Italian scientist, Torino, Italy.

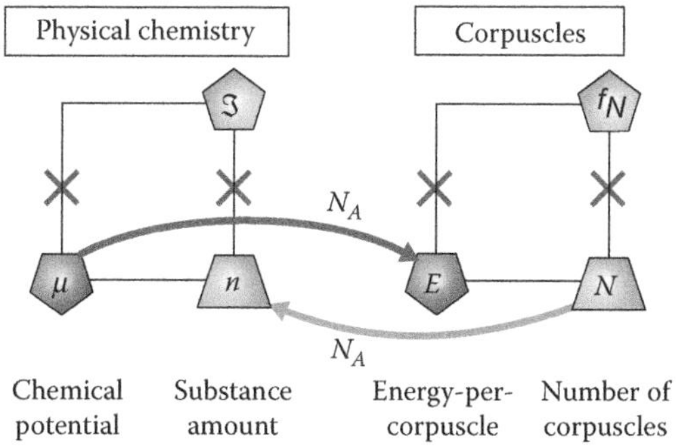

GRAPH 12.2 Formal Graph of the equivalence between physical chemical energy and corpuscular energy through the Avogadro constant. The Formal Objects are poles in this example.

The relationships featuring this equivalence are drawn in the Formal Graph in Graph 12.2.

One may ask whether this distinction between two energy varieties is a fundamental one. Obviously not, only one is enough and it is perfectly possible to use different units depending on the context, as, for instance, all non-SI units that may be used for a same quantity. However, they correspond to so markedly different worlds, macroscopic and microscopic, that it is almost impossible to confuse them without transgressing a classical and fundamental distinction. Here this custom is followed in keeping the distinction, as if they were really two different energy varieties. An advantage is that the parenthood with the chemical potential helps maintain the status of energy-per-entity of the energy-per-corpuscle, too often confused with energy itself, which is less frequent for the chemical potential.

The opposite situation exists. Some energy varieties that are interesting to individualize are not yet considered in today's practice. Case study J1 examines the case of chemical reactions involving several partners, whether belonging to the same chemical species or distributed among several species. Until now one was able to model first-order chemical reactions with the sole physical chemical energy variety (see case studies A1 and A11 in Chapter 4 and C5 in Chapter 6) but this is not possible when the order of the reaction is higher than one. By introducing a new energy variety, called *chemical reaction energy*, and by associating it through equivalence links to the physical chemical energy, this becomes feasible.

12.2.1 Translated Effort

It is interesting to establish correspondences between state variables even in the case of nonequivalent energies. When the energy variations do not match, the preceding relationship (Equation 12.4) between the differential coupling factor and efforts does not hold anymore and must be replaced by a new relation:

Nonequivalence: $$\delta\mathcal{U}_a \neq \delta\mathcal{U}_b \Leftrightarrow \frac{e_b}{e_a} \neq \lambda'_{ba} = \frac{e'_{ba}}{e_a} \quad (12.8)$$

This new relation is based on the definition of a *translated effort*, from the energy variety a to the energy variety b:

Translated effort: $$e'_{ba} \overset{def}{=} \lambda'_{ba}\, e_a \quad (12.9)$$

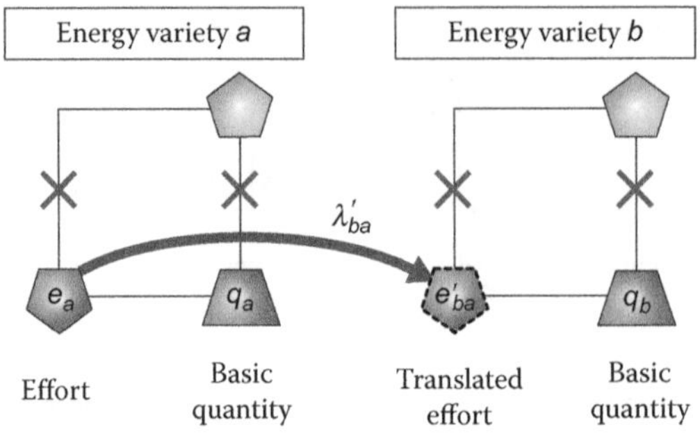

GRAPH 12.3 Formal Graph description of a translated effort from a first energy variety toward a second one, allowing considering an identical energy amount under the form of the second energy variety.

With this new variable, the energy variation in the variety *a* can be written indifferently as

$$\delta \mathcal{U}_a \underset{\exists C_a}{=} e_a \, \mathrm{d}q_a = e'_{ba} \, \mathrm{d}q_b \tag{12.10}$$

This is a way to express the energy in the variety a under the appearance of the variety b. The Formal Graph in Graph 12.3 depicts this situation by modifying the symbol of the translated effort node (dotted perimeter) in order to distinguish this relationship from the equivalence between the two energy varieties.

An illustration of the use of translated effort follows with the definition of an *apparent effort* and applications will be discussed in case studies J2 "Electrochemical Potential," J3 "Electric Forces," and J4 "Gravitational Forces."

12.2.2 Apparent Effort

When two energy varieties are not independent, that is, when a link exists between some of their state variables, algebraic models and Formal Graphs can be simplified by using apparent state variables entrusted with the representation of both energy varieties.

For example, take a system filled with two or more energy varieties (capacitive subvarieties). The total energy variation is given by a Gibbs* equation by virtue of the additive property of energy and of the First Principle of Thermodynamics:

Gibbs:
$$\mathrm{d}\mathcal{U} \underset{\exists C_i}{=} e_a \, \mathrm{d}q_a + e_b \, \mathrm{d}q_b + e_c \, \mathrm{d}q_c + \cdots \tag{12.11}$$

An upright "d" is used for notating the exact differential of the total energy because the partners may exchange energy, allowing the total energy to be independent from the followed path between two states of the system (which is an essential property of energy, as explained in Chapter 2).

Now, one assumes that the two first energy varieties are dependent through their basic quantities, which is mathematically expressed as

$$\frac{\mathrm{d}q_a}{\mathrm{d}q_b} \neq 0 \tag{12.12}$$

* Josiah Willard Gibbs (1839–1903): American physicist and chemist; New Haven, Connecticut, USA.

This dependence means that the total capacitive energy variation can be merely written as

$$\mathrm{d}\mathcal{U} \underset{\exists C_{qi}}{=} \bar{e}_b \, \mathrm{d}q_b + e_c \, \mathrm{d}q_c + \cdots \tag{12.13}$$

having defined an *apparent effort* as

Apparent effort:
$$\bar{e}_b \overset{def}{=} \left(\frac{\partial \mathcal{U}}{\partial q_b} \right)_{indep.\, qi} \tag{12.14}$$

This partial derivation must be made by maintaining constant all other basic quantities which are independent from the considered basic quantity used for the derivation. (In particular, by definition, the coupled basic quantities cannot be maintained at a constant.) It differs from the genuine definition of the effort or energy-per-entity (see Chapter 2, Section 2.1.5), which is also given by the same partial derivation but in maintaining at a constant all other basic quantities that are significant in the energy of the system. It requires complete independence between all basic quantities, that is, no coupling at all.

Effort:
$$e_b \overset{def}{=} \left(\frac{\partial \mathcal{U}}{\partial q_b} \right)_{all\, qi} \tag{12.15}$$

The dependence between the two energy varieties allows defining a coupling factor as before

Differential coupling factor:
$$\lambda'_{ba} \overset{def}{=} \frac{\mathrm{d}q_a}{\mathrm{d}q_b} \tag{12.16}$$

which, once introduced in Equation 12.11, leads to the following expressions of the apparent effort

$$\bar{e}_b = e_b + e_{ba} = e_b + \lambda'_{ba} \, e_a \tag{12.17}$$

These equations are represented in Graph 12.4 in the language of Formal Graphs.

The definition in Equation 12.14 of the apparent effort is very general; it may work in case of more than two dependent energy varieties in the system by providing directly the right expression for the apparent effort

$$\bar{e}_b = e_b + \lambda'_{ba1} \, e_{a1} + \lambda'_{ba2} \, e_{a2} + \cdots \tag{12.18}$$

It has also the noticeable advantage of avoiding being prisoner of a rigid definition (in terms of summation of several contributions, for instance) which may have a restricted domain of validity. This advantage will be used in case study J2 about the notion of the electrochemical potential, classically defined as

$$\bar{\mu} = \mu + zF\,V \tag{12.19}$$

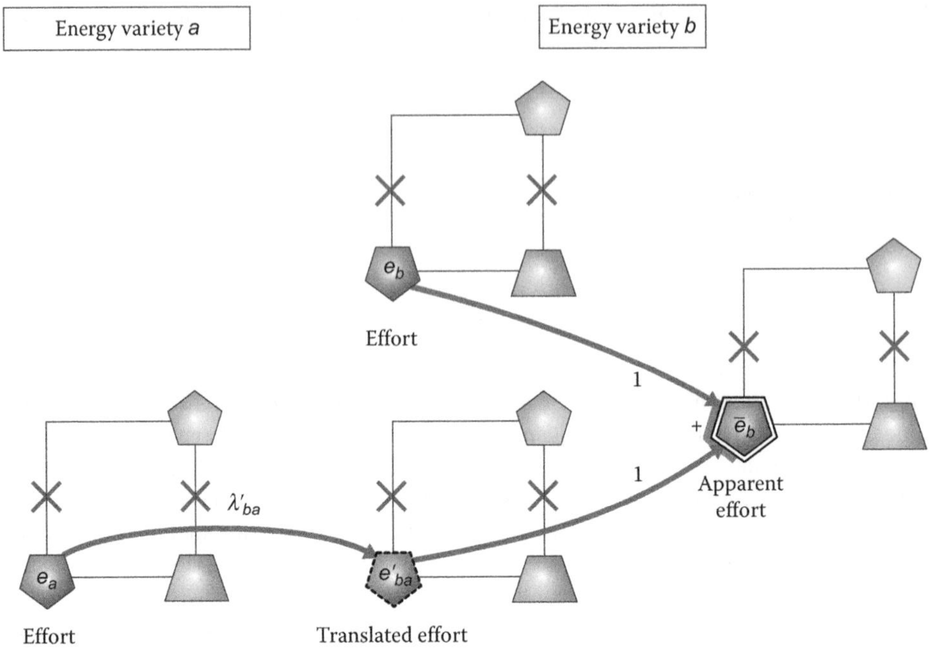

GRAPH 12.4 Formal Graph showing the construction of an apparent effort between two energy varieties (capacitive subvariety). For being incorporated into the apparent effort in energy variety *b* (on the right), the effort in the energy variety *a* (on the left) must be translated before. All objects are poles in this construction.

With the general definition in Equation 12.14, it becomes possible to extend the use of the electrochemical potential in kinetics, which is not possible with the classical definition limited to simple coupling situations in fact. This point will be detailed in case study J2.

Although the primary purpose of the notion of apparent effort is not to determine the conditions for establishing a minimum of energy, it can be used for this aim. From the definition in Equation 12.14 of the apparent effort and its expression in Equation 12.17, a relationship between efforts is deduced when the capacitive energy is extremum

$$\text{Energy extremum:} \quad \left(\frac{\partial \mathcal{U}}{\partial q_b}\right)_{indep.\,qi} = 0 \Leftrightarrow \bar{e}_b = 0 \Leftrightarrow e_b = -\lambda'_{ba}\, e_a \tag{12.20}$$

This peculiar relationship features an energetic coupling, as discussed now.

12.3 ENERGETIC COUPLING

As stated previously, the existence of several energy varieties in a system is taken into account by a sum that says nothing about the possible interactions between them. The only thing, apart from the list of significant energies, that stems from a writing of the variation of the total energy in a system is that the differential equation must be an exact one if more than one variety is present.

$$d\mathcal{E} = \delta\mathcal{E}_a + \delta\mathcal{E}_b + \delta\mathcal{E}_c + \cdots \tag{12.21}$$

Some additional properties or constraints must be brought for making progress. A first constraint is the requirement of delimitating the system for being able to impose the isolation of the system from other systems, thus restricting the energy exchanges within the listed varieties.

System isolation: $$d\mathcal{E} = 0 \tag{12.22}$$

Unhappily, there is a way for the system to fulfill this condition which is to freeze all exchanges, which is written as

Frozen exchanges: $$\delta\mathcal{E}_i = 0 \quad \text{for all varieties } i \tag{12.23}$$

In that case there is no rule for knowing the distribution of energy among the varieties and the modeling of the system is merely made by superimposing independent subsystems based on the whole preceding theory (i.e., the Formal Graph theory restricted to a single energy variety). One discards this case in restricting the list of significant energy varieties to those that are free to vary and one turns attention to a property of energy not yet discussed, the tendency to find a minimum.

A minimum or maximum, that is, an extremum, of total energy is reached when the first derivative with respect to all free entity numbers is equal to zero:

Energy extremum: $$\frac{\partial\mathcal{E}}{\partial x_i} = 0 \quad \text{for all free varieties } i \tag{12.24}$$

An absolute requirement for allowing a minimum is the existence of unconstrained exchanges between energy varieties. The concept of energetic coupling provides the necessary ingredients for this.

The coupling is defined between two energy varieties and more precisely between two subvarieties. Two kinds of elementary couplings are distinguished:

- The **internal coupling** works between the two subvarieties within a same variety.
- The **external coupling** works between two energy varieties within a same subvariety.

The Formal Graph theory does not consider as elementary a crossed coupling between two subvarieties belonging to two different varieties, which can be viewed as composed of the two previous elementary couplings (see Figure 12.2).

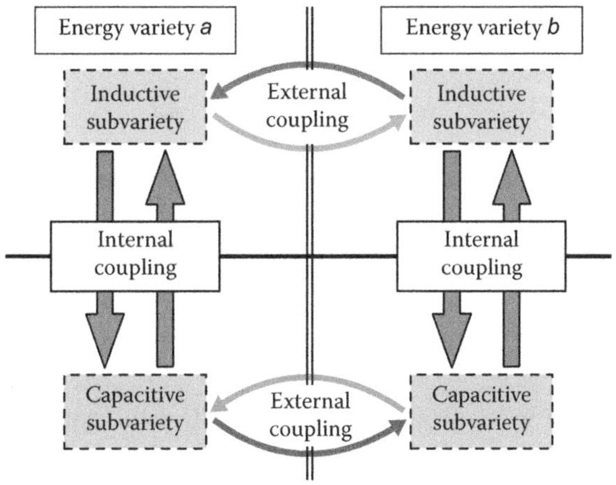

FIGURE 12.2 Scheme of the two elementary energetic couplings.

The internal coupling must not be confused with the internal conversion of energy which proceeds through evolution operators and therefore using explicitly the time, as described in Chapter 9. In the concept of coupling, no time is explicitly required.

The internal coupling, occurring in the same energy variety without conversion, has been met once in case study F8 studying the Hall effect in Chapter 9. The space–time velocity ensured the coupling between inductive and capacitive energies in the electrodynamical variety.

Timeless coupling. Normally entity numbers and energies-per-entity are linked by evolution operators. This is not the case for the internal coupling that relates the same variable families by simple scalars or vectors. These coupling factors are frequencies or velocities, closely related to time, but they are not time. This specificity applies the internal coupling essentially between timeless Formal Objects, such as the *singletons*, which model individual entities. The internal coupling is therefore a key concept in the Formal Graph theory for handling the role of energy in quantum physics. (Examples will be given in Chapter 14.)

This chapter deals only with the external coupling as it involves different energy varieties contrary to the internal coupling that works within one variety. As for the subject of energetic equivalence, we take the case of the capacitive subvariety, which is perfectly symmetric to the case of the inductive subvariety.

According to the above definition of coupling, we have to consider energy varieties in pairs for treating this subject in a first step. The extension to more than two energy varieties will be treated in the next chapter.

12.3.1 Capacitive Coupling

In the capacitive subvariety, the sum of individual energy variations is written as

$$\delta\mathcal{U} = \delta\mathcal{U}_a + \delta\mathcal{U}_b \tag{12.25}$$

It is assumed that the system contains only two capacitive energies and is isolated from the exterior and that both energies are free to vary. The following Gibbs equation expresses the total capacitive energy variation (as for the case of equivalence seen before, the Formal Objects considered must belong to the same complexity level):

Gibbs:
$$d\mathcal{U} \underset{\exists C_a, C_b}{=} e_a dq_a + e_b dq_b \underset{isol}{=} 0 \tag{12.26}$$

The status of state function of the capacitive energy when no inductive energy is present is translated into mathematical terms by stating that the Gibbs equation is an exact differential equation. A mathematical consequence is that the sum of the other terms resulting from the derivation of the products $e_i\,\Delta q_i$ must be equal to zero.

Gibbs–Duhem:
$$0 = \left(q_a - q_a^{min}\right)de_a + \left(q_b - q_b^{min}\right)de_b \tag{12.27}$$

This equation, which is a generalization of the Gibbs–Duhem* equation (Callen 1985) (originally proposed without minimums), provides a way to introduce the *integral coupling factor*,

* Pierre Duhem (1861–1916): French physicist and philosopher; Paris, France.

defined as the ratio of the relative basic quantities, for writing the Gibbs–Duhem equation in terms of ratios:

Integral coupling factor: $$\lambda_{ba} \overset{def}{=} \frac{q_a - q_a^{min}}{q_b - q_b^{min}} = -\frac{de_b}{de_a} \quad \left(q_b > q_b^{min}\right) \tag{12.28}$$

This factor has been previously defined by Equation 12.3 as a convenient proportionality factor for modeling equivalence relationships between basic quantities. Here it has also the role of relating the variations of efforts.

Now, one looks for the conditions that make the total capacitive energy minimum or maximum, that is, extremum, in this system. This is ensured when the first derivatives with respect to any basic quantities is zero:

Capacitive energy extremum: $$\frac{d\mathcal{U}}{dq_i} = 0 \quad i = a, b \tag{12.29}$$

By dividing the Gibbs equation by the variation of one of the basic quantities (the choice is arbitrary and has no consequence), one gets such a derivative:

$$\frac{d\mathcal{U}}{dq_b} \underset{\exists C_a, C_b}{=} e_a \frac{dq_a}{dq_b} + e_b \underset{\mathcal{U}\ extremum}{=} 0 \tag{12.30}$$

This expression lends naturally using again the same definition of the *differential coupling factor* as for the equivalence case, but the consequence for the efforts is opposite as now a minus sign is present:

Differential coupling factor: $$\lambda'_{ba} \overset{def}{=} \frac{dq_a}{dq_b} = -\frac{e_b}{e_a} \quad (e_a \neq 0) \tag{12.31}$$

The two double Equations 12.28 and 12.31 are the algebraic translations of the energetic coupling between two capacitive energies.

Naturally, the two coupling factors are not independent; they are related through an integration that can be viewed as a definition of the integral factor as the averaged differential factor:

$$\lambda_{ba} = \frac{1}{q_b - q_b^{min}} \int_{q_b^{min}}^{q_b} \lambda'_{ba} dq \tag{12.32}$$

The Formal Graph of a capacitive coupling between two generic Formal Objects (without specifying their nature, pole, dipole, etc.) is given in Graph 12.5.

In the general case, the two coupling factors are not equal, which corresponds to a *variable coupling*. When they are equal, which happens when the differential factor is a constant, the coupling is said to be *invariable*.

Invariable coupling: $$\lambda_{ba} = \lambda'_{ba} = C^{st}$$

Variable coupling: $$\lambda_{ba} \neq \lambda'_{ba}$$

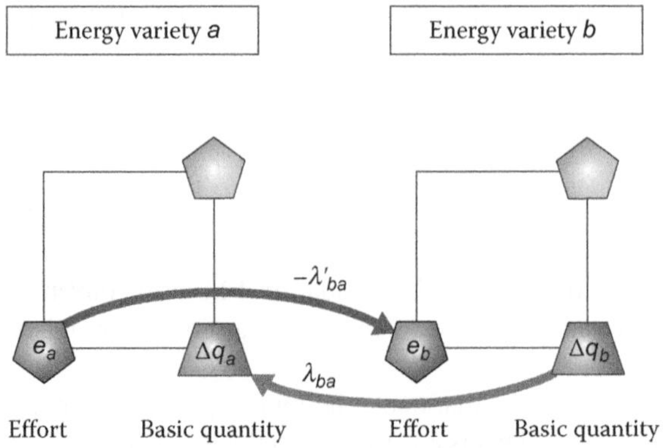

GRAPH 12.5 Formal Graph of a capacitive coupling in the general case of different factors (variable coupling). Without modifying the Formal Graph, merely by equating the two factors, the coupling becomes invariable. (The delta symbol in the basic quantity nodes means the difference with the minimum value.)

More details about these concepts will be given in Section 12.5, after having evocated the symmetrical case of inductive coupling and after all the case studies.

Generalized thermodynamical "force." A branch of thermodynamics developed by Prigogine,* called Thermodynamics of Irreversible Processes, uses the concept of generalized thermodynamical "force" for handling specifically the coupling of thermics with other energy varieties. This concept, also called "Rayleigh† force," which is defined as an effort (or an effort gradient) divided by the temperature, is not really a generalized force but merely a *differential coupling factor* in the Formal Graph theory:

Generalized thermodynamical "force": $-\lambda'_{qS} = \dfrac{e_q}{T}$ (= Differential coupling factor)

12.3.2 Inductive Coupling

It is not necessary to repeat the reasoning that leads to the definition of the energetic capacitive coupling as the same arguments are used for the inductive coupling. Consequently, the same shapes of coupling relationships work

$$\text{Integral coupling factor:} \quad \lambda_{ba} \overset{def}{=} -\frac{p_b - p_b^{min}}{p_a - p_a^{min}} = \frac{df_a}{df_b} \quad \left(p_a > p_a^{min}\right) \tag{12.33}$$

$$\text{Differential coupling factor:} \quad \lambda'_{ba} \overset{def}{=} -\frac{dp_b}{dp_a} = \frac{f_a}{f_b} \quad (f_b \neq 0) \tag{12.34}$$

The same generic notation for the coupling factors have been used because there is equality between inductive and capacitive factors, as will be demonstrated in Section 12.5.3 in the discussion

* Ilya Romanovich Prigogine (1917–2003): Russian-Belgian physical chemist; Brussels, Belgium.
† John William Strutt, 3rd Baron Rayleigh (1842–1919): English physicist; Cambridge, UK.

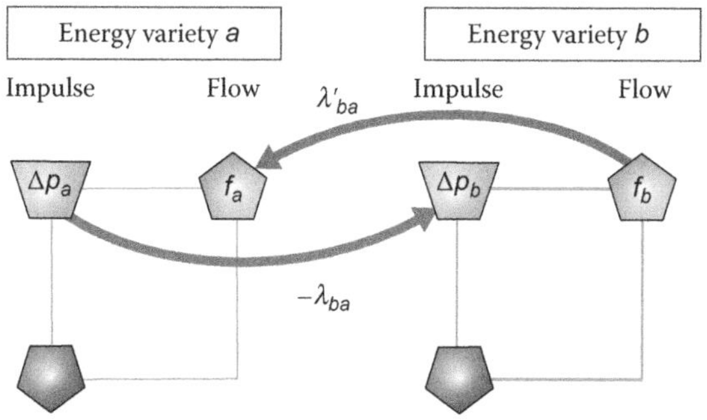

GRAPH 12.6 Formal Graph of an inductive coupling in the general case of different factors (variable coupling). Without modifying the Formal Graph, merely by equating the two factors, the coupling becomes invariable. (The delta symbol in the impulse nodes means the difference with the minimum value.)

of energy conversion. The only difference is the sign of the ratio of differences of entity numbers which is negative for the inductive coupling and positive for the capacitive one. The Formal Graph representation of an inductive coupling is given in Graph 12.6.

12.3.3 Apparent Entity Numbers

The coupling factors are defined independent of the reality of the coupling (i.e., the energies-per-entity can be independent or not). In other words, a relationship (proportionality) between two basic quantities or two impulses does not imply systematically the existence of an energy extremum, that is, an energetic coupling.

The consequence is that the possibility to use apparent variables in substitution to energies-per-entity stems from the existence of coupling factors and not necessarily from the effectiveness of the coupling. This has been exemplified with the electrochemical potential that exists even in the absence of a relationship between the chemical potential and the electric potential.

However in case of coupling, when energies-per-entity are interdependent, it becomes possible to define apparent entity numbers using the same principle.

Apparent basic quantity:

$$e_a \overset{def\rightarrow}{=} \left(\frac{\partial \mathcal{U}}{\partial \bar{q}_a}\right)_{indep.\,qi} \tag{12.35}$$

Apparent impulse:

$$f_b \overset{def\rightarrow}{=} \left(\frac{\partial \mathcal{U}}{\partial \bar{p}_b}\right)_{indep.\,pi} \tag{12.36}$$

With these definitions, one of the energy varieties can be embedded in another one, as, for instance, in the following Gibbs equations:

$$d\mathcal{U} = e_a dq_a + e_b dq_b + \cdots = e_a d\bar{q}_a + \cdots \tag{12.37}$$

$$d\mathcal{T} = f_a dp_a + f_b dp_b + \cdots = f_b d\bar{p}_b + \cdots \tag{12.38}$$

The expressions of these apparent variables are therefore:

Apparent basic quantity variation: $$\lambda'_{ba} = -\frac{e_b}{e_a} \Rightarrow d\bar{q}_a = dq_a - \lambda'_{ba}\, dq_b \tag{12.39}$$

Apparent impulse variation: $$\lambda'_{ba} = \frac{f_a}{f_b} \Rightarrow d\bar{p}_b = dp_b + \lambda'_{ba} dp_a \tag{12.40}$$

It is clear that the relationship between the variation of the apparent entity number and its integral may be quite complicated when the differential coupling factor is variable, so this notion of apparent entity number is of little use.

There is an exception, when the coupled Formal Objects are space distributed. In that case, densities along a curve or over a surface are also coupled and the previous relationship can be useful. This is the case in electrodynamics when the medium is influenced by the electric field or the magnetic field, provoking the phenomena of polarization or magnetization, respectively. In the Formal Graph theory, two specific energy varieties, polarization energy and magnetic energy,* are individualized and these two phenomena are considered as resulting from a coupling with one of these energy varieties. The models meeting the previous scheme are classically expressed as a modification of the localized entity number, electric displacement (“*electrization*”) $\boldsymbol{D}$ (represented by $\varepsilon_0\, \boldsymbol{E}$) or electromagnetic induction $\boldsymbol{B}$ (represented by $\mu_0\, \boldsymbol{H}$), by an added contribution having as result an apparent variable:

Existence of polarization: $$\boldsymbol{D} \underset{lin}{=} \varepsilon_0\, \boldsymbol{E} + \boldsymbol{P} \quad \left(= "\bar{\boldsymbol{D}}"\right) \tag{12.41}$$

Existence of magnetization: $$\boldsymbol{B} \underset{lin}{=} \mu_0\, \boldsymbol{H} + \mu_0\, \boldsymbol{M} \quad \left(= "\bar{\boldsymbol{B}}"\right) \tag{12.42}$$

The apparent variables in the left members of these equations should bear a distinctive mark, such as the upper bar, but the usage, unhappily, is not such. The Formal Graph theory cannot confuse these variables as will be seen in case studies J12 and J13.

12.4 CASE STUDIES OF ENERGETIC EQUIVALENCE, TRANSLATION, AND COUPLING

The first case study deals with the subject of energetic equivalence, used for introducing a new energy variety, the chemical reaction energy. An old notion, the affinity, is reused as effort in this variety for modeling chemical reactions having a kinetic order higher than one.

- J1: *n*th-Order chemical reaction

* Magnetic energy, or magnetism, must not be confused with the electromagnetic subvariety of electrodynamics.

The following three case studies deal with the subject of translated effort:

- J2: Electrochemical potential
- J3: Electrical force
- J4: Gravitational force

In these three cases, new definitions of the mentioned efforts are proposed, which in the case of the electrochemical potential allows one to use it out of equilibrium (in kinetics notably) and in the case of the two latter, outlines the dependence of the classical definition of the force–field relationship on the linearity of the capacitive relationship in each domain.

The following three case studies deal with the subject of coupling:

- J5: Piston
- J6: Ions distribution
- J7: Bubble

Among these, the first two case studies describe invariable coupling whereas the third one is devoted to variable coupling. The piston example demonstrates the usefulness of a thermodynamical approach in this mechanical domain and outlines the difference between a global pressure and a local pressure. In the ion distribution, the exponential function ruling the capacitive relationship in physical chemistry and corpuscular domain is exported to the electrodynamical domain. The last case study "Bubble" introduces the surface energy variety and demonstrates the Laplace law in capillarity.

The next series of case studies is concerned with the same subject of coupling between electrodynamics and thermics. This domain called thermoelectricity is an important one, owing to its numerous applications but also because reversible transformations implying entropy are at work.

- J8: Thermal junction
- J9: Seebeck* effect
- J10: Peltier† effect
- J11: Thomson‡ effect

The Seebeck effect is the production of electricity from heat whereas the Peltier effect is the reverse. All these effects are classically presented as independent ones. The Formal Graph theory shows that the Seebeck and Peltier effects correspond to invariable couplings, whereas the Thomson effect is the variable version of these couplings.

The last series of case studies is devoted to extensions of the electrodynamical energy for taking into account the modification of the distribution of energy when an electric or electromagnetic field influences a material.

- J12: Electric polarization
- J13: Magnetism

These two case studies describe the creation of new energy varieties and generalize the classical models to nonlinear ones.

---

* Thomas Johann Seebeck (1770–1831): German-Estonian physicist; Revel, Prussia; presently Tallinn, Estonia.

† Jean-Charles Athanase Peltier (1785–1845): French watchmaker and physicist; Paris, France.

‡ William Thomson (1824–1907): Lord Kelvin, British physicist; Belfast, Glasgow, and Cambridge, UK.

12.4.1 Case Study J1: *n*th-Order Chemical Reaction

Physical Chemistry
Chemical Reaction

$$\nu \mathrm{A} \xrightarrow{k}$$

$$\nu_{A1}\, \mathrm{A}_1 + \nu_{A2}\, \mathrm{A}_2 \xrightarrow{k}$$

Formal Object

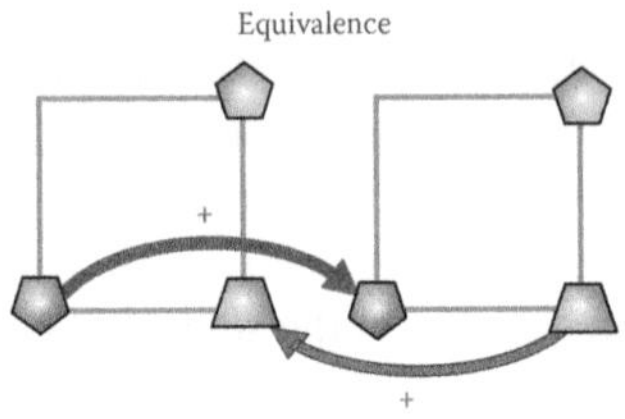

	Coupling
✓	Invariable
	Equivalence
	Conversion
✓	Poles
	Dipoles
✓	Capacitive
	Inductive
	Conductive

Formal Graph:

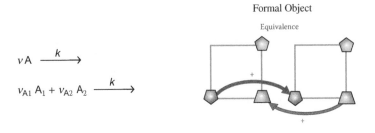

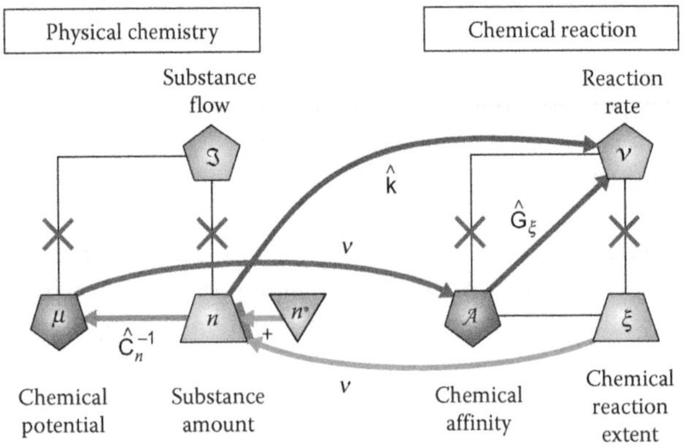

Physical chemical Capacitance:

$$\hat{\mathsf{C}}_n\, \mu = n^{\ominus} \exp\left(-\frac{\mu - \mu^{\ominus}}{\mu_\Delta}\right)$$

Reactional conductance:

$$\hat{\mathsf{G}}_\xi\, \mathcal{A} = \nu^{\ominus} \exp\left(-\frac{\mathcal{A}}{\mathcal{A}_\Delta}\right)$$

Kinetic operator:

$$\hat{\mathsf{k}} = \hat{G}_\xi\, \nu\, \hat{\mathsf{C}}_n^{-1} = k n^{\nu}$$

GRAPH 12.7

	Coupling	From	To
Variety		Physical chemical energy	Chemical reaction energy
Family	Integral factor	Basic quantity	
Name	Stoichiometric factor	Substance amount	Reaction extent
Symbol	ν	n	ξ
Unit	[-]	[mol]	[mol]
Variety		Chemical reaction energy	Physical chemical energy
Family	Differential factor	Effort	
Name	Stoichiometric factor	Affinity	Chemical potential
Symbol	ν	$\mathcal{A}$	μ
Unit	[–]	[J mol^{-1}]	[J mol^{-1}]

System Constitutive Properties		
Variety	Chemical reaction energy	Physical chemical energy
Nature	Capacitance	Capacitance
Levels	0	0
Name	Chemical reaction capacitance	Physical chemical capacitance
Symbol	C_ξ	C_n
Unit	[mol^2 J^{-1}]	[mol^2 J^{-1}]
Specificity	–	–

The *chemical reaction energy* variety has a capacitive subvariety in which the basic quantity is the *chemical reaction advance* or *chemical reaction extent* ξ with units in mole and the effort is the *chemical affinity* $\mathcal{A}$ with units in joule per mole. This latter notion has been introduced by Théophile de Donder* in 1923 (Prigogine and Defay 1962) but has disappeared now from the physical chemical landscape, being replaced by the notion of *molar free energy of reaction* notated $\Delta_r G$.

Variety	**Chemical Reaction**		
Subvariety	Capacitive (potential)		Inductive
Category	Entity number	Energy/entity	Energy/entity
Family	Basic quantity 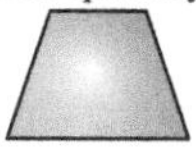	Effort 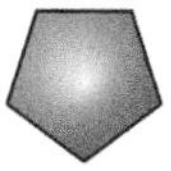	Flow 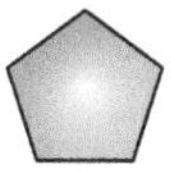
Name	Reaction extent	Affinity	Reaction rate
Symbols	ξ	$\mathcal{A}$	$\mathcal{V}$
Unit	[mol]	[J mol^{-1}]	[mol s^{-1}]

Here the notion of affinity is reintroduced, but with a slightly different acceptance, as it is taken as an energy-per-entity that may feature various Formal Objects, in contradistinction with the original concept of de Donder who associated it with a reaction, as a difference between two quantities featuring each side of the reaction. Translated into Formal Graph concepts, de Donder's affinity is a dipole affinity and is therefore the difference of two pole affinities. In this case study, only the pole affinity will be considered, as only one side of a reaction is treated.

12.4.1.1 Single Reacting Species

Let us take first the case of a single species able to react by association of several entities (or by multiple encounters in a dynamical viewpoint). The case of two species is considered later. The chemical reaction is written with its reacting side, which may represent either an irreversible reaction or only one side of a reversible reaction:

$$\nu \, \mathrm{A} \xrightarrow{k}$$

The number ν in front of the species A is called the *stoichiometric factor*. It quantifies the number of molecules that need to combine to produce another compound.

The equivalence between the two energy varieties is directly written in terms of energy variations:

$$\mathrm{d}\mathcal{U}_n = \mu \, \mathrm{d}n = \mathrm{d}\mathcal{U}_\xi = \mathcal{A} \, \mathrm{d}\xi \tag{J1.1}$$

The basic quantity ξ is the *extent of the chemical reaction*, or *chemical reaction advance*, quantifying the degree of evolution of a chemical reaction. It is counted as a substance amount, with units in mole, but it is a real number, which can be negative in case of species consumption or positive in case of production.

To determine this variable, the whole reaction must be taken into account and the assumption of a reaction mechanism in one step must be made. The reaction extent follows the increase or decrease of the substance amount of the species which has the smallest variation Δn during the reaction. This definition can be expressed mathematically in terms of absolute values, the sign

* Théophile Ernest de Donder (1872–1957): Belgian mathematician and physicist; Brussels, Belgium.

being given by the reactant ($\xi < 0$) or product ($\xi > 0$) status of the species with the minimum amplitude of variation.

$$|\xi| \overset{def}{=} min(|\Delta n_i|) \quad i = 1, \text{number of species}$$

By definition, the reaction extent is zero at the beginning of the reaction. From this definition the concept of stoichiometric factor ν is defined as the derivative of the consumed or produced substance amount with respect to the reaction extent

Stoichiometric factor:
$$\nu \overset{def}{=} \frac{\mathrm{d}n}{\mathrm{d}\xi} \tag{J1.2}$$

With the sign convention adopted for the reaction extent, the stoichiometric factor is always positive. Also it is a constant quantity, unless the reaction mechanism is more complicated than the single step assumed here. This allows identification of the stoichiometric factor with the result of the integration of its definition

$$\nu = \frac{n - n^*}{\xi} \tag{J1.3}$$

From the definition of the reaction extent, one of the stoichiometric factors is equal to 1 and the values of the others factors are equal or higher than 1. In ordinary reactions the stoichiometric factors are integers, but in more complex reactions noninteger values are frequent. An example of an ordinary reaction is the decomposition of water:

$$2H_2O \xrightarrow{k} 2H_2 + O_2$$

The species having the minimum amplitude of variation of its amount is the dioxygen molecule O_2, which is a produced species so its stoichiometric factor is 1. For water and for dihydrogen H_2 we have 2.

These definitions being established, one tackles the subject of equivalence between physical chemical energy and reaction energy in the case of a single reactant.

The definition in Equation J1.2 of the stoichiometric factor works clearly as an equivalence relationship between two basic quantities (see Section 12.2). Once introduced into the equivalence equation J1.1 between energy variations, the proportionality relation between efforts follows:

Equivalence:
$$\nu = -\frac{\mathcal{A}}{\mu} \tag{J1.4}$$

These equivalence relationships in Equations J1.3 and J1.4 are represented in the Formal Graph of the case study abstract, together with other relationships relevant to capacitive storage and kinetics. This Formal Graph uses two fundamental poles of different natures, one capacitive for the physical chemical energy variety, one conductive for the reaction energy variety. It can be remarked that when the stoichiometric factor is equal to 1 (first order reaction), the combination of the two poles into a single one is possible for producing a mixed pole, as was done in case study A11 "Reactive Chemical Species" in Chapter 4. Also, no link is drawn between the flows because it is not pertinent to use a substance flow in a purely capacitive pole (naturally, nothing forbids establishing a link by convention).

The pole in the physical chemical energy is a storage pole with a capacitance expressed, as used in the quoted case study A11 in Chapter 4 and demonstrated in Chapter 7, as

$$n = \hat{\mathsf{C}}_n \mu = n^{\ominus} \exp\left(-\frac{\mu - \mu^{\ominus}}{\mu_\Delta} \right) \tag{J1.5}$$

The pole in the reaction energy variety is a purely conductive pole endowed with a conductance, which has for expression, according to the general definition given in Chapter 8,

$$\mathcal{V} = \hat{\mathsf{G}}_\xi \mathcal{A} = \mathcal{V}^{\ominus} \exp\left(-\frac{\mathcal{A}}{\mathcal{A}_\Delta} \right) \tag{J1.6}$$

The *reaction rate* $\mathcal{V}$ is the flow and the *scaling affinity* $\mathcal{A}_\Delta$ is the scaling effort in the reaction energy variety. Both scaling efforts are identical owing to the equivalence and to the use of the same unit (J mol^{-1}):

$$\mathcal{A}_\Delta = \mu_\Delta \tag{J1.7}$$

By combining the last four equations, the usual kinetic equation giving the reaction rate as a function of the substance amount is obtained:

$$\mathcal{V} = k\, n^{\nu} \tag{J1.8}$$

The result is that the reaction order corresponds to the stoichiometric factor, which is true for ordinary reactions. The dependence of the reaction rate on the substance amount is represented in the Formal Graph in the case study abstract by a *kinetic operator* (which is not linear)

$$\hat{\mathsf{k}} = \mathsf{G}_\xi\, \mathcal{V} \hat{\mathsf{C}}_n^{-1} = k n^{\nu-} \tag{J1.9}$$

using a *kinetic constant* k expressed as

$$k = \frac{\mathcal{V}^{\ominus}}{\left(n^{\ominus}\right)^{\nu}} \exp\left(\frac{\nu\, \mu^{\ominus}}{\mu_\Delta} \right) \tag{J1.10}$$

It must be stressed that this simple link between stoichiometry and reaction order is valid only for simple reactions involving one species and occurring in one step (i.e., corresponding to the single pole description made here). For a chain reaction or for more complex mechanisms, this link is not direct and the present model does not apply for the global reaction. The case of several species reacting in one step can nevertheless be handled by generalizing the model.

12.4.1.2 Multiple Reacting Species

The case of multiple species undergoing a reaction is worth studying to show how more complex reactions can be modeled from this basic brick.

The chemical reaction is written with its reacting side, as before.

$$\nu_{A1}\, \mathrm{A}_1 + \nu_{A2}\, \mathrm{A}_2 \xrightarrow{k}$$

Each species is endowed with state variables of a capacitive pole in the chemical reaction energy variety. The two poles in the present case study are grouped in a multipole and the relationships between the poles and the multipole are

Reaction extent: $$\xi_A = \xi_{A1} = \xi_{A2} \quad \text{(J1.11)}$$

$$\mathcal{A}_A = \mathcal{A}_{A1} + \mathcal{A}_{A2} \quad \text{(J1.12)}$$

From the previous case of single reacting species, the following relationships are deduced:

Reaction extent: $$\xi_A = \frac{n_{A1} - \overset{*}{n}_{A1}}{\nu_{A1}} = \frac{n_{A2} - \overset{*}{n}_{A2}}{\nu_{A2}}; \quad (\xi_A < 0) \quad \text{(J1.13)}$$

$$\mathcal{A}_A = \nu_{A1}\mu_{A1} + \nu_{A2}\mu_{A2} \quad \text{(J1.14)}$$

The Formal Graph in Graph 12.8 contains all the Formal Objects required for modeling this system. Each species is represented by two equivalent poles, one in each energy variety, as for a single reacting species studied previously. The modeling of the group of reactants is ensured by a multipole in the chemical reaction energy variety. This architecture reflects the different roles of the energy

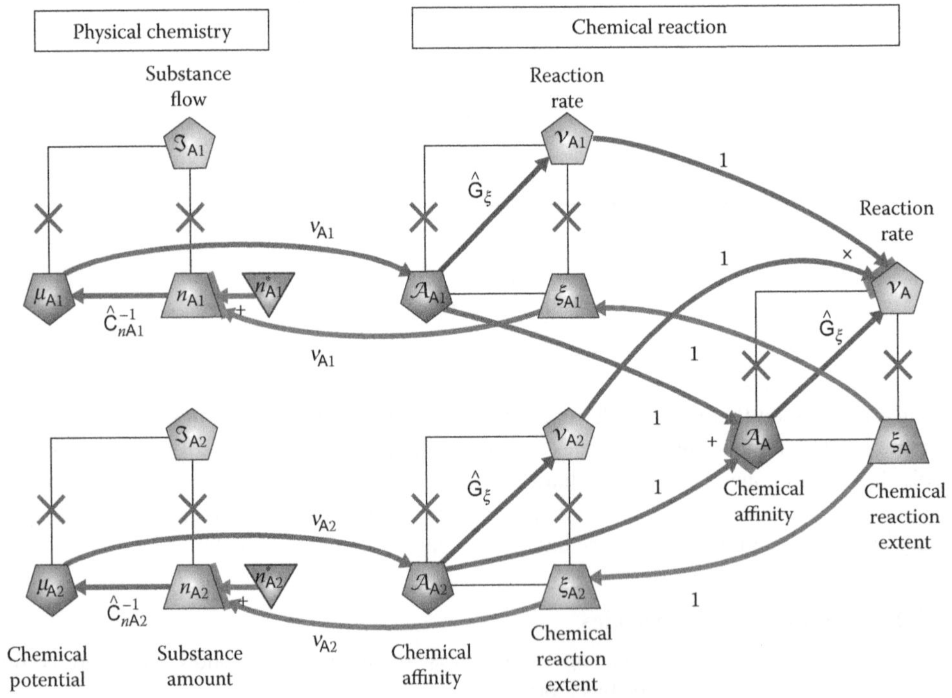

GRAPH 12.8 Complete Formal Graph of the model of two species participating in a chemical reaction. In the physical chemical energy variety are found the poles of the species and in the chemical reaction energy variety are found the equivalent poles and their association into a multipole.

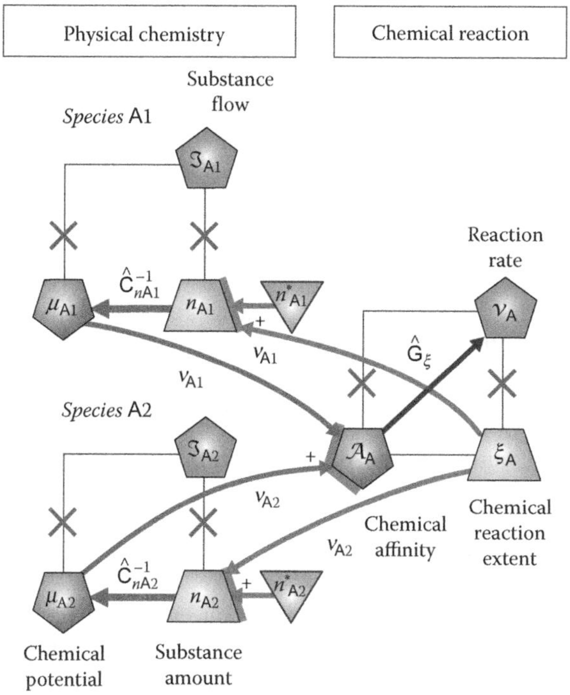

GRAPH 12.9 Formal Graph of additive energetic equivalence modeling a group of reactants with various stoichiometric factors. This Formal Graph models exactly the same system as in Graph 12.8.

varieties and the two levels of complexity of these objects. Naturally, this scheme can be extended to any number of participants to the reaction by merely adding pairs of equivalent poles as desired.

However, this Formal Graph can be simplified by removing the poles in the chemical reaction energy variety and by connecting directly the physical chemical poles to the multipole, as shown in Graph 12.9. Such a simplification corresponds to the definition of an *additive equivalence*, which extends the notion of equivalence by allowing several contributions to a more complex Formal Object.

The capacitances and the conductance having the same expressions as before, the reaction rate of the grouped reactants is deduced by using the same ingredients as for a single species, which gives

Reaction rate:
$$\mathcal{V}_A = k_A \, n_{A1}^{\nu_{A1}} n_{A2}^{\nu_{A2}} \tag{J1.15}$$

Each reaction order corresponds to the respective stoichiometric factor (provided the same requirements are met), as for a single reacting species.

Remarks

Complete reaction. Modeling a reversible reaction in the general case of any number of participants and any stoichiometry requires just one step more in the construction. In the general case, such a reaction is written as

$$\nu_{A1}\,A_1 + \nu_{A2}\,A_2 + \cdots \underset{k_B}{\overset{k_A}{\rightleftarrows}} \nu_{B1}\,B_1 + \nu_{B2}\,B_2 + \cdots$$

(continued)

It is sufficient to consider each side of the reaction as a group of reactants or products and to model each group as a multipole in the chemical reaction energy variety, as was done above in this case study. Then a dipole is built exactly on the same template as the one established in case study C5 "First-Order Chemical Reaction" in Chapter 6 devoted to dipoles. The connections between the two multipoles featuring the groups and the dipole are as follows:

$$\xi = -\xi_A = \xi_B; \quad \mathcal{A} = \mathcal{A}_B - \mathcal{A}_A; \quad \mathcal{V} = \mathcal{V}_A - \mathcal{V}_B \tag{J1.16}$$

From these relationships (and by using the same expressions for the operators) the kinetic law expressing the rate of the reaction is deduced:

$$\mathcal{V} = k_A\, n_{A1}^{\nu_{A1}} n_{A2}^{\nu_{A2}} \cdots - k_B\, n_{B1}^{\nu_{B1}} n_{B2}^{\nu_{B2}} \cdots \tag{J1.17}$$

The law of dynamic equilibrium is derived from this result by setting the reaction rate equal to zero.

Corpuscular reaction energy. The concept of reaction energy variety is not reserved for the chemical reaction and can be perfectly used for "physical" reactions as well. The *corpuscular affinity* and the *corpuscular reaction extent* are the state variables, together with the *corpuscular reaction rate*, of the reaction energy variety associated with the corpuscular energy variety (in which the number of corpuscles N is the basic quantity).

12.4.2 Case Study J2: Electrochemical Potential

Electrodynamics
Physical Chemistry

The case study abstract is given in next page.

The electrochemical potential is a concept introduced in 1929 by Guggenheim* for taking into account the influence of the charges borne by an ion and defined as (Kondepudi and Prigogine 1998)

$$\overline{\mu} = \mu + zF\,V \tag{J2.1}$$

To the chemical potential μ of the species is added a term formed by the product of z, the number of positive elementary charges (charge number) borne by the ion, F, the Faraday constant ($\cong 9.6 \times 10^4$ C mol^{-1}), and V the electric potential in the medium (also called phase).

The notation "*def*" is not indicated in the expression of the electrochemical potential because this classical definition is not recommended. It is preferable to define the electrochemical potential as an apparent effort, that is, as the partial derivative of the energy versus the variation of a basic quantity, the substance amount, as is going to be demonstrated.

$$\text{Apparent physical chemical potential: } \overline{\mu} \overset{def}{=} \left(\frac{\partial \mathcal{U}}{\partial n}\right)_{indep.\, qi} \tag{J2.2}$$

* Edward Armand Guggenheim (1901–1970): English physical chemist; Reading, UK.

J2: Electrochemical Potential

Electrodynamics
Physical Chemistry

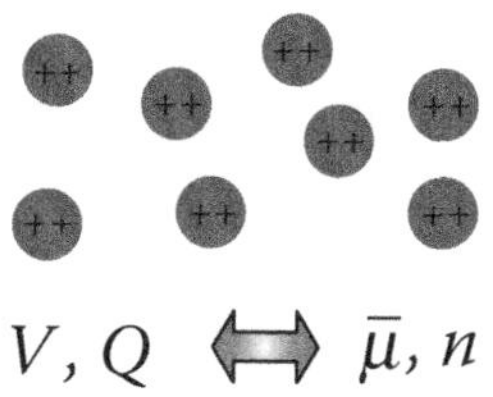

Formal Object

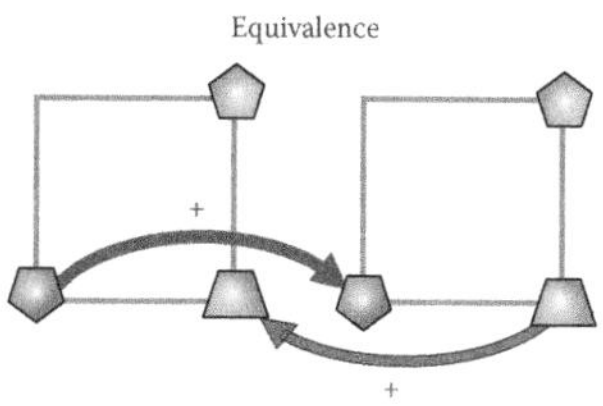

Coupling
✓ Invariable Equivalence Conversion
✓ Poles Dipoles
✓ Capacitive Inductive Conductive

Formal Graph

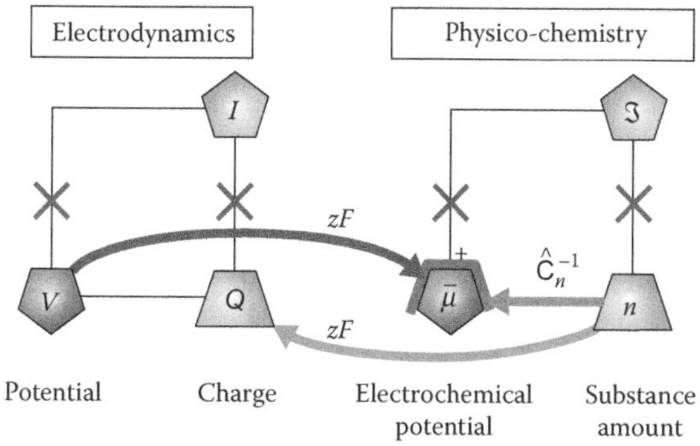

GRAPH 12.10

	Coupling	From	To
Variety		Physical chemical Energy	Electrodynamics
Family	Integral factor	Basic quantity	
Name	Charge number × Faraday	Substance amount	Charge
Symbol	zF	n	Q
Unit	[C mol^{-1}]	[mol]	[C]
Variety		Electrodynamics	Physical chemical energy
Family	Differential factor	Effort	
Name	Charge number × Faraday	Potential	Chemical potential
Symbol	zF	V	μ
Unit	[C mol^{-1}]	[V]	[J mol^{-1}]

System Constitutive Properties		
Variety	Electrodynamics	Physical chemical energy
Nature	Capacitance	Capacitance
Levels	0	0
Name	Electric capacitance	Physical chemical capacitance
Symbol	C_E	C_n
Unit	[F]	[mol^2 J^{-1}]
Specificity	–	–

This derivation defines a much more general potential than the sole electrochemical potential. This apparent physical chemical potential becomes an electrochemical potential only if the charge is the only basic quantity dependent on the substance amount. The important point in the partial derivation is the list of variables that must be kept constant. Here, all the basic quantities that are independent of the substance amount are retained. The electric charge cannot be in the list when it depends on the substance amount. As the peculiarity of an ion is to have a constant charge (in stable conditions), a collection of ions sees the overall charge Q depending directly on the substance amount through the Faraday relation

Faraday relation: $$Q = zF\,n \tag{J2.3}$$

By introducing this relation into the Gibbs equation of the system, containing any number of energy varieties for the sake of generality, one gets (in assuming the constancy of z)

Gibbs: $$\mathrm{d}\mathcal{U} = \mu\,\mathrm{d}n + V\mathrm{d}Q + \cdots = \left(\mu + zFV\right)\mathrm{d}n + \cdots \tag{J2.4}$$

From the definition of the apparent physical chemical potential, one may write

$$\mathrm{d}\mathcal{U} = \bar{\mu}\,\mathrm{d}n + \cdots \tag{J2.5}$$

and by comparison with the Gibbs equation modified by the Faraday relation the expression of the electrochemical potential is deduced as

$$\bar{\mu} = \mu + \mu_V \tag{J2.6}$$

having defined the *translated chemical potential* of an ion as

$$\mu_V \overset{def}{=} zF\,V \tag{J2.7}$$

This variable may appear superfluous at this stage of reasoning, but it corresponds to an intermediate variable used in many Formal Graph modeling coupled energy varieties. With this notion of *translated chemical potential*, the electrical energy appears under the form of physical chemical energy, with the same amount of electrical energy

$$\mathrm{d}\mathcal{U}_Q = V\mathrm{d}Q = \mu_V\,\mathrm{d}n \tag{J2.8}$$

The Formal Graph in Graph 12.11 depicts the relationships between the two energy varieties.

To model the electrochemical potential described by Equations J2.6 and J2.7, a set of four graph units is required as shown in Graph 12.12.

A simpler representation is possible using only two graph units, one for each energy variety, as shown in the Formal Graph in the case study abstract.

It relies on the expression of the chemical potential as resulting from a capacitance operator, allowing the writing of the electrochemical potential as the result of the concurring actions of two operators:

$$\bar{\mu} = \hat{\mathsf{C}}_n n + zF\,V \tag{J2.9}$$

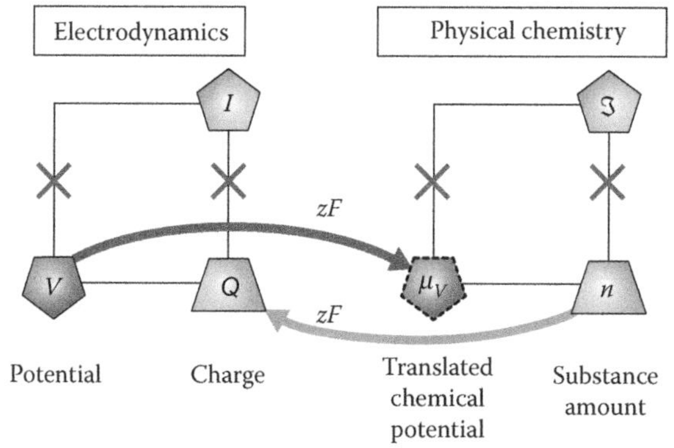

GRAPH 12.11 Formal Graph depicting how a translated chemical potential is related to the electrical potential. The two energy varieties are not equivalent, and, to avoid confusion with the case of equivalent energies, the edge of the node of the translated chemical potential is drawn with a dotted line.

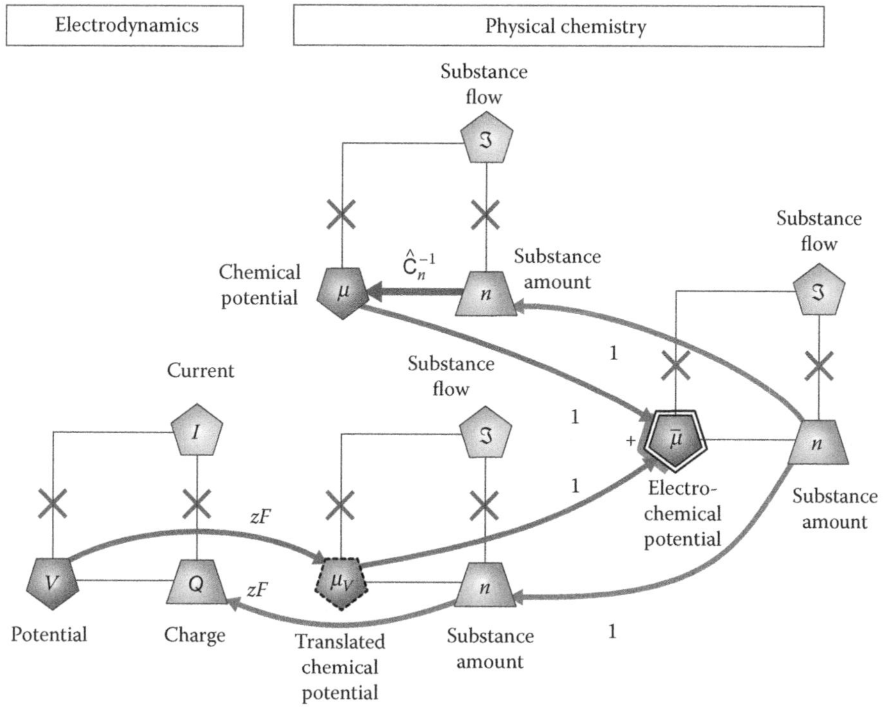

GRAPH 12.12 Construction of the electrochemical potential of an ion from the physical chemical pole and from the electrical pole translated into a physical chemical pole.

which is represented in the Formal Graph as the sum of two paths, one coming from the electrical potential and the other from the substance amount.

12.4.2.1 Electrochemical Reaction

The interest of the notion of electrochemical potential is double. First, it is used for writing equilibrium equations and determining reaction energies, that is, in the classical frame of thermodynamics restricted to static conditions. The second usage is not classical since it extends the frame of application to kinetics and dynamic conditions. The Formal Graph shown in Graph 12.13 models an

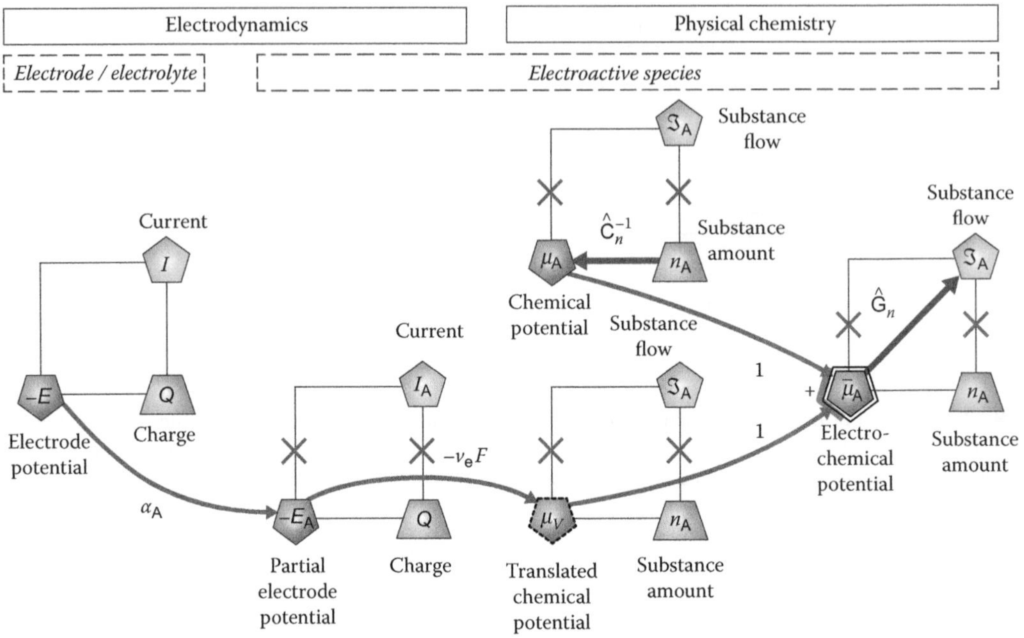

GRAPH 12.13 Formal Graph modeling the kinetics of an electroactive species. The electrical dipole on the left stands for the redox couple A/B having for effort the electrode potential $-E$. The electrical pole just besides represents the electrical energy borne by the species A with its translation into physical chemical energy on its right side. Above is the pure physical chemical pole that contributes to the total physical chemical energy of the species (right). The substance flow is then given by the conductance applied to the electrochemical potential.

electroactive species A involved in an electrochemical reaction in which ν_e is the *stoichiometric factor* of electrons, that is, the number of exchanged electrons per molecule, in counting positively this number in case of oxidation and negatively for a reduction

$$A - \nu_e e^- \rightleftarrows B$$

The two species are dissolved in an electrolyte in which an electrode is plunged. The potential difference between the electrode and the electrolyte is notated E and is merely called the *electrode potential.*

The opposite of this potential, the difference between the electrolyte and the electrode potentials, stands for the effort of a dipole modeling the electrode/electrolyte system.*

The variation of total electrical energy is therefore given by

$$\delta \mathcal{U}_Q = (-E)\mathrm{d}Q \quad \text{(J2.10)}$$

The Formal Graph theory states that owing to the charge transfer, the electrical energy $\mathcal{U}_Q$ is shared between the species A and its partner B in a manner that differs from the case in which charges are tightly bound to the molecules (as for nonelectroactive ions). The theory models this

* For historical reasons, the sign of the electrode potential is opposite to the right thermodynamical potential.

sharing by stating that the variation of electrical energy borne by a species with respect to the total electrical energy defines the *transfer factor* (also called *exchange factor*)

Transfer factor:
$$\alpha_A \overset{def}{=} \frac{\partial \mathcal{U}_{QA}}{\partial \mathcal{U}_Q} \qquad (J2.11)$$

This definition has already been given in Chapter 8, Section 8.5.6, when the theory of conduction was established and applied to conductive poles and dipoles. The model is now completed with the coupling between electrodynamics and physical chemistry for building up the electrochemical potential.

The variation of total capacitive energy for the species A is the sum of the physical chemical energy and of the electrical energy variations, this latter being expressed through the transfer factor as proportional to the variation of total electrical energy:

$$\delta \mathcal{U}_A = \delta \mathcal{U}_{nA} + \delta \mathcal{U}_{QA} = \delta \mathcal{U}_{nA} + \alpha_A \, d\mathcal{U}_Q \qquad (J2.12)$$

which gives, in using the definition in Equation J2.2 of the electrochemical potential

$$\delta \mathcal{U}_A = \bar{\mu}_A dn_A = \mu_A dn_A + \alpha_A \left(-E\right) dQ \qquad (J2.13)$$

As the charge variation in the electrical dipole is related to the substance amount variation via the Faraday relation (in which the charge number is replaced by the *stoichiometric factor* of the electrons)

$$dQ = -\nu_e F dn_A \quad \left(= \nu_e F dn_B\right) \qquad (J2.14)$$

the expression of the electrochemical potential can be written as

$$\bar{\mu}_A = \mu_A + \alpha_A \, \nu_e F E \qquad (J2.15)$$

This relationship is represented in Graph 12.13 in detailing each step in the formation of the electrochemical potential. An intermediate pole bearing the electrical energy of the species A has been drawn with a partial electrical potential $-E_A$ corresponding to the fraction α_A of the dipole potential $-E$. This variable replaces the potential V used in Graph 12.12.

12.4.2.2 Equilibrium

The dynamic equilibrium of the exchange between A and B is reached when the difference of their electrochemical potentials is equal to 0. In taking into account the necessary condition of a sum of transfer factors equal to 1, this equilibrium provides a proportionality relation between the difference of chemical potentials and the electrode potential

Equilibrium:
$$\bar{\mu}_B - \bar{\mu}_A = 0 \Leftrightarrow \bar{\mu}_B - \bar{\mu}_A = \nu_e F E \qquad (J2.16)$$

The physical chemical capacitive relationship, which has been established in Chapter 7, and the definition of a *redox potential* featuring the couple A/B

$$n_i = n^{\ominus} \exp\left(\frac{\mu_i - \mu_i^{\ominus}}{\mu_\Delta}\right); \quad E^0_{AB} \overset{def}{=} \frac{\mu_B^{\ominus} - \mu_A^{\ominus}}{\nu_e F} \qquad (J2.17)$$

lead to the following relationship, known as the Nernst[*] law (when the scaling chemical potential μ_Δ is equal to RT): (J2.18)

$$\text{Nernst law:} \qquad \frac{n_B}{n_A} = \exp\left(\frac{\mu_B - \mu_A - \mu_B^{\vartheta} + \mu_A^{\vartheta}}{\mu_\Delta}\right) = \exp\left(\frac{\nu_e F}{\mu_\Delta}\left(E - E^0\right)\right) \tag{J2.19}$$

12.4.2.3 Kinetics

To model the kinetics of the electrochemical reaction, one has to express the substance flow as a function of the electrode potential. The model of conductance is the standard exponential model established in Chapter 8 used in every energy variety, using here the electrochemical potential as effort:

$$\mathfrak{I}_A = \mathfrak{I}_{AB}^0 \exp\left(\frac{\bar{\mu}_A - \mu_{AB}^{eq}}{\mu_\Delta}\right) = \mathfrak{I}_{AB}^0 \exp\left(\frac{\mu_A - \mu_{AB}^{eq}}{\mu_\Delta}\right) \exp\left(\frac{\alpha_A \nu_e F E}{\mu_\Delta}\right) \tag{J2.20}$$

The capacitive relationship and the expression of the purely chemical kinetic constant are borrowed from case study E4 in Chapter 8:

$$k_A^0 = \frac{\mathfrak{I}_{AB}^0}{n^{\vartheta}} \exp\left(\frac{\mu_A^{\vartheta} - \mu_{AB}^{eq}}{\mu_\Delta}\right) \tag{J2.21}$$

By introducing these relationships into the expression of the substance flow, one gets a simpler expression:

$$\mathfrak{I}_A = k_A^0 \exp\left(\frac{\alpha_A \nu_e F E}{\mu_\Delta}\right) n_A \tag{J2.22}$$

The same treatment can be done for the other member of the redox couple, and, by making the difference between the two substance flows, the kinetic equation of the electrochemical reaction is obtained. The Butler[†]–Volmer[‡] (α_i constant) or more sophisticated models (α_i variable, as in the Marcus[§] theory) are easily derived from it.

Remarks

Generality of the conduction model. As explained in case study A11 "Reactive Chemical Species" in Chapter 4, the classical approach in kinetics is based on the transition state theory (Laidler and King 1998). The Formal Graph approach is based on a simpler theory of conduction which is much more general as it works in all energy varieties. For instance, the same theory is able to model electrons and holes in a p–n junction (Shockley[¶] diode) as well as molecules or enzymes involved in chemical or electrochemical reactions. Mechanical friction or viscous fluids may also be modeled with this transverse approach.

[*] Walther Hermann Nernst (1864–1941): German physical chemist; Göttingen and Berlin, Germany.
[†] Butler John Alfred Valentine (1899–1977). British electrochemist, Swansea and Edinburgh, UK.
[‡] Volmer Max (1885–1965). German physical chemist, Berlin and Hamburg, Germany.
[§] Marcus Rudolph Arthur (1923–). Canadian chemist, Ottawa, Canada and Pasadena, USA.
[¶] Shockley William Bradford (1910–1989). American physicist, Stanford, California, USA.

12.4.3 Case Study J3: Electrical Force

Translational Mechanics
Electrodynamics

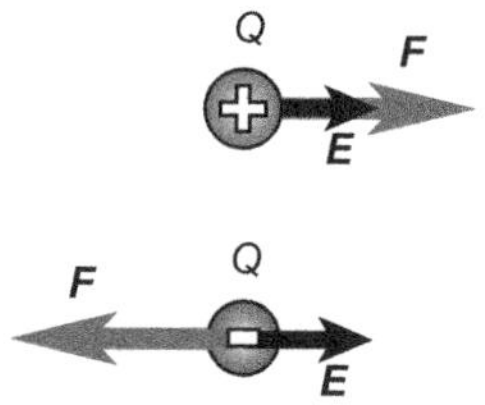

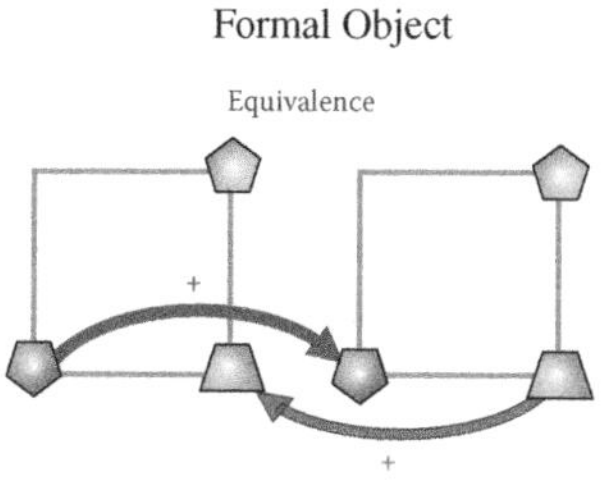

Coupling
Invariable Equivalence Conversion
✓ Poles Dipoles
✓ Capacitive Inductive Conductive

Formal Graph

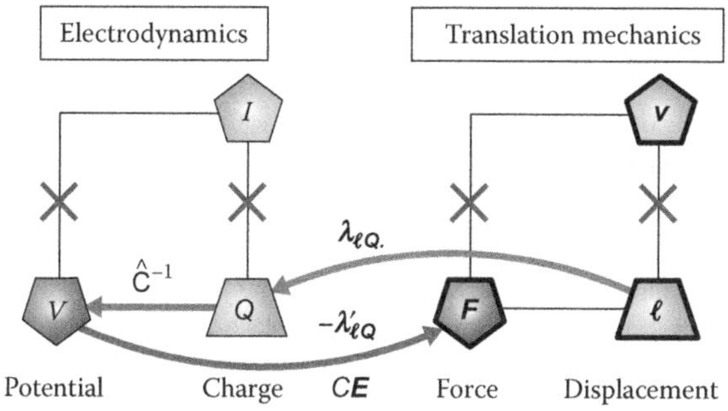

GRAPH 12.14

	Coupling	From	To
Variety		Translational mechanics	Electrodynamics
Family	Integral factor	Basic quantity	
Name		Displacement	Charge
Symbol	$\lambda_{\ell Q}$	ℓ	Q
Unit	[C m^{-1}]	[m]	[C]
Variety		Electrodynamics	Translational mechanics
Family	Differential factor	Effort	
Name		Potential	Force
Symbol	$\lambda'_{\ell Q}$	V	F
Unit	[C mol^{-1}]	[V]	[N]

System Constitutive Properties		
Variety	Electrodynamics	Translational mechanics
Nature	Capacitance	Capacitance
Levels	0	0
Name	Electric capacitance	Inverse spring constant
Symbol	C	k_e^{-1}
Unit	[F]	[m N^{-1}]
Specificity	–	–

When a charge Q is placed in an electrical field $\boldsymbol{E}$, a force $\boldsymbol{F}$ is exerted on the charge in the direction of the field when the charge is positive and in the opposite direction otherwise. It can be recalled that the charge is not necessarily a single entity but can be an electrodynamical pole, that is, a collection of entities. The relationship between field and force is the basis of many theories and notably the Coulomb law modeling the interaction through space of electrical charges (see case study D4 in Chapter 7).

The classical theory of electrodynamics states that the force $\boldsymbol{F}$ acting on the charge Q is proportional to the field strength $\boldsymbol{E}$. The proportionality factor is the charge Q itself, and this proportionality is considered as one of the fundamental laws of classical electrostatics.

$$\boldsymbol{F} \underset{lin}{=} Q\,\boldsymbol{E} \tag{J3.1}$$

The notation "*lin*" placed under the equal sign signifies that the Formal Graph theory considers this relation as valid only in the frame of linear capacitances, as will be demonstrated step by step.

The system must contain at least two capacitive energies, in the translational mechanical energy and in electrodynamical energy, which is written under the form of a Gibbs equation:

Gibbs: $$d\mathcal{U} = \boldsymbol{F} \cdot \mathbf{d}\boldsymbol{\ell} + V\,dQ + \cdots = \left(\boldsymbol{F} + V\frac{dQ}{\mathbf{d}\boldsymbol{\ell}}\right) \cdot \mathbf{d}\boldsymbol{\ell} + \cdots = \overline{\boldsymbol{F}} \cdot \mathbf{d}\boldsymbol{\ell} \tag{J3.2}$$

The two capacitive energies are coupled when charge and distance are not independent and when the apparent force, given by the partial derivative of the total capacitive energy (maintaining constant all other independent basic quantities), is equal to zero:

Coupling: $$\overline{\boldsymbol{F}} \overset{def}{=} \left(\frac{\partial \mathcal{U}}{\partial \ell}\right)_{qi} = 0 \tag{J3.3}$$

This is verified when the following equalities evidencing the differential coupling factor are satisfied:

$$\lambda'_{\ell Q} = \frac{dQ}{\mathbf{d}\boldsymbol{\ell}} = -\frac{\boldsymbol{F}}{V} \tag{J3.4}$$

The Formal Graph in the case study abstract represents the coupling with the help of this factor and with the integral one as well. In fact, the knowledge of these factors and their invariability or not is useless in the present discussion. From the second equality the force is deduced as

$$\boldsymbol{F} = -V\frac{dQ}{\mathbf{d}\boldsymbol{\ell}} \tag{J3.5}$$

When the electrical capacitance is a scalar, the validity domain is restricted to the linear case and the capacitive relationship is written as

$$Q \underset{lin}{=} C\,V \tag{J3.6}$$

and the following equality is verified:

$$V dQ \underset{lin}{=} Q\,dV \tag{J3.7}$$

so the expression of the force becomes

$$\boldsymbol{F} \underset{lin}{=} -Q\frac{dV}{\mathbf{d}\boldsymbol{\ell}} \tag{J3.8}$$

The derivative of the potential with respect to the distance corresponds to the electric field with a minus sign (in the absence of electromagnetic field)

$$\mathbf{E} = -\frac{dV}{d\ell} \tag{J3.9}$$

which leads to the classical expression in Equation J3.1 by direct substitution.

When the capacitance is a nonlinear operator, the equality in Equation J3.7 is not true anymore. It can be replaced by another equality using an *apparent charge* defined as

$$\overline{Q} \overset{def}{=} V\frac{dQ}{dV} \tag{J3.10}$$

and the general expression of the force is now written as

$$\mathbf{F} = \overline{Q}\,\mathbf{E} \tag{J3.11}$$

From the definition (Equation J3.10) of the apparent charge, it can be checked that this latter becomes equal to the actual charge when the capacitance is assumed to be a scalar, thus retrieving the classical expression in Equation J3.1. In fact, this definition can be viewed as a pseudocapacitive relationship using the differential capacitance C instead of the capacitance operator

$$\overline{Q} = CV \tag{J3.12}$$

which is equal to the integral capacitance only in case of constant differential capacitance (and therefore scalar).

Then, the differential coupling factor can be expressed as a function of this differential capacitance

$$\lambda'_{\ell Q} = -C\,\mathbf{E} \tag{J3.13}$$

allowing the general expression in Equation J3.11 to be read in the Formal Graph in the case study abstract by combining the coupling path with the capacitance path for going from Q to $\mathbf{F}$.

A step further can be made in the theory in taking the electrodynamical capacitive relationship which has been discussed in case study A5 "Electric Charges" in Chapter 4. This relationship was established for a composed pole (mix of negative and positive and written with the help of two scaling factors), Q^m for the charge and V_Δ for the potential.

$$\frac{Q+Q^m}{Q^m} = \exp\left(\frac{V}{V_\Delta}\right) \tag{J3.14}$$

The apparent charge is then deduced from the definition in Equation J3.10

$$\overline{Q} = Q^m\frac{V}{V_\Delta}\exp\left(\frac{V}{V_\Delta}\right) \tag{J3.15}$$

In comparison with the classical theory, the apparent charge can be expressed as a function of the linear charge Q_{lin} (i.e., the charge if the capacitance were a scalar), which is the asymptote at the origin and of the true charge Q given by the capacitive relationship in Equation J3.14:

$$Q_{lin} = Q^m\frac{V}{V_\Delta}; \quad \overline{Q} = Q_{lin}\left(1+\frac{Q}{Q^m}\right) \tag{J3.16}$$

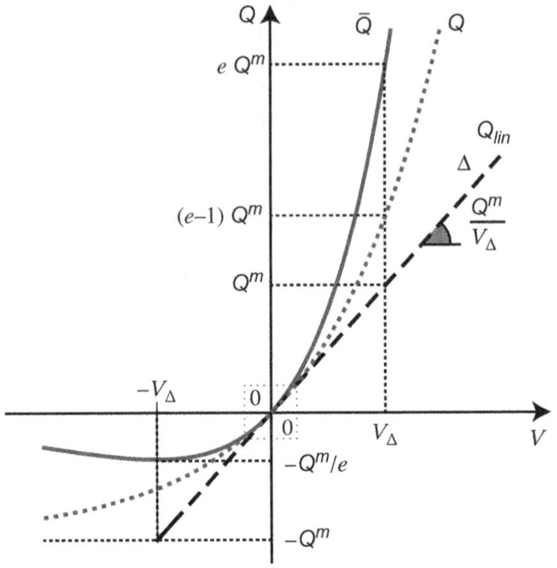

FIGURE 12.3 Apparent charge (solid curve), actual charge (dotted curve), and linear charge (dashed line) plotted versus the potential. The apparent charge is the effective one relating the electric field to the acting force. (The symbol letter e stands for the Napier constant $e = 2.71828$).*

It shows that for reaching a 1% discrepancy, the true charge must be 1% of the scaling charge Q^m. In vacuum, the scaling potential and the scaling charge are very high (in absolutely free space they are infinite, unless some coupling with another energy variety occurs), thus rendering pointless this correction of the classical law. Nevertheless, in nonlinear media the scaling charge and potential are within experimental windows and the correction is useful. Case study J6 depicts the case of ions for which the scaling potential is in the order of a few tens of millivolts.

The validity domain of the classical relationship $\boldsymbol{F} = \boldsymbol{QE}$ is schematized in Figure 12.3 approximately by the rectangle around the origin.

Piezoelectricity. When subjected to mechanical stress, some materials develop a voltage difference, called electrical polarization. The converse is also observed; the same materials exposed to an electric field are lengthened or shortened according to the direction of the field and in proportion to its strength. These behaviors are called *piezoelectric effects* from the Greek "piezein," meaning to press or to squeeze.

The *piezoelectric strain tensor* $\hat{\mathbf{d}}$ and the *piezoelectric voltage tensor* $\hat{\mathbf{g}}$ are classically used for linking the mechanical variables to the electrical ones. The *piezoelectric strain tensor* links the electric field $\boldsymbol{E}$ to the strain $\ell_{/r}$ and the *piezoelectric voltage tensor* links the stress $\boldsymbol{F}_{/A}$ to the electric field $\boldsymbol{E}$. The relationships between them and with the coupling factor are

$$\boldsymbol{F}_{/A} = \hat{\mathbf{g}}^{-1}\boldsymbol{E} = \hat{\mathrm{E}}_Y\,\hat{\mathbf{d}}\boldsymbol{E} = \frac{\mathrm{d}}{\mathrm{d}\boldsymbol{r}}\lambda_{\ell Q}^{-1}\,\boldsymbol{E} \tag{J3.17}$$

where $\hat{\mathrm{E}}_Y$ stands for the Young modulus (see case study B1 in Chapter 5).

* John Napier (or Neper) (1550–1617): Scottish mathematician and physicist; Edinburgh, UK.

12.4.4 Case Study J4: Gravitational Force

Translational Mechanics
Gravitation

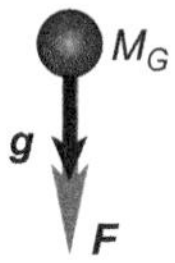

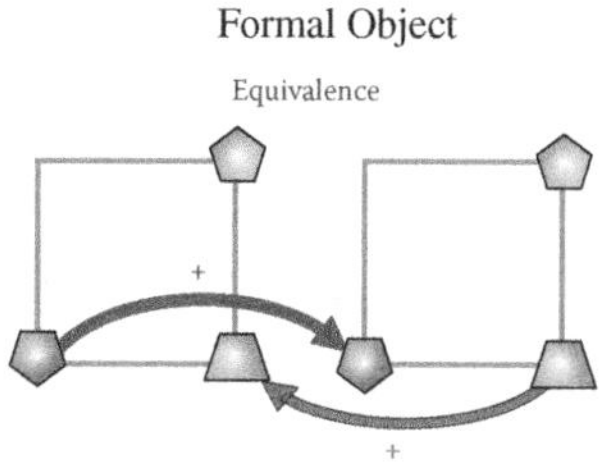

Coupling	
	Invariable
	Equivalence
	Conversion
✓	Poles
	Dipoles
✓	Capacitive
	Inductive
	Conductive

Formal Graph

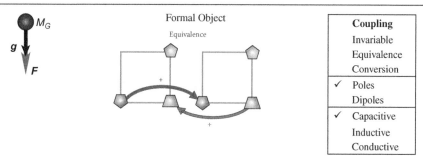

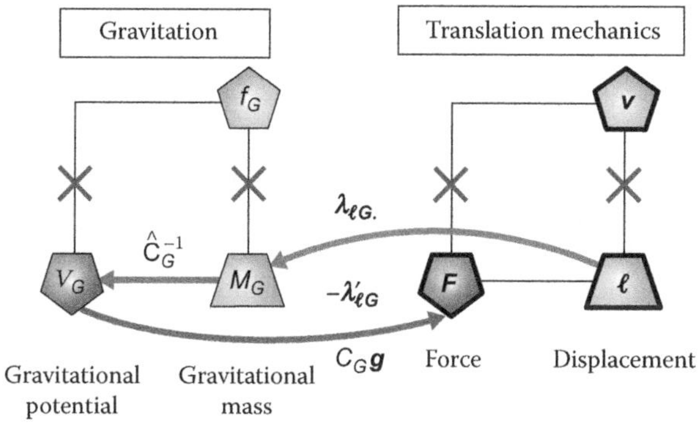

GRAPH 12.15

	Coupling	From	To
Variety		Gravitation	Electrodynamics
Family	Integral factor	Basic quantity	
Name		Displacement	Gravitational mass
Symbol	$\lambda_{\ell G}$	ℓ	M_G
Unit	[kg m^{-1}]	[m]	[kg]
Variety		Electrodynamics	Gravitation
Family	Differential factor	Effort	
Name		Gravitational potential	Force
Symbol	$\lambda'_{\ell Q}$	V_G	F
Unit	[kgm^{-1}]	[m^2 s^{-2}]	[N]

System Constitutive Properties		
Variety	Gravitation	Translational mechanics
Nature	Capacitance	Capacitance
Levels	0	0
Name	Gravitation capacitance	Inverse spring constant
Symbol	C_G	k_e^{-1}
Unit	[F]	[m N^{-1}]
Specificity	–	–

One of the important formulas in mechanics (in a broad acceptance) is the expression of a force resulting from the action of a *gravitational field* (also called *gravitational acceleration*) $\boldsymbol{g}$ on a mass:

$$\boldsymbol{F} = M_G\, \boldsymbol{g} \tag{J4.1}$$

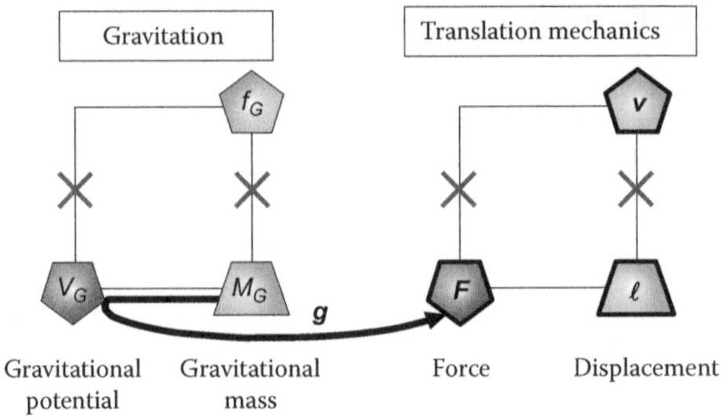

GRAPH 12.16

This relationship is one of the foundations of mechanics and of the Newtonian gravitation. It is the analogue of the expression of the force acting on an electric charge under the effect of an electric field. Its translation in Formal Graph is a composed path between two energy varieties: gravitational and translation mechanics.

The Formal Graph in Graph 12.16 is simplified to show only the composed path which uses the vector *gravitational field* $\boldsymbol{g}$ as unique operator. The complete graph with all the coupling relationships used is shown in the case study abstract. The two objects represented by the Formal Graph are two capacitive poles that are coupled. The simplified Formal Graph is not built directly from elementary properties of the poles and from the coupling between the two energy varieties but relies on a necessary hypothesis of linearity of the gravitational capacitance as it will be shown.

This is demonstrated in three steps.

1. *Coupling Gravitation and Mechanics:* The variation of capacitive energy in the system, in admitting that some other energy varieties may contribute (this possibility is indicated by suspension dots in the following equation), but without depending on the distance $\boldsymbol{\ell}$, is given by a Gibbs equation

Gibbs: $$\mathrm{d}\mathcal{U} = \boldsymbol{F}\cdot\mathbf{d}\boldsymbol{\ell} + V_G\,\mathrm{d}M_G + \cdots \qquad \text{(J4.2)}$$

By rearranging in gathering all terms that depend on space, this equation becomes

$$\mathrm{d}\mathcal{U} = \left(\boldsymbol{F} + V_G\frac{\mathrm{d}M_G}{\mathbf{d}\boldsymbol{\ell}}\right)\cdot\mathbf{d}\boldsymbol{\ell} + \cdots = \bar{\boldsymbol{F}}\cdot\mathbf{d}\boldsymbol{\ell} + \cdots \qquad \text{(J4.3)}$$

The coupling betaween the two energy varieties is assumed to be the most general one, without specifying whether or not the coupling factor is a constant. The two coupling relationships are therefore written with a *differential coupling factor* $\boldsymbol{\lambda}'_{\ell G}$ that is a vector, which allows the definition of a translated *force* $\boldsymbol{F}_G$, analogous to a translated *chemical potential* $zF\,V$, for instance (see case study J6 "Ionic Distribution").

$$\mathrm{d}M_G \overset{def\rightarrow}{=} \boldsymbol{\lambda}'_{\ell G}\cdot\mathbf{d}\boldsymbol{\ell};\quad \boldsymbol{F}_G \overset{def}{=} \boldsymbol{\lambda}'_{\ell G}\,V_G \qquad \text{(J4.4)}$$

This *differential coupling factor* is in fact the product of the differential volumic mass with the differential area, which are also coupling factors (cf. Chapter 13, Section 13.5):

$$\boldsymbol{\lambda}'_{\ell G} = \rho'_{M_G}\,\boldsymbol{A}' \qquad \text{(J4.5)}$$

One can also define, on the same scheme as the definition of an *electrochemical potential* (see case study J2), a "resulting force," which is an *apparent force* given by the sum

$$\bar{\boldsymbol{F}} \overset{def}{=} \boldsymbol{F} + \boldsymbol{F_G} \tag{J4.6}$$

Finally, when the system can minimize its energy with respect to space, this resulting force will be equal to zero. From this condition, the expression of the mechanical force $\boldsymbol{F}$ is deduced

$$\frac{\partial \mathcal{U}}{\partial \ell} = \mathbf{0} \Leftrightarrow \bar{\boldsymbol{F}} = \mathbf{0} \Leftrightarrow \boldsymbol{F} = -\boldsymbol{F_G} \Leftrightarrow \boldsymbol{F} = -V_G\, \boldsymbol{\lambda}'_{\ell G} \tag{J4.7}$$

2. *Gravitational Field:* The gravitational field is classically given by the contragradient of the gravitational potential. In fact, a more precise definition of a gravitational field states that it plays the role of factor for the variation of gravitational potential when the distance varies, which is a far better relationship for the development of the question being tackled.

$$\mathrm{d}V_G \overset{def \rightarrow}{=} -\boldsymbol{g} \cdot \mathbf{d}\boldsymbol{\ell} \tag{J4.8}$$

By introducing the gravitational field into Equation J4.4, the variation of distance can be eliminated to obtain

$$\boldsymbol{\lambda}'_{\ell G} = -\frac{\mathrm{d}M_G}{\mathrm{d}V_G}\boldsymbol{g} \tag{J4.9}$$

To obtain the same shape of equation as for Equation J4.1 to be demonstrated, an *apparent mass* can be defined according to

$$\bar{M}_G \overset{def}{=} V_G \frac{\mathrm{d}M_G}{\mathrm{d}V_G} \tag{J4.10}$$

From Equations J4.4, J4.9, and J4.10, the simple relation is deduced

$$\boldsymbol{F} = \bar{M}_G\, \boldsymbol{g} \tag{J4.11}$$

This is the most general form that can be obtained without restrictive hypotheses on the gravitational capacitance.

3. *Linear Capacitance:* Until here, the theoretical development has been carried out without particular hypotheses. Inasmuch the Formal Graphs theory strives to generalize the various particularities of each domain for attaining a common formalism; the general rule adopted for the capacitive relationships is their exponential expression and not a linear one.

Consequently, the classical capacitive relation in gravitation, which is a simple proportionality using a scalar factor C_G (equal, in the case of influence from another mass, to a distance divided by minus the gravitation constant), will be considered as a particular case corresponding to linear materials or to the vacuum.

$$M_G \underset{lin}{=} C_G\, V_G \tag{J4.12}$$

The independence of this scalar from the mass or from the gravitational potential is expressed by the equality

$$\frac{\mathrm{d}M_G}{\mathrm{d}V_G} \underset{lin}{=} \frac{M_G}{V_G} \tag{J4.13}$$

leading to a new expression of the coupling factor in Equation J4.9 (cf. the complete Formal Graph in the case study abstract)

$$\lambda'_{\ell G} \underset{lin}{=} -C_G\, \boldsymbol{g} \tag{J4.14}$$

Consequently, this independence makes the *apparent mass* in Equation J4.10 equal to the mass of the pole, and therefore to obtain the so-called fundamental law of Newtonian gravitation.

$$\boldsymbol{F} \underset{lin}{=} M_G\, \boldsymbol{g} \tag{J4.15}$$

Remarks

General relativity. In vacuum, the supporting medium by excellence of gravitation, the gravitational capacitance is linear when distances are not too long to allow the law of Newtonian gravitation to be applied without problem. On the contrary, according to Einstein's general relativity, for long distances an effect of space curvature appears and a correction is necessary.

Gravitational constant. The Formal Graph theory outlines the fact that the constancy of the gravitational constant G (leading to a linear capacitance) is a property of free space (vacuum) but that in nonlinear media this constitutive reduced property must be replaced by a more general one which is not a constant anymore. As in the case of electrodynamics for the vacuum permittivity ε_0 replaced by the permittivity ε.

Generality. The model of coupling of translation mechanics with another variety developed here is not restricted to gravitational energy. Any energy variety, in which a basic quantity depends on distance, will lead to the same model, as, for instance, hydrodynamics with the example of the piston (see case study J3). Furthermore, when the effort is spatially distributed, thus generating a field, the same relationship between force and field is obtained.

This is notably the case of electrodynamics, for which the hypothesis of linearity of the capacitance, which may seem obvious within the Newtonian framework, leads precisely to a restriction of the validity domain of classical electrostatics (or electrodynamics) to linear media (see case study J4 "Electrical Force").

12.4.5 Case Study J5: Piston

Electrodynamics
Physical Chemistry

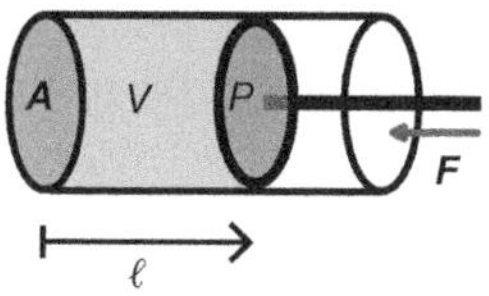

Formal Object

Coupled poles

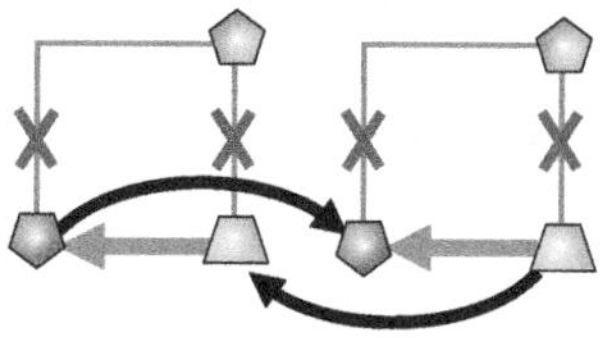

Coupling
✓ Invariable Equivalence Conversion
✓ Poles Dipoles
✓ Capacitive Inductive Conductive

Formal Graph

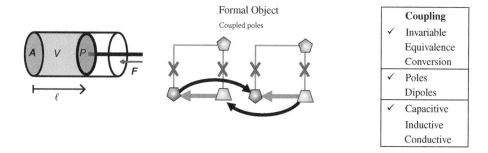

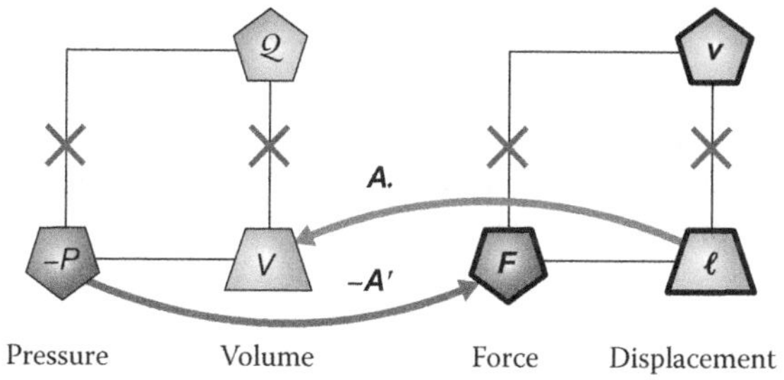

GRAPH 12.17

	Coupling	**From**	**To**
Variety		Translation mechanics	Hydrodynamics
Family	Integral factor	Basic quantity	
Name	Area	Displacement	Volume
Symbol	$\boldsymbol{A}$	$\boldsymbol{\ell}$	$\boldsymbol{V}$
Unit	[m^2]	[m]	[m^3]
Variety		Hydrodynamics	Translation mechanics
Family	Differential factor	Effort	
Name	Differential area	Pressure	Force
Symbol	$\boldsymbol{A'}$	P	$\boldsymbol{F}$
Unit	[m^2]	[Pa], [N m^{-2}]	[N]

System Constitutive Properties		
Variety	Hydrodynamics	Translational mechanics
Nature	Capacitance	Capacitance
Levels	0	0
Name	Hydrodynamical capacitance	Inverse spring constant
Symbol	C_V	k_e^{-1}
Unit	[m^5 N^{-1}]	[m N^{-1}]
Specificity	–	–

A piston is a system that couples hydrodynamic energy and mechanical energy of translation.

In hydrodynamics, in the capacitive subvariety ("volume energy"), the basic quantity is the volume V and the effort is the pressure P. In translation mechanics, still in the capacitive subvariety, the basic quantity is the displacement ℓ and the effort is the force $\boldsymbol{F}$.

In a piston, the basic quantities that are the volume of the compartment containing a compressed fluid and the displacement of the piston are coupled, because this displacement corresponds to the height (or length) of the fluid compartment. The coupling factor is therefore the area $\boldsymbol{A}$ of the piston, which in the simplest case (cylinder of constant section) is invariable:

$$V = \mathbf{A} \,.\, \ell \tag{J5.1}$$

The variation of the total capacitive energy of the system is given by a Gibbs equation, in recalling that, for historical reasons, the pressure is counted in the bad direction, thus requiring a minus sign:

$$\mathrm{d}\mathcal{U} = -P\,\mathrm{d}V + \boldsymbol{F} \,.\, \mathbf{d}\boldsymbol{\ell} \tag{J5.2}$$

The equilibrium of the system occurs by zeroing this variation, which, by taking into account the coupling between basic quantities and the constancy of the piston area, leads to the well-known linear relationship between pressure and force:

Local pressure:
$$P \underset{lin}{=} \frac{\boldsymbol{F}}{\boldsymbol{A}} \tag{J5.3}$$

The subscript "*lin*" below the equal sign means that the validity domain of this relation is restricted to the linear case in which the area is a constant. This corresponds to an **invariable coupling**, in contradistinction with a variable coupling (see below).

This coupling also works very well in the case of a varying section. It is enough that the derivative of the volume with respect to the displacement is equal to the coupling area, which is the differential definition of a varying surface. In the present theory, this coupling is called a **variable coupling** and the derivative represents a *differential coupling factor* notated $\boldsymbol{A}'$. In a variable coupling, the two coupling relationships satisfy without any restriction to the Gibbs equation J5.2 and they are written as

$$\mathrm{d}V \overset{def \rightarrow}{=} \boldsymbol{A}' \cdot \mathbf{d}\boldsymbol{\ell} \tag{J5.4}$$

$$\boldsymbol{F} = -\boldsymbol{A}'(-P) \tag{J5.5}$$

Naturally, this is not technologically as easy to conceive as in the case of a piston with a rigid disk, but it is sufficient to replace the surface of separation by an elastic film able to fit the variations of section.

It can be remarked that when the *differential coupling factor* $\boldsymbol{A}'$ is a constant, it becomes equal to the *integral coupling factor* $\boldsymbol{A}$ and the distinction between invariable and variable couplings vanishes, so the classical relation in Equation J5.4 is found again.

Remarks

Pressure definition. Classically, the pressure is defined by Equation J5.3, but this is not the case in the Formal Graph theory (and in thermodynamics), which defines the pressure as minus the effort of the subvariety in which the basic quantity is the volume, independent of any notion of force or of displacement.

$$\text{Pressure: } P \overset{def}{=} -\left(\frac{\partial \mathcal{U}}{\partial V}\right)_{q_i} \tag{J5.6}$$

This is a much more general definition because it allows one to use the concepts of pressure and volume outside of a coupling with translation mechanics (in other words, without needing a surface).

Another acceptance of the notion of pressure is met with the *local pressure*, defined as the force density over a surface, which is a localized variable of the translational mechanical variety in contradistinction with the above-defined pressure which is a state variable in the hydrodynamical energy variety.

$$\text{Local pressure: } P \overset{def}{=} \frac{\partial \boldsymbol{F}}{\partial \boldsymbol{A}} \tag{J5.7}$$

The same symbol is used in order to comply with the tradition, but it would be judicious to do otherwise.

The classical definition of pressure as the ratio of a force on an area is not always of easy application. For instance, in a gas, in a cloud of vapor, or in plasma, it is difficult to find a well-defined area unless one uses a differential definition, which amounts to the local pressure. Second, if one needs to have a surface systematically available to be able to define a pressure inside a volume, one comes up sometimes against serious difficulties, for instance, in the case of finely divided surfaces (fractal medium) or randomly varying ones (fluctuations, chaos).

Poles or dipoles? The state variables used in this case study are all relative to poles when the piston is at rest (static equilibrium). On the contrary, when the piston is asked to convert energy, by a displacement made at a given velocity (variable or not), it is necessary to complete the model for making two dipoles between the interior and the exterior of the piston for each of the two energy varieties. (Supplementary conductive dipoles may be required if mechanical friction occurs or viscous fluids are used.) Recall that it is only by using dipoles that the time may be taken into account.

12.4.6 Case Study J6: Ionic Distributions

Electrodynamics
Corpuscular Energy

The case study abstract is given in next page.

Ions are atoms or molecules possessing electric charge. The system here considered is a set (collection of entities) of ions of the same chemical species with the same charge, involving two energy varieties, electrodynamics and the corpuscular variety. An electric potential varying along the distance provokes a spatial distribution of ions, and therefore of the local electric charge, as a function of this distance and of the electric potential. Such a distribution is observed in front of electrified interfaces, as, for instance, the electric double layer in front of an electrode.

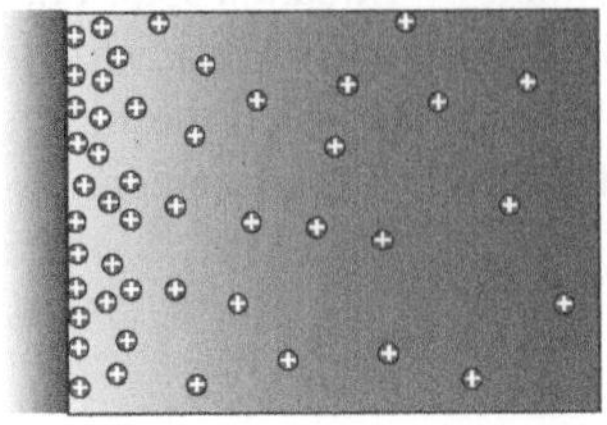

Formal Object:

Coupled poles

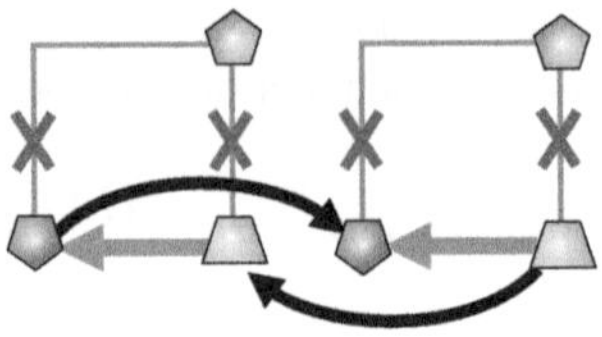

Coupling
✓ Invariable Equivalence Conversion
✓ Poles Dipoles
✓ Capacitive Inductive Conductive

Formal Graph:

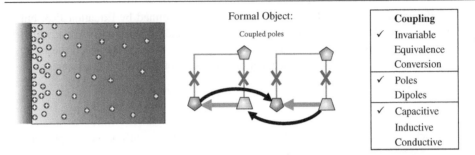

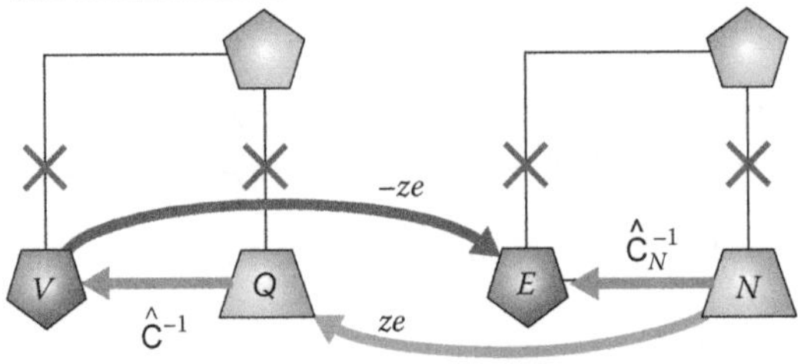

GRAPH 12.18

	Coupling	From	To
Variety		Corpuscular energy	Electrodynamics
Family	Integral factor	Basic quantity	
Name	Ionic charge	Corpuscle number	Charge
Symbol	ze	N	Q
Unit	[C corp^{-1}]	[corp]	[C]
Variety		Electrodynamics	Corpuscular energy
Family	Differential factor	Effort	
Name	Ionic charge	Potential	Energy-per-corpuscle
Symbol	ze	V	E
Unit	[C corp^{-1}]	[V]	[J corp^{-1}]

System Constitutive Properties		
Variety	Electrodynamics	Corpuscular energy
Nature	Capacitance	Capacitance
Levels	0	0
Name	Electric capacitance	Corpuscular capacitance
Symbol	C	C_N
Unit	[F]	[corp^2 J^{-1}]
Specificity	—	—

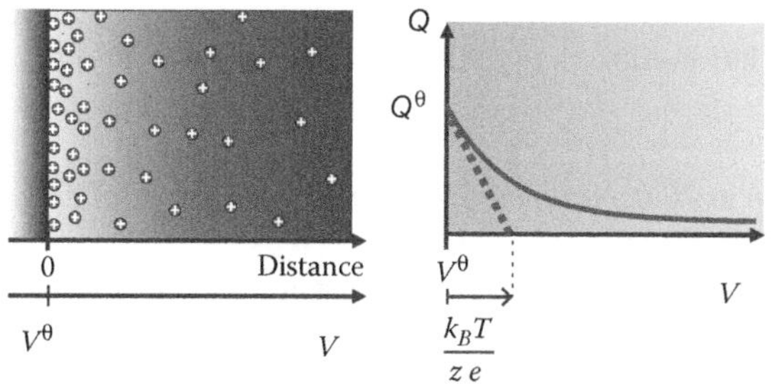

The electric potential V and the energy-per-corpuscle E are the respective efforts in electrodynamics and in physicochemistry. The respective basic quantities are the electric charge Q and the number of corpuscles N. These two last variables are coupled by the analogous of the Faraday relationship in physicochemistry, with z the electrovalency of the ion (number of charges, without dimension) and e the elementary charge (approx. 1.602×10^{-19} C corp^{-1})

$$\mathsf{Q} = ze\,N \tag{J6.1}$$

The variation of capacitive energy of the ion population is given by two terms of the Gibbs equation:

Gibbs:
$$d\mathcal{U} = E\,dN + V\,d\mathsf{Q} + \cdots \tag{J6.2}$$

Insertion of the coupling relationship into this equation, after derivation (the coupling factors are constants), and imposition of the condition of a minimum for these system energies lead to a coupling relationship between efforts:

$$E = -ze\,V \tag{J6.3}$$

The energy-per-corpuscle E in this relation is a coupled effort totally effective (it is not an apparent or virtual one) because of the equality between differential and integral coupling factors (as charges are tightly bound to the molecules, ze is a constant).

An invariable coupling like this one between two energy varieties implies that the system properties have the same shape. In corpuscular energy as well as in physical chemistry, the capacitive relationship has an exponential shape

$$N = N^{\theta}\exp\left(\frac{E - E^{\theta}}{k_B T}\right) \tag{J6.4}$$

with k_B the Boltzmann constant and T the temperature. The state variables with a zero bar as exponent correspond to a state of reference (or ground state), in standard conditions of temperature, pressure, etc., in taking as reference 1 corp. The correspondences between the origins follow the coupling relationships

$$\mathsf{Q}^{\theta} = zeN^{\theta}; \quad E^{\theta} = -ze\;V^{\theta} \tag{J6.5}$$

and replacing the corpuscular state variables with their electric homologues in the corpuscular capacitive relationship, one gets an **electrical capacitive relationship**

Electrical capacitance relationship:

$$Q = Q^{\ominus}\exp\left(\frac{-ze}{k_B T}\left(V - V^{\ominus}\right)\right) \qquad (J6.6)$$

This relation is known as *Boltzmann distribution* (or *Maxwell–Boltzmann distribution*) because it is classically established on statistical arguments.

It can be observed that this relationship is not a linear relation as $Q = CV$ which is predicted by electrostatics. Nevertheless, the Formal Graph approach shows that the structures are identical from one domain to the other and that, when an invariable coupling occurs, it should be the same for operators and therefore for the properties of a system. Two choices remain: whether the linear relationship is the general rule and an ion population is an exception, or whether the exponential shape is the rule and electrostatics is the exception.

The Formal Graph theory favors the second possibility.

Remarks

Nonlinear electrostatics. The distribution of ions in function of the distance or of the electric potential is classically called distribution of Boltzmann. It is indeed based on Boltzmann statistics, which relies on the law of large numbers for giving an exponential shape to the function describing this distribution.

The interest of this coupling is to show that a capacitive relationship between an electric charge and an electric potential is not always linear, contrary to the fundamental laws of electrostatics (which remains valid in free space).

Would it mean that electrostatics relies on statistics?

Apparent paradox. It is known that the laws of electrostatics also work with a very small number of entities, one or two, which forbids applying the approximation of large numbers normally invoked in Boltzmann statistics. The Formal Graph theory resolves this paradox by showing that the exponential shape of the capacitive relationship is not relevant to statistics but to a theory of influence between entities, as demonstrated in Chapter 7.

Electrodynamical activity. Case study A5 "Electrical charges" in Chapter 4 gives the general expression of the capacitive relationship for an electrodynamical pole in terms of generalized *activity*

$$\frac{Q - Q^{min}}{Q^0 - Q^{min}} = a_Q = \exp\frac{V}{V_\Delta} \qquad (J6.7)$$

In the case of a pole composed with charges of the same sign, the reference charge Q^0 is different from zero while the minimum charge Q^m is equal to zero. The expression J6.6 is retrieved with the following correspondences:

$$Q^0 = Q^{\ominus}\exp\left(\frac{-ze}{k_B T}V^{\ominus}\right); \quad V_\Delta = \frac{-k_B T}{ze} \qquad (J6.8)$$

12.4.7 Case Study J7: Bubble

Surface Energy
Hydrodynamics

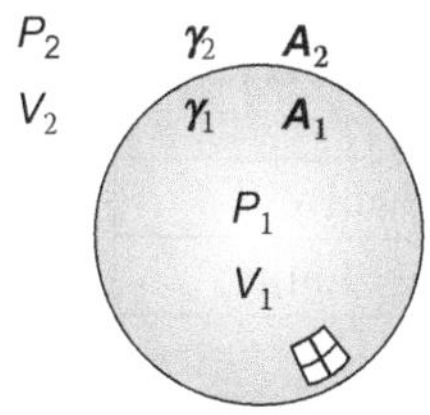

Formal Object

Coupled dipoles

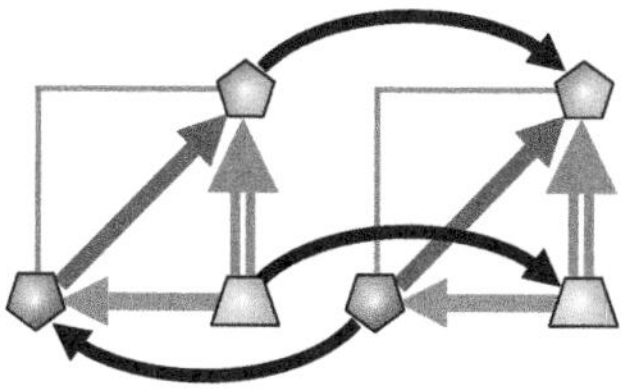

Coupling
Invariable Equivalence Conversion
Poles ✓ Dipoles
✓ Capacitive Inductive Conductive

Formal Graph

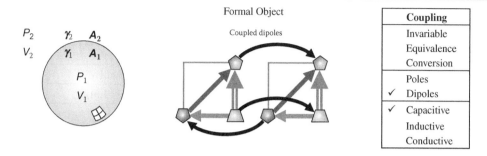

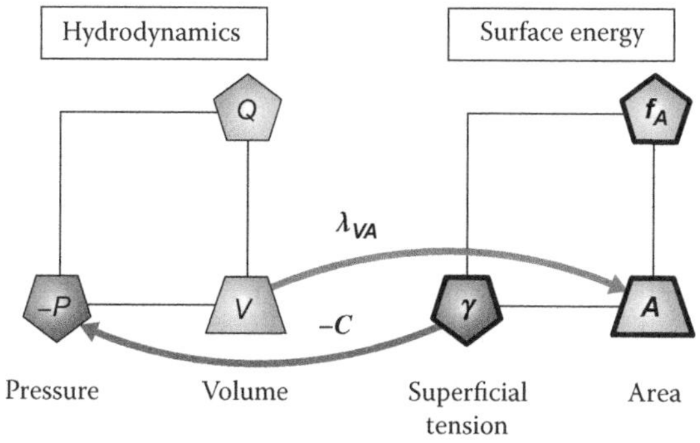

GRAPH 12.19

	Coupling	From	To
Variety		Hydrodynamics	Surface
Family	Integral factor	Basic quantity	
Name	Integrated curvature	Volume	Area
Symbol	λ_{VA}	V	A
Unit	[m^{-1}]	[m^3]	[m^3]
Variety		Surface	Hydrodynamics
Family	Differential factor	Effort	
Name	Curvature	Superficial tension	Pressure
Symbol	C	γ	P
Unit	[m^{-1}]	[N m^{-1}]	[N m^{-2}]

System Constitutive Properties		
Variety	Hydrodynamics	Surface energy
Nature	Capacitance	Capacitance
Levels	0	0
Name	Hydrodynamical capacitance	Surface capacitance
Symbol	C_V	C_A
Unit	[m^5 N^{-1}]	[m^3 N^{-1}]
Specificity	—	—

A bubble of liquid, or an ideal soap bubble, without thickness of the walls, and more generally any phase (liquid, gaseous, or solid) with a closed shape (generally spherical) in another phase, is a system involving the coupling between two dipoles, one belonging to the hydrodynamical variety, the other to the variety of surface energy.

Variety	**Surface Energy**		
Sub-variety	Capacitive (potential)		Inductive
Category	Entity Nr	Energy/Entity	Energy/Entity
Family	Basic quantity 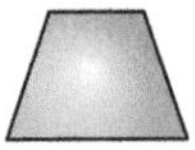	Effort 	Flow 
Name	Area	Surface tension	Surface expansion rate
Symbols	$\boldsymbol{A}$	γ	$\boldsymbol{f_A}$
Unit	[m^2]	[kg s^{-2}]	[m^2 s^{-1}]

The surface energy is a variety in the same family as the varieties of translation mechanics and hydrodynamics. Under its capacitive subvariety, it possesses as basic quantity the area $\boldsymbol{A}$ (in m^2) and as effort, the *surface tension* $\boldsymbol{\gamma}$ (in kg s^{-2} or J m^{-2}). The flow is called *rate of surface expansion*, often notated with the $\boldsymbol{A}$ surmounted by a point for indicating a temporal derivative. As it cannot be a definition for a flow, the adopted symbol in the Formal Graph theory is the generic symbol $\boldsymbol{f_A}$, with SI units in m^2 s^{-1}. All these state variables are vectors because they are oriented with respect to the surface.

The system, exemplified by the bubble, consists of two dipoles. The one belonging to the surface energy has its effort (i.e., its surface tension) that depends on the efforts of its poles according to

$$\gamma = \gamma_1 - \gamma_2 \tag{J7.1}$$

and the conservation of basic quantities that are the areas is expressed by an equality between variations

$$\mathbf{d}\boldsymbol{A} = \mathbf{d}\boldsymbol{A}_1 - \mathbf{d}\boldsymbol{A}_2 \tag{J7.2}$$

For the dipole belonging to hydrodynamics, the relation between efforts, which are the pressures, is written in the same way

$$P = P_1 - P_2 \tag{J7.3}$$

and the conservation of basic quantities, the volumes, is also written differentially

$$\mathrm{d}V = \mathrm{d}V_1 = -\mathrm{d}V_2 \tag{J7.4}$$

The variation of capacitive energy of the assembly of two dipoles is

Gibbs:
$$\mathrm{d}\mathcal{U} = -P\mathrm{d}V + \gamma \cdot \mathbf{dA} \tag{J7.5}$$

Now, one defines the differential coupling factor, called *surface curvature*, as

$$C \overset{def}{=} \frac{\mathrm{d}A}{\mathrm{d}V} \quad \text{(J7.6)}$$

It can be remarked that when the hypothesis of a circular surface and of a spherical volume is made for the phase included in the other, one finds a proportionality with the inverse of the radius, which is often (and improperly, because it is restricted to this peculiar case) given as a definition of the curvature

$$C = \frac{8\pi r \mathrm{d}r}{4\pi r^2 \mathrm{d}r} = \frac{2}{r} \quad \text{(J7.7)}$$

Incorporation of the definition of the surface curvature into the equation of the capacitive energy variation in the system and the condition of nullity of this one when the system is at equilibrium (energy minimum) lead to the *law of Laplace* (here expressed in general, without the hypothesis of sphericity)

Laplace's law: $$P = C \cdot \gamma \quad \text{(J7.8)}$$

which therefore is viewed as a coupling relation between dipolar efforts. Its symmetric relationship is the coupling relation between variations of basic quantities, utilized as a definition of the curvature. The Formal Graph shows the coupling relationship between basic quantities, defined as the ratio of the area on the volume

$$\lambda_{VA} \overset{def}{=} \frac{A}{V} \quad \text{(J7.9)}$$

Remarks

Laplace law. The "big" law in these domains, which span from capillarity to wetting, through diphasic systems, is Laplace's law. This law is demonstrated here in the most general case, with a curvature of the interface separating the two phases that can be of any shape. Physically speaking, this law assumes a simple coupling relationship between efforts.

Lagrangian and coupling. In classical mechanics, the demonstration of Laplace's law relies on the utilization of the Lagrangian* and on the determination of conditions minimizing the energy of the system. The Formal Graph theory does not use explicitly the Lagrangian, but develops the more general and powerful tool of coupling between energy varieties, which is simpler and conceptually more meaningful.

* Joseph-Louis Lagrange (1736–1813): French mathematician and physicist; Paris, France.

12.4.8 Case Study J8: Thermal Junction

Electrodynamics
Thermics

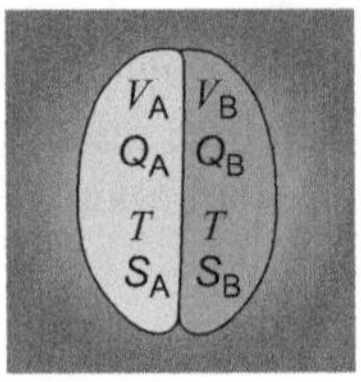

Formal Object

Coupled dipoles

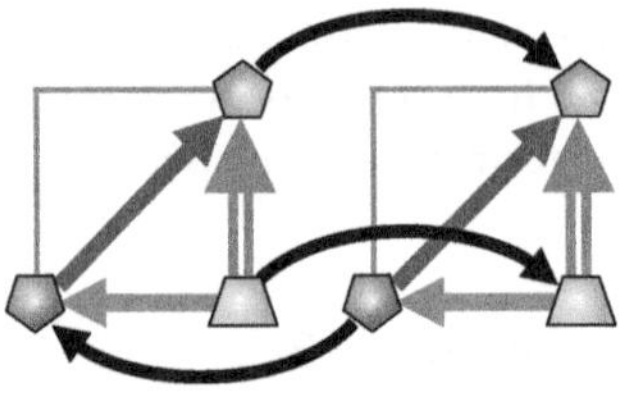

Coupling
✓ Invariable
Equivalence
Conversion
Poles
✓ Dipoles
✓ Capacitive
Inductive
Conductive

Formal Graph

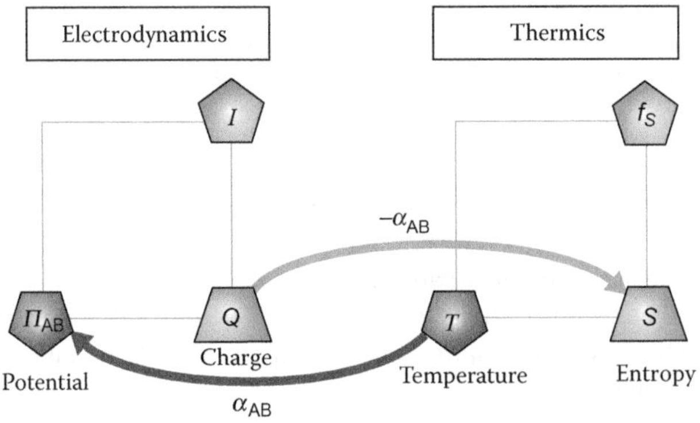

GRAPH 12.20

	Coupling	From	To
Variety		Electrodynamics	Thermics
Family	Integral factor	Basic quantity	
Name	Seebeck coefficient	Charge	Entropy
Symbol	$-\alpha_{AB}$	Q	S
Unit	[J C^{-1} K^{-1}]	[C]	[J K^{-1}]
Variety		Thermics	Electrodynamics
Family	Differential factor	Effort	
Name	Differential Seebeck coefficient	Temperature difference	Peltier potential difference
Symbol	$-\alpha'_{AB}$	T	Π_{AB}
Unit	[J C^{-1} K^{-1}]	[K]	[V]

System Constitutive Properties		
Variety	Electrodynamics	Thermics
Nature	Capacitance	Capacitance
Levels	0	0
Name	Electric capacitance	Thermal capacitance
Symbol	C	C_S
Unit	[F]	[J K^{-2}]
Specificity	—	—

A thermoelectric junction consists of a soldered joint of two materials able to contain simultaneously electric energy and thermal energy. When the soldered joint is raised to a uniform temperature, a difference of electric potential appears on both sides of the interface between the two materials. This potential difference is called *Peltier tension*, or thermoelectric electromotive force, but here the phenomenon of generation of electric energy from thermal energy is called *Seebeck effect*. Another thermoelectric effect, the Peltier effect, consists of the reverse phenomenon of generation of heat from electric energy. In fact, a thermoelectric junction utilizes the two effects in a concomitant way, and when isolation from the exterior is ensured, the two energy varieties are in equilibrium.

The scheme in the case study abstract describes the physical principle of a thermoelectric junction with two materials A and B that are able each to develop a different electric potential V_A and V_B and to contain electric charges Q_A and Q_B, respectively. The difference of thermal energy between the two sides is ensured by distinct entropies S_A and S_B as the temperature T is the same on both sides. The junction is isolated from the external medium, which means that no current nor heat flow circulates between the soldered joint and the exterior, which in electrical terms corresponds to an *open circuit*.

The variation of capacitive energy of these four poles (an electrodynamical one and a thermal one for each side) is given by the following Gibbs equation:

Gibbs:
$$d\mathcal{U} = V_A dQ_A + V_B dQ_B + T dS_A + T dS_B \tag{J8.1}$$

To transpose these polar variables into dipolar variables, it is convenient to define the Peltier potential difference, notated with an uppercase *pi* (Π), as the dipolar potential

Peltier potential:
$$\Pi_{AB} = V_A - V_B \tag{J8.2}$$

As there is globally no creation of electric charges (i.e., charge conservation), the sum of all charges is zero, which allows the dipolar charge to be equal to the polar charges and to the opposite of the other one, and to deduce the variation of dipolar charge

$$Q = Q_A = -Q_B \Rightarrow dQ = dQ_A = -dQ_B \tag{J8.3}$$

For the other energy variety, there is also conservation of the basic quantity, the entropy, but without the previous symmetry because this latter is ensured by the equality of temperatures, which amounts to saying that the total entropy is not zero

$$S_A + S_B = S \Rightarrow dS = dS_A + dS_B \tag{J8.4}$$

The previous relationships brought into Gibbs equation J8.1 provide a variation of the capacitive energy of the two dipoles. The isolation of the system corresponds to an extremum of the energy, which is expressed by zeroing this variation

$$d\mathcal{U} = \Pi_{AB} dQ + T dS \underset{isol}{=} 0 \tag{J8.5}$$

By assuming the coupling invariable, the coupling relationships involve only one factor

$$\alpha_{AB} \overset{def}{=} \frac{d\Pi_{AB}}{dT} = -\frac{S}{Q}; \quad d\alpha_{AB} = 0 \Rightarrow \alpha_{AB} = \frac{\Pi_{AB}}{T} = -\frac{dS}{dQ} \tag{J8.6}$$

The integral coupling factor is called the *Seebeck coefficient*. It is generally not a constant and depends on the temperature, which means that the invariable case treated here is an *ideal case*. The real case is modeled by involving a supplementary effect called *Thomson effect*. This case is treated in case study J11 "Thomson Effect."

The Formal Graph in the case study abstract (Graph 12.20) represents the energetic coupling between the two electrodynamical and thermal dipoles with an invariable coupling.

Graph 12.20 models a more complex level of organization than the level of poles, as they are omitted by virtue of the embedding principle. It is worth going down to the polar level to show the couplings at this level by expanding the Formal Graph, as shown in Graph 12.21.

The couplings between poles work in a similar way to the coupling between dipoles, which is shown by the coupling relationships given for the A pole. (They are the same for the B pole by substitution of subscripts.)

$$\alpha_{A} \overset{def}{=} \frac{dV_{A}}{dT_{A}} = -\frac{S_{A}}{Q_{A}}; \quad d\alpha_{A} = 0 \Rightarrow \alpha_{A} = \frac{V_{A}}{T_{A}} = -\frac{dS_{A}}{dQ_{A}} \tag{J8.7}$$

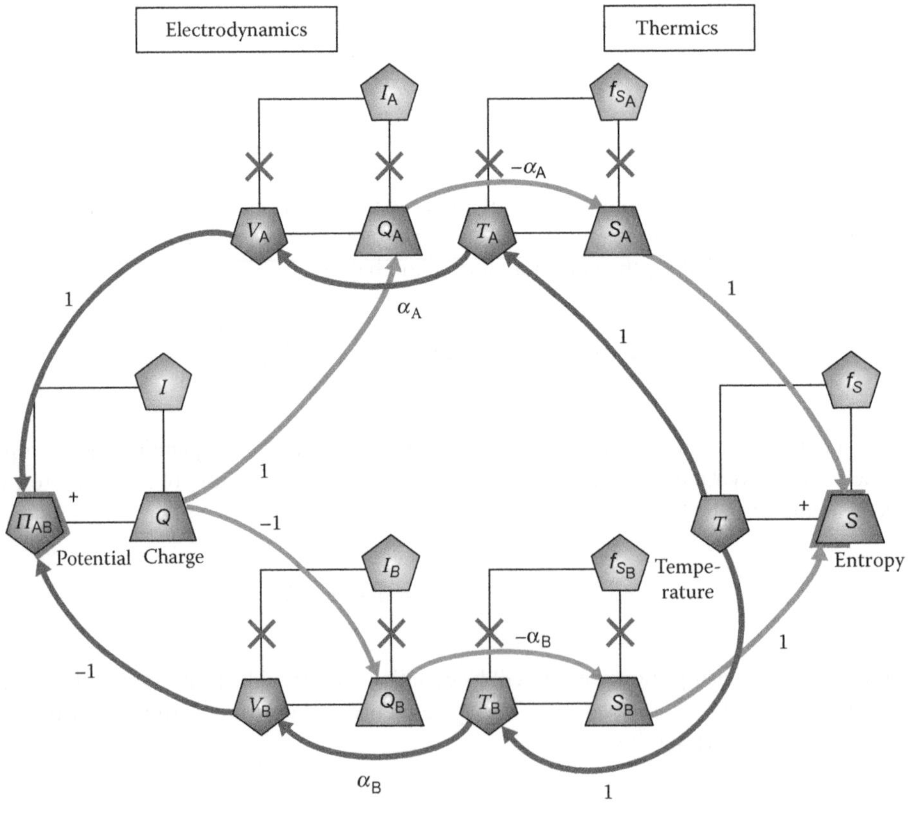

GRAPH 12.21

In the classical language, which does not rely on the pole–dipole distinction, the Seebeck coefficients defined at this level are called *absolute coefficients*. This logically leads to attributing the adjective *relative* to the dipolar Seebeck coefficient. The link between these coupling factors is, as one may read in the expanded Formal Graph (Graph 12.21),

$$\alpha_{AB} = \alpha_{A} - \alpha_{B} \tag{J8.8}$$

Remarks

Ideal case. The thermoelectric junction constitutes the basic system in the domain of thermoelectricity. It is not utilized alone as in the case study treated here but with another junction for making up a thermocouple, each junction being raised to a different temperature. The reason is quite logical: In order to electrically connect the two materials of a junction with the exterior and make an electric circuit, it is compulsory to use somewhere at least another junction, if only for closing the circuit with the same materials used in the junction.

The case study treated here is all the less realistic as the junction is assumed to be ideal, that is, with an invariable coupling factor.

Dynamic case. It can be remarked that no utilization of electric current and of entropic flow (heat "flux") is made in this model, although devoted to dipoles that are endowed with the possibility of using them. This is because of the absence of an external circuit for allowing these flows to circulate. It does not mean that it is impossible to utilize a junction in an electric and thermal circuit and to make these flows cross it, while keeping the same model of coupling (see case studies J9 "Seebeck Effect" and J10 "Peltier Effect").

12.4.9 Case Study J9: Seebeck Effect

Thermics
Electrodynamics

The case study abstract is given in next page.

The Seebeck effect corresponds to the electricity production from a difference of temperature. This effect can be reversible and is the inverse of the Peltier effect, which is the phenomenon of conversion of electric energy into thermal energy (heat). These effects can be superimposed onto the dissipative processes of transport by conduction of electric charges (Joule effect) and to the transport of heat (Fourier equation) which are both irreversible processes.

Effect	Conversion	Reversibility	Coupling
Joule	Electrodynamics $\rightarrow$ Thermics	Irreversible	Invariable
Seebeck	Thermics $\Rightarrow$ Electrodynamics	Reversible or irreversible	Invariable
Peltier	Electrodynamics $\Rightarrow$ Thermics	Reversible or irreversible	Invariable
Thomson	Electrodynamics $\Leftrightarrow$ Thermics	Reversible or irreversible	Variable

The Thomson effect is the generalization of Seebeck–Peltier effects to the case of variable coupling.

J9: Seebeck Effect

Thermics
Electrodynamics

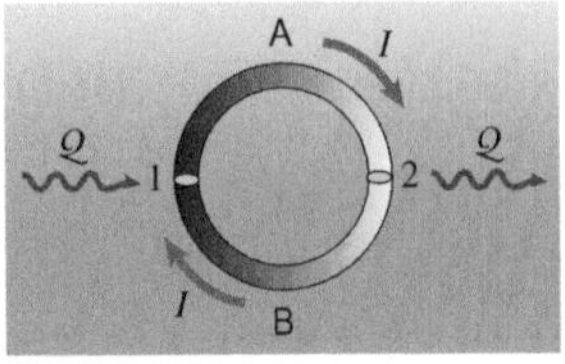

Formal Object

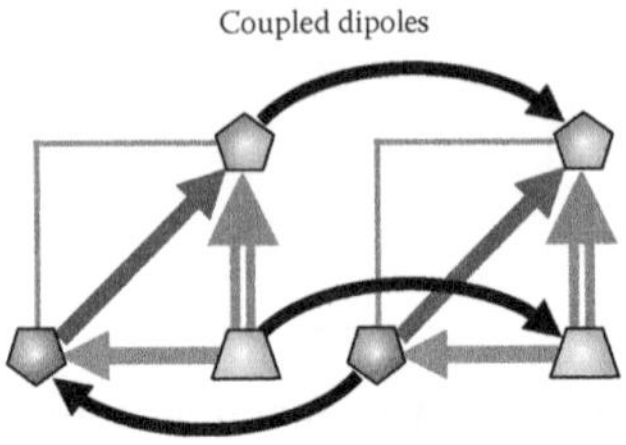

Coupling
✓ Invariable
Equivalence
✓ Conversion
Poles
✓ Dipoles
✓ Capacitive
Inductive
Conductive

Formal Graph

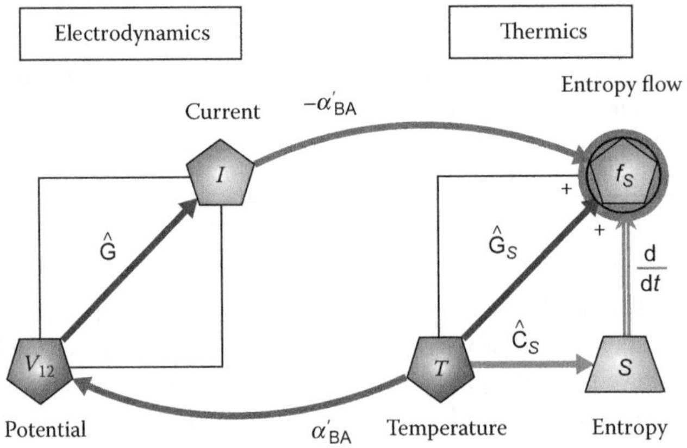

GRAPH 12.22

	Coupling	From	To
Variety		Electrodynamics	Thermics
Family	Integral factor	Basic quantity	
Name	Seebeck coefficient	Charge	Entropy
Symbol	$-\alpha_{AB}$	Q	S
Unit	[J C^{-1} K^{-1}]	[C]	[J K^{-1}]
Variety		Thermics	Electrodynamics
Family	Differential factor	Effort	
Name	Differential Seebeck coefficient	Temperature difference	Peltier potential difference
Symbol	$-\alpha'_{AB}$	T	Π_{AB}
Unit	[J C^{-1} K^{-1}]	[K]	[V]

System Constitutive Properties		
Variety	Electrodynamics	Thermics
Nature		Capacitance
Levels		0
Name		Thermal capacitance
Symbol		C_S
Unit		[J K^{-2}]
Specificity	–	–
Nature	Conductance	Conductance
Levels	0	0
Name	Conductance	Thermal conductance
Symbol	G	G_S
Unit	[S]	[JK^{-2} s^{-1}]
Specificity	—	—

The circuits schematized below show two possible utilizations of the Seebeck effect, one (left) *in closed circuit* (and therefore with a potential difference equal to zero) producing a current from thermal energy and the other (right) *in open circuit*, therefore in the absence of current, called "thermocouple" and used for measuring temperature differences. These circuits are both made up of two soldered joints of two materials having distinct thermoelectric properties. Case study J8 is devoted to the description and modeling of the thermocouple, also called thermoelectric junction, which is recalled here for comparison.

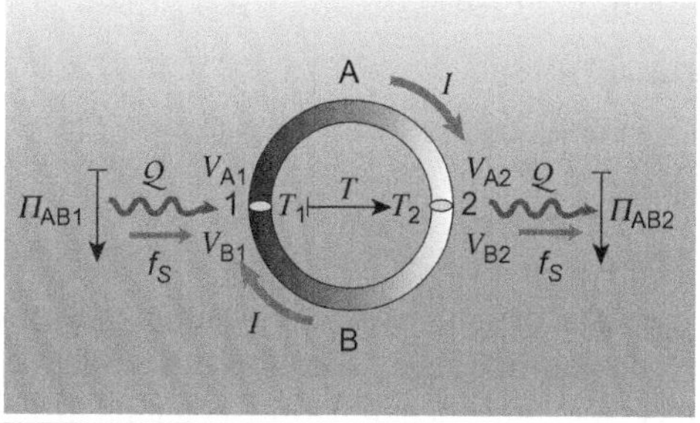

Seebeck effect: A current circulates in the loop formed with two different materials under the effect of a temperature difference between the two soldered joints. The electric circuit is **closed** (electrically isolated) and the global potential difference is zero. The thermal circuit is powered by a heat source (providing an entropy flow f_S).

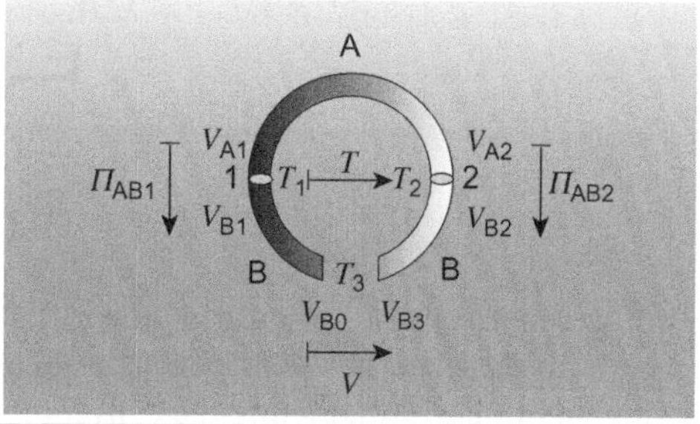

Thermocouple: A potential difference appears at the terminals of an open circuit made up of two different materials having their soldered joints brought to two different temperatures. The electric circuit is **open** and the electric current is zero. (The temperature is the same for each terminal B0, B3 of the circuit.)

The soldered joints are identical thermoelectric junctions raised to different temperatures T_1 and T_2. Each junction develops a potential difference, the Peltier tension, proportional to the temperature through the same *relative Seebeck coefficient* α_{BA} (which in the Formal Graph language is a coupling factor between dipoles) as the junctions are made of the two same materials A and B. When the junctions are assumed to be ideal, which means that the coupling is invariable, a single coupling factor, the integral factor, is utilized.

Ideal junction: $$V_{B1} - V_{A1} = \Pi_{BA1} = \alpha_{BA} T_1; \quad V_{B2} - V_{A2} = \Pi_{BA2} = \alpha_{BA} T_2 \qquad \text{(J9.1)}$$

When the junctions are not ideal but real, the coupling is a variable one, and two coupling factors are necessary, one integral factor α_{BA} for linking the basic quantities and one differential factor α'_{BA} for linking the efforts. This case corresponds to the inclusion of the Thomson effect and is treated in case study J11 "Thomson Effect."

Real junction: $$V_{B1} - V_{A1} = \Pi_{BA1} = \alpha'_{BA} \mathrm{T}_1; \quad V_{B2} - V_{A2} = \Pi_{BA2} = \alpha'_{BA} T_2 \qquad \text{(J9.2)}$$

In the general case of multiple potential differences, as in the case of the open circuit (also the case of a closed circuit when imposing a zero sum), one has

$$V = V_{A0} - V_{A3} = \left(V_{A0} - V_{A1}\right) + \left(V_{A1} - V_{B1}\right) + \left(V_{B1} - V_{B2}\right) + \left(V_{B2} - V_{A2}\right) + \left(V_{A2} - V_{A3}\right) \qquad \text{(J9.3)}$$

The Seebeck effect occurring in a closed electric circuit with a nonzero current lends to considering the way the generated electric energy is utilized and how a heat supply for making the conversion is taken into account.

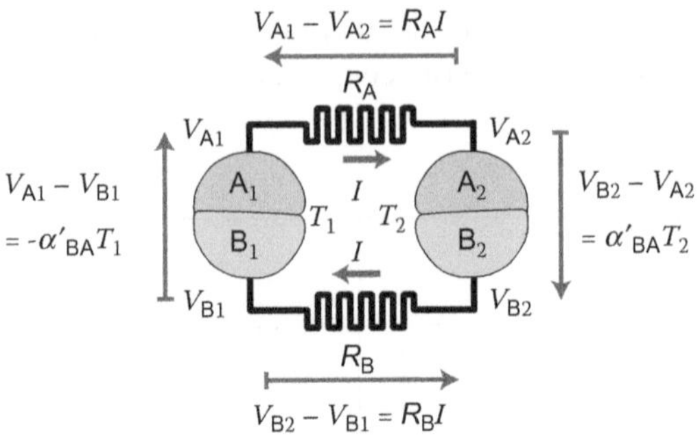

In the simple circuit (no external device) sketched above, the only way of consuming electric energy is to dissipate by Joule effect through transformation into heat. Therefore, two relations obeying Ohm's law, by adopting the convention of a receptor component (opposite potential difference and current direction), are written as

$$V_{A1} - V_{A2} = R_A I; \quad V_{B2} - V_{B1} = R_B I \tag{J9.4}$$

As the electric circuit is closed, the sum of all potential differences given by Equation J9.3 is equal to zero and using Equations J9.2 and J9.4 one finds

$$(R_A + R_B)I - \alpha'_{BA}(T_2 - T_1) = 0 \tag{J9.5}$$

This equality allows one to define a virtual potential as the common value of both members

$$V_{12} \overset{def}{=} (R_A + R_B)I = \alpha'_{BA}(T_2 - T_1) \tag{J9.6}$$

These considerations were based on the conduction of electrodynamical energy. Now, one turns attention to the thermal energy. The conduction of heat in the same circuit is possible when all the materials have a nonzero thermal conductance; then the Fourier equation allows the following equations to be written:

$$f_{SA} = \hat{G}_{SA}(T_1 - T_2); \quad f_{SB} = \hat{G}_{SB}(T_2 - T_1) \tag{J9.7}$$

From the thermal circuit viewpoint, the entropic flow is assumed to enter the circuit at the first soldered joint (T_1) and to leave it at the second soldered joint (T_2), after having circulated through the two parallel branches A and B. The entropic flow is therefore the sum of individual entropic flows:

$$f_{SG} = f_{SA} + f_{SB} = \hat{G}_S T; \quad \hat{G}_S = (\hat{G}_{SA} T + \hat{G}_{SB} T) T^{-1} \tag{J9.8}$$

The total entropy flow, entering and leaving the device, is the sum of the conductive flow and the flow converted from the current:

$$f_S = f_{SG} + f_{SBA} \tag{J9.9}$$

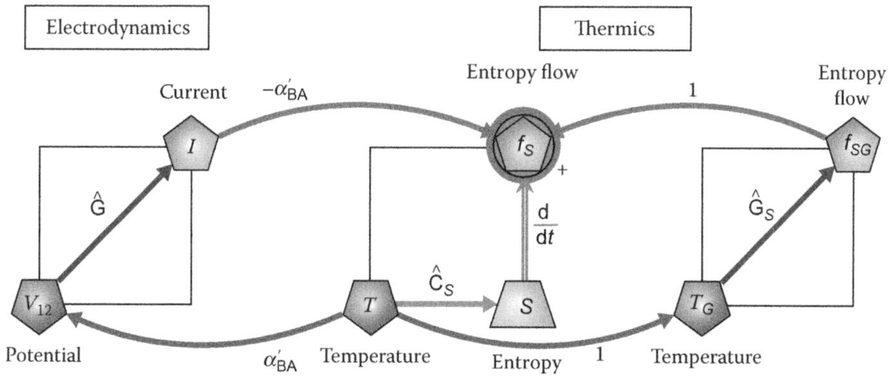

GRAPH 12.23

The energy conversion at the level of the soldered joints is done without losses and obeys a coupling (variable in the case of real junctions) that is the same for each soldered joint (as it has been assumed for writing Equation J9.2):

$$\alpha'_{BA} = \frac{\Pi_{BA1}}{T_1} = \frac{\Pi_{BA2}}{T_2} = -\frac{f_{SBA}}{I} \tag{J9.10}$$

The Formal Graph in Graph 12.23 represents the electrodynamical and thermal dipoles connected by the couplings previously described by the relations in Equation J9.8 (dropping the subscripts AB for convenience) for the summation on the node of entropic flow and by using the relation in Equation J9.6 for the virtual electric potential. In addition, a capacitive dipole in the thermal energy variety representing the heat storage in the materials has been drawn. This feature is not intrinsically included in the Seebeck effect but the Formal Graph approach requires it for modeling the dissipation through the electric resistances, according to the discussion in Chapter 11 (and also because practically all materials have a heat capacitance or "capacity").

The Formal Graph in Graph 12.23 is decomposed into dipoles, one electrical (conductive) and two thermal ones (capacitive and conductive). Another Formal Graph combining the two thermal dipoles into a dipole assembly is drawn in the case study abstract. Both graphs are strictly equivalent.

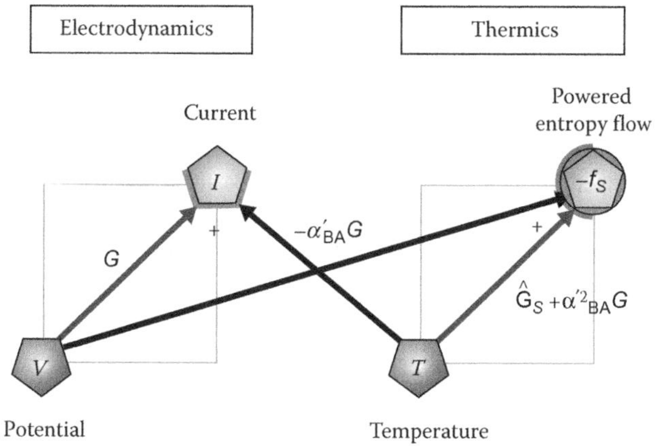

GRAPH 12.24

The relationships represented in Graph 12.23 can be represented in a different way by using crossed relations between efforts and flows instead of coupling relations between state variables of the same family. The Formal Graph in Graph 12.24 is strictly equivalent to the previous one in noticing two changes: in using the reciprocal of the resistance R, which is the conductance G, and in putting in the node of the potential one of the two potentials that make the total potential, in order to avoid a zero node.

It is worth noting that a direct translation of this graph into algebra provides a peculiar case of the Onsager* relations used by a branch of thermodynamics called Thermodynamic of Irreversible Processes developed by Prigogine* (Kondepudi and Prigogine 1998). However, a limitation of this "crossed" approach appears immediately: it can be used only when some system properties internal to the dipoles are implied (here the conductances). In contradistinction, the Formal Graph theory, in considering separately the couplings from the system properties, is able either to model energy conversions without losses (couplings alone) or to model systems without coupling, or both.

Remarks

Reversibility? Traditionally, the Seebeck effect is said to be a reversible process. In this case study, this is not so because the electric resistances of the materials dissipate the produced energy by conversion from electricity to heat. However, another setting is possible in using materials conducting perfectly electricity and electrical capacitors for storing the converted energy. In that ideal case the conversion from heat to electricity is reversible.

This is indeed the simple possibility of reversibility that justifies from the classical viewpoint the adjective "reversible," in contrast to the always irreversible dissipation.

In the Formal Graph approach, the Seebeck effect is not considered as reversible per se. It depends in fact on the accompanying processes. Chapter 11 gives a rigorous definition of a reversible process in terms of parity of the number of evolution operators in every loop in a graph. According to this criterion, a conversion process cannot be identified as reversible or irreversible until the whole process (source and receiver) is taken into consideration.

12.4.10 Case Study J10: Peltier Effect

Electrodynamics
Thermics

The case study abstract is given in next page.

The Peltier effect is a conversion of electrical energy into heat that has the remarkable property to be potentially reversible, unlike the Joule effect that also converts into heat, but by completely irreversible degradation (dissipation). The reverse process of the Peltier effect is the Seebeck effect.

The subject that matters here is the conversion with *invariable coupling* without dissipation, that is, by means of only the Peltier and Seebeck effects. (Any reversible conversion between electrical and thermal energy is based on these two complementary effects and not on only the Peltier effect, as might be suggested by the only reference to this effect when invoking heat converters.)

The diagram below (left) shows a device formed by a series of three materials A, B, and C for transferring heat from one side to another, called Peltier effect converter or "Peltier cooler/heater." The metal plates made of material C ensure a good mechanical behavior and a good contact with external heat sources/sinks and with the materials A and B, while allowing optimal thermal and electrical conduction. Between these plates, three in number, are the materials A and B to form two sequences of thermocouples ACC and CBC in series (from the electrical circuit viewpoint), each of these pairs using two

* Lars Onsager (1903–1976): Norwegian-American physical chemist; New Haven, Connecticut, USA.

J10: Peltier Effect

Electrodynamics
Thermics

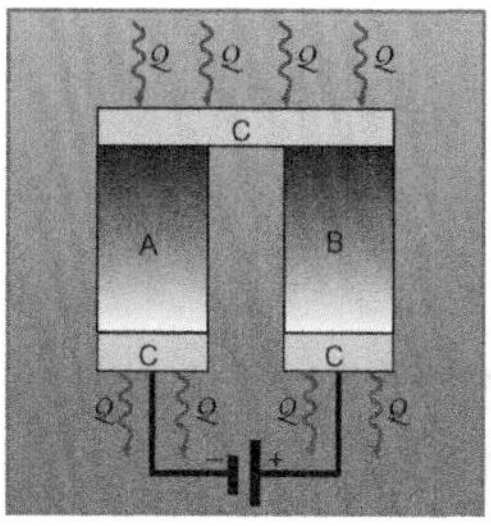

Formal Object

Coupled dipoles

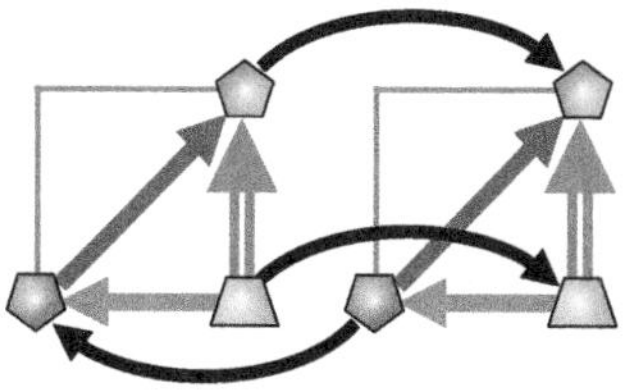

Coupling
✓ Invariable
Equivalence
✓ Conversion
Poles
✓ Dipoles
✓ Capacitive
Inductive
Conductive

Formal Graph

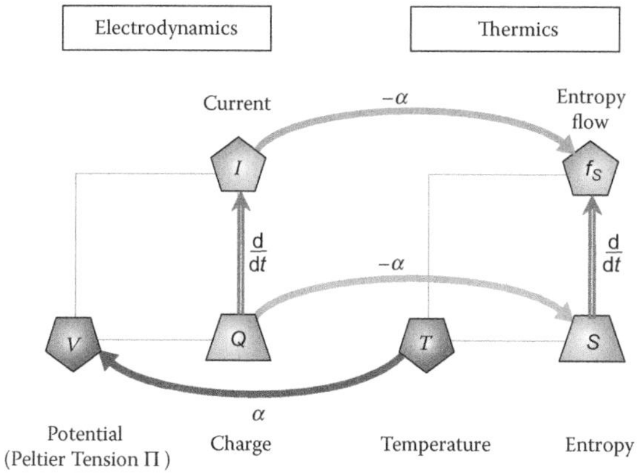

GRAPH 12.25

	Coupling	From	To
Variety		Electrodynamics	Thermics
Family	Integral factor	Basic quantity	
Name	Seebeck coefficient	Charge	Entropy
Symbol	$-\alpha_{AB}$	Q	S
Unit	[J C^{-1} K^{-1}]	[C]	[J K^{-1}]
Variety		Thermics	Electrodynamics
Family	Differential factor	Effort	
Name	Differential Seebeck coefficient	Temperature difference	Peltier potential difference
Symbol	$-\alpha'_{AB}$	T	Π_{AB}
Unit	[J C^{-1} K^{-1}]	[K]	[V]

System Constitutive Properties		
Variety	Electrodynamics	Thermics
Nature	Capacitance	Capacitance
Levels	0	0
Name	Electric capacitance	Thermics
Symbol	C	C_S
Unit	[F]	[J K^{-2}]
Specificity	—	—

thermoelectric junctions. The materials A and B are chosen with Seebeck junction factors α that are opposed, one positive and the other negative, which is the case of semiconductor materials p-doped and n-doped, respectively. This allows the two flows of entropy (heat flux) and charge (current) to run in opposite directions in one thermocouple and in the same direction in the other one, so as to add the entropic flows by operating the two thermocouples in parallel from a thermal circuit viewpoint.

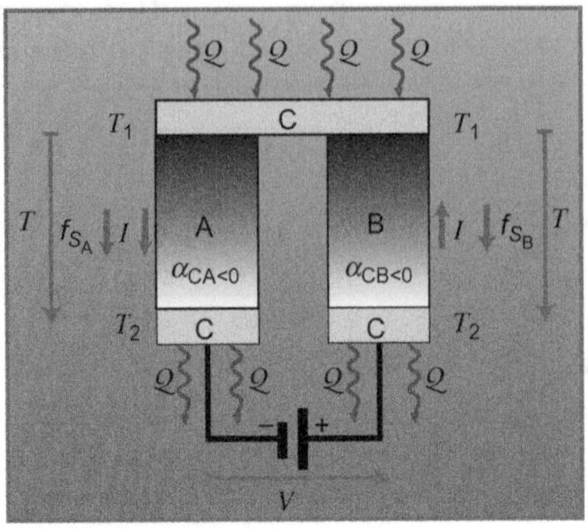

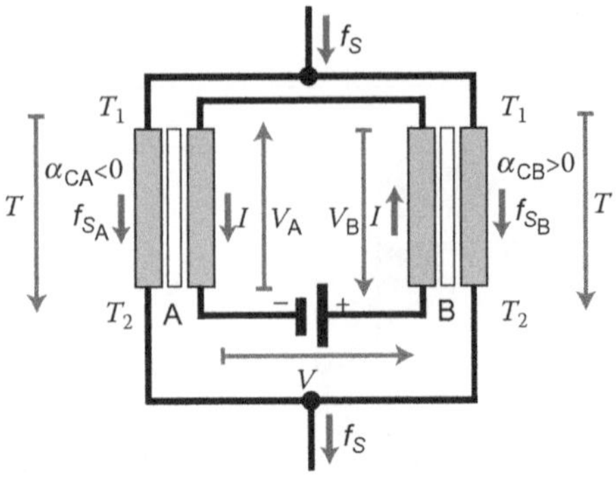

The diagram above (right) shows the two electric and thermal circuits, the latter being an electrical equivalent. Both transformers are thermocouples that are designed to convert electrical energy (each transformer plays the role of an electric receiver) into heat (each transformer plays the role of a heat generator). The amplification factor in potential, the Seebeck coefficient, is negative for the material A in order to guide the flow of charge (current) and heat in the same direction, whereas it is positive for the material B in order to steer in the opposite direction.

The thermocouples are assumed to be ideal, that is, with an invariable coupling (without Thomson effect) and without dissipation, that is, no Joule effect or heat transfer according to Fourier's equation. The potential difference of each thermocouple is proportional to the temperature difference through a constant Seebeck coefficient:

$$V_A = \alpha_{CA} T; \quad V_B = \alpha_{CB} T; \quad T = T_2 - T_1 \tag{J10.1}$$

The energy conversions take place without losses, which results in a conservation of power between receiver and generator for each thermocouple transformer.

$$0 = V_A I + T f_{SA} = V_B I + T f_{SB} \tag{J10.2}$$

The two coupling relations between flows are deduced as

$$f_{SA} = -\alpha_{CA} I; \quad f_{SB} = -\alpha_{CB} I \tag{J10.3}$$

In the thermal circuit, thermocouples operate in parallel, whereas in the electrical circuit they are in series, which is expressed by the sums

$$f_S = f_{SA} + f_{SB}; \quad V = V_A + V_B \tag{J10.4}$$

This leads to the following proportionalities for the whole assembly:

$$f_S = -\alpha I; \quad V = \alpha T \tag{J10.5}$$

having defined a global Seebeck coefficient

$$\alpha = \alpha_{CA} + \alpha_{CB} \tag{J10.6}$$

The Formal Graph in the case study abstract represents both electrodynamical and thermal dipoles connected by the coupling previously described by the relations in Equation J10.5, adding the coupling between basic quantities.

This coupling between basic quantities is derived from capacitive energy conservation expressed by the nullity of the Gibbs equation:

Gibbs:
$$d\mathcal{U} = V\,dQ + T\,dS \underset{isol}{=} 0 \tag{J10.7}$$

As the coupling factor is a constant, the same coupling factor between efforts and between basic quantities is found (except for the signs).

Remarks

Discovery. The discovery of the effect that bears his name by Jean-Charles Peltier goes back to 1834. Peltier observed a release of heat at the junction of two different metals crossed by a current. The phenomenon was interpreted by Lenz* in 1838 who concluded that the phenomenon could be reversed on changing the direction of the current (Joffe 1957).

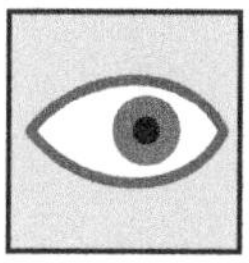

Ideal case. Note that the majority of materials do not present a constant Seebeck coefficient, which can, in principle, be translated by the indication on the previous graph of a second factor, called *differential factor of coupling*, for connecting the energies-per-entity (efforts and flows) as appropriate for a conversion with variable coupling. The coupling factor in this case can be seen, physically speaking, as a manifestation of an additional effect, called "Thomson effect," treated in case study J11.

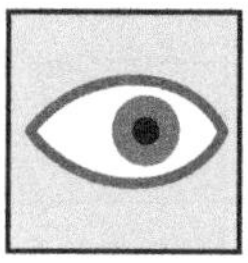

Many thermoelectric effects. The *Peltier effect* consists of the generation of an entropic flow (heat flow) at passing of a current, whereas the *Seebeck effect* is the reverse phenomenon, allowing the production of electricity from a difference of temperature.

These effects are reversible and can be superimposed onto dissipative processes that are transported by conduction of electrical charges (*Joule effect*) and *heat transfer* (Fourier equation), which are irreversible (see case study J9 "Seebeck Effect").

A third reversible effect, the *Thomson effect*, may also act as a supplement when the coupling factor between the two varieties of energy is not constant but depends on the temperature (see case study J11 "Thomson Effect").

* Heinrich Friedrich Emil Lenz (1804–1865): Russian physicist; Dorpat and Saint-Petersburg, Russia.

12.4.11 Case Study J11: Thomson Effect

Electrodynamics
Thermics

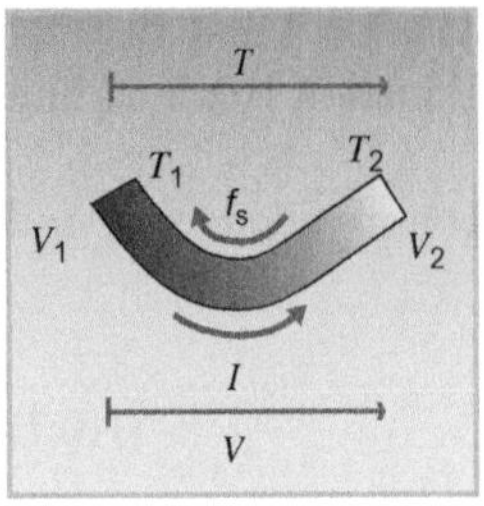

Formal Object

Coupled dipoles

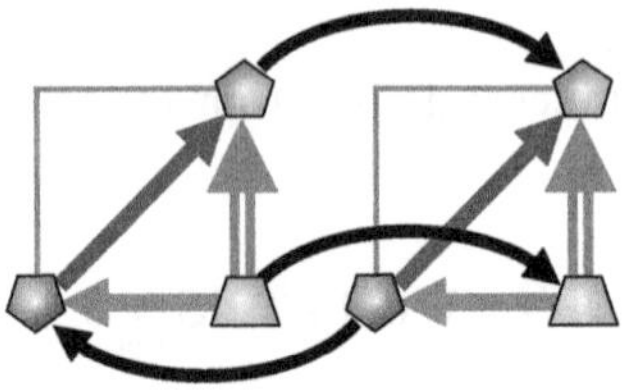

Coupling
Invariable
Equivalence
✓ Conversion
Poles
✓ Dipoles
✓ Capacitive
Inductive
Conductive

Formal Graph

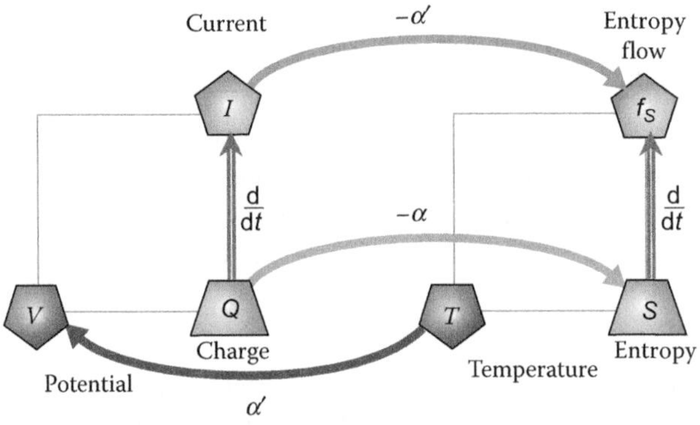

GRAPH 12. 26

	Coupling	From	To
Variety		Electrodynamics	Thermics
Family	Integral factor	Basic quantity	
Name	Seebeck coefficient	Charge	Entropy
Symbol	$-\alpha_{AB}$	Q	S
Unit	[J C^{-1} K^{-1}]	[C]	[J K^{-1}]
Variety		Thermics	Electrodynamics
Family	Differential factor	Effort	
Name	Differential See back coefficient	Temperature difference	Peltier potential difference
Symbol	$-\alpha'_{AB}$	T	Π_{AB}
Unit	[J C^{-1} K^{-1}]	[K]	[V]

System Constitutive Properties		
Variety	Electrodynamics	Thermics
Nature	Capacitance	Capacitance
Levels	0	0
Name	Electric capacitance	Thermal capacitance
Symbol	C	C_S
Unit	[F]	[J K^{-2}]
Specificity	—	—

The Thomson effect is one of three reversible thermoelectric effects that are part of conversion processes between the electric energy and the thermal energy:

- The Peltier effect consists of the generation of an entropic flow (heat flux) caused by the passing of a current.
- The Seebeck effect consists of the production of electricity as the result of a difference in temperature.

The Thomson effect, a conversion process as the first two effects, works as well in one way as in the opposite one and is the consequence of the dependence of coupling factors with the temperature, as is shown by the Formal Graph theory.

The first two effects have been treated in the case studies J9 "Seebeck Effect" and J10 "Peltier Effect."

These three effects are potentially reversible processes, in contradistinction with the Joule effect (dissipation of electric energy into heat) and with heat transfer (Fourier equation) that are irreversible processes. Nevertheless, all these processes may coexist in the same material or system. When the coupling factor is a constant, it takes the name of *Seebeck coefficient*, and only the Peltier and Seebeck effects are observed.

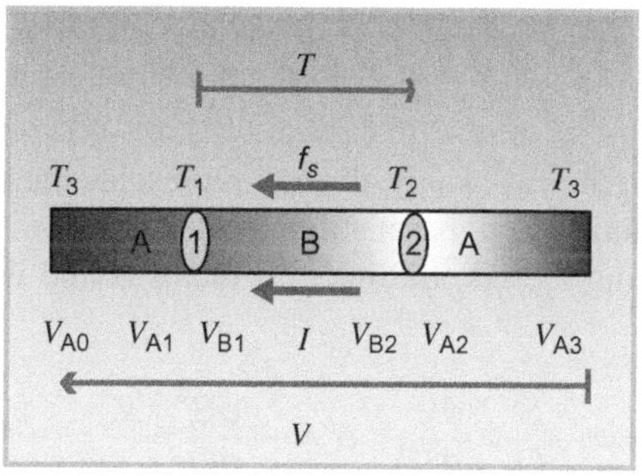

On the contrary, when the coupling factor depends on the temperature, the coupling becomes variable, and the difference of behavior with the invariable case is classically attributed to an additional effect, the Thomson effect. The same system as for the invariable coupling is taken as case study and the scheme on the left shows the three dipoles of a thermocouple sequence with the two soldered joints that are brought to temperatures T_1 and T_2 and with the two ends that are brought to the same temperature T_3 that can be of any value, as in the case of the thermocouple in case study J9 "Seebeck Effect."

The potential difference of the dipole assembly, electrically in series, is the sum of all the potential differences, whether they are due to the Peltier effect at the junctions or due to the Thomson effect in the three dipoles of materials A, B, and C

$$V = V_{A0} - V_{A3} = (V_{A0} - V_{A1}) + (V_{A1} - V_{B1}) + (V_{B1} - V_{B2}) + (V_{B2} - V_{A2}) + (V_{A2} - V_{A3}) \quad \text{(J11.1)}$$

The Peltier potential differences at the two junctions are:

$$V_{A1} - V_{B1} = \Pi_{AB1} = (\alpha_A - \alpha_B)T_1; \quad V_{B2} - V_{A2} = \Pi_{BA2} = (\alpha_B - \alpha_A)T_2 \quad \text{(J11.2)}$$

Each dipole between the junctions or ends sees a potential difference due to the Thomson effect:

$$V_{A0} - V_{A1} = \tau'_A (T_3 - T_1); \quad V_{B1} - V_{B2} = \tau'_B (T_1 - T_2); \quad V_{A2} - V_{A3} = \tau'_A (T_2 - T_3) \qquad \text{(J11.3)}$$

Insertion of all these expressions of potential differences into the sum in Equation J11.1 gives

$$V = \tau'_A (T_3 - T_1) + (\alpha_A - \alpha_B) T_1 + \tau'_B (T_1 - T_2) + (\alpha_B - \alpha_A) T_2 + \tau'_A (T_2 - T_3) \qquad \text{(J11.4)}$$

By choosing factors combining the Seebeck and Thomson effects, and by defining a *dipole temperature* as the difference of temperature of two poles,

$$\alpha'_A = \alpha_A - \tau'_A; \quad \alpha'_B = \alpha_B - \tau'_B; \quad T = T_2 - T_1 \qquad \text{(J11.5)}$$

Then, knowing that the relations between polar and dipolar factors are as follows:

$$\alpha' = \alpha'_B - \alpha'_A; \quad \alpha = \alpha_B - \alpha_A; \quad \tau' = \tau'_B - \tau'_A \qquad \text{(J11.6)}$$

the relation between temperature and potential of the dipole assembly is written as

$$V = \alpha' T = (\alpha - \tau') T \qquad \text{(J11.7)}$$

The Formal Graph given in the case study abstract represents the energetic conversion between two assemblies of electrodynamical and thermal dipoles through a variable coupling, which utilizes therefore two different coupling factors, the integral factor α and the differential factor α'. The coupling relations are then

$$\alpha = -\frac{S}{Q} = \frac{dV}{dT}; \quad \alpha' = -\frac{dS}{dQ} = \frac{V}{T} = -\frac{f_S}{I} \qquad \text{(J11.8)}$$

This Formal Graph holds everything that is necessary for describing the general model of Peltier, Seebeck, and Thomson thermoelectric effects. Nothing more in principle is requested. However, as the classical theory attributes different factors as parameters of the model, it is proper to decompose the preceding graph into two parallel couplings.

This decomposition relies on the relationship linking the two factors of a variable coupling that is written under the form of a difference in accordance with the coupling relations established in Equation J11.7:

$$\alpha' = \alpha - \tau' \qquad \text{(J11.9)}$$

with this difference defined as depending on the derivative of the differential coupling factor according to

$$\tau' \overset{\text{def}}{=} T \frac{d\alpha'}{dT} = \frac{dV}{dT} - \frac{V}{T} \qquad \text{(J11.10)}$$

as it can be verified from Equations J11.8 and J11.9. When this dependence is null, the two coupling factors become equal, making invariable the coupling as in the case of the pure Peltier–Seebeck effect.

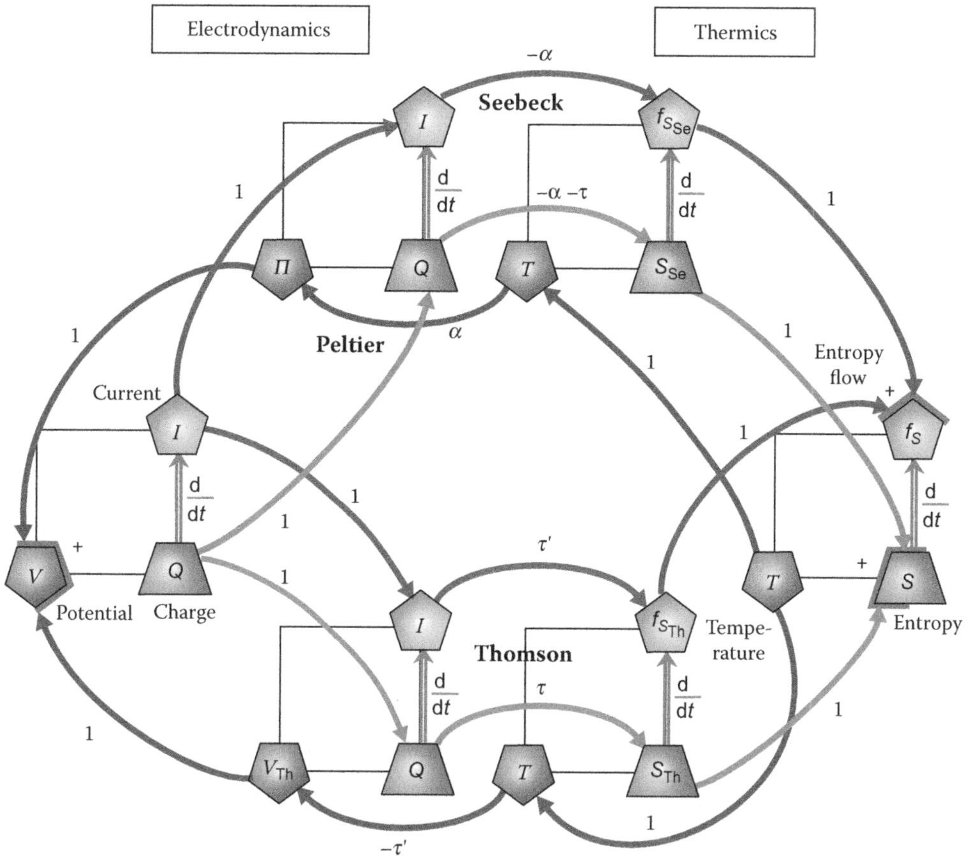

GRAPH 12.27

The Formal Graph in Graph 12.27 utilizes this difference to dissociate the global coupling into two subcouplings, the first one describing the Peltier–Seebeck effects with a differential coupling factor equal to the integral factor α of the global coupling

$$\alpha + \tau = -\frac{S_{Se}}{Q} = \frac{d\Pi}{dT}; \quad \alpha = -\frac{dS_{Se}}{dQ} = \frac{\Pi}{T} = -\frac{f_{S_{Se}}}{I} \tag{J11.11}$$

The factor τ is called the *Thomson coefficient* and it can be demonstrated from these two coupling relations, in the same manner as previously done, that it is given by the relation of definition

Thomson coefficient:
$$\tau \overset{def}{=} T\frac{d\alpha}{dT} = \frac{d\Pi}{dT} - \frac{\Pi}{T} \tag{J11.12}$$

This definition is called *Kelvin relation*. The second coupling describes the Thomson effect strictly speaking, with an integral coupling factor τ and a differential factor τ'

$$\tau = \frac{S_{Th}}{Q} = -\frac{dV_{Th}}{dT}; \quad \tau' = \frac{dS_{Th}}{dQ} = -\frac{V_{Th}}{T} = \frac{f_{S_{Th}}}{I} \tag{J11.13}$$

These coupling relations allow one to establish a differential relationship between factors giving τ'

$$\tau' = \frac{1}{T}\int \tau \, \mathrm{d}T = \frac{1}{T}\int T \, \mathrm{d}\alpha \tag{J11.14}$$

The detailed Formal Graph in Graph 12.27 shows these two couplings in parallel, adding the relations between effort variations as for the Formal Graph of the global coupling.

The parallelism of both couplings corresponds to the decomposition of global variables over the Peltier–Seebeck effect and the Thomson effect as follows:

$$V = \Pi + V_{\mathrm{Th}}; \quad S = S_{\mathrm{Se}} + S_{\mathrm{Th}}; \quad f_S = f_{S_{\mathrm{Se}}} + f_{S_{\mathrm{Th}}} \tag{J11.15}$$

The first of these relations is used to express classically the dependence of the electric potential difference on the temperature as a function of the Seebeck coefficient α and of the Thomson coefficient τ by using Equations J11.11, J11.13, and J11.14:

$$V = \alpha T - \int \tau \, \mathrm{d}T \tag{J11.16}$$

This expression can naturally be read on the detailed Formal Graph, which, it should be reminded, is only useful for retrieving the classical concepts that rely on the perception of three distinct effects. The Formal Graph theory shows that these three effects are in fact only one, which is merely an energetic conversion through a variable coupling.

In addition, the Formal Graphs show that all the three effects can be combined into a single dipole formed with only one material that must have a variable Seebeck coefficient. This unique dipole is strictly equivalent to the classical decomposition in several dipoles that separates the Peltier and Seebeck effects with a constant factor on the junctions from the Thomson effect between the junctions and the ends.

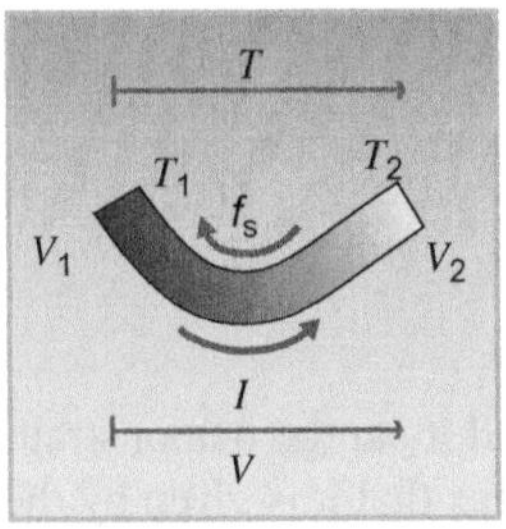

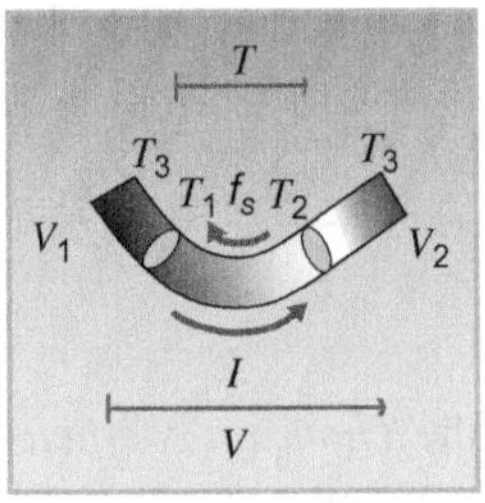

The above scheme (left) shows a dipole formed by a curved (one-dimensional) material, with its two ends being the poles, characterized by the state variables of effort that are the temperature and the electric potential. Besides, the same geometry is used to represent the three dipoles of a thermocouple sequence. The two soldered joints are brought to the same temperatures T_1 and T_2 as the ones of the ends of the system made with only one material shown on the left. On the right, the two ends are brought to the same but any temperature T_3, as in the case of the thermocouple in case study J9 "Seebeck Effect." These two systems are equivalent from the formal point of view when the Seebeck coefficient is variable (left) and when it is a constant (right) but in adding the Thomson effects.

Remarks

Thermoelectromagnetic effects. When the electromagnetic subvariety is involved in addition to the capacitive subvariety (electrostatics), other effects will be met:

- The Righi*–Leduc† effect, called thermal Hall‡ effect, is the equivalent of the Hall effect, but here it is a temperature gradient that appears under the effect of an electromagnetic field (instead of a gradient of electric potential).
- The Nernst effect consists of the apparition of an electric field under the influence of a temperature gradient in the presence of an electromagnetic field.
- The Ettingshausen§ effect consists of the apparition of a temperature gradient under the influence of an imposed current in the presence of an electromagnetic field.

These effects are potentially reversible and correspond to conversion couplings between energetic varieties, the two last effects involving three subvarieties (electrostatic, electromagnetism, thermics).

12.4.12 Case Study J12: Electric Polarization

Polarization
Electrodynamics

The case study abstract is given in next page.

The action of an electric field on a material containing electric charges is to polarize the material in displacing the charges, which amounts to modifying the distribution of energy over the various energy varieties contained in the system. A new energy variety appears, the *energy of polarization*, under a capacitive form, in which the basic quantity is the *charge of polarization* q_P and the effort is the *potential of polarization* e_P.

However, these state variables are not explicitly utilized in the classical theory, which directly models the phenomenon at the local level, in terms of vectors fields that are the *electric field* $\boldsymbol{E}$, the *electric displacement* $\boldsymbol{D}$, and the *polarization* vector $\boldsymbol{P}$, related by the classical formulas, valid in linear materials (including free space) as no operator is used for the system constitutive property.

$$\boldsymbol{D} \underset{lin}{=} \varepsilon_0 \boldsymbol{E} + \boldsymbol{P} \tag{J12.1}$$

$$\boldsymbol{P} \underset{lin}{=} \varepsilon_0 \chi_e \boldsymbol{E} \tag{J12.2}$$

The two properties of this system are the *electric permittivity* ε_0 of free space and the *electric susceptibility* χ_e which are scalars in homogeneous and isotropic media and tensors otherwise. It can be observed that, when the *electric susceptibility* becomes equal to zero, the polarization also equals zero and the material is said to be a nonpolarizable one. This parameter therefore features the capability of the material to contain energy of polarization.

The Formal Graph theory models the phenomenon of electric polarization under the form of a *variable coupling* between the two energy subvarieties that are the electrostatic energy and the energy of polarization. This approach allows one to introduce the electric susceptibility as a coupling factor, which is a new concept in this domain.

* Augusto Righi (1850–1920): Italian physicist; Bologna, Italy.
† Sylvestre Anatole Leduc (1856–1937): French physicist; Paris, France.
‡ Edwin Herbert Hall (1855–1928): North American physicist; Baltimore, Maryland, USA.
§ Albert Freiherr von Ettingshausen (1850–1932): Austrian physicist; Graz, Austria.

Formal Object

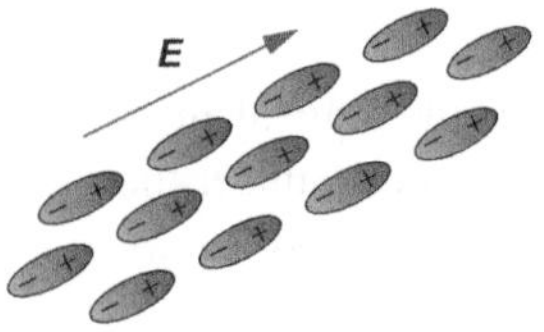

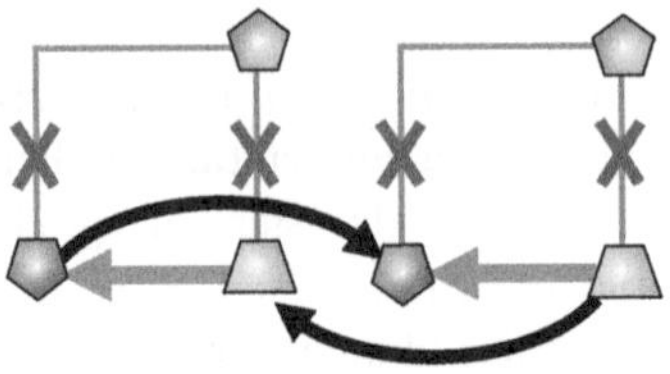

Coupling
Invariable Equivalence Conversion
✓ Poles Dipoles
✓ Capacitive Inductive Conductive

Formal Graph

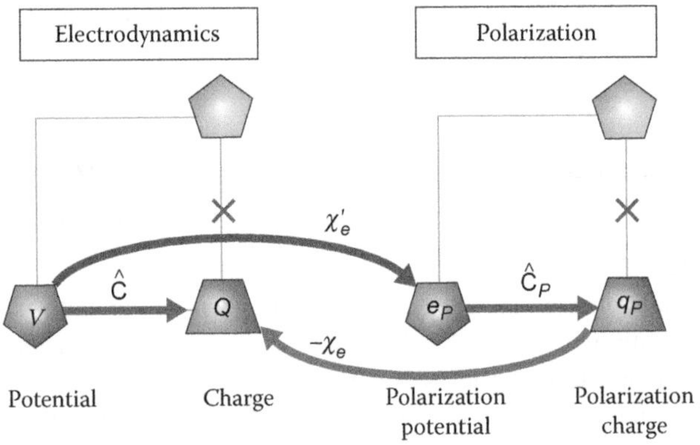

Formal Graph of the total coupling between two poles with variable coupling factors (global level, without spatial distribution).

GRAPH 12.28

	Coupling	From	To
Variety		Translation mechanics	Hydrodynamics
Family	Integral factor	Basic quantity	
Name	Electric susceptibility	Polarization charge	Charge
Symbol	$-\chi_e$	q_P	Q
Unit	[—]	[C]	[C]
Variety		Hydrodynamics	Translation mechanics
Family	Differential factor	Effort	
Name	Differential electric susceptibility	Potential	Polarization potential
Symbol	$-\chi'_e$	V	e_P
Unit	[—]	[V]	[V]

System Constitutive Properties		
Variety	Electrodynamics	Polarization
Nature	Capacitance	Capacitance
Levels	0	0
Name	Electric capacitance	Polarization capacitance
Symbol	C	C_P
Unit	[F]	[F]
Specificity	—	—
Levels	1/2	1/2
Name	Electric permittivity	Polarization permittivity
Symbol	ε	ε_P
Unit	[F m^{-1}]	[F m^{-1}]
Specificity	Three-dimensional	Three-dimensional

New energy variety. The subject of the electric polarization of a material presented here is treated in an original way. The Formal Graph theory indeed has for principle to model a system in terms of various energy varieties and to establish their mutual relationship with the help of the concept of energetic coupling. Here, it has been necessary to take recourse to a new energy variety, the *energy of polarization*, for translating the phenomenon in the Formal Graph.

Variety	**Electric Polarization**		
Subvariety	Capacitive (potential)		Inductive
Category	Entity Nr	Energy/entity	Energy/entity
Family	Basic quantity 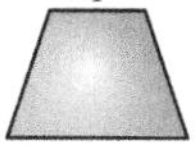	Effort 	Flow 
Name	Polarization charge	Polarization potential	Polarization current
Symbols	q_P	e_P	f_P
Unit	[C], [A s]	[V], [J C^{-1}]	[A]

The state variables of the polarization energy have exactly the same units as those belonging to electrodynamics. This is due to the historical choice of a coupling factor without unit (at that time the perception of a specific energy variety for the polarization was not clear). This is not a real problem; one could even imagine that all energy varieties were endowed with the same units for their state variables.

12.4.12.1 Capacitive Coupling

Only two varieties are supposed to be contained in the system. The variation of capacitive energy is given by the following Gibbs equation, associated with its Gibbs–Duhem equation (required for making the energy variation shown in Equation J12.3 an exact differential equation).

Gibbs:
$$\mathrm{d}\mathcal{U} = V\,\mathrm{d}Q + e_P\,\mathrm{d}q_P \tag{J12.3}$$

Gibbs–Duhem:
$$0 = Q\,\mathrm{d}V + q_P\,\mathrm{d}e_P \tag{J12.4}$$

The integral coupling factor is the *electric susceptibility* χ_e and the differential coupling factor, unknown in the classical theory, is a *differential electric susceptibility* χ'_e

$$\chi_e \overset{def}{=} -\frac{Q}{q_P} = \frac{\mathrm{d}e_P}{\mathrm{d}V}; \quad \chi'_e \overset{def}{=} -\frac{\mathrm{d}Q}{\mathrm{d}q_P} = \frac{e_P}{V} \tag{J12.5}$$

These four equalities satisfy the Gibbs and Gibbs–Duhem equations and correspond to the case of an extremum of capacitive energy. The interesting insight brought by the Formal Graph theory is to give to the electric susceptibility the status of an integral coupling factor, as many physical constants or system properties. The Formal Graph in the case study abstract uses these two coupling factors between two poles and also shows the two capacitances.

12.4.12.2 Apparent Charge

Introduction of the differential coupling factor given in Equation J12.5 into the Gibbs Equation J12.3 provides a variation of apparent electrical charge

$$\mathrm{d}\mathcal{U} = V\,\mathrm{d}\bar{Q} \tag{J12.6}$$

$$\mathrm{d}\bar{Q} = \mathrm{d}Q + \chi_e'\,\mathrm{d}q_P \tag{J12.7}$$

The interest of these relationships will become obvious when spatially reduced variables will be made explicit.

12.4.12.3 Spatial Distribution

The Formal Graph in Graph 12.29 represents the electrostatic pole together with the polarization pole, both distributed in a space with three dimensions. (Contrary to the representation of a global system given in the case study abstract, the nodes of flows are not drawn for staying in a two-dimensional representation.) In the higher part of the graph are found the relationships between global variables that are partially coupled.

12.4.12.4 Reduced Capacitances

The main local variables of a spatially distributed pole electrostatic are the *electric field* $\boldsymbol{E}$ and the *electric displacement* $\boldsymbol{D}$ (or "electrization"). These are vectors linked by the capacitive reduced

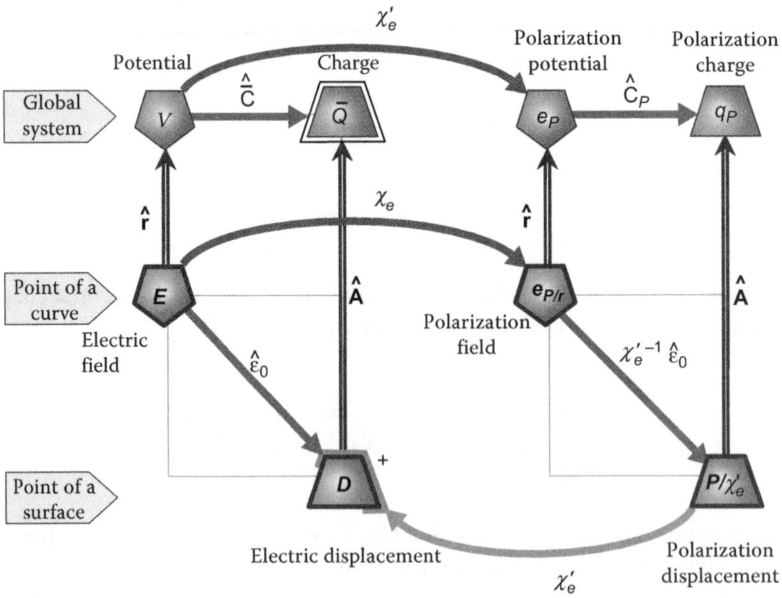

GRAPH 12.29

property called *permittivity*, which is a specific property in three-dimensional systems, represented by an operator in the Formal Graph theory for the sake of generality. In the absence of polarization, the permittivity is subscripted with a zero and no subscript is used in the presence of polarization, so the *electric displacements* in the absence ($\boldsymbol{D}_0$) and in the presence ($\boldsymbol{D}$) of polarization are given by the two relations

$$\boldsymbol{D}_0 = \hat{\varepsilon}_0\, \boldsymbol{E}; \quad \boldsymbol{D} = \hat{\varepsilon}\, \boldsymbol{E} \tag{J12.8}$$

Notation. It would have been judicious to give another symbol for the *apparent electric displacement* (with a top bar, for instance) but the tradition is to keep the bare letter $\boldsymbol{D}$, obliging to notate differently the *genuine electric displacement* $\boldsymbol{D}_0$ (which returns to the notation without subscript in every situation without possibility of polarization).

The main local variables of a spatially distributed pole in the energy of polarization are the *polarization field* $\boldsymbol{e}_{P/r}$ and the vector of the *surface density of polarization charge* $\boldsymbol{q}_{P/A}$ sometimes called *polarization displacement*, which is equal to $\boldsymbol{P}/\chi'_e$, that is, the polarization vector divided by the differential susceptibility. These vectors are linked by the polarization permittivity that is the combination of the *electric permittivity* and the reciprocal of the *differential electric susceptibility*

$$\frac{P}{\chi'_e} = \boldsymbol{q}_{P/A} = \hat{\varepsilon}_P\, \boldsymbol{e}_{P/r} = \chi_e'^{-1}\, \hat{\varepsilon}_0\, \boldsymbol{e}_{P/r} \quad \Leftrightarrow \quad P = \hat{\varepsilon}_0\, \boldsymbol{e}_{P/r} \tag{J12.9}$$

A slight difficulty in the translation of the classical theory into a Formal Graph is encountered with the nonrepresentation of the polarization vector $\boldsymbol{P}$ in a node as one could expect from the simple classical model given by Equation J12.1. The reason is that this variable is not a spatially reduced variable of a state variable, contrary to the ratio $\boldsymbol{P}/\chi'_e$ that is the surface density of polarization charge $\boldsymbol{q}_{P/A}$, but is merely the contribution of the polarization to the total electric displacement, as shown below. This difficulty is not met in the comparable example of magnetization (see case study J13), where no classical equivalent to the polarization vector is used, but an equivalent of $\boldsymbol{q}_{P/A}$ is directly used under the form of the magnetization vector $\boldsymbol{M}$. In other words, the polarization vector $\boldsymbol{P}$ is not a fundamental physical variable whereas the magnetization vector $\boldsymbol{M}$ is one, in view of the Formal Graph theory.

12.4.12.5 Reduced Couplings

The coupling between the two fields is readily obtained from the integral coupling factor in Equation J12.5 by dividing each effort variation by $-\mathbf{d}\boldsymbol{r}$ (each field is the contragradient of its effort)

$$\boldsymbol{e}_{P/r} = \chi_e\, \boldsymbol{E} \tag{J12.10}$$

By division by the surface variation $\mathbf{d}\boldsymbol{A}$, Equation J12.7 of the apparent charge variation becomes a relationship between electric and polarization displacements:

$$\frac{\mathrm{d}\bar{Q}}{\mathbf{d}\boldsymbol{A}} = \frac{\mathrm{d}Q}{\mathbf{d}\boldsymbol{A}} + \chi'_e \frac{\mathrm{d}q_P}{\mathbf{d}\boldsymbol{A}} \quad \Leftrightarrow \quad \boldsymbol{D} = \boldsymbol{D}_0 + \chi'_e \boldsymbol{q}_{P/A} \tag{J12.11}$$

This electric displacement represents the contribution of the polarization to the total electric displacement, which is written according to relationships J12.8 and J12.9:

$$\boldsymbol{D} = \hat{\varepsilon}_0\, \boldsymbol{E} + \boldsymbol{P} \tag{J12.12}$$

By restricting to the case of free space (or linear materials), the relation in Equation J12.1 of the classical model with a scalar permittivity is retrieved.

Remarks

More than two energy varieties. When several interdependent energy varieties are present in the system, the model is easily modified by extending the previous reasoning, which assumes that the apparent charge, and therefore the apparent electric displacement, results from several contributions:

$$\mathrm{d}\mathcal{U} = V\mathrm{d}Q + e_P \mathrm{d}q_P + \cdots \tag{J12.13}$$

$$\mathrm{d}\bar{Q} = \mathrm{d}Q + \chi_e' \mathrm{d}q_P + \cdots \tag{J12.14}$$

$$\boldsymbol{D} = \hat{\varepsilon}_0\, \boldsymbol{E} + \boldsymbol{P} + \cdots \tag{J12.15}$$

Generality. This model is extremely general in two aspects: First, by using an operator for the electric permittivity, allowing the modeling of nonlinear materials; second, by using a variable coupling, allowing the electric susceptibility to depend on other state variables. When this coupling factor is a constant, the coupling becomes invariable, meaning that the two coupling factors, integral and differential, are equal and constant.

It must be recalled that the usage of operators in lieu of scalars for the constitutive properties featuring such systems is a luxury allowed by the Formal Graph, for which the exact nature of links between two nodes can be as complex as possible without harming the simplicity and the clarity of the graphic representation. Therefore, it would be a pity not to use it. Here, at one's disposal is a generalization that goes well beyond what is taught or written in basic textbooks, at no expense in terms of complexification (other than a hat over the symbols and respecting the rules of noncommutativity when translating to algebra).

12.4.13 Case Study J13: Magnetism

Magnetism
Electrodynamics

The case study abstract is given in next page.

The action of an electromagnetic field on certain materials has the effect of magnetizing them by creating magnetic domains made by magnetic entities (also called, but improperly, "magnetic charges"), which amounts to modifying the distribution of energy over the various energy varieties contained in the system. A new energy variety appears, the *magnetic energy* or *energy of magnetization*, under an inductive form, the impulse being the *magnetic impulse* p_m and the flow being the

J13: Magnetism

Magnetism
Electrodynamics

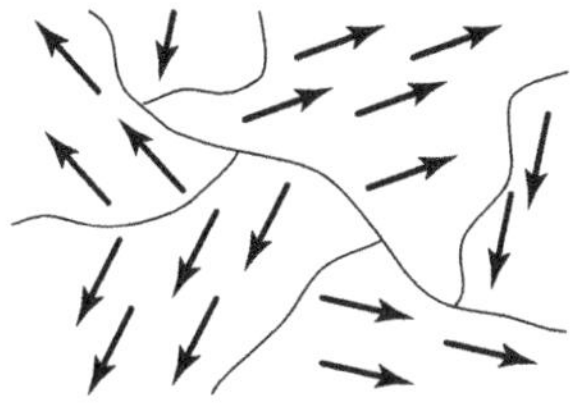

Formal Object

Coupled poles

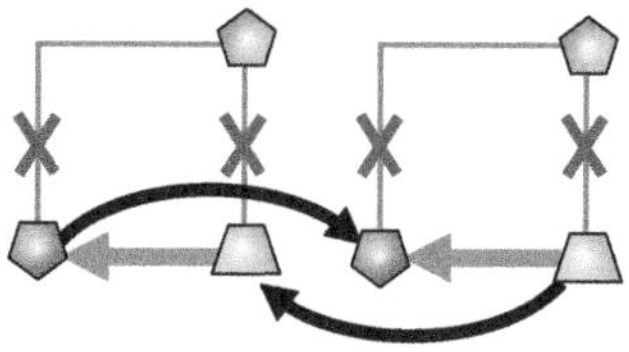

Coupling	
	Invariable
	Equivalence
	Conversion
✓	Poles
	Dipoles
	Capacitive
✓	Inductive
	Conductive

Formal Graph

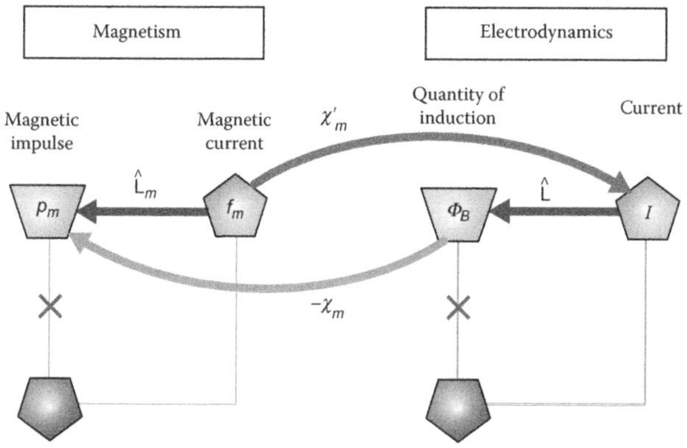

Formal Graph of the total coupling between two poles with variable coupling factors (global level, without spatial distribution).

GRAPH 12.30

	Coupling	From	To
Variety		Electrodynamics	Magnetism
Family	Integral factor	Impulse	
Name	Magnetic susceptibility	Magnetic impulse	Induction quantity (Flux)
Symbol	$-\chi_m$	p_m	$\boldsymbol{\Phi}_B$
Unit	[—]	[Wb]	[Wb]
Variety		Magnetism	Electrodynamics
Family	Differential factor	Flow	
Name	Differential magnetic susceptibility	Magnetic current	Current
Symbol	$-\chi'_m$	f_m	I
Unit	[—]	[A]	[A]

System Constitutive Properties		
Variety	Electrodynamics	Magnetism
Nature	Inductance	Inductance
Levels	0	0
Name	Inductance	Magnetic inductance
Symbol	L	L_m
Unit	[H]	[H]
Specificity	—	—
Levels	1 / 2	1 / 2
Name	Permeability	Magnetic permeability
Symbol	$\boldsymbol{\mu}$	$\boldsymbol{\mu}_m$
Unit	[H m^{-1}]	[H m^{-1}]
Specificity	Three-dimensional	Three-dimensional

magnetic current f_m (see case study A4 "Magnetization" in Chapter 4). These state variables are not explicitly utilized in the classical theory, which directly models the phenomenon at the local level, in terms of vectors fields, induction, and magnetization. The model consists of modifying the proportionality, valid in media free from magnetization, between the *electromagnetic field* $\boldsymbol{H}$ and the *electromagnetic induction* $\boldsymbol{B}$ by including a *magnetization* vector $\boldsymbol{M}$ entitled to express the modification induced by the creation of magnetic domains.

$$\boldsymbol{B} \underset{lin}{=} \mu_0 (\boldsymbol{H} + \boldsymbol{M}) \tag{J13.1}$$

$$\boldsymbol{M} = \chi_m \boldsymbol{H} \tag{J13.2}$$

According to the classical theory, the electromagnetic *free space permeability* μ_0 and the *magnetic susceptibility* χ_m are scalars or tensors, depending on the anisotropy of the medium, that relate the electromagnetic field $\boldsymbol{H}$ to the electromagnetic induction $\boldsymbol{B}$ and to the magnetization $\boldsymbol{M}$, respectively. When the *magnetic susceptibility* χ_m becomes equal to zero, the magnetization also equals zero and the material is said to be a nonmagnetizable one. This parameter therefore features the capability of the material to magnetize, that is, to contain energy of magnetization.

This model is less general than the one proposed by the Formal Graphs with the concept of *coupling* between energy varieties and with the generalization of the inductance by means of an operator, as it will be shown. The magnetic susceptibility is introduced here as a *coupling factor*, which is a new concept in this domain.

12.4.13.1 Inductive Coupling

To define this coupling, we assume that only two energy varieties are involved in an isolated system. The variation of inductive energy is given by the following Gibbs equation, associated with its Gibbs–Duhem equation (required for making the energy variation in Equation J13.3 an exact differential equation):

Gibbs:

$$\mathrm{d}\mathcal{T} = I\,\mathrm{d}\Phi_B + f_m\,\mathrm{d}p_m \underset{isol}{=} 0 \tag{J13.3}$$

Gibbs–Duhem:

$$0 = \Phi_B\,\mathrm{d}I + p_m\,\mathrm{d}f_m \tag{J13.4}$$

The integral coupling factor is the *magnetic susceptibility* χ_m and the differential coupling factor, unknown in the classical theory, is a *differential magnetic susceptibility* χ'_m.

$$\chi_m \overset{def}{=} -\frac{\Phi_B}{p_m} = \frac{\mathrm{d}f_m}{\mathrm{d}I};\quad \chi'_m \overset{def}{=} -\frac{\mathrm{d}\Phi_B}{\mathrm{d}p_m} = \frac{f_m}{I} \tag{J13.5}$$

These four equalities satisfy the Gibbs and Gibbs–Duhem equations and correspond to the case of an extremum of inductive energy, meaning that the system is isolated from any other energy variety influence. The interesting insight brought by the Formal Graph theory is to give to the magnetic susceptibility the status of a coupling factor, as many physical constants or system properties. The Formal Graph in the case study abstract uses these two coupling factors between two poles and also shows the two inductances.

12.4.13.2 Apparent Quantity of Induction

The introduction of the differential coupling factor given in Equation J13.5 into the Gibbs equation J13.3 to replace the magnetic flow provides a variation of apparent induction quantity (flux)

$$\mathrm{d}\mathcal{T} = I\,\mathrm{d}\bar{\Phi}_B + \cdots \tag{J13.6}$$

which is given by

$$d\bar{\Phi}_B = d\Phi_B + \chi'_m dp_m \tag{J13.7}$$

The interest of this relationship will become obvious when spatially reduced variables will be made explicit.

12.4.13.3 Spatial Distribution

The Formal Graph in Graph 12.31 represents the electromagnetic pole together with the magnetization pole, both distributed in a space with three dimensions. (Contrary to the representation of a global system given in the case study abstract, the nodes of efforts are not drawn for staying in a two-dimensional representation.) On the higher part of the graph are found the relationships between global variables that are partially coupled.

12.4.13.4 Reduced Inductances

The main local variables of a spatially distributed electromagnetic pole are the *electromagnetic field* ***H*** and the *electromagnetic induction* ***B***. These are vectors linked by the inductive reduced property called *electromagnetic** *permeability*, which is a specific property in a three-dimensional system. It is a scalar for an isotropic medium and a second rank tensor for an anisotropic medium and is represented by an operator in the Formal Graph theory for the sake of generality. In the absence of magnetization, the (*free*) *electromagnetic induction* and the permeability are subscripted

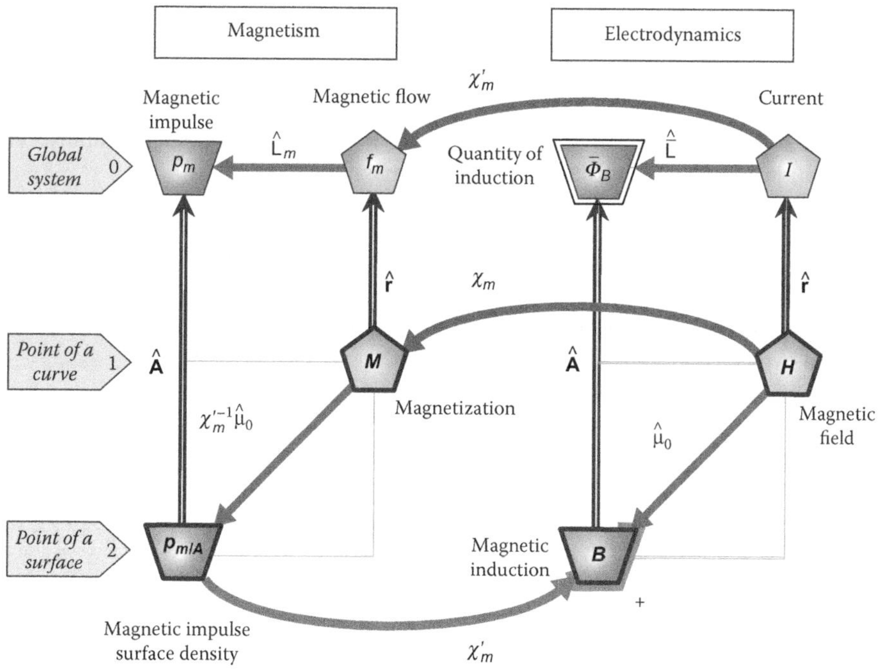

GRAPH 12.31 Formal Graph of the *partial coupling* between two poles with variable coupling factors (normally the graph is in three dimensions but the flow nodes being omitted, a two-dimensional representation of the spatial distribution is chosen for the sake of simplification).

* In fluid mechanics and earth sciences, the permeability measures the ability of a medium to transmit fluids.

with a zero. In vacuum, the permeability is a constant and takes the name of *free space permeability* or *magnetic constant*, equal to a fixed value by convention:

Free space permeability: $$\mu_0 = 4\pi \times 10^{-7}\ \mathrm{H\ m^{-1}}$$

In the presence of magnetization the *permeability* is written without a subscript, according to a widespread use (which is not the case for $\boldsymbol{B}_0$).

Reduced inductances: $$\boldsymbol{B}_0 = \hat{\mu}_0\, \boldsymbol{H}; \quad \boldsymbol{B} = \hat{\mu}\, \boldsymbol{H} \qquad \text{(J13.8)}$$

The main local variables of a spatially distributed pole in the energy of magnetization are the *magnetization field* $\boldsymbol{M}$ and the vector of the *surface density of magnetic impulse* $\mathbf{p}_{m/A}$ sometimes called *magnetic* (or *magnetization*) *induction*. These vectors are linked by the magnetic permeability that is the combination of the *electromagnetic permeability* in free space and the reciprocal of the *differential magnetic susceptibility*

$$\mathbf{p}_{m/A} = \chi_m'^{-1}\, \hat{\mu}_0\, \boldsymbol{M} \qquad \text{(J13.9)}$$

12.4.13.5 Reduced Coupling

The coupling between the two fields is readily obtained from the integral coupling factor in Equation J13.5 by dividing each flow variation by $-\mathbf{d}\boldsymbol{r}$ (each field is the contragradient of its flow)

$$-\frac{\mathrm{d}f_m}{\mathbf{dr}} = -\chi_m \frac{\mathrm{d}I}{\mathbf{dr}} = \boldsymbol{M} = \chi_m \boldsymbol{H} \qquad \text{(J13.10)}$$

By division by the surface variation $\mathbf{d}\boldsymbol{A}$, Equation J13.7 of the apparent impulse variation becomes a relationship between electromagnetic and magnetic inductions

$$\frac{\mathrm{d}\bar{\Phi}_B}{\mathbf{d}\boldsymbol{A}} = \frac{\mathrm{d}\Phi_B}{\mathbf{d}\boldsymbol{A}} + \chi_m' \frac{\mathrm{d}p_m}{\mathbf{d}\boldsymbol{A}} \Leftrightarrow \boldsymbol{B} = \boldsymbol{B}_0 + \chi_m' \mathbf{p}_{m/A} \qquad \text{(J13.11)}$$

This magnetic induction represents the contribution of magnetization to the total electromagnetic induction, which is written according to relationships J13.8 and J13.9

$$\boldsymbol{B} = \hat{\mu}_0 \left(\boldsymbol{H} + \boldsymbol{M}\right) \qquad \text{(J13.12)}$$

By restricting to the case of free space (or linear materials), the relation in Equation J13.1 of the classical model with a scalar permeability is retrieved.

Remarks

Distinction among electromagnetic inductions. The classical approach does not differentiate between these two *electromagnetic inductions*, in the absence and in the presence of magnetization, and the same symbol $\boldsymbol{B}$ is used for both. The Formal Graph approach, due to its solid basis in thermodynamics (first principle), cannot follow this usage and obliges one to distinguish the two variables. However, when the context is clear, the subscript is dropped in order to comply with the tradition.

More than two energy varieties. When several interdependent energy varieties are present in the system, the model is modified by extending the previous reasoning, which assumes that the apparent induction quantity, and therefore the induction itself, results from several contributions:

$$\mathrm{d}\mathcal{T} = I\mathrm{d}\Phi_B + f_m \mathrm{d}p_m + \cdots \tag{J13.13}$$

$$\mathrm{d}\bar{\Phi}_B = \mathrm{d}\Phi_B + \chi'_m \mathrm{d}p_m + \cdots \tag{J13.14}$$

$$\mathbf{B} = \hat{\mu}_0 \left(\mathbf{H} + \mathbf{M} + \cdots\right) \tag{J13.15}$$

Generality. This model is extremely general in two aspects: First, by systematically using an operator for the electromagnetic permeability, allowing the modeling of nonlinear materials; second, by using a variable coupling, allowing the magnetic susceptibility to depend on other state variables. When this coupling factor is a constant, coupling becomes invariable, meaning that the two coupling factors, integral and differential, are equal and constant.

12.5 PROPERTIES OF COUPLING

The two relationships (Equations 12.28 and 12.31) defining the capacitive coupling are recalled:

Integral factor:
$$\lambda_{ba} \overset{def}{=} \frac{q_a - q_a^{min}}{q_b - q_b^{min}} = -\frac{\mathrm{d}e_b}{\mathrm{d}e_a} \quad \left(q_b \neq q_b^{min}\right) \tag{12.43}$$

Differential factor:
$$\lambda'_{ba} \overset{def}{=} \frac{\mathrm{d}q_a}{\mathrm{d}q_b} = -\frac{e_b}{e_a} \quad \left(e_a \neq 0\right) \tag{12.44}$$

12.5.1 Minimum of Capacitive Energy

The minimum of capacitive energy is defined by the following two conditions:

Capacitive energy extremum:
$$\frac{\mathrm{d}\mathcal{U}}{\mathrm{d}q_i} = 0, \quad i = a, b \tag{12.45}$$

Condition for a minimum:
$$\frac{\mathrm{d}^2\mathcal{U}}{\mathrm{d}q_i^2} > 0, \quad i = a, b \tag{12.46}$$

The expression of the first derivative comes from the definition in Equation 12.44 of the differential coupling factor

$$\frac{\mathrm{d}\mathcal{U}}{\mathrm{d}q_b} = \lambda'_{ba}\, e_a + e_b \tag{12.47}$$

By proceeding to a second derivation, one gets the second condition for having a minimum of energy

$$\frac{d^2\mathcal{U}}{dq_b^2} \underset{\exists C_a, C_b}{=} \lambda_{ba}'^2 \frac{de_a}{dq_a} + \frac{de_b}{dq_b} + e_a \frac{d\lambda_{ba}'}{dq_b} \underset{\mathcal{U}\ minimum}{>} 0 \tag{12.48}$$

In this expression are found two differential elastances, one for each energy variety, which are always positive quantities.

Positive capacitances. Chapter 7 established the capacitive relationship of a pole, which has been demonstrated as given by an exponential function.

$$\text{Pole capacitance:} \quad q_i = q_i^{min} + \left(q_i^{\theta} - q_i^{min}\right) \exp\left(\frac{e_{qi} - e_{qi}^{\theta}}{e_{q\Delta}}\right) \tag{12.49}$$

The differential capacitance, or preferably for this discussion, the differential elastance, is therefore given by the ratio of two positive quantities:

$$\frac{de_i}{dq_i} = \frac{e_{q\Delta}}{q_i - q_i^{min}} > 0 \quad i = a, b \tag{12.50}$$

In the linear case, the capacitance or the elastance is a constant also given by a ratio of two positive quantities (which is a normal consequence of the above result):

$$\frac{de_i}{dq_i} \underset{lin}{=} C_{qi}^{-1} = \frac{e_{q\Delta}}{q_i^{\theta} - q_i^{min}} > 0 \quad i = a, b \tag{12.51}$$

Whatever the chosen model, general (nonlinear) or linear, a pole capacitance or elastance is always positive.

Consequently, when the differential coupling factor is a constant, the condition for having a minimum of energy is always met:

$$\frac{d\lambda_{ba}'}{dq_b} = 0 \Rightarrow \frac{d^2\mathcal{U}}{dq_b^2} \underset{\exists C_a, C_b}{=} \lambda_{ba}'^2 \frac{de_a}{dq_a} + \frac{de_b}{dq_b} \underset{\mathcal{U}\ minimum}{>} 0 \tag{12.52}$$

When the differential coupling factor is not a constant, which occurs frequently, the requirement for having a minimum is less obvious to comprehend:

$$-e_a \frac{d\lambda_{ba}'}{dq_b} \underset{\mathcal{U}\ minimum}{<} \lambda_{ba}'^2 \frac{de_a}{dq_a} + \frac{de_b}{dq_b} \tag{12.53}$$

The case of variable coupling factors requires a deeper analysis exceeding the present possibilities. For the moment, a pragmatic approach consists of noting that the cases of maximum energy are metastable situations, which would contradict the general observation that most systems are stable when at rest. It is assumed in the following sections that the condition in Equation 12.48 is always met in the studied systems, that is, that an energetic coupling always ensures a minimum for energy.

12.5.2 Relationship between Coupling Factors

From Equation 12.44 of the capacitive coupling using the differential coupling factor, rewritten as a proportionally relationship between efforts,

$$e_b = -\lambda'_{ba}\, e_a \tag{12.54}$$

is deduced the variation of the effort e_b,

$$de_b = -\lambda'_{ba}\, de_a - e_a d\lambda'_{ba} \tag{12.55}$$

Then, by invoking Equation 12.43 with the integral factor, a relationship between the two factors is obtained:

$$\lambda_{ba} = \lambda'_{ba} + e_a \frac{d\lambda'_{ba}}{de_a} \tag{12.56}$$

This equation is interesting not only for evidencing the relationship between the invariability of the differential factor and the equality of the two factors, but also for defining an increment of the differential factor in the following way:

$$\Delta\lambda'_{ba} \overset{def}{=} e_a \frac{d\lambda'_{ba}}{de_a} = \frac{e_b}{e_a} - \frac{de_b}{de_a} \tag{12.57}$$

This is exactly the equation that has been met in case study J11 "Thomson Effect" for defining the *Thomson coefficient*

Thomson coefficient:
$$\tau \overset{def}{=} T\frac{d\alpha}{dT} = \frac{d\Pi}{dT} - \frac{\Pi}{T} \tag{J11.17}$$

in which the Peltier potential difference Π and the temperature T are the efforts and α is the Seebeck coefficient, playing the role of a differential coupling factor. When the Seebeck coefficient is a constant, coupling becomes invariable and the Thomson effect disappears in order to leave the Seebeck and Peltier effects alone.

12.5.3 Energy Conversion

The conversion of energy requires time as explained in Chapter 9 devoted to dipole assemblies, in which was introduced the *internal conversion* between the two subvarieties in the same energy variety. The conversion of energy between two energy varieties is called *external conversion* and works by coupling energies-per-entity between two dipoles or dipole assemblies. The process of external conversion consists of changing (transferring) the energy from one variety to another. In itself, there are no losses in the conversion between two energy varieties as the conservation of energy is the rule, and the notions of reversibility and yield are irrelevant. They may be invoked only in case of association of a conductive dipole which diverts part of the energy in dissipation, as discussed in Chapter 11.

The conservation of energy in a dynamic process (i.e., implying time) is expressed by a conservation of powers, between the input (in the a variety to be converted) and the output (in the converted *b* variety). This is algebraically written as follows:

Powers: $$\mathcal{P}_a = e_a\, f_a; \quad \mathcal{P}_B = e_b\, f_b \tag{12.58}$$

Total power: $$\mathcal{P} = \mathcal{P}_a + \mathcal{P}_b = 0 \tag{12.59}$$

By forming the ratios of flows and of efforts and by comparing with the coupling relationships 12.31 and 12.34 established earlier, the following conversion relationship using a unique differential coupling factor is deduced:

Differential coupling factor: $$\lambda'_{ba} = \frac{f_a}{f_b} = -\frac{e_b}{e_a} \tag{12.60}$$

Graph 12.32 depicts in the language of Formal Graphs how this relationship is used to form a loop with the internal links of the Formal Objects that are involved in the conversion.

In this Formal Graph appears a loop made up of the two coupling connections and the two internal properties of the dipoles. The circularity principle provides the following equality between the combination of operators along the loop and the identity operator:

Circularity: $$\hat{Z}_a\left(\lambda'_{ba}\right)\hat{Y}_b\left(-\lambda'_{ba}\right) = \hat{1} \tag{12.61}$$

This can be written alternatively as a relationship known as *impedance adaptation* equation:

Impedance adaptation: $$\hat{Z}_a = \frac{-1}{\lambda'_{ba}}\hat{Z}_b\frac{1}{\lambda'_{ba}} \tag{12.62}$$

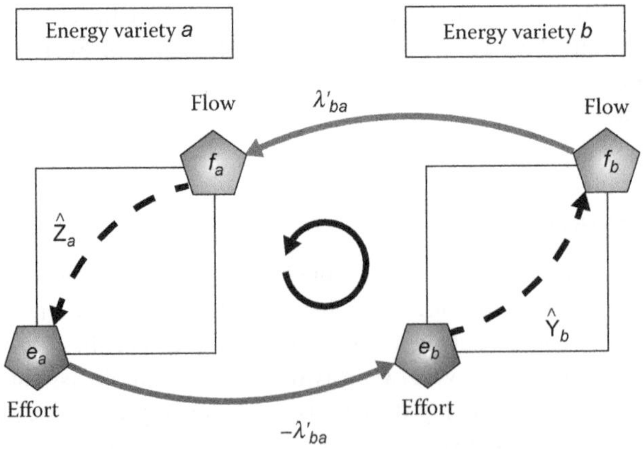

GRAPH 12.32 Formal Graph of the conversion process between two energy varieties. The Formal Objects are dipoles or dipole assemblies (not poles as time is involved) but their nature is not specified as the use of admittance and impedance exempts one from doing it. Nevertheless, the dipole in the energy variety *a* cannot be a purely conductive dipole unless the dipole in the variety *b* is the capacitive thermal dipole (i.e., implementing dissipation).

The negative sign merely corresponds to an orientation convention of potential differences and currents as explained hereafter. It can be dropped when this feature is meaningless. This is the case when AC techniques are employed with sufficiently small signal amplitudes for approximating both Fourier-transformed impedances as well as linear operators. Consequently, the admittances become proportional through the square of the coupling factor.

Linear impedance adaptation: $$\tilde{Y}_a \underset{lin}{=} \lambda'^2_{ba} \tilde{Y}_b \quad (12.63)$$

This kind of relationship is also met in systems having linear constitutive properties as will be shown in case study K5 "Rotating Bodies" in Chapter 13.

One of the classical ways used for handling the conversion of energy is the concept of generalized transformer as represented in the language of equivalent circuits shown in Figure 12.4.

This representation illustrates the role of the sign of the coupling factor which is not very important, as it merely indicates whether the flows are counted in the same direction (positive coupling factor) or in the opposite direction (negative coupling factor). In all cases, one side of the transformer is always considered as a receptor for the energy to be converted (supplied by an external part of the circuit) and the other as a generator for providing the converted energy to the rest of the circuit.

An example of conversion is worth giving owing to its ability to illustrate the principle of an electrical measurement of a phenomenon occurring in a different energy variety. The case study taken here, an electrochemical characterization of a redox couple, can be transposed to a wide range of practical situations in all domains.

The Formal Graph in Graph 12.33 (left) is issued from case study J2 about the electrochemical potential, where the graphical relationship between the physical chemical pole representing the species A and the electrodynamical dipole representing the electrode/electrolyte was established. The physical chemical dipole of the electrochemical reaction $A + \nu_e e^- \leftrightarrow B$ is built by complementing with the pole B (as explained in Chapter 6) and in Graph 12.33 only the dipoles are retained. (The transfer factors α_i disappear as their sum is equal to 1 by definition.)

The Formal Graph in Graph 12.33 (right) models the restricted case of a reversible reaction that is derived from the general case by setting the electrochemical potential equal to zero (equilibrium condition) and by dropping the conductance link (no dissipation). This procedure is not specific to the electrochemical case treated here but follows the general rule of setting to zero the effort in a dipole for describing an equilibrium, static or dynamic. Applied to an apparent effort, this equilibrium condition amounts to linking directly the components of the apparent effort. Thus, the chemical

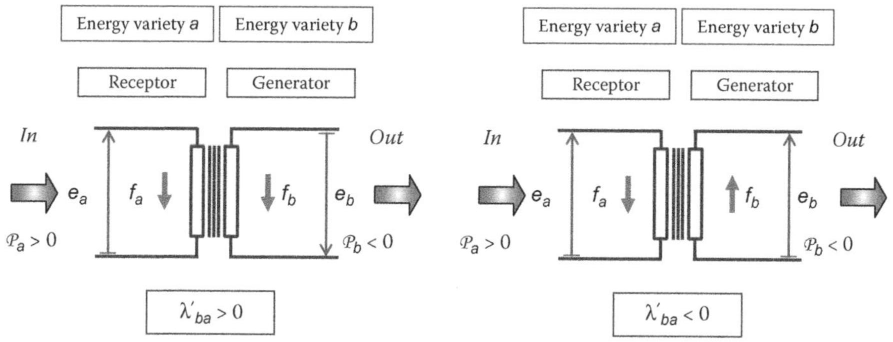

FIGURE 12.4 Two equivalent circuits representing a generalized transformer with different signs for the amplification factor (i.e., the differential coupling factor): Flows are parallel with a positive sign (left) and antiparallel with a negative sign (right).

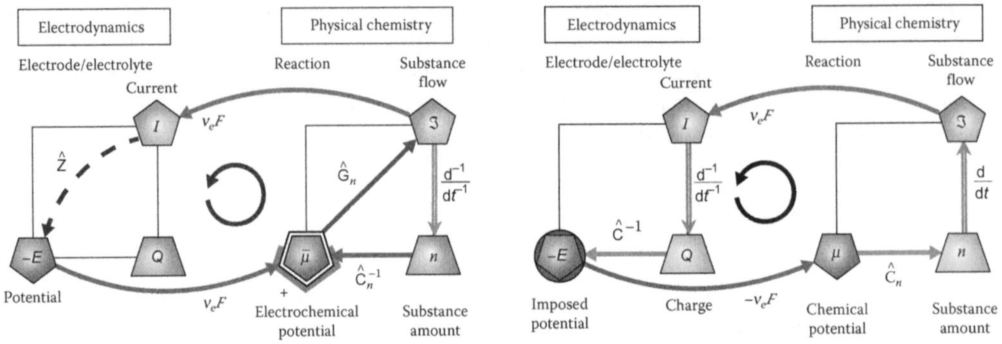

GRAPH 12.33 Two Formal Graphs of an electrochemical reaction (exchanging ν_e electrons) occurring in a thin layer deposited on an electrode (no substance transport) characterized by electrical measurement. Left: General case of reversible or irreversible reaction. Right: Reversible reaction only, with imposed electrode potential.

potential appears in the model of the physical chemical dipole at equilibrium. This equilibrium is supposed to be enforced in all conditions, notably in a dynamic regime, according to the definition of a reversible behavior for an exchange or a reaction.

When a linear ramp of potential is imposed onto the electrode, that is, an electrode potential varying with a constant slope v as a function of time,

Potential scan rate: $$v = \frac{dE}{dt} \tag{12.64}$$

the current flowing through the electrode is merely the derivative of the electrical capacitance, which is proportional to the physical chemical capacitance (see case study C5 "First-Order Chemical Reaction" in Chapter 6 for the expression of this dipole capacitance).

$$\frac{I}{I^{max}} = V_\Delta \frac{d}{dE} \frac{4}{1+\exp\left(-\frac{E-E^0}{V_\Delta}\right)} = \frac{4\exp\left(-\frac{E-E^0}{V_\Delta}\right)}{\left(1+\exp\left(-\frac{E-E^0}{V_\Delta}\right)\right)^2} \tag{12.65}$$

with the scaling potential and the scaling current given by

$$V_\Delta = \frac{RT}{\nu_e F}; \quad I^{max} = \frac{\nu_e F n^* v}{4V_\Delta} \tag{12.66}$$

It can be noted that the scaling potential differs from the expression given for an ion (cf. case study J6) as the charge number z is replaced by the electron stoichiometric factor ν_e. In Figure 12.5 are plotted the variation of the current as a function of the imposed electrode potential and the physical chemical capacitance of the reaction dipole.

The analytical technique described here is called linear sweep voltammetry or cyclic voltammetry (effecting only one half the cycle that normally returns to the initial potential) (Bard and Faulkner 2001).

The interest of this case study is to show that the notion of conversion is much wider than the notion of transformation of energy from one variety to another one for production purpose. Each time a perturbation of one of the state variables of a system is imposed, it imparts changes in the

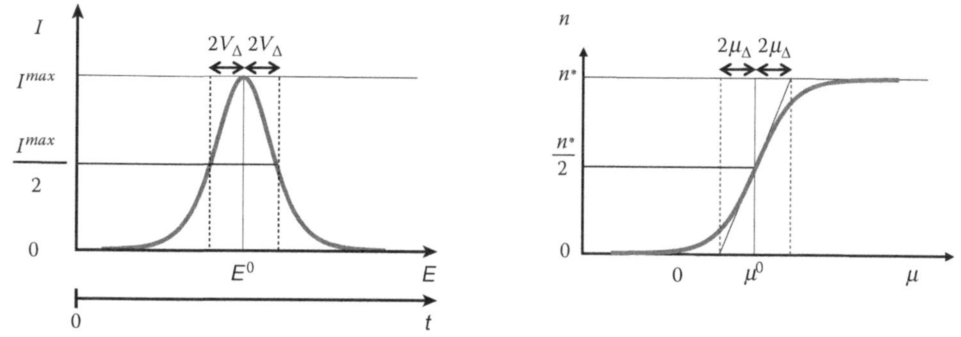

FIGURE 12.5 Plots of the electric admittance $I = f(E)$ in response to a linear ramp of electrode potential E imposed to a thin layer of electroactive substance (left) and of the physical chemical capacitance $n = f(\mu)$ of the redox couple (right).

state variables of the other energy varieties present in the system (provided energy exchanges occur), through energy conversion as discussed here. This is a universal means for characterizing phenomena that occur in a different energy variety from the "measuring" variety, that is, which possesses state variables easily connected to the exterior of the system. This is obviously one of the features of the electrodynamical energy and explains why electrical measurements are widely used in analytical or controlling techniques.

Bell shape. The plot of the electrodynamical admittance resulting from the electric measurement of a nonelectrodynamical process shown above is well known in many domains and is referred to as a bell-shaped curve. It reveals the exponential nature of the pole capacitive relationship underlying the nonelectrodynamical process. However, it should not be confused with the Gaussian* curve met in statistics, which has the same algebraic expression but with a squared argument of the exponential function, roughly producing the same bell-shaped curve (but narrower).

Pseudocapacitance. Due to its nonconstant feature, the electric capacitance given by such measurements is often called "pseudocapacitance," as if a nonlinear capacitance could not be an electrical one. The Formal Graph theory entitles this property with the status of true electrodynamical capacitance, the constant case corresponding to the "narrow" range around the summit of the curve. This appreciation of "narrow" is relative to the value of the scaling potential, which depends on the coupled energy variety. In the absence of coupling, the scaling potential is quasi-infinite, extending the "narrow" range to practically all accessible values of potential (thus retrieving the classical laws of electrostatics).

* Carl Friedrich Gauss (1777–1855): German mathematician and physicist; Brunswick and Göttingen, Germany.

12.6 IN SHORT

COUPLING FACTORS
Coupling factors express interdependence between energy varieties. They are defined between two identical subvarieties and link their entity numbers. An integral factor links the state variables whereas a differential one links their variations. These factors are not always constant and may depend on state variables. Integral and differential factors are equal when the differential factor is a constant. The existence of a coupling factor does not entail necessarily the coupling of two energy varieties, which requires in addition interdependence between energies-per-entity for allowing existence of an energy extremum.

Integral Coupling Factor	*Differential* Coupling Factor
Inductive: $\lambda_{ba} \overset{def}{=} -\dfrac{p_b - p_b^{min}}{p_a - p_a^{min}} \quad (p_a \neq p_a^{min})$ (12.33)	Inductive: $\lambda'_{ba} \overset{def}{=} -\dfrac{dp_b}{dp_a}$ (12.34)
Capacitive: $\lambda_{ba} \overset{def}{=} \dfrac{q_a - q_a^{min}}{q_b - q_b^{min}} \quad (q_b \neq q_b^{min})$ (12.3)	Capacitive: $\lambda'_{ba} \overset{def}{=} \dfrac{dq_a}{dq_b}$ (12.2)
$\lambda_{ba} = \dfrac{1}{p_b - p_b^{min}} \displaystyle\int_{p_b^{min}}^{p_b} \lambda'_{ba} dp$ (12.32) $\lambda_{ba} = \dfrac{1}{q_b - q_b^{min}} \displaystyle\int_{q_b^{min}}^{q_b} \lambda'_{ba} dp$ (12.67)	

Inductive Equivalence

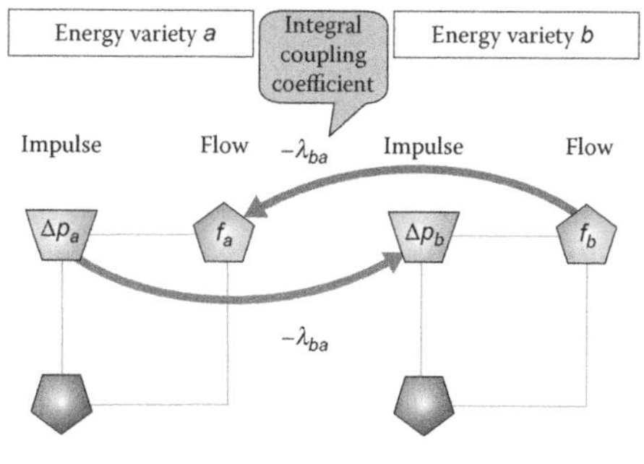

GRAPH 12.34

Two energy subvarieties are **equivalent** when their energy variations are equal.

$$\delta\mathcal{T}_a = f_a\,\mathrm{d}p_a \quad = \quad \delta\mathcal{T}_b = f_b\,\mathrm{d}p_b \qquad (12.68)$$

A common value (the differential coupling factor) exists between the ratios

$$\lambda'_{ba} \overset{def}{=} -\frac{\mathrm{d}p_b}{\mathrm{d}p_a} = -\frac{f_a}{f_b} \qquad (12.69)$$

The difference with the coupling case lies in the sign of the flows ratio, which is *negative* in case of inductive equivalence.

Inductive Coupling

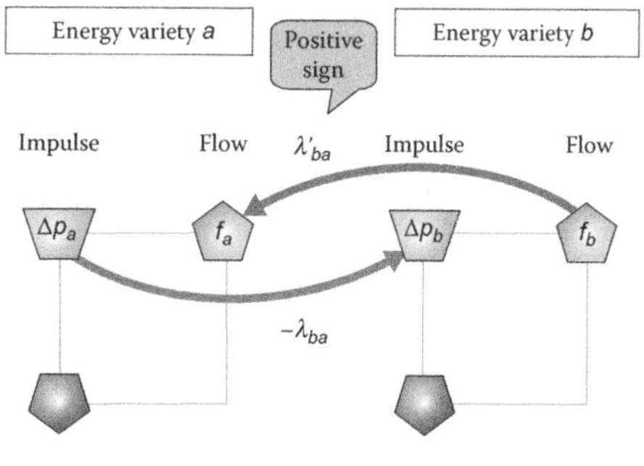

GRAPH 12.35

Two energy subvarieties are **coupled** when an extremum of energy is reached as resulting from energy exchange.

Extremum: $\quad \frac{\mathrm{d}\mathcal{T}}{\mathrm{d}p_i} = 0 \quad i = a, b$

In these coupling relationships, the sign of the flows ratio is *positive.*

$$\lambda_{ba} \overset{def}{=} -\frac{p_b - p_b^{min}}{p_a - p_a^{min}} = \frac{\mathrm{d}f_a}{\mathrm{d}f_b} \qquad (12.33)$$

$$\lambda'_{ba} \overset{def}{=} -\frac{\mathrm{d}p_b}{\mathrm{d}p_a} = +\frac{f_a}{f_b} \qquad (12.34)$$

Capacitive Equivalence

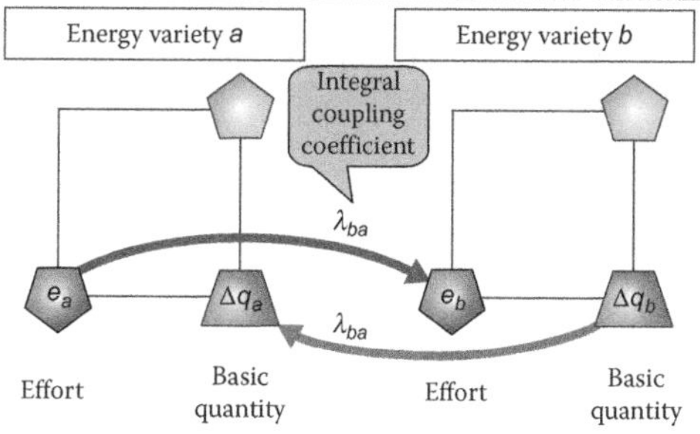

GRAPH 12.36

Capacitive Coupling

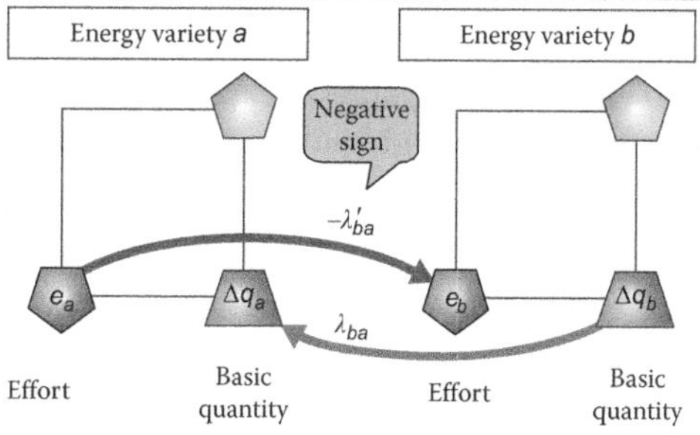

GRAPH 12.37

(continued)

Two energy subvarieties are **equivalent** when their energy variations are equal.

$$\delta \mathcal{U}_a = e_a \, dq_a = \delta \mathcal{U}_b = e_b \, dq_b \quad (12.70)$$

A common value must exist (the differential coupling factor) between the following ratios:

$$\lambda'_{ba} \overset{def}{=} \frac{dq_a}{dq_b} = + \frac{e_b}{e_a} \quad (12.71)$$

The difference with the coupling case lies in the sign of the efforts ratio, which is *positive* in case of capacitive equivalence.

Two energy subvarieties are **coupled** when an extremum of energy is reached as resulting from energy exchange.

$$\text{Extremum:} \quad \frac{d\mathcal{U}}{dq_i} = 0 \quad i = a, b \quad (12.45)$$

In these coupling relationships, the sign of the efforts ratio is *negative*.

$$\lambda_{ba} \overset{def}{=} \frac{q_a - q_a^{min}}{q_b - q_b^{min}} = - \frac{de_b}{de_a} \quad (12.28)$$

$$\lambda'_{ba} \overset{def}{=} \frac{dq_a}{dq_b} = - \frac{e_b}{e_a} \quad (12.31)$$

Translated Flow

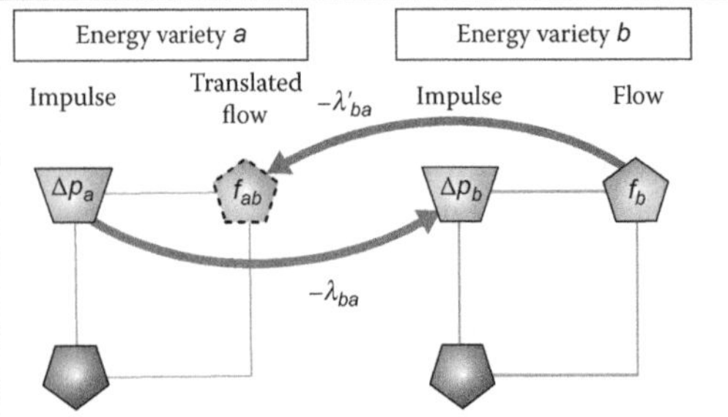

GRAPH 12.38

A **translated flow** allows one to express the amount of inductive energy in one variety in terms of state variables of another variety. For instance, the variation of inductive energy in the b variety can be expressed as the product of an a-flow times an a-impulse variation:

$$\delta \mathcal{T}_b = f_b \, dp_b = f_{ab} \, dp_a \quad (\neq \delta \mathcal{T}_a) \quad (12.72)$$

with a translated flow (from b to a)

$$f_{ab} \overset{def}{=} - \lambda'_{ba} f_b \quad (12.73)$$

Apparent Flow

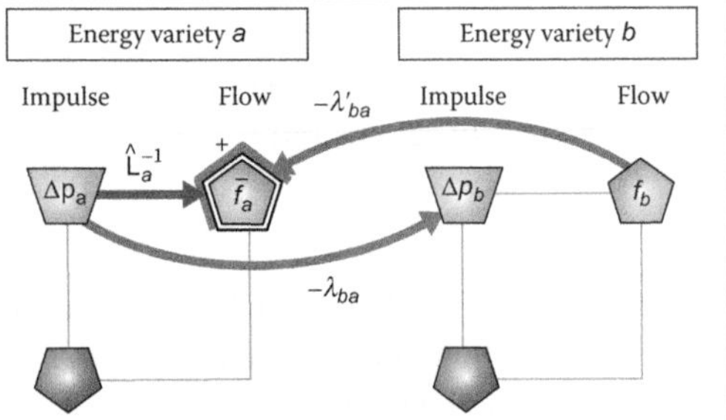

GRAPH 12.39

When two impulses are interdependent, a single **apparent flow** may be used to express the total energy variation of both varieties.

For instance, the apparent flow in the a variety is

$$\bar{f}_a \overset{def}{=} \left(\frac{\partial \mathcal{T}}{\partial p_a} \right)_{indep.\, pi} \quad (12.74)$$

Consequently, it is expressed as:

$$\bar{f}_a = f_a + f_{ab} = f_a - \lambda'_{ba} f_b \quad (12.75)$$

Translated Effort

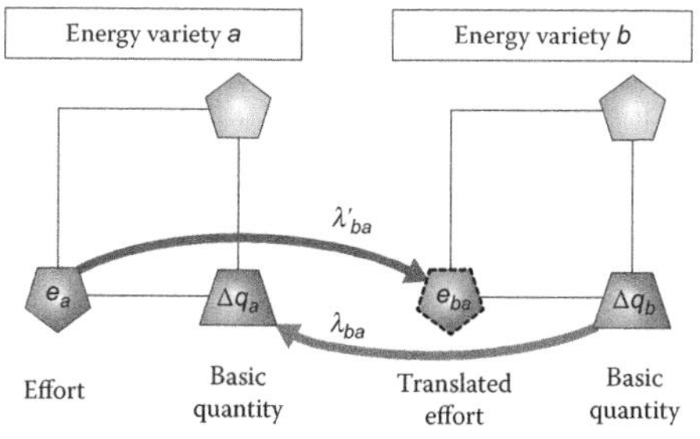

GRAPH 12.40

A **translated effort** allows one to express the amount of capacitive energy in one variety in terms of state variables of another variety. For instance, the variation of capacitive energy in the a variety can be expressed as the product of a b-effort times a b-basic quantity variation:

$$\delta \mathcal{U}_a = e_a \, dq_a = e_{ba} \, dq_b \quad (\neq \delta \mathcal{U}_b) \tag{12.76}$$

with a translated effort (from a to b)

$$e_{ba} \overset{def}{=} \lambda'_{ba} \, e_a \tag{12.77}$$

Apparent Effort

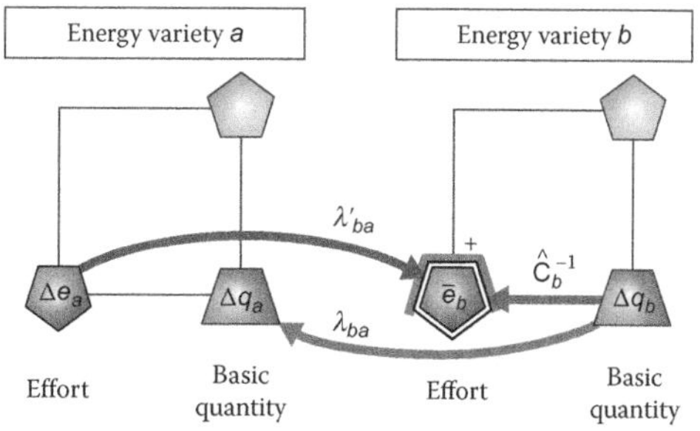

GRAPH 12.41

When two basic quantities are interdependent, a single **apparent effort** may be used to express the total energy variation of both varieties.

For instance, the apparent effort in the b variety is

$$\bar{e}_b \overset{def}{=} \left(\frac{\partial \mathcal{U}}{\partial q_b} \right)_{indep.\, qi} \tag{12.78}$$

Consequently, it is expressed as:

$$\bar{e}_b = e_b + e_{ba} = e_b + \lambda'_{ba} \, e_a \tag{12.79}$$

Energy Conversion

In an isolated system, the total power is zero:

$$\mathcal{P} = e_a f_a + e_b f_b \underset{isol}{=} 0 \tag{12.80}$$

Differential coupling factor:

$$\lambda'_{ba} = \frac{f_a}{f_b} = -\frac{e_b}{e_a} \tag{12.81}$$

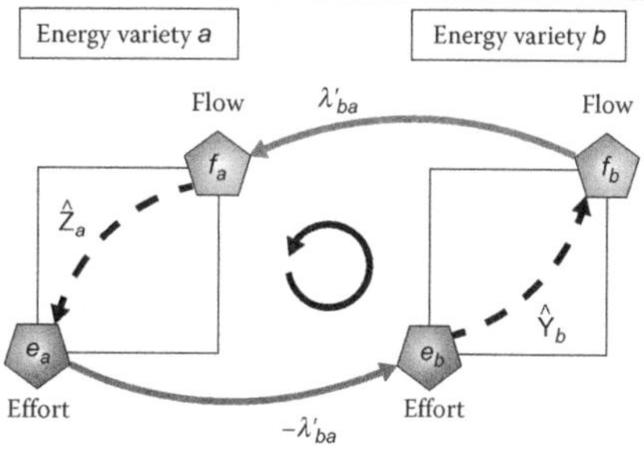

GRAPH 12.42

<table>
<tr><th colspan="3">CONVERSION AND REVERSIBILITY</th></tr>
<tr><td colspan="3">This table lists all possibilities of energy conversion from a storing dipole in the energy variety a to any dipole in the variety b. Both varieties can be thermal except in the case of conversion toward a conductive dipole, which is relevant of the dissipation process seen in Chapter 11.</td></tr>
<tr><th>From inductive subvariety</th><th>From capacitive subvariety</th><th>Reversibility</th></tr>
<tr><td>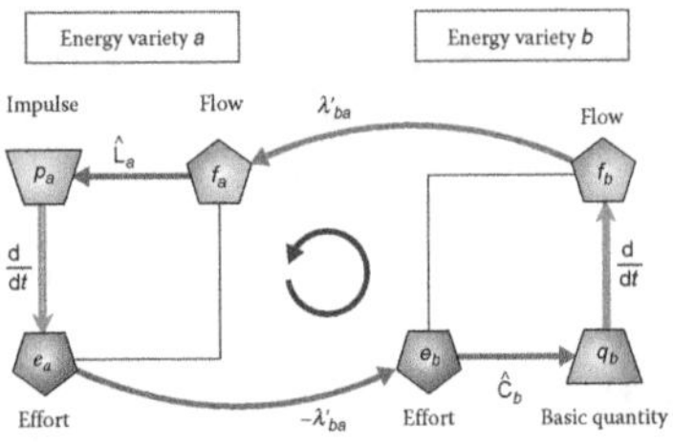

GRAPH 12.43</td><td>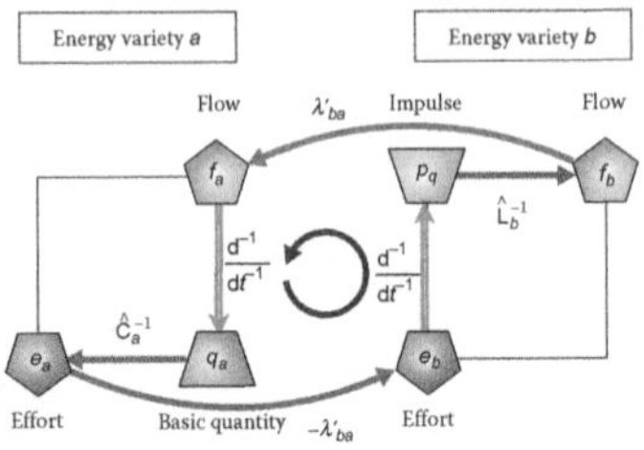

GRAPH 12.44</td><td>REVERSIBLE CONVERSION
Two evolution operators
(DIFFERENT SUBVARIETIES)</td></tr>
<tr><td>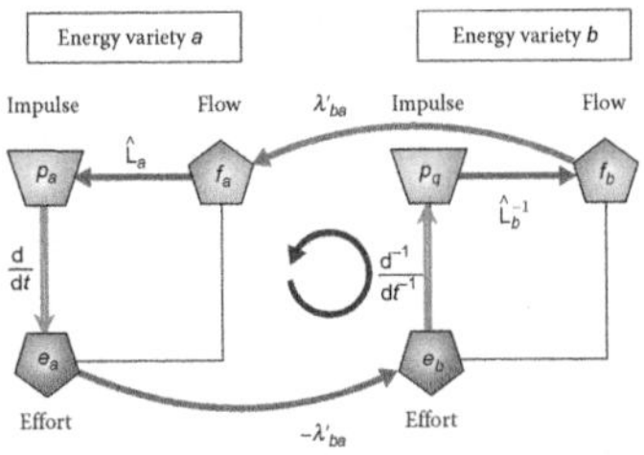

GRAPH 12.45</td><td>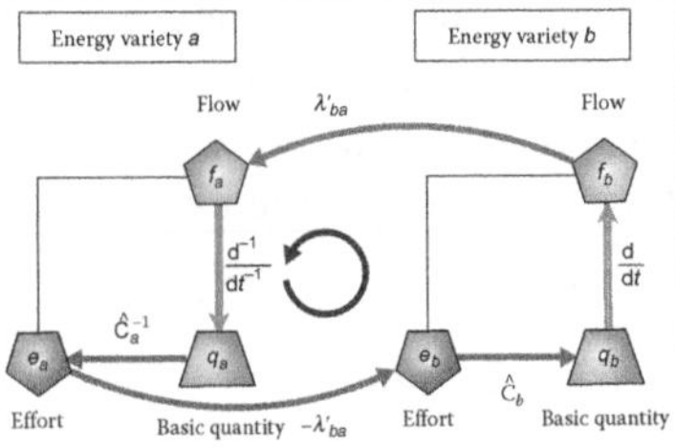

GRAPH 12.46</td><td>REVERSIBLE CONVERSION
Two evolution operators
(SAME SUBVARIETIES)</td></tr>
<tr><td>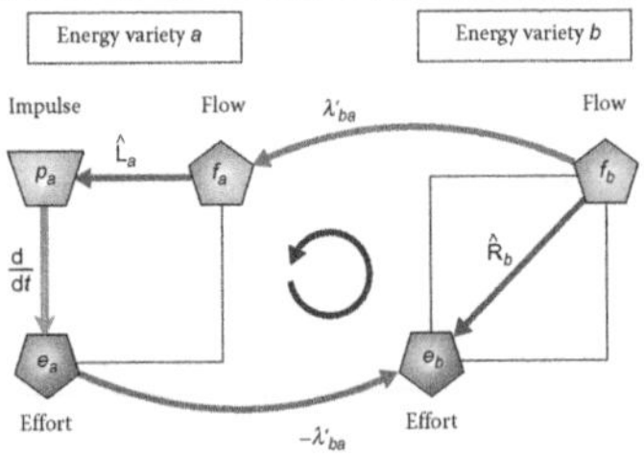

GRAPH 12.47</td><td>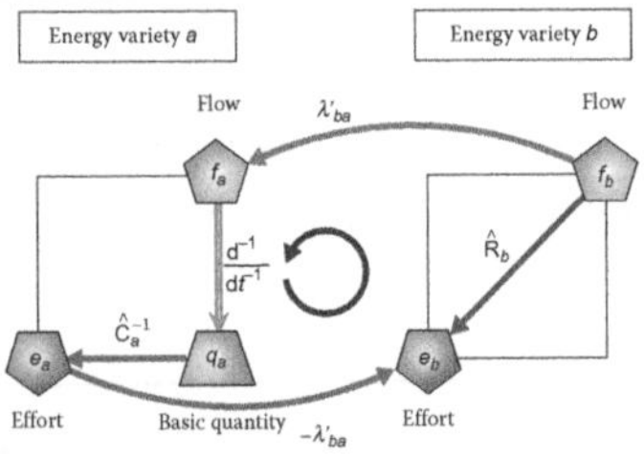

GRAPH 12.48</td><td>IRREVERSIBLE CONVERSION
One evolution operator
(Varieties # Thermics)</td></tr>
</table>

13 Multiple Couplings

CONTENTS

After having studied the coupling between pairs of energy variety, this chapter examines the case of several varieties contained in a system. The ideal gas, which is of paramount interest as an archetypal model in thermodynamics and many domains, will be modeled to ensure the coupling between three varieties, hydrodynamics, thermics, and physical chemistry. The addition of a fourth energy variety will be studied first with the case of an ideal gas placed in a gravitational field and second for handling the kinetic of gases with the internal coupling within translational mechanics.

Beyond the interest to model these important systems with a Formal Graph, this study will develop the concept of *energy of coupling*, which accounts for the part of energy shared by the coupled varieties. This will allow, at least, to demonstrate the expression of the *scaling chemical potential* μ_Δ, until now identified with the product RT without any justification other than to comply with classical models. As for any demonstration of a postulate, this allows one to determine the validity domain of the concept and opens a door for introducing variants or modifications on a rigorous basis.

This study will end with a synthesis of the various energy couplings encountered in this book, which offers an instructive picture of this part of the physical world termed "macroscopic."

Table 13.1 lists the case studies of coupling given in this chapter. (See also Figure 13.1 for the position of the coupled assemblies.)

13.1 IDEAL GAS

The concept of ideal or perfect gas is among the most important ones in physical chemistry and thermodynamics. From its properties are derived a certain number of models such as the concept of ideal substance and its counterpart, the real substance. The Formal Graph representation of this system is introduced here directly, and the classical model of the ideal gas is discussed in detail in case study K2.

TABLE 13.1
List of Case Studies of Multiple Coupling

	Name	Energy Varieties	Coupling	Page Number
$\bar{\mu}$	K1: Hydrochemical Potential	Hydrodynamics Physical chemistry	Equivalence	689
$PV = nRT$	K2: Ideal Gas	Hydrodynamics Physical chemistry Thermics	External	692
	K3: Barometric Equation	Gravitation Hydrodynamics Physical chemistry Thermics	External	698
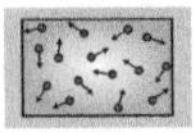	K4: Kinetic of Gases	Translational mechanics Hydrodynamics	Internal External	703
	K5: Rotating Bodies	Rotational mechanics Translational mechanics	External	711
$\rho_M \frac{\partial}{\partial t}\mathbf{v} + \rho_M(\mathbf{v}\cdot\nabla)\mathbf{v} = -\nabla P + \rho_M \mathbf{g}$	K6: Euler Equation	Gravitation Hydrodynamics	External Spatial Dipoles	715

Let us begin with the simplest Formal Graph modeling the ideal gas. This model considers two energy varieties, hydrodynamics and thermics, that are coupled through an *integral coupling factor* combining the gas constant R and the *molar volume* of the gas.

The *standard molar volume* is defined as the volume occupied by 1 mol of gas in standard state conditions ($P = 10^5$ Pa and $T = 273.15$ K) and is a constant for an ideal gas approximately equal to 22.711 08(19) $dm^3\ mol^{-1}$. The more general concept of *molar volume* V_m is used here, which is not qualified as standard and is defined as the ratio of the occupied volume on the substance amount, in any condition. For a real gas, this quantity is not a constant (i.e., depends on the substance amount).

Molar volume:
$$V_m \stackrel{def}{=} \frac{V}{n} \tag{13.1}$$

Both molar volumes are naturally equal in standard conditions of pressure and temperature. Graph 13.1 depicts the Formal Graph encoding the model of the ideal gas.

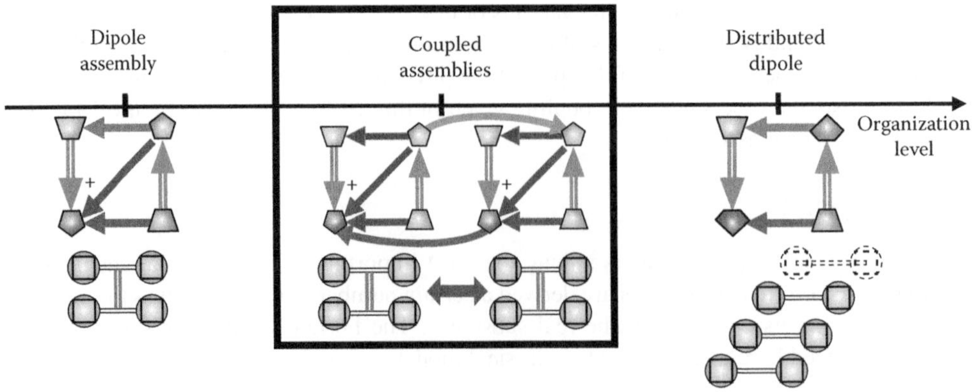

FIGURE 13.1 Position of the coupled assemblies along the complexity scale of Formal Objects.

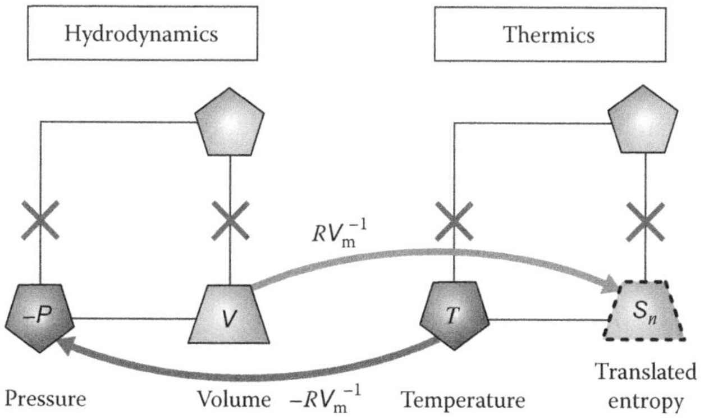

GRAPH 13.1 Simplified Formal Graph of the ideal gas using only two energy varieties. The coupling factor is the ratio of the gas constant R upon the molar volume V_m.

The algebraic translation of the coupling between efforts provides the following equation:

$$\frac{-P}{T} = -\frac{R}{V_m} \tag{13.2}$$

and the replacement of the molar volume by its definition in Equation 13.1 yields immediately the well-known state equation of the ideal gas:

Ideal gas state equation: $$PV = nRT \tag{13.3}$$

The Formal Graph contains another information about the coupling between basic quantities, using a *translated entropy* S_n and not the entropy S of the gas, but this coupling is not important to be considered for the moment.

However, such a modeling is not entirely satisfactory from the physical viewpoint because it hides the molecular nature of a gas by omitting the physical chemical (or corpuscular) energy variety. This model is not as elementary as it should be when one strives to take into account all significant energy varieties in the system. Notwithstanding this default, it corresponds to a widespread habit in thermodynamics to treat the case of 1 mol of gas instead of any amount. The next Formal Graph (Graph 13.2) improves the model by taking into consideration the three energy varieties that are required.

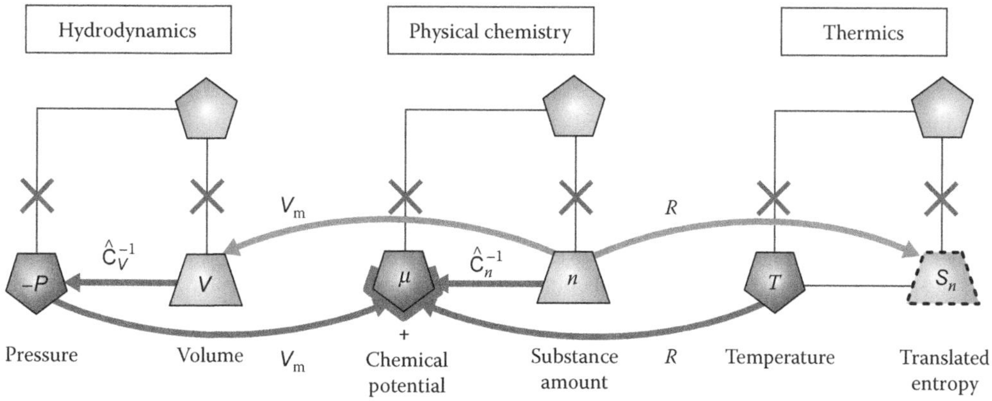

GRAPH 13.2 The full model of an ideal gas according to the requirement in the Formal Graph theory to consider all significant energy varieties.

In this graph the effort node of the physical chemical pole receives two contributions from the other efforts, without having the characteristic of an apparent effort. The node drawing, the algebraic symbol, and the name of the variable are those of an ordinary effort, the chemical potential. In other words, the apparent chemical potential is equal to the chemical potential:

Apparent chemical potential: $$\bar{\mu} = \mu \quad (13.4)$$

To obtain this equality, the sum of the two contributions must be equal to zero. These are made up of the opposite of the pressure multiplied by the *molar volume* V_m and with the temperature multiplied by the *gas constant R*.

Translated efforts: $$-V_m P + RT = 0 \quad (13.5)$$

The substitution of the molar volume by its definition in Equation 13.1 gives again the state Equation 13.3 of the ideal gas.

What defines this model is that there is a balance between the two contributions of hydrodynamics and thermics and one may suspect that this constitutes a distinctive feature of the energetic behavior of an ideal gas.

Notwithstanding the better information brought by these Formal Graphs compared to the algebraic formula, they do not provide much insight into the properties of an ideal gas.

Before tackling the ideal gas, the simpler case of a substance that contains only two energy varieties, hydrodynamical and physicochemical energy, is treated in the first case study. It allows one to present the bases of energy coupling and the various ways to link state variables in a graph.

13.2 CASE STUDIES OF MULTIPLE COUPLINGS

The first four case studies are devoted to the same physical system, a gas, with an increasing number of energy varieties participating in the system:

- K1: Hydrochemical potential
- K2: Ideal gas
- K3: Barometric equation
- K4: Kinetic of gases

The hydrodynamical and physical chemistry varieties are constitutive of every gas, as a gas is obligatorily composed of molecules possessing volume energy, and they are found in these four case studies.

In the second case study, the thermal energy is added and the conditions for having an ideal gas are discussed.

In the third case study, the gravitational energy is added, allowing the demonstration of the classical equation giving the variation of pressure with the altitude.

The fourth case study is devoted to the replacement of gravitation by translational mechanics in the system. In fact, the system is first studied with two energy varieties, hydrodynamics and translational mechanics, allowing the modeling of the equipartition of energy among the three orientations of space. Both internal and external couplings are involved for this purpose. Then, the physical chemical and thermal energy varieties are algebraically incorporated into the model through the ideal gas state equation.

The last two case studies are not directly linked to the gaseous system but they participate in the general picture of multiple couplings as enlightening examples.

- K5: Rotating bodies

The fifth case study describes the interplay between pole composition and coupling, which was used in case study K4 and now developed between translational and rotational mechanics.

- K6: Euler equation

All previous case studies were concerned with poles considered at the global level without any explicit role given to space–time (with the exception of its implicit role in the internal coupling in case study K4). This last case study deals with two space dipoles, one in gravitation and the other in hydrodynamics, for modeling the well-known Euler equation ruling the dynamic behavior of a fluid submitted to gravity.

13.2.1 Case Study K1: Hydrochemical Potential

Hydrodynamics
Physical Chemistry

$$\bar{\mu} = \mu - V'_{\mathrm{m}} P$$

Formal Objects:

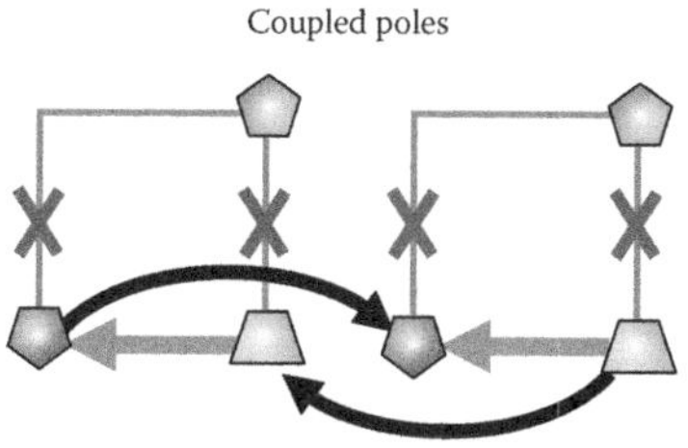

Formal Graph:

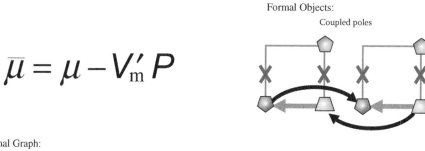

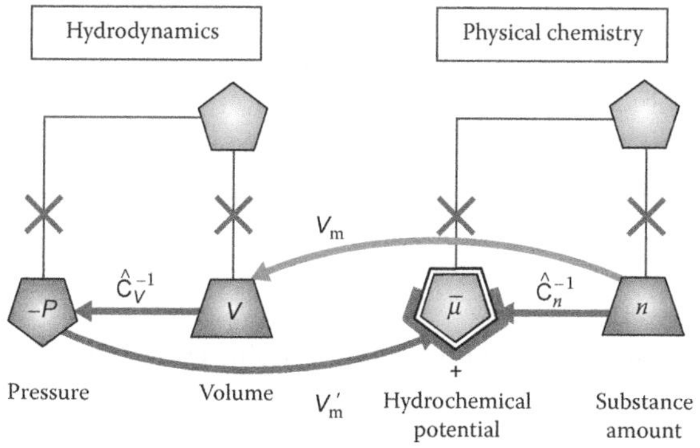

GRAPH 13.3 Formal Graph of the coupling between hydrodynamics and physical chemistry.

Coupling Factor

	Coupling	**From**	**To**
Variety		Physical chemistry	Hydrodynamics
Family	Integral factor	Basic quantity	
Name	Molar volume	Substance amount	Volume
Symbol	V_{m}	n	V
Unit	[m^3 mol^{-1}]	[mol]	[m^3]
Variety		Hydrodynamics	Physical chemistry
Family	Differential factor	Effort	
Name	Differential molar volume	Pressure	Chemical potential
Symbol	V'_{m}	P	μ
Unit	[m^3 mol^{-1}]	[Pa], [N m^{-2}]	[J mol^{-1}]

This case study deals with systems that accommodate among others varieties the hydrodynamical and physical chemical energy varieties, but without involving thermal energy. Such systems are generally made with immobilized molecules that are not subject to a significant thermal agitation but that are able to see the volume of their container vary under the effect of an external (or internal) pressure.

In thermodynamical language, the term *adiabatic*, which means "not to go through," specifies a system or a process in which no variation or exchange of heat occurs. A gas, which normally is submitted to thermal agitation, may undergo adiabatic transformation when it is thermally isolated from outside. It may behave as if no thermal energy were contained in the system, and therefore is relevant in the present case. Systems in which thermal agitation plays a significant role will be treated in case study K2 "Ideal Gas."

The variation of the total capacitive energy is given by a Gibbs* equation summing the variations of hydrodynamical energy and of physical chemical energy plus any other independent energy varieties (accounted for by a continuation mark "..."):

$$\mathrm{d}\mathcal{U} = -P\,\mathrm{d}V + \mu\,\mathrm{d}n + \cdots \qquad \text{(K1.1)}$$

The two basic quantities of these dependent energy varieties are related through the definition of the *molar volume*, playing the role of an *integral coupling factor*:

Integral coupling factor: $$V_{\mathrm{m}} \overset{def}{=} \frac{V}{n} \qquad \text{(K1.2)}$$

To this first factor is associated a second factor called *differential coupling factor*, as explained in Chapter 12:

Differential coupling factor: $$V'_{\mathrm{m}} \overset{def}{=} \frac{\mathrm{d}V}{\mathrm{d}n} \qquad \text{(K1.3)}$$

With the help of the previous equation, the Gibbs equation can be rewritten as

$$\mathrm{d}\mathcal{U} = \left(-PV'_{\mathrm{m}} + \mu\right)\mathrm{d}n + \cdots = \bar{\mu}\,\mathrm{d}n + \cdots \qquad \text{(K1.4)}$$

The grouping of the related efforts into a single variable corresponds to the use of an apparent chemical potential defined as a partial derivative:

Apparent chemical potential: $$\bar{\mu} \overset{def}{=} \left(\frac{\partial \mathcal{U}}{\partial n}\right)_{indep.\,qi} \Rightarrow \bar{\mu} = \mu - V'_{\mathrm{m}}P \qquad \text{(K1.5)}$$

In most systems, the differential molar volume is dependent on the substance amount (nonlinearity) and when it is not the case, the differential and the integral coupling factors are equal and constant and the system is said to behave ideally (hydrodynamically speaking).

Hydrodynamic ideality: $$\frac{\mathrm{d}V'_{\mathrm{m}}}{\mathrm{d}n} = 0 \Leftrightarrow V'_{\mathrm{m}} = V_{\mathrm{m}} = C^{st} \qquad \text{(K1.6)}$$

In that ideal case, the expression of the apparent chemical potential becomes

$$\bar{\mu} = \mu - V_{\mathrm{m}}P = \mu + \mu_P \qquad \text{(K1.7)}$$

* Josiah Willard Gibbs (1839–1903): American physicist and chemist; New Haven, Connecticut, USA.

The analogy with the *electrochemical potential* discussed in case study J2 in Chapter 12 leads to defining a *translated chemical potential* from hydrodynamics:

Translated chemical potential: $$\mu_P \overset{def}{\underset{ideal}{=}} -V_m P \tag{K1.8}$$

This analogy also induces the name *hydrochemical potential* for the apparent chemical potential, forged on the same pattern as the adjective electrochemical.

When the hydrodynamical and the physical chemical energy varieties are the only ones contained in the system, this ideality corresponds to an *invariable coupling*, featured by

$$\bar{\mu} = 0 \Leftrightarrow \mu = -\mu_P \underset{ideal}{=} V_m P \tag{K1.9}$$

which can be written as an invariable coupling relationship

Invariable coupling: $$V_m \overset{def}{=} \frac{V}{n} \underset{ideal}{=} -\frac{\mu}{-P} \tag{K1.10}$$

or, alternatively, as an equation of state

State equation: $$PV \underset{ideal}{=} n\mu \tag{K1.11}$$

It can be incidentally remarked that the inclusion of the thermal energy can be modeled by setting $\mu = RT$ in this equation to retrieve the state equation of an ideal gas. Notwithstanding, this is a purely heuristic remark that is not based on a rigorous approach, as considered in case study K2 "Ideal Gas."

However, an additional insight into the way the energy is shared between the two energy varieties can be given by attributing the role of representing the *coupling energy* to the common value of the two products in the above state equation (as will be seen in Section 13.3):

Coupling energy: $$\mathcal{U}_{nV} \overset{def}{=} PV \underset{ideal}{=} n\mu \tag{K1.12}$$

This energy is not the total energy of the system; it accounts for only a part, and it must be said that it is not a classical concept. This topic will be discussed in Section 13.3 in the more general case of more than two coupled energy varieties.

Adiabatic behavior. The relationship between amounts of coupling energy $\mathcal{U}_{ba}$ and of energy $\mathcal{U}_a$ of a given energy variety a can be modeled by introducing an *energetic ratio* acting as a proportionality factor. This does not imply a linear relationship because the energetic ratio may depend on the coupling energy, which is taken into account through a differential definition:

Differential energetic ratio: $$\alpha'_a \overset{def}{=} \frac{d\mathcal{U}_a}{d\mathcal{U}_{ba}} \tag{K1.13}$$

In the case of volume energy and in the present case study, this gives

$$\alpha'_V \overset{def}{=} \frac{d\mathcal{U}_V}{d\mathcal{U}_\lambda} = \frac{d\mathcal{U}_V}{dPV} \tag{K1.14}$$

As the variation of $\mathcal{U}_V$ is $-P\,dV$, one obtains the following differential equation:

$$(1+\alpha'_V)P\,dV = \alpha'_V V\,dP \tag{K1.15}$$

(continued)

which, once integrated in assuming a constant energetic ratio, gives the following adiabatic equation:

Adiabatic equation: $$PV^{\gamma} = C^{st} \quad \text{with}\ \gamma = \frac{1+\alpha'_V}{\alpha'_V} = C^{st}\ \text{(ideal system)} \qquad \text{(K1.16)}$$

This classical equation models the behavior of an ideal population of molecules when no thermal energy variation occurs, that is, in adiabatic conditions.

The significance of the *energetic ratio* will be detailed in case study K2 "Ideal Gas" and in case study K4 "Kinetic of Gases."

13.2.2 Case Study K2: Ideal Gas

Hydrodynamics
Physical Chemistry
Thermics

The case study abstract is given in next page.

The ideal gas equation, $PV = nRT$, will be established using the classical tools of thermodynamics that rely extensively on equivalences of partial derivatives obtained by maintaining at a constant some other quantities. It should be clear that this rather mathematical and cumbersome derivation is not part of the Formal Graph approach, but only a demonstration, made once and for all, establishing more interesting and powerful tools than a set of differential equations. Once these new concepts and rules are carefully defined, they can be used independently from their mathematical and thermodynamical roots, allowing a more physical understanding of models.

The starting point as usual is the writing of a Gibbs equation expressing the variation of capacitive energy as the sum of the variations of three energy varieties: hydrodynamics, physical chemistry, and thermics:

Gibbs: $$d\mathcal{U} = -P\,dV + \mu\,dn + T\,dS \qquad \text{(K2.1)}$$

To model a system containing three energy varieties, a second equation is required for linking the state variables differently. This is achieved through the variation of another form of energy called *enthalpy*, notated $\mathcal{H}$ and defined as follows:

$$\mathcal{H} \overset{def}{=} \mathcal{U} + PV \qquad \text{(K2.2)}$$

The word *enthalpy*, meaning in Greek "to put heat into," has been given by Kamerlingh Onnes* to this form of energy. It is widely used in many chemical, biological, and physical measurements because it simplifies equations modeling the energy behavior during processes involving a change of volume at constant pressure. The above operation leading to the enthalpy is known mathematically as a Legendre† transformation (see below) and is used in thermodynamics for creating several other state functions, such as the *free energy* $\mathcal{F}$ or the *free enthalpy* $\mathcal{G}$, and these forms of energy, classically called *thermodynamic potentials*, should not be confused with energy varieties.

By deriving the definition of enthalpy and replacing the variation of capacitive energy with its Gibbs expression, the following variation of enthalpy is obtained:

$$d\mathcal{H} = V\,dP + \mu\,dn + T\,dS \qquad \text{(K2.3)}$$

* Heike Kamerlingh Onnes (1853–1926): Dutch physicist; Leiden, Netherlands.
† Adrien Marie Legendre (1752–1833): French mathematician; Paris, France.

K2: Ideal Gas

Hydrodynamics
Physical Chemistry
Thermics

$PV = nRT$

Formal Objects:

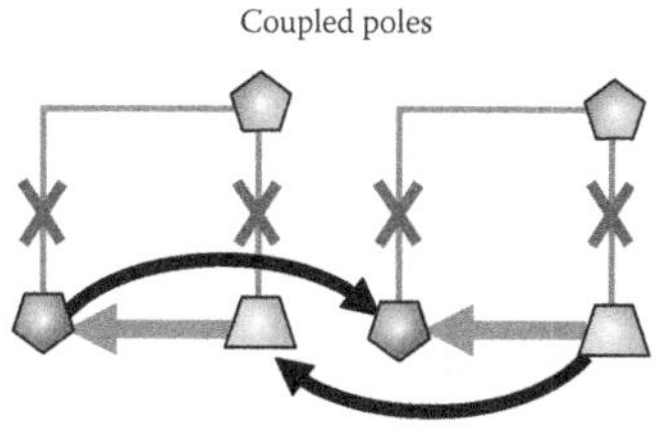

Formal Graph:

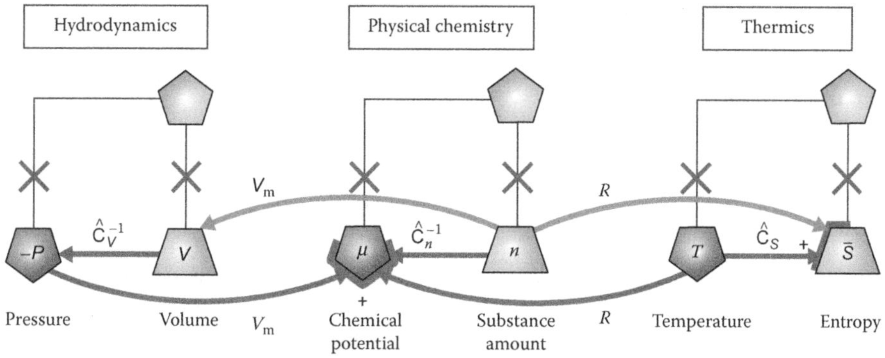

GRAPH 13.4

Coupling Factor

	Coupling	From	To
Variety		Physical chemistry	Hydrodynamics
Family	Integral factor	Basic quantity	
Name	Molar volume	Substance amount	Volume
Symbol	V_m	n	V
Unit	[m^3 mol^{-1}]	[mol]	[m^3]
Variety		Hydrodynamics	Physical chemistry
Family	Differential factor	Effort	
Name	Differential molar volume	Pressure	Chemical potential
Symbol	V_m'	P	μ
Unit	[m^3 mol^{-1}]	[Pa], [N m^{-2}]	[J mol^{-1}]

Coupling Factor

	Coupling	From	To
Variety		Physical chemistry	Thermics
Family	Integral factor	Basic quantity	
Name	Gas constant	Substance amount	Entropy
Symbol	R	n	S
Unit	[J K^{-1} mol^{-1}]	[mol]	[J K^{-1}]
Variety		Thermics	Physical chemistry
Family	Differential factor	Effort	
Name	Differential gas "constant"	Temperature	Chemical potential
Symbol	R'	T	μ
Unit	[J K^{-1} mol^{-1}]	[K]	[J mol^{-1}]

This provides the second equation sought. However, the presence of the entropy in these equations is a problem owing to the fact that a variation of entropy is not an experimentally measurable quantity, at least directly, contrary to temperature. Another Legendre transformation using the temperature/entropy couple would solve this issue (leading to the *free enthalpy* $\mathcal{G}$, also called *Gibbs free enthalpy*), but a more convenient way is possible without introducing a new energy form. It consists of writing down an *a priori* model of dependence of the energy forms on the three state variables, V, n, and T, by using unknown coefficients:

$$\mathrm{d}\mathcal{U} = -P_{T,n}\,\mathrm{d}V + \mu_{T,V}\,\mathrm{d}n + S_{V,n}\mathrm{d}T \tag{K2.4}$$

$$\mathrm{d}\mathcal{H} = V_{T,n}\,\mathrm{d}P + \mu_{T,P}\,\mathrm{d}n + S_{P,n}\mathrm{d}T \tag{K2.5}$$

Each variable playing the role of a coefficient for the variations of the state variables is defined as the partial derivative of $\mathcal{U}$ or $\mathcal{H}$ with respect to the considered state variable, by maintaining at a constant the other state variables.

Legendre transformation. The purpose of the Legendre transformation is to modify the list of independent parameters on which a function is dependent. The new parameter Y' is the derivative of the function with respect to the parameter X to be replaced and the transformed function is obtained through the following operation:

$$Y \xrightarrow{\text{Legendre}} Y - X\frac{\mathrm{d}Y}{\mathrm{d}X} \tag{K2.6}$$

Applied to the capacitive energy $\mathcal{U}$, which depends in the case of an ideal gas on the three basic quantities V, n, and S, another form of energy can be generated by replacing any of these basic quantities. When the volume V is replaced by the derivative of $\mathcal{U}$ with respect to V, which is the negative pressure $-P$ by definition:

$$-P \overset{def}{=} \left(\frac{\partial \mathcal{U}}{\partial V}\right)_{S,n} \tag{K2.7}$$

one gets the definition of the enthalpy $\mathcal{H}$:

$$\mathcal{U} \xrightarrow{\text{Legendre}} \mathcal{U} - V\left(\frac{\partial \mathcal{U}}{\partial V}\right)_{S,n} \tag{K2.8}$$

This transformation conserves all properties of energy, the most important being to be a state function and the list of independent parameters is now P, n, and S.

There are three conditions for the ideality of a gas. The first two are issued from the Joule–Thomson experiments (Planck 1922) and the third one will be enunciated shortly after. The first one states that in an ideal gas the capacitive (internal) energy is invariant with the volume at constant temperature and substance amount. This is the first Joule's condition:

First Joule's condition:

$$-P_{T,n} \overset{def}{=} \left(\frac{\partial \mathcal{U}}{\partial V}\right)_{T,n} \underset{ideal}{=} 0 \tag{K2.9}$$

The second Joule's condition, also enunciated by Joule, is the invariance of enthalpy with pressure at constant temperature and substance amount for an ideal gas:

Second Joule's condition:

$$V_{T,n} \overset{def}{=} \left(\frac{\partial \mathcal{H}}{\partial P}\right)_{T,n} \underset{ideal}{=} 0 \tag{K2.10}$$

These conditions considerably simplify the two models of variation of $\mathcal{U}$ and $\mathcal{H}$ by reducing the number of state variables to the substance amount n and the temperature T for both of them. The resemblance between the two equations offers the possibility of combining them in a single one and the interesting operation is to determine their difference for expressing the product PV as given by the enthalpy definition in Equation K2.2:

$$\mathrm{d}(PV) = \mathrm{d}(\mathcal{H} - \mathcal{U}) \underset{ideal}{=} \mu_T \, \mathrm{d}n + S_n \, \mathrm{d}T \quad \text{(K2.11)}$$

The two coefficients are the *chemical potential at constant temperature* μ_T

$$\mu_T \overset{def}{=} \mu_{T,P} - \mu_{T,V} = \left(\frac{\partial (PV)}{\partial n} \right)_T \quad \text{(K2.12)}$$

and the *entropy at constant substance amount* S_n

$$S_n \overset{def}{=} S_{P,n} - S_{V,n} = \left(\frac{\partial (PV)}{\partial T} \right)_n \quad \text{(K2.13)}$$

This product PV has the status of a state function (it corresponds to a peculiar energy as will be discussed later) and therefore its variation must be given by an exact differential equation, thus requiring equality of the second derivatives (as explained earlier in Section 2.1.5 of Chapter 2). This defines their common value which is a second derivative of energy notated R'' and has the unit J mol^{-1} K^{-1}:

$$R'' \overset{def}{=} \left(\frac{\partial \mu_T}{\partial T} \right)_n \overset{def}{=} \left(\frac{\partial S_n}{\partial n} \right)_T \quad \text{(K2.14)}$$

It is here that the third condition intervenes to make a gas satisfying these conditions an ideal gas, which is the invariance of this common value with the temperature and the substance amount:

Third condition: $$\frac{\partial R''}{\partial n} \underset{ideal}{=} 0; \quad \frac{\partial R''}{\partial T} \underset{ideal}{=} 0 \quad \text{(K2.15)}$$

In this condition, this second derivative of energy receives the name *gas constant* with the symbol R (sometimes called Regnault* constant):

Gas constant: $$R \overset{def}{=} R'' \underset{ideal}{=} C^{st} \quad \text{(K2.16)}$$

Graph 13.5 depicts the links between the variables resulting from the status of the state function of the PV product for an ideal gas under the form of a differential Formal Graph, which has been presented in Chapter 2.

* Henri Victor Regnault (1810–1878): French physical chemist; Paris, France.

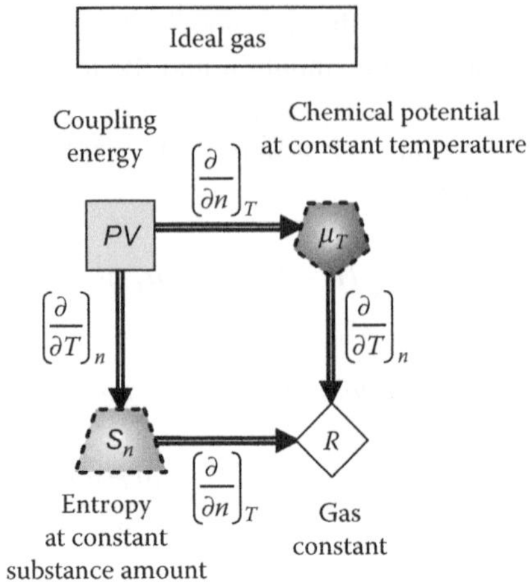

GRAPH 13.5 Differential Formal Graph of the state function status of the product of the pressure times the volume in the case of an ideal gas.

Heat capacities. In the demonstration of the ideal gas equation, the variations of the capacitive (internal) energy $\mathcal{U}$ and the enthalpy $\mathcal{H}$ have been merged into a single equation describing the variation of the PV product. If these differential equations had been kept separate and the same procedure was followed for defining the gas constant as the constant common value of two second derivatives, one would have found the definitions of the *molar heat capacities* (also called *specific heats*). Depending on the energy form, there are two heat capacities, at constant volume and at constant pressure:

$$C_V'' \overset{def}{=} \left(\frac{\partial \mu_{V,T}}{\partial T}\right)_{V,n} \overset{def}{=} \left(\frac{\partial S_{V,n}}{\partial n}\right)_{T,V} \underset{ideal}{=} C_V = C^{st} \tag{K2.17}$$

$$C_P'' \overset{def}{=} \left(\frac{\partial \mu_{P,T}}{\partial T}\right)_{P,n} \overset{def}{=} \left(\frac{\partial S_{P,n}}{\partial n}\right)_{T,P} \underset{ideal}{=} C_P = C^{st} \tag{K2.18}$$

The case of the ideal gas corresponds to the constancy of these *molar heat capacities*, allowing one to drop the double prime in their symbols. The adjective *molar* should not be omitted; otherwise it can be confused with the entropies at constant substance amount and constant volume $S_{V,n}$ or pressure $S_{P,n}$ (which occurs regularly in classical thermodynamics when one treats 1 mol of gas!).

This constancy of the *molar heat capacities* leads to simple expressions for the partial derivatives with respect to temperature:

$$S_{V,n} \underset{ideal}{=} nC_V; \quad S_{P,n} \underset{ideal}{=} nC_P \tag{K2.19}$$

The classical expressions of the energy variations are deduced from the differential Equations K2.4 and K2.5 by assuming a constant substance amount

$$d\mathcal{U} \underset{ideal}{\overset{n=C^{st}}{=}} n\,C_V dT; \quad d\mathcal{H} \underset{ideal}{\overset{n=C^{st}}{=}} n\,C_P dT \tag{K2.20}$$

(continued)

From the definition in Equation K2.12 of the chemical potential at constant temperature as a difference between the two partial chemical potentials used in the previous definitions of the molar heat capacities, their relationship with the gas constant ensues

Robert Mayer relationship: $$R \underset{ideal}{=} C_P - C_V \tag{K2.21}$$

This relationship was proposed by Robert Mayer* (Bruhat 1962).

The constancy of the second derivative of the PV product has interesting consequences with regard to the first derivatives, which become proportional to the temperature or to the substance amount through the gas constant as follows:

$$\mu_T \underset{ideal}{=} RT; \quad S_n \underset{ideal}{=} Rn \tag{K2.22}$$

Introducing these expressions into the variation in Equation K2.11 of the PV product simplifies the differential equation

$$\mathrm{d}(PV) \underset{ideal}{=} R(T\,\mathrm{d}n + n\,\mathrm{d}T) = R\,\mathrm{d}(Tn) \tag{K2.23}$$

which is easily integrated to give the ideal gas equation

Ideal gas state equation (molar): $$PV \underset{ideal}{=} RTn \tag{K2.24}$$

If one writes this well-known equation under an unusual order for the right member, it is to outline the peculiarity of the classical demonstration that first establishes the model for 1 mol of substance and then asserts that it should be valid for any substance amount in following a simple homothecy. This confers to the substance amount a simple role of coefficient (justifying its position in first place) thus ignoring its status of state variable, and by the way depriving the physical chemical energy variety from any role in the concept of ideal gas.

Naturally, the state equation can be expressed in corpuscular terms as

Ideal gas state equation (corpuscular): $$PV \underset{ideal}{=} N k_B T \tag{K2.25}$$

which is perfectly equivalent to the molar state Equation K2.24.

Heat capacity ratios. In a remark in case study K1 "Hydrochemical Potential," the notion of *differential energetic ratio* relating energy with coupling energy has been introduced, which in the case of coupled volume energy is defined as follows:

$$\alpha'_V \overset{def}{=} \frac{\mathrm{d}\mathcal{U}_V}{\mathrm{d}\mathcal{U}_{nV}} = \frac{\mathrm{d}\mathcal{U}_V}{\mathrm{d}PV} \tag{K1.14}$$

By replacing the product PV of an ideal gas by its expression (see Equation K2.23) as a function of temperature, one finds

$$\mathrm{d}\mathcal{U}_V = \alpha'_V R\,\mathrm{d}(Tn) \tag{K2.26}$$

* Julius Robert von Mayer (1814–1878): German physician and physicist; Tübingen, Germany.

Now, by comparing this energy variation with the variation of capacitive energy given in Equation K2.20 established by assuming a constant n, the *differential energetic ratio* can be identified with ratios between heat capacities and gas constant [the Mayer relationship (Equation K2.21) has been used for the second equality]:

$$d\mathcal{U}_V = d\mathcal{U} \Rightarrow \qquad \alpha'_V \overset{n=C^{st}}{\underset{ideal}{=}} \frac{C_V}{R}; \quad 1+\alpha'_V \overset{n=C^{st}}{\underset{ideal}{=}} \frac{C_P}{R} \tag{K2.27}$$

The exponent γ of the volume in the adiabatic equation in the mentioned remark, which was arbitrarily introduced for convenience, now receives a more physical meaning of ratio between heat capacities:

$$\text{Heat capacity ratio:} \qquad \gamma = \frac{1+\alpha'_V}{\alpha'_V} = \frac{C_P}{C_V} \tag{K2.28}$$

The interest of all these ratios is to relate coupling energies to more experimentally accessible variables that are the heat capacities. Furthermore, in case study K4 "Kinetic of Gases," it will be shown that these ratios can be expressed as a function of kinetic properties of gases that are independent from the intrinsic properties of the molecules (except their symmetry).

13.2.3 Case Study K3: Barometric Equation

Gravitation
Hydrodynamics
Physical Chemistry
Thermics

The case study abstract is given in next page.

A column of gas set vertically and submitted to gravity involves a minimum of four energy varieties: gravitational, hydrodynamical, physical chemical, and thermal. A real gas would involve in addition one or several other energy varieties (mechanical, superficial, etc.). The *barometric equation* gives the pressure of gas as a function of the height of the column. By taking into account simplifying hypotheses of a constant gravitational field, a constant temperature, and a behavior of perfect gas, this equation is expressed as a simple exponential function:

$$P = P^0 \exp\left(-\frac{M g}{RT} z\right) \tag{K3.1}$$

P^0 is the pressure at the bottom (ground) of the column (altitude $z = 0$), the other symbols being described hereafter.

Three couplings are found in the Formal Graph shown in the case study abstract:

1. The gravitation–physical chemistry coupling utilizes the *molar mass M* as an *integral coupling factor* and this coupling is algebraically translated into the two relations:

$$M_G = Mn$$
$$\mu = -MV_G \tag{K3.2}$$

K3: Barometric Equation

Gravitation
Hydrodynamics
Physical Chemistry
Thermics

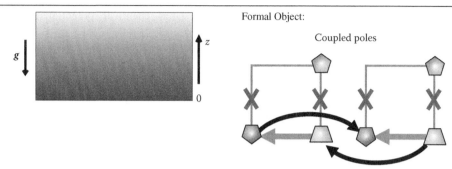

Formal Object:

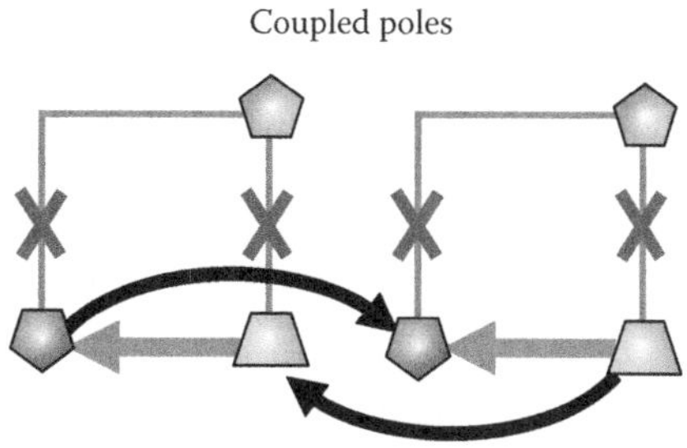

Formal Graph:

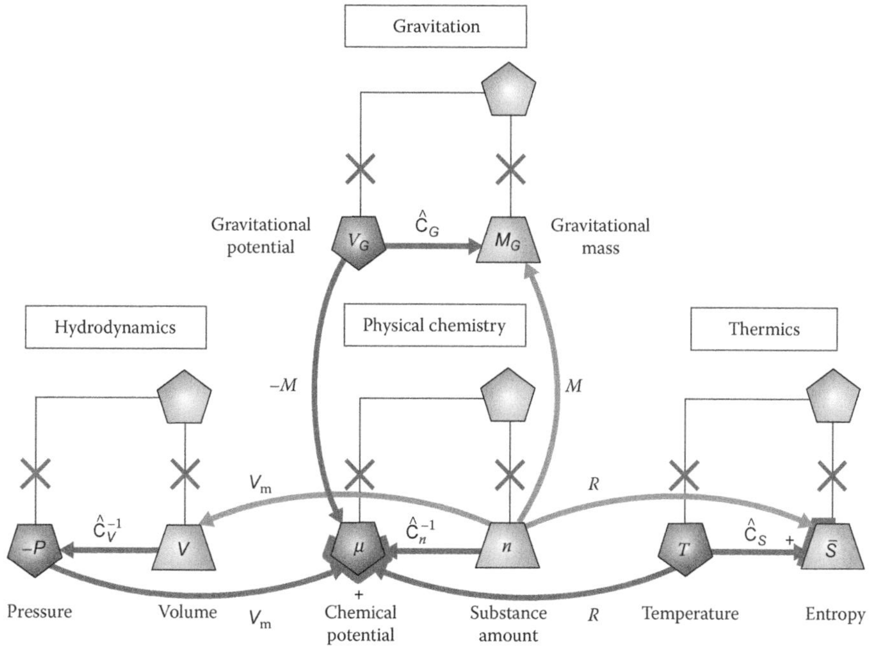

GRAPH 13.6

Coupling Factor (Integral)

	Coupling	From	To
Variety		Physical chemistry	Hydrodynamics
Family	Integral factor	Basic quantity	
Name	Molar volume	Substance amount	Volume
Symbol	V_m	n	V
Unit	[m^3 mol^{-1}]	[mol]	[m^3]

Coupling Factor (Integral)

	Coupling	From	To
Variety		Physical chemistry	Gravitation
Family	Integral factor	Basic quantity	
Name	Molar mass	Substance amount	Gravitational mass
Symbol	M	n	M_G
Unit	[kg mol^{-1}]	[mol]	[kg]

having notated V_G the gravitational potential, which is the effort in the domain of gravitation. (See case study D5 "Newtonian Gravitation" in Chapter 7.) As the molar mass is a constant, the same integral coupling factor is used between the two efforts.

2. The physical chemical–hydrodynamic coupling uses the *molar volume* as a coupling factor, supposed to be a constant, and a translated chemical potential from hydrodynamics:

$$\begin{aligned} V &= V_{\text{m}}\, n \\ \mu_P &\underset{ideal}{\overset{def}{=}} -V_{\text{m}} P \end{aligned} \tag{K3.3}$$

3. The physical chemical–thermal coupling uses the *gas constant* as a coupling factor, which means that the gas is ideal, and a translated chemical potential from thermics:

$$\begin{aligned} S_n &= Rn \\ \mu_T &\underset{ideal}{\overset{def}{=}} RT \end{aligned} \tag{K3.4}$$

These last two couplings act in the same manner as for an ideal gas, that is, by having opposite and balanced effects on the chemical potential.

Opposite translated efforts: $$\mu_P + \mu_T = 0 \tag{K3.5}$$

which consequently makes the apparent chemical potential and the chemical potential equal:

Apparent chemical potential: $$\bar{\mu} = \mu + \mu_P + \mu_T = \mu \tag{K3.6}$$

From these various couplings, the expression of the pressure as a function of the gravitational mass is derived:

$$P = \frac{RT}{V_{\text{m}}} = \frac{RT}{V} n = \frac{RT}{VM} M_G \tag{K3.7}$$

One now turns attention to the capacitive relationships in physical chemistry and in gravitation, which have the same exponential shape as it is a postulate of the Formal Graph theory.

The physical chemical capacitive relationship is (see Chapter 7)

$$n = \hat{C}_n \mu \Leftrightarrow n = n^0 \exp\frac{\mu}{\mu_\Delta} \tag{K3.8}$$

As demonstrated in Section 13.4.3, the *scaling chemical potential* is given by

$$\mu_\Delta = RT \tag{K3.9}$$

The gravitational capacitive relationship is

$$M_G = \hat{C}_G V_G \Leftrightarrow M_G = M_G^0 \exp\left(\frac{V_G}{V_{G\Delta}}\right) \tag{K3.10}$$

with a *scaling gravitational potential* $V_{G\Delta}$ related to the *scaling chemical potential* in exactly the same manner as the efforts in Equation K3.2

$$\mu_\Delta = -MV_{G\Delta} \tag{K3.11}$$

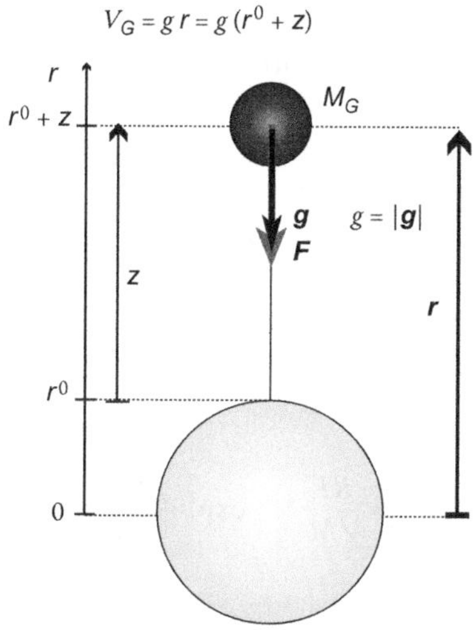

In the gravitational energy variety, the spatial distribution of gravitational potential is expressed by means of the *gravitational field* **g** (its modulus g is generally called *acceleration of gravity*), defined as the contragradient of the gravitational potential, with respect to the space coordinate that is oriented upward, which is the vector $\boldsymbol{r}$ representing the distance between the two bodies (mass centers) that are mutually attracted due to gravitation:

$$\mathbf{g} \overset{def}{=} -\frac{\mathrm{d}V_G}{\mathbf{dr}} \tag{K3.12}$$

This relation expresses the fact that, as the vector of the gravitational field is oriented downward, the gravitational potential grows with altitude. When the gravitation is assumed to obey the Newtonian model, which relies on a *linear medium* (constant gravitational permittivities), it means that the potential varies according to $1/r$ (by notating r the modulus of the vector $\boldsymbol{r}$). Concomitantly, the *acceleration of gravity* defined as the modulus g of the gravitational field varies in $1/r^2$ (see case study D5 "Newtonian Gravitation" in Chapter 7). The gravitational potential is then expressed as the product of the *gravity acceleration* g by the modulus r of distance:

$$V_G \underset{lin}{=} g(r)\, r \tag{K3.13}$$

Contrary to what is often asserted, the *gravity acceleration* does not need to be a constant to allow one to use this relationship. However, it may be the case, as integration of Equation K3.12 provides the same relation as in Equation K3.13, but the dependence on the inverse of the squared distance also works well. The gravitational capacitive relation in Equation K3.10 is then written as

$$M_G \underset{lin}{=} M_G^0 \exp\left(-\frac{M\,g\left(r^0+z\right)}{RT}\left(r^0+z\right)\right) \qquad \text{(K3.14)}$$

In this formula, the distance r between the mass centers has been replaced by the sum of the radius r^0 of the reference body (earth) and of the altitude z. The altitude origin ($z = 0$) corresponds to the surface of this body. Equation K3.14 is therefore a general model of a column of material in a Newtonian (linear medium) gravity field. In order to retrieve the barometric Equation K3.1, a certain number of approximations must be made. The first one is to assume a constant gravity field, thus allowing the writing of the gravitational capacitive relationship in Equation K3.14 as

$$M_G \overset{g=C^{st}}{\underset{lin}{=}} M_G^0 \exp\left(-\frac{M\,g}{RT}r^0\right)\exp\left(-\frac{M\,g}{RT}z\right) \qquad \text{(K3.15)}$$

The insertion of this expression into Equation K3.7 leads directly to the barometric Equation K3.1 sought, by setting for the ground pressure

$$P^0 = \frac{RT}{VM} M_G^0 \exp\left(-\frac{M\,g}{RT}r^0\right) \qquad \text{(K3.16)}$$

The second approximation is that this quantity must be a constant, which implies that the temperature is a constant (which is far from reality in the earth's atmosphere) and that the pressure is determined within layers of constant volume V.

Coupling gravitation and hydrodynamics. In this system, the choice to couple the gravitational and physical chemistry energy varieties is not compulsory. Any other combination among the four energy varieties can be used. For instance, the gravitation can be coupled with hydrodynamics according to the following coupling relationship:

Variable coupling:

$$\rho_{MG} = \frac{M_G}{V} = \frac{-P_{MG}}{V_G} \qquad \text{(K3.17)}$$

in which the *gravitational volumic mass* ρ_{MG} plays the role of an integral coupling factor. As this factor is variable, the coupling is a variable one and it has been necessary in the previous relation to use a *translated pressure* from gravitation, defined as

$$P_{MG} \overset{def}{=} -\rho_{MG} V_G = V_m^{-1}\mu \qquad \text{(K3.18)}$$

(The gravitational volumic mass ρ_{MG} is the ratio of the molar mass M on the molar volume V_m.)

13.2.4 Case Study K4: Kinetic of Gases

Translational Mechanics
Hydrodynamics

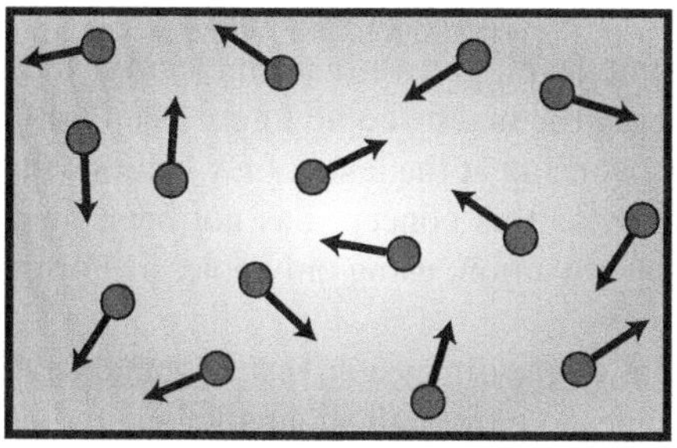

Formal Object:

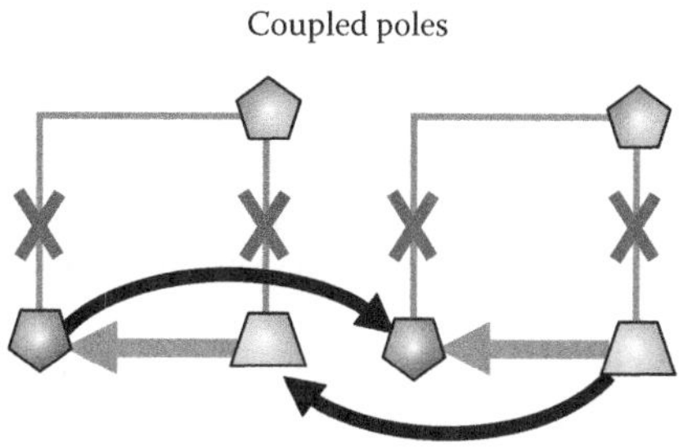

Formal Graph:

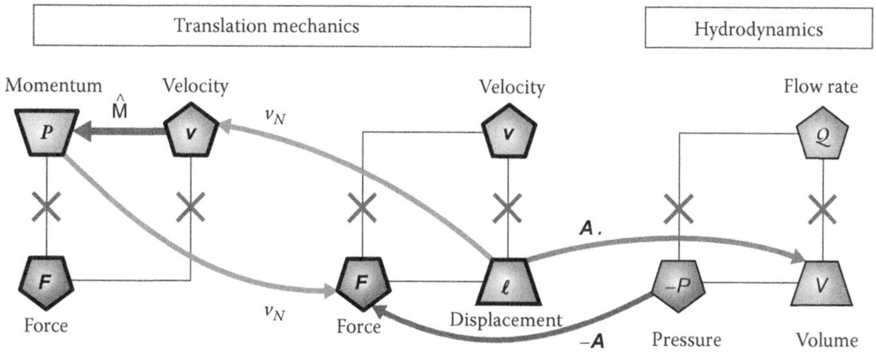

GRAPH 13.7 Formal Graph of a gas in one dimension (in a tube) using the internal coupling within translational mechanics and the external coupling with hydrodynamics.

Coupling Factor

	Coupling	From	To
Variety		Translation mechanics	Hydrodynamics
Family	Integral factor	Basic quantity	
Name	Area	Displacement	Volume
Symbol	$\boldsymbol{A}$	ℓ	V
Unit	[m²]	[m]	[m³]
Variety		Hydrodynamics	Translation mechanics
Family	Differential factor	Effort	
Name	Differential area	Pressure	Force
Symbol	$\boldsymbol{A}'$	P	$\boldsymbol{F}$
Unit	[m²]	[Pa], [N m⁻²]	[N]

Coupling Factor

	Coupling	From	To
Variety		Translation mechanics	
Family	Internal	Basic quantity	Flow
Name	Coupling frequency	Displacement	Velocity
Symbol	ν_N	ℓ	$\boldsymbol{v}$
Unit	[Hz]	[m]	[m s⁻¹]
Variety		Translation mechanics	
Family	Internal	Impulse	Effort
Name	Coupling frequency	Momentum	Force
Symbol	ν_N	P	$\boldsymbol{F}$
Unit	[Hz]	[N s]	[N]

The kinetic theory of gases is a classical model of relationship between the kinetic energy of the molecules composing a gas and its hydrodynamical energy (volume energy). This model is based on the interaction between a wall and molecules colliding elastically with it (without dissipation). During a collision, a molecule exerts a force on the wall and this force is translated into a pressure within the container of the gas.

Traditionally, the collision is treated at the microscopic level, generating a molecular force that is multiplied by the number of molecules for producing the macroscopic force. The Formal Graph theory allows such modeling with Formal Objects working at the level of few entities that are the *singletons* (cf. Chapter 4, Section 4.4.4). However, as this concept has not been much developed in this book, this system is modeled in terms of collection, using only poles as Formal Objects.

The Formal Graph in the case study abstract models a gas in one dimension, that is, contained in a linear tube closed at its two ends and allowing the molecules to move only along the longitudinal axis (no transverse movement). The case of a three-dimensional gas will be tackled shortly after. In order to focus on the main points, not all energy varieties present in a gas are taken into consideration, the thermal and corpuscular (or physical chemical) energy varieties being neglected for the moment. In fact, they do not impact on the core model of gas kinetics and it is sufficient to mention how they can be reintroduced for achieving a complete model.

Let us begin by analyzing the first coupling on the left of the Formal Graph in the case study abstract. It relates an inductive (kinetic) pole to a capacitive pole in the same energy variety, translational mechanics, which means that the coupling is *internal* to this variety. The coupling factor is a scalar, having the unit of a reciprocal of time, the hertz, called the *coupling frequency*, notated ν_N. (The subscript N refers to the property of this factor to be related to the number of corpuscles, but this is not important for the present discussion.) The internal coupling relationships are written as equality between ratios of state variables:

Internal coupling:
$$\nu_N = \frac{\boldsymbol{v}}{\boldsymbol{\ell}} = \frac{\boldsymbol{F}}{\boldsymbol{P}} \tag{K4.1}$$

The inductive (kinetic) pole is coupled through two state variables, the quantity of movement (momentum) $\boldsymbol{P}$ of the inductive pole (collection of corpuscles) and the velocity $\boldsymbol{v}$ of the pole (which can be viewed as an average of the velocities of the corpuscles). The two state variables participating in the internal coupling are the force $\boldsymbol{F}$ acting on the tube ends (i.e., the walls of the container) and the length $\boldsymbol{\ell}$ of the tube.

The reciprocal of the *coupling frequency* can be roughly viewed as corresponding to the time interval that would be required, if a conversion of energy were possible, for a molecule with velocity $\boldsymbol{v}$ to cross the tube from end to end. Notwithstanding, this image is not really adequate because there is no time in a pole and this system is in an equilibrium state. (It can also be remarked that the number of molecules able to cross the entire tube without colliding with other molecules must not be very high.) This frequency corresponds to the number per unit of time, always by admitting that time can be a pertinent notion here, of collisions against the walls that convert the momentum into a force.

Whatever the exact definition of this *coupling frequency*, the net result of the internal coupling is the equality between the energy-per-entity of the products times entity number of each energy subvariety (for the sake of generality, dot products are used, although these state variables are collinear vectors):

Internal coupling:
$$\boldsymbol{v} \cdot \boldsymbol{P} = \boldsymbol{F} \cdot \boldsymbol{\ell} \tag{K4.2}$$

Microscopic viewpoint. Instead of using the concept of internal coupling, the classical reasoning for establishing equality K4.2 uses a time interval Δt during which a shift of particle momentum $\boldsymbol{p}$ is translated into a particle force $\boldsymbol{F}/N$:

$$\frac{\boldsymbol{F}}{N} = \frac{\Delta \boldsymbol{p}}{\Delta t} \quad \text{with } \Delta \boldsymbol{p} = \boldsymbol{p}_{initial} - \boldsymbol{p}_{final} = 2\boldsymbol{p} = 2\frac{\boldsymbol{P}}{N} \quad \left(\underset{lin}{=} m\boldsymbol{v} - m(-\boldsymbol{v})\right) \tag{K4.3}$$

This shift is provoked by the collision of a molecule changing its velocity from $\boldsymbol{v}$ to $-\boldsymbol{v}$ due to the elastic bouncing against the wall. The time interval is determined by the frequency of collisions, related to the time required for traveling twice (forth and back) the length ℓ:

$$\Delta t = \frac{2\ell}{\boldsymbol{v}} \tag{K4.4}$$

It works, but this time interval is defined at the macroscopic level and the travel from one wall to the other (and conversely) takes much more time than the duration of one collision. So the physical meaning of the relationship in Equation K4.3 between momentum and force is not really more obvious than the concept of internal coupling discussed here.

Analysis of the Formal Graph is pursued with the coupling on the right which is an *external* coupling between translational mechanics and hydrodynamics. This coupling has already been presented in Chapter 12 with the case study J5 describing a piston, which is exactly the way the actual system works. The coupling factor is the *cross-sectional area* $\boldsymbol{A}$ of the tube, which is a constant vector, so the external coupling relationships are written as

External coupling: $$\boldsymbol{F} = \boldsymbol{A}P; \quad V = \boldsymbol{A}\cdot\boldsymbol{\ell} \tag{K4.5}$$

As the cross-sectional area is a vector, the *volume* of the tube is obtained by a dot product with the tube *length*, and the *pressure*, which is a scalar, is multiplied by this vector in order to obtain the *force* (in the Formal Graph, the pressure is counted negatively for complying with all other energy varieties and the coupling factor between efforts is also negative). By replacing the area in the expression of the volume by the ratio $\boldsymbol{F}/P$, one gets

External coupling: $$PV = \boldsymbol{F}\cdot\boldsymbol{\ell} \tag{K4.6}$$

It remains to assemble the two equalities (Equations K4.2 and K4.6) featuring the couplings for obtaining the relationship between kinetics and hydrodynamics:

Kinetics–hydrodynamics coupling: $$\boldsymbol{v}\cdot\boldsymbol{P} = PV \tag{K4.7}$$

In the classical theory, a model is developed in the frame of linear constitutive properties that allows translating this relationship in terms of energies. When the inertial mass is assumed to be a constant, the kinetic energy is given by

Kinetic energy: $$\mathcal{T}_\ell \underset{lin}{=} \frac{1}{2}\boldsymbol{v}\cdot\boldsymbol{P} \underset{lin}{=} \frac{1}{2}M\boldsymbol{v}^2 \tag{K4.8}$$

recalling that the inertial mass M of the collection is the particle number N times the inertial mass m of one particle (cf. Chapter 3, Section 3.1 for the notation)

$$M = Nm \tag{K4.9}$$

In addition, when the thermal and corpuscular energies are taken into account, the state equation of an ideal gas applies to this system (see case study K2 "Ideal Gas"):

Ideal gas state equation (corpuscular): $$PV \underset{ideal}{=} N k_B T \tag{K4.10}$$

By merging the last four equations, the average kinetic energy per molecule is obtained:

Corpuscular kinetic energy (one dimension): $$\frac{\mathcal{T}_\ell}{N} \underset{lin}{=} \frac{1}{2} m\mathbf{v}^2 \underset{ideal}{=} \frac{1}{2} k_B T \tag{K4.11}$$

It must be recalled that this result is for a one-dimensional system such as the tube filled with an ideal gas described above.

The case of a three-dimensional system is represented in the Formal Graph shown in Graph 13.8.

In a three-dimensional space, the simplest choice is to take a parallelepiped as container (not necessarily cubic) for the gas and then the choice of a Cartesian* system of coordinates using three orthogonal axes x, y, and z is adequate for decomposing the system in three one-dimensional subsystems. This number of subsystems corresponds to molecules having a spherical symmetry, that is, no privileged orientation in space (monoatomic molecules). If it is not the case, more subsystems are required depending on the symmetry of the molecule (e.g., five subsystems for diatomic molecules), but this does not change the reasoning made on three subsystems.

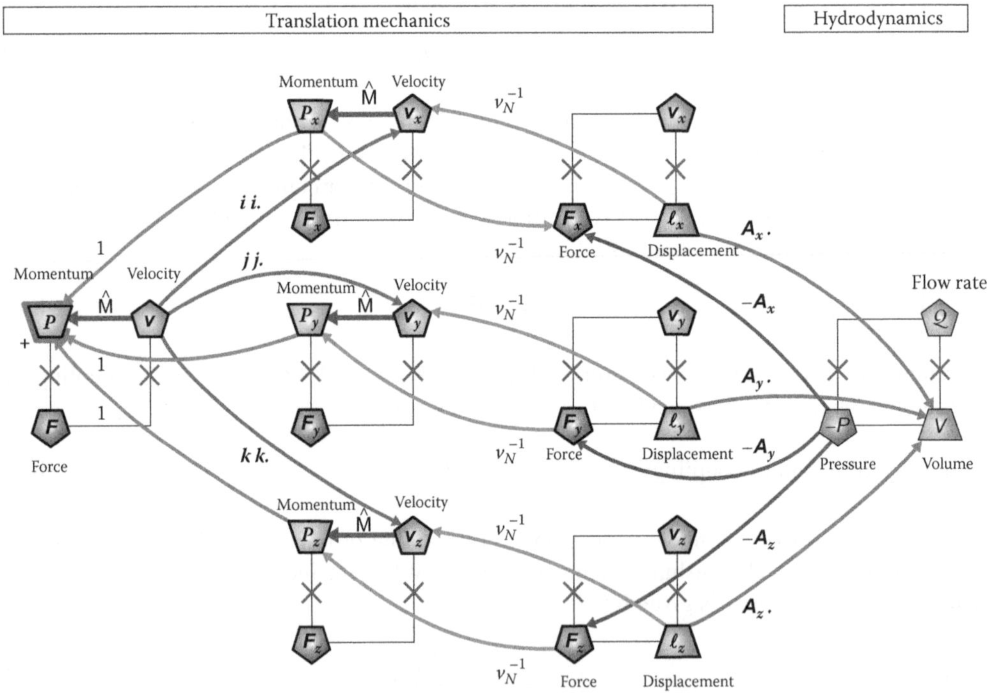

GRAPH 13.8 Formal Graph of the various couplings between translational mechanics and hydrodynamics modeling a gas with 3 degrees of freedom (one for each direction of space).

* René Descartes (1596–1650): French philosopher, mathematician, scientist, and writer; Paris, France and the Netherlands.

The Formal Graph uses therefore three pairs of individual poles for translational mechanics, each pair being made up of an inductive (kinetic) and a capacitive pole. The three inductive poles form a *composed pole* (cf. Chapter 8) featured by the momentum $\boldsymbol{P}$ and the velocity $\boldsymbol{v}$ of the gas. The coding used for assembling the composed pole relies on the algebraic decomposition of a vector into components along each axis. By using three unit vectors $\boldsymbol{i}, \boldsymbol{j}$, and $\boldsymbol{k}$, one along each axis, these two vectors can be decomposed into component vectors or component moduli:

$$\boldsymbol{P} = \boldsymbol{P}_x + \boldsymbol{P}_y + \boldsymbol{P}_z = \boldsymbol{i}P_x + \boldsymbol{j}P_y + \boldsymbol{k}P_z \quad \text{(K4.12)}$$

$$\boldsymbol{v} = \boldsymbol{v}_x + \boldsymbol{v}_y + \boldsymbol{v}_z = \boldsymbol{i}v_x + \boldsymbol{j}v_y + \boldsymbol{k}v_z \quad \text{(K4.13)}$$

Each component modulus is given by the dot product of the corresponding unit vector and the considered vector, which is written for the first components, for instance, as

$$P_x = |\boldsymbol{P}_x| = \boldsymbol{i} \cdot \boldsymbol{P}; \quad v_x = |\boldsymbol{v}_x| = \boldsymbol{i} \cdot \boldsymbol{v} \quad \text{(K4.14)}$$

The rule used in these equations results from the orthogonality of the unit vectors, meaning that the dot products of identical vector units is equal to one (identity) and equal to zero otherwise, as shown below for the case of the first unit vector:

$$\boldsymbol{i} \cdot \boldsymbol{i} = 1; \quad \boldsymbol{i} \cdot \boldsymbol{j} = 0; \quad \boldsymbol{i} \cdot \boldsymbol{k} = 0 \quad \text{(K4.15)}$$

In the Formal Graph, the relationship between the momenta directly follows this decomposition in summing the three components on the gas momentum node. In contradistinction, the relationship between velocities is coded in the reverse way as three operations transforming the vector $\boldsymbol{v}$ into its components. Each transformation consists of a double operation: the first one is the dot product between the vector and the unit vector corresponding to the component, say $\boldsymbol{i}$ for the component along the x-axis, and the second one is the ordinary multiplication by the same unit vector $\boldsymbol{i}$:

$$\boldsymbol{v}_x = \boldsymbol{i}v_x = \boldsymbol{i}(\boldsymbol{i} \cdot \boldsymbol{v}) = \boldsymbol{i}\,\boldsymbol{i} \cdot \boldsymbol{v} \quad \text{(K4.16)}$$

This achieves the algebraic description of the association between inductive subsystems that constitute the translational mechanical composed pole.

The internal couplings between the two translational mechanical subvarieties are ruled by the coupling relationships already seen for the one-dimensional gas:

Internal coupling: $$v_N = \frac{\boldsymbol{v}_w}{\boldsymbol{\ell}_w} = \frac{\boldsymbol{F}_w}{\boldsymbol{P}_w} \quad (w = x, y, z) \quad \text{(K4.17)}$$

Whether the coupling frequency is identical or different for each subsystem has no consequence on the following equalities between dot products:

Internal coupling: $$\boldsymbol{v}_w \cdot \boldsymbol{P}_w = \boldsymbol{F}_w \cdot \boldsymbol{\ell}_w \quad (w = x, y, z) \quad \text{(K4.18)}$$

However, there is no reason for having differences as the gas is the same whatever the direction of space (spherical shape of molecules and same number of molecules notably). Let us now turn our attention to the external coupling between the two capacitive subvarieties in translational mechanics and in hydrodynamics.

Owing to the symmetry of the parallelepiped chosen for containing the gas, the volume and the pressure are common to all three subsystems and they are given by coupling relationships using the cross-sectional areas as coupling factors:

$$V = \boldsymbol{A}_x \cdot \ell_x = \boldsymbol{A}_y \cdot \ell_y = \boldsymbol{A}_z \cdot \ell_z \tag{K4.19}$$

$$-P = \frac{\boldsymbol{F}_x}{-\boldsymbol{A}_x} = \frac{\boldsymbol{F}_y}{-\boldsymbol{A}_y} = \frac{\boldsymbol{F}_z}{-\boldsymbol{A}_z} \tag{K4.20}$$

By eliminating the cross-sectional areas between these last equations, the product PV appears as equivalently expressed by one of the dot products *force* · *length*:

$$PV = \boldsymbol{F}_x \cdot \ell_x = \boldsymbol{F}_y \cdot \ell_y = \boldsymbol{F}_z \cdot \ell_z \tag{K4.21}$$

The dot product of the *velocity* · *momentum* for the overall system is derived from the two component sums in Equations K4.12 and K4.13 and coupling equalities in Equation K4.18:

$$\boldsymbol{v} \cdot \boldsymbol{P} = \boldsymbol{v}_x \cdot \boldsymbol{P}_x + \boldsymbol{v}_y \cdot \boldsymbol{P}_y + \boldsymbol{v}_z \cdot \boldsymbol{P}_z = \boldsymbol{F}_x \cdot \ell_x + \boldsymbol{F}_y \cdot \ell_y + \boldsymbol{F}_z \cdot \ell_z \tag{K4.22}$$

As each dot product *force.length* is equal to the PV product by virtue of the symmetry of the container, as given by Equation K4.21, the previous expression of the dot product *velocity* · *momentum* becomes

$$\boldsymbol{v} \cdot \boldsymbol{P} = 3\,PV \tag{K4.23}$$

In the frame of linear constitutive properties for the translational mechanical energy variety, the kinetic energy is one half the dot product $\boldsymbol{v} \cdot \boldsymbol{P}$, as in the previous one-dimensional case, thus giving the coefficient 3/2 for the PV product:

Kinetic energy (three dimensions):
$$\mathcal{T}_\ell \underset{lin}{=} \frac{1}{2} M \boldsymbol{v}^2 \underset{lin}{=} \frac{1}{2} \boldsymbol{v} \cdot \boldsymbol{P} = \frac{3}{2} PV \tag{K4.24}$$

Again, when the gas is assumed to be ideal, using the state Equation K4.10 for substituting the PV product provides the kinetic energy per corpuscle as a function of temperature:

Corpuscular kinetic energy (three dimensions):
$$\frac{\mathcal{T}_\ell}{N} \underset{lin}{=} \frac{1}{2} m \boldsymbol{v}^2 \underset{ideal}{=} \frac{3}{2} k_B T \tag{K4.25}$$

The fact that one obtains exactly three times the thermal energy of a one-dimensional system is known as the *equipartition theorem*, expressing the fact that energy is equally distributed in all three subsystems composing this ideal gas.

Generalization. The numerator and the denominator of the 3/2 coefficient in the classical formula results from distinct assumptions. The numerator comes from the geometry of the container and from the molecule symmetry whereas the denominator comes from the linearity of the constitutive properties in translational mechanics. Both parameters can be generalized to more complex systems.

Value 3 of the numerator comes for one part from the equivalence between all space directions for determining the volume and the pressure of the gas:

$$V = \boldsymbol{A}_w \cdot \ell_w; \quad -P = \frac{\boldsymbol{F}_w}{-\boldsymbol{A}_w} \quad (w = x, y, z) \tag{K4.26}$$

There is no need to invoke assumptions regarding a molecular chaos or a statistical equivalence between average velocities for establishing this *equipartition* between capacitive subvarieties.

For the other part it stems independently from the number of independent subsystems, or degrees of freedom, required for taking into account the symmetry of the molecules. A single parameter d° quantifying the degrees of freedom may stand in the numerator for generalizing the model

Monoatomic: $d^\circ = 3$; Diatomic: $d^\circ = 5$; Triatomic: $d^\circ = 7$

Denominator 2 is valid only in the frame of linear mechanics which makes the kinetic energy equal to one half the product $\boldsymbol{v} \cdot \boldsymbol{P}$ or $M\,\boldsymbol{v}^2$. In the nonlinear case, this denominator can be replaced by the reciprocal of an *inductive energetic ratio* β_q, here a *kinetic ratio* β_ℓ, defined as follows:

Kinetic ratio: $$\beta_\ell \overset{def}{=} \frac{\mathcal{T}_\ell}{\boldsymbol{v} \cdot \boldsymbol{P}} \underset{lin}{=} \frac{1}{2} \tag{K4.27}$$

which becomes only equal to 1/2 in the linear case (it depends on the velocity otherwise, i.e., in the relativistic case). Consequently, the general expression of the molecular kinetic energy as a function of the temperature can now be written

Corpuscle kinetic energy (general): $$\frac{\mathcal{T}_\ell}{N} \underset{ideal}{=} d^\circ \beta_\ell \, k_B T \tag{K4.28}$$

Remarks

Energetic coupling ratios. In the first case study about the hydrodynamical potential, a *differential energetic coupling ratio* was defined as the derivative of the volume energy with respect to the coupling energy:

Differential energetic coupling ratio: $$\alpha'_V \overset{def}{=} \frac{d\mathcal{U}_V}{d\mathcal{U}_\lambda} = \frac{d\mathcal{U}_V}{dPV} \tag{K1.14}$$

As one has regularly done in case of differential definition of a variable, one has associated to it an integral variable defined by simple proportionality:

Integral energetic coupling ratio: $$\alpha_V \overset{def}{=} \frac{\mathcal{U}_V}{\mathcal{U}_\lambda} = \frac{\mathcal{U}_V}{PV} \tag{K4.29}$$

(continued)

By using the definition in Equation K4.27 of the *kinetic ratio* and by stating the equality between kinetic and volume energy, the following relationship between energetic ratios is obtained:

Integral energetic coupling ratio:
$$\frac{\alpha_V}{\beta_\ell} = d^\circ \qquad \text{(K4.30)}$$

Applied to the various degrees of freedom already met, this formula gives

Monoatomic: $d^\circ = 3$; $\alpha_V = 3/2$,
Diatomic: $d^\circ = 5$; $\alpha_V = 5/2$,
Triatomic: $d^\circ = 7$; $\alpha_V = 7/2$

When the *differential energetic ratio* is a constant, which happens in the case of an ideal gas, both ratios are identical, so the previous values found for the integral ratio also apply to the differential one:

Constant case:
$$\frac{d\alpha'_V}{dPV} = 0 \quad \Leftrightarrow \quad \alpha'_V \underset{ideal}{=} \alpha_V = d^\circ \beta_\ell \qquad \text{(K4.31)}$$

In case study K2 "Ideal Gas," the *differential energetic ratio* was identified with C_V/R, which is therefore theoretically predicted by the above reasoning.

Force surface density and pressure. The notion of pressure is traditionally thought of as a surface density of force and therefore more as a mechanical concept than a thermodynamic one. The classical development of the kinetic theory of gases uses this concept of force surface density, leading through the previously reasoning made for linking momentum and force to the following expression of the "pressure":

Local pressure:
$$F_{/A} = \frac{\mathbf{F}}{\mathbf{A}} = \frac{2}{\Delta t}\frac{\boldsymbol{P}}{\mathbf{A}} \qquad \text{(K4.32)}$$

In the Formal Graph theory, this density of force is called *superficial force* or *local pressure* and notated $F_{/A}$ (cf. case study B2 in Chapter 5) and the *pressure* P (which is a state variable) is defined as an energy per volume, that is the partial (negative) derivative of energy with respect to the volume:

Pressure:
$$P \overset{def}{=} -\left(\frac{\partial \mathcal{U}}{\partial V}\right)_{qi} \qquad \text{(K4.33)}$$

What makes it easier to confuse is the situation of coupling between hydrodynamics and translational mechanics, in which the two efforts are proportional:

$$\mathbf{F} = \mathbf{A}\,P \qquad \text{(K4.34)}$$

Notwithstanding, the distinction between these two variables allows one to think of the notion of pressure as independent from any mechanical considerations, notably without referring to particles and their motions, as is the case in solid materials with immobilized corpuscles.

Mean gas velocity. From Equation K4.24 of the kinetic energy in the Newtonian frame and from the state equation K4.10 of the ideal gas, the expression of the square of the gas velocity can be obtained for any number of degrees of freedom:

Gas velocity: $$\boldsymbol{v}^2_{ideal} = d^o \frac{RT}{M} \tag{K4.35}$$

Depending on the statistical model of distribution chosen this expression can be considered to give the root mean square of the velocities of the corpuscles.

13.2.5 Case Study K5: Rotating Bodies

Rotational Mechanics
Translational Mechanics

The case study abstract is given in next page.

This case study offers the opportunity to study the interdependence between coupling and pole composition when several subsystems are coupled differently. The composition of inductive poles has been tackled in case study K4 “Kinetic of Gases,” where the total momentum was the sum of individual momenta featuring the entity numbers of independent subsystems. Here, these subsystems are not attributed to each space dimension but to distinct objects that can be in any number. The inductive subvariety of rotational mechanics has been introduced in Chapter 4 with the case study A2 “Inertia.”

Let us begin by discussing the coupling in the case of a single body before addressing the case of many bodies.

13.2.5.1 Single Rotating Body

The figure and the Formal Graph in the case study abstract represent a rotating body featured by its *rotational inertia*, or *moment of inertia* J, which is the inductance in rotational mechanics, and by its *inertial mass* M, which is the inductance in translational mechanics. In the figure, the two vectors *angular momentum* $\boldsymbol{L}$ and *angular velocity* $\boldsymbol{\Omega}$ are perpendicular to the rotation plane, colinear with the rotation axis and oriented upward. Both *momentum* $\boldsymbol{P}$ and *translation velocity* $\boldsymbol{v}$ vectors are also colinear but belong to the rotation plane, being tangent to the circular trajectory in the direction of the movement. These vectors are the four state variables in the inductive subvariety in rotational and translational mechanics. Their colinearity is assumed for drawing the figure but is not the general case as the inductances can be more complicated properties than simple scalars (i.e., tensors for taking into account material heterogeneities). The coupling relationships are written in terms of cross-products between vectors in the following way:

$$\boldsymbol{L} = \boldsymbol{R} \times \boldsymbol{P} \tag{K5.1}$$

$$\boldsymbol{v} = -\boldsymbol{R} \times \boldsymbol{\Omega} = \boldsymbol{\Omega} \times \boldsymbol{R} \tag{K5.2}$$

The cross-product being not commutative, a minus sign is used when the order of the multiplied vectors is reversed, which is the case for the coupling factor relating the angular velocity $\boldsymbol{\Omega}$ to the translation velocity $\boldsymbol{v}$. In these relationships, the vector $\boldsymbol{R}$ is the *radius of gyration* quantifying the distance between the *center of mass* of the body and the rotation axis. Note that the unit of a *gyration radius* is the m rad^{-1}. The *center of mass* is defined as the point that would concentrate the whole mass of the body if no spatial extension were allowed. The size of the body must not be too large in the direction of the radius in order to allow a unique radius to feature the distance from the axis (a thin circular or toroidal object at constant radius is permitted). The link with the axis is assumed to be rigid, maintaining constant the radius $\boldsymbol{R}$ (no elasticity), so this variable behaves as an *integral coupling factor*. A graphical representation of the disposition of the involved vectors is given in Figure 13.2.

$$\boldsymbol{L} = \boldsymbol{R} \times \widehat{\mathrm{M}}\boldsymbol{v} = \boldsymbol{R} \times \widehat{\mathrm{M}}(-\boldsymbol{R}) \times \boldsymbol{\Omega} \tag{K5.3}$$

K5: Rotating Bodies

Rotational Mechanics
Translational Mechanics

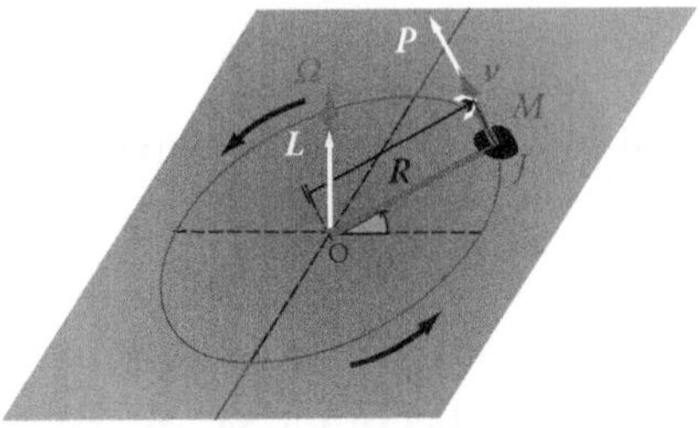

Formal Object:

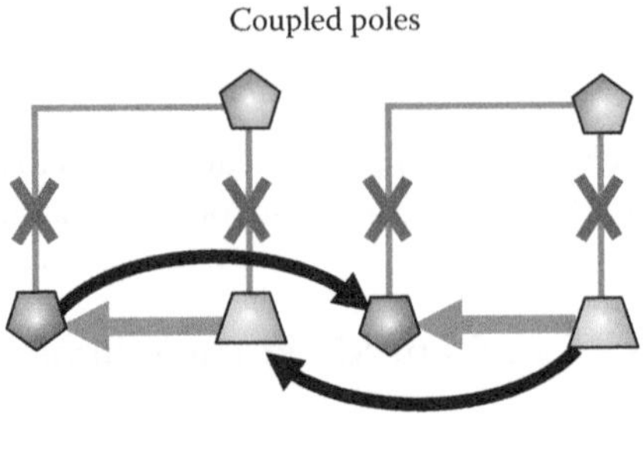

Formal Graph (see Graph 13.9):

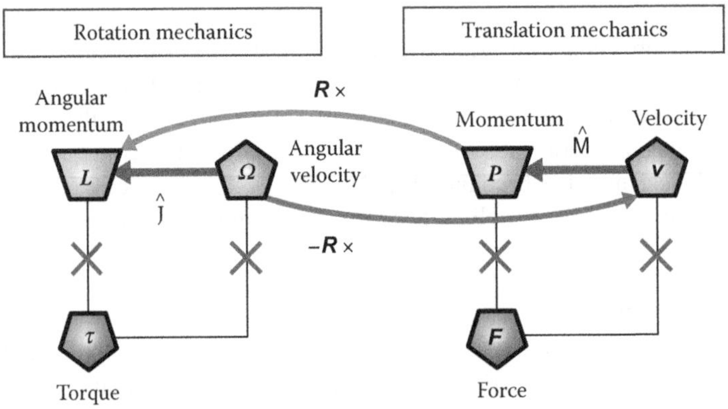

GRAPH 13.9 Formal Graph of the coupling between rotational mechanics and translational mechanics. The coupling factor is the cross-product with the radius $\boldsymbol{R}$ (distance from the rotation axis).

Coupling Factor

	Coupling	From	To
Variety		Translation mechanics	Rotation mechanics
Family	Integral factor	Impulse	
Name	Radius	Momentum	Angular momentum
Symbol	$\boldsymbol{R}$	$\boldsymbol{P}$	$\boldsymbol{L}$
Unit	[m]	[N s]	[N m s]
Variety		Rotation mechanics	Translation mechanics
Family	Differential factor	Flow	
Name	Differential radius	Angular velocity	Velocity
Symbol	$\boldsymbol{R'}$	$\boldsymbol{\Omega}$	$\boldsymbol{v}$
Unit	[m]	[rad s^{-1}]	[m s^{-1}]

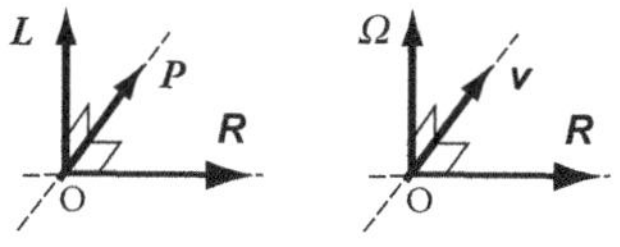

FIGURE 13.2 Three-dimensional representation of the spatial relationship between vectors belonging to rotational mechanics and translational mechanics.

The *inertia* operator is immediately deduced as a sequence of operations

$$\hat{\mathrm{J}} = \boldsymbol{R} \times \hat{\mathrm{M}}(-\boldsymbol{R}) \times \tag{K5.4}$$

which reduces to the well-known expression in case of linearity of the mass operator, as in Newtonian mechanics

$$J \underset{lin}{=} M\boldsymbol{R}^2 \tag{K5.5}$$

It can be remarked that the shape of this expression corresponds to the already discussed expression of the relationship between two linear admittances belonging to different energy varieties (see Chapter 12, Section 12.5.3).

13.2.5.2 Many Rotating Bodies

The case of several bodies rotating at the same angular velocity around a same axis is exemplified in Figure 13.3 with the cases of two and three bodies.

Again, the physical links between the bodies and the rotation axis are supposed to be constant, which features the indeformability of the system and ensures the same angular velocity for all parts. The composition of *N* poles into a single pole is based on this unique angular velocity (energy-per-entity) and on the addition of angular momenta (entity numbers):

$$\boldsymbol{\Omega} = \boldsymbol{\Omega}_1 = \boldsymbol{\Omega}_2 = \cdots = \boldsymbol{\Omega}_N \tag{K5.6}$$

$$\boldsymbol{L} = \boldsymbol{L}_1 + \boldsymbol{L}_2 + \cdots + \boldsymbol{L}_N \tag{K5.7}$$

On expressing these angular momenta as functions of the angular velocity, one writes the general expression

$$\boldsymbol{L} = \sum_{i=1}^{N} \hat{\mathrm{J}}_i \, \boldsymbol{\Omega} = \hat{\mathrm{J}} \, \boldsymbol{\Omega} \tag{K5.8}$$

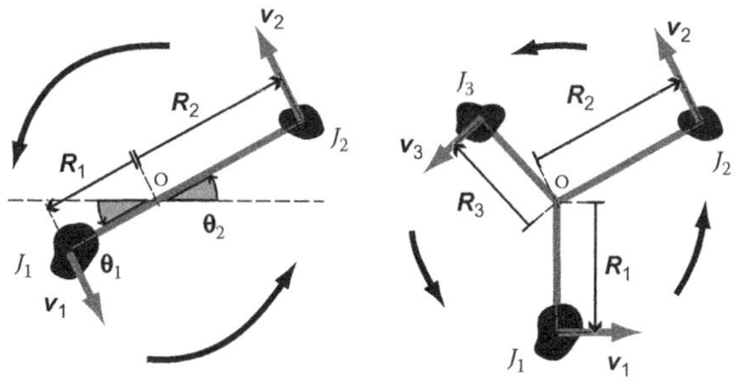

FIGURE 13.3 When several bodies are rigidly bound and rotate around a common axis, their angles and their angular velocities are identical.

When the inertial mass is a linear operator, the expression of the total rotational inertia simplifies into the classical formula for N bodies:

$$J \underset{lin}{=} \sum_{i=1}^{N} M_i \, \boldsymbol{R}_i^2 \tag{K5.9}$$

The Formal Graph modeling N rotating bodies is given in Graph 13.10.

This discussion was about objects that can be distinguished and sufficiently small for considering a constant radius as a coupling factor. They could form a single solid in being contiguous as in the case of material grains stuck together. The case of a continuum of homogeneous matter is modeled in the Newtonian frame by replacing the summation of angular momenta by an integration over all volume elements:

$$J \underset{lin}{=} \int \rho_M(V)\, R(V)^2 \, dV \tag{K5.10}$$

with ρ_M the volumic (inertial) mass which may depend on the volume element. A Formal Graph in this case would use space poles instead of global ones.

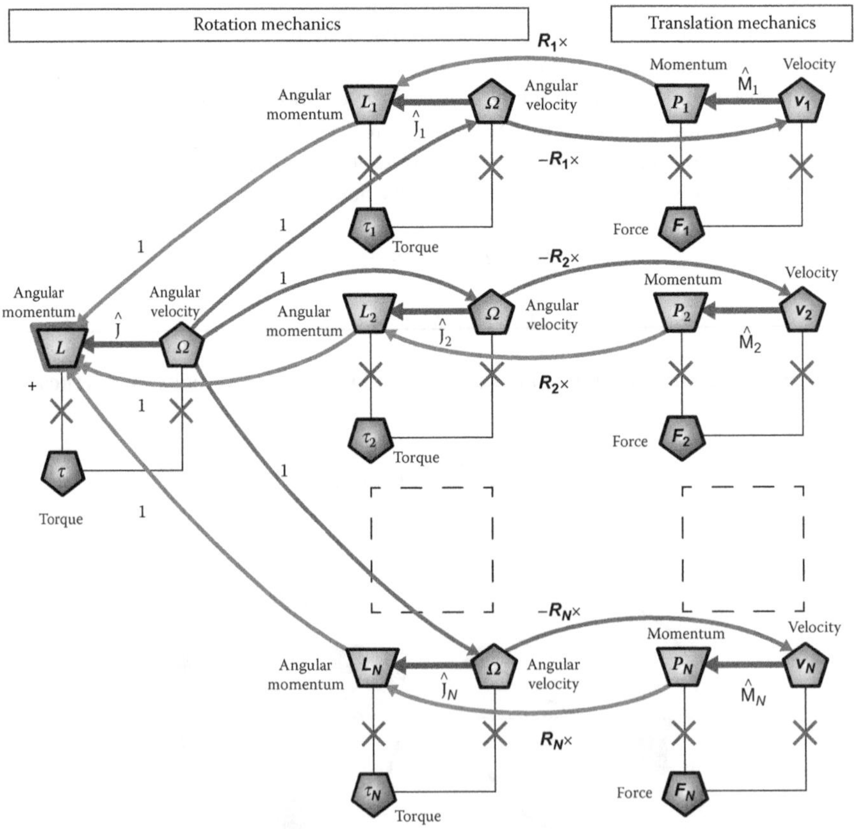

GRAPH 13.10 Formal Graph of the simultaneous rotation around a common axis of N massive bodies forming an indeformable system. All Formal Objects are inductive poles coupled by pair and are forming a composed pole.

13.2.6 Case Study K6: Euler Equation

Gravitation
Hydrodynamics

$$\rho_M \frac{\partial}{\partial t}\boldsymbol{v} + \rho_M (\boldsymbol{v} \cdot \nabla)\boldsymbol{v} = -\nabla P + \rho_M \boldsymbol{g}$$

Formal Object:

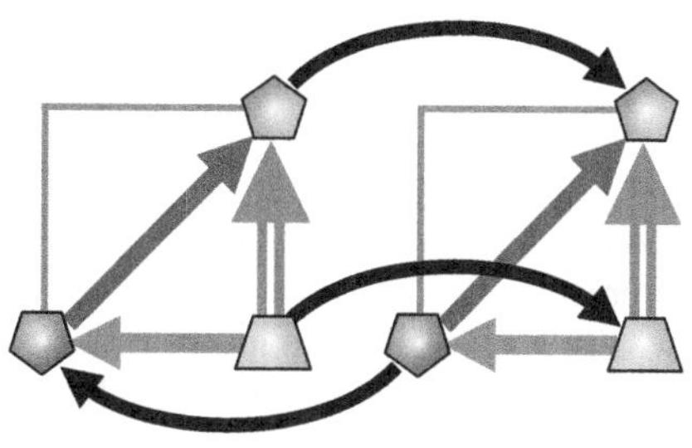

Formal Graph:

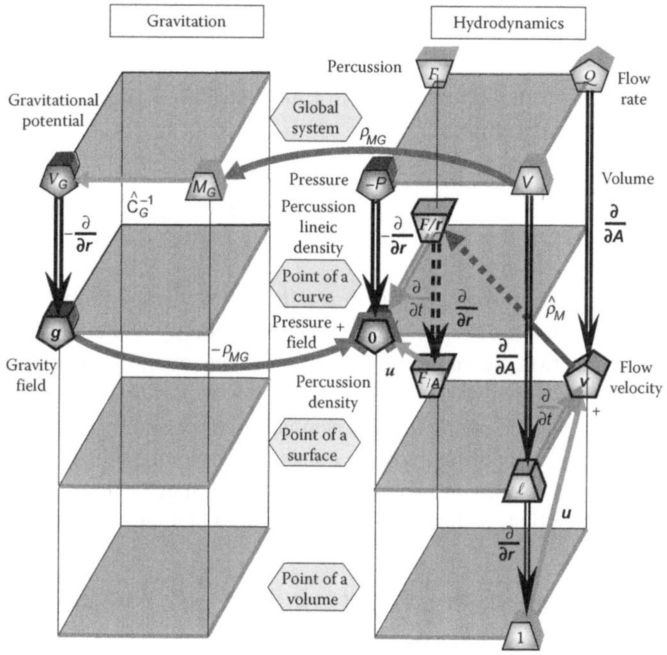

GRAPH 13.11 Formal Graph of the coupling between gravitation and hydrodynamics. The coupling factor is the volumic (gravitational) mass.

Coupling Factor

	Coupling	From	To
Variety		Hydrodynamics	Gravitation
Family	Integral factor	Basic quantity	
Name	Volumic inertial mass	Volume	Gravitational mass
Symbol	ρ_{MG}	V	M_G
Unit	[kg m^{-3}]	[m^3]	[kg]

System Constitutive Properties

Variety	Hydrodynamics	Gravitation
Nature	Inductance	Capacitance
Levels	2/1	0
Name	Volumic inertial mass	Gravitational capacitance
Symbol	ρ_{MG}	C_G
Unit	[kg m^{-3}]	[kg s^2 m^{-2}]
Specificity	—	—

In hydrodynamics, the Euler* equation holds a central position in the modeling of a moving fluid submitted to gravitational field. It therefore involves two energy varieties, hydrodynamics and gravitation. This case study is interesting from the Formal Graph viewpoint as it shows a coupling between space dipoles, using therefore a three-dimensional representation.

The Euler equation is classically written as

Euler: $$\rho_M \frac{\partial}{\partial t}\boldsymbol{v} + \rho_M (\boldsymbol{v} \cdot \nabla)\boldsymbol{v} = -\nabla P + \rho_M \boldsymbol{g} \tag{K6.1}$$

Its representation in a Formal Graph sheds some light on the physical meaning of this equation in defining precisely the role of the various variables, helping one to understand better what is in the shadow with algebra, as here it is not possible to ignore some variables. The core of this model relies on the triple dependence of the variable called "pressure field" $\boldsymbol{P}_{/r}$ which is the *lineic density of pressure* (pressure by unit of length) in a point of a curve (line) of fluid flow (fluid current):

$$\boldsymbol{P}_{/r} = \boldsymbol{P}_{/rP} + \boldsymbol{P}_{/rL} + \boldsymbol{P}_{/rg} \tag{K6.2}$$

The first term results from the inhomogeneity of the distribution of pressure along the flow line; it is given by the gradient of pressure

$$\boldsymbol{P}_{/rP} = \frac{\partial}{\partial \boldsymbol{r}} P = \nabla P \tag{K6.3}$$

The second term results from the dynamic of the fluid movement and expresses the role of time and of space (space–time) in the evolution of the inductive energy (kinetic) of a fluid element. The pertinent variable in translation mechanics is the *volumic concentration of momentum* ("concentration of movement"), which in hydrodynamics corresponds to the *lineic density of percussion* (or *percussion field*) $\boldsymbol{F}_{/r}$. Its influence on the fluid movement is expressed by means of a total derivative, called *particular derivative*, which involves the convection velocity $\boldsymbol{u}$ of the medium (see Chapter 10, Section 10.5 about the modeling of convection). This space–time velocity $\boldsymbol{u}$ coincides in Newtonian mechanics with the translation velocity $\boldsymbol{v}$ of fluid elements.

$$\boldsymbol{P}_{/rL} = \frac{\mathrm{d}}{\mathrm{d}t}\boldsymbol{F}_{/r} = \frac{\partial}{\partial t}\boldsymbol{F}_{/r} + \boldsymbol{u} \times \frac{\partial}{\partial \boldsymbol{r}}\boldsymbol{F}_{/r} = \frac{\partial}{\partial t}\boldsymbol{F}_{/r} + \boldsymbol{u} \times \boldsymbol{F}_{/A} \tag{K6.4}$$

The lineic density of percussion $\boldsymbol{F}_{/r}$ is linked to the translation velocity $\boldsymbol{v}$ by an inductive relationship with the volumic mass ρ_M as reduced inductance, which in the general case is an operator

$$\boldsymbol{F}_{/r} = \hat{\rho}_M \boldsymbol{v} \tag{K6.5}$$

It is important to recall the generality expressed by this nonlinear relationship to help remember that the Euler equation (see Equation K6.1) is valid only in the peculiar case of linear media. From now, the relation in Equation K6.5 will be utilized in this restricted case with a scalar and constant volumic mass ρ_M. The *surface density of percussion* $\boldsymbol{F}_{/A}$ is linked to the previous variable according to the specificity of the *particular derivative*, analogous to the expression of the Lorentz force in electrodynamics (see case study F8 "Hall Effect" in chapter 9), and its expression can also be given in the linear case in extracting the volumic mass from the scope of the operators:

* Leonhardt Euler (1707–1783): Swiss mathematician; Basel, Switzerland and St. Petersburg, Russia.

$$u \times F_{IA} = \left(u \cdot \frac{\partial}{\partial r} \right) F_{Ir} \underset{lin}{=} \rho_M (u \cdot \nabla) v \tag{K6.6}$$

The third contribution to the pressure field comes from the coupling with the gravitational energy. The Formal Graph in Graph 13.12 shows the total coupling, meaning that an extremum is reached by the energy of the isolated system, between gravitation and hydrodynamics. This graph can be viewed as a subset of the general Formal Graph shown previously that does not take into account the other contributions to the pressure field.

Graph 13.12, which is not a canonical Formal Graph but a differential one, represents the following Gibbs–Duhem* equation, which is issued from the variational Gibbs equation by imposing to this latter to be an exact differential equation (cf. Chapter 12, Section 12.3):

Gibbs–Duhem:

$$0 = M_G \, dV_G - V \, dP \tag{K6.7}$$

By defining a *volumic gravitational mass* ρ_{MG} as the integral coupling factor between hydrodynamics and gravitation, that is, as the ratio of the gravitational mass on the hydrodynamic volume, and then equating this factor to the ratio of the variations of the respective efforts, by virtue of the previous Gibbs–Duhem equation, one obtains

$$\rho_{MG} \overset{def}{=} \frac{M_G}{V} = \frac{dP}{dV_G} = -\frac{P_{Irg}}{g} \tag{K6.8}$$

The last equality, which stems from the identifications of the pressure field (with a subscript G indicating its gravitational origin through the coupling) and of the gravitational field g with the

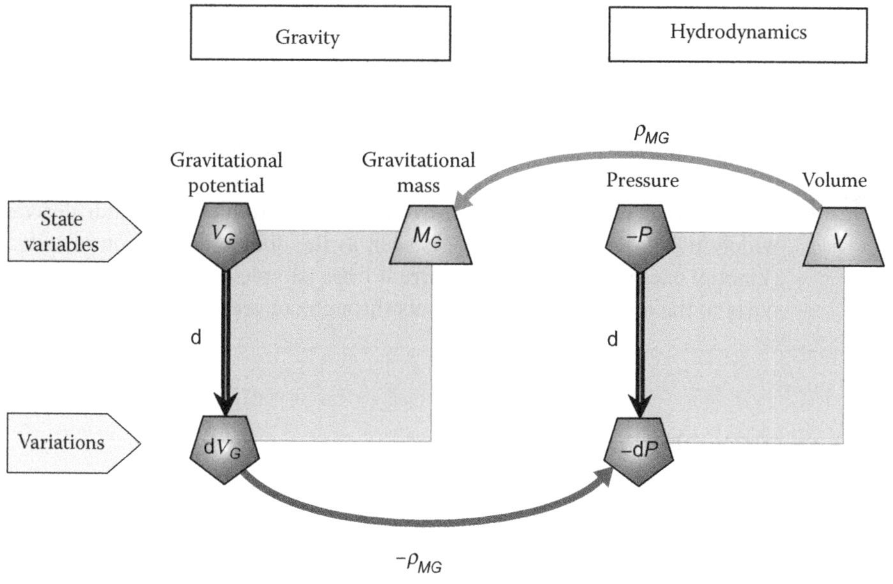

GRAPH 13.12 Differential Formal Graphs modeling the constant coupling between gravitation and hydrodynamics.

* Pierre Duhem (1861–1916): French physicist and philosopher; Paris, France.

contragradients of minus the pressure $-P$ and of the gravitational potential V_G (see case study K3 "Barometric Equation"), respectively, can be rewritten as a coupling relationship between reduced variables

$$\boldsymbol{P}_{lrg} \underset{lin}{=} -\rho_{MG}\,\boldsymbol{g} \tag{K6.9}$$

This volumic gravitational mass ρ_{MG} is equal to the volumic inertial mass ρ_M according to Einstein's postulate of equivalence between inertial and gravitational masses. However, in the Formal Graph theory, this equivalence is not seen as an identity as the gravitational mass M_G is a state variable and the inertial mass M is a system constitutive property (inductance in translational mechanics).

It remains to express that the system works according to a serial mounting, in this case an additive coupling, and in being isolated from the exterior, which amounts to balancing all contributions to the reduced effort,

$$\boldsymbol{P}_{lr} = 0 \tag{K6.10}$$

for finding again the Euler equation (see Equation K6.1), written now with more correct variables from the Formal Graph viewpoint

$$\rho_M \frac{\partial}{\partial t}\boldsymbol{v} + \rho_M(\boldsymbol{u}\cdot\nabla)\boldsymbol{v} \underset{lin}{=} -\nabla P + \rho_{MG}\,\boldsymbol{g} \tag{K6.11}$$

The fact remains that the Formal Graph in the case study abstract is a much more general model than this algebraic equation, allowing notably complicated dependence of the volumic inertial mass.

Particular derivative. Here is an example that is especially enlightening about the supplement of physical meaning brought by the graphic representation of this rather complex equation. Notably, the notion of particular derivative, widely used in fluid mechanics, is seen as the double conversion of inductive (kinetic) energy, at the level of current lines (simple evolution) and simultaneously at the level of surface elements through convection.

13.3 THE ENERGY OF COUPLING

A coupling between two energy varieties is defined as relating the state variables in pairs, as explained in Chapter 12. To be represented in a Formal Graph, the coupling relationships use a coupling factor, which can be integral (when relating entity numbers) or differential (when relating energies-per-entity). For a capacitive coupling, one obtains (the case of the inductive coupling is similar):

Integral coupling factor: $$\lambda_{ba} \overset{def}{=} \frac{q_a - q_a^{min}}{q_b - q_b^{min}} = -\frac{de_b}{de_a} = \frac{e_{ba}}{e_a} \qquad (q_b > q_b^{min}) \tag{13.6}$$

Differential coupling factor: $$\lambda'_{ba} \stackrel{def}{=} \frac{dq_a}{dq_b} = -\frac{e_b}{e_a} \quad (e_a \neq 0) \tag{13.7}$$

These relationships use the *translated effort* e_{ba} that allows one to represent graphically the relationship between efforts with the sole integral coupling factor. However, this is not necessary when the differential coupling factor is a constant, leading to equality between both coupling factors (the coupling is termed *invariable* in this peculiar case).

Table 13.2 shows the relationships using the *integral coupling factor* for some of the couplings studied until now.

It must be outlined that the variability or invariability of a coupling is not intrinsically attached to the couple of energy varieties. The coupling factors that determine the variability depend on the system and eventually on the energy content (variable case) but not on the nature of energy. For example, the hydrodynamical and translational mechanical energies may be related through an invariable coupling when the cross-sectional area is a constant, as in the case of a cylindrical piston (case study J5 in Chapter 12), and through a variable coupling otherwise. Another example is the thermoelectric coupling which is invariable in case of an ideal junction (case studies J8 through J10 about the thermal junction and Seebeck and Peltier effects, respectively) and is variable for a real junction (case study J11 "Thomson Effect").

TABLE 13.2
Nonexhaustive List of Couplings Described in the Case Studies in Chapters 12 and 13

Case Study	Coupling Relationship	Coupling Energy	Characteristics
J8: Thermal Junction	$-\alpha = \frac{S}{Q} = -\frac{V}{T}$	$\mathcal{U}_{SQ} = VQ = -TS$ Electrodynamics/thermics	Invariable (ideal junction)
J6: Ionic Distribution	$zF = \frac{Q}{n} = -\frac{\mu}{V}$	$\mathcal{U}_{nQ} = VQ = -\mu n$ Electrodynamics/physical chemistry	Invariable
K3: Barometric Equation	$M = \frac{M_G}{n} = -\frac{\mu}{V_G}$	$\mathcal{U}_{nM_G} = V_G M_G = -\mu n$ Gravitation/physical chemistry	Invariable
J5: Piston	$\mathbf{A} = \frac{V}{\ell} = -\frac{\mathbf{F}}{-P}$	$\mathcal{U}_{V\ell} = \mathbf{F} \cdot \ell = PV$ Translational mechanics/ hydrodynamics	Invariable (constant section)
K3: Barometric Equation	$\rho_M = \frac{M_G}{V} = \frac{-P_{MG}}{V_G}$	$\mathcal{U}_{VM_G} = V_G M_G = -P_{MG} V$ Gravitation/hydrodynamics	Variable
K1: Hydrochemical Potential	$V_m = \frac{V}{n} = \frac{\mu_P}{-P}$	$\mathcal{U}_{nV} = PV = -\mu_P n$ Hydrodynamics/physical chemistry	Variable
K2: Ideal Gas	$R = \frac{S_n}{n} = \frac{\mu_T}{T}$	$\mathcal{U}_{nS} = TS_n = \mu_T n$ Thermics/physical chemistry	Variable
K4: Kinetic of Gases	$v_N = \frac{\mathbf{v}}{\ell} = \frac{\mathbf{F}}{\mathbf{P}}$	$\mathcal{E}_{\ell\ell} = \mathbf{v} \cdot \mathbf{P} = \mathbf{F} \cdot \ell$ Translational mechanics	Internal

Peculiarity of the thermal agitation. In an ideal gas, the coupling between the thermal and physical chemical, or corpuscular, energies exhibits a peculiarity compared to the other couplings discussed in this chapter and in Chapter 12. Instead of involving two true state variables for both products giving the coupling energy, only one is really a state variable and the other is a *translated* one.

$$\mathcal{U}_{nS} = TS_n = \mu_T n; \quad \mathcal{U}_{NS} = TS_N = E_T N$$

In other words, the coupling energy should be equal to one of the products TS or $\mu\, n$ (or $E\, N$) as is the case for the other couplings. In fact, this exception allows one to consider this coupling as variable despite the constancy of the (differential) coupling factor R. (This makes a nuance to be brought to the definition of the variability of a coupling.)

The reason stems from the specificity of an ideal gas which is to enforce equality between two coupling energies, the hydrodynamical–physical chemical coupling energy and the thermal–physical chemical one. This holds for the ideal gas but not for a real gas and it is not a standard property of the thermal energy: the thermoelectric coupling, for instance, does not enter in this category.

Table 13.2 also shows the results of the cross-multiplications issued from the coupling relationships, which provide an alternative way for modeling a coupling. The common value of these products is an energetic variable that is identified as the energy involved in the coupling. In distinguishing between the capacitive and inductive couplings, hereafter, the generic expressions of these common values are given, first for the invariable coupling energies:

$$\left(\lambda'_{ba} = C^{st}\right): \qquad \mathcal{U}_{ba} \overset{def}{=} \pm e_a\left(q_a - q_a^{min}\right) = \mp e_b\left(q_b - q_b^{min}\right) \tag{13.8}$$

$$\left(\lambda'_{ba} = C^{st}\right): \qquad \mathcal{T}_{ba} \overset{def}{=} \pm f_a\left(p_a - p_a^{min}\right) = \mp f_b\left(p_b - p_b^{min}\right) \tag{13.9}$$

and second for the variable coupling energies

$$\left(\lambda'_{ba} \neq C^{st}\right): \qquad \mathcal{U}_{ba} \overset{def}{=} \pm e_a\left(q_a - q_a^{min}\right) = \pm e_{ba}\left(q_b - q_b^{min}\right) \tag{13.10}$$

$$\left(\lambda'_{ba} \neq C^{st}\right): \qquad \mathcal{T}_{ba} \overset{def}{=} \pm f_a\left(p_a - p_a^{min}\right) = \pm f_{ba}\left(p_b - p_b^{min}\right) \tag{13.11}$$

The case of the internal coupling energy can be also modeled in this way

$$\mathcal{E}_{aa} \overset{def}{=} \pm f_a\left(p_a - p_a^{min}\right) = \pm e_a\left(q_a - q_a^{min}\right) \tag{13.12}$$

The choice of the sign must make the coupling energy positive. Such a coupling energy must not be confused with the energy of the system, which is generally higher. It stands for the part of the energy that is shared between the coupled energy varieties, which do not necessarily pool all their energy.

The relationship between energy of a given variety and the coupling energy has been tackled with the gaseous system when an *energetic coupling ratio* α_V was defined as the ratio of hydrodynamic energy on the coupled energy (Equation K4.29 in case study K4 "Kinetic of Gases"). In the general case, these ratios are defined separately for the capacitive and inductive energies:

Integral energetic coupling ratios:
$$\alpha_a \overset{def}{=} \frac{\mathcal{U}_a}{\mathcal{U}_{ba}}; \quad \beta_a \overset{def}{=} \frac{\mathcal{T}_a}{\mathcal{T}_{ba}} \tag{13.13}$$

Prior to the definition of these integral ratios, the differential version of the energetic ratio was used in case studies K1 "Hydrochemical Potential" and K2 "Ideal Gas," which is defined in the general case as

Differential energetic coupling ratios:
$$\alpha'_a \overset{def}{=} \frac{d\mathcal{U}_a}{d\mathcal{U}_{ba}}; \quad \beta'_a \overset{def}{=} \frac{d\mathcal{T}_a}{d\mathcal{T}_{ba}} \tag{13.14}$$

As the expression of energy in a pole depends on the constitutive property that supports the energy subvariety, the expression of this property must be known for evaluating the energy amount. The simplest case is the linear relationship, corresponding to the equivalence between differential terms as follows in the capacitive case

Linear capacitance:
$$d\mathcal{U}_a = e_a\, dq_a \underset{lin}{=} \left(q_a - q_a^{min}\right) de_a \tag{13.15}$$

$$d\mathcal{U}_a \underset{lin}{=} \frac{e_a\, dq_a + \left(q_a - q_a^{min}\right) de_a}{2} = \frac{1}{2} d\mathcal{U}_{ba} \Rightarrow \alpha'_a \underset{lin}{=} \alpha_a \underset{lin}{=} \frac{1}{2} \tag{13.16}$$

As an example, let us take the inductive case with the kinetic energy in the Newtonian frame

Linear mass:
$$d\mathcal{T}_\ell = \mathbf{v} \cdot \mathbf{d}\boldsymbol{P} \underset{lin}{=} \boldsymbol{P} \cdot \mathbf{dv} \Rightarrow \mathcal{T}_\ell \underset{lin}{=} \frac{1}{2} \mathbf{v} \cdot \boldsymbol{P} \underset{lin}{=} \frac{1}{2} M\mathbf{v}^2 \tag{13.17}$$

for which there is the well-known one-half coefficient

$$\beta'_\ell \underset{lin}{=} \beta_\ell \underset{lin}{=} \frac{1}{2} \tag{13.18}$$

Expressions for energy. This book insists on the need to make a necessary disconnection between the one-half coefficient and expressions of energy. It is not a good practice to retain such formulas as $1/2CV^2$ or $1/2Mv^2$ as definitions of the electrostatic or kinetic energies because they are specific to the linear

(continued)

case, which is only a part of the whole (huge) range of the physical world. The practical expressions (which are not definitions) to retain are

$$\mathcal{U}_Q = \alpha_Q V \hat{C} V; \quad \mathcal{T}_\ell = \beta_\ell \mathbf{v} \hat{M} \mathbf{v} \tag{13.19}$$

They correspond to the general expressions

$$\mathcal{U}_q = \alpha_q e_q \hat{C}_q e_q; \quad \mathcal{T}_q = \beta_q f_q \hat{L}_q f_q \tag{13.20}$$

(The subscripts a and b are used when more than one energy variety is considered, otherwise one returns to subscript q.)

Keep in mind that the α_q or β_q coefficients are nearer the value 1/2 the energy amounts are small, and, above all, that they can be *unknown variables* to be experimentally determined (one finds here a leitmotif as this idea has been discussed in Chapter 10 when the "unknown" mass transfer operator was introduced).

Among the various case studies of energy coupling presented, it is worth discussing the case of multiple couplings in the same system, as in the case of the ideal gas (case study K2) and of the barometric equation (case study K3). In the case of the ideal gas, there is equality between the two coupling energies:

Weightless ideal gas:
$$\mathcal{U}_{nS} = \mathcal{U}_{nV} \tag{13.21}$$

which entails the following equalities:

$$TS_n = \mu_T n = -\mu_P n = PV \tag{13.22}$$

The fact that the hydrodynamical variety is the only one to participate with its two state variables (full variables and not translated ones) can be understood by the fact that the products of state variables in the other varieties ($T\,S$ and $\mu\, n$) are higher than the product PV and that, consequently, they cannot account for the coupling energy. This can be expressed by a rule of minimum saying that the coupling energy is equal to the minimum of the products of state variables.

In the case of the barometric equation, which is based on an ideal gas, the same equality as in Equation 13.22 is met, but without including the gravitational energy which is independently coupled to the physical chemical energy:

Weighting ideal gas:
$$\mathcal{U}_{nM_G} \neq \mathcal{U}_{nS} = \mathcal{U}_{nV} \tag{13.23}$$

The corresponding relationships between state (and translated) variable products are then

$$V_G M_G = -\mu n \neq TS_n = \mu_T n = -\mu_P n = PV$$

If the reason for this situation is not completely understood, the concept of energy coupling brings a supplementary insight into the question of the distribution of energy into a system. This is a somewhat opaque subject to know how the energy amount is spread among the various varieties existing in a system. It is easier to understand that the existence of some containers is conditioned by the physics of a system, but once these containers are recognized, the problem is to understand the mechanism that distributes the energy unequally (or not) among them, and how it varies on perturbation of the system.

13.4 THE SCALING CHEMICAL POTENTIAL

Let us now tackle a subject which has been left since Chapter 7 because one could not discuss it without having studied the theory of energetic coupling.

13.4.1 Scaling Effort

The *scaling effort* $e_{q\Delta}$ is a parameter that has been introduced in Chapter 7, Section 7.4.2 when the capacitive relationship was demonstrated in the generic energy variety q. The theory of influence developed at that time was based on the interaction between poles in a chain of poles (forming a sequence of dipoles also called *multipole*). This interaction was quantified by the notion of *gain*, defined as the ratio of the *pole activities*

Gain $i \rightarrow j$
$$g_{qji} \overset{def}{=} \frac{a_{qj}}{a_{qi}} \tag{13.24}$$

The theory established a relationship between this gain in activity and the variation of dipole effort, relying on a multiplicative transmission of the gains from pole to pole and an additive transmission of the efforts. In addition, complete independency of the gain and of the effort of a dipole on the other dipoles in the chain was assumed. This relationship was written as proportionality between the relative gain variation and the effort variation in a dipole:

$$de_{qji} = \pm e_{q\Delta} \frac{dg_{qji}}{g_{qji}} \tag{13.25}$$

The proportionality factor was notated $e_{q\Delta}$ and called *scaling effort*. The choice of the sign depended on the nature of the interaction, whether by mutual influence (between the two poles of a dipole) or by self-influence (within a pole). In this latter case, the positive sign was chosen and the decomposition of the previous relationship into individual relationships for each pole was written as

$$de_{qi} = e_{q\Delta} \frac{da_{qj}}{a_{qj}} \tag{13.26}$$

No more precision could be given at that time as this *scaling effort* was arbitrarily introduced. However, comparisons with classical models of capacitive relationships provided expressions for this coefficient according to the considered energy variety and three examples were given:

Physical chemistry:
$$\mu_\Delta = RT \tag{13.27}$$

Corpuscular energy:
$$E_{\Delta} = k_B T \tag{13.28}$$

Electrodynamics:
$$V_{\Delta} = \frac{k_B T}{e} = \frac{RT}{F} \tag{13.29}$$

These expressions are pure coincidences, from the methodological point of view, as the first two ensue from matching results stemming from different approaches, the Boltzmann statistics and the Formal Graph theory, and they are not produced by this latter. This comparison is a heuristic approach and one does not know the conditions of validity for these expressions (except in having recourse to the validity conditions of the Boltzmann model, which are very restrictive as analyzed in Chapter 7). The influence theory is now achieved by demonstrating the expression of the scaling chemical potential, which will bring to light these conditions.

13.4.2 Scaling Chemical Potential

The *scaling chemical potential* μ_{Δ} is the parameter normalizing the effort in all capacitive relationships in the physical chemical energy variety. Its expression is obtained from the proportionality relation between activity and chemical potential in a pole (according to relation 13.26):

$$\mu_{\Delta} = \mathsf{a}\frac{\mathsf{d}\mu}{\mathsf{da}} \tag{13.30}$$

As the relationship between activity and substance amount is a linear relation, this expression can be written identically with the substance amount. However, one must make provision for the presence of other energy varieties in the system by substituting simple derivations with partial derivations effected in maintaining constant all other basic quantities q_i:

$$\mu_{\Delta} = n\left(\frac{\partial \mu}{\partial n}\right)_{q_i} \tag{13.31}$$

The presence of several energy varieties is stated by the following Gibbs equation:

Gibbs:
$$\mathsf{d}\mathcal{U} = e_a\,\mathsf{d}q_a + \mu\,\mathsf{d}n + e_i\,\mathsf{d}q_i \tag{13.32}$$

where the first term stands for an energy variety supposed to be coupled with the physical chemical energy variety and the last term stands for all other present varieties except the physical chemical one and the first one. (Whether or not these varieties are coupled with the physical chemical variety does not matter.) The fundamental property of energy to be a state function and the assumption of inexistence of the inductive subvariety in the system are expressed by the Gibbs–Duhem equation:

Gibbs–Duhem:
$$0 = q_a\,\mathsf{d}e_a + n\,\mathsf{d}\mu + q_i\,\mathsf{d}e_i \tag{13.33}$$

This equation is appropriate for relating the term $n\,d\mu$ to other variables in order to find the scaling chemical potential, in completing it by saying that the *coupling energy* is chosen as given by the negative product of the state variables of the first energy variety:

Coupling energy: $$\mathcal{U}_{a,n,i} \overset{def}{=} -e_a q_a \tag{13.34}$$

This choice can be justified by asserting that the choice of the energy variety a is determined by the above definition, that is, this variety is selected and placed in the first position in the list because it corresponds to the coupling energy. This was the case, for instance, for the hydrodynamic variety in case study K2 "Ideal Gas," where the product PV played the role of a coupling energy. By derivation and by introducing the Gibbs–Duhem equation, one finds an expression for the variation of coupling energy:

$$d\mathcal{U}_{q_a,n,q_i} = -e_a dq_a - q_a de_a = -e_a dq_a + n\,d\mu + q_i de_i \tag{13.35}$$

This differential equation allows one to express the substance amount as the partial derivative of the coupling energy in keeping constant the basic quantity q_a and the state variables of all other varieties:

$$n = \left(\frac{\partial \mathcal{U}_{q_a,n,q_i}}{\partial \mu}\right)_{q_a,\,q_i,\,e_i} \tag{13.36}$$

Now, by introducing this expression into Equation 13.31, one gets

$$\mu_\Delta = \left(\frac{\partial \mathcal{U}_{q_a,n,q_i}}{\partial \mu}\right)_{q_a,\,q_i,\,e_i} \left(\frac{\partial \mu}{\partial n}\right)_{q_i} \tag{13.37}$$

and, by merging the two partial derivatives, the definition of the scaling chemical potential is obtained:

$$\mu_\Delta \overset{def}{=} \left(\frac{\partial \mathcal{U}_{q_a,n,q_i}}{\partial n}\right)_{q_a,\,q_i,\,e_i} \tag{13.38}$$

This is a general definition of the scaling chemical potential, valid for any number of energy varieties in the system.

13.4.3 Thermal Scaling Chemical Potential

When the physical chemical energy is coupled with the thermal energy variety, which occurs in case of a sufficient mobility of molecules for allowing the thermal agitation to represent a significant part of the total energy of the system, the scaling potential depends on the temperature.

In case study K2 "Ideal Gas," the variation of energy, either capacitive (internal) or enthalpic, was explicitly expressed as a linear differential equation with specific coefficients for the individual variations of the significant variables. This procedure can be repeated for the coupling energy by writing:

$$d\mathcal{U}_{q_a,n,S} = e_{a,n,T}\, dq_a + \mu_{T,q_a}\, dn + S_{n,q_a}\, dT \tag{13.39}$$

Each coefficient is the partial derivative of the coupling energy with respect to the considered varying variable by maintaining at a constant all other variables.

The status of the state function for the coupling energy implies that all pairs of second derivatives must be equal, which means for the physical chemical/thermal pair that the following common value can be defined as

$$R''_a \overset{def}{=} \left(\frac{\partial \mu_{T,q_a}}{\partial T}\right)_{n,q_a} \overset{def}{=} \left(\frac{\partial S_{n,q_a}}{\partial n}\right)_{T,q_a} \tag{13.40}$$

This common value is not necessarily a constant and may depend on the temperature in the general case. It can nevertheless be integrated for providing a differential coupling factor

$$R'_a \overset{def}{=} \frac{1}{T}\int_0^T R''_a \,\mathrm{d}T \tag{13.41}$$

so as to write the coupling relationship between efforts:

$$\mu_{T,q_a} = R'_a T \tag{13.42}$$

Finally, as the definition (according to Equation 13.39) of this chemical potential coefficient corresponds to the definition (according to Equation 13.38) of the scaling chemical potential, one finds for this latter

Real substance:
$$\mu_\Delta = \mu_{T,q_a} = R'_a T \tag{13.43}$$

This expression gives the scaling chemical potential in case of coupling with the thermal energy without having to restrict to the case of an ideal substance and is therefore of wide generality.

Naturally, the case of an ideal substance can be derived from the previous model by assuming a constant common value for the second derivatives

Ideal substance:
$$\frac{\partial R''_a}{\partial T}\underset{ideal}{=} 0 \tag{13.44}$$

This constant common value is identified with the gas constant R because this model must also apply to the case of the ideal gas:

Gas constant:
$$R \overset{def}{=} R'_a \underset{ideal}{=} R''_a \underset{ideal}{=} C^{st} \tag{13.45}$$

Consequently, the expression of the scaling chemical potential for an ideal substance becomes

Ideal substance:
$$\mu_\Delta = RT \tag{13.46}$$

The classical proportionality to the temperature through the gas constant is found again, but having considerably increased its validity range, as the influence theory has the great advantage of not requiring large numbers of objects, contrary to statistical methods.

However, the most interesting result is the possibility of using another expression for nonideal substances, which opens numerous possibilities, not only for the physical chemical domain but also for the other domains that can be coupled, as their scaling efforts are also coupled through the coupling factors. For instance, the following scaling efforts, among others, is directly coupled with the scaling chemical potential:

Scaling energy-per-corpuscle: $$E_{\Delta} = \frac{\mu_{\Delta}}{N_A} \quad \text{(Corpuscular energy)} \tag{13.47}$$

Scaling affinity: $$\mathcal{A}_{\Delta} = \frac{\mu_{\Delta}}{\nu} \quad \text{(Chemical reaction energy)} \tag{13.48}$$

Scaling pressure: $$P_{\Delta} = -\frac{\mu_{\Delta}}{V_{\mathrm{m}}} \quad \text{(Hydrodynamics)} \tag{13.49}$$

Scaling electrical potential: $$V_{\Delta} = \frac{\mu_{\Delta}}{F} \quad \text{(Electrodynamics)} \tag{13.50}$$

Scaling gravitational potential: $$V_{G\Delta} = -\frac{\mu_{\Delta}}{M} \quad \text{(Gravitation)} \tag{13.51}$$

This list is not exhaustive; in fact, in every energy variety that may be coupled with one of these energy varieties, a scaling effort exists.

Special treatment is required for the thermal energy variety. In this variety, the capacitive relationship is reversed as explained in case study A9 in Chapter 4 describing the thermal pole. The thermal capacitive relation does not use a scaling temperature T_{Δ} but a scaling entropy S_{Δ} identified with the Boltzmann constant

Scaling entropy: $$S_{\Delta} = k_B \quad \text{(Thermics)} \tag{13.52}$$

This enlightens the role of the Boltzmann constant as a scaling parameter throughout physics. In the next section is discussed the picture constituted by all couplings met until now.

13.5 MAP OF ENERGETIC COUPLINGS

The various case studies given in the last chapters have examined several couplings between energy varieties.

The simplified Formal Graph in Graph 13.3 synthesizes these relationships in a single map in sketching the energy varieties with only the entity numbers. The significance of the coupling factors is given in Table 13.3.

Table 13.3 mentions the coupling factors represented in Graph 13.13 and some others.

Some of these coupling factors can be considered to be composed of more elementary factors.

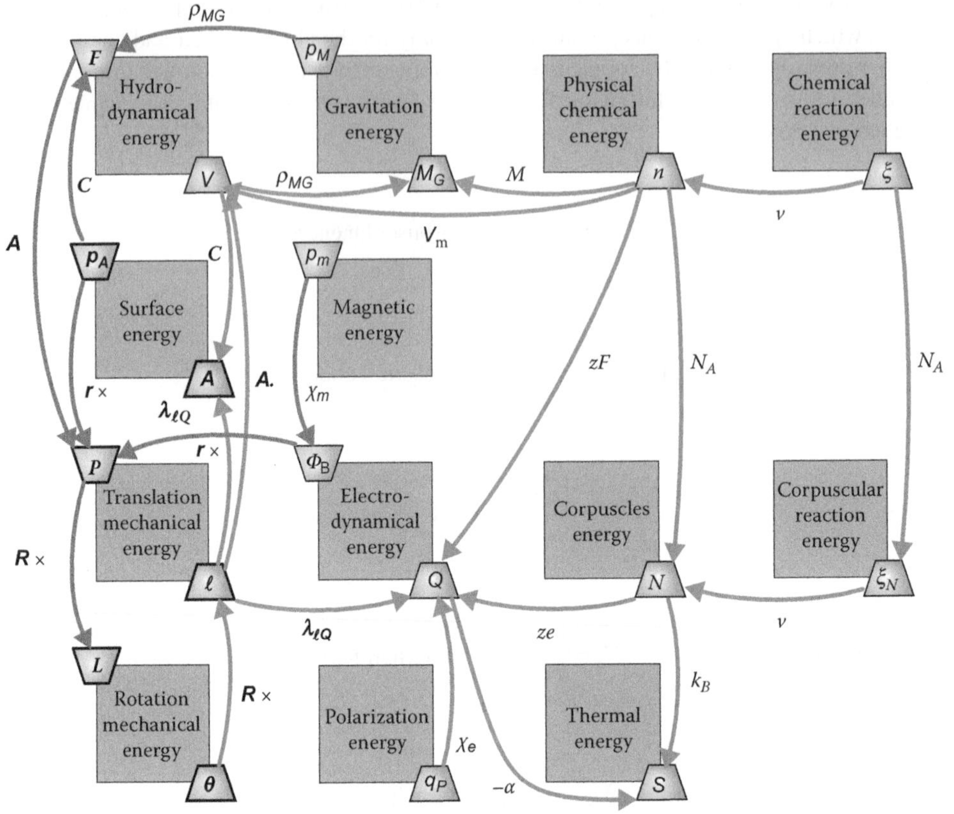

GRAPH 13.13 Mapping of the usual couplings and equivalences between energy varieties.

The coupling factor $\lambda_{\ell Q}$ between electrodynamics and translational mechanics is not classically used as such but as a *piezoelectric voltage coefficient* g (in $m^2\ C^{-1}$) divided by a characteristic length. In an anisotropic three-dimensional material, this coefficient is a tensor that links the stress $F_{/A}$ to the electric field E and is equivalent to the multiplication of the coupling factor $\lambda_{\ell Q}$ with the spatial integration of the stress (i.e., the lineic density of the force):

Piezoelectric voltage coefficient:
$$E = \hat{g}\, F_{/A} = \lambda_{\ell Q}\, F_{/r} = \lambda_{\ell Q}\, \hat{R}\, F_{/A} \qquad (13.53)$$

The coupling factor $\lambda_{\ell Q}$ was used in case study J3 "Electrical Force" in Chapter 12 and is represented as such in Graph 13.13. This is a coupling factor that can be expressed as a function of other coupling factors:

$$\lambda'_{\ell Q} = zF\frac{A'}{V'_m}; \quad \lambda_{\ell Q} = zF\frac{A}{V_m} \qquad (13.54)$$

The coupling factor $\lambda_{\ell G}$ between gravity and translational mechanics, which is similar to a lineic mass, was used in case study J4 "Gravitational Force" in Chapter 12. It can be also expressed as a function of other coupling factors.

$$\lambda'_{\ell G} = \rho'_{M_G}\, A'; \quad \lambda_{\ell G} = \rho_{M_G}\, A \qquad (13.55)$$

This factor is not represented in Graph 13.13 (for clarity).

TABLE 13.3
List of Coupling Factors between Energy Varieties

Energy Varieties					
From	**To**	**Coupling Factor**	**Name**	**Unit**[a]	**Case Study/ Chapter**
Rotational mechanics	Translational mechanics	$R\times$	Radius of gyration	m rad^{-1}	K5
Translational mechanics	Surface energy	$r\times$	Distance, height	m	
Hydrodynamics	Surface energy	C	Curvature	m^{-1}	J7
Translational mechanics	Hydrodynamics	$A.$	Cross-sectional area	m^2	J5
Hydrodynamics	Gravitation	ρ_M	Volumic mass	kg m^{-3}	K3
Electrodynamics	Translational mechanics	$\lambda_{\ell Q}$	Piezoelectric coefficient	C m^{-1}	J3
Gravitation	Translational mechanics	$\lambda_{\ell G}$	Lineic mass	kg m^{-1}	J4
Polarization energy	Electrodynamics	χ_e	Electric susceptibility	—	J12
Magnetism	Electrodynamics	χ_m	Magnetic susceptibility	—	J13
Physical chemistry	Electrodynamics	zF	Charge number × Faraday	C mol^{-1}	J2, J6
Corpuscular energy	Electrodynamics	ze	Charge number × elementary charge	C $corp^{-1}$	J6
Physical chemistry	Corpuscular energy	N_A	Avogadro constant	corp mol^{-1}	Chapter 12
Physical chemistry	Hydrodynamics	V_m	Molar volume	m^3 mol^{-1}	K1
Physical chemistry	Gravitation	M	Molar mass	kg mol^{-1}	K3
Corpuscular energy	Hydrodynamics	V_N	Corpuscular volume	m^3 $corp^{-1}$	
Corpuscular energy	Gravitation	m_N	Corpuscular gravitational mass	kg $corp^{-1}$	
Chemical reaction	Physical chemistry	ν	Stoichiometry	—	J1
Corpuscular Reaction	Corpuscular energy	ν	Stoichiometry	—	J1
Physical Chemistry	Thermics	R	Gas constant	J K^{-1} mol^{-1}	K2
Corpuscular Energy	Thermics	k_B	Boltzmann constant	J K^{-1} $corp^{-1}$	K2
Electrodynamics	Thermics	$-\alpha$	Seebeck coefficient	V K^{-1}	J9, J11

Note: The factors are given in their integral form for the relationships between basic quantities.
[a] All are SI units with the exception of corp, which is the proposed unit for counting corpuscles.

In Graph 13.13, it can be observed that the Seebeck coefficient is also a nonelementary factor as it can be expressed, when it is a constant, as the ratio of the gas constant (or the Boltzmann constant) on the ionic molar (respectively molecular) charge

$$-\alpha = \frac{R}{zF} = \frac{k_B}{ze} \tag{13.56}$$

These decompositions into more elementary factors are purely virtual ones. They do not imply the existence of real areas or true ions in the systems. A force may exist independent of any surface and there is no absolute requirement of charged carriers in a thermoelectric junction which, in principle, may work with neutral carriers.

Most of These Coupling Factors are Material or System Dependent	
Radius (of gyration)	$\boldsymbol{R}$
Distance, height	$\boldsymbol{r}$
Curvature	C
Cross-sectional area	$\boldsymbol{A}$
Volumic (gravitational) mass	ρ_{MG}
Ionic charge	zF, ze
Susceptibilities	χ_e, χ_m
Molar, corpuscular volume	V_m, V_N
Stoichiometric coefficient	ν
Seebeck coefficient	α
A Minority of them are Universal Constants	
Faraday constant:	F $(= N_A\, e)$
Elementary charge	e
Avogadro constant	N_A
Gas constant	R $(= N_A\, k_B)$
Boltzmann constant	k_B

Appendix 2 gives the values of these physical constants.

Conversely, not all universal constants are coupling factors; some are used as constitutive properties (e.g., the gravitational constant G) or space–time properties (e.g., the speed of light c in free space). The quantum world has not been tackled, but some of the universal constants met in this domain are also coupling constants, notably the most important one being the Planck constant h.

13.6 IN SHORT

ENERGY OF COUPLING

Part of the total energy of a system is involved in the coupling. This energy is a positive quantity given by the minimum of the products of state variables. The sign must be adapted for having a positive coupling energy.

EXTERNAL Capacitive

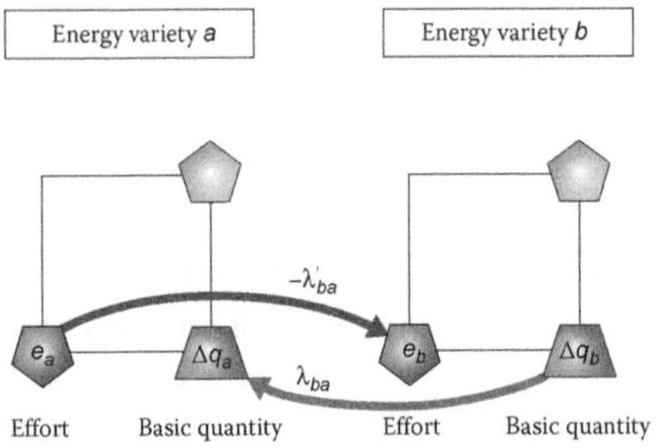

GRAPH 13.14

Invariable $\left(\lambda'_{ba} = C^{st} = \lambda_{ba}\right)$

$$\pm\mathcal{U}_{ba} \overset{def}{=} e_a\left(q_a - q_a^{min}\right) = -e_b\left(q_b - q_b^{min}\right) \quad (13.8)$$

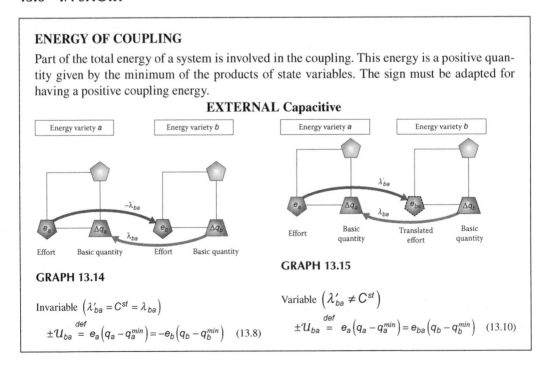

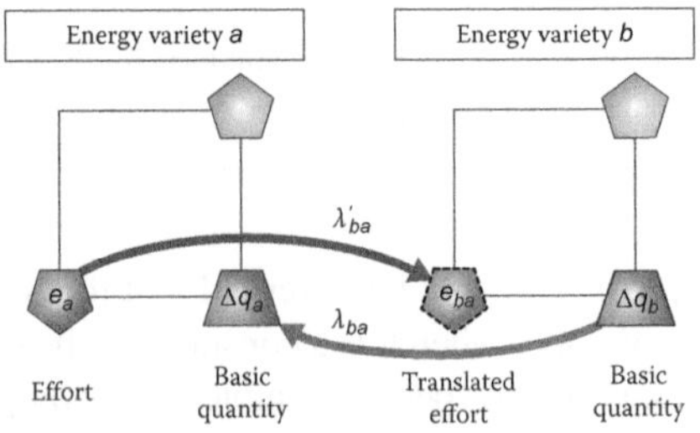

GRAPH 13.15

Variable $\left(\lambda'_{ba} \neq C^{st}\right)$

$$\pm\mathcal{U}_{ba} \overset{def}{=} e_a\left(q_a - q_a^{min}\right) = e_{ba}\left(q_b - q_b^{min}\right) \quad (13.10)$$

EXTERNAL Inductive

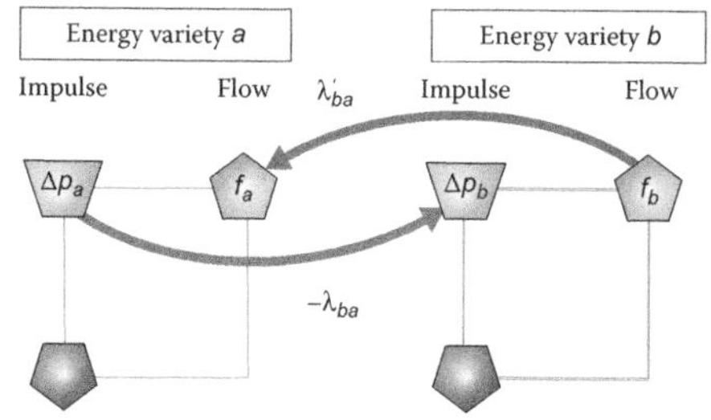

GRAPH 13.16

Invariable $\left(\lambda'_{ba} = C^{st} = \lambda_{ba}\right)$

$$\pm\mathcal{T}_{ba} \overset{def}{=} f_a\left(p_a - p_a^{min}\right) = -f_b\left(p_b - p_b^{min}\right) \quad (13.9)$$

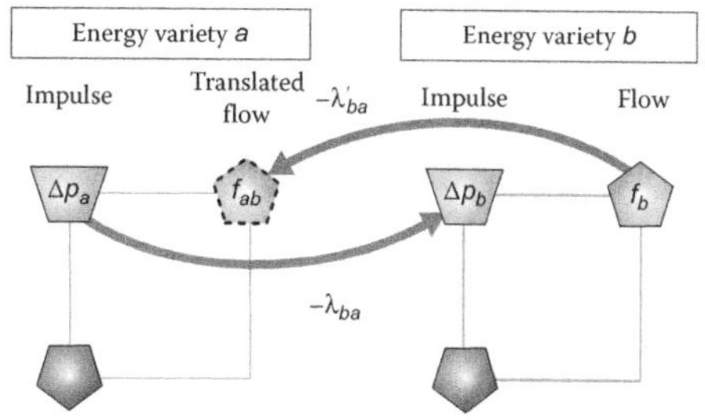

GRAPH 13.17

Variable $\left(\lambda'_{ba} \neq C^{st}\right)$

$$\pm\mathcal{T}_{ba} \overset{def}{=} f_a\left(p_a - p_a^{min}\right) = f_{ba}\left(p_b - p_b^{min}\right) \quad (13.11)$$

INTERNAL Inductive–Capacitive

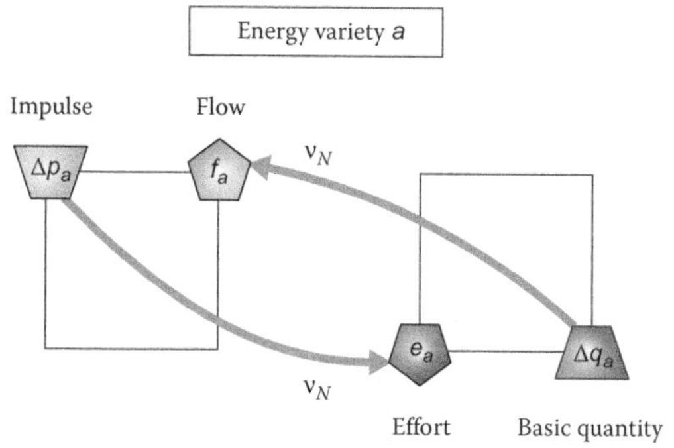

GRAPH 13.18

$$\pm\mathcal{E}_{aa} \overset{def}{=} f_a\left(p_a - p_a^{min}\right) = e_a\left(q_a - q_a^{min}\right) \quad (13.12)$$

ENERGETIC RATIOS

This is the generalization of the 1/2 coefficient found in energy linear expressions.

Subvariety	Capacitive	Inductive
Integral	$\alpha_a \overset{def}{=} \dfrac{\mathcal{U}_a}{\mathcal{U}_{ba}}$ (13.13)	$\beta_a \overset{def}{=} \dfrac{\mathcal{T}_a}{\mathcal{T}_{ba}}$ (13.13)
Differential	$\alpha'_a \overset{def}{=} \dfrac{d\mathcal{U}_a}{d\mathcal{U}_{ba}}$ (13.14)	$\beta'_a \overset{def}{=} \dfrac{d\mathcal{T}_a}{d\mathcal{T}_{ba}}$ (13.14)
Linear frame	$\alpha'_a \underset{lin}{=} \alpha_a \underset{lin}{=} \dfrac{1}{2}$ (13.16)	$\beta'_\ell \underset{lin}{=} \beta_\ell \underset{lin}{=} \dfrac{1}{2}$ (13.18)
Energy	$\mathcal{U}_q = \alpha_q\, e_q \hat{C}_q\, e_q$ (13.19)	$\mathcal{T}_q = \beta_q\, f_q \hat{L}_q\, f_q$ (13.19)
Examples	Electrodynamics (electrostatic energy) $\mathcal{U}_Q = \alpha_Q\, V \hat{C}\, V$ (13.20) $\mathcal{U}_Q \underset{lin}{=} \dfrac{1}{2} C V^2$	Translational mechanics (kinetic energy) $\mathcal{T}_\ell = \beta_\ell\, \mathbf{v} \hat{M} \mathbf{v}$ (13.20) $\mathcal{T}_\ell \underset{lin}{=} \dfrac{1}{2} M \mathbf{v}^2$

SCALING EFFORT				
The *scaling effort* is the variable normalizing the effort in a capacitive relationship. It is specific to each energy variety but strongly depends on the existence of coupling with another variety. In the absence of coupling the scaling effort is quasi-infinite, thus linearizing the capacitance in the practical range.				
Definition $\mathcal{U}_{abi}$: Coupling energy a: Coupled variety b: The considered variety i: Other varieties (eventually)	$e_{b\Delta} \overset{def}{=} \left(\frac{\partial \mathcal{U}_{abi}}{\partial q_b}\right)_{q_a, q_i, e_i}$ (generalization of Equation 13.38)			
Energy Variety	Real Case		Ideal Case	
Physical chemistry	$\mu_\Delta = R'_a T$	(13.43)	$\mu_\Delta = RT$	(13.27)
Corpuscles	$E_\Delta = \frac{\mu_\Delta}{N_A}$	(13.47)	$E_\Delta = k_B T$	(13.28)
Chemical reaction	$\mathcal{A}_\Delta = \frac{\mu_\Delta}{\nu}$	(13.48)	$\mathcal{A}_\Delta = \frac{RT}{\nu}$	
Corpuscular reaction	$\mathcal{A}_{N\Delta} = \frac{E_\Delta}{\nu}$		$\mathcal{A}_{N\Delta} = \frac{k_B T}{\nu}$	
Hydrodynamics	$P_\Delta = -\frac{\mu_\Delta}{V_m}$	(13.49)	$P_\Delta = -\frac{RT}{V_m} = -\frac{k_B T}{V_N}$	
Electrodynamics	$V_\Delta = \frac{\mu_\Delta}{F}$	(13.50)	$V_\Delta = \frac{RT}{F} = \frac{k_B T}{e}$	(13.29)
Gravitation	$V_{G\Delta} = -\frac{\mu_\Delta}{M}$	(13.51)	$V_{G\Delta} = -\frac{RT}{M} = -\frac{k_B T}{m_N}$	

14 Conclusion and Perspectives

CONTENTS

The Formal Graph theory is made up of two things: Formal Graphs and Formal Objects.

They are indissociable in this theory, although one may perfectly conceive a different language from a graph for modeling the properties of Formal Objects. The features of Formal Graphs have been thoroughly described in gross and details to avoid being recalled. The numerous examples and case studies give a practical view in a large number of disciplines, in formulating regrets that in formulating regrets that it was necessary to make selection and that all interesting domains could not be tackled. Not all energy varieties have been explored and no more complex systems have been modeled.

To conclude this presentation of the theory, the main elements—the quintessence of the Formal Graph theory—are synthesized. Then some perspectives on the subjects of relativity and quantum physics are offered.

14.1 CHARACTERISTICS OF THE THEORY

Four salient characteristics contributing to the originality of the theory are briefly outlined: the structure, the computing ability, the way space–time is handled, and the proposal of transverse models for constitutive properties.

14.1.1 Structure

The rational construction and use of a Formal Graph rely on a rigorous classification and on precise rules (grammar) that respect the principles of thermodynamics. The structure of the theory can be summarized as follows:

- One concept of energy is subdivided into several energy varieties.
- An energy variety is divided into two subvarieties:
  - **Inductive** (kinetic, electromagnetic, magnetic, etc.) (*not always supported*).
  - **Capacitive** (internal, electrostatic, elastic, etc.).

- State variables are classified into four categories (Helmholtz):
  - **Basic quantity:** Entity number in the capacitive subvariety (charge, volume, substance amount, etc.).
  - **Effort:** Energy-per-entity in the capacitive subvariety (potential, pressure, chemical potential, etc.).
  - **Impulse:** Entity number in the inductive subvariety (momentum, induction quantity (or flux), magnetic impulse, etc.).
  - **Flow:** Energy-per-entity in the inductive subvariety (velocity, current, magnetic current, etc.).
- Four categories of links can be distinguished:
  - **Constitutive properties** of the system (capacitance, inductance, conductance).
  - **Space–time properties** (evolution—time derivation, space distribution—space derivation and space–time velocity).
  - **Coupling factors** (intervariety of energy coefficients).
  - **Multiplicity coefficients** (generalized stoichiometric coefficients, connecting Formal Objects).
- Several Formal Objects are organized along a scale of complexity:
  - **Singleton** (no collective behavior).
  - **Pole** (collection of identical entities).
  - **Dipole** (two poles).
  - **Multipole** (more than two poles).
  - **Dipole assembly** (several different dipoles).
  - **Dipole distribution** (several identical dipoles distributed over energy-per-entity levels).

More details are given in Section 14.4 in the form of tables.

14.1.2 Computing Ability

Direct application to computing is one of the interesting features of the theory. It has been outlined that a Formal Graph is a peculiar implementation of neural networks. [See Haykin (2009) for the numerous examples and references therein.] This concept, borrowed from neural science, is a practical tool based on the general graph theory allowing modeling and simulating a wide variety of systems. It is more frequently being used by information technology and engineer communities, and many softwares, free or commercial (mainly as plug-ins of standard computing software), are available for computation. No special adaptation of a Formal Graph is required for using these softwares.

One may remark that this is not the case for algebraic equations that require compilation preprocessing before feeding any simulation software. This is a remarkable feature of a Formal Graph to reduce the distance between the mind and numeric results or graphical curves.

14.1.3 Time and Space

This is one of the original aspects of the theory, not to assume in an implicit manner the existence of time before to develop the main concepts and models. Time is defined only when required, when conversion between subvarieties occurs. As long as such a conversion is absent from a system, time may exist but is not involved in the various mechanisms depicting the properties and behavior of the system.

The definition of time as a variable following (or controlling) the extent of an energy conversion is a local definition, valid in the considered system, even if the same variable is used for all conversions whatever the involved subvarieties. This opens to other conceptions of time, as for instance in Einstein's theory of spacetime, in which time depends on the localization in space. Also, the Formal Graph theory allows considering some possible discontinuity or a granular structure of time through the use of evolution operators that may differ from temporal derivations.

The same granularity or difference from derivation operators is allowed for space (which is not original). Furthermore, the same approach as the one used for conditioning the role of time to the existence of energy conversion is followed for space. Space is not required as long as energy distributions do not occur.

The parallelism between time and space is pursued in applying the same concept of reversibility to both. A space distribution of energy can be a source of irreversibility as demonstrated in case study H8 "Attenuated Propagation" in Chapter 11 dealing with attenuated propagation due to radiation absorption. This is a property of space which is made clearer than in the classical theories.

14.1.4 Nonlinearity of Constitutive Properties

It can be assessed that the nonlinearity of operators is an essential and recurrent postulate of the Formal Graph theory. The linearity comes as a restricted case of validity of a more general behavior which is transverse to all energy varieties.

The trilogy pole–dipole–multipole has given rise to the possibility of developing two theories, one for **influence** and the other for **conduction**. These theories have provided two models for the system constitutive properties: capacitance and conductance.

Both of them appear quite similar because they involve an exponential function with an effort as argument. This similarity must not hinder their fundamental difference, which is that an intermediate variable, the *activity*, is used for capacitance, while conductance is a direct relationship. Furthermore, the introduction of the notion of *gain* for relating activities of different poles is not straightforward; it is clearly a postulate in the theory of influence. On the contrary, the notion of *reversibility variable* stems quasi-naturally from the principle of an exchange using forward and backward flows and from the concept of chain multipoles.

The physics behind an exchange is more intuitive than in the case of influence, which is not an easy question, especially when generalized to domains other than electrodynamics or gravitation, that is, to other domains that do not use the concept of field or action at a distance. This is also the case when no space is involved, as is the case tackled in considering the global level of the systems. It means that corpuscles, for instance, influence each other without needing to think in terms of localization or space distribution.

The situation is completely different for the third system constitutive property—inductance. As inferred from the given case studies, inductive relationships do not involve exponential functions, but are rather based on quadratic functions. These relationships have not been demonstrated because it requires more elements than have been exposed until now. In particular, some extra postulates must be added about the conservation of impulses and effort through space as done in mechanics through Einstein's theory of special relativity.

It must be repeated that the exponential shape of the capacitive and conductive relationships is only a proposal, what we call the *standard shape* of these relationships. The mathematical expressions of these properties can be modified and improved at will for taking into account the influence of other energy varieties.

14.1.5 Why Two Subvarieties?

It is tempting to see the justification of the existence of the two subvarieties in the conversion process which requires a substrate energy and a product, but it would be a circular reasoning requiring *a priori* the existence of time!

With the differences between the mathematical expressions of the two operators, capacitance and inductance, a first reason emerges for the existence of two energy subvarieties. Besides the shapes, the difference in their relationship with space–time is essential: one is space–time independent (i.e., the capacitive subvariety) whereas the other (i.e., the inductive subvariety) can be strongly

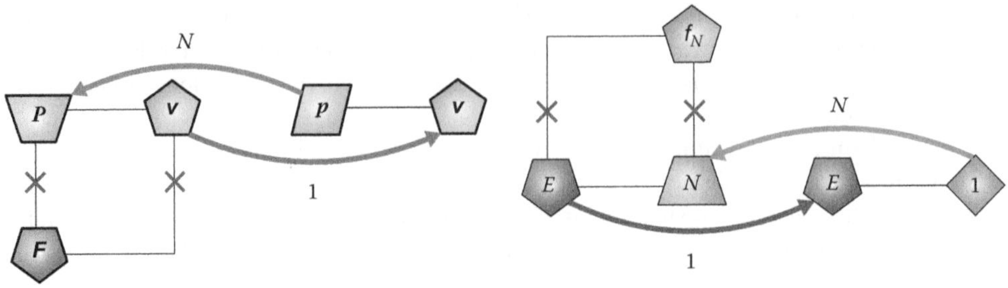

GRAPH 14.1 Connections between singletons and their collections (poles) of N corpuscles. The left Formal Graph shows the inductive singleton in the translational mechanical energy variety and the right one shows the capacitive singleton in the corpuscular energy variety.

entangled with time and space. This difference sheds some light on the reasons why some energy varieties possess the two subvarieties while others do not.

This first reason is based on macroscopic arguments because only collections are considered. When considering the lower level of singletons, one finds another reason.

In Graph 14.1 are repeated the two Formal Graphs already given in Chapter 4, making explicit the relationships between singletons and poles, but adapted to translational mechanics for the inductive subvariety and to the capacitive subvariety for corpuscular energy.

The fundamental difference between these singletons lies in their entity numbers. Both result from the division by the number of corpuscles N of the respective pole entity numbers, but in doing so the capacitive entity number becomes a constant, equal to 1, while the inductive entity number, the corpuscular momentum $\boldsymbol{p}$, remains a variable. This difference is observed for all energy varieties, the capacitive entity numbers of all singletons always being constants. Two examples will be given in the next section: the thermal singleton whose entity number is the Boltzmann constant and the singleton in the oscillation energy variety whose entity number is the Planck constant.

This is the main argument lending credence to the necessity of distinguishing between two subvarieties. Now, the question of knowing why the inductive subvariety is not supported by all systems in all energy varieties remains to be elucidated.

14.2 PERSPECTIVES

Due to the necessity of producing a condensed volume, the content of this book has been devoted to a limited part of physics and physical chemistry that excludes more complicated systems. More sophisticated Formal Objects that are the distributed dipoles and systems with assemblies of these objects have not been studied further. These concepts enable us to handle systems with several energy levels that are of paramount importance in condensed matter physics and in many other fields. The Maxwell distribution of velocities in a gas is one of the important systems that are modeled with this Formal Object.

It would have been interesting also to explore many more possibilities offered by the assembling game that consists of coupling many more energy varieties for modeling real systems instead of ideal ones as in this study. The subtle mechanisms that distribute energy over several energy varieties have not been studied and the question of knowing why some varieties receive more energy in some conditions and how this distribution of energy evolves upon perturbation of the system is still unclear.

The track beyond the open door brought by the concept of *scaling effort* has not been explored either. The possibility of modification of this variable for taking into account some influent energy varieties is a fruitful way for modeling more sophisticated constitutive relationships than the simple ones used in this study. For instance, the concept of *free volume* used in many models of conduction can be easily incorporated under the form of hydrodynamical or surface energy interactions in addition to thermal agitation.

This volume constraint has not allowed one to tackle more elaborate theories such as Einstein's relativity and quantum physics. The natural consequence of this choice is to induce the feeling that the Formal Graph theory is valid only in the classical frame of a nonrelativistic and macroscopic world. This would be a false judgment that would unfairly diminish the virtues of the proposed tools that are much more powerful than explained.

In order not to leave the reader with too brief a view, possible uses in relativity and quantum physics are examined here concisely. Notably, one will demonstrate how the Schrödinger* equation logically stems from the Formal Graph theory. These topics are not discussed for understanding the underlying physics but merely as illustrations.

14.2.1 Relativity

Extension to relativity is not difficult and does not require any modification of the Formal Graphs. Everything has been set in place to allow various improvements in Newton's theory. The first provision made for escaping from the classical frame was the decomposition of Newton's second law of motion into three operators instead of using the concept of acceleration (cf. case study F1 "Centripetal Force" in Chapter 9). The second provision was the systematic consideration of the nonlinearity of the system constitutive properties for all energy varieties and subvarieties, allowing namely the dependence of the inertial mass upon the velocity as is the case in Einstein's theory of special relativity.

Throughout this book, the properties of space and time have been presented in mentioning that the traditional derivations were not compulsory. The possibility of considering discrete space and time was evocated and it is obvious that the use of general operators $\hat{\mathbf{R}}$ and $\hat{\mathsf{T}}$ allows extension to any theory of space–time such as the theory of general relativity. This is the power of a graph to keep the same structure in allowing the links to adapt to various operator shapes.

The implementation of relativistic concepts, such as the quadrivectors and their invariance, remains to be described. Contrary to the most used approach in this domain, quantum concepts are mixed deliberately with Formal Objects allowing modeling of single entities or singletons, briefly described in Chapter 4 when discussing poles.

Graph 14.2 shows two singletons, a corpuscle capacitive one and a corpuscle inductive one, linked by two space–time velocities $\boldsymbol{u}$ and $\boldsymbol{v}$, the latter corresponding to corpuscle velocity. The inductive singleton is related to a collection of N corpuscles (inductive pole) through the following relationships

$$\boldsymbol{P} = N\boldsymbol{p}; \quad M = Nm \tag{14.1}$$

which means that the unit of the corpuscular inertial mass m is kilogram per corpuscle.

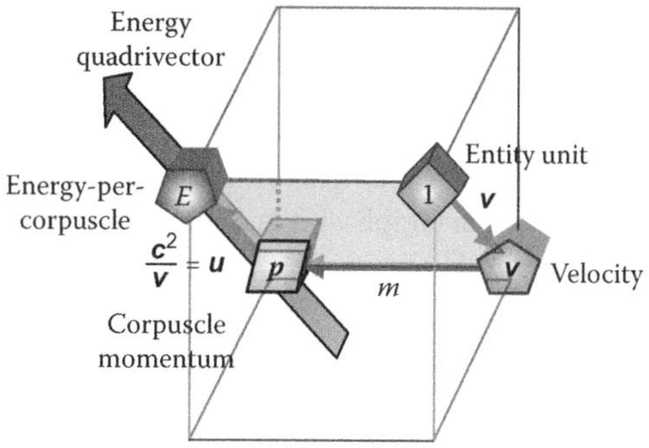

GRAPH 14.2 Representation with a Formal Graph of the concept of quadrivector and of the association through space–time velocities of two singletons, one capacitive (front) and the other inductive (rear). The well-known expression $E = mc^2$ can be read by running from the entity number 1 to the energy-per-entity E.

* Erwin Rudolf Josef Alexander Schrödinger (1887–1961): Austrian physicist; Vienna, Austria and many other European places.

The corpuscle capacitive singleton is peculiar since its entity number is one corpuscle, encoded with a "1" in its node, which is called *entity unit.*

The concept of quadrivector, which takes as components a vector in three-dimensional space (here the particle momentum $\boldsymbol{p}$) and a scalar (here the energy-per-entity E), is encoded by a thick arrow linking these two nodes on the Formal Graph. Note that E is considered as "energy" in Einstein's theory, but it amounts to the same (in taking the product NE) when a collection of singletons forming a pole (body) is considered. When the theory of relativity is applied to quantum objects, as is the case here, the distinction between energy and energy-per-entity is useful.

The hypotheses required to relate the capacitive energy-per-corpuscle to the inertial mass are expressed by the following four equations:

Inductive (kinetic) energy-per-corpuscle: $$\frac{\mathrm{d}\mathcal{T}_\ell}{N} = \mathbf{v} \cdot \mathbf{d}\boldsymbol{p} \quad (14.2)$$

Capacitive (internal) energy-per-corpuscle: $$\frac{\mathrm{d}\mathcal{U}_N}{N} = \mathrm{d}E \quad (14.3)$$

Energy equivalence: $$\mathrm{d}\mathcal{U}_N = \mathrm{d}\mathcal{T}_\ell \quad (14.4)$$

Quadrivector modulus invariance: $$E^2 - c^2\boldsymbol{p}^2 = E_0^2 \quad (14.5)$$

The first two equations correspond to the energy variations of singletons that follow the rule of the First Principle of Thermodynamics for the inductive singleton (when the entity number may vary) and that deviate from this principle in the case of the capacitive singleton as its entity number cannot vary. The energies are divided by N as we consider one corpuscle; the unit of the energy-per-corpuscle is Joules per corpuscle and not the Joule.

The third equation merely expresses the equivalence between the kinetic (inductive) energy and the internal (capacitive) energy in terms of variations, as postulated by Einstein's theory of special relativity. We shall see that a slightly different equivalence is required for establishing the Schrödinger equation (in other words, it is strongly dependent upon the energy varieties taken into account in a system).

The fourth equation states the invariance of the energy quadrivector, with the energy-per-entity E_0 at rest ($\boldsymbol{p} = 0$). This is the main hypothesis which ascribes to the space–time the role of ensuring this invariance. The scope of this book does not allow one to examine this point in detail. (For more details, see Feynman (1964).)

This set of equations leads, in a purely mathematically way, to the following expression of the energy-per-entity, which is represented in Graph 14.2 as a space–time velocity between the corpuscle momentum $\boldsymbol{p}$ and the energy-per-entity E:

$$E = \mathbf{u} \cdot p = \frac{c^2}{\mathbf{v}}\, p \quad (14.6)$$

(This provides the expression for the space–time velocity $\mathbf{u}$.) Finally, the attribution of a mass to the corpuscle is taken into account by setting the inductive relationship:

$$p = m\mathbf{v} \quad (14.7)$$

with an inertial mass m that is a variable property depending on the particle velocity (cf. case study A1 "Moving Body" in Chapter 4 which describes this relationship for a pole). These relationships are explicitly represented in Graph 14.2 together with the space–time velocity relating the entity unit to the particle velocity. By combining the two previous relationships, the well-known expression of the energy-per-entity, as proposed by Einstein, is obtained:

Einstein: $$E = mc^2 \tag{14.8}$$

The Formal Graph representation of this expression is ensured by the combination of the three links starting from the entity unit, going through the particle velocity and momentum and arriving at the energy-per-entity. These three links can be viewed as the apparent link between the entity unit and the energy-per-entity, as if the particle possessed a capacitance.

The Formal Graph puts into evidence the distinctive role played by space–time, which, on the one hand, makes a space–time velocity correspond to the particle velocity $\boldsymbol{v}$ and, on the other, uses another velocity $\boldsymbol{u}$ for relating the particle momentum $\boldsymbol{p}$ to the energy-per-entity E. In fact, the constancy of the product $\boldsymbol{u}.\boldsymbol{v} = c^2$ is another way of expressing the invariance of the energy quadrivector.

The two velocities $\boldsymbol{u}$ and $\boldsymbol{v}$ become equal when the corpuscle velocity $\boldsymbol{v}$ reaches the maximum velocity c. In this case, the two singletons are equivalent by internal coupling, which features the behavior of a photon in free space (in specifying that, in this case, no inertial mass can be attributed).

Obviously, more explanations are required to justify this model, which is a very basic one, but it is not the purpose of this brief example.

14.2.2 Quantum Physics

As for the theory of relativity, many ingredients that allow taking into account the quantum theory have been designed along with the development of the Formal Graph theory. The most important one is the clear distinction between a collective behavior of several objects, modeled by a *pole*, and the individual behavior of corpuscles (that can be several eventually), modeled by the concept of a *singleton*. This distinction is the translation into the Formal Graph language of the traditional distinction between macroscopic and microscopic. The only difference is that the nature of influence between objects, and not the size of objects, is the criterion for distinguishing in the Formal Graph theory. However, both approaches reveal identical results when considering many interdependent objects.

A second ingredient, not so important in terms of concept but essential for understanding the notion of oscillation or wave, is the notion of *wave function* ψ from the properties of oscillators, as discussed in Chapters 9 and 11. These discussions have brought out new state variables, different from those in other energy varieties considered until now, that quantify the energy contained in waves. These are the *wave pulsation* ω and the *wave vector* $\boldsymbol{k}$, which we consider state variables of the *wave energy*; from these state variables, the *frequency* ν and the *reciprocal wavelength* $\boldsymbol{1/\lambda}$ were derived by division by 2π (cf. Chapter 9, Sections 9.5 and 9.6). Now, these last state variables can be related to a variety of energies known as *oscillation energy*, as opposed to *wave energy*.

14.2.2.1 Particle/Wave Duality

The concept of energy equivalence, introduced in Chapter 12, using the full angle 2π as a coupling factor for making equivalent the *wave energy* and the *oscillation energy*, can be applied to a much more significant equivalence that may occur between a corpuscle and an oscillation, also known as the *particle–wave duality* (Pauli 1973).

The coupling factor is nothing other than the Planck constant h that relates the frequency ν and the *reciprocal wavelength* $\boldsymbol{1/\lambda}$ to the energy-per-corpuscle E and the corpuscle momentum $\boldsymbol{p}$,

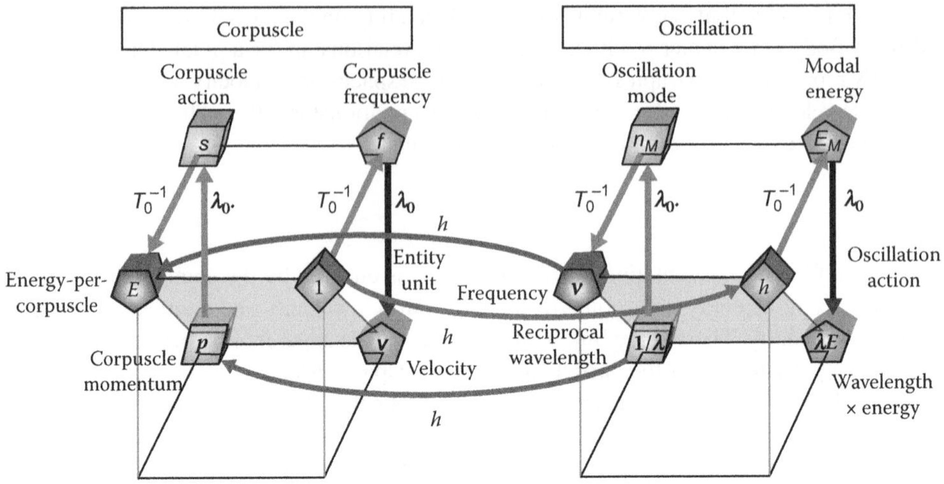

GRAPH 14.3 Equivalence between two representations of the same physical being, a corpuscle on the left and an oscillation on the right. The coupling factor is the Planck constant h. Only the significant connections are shown that model the Planck and de Broglie relationships and relate the entity numbers "1" and "h" (establishing the energetic equivalence or duality).

respectively, according to the Planck* (Equation 14.9) and de Broglie† (Equation 14.10) relationships as shown below:

Planck:

$$E = h\nu = \hbar\omega \tag{14.9}$$

de Broglie:

$$p = \frac{h}{\lambda} = \hbar\,\boldsymbol{k} \tag{14.10}$$

The Formal Graph modeling these relationships is given in Graph 14.3. It is built with only two blocks (i.e., without the block of the wave energy for simplicity), one for the corpuscle and the other for the oscillation. Each block is made of three singletons, one capacitive and two inductive. The left block was present in Graph 14.2 without the action–frequency (s, f) singleton (which has for variables, as mentioned, the *action* s of the corpuscle and its *frequency* f). Here, this singleton is internally coupled to the corpuscle capacitive singleton (1, E) through a coupling factor T_0, a scalar that can be viewed as the eigen function of the evolution operator. It is also spatially coupled to the corpuscle inductive singleton ($\boldsymbol{p}$, $\boldsymbol{v}$) through a fundamental wavelength λ_0. These coupling factors feature the time interval and space extension occupied or allowed for the system (confined object). Again, more explanations about these factors and the underlying fundamental physical processes cannot be discussed in detail. The right block exactly reproduces the same configuration and the same coupling factors work on both sides owing to the existence of the energetic equivalence.

In Chapter 9, the *phase angle* φ was shown to feature the properties of a wave, and the corresponding variable obtained by division by 2π is the *oscillation mode number* n_M. This variable without dimension quantifies a number of rotations ("turns") of the wave function and it is placed in the Formal Graph in the node of the entity number of the oscillation inductive singleton. According to the quantum theory, this number is always an integer, implying that, in a restricted time window and/or space extension (confined object), only fundamental and harmonic frequencies could be present (i.e., only complete turns of the wave function are possible). This is the fundamental principle

* Max Karl Ernst Ludwig Planck (1858–1947): German physicist; Berlin, Germany.
† Louis Victor Pierre Raymond de Broglie (1892–1987): French physicist; Paris, France.

establishing the quanticization of energy and this number is called the *principal quantum number* (Griffiths 2004).

14.2.2.2 Schrödinger Equation

All ingredients for establishing the well-known Schrödinger equation have been designed in previous chapters. The only hypothesis required is the assumption of a corpuscle possessing kinetic energy and some capacitive energy, from an energy variety q that does not need to be specified:

Energy equivalence:
$$\mathcal{U}_N = \mathcal{T}_\ell + \mathcal{U}_q \tag{14.11}$$

This equivalence slightly differs from Einstein's equivalence between corpuscular energy and kinetic energy variations, expressed by Equation 14.4, by the inclusion of a capacitive energy term. However, the two approaches amount to the same in case of constancy of the added capacitive energy.

Integrating the differential Equation 14.3, one gets

$$\frac{\mathcal{U}_N}{N} = E \tag{14.12}$$

and from the expression of the kinetic energy given in Chapter 13, Section 13.3 in the linear approximation and adapted to one corpuscle, one finds

$$\frac{\mathcal{T}_\ell}{N} \underset{lin}{=} \frac{1}{2} m\mathbf{v}^2 \underset{lin}{=} \frac{1}{2m} \mathbf{p}^2 \tag{14.13}$$

The capacitive energy for one corpuscle can be written under the form of a *corpuscle potential* as follows:

$$\frac{\mathcal{U}_q}{N} = U_q \tag{14.14}$$

The last three equations contain all the physics needed in energetic terms to establish the Shrödinger equation. What is needed in addition is to involve a wave featured by a wave function Ψ. Chapter 9, Section 9.7.4, discusses how a wave function can be built from an oscillator generating a wave. The propagation through time and space was modeled by the following two differential equations:

$$i\frac{\partial}{\partial t}\psi = \omega\,\psi \tag{14.15}$$

$$-i\frac{\partial}{\partial \mathbf{r}}\psi = \mathbf{k}\,\psi \tag{14.16}$$

Subscript q for the wave function is dropped because no reference to the energy variety of the oscillator at the origin of the wave (if any) is necessary here. Such a wave function describes the state of a wave and its interpretation in terms of probability is optional and not inherent to the demonstration.

Multiplication by the *reduced Planck constant* $\hbar$, also called *Dirac** *constant*, which is $h/2\pi$, allows involving the corpuscular energy-per-entity E and the momentum $\mathbf{p}$ by using the Planck and

* Paul Adrien Maurice Dirac (1902–1984). British physicist; Bristol, UK and Tallahassee, Florida, USA.

de Broglie relationships (Equations 14.9 and 14.10). The previous equations are rewritten with these state variables, in applying twice the space derivation in the case of the second equation for obtaining the square of the corpuscle momentum:

$$i\hbar\frac{\partial}{\partial t}\psi = \hbar\omega\,\psi = E\,\psi \tag{14.17}$$

$$-\hbar^2\frac{\partial^2}{\partial \boldsymbol{r}^2}\psi = \hbar^2\boldsymbol{k}^2\,\psi = \boldsymbol{p}^2\,\psi \tag{14.18}$$

A simple mathematical elimination of state variables between these equations and the relationships shown in Equations 14.12 through 14.14 leads to the Schrödinger equation:

Schrödinger:
$$i\hbar\frac{\partial}{\partial t}\psi = \widehat{\mathrm{H}}\psi \tag{14.19}$$

in having used the equivalence relationship in Equation 14.11 for grouping in a single operator the *kinetic operator* (as it is usually termed) and the *corpuscle potential*:

Hamiltonian:
$$\widehat{\mathrm{H}} \underset{lin}{=} \left\{-\frac{\hbar^2}{2m}\frac{\partial^2}{\partial \boldsymbol{r}^2} + U_q\right\} \tag{14.20}$$

This operator is known as the Hamiltonian* operator. It encompasses all the corpuscle energies-per-entities that feature the system and its expression may therefore vary according to the various energy varieties possibly involved.

It may also differ from the above expression in case of variability (nonlinearity) of the inertial mass obliging us to use a more general expression of the kinetic energy, as was proposed in Chapter 13, Section 13.3, with an *energetic ratio* β_ℓ that generalized the one-half coefficient of the classical theory.

As the Schrödinger equation has been established from scratch with a certain number of hypotheses rectricting its validity to the linear frame, the advantage is that the method can be modified and improved for modeling more complex systems.

14.2.3 Quantum Chemistry

It is worth discussing a last example owing to its ability to illustrate the powerfulness of the Formal Graph approach, and also because it demonstrates a result that was merely cited in Chapter 4. In case study A11 (Chapter 4) of a reactive chemical species modeled by a pole, the expression of the rate constant as a function of the *molar free enthalpy of reaction* of the reaction intermediate was given as follows:

Arrhenius:
$$k = \frac{k_B T}{h}\,exp\left(-\frac{\Delta_r G_{\neq}}{RT}\right) \tag{14.21}$$

* William Rowan Hamilton (1805–1865): Irish physicist, astronomer, and mathematician; Dublin, Ireland.

The pre-exponential factor of this Arrhenius* equation was given without demonstration as a known result of the theory of the transition state established by Wigner,† Eyring,‡ Polanyi,§ and Evans¶ in 1930 (Laidler and King 1998).

This pre-exponential factor is in fact an *intrinsic rate constant* k^0 as explained in case study A11 "Reactive Chemical Species" (Chapter 4) which can be identified with the rate constant of the intermediate species in the transition state theory. Modeling this property with the Formal Graph tools requires one pole (in the physical chemical or corpuscular energy) and five singletons. The corpuscle capacitive singleton is connected to the corpuscular pole, as explained in Chapter 4, Section 4.4, through number of corpuscles N and this singleton is coupled with the corpuscle inductive singleton and with the singleton representing the *elementary thermal energy*, having for entity the Boltzmann constant k_B. This latter also plays the role of the coupling factor. The duality corpuscle–oscillation is represented by an *energetic equivalence* between the respective singletons through the Planck constant, as discussed previously. This model is shown in Graph 14.4.

In Graph 14.4 can be read the following sequence of equalities that implies the temporal coupling between singletons and the equality between paths leading to the energy-per-corpuscle E of the corpuscle capacitive singleton:

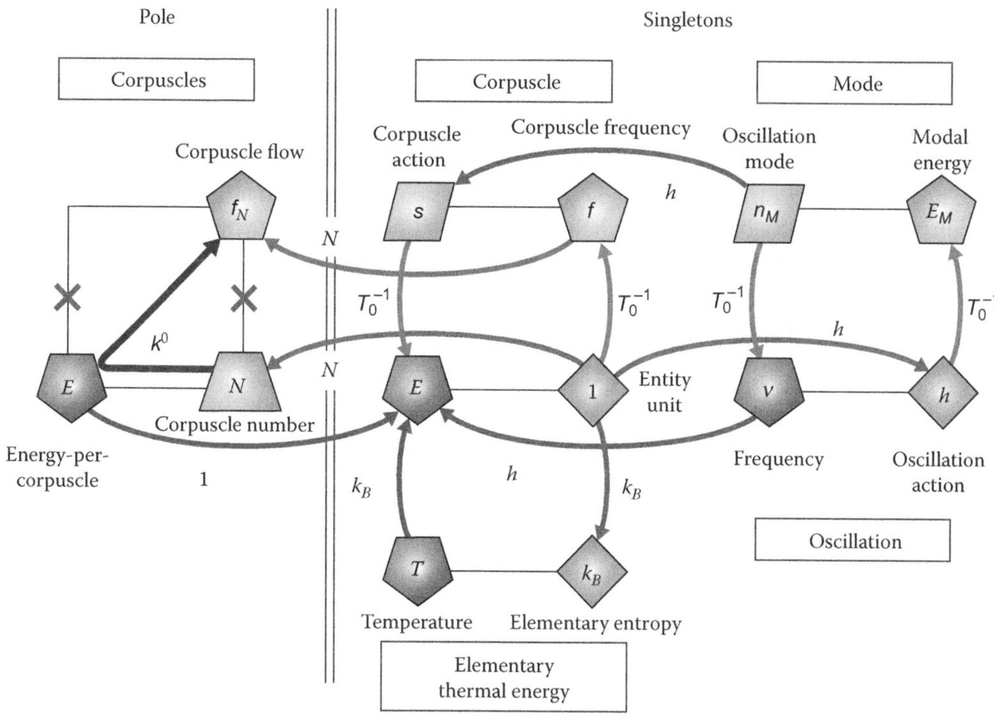

GRAPH 14.4 Formal graph of the various couplings among singletons modeling the intrinsic rate constant k^0 of a chemical or corpuscular reaction. On the left is the "macroscopic" world of poles (here a collection of N singletons) and on the right is the "microscopic" world of quantum objects.

* Svante August Arrhenius (1859–1927): Swedish chemist; Stockholm, Sweden.
† Eugene Paul Wigner (1902–1995): Hungarian-American physicist and mathematician; Princeton, NJ, USA.
‡ Henry Eyring (1901–1981): American chemist; Salt Lake City, Utah, USA.
§ Michael Polanyi (1891–1976): Hungarian-British physical chemist; Manchester, UK.
¶ Meredith Gwynne Evans (1904–1952): British physical chemist; Manchester, UK.

$$k^0 = \frac{f_N}{N} = \frac{f}{1} = \frac{\nu}{n_M} \quad (14.22)$$

$$E = h\nu = k_B T \quad (14.23)$$

By eliminating the frequency ν between these two expressions, the *intrinsic rate constant* k^0 is found:

$$k^0 = \frac{k_B T}{n_M h} \quad (14.24)$$

The classical expression is retrieved when the *number of oscillation modes* is equal to 1 (i.e., fundamental harmonic), which is the highest possible value for a reaction rate.

The physical meaning of this model is that all the reaction intermediate corpuscles behave identically (as expressed by the singleton–pole connection) as thermally agitated corpuscles in resonance with an oscillation in its fundamental mode. The transition state is featured by these couplings that are not necessarily working for the other participants in the reaction. This is a short statement, certainly too short to be clear enough, that does not enter into the details of the interesting physics involved. It can nevertheless be understood that it is a very basic model and that more complicated ones can be built by adding other couplings and energy varieties at will.

We have reached a point where the basic kinetic model of a chemical reaction has been fully demonstrated from scratch, in having based the proposed theory upon two pillars of the Formal Graph theory, the *influence theory* developed in Chapter 7 and the *conduction theory* developed in Chapter 8, without forgetting the *scaling effort* concept justifying the presence of the *RT* product in the denominator of the argument of the exponential. It can be added that all the restrictions introduced along our demonstration can be removed to access more elaborated models without too much difficulty. The generality of the approach must also be outlined since all elements used demonstrate the kinetic Arrhenius model have been established for any energy variety.

14.3 CONCLUSION

The value of a Formal Graph as a pedagogical tool is manifest and should convince many teachers to use it. The possibility offered to students and foreigners to a discipline to enter into new ones should be facilitated by using this common language at the small (?) expense of having to acquire it. The many generalizations made throughout this presentation should incite the researcher to go beyond his simplified models. The comparisons between disciplines allowed by such a unified view should be a valuable tool for people interested in scientific ideas. At last, the many demonstrations of empirical laws and postulates in the classical theories are a little contribution to the advance of science. However, there is more in this new approach.

The word "unification" is too heavily charged with the meaning of a huge attempt mobilizing generations of physicists to characterize a modest theory such as this one, proposed by a physical chemist whose latent defect is not to be a true physicist.

However, the idea is there, in proposing a transverse tool able to model systems through the main domains of physics and physical chemistry. The Formal Graph theory is a *general frame*, a cradle able to host most of the existing theories, certainly not all, because we have not checked the compatibility of all known theories, which is a long and tedious task.

The history of science has always shown that science is in perpetual evolution and that the notion of definite theory is contrary to the quintessence of science which is to progress by invalidating ["falsifying" as termed by Karl Popper* (see Popper 2002)] admitted theories in order to propose new and improved ones.

* Karl Raimund Popper (1902–1994): Austro-British philosopher of science; London, UK.

We hope that many researchers will pursue this work and will find some discrepancies and defaults lending to fruitful improvements.

14.4 IN SHORT

A brief summary recalling the structure of the theory is given in Tables 14.1 through 14.6.

TABLE 14.1
State Variables and Energy Subvarieties

	Capacitive Subvariety	Inductive Subvariety
Entity number	Basic quantity q	Impulse p_q
Energy-per-entity	Effort e_q	Flow f_q
Energy	$\mathcal{U}_q$	$\mathcal{T}_q$
Energy variation (for collections)	$\delta\mathcal{U}_q \underset{\exists C_q}{=} e_q\, dq$	$\delta\mathcal{T}_q \underset{\exists L_q}{=} f_q\, dp_q$
Total energy in each subvariety	$\mathcal{U} = \sum_i \mathcal{U}_{qi}$	$\mathcal{T} = \sum_i \mathcal{T}_{qi}$
Total energy	$\mathcal{E}_q = \mathcal{T}_q + \mathcal{U}_q;\quad \mathcal{E} = \mathcal{T} + \mathcal{U} = \sum_i \mathcal{E}_{qi}$	

TABLE 14.2
Distribution through Space: Depths (or Levels) of Spatial Localization

Depth n_R	Localization	Variable	Generic Symbol	Generic Name
0	Whole system	Global quantity	u	State variable or Gate variable
1	Point of a curve	Global quantity/Length	$u_{/r}$	Lineic density, field strength, gradient, etc.
2	Point of a surface	Global quantity/Area	$u_{/A}$	(Surface) Density
3	Point of a volume	Global quantity/Volume	$u_{/V}$	Concentration (volumic "density")

TABLE 14.3
Four Categories of Elementary Links in a Formal Graph

Link Category	Elementary Path	Operator	Physical Meaning
System constitutive properties	Inductance	$\hat{L}_q$	Inductive energy storage
	Capacitance	$\hat{C}_q$	Capacitive energy storage
	Conductance	$\hat{G}_q$	Energy dissipation
Space–time properties	Evolution	$\hat{T}^{-1}, \frac{d}{dt}$	Energy conversion
	Space distribution	$\hat{R}^{-1}, \frac{d}{d\mathbf{r}}$	Energy distribution/spreading (through space)
Energetic coefficients	Energy coupling	λ (scalar, vector)	Association of energy varieties
Connections	Multiplicity (stoichiometry)	ν (scalar, vector)	Association of Formal Objects

Note: Space–time is not necessarily considered as continuous. Discrete operators may replace derivations.

TABLE 14.4
Levels of Complexity of the Various Formal Objects

Formal Objects	Description
Lower levels	May come before if required (subparticles, quarks, etc.).
Singleton	Featured by a pair of *energy-per-entity* and *entity number*, in one *energy subvariety*. No collective behavior in a group of *singletons*. The *entity number* is invariable in a capacitive singleton.
Pole	Collection of *entities* of the same *subvariety*, in variable *number*, with a common *energy-per-entity*, which requires a collective behavior. A pole can be composed of several identical subpoles (with the same *energy-per-entity* shared by all).
Dipole	Association of two *poles* of the same *nature* (same *energy subvariety* and same *system constitutive properties*). Each pole has its own *energy-per-entity*.
Multipole	Several *poles* of the same nature. They may form a chain (in serial), or ladder (in parallel), which is an extension of the notion of *dipole*.
Dipole assembly	Association of *dipoles* of the same *energy variety* but different *natures*.
Dipole distribution	Ensemble of *dipoles* of the same nature but having different *energies-per-entities*.
Higher levels	May follow if required.

Note: Formal Objects are made up of nodes (variables) and links (operators). Several levels are distinguished in the complexity of Formal Objects. All Formal Objects can be *associated* (same energy variety) or *coupled* (different energy varieties) at all levels.

TABLE 14.5
Different Status Taken by a Variable in a Node

	Capacitive Subvariety	Inductive Subvariety
State variable	$e_q \overset{def}{=} \left(\frac{\partial \mathcal{U}}{\partial q}\right)_{q_i \neq q}$	$f_q \overset{def}{=} \left(\frac{\partial \mathcal{T}}{\partial p_q}\right)_{p_{qi} \neq p_q}$
Gate variable	$f_q = \hat{T}^{-1}\, q$ (q: State variable)	$e_q = \hat{T}^{-1}\, p_q$ (p_q: State variable)
Count variable	$q = \hat{T}\, f_q$ (f_q: State variable)	$p_q = \hat{T}\, e_q$ (e_q: State variable)
Localized variable	$q_{/r} = \hat{\mathbf{R}}^{-1}\, q$; $e_{q/r} = \hat{\mathbf{R}}^{-1}\, e_q$	$p_{q/r} = \hat{\mathbf{R}}^{-1}\, p_q$; $f_{q/r} = \hat{\mathbf{R}}^{-1}\, f_q$
	$\mathbf{q}_{/A} = \hat{\mathbf{R}}^{-2}\, q$; $\mathbf{e}_{q/A} = \hat{\mathbf{R}}^{-2}\, e_q$	$\mathbf{p}_{q/A} = \hat{\mathbf{R}}^{-2}\, p_q$; $\mathbf{e}_{q/A} = \hat{\mathbf{R}}^{-2}\, e_q$
	$q_{/V} = \hat{\mathbf{R}}^{-3}\, q$; $e_{q/V} = \hat{\mathbf{R}}^{-3}\, e_q$	$p_{q/V} = \hat{\mathbf{R}}^{-3}\, p_q$; $f_{q/V} = \hat{\mathbf{R}}^{-3}\, f_q$

Note: Translated or apparent variables are not listed.

TABLE 14.6
Capacitive and Inductive Relationships: Two Constitutive Properties Allowing a System to Store Energy and the Standard Relationships Proposed in the Formal Graph Theory

Energy Subvariety	Capacitive Relationship	Inductive Relationship
In General	**Energy Variety** q	**Energy Variety** q
General frame	$q = q^{min} + \left(q^{\theta} - q^{min}\right)\exp\left(\frac{e_q - e_q^{\theta}}{e_{q\Delta}}\right)$	$p_q = p_q^{min} + \frac{L_{q0}\, f_q}{\sqrt{1 + \frac{f_q^2}{f_{q\Delta}^2}}}$
Linear frame	$q \underset{lin}{=} C_{q0}\, e_q$	$p_q \underset{lin}{=} L_{q0}\, f_q$
Exception (Permutation *entity number/ energy-per-entity*)	Thermics (thermal energy) $T = T^0 \exp\left(\frac{S - S^0}{S_\Delta}\right)$	Translational mechanics $\boldsymbol{P} = \frac{M_0\, \boldsymbol{v}}{\sqrt{1 - \frac{v^2}{c^2}}}$

Appendix 1: Glossary

Entry	Description
Activity (a_q)	The role of this variable is to quantify the availability of entities in an *influence* process. It is defined at the level of a *pole*.
Apparent variable ($\bar{y}$)	When a *state variable* is modified by contributions from other state variables (in the same family), it becomes an apparent state variable. For example, the electrochemical potential is a chemical potential to which has been added an electric contribution under the form of proportionality with the electrical potential (as it happens with an ion).
Basic quantity (q)	One of the four families of *state variables* in the *canonical scheme*. It is the *entity number* in the *capacitive energy subvariety*. Its symbol is used for identifying the other variables belonging to the same *energy variety*.
Canonical scheme	Organization established by Helmholtz distributing the state variables into four families: *basic quantity, effort, impulse*, and *flow*.
Capacitance ($\hat{C}_q$)	*Constitutive property* enabling a system to store *capacitive energy*. A capacitance is an operator linking an *effort* to a *basic quantity*. Its reciprocal is the *elastance*.
Capacitive energy subvariety ($\mathcal{U}_q$)	One of the two *energy subvarieties* in an *energy variety*, the other being the *inductive subvariety*. Examples are: internal energy, elastic energy, potential energy, electrostatics, hydrostatics, etc.
Circularity principle	For respecting the First Principle of Thermodynamics, the Formal Graph using a connection between two *Formal Objects* must implement a loop with the connecting links and the internal links (constitutive and space–time properties) of the objects.
Collection of entities	Ensemble of identical *entities* having the same *energy-per-entity*. A pole consists of a collection of entities. By contrast, a *singleton* is a single *entity*.
Common effort	Two (or several) *dipoles* sharing the same *effort* are mounted in parallel in the equivalent circuit language.
Common flow	Two (or several) *dipoles* sharing the same *flow* are mounted in series in the equivalent circuit language.
Conductance ($\hat{G}_q$)	*Constitutive property* enabling a system to dissipate energy. A conductance is an operator linking an *effort* to a *flow*. Its reciprocal is the *resistance*.
Constitutive property	See *System constitutive property*.
Corpuscle	Usually particle and corpuscle refer to the same notion of a mechanical object, localized in space. In the Formal Graph approach, the corpuscle is distinct from the particle because it does not belong to the mechanical energy variety but to its own variety. A corpuscle has no localization in space by definition, unless a coupling with mechanics is at work.
Counted variable	Variable similar to an entity number resulting from a counting (discrete case) of circulating entities or from integration (continuous case) of the energy-per-entity. Examples of counted variables are the charge passed in an electric circuit and the distance covered by a moving object.
Coupling	Internal coupling involves two subvarieties of the same energy variety, establishing synchronization of two space–time velocities (see example F8 "Hall effect") or two frequencies (see example K4 "Kinetic of gases"). External coupling refers to the association of two energy varieties (or more) in the same energy subvariety for sharing and minimizing energy in a system. An external coupling is invariable when the two *coupling coefficients* (integral and differential) are equal and constant.

continued

Entry	Description
Coupling factor (λ_{ba})	Scalar or vector linking two *state variables* for establishing an *external coupling* or an *energetic equivalence*. The linked state variables must belong to the same *family* but in two different *energy varieties*. The integral coupling coefficient links two *entity numbers* whereas the differential coupling coefficient links their variations. A coupling coefficient may be used for establishing equivalences between *state variables* (without coupling energies) or for coupling two *energy subvarieties*. In this case the differential coupling coefficient also links the two *energies-per-entity* (with a minus sign).
Dipole	Association of two *poles* belonging to the same *energy subvariety*. A dipole is fundamental when only one system constitutive property exists, and mixed when the conductance is present with one of the two other properties (inductance or capacitance).
Dipole assembly	Association of several *dipoles* belonging to the same *energy variety*. Any number of dipoles may be assembled in different mountings, but an *elementary dipole assembly* is made up only with dipoles sharing the same energy-per-entity (common flow or common effort but not both).
Dipole distribution	*Formal Object* made up with several *dipoles* having various reference *energies-per-entity*.
Effort (e_q)	One of the four families of *state variables* in the *canonical scheme*. It is the *energy-per-entity* in the *capacitive subvariety of energy*.
Elastance ($\hat{C}_q^{-1}$)	*Constitutive property* enabling a system to store *capacitive energy*. An elastance is an operator linking a *basic quantity* to an *effort*. Its reciprocal is the *capacitance*.
Embedding principle	Whatever the level of the *Formal Objects* considered in modeling a system, the energy of the system is independent from the way to express it. For example, the energy of the two poles forming a dipole can be equivalently expressed as the sum of the pole energies or as the energy of the sole dipole. This principle also applies to the Formal Graph modeling: A Formal Graph of two poles can be replaced by a Formal Graph of the dipole. This simplifies considerably the drawing of models when the complexity of the system increases.
Energetic equivalence	Energetic equivalence means that the unit used four counting entity numbers has no effect on the amount of energy. Two energy subvarieties are equivalent when their entity numbers are identical in nature but are counted differently (e.g., moles and corpuscles).
Energetic nature	Describes a *Formal Object* in terms of storage and/or dissipation ability relatively to a given energy subvariety. The various energetic natures are called from the actual *system constitutive properties*, that is, inductive, capacitive, or conductive for fundamental objects, and for mixed ones, inductive–conductive or capacitive–conductive.
Energy-per-entity	*State variable* quantifying the energy amount in an *entity*. *Efforts* and *flows* are energies-per-entity.
Energy subvariety	An *energy variety* is composed of two subvarieties, one capacitive, the other inductive. Energy can be converted between these two subvarieties. The amount of energy of the subvariety is determined by a pair [*entity number, energy-per-entity*] of *state variables*.
Energy variety	Form of energy contained in two different kinds of containers (each kind defining an *energy subvariety*) that are able to convert energy between them.
Entity	Container of energy in a given *energy subvariety*. The nature of the container can be diverse and defines the *energy subvariety*.
Entity number	*State variable* counting the number of *entities* in a given *energy subvariety*. *Basic quantities* and *impulses* are entity numbers.
Evolution	Operation in which the *entity number* is converted into an *energy-per-entity* in the other *energy subvariety*. Evolution occurs in an energy conversion process which is featured by the variable time, playing the role of extent or advance. It is present in fundamental *dipoles* that store energy (inductive or capacitive dipoles) as a link between energies-per-entity and entity numbers belonging to different subvarieties. In a continuous space–time, the evolution operator is the derivation with respect to time. By opposition, the inverse evolution is the integration with respect to time.

Entry	Description
Exchange of entities	Process by which two *poles* equilibrate their number of entities (basic quantities or impulses) for sharing energy between them. The vector of the exchange is the energy-per-entity of the other energy subvariety (gate variable). There are two categories: exchange by flows of basic quantities and exchange by efforts of impulses.
Exchange reversibility ($r_{qi,j}$)	The *exchange reversibility* is an always positive variable that quantifies the reversibility of an *exchange of entities* between two poles labeled i and j. It ranges from zero, meaning total irreversibility in the $i \rightarrow j$ direction, to infinite, meaning total irreversibility in the opposite direction. The peculiar value 1 corresponds to an exchange in dynamic equilibrium (equal flows or efforts in both directions) and a tolerance around this value delimitates a reversible regime (or *quasireversible* when tolerance is large). It is a multiplicative variable from pole to pole when modeling a *multipole* chain.
Family	Used for describing a category of state variables having the same role for determining the energy in a system. According to Helmholtz's *canonical scheme*, there are four families named after the generic names of the state variables: *basic quantities, efforts, impulses,* and *flows*.
Flow (f_q)	One of the four families of *state variables* in the *canonical scheme*. It is the *energy-per-entity* in the *inductive subvariety of energy*.
Formal Object	The physical system represented by a Formal Graph. Formal Objects are organized by levels along a complexity scale: singletons, poles, dipoles, dipole assemblies, energy distributed dipoles, etc.
Gain ($g_{qi,j}$)	Variable quantifying the amplification of *activities* between two interacting *poles*. It is a multiplicative variable from pole to pole when modeling a *multipole* chain.
Gate variable	Variable in a node used as gate or port for exchanging energy (through entities) with another *Formal Object*. A gate variable is an *energy-per-entity* (*effort of flow*), which may work in conjunction with or without its *entity number* (it has the status of *state variable* in this case).
Ideal system	There is no absolute definition of ideality. Ideal systems are generally those having a minimum number of energy varieties participating in the system and featured by constant coupling factors. It becomes real when one more variety is added. The ideal gas is the archetype of this concept. The ideal substance is derived from it, in referring to the coupling with thermics or hydrodynamics which are not always made explicit.
Impulse (p_q)	One of the four families of *state variables* in the *canonical scheme*. It is the *entity number* in the *inductive subvariety of energy*.
Inductance ($\hat{L}_q$)	*Constitutive property* enabling a system to store *inductive energy*. An inductance is an operator linking a *flow* to an *impulse*. Its reciprocal is the *reluctance*.
Inductive energy subvariety ($\mathcal{T}_q$)	One of the two *energy subvarieties* in an *energy variety*, the other being the *capacitive subvariety*. Examples are: kinetic energy, electromagnetic energy, magnetic energy, etc.
Influence	Process in which some pole entities are acting on their own *energy-per-entity* (self-influence) or on the *energy-per-entity* of another *pole* (mutual influence). The support of the influence is a *system constitutive property*.
Localized variable	Variable defined at a point in space, which belongs to a line (curve), a surface of a volume, of the system. A localized variable results from a *state variable* by application of a space operator.
Multipole	Ensemble of several *poles* (more than two) of the same *energetic nature* (i.e., same *system constitutive properties*, same *energy subvariety*) having a common *gate variable* (effort or flow). Poles can be grouped in chains (common flow, i.e., in serial) or in ladders (common effort, i.e., in parallel) or in a mix of arrangements.
Nature	See *Energetic nature*.
Natural pulsation (ω_q)	System property featuring a temporal oscillator in a given energy variety. (The pulsation ω of the associated wave is equal to this variable only in case of conservative (nondamped) oscillator.)

continued

Entry	Description
Natural space vector ($\boldsymbol{k}_q$)	System property featuring a spatial oscillator in a given energy variety. [The space vector $\boldsymbol{k}$ of the associated spatial wave is equal to this variable only in case of conservative (nondamped) oscillator.]
Organization level	The complexity scale of *Formal Objects* is divided into levels. In this book, the range goes from *singletons to dipole distributions*.
Pole	*Formal Object* made up with a *collection of entities*. Depending on the *constitutive properties* possessed, three fundamental types of poles are defined: inductive, capacitive, and conductive, and two mixed types: inductive–conductive and capacitive–conductive.
Pole nature	The *pole nature* refers to its *energy subvariety* and to its *system constitutive properties*.
Reduced property	*Constitutive property* of a system linking two *localized variables* or one *localized variable* and a variable at the global level.
Reluctance ($\hat{L}_q^{-1}$)	*Constitutive property* enabling a system to store *inductive energy*. A reluctance is an operator linking a *flow* to an *impulse*. Its reciprocal is the *inductance*.
Resistance ($\hat{G}_q^{-1}, \hat{R}_q$)	*Constitutive property* enabling a system to dissipate energy (whatever the *energy subvariety*). A resistance is an operator linking a *flow* to an *effort*. Its reciprocal is the *conductance*.
Reversibility	Applies to a process, either for describing a potentiality to work in two opposite directions, or for featuring a behavior in which both directions are quantitatively equivalent. For instance, a reversible reaction means that the forward and the backward reaction may exist, independently of their magnitude. A reversible kinetic means that forward and backward flows are equal (dynamic equilibrium) or almost (near equilibrium). In the Formal Graph theory, the *exchange reversibility* is a variable quantifying this behavior.
Semienergy (A_q, B_q)	Energetic variable used in the Formal Graph model of oscillators, chosen as intermediate between state variables (halfway in the peculiar case of *linear system constitutive properties*). Its unit is $J^{1/2}$.
Singleton	*Formal Object* comprising a single *entity* (in contradistinction with a collection that comprises several entities).
State variable	Generic name for the *entity number* and the *energy-per-entity*. When both variables, belonging to the same *energy subvariety*, are present in a *Formal Object*, the energy amount in this subvariety can be determined, which in turn defines the state of the system.
System constitutive property	Property possessed by a system allowing to store or to dissipate (convert into heat) energy. There are three properties: *inductance, capacitance*, and *conductance*.
Translated variable	A *state variable* converted into the state variable in the same family but belonging to another *energy variety*, by multiplication by a *coupling factor*.

Appendix 2: Symbols and Constants

CONTENTS

A2.1 SYMBOLS

A2.1.1 OPERATORS

$\hat{1}$	Identity operator
$\widehat{\mathbf{A}}$	Surface collect (integration over a surface) [space–time]
d	Differentiation, exact when applied to state functions (energy)
δ	Nonexact differentiation
∂	Partial differentiation
Δ	Laplacian (= ∇^2) (also Δ means *difference, shift*)
$\hat{C}_q$	Capacitance (scalar or vector operator) [variety q]
$\hat{D}$	Diffusivity tensor [physical chemistry]
$\hat{E}_q$	Elastance (scalar or vector operator) [variety q]
$\hat{G}_q$	Conductance (scalar or vector operator) [variety q]
$\hat{H}$	Hamiltonian [several varieties]
$\hat{I}_q$	Influence operator [variety q]
$\hat{\mathbf{k}}$	Space–vector operator for a wave function [space–time]
$\hat{k}_e$	Rigidity, stiffness [translational mechanics]
$\hat{k}_f$	Viscous (friction) resistance [translational mechanics]
$\hat{k}_p$	Viscous inertia [translational mechanics]
$\hat{k}_q$	Natural relaxation operator [variety q]

* The mention "*scalar* or *vector* operator" applies to the result of the operator.

$\hat{L}_q$	Inductance (scalar or vector operator) [variety q]
$\hat{M}$	Inertial mass or global mass transfer operator [translational mechanics or physical chemistry]
$\hat{M}_q$	Kinetic operator (basic quantity transfer) (global) [variety q]
$\hat{\mathbf{m}}$	Reduced mass transfer operator [physical chemistry]
$\hat{\mathbf{m}}_q$	Reduced kinetic operator (basic quantity transfer) [variety q]
$\hat{N}_q$	Activation operator [variety q]
$\hat{N}_q$	Kinematic operator (impulse transfer) (global) [variety q]
$\hat{\mathbf{n}}_q$	Reduced kinematic operator (impulse transfer) [variety q]
$\hat{O}$	Operator
$\hat{\mathbf{R}}$	Space distribution or spreading (integration along a distance) [space–time]
$\hat{R}_q$	Resistance (scalar or vector operator) [variety q]
$\hat{\mathfrak{R}}_q$	Reluctance (scalar or vector operator) [variety q]
$\hat{T}$	Evolution (general case) [space–time]
$\hat{\mathbf{U}}$	Space–time velocity [space–time]
$\hat{\mathbf{u}}$	Velocity operator for a wave function [space–time]
$\hat{V}$	Integration over a volume [space–time]
$\hat{X}_a$	Integration with respect to variable x_a (continuous case)
$\hat{Y}_q$	Admittance [variety q]
$\hat{Z}_q$	Impedance [variety q]

A2.1.2 Greek and Other Symbols

$\hat{\gamma}_q$	Temporal damping operator of an oscillator [variety q]
$\hat{\zeta}_q$	Spatial damping operator of an oscillator [variety q]
$\hat{\boldsymbol{\nu}}$	Kinetic viscosity tensor [translational mechanics]
$\hat{\tau}_{tL}$	Viscous inertia relaxation [translational mechanics]
$\hat{\omega}_q$	Natural pulsation operator of an oscillator [variety q]
$\hat{\omega}$	Pulsation operator of a wave function [space–time]
$\frac{d}{dt}$	Evolution, derivation/time (continuous case) [space–time]
$\frac{d^{-1}}{dt^{-1}}$	Inverse evolution, integration over [0, t] (continuous case) [space–time]
$\frac{d}{d\mathbf{r}}$	Space distribution, derivation/space (continuous case) [space–time]
∇	Derivation/space ("Nabla" or "Del") (continuous case) [space–time]
$\frac{d}{d\mathbf{w}}$	Space–time operator (continuous case) [space–time]
$\left(\frac{\partial y}{\partial x_i}\right)_{x_j}$	Partial derivation of y with respect to x_i in maintaining all other x_j constant

A2.1.3 Dual Operators

×	Multiplication between operator results
×	Curl product (vector product)
•	Dot product (scalar product)
*	Convolution product

A2.1.4 Variables and Constants

A

$\mathcal{A}$	Affinity [chemical reaction energy]
A	Exponential prefactor
A_q	Semienergy of an oscillator (in $J^{1/2}$) [capacitive subvariety q]
a	Activity [physical chemistry]
a_q	Generalized activity [variety q]
$\boldsymbol{A}$	Potential vector [electrodynamics]
$\mathbf{A}$	Area (vector) [surface energy]
$\mathbf{a}$	Acceleration (vector) [translational mechanics]

B

B_q	Semi-energy of an oscillator (in $J^{1/2}$) [inductive subvariety q]
b_q	Exchange activity [variety q]
$\mathbf{B}$	Electromagnetic induction (vector) [electrodynamics]

C

C	Electric capacitance (scalar) [electrodynamics]
C_q	Generalized capacitance (scalar) [variety q]
C_θ	Torsion constant (scalar) [rotational mechanics]
c_n	Concentration of species (in mol. m^{-3}) [physical chemistry]
c_N	Concentration of corpuscles (in corp. m^{-3}) [corpuscular energy]
c	Maximum speed constant (speed light in vacuum) [space–time]
$\boldsymbol{C}$	Curvature of an interface (vector) [surface energy + hydrodynamics]
C_V, C_P	Molar heat capacities at constant volume or pressure [thermics + physical chemistry]

D

D	Number of dimensions of a Formal Graph
D	Diffusivity [physical chemistry and corpuscular energy]
$\mathbf{D}$	"Electrization," electric displacement [electrodynamics]
$\mathcal{D}$	Density of energy states [variety q]
d	Number of space dimensions
d°	Degree of freedom of a system

E

E	Energy-per-corpuscle [corpuscular energy]
E_F	Fermi level [corpuscular energy]
E_C	Energy-per-corpuscle (level) of the conduction band [corpuscular energy]
E_V	Energy-per-corpuscle (level) of the valence band [corpuscular energy]
E_Y	Young modulus [translational mechanics]
E	Electrode potential (electrochemistry) [electrodynamics]
E	Electric Elastance (scalar) [electrodynamics]
E_a	Molar energy of activation [physical chemistry]
$\mathcal{E}_q$	Energy [variety q]
e	Elementary charge [electrodynamics]
e_m	Magnetic potential [magnetism]
e_P	Polarization potential [electric polarization energy]

e_q	Effort [variety q]
e_{qC}	Effort involved in capacitive storage [variety q]
e_{qG}	Effort involved in conduction [variety q]
e_{qL}	Effort resulting from evolution of the impulse [variety q]
E_q	Generalized elastance (reciprocal of capacitance) [variety q]
$\mathbf{E}$	Electric field [electrodynamics]
$€_q$	Generalized elastance (reciprocal of capacitance) [variety q]

F

F	Faraday constant [physical chemistry + electrodynamics]
F	Percussion [hydrodynamics]
f_q	Function of distribution [variety q]
f_M	Mass flow [gravitation]
f_m	Magnetic current [magnetism]
f_N	Corpuscle flow [corpuscular energy]
f_P	Polarization current [electric polarization energy]
f_q	Flow [variety q]
f_{qC}	Flow resulting from evolution of the basic quantity [variety q]
f_{qG}	Flow involved in conduction [variety q]
f_{qL}	Flow involved in inductive storage [variety q]
f_S	Entropy flow [thermics]
$\mathbf{F}$	Force (vector) [translational mechanics]
$\mathbf{f}_A$	Surface expansion velocity (vector) [surface energy]

G

G	Gravitational constant [gravitation]
G_γ	Shear modulus [translational mechanics]
g	Gravitational field modulus (scalar) [gravitation]
g_q	Gain of activity [variety q]
$\mathbf{G}$	"Gravitization," gravitational displacement (vector) [gravitation]
$\mathbf{g}$	Gravitational field (vector) [gravitation]

H

h	Planck's constant [corpuscular energy]
$\hbar$	Dirac's constant (= $h/2\pi$) [corpuscular energy]
$\mathbf{H}$	Electromagnetic field (vector) [electrodynamics]
$\mathcal{H}$	Enthalpy
$\mathcal{H}(t)$	Heaviside function (step function)

I

I	Electric current [electrodynamics]
I_Δ	Scaling electric current [electrodynamics]
I_S	Saturation current [electrodynamics]
$\boldsymbol{I}_{\lambda q}$	Irradiance (or radiation intensity) at wavelength λ (vector) [variety q]
$\mathfrak{I}$	Substance flow [physical chemistry] (*I Fraktur*)
i	Imaginary number [*Maths*]

J

J Rotational inertia (or moment of) (scalar or tensor) [rotational mechanics]
$\boldsymbol{J}$ Substance flow density (vector) [physical chemistry]
$\boldsymbol{j}$ Current density (vector) [electrodynamics]

K

k_B Boltzmann constant [corpuscular energy + thermics]
k, k_A, k_B Reaction specific rates ("rate constants") [physical chemistry]
k_e Spring constant, elastance (elasticity) [translational mechanics]
k_q Kinetic constant [variety q]
k_{qC} Capacitive relaxation constant [variety q]
k_{qL} Inductive relaxation constant [variety q]
k_θ Rotational friction coefficient [rotational mechanics]
k_f Friction resistance [translational mechanics]
$\boldsymbol{k}$ Wave vector of a wave (vector, in rad m^{-1}) [space wave energy]
$\boldsymbol{k}_q$ Natural wave vector of an oscillator (vector, in rad m^{-1}) [variety q]

L

L Electric inductance (scalar) [electrodynamics]
L_q Generalized inductance (scalar) [variety q]
L_m Magnetic inductance (scalar) [magnetism]
$\boldsymbol{L}$ Angular momentum (vector) (or kinetic momentum) [rotational mechanics]
$\boldsymbol{\ell}$ Displacement, position, distance (vector) [translational mechanics]
$\boldsymbol{\ell}^0$ Origin of displacement, of position (vector) [translational mechanics]
$\boldsymbol{\ell}^{min}$ Minimum distance (vector) [translational mechanics]

M

M Inertial mass (scalar) [translational mechanics]
M_G Gravitational mass [gravitation]
M Molar mass [gravitation + physical chemistry]
M_Q "Memristance" [electrodynamics]
m Mass of a corpuscle [translational mechanics]
$\boldsymbol{M}$ Magnetization (vector) [magnetism]

N

N Number of objects
N Number of corpuscles [corpuscular energy]
N_A Avogadro number [corpuscular energy + physical chemistry]
N_C Number of corpuscles in the conduction band [corpuscular energy]
N_g Energy state degeneracy (maximum number of corpuscles per energy state) [corpuscular]
N_V Number of corpuscles in the valence band [corpuscular energy]
$\mathcal{N}$ Number of energy states [corpuscular energy]
n Substance amount, number of moles [physical chemistry]
n_R Spatial reduction depth (from 0 = global to 3 = volume)
n_λ Index of refraction at wavelength λ [space–time]

P

P Pressure [hydrodynamics]
$\mathcal{P}_q$ Power [variety q]

p	Evolution mode, or conservation yield, of a transfer
p_M	Gravitational impulse [gravitation]
p_m	Magnetic impulse [magnetism]
p_N	Corpuscular impulse [corpuscular energy]
p_n	Substance impulse [physical chemistry]
p_q	Impulse [variety q]
p_S	Entropy impulse [thermics]
p_ξ	Reaction impulse [chemical reaction energy]
$\boldsymbol{P}$	Momentum, quantity of movement (vector) [translational mechanics]
$\boldsymbol{p}$	Particle momentum (vector) [translational mechanics]
$\mathbf{P}$	Polarization vector [polarization]
$\mathbf{p}_A$	Superficial impulse (vector) [surface energy]

Q

Q	Heat (thermal energy) [thermics]
Q	Electric charge [electrodynamics]
$\mathcal{Q}$	Volume flow [hydrodynamics]
Q^m	Minimum electric charge [electrodynamics]
Q_P	Passed electric charge [electrodynamics]
q	Basic quantity [variety q]
q_P	Polarization charge [electric polarization energy]

R

R	Gas constant [physical chemistry + thermics]
R	Electric resistance (scalar) [electrodynamics]
R_q	Generalized resistance (scalar) [variety q]
R_V	Hydraulic resistance (scalar) [hydrodynamics]
$\mathbf{R}$	Radius (vector) [space–time]
$\mathcal{R}_H$	Hall coefficient (=inverse of charge concentration) [electrodynamics]
r_q	Reversibility of an exchange [variety q]
$\boldsymbol{r}$	Position, distance (vector) [space–time]
$\mathfrak{R}_q$	Generalized reluctance (reciprocal of inductance) [variety q]

S

S	Action
S	Entropy [thermics]
s_D	Separability factor [variety q]
s_q	Sink [variety q]

T

T	Temperature [thermics]
T_q	Transmittance of a variable [variety q]
$T_{\lambda q}$	Transmittance of a radiation intensity at a given wavelength [variety q]
T	Period of a wave (in s) [space–time]
$\mathcal{T}_q$	Inductive energy (kinetic, electromagnetic, etc.) [variety q]
t	Time [space–time]
$\boldsymbol{T}_\ell$	Mechanical tension (vector) [translational mechanics]

U

$\mathcal{U}_q$	Capacitive energy (potential, internal, etc.) [variety q]
U_q	Corpuscular potential [variety q]
U	Space–time volume rate (in $m^3\ s^{-1}$) [space–time]
u_q	Mobility [variety q]
$\boldsymbol{u}$	Space–time velocity (vector, in $m\ s^{-1}$) [space–time]

V

V	Electric potential [electrodynamics]
V_G	Gravitational potential [gravitation]
V_τ	Thomson potential [electrodynamics + thermics]
V	Volume [hydrodynamics]
V_m	Molar volume [hydrodynamics + physical chemistry]
v	Potential scan rate [electrodynamics]
$\mathcal{V}$	Reaction rate [chemical reaction energy]
$\boldsymbol{v}$	Translation velocity (vector) [translational mechanics]

W

$\boldsymbol{w}$	Space–time vector (in $m\ s^{-1}$) [space–time]

X

x_q	Entity number [variety q]
x	Any variable
x	Space coordinates

Y

y_q	Energy-per-entity [variety q]
y	Any variable
y	Space coordinates

Z

z	Charge number [electrodynamics]
z	Altitude [space–time]
z	Space coordinates [space–time]

A2.1.5 Greek Symbols

α	Alpha	Seebeck's coefficient [electrodynamics + thermics]
α_i		Transfer factor of species i [electrodynamics + physical chemistry]
α_q		Energetic coupling ratio (= $\mathcal{U}_q$ / $\mathcal{U}_\lambda$) [variety q]
$\boldsymbol{\alpha_q}$		Spatial energy-damping factor (vector, in m^{-1})
$\alpha_{qi:j}$		Exchange or transfer coefficient from pole i to pole j [Variety q]
β_{Cq}	Beta	Capacitive energetic ratio (= $e_q\ q/\mathcal{U}_q$) [variety q]
β_{Lq}		Inductive energetic ratio (= $f_q\ p_q/\mathcal{T}_q$) [variety q]
$\Delta_r G$		Molar free enthalpy of reaction [physical chemistry]
ε	Epsilon	Electric (dielectric) permittivity [electrodynamics]
ε_ℓ		Relative elongation [translational mechanics]
ε_M		Gravitational permittivity [gravitation]
ζ_q	Zeta	Spatial damping factor (vector) of an oscillator (system property, in m^{-1})

η	Eta	Dynamic viscosity [translational mechanics]
η		Overpotential [electrodynamics]
θ	Theta	Angle (vector) [rotational mechanics]
Φ_B	Phi	Quantity of induction, induction flux [electrodynamics]
Φ_{BN}		Multiple of quantity of induction [electrodynamics]
Φ_S		Electromagnetic saturation [electrodynamics]
ϕ	Phi	Phase shift (in rad)
φ		Phase angle of an oscillator or wave (in rad) [phase energy]
$\Gamma(p)$	Gamma	Euler's function
γ	Gamma	Heat capacities ratio [thermics + physical chemistry]
$\boldsymbol{\gamma}$		Superficial tension (vector) [surface energy]
γ_q		Temporal damping factor of an oscillator (system property, in s^{-1})
κ	Kappa	Thermal conductivity [thermics]
κ_n		Molar specific absorbance of a species (absolute, or Naperian) (in $m^2\ mol^{-1}$)
κ_N		Specific absorbance of corpuscles (absolute, or Naperian) (in $m^2\ corp^{-1}$)
κ_λ		Imaginary part of index of refraction at wavelength λ [space–time]
λ_{ba}	Lamda	Integral coupling coefficient [varieties q_a and q_b]
λ'_{ba}		Differential coupling coefficient [varieties q_a and q_b]
$\boldsymbol{\lambda}$		Wavelength of an oscillator or wave (vector, in m)
μ	Mu	Chemical potential [physical chemistry]
μ		Electromagnetic permeability [electrodynamics]
$\bar{\mu}$		Electrochemical potential [physical chemistry]
$\bar{\mu}$		Apparent chemical potential [physical chemistry]
$\boldsymbol{\mu_M}$		Lineic mass [translational mechanics]
ν	Nu	Frequency (in Hz)
ν		Kinematic viscosity [translational mechanics]
ν_N		Coupling frequency (in Hz)
Π	Pi	Peltier potential [electrodynamics + thermics]
ρ	Rho	Charge concentration (volumic density) [electrodynamics]
ρ_M		Volumic inertial mass [translational mechanics]
ρ_{MG}		Volumic gravitational mass [gravitation]
σ	Sigma	Electric conductivity [electrodynamics]
σ_q		Generalized conductivity [variety q]
τ	Tau	Thomson coefficient [electrodynamics + thermics]
$\boldsymbol{\tau}$		Torque, couple (vector) [rotational mechanics]
τ_{qC}		Capacitive relaxation time constant [variety q]
τ_{qL}		Inductive relaxation time constant [variety q]
ξ	Xi	Extent, advance [chemical reaction energy]
χ	Chi	Compressibility [hydrodynamics]
χ_e		Electric susceptibility [electrodynamics + polarization]
χ_m		Magnetic susceptibility [electrodynamics + magnetization]
Ψ_q	Psi	Wave function of an oscillator (in $J^{1/2}$) [variety q]
$\boldsymbol{\Omega}$	Omega	Angular velocity (vector, in rad s^{-1}) [rotational mechanics]
ω		Pulsation (angular frequency) of a wave (in rad s^{-1}) [phase energy]
ω_q		Natural pulsation of an oscillator (system property, in rad s^{-1}) [variety q]

A2.1.6 Superscripts

0	Reference
θ	Standard
eq	At equilibrium
m	Minimum or maximum (extremum)
max	Maximum
min	Minimum
mid	Middle
T	Transposed matrix or operator
tot	Total quantity
$*$	Separated (or initial) state
$'$	(*prime*) Differential quantity or coefficient or real part
$''$	(*double prime*) Partial derivative or second order derivative or imaginary part

A2.1.7 Subscripts

a	Activation or energy variety q_a
b	Energy variety q_b
m	Molar quantity (i.e., divided by the number of moles)
q	Energy variety q
Δ	Scaling
$/r$	Lineic density
$/A$	Surface density
$/V$	Volume density
$=$	Means that the order of subscripts can be any
$:$	Used for separating subscripts in indicating that their order matters

A2.1.8 Mathematical Symbols

δ	Dirac function
e	Napier constant (Exponential constant) (= 2.71828...)
i	Imaginary number $(-i^2 = 1)$
π	Archimedes constant *pi* (=3.14159...)
$\vert...\vert$	Absolute value
$\Vert...\Vert$	Modulus
∞	Infinite
$\pm$	Plus or minus
$=$	Is equal to
$\equiv$	Is similar to, or is congruent to
$\neq$	Is different from
$\sim$	Is equivalent to
$\approx$	Is approximately equal to
$\propto$	Is proportional to
$\rightarrow$	Tends toward
$<, >$	Is less than, is greater than
$\leq, \geq$	Is less than or equal to, is greater than or equal to
$\ll, \gg$	Is much less than, is much greater than
$\Rightarrow$	Implies
$\Leftrightarrow$	Implies and reciprocally

$\exists$	Exists
$\Re e$	Real part of a complex variable
$\Im m$	Imaginary part of a complex variable

A2.1.9 Conditions and Definitions in Equalities

$\overset{def}{=}$	Definition equation of the left member (this is the normal use)
$\overset{def\to}{=}$	Definition equation of the right member
$\underset{condition}{=}$	Equality valid under specified condition
$\underset{lin}{=}$	Equality valid in linear domain
$\underset{hom}{=}$	Equality valid in spatially homogeneous systems
$\underset{hom\ time}{=}$	Equality valid in temporally homogeneous systems
$\underset{homot}{=}$	Equality valid in case of homothetic system properties (proportionality)
$\underset{cont}{=}$	Equality valid in continuous space–time
$\underset{quad}{=}$	Equality valid in case of quadratic dependence
$\underset{\exists X}{=}$	Equality valid only if the property X exists
$\underset{rec}{=}$	Equality according to the *receptor* convention (effort opposite to flow)
$\underset{gen}{=}$	Equality according to the *generator* convention (effort oriented as flow)
$\underset{reg}{=}$	Equality valid for a regular multipole (identical dipoles)
$\underset{isol}{=}$	Equality valid for isolated systems (neither exchange nor interaction)
$\underset{node}{=}$	Provides a definition owing to the node location in a Formal Graph

A2.1.10 Diacritical Marks

$\bar{X}$	(*macron*) Apparent variable or complex variable
$\hat{X}$	(*circumflex*) Operator
$\tilde{X}$	(*tilde*) Fourier transformed variable or operator

A2.2 CONSTANTS*

Avogadro constant	$N_A \cong 6.022\ 141\ 79(30) \times 10^{23}\ \text{mol}^{-1}$	(see CGPM 2007)
Boltzmann constant	$k_B \cong 1.380\ 650\ 4(24) \times 10^{-23}\ \text{J K}^{-1}$	
Elementary charge	$e \cong 1.602\ 176\ 487(40) \times 10^{-19}\ \text{C}$	
Faraday constant	$F \cong 9.648\ 533\ 83(83) \times 10^{4}\ \text{C mol}^{-1}$	
Gas constant	$R \cong 8.314\ 472(15)\ \text{J mol}^{-1}\ \text{K}^{-1}$	(see CODATA 2006)
Gravitation constant	$G \cong 6.674\ 28(67) \times 10^{-11}\ \text{m}^3\ \text{kg}^{-1}\ \text{s}^{-2}$	
Permeability of free space	$\mu_0 = 4\pi \times 10^{-7}\ \text{V s A}^{-1}\ \text{m}^{-1}$	(by definition) (see CGPM 1983)
Permittivity of free space	$\varepsilon_0 = 8.854\ 187\ 817... \times 10^{-12}\ \text{A s V}^{-1}\ \text{m}^{-1}$	
Planck constant	$h \cong 6.626\ 068\ 96(33) \times 10^{-34}\ \text{J s}$	
Speed of light in free space	$c = 2.997\ 924\ 58 \times 10^{8}\ \text{m s}^{-1}$	(by definition) (see CGPM 1983)

* The number in brackets is the standard uncertainty.

Appendix 3: Formal Graph Encoding

CONTENTS

The rules for drawing Formal Graphs are summarized in this appendix.

A3.1 RULES FOR COLORING GRAPHS (WHEN POSSIBLE)

A3.1.1 Variables (Nodes)

The various forms of energy (including *coupling energies*, *Legendre's transformed* ones, etc.) are colored in yellow, but subvarieties have their own coloration (inductive is represented by light brown and capacitive by light green). One color for each variable family: basic quantities (light green), efforts (red), impulses (light blue), and flows (purple). Activity and counting variables are in white.

A3.1.2 Operators (Links)

Inductances are in brown, capacitances in dark green, and conductances in dark blue. Temporal operators are in orange (and space–time velocities) whereas purely spatial operators are in black. Gray indicates localization levels (spatial reduction depth) (see Figure A3.1).

A3.2 NODES

A3.2.1 Node Shapes

In a Formal Graph, nodes are variables of the system. By default, a disc (round) is used for any variable, when its nature or status is not specified.

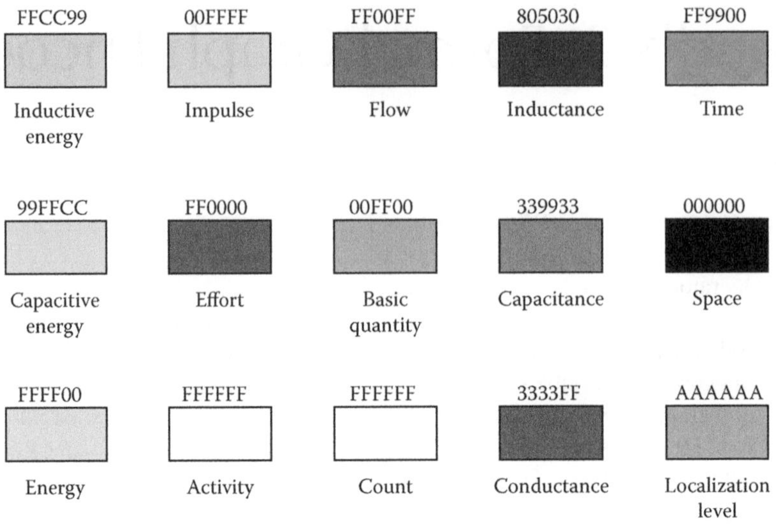

FIGURE A3.1 Colors used in a Formal Graph. The color is given by its hexadecimal code in the RGB system.

A3.2.2 Categories of Variables

Other shapes are used for well-defined categories of variables. The graphical distinction made according to the status of the variable is the following: trapezoid for an entity number, pentagonal for an energy-per-entity, square for a state function, rounded square for a counting variable or activity, and diamond for a coefficient, which can be a constant, a coupling factor, etc. (see Graph A3.1).

In the above shapes, the entity number and the energy-per-entity are represented for the capacitive subvariety. The shapes are turned upside down (horizontal symmetry) for representing the inductive subvariety.

In a *canonical* Formal Graph, the only authorized nodes are state variables or gate variables of the considered system, belonging to the four families described: basic quantities, efforts, impulses, and flows. Localized variables of state variables are naturally authorized.

A3.2.3 Mathematical Structures of Variables

Nodes for a scalar variable have a thin perimeter, for a vector have a medium bold one, and for a tensor or a matrix have a thick bold one (see Graph A3.2).

A3.2.4 Node Drawings

The graphical representation of a node is a combination of shape and perimeter style and optionally of color. All nodes are placed at the corners of a square pattern or a segment (case of *singletons*), with the exception of the *activity* (cf. Chapter 7).

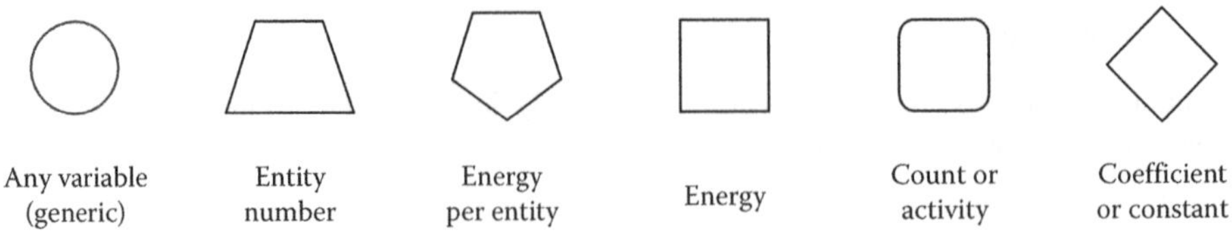

GRAPH A3.1 Convention for representing variables with node shapes.

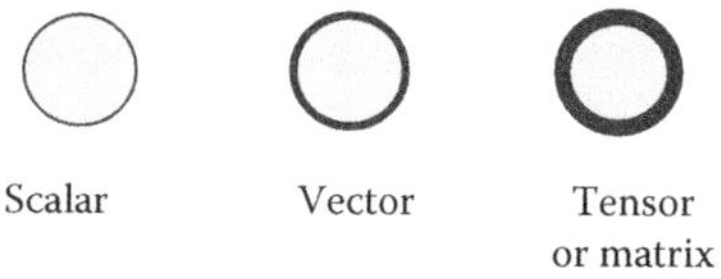

GRAPH A3.2 Convention for representing nodes according to the mathematical structure of variables (i.e., scalars, vectors, and tensors). The circle is used as a generic drawing.

A3.2.5 State, Apparent, and Translated Variables

A *translated variable* results from the multiplication of a state variable by a coupling factor, thus producing a variable in another energy variety. It amounts to expressing the state variable in the units of the other energy variety (e.g., a translated electrical potential into physical chemical energy, by multiplication with the Faraday constant, has the Joule per mole as a unit) (cf. Chapter 12).

An *apparent variable* results from the contribution of several other variables from the same family. Generally, it corresponds to the sum of a state variable with translated variables (see Graph A3.3).

These variations in line style are compatible with the variations of line thickness used for encoding the mathematical structure of variables (e.g., a translated vector is drawn with a dotted and bold perimeter).

A3.2.6 Other Variables

When the supply of energy to the system is made by external generators that impose either the flow (current generators) or the effort (potential generators), the *powered energy-per-entity* is indicated by a disc in the background of the node drawing.

An ensemble of identical dipoles but with various values of energy-per-entity (i.e., a distribution of values over a range of discrete values) constitutes a Formal Object called *dipole distribution*. The energy-per-entity node of this Formal Object is drawn with a flattened pentagon (i.e., a regular pentagon deformed by moving its horizontal side toward the center).

The entity number of the Formal Object called *singleton* is drawn with a diamond (lozenge) in the case of the capacitive singleton and with a slanted square in the case of the inductive one. As a singleton is featured by only two state variables, its representation does not use a square to place its nodes at the summit, but a segment (see Graph A3.4).

Naturally, the same variability in perimeter styles and thicknesses as previously indicated is available for these nodes.

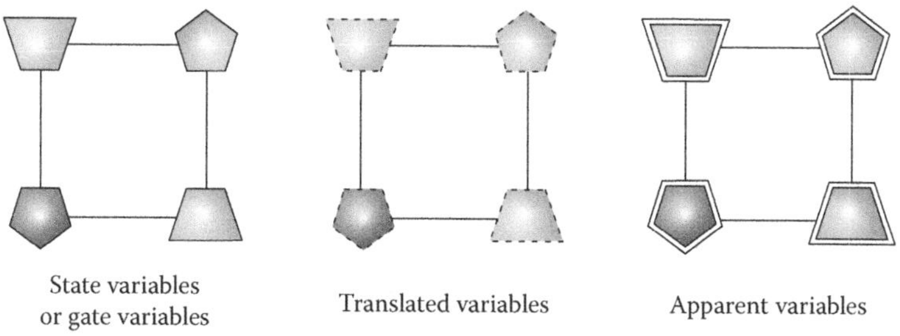

GRAPH A3.3 Various line styles for the node perimeter according to the status of the variable.

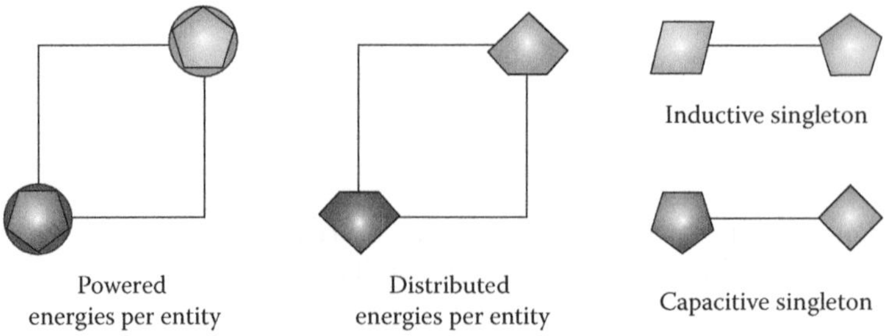

GRAPH A3.4 Various shapes for the nodes used for representing energies-per-entity powered by external sources (left) or distributed over discrete values (center). On the right are shown the simplest Formal Objects (on the complexity scale), the inductive and the capacitive singletons.

A3.3 LINKS

A3.3.1 Elementary Links

The general rule is to encode an operator with an arrow in placing nearby its algebraic symbol. When a binary operation involving another variable (addition, subtraction, exponentiation, etc.) needs to be made explicit, an oval is positioned over the arrow (eventually drawn with a filled form) with the indication of the operation (in case of addition, the "+" sign can be omitted) (see Graph A3.5).

A path defines the way state variables act on each other through a certain number of connections (and operators) in a precise order. *Elementary paths* are represented by single graph connections called links. *Composed paths* are defined sequences of elementary links. *Apparent paths* are virtual paths, which can be materialized by one amidst several possible paths.

Elementary links are used to represent the properties of the energy container through an operator. In a canonical Formal Graph, two kinds of links are possible: those belonging to the system itself, such as capacitance, inductance, and conductance, and those supported by the space–time, such as the evolution property represented by the time operator or the spatial distribution represented by a spatial operator. When the operator is a purely differential one (integration or derivation), a double line is used, otherwise (space–time velocity, coupling frequency, mass transfer operator, etc.) a simple but thicker line is used. In a Differential Formal Graph, only partial derivatives with respect to a variable are allowed.

The interdiction of link is stipulated by a cross on the edge where a connection may exist in other Formal Objects (see Graph A3.6).

The constitutive properties of a system are normally represented by simple straight arrows when they are elementary (i.e., composed by a single direct link). However, curved lines may be used when parallel paths need to be represented.

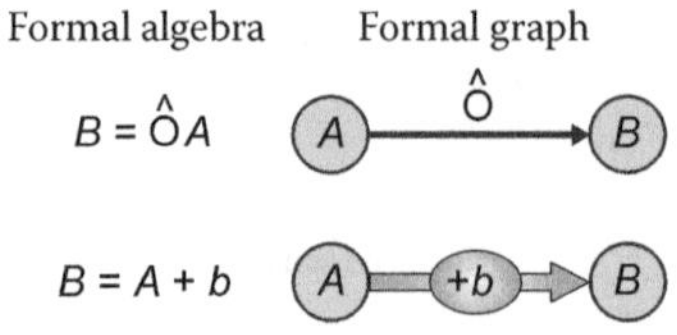

GRAPH A3.5 Basic encoding of algebraic operators in a Formal Graph.

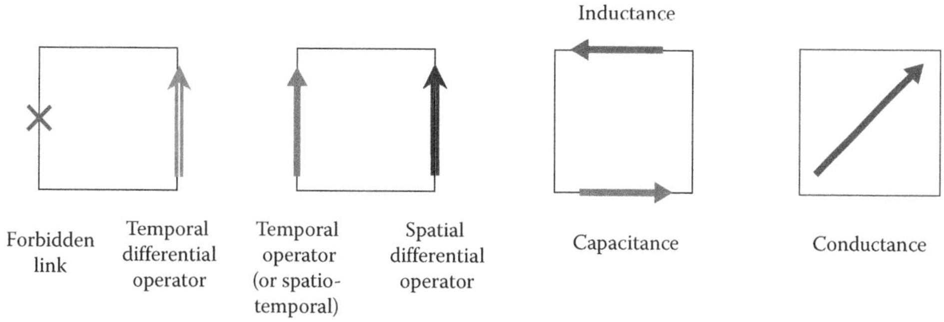

GRAPH A3.6 Various encodings of internal links.

A3.3.2 Composed and Apparent Paths

When paths are composed, a broken line is used. An apparent path is one that connects two nodes without indication of the links composing the path (see Graph A3.7). A dashed line is used for it.

A3.3.3 External Links

External links include several categories of connectors that have in common the property to use scalars as operators. These include:

- Building connectors that relate nodes of the same family but different Formal Objects. Their scalar is plus or minus the identity operator (i.e., +1 or −1).
- External coupling links that relate nodes of the same family and same Formal Objects but different energy variety. Their scalar is the coupling factor that makes a change of scale.
- Internal coupling links working between the two subvarieties within a same Formal Object. They are space–time scalars, such as a frequency or a velocity. They are internal with respect to the energy variety but external with respect to the Formal Objects.

These various links are sketched in Graph A3.8.

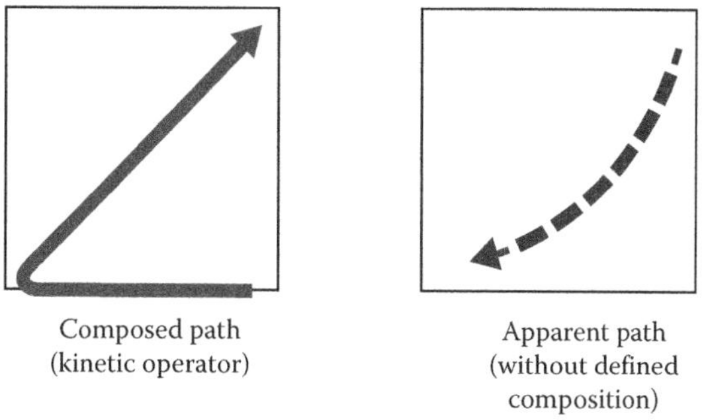

GRAPH A3.7 Two versions of a nonelementary path: the composed path on the left, built by chaining several elementary links (here a capacitance, then a conductance), and the apparent path on the right, with no defined composition (here a generalized impedance).

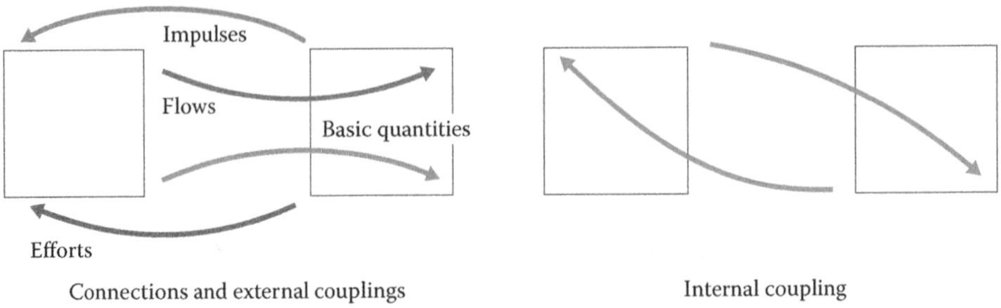

GRAPH A3.8 Encoding of connectors used for constructing Formal Objects and external coupling (left) and for ensuring internal coupling (right).

A3.4 OPERATIONS WITH PATHS

When a node is linked by several paths, it may have various consequences for its variable depending on the way they are connected. When only one path is arriving and the others are leaving the node, the variable plays the passive role of relay in a paths chain; this corresponds to the *path transmission*. When several paths are converging on a node without indication of operation, they are equivalent: whatever the followed path, the action on the variable is the same. When the operation is specified with a plus sign, the effect of the paths is cumulative; this is translated under the form of an addition of paths called *summation*. At last, *complemented paths* are paths sharing each one a varying part of the action on a variable.

The coding used for each kind of path contribution on a node is listed in Graph A3.9.

These combinations are not exclusive; they can be superimposed on the same node, as for instance in Graph A3.10.

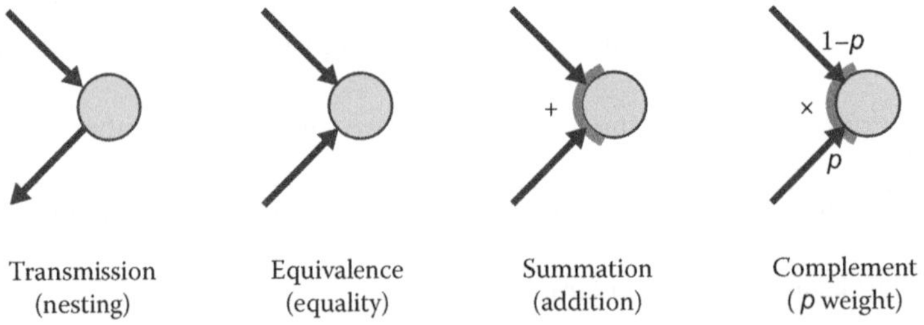

GRAPH A3.9 The four basic combination rules applying to paths on the same node in Formal Graphs.

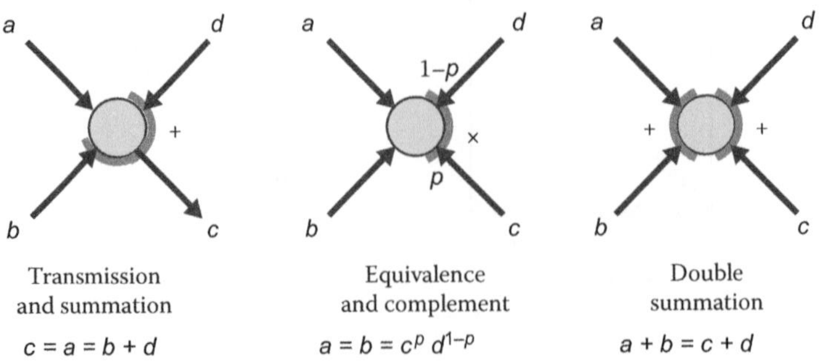

GRAPH A3.10 Examples of multiple combinations of paths on the same node. The value of the variable in the node is the common value of the equalities.

The important detail to take into account is the extension of the gathering curve which partially surrounds the node: Only paths arriving onto this curve are involved in the indicated operation.

A3.5 FORMAL GRAPHS IN THREE DIMENSIONS

For a canonical Formal Graph, the convention used in this book is to place the global level (nonlocalized variables) at the top of the graph and then to go down with the successive localized levels:

- Global level: Depth 0 State variables
- Curve level: Depth 1 Lineic densities
- Surface level: Depth 2 Surface densities
- Volume level: Depth 3 Volumic concentrations

The same spatial operators are used for going from one level to the other. The four corners of the squares at each level use the same state variable configuration as in a two-dimensional Formal Graph (i.e., a vertical projection from above restitutes the usual canonical two-dimensional graph). This convention constitutes the capacitive variables occupying the front plane whereas the inductive variables are in the rear.

Appendix 4: List of Examples and Case Studies

CONTENTS

A4.1 EXAMPLES

Subjects that are treated outside the case studies (or within but devoted to another topic) are given as examples. The chapter number is indicated before the case study number in the first column.

	Translational Mechanics (ℓ, F) + (P, v)
6-C2	Frame of reference
9-F1	Centripetal force
14	$E = mc^2$
	Electrodynamics (Q, V) + (Φ_B, I)
7-4.1 and 11-H8	Lambert–Beer–Bouguer law (radiation absorption)
8-5.6	Shockley diode
8-1.2	Bipolar junction transistor
9-F8	Lorentz force
11-H8	Index of refraction
12-J3	Piezoelectricity
	Physical Chemistry/Corpuscular Energy (n, μ) + (. . .)/(N, E) + (. . .)
6-C9	Energy levels distribution
8-5.8	Arrhenius law
8-5.6	Electrochemical kinetics (Butler–Volmer)
8-E5	Equivalence reaction/diffusion
8-5.7	First-order kinetic law (demonstration)
10-6.1	Randle's circuit in electrochemistry
13-K2	Heat capacity
14	Kinetic constant in quantum physics
	General
7-4.3	Statistics versus influence
8-5.10	Comparison with statistics
9-5.3	Introduction of the imaginary number i
9-5.4	Construction of the exponential function and Euler's formula
9-5 and 11-5	Demonstration (construction) of the wave function
9-2.1	Definition of time
10-1.4	Fractional derivation/integration
11-1.2	Reversibility definition (temporal)
11-H8	Irreversibility due to spatial distribution
11-4.3	Convolution, Fourier, and Laplace transforms
11-4.4	Complex transfer functions
13-K6	Particular derivative
13-5	Coupling factors and universal constants
14	Wave/particle duality

A4.2 CASE STUDIES

The chapter number is indicated before the case study number in the first column. In the last column, L represents inductive energy, C represents capacitive energy, and G represents conductive energy.

		Translational Mechanics (ℓ, F) + (P, v)	
4-A1	Moving Body	Pole	L
4-A8	Spring End	Pole	C
4-A10	Motion with Friction	Pole	LG
5-B1	Elastic Element	One-dimensional space pole	C
5-B2	Elasticity of Solids	Three-dimensional space pole	C
6-C1	Colliding Bodies	Dipole (conservative)	L
6-C2	Relative Motion	Dipole (Nonconservative)	L
6-C3	Viscous Layers	Dipole	LG
6-C8	Spring	Dipole (inseparable)	C
8-E2	Concurring Forces	Multipole	C
9-F1	Accelerated Motion	Dipole assembly (powered)	L
9-F4	Vibrating String	One-dimensional dipole assembly (isolated)	LC
10-G4	Newton's Law of Viscosity	Three-dimensional pole	LG
11-H2	Slowed-down Motion	Three-dimensional pole	LG
11-H6	Viscoelastic Relaxation	Three-dimensional pole	CG
11-H7	Damped Mechanical Oscillator	Dipole assembly	L + C + G
12-J3	Electrical Force	Coupling with electrodynamics	C
12-J4	Gravitational Force	Coupling with gravitation	C
12-J5	Piston	Coupling with hydrodynamics	C
13-K4	Kinetic of Gases	Coupling with hydrodynamics	LC
13-K5	Rotating Bodies	Coupling with rotational mechanics	L
		Rotational Mechanics (θ, τ) + (L, Ω)	
4-A2	Moment of Inertia	Pole	L
8-E3	Levers and Balance	Chain multipole (regular)	C
9-F3	Torsion Pendulum	Dipole assembly (isolated)	L + C
13-K5	Rotating Bodies	Coupling with translational mechanics	L
		Electrodynamics (Q, V) + (Φ_B, I)	
4-A3	Current Loop	Pole	L
4-A5	Electric Charges	Pole	C
5-B3	Electric Space Charges	Three-dimensional pole	C
5-B4	Poisson Equation	Three-dimensional pole	C
5-B5	Gauss Theorem	Three-dimensional pole	C
6-C7	Electric Capacitor	Dipole	C
7-D1	Electromagnetic Influence	Dipole	L
7-D2	Electrostatic Influence	Dipole	C
7-D4	Coulomb Law	Three-dimensional dipole	C
8-E1	Parallel Capacitors	Ladder multipole (irregular)	C
8-E6	Solenoid	Chain multipole (regular)	L
9-F2	LC Circuit	Dipole assembly (isolated)	L + C
9-F6	Light Propagation	Three-dimensional dipole assembly (isolated)	L + C
9-F7	Maxwell Equations	Three-dimensional dipole assembly (isolated)	L + C
9-F8	Hall Effect	Dipole assembly (powered)	L + C
10-G1	Reduced Ohm's Law	Three-dimensional pole	G
10-G5	Capacitive Transmission Line	Dipole	C + G
10-G6	Constant Phase Element	Dipole	C + G and L + G

11-H1	Electrical Joule Effect	Dipole assembly	C + G
11-H4	RC Circuit	Dipole assembly	C + G
11-H5	Dielectric Relaxation	Three-dimensional dipole assembly	C + G
11-H8	Attenuated Propagation	Three-dimensional dipole	L + C + G
12-J2	Electrochemical Potential	Coupling with physical chemistry	C
12-J3	Electrical Force	Coupling with translational mechanics	C
12-J8	Thermal Junction	Coupling with thermics	C
12-J9	Seebeck Effect	Coupling with thermics	C
12-J10	Peltier Effect	Coupling with thermics	C
12-J11	Thomson Effect	Coupling with thermics	C
12-J12	Electric Polarization	Coupling with electric polarization	C
12-J13	Magnetism	Coupling with magnetism	L
		Electric Polarization (q_P, e_P) + (. . .)	
12-J12	Electric Polarization	Coupling with electrodynamics	C
		Magnetism (. . .) + (p_m, f_m)	
4-A4	Magnetization	Pole	L
7-D6	Magnetic Interaction	Three-dimensional dipole	L
12-J13	Magnetism	Coupling with electrodynamics	L
		Physical Chemistry (n, μ) + (. . .)	
4-A6	Chemical Species	Pole	C
4-A11	Reactive Chemical Species	Pole	CG
6-C5	First-Order Chemical Reaction	Dipole	CG
6-C6	Physical Chemical Interface	Dipole	C
7-D3	Physical Chemical Influence	Dipole	C
8-E4	First-Order Chemical Reactions in Series	Chain multipole (irregular)	CG
10-G3	Fick's Law of Diffusion	Three-dimensional pole	CG
10-G7	Transient Diffusion	Three-dimensional dipole	CG
10-G8	Anomalous Diffusion	Three-dimensional dipole	CG
10-G9	Generalized Mass Transfer	Three-dimensional dipole	CG
11-H3	Occurring Reaction	Dipole assembly	CG
12-J1	The *n*th-Order Chemical Reaction	Coupling with chemical reaction energy	C
12-J2	Electrochemical Potential	Coupling with electrodynamics	C
13-K1	Hydrochemical Potential	Coupling with hydrodynamics	C
13-K2	Ideal Gas	Coupling with hydrodynamics and thermics	C
13-K3	Barometric Equation	Coupling with hydrodynamics, thermics, and gravitation	C
13-K4	Kinetic of Gases	Coupling with translational mechanics, hydrodynamics, and thermics	C
		Corpuscles Groups (Corspuscular Energy) (N, E_N) + (. . .)	
4-A7	Group of Corpuscles	Pole	C
6-C9	Electrons and Holes	Dipole (inseparable)	C
8-E5	Diffusion through Layers	Chain multipole (regular)	CG
12-J6	Ions Distribution	Coupling with electrodynamics	C
		Thermics (Heat or Thermal Energy) (S, T) + (. . .)	
4-A9	Thermal Pole	Pole	C
10-G2	Fourier Equation (Heat Transfer)	Three-dimensional pole	G
12-J8	Thermal Junction	Coupling with electrodynamics	C
12-J9	Seebeck Effect	Coupling with electrodynamics	C
12-J10	Peltier Effect	Coupling with electrodynamics	C

continued

12-J11	Thomson Effect	Coupling with electrodynamics	C
13-K2	Ideal Gas	Coupling with physical chemistry and hydrodynamics	C
13-K3	Barometric equation	Coupling with physical chemistry, hydrodynamics, and gravitation	C
13-K4	Kinetic of Gases	Coupling with translational mechanics, hydrodynamics, and physical chemistry	C
		Hydrodynamics (V, P) + (F, $\mathcal{Q}$)	
6-C4	Pipe	Dipole (nonconservative)	G
9-F5	Sound Propagation	Three-dimensional dipole assembly (spatiotemporal, isolated)	LC
12-J5	Piston	Coupling with translational mechanics	C
12-J7	Bubble	Coupling with surface energy	C
13-K1	Hydrochemical Potential	Coupling with physical chemistry	C
13-K2	Ideal Gas	Coupling with physical chemistry and thermics	C
13-K3	Barometric Equation	Coupling with physical chemistry, gravitation, and thermics	C
13-K4	Kinetic of Gases	Coupling with translational mechanics, physical chemistry, and thermics	C
13-K6	Euler Equation	Coupling with gravitation	LC
		Gravitation (M_G, V_G) + (p_G, f_G)	
7-D5	Newtonian Gravitation	Three-dimensional dipole	C
12-J4	Gravitational Force	Coupling with translational mechanics	C
13-K3	Barometric Equation	Coupling with physical chemistry, hydrodynamics, and thermics	C
13-K6	Euler Equation	Coupling with hydrodynamics	C
		Chemical Reaction Energy (ξ, $\mathcal{A}$) + (. . .)	
12-J1	The *n*th-Order Chemical Reaction	Coupling with physical chemistry	C
		Surface Energy (A, γ) + (p_A, f_A)	
12-J7	Bubble	Coupling with hydrodynamics	C

Appendix 5: CD-Rom Content

CONTENTS

The companion CD-Rom contains the following:

- The "Circuit-to-Graph" software.
- All Formal Graphs of the book in bitmap color files.

No installation is required and the software is platform independent.*

A5.1 CIRCUIT-TO-GRAPH TRANSLATOR

Open the file "translator.htm" in a web browser to access the Circuit-to-Graph translator.

The Flash player plug-in must be installed, with a version 9.0 or higher, and the browser must be enabled for executing scripts. If the Flash player is not installed, the browser will ask to download it from the Adobe site (http://www.adobe.com/go/getflashplayer).

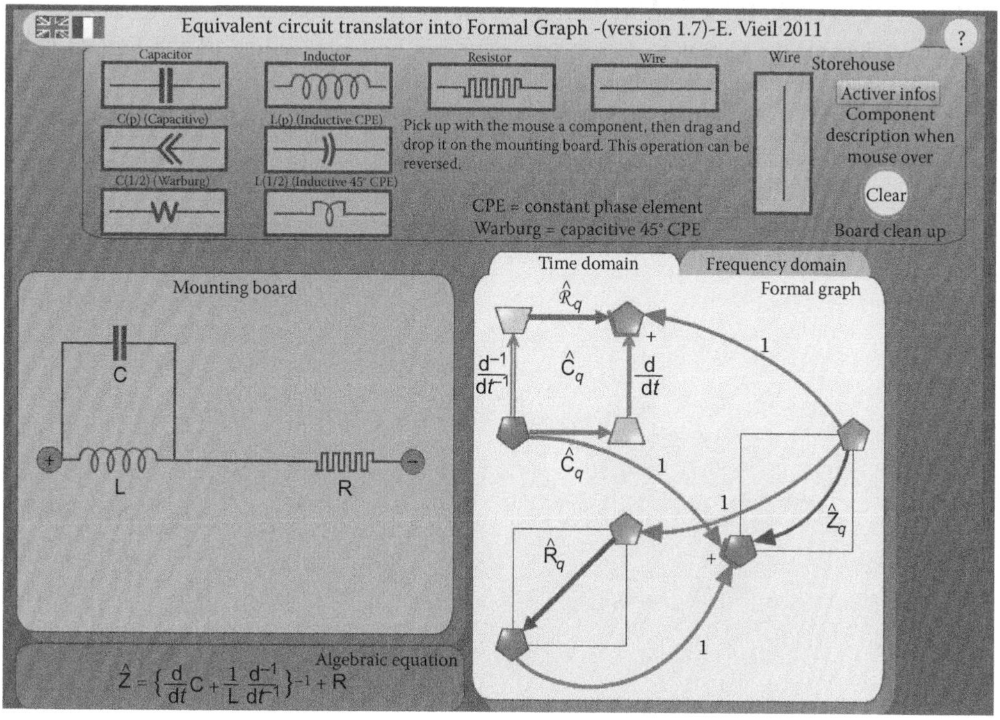

* Tested with Mac OS 10.6 and Windows XP, Vista, Seven.

Sounds are used as command feedback.

English/French languages can be selected by clicking on the corresponding flag (upper left corner).

Help and explanations are available by clicking on the question mark round button (upper right corner).

The circuit is mounted by picking up with the mouse components in the storehouse (upper zone). Components are dragged and dropped on the mounting board (left zone). This operation can be reversed.

The Formal Graph appears in the right zone, either in temporal representation (time domain) or in imaginary pulsation representation (frequency domain), depending on the selected tab.

The corresponding algebraic equation appears in the lower box.

A5.1.1 Limitations

- Only one energy variety.
- No more than two system properties with the same nature, capacitive, inductive, or conductive (e.g., two resistors, two resistors and two capacitors, but not three resistors, etc.).
- Nesting between parallel and serial branches limited to two levels.

A5.2 FORMAL GRAPH FILES

All Formal Graphs of the book are reproduced in color (PNG format, Portable Network Graphics) in the "Formal_Graphs" directory. They are grouped by chapter.

A5.3 FORMAL GRAPHS WEBSITE

Updated versions of the Circuit-to-Graph translator can be found online on the Formal Graph website: http://edmgp-graphes.lepmi.grenoble-inp.fr/formal/

References

CHAPTER 1

C. Berge, *The Theory of Graphs*, Dover, New York, USA, 1962.
A. Carling, *Introducing Neural Networks*, Sigma Press, Wilmslow, UK, 1992.
J. N. Shive, R. L. Weber, *Similarities in Physics*, Adam Hilger Ltd., Bristol, UK, 1982.
L. Smolin, *The Trouble with Physics*, Mariner Books, Boston, USA, 2007.
E. Vieil, A new graph language for representing the macroscopic formalism in physico-chemistry and physics, *ECS Trans.*, 1, 2006, 37–48, doi: 10.1149/1.2214633.
E. Vieil, Introduction to Formal Graphs: A new approach of the classical formalism, *Phys. Chem. Chem. Phys.*, 9, 2007, 3877–3896, doi:10.1039/b700797c.

CHAPTER 2

L. de Broglie, Some questions about mechanics and thermodynamics in the classical and relativistic frames, [Diverses questions de mécanique et de thermodynamique classiques et relativistes (French)], G. Lochak, ed., *Lecture Notes in Physics M 32*, Springer-Verlag, Berlin, 1995.
G. F. Oster, A. Perelson, A. Katchalsky, Network thermodynamics, *Nature*, 234, 1971, 393–399.
G. F. Oster, A. Perelson, A. Katchalsky, Network thermodynamics, dynamic modeling of biophysical systems, *Q. Rev. Biophys.*, 6(I), 1973, 1–134.
H. M. Paynter, *Analysis and Design of Engineering Systems*, MIT Press, Cambridge, MA, 1961.

CHAPTER 3

L. O. Chua, Memristor—The missing circuit element, *IEEE Trans. Circuit Theor.*, 18, 1971, 507–519.
J. M. Tour, T. He, The fourth element, *Nature* 453, 2008, 42–43, doi:10.1038/453042a.

CHAPTER 4

P. W. Atkins, *Physical Chemistry*, 6th ed., Oxford University Press, Oxford, UK, 1998.
K. Laidler, C. King, A lifetime of transition-state theory, *Chem. Intell.*, 4(3), 1998, 39.

CHAPTER 5

R. W. R. Darling, *Differential Forms and Connections*, Cambridge University Press, Cambridge, UK, 1994.
G. A. Deschamps, Electromagnetics and differential forms, *IEEE Proc.*, 69, 1981, 676–696.
T. Frankel, *The Geometry of Physics*, 2nd ed., Cambridge University Press, Cambridge, UK, 2004.
G. Kron, Equivalent circuits to represent the electromagnetic field equations, *Phys. Rev.*, 64, 1943, 126–128.
N. Schleifer, Differential forms as a basis for vector analysis—With applications to electrodynamics, *Am. J. Phys.*, 51, 1983, 1139–1146.
E. Tonti, On the mathematical structure of a large class of physical theories, *Rend. Acc. Lincei*, 52, 1972, 48–56.
E. Tonti, The reason for analogies between physical theories, *Appl. Math. Modelling*, 1, 1976, 37–50.
E. Tonti, On the geometrical structure of electromagnetism, In G. Ferrarese, ed., *Gravitation, Electromagnetism and Geometrical Structure*, Pitagora Editrice, Bologna, 1995, pp. 281–308.
B. Schutz, *Geometrical Methods of Mathematical Physics*, Cambridge University Press, Cambridge, UK, 1980.
K. F. Warnick, R. H. Selfridge, D. V. Arnold, Teaching electromagnetic field theory using differential forms, *IEEE Trans. Educ.*, 40, 1997, 53–68.

CHAPTER 6

C. Berge, *The Theory of Graphs*, Dover, New York, 1962.

C. Berge, *La théorie des graphes et ses applications* (French), Dunod, Paris, 1958.

A. Einstein, *Relativity: The Special and General Theory*, Three Rivers Press, Crown Publishing, New York, USA, 1961.

A. Einstein, *Théorie de la relativité restreinte et générale* (French), Dunod, Paris, 2004.

A. K. Jonscher, *Universal Relaxation Law*, Chelsea Dielectrics Press, London, UK, 1996.

J. Magueiro, *Faster than the Speed of Light*, Basic Books, Perseus Publishing, New York, USA, 2003.

J. Magueiro, *Plus vite que la lumière* (French), Dunod, Paris, 2009.

J. W. Moffat, *Reinventing Gravity*, Smithsonian Books, Collins, New York, USA, 2008.

CHAPTER 7

P. W. Atkins, *Physical Chemistry*, 6th ed., Oxford University Press, Oxford, UK, 1998.

P. Atkins, J. De Paula, *Physical Chemistry*, 9th ed., Oxford University Press, Oxford, UK, 2010.

E. Hecht, *Physics. Calculus*, Brooks/Cole, International Thomson Publishing, Pacific Grove, CA, USA, 1996.

E. Hecht, *Physique* (French), 5th ed., De Boeck, International Thomson Publishing, Brussels, Belgium, 2007.

L. D. Landau, E. F. Lifchitz, *Statistical Physics (Vol. 5 of Course of Theoretical Physics)*, Pergamon, New York, USA, 1958.

K. F. Riley, M. P. Hobson, S. J. Bence, *Mathematical Methods for Physics and Engineering*, 3rd ed., Cambridge University Press, Cambridge, UK, 2006.

CHAPTER 8

A. J. Bard, L. R. Faulkner, *Electrochemical Methods. Fundamentals and Applications*, 2nd ed., John Wiley & Sons, New York, USA, 2001.

R. A. Smith, *Semiconductors*, Cambridge University Press, Cambridge, UK, 1959.

CHAPTER 9

Conférence Générale des Poids et Mesures, 17th Meeting, 1983. http://www.bipm.org/en/CGPM/db/17/1/

R. P. Feynman, *The Feynman Lectures on Physics*, Addison-Wesley, Reading, MA, USA, 1964.

E. Hecht, *Physics. Calculus*, Brooks/Cole, International Thomson Publishing, Pacific Grove, CA, USA, 1996.

E. Hecht, *Physique* (French), 5th ed., De Boeck, International Thomson Publishing, Brussels, Belgium, 2007.

G. S. Nolas, J. Sharp, H. J. Goldsmid, *Thermoelectrics: Basic Principles and New Materials Developments*, Springer Verlag, Berlin, 1962.

B. D. H. Tellegen, A general network theorem, with applications, *Philips Res. Rep.*, 7, 1952, 259–269.

G. Woan, *The Cambridge Handbook of Physics Formulas*, Cambridge University Press, Cambridge, UK, 1983.

CHAPTER 10

A. J. Bard, L. R. Faulkner, *Electrochemical Methods. Fundamentals and Applications*, 2nd ed., John Wiley & Sons, New York, USA, 2001.

C. Berge, *The Theory of Graphs*, Dover, New York, USA, 1962.

P. A. Kullstam, Heaviside's Operational Calculus: Oliver's Revenge, *IEEE Trans. Educ.*, 34, 1991, 155–156.

A. Le Méhauté, G. Crépy, Introduction to transfer and motion in fractal media: The geometry of kinetics, *Solid State Ionics* 9–10, 1983, 17–30, doi:10.1016/0167-2738(83)90207-2.

F. Miomandre, E. Vieil, Reversible charge-transfer with unknown mass-transfer. Characterization and modeling by using a mass-transfer operator determined from experimental data, *J. Electroanal. Chem.*, 375, 1994, 275–292.

K. B. Oldham, J. Spanier, *The Fractional Calculus*, Academic Press, New York, USA, 1974.

N. Phan-Tien, *Understanding Rheology*, Springer-Verlag, Berlin, 2002.

S. G. Samko, B. Ross, Integration and differentiation to a variable fractional order, *Integral Transforms and Special Functions*, 1, 1993, 277–300.

E. Vieil, The mass transfer rate in electrochemistry, *J. Electroanal. Chem.*, 297, 1991, 61–92, doi:10.1016/0022-0728(91)85359-W.
E. Vieil, Quantification of the mass-transfer mode, *J. Electroanal. Chem.*, 318, 1991, 61–68, doi: 10.1016/0022-0728(91)85294-Y.
E. Vieil, F. Miomandre, Non-reversible charge-transfer and mass-transfer. Methodology for an experimental characterization without theoretical Models, *J. Electroanal. Chem.*, 395, 1995, 15–27.
E. Vieil, Simple and direct interpretation of phase angles or derivation degrees in term of energy conservation vs. dissipation with Formal Graphs, *J. Solid State Electrochem.*, 15, 2011, 955–969, doi: 10.1007/s10008-011-1308-9.

CHAPTER 11

H. B. Callen, *Thermodynamics and an Introduction to Thermostatistics*, John Wiley & Sons, New York, USA, 1985.
R. P. Feynman, R. B. Leighton, M. Sands, *The Feynman Lectures on Physics*, Definitive ed., Addison-Wesley, Reading, MA, USA, 2006, vol. II, chapter 32.
R. Giles, *Mathematical Foundations of Thermodynamics*, Pergamon, Oxford, UK, 1964.
J. D. Jackson, *Classical Electrodynamics*, 3rd ed., John Wiley & Sons, New York, USA, 1998, chapter 7.
A. K. Jonscher, *Universal Relaxation Law*, Chelsea Dielectrics Press, London, UK, 1996.
L. D. Landau, E. F. Lifchitz, *Statistical Physics (Vol. 5 of Course of Theoretical Physics)*, Pergamon, New York, USA, 1958.
P. Oswald, *Rhéophysique*, Belin, Paris, 2005.
N. Phan-Tien, *Understanding Rheology*, Springer-Verlag, Berlin, 2002.
M. Planck, *Treatise on Thermodynamics*, 7th ed., Dover Publications, New York, USA, 1922, republished 1990.
G. Woan, *The Cambridge Handbook of Physics Formulas*, Cambridge University Press, Cambridge, UK, 2003.

CHAPTER 12

A. J. Bard, L. R. Faulkner, *Electrochemical Methods. Fundamentals and Applications*, 2nd ed., John Wiley & Sons, New York, USA, 2001.
H. B. Callen, *Thermodynamics and an Introduction to Thermostatistics*, 2nd ed., John Wiley & Sons, New York, USA, 1985.
Conférence Générale des Poids et Mesures, 23rd Meeting, 2007. http://www.bipm.org/en/CGPM/db/23/12/
The IUPAC Green Book, *Quantities, Units and Symbols in Physical Chemistry*, 3rd ed., RSC Publishing, Cambridge, UK, 2007, p. 4.
A. F. Joffe, *Semiconductor Thermoelements and Thermoelectric Cooling*, Infosearch, London, 1957.
D. Kondepudi, I. Prigogine, *Modern Thermodynamics*, John Wiley & Sons, New York, USA, 1998.
K. Laidler, C. King, A lifetime of transition-state theory, *Chem. Intell.*, 4(3), 1998, 39.
M. S. Longair, *Theoretical Concepts in Physics*, 2nd ed., Cambridge University Press, Cambridge, UK, 2003.
I. Prigogine, R. Defay, *Thermodynamique chimique conformément aux méthodes de Gibbs et de Donder* (French), Desoer, Liège (Belgium), 1944–1946 (2 Tomes).
I. Prigogine, R. Defay, *Chemical Thermodynamics*, translated D. H. Everett, Longmans Green, London, UK, 1962.

CHAPTER 13

G. Bruhat, *Thermodynamique*, (French), Masson, Paris, 1962.
P. J. Mohr, B. N. Taylor, D. B. Newell, Recommended values of the fundamental physical constants, Committee on Data for Science and Technology (CODATA 2006), *Rev. Mod. Phys.*, 80, 2008, 633.
M. Planck, *Treatise on Thermodynamics*, 7th ed., Dover Publications, New York, USA, 1922, republished 1990.

CHAPTER 14

R. P. Feynman, *The Feynman Lectures on Physics*, Addison-Wesley, Reading, MA, USA, 1964.
D. J. Griffiths, *Introduction to Elementary Particles*, Wiley-VCH, Weinheim, Germany, 2004.

S. Haykin, *Neural Networks and Learning Machines*, 3rd ed., Pearson Education Inc., Upper Saddle River, NJ, USA, 2009.
K. Laidler, C. King, A lifetime of transition-state theory, *Chem. Intell.*, 4(3), 1998, 39.
W. Pauli, *Wave Mechanics*, *Pauli Lectures on Physics*, Vol. 5, Dover Publications, New York, USA, 1973.
K. Popper, *Unended Quest*, (re-ed. of 1976), Taylor & Francis, Boca Raton, FL, USA, 2002.
K. Popper, *La quête inachevée*, (French), Calmann_Lévy, Paris, 1981.

Index

Note: n = footnote

C

D

J

K

Q

R

T

U

V

W

Y

Printed and bound by CPI Group (UK) Ltd, Croydon, CR0 4YY
21/10/2024
01777095-0020